EQUATIONS OF CHANGE IN TERMS OF THE TOTAL FLUXES

These equations are valid only for systems in which gravity is the sole external force. More information can be found in §19.2.

Momentum:

$$\frac{\partial}{\partial t}\rho\mathbf{v} = -[\nabla \cdot \boldsymbol{\phi}] + \rho\mathbf{g}$$

Energy:

$$\frac{\partial}{\partial t}\left(\rho\hat{U} + \frac{1}{2}\rho v^2\right) = -(\nabla \cdot \mathbf{e}) + \rho(\mathbf{v} \cdot \mathbf{g})$$

Mass:

$$\frac{\partial}{\partial t}\rho\omega_A = -(\nabla \cdot \mathbf{n}_A) + r_A$$

EQUATIONS OF CHANGE (RESTRICTED FORMS)

Momentum (for Newtonian fluids with constant ρ and μ):

$$\rho\frac{D\mathbf{v}}{Dt} \equiv \rho\left(\frac{\partial\mathbf{v}}{\partial t} + [\mathbf{v} \cdot \nabla\mathbf{v}]\right) = -\nabla p + \mu\nabla^2\mathbf{v} + \rho\mathbf{g}$$

Energy (for Newtonian fluids with constant ρ and k):

$$\rho\hat{C}_p\frac{DT}{Dt} \equiv \rho\hat{C}_p\left(\frac{\partial T}{\partial t} + (\mathbf{v} \cdot \nabla T)\right) = k\nabla^2 T + \mu\Phi_v$$

Mass (for binary mixtures of A and B with constant $\rho\mathscr{D}_{AB}$ and no cross effects):

$$\rho\frac{D\omega_A}{Dt} \equiv \rho\left(\frac{\partial\omega_A}{\partial t} + (\mathbf{v} \cdot \nabla\omega_\mathbf{A})\right) = \rho\mathscr{D}_{AB}\nabla^2\omega_\mathbf{A} + r_A$$

DIMENSIONLESS GROUPS

(l_0 and v_0 are a characteristic length and velocity, respectively)

$\mathrm{Re} = l_0 v_0 \rho/\mu$	$\mathrm{Pr} = \hat{C}_p\mu/k$	$\mathrm{Sc} = \mu/\rho\mathscr{D}_{AB}$
$\mathrm{Ra} = \mathrm{GrPr}$	$\mathrm{Gr} = g\beta l_0^3\Delta T/\nu^2$	$\mathrm{Gr}_\omega = g\zeta l_0^3\Delta\omega_A/\nu^2$
$\mathrm{Nu} = hl_0/k$	$\mathrm{Pé} = \mathrm{RePr}$	$\mathrm{Pé}_{AB} = \mathrm{ReSc}$
$\mathrm{Sh} = k_c l_0/\mathscr{D}_{AB}$	$j_H = \mathrm{Nu}/\mathrm{RePr}^{1/3}$	$j_D = \mathrm{Sh}/\mathrm{ReSc}^{1/3}$

Sphere: $Nu_m = 2 + 0.6\,Re^{\frac{1}{2}}\,Pr^{\frac{1}{3}}$

$A = 4\pi R^2$

$Nu = 2$ $\quad h_m = \frac{Nu(k)}{D}$

Introductory Transport Phenomena

R. Byron Bird
Warren E. Stewart
Edwin N. Lightfoot
Daniel J. Klingenberg

WILEY

VP & Executive Publisher: Don Fowley
Executive Editor: Dan Sayre
Product Designer: Jenny Welter
Marketing Manager: Christopher Ruel
Operations Manager: Yana Mermel
Editorial Assistant: Francesca Baratta
Associate Production Manager: Joyce Poh
Designer: Kenji Ngieng

This book was set in Palatino by Laserwords Private Limited.

Founded in 1807, John Wiley & Sons, Inc. has been a valued source of knowledge and understanding for more than 200 years, helping people around the world meet their needs and fulfill their aspirations. Our company is built on a foundation of principles that include responsibility to the communities we serve and where we live and work. In 2008, we launched a Corporate Citizenship Initiative, a global effort to address the environmental, social, economic, and ethical challenges we face in our business. Among the issues we are addressing are carbon impact, paper specifications and procurement, ethical conduct within our business and among our vendors, and community and charitable support. For more information, please visit our website: www.wiley.com/go/citizenship.

ISBN 978-1-118-77552-3 (cloth)

Printed in the United States of America

SKY10051796_072523

Preface

Transport phenomena is now generally recognized as a key scientific subject supporting study of several branches of engineering, agriculture, meteorology, medicine, and environmental studies. Historically, the subject had its origins in Europe, and some of the key original papers and books were written in German and French, as well as in English. In the years following World War II, interest in the teaching of transport phenomena in U.S. universities led to the rapid development of courses and textbooks in this area, first for background in engineering curricula, with later textbooks in various specialized areas within engineering as well as outside.

So what exactly is meant by the term "transport phenomena"? Very simply put, it includes the transport of *momentum* (or "fluid mechanics"), the transport of *energy* (or "heat transfer"), and the transport of *chemical species* (or "mass transfer"). Of particular importance are then the transport coefficients—the *viscosity* (describing the transport of momentum), the *thermal conductivity* (describing the transport of energy), and the *diffusivity* (describing the transport of chemical species). Luckily there are "conservation laws" that apply to the three entities being transported, and we will pay considerable attention to these conservation laws throughout this textbook.

Understandably, the book divides itself quite naturally into three parts, one for each of the entities being transported. Much emphasis will be placed on the similarities—and differences—between the development of the three types of conservation equations and their associated transport coefficients. Each of the three areas just described may be studied at one of three different levels: the *molecular* level (where we try to understand the transport coefficients in terms of the molecular interactions); the *microscopic* level (where we regard the materials as continua); and the *macroscopic* level (where we examine large systems—such as pieces of equipment or biological organs). It is important, we feel, for those first making acquaintance with the subject of transport phenomena to understand the connection between the three levels of the subject, as well as that between the three quantities that are transported. Since there are these various connections, we can lay out the general plan of this textbook as in the table shown in Chapter 0—the subject material just organizes itself! Despite this apparently obvious self-organization, there are in each of the three main parts of the book some topics that do not fit into this tidy scheme. Each of the three parts, therefore, has an "eighth chapter," in which these oddball topics—but very important topics—are discussed.

Due to the arrangement of the topics to be discussed, it is possible for teaching purposes to consider two ways of proceeding. In the first way, we can teach the material "by columns"—that is, by teaching the chapters in the order 1, 2, 3, and so on. Clearly this is the best course of action for instruction of undergraduates. As for graduate students, teaching "by rows"—that is, by teaching the chapters in the sequence 1, 9, 17, 2, 10, 18, and so on—may be preferable.

The chapters are provided with a summary section as well as questions for discussion, and these may be helpful in suggesting ways to organize "problems sessions" or "quiz sections." Each chapter also has, at the end, a number of problems, which have been grouped into various classes: Class A, illustrating direct numerical applications of material in the text; Class B, involving elementary analysis of physical situations; and Class C, requiring more mature analysis or material from several chapters.

Depending on the number of credit hours allotted to this subject, it may not be possible for all the material in the book to be covered. Therefore, in the chapter outlines at the beginnings of the chapters, as an aid to instructors, we designate optional sections by ($^{\circ}$).

Omission of the topics so designated, will still leave enough material for a well-balanced introductory course.

The subject of transport phenomena has long been regarded as a rather mathematical subject. However, it should be taught as a topic in applied physics. Emphasis should be placed on the visualization of physical systems, the physical interpretation of the results of problem solving, and, of course, the choice of the physical laws needed. Mathematics will inevitably have to be used in order to solve the problems, but the mathematics is secondary in importance to the physics. Every effort is made in this introductory text to explain how the mathematics is used, and to include sufficient intermediate steps in derivations so that the text will be useful for self-study. Much emphasis should be placed on checking the mathematics at each step in a development and learning to ask whether the equations are giving results in agreement with physical intuition.

Of importance to teachers will be the difference between BSL (*Transport Phenomena*, by Bird, Stewart, and Lightfoot) and the present textbook, which we will designate by BSLK. Readers familiar with the former book will recognize that the overall organization of the contents is similar, and that there is considerable overlap in the contents of the two books. You will find, however, that considerable more space has been devoted to filling in missing steps in mathematical derivations and to fuller explanations of mathematical developments, including an enlargement of the appendix devoted to mathematical topics; in addition, much material has been removed that we felt was beyond the level of mathematical preparation of most undergraduates. Our intention was to make this book reflect the topics covered in our undergraduate course, with the exception of a few advanced topics included for the brightest students. Finally, we removed the chapters dealing with two independent variables and replaced them with chapters on dimensional analysis, including the much-used Buckingham pi theorem.

We wish to thank the people who have influenced the final form of this textbook, and at the top of the list we must put Olaf Andreas Hougen, who guided our department and emphasized the scientific background of chemical engineering. Also, he felt very strongly that no topic should be taught to engineering students unless there are clear indications that it can be used in applications. Rarely did a week go by that we did not get questions from students or professors at other institutions about unclear portions of the text in the predecessors of this volume. Responding to these queries has been a very helpful exercise for the coauthors in planning future printings of the book and for better understanding the pedagogical problems that the readers have had to cope with. Everyone should feel free to contact us in connection with the present volume, and we encourage you to point out errors and inconsistencies in the text. Next we would like to acknowledge our own students and colleagues from whom we have learned so much; conversations with Professors Edwin N. Lightfoot, Michael D. Graham, Eric V. Shusta, A. Jeffrey Giacomin, and Ross E. Swaney have been particularly helpful. Each and every one of them has in some way contributed to our knowledge of the subject and the solving of pedagogical problems encountered. Special thanks go to Professor Carlos Ramirez of the University of Puerto Rico for the time and trouble that he took in supplying us with copious corrigenda for the second edition of BSL, and to Dr. M. I. Hill of Columbia University for offering critiques of several chapters for BSLK.

Transport phenomena is an evolving subject. Each month, new applications appear in the technical journals, and new techniques for problem solving are presented. We are acutely aware that we cannot possibly cover all the newest of theoretical and experimental developments in a beginning textbook, but we hope that we have provided a springboard from which the readers can launch themselves into the new areas.

And now, we point out that the chief coauthors of this BSLK book are RBB and DJK, who have tried to carry the torch forward; DJK has taught the undergraduate and graduate transport phenomena courses for about twenty years, and his understanding of students' challenges has been very helpful. Regretfully ENL was not able to participate fully in this rewrite of the *Transport Phenomena* textbook, although his contributions to the earlier BSL editions—particularly some of the more challenging examples and problems—have been retained. The untimely passing of WES has been keenly felt by our "team," and to him we also owe generous thanks for his organizational skills, his passion for accuracy, and his wonderful sense of humor.

RBB
ENL
WES
DJK

Madison, Wisconsin
Summer 2013

Contents

Chapter 0

The Subject of Transport Phenomena

The purpose of this introductory chapter is to describe the scope, aims, and methods of the subject of transport phenomena. It is important to have some idea about the structure of the field before plunging into the details; without this perspective it is not possible to appreciate the unifying principles of the subject and the interrelation of the various individual topics. A good grasp of transport phenomena is essential for understanding many processes in engineering, agriculture, meteorology, physiology, biology, analytical chemistry, materials science, pharmacy, and other areas. Transport phenomena is a well-developed and eminently useful branch of physics that pervades many areas of applied science and engineering.

§0.1 WHAT ARE THE TRANSPORT PHENOMENA?

The subject of transport phenomena includes three closely related topics: fluid dynamics, heat transfer, and mass transfer. Fluid dynamics involves the transport of *linear momentum* and *angular momentum*, heat transfer deals with the transport of *energy*, and mass transfer is concerned with the transport of *mass* of various chemical species. These three transport phenomena should, at the introductory level, be studied together for the following reasons:

- They frequently occur simultaneously in industrial, biological, agricultural, and meteorological problems; in fact, the occurrence of any one transport process by itself is the exception rather than the rule.
- The basic equations that describe the three transport phenomena are closely related. The similarity of the equations under simple conditions is the basis for solving problems "by analogy."
- The mathematical tools needed for describing these phenomena are very similar. Although it is not the aim of this book to teach mathematics, the student may find it necessary to review various mathematical topics as the development unfolds.

Learning how to use mathematics may be a very valuable by-product of studying transport phenomena.

- The molecular mechanisms underlying the various transport phenomena are very closely related. All materials are made up of molecules, and the same molecular motions and interactions are responsible for viscosity, thermal conductivity, and diffusivity.

The main aim of this book is to give a balanced overview of the field of transport phenomena, present the fundamental equations of the subject, and illustrate how to use them to solve problems.

There are many excellent treatises on fluid dynamics, heat transfer, and mass transfer. In addition, there are many research and review journals devoted to these individual subjects and even to specialized subfields. The reader who has mastered the content of this book should find it possible to consult the treatises and journals and go more deeply into other aspects of the theory, experimental techniques, empirical correlations, design methods, and applications. That is, this book should not be regarded as the complete presentation of the subject, but rather as a stepping stone to a wealth of knowledge that lies beyond.

§0.2 THREE LEVELS FOR THE STUDY OF TRANSPORT PHENOMENA

In Fig. 0.2-1 we show a schematic diagram of a large system—for example, a large piece of equipment through which a fluid mixture is flowing. We can describe the transport of mass, momentum, energy, and angular momentum at three different levels.

At the *macroscopic level* (Fig. 0.2-1*a*), we can write down a set of equations called the "macroscopic balances," which describe how the mass, momentum, energy, and angular momentum in the system change because of the introduction and removal of these entities via the entering and leaving streams, and because of various other inputs to the system from the surroundings. No attempt is made to understand all the details of the system. In studying an engineering or biological system, it is a good idea to start with this macroscopic description in order to make a global assessment of the problem; in some instances, it is only this overall view that is needed.

At the *microscopic level* (Fig. 0.2-1*b*), we examine what is happening to the fluid mixture in a tiny region within the equipment. We write down a set of equations called the "equations of change," which describe how the mass, momentum, energy, and angular momentum change within this tiny region. The aim here is to get information about

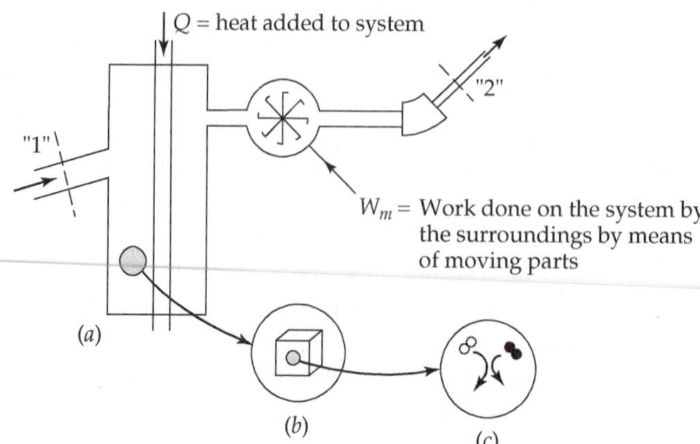

Fig. 0.2-1. (*a*) A macroscopic flow system containing N_2 and O_2; (*b*) a microscopic region within the macroscopic system containing N_2 and O_2, which are in a state of flow; (*c*) a collision between a molecule of N_2 and a molecule of O_2.

velocity, temperature, pressure, and concentration profiles within the system. This more detailed information may be required for the understanding of some processes.

At the *molecular level* (Fig. 0.2-1*c*), we seek a fundamental understanding of the mechanisms of momentum, energy, and mass transport in terms of molecular structure and intermolecular forces. Generally this belongs in the realm of the theoretical physicist or physical chemist, but occasionally engineers and applied scientists have to get involved at this level. This is particularly true if the processes being studied involve complex molecules, extreme ranges of temperature and pressure, or chemically reacting systems.

It should be evident that these three levels of description involve different "length scales": for example, in a typical industrial problem, at the macroscopic level the dimensions of the flow systems may be of the order of centimeters or meters; the microscopic level involves what is happening in the micron to the centimeter range; and molecular level problems involve ranges of about 1 to 1000 nanometers.

This book is divided into three parts:

- Chapters 1–8: Flow of pure fluids at constant temperature (with emphasis on viscous and convective momentum transport)
- Chapters 9–16: Flow of pure fluids with varying temperature (with emphasis on conductive, convective, and radiative energy transport)
- Chapters 17–24: Flow of binary fluid mixtures with varying composition (with emphasis on diffusive and convective mass transport), as well as simultaneous energy and mass transport

That is, we build from the simpler to the more difficult problems. Within each of these parts, we start with an initial chapter dealing with some results of the *molecular* theory of the transport properties (viscosity, thermal conductivity, and diffusivity). Then we proceed to the *microscopic* level and learn how to determine the velocity, temperature, and concentration profiles in various kinds of systems. The discussion then focuses on the *macroscopic* level and the description of large systems. Each part concludes with a chapter that introduces special topics that are somewhat outside the main story.

As the discussion unfolds, the reader will appreciate that there are many connections between the levels of description. The transport properties that are described by molecular theory are used at the microscopic level. Furthermore, the equations developed at the microscopic level are needed in order to provide some input into problem solving at the macroscopic level.

There are also many connections between the three areas of momentum, energy, and mass transport. By learning how to solve problems in one area, one also learns the techniques for solving problems in another area. The similarities of the equations in the three areas mean that in many instances one can solve a problem "by analogy"—that is, by taking over a solution directly from one area and, by changing the symbols in the equations, immediately writing down the solution to a problem in another area.

The student will find that these connections—among levels, and among the various transport phenomena—reinforce the learning process. As one goes from the first part of the book (momentum transport) to the second part (energy transport) and then on to the third part (mass transport), the story will be very similar but the "names of the players" will change.

Table 0.2-1 shows the arrangement of the chapters in the form of a 3×8 "matrix." Just a brief glance at the matrix will make it abundantly clear what kinds of interconnections can be expected in the course of the study of the book. We have traditionally taught our undergraduate introductory transport phenomena course by columns, but a course may be taught by rows as well.

Table 0.2-1. Organization of Topics

Type of transport		Momentum		Energy		Mass
Transport mechanisms	1	Viscosity and the momentum flux vector	9	Thermal conductivity and the heat flux vector	17	Diffusivity and the mass flux vector
Transport in one dimension (shell balance methods)	2	Shell momentum balances and velocity distributions	10	Shell energy balances and temperature distributions	18	Shell mass balances and concentration distributions
Transport in arbitrary continua	3	Equations of change [isothermal]	11	Equations of change [nonisothermal]	19	Equations of change [binary mixtures]
Transport in turbulent flow	4	Turbulent momentum transport	12	Turbulent energy transport	20	Turbulent mass transport
Dimensional analysis	5	Dimensional analysis of momentum transport	13	Dimensional analysis of energy transport	21	Dimensional analysis of mass transport
Transport across phase boundaries	6	Friction factors; use of empirical correlations	14	Heat-transfer coefficients; use of empirical correlations	22	Mass-transfer coefficients; use of empirical correlations
Transport in large systems, such as pieces of equipment or parts thereof	7	Macroscopic balances [isothermal]	15	Macroscopic balances [nonisothermal]	23	Macroscopic balances [mixtures]
Transport by other mechanisms	8	Momentum transport in complex fluids	16	Energy transport by radiation	24	Cross effects; multicomponent systems

At all three levels of description—molecular, microscopic, and macroscopic—the *conservation laws* play a key role. The derivation of the conservation laws for molecular systems is straightforward and instructive. With elementary physics and a minimum of mathematics, we can illustrate the main concepts and review key physical quantities that will be encountered throughout this book. That is the topic of the next section.

§0.3 THE CONSERVATION LAWS: A MOLECULAR COLLISION EXAMPLE

The system we consider here is that of two colliding diatomic molecules. For simplicity we assume that the molecules do not interact chemically and that each molecule is homonuclear—that is, that its atomic nuclei are identical. The molecules are in a

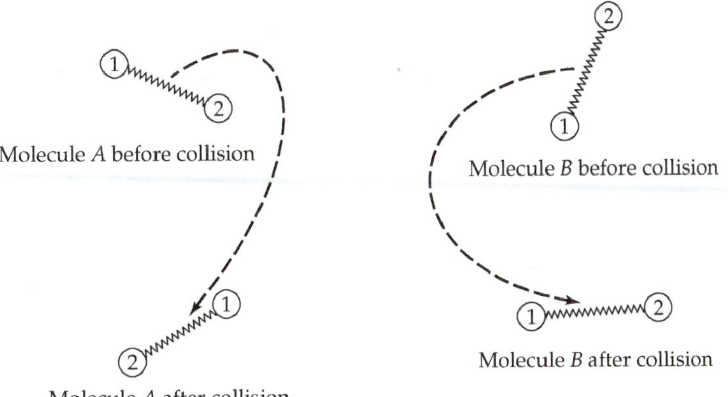

Fig. 0.3-1. A collision between identical, homonuclear diatomic molecules, such as N_2 and O_2. Molecule A is made up of two atoms $A1$ and $A2$. Molecule B is made up of two atoms $B1$ and $B2$. From R. B. Bird, *Korean J. Chem. Eng.*, **15**, 105–123 (1998).

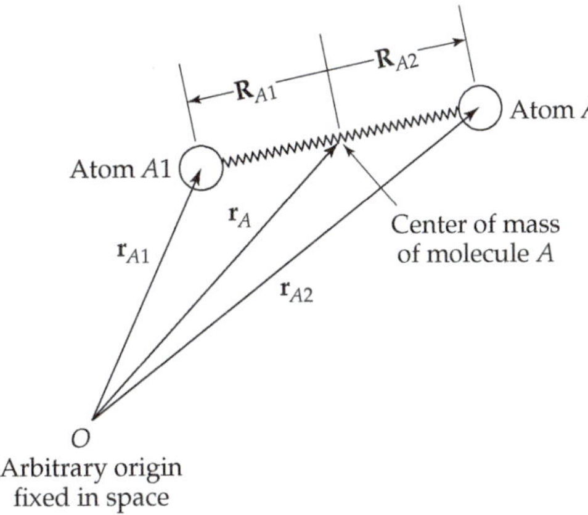

Fig. 0.3-2. Position vectors for the atoms $A1$ and $A2$ in molecule A. From R. B. Bird, *Korean J. Chem. Eng.*, **15**, 105–123 (1998).

low-density gas, so that we need not consider interactions with other molecules. In Fig. 0.3-1 we show the collision between the two homonuclear diatomic molecules, A and B, and in Fig. 0.3-2 we show the notation for specifying the locations of the two atoms of one molecule by means of position vectors drawn from an arbitrary origin.

Actually the description of events at the atomic and molecular level should be made by using quantum mechanics. However, except for the lightest molecules (H_2 and He) at temperatures lower than 50 K, the kinetic theory of gases can be developed quite satisfactorily by use of classical mechanics.

Several relations must hold between quantities before and after a collision. Both before and after the collision, the molecules are presumed to be sufficiently distant that the two molecules cannot "feel" the intermolecular force between them; beyond a distance of about 5 molecular diameters, the intermolecular force is known to be negligible. Quantities after the collision are indicated with primes.

(a) According to the *law of conservation of mass*, the total mass of the molecules entering and leaving the collision must be equal:

$$m_A + m_B = m'_A + m'_B \tag{0.3-1}$$

Here m_A and m_B are the masses of molecules A and B. Since there are no chemical reactions, the masses of the individual species will also be conserved, so that

$$m_A = m'_A \text{ and } m_B = m'_B \tag{0.3-2}$$

(b) According to the *law of conservation of momentum*, the sum of the momenta of all the atoms before the collision must be equal to that after the collision, so that

$$m_{A1}\mathbf{v}_{A1} + m_{A2}\mathbf{v}_{A2} + m_{B1}\mathbf{v}_{B1} + m_{B2}\mathbf{v}_{B2} = m'_{A1}\mathbf{v}'_{A1} + m'_{A2}\mathbf{v}'_{A2} + m'_{B1}\mathbf{v}'_{B1} + m'_{B2}\mathbf{v}'_{B2} \tag{0.3-3}$$

in which $\mathbf{v}_{A1} = d\mathbf{r}_{A1}/dt$ is the velocity of atom 1 of molecule A before the collision, and $\mathbf{v}'_{A1} = d\mathbf{r}_{A1}/dt$ its velocity after the collision. We now write $\mathbf{r}_{A1} = \mathbf{r}_A + \mathbf{R}_{A1}$, so that $\mathbf{r}_{A1}$ is written as the sum of the position vector for the center of mass of molecule A and the position vector of the atom 1 with respect to the center of mass of molecule A. We recognize that $\mathbf{R}_{A2} = -\mathbf{R}_{A1}$, and, by taking the time derivative of this relation, that $d\mathbf{R}_{A2}/dt = -d\mathbf{R}_{A1}/dt$, or $\mathbf{V}_{A2} = -\mathbf{V}_{A1}$.

In Eq. 0.3-3, we replace $\mathbf{r}_{A1}$ by $\mathbf{r}_A + \mathbf{R}_{A1}$ and make analogous replacements for $\mathbf{r}_{A2}$, $\mathbf{r}_{B1}$, and $\mathbf{r}_{B2}$. We also replace $\mathbf{v}_{A1}$ by $\mathbf{v}_A + \mathbf{V}_{A1}$ and make similar replacements for $\mathbf{v}_{A2}$, $\mathbf{v}_{B1}$, and $\mathbf{v}_{B2}$. We treat the primed quantities similarly. We also let $m_{A1} = m_{A2} = \frac{1}{2}m_A$. With these substitutions along with Eq. 0.3-2, we then get

$$\begin{aligned}\frac{1}{2}m_A(\mathbf{v}_A + \mathbf{V}_{A1}) + \frac{1}{2}m_A(\mathbf{v}_A + \mathbf{V}_{A2}) + \frac{1}{2}m_B(\mathbf{v}_B + \mathbf{V}_{B1}) + \frac{1}{2}m_B(\mathbf{v}_B + \mathbf{V}_{B2}) = \\ \frac{1}{2}m_A(\mathbf{v}'_A + \mathbf{V}'_{A1}) + \frac{1}{2}m_A(\mathbf{v}'_A + \mathbf{V}'_{A2}) + \frac{1}{2}m_B(\mathbf{v}'_B + \mathbf{V}'_{B1}) + \frac{1}{2}m_B(\mathbf{v}'_B + \mathbf{V}'_{B2})\end{aligned} \tag{0.3-4}$$

Then using $\mathbf{V}_{A2} = -\mathbf{V}_{A1}$ and $\mathbf{V}_{B2} = -\mathbf{V}_{B1}$, and similar relations for the primed quantities, we get finally

$$m_A\mathbf{v}_A + m_B\mathbf{v}_B = m_A\mathbf{v}'_A + m_B\mathbf{v}'_B \tag{0.3-5}$$

It is important to notice that the *atomic* masses and velocities have completely disappeared in this final result, so that the law of conservation of momentum is written in terms of the *molecular* masses and velocities alone. This does not happen for energy and angular momentum, as we shall see presently.

(c) According to the *law of conservation of energy*, the energy of the colliding pair of molecules must be the same before and after the collision. The energy of an isolated molecule is the sum of the kinetic energies of the two atoms and the interatomic potential energy, ϕ_A, which describes the force of the chemical bond joining the two atoms 1 and 2 of molecule A, and is a function of the interatomic distance $|\mathbf{r}_{A2} - \mathbf{r}_{A1}|$. Therefore, energy conservation leads to

$$\begin{aligned}\left(\frac{1}{2}m_{A1}v_{A1}^2 + \frac{1}{2}m_{A2}v_{A2}^2 + \phi_A\right) + \left(\frac{1}{2}m_{B1}v_{B1}^2 + \frac{1}{2}m_{B2}v_{B2}^2 + \phi_B\right) = \\ \left(\frac{1}{2}m'_{A1}v'^2_{A1} + \frac{1}{2}m'_{A2}v'^2_{A2} + \phi'_A\right) + \left(\frac{1}{2}m'_{B1}v'^2_{B1} + \frac{1}{2}m'_{B2}v'^2_{B2} + \phi'_B\right)\end{aligned} \tag{0.3-6}$$

Note that we have used the abbreviated notation $v_{A1}^2 = (\mathbf{v}_{A1} \cdot \mathbf{v}_{A1})$.

We may now replace $\mathbf{r}_{A1}$ by $\mathbf{r}_A + \mathbf{R}_{A1}$ and $\mathbf{v}_{A1}$ by $\mathbf{v}_A + \mathbf{V}_{A1}$ as above in (b). We also let $m_{A1} = m_{A2} = \frac{1}{2}m_A$. Then Eq. 0.3-6 becomes

$$\begin{aligned}
&\frac{1}{2}\frac{1}{2}m_A\left[(\mathbf{v}_A\cdot\mathbf{v}_A) + \underline{2(\mathbf{v}_A\cdot\mathbf{V}_{A1})} + (\mathbf{V}_{A1}\cdot\mathbf{V}_{A1})\right] \\
&\quad + \frac{1}{2}\frac{1}{2}m_A\left[(\mathbf{v}_A\cdot\mathbf{v}_A) + \underline{\underline{2(\mathbf{v}_A\cdot\mathbf{V}_{A2})}} + (\mathbf{V}_{A2}\cdot\mathbf{V}_{A2})\right] + \phi_A \\
&\quad + \frac{1}{2}\frac{1}{2}m_B\left[(\mathbf{v}_B\cdot\mathbf{v}_B) + \underline{2(\mathbf{v}_B\cdot\mathbf{V}_{B1})} + (\mathbf{V}_{B1}\cdot\mathbf{V}_{B1})\right] \\
&\quad + \frac{1}{2}\frac{1}{2}m_B\left[(\mathbf{v}_B\cdot\mathbf{v}_B) + \underline{\underline{2(\mathbf{v}_B\cdot\mathbf{V}_{B2})}} + (\mathbf{V}_{B2}\cdot\mathbf{V}_{B2})\right] + \phi_B \\
&= \frac{1}{2}\frac{1}{2}m_A\left[(\mathbf{v}'_A\cdot\mathbf{v}'_A) + \underline{2(\mathbf{v}'_A\cdot\mathbf{V}'_{A1})} + (\mathbf{V}'_{A1}\cdot\mathbf{V}'_{A1})\right] \\
&\quad + \frac{1}{2}\frac{1}{2}m_A\left[(\mathbf{v}'_A\cdot\mathbf{v}'_A) + \underline{2(\mathbf{v}'_A\cdot\mathbf{V}'_{A2})} + (\mathbf{V}'_{A2}\cdot\mathbf{V}'_{A2})\right] + \phi'_A \\
&\quad + \frac{1}{2}\frac{1}{2}m_B\left[(\mathbf{v}'_B\cdot\mathbf{v}'_B) + \underline{2(\mathbf{v}'_B\cdot\mathbf{V}'_{B1})} + (\mathbf{V}'_{B1}\cdot\mathbf{V}'_{B1})\right] \\
&\quad + \frac{1}{2}\frac{1}{2}m_B\left[(\mathbf{v}'_B\cdot\mathbf{v}'_B) + \underline{\underline{2(\mathbf{v}'_B\cdot\mathbf{V}'_{B2})}} + (\mathbf{V}'_{B2}\cdot\mathbf{V}'_{B2})\right] + \phi'_B
\end{aligned} \tag{0.3-7}$$

In Eq. 0.3-7, each single-underlined term exactly cancels the doubly underlined term in the following line, because $\mathbf{V}_{A1} = -\mathbf{V}_{A2}$ and $\mathbf{V}_{B1} = -\mathbf{V}_{B2}$. Hence we get

$$\begin{aligned}
&\frac{1}{2}m_A(\mathbf{v}_A\cdot\mathbf{v}_A) + \frac{1}{2}m_{A1}(\mathbf{V}_{A1}\cdot\mathbf{V}_{A1}) + \frac{1}{2}m_{A2}(\mathbf{V}_{A2}\cdot\mathbf{V}_{A2}) + \phi_A \\
&\quad + \frac{1}{2}m_B(\mathbf{v}_B\cdot\mathbf{v}_B) + \frac{1}{2}m_{B1}(\mathbf{V}_{B1}\cdot\mathbf{V}_{B1}) + \frac{1}{2}m_{B2}(\mathbf{V}_{B2}\cdot\mathbf{V}_{B2}) + \phi_B \\
&= \frac{1}{2}m_A(\mathbf{v}'_A\cdot\mathbf{v}'_A) + \frac{1}{2}m_{A1}(\mathbf{V}'_{A1}\cdot\mathbf{V}'_{A1}) + \frac{1}{2}m_{A2}(\mathbf{V}'_{A2}\cdot\mathbf{V}'_{A2}) + \phi'_A \\
&\quad + \frac{1}{2}m_B(\mathbf{v}'_B\cdot\mathbf{v}'_B) + \frac{1}{2}m_{B1}(\mathbf{V}'_{B1}\cdot\mathbf{V}'_{B1}) + \frac{1}{2}m_{B2}(\mathbf{V}'_{B2}\cdot\mathbf{V}'_{B2}) + \phi'_B
\end{aligned} \tag{0.3-8}$$

In the first line of the equation above, the terms have the following significance: term 1 is the kinetic energy of molecule A in a fixed coordinate system; term 2 is the kinetic energy of atom $A1$ in a coordinate system fixed at the center of mass of molecule A; term 3 is the kinetic energy of atom $A2$ in a coordinate system fixed at the center of mass of molecule A; term 4 is the potential energy of molecule A, which is a function of $|\mathbf{r}_{A2} - \mathbf{r}_{A1}|$, the separation of the two atoms in molecule A. Equation 0.3-8 may be rewritten

$$\left(\frac{1}{2}m_A v_A^2 + u_A\right) + \left(\frac{1}{2}m_B v_B^2 + u_B\right) = \left(\frac{1}{2}m_A v'^2_A + u'_A\right) + \left(\frac{1}{2}m_B v'^2_B + u'_B\right) \tag{0.3-9}$$

in which the "internal energy" $u_A \equiv \frac{1}{2}m_{A1}V_{A1}^2 + \frac{1}{2}m_{A2}V_{A2}^2 + \phi_A$ is the sum of the kinetic energies of the atoms, referred to the center of mass of molecule A, and the interatomic potential energy of molecule A. It is important to note that the sum of the kinetic energies of the molecules $\frac{1}{2}m_A v_A^2 + \frac{1}{2}m_B v_B^2$ is not equal to $\frac{1}{2}m_A v'^2_A + \frac{1}{2}m_B v'^2_B$ and is hence not conserved; the same may be said of the sum of the internal energies $u_A + u_B$. Thus, it is only the sum of kinetic and internal energies that is conserved, that is, there may in general be an interchange between the kinetic and internal energies during a collision. In Chapter 11 we shall see again the interconversion among different forms of energy for flowing fluids.

This discussion of the collision between two diatomic molecules is interesting, for several reasons. It shows how the idea of "internal energy" arises in a very simple system. We encounter this concept later in §11.1 where the terms "kinetic energy" and "internal energy" are used in connection with a fluid regarded as a continuum. When the fluid is regarded as a continuum, it may be difficult to understand how one goes about splitting the energy of a fluid into kinetic energy and internal energy, and how to define the latter. In considering the collision between two diatomic molecules, however, the splitting is quite straightforward.

(d) Finally, the *law of conservation of angular momentum* may be applied to a collision to give

$$([\mathbf{r}_{A1} \times m_{A1}\mathbf{v}_{A1}] + [\mathbf{r}_{A2} \times m_{A2}\mathbf{v}_{A2}]) + ([\mathbf{r}_{B1} \times m_{B1}\mathbf{v}_{B1}] + [\mathbf{r}_{B2} \times m_{B2}\mathbf{v}_{B2}]) = ([\mathbf{r}'_{A1} \times m'_{A1}\mathbf{v}'_{A1}] + [\mathbf{r}'_{A2} \times m'_{A2}\mathbf{v}'_{A2}]) + ([\mathbf{r}'_{B1} \times m'_{B1}\mathbf{v}'_{B1}] + [\mathbf{r}'_{B2} \times m'_{B2}\mathbf{v}'_{B2}]) \quad (0.3\text{-}10)$$

in which $\times$ is used for the cross product of two vectors. Next we introduce the center of mass and relative position vectors and velocity vectors exactly as we did in parts (b) and (c) above. Without giving the intermediate steps we get

$$([\mathbf{r}_A \times m_A\mathbf{v}_A] + \mathbf{L}_A) + ([\mathbf{r}_B \times m_B\mathbf{v}_B] + \mathbf{L}_B) = ([\mathbf{r}'_A \times m_A\mathbf{v}'_A] + \mathbf{L}'_A) + ([\mathbf{r}'_B \times m_B\mathbf{v}'_B] + \mathbf{L}'_B) \quad (0.3\text{-}11)$$

in which $\mathbf{L}_A = [\mathbf{R}_{A1} \times m_{A1}\mathbf{V}_{A1}] + [\mathbf{R}_{A2} \times m_{A2}\mathbf{V}_{A2}]$ is the sum of the angular momenta of the atoms with respect to the center of mass of the molecule—that is, the "internal angular momentum."

Here again there is the possibility for interchange between the angular momentum of the molecules (with respect to the origin of coordinates) and the internal angular momentum (with respect to the centers of mass of the molecules). If, however, the **L**'s are negligibly small, then there is virtually no interchange. Exchange of angular momentum will be referred to later in Chapter 3, where we discuss the equation of change for angular momentum in fluids.

Much of this book is concerned with establishing the conservation laws for fluids, regarded as "continua," and for large pieces of equipment or parts thereof. The above discussion gives a good perspective for this adventure.

§0.4 FROM MOLECULES TO CONTINUA

We know that gases and liquids consist of molecules and that the molecular motions ultimately determine how fluids flow, how energy moves around in the fluids, and how diffusion in mixtures occurs. In the foregoing section, we have illustrated the application of the fundamental conservation laws to the collisions between diatomic molecules. However, for engineering applications, we prefer to think of fluids as being "continua." After all, when we look at a fluid, we do not see the individual molecules, but instead a smoothed-out material—a continuum—rather than something of evident corpuscular nature. In this section, we set forth some ideas that will be used throughout the rest of this book in discussions of the conservation principles.

Instead of talking about masses of atoms and molecules, we shall talk about the mass of material inside a "tiny" region, $\Delta x \Delta y \Delta z$, containing N molecules, as shown in Fig. 0.2-1(b). We may take the edges of this box to be of length of the order of 100 nm, so that it is many times larger than molecular dimensions (1 nm), but much smaller than most systems of interest (including biological cells of about 1000 nm). Then the fluid within this volume will have a *density* (for a pure fluid in which all N molecules have the same mass m_i)

$$\rho = \frac{\sum_i^N m_i}{\Delta x \Delta y \Delta z} \; [=] \; \frac{M}{L^3} \quad (0.4\text{-}1)^1$$

[1]Here we have introduced the symbol "[=]" to mean "has dimensions of," where M, L, t, and T mean *mass, length, time,* and *temperature*. The symbol "[=]" may also be used to mean "has units of" in this text.

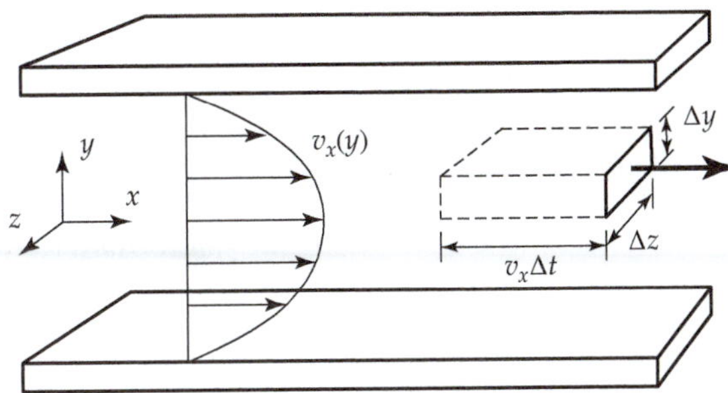

Fig. 0.4-1. Schematic diagram illustrating the volume of fluid that passes through the area element $\Delta y\Delta z$ in a time Δt.

The fluid density will in general be a function of position and time, so that we write $\rho(x,y,z,t)$. For many situations, the density may be considered to be constant, and then we speak of an "incompressible fluid."

The fluid will also have a *velocity* given by the average of the individual molecular velocities within the tiny region

$$\mathbf{v} = \frac{\sum_i^N \mathbf{v}_i}{N} \; [=] \; \frac{L}{t} \tag{0.4-2}$$

The velocity vector may be observed by putting tiny neutrally buoyant particles in the fluid and observing how they move. In this way, we can see how the velocity of the fluid changes with time at every point in the fluid, and thus obtain the functions $v_x(x,y,z,t)$, $v_y(x,y,z,t)$, and $v_z(x,y,z,t)$. We can reserve the term *speed* for the scalar $v = \sqrt{v_x^2 + v_y^2 + v_z^2}$, which is the magnitude of the velocity vector.

For the time-independent flow in Fig. 0.4-1, with just one velocity component v_x, in a time interval Δt, a volume $(v_x\Delta t)\Delta y\Delta z$ of fluid will pass through the tiny area element $\Delta y\Delta z$. Then the *volume rate of flow* through $\Delta y\Delta z$ is

$$\text{Rate of volume flow} = \frac{\text{volume}}{\text{time}} = v_x\Delta y\Delta z \; [=] \; \frac{L^3}{t} \tag{0.4-3}$$

Because the fluid is flowing, it carries with it all of its properties (mass, momentum, kinetic energy, etc.). That is, the fluid motion provides a mechanism for transport of the various quantities. We call this type of transport *convective*.

Knowing the volume rate of flow of the fluid, we are now in a position to get expressions for the rate of convective transport of various other quantities through the area element $\Delta y\Delta z$ by being swept along by the fluid. For example, since the mass density ρ is the mass per unit volume, the *convective rate of flow of mass* through $\Delta y\Delta z$ is

$$\text{Rate of mass flow} = \frac{\text{mass}}{\text{volume}} \times \frac{\text{volume}}{\text{time}} = \rho(v_x\Delta y\Delta z) \; [=] \; \frac{M}{t} \tag{0.4-4}$$

Since ρv_x is the x momentum per unit volume, the convective *rate of x momentum flow* through $\Delta y\Delta z$ is

$$\begin{aligned}\text{Rate of momentum flow} &= \frac{x\text{ momentum}}{\text{volume}} \times \frac{\text{volume}}{\text{time}} \\ &= \rho v_x(v_x\Delta y\Delta z) \; [=] \; \frac{ML}{t^2}\end{aligned} \tag{0.4-5}$$

Since $\frac{1}{2}\rho v_x^2$ is the kinetic energy per unit volume, the *convective rate of kinetic energy flow* through $\Delta y \Delta z$ is

$$\text{Rate of kinetic energy flow} = \frac{\text{kinetic energy}}{\text{volume}} \times \frac{\text{volume}}{\text{time}}$$
$$= \frac{1}{2}\rho v_x^2 (v_x \Delta y \Delta z) \; [=] \; \frac{ML^2}{t^3} \tag{0.4-6}$$

Similarly, if $\hat{U}$ is the internal energy per unit mass, then $\rho\hat{U}$ is the internal energy per unit volume, and the *convective rate of flow of internal energy* through $\Delta y \Delta z$ is

$$\text{Rate of internal energy flow} = \frac{\text{internal energy}}{\text{volume}} \times \frac{\text{volume}}{\text{time}}$$
$$= \rho\hat{U}(v_x \Delta y \Delta z) \; [=] \; \frac{ML^2}{t^3} \tag{0.4-7}$$

Thus, we have seen that the *convective rates of flow* of mass, momentum, kinetic energy, and internal energy are all evaluated in the same way.

Now we divide each of these rates of flow by the area $\Delta y \Delta z$ to get the corresponding *fluxes* (for a flow in which the fluid is flowing in the x direction only)

$$\text{Flux of mass:} \qquad \rho v_x \; [=] \; \frac{M}{L^2 t} \tag{0.4-8}$$

$$\text{Flux of momentum:} \qquad \rho v_x v_x \; [=] \; \frac{M}{L t^2} \tag{0.4-9}$$

$$\text{Flux of kinetic energy:} \qquad \frac{1}{2}\rho v_x^2 v_x \; [=] \; \frac{M}{t^3} \tag{0.4-10}$$

$$\text{Flux of internal energy:} \qquad \rho\hat{U} v_x \; [=] \; \frac{M}{t^3} \tag{0.4-11}$$

These are referred to as *convective fluxes*, inasmuch as they describe how the various entities are swept along (i.e., convected) with the fluid. In the next chapter, we describe another mechanism for transport, namely that caused by molecular motion and interactions. The corresponding *molecular fluxes*, along with the convective fluxes, will play a prominent role in the development of the various theories in this book.

In §0.3, we showed how to apply the conservation laws to a molecular collision. In Chapters 3, 11, and 19, we apply the conservation laws to a fluid, regarded as a continuum, in a tiny region $\Delta x \Delta y \Delta z$ fixed in space. This will lead to a set of partial differential equations of great generality—the *equations of change*. We will then show how these equations can be used to get the velocity distribution, the temperature distribution (the temperature being related to the internal energy), and the concentration distribution for binary mixtures. These equations contain the *transport properties*—the viscosity, the thermal conductivity, and the diffusivity—that will appear in the expressions for the fluxes (see Chapters 1, 9, and 17).

Finally, we point out in Chapters 7, 15, and 23 that the equations of change can be integrated over large systems of industrial or biological interest. These *macroscopic balances* are in essence the laws of conservation of mass, momentum, energy, and angular momentum applied to large, complex systems. Thus, the subject of this book starts with the *molecular* level (atoms and molecules), then moves to the *microscopic* level ($\Delta x \Delta y \Delta z$), and finally to the *macroscopic* level (large systems).

§0.5 CONCLUDING COMMENTS

To use the macroscopic balances intelligently, one needs information about interphase transport obtainable from the equations of change. To use the equations of change, we need the transport properties, which are described by various molecular theories.

Therefore, from a teaching point of view, it seems best to start at the molecular level and work upward toward the larger systems.

All the discussions of theory are accompanied by examples to illustrate how the theory is applied to problem solving. Then, at the end of each chapter, there are problems to provide extra experience in using the ideas given in the chapter. The problems are grouped into three classes:

Class A: Numerical problems that are designed to highlight important equations in the text and to give a feeling for the orders of magnitude.

Class B: Analytical problems that require doing elementary derivations using ideas mainly from the chapter.

Class C: Somewhat more advanced analytical problems.

Many of the problems and illustrative examples are rather elementary in that they involve oversimplified systems or very idealized models. It is, however, advisable to start with these elementary problems to understand how the theory works and to develop confidence in using it. In addition, some of these elementary examples can be very useful in making order-of-magnitude estimates in complex problems.

Here are a few suggestions for studying the subject of transport phenomena:

- Always read the text with pencil and paper in hand; work through the details of the mathematical developments and supply any missing steps.
- Whenever necessary, go back to the mathematics textbooks to brush up on calculus, differential equations, vectors, etc. This is an excellent time to review the mathematics that was learned earlier (but possibly not as carefully as it should have been).
- Make it a habit of giving a physical interpretation of key results; that is, get in the habit of relating the physical ideas to the equations.
- Always ask whether the results seem reasonable. If the results do not agree with intuition, it is important to find out which is incorrect.
- Check the dimensions of all results. This is one very good way of avoiding errors in your work.

We conclude with Table 0.5-1, which emphasizes the role of the three conservation laws (mass, momentum, and energy) at the three levels of description (molecular, microscopic, and macroscopic).

We hope that the reader will share our enthusiasm for the subject of transport phenomena. It will take some effort to learn the material, but the rewards will be worth the time and energy required.

QUESTIONS FOR DISCUSSION

1. What are the definitions of momentum, angular momentum, and kinetic energy for a single particle? What are the dimensions of these quantities?
2. What are the dimensions of velocity, angular velocity, pressure, density, force, work, and torque? What are some common units used for these quantities?
3. Verify that it is possible to go from Eq. 0.3-3 to Eq. 0.3-5.
4. Go through all the details needed to get Eq. 0.3-9 from Eq. 0.3-6.
5. Suppose that the origin of coordinates is shifted to a new position. What effect would that have on Eq. 0.3-10? Is the equation changed?
6. Compare and contrast the terms *angular velocity* and *angular momentum*.
7. What is meant by *internal energy*? *Potential energy*?
8. Is the law of conservation of mass always valid? What are the limitations?

Table 0.5-1. Summary of Conservation Laws and Systems to Which They are Applied

		CONSERVATION LAWS		
		MASS m	MOMENTUM $m\mathbf{v}$	ENERGY $\frac{1}{2}mv^2$
SYSTEM TO WHICH THE CONSERVATION LAWS ARE APPPLIED	Collision of two molecules (Fig. 0.3-1)	Conservation of mass in a binary collision (Eq. 0.3-1)	Conservation of momentum in a binary collision (Eq. 0.3-5)	Conservation of energy in a binary collision (Eq. 0.3-9)
	Region $\Delta x \Delta y \Delta z$ in a continuum (Fig. 3.1-2)	Equation of continuity (Eqs. 3.1-4 & 19.1-7)	Equation of motion (Eq. 3.2-9)	Equation of energy (Eq. 11.1-8 & Table 19.2-4)
	Piece of equipment (Fig. 7.0-1)	Macroscopic mass balance (Eqs. 7.1-3 & 23.1-1)	Macroscopic momentum balance (Eqs. 7.2-2 & 23.2-1)	Macroscopic energy balance (Eqs. 15.1-2 & 23.3-1)

Part One

Momentum Transport

Chapter 1

Viscosity and the Mechanisms of Momentum Transport

The first seven chapters of this book deal with the flow of viscous fluids of molecular weight less than about 1000 g/g-mol, such as air, water, benzene, and glycerin. Polymeric liquids of molecular weight greater than about 1000 g/g-mol, which exhibit flow properties that are quite different from the low molecular weight fluids, are discussed in Chapter 8.

Fluid dynamics involves the transport of momentum. There are two mechanisms for this transport: *convective momentum transport*, discussed in §1.1, and *molecular momentum transport*, discussed in §1.2. The combination of the two types of transport is given in §1.3. The convective transport involves just one physical property, namely the fluid density ρ. The molecular transport involves two physical properties, the viscosity μ and the dilatational viscosity κ. In §1.4, we present experimental data for viscosities of a wide variety of gases and liquids. In §1.5, we use a correlation of fluid viscosity data obtained from the principle of corresponding states to illustrate how the viscosities of gases and liquids depend on pressure and temperature, and to show how viscosity values may be estimated. In §1.6, we present the results of a rigorous molecular theory for the viscosity for gases at low density. In §1.7 and §1.8 we give a few widely used expressions for the viscosities of liquids and suspensions.

In this chapter we are primarily concerned with expressions for the *momentum flux*—the momentum flow per unit area per unit time. This quantity is needed for setting up the equations for finding the velocity distribution in flow systems, and then getting expressions for the rate of flow in terms of the system geometry and the physical properties. This will be made abundantly clear in Chapters 2 and 3.

§1.1 CONVECTIVE MOMENTUM FLUX TENSOR

The notion of convective momentum transport was introduced in §0.4. We now show how to develop the expressions for the convective momentum flux. We divide the discussion into three parts: (a) unidirectional shear flow, (b) general three-dimensional flows, and (c) flows in curvilinear coordinates.

a. Unidirectional shear flow

Let us now consider a simple flow system, namely that pictured in Fig. 1.1-1 where the fluid flows in the x direction between a pair of parallel plates. The spacing between the two planes is y_0, and v_0 is the constant speed with which the lower plate is moving to the right. As discussed in §0.4, the volume rate of flow of fluid in the x direction through an area element $\Delta y \Delta z$ perpendicular to the flow direction is $v_x \Delta y \Delta z$, and the rate of convective flow of x momentum through an element of area $\Delta y \Delta z$ is

$$\left(\frac{\text{volume of fluid}}{\text{time}}\right)\left(\frac{x\text{ momentum}}{\text{volume of fluid}}\right) = \left(\frac{x\text{ momentum}}{\text{time}}\right) = (v_x \Delta y \Delta z)(\rho v_x) \tag{1.1-1}$$

If we now divide through by $\Delta y \Delta z$, we get the *convective flux of x momentum* in the x direction

$$\pi_{xx}^{(c)} = \rho v_x v_x \tag{1.1-2}$$

which has dimensions of momentum/area · time. Throughout this book, the flux of any entity will have dimensions of entity/area · time.

b. General three-dimensional time-dependent flow

We will generally be interested in a flow system that is much more complicated than that in Fig. 1.1-1. Consider a general flow in three-dimensional space with three velocity components that depend on all three spatial coordinates as well as the time, $v_x(x,y,z,t)$, $v_y(x,y,z,t)$, and $v_z(x,y,z,t)$, such as that shown in Fig. 1.1-2. The momentum flux components can be deduced by generalizing the discussion that led to Eq. 1.1-2.

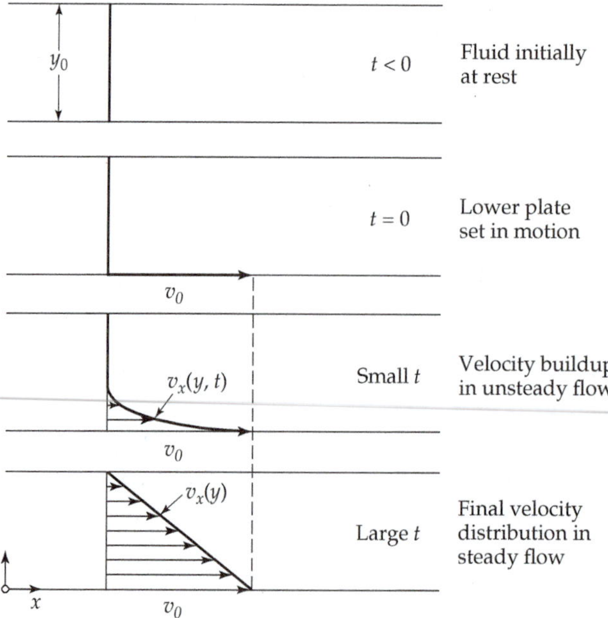

Fig. 1.1-1. The buildup to the steady, laminar velocity profile for a fluid contained between two plates. The flow is called "laminar" because the adjacent layers of fluid ("laminae") slide past one another in an orderly fashion.

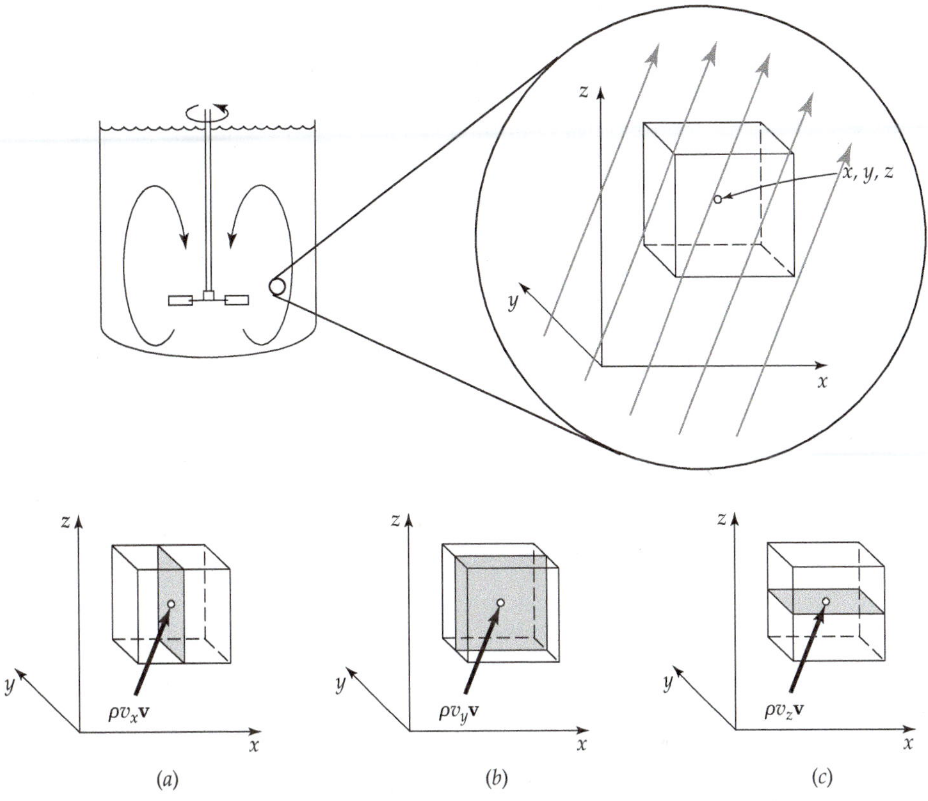

Fig. 1.1-2. The convective momentum fluxes through planes of unit area perpendicular to the coordinate directions.

Somewhere in the midst of this flow, we can imagine a tiny volume element $\Delta x \Delta y \Delta z$, fixed in space, through which the fluid is flowing. For a sufficiently small volume element, the velocity vector $\mathbf{v}$ will be approximately constant throughout. Then across the plane perpendicular to the x direction (Fig. 1.1-2(*a*)), there will be convective fluxes of x momentum, y momentum, and z momentum given by the three components of $v_x(\rho\mathbf{v})$

$$\pi_{xx}^{(c)} = \rho v_x v_x \qquad \pi_{xy}^{(c)} = \rho v_x v_y \qquad \pi_{xz}^{(c)} = \rho v_x v_z \tag{1.1-3}$$

Similarly across the plane perpendicular to the y direction (Fig. 1.1-2(*b*)), there will be the three components of $v_y(\rho\mathbf{v})$

$$\pi_{yx}^{(c)} = \rho v_y v_x \qquad \pi_{yy}^{(c)} = \rho v_y v_y \qquad \pi_{yz}^{(c)} = \rho v_y v_z \tag{1.1-4}$$

and across the plane perpendicular to the z direction (Fig. 1.1-2(*c*)), there will be the three components of $v_z(\rho\mathbf{v})$

$$\pi_{zx}^{(c)} = \rho v_z v_x \qquad \pi_{zy}^{(c)} = \rho v_z v_y \qquad \pi_{zz}^{(c)} = \rho v_z v_z \tag{1.1-5}$$

Note that the first subscript on $\pi_{ij}^{(c)}$ and $\rho v_i v_j$ gives the direction of transport and the second gives the component of the momentum being transported. Quantities with components such as $\rho v_i v_j$, which have two subscripts associated with the coordinate directions, are referred to as "tensors," just as quantities (such as velocity) that have one subscript associated with the coordinate directions are called "vectors." A discussion of vectors and tensors can be found in Appendix A.

Table 1.1-1. Summary of the Convective Momentum Flux Components

Direction normal to shaded surface	Flux of momentum through the shaded surface	Convective momentum flux components		
		x component	y component	z component
x	$\rho v_x \mathbf{v}$	$\rho v_x v_x$	$\rho v_x v_y$	$\rho v_x v_z$
y	$\rho v_y \mathbf{v}$	$\rho v_y v_x$	$\rho v_y v_y$	$\rho v_y v_z$
z	$\rho v_z \mathbf{v}$	$\rho v_z v_x$	$\rho v_z v_y$	$\rho v_z v_z$

The set of nine equations for $\pi_{ij}^{(c)}$ may be written symbolically as

$$\boldsymbol{\pi}^{(c)} = \rho\mathbf{v}\mathbf{v} \tag{1.1-6}$$

where $\boldsymbol{\pi}^{(c)}$ is the convective momentum flux tensor (with nine components $\pi_{xx}^{(c)}$, $\pi_{xy}^{(c)}$, etc.), and $\mathbf{vv}$ is the dyadic product of the vector velocity with itself (with nine components v_xv_x, v_xv_y, etc.). These convective tensor components are summarized in Table. 1.1-1. Note that whereas vectors, such as $\mathbf{v}$, have three components, one corresponding to each coordinate direction, (second order) tensors and dyadic products, such as $\boldsymbol{\pi}^{(c)}$ and $\mathbf{vv}$, have nine components, corresponding to all possible ordered pairs of coordinate directions.

c. Flows in curvilinear coordinates

In curvilinear coordinates (e.g., cylindrical and spherical coordinates), the convective momentum flux is still defined by Eq. 1.1-6, or

$$\pi_{ij}^{(c)} = \rho v_i v_j \tag{1.1-7}$$

where, $\pi_{ij}^{(c)}$ is again interpreted as the convective flux of j momentum transported in the positive i direction, except that now i and j take on the values r,θ,z (cylindrical coordinates) or r,θ,ϕ (spherical coordinates). Thus, the components of $\boldsymbol{\pi}^{(c)}$ in cylindrical coordinates can be obtained by replacing x,y,z by r,θ,z in Table 1.1-1 (i.e., $\pi_{rr}^{(c)} = \rho v_r v_r$, $\pi_{r\theta}^{(c)} = \rho v_r v_\theta$, $\pi_{rz}^{(c)} = \rho v_r v_z$, etc.); similarly, the components of $\boldsymbol{\pi}^{(c)}$ in spherical coordinates can be obtained by replacing x,y,z by r,θ,ϕ in Table 1.1-1 (i.e., $\pi_{rr}^{(c)} = \rho v_r v_r$, $\pi_{r\theta}^{(c)} = \rho v_r v_\theta$, $\pi_{r\phi}^{(c)} = \rho v_r v_\phi$, etc.). (See Fig. A.6-1 for definitions of curvilinear coordinates.)

§1.2 MOLECULAR MOMENTUM FLUX TENSOR—NEWTON'S LAW

In the foregoing section we discussed how momentum is transported by being swept along by the bulk movement of the fluid. But there is another mechanism by which momentum may be transported, namely, by the motion, interactions, and collisions of the fluid molecules. As before, we develop the expressions for the *molecular momentum flux*, by considering first (a) unidirectional shear flow, then (b) general three-dimensional flows, and finally (c) flows in curvilinear coordinates.

a. Unidirectional shear flow

Once again we consider the simple flow of Fig. 1.1-1, in which the fluid—either a gas or a liquid—is located between a pair of parallel plates, each of area A separated by a distance y_0. The system is initially at rest. At time $t = 0$ the lower plate is set in motion in the positive x direction with a constant speed v_0. As time increases, the fluid near the lower plate gains momentum, and then successive layers of fluid in the y direction gain momentum, and ultimately the linear steady-state velocity profile shown in the figure is established. We require that the flow be laminar ("laminar" flow is the orderly type of flow that one usually observes when honey is poured, in contrast to the highly irregular "turbulent" flow that one sees in a high-speed mixer). When the final state of steady motion has been attained, a constant force in the x direction of magnitude F is required to maintain the motion of the lower plate. This force may be expressed as

$$\frac{F}{A} = \mu \frac{v_0}{y_0} \tag{1.2-1}$$

That is, the force is proportional to the area of the plates and the speed of the lower plate, and inversely proportional to the distance between the plates. The constant of proportionality μ is a property of the fluid, called the *viscosity*.

We now rewrite Eq. 1.2-1 in the notation that will be used throughout the book. First, we replace F/A by the symbol τ_{yx}, which is the force in the x direction acting on a unit area perpendicular to the y direction. It is understood that this is the force exerted by the fluid of lesser y on the fluid of greater y. Furthermore, we replace v_0/y_0 by $-dv_x/dy$ (note that $-dv_x/dy > 0$ in Fig. 1.1-1). Then Eq. 1.2-1 becomes

$$\tau_{yx} = -\mu \frac{dv_x}{dy} \tag{1.2-2}$$[1]

This equation, which states that the shearing force per unit area, or *shear stress*, is proportional to the negative of the velocity gradient, is often called *Newton's law of viscosity*.[2] Newton suggested this "law" as an empiricism—the simplest proposal that could be made for relating the shear stress and the velocity gradient. This simple linear relation has proven to be very useful for gases and low molecular weight liquids, and these fluids are referred to as *Newtonian fluids*. Many polymeric liquids, particulate suspensions, and other complex fluids are not described by Eq. 1.2-2 and are referred to as *non-Newtonian fluids*. These fluids are discussed in Chapter 8.

The quantity τ_{yx} may be interpreted in another way. In the neighborhood of the moving solid surface at $y = 0$, the fluid acquires a certain amount of x momentum. This fluid, in turn, imparts momentum to the adjacent layer of fluid through molecular motion, interactions, and collisions, causing this adjacent layer to remain in motion in the x direction. Hence, x momentum is being transmitted in the positive y direction via these molecular interactions. The term τ_{yx} can also be interpreted as the *flux of x momentum in the positive y direction* (that is, from the region of smaller y values, to the region of larger y values), where the term *flux* means "rate of transport per unit area." This picture is consistent with the molecular picture of momentum transport and the kinetic theory of gases and liquids. It is also in harmony with the analogous treatment given later for energy and mass transport. The kinetic theory of gas viscosity is discussed in §1.6.

---

[1]Some authors write Eq. 1.2-2 in the form

$$g_c \tau_{yx} = -\mu \frac{dv_x}{dy} \tag{1.2-2a}$$

in which g_c is the "gravitational conversion factor." In the English Engineering System, τ_{yx} [=] lb_f/ft^2, v_x [=] ft/s, y [=] ft, and μ [=] $\text{lb}_m/\text{ft}\cdot\text{s}$, and thus $g_c = 32.174\ \text{ft}\cdot\text{lb}_m/\text{lb}_f\cdot\text{s}^2$. In this book we will always use Eq. 1.2-2 rather than Eq. 1.2-2a.

[2]Sir **Isaac Newton** (1643–1727), a professor at Cambridge University and later Master of the Mint, was the founder of classical mechanics and contributed to other fields of physics as well. Actually Eq. 1.1-2 does not appear in Sir Isaac Newton's *Philosophiae Naturalis Principia Mathematica* (1687), but the germ of the idea is there. For illuminating comments, see D. J. Acheson, *Elementary Fluid Dynamics*, Oxford University Press, 1990, §6.1.

Table 1.2-1. Summary of Units in the SI, c.g.s., and EE Systems for Quantities Related to Eq. 1.2-2

	SI	c.g.s.	EE
τ_{yx}	Pa	dyn/cm^2	lb$_f$/ft^2
v_x	m/s	cm/s	ft/s
y	M	cm	ft
μ	Pa · s	g/cm · s = poise	lb$_m$/ft · s
ν	m^2/s	cm^2/s	ft^2/s

Note: The pascal, Pa, is the same as N/m^2, and the newton, N, is the same as kg/m · s^2. The abbreviation of "centipoise" is "cp."

The dual interpretation of τ_{yx} in Eq. 1.2-1 as a force per unit area (shear stress) and as a momentum flux—both caused by molecular forces during shear flow—leads to two interchangeable names for this quantity. We will call this quantity the *viscous shear stress* as well as the *viscous momentum flux*. Both interpretations are useful and will be discussed in more detail below.

Equation 1.2-2 may also be interpreted as stating that momentum naturally flows "downhill" from a region of high velocity to a region of low velocity—just as a sled goes downhill from a region of high elevation to a region of low elevation, or the way heat flows from a region of high temperature to a region of low temperature. The velocity gradient can therefore be thought of as a "driving force" for molecular momentum transport.

Often fluid dynamicists use the symbol ν to represent the viscosity divided by the density (mass per unit volume) of the fluid, thus:

$$\nu = \mu/\rho \tag{1.2-3}$$

This quantity is called the *kinematic viscosity*.

Next we make a few comments about the units of the quantities we have defined. If we use the symbol "[=]" to mean "has units of," then in the SI system,[3] τ_{yx} [=] N/m^2 = Pa, v_x [=] m/s, and y [=] m, so that

$$\mu = -\tau_{yx}\left(\frac{dv_x}{dy}\right)^{-1} [=] \text{Pa}\left(\frac{\text{m/s}}{\text{m}}\right)^{-1} = \text{Pa} \cdot \text{s} \tag{1.2-4}$$

We summarize the above and also give the units for the c.g.s. system and the English Engineering (EE) system in Table 1.2-1. The conversion tables in Appendix E will prove to be very useful for solving numerical problems involving diverse systems of units.

EXAMPLE 1.2-1

Calculation of Momentum Flux

Compute the steady-state momentum flux τ_{yx} in lb$_f$/ft^2 when the lower plate velocity v_0 in Fig. 1.1-1(*d*) is 1 ft/s in the positive x direction, the plate separation y_0 is 0.001 ft, and the fluid viscosity μ is 0.7 cp.

SOLUTION

Since τ_{yx} is desired in English Engineering units, we should convert the viscosity into that system of units. Thus, making use of Table E.3-4, we find μ = (0.7 cp)(2.0886 × 10^{-5} lb$_f$ · s/ft^2 · cp) =

[3]SI (Système Internationale d'Unités) has seven basic units: meter (m), kilogram (kg), second (s), ampere (A), kelvin (K), mole (mol), candela (cd). See R. J. Silbey, R. A. Alberty, and M. G. Bawendi, *Physical Chemistry*, 4th edition, Wiley, New York (2005), Appendix A.

1.46×10^{-5} $\text{lb}_f \cdot \text{s/ft}^2$. The velocity profile is linear so that

$$\frac{dv_x}{dy} = -\frac{v_0}{y_0} = -\frac{10 \text{ ft/s}}{0.001 \text{ ft}} = -1000 \text{ s}^{-1} \tag{1.2-5}$$

Substitution into Eq. 1.2-2 gives

$$\tau_{yx} = -\mu\frac{dv_x}{dy} = -\left(1.46 \times 10^{-5}\frac{\text{lb}_f \cdot \text{s}}{\text{ft}^2}\right)\left(-1000\ \frac{1}{\text{s}}\right) = 1.46 \times 10^{-2}\frac{\text{lb}_f}{\text{ft}^2} \tag{1.2-6}$$

---

b. General three-dimensional, time-dependent flow

Next we consider an arbitrary flow in which all three velocity components $v_x(x,y,z,t)$, $v_y(x,y,z,t)$, and $v_z(x,y,z,t)$ may be nonzero, and may depend on all three spatial coordinates as well as time. An example of such a flow is illustrated in Fig. 1.2-1. We ask how to relate the various viscous stresses and momentum fluxes to the velocity gradients. That is, we need to generalize Eq. 1.2-2 to arbitrary flows. This generalization of Newton's law of viscosity is associated with the names of Navier, Poisson, and Stokes—all eminent mathematicians and scientists.[4] The development took a century and a half, so the beginner need not feel that this generalization is somehow obvious—it isn't. Fortunately, it is much simpler to use the stress and momentum flux components than it is to derive them.

Consider the tiny volume element $\Delta x\Delta y\Delta z$ within a flow field, as illustrated in Fig. 1.2-1. The center of the volume element is at the position x,y,z. We can slice the volume element in such a way as to remove half of the fluid within it. As shown in the figure, we can cut the volume perpendicular to each of the three coordinate directions in turn. We can then ask what force has to be applied on the shaded surface in order to replace the force that had been exerted by the fluid that was removed. There will be two contributions to the force: the viscous forces caused by the flow, and the pressure force that remains even when the fluid is not moving.

In general, the viscous forces are neither perpendicular nor parallel to the surface element, but rather act at some angle to the surface, as illustrated in Fig. 1.2-1. In Fig. 1.2-1(*a*) we see a force per unit area $\boldsymbol{\tau}_x$ exerted on the shaded area, and in (*b*) and (*c*) we see forces per unit area $\boldsymbol{\tau}_y$ and $\boldsymbol{\tau}_z$. Each of these forces (which are vectors) has components (scalars); for example, $\boldsymbol{\tau}_x$ has components τ_{xx}, τ_{xy}, and τ_{xz}, which denote the x, y, and z components of the viscous force per unit area acting on an area element perpendicular to the x direction.

The question now is: How are the stresses (or momentum fluxes) τ_{ij} related to the velocity gradients in the fluid? In generalizing Eq. 1.2-2, it is customary to require that

- **i.** the τ_{ij} do not depend explicitly on time
- **ii.** the τ_{ij} are not affected by pure rotation of the fluid
- **iii.** the fluid is isotropic
- **iv.** in the steady shear flow system of Fig. 1.1-1, the general expression for τ_{ij} will simplify to Eq. 1.2-2

---

[4]C.-L.-M.-H. Navier, *Ann. Chimie*, **19**, 244–260 (1821); S.-D. Poisson, *J. École Polytech.*, **13**, Cahier 20, 1–174 (1831); G. G. Stokes, *Trans. Camb. Phil. Soc.*, **8**, 287–305 (1845). **Claude-Louis-Marie-Henri Navier** (1785–1836) (pronounced "Nah-vyay," with the second syllable accented) was a civil engineer whose specialty was road and bridge building; **George Gabriel Stokes** (1819–1903) taught at Cambridge University and was president of the Royal Society. Navier and Stokes are well known because of the Navier–Stokes equations (see Chapter 3). See also D. J. Acheson, *Elementary Fluid Mechanics*, Oxford University Press (1990), pp. 209–212, 218, and L. D. Landau and E. M. Lifshitz, *Fluid Mechanics*, 2nd edition, Pergamon Press, Oxford (1987), pp. 44–46. **Lev Davydovich Landau** (1908–1968) received the Nobel Prize in 1962 for his work on liquid helium and superfluid dynamics.

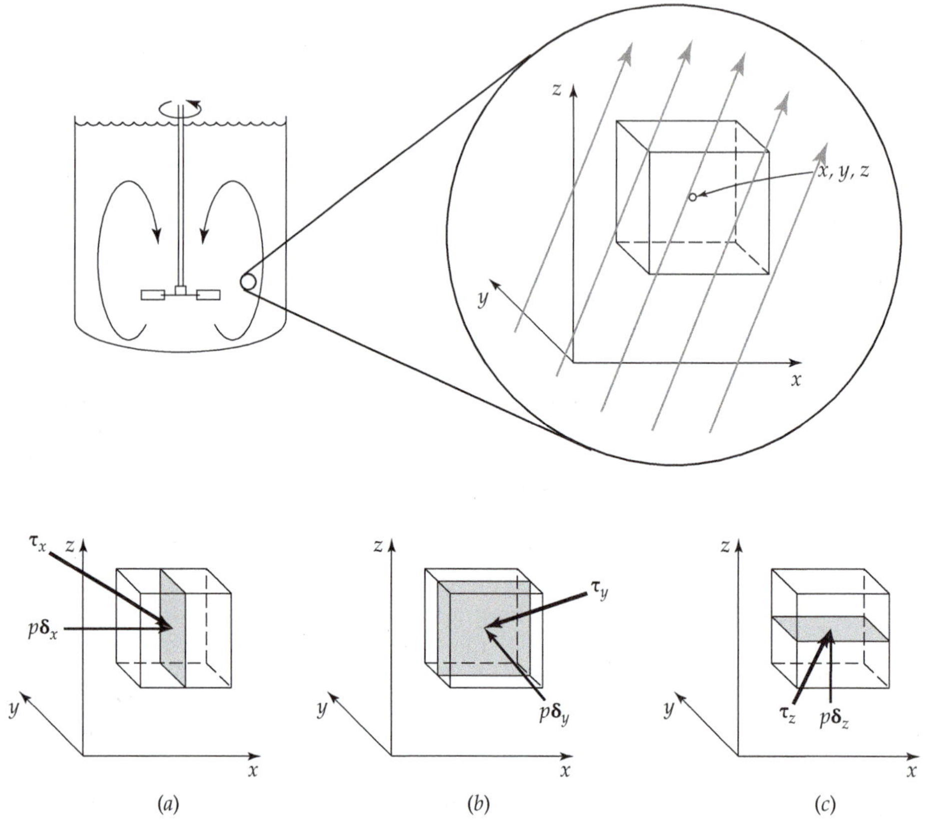

Fig. 1.2-1. Pressure and viscous forces acting on planes in the fluid perpendicular to the three coordinate directions.

Then it may be shown that the most general form for τ_{ij} in Cartesian coordinates must be

$$\tau_{xx} = -2\mu\frac{\partial v_x}{\partial x} + \left(\frac{2}{3}\mu - \kappa\right)\left(\frac{\partial v_x}{\partial x} + \frac{\partial v_y}{\partial y} + \frac{\partial v_z}{\partial z}\right) \tag{1.2-7}$$

$$\tau_{yy} = -2\mu\frac{\partial v_y}{\partial y} + \left(\frac{2}{3}\mu - \kappa\right)\left(\frac{\partial v_x}{\partial x} + \frac{\partial v_y}{\partial y} + \frac{\partial v_z}{\partial z}\right) \tag{1.2-8}$$

$$\tau_{zz} = -2\mu\frac{\partial v_z}{\partial z} + \left(\frac{2}{3}\mu - \kappa\right)\left(\frac{\partial v_x}{\partial x} + \frac{\partial v_y}{\partial y} + \frac{\partial v_z}{\partial z}\right) \tag{1.2-9}$$

$$\tau_{xy} = \tau_{yx} = -\mu\left(\frac{\partial v_y}{\partial x} + \frac{\partial v_x}{\partial y}\right) \tag{1.2-10}$$

$$\tau_{yz} = \tau_{zy} = -\mu\left(\frac{\partial v_z}{\partial y} + \frac{\partial v_y}{\partial z}\right) \tag{1.2-11}$$

$$\tau_{zx} = \tau_{xz} = -\mu\left(\frac{\partial v_x}{\partial z} + \frac{\partial v_z}{\partial x}\right) \tag{1.2-12}$$

These expressions contain two material properties: the *viscosity* μ,[5] and the *dilatational viscosity* κ. Usually, in solving fluid dynamics problems, κ can be omitted. For monatomic

---

[5] Some writers refer to μ as the *shear viscosity*, but this is inappropriate nomenclature as μ applies to nonshearing flows as well as shearing flows. The term *dynamic viscosity* is also occasionally seen, but this term has a specific meaning in the field of viscoelasticity and is an inappropriate term for μ.

gases at low density, molecular theory gives $\kappa = 0$; and for liquids that can be assumed to be incompressible, the collection of derivatives following $\left(\frac{2}{3}\mu - \kappa\right)$ is exactly zero (see §3.1). For these cases, κ does not influence the viscous stress. It is, however, important in describing sound absorption in polyatomic gases and in understanding the fluid dynamics of liquids containing gas bubbles.

As we encountered before in §1.1, quantities with components such as τ_{ij}, which have two subscripts associated with the coordinate directions, are referred to as "tensors," just as quantities (such as velocity) that have one subscript associated with the coordinate directions are called "vectors." Therefore, we will refer to $\boldsymbol{\tau}$ as the *viscous stress tensor* or the *viscous momentum flux tensor* (with components τ_{ij}). When there is no chance for confusion, the modifier "viscous" may be omitted. A discussion of vectors and tensors can be found in Appendix A.

When one looks at the components of $\boldsymbol{\tau}$ given by Eqs. 1.2-7 through 1.2-12, it is clear why people working in this field like to use a more compact notation. The viscous stress (momentum flux) tensor can be written in vector-tensor notation as

$$\boldsymbol{\tau} = -\mu \left[\nabla \mathbf{v} + (\nabla \mathbf{v})^{\dagger}\right] + \left(\frac{2}{3}\mu - \kappa\right)(\nabla \cdot \mathbf{v})\boldsymbol{\delta} \tag{1.2-13}$$

Here $\nabla \mathbf{v}$ is called the *velocity gradient tensor*, $(\nabla \mathbf{v})^{\dagger}$ is the *transpose* of the velocity gradient tensor, $\boldsymbol{\delta}$ is the *unit tensor*, and $(\nabla \cdot \mathbf{v})$ is the *divergence* of the vector $\mathbf{v}$, a scalar quantity. The unit tensor has components δ_{ij}, where δ_{ij} is the *Kronecker delta*, which is 1 if $i = j$ and 0 if $i \neq j$.

When there is no flow ($\mathbf{v} = \mathbf{0}$), the viscous stresses (momentum fluxes) τ_{ij} are all zero. The pressure force, however, is present even if the fluid is stationary. The pressure force is always perpendicular to an exposed surface, as illustrated in Fig. 1.2-1. Hence, in Fig. 1.2-1(*a*), the pressure force per unit area on the shaded surface will be a vector $p\boldsymbol{\delta}_x$, that is, the pressure (a scalar) multiplied by the unit vector $\boldsymbol{\delta}_x$ in the x direction. Similarly the pressure force per unit area on the shaded surface in (*b*) will be $p\boldsymbol{\delta}_y$, and in (*c*) the pressure force per unit area will be $p\boldsymbol{\delta}_z$.

Sometimes we will find it convenient to have a symbol that includes both types of stresses, and so we define the *molecular stresses* (or *molecular momentum fluxes)* as

$$\pi_{ij} = p\delta_{ij} + \tau_{ij} \quad \text{where } i \text{ and } j \text{ may be } x, y, \text{ or } z \tag{1.2-14}$$

Here the *Kronecker delta* δ_{ij} is used to indicate that the pressure is normal to surfaces. The term "molecular stress" (or "molecular momentum flux") is adopted because both p and τ_{ij} arise from molecular motion, interactions, and collisions. The *molecular stress (momentum flux) tensor* is written in vector-tensor notation as

$$\boldsymbol{\pi} = p\boldsymbol{\delta} + \boldsymbol{\tau} \tag{1.2-15}$$

As with the viscous stress (momentum flux) τ_{yx} in §1.2, the components π_{ij} may be interpreted in two ways:

$\pi_{ij} =$ total molecular (viscous plus pressure) force in the j direction per unit area perpendicular to the i direction, exerted *by* the fluid at lesser i *on* the fluid at greater i, and

$\pi_{ij} =$ total molecular (viscous plus pressure) flux of j momentum in the positive i direction—that is, from the region of lesser i to that of greater i.

Both interpretations are used in this book; the first one is particularly useful in describing the forces exerted by the fluid on solid surfaces. The stresses $\pi_{xx} = p + \tau_{xx}$, $\pi_{yy} = p + \tau_{yy}$, and $\pi_{zz} = p + \tau_{zz}$ are called *normal stresses*, whereas the remaining quantities, $\pi_{xy} = \tau_{xy}$, $\pi_{yz} = \tau_{yz}$, $\cdots$ are called *shear stresses*. The stresses acting on the various shaded surfaces in Fig. 1.2-1 are summarized in Table 1.2-2.

Table 1.2-2. Summary of the Components of the Molecular Stress Tensor (or Molecular Momentum Flux Tensor)

Direction normal to the shaded area	Vector force per unit area on the shaded face (momentum flux through the shaded face)	Components of the forces (per unit area) acting on the shaded face (components of the momentum flux through the shaded face)		
		x component	y component	z component
x	$\boldsymbol{\pi}_x = p\boldsymbol{\delta}_x + \boldsymbol{\tau}_x$	$\pi_{xx} = p + \tau_{xx}$	$\pi_{xy} = \tau_{xy}$	$\pi_{xz} = \tau_{xz}$
y	$\boldsymbol{\pi}_y = p\boldsymbol{\delta}_y + \boldsymbol{\tau}_y$	$\pi_{yx} = \tau_{yx}$	$\pi_{yy} = p + \tau_{yy}$	$\pi_{yz} = \tau_{yz}$
z	$\boldsymbol{\pi}_z = p\boldsymbol{\delta}_z + \boldsymbol{\tau}_z$	$\pi_{zx} = \tau_{zx}$	$\pi_{zy} = \tau_{zy}$	$\pi_{zz} = p + \tau_{zz}$

c. *Flows in curvilinear coordinates*

In curvilinear coordinates (e.g., cylindrical and spherical coordinates), the stresses and momentum fluxes retain their definitions. That is, π_{ij} can still be interpreted as the molecular force in the j direction per unit area perpendicular to the i direction, exerted by the fluid at lesser i on the fluid at greater i, except that now i and j take on the values r,θ,z (cylindrical coordinates) or r,θ,ϕ (spherical coordinates). Thus, the components of $\boldsymbol{\pi}$ in cylindrical coordinates can be obtained by replacing x,y,z with r,θ,z in Table 1.2-2 (i.e., $\pi_{rr} = p + \tau_{rr}$, $\pi_{r\theta} = \tau_{r\theta}$, etc.); similarly, the components of $\boldsymbol{\pi}$ in spherical coordinates can be obtained by replacing x,y,z with r,θ,ϕ in Table 1.2-2. Figure 1.2-2 illustrates some typical surface elements and differential forces in terms of stress tensor components that arise in fluid dynamics.

Some caution should be used, however, when expressing the components of the viscous stress (momentum flux) tensor τ_{ij} in terms of the velocity gradients. While the symbolic form of Newton's law of viscosity in Eq. 1.2-13 still applies, the components expressed in Eqs. 1.2-7 through 1.2-12 are valid only for Cartesian coordinates and cannot be converted to curvilinear coordinates by simply replacing indices. Therefore, Eq. 1.2-13 is written out in full in Cartesian (x,y,z), cylindrical (r,θ,z), and spherical (r,θ,ϕ) coordinates in Appendix B.1. We recommend that the beginning students not concern themselves with the details of the derivation of Eq. 1.2-13, but rather concentrate on using the tabulated results. Chapters 2 and 3 will give ample practice in doing this.

The shear stresses are usually easy to visualize, but the normal stresses may cause conceptual problems. For example, τ_{zz} is a force per unit area in the z direction on a plane perpendicular to the z direction. For the flow of an incompressible fluid in the convergent channel of Fig. 1.2-3 we know intuitively that v_z increases with decreasing z; hence, according to Eq. 1.2-2, there is a nonzero stress $\tau_{zz} = -2\mu(\partial v_z/\partial z)$ acting in the fluid.

Note on the Sign Convention for the Stress Tensor. We have emphasized in connection with Eq. 1.2-2 (and in the generalization in this section) that τ_{yx} is the force in the positive x direction on a plane perpendicular to the y direction, and that this is the force exerted by the fluid in the region of the *lesser* y on the fluid of *greater* y. In most fluid dynamics and elasticity books, the words "lesser" and "greater" are interchanged and Eq. 1.2-2 is written as $\tau_{yx} = +\mu(dv_x/dy)$. The advantages of the sign convention used in this book are: (a) the sign convention used in Newton's law of viscosity is consistent with that used in Fourier's law of heat conduction (Chapter 9) and Fick's law of diffusion (Chapter 17); (b) the sign convention for τ_{ij} is the same as that for the convective momentum flux

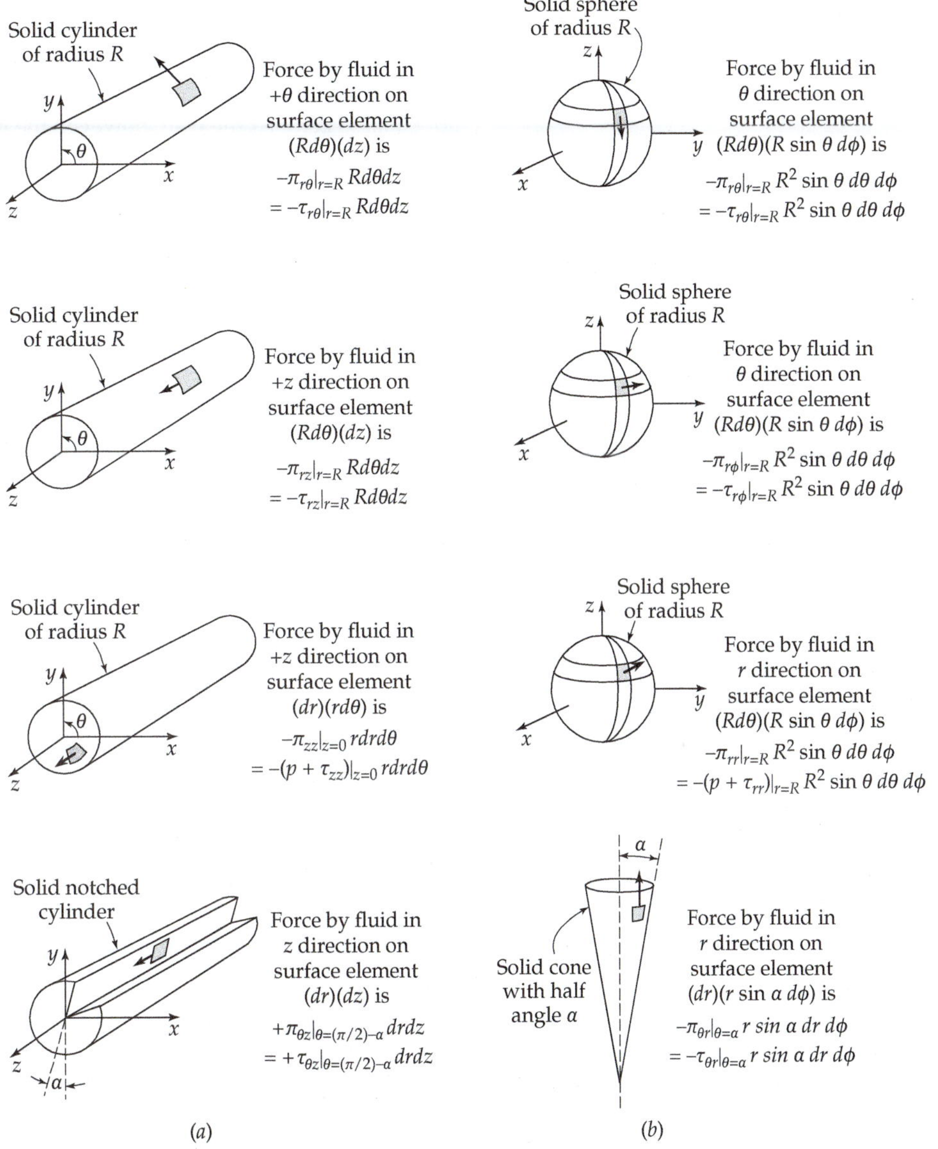

Fig. 1.2-2. (*a*) Some typical surface elements and differential forces expressed in terms of stress-tensor components in the cylindrical coordinate system. (*b*) Some typical surface elements and differential forces expressed in terms of stress-tensor components in the spherical coordinate system.

$\rho\mathbf{vv}$ (see §1.1); (c) in Eq. 1.2-14, the terms $p\delta_{ij}$ and τ_{ij} have the same sign affixed, and the terms p and τ_{ii} are both positive in compression (in accordance with common usage in thermodynamics); (d) all terms in the entropy production have the same sign (see Chapter 24). Clearly the sign convention in Eqs. 1.2-2 and 1.2-13 is arbitrary, and either sign convention can be used, provided that the physical meaning of the sign convention is clearly understood.

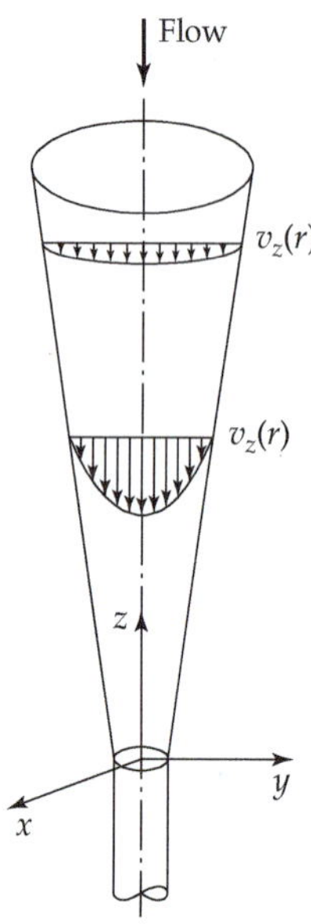

Fig. 1.2-3. The flow in a converging duct is an example of a situation in which the normal stresses are not zero. Since v_z is a function of r and z, the normal-stress component $\tau_{zz} = -2\mu(\partial v_z/\partial z)$ is nonzero. Also, since v_r depends on r and z, the normal-stress component $\tau_{rr} = -2\mu(\partial v_r/\partial r)$ is not equal to zero. At the wall, however, the normal stresses all vanish for fluids described by Eq. 1.2-13 provided that the density is constant (see Example 3.1-1).

§1.3 TOTAL MOMENTUM FLUX TENSOR

Thus far, we have defined the *convective momentum flux* in §1.1 and the *molecular momentum flux* in §1.2. In setting up shell momentum balances in Chapter 2, and in establishing the equation of motion (a general momentum balance) in Chapter 3, we will find it useful to define the total momentum flux as the sum of the convective and molecular momentum fluxes.

a. Cartesian coordinates

In Cartesian coordinates, the components of the *total momentum flux* is written as the sum of the convective and molecular momentum flux components as

$$\phi_{ij} = \rho v_i v_j + \pi_{ij} = \rho v_i v_j + p\delta_{ij} + \tau_{ij} \tag{1.3-1}$$

Here, ϕ_{ij} is the total flux of j momentum transported in the positive i direction (from the region of lesser i to that of greater i). All nine components of ϕ_{ij} in Cartesian coordinates are tabulated in Table 1.3-1. In vector-tensor notation, the *total momentum flux tensor* can be written

$$\boldsymbol{\phi} = \rho \mathbf{v}\mathbf{v} + \boldsymbol{\pi} = \rho \mathbf{v}\mathbf{v} + p\boldsymbol{\delta} + \boldsymbol{\tau} \tag{1.3-2}$$

Keep in mind that the contribution $p\boldsymbol{\delta}$ contains no velocity, only the pressure; the combination $\rho\mathbf{v}\mathbf{v}$ contains the density and products of the velocity components; and the contribution $\boldsymbol{\tau}$, for Newtonian fluids, contains the viscosities μ and κ and is linear in the velocity gradients.

b. Curvilinear coordinates

In curvilinear coordinates (e.g., cylindrical and spherical coordinates), the total momentum flux components are still defined by Eq. 1.3-1. That is, ϕ_{ij} can still be interpreted as the total flux of j momentum transported in the positive i direction, except that now i and j take on the values r,θ,z (cylindrical coordinates) or r,θ,ϕ (spherical coordinates). Thus, the components of $\boldsymbol{\phi}$ in cylindrical coordinates can be obtained by replacing x,y,z with r,θ,z in Table 1.3-1 (i.e., $\phi_{rr} = \rho v_r v_r + p + \tau_{rr}$, $\phi_{r\theta} = \rho v_r v_\theta + \tau_{r\theta}$, etc.);

Table 1.3-1. Components of the Total Momentum Flux Tensor $\boldsymbol{\phi}$ for Cartesian Coordinates

Direction momentum is being transported	Component of momentum being transported		
	x component	y component	z component
x	$\phi_{xx} = \rho v_x v_x + p + \tau_{xx}$	$\phi_{xy} = \rho v_x v_y + \tau_{xy}$	$\phi_{xz} = \rho v_x v_z + \tau_{xz}$
y	$\phi_{yx} = \rho v_y v_x + \tau_{yx}$	$\phi_{yy} = \rho v_y v_y + p + \tau_{yy}$	$\phi_{yz} = \rho v_y v_z + \tau_{yz}$
z	$\phi_{zx} = \rho v_z v_x + \tau_{zx}$	$\phi_{zy} = \rho v_z v_y + \tau_{zy}$	$\phi_{zz} = \rho v_z v_z + p + \tau_{zz}$

Note: The components of τ_{ij} are tabulated in Appendix B.1 for the various coordinate systems.

Table 1.3-2. Summary of Notation for the Momentum Fluxes

Symbol	Meaning	Reference
$\boldsymbol{\pi}^{(c)} = \rho\mathbf{vv}$	Convective momentum flux tensor	Eq. 1.1-6
$\boldsymbol{\tau}$	Viscous momentum flux tensor[a]	Eq. 1.2-13
$\boldsymbol{\pi} = p\boldsymbol{\delta} + \boldsymbol{\tau}$	Molecular momentum flux tensor[b]	Eq. 1.2-15
$\boldsymbol{\phi} = \rho\mathbf{vv} + p\boldsymbol{\delta} + \boldsymbol{\tau}$	Total momentum flux tensor	Eq. 1.3-2

[a]For viscoelastic fluids (Chapter 8), this should be called the "viscoelastic momentum flux tensor" or the "viscoelastic stress tensor."

[b]This may also be called the molecular stress tensor.

similarly, the components of $\boldsymbol{\phi}$ in spherical coordinates can be obtained by replacing x,y,z with r,θ,ϕ in Table 1.3-1. As before, the components τ_{ij} appearing in Table 1.3-1 are expressed in terms of the velocity gradients in Cartesian (x,y,z), cylindrical (r,θ,z), and spherical (r,θ,ϕ) coordinates in Appendix B.1.

The nomenclature and various symbols for the momentum fluxes discussed in this chapter are summarized in Table. 1.3-2.

§1.4 VISCOSITY DATA FROM EXPERIMENTS

The viscosities of fluids vary over many orders of magnitude, with the viscosity of air at 20°C being 1.8×10^{-5} Pa · s, that of water being 10^{-3} Pa · s, and that of glycerol being about 1 Pa · s, with many silicone oils being even more viscous. In Tables 1.4-1, 1.4-2, and 1.4-3 some experimental data[1] are given for pure fluids at 1 atm pressure. Note that for gases

[1]A comprehensive presentation of experimental techniques for measuring transport properties can be found in W. A. Wakeham, A. Nagashima, and J. V. Sengers, *Measurement of the Transport Properties of Fluids*, CRC Press, Boca Raton, FL (1991). Sources for experimental data are: *Landolt-Börnstein: Zahlenwerte und Funktionen*, Vol. II, 5, Springer (1968–1969); *International Critical Tables*, McGraw-Hill, New York (1926–1930); Y. S. Touloukian, P. E. Liley, and S. C. Saxena, *Thermophysical Properties of Matter*, Plenum Press, New York (1970); and also numerous handbooks of chemistry, physics, fluid dynamics, and heat transfer.

Table 1.4-1. Viscosity of Water and Air at 1 atm Pressure

	Water (liq.)[a]		Air[b]	
Temperature T(°C)	Viscosity μ (mPa · s)	Kinematic Viscosity ν (cm^2/s)	Viscosity μ (mPa · s)	Kinematic Viscosity ν (cm^2/s)
0	1.787	0.01787	0.01716	0.1327
20	1.0019	0.010037	0.01813	0.1505
40	0.6530	0.006581	0.01908	0.1692
60	0.4665	0.004744	0.01999	0.1886
80	0.3548	0.003651	0.02087	0.2088
100	0.2821	0.002944	0.02173	0.2298

[a]Calculated from the results of R. C. Hardy and R. L. Cottington, *J. Research Nat. Bur. Standards,* **42,** 573–578 (1949); and J. F. Swidells, J. R. Coe, Jr., and T. B. Godfrey, *J. Research Nat. Bur. Standards,* **48,** 1–31 (1952).

[b]Calculated from "Tables of Thermal Properties of Gases," *National Bureau of Standards Circular* **464** (1955), Chapter 2.

Table 1.4-2. Viscosities of Some Gases and Liquids at Atmospheric Pressure[a]

Gases	Temperature T(°C)	Viscosity μ (mPa · s)	Liquids	Temperature T(°C)	Viscosity μ (mPa · s)
i-C_4H_{10}	23	0.0076[c]	$(C_2H_5)_2O$	0	0.283
SF_6	23	0.0153		25	0.224
CH_4	20	0.0109[b]	C_6H_6	20	0.649
H_2O	100	0.01211[d]	Br_2	25	0.744
CO_2	20	0.0146[b]	Hg	20	1.552
N_2	20	0.0175[b]	C_2H_5OH	0	1.786
O_2	20	0.0204		25	1.074
Hg	380	0.0654[d]		50	0.694
			H_2SO_4	25	25.54
			Glycerol	25	934.

[a]Values taken from N. A. Lange, *Handbook of Chemistry,* McGraw-Hill, New York, 15th edition (1999), Tables 5.16 and 5.18.

[b]H. L. Johnston and K. E. McKloskey, *J. Phys. Chem.,* **44,** 1038–1058 (1940).

[c]*CRC Handbook of Chemistry and Physics,* CRC Press, Boca Raton, FL (1999).

[d]*Landolt-Börnstein: Zahlenwerte und Funktionen,* Springer (1969).

at low density, the viscosity *increases* with increasing temperature, whereas for liquids the viscosity usually *decreases* with increasing temperature. In gases, the momentum is transported by the molecules in free flight between collisions, but in liquids the transport takes place predominantly via intermolecular forces that molecules experience as they wind their way among their neighbors.

Table 1.4-3. Viscosities of Some Liquid Metals

Metal	Temperature T(°C)	Viscosity μ (mPa · s)
Li	183.4	0.5918
	216.0	0.5406
	285.5	0.4548
Na	103.7	0.686
	250	0.381
	700	0.182
K	69.6	0.515
	250	0.258
	700	0.136
Hg	−20	1.85
	20	1.55
	100	1.21
	200	1.01
Pb	441	2.116
	551	1.700
	844	1.185

Data taken from *The Reactor Handbook*, Vol. 2, Atomic Energy Commission AECD-3646, U.S. Government Printing Office, Washington, D. C. (May 1955), pp. 258 *et seq*.

§1.5 VISCOSITY DATA AND THE PRINCIPLE OF CORRESPONDING STATES

Extensive data on viscosities of pure gases and liquids are available in various science and engineering handbooks.[1] When values of viscosities are required for practical calculations, experimental data should be used. If experimental data are lacking and there is not time to obtain them, the viscosity can be estimated by empirical methods, making use of other data on the given substance. We present here a *corresponding-states correlation* that facilitates such estimates and illustrates general trends of viscosity with temperature and pressure for ordinary fluids. The principle of corresponding states, which has a sound scientific basis,[2] is widely used for correlating transport and thermodynamic properties. Discussions of this principle can be found in textbooks on physical chemistry and thermodynamics.

---

[1] J. A. Schetz and A. E. Fuhs (eds.), *Handbook of Fluid Dynamics and Fluid Machinery*, Wiley-Interscience, New York (1996), Vol. 1, Chapter 2; W. M. Rohsenow, J. P. Hartnett, and Y. I. Cho, *Handbook of Heat Transfer*, McGraw-Hill, New York, 3rd edition (1998), Chapter 2. Other sources are mentioned in footnote 1 of §1.4.

[2] J. Millat, J. H. Dymond, and C. A. Nieto de Castro (eds.), *Transport Properties of Fluids*, Cambridge University Press (1996), Chapter 11, by E. A. Mason and F. J. Uribe, and Chapter 12, by M. L. Huber and H. J. M. Hanley.

The plot in Fig. 1.5-1 gives a global view of the pressure and temperature dependence of viscosity. The reduced viscosity $\mu_r = \mu/\mu_c$ is plotted versus the reduced temperature $T_r = T/T_c$ for various values of the reduced pressure $p_r = p/p_c$. A "reduced" quantity is one that has been made dimensionless by dividing by the same quantity at the critical point. The chart shows that the viscosity of a gas approaches a limit (the low-density limit) as the pressure becomes smaller; for most gases, this limit is nearly attained at 1 atm pressure. The viscosity of a gas at low density *increases* with increasing temperature, whereas the viscosity of a liquid *decreases* with increasing temperature.

Experimental values of the critical viscosity μ_c are seldom available. However μ_c may be estimated in one of the following ways: (i) if a value of viscosity is known at a given reduced pressure and temperature, preferably at conditions near to those of interest, then μ_c can be calculated from $\mu_c = \mu/\mu_r$; or (ii) if critical p-V-T data are available, then μ_c may

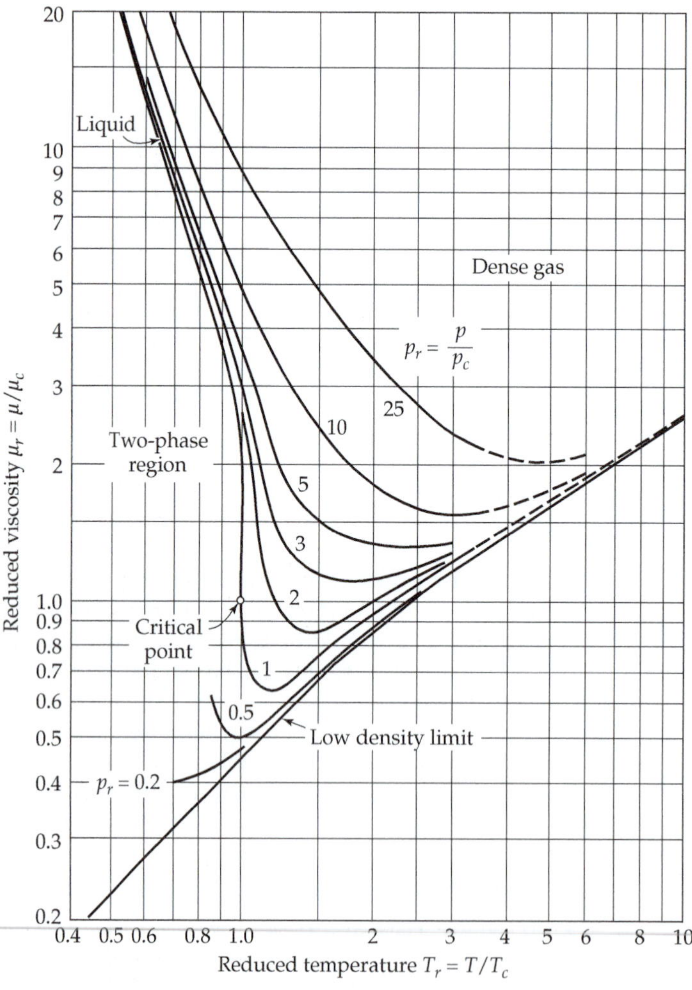

Fig. 1.5-1. Reduced viscosity $\mu_r = \mu/\mu_c$ as a function of reduced temperature for several values of the reduced pressure. [O. A. Uyehara and K. M. Watson, *Nat. Petroleum News, Tech. Section*, **36**, 764 (Oct. 4, 1944); revised by K. M. Watson (1960). A large-scale version of this graph is available in O. A. Hougen, K. M. Watson, and R. A. Ragatz, *C. P. P. Charts*, Wiley, New York, 2nd edition (1960).]

be estimated from these empirical relations:

$$\mu_c = 61.6(MT_c)^{1/2}(\tilde{V}_c)^{-2/3} \quad \text{or} \quad \mu_c = 7.70M^{1/2}p_c^{2/3}T_c^{-1/6} \tag{1.5-1a,b}$$

Here μ_c is in micropoises, M (mol. wt.) in g/g-mol, p_c in atm, T_c in K, and $\tilde{V}_c$ in cm^3/g-mol. A tabulation of critical viscosities[3] computed by method (i) is given in Appendix D.

Figure 1.5-1 can also be used for rough estimation of viscosities of mixtures. For an N-component fluid with mole fractions x_α, the "pseudocritical" properties[4] are

$$p'_c = \sum_{\alpha=1}^{N} x_\alpha p_{c\alpha} \qquad T'_c = \sum_{\alpha=1}^{N} x_\alpha T_{c\alpha} \qquad \mu'_c = \sum_{\alpha=1}^{N} x_\alpha \mu_{c\alpha} \tag{1.5-2a, b, c}$$

That is, one uses the chart exactly as for pure fluids, but with the pseudocritical properties instead of the critical properties. This empirical procedure works reasonably well unless there are chemically dissimilar substances in the mixture or the critical properties of the components differ greatly.

There are many variants on the above method, as well as a number of other empiricisms. These can be found in the extensive compilation of Poling, Prausnitz, and O'Connell.[5]

EXAMPLE 1.5-1

Estimation of Viscosity from Critical Properties

Estimate the viscosity of N_2 at 50°C and 854 atm, given $M = 28.0$ g/g-mol, $p_c = 33.5$ atm, and $T_c = 126.2$ K.

SOLUTION

Using Eq. 1.5-1b, we get

$$\begin{aligned}\mu_c &= 7.70(2.80)^{1/2}(33.5)^{2/3}(126.2)^{-1/6} \\ &= 189 \text{ micropoises} = 189 \times 10^{-6} \text{ poise}\end{aligned} \tag{1.5-3}$$

The reduced temperature and pressure are:

$$T_r = \frac{273.2 + 50 \text{ K}}{126.2 \text{ K}} = 2.56; \quad p_r = \frac{854 \text{ atm}}{33.5 \text{ atm}} = 25.5 \tag{1.5-4a, b}$$

From Fig. 1.5-1, we obtain $\mu_r = \mu/\mu_c = 2.6$. Hence, the predicted value of the viscosity is

$$\mu = \mu_c(\mu/\mu_c) = (189 \times 10^{-6} \text{ poise})(2.6) = 490 \times 10^{-6} \text{ poise} \tag{1.5-5}$$

The measured value[6] is 455×10^{-6} poise.

---

[3] O. A. Hougen and K. M. Watson, *Chemical Process Principles*, Part III, Wiley, New York (1947), p. 873. **Olaf Andreas Hougen** (pronounced "How-gen") (1893–1986) was a leader in the development of chemical engineering for four decades; together with K. M. Watson and R. A. Ragatz, he wrote influential books on thermodynamics and kinetics.

[4] O. A. Hougen, K. M. Watson, and R. A. Ragatz, *Chemical Process Principles*, Part II, Wiley, New York, 2nd edition (1959), p. 859.

[5] B. E. Poling, J. M. Prausnitz, and J. P. O'Connell, *The Properties of Gases and Liquids*, McGraw-Hill, New York, 5th edition (2001), Chapter 9.

[6] A. M. J. F. Michels and R. E. Gibson, *Proc. Roy. Soc.* (London), **A134**, 288–307 (1931).

§1.6 VISCOSITY OF GASES AND KINETIC THEORY

To get a better appreciation of the concept of molecular momentum transport, we examine this transport mechanism from the point of view of an elementary kinetic theory of gases.

We consider a pure gas composed of rigid, nonattracting spherical molecules of diameter d and mass m, and the number density (number of molecules per unit volume) is taken to be n. The concentration of gas molecules is presumed to be sufficiently small so that the average distance between molecules is many times their diameter d. In such a gas it is known[1] that, at equilibrium, the molecular velocities are randomly directed and have an average magnitude given by

$$\bar{u} = \sqrt{\frac{8\kappa T}{\pi m}} \tag{1.6-1}$$

in which κ is the Boltzmann constant (see Appendix E). The frequency of molecular bombardment per unit area on one side of any stationary surface exposed to the gas is

$$Z = \frac{1}{4}n\bar{u} \tag{1.6-2}$$

The average distance traveled by a molecule between successive collisions is the *mean-free path* λ, given by

$$\lambda = \frac{1}{\sqrt{2}\pi d^2 n} \tag{1.6-3}$$

On average, the molecules reaching a plane will have experienced their last collision at a distance a from the plane, where a is given very roughly by

$$a = \frac{2}{3}\lambda \tag{1.6-4}$$

The concept of the mean-free path is intuitively appealing, but it is meaningful only when λ is large compared to the range of intermolecular forces. The concept is appropriate for the rigid-sphere molecular model considered here.

To determine the viscosity of a gas in terms of the molecular model parameters d and m, we consider the behavior of the gas when it flows parallel to the xz plane with a velocity gradient dv_x/dy (see Fig. 1.6-1). We assume that Eqs. 1.6-1 to 1.6-4 remain valid in this nonequilibrium situation, provided that all molecular velocities are calculated relative to the average velocity $\mathbf{v}$ in the region in which the given molecule had its last collision. The flux of x momentum across any plane of constant y is found by summing the x momenta of the molecules that cross in the positive y direction and subtracting the x momenta of those that cross in the opposite direction, as follows:

$$\tau_{yx} = Zmv_x|_{y-a} - Zmv_x|_{y+a} \tag{1.6-5}$$

In writing this equation, we have assumed that all molecules have velocities representative of the region in which they last collided and that the velocity profile $v_x(y)$ is essentially linear for a distance of several mean-free paths. In view of the latter assumption, we may further write

$$v_x|_{y\pm a} = v_x|_y \pm \frac{2}{3}\lambda\frac{dv_x}{dy} \tag{1.6-6}$$

[1] The first four equations in this section are given without proof. Detailed justifications are given in books on kinetic theory, e.g., E. H. Kennard, *Kinetic Theory of Gases*, McGraw-Hill, New York (1938), Chapters II and III. Also E. A. Guggenheim, *Elements of the Kinetic Theory of Gases*, Pergamon Press, New York (1960), Chapter 7, has given a short account of the elementary theory of viscosity. For a readable summary of the kinetic theory of gases, see R. J. Silbey, R. A. Alberty, and M. G. Bawendi, *Physical Chemistry*, Wiley, New York, 4th edition (2005), Chapter 17, or R. S. Berry, S. A. Rice, and J. Ross, *Physical Chemistry*, Oxford University Press, 2nd edition (2000), Chapter 28.

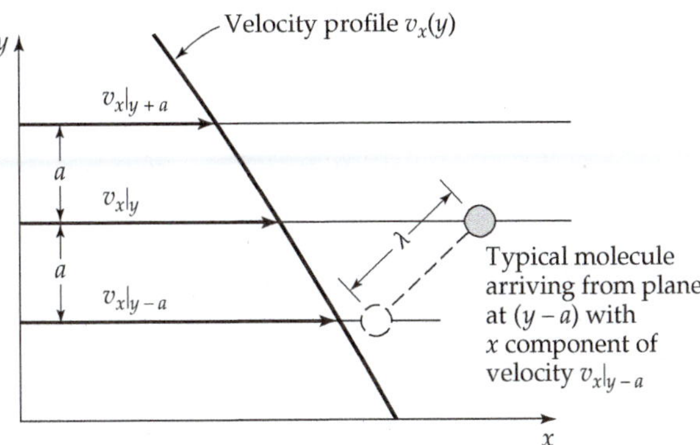

Fig. 1.6-1. Molecular transport of x momentum from the plane at $(y - a)$ to the plane at y.

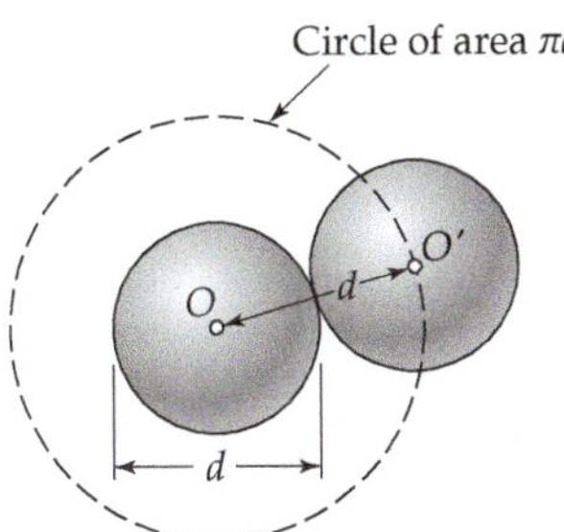

Fig. 1.6-2. When two rigid spheres of diameter d approach each other, the center of one sphere (at O') "sees" a circle of area πd^2 about the center of the other sphere (at O), on which a collision can occur. The area πd^2 is referred to as the "collision cross section."

When we combine Eqs. 1.6-2, 1.6-5, and 1.6-6, we get for the net flux of x momentum in the positive y direction

$$\tau_{yx} = -\frac{1}{3} nm\bar{u}\lambda \frac{dv_x}{dy} \tag{1.6-7}$$

This has the same form as Newton's law of viscosity given in Eq. 1.2-2. Comparing the two equations gives an equation for the viscosity

$$\mu = \frac{1}{3} nm\bar{u}\lambda = \frac{1}{3}\rho\bar{u}\lambda \tag{1.6-8}$$

or, by combining Eqs. 1.6-1, 1.6-3, and 1.6-8

$$\mu = \frac{2}{3}\frac{\sqrt{m\kappa T/\pi}}{\pi d^2} = \frac{2}{3\pi}\frac{\sqrt{\pi m\kappa T}}{\pi d^2} \tag{1.6-9}$$

This expression for the viscosity was obtained by Maxwell[2] in 1860. The quantity πd^2 is called the *collision cross section* (see Fig. 1.6-2).

The above derivation, which gives a qualitatively correct picture of momentum transfer in a gas at low density, makes it clear why we wished to introduce the term "momentum flux" for τ_{yx} in §1.2.

The prediction of Eq. 1.6-9 that μ is independent of pressure agrees with experimental data up to about 10 atm at temperatures above the critical temperature (see Fig. 1.5-1).

[2]**James Clerk Maxwell** (1831–1879) was one of the greatest physicists of all times; he is particularly famous for his development of the field of electromagnetism and his contributions to the kinetic theory of gases. In connection with the latter, see J. C. Maxwell, *Phil. Mag.*, **19**, 19, Prop. XIII; S. G. Brush, *Am. J. Phys.*, **30**, 269–281 (1962). There is some controversy concerning Eqs. 1.6-4 and 1.6-9 (see S. Chapman and T. G. Cowling, *The Mathematical Theory of Non-Uniform Gases*, Cambridge University Press, 3rd edition (1970), p. 98; R. E. Cunningham and R. J. J. Williams, *Diffusion in Gases and Porous Media*, Plenum Press, New York (1980), §6.4.)

The predicted temperature dependence is less satisfactory; data for various gases indicate that μ increases more rapidly than $\sqrt{T}$. For a better description of the temperature dependence of μ, it is necessary to replace the rigid-sphere model by one that portrays the attractive and repulsive forces more accurately. It is also necessary to abandon the mean-free path theories and use the Boltzmann equation to obtain the molecular velocity distribution in nonequilibrium systems more accurately. We present here the main results; more details are presented elsewhere.[3,4,5,6]

A rigorous kinetic theory of monatomic gases at low density was developed early in the twentieth century by Chapman in England and independently by Enskog in Sweden. The Chapman-Enskog theory gives expressions for the transport properties in terms of the *intermolecular potential energy* $\varphi(r)$, where r is the distance between a pair of molecules undergoing a collision. The intermolecular force is then given by $F(r) = -d\varphi/dr$. The exact functional form of $\varphi(r)$ is not known; however, for nonpolar molecules a satisfactory empirical expression is the *Lennard-Jones (6–12) potential*[7] given by

$$\varphi(r) = 4\varepsilon\left[\left(\frac{\sigma}{r}\right)^{12} - \left(\frac{\sigma}{r}\right)^{6}\right] \tag{1.6-10}$$

in which σ is a characteristic diameter of the molecules, often called the *collision diameter* and ε is a characteristic energy, actually the *maximum energy of attraction* between a pair of molecules. This function, shown in Fig. 1.6-3, exhibits the characteristic features of intermolecular forces: weak attractions at large separations, and strong repulsions at small separations. Values of the parameters σ and ε are known for many substances; a partial list is given in Table D.1, and a more extensive list is available elsewhere.[4] When σ and ε are not known, they may be estimated from properties of the fluid at the critical point (c), the liquid at the normal boiling point (b), or the solid at the melting point (m), by means of the

---

[3]**Sydney Chapman** (1888–1970) taught at Imperial College in London, and thereafter was at the High Altitude Observatory in Boulder, Colorado; in addition to his seminal work on gas kinetic theory, he contributed to kinetic theory of plasmas and the theory of flames and detonations. **David Enskog** (1884–1947) (pronounced, roughly, "Ayn-skohg") is famous for his work on kinetic theories of low- and high-density gases. The standard reference on the Chapman-Enskog kinetic theory of dilute gases is S. Chapman and T. G. Cowling, *The Mathematical Theory of Non-Uniform Gases*, Cambridge University Press, 3rd edition (1970); pp. 407–409 give a historical summary of the kinetic theory. See also D. Enskog, *Inaugural Dissertation*, Uppsala (1917). In addition J. H. Ferziger and H. G. Kaper, *Mathematical Theory of Transport Processes in Gases*, North-Holland, Amsterdam (1972), is a very readable account of molecular theory.

[4]The Curtiss-Hirschfelder[5] extension of the Chapman-Enskog theory to multicomponent gas mixtures, as well as the development of useful tables for computation, can be found in J. O. Hirschfelder, C. F. Curtiss, and R. B. Bird, *Molecular Theory of Gases and Liquids*, Wiley, New York, 2nd corrected printing (1964). See also C. F. Curtiss, *J. Chem. Phys.*, **49**, 2917–2919 (1968), as well as references given in Appendix D. **Joseph Oakland Hirschfelder** (1911–1990), founding director of the Theoretical Chemistry Institute at the University of Wisconsin, specialized in intermolecular forces and applications of kinetic theory. **Charles Francis ("Chuck") Curtiss** (1921–2007) of the Theoretical Chemistry Institute at the University of Wisconsin specialized in the kinetic theory of nonspherical and polyatomic molecules. He also developed a comprehensive theory for the transport properties of polymeric fluids.

[5]C. F. Curtiss and J. O. Hirschfelder, *J. Chem. Phys.*, **17**, 550–555 (1949).

[6]R. B. Bird, W. E. Stewart, and E. N. Lightfoot, *Transport Phenomena*, Revised Second Edition, Wiley, New York (2007), Appendix D.

[7]J. E. (Lennard-)Jones, *Proc. Roy. Soc.*, **A106**, 441–462, 463–477 (1924). See also R. J. Silbey, R. A. Alberty, and M. G. Bawendi, *Physical Chemistry*, Wiley, 4th edition (2005), §11.9, §16.11, and §17.8, and R. S. Berry, S. A. Rice, and J. Ross, *Physical Chemistry*, Oxford University Press, 2nd edition (2000), §10.2.

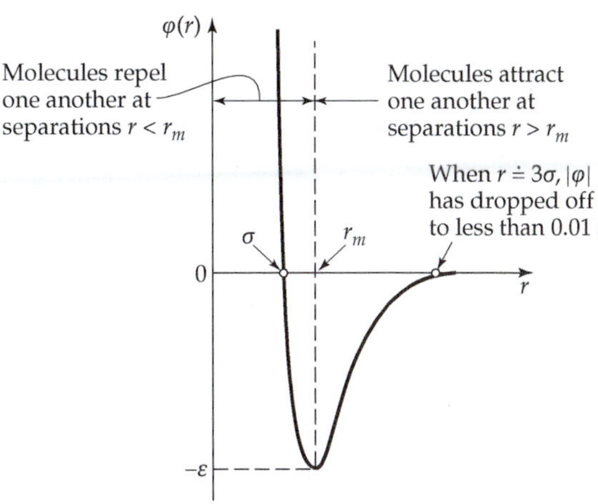

Fig. 1.6-3. Potential energy function $\varphi(r)$ describing the interaction of two spherical, nonpolar molecules. The Lennard-Jones (6–12) potential, given in Eq. 1.6-10, is one of the many empirical equations proposed for fitting this curve. For $r < r_m$ the molecules repel one another, whereas for $r > r_m$ the molecules attract one another.

following empirical relations:[4]

$$\varepsilon/\kappa = 0.77T_c \qquad \sigma = 0.841\tilde{V}_c^{1/3} \quad \text{or} \quad \sigma = 2.44(T_c/p_c)^{1/3} \tag{1.6-11a, b, c}$$

$$\varepsilon/\kappa = 1.15T_b \qquad \sigma = 1.166\tilde{V}_{b,\text{liq}}^{1/3} \tag{1.6-12a, b}$$

$$\varepsilon/\kappa = 1.92T_m \qquad \sigma = 1.222\tilde{V}_{m,\text{sol}}^{1/3} \tag{1.6-13a, b}$$

Here ε/κ and T are in K, σ is in Ångströms (1 Å = 10^{-10} m), $\tilde{V}$ is in cm^3/g-mol, and p_c is in atmospheres.

The viscosity of a pure monatomic gas of molecular weight M may be written in terms of the Lennard-Jones parameters as:

$$\mu = \frac{5}{16}\frac{\sqrt{\pi m\kappa T}}{\pi\sigma^2\Omega_\mu} \quad \text{or} \quad \mu = 2.6693 \times 10^{-5}\frac{\sqrt{MT}}{\sigma^2\Omega_\mu} \tag{1.6-14}$$

In the second form of this equation, if T [=] K and σ [=] Å, then μ [=] g/cm · s. The dimensionless quantity Ω_μ is a slowly varying function of the dimensionless temperature $\kappa T/\varepsilon$, of the order of magnitude of unity, given in Table D.2. It is called the "collision integral for viscosity," since it accounts for the details of the paths that the molecules take during a binary collision. If the gas were made up of rigid spheres of diameter σ (instead of real molecules with attractive and repulsive forces), then Ω_μ would be exactly unity. Hence, the function Ω_μ may be interpreted as describing the deviation from rigid-sphere behavior.

Although Eq. 1.6-14 is a result of the kinetic theory of monatomic gases, it has been found to be remarkably good for polyatomic gases as well. The reason for this is that, in the equation of conservation of momentum for a collision between polyatomic molecules, the center of mass coordinates are more important than the internal coordinates (see §0.3(*b*)). The temperature dependence predicted by Eq. 1.6-14 is in good agreement with that found from the low-density line in the empirical correlation of Fig. 1.5-1. The viscosity of gases at low density increases with temperature, roughly as the 0.6 to 1.0 power of the absolute temperature, and is independent of the pressure. Equation 1.6-14 will not, however, give reliable results for gases consisting of polar or highly elongated molecules because of the angle-dependent force fields that exist between such molecules. For polar vapors, such as H_2O, NH_3, CHOH, and NOCl, an angle-dependent modification of Eq. 1.6-10 has given good results.[8] For the light gases H_2 and He below about 100 K, quantum effects have to

[8] E. A. Mason and L. Monchick, *J. Chem. Phys.*, **35**, 1676–1697 (1961), **36**, 1622–1639, 2746–2757 (1962).

be taken into account.[9] Viscosities of gas mixtures can be calculated using an extension of the Chapman-Enskog theory.[4,5] A variety of empirical approaches can also be used to give satisfactory results for gas mixtures.[10,11,12]

EXAMPLE 1.6-1

Computation of the Viscosity of a Pure Gas at Low Density

Compute the viscosity of CO_2 at 200, 300, and 800 K and 1 atm.

SOLUTION

Use Eq. 1.6-14. From Table D.1, we find the Lennard-Jones parameters for CO_2 to be $\varepsilon/\kappa = 190$ K and $\sigma = 3.996$ Å. The molecular weight of CO_2 is 44.01 g/g-mol. Substitution of M and σ into Eq. 1.6-14 gives

$$\mu = 2.6693 \times 10^{-5} \frac{\sqrt{44.01T}}{(3.996)^2 \Omega_\mu} = 1.109 \times 10^{-5} \frac{\sqrt{T}}{\Omega_\mu} \tag{1.6-15}$$

in which μ [=] g/cm · s and T [=] K. The remaining calculations may be displayed in a table.

				Viscosity (g/cm · s)	
T (K)	$\kappa T/\varepsilon$	Ω_μ	$\sqrt{T}$	Predicted	Observed[13]
200	1.053	1.548	14.14	1.013×10^{-4}	1.015×10^{-4}
300	1.58	1.286	17.32	1.494×10^{-4}	1.495×10^{-4}
800	4.21	0.9595	28.28	3.296×10^{-4}	...

Experimental data are shown in the last column for comparison. The good agreement is to be expected, since the Lennard-Jones parameters of Table D.1 were derived from viscosity data.

§1.7 VISCOSITY OF LIQUIDS

Although the molecular theory has been worked out in some detail for gases at low density and is useful for making calculations, the same cannot be said of liquids, suspensions, pastes, and other fluids. In the absence of experimental data, it is necessary to use empirical formulas of varying degrees of reliability.

[9] J. O. Hirschfelder, C. F. Curtiss, and R. B. Bird, *op. cit.*, Chapter 10 by J. de Boer and R. B. Bird; H. T. Wood and C. F. Curtiss, *J. Chem. Phys.*, **41**, 1167–1173 (1964); R. J. Munn, F. J. Smith, and E. A. Mason, *J. Chem. Phys.*, **42**, 537–539 (1965); S. Imam-Rahajoe, C. F. Curtiss, and R. B. Bernstein, *J. Chem. Phys.*, **42**, 530–536 (1965).

[10] R. B. Bird, W. E. Stewart, and E. N. Lightfoot, *Transport Phenomena*, Revised Second Edition, Wiley, New York (2007), pp. 27–29.

[11] C. R. Wilke, *J. Chem. Phys.*, **18**, 517–519 (1950); see also J. W. Buddenberg and C. R. Wilke, *Ind. Eng. Chem.*, **41**, 1345–1347 (1949).

[12] B. E. Poling, J. M. Prausnitz, and J. P. O'Connell, *The Properties of Gases and Liquids*, McGraw-Hill, New York, 5th edition (2001), Chapter 9.

[13] H. L. Johnston and K. E. McCloskey, *J. Phys. Chem.*, **44**, 1038–1058 (1940).

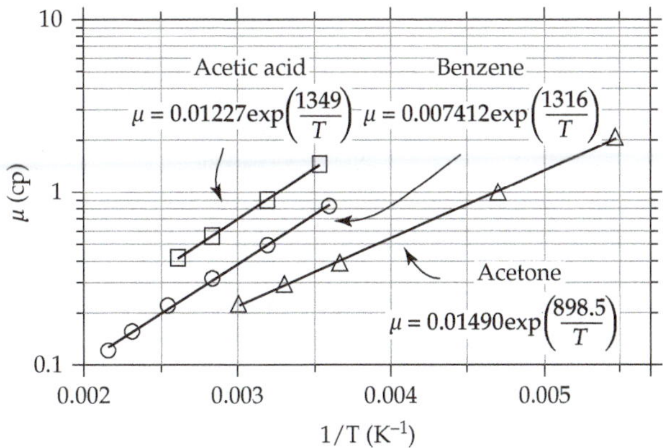

Fig. 1.7-1. Viscosity as a function of $1/T$ for three different liquids. [Data from B. E. Poling, J. M. Prausnitz, and J. P. O'Connell, *The Properties of Gases and Liquids*, McGraw-Hill, New York, 5th edition (2001), Chapter 9.]

For *simple liquids*, a useful empiricism, based on early work by Eyring[1] and coworkers, is

$$\mu = \frac{\tilde{N}h}{\tilde{V}} \exp\left(\frac{3.8T_b}{T}\right) \tag{1.7-1}$$

in which $\tilde{N}$ is Avogadro's number, h is Planck's constant, $\tilde{V}$ is the molar volume, T_b is the boiling point temperature, and T is the absolute temperature. This expression illustrates that the viscosity of a liquid decreases with increasing temperature, and agrees with the long-used empiricism $\mu = A\exp(B/T)$. An approximately exponential dependence on the inverse temperature is often observed as illustrated in Fig. 1.7-1. However, errors as large as 30% are common, and this empiricism should not be used for long slender molecules, such as n-$C_{20}H_{42}$. The approximately exponential dependence on the inverse absolute temperature is best used for interpolation between viscosity values at different temperatures (e.g., by plotting $\ln \mu$ vs $1/T$).

§1.8 VISCOSITY OF SUSPENSIONS

For very dilute *suspensions of spheres* in a suspending medium with viscosity μ_0 and volume fraction of spheres ϕ, a rigorous theory by Einstein[1] gives for the effective viscosity of the suspension ($\phi \leq 0.1$)

$$\frac{\mu_{\text{eff}}}{\mu_0} = 1 + \frac{5}{2}\phi \tag{1.8-1}$$

This expression illustrates that a dilute suspension of particles is Newtonian, and its viscosity increases with increasing concentration of the solid particles. It further indicates that, for low-volume fractions, the effective viscosity is independent of the particle size.

---

[1]S. Glasstone, K. J. Laidler, and H. Eyring, *Theory of Rate Processes*, McGraw-Hill, New York (1941), Chapter 9; H. Eyring, D. Henderson, B. J. Stover, and E. M. Eyring, *Statistical Mechanics*, Wiley, New York (1964), Chapter 16. See also R. J. Silbey, R. A. Alberty, and M. G. Bawendi, *Physical Chemistry*, Wiley, 4th edition (2005), §20.1; and R. S. Berry, S. A. Rice, and J. Ross, *Physical Chemistry*, Oxford University Press, 2nd edition (2000), Ch. 29. **Henry Eyring** (1901–1981) developed theories for the transport properties based on simple physical models; he was also a key figure in the theory of absolute reaction rates.

[1]**Albert Einstein** (1879–1955) received the Nobel Prize for his explanation of the photoelectric effect, not for his development of the theory of special relativity. His seminal work on suspensions appeared in A. Einstein, *Ann. Phys. (Leipzig)*, **19**, 289–306 (1906); erratum, *ibid.*, **34**, 591–592 (1911). In the original publication, Einstein made an error in the derivation of Eq. 1.8-1 and got ϕ instead of $\frac{5}{2}\phi$. After experiments showed that his equation did not agree with the experimental data, he recalculated the coefficient. Einstein's original derivation is quite lengthy; for a more compact development, see L. D. Landau and E. M. Lifshitz, *Fluid Mechanics*, Pergamon Press, Oxford, 2nd edition (1987), pp. 73–75.

Equation 1.8-1, however, is accurate only for relatively dilute suspensions. For concentrated suspensions of particles, the Krieger-Dougherty[2] equation is often used:

$$\frac{\mu_{\text{eff}}}{\mu_0} = \left(1 - \frac{\phi}{\phi_{\max}}\right)^{-A\phi_{\max}} \tag{1.8-2}$$

This expression illustrates that the effective viscosity of a suspension increases rapidly as ϕ approaches the maximum packing fraction $\phi_{\max}$. The constants $\phi_{\max}$ and A are typically treated as adjustable parameters that are fit to experimental data.

Suspensions can exhibit non-Newtonian behavior, particularly at large concentrations. Viscosities in these cases depend on the velocity gradient and may be different in shear and elongational flows. Therefore, equations such as Eq. 1.8-2 should be used with some caution.

§1.9 CONCLUDING COMMENTS

The main purpose of this chapter is to describe the different ways that momentum can be transported, and to define the corresponding momentum fluxes. The convective momentum flux $\boldsymbol{\pi}^{(c)}$ arises from the bulk flow of the fluid (§0.4). The molecular momentum flux $\boldsymbol{\pi}$ arises from molecular motion, collisions, and interactions, and consists of two parts: a pressure term $p\boldsymbol{\delta}$ that is present even when the fluid is static, and a viscous term $\boldsymbol{\tau}$ that arises when the fluid is flowing. The convective and molecular momentum fluxes are added together to form the total momentum flux $\boldsymbol{\phi}$, which will be used in Chapters 2 and 3 to solve flow problems.

The momentum fluxes are second-order tensors. It is important to note, however, that the treatment of transport phenomena in this text does not require the reader to understand tensor mathematics, but only to understand how the indices on the scalar tensor components are assigned. For the momentum flux tensor components, the first index represents the direction that momentum is being transported, and the second index represents the component of momentum that is being transported. In Chapters 2 and 3, we will show how the scalar tensor components are used to set up momentum balances. A variety of mathematical topics are reviewed in Appendix A, including a summary of tensor operations in the various coordinate systems.

For the viscous momentum flux, our attention will focus almost exclusively on Newton's law of viscosity, where the flux components are written in terms of derivatives of the velocity components. These relationships are written explicitly in Appendix B in Cartesian, cylindrical, and spherical coordinates. Setting up and solving the flow problems in Chapters 2 and 3 will require frequent use of the key equations in Appendix B.

Finally, we have described methods for estimating values of the viscosities of gases and liquids in the event that experimental data are not available. Appendix E contains conversion factors for converting between different systems of units. For practical calculations that appear later, this appendix will prove to be indispensable.

QUESTIONS FOR DISCUSSION

1. Compare Newton's law of viscosity and Hooke's law of elasticity. What is the origin of these "laws"?
2. Verify that "momentum per unit area per unit time" has the same dimensions as "force per unit area."
3. Compare and contrast the molecular and convective mechanisms for momentum transport.

---

[2] I. M. Krieger and T. J. Dougherty, *Trans. Soc. Rheol.*, **3**, 137–152 (1959).

4. What are the physical meanings of the Lennard-Jones parameters and how can they be determined from viscosity data? Is the determination unique?
5. How do the viscosities of liquids and low-density gases depend on the temperature and pressure?
6. The Lennard-Jones potential depends only on the intermolecular separation. For what kinds of molecules would you expect that this kind of potential would be inappropriate?
7. Sketch the potential energy function $\varphi(r)$ for rigid, nonattracting spheres.
8. Molecules differing only in their atomic isotopes have the same values of the Lennard-Jones potential parameters. Would you expect the viscosity of CD_4 to be larger or smaller than that of CH_4 at the same temperature and pressure?
9. Fluid A has a viscosity twice that of fluid B; which fluid would you expect to flow more rapidly through a horizontal tube of length L and radius R when the same pressure difference is imposed?
10. Draw a sketch of the intermolecular force $F(r)$ obtained from the Lennard-Jones function for $\varphi(r)$. Also, determine the value of r_m in Fig. 1.6-3 in terms of the Lennard-Jones parameters.
11. What main ideas are used when one goes from Newton's law of viscosity in Eq. 1.2-2 to the generalization in Eq. 1.2-13?

PROBLEMS

1A.1 Estimation of dense-gas viscosity. Estimate the viscosity of nitrogen at 68°F and 1000 psig by means of Fig. 1.5-1, using the critical viscosity from Table D.1. Give the result in units of $\text{lb}_m/\text{ft}\cdot\text{s}$. For the meaning of "psig," see Table E.3.2.

Answer: $1.4 \times 10^{-4}\ \text{lb}_m/\text{ft}\cdot\text{s}$

1A.2 Estimation of the viscosity of methyl fluoride. Use Fig. 1.5-1 to find the viscosity in Pa· s of CH_3F at 370°C and 120 atm. Use the following values[1] for the critical constants: $T_c = 4.55°\text{C}$, $p_c = 58.0$ atm, $\rho_c = 0.300\ \text{g/cm}^3$.

1A.3 Computation of the viscosities of gases at low density. Predict the viscosities of molecular oxygen, nitrogen, and methane at 20°C and atmospheric pressure, and express the results in mPa· s. Compare the results with experimental data given in this chapter.

Answers: 0.0202, 0.0172, 0.0107 mPa· s

1A.4 Estimation of liquid viscosity. Estimate the viscosity of saturated liquid water at 0°C and at 100°C by means of Eq. 1.7-1. Compare the results with the values in Table 1.4-1.

Answer: 4.0 cp, 0.95 cp

1A.5 Molecular velocity and mean-free path. Compute the mean molecular velocity $\bar{u}$ (in cm/s) and the mean-free path λ (in cm) for oxygen at 1 atm and 273.2 K. A reasonable value for d is 3 Å. What is the ratio of the mean-free path to the molecular diameter under these conditions? What would be the order of magnitude of the corresponding ratio in the liquid state?

Answers: $\bar{u} = 4.25 \times 10^4$ cm/s, $\lambda = 9.3 \times 10^{-6}$ cm

1A.6 Checking dimensions in equations. It is very important to make a habit of checking equations for dimensional consistency. Show that the following equations in the text are dimensionally consistent: Eq. 1.6-14, Eq. 1.7-1, and Eq. 1.3-2. Do this by replacing the symbols in the formulas by the appropriate dimensions. Omit any numerical factors that appear.

1B.1 Velocity profiles and the stress components τ_{ij}. For each of the following velocity distributions, draw a meaningful sketch showing the flow pattern. Then find all the components of $\boldsymbol{\tau}$ and $\rho\mathbf{vv}$ for the Newtonian fluid. The parameter b is a constant.

(a) $v_x = by,\ v_y = 0,\ v_z = 0$
(b) $v_x = by,\ v_y = bx,\ v_z = 0$

---

[1] K. A. Kobe and R. E. Lynn, Jr., *Chem. Revs.* **52**, 117–236 (1953), see p. 202.

(c) $v_x = -by, v_y = bx, v_z = 0$

(d) $v_x = -\frac{1}{2}bx, v_y = -\frac{1}{2}by, v_z = bz$

1B.2 A fluid in a state of rigid rotation.

(a) Verify that the velocity distribution (d) in Problem 1B.1 describes a fluid in a state of pure rotation; that is, the fluid is rotating like a rigid body. What is the angular velocity of rotation?

(b) For that flow pattern evaluate the symmetric and antisymmetric combinations of velocity derivatives:

i. $(\partial v_y/\partial x) + (\partial v_x/\partial y)$

ii. $(\partial v_y/\partial x) - (\partial v_x/\partial y)$

(c) Discuss the results of (b) in connection with the development in §1.2.

Chapter 2

Shell Momentum Balances and Velocity Distributions in Laminar Flow

In this chapter we show how to obtain the velocity profiles for fluids in simple laminar flows. These derivations make use of the definition of viscosity, the expressions for the convective and molecular momentum fluxes, and the concept of a momentum balance. Once the velocity profiles have been obtained, we can then get other quantities such as the maximum velocity, the average velocity, or the force on a surface. Often it is these latter quantities that are of interest in engineering problems.

In the first section we make a few general remarks about how to set up differential momentum balances. In the sections that follow, we show how to solve several classical viscous flow problems. These examples should be thoroughly understood, since we shall have frequent occasions to refer to them in subsequent chapters. Although these problems are rather simple and involve idealized systems, the solution methods and results are nonetheless often used in solving practical problems.

The systems studied in this chapter are so arranged that the reader is gradually introduced to a variety of factors that arise in the solution of viscous flow problems. In §2.2 the falling-film problem illustrates the role of gravity forces and the use of Cartesian coordinates; it also shows how to solve the problem when viscosity may be a function of position. In §2.3 the flow in a circular tube illustrates the role of pressure and gravity forces and the use of cylindrical coordinates; an approximate extension to compressible flow is also given. In §2.4 the flow in a cylindrical annulus emphasizes the role played by the boundary conditions. Then in §2.5 the question of boundary conditions is pursued further in the discussion of the flow of two adjacent immiscible liquids. In §2.6 the flow in a cone-and-plate viscometer is solved approximately to illustrate a problem in spherical coordinates. This problem also introduces the calculation of a torque exerted by the flow.

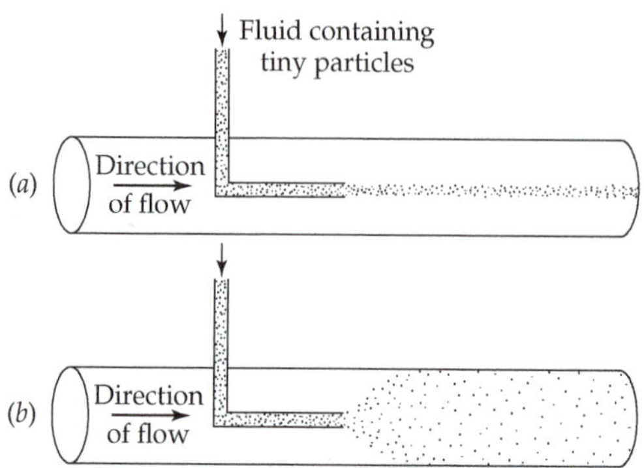

Fig. 2.0-1. (*a*) Laminar flow, in which fluid layers move smoothly over one another in the direction of flow, and (*b*) turbulent flow, in which the flow pattern is complex and time-dependent, with considerable motion perpendicular to the principal flow direction.

Finally, in §2.7 the flow around a sphere is discussed briefly to illustrate another problem in spherical coordinates and also to point out how both tangential and normal forces contribute to the total force on an object immersed in a flow.

The method of setting up momentum balances in this chapter applies only to *steady rectilinear flow*. By "steady" we mean that the pressure, density, and velocity components at each point in the fluid do not change with time. By "rectilinear" we mean that the fluid moves along in straight parallel lines. The flows in §2.6 and §2.7 are examples of nonrectilinear flows. The general equations for unsteady flow with arbitrary flow patterns are given in Chapter 3.

This chapter is also concerned only with *laminar flow*. Laminar flow is the orderly flow where thin fluid layers ("lamina") appear to slide past one another (as in Fig. 1.1-1). Laminar flow is observed, for example, in tube flow at velocities sufficiently low that tiny particles injected into the tube move along in a thin line. This is in sharp contrast with the wildly chaotic "turbulent flow" at sufficiently high velocities that the particles are flung apart and dispersed throughout the entire cross section of the tube. Turbulent flow is the subject of Chapter 4. The sketches in Fig. 2.0-1 illustrate the difference between the two flow regimes.

§2.1 SHELL MOMENTUM BALANCES; BOUNDARY CONDITIONS

The problems discussed in §2.2 through §2.5 are approached by setting up momentum balances for appropriately selected volumes of fluid within the flows. These volumes, which are fixed in space, are open systems through which the fluid is flowing. We call these volumes "shells" because they are thin in one dimension, specifically in the direction in which the fluid velocity varies. For *steady flow*, the momentum balance over a shell is

$$\left\{\begin{matrix}\text{Total rate}\\ \text{of momentum}\\ \text{transported } \textit{in}\end{matrix}\right\} - \left\{\begin{matrix}\text{Total rate}\\ \text{of momentum}\\ \text{transported } \textit{out}\end{matrix}\right\} + \left\{\begin{matrix}\text{Force of}\\ \text{gravity acting}\\ \text{on the fluid}\end{matrix}\right\} = 0 \tag{2.1-1}$$

This is a restricted statement of the law of conservation of momentum. In §2.2 through §2.5 of this chapter we apply this statement only to one component of the momentum, namely the component in the direction of flow.

Applying Eq. 2.1-1 for specific flow problems requires specifying the rates of momentum transport into and out of the shell, across each of the shell faces. These rates are simply the appropriate components of the total momentum flux $\boldsymbol{\phi}$ (tabulated in Table 1.3-1), multiplied by the corresponding face areas. Such applications to flow problems are illustrated in detail in the following sections. Keep in mind that the total momentum flux is the sum of the convective momentum flux (see Table 1.1-1) and the molecular momentum flux (see Table 1.2-1), and that the molecular momentum flux includes both the pressure and the viscous contributions.

In §2.2 to §2.5 the momentum balance is applied only to flows in which there is just one velocity component, which depends on just one spatial variable. In the next chapter the momentum balance concept is extended to unsteady-state systems and to flows with curvilinear streamlines[1] and more than one velocity component.

The procedure recommended in this chapter for setting up and solving viscous flow problems is as follows:

- Draw a sketch of the flow geometry being studied, including your best guess as to what the velocity distribution will look like.
- Identify the nonvanishing velocity component and the spatial variable on which it depends.
- Draw the "shell"—a volume with faces parallel or perpendicular to the velocity—that is thin in the direction in which the velocity varies.
- Write a momentum balance of the form of Eq. 2.1-1 for the thin shell.
- Let the thickness of the shell approach zero and make use of the definition of the first derivative to obtain a differential equation for the momentum flux.
- Integrate this equation to get the momentum-flux distribution.
- Insert the expression for the total momentum flux, including Newton's law of viscosity, and obtain a differential equation for the velocity.
- Integrate this equation to get the velocity distribution.
- Use the velocity distribution to get other quantities, such as the maximum velocity, average velocity, or forces on solid surfaces.

In the integrations mentioned above, several constants of integration appear, and these are evaluated by using "boundary conditions," that is, statements about the velocity or stress at the boundaries of the system. The most commonly used boundary conditions are:

a. At *solid-fluid* interfaces the fluid velocity equals the velocity with which the solid surface is moving; this statement is applied to both the tangential and the normal components of the velocity vector. The equality of the tangential components is referred to as the "no-slip condition."

b. At a *liquid-liquid* interfacial plane of constant x, the tangential velocity components v_y and v_z are continuous through the interface (the "no-slip condition") as are also the molecular stresses $\pi_{xx} = p + \tau_{xx}$, $\pi_{xy} = \tau_{xy}$, and $\pi_{xz} = \tau_{xz}$.

c. At a *liquid-gas* interfacial plane of constant x, the molecular stresses $\pi_{xy} = \tau_{xy}$ and $\pi_{xz} = \tau_{xz}$ are taken to be zero, provided that the gas-side velocity gradient is not too large. This is reasonable, since the viscosities of gases are much smaller than those of liquids.

In all of these boundary conditions it is presumed that there is no material passing through the interface; that is, there is no adsorption, absorption, dissolution, evaporation, melting, or chemical reaction at the surface between the two phases. In this section we have presented some guidelines for solving simple viscous flow problems. For some problems slight variations on these guidelines may prove to be appropriate.

§2.2 FLOW OF A FALLING FILM

The first example we discuss is that of the flow of a liquid down an inclined flat plate of length L and width W, as shown in Figs. 2.2-1 and 2.2-2. Such films have been studied in connection with wetted-wall towers, evaporation, and gas-absorption experiments,

---

[1] A *streamline* is a curve that is tangent to the instantaneous velocity.

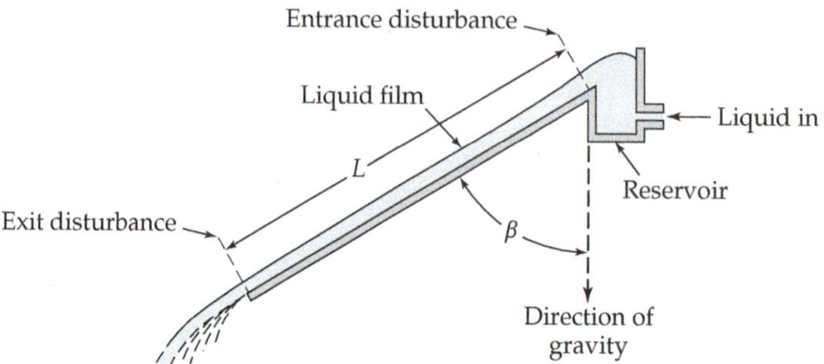

Fig. 2.2-1. Schematic diagram of the falling-film experiment, showing end effects.

and applications of coatings. We consider the viscosity and density of the fluid to be constant.

A complete description of the liquid flow is difficult because of the disturbances at the edges of the system ($z = 0, z = L, y = 0, y = W$). An adequate description can often be obtained by neglecting such disturbances, particularly if W and L are large compared to the film thickness δ. For small flow rates we expect that the viscous forces will prevent continued acceleration of the liquid down the wall, so that v_z will become independent of z in a short distance down the plate. Therefore, it seems reasonable to *postulate* that $v_z = v_z(x)$, $v_x = 0$, and $v_y = 0$, and further that $p = p(x)$.

We now select as the "system" a shell that is thin in the x direction; that is, a rectangular region of thickness Δx, bounded by the planes $z = 0$ and $z = L$, and extending a distance W in the y direction. This shell, depicted by the shaded region in Fig. 2.2-2, is shown in more detail in Fig. 2.2-3.

Next we set up a z momentum balance over this shell, using the total momentum-flux components ϕ_{xz}, ϕ_{yz}, and ϕ_{zz} to describe the flux of z momentum across the x, y, and z faces, respectively. These components, listed in the "z component" column of Table 1.3-1,

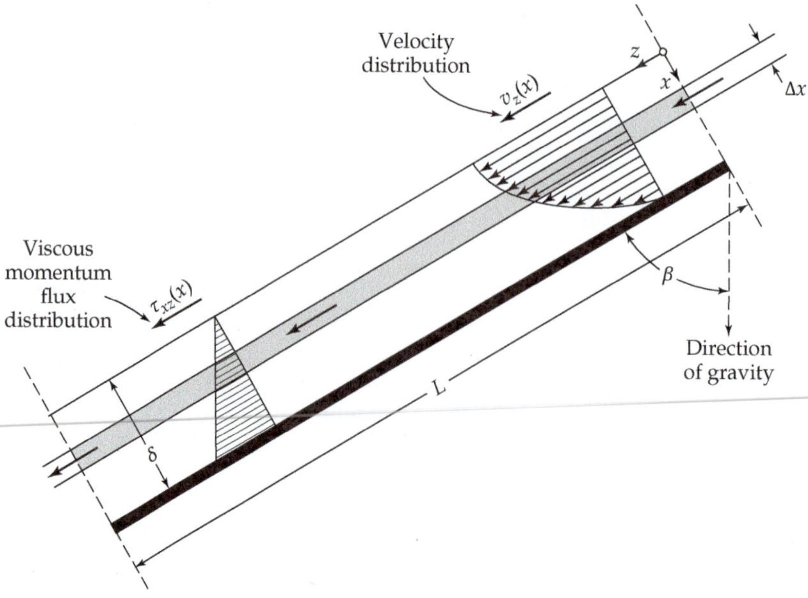

Fig. 2.2-2. Sketch of the falling-film flow, including the thin shell of thickness Δx, over which the z momentum balance is made. Also shown are the viscous momentum-flux distribution $\tau_{xz}(x)$ and the velocity distribution $v_z(x)$.

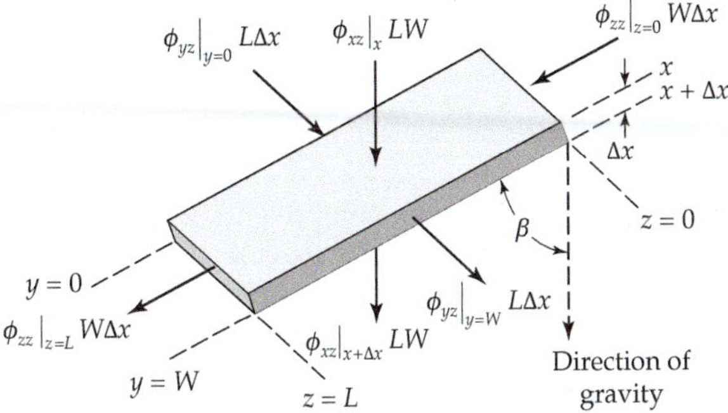

Fig. 2.2-3. Three-dimensional sketch of the shell of thickness Δx over which the z momentum balance is made. The arrows show the rates of momentum transport—momentum-flux components multiplied by the corresponding face areas—into and out of the shell. Note that the arrows always point in the directions of the positive coordinate axes.

are then multiplied by the corresponding areas of the shell faces to obtain the total rates of z momentum transported "in" and "out" of the shell across each face. The z momentum balance is illustrated pictorially in Fig. 2.2-3, where these "in" and "out" terms are depicted along with arrows. These arrows indicate that the momentum-flux components ϕ_{xz}, ϕ_{yz}, and ϕ_{zz} are defined as the fluxes of z momentum in the positive x, y, and z directions, respectively (regardless of the direction that z momentum is actually being transported). The terms in the z momentum balance are thus

Total rate of z momentum in across surface at x:	$\phi_{xz}\|_x LW$	(2.2-1)
Total rate of z momentum out across surface at x + Δx:	$\phi_{xz}\|_{x+\Delta x} LW$	(2.2-2)
Total rate of z momentum in across surface at y = 0:	$\phi_{yz}\|_{y=0} L\Delta x$	(2.2-3)
Total rate of z momentum out across surface at y = W:	$\phi_{yz}\|_{y=W} L\Delta x$	(2.2-4)
Total rate of z momentum in across surface at z = 0:	$\phi_{zz}\|_{z=0} W\Delta x$	(2.2-5)
Total rate of z momentum out across surface at z = L:	$\phi_{zz}\|_{z=L} W\Delta x$	(2.2-6)
Gravity force acting on the fluid in the z direction:	$\rho(LW\Delta x)(g\cos\beta)$	(2.2-7)

By using the quantities ϕ_{xz}, ϕ_{yz}, and ϕ_{zz} we account for the z momentum transport by both the convective and molecular mechanisms. The notation $|_{x+\Delta x}$ means "the term to the left is evaluated at $x + \Delta x$," and g is the gravitational acceleration.

When these terms are substituted into the z momentum balance of Eq. 2.1-1, we get

$$\left(\phi_{xz}|_x - \phi_{xz}|_{x+\Delta x}\right) LW + \left(\phi_{yz}|_{y=0} - \phi_{yz}|_{y=W}\right) L\Delta x + \left(\phi_{zz}|_{z=0} - \phi_{zz}|_{z=L}\right) W\Delta x + \rho(LW\Delta x)(g\cos\beta) = 0 \tag{2.2-8}$$

When this equation is divided by $LW\Delta x$, and the limit taken as Δx approaches zero, we get

$$\lim_{\Delta x\to 0}\left(\frac{\phi_{xz}|_{x+\Delta x} - \phi_{xz}|_x}{\Delta x}\right) + \frac{\phi_{yz}|_{y=W} - \phi_{yz}|_{y=0}}{W} + \frac{\phi_{zz}|_{z=L} - \phi_{zz}|_{z=0}}{L} = \rho g\cos\beta \tag{2.2-9}$$

The first term on the left side is just the definition of the derivative of ϕ_{xz} with respect to x. Therefore, Eq. 2.2-8 becomes

$$\frac{d\phi_{xz}}{dx} + \frac{\phi_{yz}|_{y=W} - \phi_{yz}|_{y=0}}{W} + \frac{\phi_{zz}|_{z=L} - \phi_{zz}|_{z=0}}{L} = \rho g \cos\beta \tag{2.2-10}$$

At this point we write explicitly the components ϕ_{xz}, ϕ_{yz}, and ϕ_{zz}, making use of the definition of $\boldsymbol{\phi}$ in Eq. 1.3-1 and Table 1.3-1. We also use Newton's law of viscosity (Appendix B.1) to write the viscous momentum-flux components τ_{xz}, τ_{yz}, and τ_{zz} in terms of derivatives of the velocity components. The components are thus written

$$\phi_{xz} = \tau_{xz} + \rho v_x v_z = -\mu\frac{\partial v_z}{\partial x} + \rho v_x v_z \tag{2.2-11}$$

$$\phi_{yz} = \tau_{yz} + \rho v_y v_z = -\mu\frac{\partial v_z}{\partial y} + \rho v_y v_z \tag{2.2-12}$$

$$\phi_{zz} = p + \tau_{zz} + \rho v_z v_z = p - 2\mu\frac{\partial v_z}{\partial z} + \rho v_z v_z \tag{2.2-13}$$

In accordance with the postulates that $v_z = v_z(x)$, $v_x = 0$, $v_y = 0$, and $p = p(x)$, we see that

(i) because $v_x = 0$, the term $\rho v_x v_z$ in Eq. 2.2-11 is zero.

(ii) because $v_z = v_z(x)$, the term $-\mu(\partial v_z/\partial y) = \tau_{yz}$ in Eq. 2.2-12 is zero.

(iii) because $v_y = 0$, the term $\rho v_y v_z$ in Eq. 2.2-12 is zero.

(iv) because $v_z = v_z(x)$, the term $-2\mu(\partial v_z/\partial z) = \tau_{zz}$ in Eq. 2.2-13 is zero.

(v) because $v_z = v_z(x)$, the term $\rho v_z v_z$ is the same at $z = 0$ and $z = L$.

(vi) because $p = p(x)$, the term p is the same at $z = 0$ and $z = L$.

Hence, Eq. 2.2-10 simplifies to

$$\frac{d\tau_{xz}}{dx} = \rho g \cos\beta \tag{2.2-14}$$

This is the differential equation for the viscous momentum flux τ_{xz}. It may be integrated to give

$$\tau_{xz}(x) = (\rho g \cos\beta)x + C_1 \tag{2.2-15}$$

The constant of integration may be evaluated by using the boundary condition at the gas-liquid interface (see §2.1)

B.C. 1: $$\text{at } x = 0,\ \tau_{xz} = 0 \tag{2.2-16}$$

Substitution of this boundary condition into Eq. 2.2-15 shows that $C_1 = 0$. Therefore, the momentum-flux distribution is

$$\boxed{\tau_{xz}(x) = (\rho g \cos\beta)x} \tag{2.2-17}$$

as shown in Fig. 2.2-2.

Next we substitute Newton's law of viscosity (Appendix B.1)

$$\tau_{xz} = -\mu\frac{dv_z}{dx} \tag{2.2-18}$$

into the left side of Eq. 2.2-17 to obtain

$$\frac{dv_z}{dx} = -\left(\frac{\rho g \cos\beta}{\mu}\right)x \tag{2.2-19}$$

which is the differential equation for the velocity distribution. It can be integrated to give

$$v_z(x) = -\left(\frac{\rho g \cos\beta}{2\mu}\right)x^2 + C_2 \tag{2.2-20}$$

The constant of integration is evaluated by using the no-slip boundary condition at the solid surface:

B.C. 2: $\quad$ at $x = \delta,\ v_z = 0$ $\quad$ (2.2-21)

Substitution of this boundary condition into Eq. 2.2-20 shows that $C_2 = (\rho g \cos\beta/2\mu)\delta^2$. Consequently the velocity distribution is

$$\boxed{v_z(x) = \frac{\rho g \delta^2 \cos\beta}{2\mu}\left[1 - \left(\frac{x}{\delta}\right)^2\right]} \tag{2.2-22}$$

This parabolic velocity distribution is shown in Fig. 2.2-2. It is consistent with the postulates made initially and must therefore be a *possible* solution. Other solutions might be possible, and experiments are normally required to tell whether other flow patterns can actually arise. We return to this point after Eq. 2.2-28.

Once the velocity distribution is known, a number of quantities can be calculated:

(i) The *maximum velocity* $v_{z,\max}$ is clearly the velocity at $x = 0$; that is

$$v_{z,\max} = \frac{\rho g \delta^2 \cos\beta}{2\mu} \tag{2.2-23}$$

(ii) The *volume rate of flow* Q is obtained by integrating the volumetric flow rate through a differential element of the cross section, $v_z dxdy$ (see §0.4), over the entire cross section

$$\begin{aligned} Q &= \int_0^W \int_0^\delta v_z(x)dxdy = W\frac{\rho g \delta^2 \cos\beta}{2\mu}\int_0^\delta \left[1 - \left(\frac{x}{\delta}\right)^2\right]dx \\ &= \frac{\rho g W \delta^3 \cos\beta}{2\mu}\int_0^1 (1 - \xi^2)d\xi \\ &= \frac{\rho g W \delta^3 \cos\beta}{3\mu} \end{aligned} \tag{2.2-24}$$

Here the dimensionless integration variable $\xi = x/\delta$ has been introduced.

(iii) The *average velocity* $\langle v_z \rangle$ over a cross section of the film is the volume rate of flow divided by the cross-sectional area

$$\langle v_z \rangle = \frac{\int_0^W \int_0^\delta v_z(x)dxdy}{\int_0^W \int_0^\delta dxdy} = \frac{Q}{W\delta} = \frac{\rho g \delta^2 \cos\beta}{3\mu} = \frac{2}{3}v_{z,\max} \tag{2.2-25}$$

(iv) The *mass rate of flow* w is obtained by integration of the convective mass flux distribution $\rho v_z(x)$ (see §0.4) over the entire cross section, or from the volume rate of flow,

$$w = \int_0^W \int_0^\delta \rho v_z(x)dxdy = \rho Q = \rho W \delta \langle v_z \rangle = \frac{\rho^2 g W \delta^3 \cos\beta}{3\mu} \tag{2.2-26}$$

(v) The *film thickness* δ may be given in terms of the average velocity or the mass rate of flow

$$\delta = \sqrt{\frac{3\mu\langle v_z \rangle}{\rho g \cos\beta}} = \sqrt[3]{\frac{3\mu w}{\rho^2 g W \cos\beta}} \tag{2.2-27}$$

(vi) The force per unit area in the z direction on a surface element perpendicular to the x direction is $+\pi_{xz} = +\tau_{xz}$ evaluated at $x = \delta$. This is the force exerted by the fluid (region of lesser x) on a unit area of the wall (region of greater x). The z component of the *force* **F** *of the fluid on the solid surface* is obtained by integrating the shear stress over the fluid-solid interface:

$$F_z = \int_0^L \int_0^W \left(\pi_{xz}\big|_{x=\delta}\right) dydz = \int_0^L \int_0^W \left(\tau_{xz}\big|_{x=\delta}\right) dydz = \int_0^L \int_0^W \left(-\mu \frac{dv_z}{dx}\bigg|_{x=\delta}\right) dydz$$
$$= (LW)(-\mu)\left(\frac{\rho g \delta \cos\beta}{\mu}\right) = \rho(LW\delta) g \cos\beta \qquad (2.2\text{-}28)$$

This is just the z component of the weight of the fluid in the entire film—as we would have expected.

Experimental observations of falling films show that there are actually three "flow regimes," and that these may be classified according to the *Reynolds number*,[1] Re, for the flow. For falling films the (dimensionless) Reynolds number is defined by $\text{Re} = 4\delta\langle v_z\rangle\rho/\mu$. The three flow regimes are

laminar flow with negligible rippling:	Re < 20
laminar flow with pronounced rippling:	20 < Re < 1500
turbulent flow:	Re > 1500

The analysis we have given above is valid only for the first regime, since the analysis was restricted by the postulates made at the outset. Ripples appear on the surface of the fluid at all Reynolds numbers. For Reynolds numbers less than about 20, the ripples are very long and grow rather slowly as they travel down the surface of the liquid; as a result, the formulas derived above are useful up to about Re = 20 for plates of moderate length. Above that value of Re, the ripple growth increases very rapidly, although the flow remains laminar. At about Re = 1500 the flow becomes irregular and chaotic, and the flow is said to be turbulent.[2,3] At this point it is not clear why the value of the Reynolds number should be used to delineate the flow regimes. We shall have more to say about this in Chapter 5.

This discussion illustrates a very important point: the theoretical analysis of a flow system is limited by the postulates that are made in setting up the problem. It is absolutely necessary to do experiments to establish the flow regimes in order to know when instabilities (e.g., spontaneous rippling) occur and when the flow becomes turbulent. Some information about the onset of instability and the demarcation of the flow regimes can be obtained by theoretical analysis, but this is an extraordinarily difficult subject. This is a result of the inherent nonlinear nature of the governing equations of fluid dynamics, as will be explained in Chapter 3. Suffice it to say, at this point, that experiments play a *very* important role in the field of fluid dynamics.

EXAMPLE 2.2-1

Calculation of Film Velocity

An oil has a kinematic viscosity of 2×10^{-4} m^2/s and a density of 0.8×10^3 kg/m^3. If we want to have a falling film of thickness of 2.5 mm on a vertical wall, what should the mass rate of flow of the liquid be?

---

[1] This dimensionless group is named for **Osborne Reynolds** (1842–1912), professor of engineering at the University of Manchester in England. He studied the laminar-turbulent transition, turbulent heat transfer, and theory of lubrication. We shall see in the Chapter 5 that the Reynolds number is the ratio of the inertial forces to the viscous forces.

[2] G. D. Fulford, *Adv. Chem. Engr.*, **5**, 151–236 (1964); S. Whitaker, *Ind. Eng. Chem. Fund.*, **3**, 132–142 (1964); V. G. Levich, *Physicochemical Hydrodynamics*, Prentice-Hall, Englewood Cliffs, NJ (1962), §135.

[3] H.-C. Chang, *Ann. Rev. Fluid Mech.*, **26**, 103–136 (1994); S.-H. Hwang and H.-C. Chang, *Phys. Fluids*, **30**, 1259–1268 (1987).

SOLUTION

According to Eq. 2.2-26, the mass rate of flow in kg/s is

$$w = \frac{\rho g \delta^3 W}{3\nu} = \frac{(0.8 \times 10^3\ \text{kg/m}^3)(9.80\ \text{m/s}^2)(2.5 \times 10^{-3}\ \text{m})^3 W}{3(2 \times 10^{-4}\ \text{m}^2/\text{s})} = (0.204\ \text{kg/m}\cdot\text{s})W \quad (2.2\text{-}29)$$

To get the mass rate of flow, one then needs to insert a value for the width of the wall W in meters. This is the desired result provided that the flow is laminar and nonrippling. To determine the flow regime we calculate the Reynolds number making use of Eq. 2.2-26

$$\begin{aligned} \text{Re} &= \frac{4\delta\langle v_z\rangle\rho}{\mu} = \frac{4w/W}{\nu\rho} \\ &= \frac{4(0.204\ \text{kg/m}\cdot\text{s})}{(2 \times 10^{-4}\ \text{m}^2/\text{s})(0.8 \times 10^3\ \text{kg/m}^3)} = 5.1 \end{aligned} \quad (2.2\text{-}30)$$

This Reynolds number is sufficiently low that rippling will not be pronounced, and therefore, using the expression for the mass rate of flow in Eq. 2.2-26 is reasonable.

EXAMPLE 2.2-2

Falling Film with Variable Viscosity

Rework the falling-film problem for a position-dependent viscosity $\mu = \mu_0 e^{-\alpha x/\delta}$, which arises when the film is nonisothermal, as in the condensation of a vapor on a wall. Here μ_0 is the viscosity at the surface of the film and α is a constant that describes how rapidly μ decreases as x increases. Such a variation could arise in the flow of a condensate down a wall with a linear temperature gradient through the film.

SOLUTION

The development proceeds as before up to Eq. 2.2-17. Substituting Newton's law with variable viscosity into Eq. 2.2-17 gives

$$-\mu_0 e^{-\alpha x/\delta}\frac{dv_z}{dx} = (\rho g \cos\beta)x \quad (2.2\text{-}31)$$

This equation can be integrated, and using the no-slip boundary condition (Eq. 2.2-21) enables us to evaluate the integration constant. The velocity profile is then

$$v_z(x) = \frac{\rho g \delta^2 \cos\beta}{\mu_0}\left[e^{\alpha}\left(\frac{1}{\alpha} - \frac{1}{\alpha^2}\right) - e^{\alpha x/\delta}\left(\frac{x}{\alpha\delta} - \frac{1}{\alpha^2}\right)\right] \quad (2.2\text{-}32)$$

As a check we evaluate the velocity distribution for the constant-viscosity problem (that is, when α is zero). However, setting $\alpha = 0$ gives $\infty - \infty$ in the two expressions within parentheses. This difficulty can be overcome if we expand the two exponentials in Taylor series (see §C.2), as follows:

$$\begin{aligned} [v_z(x)]_{\alpha=0} &= \frac{\rho g \delta^2 \cos\beta}{\mu_0}\lim_{\alpha\to 0}\left[\left(1 + \alpha + \frac{\alpha^2}{2!} + \frac{\alpha^3}{3!} + \cdots\right)\left(\frac{1}{\alpha} - \frac{1}{\alpha^2}\right)\right. \\ &\quad \left. - \left(1 + \frac{\alpha x}{\delta} + \frac{1}{2!}\frac{\alpha^2 x^2}{\delta^2} + \frac{1}{3!}\frac{\alpha^3 x^3}{\delta^3} + \cdots\right)\left(\frac{x}{\alpha\delta} - \frac{1}{\alpha^2}\right)\right] \\ &= \frac{\rho g \delta^2 \cos\beta}{\mu_0}\lim_{\alpha\to 0}\left[\begin{array}{l}\left(-\frac{1}{\alpha^2} + \frac{1}{2} + \frac{1}{3}\alpha + \cdots\right) \\ -\left(-\frac{1}{\alpha^2} + \frac{1}{2}\frac{x^2}{\delta^2} + \frac{1}{3}\frac{x^3}{\delta^3}\alpha + \cdots\right)\end{array}\right] \\ &= \frac{\rho g \delta^2 \cos\beta}{2\mu_0}\left[1 - \left(\frac{x}{\delta}\right)^2\right] \end{aligned} \quad (2.2\text{-}33)$$

which is in agreement with Eq. 2.2-22.

From Eq. 2.2-32 it may be shown that the average velocity is

$$\langle v_z\rangle = \frac{\rho g \delta^2 \cos\beta}{\mu_0}\left[e^{\alpha}\left(\frac{1}{\alpha} - \frac{2}{\alpha^2} + \frac{2}{\alpha^3}\right) - \frac{2}{\alpha^3}\right] \quad (2.2\text{-}34)$$

The reader may verify that this result simplifies to Eq. 2.2-25 when α goes to zero.

§2.3 FLOW THROUGH A CIRCULAR TUBE

The flow of fluids in circular tubes is encountered frequently in physics, chemistry, biology, and engineering. The laminar flow of fluids in circular tubes may be analyzed by means of the momentum balance described in §2.1. The only new feature introduced here is the use of cylindrical coordinates, which are the natural coordinates for describing positions in a pipe of circular cross section.

We consider then the steady-state, laminar flow of a fluid of constant density ρ and viscosity μ in a vertical tube section of length L and radius R (see Fig. 2.3-1(*a*)). The liquid flows downward under the influence of a pressure difference and gravity; the coordinate system is that shown in Fig. 2.3-1(*b*). If we wish to apply this analysis to the entire tube length, we require that the tube length be very large with respect to the tube radius, so that "end effects" will be unimportant throughout most of the tube; that is, we can ignore the fact that at the tube entrance and exit the flow will not necessarily be parallel to the tube wall.

We postulate that $v_z = v_z(r)$, $v_r = 0$, $v_\theta = 0$, and $p = p(z)$. Because we are interested in determining the velocity component v_z, we will set up and solve a z momentum balance.

We select as our system a cylindrical shell of thickness Δr and length L, located at an arbitrary radial position r. A section of this shell is illustrated as the lightly shaded region in Fig. 2.3-1(*b*).

Next we set up a z momentum balance for this shell, using the total momentum-flux components ϕ_{rz} and ϕ_{zz} to describe the flux of z momentum across the r and z faces, respectively (see Table 1.3-1; for a complete cylindrical shell, there are no faces with constant θ). These components are then multiplied by the corresponding areas of the shell faces to obtain the total rates of z momentum transported "in" or "out" of the shell across each face. The z momentum balance is illustrated pictorially in Fig. 2.3-1(*b*), where these

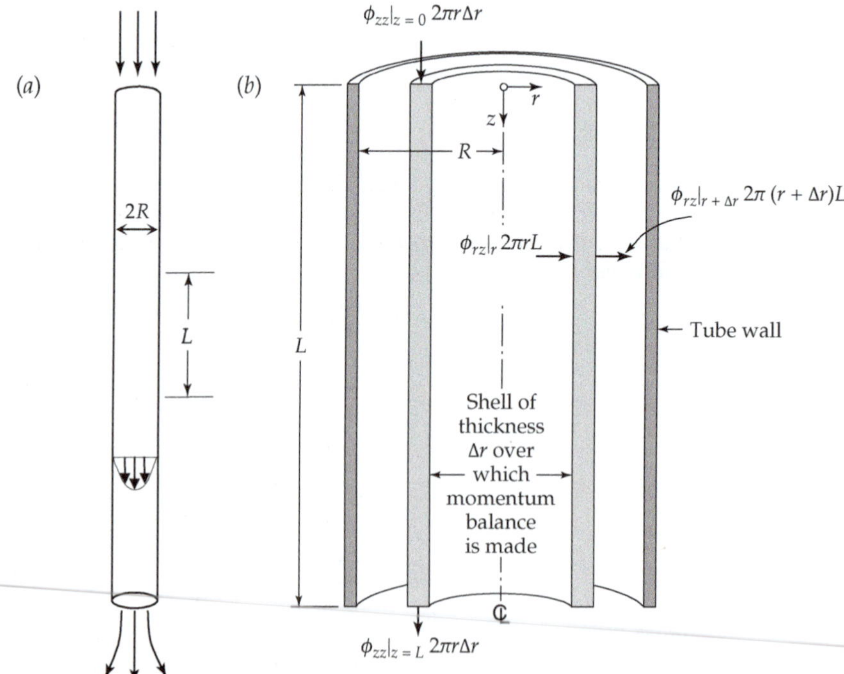

Fig. 2.3-1. Cylindrical shell of fluid over which the z momentum balance is made for axial flow in a circular tube (see Eqs. 2.3-1 to 2.3-5). The z momentum fluxes ϕ_{rz} and ϕ_{zz} are given in full in Eqs. 2.3-9 and 2.3-10.

"in" and "out" terms are depicted along with arrows. These arrows indicate that the momentum-flux components ϕ_{rz} and ϕ_{zz} are defined as the fluxes of z momentum in the positive r and z directions, respectively (regardless of the direction that z momentum is actually being transported). The terms in the z momentum balance are thus

Rate of total z momentum in across washer-shaped surface at z = 0:	$\phi_{zz}\|_{z=0} 2\pi r \Delta r$	(2.3-1)
Rate of total z momentum out across washer-shaped surface at z = L:	$\phi_{zz}\|_{z=L} 2\pi r \Delta r$	(2.3-2)
Rate of total z momentum in across cylindrical surface at r:	$\phi_{rz}\|_r 2\pi r L = (2\pi r L \phi_{rz})\|_r$	(2.3-3)
Rate of total z momentum out across cylindrical surface at r + Δr:	$\phi_{rz}\|_{r+\Delta r} 2\pi(r+\Delta r)L = (2\pi r L \phi_{rz})\|_{r+\Delta r}$	(2.3-4)
Gravity force acting in z direction on cylindrical shell:	$\rho(2\pi r \Delta r L)g$	(2.3-5)

The quantities ϕ_{zz} and ϕ_{rz} account for the momentum transport by both the convective and molecular mechanisms. In Eq. 2.3-4, $(r)|_{r+\Delta r}$ is used as an alternate way of writing $(r + \Delta r)$.

When these terms are substituted into the z momentum balance of Eq. 2.1-1, we get

$$(2\pi r L \phi_{rz})|_r - (2\pi r L \phi_{rz})|_{r+\Delta r} + \phi_{zz}|_{z=0} 2\pi r \Delta r - \phi_{zz}|_{z=L} 2\pi r \Delta r + \rho(2\pi r \Delta r L)g = 0 \tag{2.3-6}$$

When we divide Eq. (2.3-6) by $2\pi L \Delta r$ and take the limit as $\Delta r \to 0$, we get

$$\lim_{\Delta r \to 0} \left(\frac{(r\phi_{rz})|_{r+\Delta r} - (r\phi_{rz})|_r}{\Delta r} \right) = \left(\frac{\phi_{zz}|_{z=0} - \phi_{zz}|_{z=L}}{L} + \rho g \right) r \tag{2.3-7}$$

The left side is the definition of the first derivative of $r\phi_{rz}$ with respect to r. Hence, Eq. 2.3-7 may be rewritten as

$$\frac{d}{dr}(r\phi_{rz}) = \left(\frac{\phi_{zz}|_{z=0} - \phi_{zz}|_{z=L}}{L} + \rho g \right) r \tag{2.3-8}$$

Now we evaluate the components ϕ_{rz} and ϕ_{zz} using Eq. 1.3-1 or Table 1.3-1, along with Appendix B.1 to evaluate the viscous stress components τ_{rz} and τ_{zz}

$$\phi_{rz} = \tau_{rz} + \rho v_r v_z = -\mu \frac{\partial v_z}{\partial r} + \rho v_r v_z \tag{2.3-9}$$

$$\phi_{zz} = p + \tau_{zz} + \rho v_z v_z = p - 2\mu \frac{\partial v_z}{\partial z} + \rho v_z v_z \tag{2.3-10}$$

In accordance with the postulates that $v_z = v_z(r)$, $v_r = 0$, $v_\theta = 0$, and $p = p(z)$, we see that:

(i) because $v_r = 0$, the term $\rho v_r v_z$ in Eq. 2.3-9 is zero.

(ii) because $v_z = v_z(r)$, the term $\rho v_z v_z$ in Eq. 2.3-10 will be the same at $z = 0$ and $z = L$.

(iii) because $v_z = v_z(r)$, the term $-2\mu(\partial v_z/\partial z) = \tau_{zz}$ in Eq. 2.3-10 is zero.

Hence, Eq. 2.3-8 simplifies to

$$\frac{d}{dr}(r\tau_{rz}) = \left(\frac{(p_0 - \rho g 0) - (p_L - \rho g L)}{L} \right) r = \left(\frac{\mathscr{P}_0 - \mathscr{P}_L}{L} \right) r \tag{2.3-11}$$

in which $\mathscr{P} = p - \rho g z$ is a convenient abbreviation for the sum of the pressure and gravitational terms.[1] Equation 2.3-11 may be integrated to give

$$\tau_{rz}(r) = \left(\frac{\mathscr{P}_0 - \mathscr{P}_L}{2L}\right) r + \frac{C_1}{r} \tag{2.3-12}$$

The constant C_1 is evaluated by using the boundary condition

B.C. 1: at $r = 0$, τ_{rz} = finite (2.3-13)

Consequently C_1 must be zero, for otherwise the momentum flux would be infinite at the axis of the tube. Therefore, the viscous momentum-flux distribution is

$$\boxed{\tau_{rz}(r) = \left(\frac{\mathscr{P}_0 - \mathscr{P}_L}{2L}\right) r} \tag{2.3-14}$$

This distribution is shown in Fig. 2.3-2.

Newton's law of viscosity for this situation is obtained from Appendix B.1

$$\tau_{rz} = -\mu \frac{dv_z}{dr} \tag{2.3-15}$$

Substitution of this expression into Eq. 2.3-14 then gives the following differential equation for the velocity:

$$\frac{dv_z}{dr} = -\left(\frac{\mathscr{P}_0 - \mathscr{P}_L}{2\mu L}\right) r \tag{2.3-16}$$

This first-order, separable differential equation may be integrated to give

$$v_z(r) = -\left(\frac{\mathscr{P}_0 - \mathscr{P}_L}{4\mu L}\right) r^2 + C_2 \tag{2.3-17}$$

The constant C_2 is evaluated from the no-slip boundary condition

B. C. 2: at $r = R$, $v_z = 0$ (2.3-18)

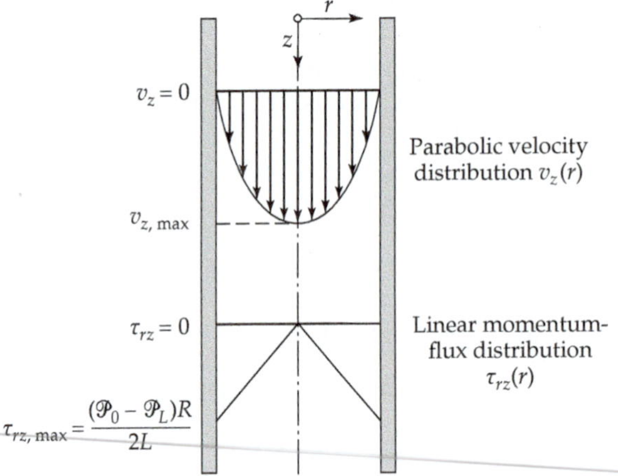

Fig. 2.3-2. The viscous momentum-flux distribution $\tau_{rz}(r)$ and velocity distribution $v_z(r)$ for the downward flow in a circular tube.

---

[1] The quantity designated by $\mathscr{P}$ is called the *modified pressure*. In general it is defined by $\mathscr{P} = p + \rho g h$, where h is the distance "upward"—that is, in the direction opposed to gravity—from some preselected reference plane. In this problem, $h = -z$.

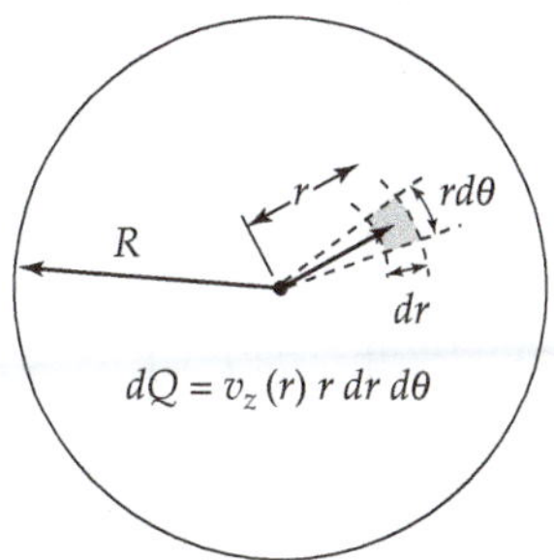

Fig. 2.3-3. Differential area element used to calculate the volumetric flow rate. This flow rate is obtained by integrating dQ over the entire cross section (i.e., as r goes from 0 to R, and θ goes from 0 to 2π).

Substitution of this boundary condition into Eq. 2.3-17 gives $C_2 = (\mathscr{P}_0 - \mathscr{P}_L)R^2/4\mu L$. Hence, the velocity distribution is

$$\boxed{v_z(r) = \frac{(\mathscr{P}_0 - \mathscr{P}_L)R^2}{4\mu L}\left[1 - \left(\frac{r}{R}\right)^2\right]} \tag{2.3-19}$$

We see that the velocity distribution for laminar, incompressible flow of a Newtonian fluid in a long tube is parabolic (see Fig. 2.3-2).

Once the velocity profile has been established, various derived quantities can be obtained:

(i) The *maximum velocity* $v_{z,\max}$ occurs at $r = 0$ and is

$$v_{z,\max} = \frac{(\mathscr{P}_0 - \mathscr{P}_L)R^2}{4\mu L} \tag{2.3-20}$$

(ii) The *volume rate of flow* Q is obtained by integrating the volumetric flow rate through an area element, $rdrd\theta$, perpendicular to the flow over the entire cross section (see Fig. 2.3-3)

$$\begin{aligned} Q &= \int_0^{2\pi}\int_0^R v_z(r)rdrd\theta = 2\pi\cdot\frac{(\mathscr{P}_0 - \mathscr{P}_L)R^2}{4\mu L}\int_0^R\left[1 - \left(\frac{r}{R}\right)^2\right]rdr \\ &= \frac{\pi(\mathscr{P}_0 - \mathscr{P}_L)R^4}{2\mu L}\int_0^1 (1-\xi^2)\xi d\xi = \frac{\pi(\mathscr{P}_0 - \mathscr{P}_L)R^4}{8\mu L} \end{aligned} \tag{2.3-21}$$

in which $\xi = r/R$. This rather famous result is called the *Hagen-Poiseuille*[2] *equation*. It is used, along with experimental data for the rate of flow and the modified pressure difference, to determine the viscosity of fluids (see Example 2.3.1) in a "capillary viscometer."

(iii) The *average velocity* $\langle v_z\rangle$ is obtained by dividing the total volumetric flow rate by the cross-sectional area

$$\langle v_z\rangle = \frac{\pi(\mathscr{P}_0 - \mathscr{P}_L)R^4}{8\mu L}\cdot\frac{1}{\pi R^2} = \frac{(\mathscr{P}_0 - \mathscr{P}_L)R^2}{8\mu L} = \frac{1}{2}v_{z,\max} \tag{2.3-22}$$

(iv) The *mass rate of flow* w is the product of the volumetric rate of flow Q and the density ρ

$$w = \frac{\pi(\mathscr{P}_0 - \mathscr{P}_L)R^4\rho}{8\mu L} \tag{2.3-23}$$

[2]G. Hagen, *Ann. Phys. Chem.*, **46**, 423–442 (1839); J. L. Poiseuille, *Comptes Rendus*, **11**, 961 and 1041 (1841). **Jean Louis Poiseuille** (1799–1869) (pronounced "Pwa-zø'-yuh," where ø is roughly the "oo" in book) was a physician interested in the flow of blood. The unit "poise" was named after Poiseuille. Although Hagen and Poiseuille established the dependence of the flow rate on the fourth power of the tube radius, Eq. 2.3-21 was first derived by E. Hagenbach, *Pogg. Annalen der Physik u. Chemie*, **108**, 385–426 (1860).

(v) The *rate of flow of kinetic energy* through the tube is obtained by integrating the volume rate of flow through an element of cross section $rdrd\theta$, namely, $v_z rdrd\theta$, multiplied by the kinetic energy per unit volume of the fluid $\frac{1}{2}\rho v_z^2$. Therefore, the total amount of kinetic energy per unit volume flowing through the tube is

$$\begin{aligned}\int_0^{2\pi}\int_0^R \left(\frac{1}{2}\rho v_z^2\right) v_z rdrd\theta &= 2\pi \int_0^R \left(\frac{1}{2}\rho v_z^2\right) v_z rdr \\ &= 2\pi \cdot \frac{1}{2}\rho \cdot R^2 \int_0^1 v_z^3 \xi d\xi \\ &= \rho\pi R^2 v_{z,\max}^3 \int_0^1 \left(1-\xi^2\right)^3 \xi d\xi \end{aligned} \tag{2.3-24}$$

Here, we have introduced the dimensionless coordinate $\xi = r/R$ and the maximum velocity $v_{z,\max} = (\mathscr{P}_0 - \mathscr{P}_L)R^2/4\mu L$. Now all that remains is to evaluate the integral. The rate of flow of kinetic energy is then

$$\begin{aligned}\rho\pi R^2 v_{z,\max}^3 \int_0^1 (1-3\xi^2+3\xi^4-\xi^6)\,\xi d\xi &= \rho\pi R^2 v_{z,\max}^3\left(\frac{1}{2}-\frac{3}{4}+\frac{3}{6}-\frac{1}{8}\right) \\ &= \rho\pi R^2 v_{z,\max}^3\left(\frac{4-6+4-1}{8}\right) \\ &= \frac{1}{8}\rho\pi R^2 v_{z,\max}^3 \\ &= \frac{1}{4}(\pi R^2 v_{z,\max})\left(\frac{1}{2}\rho v_{z,\max}^2\right)\end{aligned} \tag{2.3-25}$$

(vi) The z component of the *force, F_z, of the fluid on the wetted surface* of the pipe is just the shear stress $\pi_{rz} = \tau_{rz}$ integrated over the wetted area

$$\begin{aligned}F_z = \int_0^L \int_0^{2\pi} \tau_{rz}|_{r=R} Rd\theta dz &= (2\pi RL)\left(-\mu\frac{dv_z}{dr}\right)\bigg|_{r=R} = \pi R^2(\mathscr{P}_0 - \mathscr{P}_L) \\ &= \pi R^2(p_0 - p_L) + \pi R^2 L\rho g\end{aligned} \tag{2.3-26}$$

This result states that the viscous force F_z exerted by the fluid on the solid pipe wall is equal to the sum of the net pressure force and the gravitational force. By Newton's third law of motion, the force exerted *by the wall on the fluid* is $-F_z$. This is exactly what one would obtain from making a force balance over all of the fluid in the tube, where all of the forces acting on the fluid—the pressure force, gravity force, and (viscous) force *exerted by the wall*—must sum to zero at steady state. We will discuss such macroscopic balances in more detail in Chapter 7.

The results of this section are only as good as the postulates introduced at the beginning of the section, namely, that $v_z = v_z(r)$ and $p = p(z)$. Experiments have shown that these postulates are in fact realized for Reynolds numbers up to about 2100; above that value, the flow will be turbulent if there are any appreciable disturbances in the system—that is, wall roughness or vibrations.[3] For circular tubes the Reynolds number is defined by $\text{Re} = D\langle v_z\rangle\rho/\mu$, where $D = 2R$ is the tube diameter.

We now summarize all the assumptions that were made in obtaining the Hagen-Poiseuille equation.

(a) The flow is laminar; that is, Re must be less than about 2100.

(b) The density is constant ("incompressible flow").

---

[3] A. A. Draad [Doctoral Dissertation, Technical University of Delft (1996)] in a carefully controlled experiment, attained laminar flow up to Re = 60,000. He also studied the nonparabolic velocity profile induced by the Earth's rotation (through the Coriolis effect). See also A. A. Draad and F. T. M. Nieuwstadt, *J. Fluid Mech.*, **361**, 207-308 (1998).

(c) The flow is "steady" (i.e., it does not change with time).

(d) The fluid is Newtonian (Eq. 2.3-15 is valid).

(e) End effects are neglected. Actually an "entrance length," after the tube entrance, of the order of $L_e = 0.035D\text{Re}$, is needed for the buildup to the parabolic profile. If the section of pipe of interest includes the entrance region, a correction must be applied.[4] The fractional correction in the pressure difference or mass rate of flow never exceeds L_e/L if $L > L_e$.

(f) The fluid behaves as a continuum. This assumption is valid for most situations, but becomes invalid for very dilute gases or very narrow tubes in which the molecular mean-free path is comparable to the tube diameter (the "slip flow region") or much greater than the tube diameter (the "Knudsen flow" or "free molecule flow" regime).[5]

(g) There is no slip at the wall, so that B. C. 2 is valid; this is an excellent assumption for pure fluids under the conditions assumed in (f). See Problem 2B.10 for a discussion of wall slip.

EXAMPLE 2.3-1

Determination of Viscosity from Capillary Flow Data

Glycerol ($CH_2OH \cdot CHOH \cdot CH_2OH$) at 26.5°C is flowing through a horizontal tube 1 ft long and with 0.1 in. inside diameter. For a pressure drop of 40 psi, the volume flow rate Q is 0.00398 ft^3/min. The density of glycerol at 26.5°C is 1.261 g/cm^3. From the flow data, find the viscosity of glycerol in centipoises and in Pa · s. (For conversion of units, see Appendix E.)

SOLUTION

From the Hagen-Poiseuille equation (Eq. 2.3-21), we find

$$\begin{aligned}\mu &= \frac{\pi(\mathcal{P}_0 - \mathcal{P}_L)R^4}{8QL} \\ &= \frac{\pi\left(40\dfrac{\text{lb}_f}{\text{in.}^2}\right)\left(\dfrac{6.8947\times10^4\ \text{dyn/cm}^2}{\text{lb}_f/\text{in.}^2}\right)\left(0.05\ \text{in.}\times\dfrac{1\ \text{ft}}{12\ \text{in.}}\right)^4}{8\left(0.00398\dfrac{\text{ft}^3}{\text{min}}\times\dfrac{1\ \text{min}}{60\ \text{s}}\right)(1\ \text{ft})} \\ &= 4.92\frac{\text{g}}{\text{cm}\cdot\text{s}} = 492\ \text{cp} = 0.492\ \text{Pa}\cdot\text{s}\end{aligned} \tag{2.3-27}$$

To check whether the flow is laminar, we calculate the Reynolds number

$$\begin{aligned}\text{Re} &= \frac{D\langle v_z\rangle\rho}{\mu} = \frac{4Q\rho}{\pi D\mu} \\ &= \frac{4\left(0.00398\dfrac{\text{ft}^3}{\text{min}}\right)\left(\dfrac{2.54\ \text{cm}}{1\ \text{in.}}\times\dfrac{12\ \text{in.}}{1\ \text{ft}}\right)^3\left(\dfrac{1\ \text{min}}{60\ \text{s}}\right)\left(1.261\dfrac{\text{g}}{\text{cm}^3}\right)}{\pi\left(0.1\ \text{in.}\times\dfrac{2.54\ \text{cm}}{1\ \text{in.}}\right)\left(4.92\dfrac{\text{g}}{\text{cm}\cdot\text{s}}\right)} \\ &= 2.41(\text{dimensionless})\end{aligned} \tag{2.3-28}$$

---

[4] J. H. Perry, *Chemical Engineers Handbook*, McGraw-Hill, New York, 3rd edition (1950), pp. 388–389; W. M. Kays and A. L. London, *Compact Heat Exchangers*, McGraw-Hill, New York (1958), p. 49.

[5] **Martin Hans Christian Knudsen** (1871–1949), professor of physics at the University of Copenhagen, did key experiments on the behavior of very dilute gases. The lectures he gave at the University of Glasgow were published as M. Knudsen, *The Kinetic Theory of Gases*, Methuen, London (1934); G. N. Patterson, *Molecular Flow of Gases*, Wiley, New York (1956). See also J. H. Ferziger and H. G. Kaper, *Mathematical Theory of Transport Processes in Gases*, North-Holland, Amsterdam (1972), Chapter 15.

Hence, the flow is indeed laminar. Furthermore, the entrance length is

$$L_e = 0.035D\,\text{Re} = 0.035\left(0.1\text{ in.} \times \frac{1\text{ ft}}{12\text{ in.}}\right)(2.41) = 0.0007\text{ ft} \tag{2.3-29}$$

Hence, entrance effects are not important, and the viscosity value given above has been calculated properly.

EXAMPLE 2.3-2

Compressible Flow in a Horizontal Circular Tube[6]

Obtain an expression for the mass rate of flow w for an ideal gas in laminar flow in a long circular tube. The flow is presumed to be isothermal, so that the viscosity can be regarded as constant (because the gas is ideal, the viscosity does not depend on pressure).

SOLUTION

This problem can be solved *approximately* by assuming that the Hagen-Poiseuille equation in the form of Eq. 2.3-23 can be applied over a small length dz of the tube as follows:

$$w = \frac{\pi\rho R^4}{8\mu}\left(-\frac{dp}{dz}\right) \tag{2.3-30}$$

To eliminate ρ in favor of p, we use the ideal gas law in the form $p/\rho = p_0/\rho_0$, where p_0 and ρ_0 are the pressure and density at $z = 0$. This gives

$$w = \frac{\pi R^4}{8\mu}\frac{\rho_0}{p_0}\left(-p\frac{dp}{dz}\right) \tag{2.3-31}$$

The mass rate of flow w is the same for all z. Hence, Eq. 2.3-31 can be integrated from $z = 0$ to $z = L$ to give

$$w = \frac{\pi R^4}{16\mu L}\frac{\rho_0}{p_0}(p_0^2 - p_L^2) \tag{2.3-32}$$

Since $p_0^2 - p_L^2 = (p_0 + p_L)(p_0 - p_L)$, we get finally

$$w = \frac{\pi(p_0 - p_L)R^4\rho_{\text{avg}}}{8\mu L} \tag{2.3-33}$$

where $\rho_{\text{avg}} = (\rho_0 + \rho_L)/2$ is the average density calculated at the average pressure $p_{\text{avg}} = (p_0 + p_L)/2$.

EXAMPLE 2.3-3

Tube-Branching in Poiseuille Flow in Blood Vessels

A tube of radius R_1 branches into two tubes of equal radii R_2. Find the ratio R_2/R_1 when it is required that the wall shear stress be the same in all the tubes. It has been established that this requirement follows from a minimum principle that appears to be applicable to the bifurcation of blood vessels in the human body.[7]

SOLUTION

From the law of conservation of mass, the mass flow rate in the tube of radius R_1 must be equal to twice the mass flow rate in one of the tubes of radius R_2. That is, from Eq. 2.3-23

$$w = \frac{\rho\pi(\Delta\mathscr{P}/L)_1}{8\mu}R_1^4 = 2\frac{\rho\pi(\Delta\mathscr{P}/L)_2}{8\mu}R_2^4 \tag{2.3-34}$$

or

$$\left(\frac{R_2}{R_1}\right)^4 = \frac{1}{2}\frac{(\Delta\mathscr{P}/L)_1}{(\Delta\mathscr{P}/L)_2} \tag{2.3-35}$$

---

[6]L. Landau and E. M. Lifshitz, *Fluid Mechanics*, Pergamon, 2nd edition (1987), §17, Problem 6. A perturbation solution of this problem has been obtained by R. K. Prud'homme, T. W. Chapman, and J. R. Bowen, *Appl. Sci. Res.*, **43**, 67–74 (1986).

[7]Y. C. Fung, *Biomechanics: Circulation*, Springer, New York, 2nd edition (1997), §3.3.

The requirement that the wall shear stress in both tubes be the same is, according to Eq. 2.3-14

$$\frac{(\Delta\mathscr{P}/L)_1 R_1}{2} = \frac{(\Delta\mathscr{P}/L)_2 R_2}{2} \tag{2.3-36}$$

or

$$\left(\frac{R_2}{R_1}\right) = \frac{(\Delta\mathscr{P}/L)_1}{(\Delta\mathscr{P}/L)_2} \tag{2.3-37}$$

Dividing Eq. 2.3-35 by Eq. 2.3-37 gives

$$\left(\frac{R_2}{R_1}\right)^3 = \frac{1}{2} \qquad \text{or} \qquad \frac{R_2}{R_1} = \frac{1}{\sqrt[3]{2}} = 0.794 \tag{2.3-38}$$

This gives a reliable result for the bifurcation of blood vessels, until the size of the blood vessels becomes of capillary size.

§2.4 FLOW THROUGH AN ANNULUS

We now solve another viscous flow problem in cylindrical coordinates, namely, the steady-state axial flow of an incompressible liquid in an annular region between two coaxial circular cylinders of radii κR and R as shown in Fig. 2.4-1. The fluid is flowing upwards in the annulus, that is, in the direction opposed to gravity. We make the same postulates as in §2.3: $v_z = v_z(r)$, $v_\theta = 0$, $v_r = 0$, and $p = p(z)$.

Because the annular flow problem and its postulates are similar to those for the tube flow problem, in that both are described in cylindrical coordinates, the shell of thickness Δr for the momentum balance can be chosen similarly—a thin cylindrical shell that extends from $z = 0$ to $z = L$. Furthermore, the z momentum balance will be identical to that derived in §2.3, up to Eq. 2.3-11. Thus, the z momentum balance produces the differential equation

$$\frac{d}{dr}(r\tau_{rz}) = \left(\frac{(p_0 + \rho g 0) - (p_L + \rho g L)}{L}\right) r \equiv \left(\frac{\mathscr{P}_0 - \mathscr{P}_L}{L}\right) r \tag{2.4-1}$$

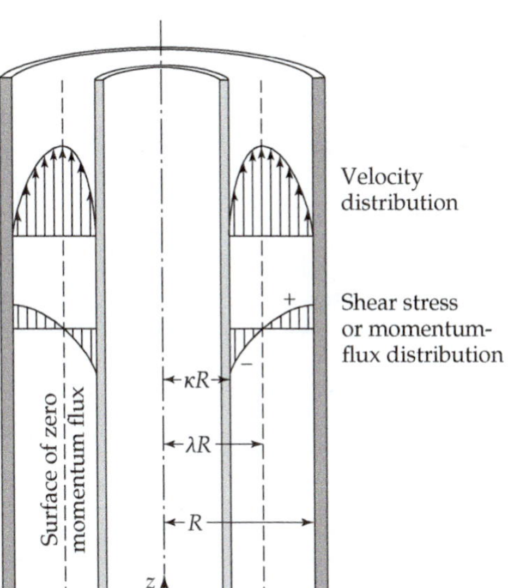

Fig. 2.4-1. The viscous momentum-flux distribution $\tau_{rz}(r)$ and velocity distribution $v_z(r)$ for the upward flow in a cylindrical annulus. Note that the momentum flux changes sign at the same value of r for which the velocity has a maximum.

This differs from Eq. 2.3-11 only in that $\mathscr{P} = p + \rho g z$ here, since the coordinate z is in the direction opposed to gravity (i.e., z is the same as the h of footnote 1 in §2.3). Integration of Eq. 2.4-1 gives

$$\tau_{rz}(r) = \left(\frac{\mathscr{P}_0 - \mathscr{P}_L}{2L}\right) r + \frac{C_1}{r} \tag{2.4-2}$$

just as in Eq. 2.3-12.

The constant C_1 cannot be determined immediately, since we have no information about the momentum flux at the fixed surfaces $r = \kappa R$ and $r = R$. We know only that there will be a maximum in the velocity curve at some (as yet unknown) cylindrical surface $r = \lambda R$ at which the momentum flux will be zero; here λ is a dimensionless quantity. That is

$$0 = \left(\frac{\mathscr{P}_0 - \mathscr{P}_L}{2L}\right) \lambda R + \frac{C_1}{\lambda R} \tag{2.4-3}$$

When we solve this equation for C_1 and substitute it into Eq. 2.4-2, we get

$$\tau_{rz}(r) = \frac{(\mathscr{P}_0 - \mathscr{P}_L)R}{2L}\left[\left(\frac{r}{R}\right) - \lambda^2\left(\frac{R}{r}\right)\right] \tag{2.4-4}$$

The only difference between this equation and Eq. 2.4-2 is that the constant of integration C_1 has been eliminated in favor of a different constant λ. The advantage of this is that we know the geometrical significance of λ.

We now substitute Newton's law of viscosity, $\tau_{rz} = -\mu(dv_z/dr)$, into Eq. 2.4-4 to obtain a differential equation for v_z

$$\frac{dv_z}{dr} = -\frac{(\mathscr{P}_0 - \mathscr{P}_L)R}{2\mu L}\left[\left(\frac{r}{R}\right) - \lambda^2\left(\frac{R}{r}\right)\right] \tag{2.4-5}$$

Integration of this first-order separable differential equation then gives

$$v_z(r) = -\frac{(\mathscr{P}_0 - \mathscr{P}_L)R^2}{4\mu L}\left[\left(\frac{r}{R}\right)^2 - 2\lambda^2 \ln\left(\frac{r}{R}\right) + C_2\right] \tag{2.4-6}$$

We now evaluate the two constants of integration λ and C_2 by using the no-slip condition on each solid boundary:

B. C. 1: $\quad$ at $r = \kappa R, \quad v_z = 0 \qquad (2.4\text{-}7)$

B. C. 2: $\quad$ at $r = R, \quad v_z = 0 \qquad (2.4\text{-}8)$

Substitution of these boundary conditions into Eq. 2.4-6 then gives two equations

$$0 = \kappa^2 - 2\lambda^2 \ln \kappa + C_2; \qquad 0 = 1 + C_2 \tag{2.4-9,10}$$

From these, the two integration constants λ and C_2 are found to be

$$C_2 = -1; \qquad 2\lambda^2 = \frac{1-\kappa^2}{\ln(1/\kappa)} \tag{2.4-11,12}$$

These expressions can be inserted into Eqs. 2.4-4 and 2.4-6 to give the viscous momentum-flux distribution and the velocity distribution[1] as follows:

$$\boxed{\tau_{rz}(r) = \frac{(\mathscr{P}_0 - \mathscr{P}_L)R}{2L}\left[\left(\frac{r}{R}\right) - \frac{1-\kappa^2}{2\ln(1/\kappa)}\left(\frac{R}{r}\right)\right]} \tag{2.4-13}$$

$$\boxed{v_z(r) = \frac{(\mathscr{P}_0 - \mathscr{P}_L)R^2}{4\mu L}\left[1 - \left(\frac{r}{R}\right)^2 - \frac{1-\kappa^2}{\ln(1/\kappa)}\ln\left(\frac{R}{r}\right)\right]} \tag{2.4-14}$$

[1] H. Lamb, *Hydrodynamics*, Cambridge University Press, 2nd edition (1895), p. 522.

Note that when the annulus becomes very thin (i.e., κ only slightly less than unity), these results simplify to those for a plane slit (see Problem 2B.6). It's always a good idea to check "limiting cases" such as these whenever possible.

The lower limit of $\kappa \to 0$ is not so simple, because the ratio $\ln(R/r)/\ln(1/\kappa)$ will always be important in a region near the inner boundary. Hence, Eq. 2.4-14 does not simplify to the parabolic distribution. However, Eq. 2.4-18 (below) for the mass rate of flow does simplify to Eq. 2.3-23 as $\kappa \to 0$.

Once we have the viscous momentum-flux and velocity distributions, it is straightforward to get other results of interest:

(i) The *maximum velocity* is

$$v_{z,\max} = v_z|_{r=\lambda R} = \frac{(\mathscr{P}_0 - \mathscr{P}_L)R^2}{4\mu L}\left[1 - \lambda^2\left(1 - \ln \lambda^2\right)\right] \tag{2.4-15}$$

where the dimensionless location of the maximum velocity λ is given in Eq. 2.4-12. It can then be shown that $\lambda < (1+\kappa)/2$; that is, the maximum velocity is at a point that is closer to the inner wall than to the outer wall.

(ii) The *volume rate of flow* Q is obtained in a manner similar to that employed for flow in a tube—by integrating the volumetric flow rate through an area element, $rdrd\theta$, perpendicular to the flow over the entire fluid cross section ($\kappa R < r < R$), to give

$$\begin{aligned} Q &= \int_0^{2\pi}\int_{\kappa R}^{R} v_z(r) r dr d\theta \\ &= 2\pi \cdot \frac{(\mathscr{P}_0 - \mathscr{P}_L)R^2}{4\mu L}\int_{\kappa R}^{R}\left[1 - \left(\frac{r}{R}\right)^2 - \frac{1-\kappa^2}{\ln(1/\kappa)}\ln\left(\frac{R}{r}\right)\right] r dr \\ &= \frac{\pi(\mathscr{P}_0 - \mathscr{P}_L)R^4}{8\mu L}\left[\left(1-\kappa^4\right) - \frac{(1-\kappa^2)^2}{\ln(1/\kappa)}\right] \end{aligned} \tag{2.4-16}$$

(iii) The *average velocity* is given by the volume rate of flow divided by the cross-sectional area

$$\langle v_z \rangle = \frac{\int_0^{2\pi}\int_{\kappa R}^{R} v_z(r) r dr d\theta}{\int_0^{2\pi}\int_{\kappa R}^{R} r dr d\theta} = \frac{Q}{\pi R^2 - \pi(\kappa R)^2} = \frac{(\mathscr{P}_0 - \mathscr{P}_L)R^2}{8\mu L}\left[\frac{1-\kappa^4}{1-\kappa^2} - \frac{1-\kappa^2}{\ln(1/\kappa)}\right] \tag{2.4-17}$$

(iv) The *mass rate of flow* can be obtained from the volume rate of flow via $w = \rho Q = \rho \cdot \pi R^2(1-\kappa^2)\langle v_z \rangle$

$$w = \frac{\pi(\mathscr{P}_0 - \mathscr{P}_L)R^4\rho}{8\mu L}\left[\left(1-\kappa^4\right) - \frac{(1-\kappa^2)^2}{\ln(1/\kappa)}\right] \tag{2.4-18}$$

In the limit that $\kappa \to 0$, Eq. 2.4-18 becomes the same as Eq. 2.3-23 for circular tubes.

(v) The *force exerted by the fluid on the solid surfaces* is obtained by summing the forces acting on the inner and outer cylinders, as follows:

$$\begin{aligned} F_z &= (2\pi\kappa RL)(-\pi_{rz}|_{r=\kappa R}) + (2\pi RL)(+\pi_{rz}|_{r=R}) \\ &= (2\pi\kappa RL)(-\tau_{rz}|_{r=\kappa R}) + (2\pi RL)(+\tau_{rz}|_{r=R}) \\ &= \pi R^2(1-\kappa^2)(\mathscr{P}_0 - \mathscr{P}_L) \end{aligned} \tag{2.4-19}$$

The reader should explain the choice of signs in front of the shear stresses above and also give an interpretation of the final result.

The equations derived above are valid only for laminar flow. The laminar-turbulent transition occurs in the neighborhood of Re = 2000, with the Reynolds number defined as $\text{Re} = 2R(1-\kappa)\langle v_z \rangle \rho/\mu$.

§2.5 FLOW OF TWO ADJACENT IMMISCIBLE FLUIDS[1]

Thus far we have considered flow situations with solid-fluid and liquid-gas boundaries. We now give one example of a flow problem with a liquid-liquid interface (see Fig. 2.5-1).

Two immiscible, incompressible liquids are flowing in the z direction in a horizontal thin slit of length L and width W under the influence of a horizontal pressure gradient $(p_0 - p_L)/L$. The fluid flow rates are adjusted so that the slit is half filled with fluid I (the more dense phase) and half filled with fluid II (the less dense phase). The fluids are flowing sufficiently slowly that no instabilities occur—that is, that the interface remains exactly planar. It is desired to find the viscous momentum-flux and velocity distributions.

A z momentum balance over a shell of thickness Δx leads to the following differential equation for the viscous momentum flux:

$$\frac{d\tau_{xz}}{dx} = \frac{p_0 - p_L}{L} \tag{2.5-1}$$

This equation is obtained for both phase I and phase II. Integration of Eq. 2.5-1 for the two regions gives

$$\tau_{xz}^{\text{I}}(x) = \left(\frac{p_0 - p_L}{L}\right)x + C_1^{\text{I}} \tag{2.5-2}$$

$$\tau_{xz}^{\text{II}}(x) = \left(\frac{p_0 - p_L}{L}\right)x + C_1^{\text{II}} \tag{2.5-3}$$

We may immediately make use of one of the boundary conditions, namely, that the momentum flux τ_{xz} is continuous through the fluid-fluid interface

B. C. 1: $\quad$ at $x = 0$, $\tau_{xz}^{\text{I}} = \tau_{xz}^{\text{II}}$ $\quad$ (2.5-4)

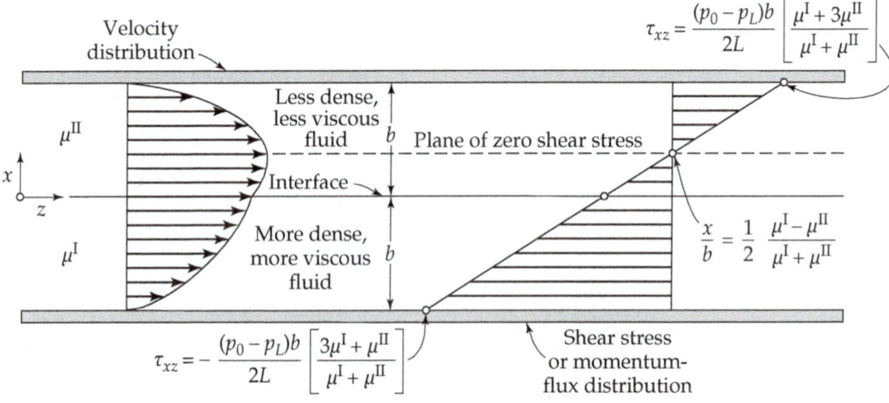

Fig. 2.5-1. Flow of two immiscible fluids between a pair of horizontal plates under the influence of a pressure gradient.

---

[1] The adjacent flow of gases and liquids in conduits has been reviewed by A. E. Dukler and M. Wicks, III, in Chapter 8 of *Modern Chemical Engineering*, Vol. **1**, "Physical Operations," A. Acrivos (ed.), Reinhold, New York (1963).

This tells us that $C_1^{\mathrm{I}} = C_1^{\mathrm{II}}$; hence, we drop the superscript and call both integration constants C_1.

When Newton's law of viscosity is substituted into Eqs. 2.5-2 and 2.5-3, we get

$$-\mu^{\mathrm{I}}\frac{dv_z^{\mathrm{I}}}{dx} = \left(\frac{p_0 - p_L}{L}\right)x + C_1 \tag{2.5-5}$$

$$-\mu^{\mathrm{II}}\frac{dv_z^{\mathrm{II}}}{dx} = \left(\frac{p_0 - p_L}{L}\right)x + C_1 \tag{2.5-6}$$

These two equations can be integrated to give

$$v_z^{\mathrm{I}}(x) = -\left(\frac{p_0 - p_L}{2\mu^{\mathrm{I}}L}\right)x^2 - \frac{C_1}{\mu^{\mathrm{I}}}x + C_2^{\mathrm{I}} \tag{2.5-7}$$

$$v_z^{\mathrm{II}}(x) = -\left(\frac{p_0 - p_L}{2\mu^{\mathrm{II}}L}\right)x^2 - \frac{C_1}{\mu^{\mathrm{II}}}x + C_2^{\mathrm{II}} \tag{2.5-8}$$

The three integration constants can be determined from the following no-slip boundary conditions:

B. C. 2: at $x = 0$, $v_z^{\mathrm{I}} = v_z^{\mathrm{II}}$ (2.5-9)

B. C. 3: at $x = -b$, $v_z^{\mathrm{I}} = 0$ (2.5-10)

B. C. 4: at $x = +b$, $v_z^{\mathrm{II}} = 0$ (2.5-11)

When these three boundary conditions are applied, we get three simultaneous equations for the integration constants:

From B. C. 2: $C_2^{\mathrm{I}} = C_2^{\mathrm{II}}$ (2.5-12)

From B. C. 3: $0 = -\left(\frac{p_0 - p_L}{2\mu^{\mathrm{I}}L}\right)b^2 + \frac{C_1}{\mu^{\mathrm{I}}}b + C_2^{\mathrm{I}}$ (2.5-13)

From B. C. 4: $0 = -\left(\frac{p_0 - p_L}{2\mu^{\mathrm{II}}L}\right)b^2 - \frac{C_1}{\mu^{\mathrm{II}}}b + C_2^{\mathrm{II}}$ (2.5-14)

From these three equations we get

$$C_1 = -\frac{(p_0 - p_L)b}{2L}\left(\frac{\mu^{\mathrm{I}} - \mu^{\mathrm{II}}}{\mu^{\mathrm{I}} + \mu^{\mathrm{II}}}\right) \tag{2.5-15}$$

$$C_2^{\mathrm{I}} = +\frac{(p_0 - p_L)b^2}{2\mu^{\mathrm{I}}L}\left(\frac{2\mu^{\mathrm{I}}}{\mu^{\mathrm{I}} + \mu^{\mathrm{II}}}\right) = C_2^{\mathrm{II}} \tag{2.5-16}$$

The resulting momentum-flux and velocity profiles are

$$\boxed{\tau_{xz}(x) = \frac{(p_0 - p_L)b}{2L}\left[\left(\frac{x}{b}\right) - \frac{1}{2}\left(\frac{\mu^{\mathrm{I}} - \mu^{\mathrm{II}}}{\mu^{\mathrm{I}} + \mu^{\mathrm{II}}}\right)\right]} \tag{2.5-17}$$

$$\boxed{v_z^{\mathrm{I}}(x) = \frac{(p_0 - p_L)b^2}{2\mu^{\mathrm{I}}L}\left[\left(\frac{2\mu^{\mathrm{I}}}{\mu^{\mathrm{I}} + \mu^{\mathrm{II}}}\right) + \left(\frac{\mu^{\mathrm{I}} - \mu^{\mathrm{II}}}{\mu^{\mathrm{I}} + \mu^{\mathrm{II}}}\right)\left(\frac{x}{b}\right) - \left(\frac{x}{b}\right)^2\right]} \tag{2.5-18}$$

$$\boxed{v_z^{\mathrm{II}}(x) = \frac{(p_0 - p_L)b^2}{2\mu^{\mathrm{II}}L}\left[\left(\frac{2\mu^{\mathrm{II}}}{\mu^{\mathrm{I}} + \mu^{\mathrm{II}}}\right) + \left(\frac{\mu^{\mathrm{I}} - \mu^{\mathrm{II}}}{\mu^{\mathrm{I}} + \mu^{\mathrm{II}}}\right)\left(\frac{x}{b}\right) - \left(\frac{x}{b}\right)^2\right]} \tag{2.5-19}$$

These distributions are shown in Fig. 2.5-1. If both viscosities are the same, then the velocity distribution is parabolic, as one would expect for a pure fluid flowing between parallel plates (see Eq. 2B.4-2).

The *average velocity* in each layer can be obtained and the results are

$$\langle v_z^{\rm I} \rangle = \frac{1}{b}\int_{-b}^{0} v_z^{\rm I}(x)dx = \frac{(\mathscr{P}_0 - \mathscr{P}_L)b^2}{12\mu^{\rm I}L}\left(\frac{7\mu^{\rm I} + \mu^{\rm II}}{\mu^{\rm I} + \mu^{\rm II}}\right) \tag{2.5-20}$$

$$\langle v_z^{\rm II} \rangle = \frac{1}{b}\int_{0}^{b} v_z^{\rm II}(x)dx = \frac{(\mathscr{P}_0 - \mathscr{P}_L)b^2}{12\mu^{\rm II}L}\left(\frac{\mu^{\rm I} + 7\mu^{\rm II}}{\mu^{\rm I} + \mu^{\rm II}}\right) \tag{2.5-21}$$

From the velocity and momentum-flux distributions given above, one can also calculate the maximum velocity, the velocity at the interface, the location of the plane of zero shear stress, and the force exerted by the fluid on the walls of the slit.

§2.6 FLOW IN A CONE-AND-PLATE VISCOMETER

A very popular type of viscometer consists of a stationary circular flat plate of radius R, and a rotating cone, which can be used to measure the viscosity of liquids (see Fig. 2.6-1). Although the very tip of the cone is typically removed, the cone is positioned so that the cone tip, if present, would just contact the center of the plate (see Fig. 2.6-2(*a*)). The angle between the cone and the plate is ψ_0, which is quite small—from 1 to 5 degrees in commercial instruments.

A liquid, whose viscosity is to be measured, is placed in the gap between the cone and the plate. The cone is made to rotate with an angular velocity Ω, and the torque T_z required

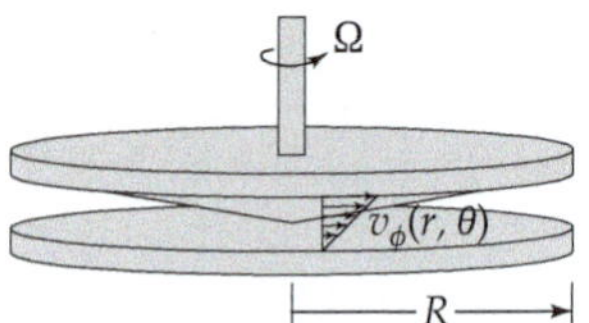

Fig. 2.6-1. Cone-and-plate viscometer. The plate remains fixed, and the cone is rotated at Ω rad/s.

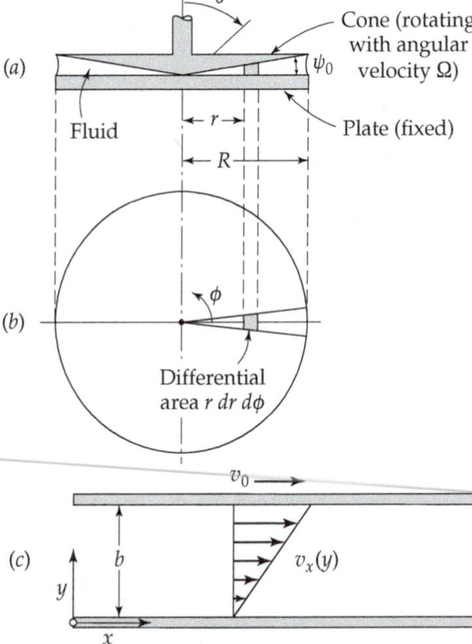

Fig. 2.6-2. Detailed illustrations of the cone-and-plate viscometer: (*a*) side view of the instrument, (*b*) top view of the cone-plate system, showing a differential element $r\,dr\,d\phi$, (*c*) an approximate velocity distribution within the differential region. To equate the systems in (*a*) and (*c*), we identify the following equivalences: $v_0 = \Omega r$ and $b = r\sin\psi_0 \approx r\psi_0$.

to turn the cone is measured. We want to find the relation between Ω, T_z, R, ψ_0 and the viscosity μ. This relation may be used to determine the liquid viscosity from experimental measurements.

The advantages of this instrument are that only a very small amount of liquid is needed, and that the velocity gradient is essentially constant throughout (which is more of an advantage for non-Newtonian fluids whose viscosities can depend on the velocity gradient; see Chapter 8). Disadvantages are that evaporation and surface tension at the outer edge can complicate the viscosity determination for some fluids, and that at high rates of rotation, the sample may be thrown outward from the instrument.

It is appropriate to use spherical coordinates to solve this problem. The variable r is the distance outward from the point of contact of the cone and plate. The variable ϕ is the direction of rotational motion of the cone, and θ is the angle downward from a line perpendicular to the plate. These variables are shown in Figs. 2.6-2(a) and (b). Thus, in spherical coordinates, the cone and the plate are both surfaces of constant θ (the cone surface is at $\theta = \frac{1}{2}\pi - \psi_0$ and the plate surface is at $\theta = \frac{1}{2}\pi$), and it is reasonable to assume that the flow is only in the ϕ direction. We will find it convenient to introduce the angle $\psi = \frac{1}{2}\pi - \theta$, which is the angle measured upward from the plate.

For the steady flow in spherical coordinates, we postulate that $v_\phi = v_\phi(r,\theta)$, $v_r = 0$, and $v_\theta = 0$. Because this flow is not rectilinear, we cannot employ the shell momentum balance described in §2.1 to obtain an exact solution to this flow problem. However, we can obtain an approximate solution for small cone angles ψ_0 because the flow at any radial position will be, locally, approximately rectilinear; that is, at any radial position, the fluid is bounded below by the stationary plate and bounded above by the nearly parallel cone, which locally appears to translate in a uniform direction. The velocity profile at any radial position can then be determined approximately by applying a momentum balance for a rectilinear flow in an appropriate Cartesian coordinate system.

To make further progress, we focus on a small volume of liquid at a position r, that extends from a surface area element $r\,dr\,d\phi$ on the plate upward to the cone as depicted in Figs. 2.6-2(a) and (b). *Locally* we can regard the flow as being very nearly that of the steady, rectilinear flow in Fig. 2.6-2(c), where the fluid is between parallel plates separated by a distance b and the top plate moves in the $+x$ direction with speed v_0. The variables in the Cartesian coordinate system of Fig. 2.6-2(c) are related to variables in the spherical coordinate system of Figs. 2.6-2(a) and (b) as described in Table 2.6-1.

The velocity profile $v_x(y)$ in Fig. 2.6-2(c) can be obtained by using a shell x momentum balance (Problem 2B.1). The resulting velocity profile is

$$v_x(y) = v_0 \frac{y}{b} \tag{2.6-1}$$

Table 2.6-1. Correspondences between the Variables Used to Describe Flow in a Cone-and-Plate Viscometer

Variable	Variable used in the local Cartesian coordinate system of Fig. 2.6-2(c)	Equivalent expression using the spherical coordinate system of Figs. 2.6-2(a) and (b)[a]
Fluid velocity	v_x	v_ϕ
Cone surface speed	v_0	Ωr
Plate separation	b	$r \sin \psi_0 \approx r\psi_0$
Vertical position within the fluid	y	$r \sin \psi \approx r\psi = r\left(\frac{1}{2}\pi - \theta\right)$

[a]The approximation $r \sin \psi \approx r\psi = r\left(\frac{1}{2}\pi - \theta\right)$ for small cone angles is used.

Using the relationships in Table 2.6-1, this velocity profile can be converted to the profile for $v_\phi(r,\theta)$ in Figs. 2.6-2(a,b)

$$v_\phi(r,\theta) = \frac{r\left(\frac{1}{2}\pi - \theta\right)}{r\psi_0}\Omega r = \frac{\left(\frac{1}{2}\pi - \theta\right)}{\psi_0}\Omega r \tag{2.6-2}$$

Now that we know the local velocity in spherical coordinates, we can evaluate the shear stress $\tau_{\theta\phi}$ from Eq. B.1-19.

$$\tau_{\theta\phi} = -\mu\frac{\sin\theta}{r}\frac{d}{d\theta}\left(\frac{v_\phi}{\sin\theta}\right) \tag{2.6-3}$$

Because θ is very nearly equal to $\frac{1}{2}\pi$, we can set $\sin\theta = 1$ in both places in Eq. 2.6-3, and write

$$\tau_{\theta\phi} = -\mu\frac{1}{r}\frac{dv_\phi}{d\theta} = \mu\frac{\Omega}{\psi_0} \tag{2.6-4}$$

This shows that the shear stress is uniform throughout the gap (when ψ_0 is small).

At steady state, the torque[1] T_z about the z axis required to rotate the cone is equal to the torque about the z axis exerted by the fluid on the lower plate. To calculate this torque, we integrate the torque exerted on a surface area element $r\,dr\,d\phi$ over the entire plate surface. As illustrated in Fig. 2.6-3, the torque about the z axis is the moment arm r multiplied by the component of the differential force that is perpendicular to the moment arm. In this case, the component of the force (exerted by the fluid on the area element) that is perpendicular to the moment arm is $dF_\phi = +\tau_{\theta\phi}|_{\theta=\pi/2}r\,dr\,d\phi$. Thus, the total torque about the z axis exerted by the fluid on the plate is given by the integral of $r\,dF_\phi$ over the plate surface

$$T_z = \int_0^{2\pi}\int_0^R r(\tau_{\theta\phi})|_{\theta=\pi/2}r\,dr\,d\phi \tag{2.6-5}$$

When the shear stress from Eq. 2.6-4 is inserted into Eq. 2.6-5 and the integration is performed, we get

$$\boxed{T_z = 2\pi\mu\frac{\Omega}{\psi_0}\int_0^R r^2 dr = \frac{2\pi\mu\Omega R^3}{3\psi_0}} \tag{2.6-6}$$

This is the standard formula used for calculating the viscosity from measurements of the torque and the angular velocity for a cone-and-plate instrument of known radius R and cone angle ψ_0.

This problem illustrates how we can obtain a satisfactory solution to some problems by approximating the flow system locally as a steady shear flow. This particular problem can actually be solved exactly in spherical coordinates to give for the velocity distribution[2]

$$\frac{v_\phi(r,\theta)}{r} = \Omega\sin\theta_0\left[\frac{\cot\theta + \frac{1}{2}\left(\ln\dfrac{1+\cos\theta}{1-\cos\theta}\right)\sin\theta}{\cot\theta_0 + \frac{1}{2}\left(\ln\dfrac{1+\cos\theta_0}{1-\cos\theta_0}\right)\sin\theta_0}\right] \tag{2.6-7}$$

where $\theta_0 = \frac{1}{2}\pi - \psi_0$. When both θ and θ_0 are very close to $\frac{1}{2}\pi$, Eq. 2.6-7 reduces to Eq. 2.6-2. The reader should be very pleased that it is not necessary to derive Eq. 2.6-7.

---

[1]Torque (a vector) is defined as

$$\mathbf{T} = [\mathbf{r} \times \mathbf{F}] \tag{2.6-4a}$$

where **r** is the lever arm and **F** is the force. Therefore, in calculating the torque in the cone-and-plate system, if we take the lever arm to be in the x direction and the force to be in the y direction, the torque will be in the z direction. This explains the "z" subscript on T.

[2]R. B. Bird, W. E. Stewart, and E. N. Lightfoot, *Transport Phenomena*, Wiley, New York, 1st edition (1960), p. 119.

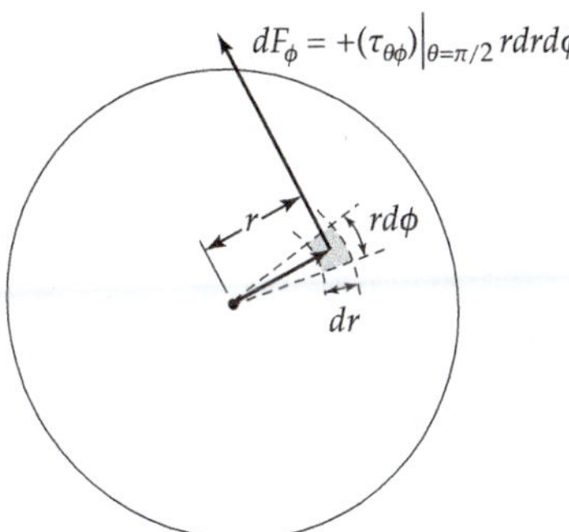

Fig. 2.6-3. Illustration of the torque about the z axis exerted by the fluid on an area element $r\,dr\,d\phi$ on the plate surface.

§2.7 FLOW AROUND A SPHERE[1,2,3,4]

In the preceding sections several elementary viscous flow problems have been solved. These have all dealt with rectilinear flows with only one nonvanishing velocity component. Since the flow around a sphere involves two nonvanishing velocity components, v_r and v_θ, it cannot be conveniently analyzed by the techniques explained at the beginning of this chapter. Nonetheless, a brief discussion of flow around a sphere is warranted here because of the importance of flow around submerged objects. In fluid dynamics textbooks,[2-4] it is shown how to obtain the velocity and pressure distributions. Here we just cite those results and show how they can be used to derive some important relations that we need in later discussions. This exercise also provides the opportunity to reinforce the readers' understanding of the definitions of stress components, and their ability to use these components to evaluate forces on objects. The problem solved here is concerned with "creeping flow," that is, very slow flow. This type of flow is also referred to as "Stokes flow" (see Eq. 3.6-3).

We consider here the flow of an incompressible fluid about a solid sphere of radius R and diameter D as shown in Fig. 2.7-1. The fluid, with density ρ and viscosity μ, approaches the fixed sphere vertically upward along the negative z axis with a uniform velocity v_∞. For this problem, "creeping flow" means that the Reynolds number $\text{Re} = Dv_\infty\rho/\mu$ is less than about 0.1. This flow regime is characterized by the absence of eddy formation downstream from the sphere.

The velocity and pressure distributions for this creeping flow are[2,4]

$$\boxed{v_r(r,\theta) = v_\infty\left[1 - \frac{3}{2}\left(\frac{R}{r}\right) + \frac{1}{2}\left(\frac{R}{r}\right)^3\right]\cos\theta} \tag{2.7-1}$$

$$\boxed{v_\theta(r,\theta) = v_\infty\left[-1 + \frac{3}{4}\left(\frac{R}{r}\right) + \frac{1}{4}\left(\frac{R}{r}\right)^3\right]\sin\theta} \tag{2.7-2}$$

$$\boxed{v_\phi = 0} \tag{2.7-3}$$

---

[1] G. G. Stokes, *Trans. Cambridge Phil. Soc.*, **9**, 8–106 (1851). For creeping flow around an object of arbitrary shape, see H. Brenner, *Chem. Engr. Sci.*, **19**, 703–727 (1964).

[2] L. D. Landau and E. M. Lifshitz, *Fluid Mechanics*, 2nd edition, Pergamon, London (1987), §20.

[3] S. Kim and S. J. Karrila, *Microhydrodynamics: Principles and Selected Applications*, Butterworth-Heinemann, Boston (1991); Dover, Mineola, NY (2005), §4.2.3; this book contains a thorough discussion of "creeping flow" problems.

[4] R. B. Bird, W. E. Stewart, and E. N. Lightfoot, *Transport Phenomena*, Revised 2nd Edition, Wiley, New York (2007), Example 4.2-1.

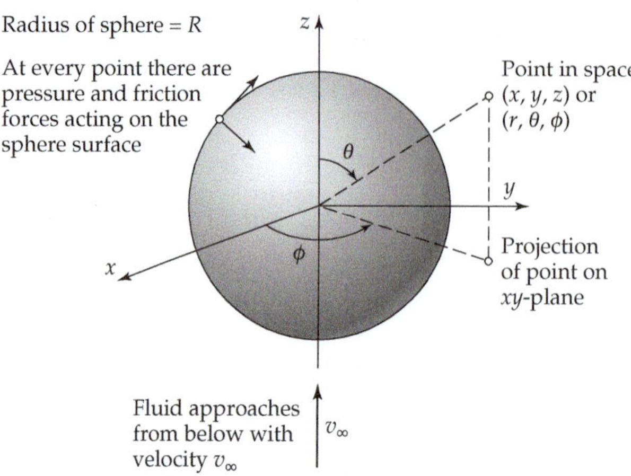

Fig. 2.7-1. Sphere of radius R around which a fluid is flowing. The coordinates r, θ, and ϕ are shown. For more information on spherical coordinates, see Fig. A.8.2.

$$p(r,\theta) = p_0 - \rho g z - \frac{3}{2}\frac{\mu v_\infty}{R}\left(\frac{R}{r}\right)^2 \cos\theta \tag{2.7-4}$$

In the last equation, the quantity p_0 is the pressure in the plane $z = 0$ far away from the sphere. The term $-\rho g z$ is the hydrostatic pressure resulting from the weight of the fluid, and the term containing v_∞ is the contribution of the fluid motion. Equations 2.7-1, 2.7-2, and 2.7-3 show that the fluid velocity is zero at the surface of the sphere. Furthermore, in the limit as $r \to \infty$, the fluid velocity is in the z direction with uniform magnitude v_∞; this follows from the fact that $v_z = v_r \cos\theta - v_\theta \sin\theta$, which can be derived by using Eq. A.6-33, and $v_x = v_y = 0$, which follows from Eqs. A.6-31 and A.6-32 (see Problem 2C.5).

The components of the viscous stress tensor $\boldsymbol{\tau}$ in spherical coordinates may be obtained from the velocity distribution above by using Table B.1. They are

$$\begin{aligned}\tau_{rr}(r,\theta) &= -2\tau_{\theta\theta}(r,\theta) = -2\tau_{\phi\phi}(r,\theta)\\ &= \frac{3\mu v_\infty}{R}\left[-\left(\frac{R}{r}\right)^2 + \left(\frac{R}{r}\right)^4\right]\cos\theta\end{aligned} \tag{2.7-5}$$

$$\tau_{r\theta}(r,\theta) = \tau_{\theta r}(r,\theta) = \frac{3}{2}\frac{\mu v_\infty}{R}\left(\frac{R}{r}\right)^4 \sin\theta \tag{2.7-6}$$

and all other components are zero. Note that the normal stresses for this flow are nonzero, except at $r = R$.

Let us now determine the force exerted by the flowing fluid on the sphere. Because of the symmetry around the z axis, the resultant force will be in the z direction. Therefore, the force can be obtained by integrating the z components of the normal and tangential forces over the sphere surface.

Integration of the Normal Force

At each point on the surface of the sphere the fluid exerts a force per unit area $-\pi_{rr}|_{r=R} = -(p + \tau_{rr})|_{r=R}$ on the solid, acting normal to the surface. Since the fluid is in the region of greater r and the sphere in the region of lesser r, we have to affix a minus sign in accordance with the sign convention established in §1.2. The z component of the force (per unit area) is $-\pi_{rr}|_{r=R}\cos\theta = -(p + \tau_{rr})|_{r=R}\cos\theta$. We now multiply this by a differential surface element $R^2 \sin\theta\, d\theta\, d\varphi$ to get the force on the surface element (see Fig. A.8.2). Then we integrate over the surface of the sphere to get the resultant normal force in the z direction

$$F^{(n)} = \int_0^{2\pi}\int_0^{\pi}\left[-(p+\tau_{rr})\Big|_{r=R}\cos\theta\right]R^2\sin\theta\,d\theta\,d\phi \tag{2.7-7}$$

According to Eq. 2.7-5, the normal stress τ_{rr} is zero[5] at $r = R$ and can be omitted in the integral in Eq. 2.7-7. The pressure distribution at the surface of the sphere is, according to Eq. 2.7-4

$$p|_{r=R} = p_0 - \rho g R\cos\theta - \frac{3}{2}\frac{\mu v_\infty}{R}\cos\theta \tag{2.7-8}$$

When this is substituted into Eq. 2.7-7 and the integration performed, we get

$$\begin{aligned} F^{(n)} &= 2\pi\int_0^{\pi}\left[-p_0 + \rho g R\cos\theta + \frac{3}{2}\frac{\mu v_\infty}{R}\cos\theta\right](\cos\theta)R^2\sin\theta\,d\theta \\ &= -2\pi p_0 R^2\int_0^{\pi}\cos\theta\,\sin\,\theta\,d\theta + 2\pi R^2\left(\rho g R + \frac{3}{2}\frac{\mu v_\infty}{R}\right)\int_0^{\pi}\cos^2\theta\,\sin\,\theta\,d\theta \\ &= 2\pi R^2\left(\rho g R + \frac{3}{2}\frac{\mu v_\infty}{R}\right)\left(\frac{2}{3}\right) \\ &= \frac{4}{3}\pi R^3\rho g + 2\pi\mu R v_\infty \end{aligned} \tag{2.7-9}$$

In the integration, the term containing p_0 gives zero, the term containing the gravitational acceleration g gives the "buoyant force," $\frac{4}{3}\pi R^3\rho g$, and the term containing the approach velocity v_∞ gives the "form drag," $2\pi\mu R v_\infty$.

Integration of the Tangential Force

At each point on the solid surface there is also a shear stress acting tangentially. The force per unit area exerted in the $-\theta$ direction by the fluid (region of greater r) on the solid (region of lesser r) is $+\pi_{r\theta}|_{r=R} = +\tau_{r\theta}|_{r=R}$. The z component of this force per unit area is $\pi_{r\theta}|_{r=R}\sin\theta = \tau_{r\theta}|_{r=R}\sin\theta$. We now multiply this by the differential surface area $R^2\sin\theta\,d\theta\,d\phi$ and integrate over the entire spherical surface. This gives the resultant force in the z direction:

$$F^{(t)} = \int_0^{2\pi}\int_0^{\pi}\left(\tau_{r\theta}|_{r=R}\sin\theta\right)R^2\sin\theta\,d\theta\,d\phi \tag{2.7-10}$$

The shear stress distribution on the sphere surface, from Eq. 2.7-6, is

$$\tau_{r\theta}|_{r=R} = \frac{3}{2}\frac{\mu v_\infty}{R}\sin\theta \tag{2.7-11}$$

Substitution of this expression into the integral in Eq. 2.7-10 gives

$$\begin{aligned} F^{(t)} &= 2\pi\int_0^{\pi}\left(\frac{3}{2}\frac{\mu v_\infty}{R}\sin\theta\right)(\sin\theta)R^2\sin\theta\,d\theta \\ &= 3\pi\mu R v_\infty\int_0^{\pi}\sin^3\theta\,d\theta = 3\pi\mu R v_\infty\left(\frac{4}{3}\right) \\ &= 4\pi\mu R v_\infty \end{aligned} \tag{2.7-12}$$

which is the "friction drag."

Hence, the total force F exerted by the fluid on the sphere in the $+z$ direction is given by the sum of Eqs. 2.7-9 and 2.7-12

$$F = \underbrace{\frac{4}{3}\pi R^3\rho g}_{\text{buoyant force}} + \underbrace{2\pi\mu R v_\infty}_{\text{form drag}} + \underbrace{4\pi\mu R v_\infty}_{\text{friction drag}} \tag{2.7-13}$$

[5] In Example 3.1-1 we show that, for incompressible, Newtonian fluids, all three of the normal stresses are zero at fixed solid surfaces in all flows.

or

$$F = F_b + F_k = \underbrace{\frac{4}{3}\pi R^3 \rho g}_{\text{buoyant force}} + \underbrace{6\pi\mu R v_\infty}_{\text{kinetic force}} \qquad (2.7\text{-}14)$$

The first term is the *buoyant force*, F_b, which would be present in a fluid at rest; it is just the mass of the displaced fluid multiplied by the gravitational acceleration. The second term, the *kinetic force*, F_k, results from the motion of the fluid. The relation

$$F_k = 6\pi\mu R v_\infty \qquad (2.7\text{-}15)$$

is known as *Stokes' law*.[1] It is used in describing the motion of colloidal particles under an electric field, in the theory of sedimentation, and in the study of the motion of aerosol particles. Stokes' law is useful only up to a Reynolds number $\text{Re} = Dv_\infty\rho/\mu$ of about 0.1. At $\text{Re} = 1$, Stokes' law predicts a force that is about 10 per cent too low. The flow behavior for larger Reynolds numbers is discussed in Chapter 6.

This problem, which could not be solved by the shell balance method, emphasizes the need for a more general method for coping with flow problems in which the streamlines are not rectilinear. That is the subject of the following chapter.

EXAMPLE 2.7-1

Determination of Viscosity from the Terminal Velocity of a Falling Sphere

Derive a relation that enables one to get the viscosity of a fluid by measuring the terminal velocity v_t of a small sphere of radius R in the fluid.

SOLUTION

If a small sphere is allowed to fall from rest in a viscous fluid, it will accelerate until it reaches a constant velocity—the *terminal velocity*. When this steady-state condition has been reached the sum of all the forces acting on the sphere must be zero. The force of gravity on the solid acts in the direction of fall, and the buoyant and kinetic forces act in the opposite direction. Thus, the z force balance is

$$\frac{4}{3}\pi R^3 \rho_s g = \frac{4}{3}\pi R^3 \rho g + 6\pi\mu R v_t \qquad (2.7\text{-}16)$$

Here ρ_s and ρ are the densities of the solid sphere and the fluid. Solving this equation for the viscosity gives

$$\mu = \frac{2}{9}\frac{R^2(\rho_s - \rho)g}{v_t} \qquad (2.7\text{-}17)$$

This result may be used only if the Reynolds number $Dv_t\rho/\mu$ is less than about 0.1.

This experiment provides an apparently simple method for determining viscosity. However, it is difficult to keep a homogeneous sphere from rotating during its descent, and if it does rotate, then Eq. 2.7-17 cannot be used. Sometimes asymmetrically weighted spheres are used in order to preclude rotation; then the left side of Eq. 2.7-16 has to be replaced by m, the mass of the weighted sphere, times the gravitational acceleration.

§2.8 CONCLUDING COMMENTS

In this chapter we have shown how to set up and solve some time-independent, rectilinear flow problems by using a shell momentum balance. We have also given two examples of flows that we could not solve by the shell balance method—namely, the flow in a cone-and-plate viscometer and the flow around a sphere—because they are not rectilinear flows (although we can obtain an approximate solution to the flow in a cone-and-plate viscometer using a local shell momentum balance). In the next chapter we will set up the general equations, the "equations of change," that will enable us to solve time-dependent problems as well as flows that are more complicated than rectilinear.

In setting up the shell momentum balances, components of the convective momentum flux $\rho \mathbf{vv}$ appeared in Eqs. 2.2-11 through 2.2-13, and also in Eqs. 2.3-9 and 2.3-10, but did not contribute to the final differential equation. This invariably happens for incompressible, rectilinear flows. Also, in §2.7, in the flow around the sphere, the $\rho \mathbf{vv}$ term does not appear, because of the creeping-flow assumption; inasmuch as the velocity $\mathbf{v}$ is assumed to be very small, the term $\rho \mathbf{vv}$ is even smaller and therefore can be safely disregarded. In Chapter 3, the $\rho \mathbf{vv}$ term is needed for the derivation of the general momentum conservation equation. The term appears in problems involving transpiration cooling, one-dimensional compressible flow, and many other problems.

QUESTIONS FOR DISCUSSION

1. Summarize the procedure used in the solution of viscous flow problems by the shell balance method. What kinds of problems can and cannot be solved by this method? How is the definition of the first derivative used in the method?
2. Which of the flow systems in this chapter can be used as a viscometer? List the difficulties that might be encountered in each.
3. How are the Reynolds numbers defined for films, tubes, and spheres? What are the dimensions of Re?
4. How can one modify the film thickness formula in §2.2 to describe a thin film falling down the interior wall of a cylinder? What restrictions might have to be replaced on this modified formula?
5. How can the results in §2.3 be used to estimate the time required for a liquid to drain out of a vertical tube that is open at both ends?
6. Contrast the radial dependence of the shear stress for the laminar flow of a Newtonian liquid in a tube and in an annulus. In the latter, why does the function change sign?
7. Show that the Hagen-Poiseuille formula (Eq. 2.3-21) is dimensionally consistent.
8. What differences are there between the flow in a circular tube of radius R and the flow in the same tube with a thin wire placed along the axis?
9. Under what conditions would you expect the analysis in §2.5 to be inapplicable?
10. In §2.6, why is the torque required to turn the cone equal to the torque exerted by the fluid on the bottom plate? If one were to rotate the plate and keep the cone fixed, what would be the magnitude of the torque exerted by the fluid on the cone?
11. Is Stokes' law valid for droplets of oil falling in water? For air bubbles rising in benzene? For tiny particles falling in air, if the particle diameters are of the order of the mean-free path of the molecules in the air?
12. Two immiscible liquids, A and B, are flowing in laminar flow between two parallel plates. Is it possible that the velocity profiles would be of the following form? Explain.

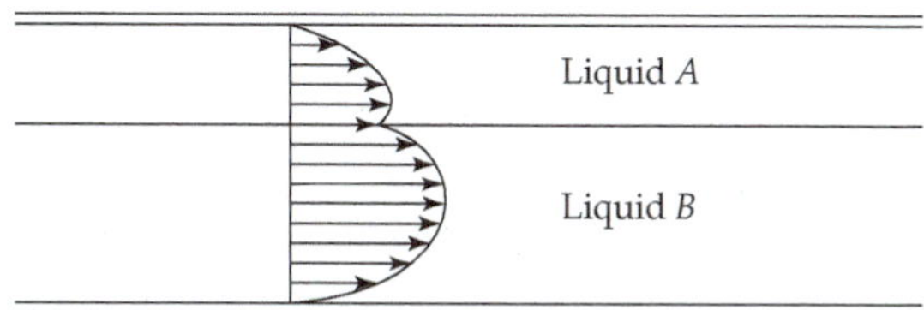

13. What is the terminal velocity of a spherical colloidal particle having an electric charge e in an electric field of strength $\mathscr{E}$? How is this used in the Millikan oil-drop experiment?

PROBLEMS

2A.1 Thickness of a falling film. Water at 20°C is flowing down a vertical wall with Re = 10. Calculate **(a)** the flow rate, in gallons per hour per foot of wall width, and **(b)** the film thickness in inches.

Answers: **(a)** 0.727 gal/hr · ft; **(b)** 0.00361 in.

2A.2 Determination of capillary radius by flow measurement. One method for determining the radius of a capillary tube is by measuring the rate of flow of a Newtonian liquid through the tube. Find the radius of a capillary from the following flow data:

Length of capillary tube	50.02 cm
Kinematic viscosity of liquid	4.03×10^{-5} m^2/s
Density of liquid	0.9552×10^3 kg/m^3
Pressure drop in the horizontal tube	4.829×10^5 Pa
Mass rate of flow through tube	2.997×10^{-3} kg/s

What difficulties may be encountered in this method? Suggest some other methods for determining the radii of capillary tubes.

2A.3 Volume flow rate through an annulus. A horizontal annulus, 27 ft in length, has an inner radius of 0.495 in. and an outer radius of 1.1 in. A 60% aqueous solution of sucrose ($C_{12}H_{22}O_{11}$) is to be pumped through the annulus at 20°C. At this temperature the solution density is 80.3 lb$_m$/ft^3 and the viscosity is 136.8 lb$_m$/ft · hr. What is the volume flow rate when the impressed pressure difference is 5.39 psi?

Answer: 0.110 ft^3/s

2A.4 Loss of catalyst particles in stack gas.

(a) Estimate the maximum diameter of microspherical catalyst particles that could be lost in the stack gas of a fluid cracking unit under the following conditions:

Gas velocity at axis of stack	= 1.0 ft/s (vertically upward)
Gas viscosity	= 0.026 cp
Gas density	= 0.045 lb$_m$/ft^3
Density of a catalyst particle	= 1.2 g/cm^3

Express the result in microns (1 micron = 10^{-6} m = 1μm).

(b) Is it permissible to use Stokes' law in (*a*)?

Answers: **(a)** 110 μm; Re = 0.93

2B.1 Simple shear flow between parallel plates. Consider the flow of an incompressible Newtonian fluid between parallel plates, such as that depicted in Fig. 2.6-2(*c*). The top plate (located at $y = b$) moves in the positive x direction with speed v_0, and the bottom plate (located at $y = 0$) remains at rest. The dimensions of the plates in the x and z directions are much larger than b, and thus, we postulate that at steady state $v_x = v_x(y)$ and $v_y = v_z = 0$. Gravity acts in the $-y$ direction.

(a) Using a shell x momentum balance, obtain the following expressions for the viscous momentum-flux and velocity distributions:

$$\tau_{yx}(y) = -\mu \frac{v_0}{b} \tag{2B.1-1}$$

$$v_x(y) = v_0 \frac{y}{b} \tag{2B.1-2}$$

(b) Obtain an expression for the volumetric rate of flow.

(c) Repeat the problem for the case where the upper plate is fixed and the lower plate is moving in the $+x$ direction with a speed v_0.

2B.2 Different choice of coordinates for the falling-film problem. Re-derive the velocity profile and the average velocity in §2.2, by replacing x by a coordinate $\bar{x}$ measured away from the wall; that is, $\bar{x} = 0$ is the wall surface, and $\bar{x} = \delta$ is the liquid-gas interface. Show that the velocity distribution is then given by

$$v_z(\bar{x}) = \frac{\rho g \delta^2 \cos\beta}{\mu}\left[\frac{\bar{x}}{\delta} - \frac{1}{2}\left(\frac{\bar{x}}{\delta}\right)^2\right] \tag{2B.2-1}$$

and then use this to get the average velocity. Show how one can get Eq. 2B.2-1 from Eq. 2.2-22 by making a change of variable.

2B.3 Alternate procedure for solving flow problems. In this chapter we have used the following procedure: (i) derive an equation for the viscous momentum flux, (ii) integrate this equation, (iii) insert Newton's law to get a first-order differential equation for the velocity, and (iv) integrate the latter to get the velocity distribution. Another method is: (i) derive an equation for the viscous momentum flux, (ii) insert Newton's law to get a second-order differential equation for the velocity profile, and (iii) integrate the latter to get the velocity distribution. Apply this second method to the falling-film problem by substituting Eq. 2.2-18 into Eq. 2.2-14 to obtain a second-order, ordinary differential equation for the velocity profile $v_z(x)$. Integrate this equation twice and apply appropriate boundary conditions for the velocity and its derivative to evaluate the constants of integration and determine the velocity profile.

2B.4 Laminar flow in a narrow slit (see Fig. 2B.4).

(a) An incompressible Newtonian fluid is in laminar flow in a narrow slit formed by two parallel walls a distance $2B$ apart. It is understood that $B \ll W$ and $B \ll L$, so that "edge effects" are unimportant. Make a shell momentum balance, and obtain the following expressions for the viscous momentum-flux and velocity distributions:

$$\tau_{xz}(x) = \left(\frac{\mathscr{P}_0 - \mathscr{P}_L}{L}\right) x \tag{2B.4-1}$$

$$v_z(x) = \frac{(\mathscr{P}_0 - \mathscr{P}_L)B^2}{2\mu L}\left[1 - \left(\frac{x}{B}\right)^2\right] \tag{2B.4-2}$$

In these expressions $\mathscr{P} = p + \rho g h = p - \rho g z$.

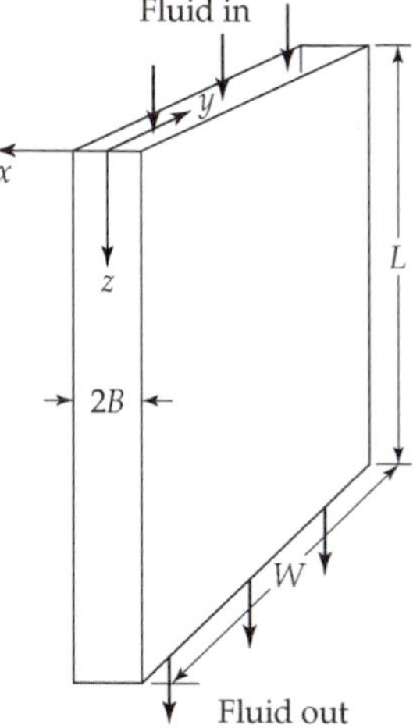

Fig. 2B.4 Flow through a slit, with $B \ll W \ll L$.

(b) What is the ratio of the average velocity to the maximum velocity for this flow?

(c) Obtain the slit analog of the Hagen-Poiseuille equation.

(d) Draw a meaningful sketch to show why the above analysis is inapplicable if $B = W$.

(e) How can the result in (b) be obtained from the results of §2.5?

Answers: **(b)** $\langle v_z \rangle / v_{z,\max} = \frac{2}{3}$;

(c) $Q = \frac{2}{3}\frac{(\mathscr{P}_0 - \mathscr{P}_L)B^3 W}{\mu L}$

2B.5 Laminar slit flow with a moving wall ("plane Couette flow"). Extend Problem 2B.4 by allowing the wall at $x = B$ to move in the positive z direction at a steady speed v_0. Obtain

(a) the shear-stress distribution, and

(b) the velocity distribution.

Draw carefully labeled sketches of these functions.

Answers: **(a)** $\tau_{xz}(x) = \left(\dfrac{\mathscr{P}_0 - \mathscr{P}_L}{L}\right)x - \dfrac{\mu v_0}{2B}$

(b) $v_z(x) = \dfrac{(\mathscr{P}_0 - \mathscr{P}_L)B^2}{2\mu L}\left[1 - \left(\dfrac{x}{B}\right)^2\right] + \dfrac{v_0}{2}\left(1 + \dfrac{x}{B}\right)$

2B.6 Interrelation of slit and annulus formulas. When an annulus is very thin, it may, to a good approximation, be considered as a thin slit. Then the results of Problem 2B.4 can be used with suitable modifications. For example, the mass rate of flow in an annulus with outer wall of radius R and inner wall of radius $(1 - \varepsilon)R$, where ε is small, may be obtained from Problem 2B.4 by replacing $2B$ by εR, and W by $2\pi\left(1 - \frac{1}{2}\varepsilon\right)R$. In this way we get for the mass rate of flow

$$w = \frac{\pi(\mathscr{P}_0 - \mathscr{P}_L)R^4\varepsilon^3\rho}{6\mu L}\left(1 - \frac{1}{2}\varepsilon\right) \tag{2B.6-1}$$

Show that this same result may be obtained from Eq. 2.4-18 by setting κ equal to $1 - \varepsilon$ everywhere in the formula and then expanding the expression for w in powers of ε. This requires using the Taylor series (see §C.2)

$$\ln(1 - \varepsilon) = -\varepsilon - \frac{1}{2}\varepsilon^2 - \frac{1}{3}\varepsilon^3 - \frac{1}{4}\varepsilon^4 - \cdots \tag{2B.6-2}$$

and then performing a long division. The first term in the resulting series will be Eq. 2B.6-1. *Caution*: In the derivation it is necessary to use the first *four* terms of the Taylor series in Eq. 2B.6-2.

2B.7 Flow of a film on the outside of a circular tube. In a gas absorption experiment, a viscous fluid flows upward through a small circular tube and then downward in laminar flow on the outside. Set up a momentum balance over a shell of thickness Δr in the film, as shown in Fig. 2B.7.

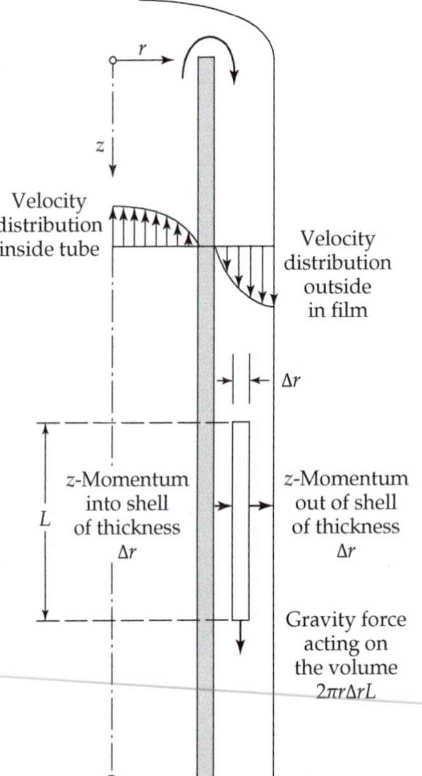

Fig. 2B.7 Velocity distribution $v_z(r)$ and z momentum balance for the flow of a falling film on the outside of a circular tube.

Note that the "momentum in" and "momentum out" arrows are always taken in the positive coordinate direction, even though in this problem the momentum is flowing in the negative r direction.

(a) Show that the velocity distribution in the falling film (neglecting end effects) is

$$v_z(r) = \frac{\rho g R^2}{4\mu}\left[1 - \left(\frac{r}{R}\right)^2 + 2a^2 \ln\left(\frac{r}{R}\right)\right] \qquad (2B.7\text{-}1)$$

(b) Obtain an expression for the mass rate of flow in the film.

(c) Show that the result in (b) simplifies to Eq. 2.2-26 if the film thickness is very small.

2B.8 Annular flow with inner cylinder moving axially. A cylindrical rod of radius κR moves axially with velocity v_0 along the axis of a cylindrical cavity of radius R as seen in Fig. 2B.8. The pressure at both ends of the cavity is the same, so that the fluid moves through the annular region solely because of the rod motion.

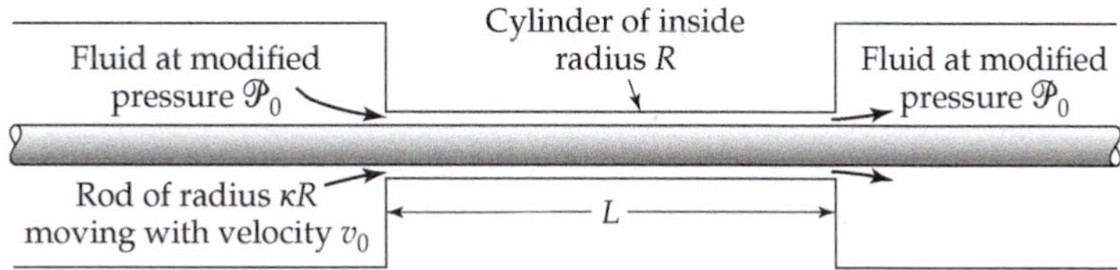

Fig. 2B.8 Annular flow with the inner cylinder moving axially.

(a) Find the velocity distribution in the narrow annular region.

(b) Find the mass rate of flow through the annular region.

(c) Obtain the viscous force acting on the rod over the length L.

(d) Show that the result in (c) can be written as a "plane slit" formula multiplied by a "curvature correction." Problems of this kind arise in studying the performance of plastics extrusion dies for wire coating.[1]

Answers: **(a)** $\dfrac{v_z(r)}{v_0} = \dfrac{\ln(r/R)}{\ln \kappa}$

(b) $w = \dfrac{\pi R^2 \rho v_0}{2}\left[\dfrac{1-\kappa^2}{\ln(1/\kappa)} - 2\kappa^2\right]$

(c) $F_z = -2\pi\mu L v_0/\ln(1/\kappa)$

(d) $F_z = -\dfrac{2\pi\mu L v_0}{\varepsilon}\left(1 - \dfrac{1}{2}\varepsilon - \dfrac{1}{12}\varepsilon^2 + \cdots\right)$

where $\varepsilon = 1 - \kappa$ (see Problem 2B.6)

2B.9 Analysis of a capillary flow meter. Determine the rate of flow (in lb_m/hr) through the capillary flow meter shown in Fig. 2B.9. The fluid flowing in the inclined tube is water at 20°C, and the manometer fluid is carbon tetrachloride (CCl_4) with density 1.594 g/cm^3. The capillary diameter is 0.010 in. *Note*: Measurements of H and L are sufficient to calculate the flow rate; θ need not be measured. Why?

---

[1] J. B. Paton, P. H. Squires, W. H. Darnell, F. M. Cash, and J. F. Carley, *Processing of Thermoplastic Materials*, E. C. Bernhardt (ed.), Reinhold, New York (1959), Chapter 4.

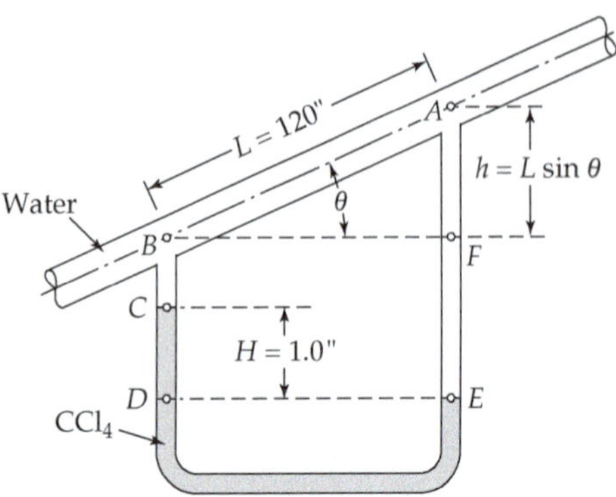

Fig. 2B.9 A capillary flow meter.

2B.10 Low-density phenomena in tube flow with compressible fluids.[2,3] As the pressure is decreased in the system studied in Example 2.3-2, deviations from Eqs. 2.3-32 and 2.3-33 arise. The gas behaves as if it slips at the tube wall. It is conventional[2] to replace the customary "no-slip" boundary condition that $v_z = 0$ at the tube wall by

$$v_z = -\zeta \frac{dv_z}{dr} \quad \text{at } r = R \tag{2B.10-1}$$

in which ζ is the *slip coefficient*. Repeat the derivation in Example 2.3-2 using Eq. 2B.10-1 as the boundary condition. Also make use of the experimental fact that the slip coefficient varies inversely with the pressure, $\zeta = \zeta_0/p$, in which ζ_0 is a constant. Show that the mass rate of flow is

$$w = \frac{\pi(p_0 - p_L)R^4\rho_{\text{avg}}}{8\mu L}\left(1 + \frac{4\zeta_0}{Rp_{\text{avg}}}\right) \tag{2B.10-2}$$

in which $p_{\text{avg}} = \frac{1}{2}(p_0 + p_L)$ and ρ_{avg} is the average density calculated at p_{avg}.

When the pressure is decreased further, a flow regime is reached in which the mean-free path of the gas molecules is large with respect to the tube radius (*Knudsen flow*). In that regime[3]

$$w = \sqrt{\frac{2m}{\pi\kappa T}}\left(\frac{4}{3}\pi R^3\right)\left(\frac{p_0 - p_L}{L}\right) \tag{2B.10-3}$$

in which m is the molecular mass and κ is the Boltzmann constant. In the derivation of this result it is assumed that all collisions of the molecules with the solid surfaces are *diffuse* and not *specular*. The results in Eqs. 2.3-33, 2B.10-2, and 2B.10-3 are summarized in Fig. 2B.10.

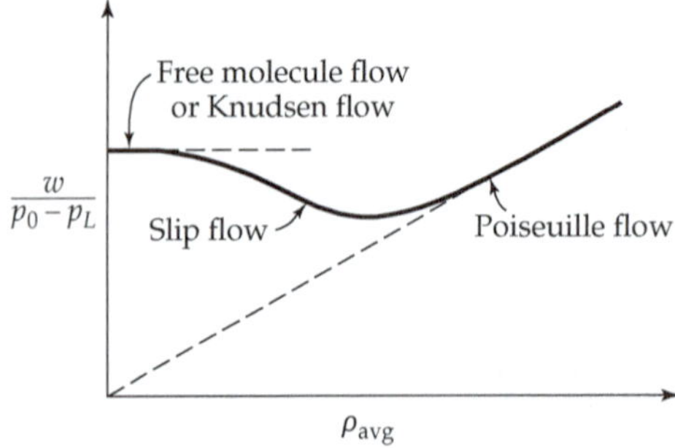

Fig. 2B.10 A comparison of the flow regimes in gas flow through a tube.

---

[2] E. H. Kennard, *Kinetic Theory of Gases*, McGraw-Hill, New York (1938), pp. 292–295, 300–306.

[3] M. Knudsen, *The Kinetic Theory of Gases*, Methuen, London, 3rd edition (1950). See also R. J. Silbey, R. A. Alberty, and M. G. Bawendi, *Physical Chemistry*, 4th edition, Wiley, New York (2005), §17.6.

2B.11 Incompressible flow in a slightly tapered tube. An incompressible fluid flows through a tube of circular cross section, for which the tube radius changes linearly from R_0 at the tube entrance to a slightly smaller value R_L at the tube exit. Assume that the Hagen-Poiseuille equation is *approximately* valid over a differential length of the tube, dz, so that the mass flow rate is

$$w = \frac{\pi[R(z)]^4\rho}{8\mu}\left(-\frac{d\mathscr{P}}{dz}\right) \tag{2B.11-1}$$

This is a differential equation for $\mathscr{P}$ as a function of z, but, when the explicit expression for $R(z)$ is inserted, integration of Eq. 2B.11-1 is cumbersome. This can be avoided by making a change of variable.

(a) Write down the expression for R as a function of z.

(b) Change the independent variable in the above equation to R, so that the equation becomes

$$w = \frac{\pi R^4\rho}{8\mu}\left(-\frac{d\mathscr{P}}{dR}\right)\left(\frac{R_L - R_0}{L}\right) \tag{2B.11-2}$$

(c) Integrate this equation, and then show that the solution can be rearranged to give

$$w = \frac{\pi(\mathscr{P}_0 - \mathscr{P}_L)R_0^4\rho}{8\mu L}\left[1 - \frac{1 + (R_L/R_0) + (R_L/R_0)^2 - 3(R_L/R_0)^3}{1 + (R_L/R_0) + (R_L/R_0)^2}\right] \tag{2B.11-3}$$

Interpret the result. The approximation used here that a flow between nonparallel surfaces may be regarded locally as flow between parallel surfaces is sometimes called the *lubrication approximation* and is widely used in the theory of lubrication. By making a careful order-of-magnitude analysis, it can be shown that, for this problem, the lubrication approximation is valid as long as[4]

$$\frac{R_L}{L}\left[1 - \left(\frac{R_L}{R_0}\right)^2\right] \ll 1 \tag{2B.11-4}$$

2B.12 Flow of a fluid in a network of tubes. A fluid is flowing in laminar flow from A to B through a network of tubes, as depicted in Fig. 2B.12. Obtain an expression for the mass flow rate w of the fluid entering at A (or leaving at B) as a function of the modified pressure drop $\mathscr{P}_A - \mathscr{P}_B$. Neglect the disturbances at the various tube junctions.

Answer: $w = \dfrac{3\pi(\mathscr{P}_A - \mathscr{P}_B)R^4\rho}{20\mu L}$

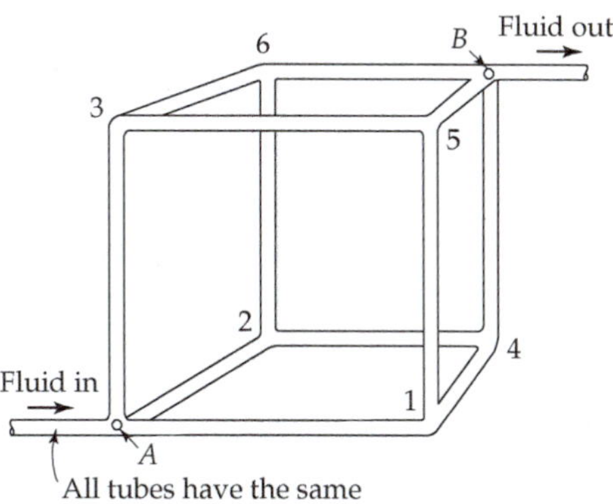

Fig. 2B.12 Flow of a fluid in a network with branching.

---

[4]R. B. Bird, R. C. Armstrong, and O. Hassager, *Dynamics of Polymeric Liquids*, Vol. **1**, Wiley-Interscience, New York, 2nd edition (1987), pp. 16–18.

2B.13 Location of maximum in the velocity distribution for annular flow. Verify the statement that the maximum in the velocity distribution for flow in an annulus is closer to the inner wall than to the outer wall. Use a numerical calculation if necessary.

2C.1 Performance of an electric dust collector.[5]

(a) A dust precipitator consists of a pair of oppositely charged plates between which dust-laden gases flow, as depicted in Fig. 2C.1. It is desired to establish a criterion for the minimum length of the precipitator in terms of the charge on the particle e, the electric field strength $\mathscr{E}$, the pressure difference $p_0 - p_L$, the particle mass m, and the gas viscosity μ. That is, for what length L will the smallest particle present (mass m) reach the bottom just before it has a chance to be swept out of the channel? Assume that the flow between the plates is laminar so that the velocity distribution is described by Eq. 2B.4-2. Assume also that the particle velocity in the z direction is the same as the fluid velocity in the z direction. Assume further that the Stokes drag on the sphere as well as the gravity force acting on the particle as it is accelerated in the $-x$ direction can be neglected.

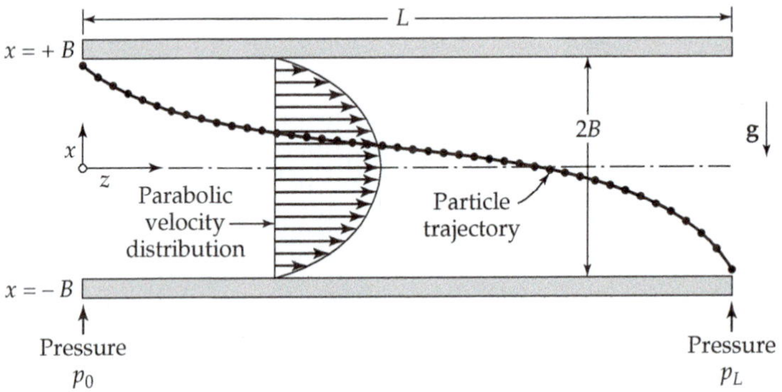

Fig. 2C.1 A possible particle trajectory in an electric dust collector. The particle that begins at $z = 0$, $x = +B$ and ends up at $x = -B$ may not necessarily travel the longest distance in the z direction.

(b) Rework the problem neglecting the particle acceleration in the x direction resulting from gravity, but including the Stokes drag.

(c) Compare the usefulness of the solutions in (a) and (b), considering that stable aerosol particles have effective diameters of about 1-10 μm and densities of about 1 g/cm^3.

Answer: **(a)** $L_{\text{min}} = [12(p_0 - p_L)^2 B^5 m / 25\mu^2 e \mathscr{E}]^{1/4}$

2C.2 Residence-time distribution in tube flow. Define the *residence-time function* $F(t)$ to be the fraction of the fluid flowing in a conduit that flows completely through the conduit in a time interval t. Also define the *mean residence time* t_m by the relation

$$t_m = \int_0^1 t\, dF \tag{2C.2-1}$$

(a) An incompressible Newtonian liquid is flowing in a circular tube of length L and radius R, and the average flow velocity is $\langle v_z \rangle$. Show that

$$F(t) = 0 \qquad \text{for } t \leq L/2\langle v_z \rangle \tag{2C.2-2}$$

$$F(t) = 1 - (L/2\langle v_z \rangle t)^2 \qquad \text{for } t \geq L/2\langle v_z \rangle \tag{2C.2-3}$$

(b) Show that $t_m = L/\langle v_z \rangle$.

[5] In the first edition of *Transport Phenomena*, by R. B. Bird, W. E. Stewart, and E. N. Lightfoot, Wiley, New York (1960), the answer given to this problem (Problem 2.M) was incorrect, as pointed out in 1970 by Nau Gab Lee of Seoul National University.

2C.3 Velocity distribution in a tube. You have received a manuscript to referee for a technical journal. The paper deals with heat transfer in tube flow. The authors state that, because they are concerned with nonisothermal flow, they must have a "general" expression for the velocity distribution, one that can be used even when the viscosity of the fluid is a function of temperature (and hence, position). The authors state that a "general expression for the velocity distribution for flow in a tube" is

$$\frac{v_z(y)}{\langle v_z \rangle} = \frac{\displaystyle\int_y^1 (\bar{y}/\mu)d\bar{y}}{\displaystyle\int_0^1 (\bar{y}^3/\mu)d\bar{y}} \tag{2C.3-1}$$

in which $y = r/R$ and $\bar{y}$ is a dummy variable of integration. The authors give no derivation, nor do they give a literature citation. As the referee you feel obliged to derive the formula and list any restrictions implied.

2C.4 Falling-cylinder viscometer.[6] A falling-cylinder viscometer consists of a long vertical cylindrical container (radius R), capped at both ends, with a solid cylindrical slug (radius κR), as illustrated in Fig. 2C.4(a). The slug is equipped with fins so that its axis is coincident with that of the tube.

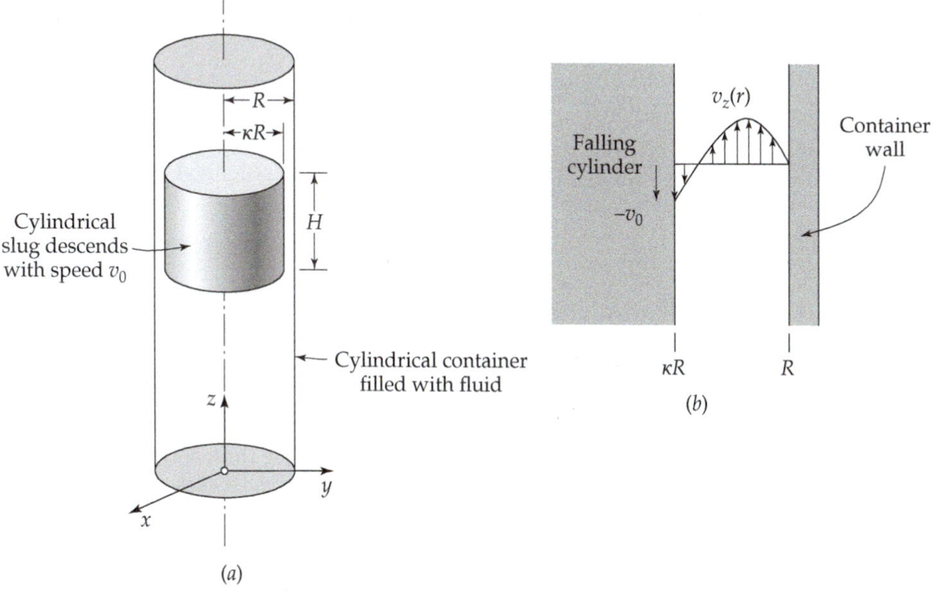

Fig. 2C.4 (a) A falling-cylinder viscometer with a tightly fitting solid cylinder moving vertically. The cylinder is usually equipped with fins to maintain centering within the tube. The fluid completely fills the tube and the top and bottom are closed. (b) Magnification of the gap between the slug and the container wall, illustrating the velocity profile $v_z(r)$.

As the cylindrical slug descends in the $-z$ direction, the (incompressible) fluid is pushed upward through the gap between the falling slug and the container walls. A sketch of the resulting fluid velocity profile is illustrated in Fig. 2C.4(b). Note that a portion of the fluid will be moving upward, while another portion of the fluid is moving downward. Note also that the pressure at the bottom of the falling slug (p_0) will be greater than the pressure at the top

[6] J. Lohrenz, G. W. Swift, and F. Kurata, *AIChE Journal*, **6**, 547–550 (1960) and **7**, 6S (1961); E. Ashare, R. B. Bird, and J. A. Lescarboura, *AIChE Journal*, **11**, 910–916 (1965); F. J. Eichstadt and G. W. Swift, *AIChE Journal*, **12**, 1179–1183 (1966); M. C. S. Chen and J. A. Lescarboura, *AIChE Journal*, **14**, 123–127 (1968).

of the falling slug (p_H). At steady state, we postulate that $v_z = v_z(r)$, $v_r = 0$, $v_\theta = 0$, and $p = p(z)$ within the gap between the slug and outer cylinder.

One can observe the rate of descent of the slug in the cylindrical container when the latter is filled with fluid. In this analysis, we will find an equation that gives the viscosity of the fluid in terms of the terminal speed of the slug v_0 and the various geometric quantities shown in the figure.

(a) Use a shell z momentum balance and the postulates above to obtain the following differential equation for the shear stress in the fluid-filled gap

$$\frac{d}{dr}(r\tau_{rz}) = r\frac{\mathscr{P}_0 - \mathscr{P}_H}{H} \tag{2C.4-1}$$

(b) One can integrate Eq. 2C.4-1 to obtain an expression for τ_{rz}, which will include a constant of integration. Inserting Newton's law of viscosity will produce a differential equation for $v_z(r)$, which can be integrated to produce a second constant of integration. Show that by applying appropriate boundary conditions at $r = \kappa R$ and $r = R$, one can obtain the velocity distribution

$$v_z(r) = \frac{\Delta\mathscr{P}}{4\mu H}R^2\left[1 - \left(\frac{r}{R}\right)^2\right] - \left[v_0 + \frac{\Delta\mathscr{P}}{4\mu H}R^2(1-\kappa^2)\right]\frac{\ln(r/R)}{\ln\kappa} \tag{2C.4-2}$$

where $\Delta\mathscr{P} = \mathscr{P}_0 - \mathscr{P}_H$.

(c) Because the slug and the fluid are both incompressible, the downward volumetric flow rate of the cylindrical slug $[\pi(\kappa R)^2 v_0]$ must be equal to the upward volumetric flow rate of the fluid. Use this constraint, with the fluid volumetric flow rate determined from the velocity profile above, to show that

$$\frac{\Delta\mathscr{P}}{4\mu H}R^2 = \frac{v_0}{(1+\kappa^2)\ln\frac{1}{\kappa} - (1-\kappa^2)} \tag{2C.4-3}$$

(d) Show that by using Eq. 2C.4-3 to eliminate $\Delta\mathscr{P}$ from Eq. 2C.4-2, the velocity distribution in the annular slit can be written

$$\frac{v_z(\xi)}{v_0} = -\frac{(1-\xi^2) - (1+\kappa^2)\ln(1/\xi)}{(1-\kappa^2) - (1+\kappa^2)\ln(1/\kappa)} \tag{2C.4-4}$$

in which $\xi = r/R$ is a dimensionless radial coordinate.

(e) Make a force balance on the cylindrical slug and obtain

$$\mu = \frac{(\rho_0 - \rho)g(\kappa R)^2}{2v_0}\left[\left(\ln\frac{1}{\kappa}\right) - \left(\frac{1-\kappa^2}{1+\kappa^2}\right)\right] \tag{2C.4-5}$$

in which ρ and ρ_0 are the densities of the fluid and the slug, respectively. To obtain the correct result, you must include both the (modified) pressure and shear stress contributions to the force exerted by the fluid on the slug.

(f) Show that, for very small slit widths, the result in (e) may be expanded in powers of $\varepsilon = 1 - \kappa$ to give

$$\mu = \frac{(\rho_0 - \rho)gR^2\varepsilon^3}{6v_0}\left(1 - \frac{1}{2}\varepsilon - \frac{13}{20}\varepsilon^2 + \cdots\right) \tag{2C.4-6}$$

See §C.2 for information on expansions in Taylor series.

2C.5 Relations needed for development of Stokes' law.

(a) To get the expression for v_z above Eq. 2.7-5, start by using Eq. A.2-16 to get $v_z = (\mathbf{v} \cdot \boldsymbol{\delta}_z)$. Then use Eq. A.6-33, relating the unit vectors in the x,y,z coordinates to those in the r,θ,ϕ coordinates, to get $v_z = (\mathbf{v} \cdot \{\boldsymbol{\delta}_r \cos\theta - \boldsymbol{\delta}_\theta \sin\theta + \boldsymbol{\delta}_\phi 0\})$. Next perform the dot product operations between $\mathbf{v}$ and the unit vectors in the spherical coordinate system, to get the desired result.

(b) Proceed as in part (a) by writing $v_x = (\mathbf{v} \cdot \boldsymbol{\delta}_x)$. Then use Eq. A.6-31 for the interrelation of the unit vectors. This should give you $v_x = v_r \sin\theta\cos\phi + v_\theta \cos\theta\cos\phi$. Next recognize that, far from the sphere, $v_r = v_\infty \cos\theta$ and $v_\theta = -v_\infty \sin\theta$, so that $v_x = 0$.

(c) Following a procedure similar to that in (b), show that $v_y = 0$.

2C.6 Falling film on a conical surface.[7] A fluid flows upward through a circular pipe and then downward on a conical surface, as illustrated in Fig. 2C.6. Find the film thickness as a function of the distance s down the cone.

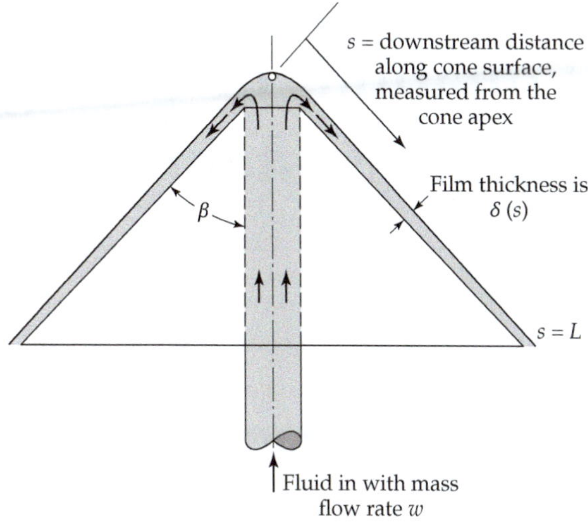

Fig. 2C.6 A falling film on a conical surface. [R. B. Bird, in *Selected Topics in Transport Phenomena*, CEP Symposium Series #58, **61**, 1–15 (1965).]

(a) Assume that the results of §2.2 apply *approximately* over any small region of the cone surface. Show that a mass balance on a ring of liquid contained between s and $s + \Delta s$ gives

$$\frac{d}{ds}(s\delta\langle v\rangle) = 0 \quad \text{or} \quad \frac{d}{ds}(s\delta^3) = 0 \tag{2C.6-1}$$

(b) Integrate this equation and evaluate the constant of integration by equating the mass rate of flow w up the central tube to that flowing down the conical surface at $s = L$. Obtain the following expression for the film thickness

$$\delta = \sqrt[3]{\frac{3\mu w}{\pi\rho^2 g L \sin 2\beta}\left(\frac{L}{s}\right)} \tag{2C.6-2}$$

[7] R. B. Bird, in *Selected Topics in Transport Phenomena*, CEP Symposium Series #58, **61**, 1–15 (1965).

Chapter 3

The Equations of Change for Isothermal Systems

In Chapter 2, velocity distributions were determined for several simple flow systems by the shell momentum balance method. The resulting velocity distributions were then used to get other quantities, such as the average velocity and drag force. The shell balance approach was used to acquaint the novice with the notion of a momentum balance. Even though we made no mention of it in Chapter 2, at several points we tacitly made use of the idea of a mass balance.

It is tedious to set up a shell balance for each problem that one encounters. Furthermore, the shell balance method cannot be applied to all flows, such as flows with curved streamlines. What we need is a general mass balance and a general momentum balance that can be applied to any problem, including problems with nonrectilinear motion. That is the main point of this chapter. The two main equations that we derive are called the *equation of continuity* (for the mass balance) and the *equation of motion* (for the momentum balance). These equations can be used as the starting point for studying all problems involving the isothermal flow of a pure fluid. In fact, all of the flow problems solved in Chapter 2 can be formulated and solved more easily using the equations derived in this chapter.

In Chapter 11, we enlarge our problem-solving capability by developing the equations needed for nonisothermal pure fluids by adding an equation for the temperature. In Chapter 19, we go even further and add equations of continuity for the concentrations of the individual species. Thus, as we go from Chapter 3 to Chapter 11 and on to Chapter 19, we are able to analyze systems of increasing complexity, using the complete set of *equations of change*. It should be evident that Chapter 3 is a very important chapter—perhaps the most important chapter in the book—and it should be mastered thoroughly.

In §3.1, the equation of continuity is developed by applying the *law of conservation of mass* to a tiny element of volume through which the fluid is flowing. Then the size of this element is allowed to go to zero (thereby treating the fluid as a continuum), and the desired partial differential equation is generated.

In §3.2, the equation of motion is developed by applying the *law of conservation of momentum* to a tiny element of volume and letting the volume element become infinitesimally small. Here again a partial differential equation is generated. This equation of motion can be used, along with some help from the equation of continuity, to set up and solve all the problems given in Chapter 2 and many more complicated ones. It is thus a key equation in transport phenomena.

In §3.3 and §3.4, we digress briefly to introduce the equations of change for mechanical energy and angular momentum. These equations are obtained from the equation of motion and hence contain no new physical information. However, they provide a convenient starting point for several applications in this book—particularly the macroscopic balances in Chapter 7.

The term "equation of change" may be used to describe how any physical quantity changes with time and position. The more restricted term "conservation equation" is reserved for an equation that is derived from a statement of a conservation law. The equations of continuity and motion are based on the conservation laws for mass and momentum. The equation of change for mechanical energy is not based on a conservation statement, and the equation for angular momentum given here does not account for the "internal angular momentum" and hence is not based on a complete conservation statement. A brief review of §0.3 at this point would not be out of place.

In §3.5, we introduce the "substantial derivative." This is the time derivative following the motion of the substance (i.e., the fluid). Because it is widely used in books on fluid dynamics and transport phenomena, we then show how the various equations of change can be rewritten in terms of the substantial derivatives.

In §3.6, some commonly used special forms of the equation of motion are given. These include the equation for inviscid fluids, the equation for incompressible, constant viscosity fluids, and the equation for situations where the acceleration terms can be neglected.

In §3.7, we discuss the solution of flow problems by use of the equations of continuity and motion. Although these are partial differential equations, we can solve many problems by postulating the form of the solution and then discarding many terms in these equations. In this way we end up with a simpler set of equations to solve. In §3.7, we solve only problems in which the general equations may be simplified to one or more ordinary differential equations.

In §3.8, we examine problems of greater complexity, in that they involve more than one velocity component or more than one independent variable. These problems are solved by using various standard methods to reduce the governing equations to ordinary differential equations. Thus, one does not need a background in partial differential equations to work any of the problems in this chapter.

At the end of §2.2, we emphasized the importance of experiments in fluid dynamics. We repeat those words of caution here and point out that visualization of actual flows have provided us with a much deeper understanding of flow problems than would be possible by theory alone.[1] Keep in mind that when one derives a flow field from the equations of change, it is not necessarily the only physically admissible solution.

---

[1]We recommend particularly M. Van Dyke, *An Album of Fluid Motion*, Parabolic Press, Stanford (1982); H. Werlé, *Ann. Rev. Fluid Mech.*, **5**, 361–382 (1973); D. V. Boger and K. Walters, *Rheological Phenomena in Focus*, Elsevier, Amsterdam (1993).

Vector and tensor notations are occasionally used in this chapter, primarily for the purpose of abbreviating otherwise lengthy expressions. The beginning student will find that only an elementary knowledge of vector and tensor notation is needed for reading this chapter and for solving flow problems. The advanced student will find Appendix A helpful for a better understanding of vector and tensor manipulations. With regard to the notation, it should be kept in mind that we use *lightface italic* symbols for scalars, **boldface Roman** symbols for vectors, and **boldface Greek** symbols for tensors. Also dot product operations enclosed in ( ) are scalars, dot and cross product operations in [ ] are vectors, and those in { } are tensors.

§3.1 THE EQUATION OF CONTINUITY

We consider here an arbitrary flow in which all three velocity components $v_x(x,y,z,t)$, $v_y(x,y,z,t)$, and $v_z(x,y,z,t)$ may be nonzero, and may depend on all three spatial coordinates as well as time (for example, such as that depicted in Fig. 1.1.2). We apply the law of conservation of mass to a tiny volume element $\Delta x \Delta y \Delta z$, *fixed in space*, through which a fluid is flowing (see Fig. 3.1-1). Stated in words, this is

$$\begin{Bmatrix} \text{rate of} \\ \text{increase} \\ \text{of mass} \end{Bmatrix} = \begin{Bmatrix} \text{rate of} \\ \text{mass} \\ \text{in} \end{Bmatrix} - \begin{Bmatrix} \text{rate of} \\ \text{mass} \\ \text{out} \end{Bmatrix} \tag{3.1-1}$$

Now we have to translate this simple conservation statement into mathematical language.

The rate of increase of mass with time is $\partial(\rho \Delta x \Delta y \Delta z)/\partial t = \Delta x \Delta y \Delta z(\partial \rho/\partial t)$. As discussed in §0.4, the fact that the fluid is flowing means that there is a convective mass flux, whose components contribute to the "in" and "out" terms in Eq. 3.1-1. Consider first the two shaded faces in Fig. 3.1-1, which are perpendicular to the x axis. The rate of mass entering the volume element through the shaded face at x is the convective mass flux in the x direction $(\rho v_x)|_x$ multiplied by the area of the shaded face $\Delta y \Delta z$, or $(\rho v_x)|_x \Delta y \Delta z$. The rate of mass leaving through the shaded face at $x + \Delta x$ is $(\rho v_x)|_{x+\Delta x} \Delta y \Delta z$. Similar expressions can be written for the other two pairs of faces. The mass balance then becomes

$$\begin{aligned} \Delta x \Delta y \Delta z \frac{\partial \rho}{\partial t} &= \Delta y \Delta z[(\rho v_x)|_x - (\rho v_x)|_{x+\Delta x}] \\ &+ \Delta z \Delta x[(\rho v_y)|_y - (\rho v_y)|_{y+\Delta y}] \\ &+ \Delta x \Delta y[(\rho v_z)|_z - (\rho v_z)|_{z+\Delta z}] \end{aligned} \tag{3.1-2}$$

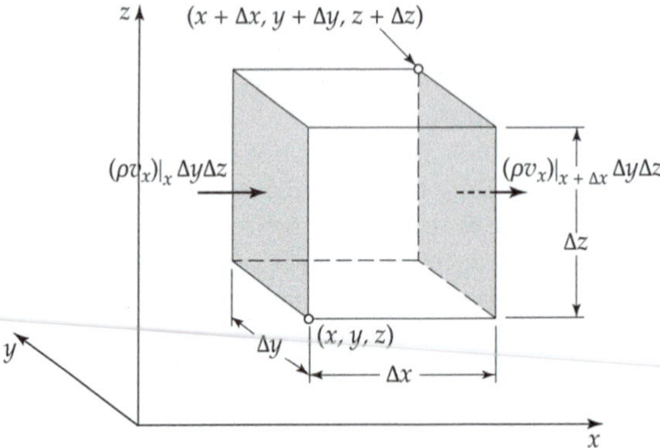

Fig. 3.1-1. Tiny fixed volume element $\Delta x \Delta y \Delta z$ through which a fluid is flowing. The arrows indicate the mass transport rates in and out of the volume through the two shaded faces located at x and $x + \Delta x$.

By dividing the entire equation by $\Delta x \Delta y \Delta z$ and taking the limits as Δx, Δy, and Δz go to zero, and then using the definitions of the partial derivatives, we get[1]

$$\frac{\partial \rho}{\partial t} = -\left(\frac{\partial}{\partial x}\rho v_x + \frac{\partial}{\partial y}\rho v_y + \frac{\partial}{\partial z}\rho v_z \right) \quad (3.1\text{-}3)$$

This is the *equation of continuity*, which describes the time rate of change of the fluid density at a fixed point in space. This equation can be written more concisely by using vector notation as follows:

$$\underset{\substack{\text{rate of}\\ \text{increase of}\\ \text{mass per}\\ \text{unit volume}}}{\frac{\partial \rho}{\partial t}} = \underset{\substack{\text{net rate of mass}\\ \text{addition per}\\ \text{unit volume}\\ \text{by convection}}}{-(\nabla \cdot \rho \mathbf{v})} \quad (3.1\text{-}4)$$

Here $(\nabla \cdot \rho \mathbf{v})$ is called the "divergence of $\rho \mathbf{v}$," sometimes written as "div$\rho \mathbf{v}$." The divergence of the mass flux vector $\rho \mathbf{v}$ has a simple meaning: it is the net rate of mass efflux per unit volume. Equation 3.1-4 is written in component form in Cartesian, cylindrical, and spherical coordinates in Appendix B.1.

A very important special form of the equation of continuity is that for a fluid of constant density, for which Eq. 3.1-4 assumes the particularly simple form

(incompressible fluid) $$(\nabla \cdot \mathbf{v}) = 0 \quad (3.1\text{-}5)$$

Of course, no fluid is truly incompressible, but frequently in engineering and biological applications, the assumption of constant density results in considerable simplification and very little error.[2,3]

EXAMPLE 3.1-1

Normal Stresses at Solid Surfaces for Incompressible Newtonian Fluids

Show that for any kind of flow pattern, the normal stresses are zero at fluid–solid boundaries, *for Newtonian fluids with constant density*. This is an important result that we shall use often.

SOLUTION

We visualize the flow of a fluid near some solid surface, which may or may not be flat. The flow may be quite general, with all three velocity components being functions of all three coordinates and time. At some point P on the surface we erect a Cartesian coordinate system with the origin

---

[1]In this book, when we write

$$\frac{\partial}{\partial x}\rho v_x \quad (3.1\text{-}3a)$$

we mean that the product of ρ and v_x is being differentiated. Parentheses in this situation are unnecessary, but will be added occasionally for clarity. The same comments apply to $(\partial/\partial t)\rho \mathbf{v}$ and ∇rs. If we write

$$\left(\frac{\partial}{\partial x}\rho\right) v_x \quad (3.1\text{-}3b)$$

then only ρ is being differentiated.

[2]L. D. Landau and E. M. Lifshitz, *Fluid Mechanics*, Pergamon Press, Oxford (1987), p. 21, point out that for steady, isentropic flows, commonly encountered in aerodynamics, the incompressibility assumption is valid when the fluid velocity is small compared to the velocity of sound (i.e., low Mach number).

[3]Equation 3.1-5 is the basis for Chapter 2 in G. K. Batchelor, An Introduction to Fluid Dynamics, *Cambridge University Press* (1967), which is a lengthy discussion of the kinematical consequences of the equation of continuity.

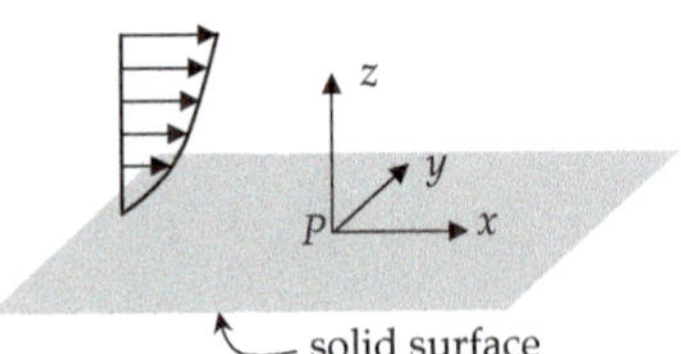

Fig. 3.1-2. Portion of a solid surface on which a Cartesian coordinate system is constructed. The x and y axes are in the plane of the surface, and the z axis is normal to the surface, pointing into the fluid.

at P with z as the normal direction, as illustrated in Fig. 3.1-2. We now ask what the normal stress τ_{zz} is at P.

According to Table B.1 or Eq. 1.2-9, $\tau_{zz} = -2\mu(\partial v_z/\partial z)$, because $(\nabla \cdot \mathbf{v}) = 0$ for incompressible fluids. Also, for constant ρ, Eq. 3.1-4 can be simplified to

$$\frac{\partial v_z}{\partial z} = -\left(\frac{\partial v_x}{\partial x} + \frac{\partial v_y}{\partial y}\right) \tag{3.1-6}$$

Substituting this into the expression for τ_{zz} gives for point P on the surface of the solid

$$\tau_{zz}|_{z=0} = -2\mu \left.\frac{\partial v_z}{\partial z}\right|_{z=0} = +2\mu \left.\left(\frac{\partial v_x}{\partial x} + \frac{\partial v_y}{\partial y}\right)\right|_{z=0} \tag{3.1-7}$$

Everywhere on the solid surface at $z = 0$, the velocity v_x is zero by the no-slip condition (see §2.1), and therefore, the derivative $\partial v_x/\partial x$ on the surface must be zero (i.e., v_x does not vary with x). Similarly, $\partial v_y/\partial y$ is also zero on the surface. Therefore, τ_{zz} is zero on the surface. One can show that τ_{xx} and τ_{yy} are also zero at the surface because of the vanishing of the derivatives at $z = 0$.

§3.2 THE EQUATION OF MOTION

To derive the equation of motion for an arbitrary flow, we again consider a tiny volume element $\Delta x \Delta y \Delta z$, illustrated in Fig. 3.2-1. Conservation of momentum for this volume element can be written in words as

$$\left\{\begin{matrix}\text{rate of} \\ \text{increase of} \\ \text{momentum}\end{matrix}\right\} = \left\{\begin{matrix}\text{rate of total} \\ \text{momentum} \\ \text{in}\end{matrix}\right\} - \left\{\begin{matrix}\text{rate of total} \\ \text{momentum} \\ \text{out}\end{matrix}\right\} + \left\{\begin{matrix}\text{external} \\ \text{force on} \\ \text{the fluid}\end{matrix}\right\} \tag{3.2-1}$$

Note that Eq. 3.2-1 is just an extension of Eq. 2.1-1 to unsteady problems. Therefore, we proceed in much the same way as in Chapter 2. However, in addition to including the unsteady term, we must allow the fluid to move through all six faces of the volume element. Remember that Eq. 3.2-1 is a vector equation with components in each of the three coordinate directions x,y, and z. We develop the x component of each term in Eq. 3.2-1; the y and z components may be treated analogously.[1]

First, we consider the rates of flow of the x component of momentum into and out of the volume element shown in Fig. 3.2-1. Momentum enters and leaves $\Delta x \Delta y \Delta z$ by two mechanisms: convective transport (see §1.1), and molecular transport (see §1.2), the sum of which is described by the total momentum flux $\boldsymbol{\phi}$ (see §1.3).

The rate at which the x component of momentum enters across the shaded face at x by all mechanisms—both convective and molecular—is $\phi_{xx}|_x \Delta y \Delta z$ and the rate at which

[1] In this book all the equations of change are derived by applying the conservation laws to a region $\Delta x \Delta y \Delta z$ fixed in space. The same equations can be obtained by using an arbitrary region fixed in space, or one moving along with the fluid.

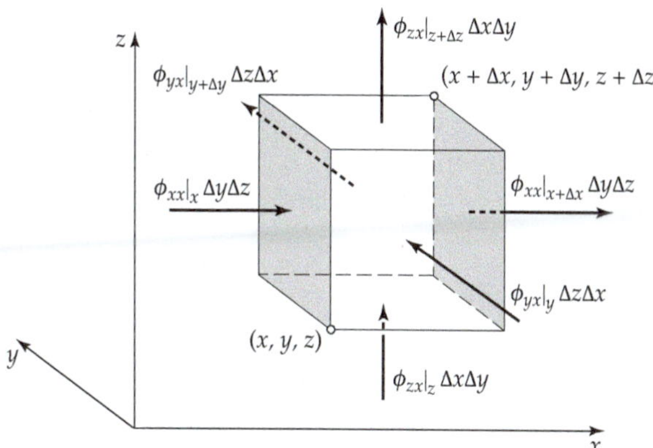

Fig. 3.2-1. Tiny fixed volume element $\Delta x\Delta y\Delta z$, with six arrows indicating the rates of x momentum transported through the surfaces by all mechanisms. The shaded faces are located at x and $x+\Delta x$.

it leaves the shaded face at $x+\Delta x$ is $\phi_{xx}|_{x+\Delta x}\Delta y\Delta z$. The rates at which x momentum enters and leaves through the faces at y and $y+\Delta y$ are $\phi_{yx}|_y\Delta z\Delta x$ and $\phi_{yx}|_{y+\Delta y}\Delta z\Delta x$, respectively. Similarly the rates at which x momentum enters and leaves through the faces at z and $z+\Delta z$ are $\phi_{zx}|_z\Delta x\Delta y$ and $\phi_{zx}|_{z+\Delta z}\Delta x\Delta y$. When these contributions are added, we get for the net rate of addition ("in" minus "out") of x momentum

$$\Delta y\Delta z(\phi_{xx}|_x - \phi_{xx}|_{x+\Delta x}) + \Delta z\Delta x(\phi_{yx}|_y - \phi_{yx}|_{y+\Delta y}) + \Delta x\Delta y(\phi_{zx}|_z - \phi_{zx}|_{z+\Delta z}) \tag{3.2-2}$$

for transport across all three pairs of faces.

Next there is the external force (typically the gravitational force) acting on the fluid in the volume element. The x component of this force is

$$\rho g_x \Delta x\Delta y\Delta z \tag{3.2-3}$$

where the x component of the acceleration of gravity g_x can be replaced by the body force per unit mass if other body forces are present (e.g., electric or magnetic forces). Equations 3.2-2 and 3.2-3 give the x components of the three terms on the right side of Eq. 3.2-1. The sum of these terms must then be equated to the rate of increase of x momentum within the volume element, $\partial(\rho v_x\Delta x\Delta y\Delta z)/\partial t = \Delta x\Delta y\Delta z\partial(\rho v_x)/\partial t$. When this is done, we have the x component of the momentum balance. When this equation is divided by $\Delta x\Delta y\Delta z$ and the limits are taken as Δx, Δy, and Δz go to zero, we get

$$\frac{\partial}{\partial t}\rho v_x = -\left(\frac{\partial}{\partial x}\phi_{xx} + \frac{\partial}{\partial y}\phi_{yx} + \frac{\partial}{\partial z}\phi_{zx}\right) + \rho g_x \tag{3.2-4}$$

where we have made use of the definitions of the partial derivatives. Similar equations can be developed for the y component of the momentum balance

$$\frac{\partial}{\partial t}\rho v_y = -\left(\frac{\partial}{\partial x}\phi_{xy} + \frac{\partial}{\partial y}\phi_{yy} + \frac{\partial}{\partial z}\phi_{zy}\right) + \rho g_y \tag{3.2-5}$$

and the z component

$$\frac{\partial}{\partial t}\rho v_z = -\left(\frac{\partial}{\partial x}\phi_{xz} + \frac{\partial}{\partial y}\phi_{yz} + \frac{\partial}{\partial z}\phi_{zz}\right) + \rho g_z \tag{3.2-6}$$

By using vector-tensor notation, these three equations can be written as follows:

$$\frac{\partial}{\partial t}\rho v_i = -[\nabla\cdot\boldsymbol{\phi}]_i + \rho g_i \qquad i = x,y,z \tag{3.2-7}$$

That is, by letting i be successively x,y, and z, Eqs. 3.2-4, 3.2-5, and 3.2-6 can be reproduced. The quantities ρv_i are the components of the vector $\rho\mathbf{v}$, which is the momentum per unit volume at a point in the fluid. Similarly, the quantities ρg_i are the components of the vector $\rho\mathbf{g}$, which is the external force per unit volume. The term $-[\nabla \cdot \boldsymbol{\phi}]_i$ is the ith component of the vector $-[\nabla \cdot \boldsymbol{\phi}]$, which represents the net rate of addition of i momentum by transport across the faces, per unit volume.

When each of the components of Eq. 3.2-7 is multiplied by the unit vector in the ith direction and then added together vectorially, we get

$$\frac{\partial}{\partial t}\rho\mathbf{v} = -[\nabla \cdot \boldsymbol{\phi}] + \rho\mathbf{g} \tag{3.2-8}$$

which is the differential statement of the law of conservation of momentum. It is the translation of Eq. 3.2-1 into mathematical symbols.

In Eq. 1.3-2 it was shown that the total momentum-flux tensor $\boldsymbol{\phi}$ is the sum of the convective momentum-flux tensor $\rho\mathbf{vv}$ and the molecular momentum-flux tensor $\boldsymbol{\pi}$, and that the latter can be written as the sum of $p\boldsymbol{\delta}$ and $\boldsymbol{\tau}$. When we insert $\boldsymbol{\phi} = \rho\mathbf{vv} + p\boldsymbol{\delta} + \boldsymbol{\tau}$ into Eq. 3.2-8, we get (using Eq. A.4-26) the following *equation of motion:*[2]

$\frac{\partial}{\partial t}\rho\mathbf{v}$	$=$	$-[\nabla \cdot \rho\mathbf{vv}]$	$-\nabla p \quad -[\nabla \cdot \boldsymbol{\tau}]$	$+\rho\mathbf{g}$	
rate of increase of momentum per unit volume		rate of momentum addition by convection per unit volume	rate of momentum addition by molecular transport per unit volume	external force on fluid per unit volume	(3.2-9)

In this equation ∇p is a vector called the "gradient of (the scalar) p," sometimes written as "grad p." The symbol $[\nabla \cdot \boldsymbol{\tau}]$ is a vector called the "divergence of (the tensor) $\boldsymbol{\tau}$" and $[\nabla \cdot \rho\mathbf{vv}]$ is a vector called the "divergence of (the dyadic product) $\rho\mathbf{vv}$."

In the next two sections we give some formal results that are based on the equation of motion. The equation of change for angular momentum is not used for problem solving in this chapter, but will be referred to in Chapter 7. Beginners are advised to skim this section on first reading and to refer to it later as the need arises.

§3.3 THE EQUATION OF CHANGE FOR MECHANICAL ENERGY

It was shown in §0.3 that for binary collisions between two molecules, although total energy is conserved, kinetic energy is not. Similarly, in a flow system, kinetic energy is not conserved, but that does not prevent us from developing an equation of change for this quantity. In fact, during the course of this book, we will obtain equations of change for a number of nonconserved quantities, such as internal energy, enthalpy, and temperature. An equation of change for *mechanical* energy, which involves kinetic energy and other mechanical terms, may be derived from the equation of motion of §3.2. The resulting equation is referred to in several places later in the book.

We take the dot product of the velocity vector $\mathbf{v}$ with the equation of motion in Eq. 3.2-9 and then do some rather lengthy rearranging, making use of the equation of continuity, Eq. 3.1-4 (see Problem 3C.4). We also split up each of the terms containing p and $\boldsymbol{\tau}$ into two parts. The final result is the *equation of change for kinetic energy,* for

[2]This equation is attributed to A.-L. Cauchy, *Ex. de math.*, **2**, 108–111 (1827). **(Baron) Augustin-Louis Cauchy** (1789–1857) (pronounced "Koh-shee" with the accent on the second syllable), originally trained as an engineer, made great contributions to theoretical physics and mathematics, including the calculus of complex variables.

symmetric $\boldsymbol{\tau}$:[1]

$$\frac{\partial}{\partial t}\left(\tfrac{1}{2}\rho v^2\right) = -\left(\nabla \cdot \tfrac{1}{2}\rho v^2 \mathbf{v}\right) - (\nabla \cdot p\mathbf{v}) - p(-\nabla \cdot \mathbf{v}) - (\nabla \cdot [\boldsymbol{\tau} \cdot \mathbf{v}]) - (-\boldsymbol{\tau} : \nabla \mathbf{v}) + \rho(\mathbf{v} \cdot \mathbf{g}) \tag{3.3-1}$$

$\frac{\partial}{\partial t}\left(\tfrac{1}{2}\rho v^2\right)$	rate of increase of kinetic energy per unit volume
$-\left(\nabla \cdot \tfrac{1}{2}\rho v^2 \mathbf{v}\right)$	rate of addition of kinetic energy by convection per unit volume
$-(\nabla \cdot p\mathbf{v})$	rate of work done by pressure of surroundings on the fluid per unit volume
$-p(-\nabla \cdot \mathbf{v})$	rate of *reversible* conversion of kinetic energy into internal energy per unit volume
$-(\nabla \cdot [\boldsymbol{\tau} \cdot \mathbf{v}])$	rate of work done by viscous force on the fluid per unit volume
$-(-\boldsymbol{\tau} : \nabla \mathbf{v})$	rate of *irreversible* conversion from kinetic to internal energy per unit volume
$+\rho(\mathbf{v} \cdot \mathbf{g})$	rate of work done by external force on the fluid per unit volume

At this point it is not clear why we have attributed the indicated physical significance to the terms $p(\nabla \cdot \mathbf{v})$ and $(\boldsymbol{\tau} : \nabla \mathbf{v})$. Their meaning cannot be properly appreciated until one has studied the energy balance in Chapter 11. There it will be seen how these same two terms appear with the opposite sign in the equation of change for internal energy.

We now introduce the *potential energy*[2] (per unit mass) $\hat{\Phi}$, defined by $\mathbf{g} = -\nabla\hat{\Phi}$. Then the last term in Eq. 3.3-1 may be rewritten as $-\rho(\mathbf{v} \cdot \nabla\hat{\Phi}) = -(\nabla \cdot \rho\mathbf{v}\hat{\Phi}) + \hat{\Phi}(\nabla \cdot \rho\mathbf{v})$. The equation of continuity (Eq. 3.1-4) may now be used to replace $+\hat{\Phi}(\nabla \cdot \rho\mathbf{v})$ by $-\hat{\Phi}(\partial\rho/\partial t)$. The latter may be written as $-\partial(\rho\hat{\Phi})/\partial t$, if the potential energy is independent of the time. This is true for the gravitational field for systems that are located on the surface of the earth; then $\hat{\Phi} = gh$ where g is the (scalar constant) gravitational acceleration and h is the elevation coordinate in the gravitational field.

With the introduction of the potential energy, Eq. 3.3-1 assumes the following form:

$$\frac{\partial}{\partial t}\left(\tfrac{1}{2}\rho v^2 + \rho\hat{\Phi}\right) = -\left(\nabla \cdot \left(\tfrac{1}{2}\rho v^2 + \rho\hat{\Phi}\right)\mathbf{v}\right) - (\nabla \cdot p\mathbf{v}) - p(-\nabla \cdot \mathbf{v}) - (\nabla \cdot [\boldsymbol{\tau} \cdot \mathbf{v}]) - (-\boldsymbol{\tau} : \nabla \mathbf{v}) \tag{3.3-2}$$

This is an *equation of change for kinetic-plus-potential energy.* Since Eqs. 3.3-1 and 3.3-2 contain only mechanical terms, they may both be called *equations of change for mechanical energy.*

The term $p(\nabla \cdot \mathbf{v})$ may be either positive or negative, depending on whether the fluid is undergoing *expansion* or *compression*. The resulting temperature changes can be rather large for gases in compressors, turbines, and shock tubes. The term $(-\boldsymbol{\tau} : \nabla \mathbf{v})$ is always positive for *Newtonian* fluids,[3] because it may be written as a sum of squared terms:

$$\begin{aligned}(-\boldsymbol{\tau} : \nabla \mathbf{v}) &= \frac{1}{2}\mu \sum_i \sum_j \left[\left(\frac{\partial v_i}{\partial x_j} + \frac{\partial v_j}{\partial x_i}\right) - \frac{2}{3}(\nabla \cdot \mathbf{v})\delta_{ij}\right]^2 + \kappa(\nabla \cdot \mathbf{v})^2 \\ &\equiv \mu\Phi_v + \kappa\Psi_v\end{aligned} \tag{3.3-3}$$

---

[1]The interpretation under the $(\boldsymbol{\tau} : \nabla \mathbf{v})$ term is correct only for inelastic fluids, such as Newtonian fluids; for viscoelastic fluids, such as polymers, this term may include reversible conversion to elastic energy.

[2]If $\mathbf{g} = -\boldsymbol{\delta}_z g$ is a vector of magnitude g in the $-z$ direction, then the potential energy per unit mass is $\hat{\Phi} = gz$, where z is the elevation in the gravitational field.

[3]An amusing consequence of the viscous dissipation for air is the study by H. K. Moffatt [*Nature*, **404**, 833–834 (2000)] of the way in which a spinning coin comes to rest on a table.

which serves to define the two quantities Φ_v and Ψ_v. The viscous dissipation function Φ_v is tabulated in component form for Cartesian, cylindrical, and spherical coordinates in Appendix B. The function Ψ_v is zero for incompressible fluids, and thus is often omitted.

The quantity $(-\boldsymbol{\tau} : \nabla\mathbf{v})$ describes the degradation of mechanical energy into thermal energy that occurs in all flow systems (sometimes called *viscous dissipation*).[4] This thermal energy production can result in considerable temperature rises in systems with large viscosity and large velocity gradients, as in lubrication, rapid extrusion, and high-speed flight. (Another example of conversion of mechanical energy into heat is the rubbing of two sticks together to start a fire.)

When we speak of "isothermal systems," we mean systems in which there are no externally imposed temperature gradients and no appreciable temperature change resulting from expansion, contraction, or viscous dissipation.

The most important uses of Eq. 3.3-2 are for the development of the macroscopic mechanical energy balance (or engineering Bernoulli equation), and for developing the internal energy equation in §11.2.

EXAMPLE 3.3-1

The Bernoulli Equation for the Steady Flow of Inviscid Fluids

The Bernoulli equation for steady flow of inviscid fluids (fluids with $\mu \approx 0$) is one of the most famous equations of classical fluid dynamics.[5] Show how it is obtained from the equation of change for mechanical energy, given in Eq. 3.3-2.

SOLUTION

If we assume steady, inviscid flow, then Eq. 3.3-2 can be rewritten as

$$\left(\nabla \cdot \left(\frac{1}{2}\rho v^2 + \rho\hat{\Phi}\right)\mathbf{v}\right) = -(\mathbf{v} \cdot \nabla p) \tag{3.3-4}$$

where the inviscid flow assumption has been used to eliminate the terms containing $\boldsymbol{\tau}$. Next we use the appropriate rule for differentiating the product of a scalar and a vector (Eq. A.4-19) to expand the left side of the equation

$$\left(\rho\mathbf{v} \cdot \nabla\left(\frac{1}{2}v^2 + \hat{\Phi}\right)\right) + \left(\frac{1}{2}v^2 + \hat{\Phi}\right)(\nabla \cdot \rho\mathbf{v}) = -(\mathbf{v} \cdot \nabla p) \tag{3.3-5}$$

The second term on the left side is zero because the continuity equation (Eq. 3.1-4) gives $(\nabla \cdot \rho\mathbf{v}) = 0$ at steady state. After expanding the first term on the left side, the equation of change for mechanical energy is written as

$$\rho\left(\mathbf{v} \cdot \nabla\frac{1}{2}v^2\right) + \rho(\mathbf{v} \cdot \nabla\hat{\Phi}) = -(\mathbf{v} \cdot \nabla p) \tag{3.3-6}$$

Then we write the potential energy term using $\nabla\hat{\Phi} = g\nabla h$, where h is the coordinate in the direction opposite to the direction of gravity. This gives

$$\left(\mathbf{v} \cdot \nabla\frac{1}{2}v^2\right) + (\mathbf{v} \cdot g\nabla h) + \frac{1}{\rho}(\mathbf{v} \cdot \nabla p) = 0 \tag{3.3-7}$$

We can now divide each term by $|\mathbf{v}|$ and introduce the unit vector $\mathbf{t} = \mathbf{v}/|\mathbf{v}|$, and then recognize that $(\mathbf{t} \cdot \nabla) = d/ds$, where s is the distance along a streamline. That is, $(\mathbf{t} \cdot \nabla)$ is the derivative in the direction of flow. Equation 3.3-7 may now be written as

$$\frac{d}{ds}\left(\frac{1}{2}v^2\right) + g\frac{dh}{ds} + \frac{1}{\rho}\frac{dp}{ds} = 0 \tag{3.3-8}$$

---

[4]G. G. Stokes, *Trans. Camb. Phil. Soc.*, **9**, 8–106 (1851), see pp. 57–59.

[5]**Daniel Bernoulli** (1700–1782) was one of the early researchers in fluid dynamics and also the kinetic theory of gases. His hydrodynamical ideas were summarized in D. Bernoulli, *Hydrodynamica sive de viribus et motibus fluidorum commentarii*, Argentorati (1738), however, he did not actually give Eq. 3.3-9. The credit for the derivation of Eq. 3.3-9 goes to L. Euler, *Histoires de l'Académie de Berlin* (1755).

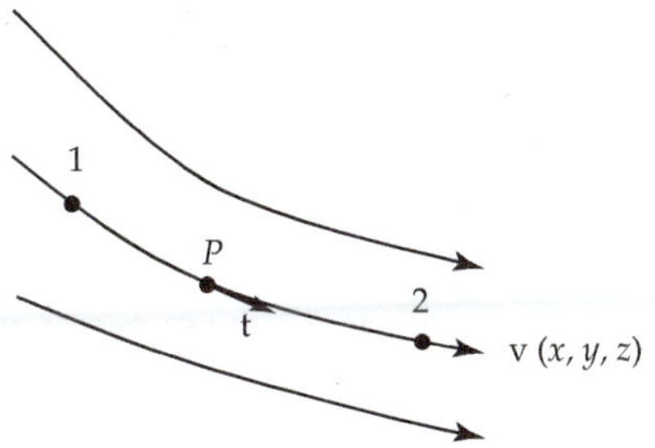

Fig. 3.3-1. Streamlines in a steady flow. Recall that a streamline is a curve that is tangent to the instantaneous velocity. The unit tangent vector at point P is shown.

When this equation is integrated along a streamline from point 1 to point 2 (see Fig. 3.3-1), we get

$$\frac{1}{2}(v_2^2 - v_1^2) + \int_{p_1}^{p_2} \frac{1}{\rho} dp + g(h_2 - h_1) = 0 \tag{3.3-9}$$

which is the *Bernoulli equation*. It relates the velocity, pressure, and elevation at two points along a streamline in a fluid for steady flow. It is used in situations where it can be assumed that viscosity plays a rather minor role (see Example 3.3-2 below).

EXAMPLE 3.3-2

Torricelli's Equation for Efflux from a Tank

A large uncovered tank is filled with an incompressible liquid to a height h. Near the bottom of the tank, there is a hole that allows the fluid to exit to the atmosphere. Apply Bernoulli's equation to a streamline that extends from the liquid surface at the top to a point in the stream just outside the vessel (see Fig. 3.3-2) to obtain an expression for the *efflux velocity*.

SOLUTION

At the top liquid surface (point 1), the pressure is p_{atm}, and the velocity is negligible. In the stream just outside the vessel (point 2), the pressure is also p_{atm} and the velocity magnitude is v_{efflux}. From Eq. 3.3-9 we get

$$\frac{1}{2}(v_{\text{efflux}}^2 - 0) + \frac{1}{\rho}(p_{\text{atm}} - p_{\text{atm}}) + g(0 - h) = 0 \tag{3.3-10}$$

This leads to

$$v_{\text{efflux}} = \sqrt{2gh} \tag{3.3-11}$$

This is known as *Torricelli's equation*.

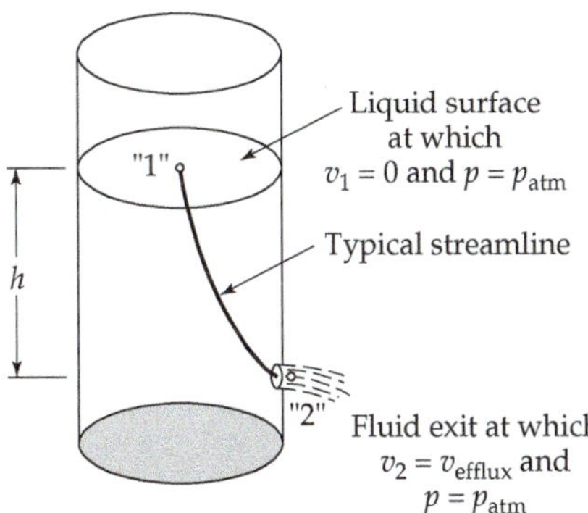

Fig. 3.3-2. Fluid draining from a tank. Points 1 and 2 are on the same streamline. The vertical distance between the surface and the exit hole is h.

Note that in this problem, we have used a formula that is valid at steady state (Eq. 3.3-9) to a situation in which is not actually at steady state (i.e., the liquid height in the tank decreases slowly with time). The assumption that a steady-state equation can be applied to an unsteady system at each instant is called the *quasi-steady-state* assumption.

§3.4 THE EQUATION OF CHANGE FOR ANGULAR MOMENTUM

An equation of change for angular momentum can be obtained from the equation of motion by forming the cross product of the position vector $\mathbf{r}$ (which has Cartesian components x,y,z) with Eq. 3.2-9. When the cross product is formed, we get—after some vector-tensor manipulations—the following *equation of change for angular momentum*:

$$\frac{\partial}{\partial t}\rho[\mathbf{r}\times\mathbf{v}] = -[\nabla\cdot\rho\mathbf{v}[\mathbf{r}\times\mathbf{v}]] - [\nabla\cdot\{\mathbf{r}\times p\boldsymbol{\delta}\}^{\dagger}]$$
$$-[\nabla\cdot\{\mathbf{r}\times\boldsymbol{\tau}^{\dagger}\}^{\dagger}] + [\mathbf{r}\times\rho\mathbf{g}] - [\boldsymbol{\varepsilon}:\boldsymbol{\tau}] \qquad (3.4\text{-}1)$$

Here the superscript † denotes the transpose, and $\boldsymbol{\varepsilon}$ is a third-order tensor with components ε_{ijk} (the permutation symbol defined in §A.2). If the stress tensor $\boldsymbol{\tau}$ is symmetric, as for Newtonian fluids, the last term is zero. According to the kinetic theories of dilute gases, monatomic liquids, and polymers, the tensor $\boldsymbol{\tau}$ is symmetric, in the absence of electric and magnetic torques.[1] If, on the other hand, $\boldsymbol{\tau}$ is asymmetric, then the last term describes the rate of interconversion of bulk angular momentum and internal angular momentum.

The assumption of a symmetric stress tensor, then, is equivalent to an assertion that there is no interconversion between bulk angular momentum and internal angular momentum and that the two forms of angular momentum are conserved separately. This corresponds, in Eq. 0.3-11, to equating the cross-product terms and the internal angular momentum terms separately.

Eq. 3.4-1 will be referred to only in Chapter 7, where we indicate that the macroscopic angular momentum balance can be obtained from it.

§3.5 THE EQUATIONS OF CHANGE (SUBSTANTIAL DERIVATIVE FORM)

Before proceeding, we point out that several different time derivatives may be encountered in transport phenomena. We illustrate these by a homely example—namely, the observation of the concentration of fish in the St. Croix River. Because fish swim around, the fish concentration c will in general be a function of position (x,y,z) and time (t).

The Partial Time Derivative $\partial/\partial t$

Suppose we stand on a bridge and record the concentration of fish just below us as a function of time. From this record of the fish concentration as a function of time at the fixed location x,y,z, we can then calculate the time rate of change of the fish concentration at a fixed location using

$$\left(\frac{\partial c}{\partial t}\right)_{x,y,z} \equiv \lim_{\Delta t\to 0}\frac{c(x,y,z,t+\Delta t) - c(x,y,z,t)}{\Delta t} \qquad (3.5\text{-}1)$$

This is the familiar *partial derivative* of c with respect to t, at constant x,y, and z.

[1] J. S. Dahler and L. E. Scriven, *Nature*, **192**, 36–37 (1961). S. R. de Groot and P. Mazur, *Nonequilibrium Thermodynamics*, North Holland, Amsterdam (1962), Chapter XII. A literature review can be found in G. D. C. Kuiken, *Ind. Eng. Chem. Res.*, **34**, 3568–3572 (1995).

The Total Time Derivative d/dt

Now suppose that we jump into a motor boat and speed around on the river, sometimes going upstream, sometimes downstream, and sometimes across the current. All the time we are recording the fish concentration as a function of time and the current boat location. The time rate of change of the observed fish concentration can be calculated at any instant from this data using

$$\frac{dc}{dt} \equiv \lim_{\Delta t \to 0} \frac{c(x + u_x \Delta t, y + u_y \Delta t, z + u_z \Delta t, t + \Delta t) - c(x,y,z,t)}{\Delta t} \tag{3.5-2}$$

where u_x, u_y, and u_z are the x,y, and z components of the velocity of the boat. Equation 3.5-2 differs from the partial derivative because the "observer" is not fixed in space, but moves with velocity $\mathbf{u}$. This derivative is called the *total derivative*. By expanding the first term in the numerator of Eq. 3.5-2 in a Taylor series (§C.2) about x,y, and z, the total derivative can be rewritten

$$\begin{aligned} \frac{dc}{dt} &= \left(\frac{\partial c}{\partial t}\right)_{x,y,z} + u_x \left(\frac{\partial c}{\partial x}\right)_{y,z,t} + u_y \left(\frac{\partial c}{\partial y}\right)_{z,x,t} + u_z \left(\frac{\partial c}{\partial z}\right)_{x,y,t} \\ &= \frac{\partial c}{\partial t} + (\mathbf{u} \cdot \nabla c) \end{aligned} \tag{3.5-3}$$

Equation 3.5-3 can be applied generally to describe the time rate of change of any quantity as the point of observation moves with velocity $\mathbf{u}$.

The Substantial Time Derivative D/Dt

Next we climb into a canoe, and not feeling energetic, we just float along with the current, recording the fish concentration. In this situation the velocity of the observer $\mathbf{u}$ is the same as the velocity $\mathbf{v}$ of the stream, which has components v_x, v_y, and v_z. If at any instant we report the time rate of change of fish concentration, we are then giving

$$\begin{aligned} \frac{Dc}{Dt} &= \frac{\partial c}{\partial t} + v_x \frac{\partial c}{\partial x} + v_y \frac{\partial c}{\partial y} + v_z \frac{\partial c}{\partial z} \\ &= \frac{\partial c}{\partial t} + (\mathbf{v} \cdot \nabla c) \end{aligned} \tag{3.5-4}$$

The special operator $D/Dt = \partial/\partial t + \mathbf{v} \cdot \nabla$ is called the *substantial derivative* (meaning that the time rate of change is reported as the observer moves with the "substance"). The terms *material derivative, hydrodynamic derivative*, and *derivative following the motion* are also used. Equation 3.5-4 can be utilized in any flow problem to describe the time rate of change of a quantity as the point of observation moves with the local fluid velocity $\mathbf{v}$.

Now we need to know how to convert equations expressed in terms of $\partial/\partial t$ into equations written with D/Dt. For any scalar function $f(x,y,z,t)$, we can do the following manipulations:

$$\begin{aligned} &\frac{\partial}{\partial t}\rho f + \frac{\partial}{\partial x}\rho v_x f + \frac{\partial}{\partial y}\rho v_y f + \frac{\partial}{\partial z}\rho v_z f \\ &= \rho\left(\frac{\partial f}{\partial t} + v_x \frac{\partial f}{\partial x} + v_y \frac{\partial f}{\partial y} + v_z \frac{\partial f}{\partial z}\right) + f\left(\frac{\partial}{\partial t}\rho + \frac{\partial}{\partial x}\rho v_x + \frac{\partial}{\partial y}\rho v_y + \frac{\partial}{\partial z}\rho v_z\right) \\ &= \rho \frac{Df}{Dt} \end{aligned} \tag{3.5-5}$$

The quantity in the second parentheses in the second line is zero according to the equation of continuity. Equation 3.5-5 can be written in vector form as:

$$\frac{\partial}{\partial t}\rho f + (\nabla \cdot \rho \mathbf{v} f) = \rho \frac{Df}{Dt} \tag{3.5-6}$$

Table 3.5-1. The Equations of Change for Isothermal Systems in the D/Dt-Form[a].
Note: At the left are given the equation numbers for the $\partial/\partial t$ forms.

(3.1-4)	$\frac{D\rho}{Dt} = -\rho(\nabla \cdot \mathbf{v})$	(A)
(3.2-9)	$\rho\frac{D\mathbf{v}}{Dt} = -\nabla p - [\nabla \cdot \boldsymbol{\tau}] + \rho\mathbf{g}$	(B)
(3.3-1)	$\rho\frac{D}{Dt}\left(\frac{1}{2}v^2\right) = -(\mathbf{v} \cdot \nabla p) - (\mathbf{v} \cdot [\nabla \cdot \boldsymbol{\tau}]) + \rho(\mathbf{v} \cdot \mathbf{g})$	(C)
(3.4-1)	$\rho\frac{D}{Dt}[\mathbf{r} \times \mathbf{v}] = -[\nabla \cdot \{\mathbf{r} \times p\boldsymbol{\delta}\}^{\dagger}] - [\nabla \cdot \{\mathbf{r} \times \boldsymbol{\tau}\}^{\dagger}] + [\mathbf{r} \times \rho\mathbf{g}]$	(D)

[a]Equations (A) and (B) are obtained from Eqs. 3.1-4 and 3.2-9 with *no assumptions*. Whereas Eq. 3.3-1 is restricted to symmetric $\boldsymbol{\tau}$, Eq. (C) is valid for any $\boldsymbol{\tau}$. Equation (D) is valid for symmetric $\boldsymbol{\tau}$ only. Note that $\boldsymbol{\tau}$ is symmetric for Newtonian fluids.

Similarly for any vector function $\mathbf{f}(x,y,z,t)$

$$\frac{\partial}{\partial t}\rho\mathbf{f} + [\nabla \cdot \rho\mathbf{v}\mathbf{f}] = \rho\frac{D\mathbf{f}}{Dt} \tag{3.5-7}$$

These equations can be used to rewrite the equations of change given in §3.1 to §3.4 in terms of the substantial derivative as shown in Table 3.5-1.

Equation (A) in Table 3.5-1 tells how the density is decreasing or increasing as one moves along with the fluid, because of the compression $[(\nabla \cdot \mathbf{v}) < 0]$ or expansion of the fluid $[(\nabla \cdot \mathbf{v}) > 0]$. Equation (B) can be interpreted as (mass) $\times$ (acceleration) = the sum of the pressure forces, viscous forces, and the external force (all per unit volume). In other words, Eq. 3.2-9 is equivalent to Newton's second law of motion applied to a tiny blob of fluid whose envelope moves with the local fluid velocity $\mathbf{v}$. Equation (B) is tabulated in component form in Cartesian, cylindrical, and spherical coordinates in §B.5.

§3.6 COMMON SIMPLIFICATIONS OF THE EQUATION OF MOTION

We now discuss briefly the three most common simplifications of the equation of motion.

(i) For *constant* ρ *and* μ, insertion of the Newtonian expression for $\boldsymbol{\tau}$ from Eq. 1.2-13 into the equation of motion leads to the very famous *Navier–Stokes equation*, first developed from molecular arguments by Navier and from continuum arguments by Stokes[1]:

$$\boxed{\rho\frac{D}{Dt}\mathbf{v} = -\nabla p + \mu\nabla^2\mathbf{v} + \rho\mathbf{g}} \tag{3.6-1}$$

Equation 3.6-1 is a standard starting point for describing isothermal flows of incompressible Newtonian fluids. This equation is written in component form in Cartesian, cylindrical, and spherical coordinates in §B.6.

Equation 3.6-1 can be written in terms of the modified pressure by replacing p with $\mathscr{P} - \rho g h$, to give

$$\rho\frac{D}{Dt}\mathbf{v} = -\nabla\mathscr{P} + \mu\nabla^2\mathbf{v} \tag{3.6-2}$$

[1]C.-L.-M.-H. Navier, *Mémoires de l'Académie Royale des Sciences*, **6**, 389–440 (1827); G. G. Stokes, *Proc. Cambridge Phil. Soc.*, **8**, 287–319 (1845). The name Navier is pronounced "Nah-vyay."

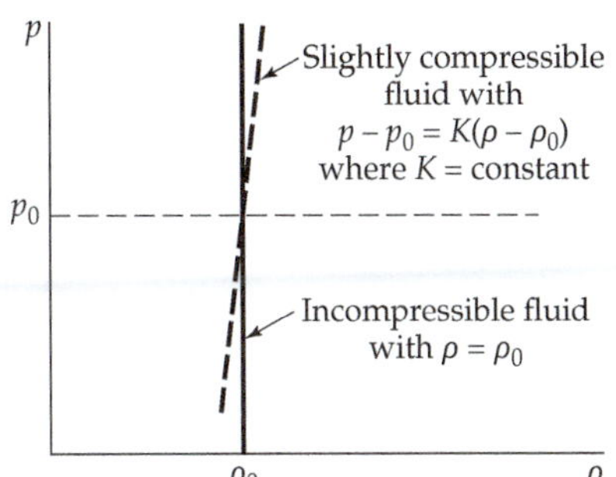

Fig. 3.6-1. The equation of state for a slightly compressible fluid and an incompressible fluid when T is constant.

Appendix B.6 can also be used to obtain the component forms of this equation in Cartesian, cylindrical, and spherical coordinates. One need only replace p in the tables in §B.6 by $\mathscr{P}$, and omit all terms containing the acceleration of gravity.

It must be kept in mind that, when constant ρ is assumed, the equation of state (at constant T) is a vertical line on a plot of p vs. ρ (see Fig. 3.6-1). Thus, the absolute pressure is no longer determinable from ρ and T, although pressure gradients and instantaneous differences remain determinate by Eq. 3.6-1 or Eq. 3.6-2. Absolute pressures are also obtainable if p is known at some point in the system.

(ii) When the *acceleration terms* in the Navier–Stokes equation are neglected—that is, when $\rho(D\mathbf{v}/Dt) = \mathbf{0}$—we get

$$\boxed{\mathbf{0} = -\nabla p + \mu\nabla^2\mathbf{v} + \rho\mathbf{g}} \tag{3.6-3}$$

which is called the *Stokes equation*. It is sometimes called the *creeping flow equation*, because the term $\rho[\mathbf{v}\cdot\nabla\mathbf{v}]$, which is quadratic in the velocity, can be discarded when the flow is extremely slow. For some flows, such as the Hagen-Poiseuille tube flow, the term $\rho[\mathbf{v}\cdot\nabla\mathbf{v}]$ drops out, and a restriction to slow flow is not implied. The Stokes flow equation is important in lubrication theory, the study of particle motions in suspension, flow through porous media, and swimming of microbes. There is a vast literature on this subject.[2]

(iii) When *viscous forces are neglected*—that is, $[\nabla\cdot\boldsymbol{\tau}] = \mathbf{0}$—the equation of motion becomes

$$\boxed{\rho\frac{D\mathbf{v}}{Dt} = -\nabla p + \rho\mathbf{g}} \tag{3.6-4}$$

which is known as the *Euler equation* for "inviscid" fluids.[3] Of course, there are no truly "inviscid" fluids, but there are many flows in which the viscous forces are relatively unimportant. Examples are the flow around airplane wings (except near the solid boundary), flow of rivers around the upstream surfaces of bridge abutments, some problems in compressible gas dynamics, and flow of ocean currents.[4]

---

[2] J. Happel and H. Brenner, *Low Reynolds Number Hydrodynamics*, Martinus Nijhoff, The Hague (1983); S. Kim and S. J. Karrila, *Microhydrodynamics: Principles and Selected Applications*, Dover, NY (2005).

[3] L. Euler, *Mém. Acad. Sci. Berlin*, **11**, 217–273, 274–315, 316–361 (1755). The Swiss-born mathematician **Leonhard Euler** (1707–1783) (pronounced "Oiler") taught in St. Petersburg, Basel, and Berlin and published extensively in many fields of mathematics and physics.

[4] See, for example, D. J. Acheson, *Elementary Fluid Mechanics*, Clarendon Press, Oxford (1990), Chapters 3–5; and G. K. Batchelor, *An Introduction to Fluid Dynamics*, Cambridge University Press (1967), Chapter 6.

§3.7 THE EQUATIONS OF CHANGE AND SOLVING STEADY-STATE PROBLEMS WITH ONE INDEPENDENT VARIABLE

For most applications of the equation of motion, we have to insert the expression for $\boldsymbol{\tau}$ from Eq. 1.2-13 into Eq. 3.2-9 (or, equivalently, the components of $\boldsymbol{\tau}$ from Eqs. 1.2-7 through 1.2-12 or from §B.1 into Eqs. 3.2-4, 3.2-5, and 3.2-6). Then to describe the flow of a Newtonian fluid at constant temperature, we need in general

The equation of continuity	Eq. 3.1-4
The equation of motion	Eq. 3.2-9
The components of $\boldsymbol{\tau}$	Eq. 1.2-13
The equation of state	$p = p(\rho)$
The equations for the viscosities	$\mu = \mu(\rho)$, $\kappa = \kappa(\rho)$

These equations, along with the necessary boundary and initial conditions, determine completely the pressure, density, and velocity distributions in the fluid. However, they are seldom used in their complete form to solve fluid dynamics problems. Usually restricted forms are used for convenience, as in this chapter.

If it is appropriate to assume *constant density and viscosity*, then we use

The equation of continuity	Eq. 3.1-4 (or §B.4)
The Navier–Stokes equation	Eq. 3.6-1 (or §B.6)

along with initial and boundary conditions. From these, one determines the pressure (within an additive constant; see §3.6) and velocity distributions.

In Chapter 1 we gave the components of the stress tensor in Cartesian coordinates, and in this chapter we have derived the equations of continuity and motion in Cartesian coordinates. In §B.1, §B.4, §B.5, and §B.6, we have summarized these key equations in three much used coordinate systems: Cartesian (x,y,z), cylindrical (r,θ,z), and spherical (r,θ,ϕ) coordinates. Beginning students need not concern themselves with the derivation of these equations, but they should be very familiar with the tables in Appendix B and be able to use them for setting up fluid dynamics problems.

In this section we illustrate how to set up and solve some problems involving the steady, isothermal, laminar flow of Newtonian fluids. The relatively simple analytical solutions given here are not to be regarded as ends in themselves, but rather as a preparation for moving on to the analytical or numerical solution of more complex problems, the use of various approximate methods, or the use of dimensional analysis.

The complete solution of viscous flow problems, including proofs of uniqueness and criteria for stability, is a formidable task. Indeed the attention of some of the world's best applied mathematicians has been devoted to the challenge of solving the equations of continuity and motion. The beginner may well feel inadequate when faced with these equations for the first time. All we attempt to do in the illustrative examples in this section is to solve a few simple problems for stable flows that are known to exist. In each case we begin by making some *postulates* about the form for the pressure and velocity distributions—that is, we make *informed guesses* about how p and $\mathbf{v}$ should depend on position in the problem being studied. Then we discard all the terms in the equations of continuity and motion that are zero or negligible according to the postulates made. For example, if one postulates that v_x is a function of y alone, then terms like $\partial v_x/\partial x$ and $\partial^2 v_x/\partial z^2$ are zero and can be eliminated from the governing equations. When all the appropriate terms have been eliminated, one is frequently left with a small number of relatively simple equations; and if the problem is sufficiently simple, an analytical solution can be obtained.

It must be emphasized that in listing the postulates, one makes use of *intuition*. The latter is based on our daily experience with flow phenomena. Our intuition often tells us that a flow will be symmetric about an axis, or that some component of the velocity is zero. Having used our intuition to make such postulates, we must remember that the final solution is correspondingly restricted. However, by starting with the equations of change, when we have finished the "discarding process," we do at least have a complete listing of all the assumptions used in the solution. In some instances, it is possible to go back and remove some of the assumptions and get a better solution.

In several examples to be discussed, we will find one solution to the fluid dynamical equations. However, because the full equations are nonlinear, there may be other solutions to the problem. Thus, a complete solution to a fluid dynamics problem requires the specification of the limits on the stable flow regimes as well as any ranges of unstable behavior. That is, we have to develop a "map" showing the various flow regimes that are possible. Usually analytical solutions can be obtained for only the simplest flow regimes; the remainder of the information is generally obtained by experiment or by very detailed numerical solutions. In other words, although we know the differential equations that govern the fluid motion, much is yet unknown about how to solve them. This is a challenging area of applied mathematics, well above the level of an introductory textbook.

When difficult problems are encountered, a search should be made through some of the advanced treatises on fluid dynamics.[1]

We now turn to the illustrative examples. The first two are problems that were discussed in the preceding chapter; we rework these just to illustrate the use of the equations of change. Then we consider some other problems that would be difficult to set up by the shell balance method of Chapter 2.

EXAMPLE 3.7-1

Steady Flow in a Long Circular Tube

Rework the tube-flow problem of §2.3 using the equations of continuity and motion. This illustrates the use of the tabulated equations for constant viscosity and density in cylindrical coordinates, given in Appendix B.

SOLUTION

We postulate that $\mathbf{v} = \boldsymbol{\delta}_z v_z(r,z)$. This postulate implies that there is no radial flow ($v_r = 0$) and no tangential flow ($v_\theta = 0$), and that v_z does not depend on θ. Consequently we can discard many terms from the tabulated equations of change, leaving

equation of continuity:
$$\frac{\partial v_z}{\partial z} = 0 \tag{3.7-1}$$

r equation of motion:
$$0 = -\frac{\partial \mathscr{P}}{\partial r} \tag{3.7-2}$$

θ equation of motion:
$$0 = -\frac{\partial \mathscr{P}}{\partial \theta} \tag{3.7-3}$$

z equation of motion:
$$0 = -\frac{\partial \mathscr{P}}{\partial z} + \mu \frac{1}{r}\frac{\partial}{\partial r}\left(r\frac{\partial v_z}{\partial r}\right) \tag{3.7-4}$$

The process of eliminating terms in the z component of the equation of motion to arrive at Eq. 3.7-4 is illustrated in Fig. 3.7-1. Equation 3.7-1 indicates that v_z depends only on r; hence, the partial derivatives in the second term on the right side of Eq. 3.7-4 can be replaced by ordinary derivatives. By using the modified pressure $\mathscr{P} = p + \rho g h$ (where h is the height above some arbitrary datum plane), we avoid the necessity of calculating the components of **g** in cylindrical coordinates, and we obtain a solution valid for any orientation of the tube.

---

[1] R. Berker, *Handbuch der Physik*, Volume **VIII-2**, Springer, Berlin (1963), pp. 1–384 (in French); G. K. Batchelor, *An Introduction to Fluid Dynamics*, Cambridge University Press (1967); L. Landau and E. M. Lifshitz, *Fluid Mechanics*, Pergamon Press, Oxford, 2nd edition (1987); J. A. Schetz and A. E. Fuhs (eds.), *Handbook of Fluid Dynamics and Fluid Machinery*, Wiley-Interscience, New York (1996); R. W. Johnson (ed.), *The Handbook of Fluid Dynamics*, CRC Press, Boca Raton, Fla. (1998); C. Y. Wang, *Ann. Revs. Fluid Mech.*, **23**, 159–177 (1991).

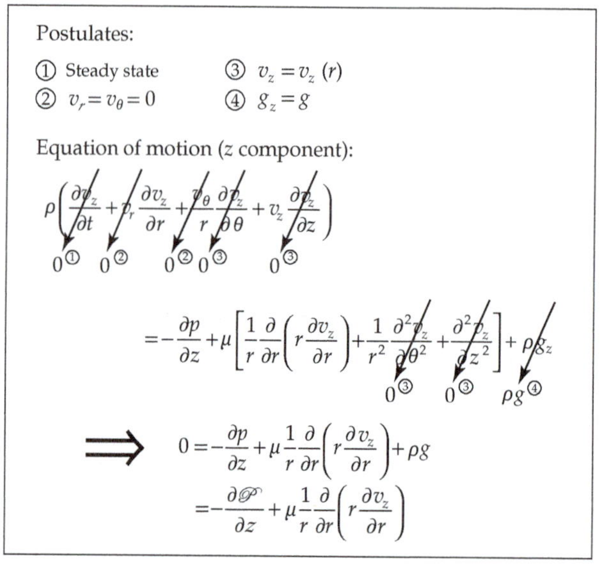

Postulates:

① Steady state ③ $v_z = v_z(r)$

② $v_r = v_\theta = 0$ ④ $g_z = g$

Equation of motion (z component):

$$\rho\left(\frac{\partial v_z}{\partial t} + v_r\frac{\partial v_z}{\partial r} + \frac{v_\theta}{r}\frac{\partial v_z}{\partial \theta} + v_z\frac{\partial v_z}{\partial z}\right)$$

$$0^{①} \quad 0^{②} \quad 0^{②}\, 0^{③} \quad 0^{③}$$

$$= -\frac{\partial p}{\partial z} + \mu\left[\frac{1}{r}\frac{\partial}{\partial r}\left(r\frac{\partial v_z}{\partial r}\right) + \frac{1}{r^2}\frac{\partial^2 v_z}{\partial \theta^2} + \frac{\partial^2 v_z}{\partial z^2}\right] + \rho g_z$$

$$0^{③} \quad 0^{③} \quad \rho g^{④}$$

$$\Rightarrow \quad 0 = -\frac{\partial p}{\partial z} + \mu\frac{1}{r}\frac{\partial}{\partial r}\left(r\frac{\partial v_z}{\partial r}\right) + \rho g$$

$$= -\frac{\partial \mathcal{P}}{\partial z} + \mu\frac{1}{r}\frac{\partial}{\partial r}\left(r\frac{\partial v_z}{\partial r}\right)$$

Fig. 3.7-1. Illustration of the process of writing postulates and using them to eliminate terms in the z component of the equation of motion.

Equations 3.7-2 and 3.7-3 show that $\mathcal{P}$ is a function of z alone, and therefore, the partial derivative in the first term of Eq. 3.7-4 may be replaced by an ordinary derivative. Equation 3.7-4 thus contains two nonzero terms: a pressure term that can be a function only of z, and a velocity term that can only be a function of r. The only way that we can have a function of r plus a function of z equal to zero is for each term individually to be a constant—say C_0—so that

$$\mu\frac{1}{r}\frac{d}{dr}\left(r\frac{dv_z}{dr}\right) = C_0 = \frac{d\mathcal{P}}{dz} \tag{3.7-5}$$

The equation for $\mathcal{P}$ can be integrated at once to give

$$\mathcal{P}(z) = C_0 z + C_1 \tag{3.7-6}$$

To integrate the equation for v_z, one should not try to simplify the left side of Eq. 3.7-5 by using the product rule for derivatives, but instead separate variables and integrate twice. First, multiply by r/μ to obtain

$$\frac{d}{dr}\left(r\frac{dv_z}{dr}\right) = \frac{C_0}{\mu}r \tag{3.7-7}$$

Integrate this equation to get

$$r\frac{dv_z}{dr} = \frac{C_0}{2\mu}r^2 + C_2 \tag{3.7-8}$$

Multiply this equation by $1/r$ and integrate again to obtain

$$v_z(r) = \frac{C_0}{4\mu}r^2 + C_2 \ln r + C_3 \tag{3.7-9}$$

The three constants of integration and C_0 can be obtained by applying the following boundary conditions to Eqs. 3.7-6 and 3.7-9:

B. C. 1	at $z = 0$,	$\mathcal{P} = \mathcal{P}_0$	(3.7-10)
B. C. 2	at $z = L$,	$\mathcal{P} = \mathcal{P}_L$	(3.7-11)
B. C. 3	at $r = R$,	$v_z = 0$	(3.7-12)
B. C. 4	at $r = 0$,	$v_z = \text{finite}$	(3.7-13)

The resulting solutions are:

$$\mathcal{P}(z) = \mathcal{P}_0 - (\mathcal{P}_0 - \mathcal{P}_L)\frac{z}{L} \tag{3.7-14}$$

$$v_z(r) = \frac{(\mathscr{P}_0 - \mathscr{P}_L)R^2}{4\mu L}\left[1 - \left(\frac{r}{R}\right)^2\right] \tag{3.7-15}$$

Equation 3.7-15 is the same as Eq. 2.3-19. The pressure profile in Eq. 3.7-14 was not obtained in §2.3, but was tacitly postulated; we could have done that here, too, but we chose to work with a minimal number of postulates.

As pointed out in §2.3, Eq. 3.7-15 is valid only in the laminar-flow regime, and at locations not too near the tube entrance and exit. For Reynolds numbers above about 2100, a turbulent-flow regime exists downstream of the entrance region, and Eq. 3.7-15 is no longer valid.

EXAMPLE 3.7-2

Falling Film with Variable Viscosity

Set up the problem in Example 2.2-2 by using the equations of Appendix B. This illustrates the use of the equation of motion in terms of $\boldsymbol{\tau}$.

SOLUTION

As in Example 2.2-2 we postulate a steady flow with constant density, but with viscosity depending on x. We postulate, as before, that the x and y components of the velocity are zero and that $v_z = v_z(x)$. With these postulates, the equation of continuity is identically satisfied. According to §B.1, the only nonzero components of $\boldsymbol{\tau}$ are $\tau_{xz} = \tau_{zx} = -\mu(dv_z/dx)$. The components of the equation of motion in terms of $\boldsymbol{\tau}$ are, from §B.5,

$$0 = -\frac{\partial p}{\partial x} + \rho g \sin\beta \tag{3.7-16}$$

$$0 = -\frac{\partial p}{\partial y} \tag{3.7-17}$$

$$0 = -\frac{\partial p}{\partial z} - \frac{d}{dx}\tau_{xz} + \rho g \cos\beta \tag{3.7-18}$$

where β is the angle shown in Figs. 2.2-1 and 2.2-2.

Integration of Eq. 3.7-16 gives

$$p = \rho g x \sin\beta + f(y,z) \tag{3.7-19}$$

in which $f(y,z)$ is an arbitrary function of integration (i.e., the most general term that can be added to the pressure and still satisfy Eq. 3.7-16). Equation 3.7-17 shows that f cannot be a function of y. We next recognize that the pressure in the gas phase is very nearly constant at the prevailing atmospheric pressure p_{atm}. Therefore, all along the gas–liquid interface $x = 0$, the pressure is also constant at the value p_{atm}. Consequently f is also independent of z, and can be set equal to p_{atm}. We obtain finally

$$p = \rho g x \sin\beta + p_{\text{atm}} \tag{3.7-20}$$

Equation 3.7-18 then becomes

$$\frac{d}{dx}\tau_{xz} = \rho g \cos\beta \tag{3.7-21}$$

which is the same as Eq. 2.2-14. The remainder of the solution is the same as in Example 2.2-2.

EXAMPLE 3.7-3

Operation of a Couette Viscometer

It has been mentioned earlier that the measurement of pressure difference vs. mass flow rate through a cylindrical tube is the basis for the determination of viscosity in commercial capillary viscometers. The viscosity may also be determined by measuring the torque required to turn a solid object in contact with a fluid. The forerunner of all rotational viscometers is the Couette instrument, a common version of which is illustrated in Figure 3.7-2.

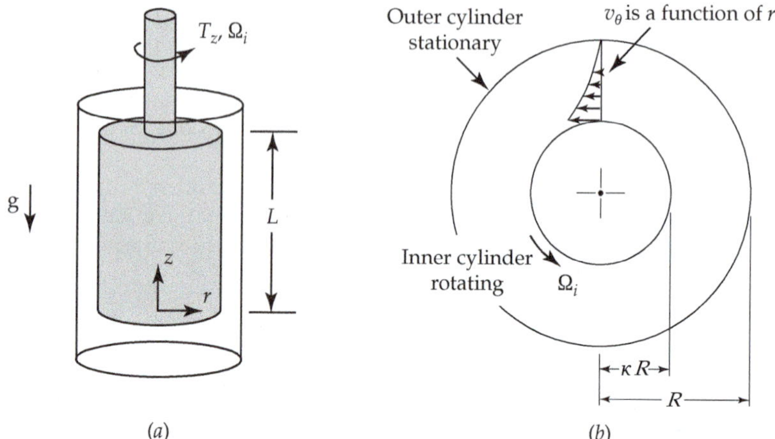

Fig. 3.7-2. (*a*) A diagram of a Couette viscometer. One measures the angular velocity Ω_i of the bob that results from an external torque T_z applied to the bob. Equation 3.7-33 gives the viscosity μ, in terms of Ω_i and the torque T_z. (*b*) Tangential laminar flow of an incompressible fluid in the space between two cylinders, the inner one of which is moving with an angular velocity Ω_i.

The fluid is placed in the outer cylinder or "cup," and the inner cylinder or "bob" is then made to rotate by applying a specified torque T_z about the z axis with a motor placed above. The applied torque causes the suspended bob to rotate with angular speed Ω_i that is measured by the instrument (the subscript "*i*" stands for "inner"). A steady flow is obtained when the externally applied torque is balanced by the torque exerted by the fluid on the inner cylinder. By deriving an expression for the relationship between the torque and the rotation rate, the viscosity of the liquid can be obtained. In this analysis, end effects over the region including the bob height L are neglected. This is reasonable, provided that the bob height is large compared to the gap between the cylinders.

To analyze this measurement, we apply the equations of continuity and motion for a liquid of constant ρ and μ in cylindrical coordinates to the tangential flow in the annular region between the bob and the outer cylinder. Ultimately we want an expression for the viscosity in terms of (the z component of) the torque on the inner cylinder T_z, the angular velocity of the rotating bob Ω_i, the bob height L, and the radii κR and R of the bob and cup.

SOLUTION

In the portion of the annulus under consideration, the fluid moves in a circular pattern. Reasonable postulates for the velocity and pressure are: $v_\theta = v_\theta(r)$, $v_r = 0$, $v_z = 0$, and $p = p(r,z)$. We expect p to depend on z because of gravity, and on r because of the centrifugal force.

For these postulates all the terms in the equation of continuity are zero, and the components of the equation of motion in Eqs. B.6-4 to B.6-6 simplify to

r component:
$$-\rho\frac{v_\theta^2}{r} = -\frac{\partial p}{\partial r} \tag{3.7-22}$$

θ component:
$$0 = \frac{d}{dr}\left(\frac{1}{r}\frac{d}{dr}(rv_\theta)\right) \tag{3.7-23}$$

z component:
$$0 = -\frac{\partial p}{\partial z} - \rho g \tag{3.7-24}$$

The second equation can be used to obtain the velocity distribution. The third equation gives the effect of gravity on the pressure (the hydrostatic effect), and the first equation tells how the centrifugal force affects the pressure. For the problem at hand we need only the θ component of the equation of motion.[2]

Equation 3.7-23 can be integrated once directly to give

$$\frac{1}{r}\frac{d}{dr}(rv_\theta) = C_1 \tag{3.7-25}$$

Multiplication by r yields

$$\frac{d}{dr}(rv_\theta) = C_1 r \tag{3.7-26}$$

which can then be integrated to give

$$rv_\theta = \frac{1}{2}C_1 r^2 + C_2 \tag{3.7-27}$$

or

$$v_\theta(r) = \frac{1}{2}C_1 r + \frac{C_2}{r} \tag{3.7-28}$$

The boundary conditions are that the fluid does not slip at the two cylindrical surfaces:

B. C. 1 at $r = \kappa R$, $v_\theta = \Omega_i \kappa R$ (3.7-29)

B. C. 2 at $r = R$, $v_\theta = 0$ (3.7-30)

These boundary conditions can be used to get the constants of integration $C_1 = -2\Omega_i\kappa[(1/\kappa) -\kappa]^{-1}$ and $C_2 = +\Omega_i\kappa R^2[(1/\kappa) - \kappa]^{-1}$. When these are inserted in Eq. 3.7-28, we get

$$v_\theta(r) = \Omega_i \kappa R \frac{\left(\dfrac{R}{r} - \dfrac{r}{R}\right)}{\left(\dfrac{1}{\kappa} - \kappa\right)} \tag{3.7-31}$$

By writing the result in this form, with similar terms in the numerator and denominator, it is clear that both boundary conditions are satisfied and that the equation is dimensionally consistent.

From the velocity distribution we can find the shear stress (momentum flux) by using §B.1:

$$\tau_{r\theta}(r) = -\mu r \frac{d}{dr}\left(\frac{v_\theta}{r}\right) = 2\mu\Omega_i \left(\frac{\kappa^2}{1-\kappa^2}\right)\left(\frac{R}{r}\right)^2 \tag{3.7-32}$$

The torque about the z axis exerted by the inner cylinder on the fluid is then given by the product of the force per unit area exerted by the cylinder on the fluid in the θ direction ($+\tau_{r\theta}|_{r=\kappa R}$), the surface area of the cylinder ($2\pi\kappa RL$), and the lever arm (κR),

$$T_z = (+\tau_{r\theta})|_{r=\kappa R} \cdot 2\pi\kappa RL \cdot \kappa R = 4\pi\mu\Omega_i R^2 L \left(\frac{\kappa^2}{1-\kappa^2}\right) \tag{3.7-33}$$

Therefore, measurement of the angular velocity of the bob for a known torque makes it possible to determine the viscosity. The same kind of analysis is available for other rotational viscometers.[3]

---

[2]See R. B. Bird, C. F. Curtiss, and W. E. Stewart, *Chem. Eng. Sci.*, **11**, 114–117 (1959) for a method of getting $p(r,z)$ for such systems. The time-dependent buildup to the steady-state profiles is given by R. B. Bird and C. F. Curtiss, *Chem. Eng. Sci.*, **11**, 108–113 (1959).

[3]J. R. VanWazer, J. W. Lyons, K. Y. Kim, and R. E. Colwell, *Viscosity and Flow Measurement*, Wiley, New York (1963); K. Walters, *Rheometry*, Chapman and Hall, London (1975).

One might ask what happens if we hold the inner cylinder fixed and cause the outer cylinder to rotate with an angular velocity Ω_o (the subscript "o" stands for outer). Then the velocity distribution may be shown to be

$$v_\theta(r) = \Omega_o R \frac{\left(\frac{r}{\kappa R} - \frac{\kappa R}{r}\right)}{\left(\frac{1}{\kappa} - \kappa\right)} \tag{3.7-34}$$

This is obtained by making the same postulates (see before Eq. 3.7-22) and solving the same differential equation (Eq. 3.7-23), but with a different set of boundary conditions. The torque required to rotate the outer cylinder is still given by Eq. 3.7-33, but with Ω_i replaced by Ω_o (the case of both cylinders rotating is addressed in Problem 3B.1).

Equation 3.7-31 describes the flow accurately for small values of Ω_i. However, when Ω_i reaches a critical value ($\Omega_{i,\text{crit}} \approx 41.3(\mu/R^2(1-\kappa)^{3/2}\rho)$ for $\kappa \approx 1$), the fluid develops a secondary flow, which is superimposed on the primary (tangential) flow and which is periodic in the axial direction. A very neat system of toroidal vortices, called *Taylor vortices*, is formed, as depicted in Figs. 3.7-3 and 3.7-4(b). The loci of the centers of these vortices are circles, whose centers are located on the common axis of the cylinders. This is still laminar motion—but certainly inconsistent with the postulates made at the beginning of the problem. When the angular velocity Ω_i is increased further, the loci of the centers of the vortices become traveling waves; that is, the flow becomes, in addition, periodic in the tangential direction (see Fig. 3.7-4(c)). Furthermore the angular velocity of the traveling waves is approximately $\frac{1}{3}\Omega_i$. When the angular velocity Ω_i is further increased, the flow becomes turbulent. Figure 3.7-5 shows the various flow regimes, with the inner and outer cylinders both rotating, determined for a specific apparatus and a specific fluid. This diagram demonstrates how complicated this apparently simple system is.

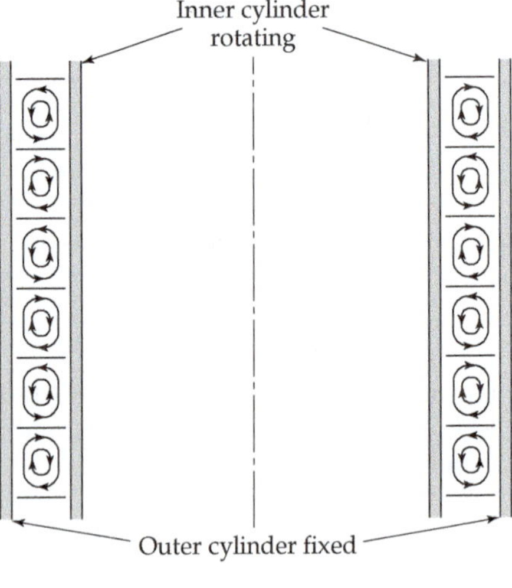

Fig. 3.7-3. Counter-rotating toroidal vortices, called *Taylor vortices*, observed in the annular space between two cylinders. The streamlines have the form of helices, with the axes wrapped around the common axis of the cylinders. This corresponds to Fig. 3.7-4(b). **Geoffrey Ingram Taylor** (1886–1975) is famous for Taylor dispersion, Taylor vortices, and his work on the statistical theory of turbulence; he attacked many complex problems in ingenious ways that made maximum use of the physical processes involved.

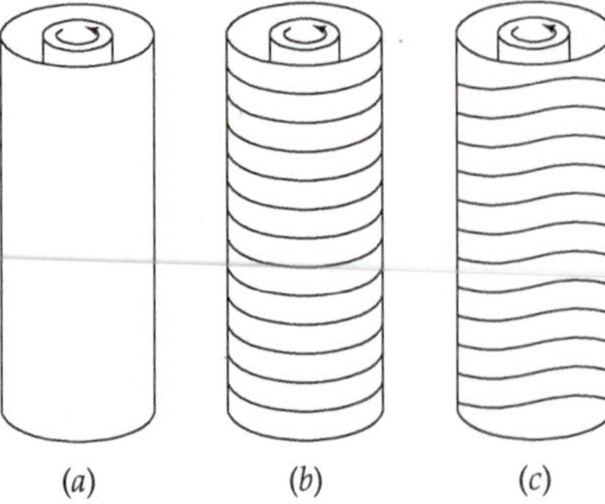

Fig. 3.7-4. Sketches showing the phenomena observed in the annular space between two cylinders: (a) purely tangential flow; (b) singly periodic flow (Taylor vortices); and (c) doubly periodic flow in which an undulatory motion is superposed on the Taylor vortices.

Further details may be found elsewhere.[4,5] The critical Reynolds number $(\Omega_o R^2 \rho/\mu)_{\text{crit}}$, above which the system becomes turbulent, is shown in Fig. 3.7-6 as a function of the radius ratio κ for the case of the outer cylinder rotating with the inner cylinder fixed.

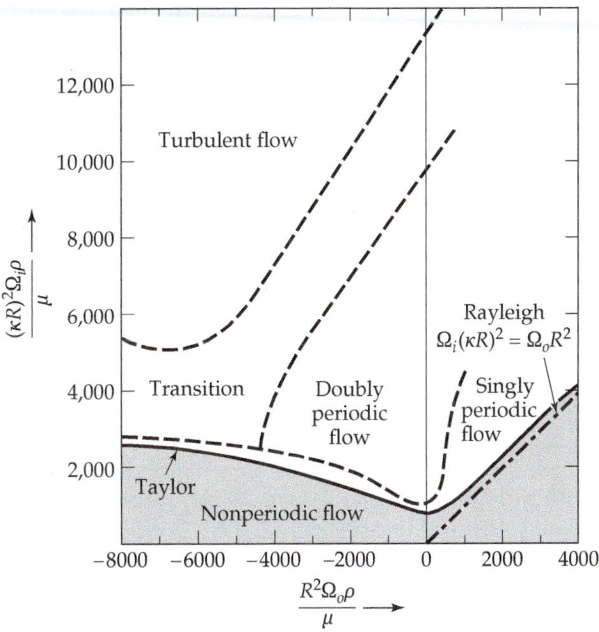

Fig. 3.7-5. Flow-regime diagram for the flow between two coaxial cylinders. The straight line, labeled "Rayleigh" is Lord Rayleigh's analytic solution for an inviscid fluid. [See D. Coles, *J. Fluid Mech.*, **21**, 385–425 (1965).]

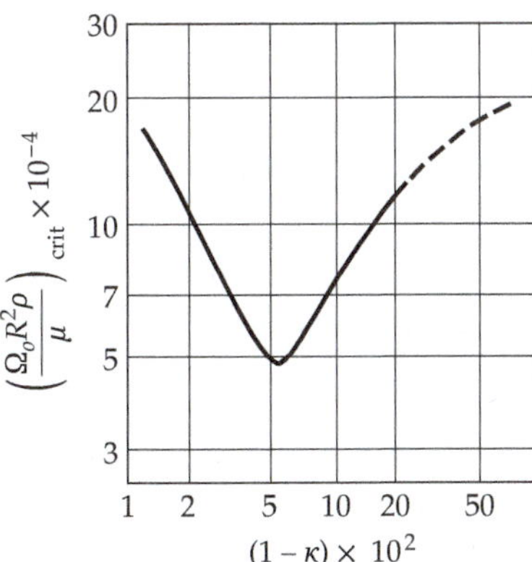

Fig. 3.7-6. Critical Reynolds number for the tangential flow in an annulus, with the outer cylinder rotating and the inner cylinder stationary. [H. Schlichting, *Boundary Layer Theory*, McGraw-Hill, New York (1955), p. 357.]

[4]The initial work on this subject was done by **John William Strutt (Lord Rayleigh)** (1842–1919), who established the field of acoustics with his *Theory of Sound*, written on a houseboat on the Nile River. Some original references on Taylor instability are: J. W. Strutt (Lord Rayleigh), *Proc. Roy. Soc.*, **A93**, 148–154 (1916); G. I. Taylor, *Phil. Trans.*, **A223**, 289–343 (1923) and *Proc. Roy. Soc.* **A157**, 546–564 (1936); P. Schultz-Grunow and H. Hein, *Zeits. Flugwiss.*, **4**, 28–30 (1956); D. Coles, *J. Fluid Mech.* **21**, 385–425 (1965). See also R. P. Feynman, R. B. Leighton, and M. Sands, *The Feynman Lectures in Physics*, Addison-Wesley, Reading, MA (1964), §41-6.

[5]Other references on Taylor instability, as well as instability in other flow systems, are: L. D. Landau and E. M. Lifshitz, *Fluid Mechanics*, Pergamon, Oxford, 2nd edition (1987), pp. 99–106; S. Chandrasekhar, *Hydrodynamic and Hydromagnetic Stability*, Oxford University Press (1961), pp. 272–342; H. Schlichting and K. Gersten, *Boundary-Layer Theory*, Springer-Verlag, Berlin, 8th edition (2000), Chapter 15; P. G. Drazin and W. H. Reid, *Hydrodynamic Stability*, Cambridge University Press (1981); M. Van Dyke, *An Album of Fluid Motion*, Parabolic Press, Stanford (1982).

The preceding discussion should serve as a stern warning that intuitive postulates may be misleading. Most of us would not think about postulating the singly and doubly periodic solutions just described. Nonetheless, this information *is* contained in the Navier–Stokes equations! However, since problems involving instability and transitions between several flow regimes are extremely complex, we are forced to use a combination of theory and experiment to describe them. Theory alone cannot yet give us all the answers, and carefully controlled experiments will be needed for years to come.

EXAMPLE 3.7-4

Shape of the Surface of a Rotating Liquid

A liquid of constant density and viscosity is in a cylindrical container of radius R as shown in Fig. 3.7-7. The container is caused to rotate about its own axis at an angular velocity Ω. The cylinder axis is vertical, so that in cylindrical coordinates $g_r = 0$, $g_\theta = 0$, and $g_z = -g$, in which g is the magnitude of the gravitational acceleration. Find the shape of the free surface of the liquid at steady state.

SOLUTION

Cylindrical coordinates are appropriate for this problem, and the equations of change are given in §B.4 and §B.6. At steady state we postulate that v_r and v_z are both zero and that v_θ depends only on r. We also postulate that p depends on z because of the gravitational force, and on r because of the centrifugal force.

These postulates give $0 = 0$ for the equation of continuity, and the equation of motion gives:

r component:
$$-\rho\frac{v_\theta^2}{r} = -\frac{\partial p}{\partial r} \tag{3.7-35}$$

θ component:
$$0 = \frac{d}{dr}\left(\frac{1}{r}\frac{d}{dr}(rv_\theta)\right) \tag{3.7-36}$$

z component:
$$0 = -\frac{\partial p}{\partial z} - \rho g \tag{3.7-37}$$

The θ component of the equation of motion can be integrated twice to give

$$v_\theta(r) = \frac{1}{2}C_1 r + \frac{C_2}{r} \tag{3.7-38}$$

in which C_1 and C_2 are constants of integration. Because v_θ cannot be infinite at $r = 0$, the constant C_2 must be zero. At $r = R$ the velocity v_θ is $R\Omega$. Hence, $C_1 = 2\Omega$ and

$$v_\theta(r) = \Omega r \tag{3.7-39}$$

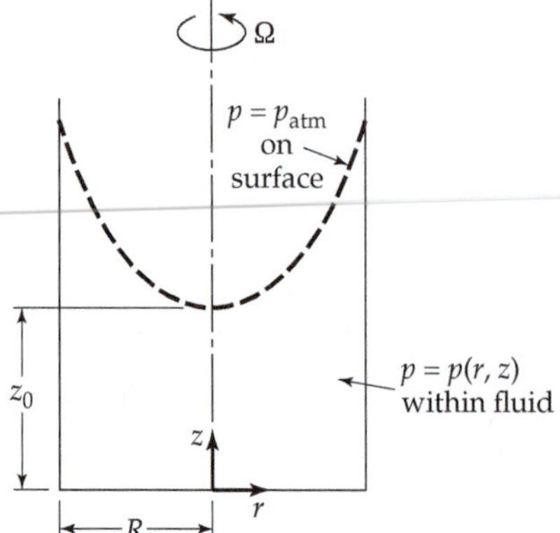

Fig. 3.7-7. Rotating liquid with a free surface, the shape of which is a paraboloid of revolution.

This states that each element of the rotating liquid moves as an element of a rigid body. When the result in Eq. 3.7-39 is substituted into Eq. 3.7-35, we then have the following two equations for the pressure derivatives:

$$\frac{\partial p}{\partial r} = \rho \Omega^2 r \qquad \text{and} \qquad \frac{\partial p}{\partial z} = -\rho g \tag{3.7-40, 41}$$

Each of these equations can be integrated to give

$$p(r,\theta,z) = \frac{1}{2}\rho \Omega^2 r^2 + f_1(\theta,z) \qquad \text{and} \qquad p(r,\theta,z) = -\rho g z + f_2(r,\theta) \tag{3.7-42, 43}$$

where $f_1(\theta,z)$ and $f_2(r,\theta)$ are arbitrary functions of integration. Since we have postulated that p does not depend on θ, we can choose $f_1(z) = -\rho g z + C$ and $f_2(r) = \frac{1}{2}\rho \Omega^2 r^2 + C$, where C is a constant, to satisfy Eqs. 3.7-40 and 3.7-41. Thus, the solution to Eqs. 3.7-40 and 3.7-41 is

$$p(r,z) = -\rho g z + \frac{1}{2}\rho \Omega^2 r^2 + C \tag{3.7-44}$$

The constant C may be determined by requiring that $p = p_{\text{atm}}$ at $r = 0$ and $z = z_0$, the latter being the elevation of the liquid surface at $r = 0$. When C is obtained in this way, we get

$$p(r,z) - p_{\text{atm}} = -\rho g (z - z_0) + \frac{1}{2}\rho\, \Omega^2 r^2 \tag{3.7-45}$$

This equation gives the pressure at all points within the liquid. Right at the liquid-air interface, $p = p_{\text{atm}}$, and with this substitution Eq. 3.7-45 gives the shape of the liquid-air interface

$$z(r) - z_0 = \left(\frac{\Omega^2}{2g}\right) r^2 \tag{3.7-46}$$

This is the equation for a parabola. The reader can verify that the free surface of a liquid in a rotating annular container obeys a similar relation.

§3.8 THE EQUATIONS OF CHANGE AND SOLVING PROBLEMS WITH TWO INDEPENDENT VARIABLES

In this section, we show how some problems with more than one independent variable can be solved. For problems with two independent variables, the equations of change produce partial differential equations that can be quite challenging to solve. In the examples below, the problems are solved by using a few well-known methods. There are numerous other methods that can be exploited to solve flow problems, which are beyond the scope of this text. The reasons for including some examples here are (1) to provide additional examples where the equations of change are reduced using postulates and boundary (and initial) conditions are used to evaluate unknown constants, (2) to illustrate some useful and interesting transport phenomena, and (3) to show the reader that with a little more experience with mathematics, many more complicated flow problems can be solved.

Example 3.8-1 illustrates a rectilinear flow problem in which the only nonzero velocity component depends on both position and time. This problem also introduces "similarity solution" methods, as well as the error function and complementary error function, which appear in numerous problems in transport phenomena. Example 3.8-2 illustrates a creeping flow problem in spherical coordinates. Example 3.8-3 illustrates a creeping flow problem in cylindrical coordinates, and also demonstrates how the equation of continuity can be used in solving flow problems.

EXAMPLE 3.8-1

Flow near a Wall Suddenly Set in Motion

A semi-infinite body of liquid with constant density and viscosity is bounded below by a horizontal surface (the xz plane). Initially the fluid and the solid are at rest. Then at time $t = 0$, the solid surface is set in motion in the positive x direction with speed v_0 as shown in Fig. 3.8-1. Find the velocity v_x as a function of y and t. There is no pressure gradient or gravity force in the x direction, and the flow is presumed to be laminar.

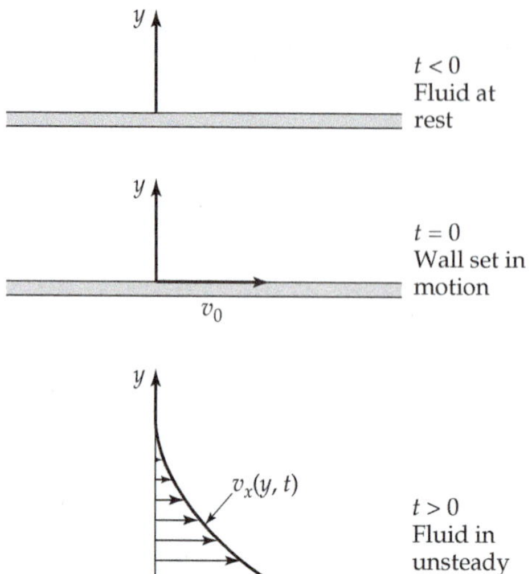

Fig. 3.8-1. Viscous flow of a fluid near a wall suddenly set in motion.

SOLUTION

For this system, $v_x = v_x(y,t)$, $v_y = 0$, and $v_z = 0$. Then from §B.4 we find that the equation of continuity is satisfied directly, and from §B.6 we get

$$\frac{\partial v_x}{\partial t} = \nu \frac{\partial^2 v_x}{\partial y^2} \tag{3.8-1}$$

in which $\nu = \mu/\rho$ is the kinematic viscosity. The initial and boundary conditions are

I. C.:	at $t \leq 0$,	$v_x = 0$	for all y	(3.8-2)
B. C. 1:	at $y = 0$,	$v_x = v_0$	for all $t > 0$	(3.8-3)
B. C. 2:	as $y \to \infty$,	$v_x \to 0$	for all $t > 0$	(3.8-4)

Next we introduce a dimensionless velocity $\phi = v_x/v_0$, so that Eq. 3.8-1 becomes

$$\frac{\partial \phi}{\partial t} = \nu \frac{\partial^2 \phi}{\partial y^2} \tag{3.8-5}$$

with the initial condition $\phi(y,0) = 0$, and boundary conditions $\phi(0,t) = 1$ and $\phi(\infty,t) = 0$. Because ϕ is a dimensionless function and the initial and boundary conditions contain only pure (dimensionless) numbers, the quantities y,t, and ν must always appear in the function $\phi(y,t;\nu)$ in a dimensionless combination. The only dimensionless combinations of these three quantities are $y/\sqrt{\nu t}$, or powers or multiples thereof. We therefore conclude that

$$\phi = \phi(\eta) \qquad \text{where} \qquad \eta = \frac{y}{\sqrt{4\nu t}} \tag{3.8-6}$$

Combining variables in this way is known as the "similarity method" or the "method of combination of (independent) variables." The "4" is included so that the final result in Eq. 3.8-15 will look neater; we know to do this only after solving the problem without it. The form of the solution in Eq. 3.8-6 is possible essentially because there is no characteristic length or time in the physical system.

We now convert the derivatives in Eq. 3.8-5 into derivatives with respect to the "combined variable" η as follows:

$$\frac{\partial \phi}{\partial t} = \frac{d\phi}{d\eta}\frac{\partial \eta}{\partial t} = -\frac{1}{2}\frac{y}{\sqrt{4\nu}}\frac{1}{t^{3/2}}\frac{d\phi}{d\eta} = -\frac{1}{2}\frac{\eta}{t}\frac{d\phi}{d\eta} \tag{3.8-7}$$

$$\frac{\partial \phi}{\partial y} = \frac{d\phi}{d\eta}\frac{\partial \eta}{\partial y} = \frac{1}{\sqrt{4\nu t}}\frac{d\phi}{d\eta} \tag{3.8-8}$$

$$\frac{\partial^2 \phi}{\partial y^2} = \frac{1}{4\nu t}\frac{d^2\phi}{d\eta^2} \tag{3.8-9}$$

Substitution of Eqs. 3.8-7 and 3.8-9 into Eq. 3.8-5 then gives

$$\frac{d^2\phi}{d\eta^2} + 2\eta\frac{d\phi}{d\eta} = 0 \tag{3.8-10}$$

This is an ordinary differential equation of the type given in Eq. C.1-8, and the accompanying boundary conditions are

B. C. 1: at $\eta = 0, \quad \phi = 1$ (3.8-11)

B. C. 2: as $\eta \to \infty, \quad \phi \to 0$ (3.8-12)

The first of these boundary conditions is related to Eq. 3.8-3, and the second includes Eqs. 3.8-2 and 3.8-4. If now we let $d\phi/d\eta = \psi$, we get a first-order separable equation for ψ, which may be solved to give

$$\psi(\eta) = \frac{d\phi}{d\eta} = C_1 \exp(-\eta^2) \tag{3.8-13}$$

A second integration then gives

$$\phi(\eta) = C_1 \int_0^{\eta} \exp(-\bar{\eta}^2)d\bar{\eta} + C_2 \tag{3.8-14}$$

The choice of 0 for the lower limit of the integral is arbitrary; another choice would lead to a different value of C_2, which is still undetermined. Note that we have been careful to use an overbar for the dummy variable of integration ($\bar{\eta}$) to distinguish it from the η in the upper limit.

Application of the two boundary conditions makes it possible to evaluate the two integration constants, and we get finally

$$\phi(\eta) = 1 - \frac{\int_0^{\eta} \exp(-\bar{\eta}^2)d\bar{\eta}}{\int_0^{\infty} \exp(-\bar{\eta}^2)d\bar{\eta}} = 1 - \frac{2}{\sqrt{\pi}}\int_0^{\eta} \exp(-\bar{\eta}^2)d\bar{\eta}$$
$$= 1 - \text{erf}\,\eta = \text{erfc}\,\eta \tag{3.8-15}$$

where erf η and erfc η are known as the "error function" and the "complementary error function," respectively (see §C.6). They are well-known functions, available in mathematics handbooks and computer software. When Eq. 3.8-15 is rewritten in the original variables, it becomes

$$\frac{v_x(y,t)}{v_0} = 1 - \text{erf}\frac{y}{\sqrt{4\nu t}} = \text{erfc}\frac{y}{\sqrt{4\nu t}} \tag{3.8-16}$$

Plots of Eq. 3.8-16 are given in Fig. 3.8-2(*a*) and (*b*). Note that by plotting the result in terms of dimensionless quantities, only one curve is needed to describe the velocity profile at all times and for all fluids.

The complementary error function erfc η is a monotone decreasing function that goes from 1 to 0 and drops to 0.01 when η is about 2.0. We can use this fact to define a "boundary-layer thickness" δ as that distance y for which $\phi = v_x/v_0$ has dropped to a value of 0.01. This gives $\delta = 4\sqrt{\nu t}$ as a natural length scale for the diffusion of momentum. This distance is a measure of the extent to which momentum has "penetrated" into the body of the fluid. Note that this boundary-layer thickness is proportional to the square root of the elapsed time. The boundary-layer thickness also quantifies the notion of "similarity" of the velocity profiles at different times—when the position y is divided by the boundary-layer thickness, the scaled velocity profiles $\phi = v_x/v_0$ superpose onto a single curve, as illustrated in Fig. 3.8-2(*b*).

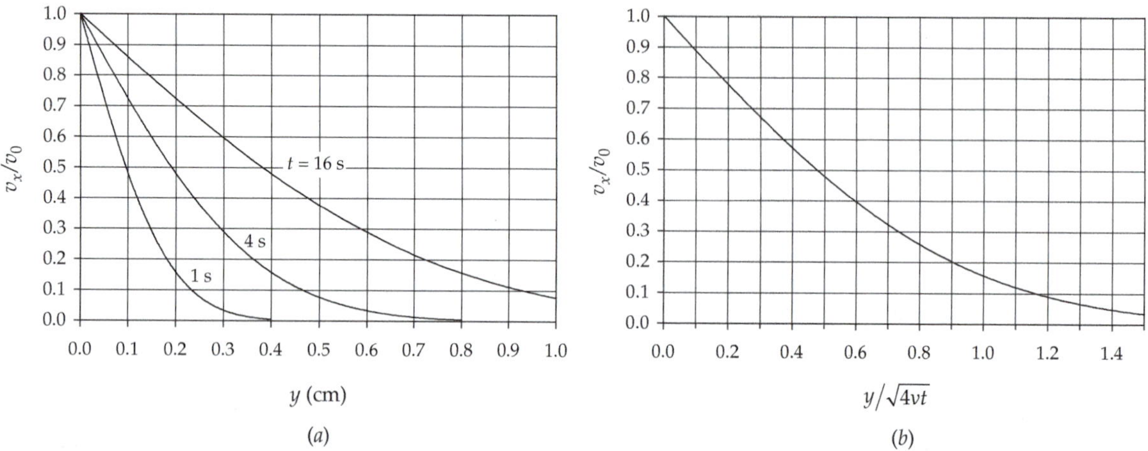

Fig. 3.8-2. (*a*) Velocity distribution at various times for the flow of water at 20°C ($\nu = 10^{-6}$ m^2/s) in the neighborhood of a wall suddenly set in motion. (*b*) Velocity distribution in dimensionless form for flow in the neighborhood of a wall suddenly set in motion.

EXAMPLE 3.8-2

Flow near a Slowly Rotating Sphere

A solid sphere of radius R is rotating slowly at a constant angular velocity Ω in a large body of quiescent fluid (see Fig. 3.8-3). Develop expressions for the pressure and velocity distributions in the fluid and for the torque T_z required to maintain the motion. It is assumed that the sphere rotates sufficiently slowly that it is appropriate to use the *creeping flow* version of the equation of motion in Eq. 3.6-3. This problem illustrates setting up and solving a problem in spherical coordinates.

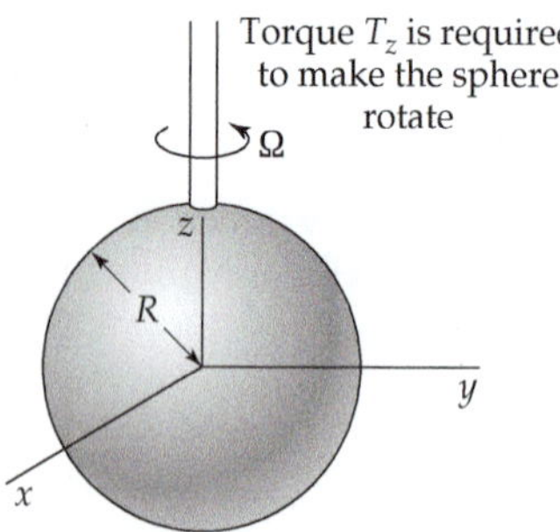

Fig. 3.8-3. A slowly rotating sphere of radius R in an infinite expanse of fluid. The primary flow is $v_\phi = \Omega R(R/r)^2 \sin\theta$.

SOLUTION

The equations of continuity and motion in spherical coordinates are given in §B.4 and §B.6, respectively. We postulate that, for steady creeping flow, the velocity distribution will have the general form $\mathbf{v} = \boldsymbol{\delta}_\phi v_\phi(r,\theta)$, and that the modified pressure will be of the form $\mathscr{P} = \mathscr{P}(r,\theta)$. Since the solution is expected to be symmetric about the z axis, there is no dependence on the angle ϕ.

With these postulates, the equation of continuity is exactly satisfied, and the components of the creeping flow equation of motion become

r component: $$0 = -\frac{\partial \mathscr{P}}{\partial r} \tag{3.8-17}$$

θ component: $$0 = -\frac{1}{r}\frac{\partial \mathscr{P}}{\partial \theta} \tag{3.8-18}$$

ϕ component: $$0 = \frac{1}{r^2}\frac{\partial}{\partial r}\left(r^2 \frac{\partial v_\phi}{\partial r}\right) + \frac{1}{r^2}\frac{\partial}{\partial \theta}\left(\frac{1}{\sin\theta}\frac{\partial}{\partial \theta}(v_\phi \sin\theta)\right) \tag{3.8-19}$$

The boundary conditions may be summarized as

B. C. 1: $$\text{at } r = R, \quad v_r = 0, \quad v_\theta = 0, \quad v_\phi = R\Omega \sin\theta \tag{3.8-20}$$

B. C. 2: $$\text{as } r \to \infty, \quad v_r \to 0, \quad v_\theta \to 0, \quad v_\phi \to 0 \tag{3.8-21}$$

B. C. 3: $$\text{as } r \to \infty, \quad \mathscr{P} \to p_0 \tag{3.8-22}$$

where $\mathscr{P} = p + \rho g z$, and p_0 is the fluid pressure far from the sphere at the plane $z = 0$.

Equation 3.8-19 is a partial differential equation for $v_\phi(r,\theta)$. To solve this, we try a solution of the form $v_\phi = f(r)\sin\theta$. This is just a guess, motivated by the form of the boundary condition in Eq. 3.8-20. When this form for the velocity distribution is inserted into Eq. 3.8-19, we get the following ordinary differential equation for $f(r)$:

$$\frac{d}{dr}\left(r^2 \frac{df}{dr}\right) - 2f = 0 \tag{3.8-23}$$

This is an "equidimensional equation" (or "Cauchy-Euler equation"), which may be solved by assuming a solution of the form $f(r) = r^n$ (see Eq. C.1-14). Substitution of this into Eq 3.8-23 shows that the equation can be satisfied with $n = 1$ or -2. The general solution of Eq. 3.8-23 is then

$$f(r) = C_1 r + \frac{C_2}{r^2} \tag{3.8-24}$$

so that

$$v_\phi(r,\theta) = \left(C_1 r + \frac{C_2}{r^2}\right)\sin\theta \tag{3.8-25}$$

Application of the first two boundary conditions shows that $C_1 = 0$ and $C_2 = \Omega R^3$. Therefore, the final expression for the velocity distribution is

$$v_\phi(r,\theta) = \Omega R\left(\frac{R}{r}\right)^2 \sin\theta \tag{3.8-26}$$

Next we evaluate the torque needed to maintain the rotation of the sphere. This will be the integral, over the entire sphere, of the tangential force $(\tau_{r\phi}|_{r=R})R^2 \sin\theta d\theta d\phi$ exerted on the fluid by a solid-surface element, multiplied by the lever arm $R\sin\theta$ for that element

$$\begin{aligned} T_z &= \int_0^{2\pi}\int_0^{\pi} (\tau_{r\phi})|_{r=R}(R\sin\theta)R^2 \sin\theta d\theta d\phi \\ &= \int_0^{2\pi}\int_0^{\pi} (3\mu\Omega\sin\theta)(R\sin\theta)R^2\sin\theta d\theta d\phi \\ &= 6\pi\mu\Omega R^3 \int_0^{\pi} \sin^3\theta d\theta \\ &= 8\pi\mu\Omega R^3 \end{aligned} \tag{3.8-27}$$

In going from the first to the second line, we have used §B.1, and in going from the second to the third line, we have done the integration over ϕ. The integral in the third line gives $\frac{4}{3}$.

As the angular velocity increases, deviations from the "primary flow" of Eq. 3.8-26 occur. The centrifugal force causes fluid to be pulled in toward the poles of the sphere and pushed outward along the equator as shown in Fig. 3.8-4. To describe this "secondary flow," one has to include the $[\mathbf{v} \cdot \nabla\mathbf{v}]$ term in the equation of motion. This can be done by a "stream function" method.[1]

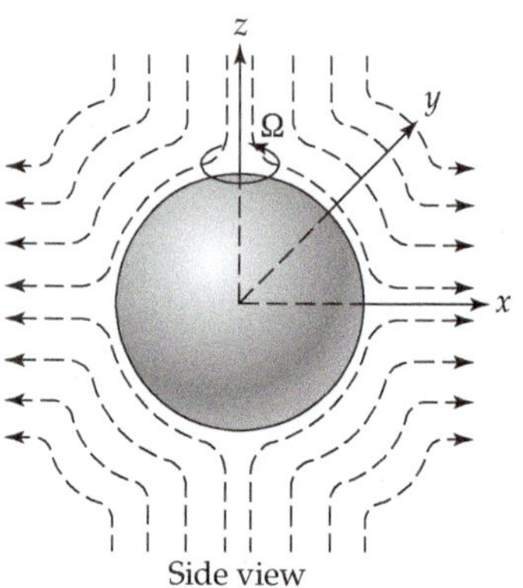

Fig. 3.8-4. Rough sketch showing the secondary flow that appears around a rotating sphere as the Reynolds number is increased.

EXAMPLE 3.8-3

Creeping Flow toward a Slot

Next we examine the flow pictured in Fig. 3.8-5. An incompressible Newtonian liquid is flowing very slowly at steady state into a thin slot of thickness $2B$ (in the y direction) and width W in the z direction ($W \gg B$). The mass rate of flow through the slot is w. The fluid outside the slot is of semi-infinite extent. We want to find the velocity and pressure distributions of the fluid as it approaches the slot ($x > 0$). We postulate that, in cylindrical coordinates, $v_\theta = 0$, $v_z = 0$, and $v_r = v_r(r,\theta)$, and further that $\mathscr{P} = \mathscr{P}(r,\theta)$. Furthermore, we assume that the flow is sufficiently slow that the acceleration terms can be neglected in the equations of motion.

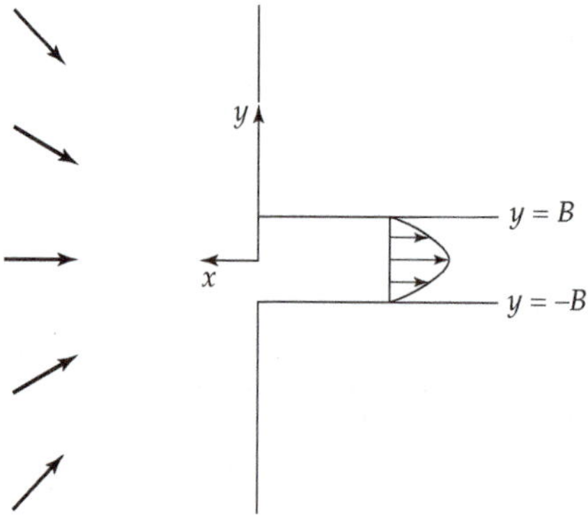

Fig. 3.8-5. Converging flow of a liquid into a slot from a semi-infinite region $x > 0$.

SOLUTION

The equation of continuity is given by Eq. B.4-2, which for the postulated velocity distribution reduces to

$$\frac{1}{r}\frac{\partial}{\partial r}(rv_r) = 0 \tag{3.8-28}$$

---

[1]See, for example, the development by O. Hassager in R. B. Bird, R. C. Armstrong, and O. Hassager, *Dynamics of Polymeric Liquids*, Vol 1, Wiley-Interscience, New York, 2nd edition (1987), pp. 31–33. See also L. Landau and E. M. Lifshitz, *Fluid Mechanics*, Pergamon, Oxford, 2nd edition (1987), p. 65; and L. G. Leal, *Laminar Flow and Convective Transport Processes*, Butterworth-Heinemann, Boston (1992), pp. 180–181.

from which it follows that

$$v_r = \frac{1}{r}f(\theta) \tag{3.8-29}$$

Since the flow is symmetric about $\theta = 0$ (i.e., the x axis), we must have $df/d\theta = 0$ at $\theta = 0$; furthermore, since the fluid velocity is zero at $\theta = \pm\frac{1}{2}\pi$ (i.e., at the solid wall), it follows that $f = 0$ at $\theta = \pm\frac{1}{2}\pi$.

The r and θ components of the equation of motion are given by Eqs. B.6-4 and B.6-5, which for the postulated pressure distribution reduce to

$$0 = -\frac{\partial \mathscr{P}}{\partial r} + \mu\frac{1}{r^3}\frac{d^2 f}{d\theta^2} \qquad 0 = -\frac{\partial \mathscr{P}}{\partial \theta} + 2\mu\frac{1}{r^2}\frac{df}{d\theta} \tag{3.8-30, 31}$$

When the r component of the equation of motion is differentiated with respect to θ, and the θ component is differentiated with respect to r, the two expressions for $\partial^2\mathscr{P}/\partial r\partial\theta$ may be equated to give

$$\frac{d^3 f}{d\theta^3} + 4\frac{df}{d\theta} = 0 \tag{3.8-32}$$

This equation can be integrated once to give

$$\frac{d^2 f}{d\theta^2} + 4f = C_1 \tag{3.8-33}$$

This equation has a particular integral $f_{\text{PI}} = \frac{1}{4}C_1$, and the complementary function is $f_{\text{CF}} = C_2 \cos 2\theta + C_3 \sin 2\theta$ (according to Eq. C.1-3). Therefore, the solution of Eq. 3.8-33 is the sum of these two functions

$$f(\theta) = \frac{1}{4}C_1 + C_2 \cos 2\theta + C_3 \sin 2\theta \tag{3.8-34}$$

The boundary conditions on $f(\theta)$ are given immediately after Eq. 3.8-29. It is found that $C_3 = 0$, from the condition on $df/d\theta$, and that $\frac{1}{4}C_1 = C_2$, from the condition on f at the solid walls. Thus, we have

$$f(\theta) = C_2(1 + \cos 2\theta) = 2C_2\cos^2\theta \tag{3.8-35}$$

The remaining constant can be determined by requiring that the total mass flow rate across any cylindrical surface must be w, that is

$$\int_0^W \int_{-\pi/2}^{+\pi/2} (-\rho v_r) r\,d\theta\,dz = \rho \int_0^W \int_{-\pi/2}^{+\pi/2} \left(-\frac{1}{r}\cdot 2C_2\cos^2\theta\right) r\,d\theta\,dz = w \tag{3.8-36}$$

When the integral is performed, we get $C_2 = -w/\pi\rho W$, so that

$$v_r(r,\theta) = -\frac{2w}{\pi\rho W r}\cos^2\theta \tag{3.8-37}$$

This is the velocity distribution in the region of positive x. Of course, it will not be valid for very small values of r, inasmuch as there will be a disturbance in the region where the fluid enters the slit.

Next we turn to the determination of the modified pressure distribution $\mathscr{P}(r,\theta)$. We substitute $f(\theta)$ from Eq. 3.8-35, with $C_2 = -w/\pi\rho W$, into Eqs. 3.7-30 and 3.7-31 to get

$$\frac{\partial \mathscr{P}}{\partial r} = \frac{\mu}{r^3}\left(\frac{4w}{\pi\rho W}\right)(\cos^2\theta - \sin^2\theta) \tag{3.8-38}$$

$$\frac{\partial \mathscr{P}}{\partial \theta} = \frac{2\mu}{r^2}\frac{d}{d\theta}\left(-\frac{2w}{\pi W\rho}\cos^2\theta\right) \tag{3.8-39}$$

Each of these equations can be integrated. From Eq. 3.8-38 we obtain

$$\mathscr{P} = -\frac{\mu}{r^2}\left(\frac{2w}{\pi\rho W}\right)(\cos^2\theta - \sin^2\theta) + F(\theta) \tag{3.8-40}$$

and from Eq. 3.8-39

$$\mathscr{P} = -\frac{2\mu}{r^2}\left(\frac{2w}{\pi\rho W}\right)\cos^2\theta + G(r) \tag{3.8-41}$$

Equation 3.8-41 can be rewritten as

$$\begin{aligned}\mathscr{P} &= -\frac{\mu}{r^2}\left(\frac{2w}{\pi\rho W}\right)(\cos^2\theta + \cos^2\theta) + G(r) \\ &= -\frac{\mu}{r^2}\left(\frac{2w}{\pi\rho W}\right)(\cos^2\theta + (1 - \sin^2\theta)) + G(r) \\ &= -\frac{\mu}{r^2}\left(\frac{2w}{\pi\rho W}\right)(\cos^2\theta - \sin^2\theta) + H(r)\end{aligned} \tag{3.8-42}$$

In comparing Eqs. 3.8-40 and 3.8-42, we see that the two expressions for $\mathscr{P}$ are the same except for the functions F and H. Since the first is a function of θ alone, and the second a function of r alone, they must both be equal to a constant, which we call $\mathscr{P}_\infty$. This is the modified pressure in the limit as $r \to \infty$.

Next we get the total normal stress at the wall at $\theta = \pm\frac{1}{2}\pi$. If we use the result of Example 3.1-1 (i.e., $\tau_{\theta\theta}|_{\theta=\pm\frac{1}{2}\pi} = 0$), we find

$$\begin{aligned}\pi_{\theta\theta}|_{\theta=\pm\frac{1}{2}\pi} &= (p + \tau_{\theta\theta})|_{\theta=\pm\frac{1}{2}\pi} = (\mathscr{P} - \rho g h)|_{\theta=\pm\frac{1}{2}\pi} \\ &= \mathscr{P}_\infty + \frac{2w\mu}{\pi\rho W r^2} - \rho g h = p_\infty + \frac{2w\mu}{\pi\rho W r^2}\end{aligned} \tag{3.8-43}$$

And finally we get the shear stress acting on the walls. From Eq. B.1-11

$$\tau_{\theta r}|_{\theta=\pm\frac{\pi}{2}} = -\mu\frac{1}{r}\frac{\partial v_r}{\partial\theta}\bigg|_{\theta=\pm\frac{\pi}{2}} = -\mu\left(-\frac{2w}{\pi W\rho r^2}\right)(-2\cos\theta\sin\theta)\bigg|_{\theta=\pm\frac{\pi}{2}} = 0 \tag{3.8-44}$$

Although the fluid is flowing along the wall, there is no tangential force on the wall, because the velocity gradient is zero! This is an atypical result, and should not be automatically applied to other problems involving flow along a solid wall.

§3.9 CONCLUDING COMMENTS

In Chapter 2 it was shown how to set up momentum conservation statements by using shell-balance arguments. The method was, however, restricted to incompressible, rectilinear flows.

In Chapter 3, we have seen how to set up quite general mass and momentum conservation equations—the equations of continuity and motion—that enable us to formulate problems in much more general flows, using several different coordinate systems. Then we illustrated problem solving by working through the solutions to seven laminar-flow problems of varying degrees of complexity. Of course, there are many more fluid dynamics problems for which the solutions can be found in advanced fluid mechanics texts and handbooks. One's ability to solve these more complex problems depends on one's facility with mathematics. These problems, as well as more complicated ones, can also be solved numerically by using a variety of commercial software packages. The problems that have been solved analytically in this chapter can be used to verify the user's ability to employ those software packages.

In addition, you are now prepared to approach problem solving in turbulent flow (Chapter 4) and to examine problems by dimensional analysis (Chapter 5). Also, in Part II and Part III of this book, where we look at nonisothermal problems and flow of binary mixtures, you will see the virtue of problem solving by using equations of change for flows with temperature and concentration gradients.

QUESTIONS FOR DISCUSSION

1. What is the physical meaning of the term $\Delta x \Delta y(\rho v_z)|_z$ in Eq. 3.1-2? What is the physical meaning of $(\nabla \cdot \mathbf{v})$? of $(\nabla \cdot \rho\mathbf{v})$?
2. By making a mass balance over a volume element $(\Delta r)(r\Delta\theta)(\Delta z)$, derive the equation of continuity in cylindrical coordinates. Compare your result with the appropriate entry in Appendix B.
3. What is the physical meaning of the term $\Delta x \Delta y(\phi_{zx})|_z$ in Eq. 3.2-2? What is the physical meaning of $[\nabla \cdot \boldsymbol{\phi}]$?
4. What happens when f is set equal to unity in Eq. 3.5-6?
5. Equation (B) in Table 3.5-1 is *not* restricted to fluids with constant density, even though ρ is to the left of the substantial derivative. Explain.
6. In the tangential annular flow problem in Example 3.7-3, would you expect velocity profiles relative to the inner cylinder and the torques on the inner cylinder to be the same in the following two situations: (i) the inner cylinder is fixed and the outer cylinder rotates with an angular velocity Ω; (ii) the outer cylinder is fixed and the inner cylinder rotates with an angular velocity $-\Omega$? Both flows are presumed to be laminar and stable. How are the torques exerted by the fluid on the outer cylinder related to the torques on the inner cylinder in these two cases?
7. Suppose that, in Example 3.7-4, there were two immiscible liquids in the rotating beaker. What would be the shape of the interface between the two liquid regions?
8. Would the system discussed in Example 3.8-2 be useful as a viscometer?
9. In Eq. 3.8-27, explain by means of a carefully drawn sketch the choice of limits in the integration and the meaning of each factor in the first integrand.

PROBLEMS

3A.1 Torque required to turn a friction bearing. Calculate the required torque in $\text{lb}_f \cdot \text{ft}$ and power consumption in horsepower to turn the shaft in the friction bearing shown in Fig. 3A.1. The length of the bearing surface on the shaft is 2 in., and the shaft is rotating at 200 rpm. The viscosity of the lubricant is 200 cp, and its density is 50 lb_m/ft^3. Neglect the effect of eccentricity.

Answers: 0.32 $\text{lb}_f \cdot \text{ft}$; 0.012 hp = 0.009 kW

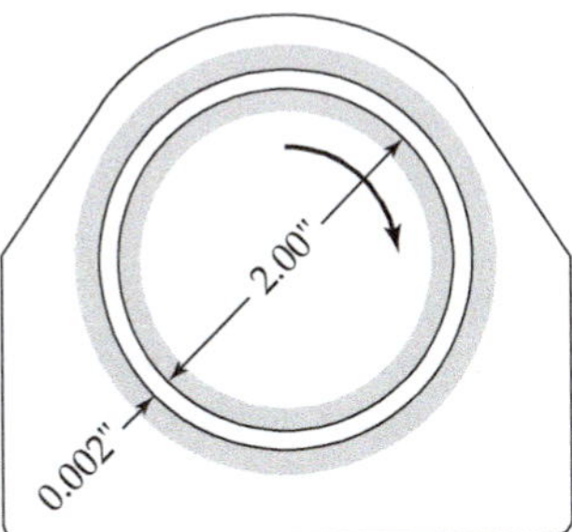

Fig. 3A.1 Friction bearing.

3A.2 Friction loss in bearings.[1] Each of two screws on a large motor-ship is driven by a 4000-hp engine. The shaft that connects the motor and the screw is 16 in. in diameter and rests in a series of sleeve bearings that give a 0.005 in. clearance. The shaft rotates at 50 rpm, the lubricant has a viscosity of 5000 cp, and there are 20 bearings, each 1 ft in length. Estimate the fraction of engine power expended in rotating the shafts in their bearings. Neglect the effect of the eccentricity.

Answer: 0.116

[1]This problem was contributed by Prof. E. J. Crosby, University of Wisconsin, author of *Experiments in Transport Phenomena*, Wiley, New York (1961).

3A.3 Effect of altitude on air pressure. When standing at the mouth of the Ontonagon River on the south shore of Lake Superior (602 ft above mean sea level), your portable barometer indicates a pressure of 750 mm Hg. Use the equation of motion to estimate the barometric pressure at the top of Government Peak (2023 ft above mean sea level) in the nearby Porcupine Mountains. Assume the temperature at lake level is 70°F and that the temperature decreases with increasing altitude at a steady rate of 3°F per 1000 feet. The gravitational acceleration at the south shore of Lake Superior is about 32.19 ft/s^2, and its variation with altitude may be neglected in this problem.

Answer: 713 mm Hg $= 9.49 \times 10^4$ N/m^2(if $\rho = \rho(p)$)

3A.4 Viscosity determination with a rotating-cylinder viscometer. It is desired to measure the viscosities of sucrose solutions of about 60% concentration by weight at about 20°C with a rotating-cylinder viscometer such as that shown in Fig. 3.7-2. This instrument has an inner cylinder 4.000 cm in diameter surrounded by a rotating concentric cylinder 4.500 cm in diameter. The length L is 4.00 cm. The viscosity of a 60% sucrose solution at 20°C is about 57 cp, and its density is about 1.29 g/cm^3.

On the basis of past experience, it seems possible that end effects will be important, and it is therefore decided to calibrate the viscometer by measurements on some known solutions of approximately the same viscosity as those of the unknown sucrose solutions.

Determine a reasonable value for the applied torque to be used in calibration if the torque measurements are reliable within 100 dyne/cm and the angular velocity can be measured within 0.5%. What will be the resultant angular velocity?

3A.5 Fabrication of a parabolic mirror. It is proposed to make a backing for a parabolic mirror, by rotating a pan of slow-hardening plastic resin at constant speed until it hardens (see Fig. 3.7-7). Calculate the rotational speed required to produce a mirror of focal length $f = 100$ cm. The focal length is one-half the radius of curvature at the axis, which in turn is given by

$$r_c = \left[1 + \left(\frac{dz}{dr}\right)^2\right]^{3/2} \left(\frac{d^2z}{dr^2}\right)^{-1} \tag{3A.5-1}$$

Answer: 21.1 rpm

3B.1 Flow between coaxial cylinders and concentric spheres.

(a) The space between two coaxial cylinders is filled with an incompressible fluid at constant temperature. The radii of the inner and outer wetted surfaces are κR and R, respectively. The angular velocities of rotation of the inner and outer cylinders are Ω_i and Ω_o. Determine the velocity distribution in the fluid, and the torques exerted by the fluid on the two cylinders needed to maintain the motion.

(b) Repeat part (a) for two concentric spheres.

Answers: **(a)** $v_\theta(r) = \dfrac{\kappa R}{1-\kappa^2}\left[(\Omega_o - \Omega_i\kappa^2)\left(\dfrac{r}{\kappa R}\right) + (\Omega_i - \Omega_o)\left(\dfrac{\kappa R}{r}\right)\right]$

(b) $v_\phi(r,\theta) = \dfrac{\kappa R}{1-\kappa^3}\left[(\Omega_o - \Omega_i\kappa^3)\left(\dfrac{r}{\kappa R}\right) + (\Omega_i - \Omega_o)\left(\dfrac{\kappa R}{r}\right)^2\right]\sin\theta$

3B.2 Laminar flow in a triangular duct.[2] One type of compact heat exchanger is shown in Fig. 3B.2(*a*). To analyze the performance of such an apparatus, it is necessary to understand the flow in a duct whose cross section is an equilateral triangle. This is done most easily by installing a coordinate system as shown in Fig. 3B.2(*b*).

[2] An alternative formulation of the velocity profile is given by L. D. Landau and E. M. Lifshitz, *Fluid Mechanics*, Pergamon, Oxford, 2nd edition (1987), p. 54.

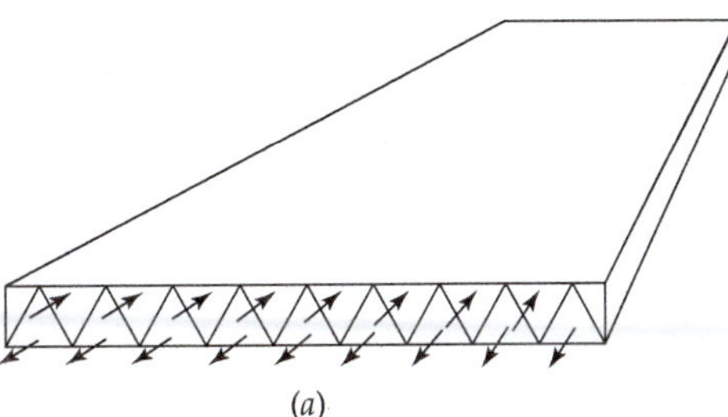

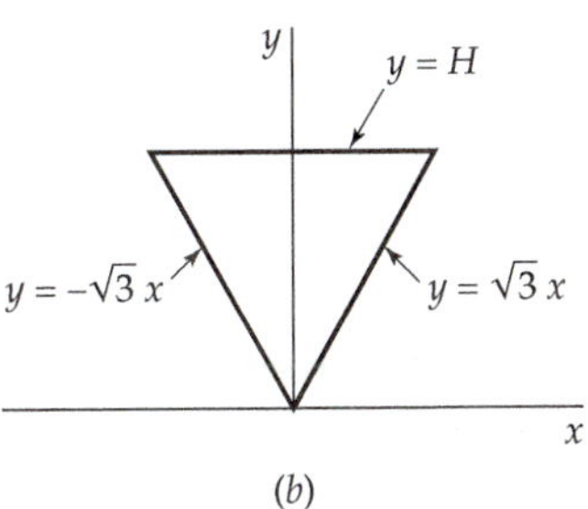

Fig. 3B.2 (*a*) Compact heat-exchanger element, showing channels of a triangular cross section; (*b*) coordinate system for an equilateral-triangular duct.

(a) Verify that the velocity distribution for the laminar flow of a Newtonian fluid in a duct of this type is given by

$$v_z(x,y) = \frac{(\mathscr{P}_0 - \mathscr{P}_L)}{4\mu LH}(y - H)(3x^2 - y^2) \tag{3B.2-1}$$

(b) From Eq. 3B.2-1, find the average velocity, maximum velocity, and mass flow rate.

Answers: **(b)** $\langle v_z \rangle = \dfrac{(\mathscr{P}_0 - \mathscr{P}_L)H^2}{60\mu L} = \dfrac{9}{20}v_{z,\max}$;

$$w = \frac{\sqrt{3}(\mathscr{P}_0 - \mathscr{P}_L)H^4\rho}{180\mu L}$$

3B.3 Laminar flow in a square duct.

(a) A straight duct extends in the z direction for a length L and has a square cross section, bordered by the lines $x = \pm B$ and $y = \pm B$. A colleague has told you that the velocity distribution is given by

$$v_z(x,y) = \frac{(\mathscr{P}_0 - \mathscr{P}_L)B^2}{4\mu L}\left[1 - \left(\frac{x}{B}\right)^2\right]\left[1 - \left(\frac{y}{B}\right)^2\right] \tag{3B.3-1}$$

To be helpful, you feel obliged to check the result. Does it satisfy the relevant boundary conditions and the relevant differential equation?

(b) According to the review article by Berker,[3] the mass rate of flow in a square duct is given by:

$$w = \frac{0.563(\mathscr{P}_0 - \mathscr{P}_L)B^4\rho}{\mu L} \tag{3B.3-2}$$

Compare the numerical coefficient in this expression with the coefficient that one obtains from Eq. 3B.3-1.

3B.4 Creeping flow between two concentric spheres. A very viscous Newtonian fluid flows in the space between two concentric spheres, as shown in Fig. 3B.4. It is desired to find the rate of

[3] R. Berker, *Handbuch der Physik*, Vol. **VIII/2**, Springer, Berlin (1963); see pp. 67–77 for laminar flow in conduits of noncircular cross sections. See also W. E. Stewart, *AIChE Journal*, **8**, 425–428 (1962).

flow in the system as a function of the imposed pressure difference. Neglect end effects and postulate that v_θ depends only on r and θ with the other velocity components zero.

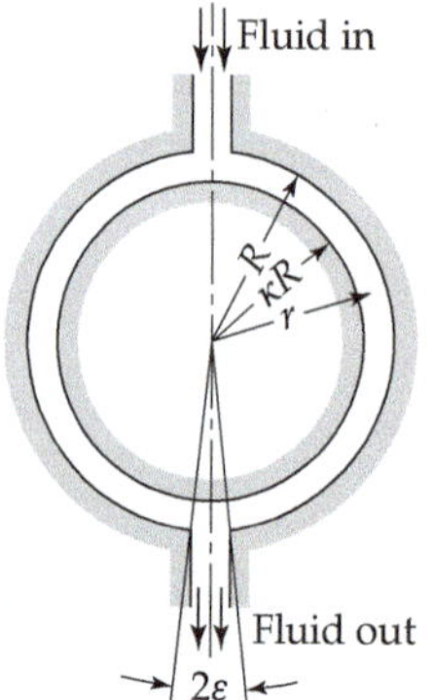

Fig. 3B.4 Creeping flow in the region between two stationary concentric spheres.

(a) Using the equation of continuity, show that $v_\theta(r,\theta)\sin\theta = u(r)$, where $u(r)$ is a function of r to be determined.

(b) Write the θ component of the equation of motion for this system, assuming the flow to be slow enough that the $[\mathbf{v}\cdot\nabla\mathbf{v}]$ term is negligible. Show that this gives

$$0 = -\frac{1}{r}\frac{\partial\mathscr{P}}{\partial\theta} + \mu\left[\frac{1}{\sin\theta}\frac{1}{r^2}\frac{d}{dr}\left(r^2\frac{du}{dr}\right)\right] \tag{3B.4-1}$$

(c) Separate this into two equations

$$\sin\theta\frac{d\mathscr{P}}{d\theta} = B; \quad \frac{\mu}{r}\frac{d}{dr}\left(r^2\frac{du}{dr}\right) = B \tag{3B.4-2,3}$$

where B is the separation constant, and solve the two equations to get

$$B = \frac{\mathscr{P}_2 - \mathscr{P}_1}{2\ln\cot(\varepsilon/2)} \tag{3B.4-4}$$

$$u(r) = \frac{(\mathscr{P}_1 - \mathscr{P}_2)R}{4\mu\ln\cot(\varepsilon/2)}\left[\left(1-\frac{r}{R}\right) + \kappa\left(1-\frac{R}{r}\right)\right] \tag{3B.4-5}$$

where $\mathscr{P}_1$ and $\mathscr{P}_2$ are the values of the modified pressure at $\theta = \varepsilon$ and $\theta = \pi - \varepsilon$, respectively.

(d) Use the results above to get the mass rate of flow

$$w = \frac{\pi(\mathscr{P}_1 - \mathscr{P}_2)R^3(1-\kappa)^3\rho}{12\mu\ln\cot(\varepsilon/2)} \tag{3B.4-6}$$

3B.5 Parallel-disk viscometer. A fluid, whose viscosity is to be measured, is placed in the gap of thickness B between the two disks of radius R (Fig. 3B.5). One measures the torque T_z required to turn the upper disk at an angular velocity Ω. Develop the formula for deducing the viscosity from these measurements. Assume creeping flow.

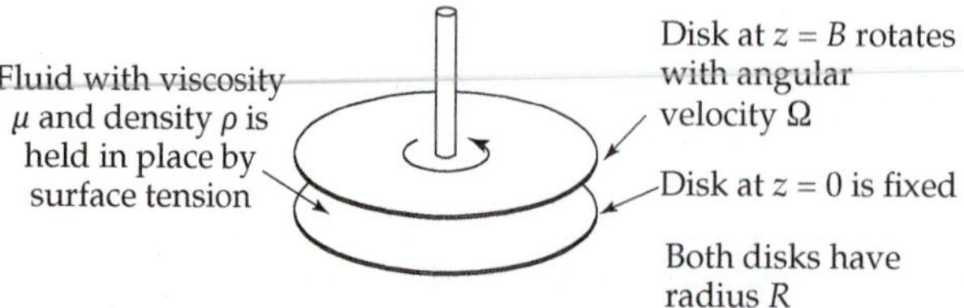

Fig. 3B.5 Parallel-disk viscometer.

(a) Postulate that for small values of Ω the velocity profiles have the form $v_r = 0$, $v_z = 0$, and $v_\theta = rf(z)$; why does this form for the tangential velocity seem reasonable? Postulate further that $\mathscr{P} = \mathscr{P}(r,z)$. Write down the resulting simplified equations of continuity and motion.

(b) From the θ component of the equation of motion, obtain a differential equation for $f(z)$. Solve the equation for $f(z)$ and evaluate the constants of integration. This leads ultimately to the result $v_\theta(r,z) = \Omega r(z/B)$. Could you have guessed this result?

(c) Show that the desired working equation for deducing the viscosity is $\mu = 2BT_z/\pi\Omega R^4$.

(d) Discuss the advantages and disadvantages of this instrument.

3B.6 Circulating axial flow in an annulus. A rod of radius κR moves upward with a constant speed v_0 through a cylindrical container of inner radius R containing a Newtonian liquid (Fig. 3B.6). The liquid circulates in the cylinder, moving upward along the moving central rod and moving downward near the fixed container wall. Find the velocity distribution in the annular region, far from the end disturbances. Flows similar to this occur in the seals of some reciprocating machinery, for example, in the annular space between piston rings.

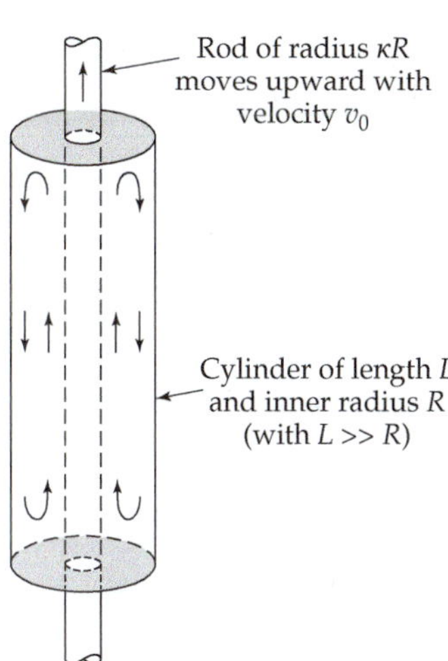

Fig. 3B.6 Circulating flow produced by an axially moving rod in a closed annular region.

(a) First, consider the problem where the annular region is quite narrow—that is, where κ is just slightly less than unity. In that case, the annulus may be approximated by a thin plane slit and the curvature can be neglected. Show that in this limit, the velocity distribution is given by

$$\frac{v_z(\xi)}{v_0} = 3\left(\frac{\xi-\kappa}{1-\kappa}\right)^2 - 4\left(\frac{\xi-\kappa}{1-\kappa}\right) + 1 \tag{3B.6-1}$$

where $\xi = r/R$.

(b) Next, work the problem without the thin-slit assumption. Show that the velocity distribution is given by

$$\frac{v_z(\xi)}{v_0} = \frac{(1-\xi^2)\left(1 - \dfrac{2\kappa^2}{1-\kappa^2}\ln\dfrac{1}{\kappa}\right) - (1-\kappa^2)\ln\dfrac{1}{\xi}}{(1-\kappa^2) - (1+\kappa^2)\ln\dfrac{1}{\kappa}} \tag{3B.6-2}$$

3B.7 Comparison of tube flows for several cross sections. The laminar flow in a circular tube with radius R is discussed in §2.3, and the laminar flow in tubes with equilateral triangular cross section of height H is described in Problem 3B.2. Both tubes have the same length, L. We want to compare these two flow problems.

(a) Compare the mass rates of flow for the two tubes when their cross-sectional areas are the same.

(b) Compare the mass rates of flow for the two tubes when the perimeters of their cross sections are the same.

Answers: **(a)** $\dfrac{w_\Delta}{w_\bigcirc} = \dfrac{3\sqrt{3}}{180} \cdot 8\pi = 0.726$

(b) $\dfrac{w_\Delta}{w_\bigcirc} = \dfrac{\sqrt{3}}{(2\sqrt{3})^4(180)} \cdot \dfrac{(8)(2\pi)^4}{\pi} = 0.265$

For the square cross section (see Problem 3B.3), the ratios corresponding to (a) and (b) may be found to be

$$\frac{w_\square}{w_\bigcirc} = 0.884 \text{ (same cross-sectional areas)}$$

$$\frac{w_\square}{w_\bigcirc} = 0.545 \text{ (same perimeters)}$$

What, if any, conclusions can you draw from this problem?

3B.8 Flow near a wall suddenly set in motion. When you look at Fig. 3.8-2(*b*), you might wonder whether the slope at $y = 0$ is -1. You can answer that question by differentiating Eq. 3.8-15 with respect to η. Show that

$$\left.\frac{d}{d\eta} \operatorname{erfc} \eta\right|_{\eta=0} = -1.1287 \tag{3B.8-1}$$

To do this, use the Leibniz formula for differentiating integrals in §C.3 and the definition of the complementary error function in §C.6. As may be seen from Fig. 3.8-2(*b*), the slope is somewhat steeper than minus 1.

3B.9 Slow transverse flow around a cylinder. An incompressible Newtonian fluid approaches a stationary cylinder with a uniform, steady velocity v_∞ in the positive x direction (Fig. 3B.9). When the equations of change are solved for creeping flow, the following expressions[4] are found for the pressure and velocity in the immediate vicinity of the cylinder (they are *not* valid at large distances):

$$p(r,\theta) = p_\infty - C\mu \frac{v_\infty \cos\theta}{r} - \rho g r \sin\theta \tag{3B.9-1}$$

$$v_r(r,\theta) = Cv_\infty \left[\frac{1}{2}\ln\left(\frac{r}{R}\right) - \frac{1}{4} + \frac{1}{4}\left(\frac{R}{r}\right)^2\right]\cos\theta \tag{3B.9-2}$$

$$v_\theta(r,\theta) = -Cv_\infty \left[\frac{1}{2}\ln\left(\frac{r}{R}\right) + \frac{1}{4} - \frac{1}{4}\left(\frac{R}{r}\right)^2\right]\sin\theta \tag{3B.9-3}$$

Here p_∞ is the pressure far from the cylinder at $y = 0$ and

$$C = \frac{2}{\ln(7.4/\text{Re})} \tag{3B.9-4}$$

with the Reynolds number defined as $\text{Re} = 2Rv_\infty \rho/\mu$.

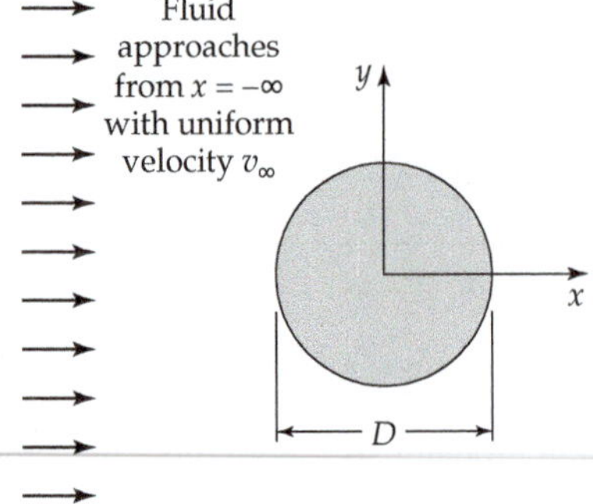

Fig. 3B.9 Transverse flow around a cylinder.

[4]See G. K. Batchelor, *An Introduction to Fluid Dynamics*, Cambridge University Press (1967), pp. 244–246, 261.

(a) Use these results to get the pressure p, the shear stress $\tau_{r\theta}$, and the normal stress τ_{rr} at the surface of the cylinder.

(b) Show that the x component of the force per unit area exerted by the liquid on the cylinder is

$$-p|_{r=R}\cos\theta + \tau_{r\theta}|_{r=R}\sin\theta \tag{3B.9-5}$$

(c) Obtain the force $F_x = 2C\pi L\mu v_\infty$ exerted by the fluid in the x direction on a length L of the cylinder.

3B.10 Radial flow between parallel disks. A part of a lubrication system consists of two circular disks between which a lubricant flows radially, as illustrated in Fig. 3B.10. The flow takes place because of a modified pressure difference $\mathscr{P}_1 - \mathscr{P}_2$ between the inner and outer radii r_1 and r_2, respectively.

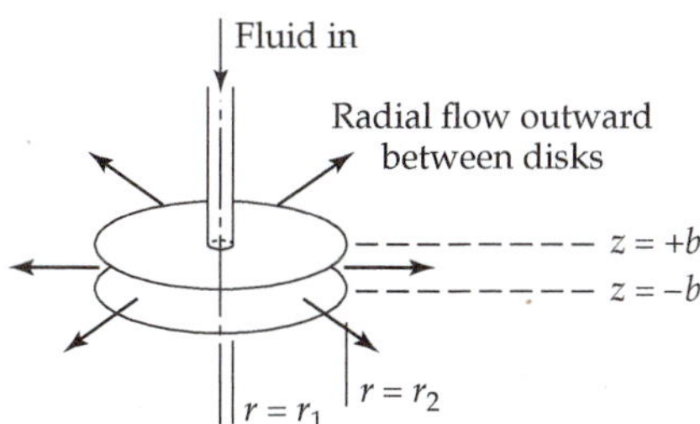

Fig. 3B.10 Outward radial flow in the space between two parallel, circular disks.

(a) Write the equations of continuity and motion for this flow system, assuming steady-state, laminar, incompressible Newtonian flow. Consider only the region $r_1 \le r \le r_2$ and a flow that is radially directed.

(b) Show how the equation of continuity enables one to simplify the equation of motion to give

$$-\rho\frac{\phi^2}{r^3} = -\frac{d\mathscr{P}}{dr} + \mu\frac{1}{r}\frac{d^2\phi}{dz^2} \tag{3B.10-1}$$

in which $\phi = rv_r$ is a function of z only. Why is ϕ independent of r?

(c) It can be shown that no solution exists for Eq. 3B.10-1 unless the nonlinear term containing ϕ is omitted. Omission of this term corresponds to the "creeping flow assumption." Show that for creeping flow, Eq. 3B.10-1 can be integrated with respect to r to give

$$0 = (\mathscr{P}_1 - \mathscr{P}_2) + \left(\mu\ln\frac{r_2}{r_1}\right)\frac{d^2\phi}{dz^2} \tag{3B.10-2}$$

(d) Show that further integration with respect to z gives

$$v_r(r,z) = \frac{(\mathscr{P}_1 - \mathscr{P}_2)b^2}{2\mu r\ln(r_2/r_1)}\left[1 - \left(\frac{z}{b}\right)^2\right] \tag{3B.10-3}$$

(e) Show that the mass flow rate is

$$w = \frac{4\pi(\mathscr{P}_1 - \mathscr{P}_2)b^3\rho}{3\mu\ln(r_2/r_1)} \tag{3B.10-4}$$

(f) Sketch the curves $\mathscr{P}(r)$ and $v_r(r,z)$.

3B.11 Radial flow between two coaxial cylinders. Consider an incompressible fluid, at constant temperature, flowing radially between two porous cylindrical shells with inner and outer radii κR and R.

(a) Show that the equation of continuity leads to $v_r = C/r$ where C is a constant.

(b) Simplify the components of the equation of motion to obtain the following expressions for the modified pressure distribution:

$$\frac{d\mathscr{P}}{dr} = -\rho v_r\frac{dv_r}{dr}, \quad \frac{d\mathscr{P}}{d\theta} = 0, \quad \frac{d\mathscr{P}}{dz} = 0 \tag{3B.11-1}$$

(c) Integrate the expression for $d\mathscr{P}/dr$ above to get

$$\mathscr{P}(r) - \mathscr{P}(R) = \frac{1}{2}\rho[v_r(R)]^2\left[1 - \left(\frac{R}{r}\right)^2\right] \tag{3B.11-2}$$

(d) Write out all the nonzero components of $\boldsymbol{\tau}$ for this flow.

(e) Repeat the problem for concentric spheres.

3B.12 Pressure distribution in incompressible fluids. Penelope is staring at a beaker filled with a liquid, which for all practical purposes can be considered as incompressible; let its density be ρ_0. She tells you she is trying to understand how the pressure in the liquid varies with depth. She has taken the origin of coordinates at the liquid-air interface, with the positive z axis pointing away from the liquid. She says to you:

"If I simplify the equation of motion for an incompressible liquid at rest, I get $0 = -dp/dz - \rho_0 g$. I can solve this and get $p = p_{atm} - \rho_0 g z$. That seems reasonable—the pressure increases with increasing depth.

"But, on the other hand, the equation of state for any fluid is $p = p(\rho, T)$, and if the system is at constant temperature, this just simplifies to $p = p(\rho)$. And, since the fluid is incompressible, $p = p(\rho_0)$, and p must be a constant throughout the fluid! How can that be?"

Clearly Penelope needs help. Provide a useful explanation.

3B.13 Flow of a fluid through a sudden contraction.

(a) An incompressible liquid flows through a sudden contraction from a pipe of diameter D_1 into a pipe of smaller diameter D_2. What does the Bernoulli equation predict for $\mathscr{P}_1 - \mathscr{P}_2$, the difference between the modified pressures upstream and downstream of the contraction? Does this result agree with experimental observations?

(b) Repeat the derivation for the isothermal horizontal flow of an ideal gas through a sudden contraction.

3B.14 Normal stresses at solid surfaces for compressible fluids. Extend Example 3.1-1 to compressible fluids. Show that

$$\tau_{zz}|_{z=0} = \left(\frac{4}{3}\mu + \kappa\right)(\partial \ln \rho/\partial t)|_{z=0} \tag{3B.14-1}$$

Discuss the physical significance of this result.

3B.15 Shape of free surface in tangential annular flow.

(a) A liquid is in the annular space between two vertical cylinders of radii κR and R, and the liquid is open to the atmosphere at the top.

Show that when the inner cylinder rotates with an angular velocity Ω_i, and the outer cylinder is fixed, the free liquid surface has the shape $z(\xi)$ given by

$$z_R - z(\xi) = \frac{1}{2g}\left(\frac{\kappa^2 R\Omega_i}{1-\kappa^2}\right)^2\left(\frac{1}{\xi^2} + 4\ln\xi - \xi^2\right) \tag{3B.15-1}$$

in which z_R is the height of the liquid at the outer-cylinder wall, and $\xi = r/R$.

(b) Repeat (a) but with the inner cylinder fixed and the outer cylinder rotating with an angular velocity Ω_o. Show that the shape of the liquid surface is

$$z_R - z(\xi) = \frac{1}{2g}\left(\frac{\kappa^2 R\Omega_o}{1-\kappa^2}\right)^2\left[\left(\frac{1}{\xi^2} - 1\right) + \frac{4}{\kappa^2}\ln\xi - \frac{1}{\kappa^4}(\xi^2 - 1)\right] \tag{3B.15-2}$$

(c) Draw a sketch comparing these two liquid-surface shapes.

3B.16 Flow in a slit with uniform cross flow. A fluid flows in the positive x direction through a long flat duct of length L, width W, and thickness B, where $L \gg W \gg B$ (see Fig. 3B.16). The duct has porous walls at $y = 0$ and $y = B$, so that a constant cross flow can be maintained, with $v_y = v_0$, a constant, everywhere. Flows of this type are important in connection with separations processes using the sweep-diffusion effect. By carefully controlling the cross flow, one can concentrate the larger constituents (molecules, dust particles, etc.) near the upper wall.

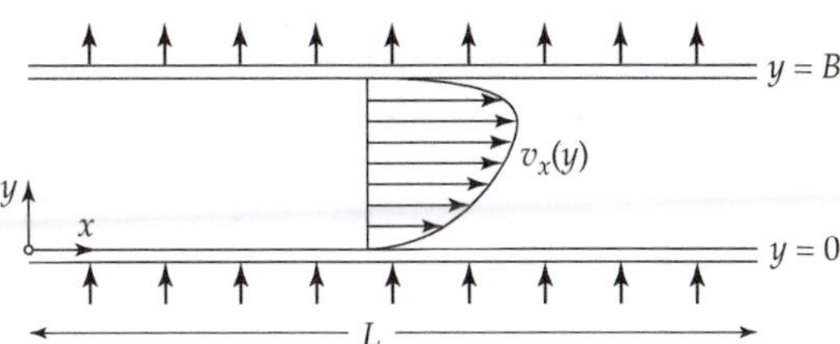

Fig. 3B.16 Flow in a slit of length L, width W, and thickness B. The walls at $y = 0$ and $y = B$ are porous, and there is a flow of the fluid in the y direction, with a uniform velocity $v_y = v_0$.

(a) Show that the velocity profile for the system is given by

$$v_x(y) = \frac{(\mathscr{P}_0 - \mathscr{P}_L)B^2}{\mu L}\frac{1}{A}\left(\frac{y}{B} - \frac{e^{Ay/B} - 1}{e^A - 1}\right) \tag{3B.16-1}$$

in which $A = Bv_0\rho/\mu$.

(b) Show that the mass flow rate in the x direction is

$$w = \frac{(\mathscr{P}_0 - \mathscr{P}_L)B^3 W\rho}{\mu L}\frac{1}{A}\left(\frac{1}{2} - \frac{1}{A} + \frac{1}{e^A - 1}\right) \tag{3B.16-2}$$

(c) Verify that the above results simplify to those of Problem 2B.4 in the limit that there is no cross flow at all (that is, as $A \to 0$).

(d) A colleague has also solved this problem, but taking a coordinate system with $y = 0$ at the midplane of the slit, with the porous walls located at $y = \pm b$. His answer to part (a) above is

$$\frac{v_x(\eta)}{\langle v_x \rangle} = \frac{e^{a\eta} - \eta \sinh a - \cosh a}{(1/a)\sinh a - \cosh a} \tag{3B.16-3}$$

in which $a = bv_0\rho/\mu$ and $\eta = y/b$. Is this result equivalent to Eq. 3B.16-1?

3C.1 Flow near a wall suddenly set in motion (approximate solution). This problem was solved in Example 3.8-1 by the method of combination of independent variables. We now want to show how the same problem can be solved by an approximate method.

First, we integrate Eq. 3.8-1 over all y from $y = 0$ to $y = \infty$ to get

$$\int_0^\infty \frac{\partial v_x}{\partial t} dy = \nu \left.\frac{\partial v_x}{\partial y}\right|_0^\infty \quad \text{or} \quad \frac{d}{dt}\int_0^\infty v_x dy = \nu \left.\frac{\partial v_x}{\partial y}\right|_0^\infty \tag{3C.1-1}$$

Now, instead of the true velocity profiles in Fig. 3C.1(*a*), let us approximate the profiles as in Fig 3C.1(*b*). That is, we say that at each value of the time t, the fluid will be in motion out to a distance $\delta(t)$, beyond which the velocity will be zero. The distance $\delta(t)$ will indicate the "boundary layer" within which the fluid is in motion.

We can make a reasonably good guess as to the shape of the velocity profile within the boundary layer. For example, the following function seems reasonable:

$$\frac{v_x(y,t)}{v_0} = 1 - \frac{3}{2}\frac{y}{\delta(t)} + \frac{1}{2}\left(\frac{y}{\delta(t)}\right)^3 \quad \text{for} \quad 0 \le y \le \delta(t) \tag{3C.1-2}$$

and $v_x(y,t)/v_0 = 0$ for $y \ge \delta(t)$. Equation 3C.1-2 satisfies the imposed requirements of $v_x = v_0$ at $y = 0$ and $v_x = 0$ at $y = \delta(t)$; this equation also gives $\partial v_x/\partial y = 0$ at $y = \delta(t)$, which ensures that the shear stress τ_{yx} is continuous.

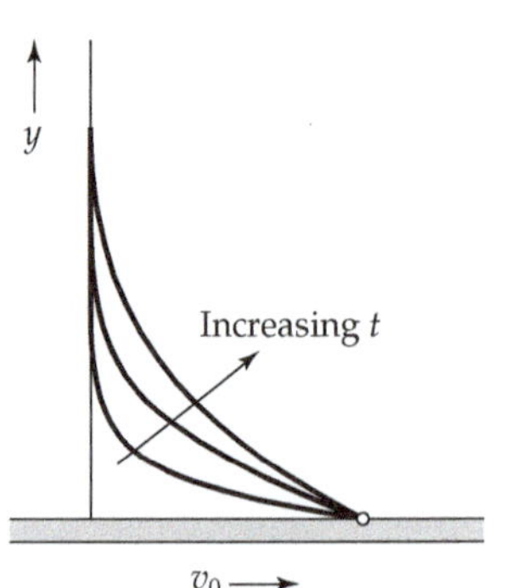

(a) True solution

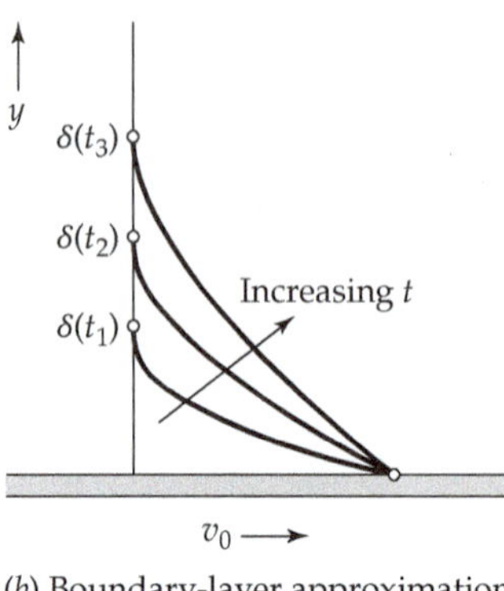

(b) Boundary-layer approximation

Fig. 3C.1 Comparison of true and approximate velocity profiles near a wall suddenly set in motion with velocity v_0.

(a) Substitute this guessed velocity profile into Eq. 3C.1-1 and get, with $\eta = y/\delta(t)$,

$$\frac{d}{dt}\delta(t)\int_0^1\left(1-\frac{3}{2}\eta+\frac{1}{2}\eta^3\right)d\eta = -\nu\left(-\frac{3}{2}+\frac{3}{2}\eta^2\right)\Bigg|_{\eta=0}\frac{1}{\delta(t)} \tag{3C.1-3}$$

(b) Evaluate the integral and obtain the following differential equation for the boundary-layer thickness:

$$\delta\frac{d\delta}{dt} = 4\nu \tag{3C.1-4}$$

(c) Solve this and get $\delta(t) = \sqrt{8\nu t}$. Substitution of this into Eq. 3C.1-2 then gives an approximation to the velocity profile.

(d) Compare the approximate velocity profiles with the exact results of Example 3.8-1.

3C.2 Boundary-layer development for flow past a flat plate. Having seen the success of the approximate time-dependent boundary-layer development in Problem 3C.1, we now apply the same ideas to the space-dependent boundary-layer for flow past a flat plate of negligible thickness, as shown in Fig. 3C.2.

The differential equations describing the flow field will be the steady-state, two-dimensional equation of continuity and the x component of the equation of motion (intuition tells us that we do not need the y component because the motion in the y direction is expected to be very small):

$$\frac{\partial v_x}{\partial x}+\frac{\partial v_y}{\partial y} = 0 \tag{3C.2-1}$$

$$v_x\frac{\partial v_x}{\partial x}+v_y\frac{\partial v_x}{\partial y} = \nu\left(\frac{\partial^2 v_x}{\partial x^2}+\frac{\partial^2 v_x}{\partial y^2}\right) \tag{3C.2-2}$$

Order of magnitude arguments can be given that suggest that the term $\nu\partial^2 v_x/\partial x^2$ is small compared to $v_x\partial v_x/\partial x$.

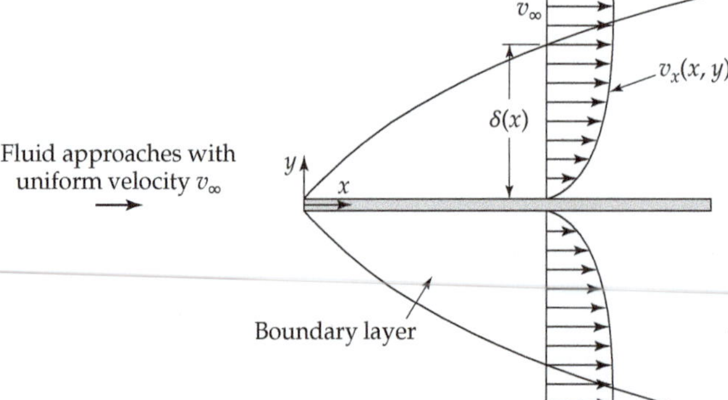

Fig. 3C.2 Boundary-layer development near a flat plate of negligible thickness.

(a) Solve Eq. 3C.2-1 for v_y using the boundary condition that $v_y = 0$ at $y = 0$, and substitute the result to Eq. 3C.2-2 to get

$$v_x \frac{\partial v_x}{\partial x} - \left(\int_0^y \frac{\partial v_x}{\partial x} dy \right) \frac{\partial v_x}{\partial y} = \nu \frac{\partial^2 v_x}{\partial y^2} \tag{3C.2-3}$$

(b) Eq. 3C. 2–3 can be solved approximately by the same method used in Problem 3C.1. We let the velocity profile be of the form

$$\frac{v_x(x,y)}{v_\infty} = \phi(\eta) \quad \text{where} \quad \eta = \frac{y}{\delta(x)} \tag{3C.2-4}$$

in which $\delta(x)$ is the boundary-layer thickness at a distance x from the leading edge of the plate. Also, for $y > \delta(x)$, $v_x = v_\infty$. Then show that the derivatives in Eq. 3C.2-3 are

$$\frac{\partial v_x}{\partial x} = v_\infty \phi' \left(-\frac{\eta}{\delta} \right) \frac{d\delta}{dx}; \quad \frac{\partial v_x}{\partial y} = v_\infty \phi' \frac{1}{\delta}; \quad \frac{\partial^2 v_x}{\partial y^2} = v_\infty \phi'' \frac{1}{\delta^2} \tag{3C.2-5}$$

where primes denote differentiation with respect to η.

(c) Show next that, when the results of (b) are substituted into Eq. 3C.2-3, and when the resulting equation is integrated with respect to η, the following equation for the boundary-layer thickness is obtained:

$$(B - A)\delta \frac{d\delta}{dx} = \frac{\nu}{v_\infty} C \tag{3C.2-6}$$

in which the constants A,B, and C are

$$A = \int_0^1 \phi\phi'\eta d\eta; \; B = \int_0^1 \phi' \left(\int_0^{\bar{\eta}} \phi'\eta d\eta \right) d\bar{\eta} = -A + \int_0^1 \phi'\eta d\eta$$

$$C = \int_0^1 \phi'' d\eta = \phi'|_0^1 \tag{3C.2-7}$$

Then integration of Eq. 3C.2-6 gives

$$\delta(x) = \sqrt{2 \left(\frac{C}{B - A} \right) \frac{\nu x}{v_\infty}} \tag{3C.2-8}$$

This completes the derivation of the expression for the boundary-layer thickness for any assumed velocity profile $\phi(\eta)$ (other than determining the constants A,B, and C explicitly). Note that this analysis predicts that $\delta(x) \propto \sqrt{x}$, which is consistent with the exact solution to this problem.[5]

(d) Finally select the approximate velocity profile $\phi(\eta) = \frac{3}{2}\eta - \frac{1}{2}\eta^3$. Why is this "reasonable"? Using this velocity profile, obtain $\delta(x) = 4.64\sqrt{\nu x / v_\infty}$, and then write down the expression for the velocity distribution in terms of dimensional variables.

(e) Show that the $v_x(x,y)$, thus obtained, leads to the following expression for the drag force over the flat plate, wetted on both sides:

$$F_x = 2\int_0^W \int_0^L \left(+\mu \frac{\partial v_x}{\partial y} \right) \bigg|_{y=0} dx dz = 1.293\sqrt{\rho\mu L W^2 v_\infty^3} \tag{3C.2-9}$$

A more accurate treatment[6] leads to a numerical constant 1.328.

---

[5]See Eq. 4.4-25 of R. B. Bird, W. E. Stewart, and E. N. Lightfoot, *Transport Phenomena,* Revised 2nd Edition, Wiley, New York (2007).

[6]H. Blasius, *Z. Math. Phys.*, **56**, 1–37 (1908).

3C.3 Parallel-disk compression viscometer.[7] An incompressible fluid fills completely the region between two circular disks of radius R (see Fig. 3C.3). The bottom disk is fixed, and the upper disk is made to approach the lower one very slowly with a constant speed v_0, starting from a height H_0 (and $H_0 \ll R$). The instantaneous height of the upper disk is $H(t)$. It is desired to find the force needed to maintain the speed v_0. This kind of viscometer has been found to be useful for extremely viscous fluids, such as tars and cheeses.

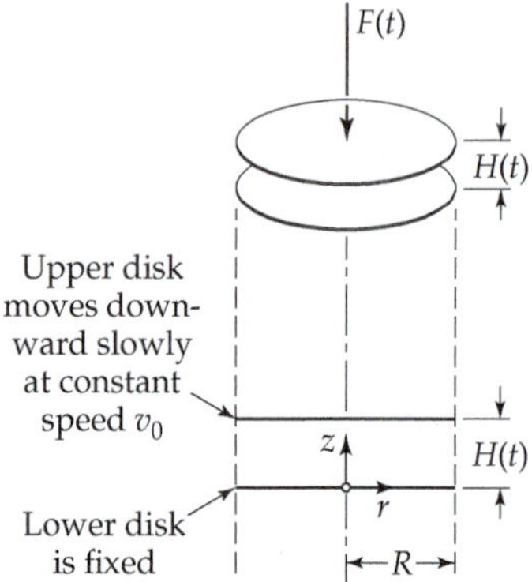

Fig. 3C.3 Squeezing flow in a parallel-disk compression viscometer.

This problem is inherently a rather complicated unsteady flow problem. However, a useful approximate solution can be obtained by making two simplifications in the equations of change: (i) we assume that the speed v_0 is so small that all terms containing time derivatives can be omitted; this is the so-called "quasi-steady-state" assumption; (ii) we use the fact that $H_0 \ll R$ to neglect quite a few terms in the equations of change by order-of-magnitude arguments. Note that the rate of decrease of the fluid volume between the disks is $\pi R^2 v_0$, and that this must equal the rate of outflow from between the disks, which is $2\pi RH\langle v_r\rangle|_{r=R}$. Hence,

$$\langle v_r\rangle|_{r=R} = \frac{Rv_0}{2H(t)} \tag{3C.3-1}$$

We now argue that $v_r(r,z)$ will be of the order of magnitude of $\langle v_r\rangle|_{r=R}$ and that $v_z(r,z)$ is of the order of magnitude of v_0, so that

$$v_r \approx (R/H)v_0; \quad v_z \approx -v_0 \tag{3C.3-2, 3}$$

and hence, $|v_z| \ll v_r$ because $R/H \gg 1$. We may now estimate the order of magnitude of various derivatives as follows: as r goes from 0 to R, the radial velocity v_r goes from zero to approximately $(R/H)v_0$. By this kind of reasoning we get

$$\frac{\partial v_r}{\partial r} \approx \frac{(R/H)v_0 - 0}{R - 0} = \frac{v_0}{H} \tag{3C.3-4}$$

$$\frac{\partial v_z}{\partial z} \approx \frac{(-v_0) - 0}{H - 0} = -\frac{v_0}{H}, \text{ etc.} \tag{3C.3-5}$$

By the above-outlined order-of-magnitude analysis, it can be shown that the continuity equation and the r component of the equation of motion become (with g_z neglected)

continuity: $$\frac{1}{r}\frac{\partial}{\partial r}(rv_r) + \frac{\partial v_z}{\partial z} = 0 \tag{3C.3-6}$$

motion: $$0 = -\frac{dp}{dr} + \mu\frac{\partial^2 v_r}{\partial z^2} \tag{3C.3-7}$$

---

[7] J. R. Van Wazer, J. W. Lyons, K. Y. Kim, and R. E. Colwell, *Viscosity and Flow Measurement*, Wiley-Interscience, New York (1963), pp. 292–295.

with the boundary conditions

B. C. 1:	at $z = 0,\quad v_r = 0,\quad v_z = 0$	(3C.3-8)
B. C. 2:	at $z = H(t),\quad v_r = 0,\quad v_z = -v_0$	(3C.3-9)
B. C. 3:	at $r = R,\quad p = p_{\text{atm}}$	(3C.3-10)

(a) By integrating Eq. 3C.3-7 and using the boundary conditions in Eqs. 3C.3-8 and 3C.3-9, show that

$$v_r = \frac{1}{2\mu}\left(\frac{dp}{dr}\right)z(z - H) \tag{3C.3-11}$$

(b) Next, integrate Eq. 3C.3-6 with respect to z from 0 to H and substitute the result from Eq. 3C.3-11 to get

$$v_0 = -\frac{H^3}{12\mu}\frac{1}{r}\frac{d}{dr}\left(r\frac{dp}{dr}\right) \tag{3C.3-12}$$

(c) Integrate Eq. 3C.3-12 to get the pressure distribution

$$p = p_{\text{atm}} + \frac{3\mu v_0 R^2}{H^3}\left[1 - \left(\frac{r}{R}\right)^2\right] \tag{3C.3-13}$$

(d) Integrate $[(p + \tau_{zz}) - p_{\text{atm}}]$ over the moving-disk surface to find the total force needed to maintain the disk motion:

$$F(t) = \frac{3\pi\mu v_0 R^4}{2[H(t)]^3} \tag{3C.3-14}$$

This result can be used to obtain the viscosity from the force and velocity measurements.

(e) Repeat the analysis for a viscometer that is operated with constant applied force, F_0. The speed of the upper plate in the z direction, $v_0 = -dH/dt$, is not constant, but Eq. 3C.3-14 can be integrated to obtain

$$\frac{1}{[H(t)]^2} = \frac{1}{H_0^2} + \frac{4F_0 t}{3\pi\mu R^4} \tag{3C.3-15}$$

The viscosity may then be determined by measuring the height of the upper plate H as a function of the time that has elapsed since the beginning of the experiment, t.

(f) Repeat the analysis for a viscometer that is operated in such a way that a centered, circular glob of liquid never completely fills the space between the two disks. Let the volume of the fluid sample be V, and show that

$$F(t) = \frac{3\mu v_0 V^2}{2\pi[H(t)]^5} \tag{3C.3-16}$$

3C.4 **Derivation of the equation of change for mechanical energy.** Show how Equation 3.3-1 is derived by forming the dot product of the flow velocity with the equation of motion in Eq. 3.2-9. This development uses only the equation of continuity (Eq. 3.1-4) and several identities from Appendix A. We attack the problem by dotting $\mathbf{v}$ into Eq. 3.2-9 term by term.

(a) Form the scalar product of $\mathbf{v}$ with the external force term to obtain directly

$$(\mathbf{v} \cdot \rho\mathbf{g}) = \rho(\mathbf{v} \cdot \mathbf{g}) \tag{3C.4-1}$$

(b) Take the dot product of $\mathbf{v}$ with the $[\nabla \cdot \boldsymbol{\tau}]$ term. Rearrange using Eq. A.4-29 in Example A.4-1 to obtain

$$-(\mathbf{v} \cdot [\nabla \cdot \boldsymbol{\tau}]) = -(\nabla \cdot [\boldsymbol{\tau} \cdot \mathbf{v}]) + (\boldsymbol{\tau} : \nabla\mathbf{v}) \tag{3C.4-2}$$

(c) Treat the term containing ∇p similarly by using Eq. A.4-19 to get

$$-(\mathbf{v} \cdot \nabla p) = -(\nabla \cdot p\mathbf{v}) + p(\nabla \cdot \mathbf{v}) \tag{3C.4-3}$$

(d) Combine the dot products of the remaining two terms with $\mathbf{v}$, and then differentiate the products, using Eq. A.4-24 to expand the gradient term

$$\left(\mathbf{v}\cdot\frac{\partial}{\partial t}\rho\mathbf{v}\right)+(\mathbf{v}\cdot[\nabla\cdot\rho\mathbf{v}\mathbf{v}])=\left(\rho\mathbf{v}\cdot\frac{\partial}{\partial t}\mathbf{v}\right)+\underline{(\mathbf{v}\cdot\mathbf{v})\frac{\partial}{\partial t}\rho}+(\mathbf{v}\cdot[\rho\mathbf{v}\cdot\nabla\mathbf{v}])+\underline{(\mathbf{v}\cdot\mathbf{v})(\nabla\cdot\rho\mathbf{v})} \quad (3C.4\text{-}4)$$

The underlined terms then sum to zero by using the equation of continuity.

Then on the right side of Eq. 3C.4-4, rewrite the first term as two terms, and rearrange the third term, so that Eq. 3C.4-4 becomes

$$\left(\mathbf{v}\cdot\frac{\partial}{\partial t}\rho\mathbf{v}\right)+(\mathbf{v}\cdot[\nabla\cdot\rho\mathbf{v}\mathbf{v}])=\frac{\partial}{\partial t}\left(\frac{1}{2}\rho(\mathbf{v}\cdot\mathbf{v})\right)-\frac{1}{2}(\mathbf{v}\cdot\mathbf{v})\frac{\partial}{\partial t}\rho+(\rho\mathbf{v}\cdot[\mathbf{v}\cdot\nabla\mathbf{v}]) \quad (3C.4\text{-}5)$$

Again, use the equation of continuity to rewrite the second term on the right side of Eq. 3C.4-5 to get

$$\left(\mathbf{v}\cdot\frac{\partial}{\partial t}\rho\mathbf{v}\right)+(\mathbf{v}\cdot[\nabla\cdot\rho\mathbf{v}\mathbf{v}])=\frac{\partial}{\partial t}\left(\frac{1}{2}\rho(\mathbf{v}\cdot\mathbf{v})\right)+\frac{1}{2}(\mathbf{v}\cdot\mathbf{v})(\nabla\cdot\rho\mathbf{v})+(\rho\mathbf{v}\cdot[\mathbf{v}\cdot\nabla\mathbf{v}]) \quad (3C.4\text{-}6)$$

Now combine the second and third terms by using Eq. A.4-19 with s replaced by $\frac{1}{2}(\mathbf{v}\cdot\mathbf{v})$ and $\mathbf{v}$ replaced by $\rho\mathbf{v}$. This gives finally

$$\left(\mathbf{v}\cdot\frac{\partial}{\partial t}\rho\mathbf{v}\right)+(\mathbf{v}\cdot[\nabla\cdot\rho\mathbf{v}\mathbf{v}])=\frac{\partial}{\partial t}\left(\frac{1}{2}\rho v^2\right)+\left(\nabla\cdot\frac{1}{2}\rho v^2\mathbf{v}\right) \quad (3C.4\text{-}7)$$

Combination of Eqs. 3C.4-1, 3C.4-2, 3C.4-3, and 3C.4-7 gives Equation 3.3-1.

3C.5 Deformation of a fluid line. A fluid is contained in the annular space between two cylinders of radii κR and R (see Fig. 3C.5). The inner cylinder is made to rotate with a constant angular velocity of Ω_i. Consider a line of fluid particles in the plane $z = 0$ extending from the inner cylinder to the outer cylinder and initially located at $\theta = 0$, normal to the two surfaces. How does this fluid line deform into a curve $\theta(r,t)$? What is the length, l, of the curve after N revolutions of the inner cylinder? Use Eq. 3.7-31.

Answer: $\dfrac{l}{R}=\displaystyle\int_\kappa^1\sqrt{1+\frac{16\pi^2N^2}{[(1/\kappa)^2-1]^2\xi^4}}\,d\xi$

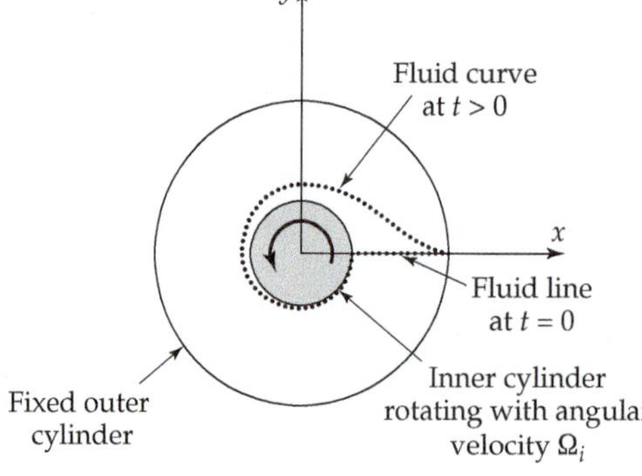

Fig. 3C.5 Deformation of a fluid line in Couette flow.

Chapter 4

Velocity Distributions in Turbulent Flow

In Chapter 3 we discussed laminar-flow problems only. We have seen that the differential equations describing laminar flow are well understood and that, for a number of simple systems, the velocity distribution and various derived quantities can be obtained in a straightforward fashion. The limiting factor in applying the equations of change is the mathematical complexity that one encounters in problems for which there are several velocity components that are functions of several variables. Even there, with the rapid development of computational fluid dynamics, such problems are gradually yielding to numerical solution.

In this chapter we turn our attention to turbulent flow. Whereas laminar flow is orderly, turbulent flow is chaotic. It is this chaotic nature of turbulent flow that poses all sorts of difficulties. In fact, one might even question whether or not the equations of change given in Chapter 3 are even capable of describing the violently fluctuating motions in turbulent flow. Since the sizes of the turbulent eddies are several orders of magnitude larger than the mean-free path of the molecules of the fluid, the equations of change *are* applicable. Numerical solutions of these equations are obtainable and can be used for studying the details of the turbulence structure. For many purposes, however, we are not interested in having such detailed information, in view of the computational effort required. Therefore, in this chapter we shall concern ourselves primarily with methods that enable us to describe the time-smoothed velocity and pressure profiles.

In §4.1 we start by comparing the experimental results for laminar and turbulent flows in several simple systems. In this way we can get some qualitative ideas about the main differences between laminar and turbulent motions. These experiments help to define some of the challenges that face the fluid dynamicist.

In §4.2 we define several *time-smoothed* quantities, and show how these definitions can be used to time-average the equations of change over a short time interval. These equations describe the behavior of the time-smoothed velocity and pressure. The time-smoothed equation of motion, however, contains the *turbulent momentum flux*.

This flux cannot be simply related to velocity gradients in the way that the viscous momentum flux is given by Newton's law of viscosity in Chapter 1. At the present time the turbulent momentum flux is usually estimated experimentally or else modeled by some type of empiricism based on experimental measurements.

Fortunately, for turbulent flow near a solid surface, there are several rather general results that are very helpful in fluid dynamics and transport phenomena: the Taylor-series development for the velocity near the wall, and the logarithmic and power-law velocity profiles for regions further from the wall, the latter being obtained by dimensional reasoning. These expressions for the time-smoothed velocity distribution are given in §4.3.

In the following section, §4.4, we present a few of the empiricisms that have been proposed for the turbulent momentum flux. These empiricisms are of historical interest and have also been widely used in engineering calculations. When applied with proper judgment, these empirical expressions can be useful.

The remainder of the chapter is devoted to a discussion of two types of turbulent flows: flows in closed conduits (§4.5) and flows in jets (§4.6). These flows are illustrative of the two classes of flows that are usually discussed under the headings of *wall turbulence* and *free turbulence*.

In this brief introduction to turbulence we deal primarily with the description of the fully developed turbulent flow of an incompressible fluid. We do not consider at all the theoretical methods for predicting the inception of turbulence, nor the experimental techniques devised for probing the structure of turbulent flow. We also give no discussion of the statistical theories of turbulence and the way in which the turbulent energy is distributed over the various modes of motion. For these and other interesting topics, the reader should consult some of the standard books on turbulence.[1–6] There is a growing literature on experimental and computational evidence for "coherent structures" (vortices) in turbulent flows.[7]

Turbulence is an important subject. In fact, most flows encountered in engineering are turbulent and not laminar! Although our understanding of turbulence is far from satisfactory, it is a subject that must be studied and appreciated. For the solution to many industrial problems we cannot get neat analytical solutions, and, for the most part, such problems are attacked by using a combination of dimensional analysis and experimental data. These methods are discussed in Chapters 5 and 6.

§4.1 COMPARISONS OF LAMINAR AND TURBULENT FLOWS

Before discussing any theoretical ideas about turbulence, it is important to illustrate the differences between laminar and turbulent flows in several simple systems. Specifically we consider the flow in conduits of circular and triangular cross section, flow along a flat

---

[1]S. Corrsin, "Turbulence: Experimental Methods," in *Handbuch der Physik*, Springer, Berlin (1963), Vol. VIII/2. **Stanley Corrsin** (1920–1986), a professor at The Johns Hopkins University, was an excellent experimentalist and teacher; he studied the interaction between chemical reactions and turbulence and the propagation of the double temperature correlations.

[2]A. A. Townsend, *The Structure of Turbulent Shear Flow*, Cambridge University Press, 2nd edition (1976); see also A. A. Townsend in *Handbook of Fluid Dynamics* (V. L. Streeter, ed.), McGraw-Hill (1961) for a readable survey.

[3]J. O. Hinze, *Turbulence*, McGraw-Hill, New York, 2nd edition (1975).

[4]H. Tennekes and J. L. Lumley, *A First Course in Turbulence*, MIT Press, Cambridge, MA (1972); Chapters 1 and 2 of this book provide an introduction to the physical interpretations of turbulent flow phenomena.

[5]M. Lesieur, *La Turbulence*, Presses Universitaires de Grenoble (1994); this book contains beautiful color photographs of turbulent flow systems.

[6]W. D. McComb, *The Physics of Fluid Turbulence*, Oxford University Press (1990).

[7]P. Holmes, J. L. Lumley, and G. Berkooz, *Turbulence, Coherent Structures, Dynamical Systems, and Symmetry*, Cambridge University Press (1996); F. Waleffe, *Phys. Rev. Lett.*, **81**, 4140–4148 (1998).

plate, and flows in jets. The first two of these have been considered for laminar flow in §2.3 and Problem 3B.2.

a. Circular tubes

For the steady, fully developed, laminar flow in a circular tube of radius R we know that the velocity distribution and the average velocity are given by

$$\frac{v_z(r)}{v_{z,\max}} = 1 - \left(\frac{r}{R}\right)^2 \quad \text{and} \quad \frac{\langle v_z \rangle}{v_{z,\max}} = \frac{1}{2} \quad (\text{Re} < 2100) \tag{4.1-1,2}$$

and that the pressure drop and mass flow rate w are linearly related:

$$\mathscr{P}_0 - \mathscr{P}_L = \left(\frac{8\mu L}{\pi\rho R^4}\right) w \quad (\text{Re} < 2100) \tag{4.1-3}$$

For turbulent flow, on the other hand, the velocity is fluctuating with time chaotically at each point in the tube. We can measure a "time-smoothed velocity" at each point with, say, a Pitot tube. This type of instrument is not sensitive to rapid velocity fluctuations, but senses the velocity averaged over several seconds. The time-smoothed velocity (which is defined in the next section) will have a z component represented by $\bar{v}_z$, and its shape and average value will be given very roughly by[1]

$$\frac{\bar{v}_z(r)}{\bar{v}_{z,\max}} \approx \left(1 - \frac{r}{R}\right)^{1/7} \quad \text{and} \quad \frac{\langle \bar{v}_z \rangle}{\bar{v}_{z,\max}} \approx \frac{4}{5} \quad \left(10^4 < \text{Re} < 10^5\right) \tag{4.1-4,5}$$

This $\frac{1}{7}$-power expression for the velocity distribution is too crude to give a realistic velocity derivative at the wall. For example, it gives an infinite velocity gradient at $r = R$. The laminar and turbulent velocity profiles are compared in Fig. 4.1-1.

Over the same range of Reynolds numbers, the pressure drop is no longer proportional to the mass flow rate, but obeys roughly the following relation:

$$\mathscr{P}_0 - \mathscr{P}_L \approx 0.198\left(\frac{2}{\pi}\right)^{7/4} \left(\frac{\mu^{1/4} L}{\rho R^{19/4}}\right) w^{7/4} \quad \left(10^4 < \text{Re} < 10^5\right) \tag{4.1-6}$$

The stronger dependence of pressure drop on mass flow rate for turbulent flow results from the fact that more energy has to be supplied in order to maintain the violent eddy motion in the fluid.

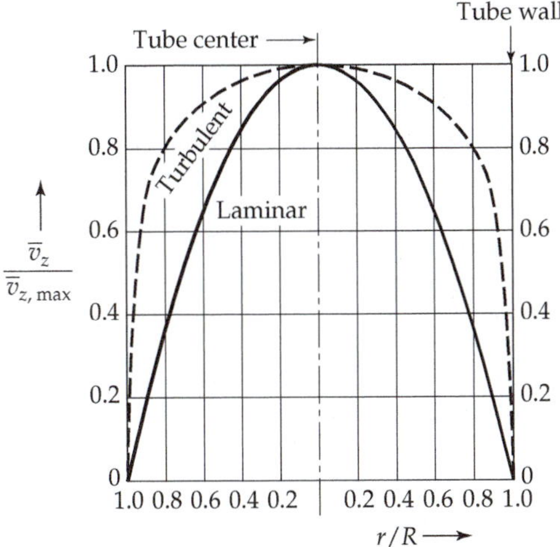

Fig. 4.1-1. Qualitative comparison of laminar and turbulent velocity profiles. For a more detailed description of the turbulent velocity distribution near the wall, see Fig. 4.5-3.

[1] H. Schlichting, *Boundary-Layer Theory*, McGraw-Hill, New York, 7th edition (1979), Chapter XX (tube flow), Chapters VII and XXI (flat plate flow), Chapters IX and XXIIV (jet flows).

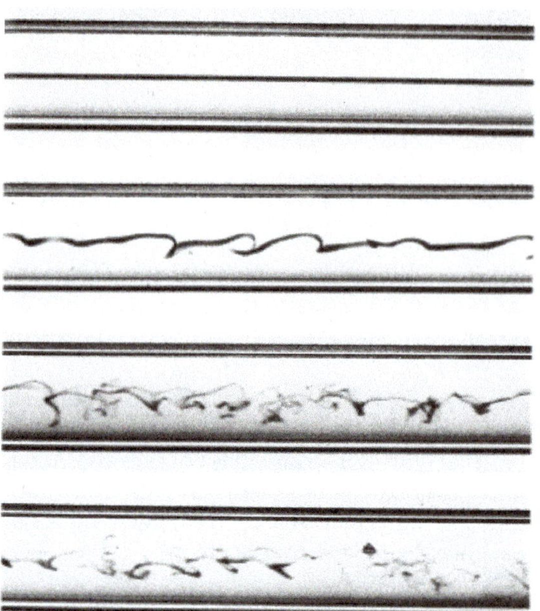

Fig. 4.1-2. Behavior of a colored stream of liquid injected into the flow of water in a glass tube. In the top photograph, the flow is laminar and the colored liquid stream remains thin. The Reynolds number is increased successively in the lower photographs. As the flow becomes more turbulent, the colored liquid disperses more quickly in the tube. [Experiments performed by N. H. Johannesen and C. Lowe with O. Reynolds's original equipment; photograph from M. van Dyke, *An Album of Fluid Motion*, Parabolic Press, Stanford (1982).]

The laminar-turbulent transition in circular pipes normally occurs at a *critical Reynolds number* of roughly 2100, although this number may be higher if extreme care is taken to eliminate vibrations in the system.[2] The transition from laminar flow to turbulent flow can be demonstrated by the simple experiment originally performed by Reynolds. One sets up a long transparent tube equipped with a device for injecting a small amount of colored liquid into the stream along the tube axis. Results from one such study are illustrated in Fig. 4.1-2. When the flow is laminar, the colored liquid moves downstream as a straight, coherent filament. For turbulent flow, on the other hand, the colored liquid spreads quickly over the entire cross section. This *dispersion* is similar to the motion of particles in turbulent flow illustrated in Fig. 2.0-1(*b*); in both cases, the displacement of material perpendicular to the flow direction is caused by eddy motion (turbulent diffusion). This turbulent mixing is important in heat and mass transport.

b. Noncircular tubes

For developed laminar flow in the triangular duct shown in Fig. 3B.2(*b*), the fluid particles move rectilinearly in the z direction, parallel to the walls of the duct. By contrast, in turbulent flow there is superposed on the time-smoothed flow in the z direction (the *primary flow*), a time-smoothed motion in the xy plane (the *secondary flow*). The secondary flow is much weaker than the primary flow and manifests itself as a set of six vortices arranged in a symmetric pattern around the duct axis (see Fig. 4.1-3). Other noncircular tubes also exhibit secondary flows.

c. Flat plate

For the laminar flow around a flat plate, wetted on both sides and aligned with the flow, the Blasius solution of the boundary-layer equations gives the drag-force expression (see Problem 3C.2)

$$F = 1.328\sqrt{\rho\mu L W^2 v_\infty^3} \quad (\text{laminar; } 0 < \text{Re}_L < 5 \times 10^5) \tag{4.1-7}$$

[2]O. Reynolds, *Phil. Trans. Roy. Soc.*, **174**, Part III, 935–982 (1883). See also A. A. Draad and F. M. T. Nieuwstadt, *J. Fluid Mech.*, **361**, 297–308 (1998).

Fig. 4.1-3. Sketch showing the secondary flow patterns for turbulent flow in a tube of triangular cross section. [H. Schlichting, *Boundary-Layer Theory*, McGraw-Hill, New York, 7th edition (1979), p. 613.]

in which $\text{Re}_L = Lv_\infty\rho/\mu$ is the Reynolds number for a plate of length L in the flow direction; the plate width is W, and the approach velocity of the fluid is v_∞.

For turbulent flow, on the other hand, the dependence on the geometrical and physical properties is quite different:[1]

$$F \approx 0.074\sqrt[5]{\rho^4\mu L^4 W^5 v_\infty^9} \qquad (\text{turbulent; } 5\times 10^5 < \text{Re}_L < 10^7) \tag{4.1-8}$$

Thus, the force is proportional to the $\frac{3}{2}$-power of the approach velocity for laminar flow, but to the $\frac{9}{5}$-power for turbulent flow. The stronger dependence on the approach velocity reflects the extra energy needed to maintain the irregular eddy motions in the fluid.

d. Circular and plane jets

Next, we examine the behavior of jets that emerge from a flat wall, which is taken to be the xy plane (see Fig. 4.6-1). The fluid emerges from a circular tube or a long narrow slot, and flows into a large body of the same fluid. Various observations on the jets can be made: the width of the jet, the center-line velocity of the jet, and the mass flow rate through a cross section parallel to the xy plane. All these properties can be measured as functions of the distance z from the wall. In Table 4.1-1 we summarize the properties of the circular and two-dimensional jets for laminar and turbulent flow.[1] It is curious that, for the circular jet, the jet width, center-line velocity, and mass flow rate have exactly the same dependence on z in both laminar and turbulent flow. We shall return to this point later in §4.6.

The above examples should make it clear that the gross features of laminar and turbulent flow are generally quite different. One of the many challenges in turbulence theory is to try to explain these differences.

Table 4.1-1. Dependence of Jet Parameters on Distance z from Wall

	Laminar flow			Turbulent flow		
	Width of jet	Center-line velocity	Mass flow rate	Width of jet	Center-line velocity	Mass flow rate
Circular jet	z	z^{-1}	z	z	z^{-1}	z
Plane jet	$z^{2/3}$	$z^{-1/3}$	$z^{1/3}$	z	$z^{-1/2}$	$z^{1/2}$

§4.2 TIME-SMOOTHED EQUATIONS OF CHANGE FOR INCOMPRESSIBLE FLUIDS

We begin by considering a turbulent flow in a tube with a constant imposed pressure gradient. If at a particular point in the fluid we observe one component of the velocity as a function of time, we find that it is fluctuating in a chaotic fashion as shown in

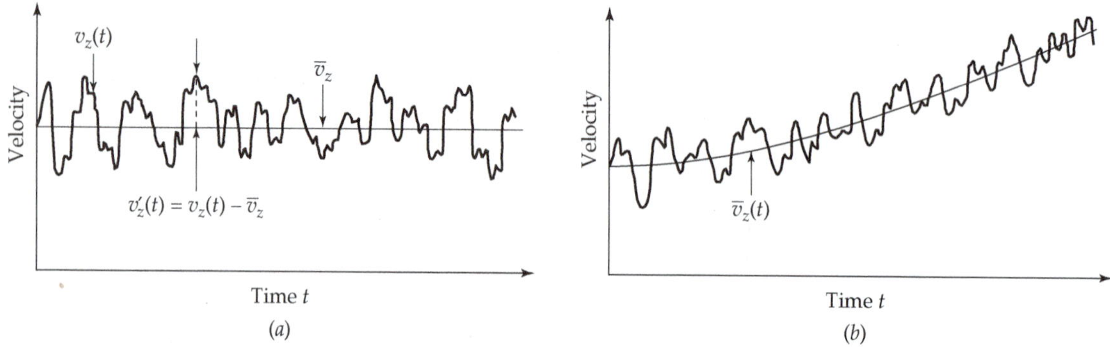

Fig. 4.2-1. Sketch showing the velocity component v_z, as well as its time-smoothed value $\bar{v}_z$, and its fluctuation v_z' in turbulent flow (*a*) for "steadily driven turbulent flow" in which $\bar{v}_z$ does not depend on time, and (*b*) for a situation in which $\bar{v}_z$ does depend on time.

Fig. 4.2-1(*a*). The fluctuations are irregular deviations from a mean value. The actual velocity can be regarded as the sum of the mean value (designated by an overbar) and the fluctuation (designated by a prime). For example, for the z component of the velocity we write

$$v_z(t) = \bar{v}_z + v_z'(t) \tag{4.2-1}$$

which is sometimes called the *Reynolds decomposition*. The mean value is obtained from $v_z(t)$ by making a time average over a large number of fluctuations

$$\bar{v}_z = \frac{1}{t_0}\int_{t-\frac{1}{2}t_0}^{t+\frac{1}{2}t_0} v_z(s)ds \tag{4.2-2}$$

the period t_0 being long enough to give a smooth averaged function. For the system at hand, the quantity $\bar{v}_z$, which we call the *time-smoothed velocity*, is independent of time, but of course depends on position. When the time-smoothed velocity does not depend on time, we speak of *steadily driven turbulent flow*. The same comments we have made for velocity can also be made for the pressure.

Next we consider turbulent flow in a tube with a time-dependent pressure gradient. For such a flow one can define time-smoothed quantities as above, but one has to understand that the period t_0 must be small with respect to the changes in the pressure gradient, but still large with respect to the periods of fluctuations. For such a situation, the time-smoothed velocity and the actual velocity are illustrated in Fig. 4.2-1(*b*).[1]

According to the definition in Eq. 4.2-2, one can show that the following relations are true:

$$\overline{v_z'} = 0; \quad \overline{\bar{v}_z} = \bar{v}_z; \quad \overline{\bar{v}_z v_z'} = 0; \quad \overline{\frac{\partial}{\partial x} v_z} = \frac{\partial}{\partial x}\bar{v}_z; \quad \overline{\frac{\partial}{\partial t} v_z} = \frac{\partial}{\partial t}\bar{v}_z \tag{4.2-3}$$

The quantity $\overline{v_z'^2}$ will not, however, be zero, and in fact, the ratio $\sqrt{\overline{v_z'^2}}/\langle\bar{v}_z\rangle$ can be taken to be a measure of the magnitude of the turbulent fluctuations. This quantity, known as

[1]One can also define the "overbar" quantities in terms of an "ensemble average." For most purposes, the results are equivalent or are assumed to be so. See, for example, A. A. Townsend, *The Structure of Turbulent Shear Flow*, Cambridge University Press, 2nd edition (1976). See also P. K. Kundu, *Fluid Mechanics*, Academic Press, New York (1990), p. 421, regarding the last of the formulas given in Eq. 4.2-3.

the *intensity of turbulence*, may have values from 1% to 10% in the main part of a turbulent stream and values of 25% or higher in the neighborhood of a solid wall. Hence, it must be emphasized that we are not necessarily dealing with tiny disturbances; sometimes the fluctuations are actually quite violent and large.

Quantities such as $\overline{v'_x v'_y}$ are also nonzero. The reason for this is that the local motions in the x and y directions are *correlated*. In other words, the fluctuations in the x direction are not independent of the fluctuations in the y direction. We shall see presently that these time-smoothed values of the products of fluctuating properties have an important role in turbulent momentum transfer. Later we shall find similar correlations arising in turbulent energy and mass transport.

Having defined the time-smoothed quantities and discussed some of the properties of the fluctuating quantities, we can now move on to the time-smoothing of the equations of change. To keep the development as simple as possible, we consider here only the equations for a fluid of constant density and viscosity. The equation of continuity is then just $(\nabla \cdot \mathbf{v}) = 0$. We then write the equations of continuity and motion with $\mathbf{v}$ replaced by its equivalent $\overline{\mathbf{v}} + \mathbf{v}'$ and p by its equivalent $\overline{p} + p'$. We write the x component of the equation of motion as the Navier–Stokes equation, Eq. 3.6-1, in the $\partial/\partial t$ form by using Eq. 3.5-7:

$$\frac{\partial}{\partial x}\left(\overline{v}_x + v'_x\right) + \frac{\partial}{\partial y}\left(\overline{v}_y + v'_y\right) + \frac{\partial}{\partial z}\left(\overline{v}_z + v'_z\right) = 0 \tag{4.2-4}$$

$$\begin{aligned}\frac{\partial}{\partial t}\rho\left(\overline{v}_x + v'_x\right) = &-\frac{\partial}{\partial x}\left(\overline{p} + p'\right) - \left(\frac{\partial}{\partial x}\rho\left(\overline{v}_x + v'_x\right)\left(\overline{v}_x + v'_x\right)\right.\\ &\left.+ \frac{\partial}{\partial y}\rho\left(\overline{v}_y + v'_y\right)\left(\overline{v}_x + v'_x\right) + \frac{\partial}{\partial z}\rho\left(\overline{v}_z + v'_z\right)\left(\overline{v}_x + v'_x\right)\right)\\ &+ \mu\nabla^2\left(\overline{v}_x + v'_x\right) + \rho g_x\end{aligned} \tag{4.2-5}$$

The y and z components of the equation of motion can be similarly written. We next time-smooth these equations, making use of the relations given in Eq. 4.2-3. This gives

$$\frac{\partial}{\partial x}\overline{v}_x + \frac{\partial}{\partial y}\overline{v}_y + \frac{\partial}{\partial z}\overline{v}_z = 0 \tag{4.2-6}$$

$$\begin{aligned}\frac{\partial}{\partial t}\rho\overline{v}_x = &-\frac{\partial}{\partial x}\overline{p} - \left(\frac{\partial}{\partial x}\rho\overline{v}_x\overline{v}_x + \frac{\partial}{\partial y}\rho\overline{v}_y\overline{v}_x + \frac{\partial}{\partial z}\rho\overline{v}_z\overline{v}_x\right)\\ &\underline{-\left(\frac{\partial}{\partial x}\rho\overline{v'_x v'_x} + \frac{\partial}{\partial y}\rho\overline{v'_y v'_x} + \frac{\partial}{\partial z}\rho\overline{v'_z v'_x}\right)} + \mu\nabla^2\overline{v}_x + \rho g_x\end{aligned} \tag{4.2-7}$$

with similar relations for the y and z components of the equation of motion. These are then the *time-smoothed equations of continuity and motion* for a fluid with constant density and viscosity. By comparing them with the corresponding equations in Eq. 3.1-5 and Eq. 3.6-1 (the latter rewritten in terms of $\partial/\partial t$), we conclude that:

- **a.** The equation of continuity is the same as we had previously, except that $\mathbf{v}$ is now replaced by $\overline{\mathbf{v}}$.
- **b.** The equation of motion now has $\overline{\mathbf{v}}$ and $\overline{p}$ where we previously had $\mathbf{v}$ and p. In addition there appears the dashed-underlined term, which describes the momentum transport associated with the turbulent fluctuations.

We may rewrite Eq. 4.2-7 by introducing the *turbulent momentum flux tensor* $\overline{\boldsymbol{\tau}}^{(t)}$ with components

$$\overline{\tau}^{(t)}_{xx} = \rho\overline{v'_x v'_x}; \qquad \overline{\tau}^{(t)}_{xy} = \rho\overline{v'_x v'_y}; \qquad \overline{\tau}^{(t)}_{xz} = \rho\overline{v'_x v'_z} \qquad \text{etc.} \tag{4.2-8}$$

These quantities are usually referred to as the *Reynolds stresses*. We may also introduce a symbol $\overline{\boldsymbol{\tau}}^{(v)}$ for the time-smoothed viscous momentum flux. The components of this tensor have the same appearance as the expressions given in §B.1, except that the time-smoothed velocity components appear in them:

$$\bar{\tau}_{xx}^{(v)} = -2\mu \frac{\partial \bar{v}_x}{\partial x}; \quad \bar{\tau}_{xy}^{(v)} = -\mu \left(\frac{\partial \bar{v}_y}{\partial x} + \frac{\partial \bar{v}_x}{\partial y} \right); \quad \text{etc.} \tag{4.2-9}$$

This enables us then to write the equations of change in vector-tensor form as

$$(\nabla \cdot \bar{\mathbf{v}}) = 0 \quad \text{and} \quad (\nabla \cdot \mathbf{v}') = 0 \tag{4.2-10,11}$$

$$\frac{\partial}{\partial t}\rho\bar{\mathbf{v}} = -\nabla\bar{p} - [\nabla \cdot \rho\overline{\mathbf{v}\mathbf{v}}] - \left[\nabla \cdot \left(\bar{\boldsymbol{\tau}}^{(v)} + \bar{\boldsymbol{\tau}}^{(t)}\right)\right] + \rho\mathbf{g} \tag{4.2-12}$$

Equation 4.2-11 is an extra equation obtained by subtracting Eq. 4.2-10 from the original equation of continuity.

The principal result of this section is that the equation of motion in terms of the stress tensor, summarized in §B.5, can be adapted for time-smoothed turbulent flow by changing all v_i to $\bar{v}_i$, p to $\bar{p}$, and τ_{ij} to $\bar{\tau}_{ij} = \bar{\tau}_{ij}^{(v)} + \bar{\tau}_{ij}^{(t)}$ in any of the coordinate systems given.

We have now arrived at the main stumbling block in the theory of turbulence. The Reynolds stresses $\bar{\tau}_{ij}^{(t)}$ above are not related to the velocity gradients in a simple way as are the time-smoothed viscous stresses $\bar{\tau}_{ij}^{(v)}$ in Eq. 4.2-9. They are, instead, complicated functions of the position and the turbulence intensity. To solve flow problems, we must have experimental information about the Reynolds stresses or else resort to some empirical expression. In §4.4 we discuss some of the empiricisms that are available.

Actually one can also obtain equations of change for the Reynolds stresses. However, these equations contain quantities like $\overline{v_i'v_j'v_k'}$. Similarly, the equations of change for the $\overline{v_i'v_j'v_k'}$ contain the next higher-order correlation $\overline{v_i'v_j'v_k'v_l'}$, and so on. That is, there is a never-ending hierarchy of equations that must be solved. To attack flow problems, one has to "truncate" this hierarchy by introducing empiricisms. If we use empiricisms for the Reynolds stresses, we then have a "first-order" theory. If we introduce empiricisms for the $\overline{v_i'v_j'v_k'}$, we then have a "second-order theory," and so on. The problem of introducing empiricisms to get a closed set of equations that can be solved for the velocity and pressure distributions is referred to as the "closure problem." The discussion in §4.4 deals with closure at the first order. At the second order the "k-ε empiricism" has been extensively studied and widely used in computational fluid mechanics.[2]

§4.3 THE TIME-SMOOTHED VELOCITY PROFILE NEAR A WALL

Before we discuss the various empirical expressions used for the Reynolds stresses, we present here several developments that do not depend on any empiricisms. We are concerned here with the fully developed, time-smoothed velocity distribution in the neighborhood of a wall. We discuss two results: (a) the Taylor expansion of the velocity near the wall, and (b) the universal logarithmic velocity distribution a little further out from the wall.

The flow near a flat surface is shown in Fig. 4.3-1. It is usual to distinguish—somewhat arbitrarily—four regions of flow:

- the *viscous sublayer* very near the wall, in which viscosity plays a key role
- the *buffer layer* in which the transition occurs between the viscous and inertial sublayers

[2]J. L. Lumley, *Adv. Appl. Mech.*, **18**, 123–176 (1978); C. G. Speziale, *Ann. Revs. Fluid Mech.*, **23**, 107–157 (1991); H. Schlichting and K. Gersten, *Boundary-Layer Theory*, Springer, Berlin, 8th edition (2000), pp. 560–563.

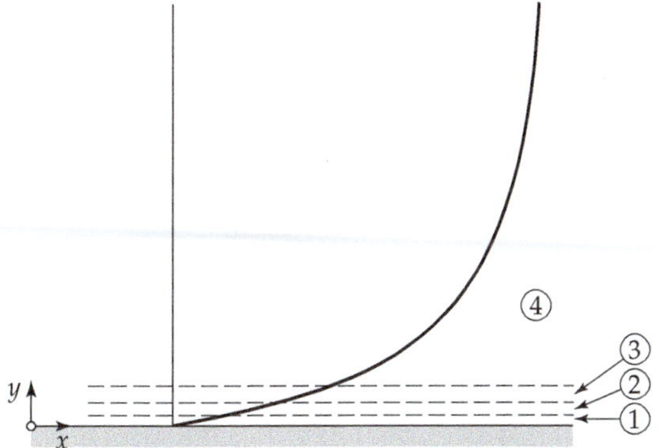

Fig. 4.3-1. Flow regions for describing turbulent flow near a wall: (1) viscous sublayer, (2) buffer layer, (3) inertial sublayer, (4) main turbulent stream.

- the *inertial sublayer* at the beginning of the main turbulent stream, in which viscosity plays at most a minor role
- the *main turbulent stream* in which the time-smoothed velocity distribution is nearly flat and viscosity is unimportant

It is only in the viscous sublayer that we can say anything very precise, although a dimensional analysis treatment can give some information about the velocity distribution in the inertial sublayer.

a. Taylor-series development in the viscous sublayer

We start by writing the Taylor series (see §C.2) for $\bar{v}_x$ as a function of y, about $y = 0$

$$\bar{v}_x(y) = \bar{v}_x(0) + \left.\frac{\partial \bar{v}_x}{\partial y}\right|_{y=0} y + \frac{1}{2!}\left.\frac{\partial^2 \bar{v}_x}{\partial y^2}\right|_{y=0} y^2 + \frac{1}{3!}\left.\frac{\partial^3 \bar{v}_x}{\partial y^3}\right|_{y=0} y^3 + \cdots \tag{4.3-1}$$

To evaluate the terms in this series, we need the expression for the time-smoothed shear stress in the vicinity of a wall. For the special case of the steadily driven flow in a slit of thickness $2B$, the shear stress will be of the form $\bar{\tau}_{yx} = \bar{\tau}_{yx}^{(v)} + \bar{\tau}_{yx}^{(t)} = -\tau_0\left[1 - (y/B)\right]$. Here, $\tau_0 \equiv -\bar{\tau}_{yx}|_{y=0}$ is the time-smoothed shear stress acting on the wall at $y = 0$. Then from Eqs. 4.2-8 and 4.2-9, we have

$$+\mu\frac{\partial \bar{v}_x}{\partial y} - \rho\overline{v_x' v_y'} = \tau_0\left(1 - \frac{y}{B}\right) \tag{4.3-2}$$

Now we examine the terms that appear in Eq. 4.3-1 one by one:[1]

(i) The first term is zero by the no-slip condition.

(ii) The coefficient of the second term can be obtained from Eq. 4.3-2, recognizing that both v_x' and v_y' are zero at the wall so that

$$\left.\frac{\partial \bar{v}_x}{\partial y}\right|_{y=0} = \frac{\tau_0}{\mu} \tag{4.3-3}$$

(iii) The coefficient of the third term involves the second derivative, which may be obtained by differentiating Eq. 4.3-2 with respect to y and then setting $y = 0$, as follows:

$$\left.\frac{\partial^2 \bar{v}_x}{\partial y^2}\right|_{y=0} = \frac{\rho}{\mu}\left.\overline{\left(v_x'\frac{\partial v_y'}{\partial y} + v_y'\frac{\partial v_x'}{\partial y}\right)}\right|_{y=0} - \frac{\tau_0}{\mu B} = -\frac{\tau_0}{\mu B} \tag{4.3-4}$$

since both v_x' and v_y' are zero at the wall.

[1] A. A. Townsend, *The Structure of Turbulent Shear Flow*, Cambridge University Press, 2nd edition (1976), p. 163.

(iv) The coefficient of the fourth term involves the third derivative, which may be obtained from Eq. 4.3-4, and this is

$$\left.\frac{\partial^3 \bar{v}_x}{\partial y^3}\right|_{y=0} = \frac{\rho}{\mu}\left.\left(\overline{v'_x \frac{\partial^2 v'_y}{\partial y^2} + 2\frac{\partial v'_y}{\partial y}\frac{\partial v'_x}{\partial y} + v'_y \frac{\partial^2 v'_x}{\partial y^2}}\right)\right|_{y=0}$$

$$= -\frac{\rho}{\mu}\left.\left(\overline{+2\left(\frac{\partial v'_x}{\partial x} + \frac{\partial v'_z}{\partial z}\right)\frac{\partial v'_x}{\partial y}}\right)\right|_{y=0} = 0 \qquad (4.3\text{-}5)$$

Here Eq. 4.2-10,11 has been used.

There appears to be no reason to set the next coefficient equal to zero, so that we find that the Taylor series, in dimensionless quantities, has the form

$$\frac{\bar{v}_x(y)}{v_*} = \frac{yv_*}{\nu} - \frac{1}{2}\left(\frac{\nu}{v_* B}\right)\left(\frac{yv_*}{\nu}\right)^2 + C\left(\frac{yv_*}{\nu}\right)^4 + \cdots \qquad (4.3\text{-}6)$$

where the quantity $v_* \equiv \sqrt{\tau_0/\rho}$ has dimensions of velocity, and is called the *friction velocity*. The coefficient C has been obtained experimentally,[2] and so we have the final result:

$$\frac{\bar{v}_x(y)}{v_*} = \frac{yv_*}{\nu}\left[1 - \frac{1}{2}\left(\frac{\nu}{v_* B}\right)\left(\frac{yv_*}{\nu}\right) - \frac{1}{4}\left(\frac{yv_*}{14.5\nu}\right)^3 + \cdots\right] \qquad 0 < \frac{yv_*}{\nu} < 5 \qquad (4.3\text{-}7)$$

The second term in the brackets is negligible for turbulent flow as long as B is not too small (i.e., the length scale ν/v_* is typically much less than B). The y^3 term in the brackets will turn out to be very important in connection with turbulent energy- and mass-transfer correlations in Chapters 12, 14, 20, and 22.

Equation 4.3-7 is valid in the viscous sublayer, defined approximately by the region $0 < yv_*/\nu < 5$. For the region $5 < yv_*/\nu < 30$ no simple analytical derivations are available, and empirical curve fits are sometimes used. One of these is shown in Fig. 4.5-3 for circular tubes.

b. The logarithmic velocity profile in the inertial sublayer[3–6]

Next we consider an analysis appropriate for the inertial sublayer ($yv_*/\nu > 30$). Here the shear stress will not be very different from the value τ_0. We now ask: on what quantities will the time-smoothed velocity gradient $d\bar{v}_x/dy$ depend? It should not depend on the viscosity, since, out beyond the buffer layer, momentum transport should depend primarily on the velocity fluctuations (loosely referred to as "eddy motion"). It may depend on the density ρ, the wall shear stress τ_0, and the distance y from the wall. The only combination

---

[2] C. S. Lin, R. W. Moulton, and G. L. Putnam, *Ind. Eng. Chem.*, **45**, 636–640 (1953); the numerical coefficient was determined from mass transfer experiments in circular tubes. The importance of the y^4 term in heat and mass transfer was recognized earlier by E. V. Murphree, *Ind. Eng. Chem.*, **24**, 726–736 (1932). **Eger Vaughn Murphree** (1898–1962) was captain of the University of Kentucky football team in 1920 and became president of the Standard Oil Development Company.

[3] L. Landau and E. M. Lifshitz, *Fluid Mechanics*, Pergamon, Oxford, 2nd edition (1987), pp. 172–178.

[4] H. Schlichting and K. Gersten, *Boundary-Layer Theory*, Springer-Verlag, Berlin, 8th edition (2000), §17.2.3.

[5] T. von Kármán, *Nachr. Ges. Wiss. Göttingen, Math-Phys. Klasse* Series 5, 58–76 (1930); L. Prandtl, *Ergeb. Aerodyn. Versuch.*, Göttingen, Series 4, 18–29 (1932).

[6] G. I. Barenblatt and A. J. Chorin, *Proc. Nat. Acad. Sci. USA*, **93**, 6749–6752 (1996) and *SIAM Rev.*, **40**, 265–291 (1981); G. I. Barenblatt, A. J. Chorin, and V. M. Prostokishin, *Proc. Nat. Acad. Sci. USA*, **94**, 773–776 (1997). See also G. I. Barenblatt, *Scaling, Self-Similarity, and Intermediate Asymptotics*, Cambridge University Press (1992), §10.2.

of these three quantities that has the dimensions of a velocity gradient is $\sqrt{\tau_0/\rho}/y$. Hence, we write

$$\frac{d\bar{v}_x}{dy} = \frac{1}{\kappa}\sqrt{\frac{\tau_0}{\rho}}\frac{1}{y} \tag{4.3-8}$$

in which κ is a dimensionless constant, which must be determined experimentally. When Eq. 4.3-8 is integrated we get (using $v_* = \sqrt{\tau_0/\rho}$)

$$\bar{v}_x(y) = \frac{v_*}{\kappa}\ln y + \lambda' \tag{4.3-9}$$

λ' being an integration constant. In order to use dimensionless groupings, we rewrite Eq. 4.3-9 as

$$\frac{\bar{v}_x(y)}{v_*} = \frac{1}{\kappa}\ln\left(\frac{yv_*}{\nu}\right) + \lambda \tag{4.3-10}$$

in which λ is a constant simply related to λ'; the kinematic viscosity ν was included in order to construct the dimensionless argument of the logarithm. Experimentally it has been found that reasonable values of the constants[4] are $\kappa = 0.4$ and $\lambda = 5.5$, giving

$$\frac{\bar{v}_x}{v_*} = 2.5\ln\left(\frac{yv_*}{\nu}\right) + 5.5 \qquad \frac{yv_*}{\nu} > 30 \tag{4.3-11}$$

This is called the *von Kármán-Prandtl universal logarithmic velocity distribution;*[5] it is intended to apply only in the inertial sublayer. Later we shall see (in Fig. 4.5-3) that this function describes moderately well the experimental data somewhat beyond the inertial sublayer.

§4.4 EMPIRICAL EXPRESSIONS FOR THE TURBULENT MOMENTUM FLUX

We now return to the problem of using the time-smoothed equations of change in Eqs. 4.2-10, 4.2-11, and 4.2-12 to obtain the time-smoothed velocity and pressure distributions. As pointed out in the preceding section, some information about the velocity distribution can be obtained without having a specific expression for the turbulent momentum flux $\bar{\boldsymbol{\tau}}^{(t)}$. However, it has been popular among engineers to use various empiricisms for $\bar{\boldsymbol{\tau}}^{(t)}$ that involve velocity gradients. We mention a few of these, and many more can be found in the turbulence literature.

a. The eddy viscosity of Boussinesq

By analogy with Newton's law of viscosity, Eq. 1.2-2, one may write for a turbulent shear flow[1]

$$\bar{\tau}_{yx}^{(t)} = -\mu^{(t)}\frac{d\bar{v}_x}{dy} \tag{4.4-1}$$

in which $\mu^{(t)}$ is the *turbulent viscosity* (often called the *eddy viscosity*, and given the symbol ε). As one can see from Table 4.1-1, for at least one of the flows given there, the circular jet, one might expect Eq. 4.4-1 to be useful. Usually, however, $\mu^{(t)}$ is a strong function of position and the intensity of turbulence. In fact, for some systems[2] $\mu^{(t)}$ may even be negative in

[1] J. Boussinesq, *Mém. prés. par div. savants à l'acad. sci. de Paris*, **23**, #1, 1–680 (1877), **24**, #2, 1–64 (1877). **Joseph Valentin Boussinesq** (1842–1929), university professor in Lille, France, wrote a two-volume treatise on heat, and is famous for the "Boussinesq approximation" and the idea of "eddy viscosity."

[2] J. O. Hinze, *Appl. Sci. Res.*, **22**, 163–175 (1970); V. Kruka and S. Eskinazi, *J. Fluid Mech.*, **20**, 555–579 (1964).

some regions. It must be emphasized that the viscosity μ is a property of the *fluid*, whereas the turbulent viscosity $\mu^{(t)}$ is primarily a property of the *flow*.

For two kinds of turbulent flows (i.e., flows along surfaces, and flows in jets and wakes), special expressions for $\mu^{(t)}$ are available:

(i) Wall turbulence:

$$\mu^{(t)} = \mu\left(\frac{yv_*}{14.5\nu}\right)^3 \qquad 0 < \frac{yv_*}{\nu} < 5 \tag{4.4-2}$$

This expression, derivable from Eq. 4.3-7, is valid only very near the wall. It is of considerable importance in the theory of turbulent heat and mass transfer at fluid–solid interfaces.[3]

(ii) Free turbulence:

$$\mu^{(t)} = \rho\kappa_0 b\left(\bar{v}_{z,\max} - \bar{v}_{z,\min}\right) \tag{4.4-3}$$

in which κ_0 is a dimensionless coefficient to be determined experimentally, b is the width of the mixing zone at a downstream distance z, and the quantity in parentheses represents the maximum difference in the z component of the time-smoothed velocities at that distance z. Prandtl[4] found Eq. 4.4-3 to be a useful empiricism for jets and wakes.

b. The mixing length of Prandtl

By assuming that eddies move around in a fluid very much as molecules move around in a low-density gas (not a very good analogy!), Prandtl[5] developed an expression for momentum transport in a turbulent fluid. The "mixing length" l plays roughly the same role as the mean-free path in kinetic theory (see §1.6). This kind of reasoning led Prandtl to the following relation:

$$\bar{\tau}_{yx}^{(t)} = -\rho l^2 \left|\frac{d\bar{v}_x}{dy}\right| \frac{d\bar{v}_x}{dy} \tag{4.4-4}$$

If the mixing length were a universal constant, Eq. 4.4-4 would be very attractive, but in fact l has been found to be a function of position. Prandtl proposed the following expressions for l:

(i) Wall turbulence:

$$l = \kappa_1 y \qquad (y = \text{distance from wall}) \tag{4.4-5}$$

(ii) Free turbulence:

$$l = \kappa_2 b \qquad (b = \text{width of mixing zone}) \tag{4.4-6}$$

in which κ_1 and κ_2 are constants. A result similar to Eq. 4.4-4 was obtained by Taylor[6] by his "vorticity transport theory" some years prior to Prandtl's proposal.

c. The modified van Driest equation

There have been numerous attempts to devise empirical expressions that can describe the turbulent shear stress all the way from the wall to the main turbulent stream. Here we give a modification of the equation of van Driest.[7] This is a formula for the mixing length

---

[3] C. S. Lin, R. W. Moulton, and G. L. Putnam, *Ind. Eng. Chem.*, **45**, 636–640 (1953).

[4] L. Prandtl, *Zeits. f. angew. Math. u. Mech.*, **22**, 241–243 (1942). **Ludwig Prondtl** (pronounced "Prahnt'l"), who taught in Hannover and Göttingen and later served as the director of the Kaiser Wilhelm Institute for Fluid Dynamics, was one of the people who shaped the future of his field at the beginning of the twentieth century; he made contributions to turbulent flow, heat transfer, and boundary-layer theory.

[5] L. Prandtl, *Zeits. f. angew. Math. u. Mech.*, **5** , 136–139 (1925).

[6] G. I. Taylor, *Phil. Trans.* **A215**, 1–26 (1915), *Proc. Roy. Soc. (London)*, **A135**, 685–701 (1932).

[7] E. R. van Driest, *J. Aero. Sci.*, **23**, 1007–1011 and 1036 (1956). Van Driest's original equation did not have the square root in the denominator. This modification was made by O. T. Hanna, O. C. Sandall, and P. R. Mazet, *AIChE Journal*, **27**, 693–697 (1981) so that the turbulent viscosity would be proportional to y^3 as $y \to 0$.

of Eq. 4.4-4:

$$l = 0.4y\frac{1 - \exp\left(-yv_*/26\nu\right)}{\sqrt{1 - \exp\left(-0.26yv_*/\nu\right)}} \tag{4.4-7}$$

This relation has been found to be useful for predicting heat and mass transfer rates in flow in channels.

In the next two sections and in several problems at the end of the chapter, the use of the above empiricisms is illustrated. It should be kept in mind that these expressions for the Reynolds stresses are little more than crutches that can be used for the representation of experimental data, or for solving problems that fall into rather special classes.

§4.5 TURBULENT FLOW IN DUCTS

We start this section with a short discussion of experimental measurements for turbulent flow in rectangular ducts, in order to give some impressions about the Reynolds stresses. In Figs. 4.5-1 and 4.5-2 are shown some experimental measurements of the time-smoothed quantities $\overline{v_z'^2}$, $\overline{v_x'^2}$, and $\overline{v_x'v_z'}$ for the flow in the z direction in a rectangular duct.

In Fig. 4.5-1, note that quite close to the wall, $\sqrt{\overline{v_z'^2}}$ is about 13% of the time-smoothed centerline velocity $\bar{v}_{z,\max}$, whereas $\sqrt{\overline{v_x'^2}}$ is about 5%. This means that, near the wall, the velocity fluctuations in the flow direction are appreciably greater than those in the transverse direction. Near the center of the duct, the two fluctuation amplitudes are nearly equal and we say that the turbulence is nearly *isotropic* there.

In Fig. 4.5-2, the turbulent shear stress $\bar{\tau}_{xz}^{(t)} = \rho\overline{v_x'v_z'}$ is compared with the total shear stress $\bar{\tau}_{xz} = \bar{\tau}_{xz}^{(v)} + \bar{\tau}_{xz}^{(t)}$ across the duct. It is evident that the turbulent contribution is the more important over most of the cross section, and that the viscous contribution is important only in the vicinity of the wall. This is further illustrated in Example 4.5-3. Analogous behavior is observed in tubes of circular cross section.

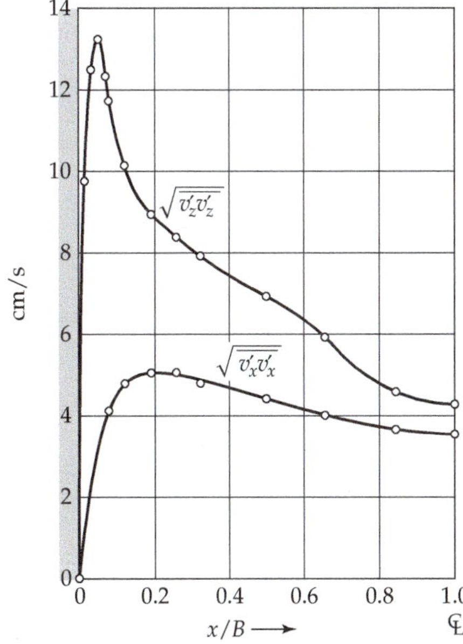

Fig. 4.5-1. Measurements of H. Reichardt [*Naturwissenschaften*, 404 (1938), *Zeits. f. angew. Math. u. Mech.*, **13**, 177–180 (1933), **18**, 358–361 (1938)] for the turbulent flow of air in a rectangular duct with $\bar{v}_{z,\max} = 100$ cm/s. Here the quantities $\sqrt{\overline{v_x'v_x'}}$ and $\sqrt{\overline{v_z'v_z'}}$ are shown.

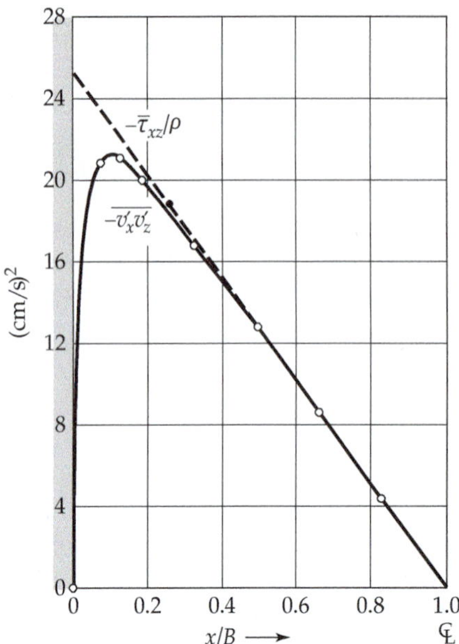

Fig. 4.5-2. Measurements of Reichardt (see Fig. 4.5-1) for the quantity $\overline{v'_x v'_z}$ in a rectangular duct. Note that this quantity differs from $\bar{\tau}_{xz}/\rho$ only near the duct wall.

EXAMPLE 4.5-1

Estimation of Average Velocity in Circular Tube

Apply the results of §4.3 to obtain the average velocity for turbulent flow in a circular tube.

SOLUTION

We can use the time-smoothed velocity distribution in the caption to Fig. 4.5-3. To get the average velocity in the tube, one should integrate over four regions: the viscous sublayer $(y^+ < 5)$, the buffer zone $(5 < y^+ < 30)$, the inertial sublayer, and the main turbulent stream, which is roughly parabolic in shape. One can certainly do this, but it has been found that integrating the logarithmic profile of Eq. 4.3-11 over the entire cross section gives results that are roughly of the right form, thus

$$\frac{\langle \bar{v}_z \rangle}{v_*} = 2.5 \ln \left(\frac{R v_*}{\nu} \right) + 1.75 \tag{4.5-1}$$

If this is compared with experimental data on flow rate vs. pressure drop, it is found that good agreement can be obtained by changing 2.5 to 2.45 and 1.75 to 2.0. This "fudging" of the constants would probably not be necessary if the integration over the cross section had been done by using the local expression for the velocity in the various layers. On the other hand, there is some virtue in having a simple logarithmic relation such as Eq. 4.5-1 to describe pressure drop vs. flow rate.

EXAMPLE 4.5-2

Application of Prandtl's Mixing-Length Formula to Turbulent Flow in Circular Tubes

Show how Eqs. 4.4-4 and 4.4-5 can be used to describe turbulent flow in a circular tube.

SOLUTION

From Eq. 4.2-12 we get for the equation of motion for flow in a circular tube:

$$0 = \frac{\mathcal{P}_0 - \mathcal{P}_L}{L} - \frac{1}{r}\frac{d}{dr}(r\bar{\tau}_{rz}) \tag{4.5-2}$$

in which $\bar{\tau}_{rz} = \bar{\tau}_{rz}^{(v)} + \bar{\tau}_{rz}^{(t)}$. Over most of the tube the viscous contribution is quite small; here we neglect it entirely. Integration of Eq. 4.5-2 then gives

$$\bar{\tau}_{rz}^{(t)} = \frac{(\mathcal{P}_0 - \mathcal{P}_L) r}{2L} = \tau_0 \left(1 - \frac{y}{R}\right) \tag{4.5-3}$$

where τ_0 is the wall shear stress, and $y = R - r$ is the distance from the tube wall.

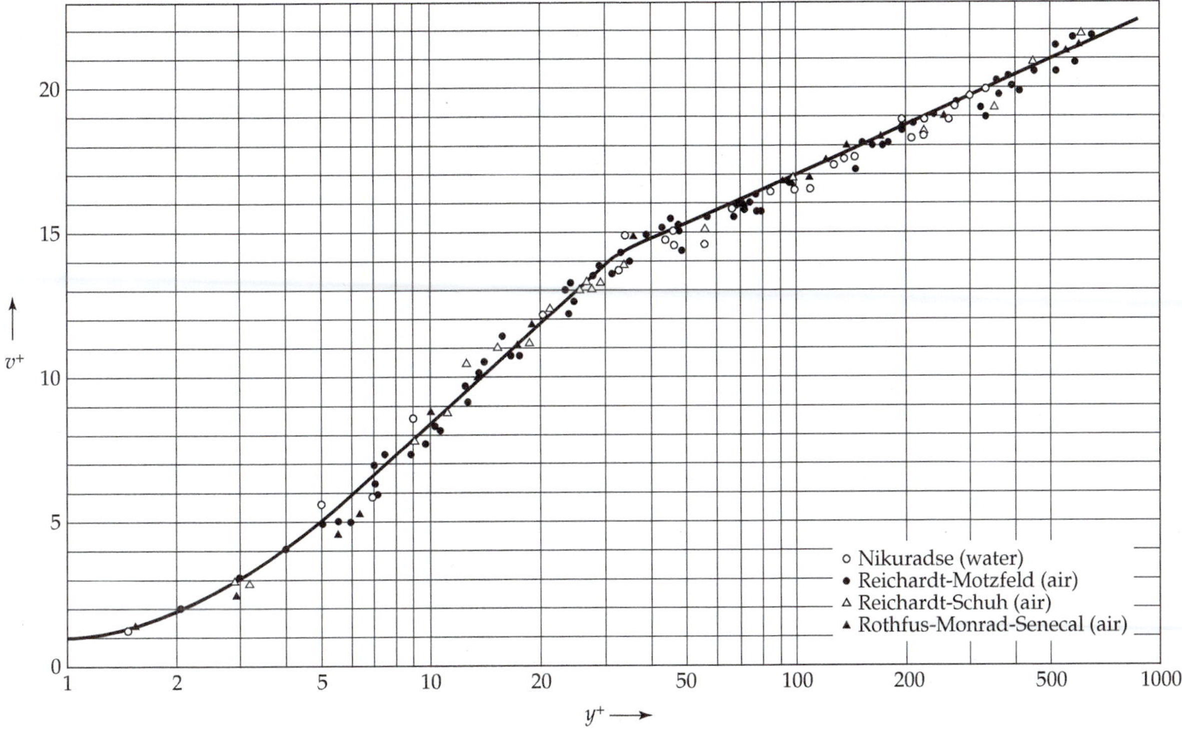

Fig. 4.5-3. Dimensionless velocity distribution for turbulent flow in circular tubes, presented as $v^+ = \bar{v}_z/v_*$ vs. $y^+ = yv_*\rho/\mu$, where $v_* = \sqrt{\tau_0/\rho}$ and τ_0 is the wall shear stress. The solid curves are those suggested by Lin, Moulton, and Putnam [*Ind. Eng. Chem.*, **45**, 636–640 (1953)]:

$$\begin{aligned} 0 < y^+ < 5: &\quad v^+ = y^+\left[1 - \tfrac{1}{4}(y^+/14.5)^3\right] \\ 5 < y^+ < 30: &\quad v^+ = 5\ln\left(y^+ + 0.205\right) - 3.27 \\ 30 < y^+: &\quad v^+ = 2.5\ln y^+ + 5.5 \end{aligned}$$

The experimental data are those of: J. Nikuradse for water (o) [*VDI Forschungsheft*, **H356** (1932)]; Reichardt and Motzfeld for air (•); Reichardt and Schuh (Δ) for air [H. Reichardt, NACA, Tech. Mem. 1047 (1943)]; and R. R. Rothfus, C. C. Monrad, and V. E. Senecal for air (▲) [*Ind. Eng. Chem.*, **42**, 2511–2520 (1950)].

According to the mixing-length theory in Eq. 4.4-4, with the empirical expression in Eq. 4.4-5, we have for $d\bar{v}_z/dr < 0$

$$\bar{\tau}_{rz}^{(t)} = -\rho l^2 \left|\frac{d\bar{v}_z}{dr}\right| \frac{d\bar{v}_z}{dr} = +\rho(\kappa_1 y)^2 \left(\frac{d\bar{v}_z}{dy}\right)^2 \tag{4.5-4}$$

Substitution of this into Eq. 4.5-3 gives a differential equation for the time-smoothed velocity. If we follow Prandtl and extrapolate the inertial sublayer to the wall, then in Eq. 4.5-4 it is appropriate to replace $\bar{\tau}_{rz}^{(t)}$ by τ_0. When this is done, Eq. 4.5-4 can be integrated to give

$$\bar{v}_z = \frac{v_*}{\kappa_1}\ln y + \text{constant} \tag{4.5-5}$$

Thus, a logarithmic profile is obtained, and hence, the results from Example 4.5-1 can be used; that is, one can apply Eq. 4.5-5 as a very rough approximation over the entire tube cross section.

EXAMPLE 4.5-3

Relative Magnitude of Viscosity and Eddy Viscosity

Determine the ratio $\mu^{(t)}/\mu$ at $y = R/2$ for water flowing at a steady rate in a long, smooth, round tube under the following conditions:

$$R = \text{tube radius} = 3 \text{ in.} = 7.62 \text{ cm}$$

$$\tau_0 = \text{wall shear stress} = 2.36 \times 10^{-5} \text{ lb}_f/\text{in.}^2 = 0.163 \text{ Pa}$$

$$\rho = \text{density} = 62.4 \text{ lb}_m/\text{ft}^3 = 1000 \text{ kg/m}^3$$

$$\nu = \text{kinematic viscosity} = 1.1 \times 10^{-5} \text{ ft}^2/\text{s} = 1.02 \times 10^{-6} \text{ m}^2/\text{s}$$

SOLUTION

The expression for the time-smoothed momentum flux is

$$\bar{\tau}_{rz}^{(t)} = -\mu \frac{d\bar{v}_z}{dr} - \mu^{(t)} \frac{d\bar{v}_z}{dr} \tag{4.5-6}$$

This result may be solved for $\mu^{(t)}/\mu$ and the result can be expressed in terms of dimensionless variables:

$$\begin{aligned}\frac{\mu^{(t)}}{\mu} &= \frac{1}{\mu} \frac{\bar{\tau}_{rz}}{d\bar{v}_z/dy} - 1 = \frac{1}{\mu} \frac{\tau_0 \left[1 - (y/R)\right]}{d\bar{v}_z/dy} - 1 \\ &= \frac{\left[1 - (y/R)\right]}{dv^+/dy^+} - 1\end{aligned} \tag{4.5-7}$$

where $y^+ = yv_*\rho/\mu$ and $v^+ = \bar{v}_z/v_*$. When $y = R/2$, the value of y^+ is

$$y^+ = \frac{yv_*\rho}{\mu} = \frac{(R/2)\sqrt{\tau_0/\rho\rho}}{\mu} = 477 \tag{4.5-8}$$

For this value of y^+, the logarithmic distribution in the caption of Fig. 4.5-3 gives:

$$\frac{d\bar{v}_z^+}{dy^+} = \frac{2.5}{477} = 0.00524 \tag{4.5-9}$$

Substituting this into Eq. 4.5-7 gives

$$\frac{\mu^{(t)}}{\mu} = \frac{1/2}{0.00524} - 1 = 94 \tag{4.5-10}$$

This result emphasizes that, far from the tube wall, molecular momentum transport is negligible in comparison with eddy transport.

§4.6 TURBULENT FLOW IN JETS

In the preceding section we discussed the flow in ducts, such as circular tubes; such flows are examples of *wall turbulence*. Another main class of turbulent flows is *free turbulence*, and the main examples of these flows are jets and wakes. The time-smoothed velocity in these types of flows can be described adequately by using Prandtl's expression for the eddy viscosity in Eq. 4.4-3, or by using Prandtl's mixing-length theory with the empiricism given in Eq. 4.4-6. The former method is simpler, and hence, we use it in the following illustrative example.

EXAMPLE 4.6-1

Time-Smoothed Velocity Distribution in a Circular Wall Jet[1–4]

A jet of fluid emerges from a circular hole into a semi-infinite reservoir of the same fluid as depicted in Fig. 4.6-1. In the same figure we show roughly what we expect the profiles of the z component of the velocity to look like. We would expect that for various values of z, the profiles will be similar in shape, differing only by a scale factor for distance and velocity. We also can imagine that as the jet moves outward, it will create a net radial inflow so that some of the surrounding fluid will be dragged along. We want to find the time-smoothed velocity distribution in the jet and also the amount of fluid crossing each plane of constant z. Before working through the solution, it may be useful to review the information on jets in Table 4.1-1.

SOLUTION

In order to use Eq. 4.4-3, it is necessary to know how b and $\bar{v}_{z,\max} - \bar{v}_{z,\min}$ vary with z for the circular jet. We know that the total rate of flow of z momentum J will be the same for all values of z. We presume that the convective momentum flux is much greater than the viscous momentum flux. This permits us to postulate that the jet width b depends on J, on the density ρ and the kinematic viscosity ν of the fluid, and on the downstream distance z from the wall. The only combination of these variables that has the dimensions of length is $b \propto Jz/\rho\nu^2$, so that the jet width is proportional to z.

We next postulate that the velocity profiles are "similar," that is

$$\frac{\bar{v}_z}{\bar{v}_{z,\max}} = f(\xi) \quad \text{where} \quad \xi = \frac{r}{b(z)} \tag{4.6-1}$$

which seems like a plausible proposal; here $\bar{v}_{z,\max}$ is the velocity along the centerline. When Eq. 4.6-1 is substituted into the expression for the rate of momentum flow in the jet (neglecting the contribution from $\bar{\tau}_{zz}$)

$$J = \int_0^{2\pi}\int_0^{\infty} \rho\bar{v}_z^2 r\,dr\,d\theta \tag{4.6-2}$$

it is found that

$$J = 2\pi\rho b^2\bar{v}_{z,\max}^2 \int_0^{\infty} f^2\xi\,d\xi = \text{constant} \times \rho b^2\bar{v}_{z,\max}^2 \tag{4.6-3}$$

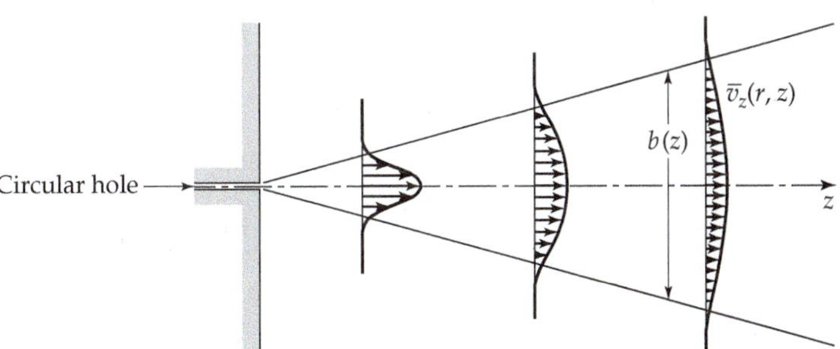

Fig. 4.6-1. Circular jet emerging from a plane wall. [H. Schlichting, *Boundary-Layer Theory*, McGraw-Hill, New York, 7th edition (1979), Fig. 24.1(b).]

---

[1] H. Schlichting, *Boundary-Layer Theory*, McGraw-Hill, New York, 7th edition (1979), pp. 747–750.

[2] A. A. Townsend, *The Structure of Turbulent Shear Flow*, Cambridge University Press, 2nd edition (1976), Chapter 6.

[3] J. O. Hinze, *Turbulence*, McGraw-Hill, New York, 2nd edition (1975), Chapter 6.

[4] S. Goldstein, *Modern Developments in Fluid Dynamics*, Oxford University Press (1938), Dover Reprint (1965), pp. 592–597.

Since J does not depend on z, and since b is proportional to z, then $\bar{v}_{z,\max}$ has to be inversely proportional to z.

The $\bar{v}_{z,\min}$ in Eq. 4.4-3 occurs at the outer edge of the jet and is zero. Therefore, because $b \propto z$ and $\bar{v}_{z,\max} \propto z^{-1}$, we find from Eq. 4.4-3 that $\mu^{(t)}$ is a constant. Therefore, we can use the equations of motion for laminar flow and just replace the viscosity μ by the eddy viscosity $\mu^{(t)}$, or ν by $\nu^{(t)}$.

In the jet the main motion is in the z direction, i.e., $|\bar{v}_r| \ll |\bar{v}_z|$. Hence, we can use a boundary-layer approximation[5] for the time-smoothed equations of change and write

$$\textit{continuity:} \qquad \frac{1}{r}\frac{\partial}{\partial r}(r\bar{v}_r) + \frac{\partial \bar{v}_z}{\partial z} = 0 \tag{4.6-4}$$

$$\textit{motion:} \qquad \bar{v}_r\frac{\partial \bar{v}_z}{\partial r} + \bar{v}_z\frac{\partial \bar{v}_z}{\partial z} = \nu^{(t)}\frac{1}{r}\frac{\partial}{\partial r}\left(r\frac{\partial \bar{v}_z}{\partial r}\right) \tag{4.6-5}$$

These equations are to be solved with the following boundary conditions:

$$\text{B.C.1:} \qquad \text{at } r = 0 \qquad \bar{v}_r = 0 \tag{4.6-6}$$

$$\text{B.C.2:} \qquad \text{at } r = 0 \qquad \partial\bar{v}_z/\partial r = 0 \tag{4.6-7}$$

$$\text{B.C.3:} \qquad \text{as } z \to \infty; \qquad \bar{v}_z \to 0 \tag{4.6-8}$$

The last boundary condition is automatically satisfied, inasmuch as we have already found that $\bar{v}_{z,\max}$ is inversely proportional to z. We now seek a solution to Eq. 4.6-5 of the form of Eq. 4.6-1 with $b = z$. The equations above, along with their boundary conditions, may be solved to give[6]

$$\bar{v}_z = \frac{\nu^{(t)}}{z}\frac{2C_3^2}{\left[1 + \frac{1}{4}(C_3 r/z)^2\right]^2} \tag{4.6-9}$$

$$\bar{v}_r = \frac{C_3\nu^{(t)}}{z}\frac{(C_3 r/z) - \frac{1}{4}(C_3 r/z)^3}{\left[1 + \frac{1}{4}(C_3 r/z)^2\right]^2} \tag{4.6-10}$$

A measurable quantity in jet flow is the radial position corresponding to an axial velocity one-half the centerline value; we call this half-width $b_{1/2}$. From Eq. 4.6-9 we then obtain:

$$\frac{\bar{v}_z(b_{1/2}, z)}{\bar{v}_{z,\max}(z)} = \frac{1}{2} = \frac{1}{\left[1 + \frac{1}{4}(C_3 b_{1/2}/z)^2\right]^2} \tag{4.6-11}$$

Experiments indicate[7] that $b_{1/2} = 0.0848z$. When this is inserted into Eq. 4.6-11, it is found that $C_3 = 15.1$.

Figure 4.6-2 gives a comparison of the above axial velocity profile with experimental data. The calculated curve obtained from the Prandtl mixing-length theory is also shown.[8] Both methods appear to give reasonably good curve fits of the experimental profiles. The eddy viscosity method seems to be somewhat better in the neighborhood of the maximum, whereas the mixing-length results are better at the outer edge of the jet.

Once the velocity profiles are known, the streamlines can be obtained. From the streamlines, shown in Fig. 4.6-3, it can be seen how the jet draws in fluid from the surrounding mass of fluid. Hence, the mass of fluid carried by the jet increases with the distance from the source. This mass rate of flow is

$$w = \int_0^{2\pi}\int_0^{\infty} \rho\bar{v}_z r\,dr\,d\theta = 8\pi\rho\nu^{(t)}z \tag{4.6-12}$$

This result corresponds to an entry in Table 4.1-1.

---

[5]See R. B. Bird, W. E. Stewart, and E. N. Lightfoot, *Transport Phenomena*, Revised Second Edition, Wiley, New York (2007), §4.4.

[6]See R. B. Bird, W. E. Stewart, and E. N. Lightfoot, *Transport Phenomena*, Revised Second Edition, Wiley, New York (2007), §5.6.

[7]H. Reichardt, *VDI Forschungsheft*, **414** (1942).

[8]W. Tollmien, *Zeits. f. angew. Math. u. Mech.*, **6**, 468–478 (1926).

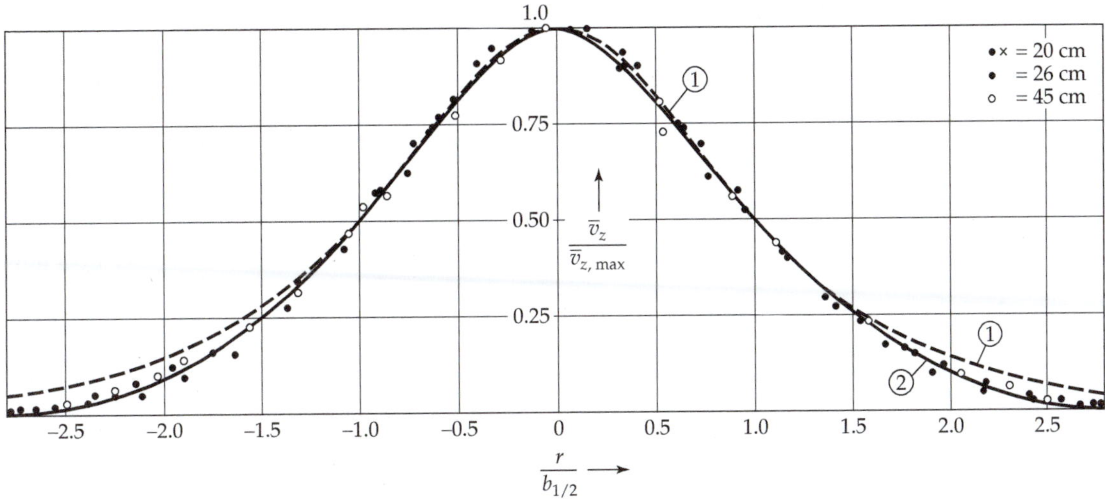

Fig. 4.6-2. Velocity distribution in a circular jet in turbulent flow [H. Schlichting, *Boundary Layer Theory*, McGraw-Hill, New York, 7th edition (1979), Fig. 24.9]. The eddy viscosity calculation (curve 1) and the Prandtl mixing-length calculation (curve 2) are compared with the measurements of H. Reichardt [*VDI Forschungsheft*, 414 (1942), 2nd edition (1951)]. Further measurements by others are cited by S. Corrsin ["Turbulence: Experimental Methods," in *Handbuch der Physik*, Vol. VIII/2, Springer, Berlin (1963)].

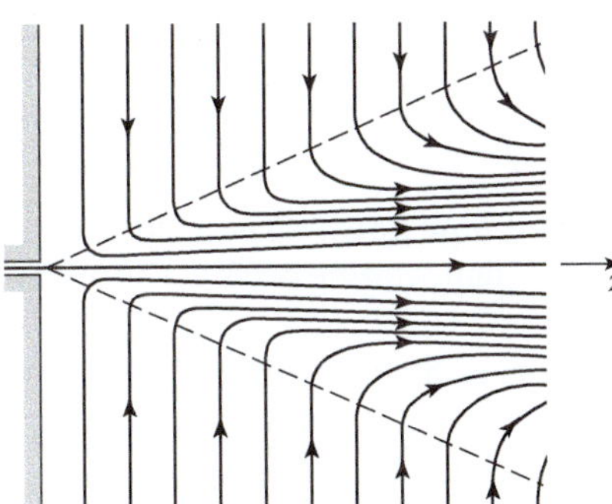

Fig. 4.6-3. Streamline pattern in a circular jet in turbulent flow. [H. Schlichting, *Boundary Layer Theory*, McGraw-Hill, New York, 7th edition (1979), Fig. 24.10.]

The two-dimensional jet issuing from a thin slot may be analyzed similarly. In that problem, however, the turbulent viscosity is a function of position.

§4.7 CONCLUDING COMMENTS

This chapter began by comparing and contrasting laminar flows and turbulent flows in several simple systems. It has also introduced some important concepts and definitions that will be needed in later chapters: time smoothing, Reynolds decomposition, correlation, eddy motion, turbulent mixing, turbulent momentum fluxes (or Reynolds stresses), and the regions in the neighborhood of the wall–viscous sublayer, buffer layer, and inertial sublayer.

In addition, we have given several important semi-empirical results, such as the logarithmic velocity profile, the eddy viscosity, and Prandtl's mixing-length hypothesis.

Then we have shown how these ideas can be used for describing two important turbulent flows, namely, the flow in a circular tube (where wall turbulence plays an important role), and the flow in a jet (where free turbulence is the key mechanism).

Much of this material will be referred to later in the chapters on turbulent heat transfer (Chapter 12) and turbulent mass transfer (Chapter 20).

Keep in mind that this chapter has merely scratched the surface of a very large and rapidly expanding subject.

QUESTIONS FOR DISCUSSION

1. Compare and contrast the procedures for solving laminar flow problems and turbulent flow problems.
2. Why must Eq. 4.1-4 *not* be used for evaluating the velocity gradient at the solid boundary?
3. What does the logarithmic profile of Eq. 4.3-11 give for the fluid velocity at the wall? Why doesn't this create a problem in Example 4.5-1 when the logarithmic profile is integrated over the tube cross section?
4. Discuss the physical interpretation of each term in Eq. 4.2-12.
5. Why is the absolute value sign used in Eq. 4.4-4? How is it eliminated in Eq. 4.5-5?
6. In Example 4.6-1, how do we know that the momentum flow through any plane of constant z is a constant? Can you imagine a modification of the jet problem in which that would not be the case?
7. In Eq. 4.3-8, why do we investigate the functional dependence of the velocity gradient rather than the velocity itself?
8. Why is turbulence such a difficult topic?

PROBLEMS

4A.1 Pressure drop needed for laminar-turbulent transition. A fluid with viscosity 18.3 cp and density 1.32 g/cm^3 is flowing in a long horizontal tube of radius 1.05 in. (2.67 cm). For what pressure gradient will the flow become turbulent?

Answer: 26 psi/mile $(1.12 \times 10^5 \text{ Pa/km})$

4A.2 Velocity distribution in turbulent pipe flow. Water is flowing through a long, straight, level run of smooth 6.00 in. i.d. pipe, at a temperature of 68°F. The pressure gradient along the length of the pipe is 1.0 psi/mile.

(a) Determine the wall shear stress τ_0 in psi (lb$_f$/in.^{2}) and Pa.

(b) Assume the flow to be turbulent and determine the radial distances from the pipe wall at which $\bar{v}_z/\bar{v}_{z,\max} = 0.0, 0.1, 0.2, 0.4, 0.7, 0.85, 1.0$.

(c) Plot the complete velocity profile, $\bar{v}_z/\bar{v}_{z,\max}$ vs. $y = R - r$.

(d) Is the assumption of turbulent flow justified?

(e) What is the mass flow rate?

4B.1 Average flow velocity in turbulent tube flow.

(a) For the turbulent flow in smooth circular tubes, the function[1]

$$\frac{\bar{v}_z(r)}{\bar{v}_{z,\max}} = \left(1 - \frac{r}{R}\right)^{1/n} \tag{4B.1-1}$$

is sometimes useful for curve-fitting purposes: near Re $= 4 \times 10^3$, $n = 6$; near Re $= 1.1 \times 10^5$, $n = 7$; and near 3.2×10^6, $n = 10$. Show that the ratio of average to maximum velocity is

$$\frac{\langle \bar{v}_z \rangle}{\bar{v}_{z,\max}} = \frac{2n^2}{(n+1)(2n+1)} \tag{4B.1-2}$$

and verify the result in Eq. 4.1-4.

[1] H. Schlichting, *Boundary-Layer Theory*, McGraw-Hill, New York, 7th edition (1979), pp. 596–600.

(b) Sketch the logarithmic profile in Eq. 4.3-11 as a function of r when applied to a circular tube of radius R. Then show how this function may be integrated over the tube cross section to get Eq. 4.5-1. List all the assumptions that have been made to get this result.

4B.2 Mass flow rate in a turbulent circular jet.

(a) Verify that the velocity distributions in Eqs. 4.6-9 and 4.6-10 do indeed satisfy the differential equations and boundary conditions.

(b) Verify that Eq. 4.6-12 follows from Eq. 4.6-9.

4B.3 The eddy viscosity expression in the viscous sublayer. Verify that Eq. 4.4-2 for the eddy viscosity comes directly from the Taylor-series expression in Eq. 4.3-7.

4C.1 Axial turbulent flow in an annulus. An annulus is bounded by cylindrical walls at $r = aR$ and $r = R$ (where $a < 1$). Obtain expressions for the turbulent velocity profiles and the mass flow rate. Apply the logarithmic profile of Eq. 4.3-10 for the flow in the neighborhood of each wall. Assume that the location of the maximum in the velocity occurs on the same cylindrical surface $r = bR$ found for laminar annular flow:

$$b = \sqrt{\frac{1 - a^2}{2 \ln (1/a)}} \tag{4C.1-1}$$

Measured velocity profiles suggest that this assumption for b is reasonable, at least for high Reynolds numbers.[2] Assume further that κ in Eq. 4.3-10 is the same at the inner and outer walls.

(a) Show that direct application of Eq. 4.3-10 leads immediately to the following velocity profiles[3] in the region $r < bR$ (designated by $<$) and $r > bR$ (designated by $>$):

$$\frac{\bar{v}_z^<(r)}{v_*^<} = \frac{1}{\kappa} \ln \left(\frac{(r - aR)\, v_*^<}{\nu} \right) + \lambda^< \quad \text{where} \quad v_*^< = v_{**}\sqrt{\frac{b^2 - a^2}{a}} \tag{4C.1-2}$$

$$\frac{\bar{v}_z^>(r)}{v_*^>} = \frac{1}{\kappa} \ln \left(\frac{(r - aR)\, v_*^>}{\nu} \right) + \lambda^> \quad \text{where} \quad v_*^> = v_{**}\sqrt{1 - b^2} \tag{4C.1-3}$$

in which $v_{**} = \sqrt{(\mathscr{P}_0 - \mathscr{P}_L)\, R/2L\rho}$.

(b) Obtain a relation between the constants $\lambda^<$ and $\lambda^>$ by requiring that the velocity be continuous at $r = bR$.

(c) Use the results of (b) to show that the mass flow rate through the annulus is

$$w = \pi R^2 \rho v_{**} \left\{ \sqrt{1 - b^2}\,(1 - a^2) \left[\frac{1}{\kappa} \ln \frac{R(1 - b)\sqrt{1 - b^2}\, v_{**}}{\nu} + \lambda^> \right] - B \right\} \tag{4C.1-4}$$

in which B is:

$$B = \frac{(b^2 - a^2)^{3/2}}{\kappa \sqrt{a}} \left(\frac{a}{a + b} + \frac{1}{2} \right) + \frac{(1 - b^2)^{3/2}}{\kappa} \left(\frac{1}{1 + b} + \frac{1}{2} \right) \tag{4C.1-5}$$

---

[2] J. G. Knudsen and D. L. Katz, *Fluid Dynamics and Heat Transfer*, McGraw-Hill, New York (1958); R. R. Rothfus (1948), J. E. Walker (1957), and G. A. Whan (1956), Doctoral Theses, Carnegie Institute of Technology (now Carnegie-Mellon University), Pittsburgh PA.

[3] W. Tiedt, *Berechnung des laminaren u. turbulenten Reibungswiderstandes konzentrischer u. exzentrischer Ringspalten*, Technischer Bericht Nr. 4, Inst. f. Hydraulik u. Hydraulogie, Technische Hochschule, Darmstadt (1968); D. M. Meter and R. B. Bird, *AIChE Journal*, **7**, 41–45 (1961) did the same analysis using the Prandtl mixing-length theory.

Chapter 5

Dimensional Analysis for Isothermal Systems

Many flow problems are far too complicated to be solved analytically. Common examples of such situations include determining time-dependent velocity profiles in turbulent flow, or solving for laminar flow in a geometry in which the bounding surfaces cannot be simply described in a convenient coordinate system. Consequently we frequently resort to various kinds of dimensional analyses in order to get useful information.

In this chapter we start by nondimensionalizing the equations of change. This leads to dimensionless groups of variables that will be useful for correlating experimental data. One such important group that will arise naturally is the Reynolds number, which appeared several times in Chapters 2, 3, and 4. We will also see how this dimensional analysis typically reduces the number of parameters on which solutions depend. The dimensionless equations are then used to obtain useful information about various problems, without ever solving the equations of change.

Finally we show an alternative dimensional analysis method using the "Buckingham pi theorem," which has been popular in engineering circles for many decades. To use this theorem, one does not need to know the governing equations for a process or phenomenon, such as the equations of change. One does need to have some intuition, however, and also must know all of the important parameters and their dimensions.

§5.1 DIMENSIONAL ANALYSIS OF THE EQUATIONS OF CHANGE FOR A PURE ISOTHERMAL FLUID

Suppose that we have taken experimental data on, or captured images of, the flow through some system that cannot be analyzed by solving the equations of change analytically. An example of such a system is the flow of a fluid through an orifice meter in a pipe (this consists of a disk with a centered hole in it, placed in the tube, with pressure-sensing devices upstream and downstream of the disk). Suppose now that we want to scale up (or down) the experimental system, in order to build a new one in which exactly the same flow patterns occur (but appropriately scaled up (or down)). First of all, we need to have *geometric similarity*—that is, the ratios of all dimensions of the pipe and orifice plate in the original system and in the scaled-up (or scaled-down) system must be the same. In addition, we must have *dynamic similarity*—that is, the dimensionless groups in the relevant differential equations and boundary conditions (such as the Reynolds number) must be the same. The study of dynamic similarity is best understood by writing the equations of change, along with boundary and initial conditions, in dimensionless form.[1,2]

For simplicity we restrict the discussion here to fluids of constant density and viscosity, for which the equations of change are Eqs. 3.1-5 and 3.6-2 (and using Eq. 3.5-4 for the substantial derivative),

$$(\nabla \cdot \mathbf{v}) = 0 \tag{5.1-1}$$

$$\rho\left(\frac{\partial \mathbf{v}}{\partial t} + [\mathbf{v} \cdot \nabla \mathbf{v}]\right) = -\nabla \mathscr{P} + \mu \nabla^2 \mathbf{v} \tag{5.1-2}$$

In most flow systems one can identify the following "scale factors": a characteristic length scale l_0, a characteristic velocity scale v_0, and a characteristic pressure scale ρv_0^2. These scales represent the magnitudes of the variables in the system. For example, for flow in a tube, we could choose the tube diameter D as the length scale, because all r values will be of the same magnitude as the diameter. We could choose the average velocity $\langle \overline{v}_z \rangle$ as the velocity scale, because the axial velocities at different radial positions will be of the same magnitude as the average velocity. The pressure scale ρv_0^2 is selected because the parameter ρ is already present in the governing equation, and when multiplied by v_0^2, gives a quantity with dimensions of pressure, which makes the creation of a new pressure scale unnecessary (this scale is appropriate for large Reynolds number; in Example 5.1-1, this dimensional analysis is repeated for a different choice for the pressure scale).

Next we define dimensionless variables and differential operators (indicated by a "˘" (˘) over the symbols) as follows:

$$\breve{x} \equiv \frac{x}{l_0}; \quad \breve{y} \equiv \frac{y}{l_0}; \quad \breve{z} \equiv \frac{z}{l_0}; \quad \breve{t} \equiv \frac{v_0 t}{l_0} \tag{5.1-3}$$

$$\breve{\mathbf{v}} \equiv \frac{\mathbf{v}}{v_0}; \quad \breve{\mathscr{P}} \equiv \frac{\mathscr{P} - \mathscr{P}_0}{\rho v_0^2} \tag{5.1-4}$$

$$\breve{\nabla} \equiv l_0 \nabla = \boldsymbol{\delta}_x \left(\frac{\partial}{\partial \breve{x}}\right) + \boldsymbol{\delta}_y \left(\frac{\partial}{\partial \breve{y}}\right) + \boldsymbol{\delta}_z \left(\frac{\partial}{\partial \breve{z}}\right) \tag{5.1-5}$$

$$\breve{\nabla}^2 \equiv l_0^2 \nabla^2 = \left(\frac{\partial^2}{\partial \breve{x}^2}\right) + \left(\frac{\partial^2}{\partial \breve{y}^2}\right) + \left(\frac{\partial^2}{\partial \breve{z}^2}\right) \tag{5.1-6}$$

A constant $\mathscr{P}_0$ is typically used in the dimensionless pressure for convenience. When the equations of change in Eqs. 5.1-1 and 5.1-2 are rewritten in terms of the dimensionless quantities, they become

$$(\breve{\nabla} \cdot \breve{\mathbf{v}}) = 0 \tag{5.1-7}$$

$$\frac{\partial \breve{\mathbf{v}}}{\partial \breve{t}} + [\breve{\mathbf{v}} \cdot \breve{\nabla} \breve{\mathbf{v}}] = -\breve{\nabla} \breve{\mathscr{P}} + \left[\!\left[\frac{\mu}{l_0 v_0 \rho}\right]\!\right] \breve{\nabla}^2 \breve{\mathbf{v}} \tag{5.1-8}$$

In these dimensionless equations, the four dimensional parameters l_0, v_0, ρ, and μ appear in one dimensionless group. The reciprocal of this group is named after a famous fluid

---

[1] G. Birkhoff, *Hydrodynamics*, Dover, New York (1955), Chapter IV. Our dimensional analysis procedure corresponds to Birkhoff's "complete inspectional analysis."

[2] R. W. Powell, *An Elementary Text in Hydraulics and Fluid Mechanics*, Macmillan, New York (1951), Chapter VIII, and H. Rouse and S. Ince, *History of Hydraulics*, Dover, New York (1963) have interesting historical material regarding the dimensionless groups and the persons for whom they were named.

dynamicist, Osborne Reynolds[3]

$$\text{Re} = \left[\!\left[\frac{l_0 v_0 \rho}{\mu} \right]\!\right] = \textit{Reynolds number} \tag{5.1-9}$$

The magnitude of this dimensionless group gives an indication of the relative importance of inertial and viscous forces in the fluid system. This feature is elaborated on further following Example 5.1-1 below.

Additional dimensionless groups may arise in alternate formulations, or in the initial and boundary conditions; two that appear in problems with fluid-fluid interfaces are[4,5]

$$\text{Fr} = \left[\!\left[\frac{v_0^2}{l_0 g} \right]\!\right] = \textit{Froude number} \tag{5.1-10}$$[4]

$$\text{We} = \left[\!\left[\frac{l_0 v_0^2 \rho}{\sigma} \right]\!\right] = \textit{Weber number} \tag{5.1-11}$$[5]

The first of these contains the gravitational acceleration g, and the second contains the interfacial tension σ, which may enter into a boundary condition (see Problem 5C.1). Still other groups may appear, such as ratios of lengths in the flow system (for example, the ratio of the tube diameter to the diameter of the hole in an orifice meter).

EXAMPLE 5.1-1

Dimensional Analysis with a Different Pressure Scale

For flows in which viscous forces play a dominant role, a more appropriate pressure scale is $\mu v_0/l_0$. Nondimensionalize the equations of change using this pressure scale, in addition to the length and velocity scales l_0 and v_0.

SOLUTION

The dimensionless coordinates, time, velocity, and gradient operator that are defined above can still be used here. A different dimensionless pressure is now defined

$$\breve{\mathscr{P}} \equiv \frac{\mathscr{P} - \mathscr{P}_0}{\mu v_0 / l_0} \tag{5.1-12}$$

To nondimensionalize the governing equations, we first rearrange all of the definitions to express the dimensional quantities in terms of the dimensionless quantities

$$x = l_0 \breve{x}, \quad y = l_0 \breve{y}, \quad z = l_0 \breve{z}, \quad t = \frac{l_0}{v_0} \breve{t} \tag{5.1-13}$$

$$\mathbf{v} = v_0 \breve{\mathbf{v}}, \quad \nabla = \frac{1}{l_0} \breve{\nabla}, \quad \nabla^2 = \frac{1}{l_0^2} \breve{\nabla}^2 \tag{5.1-14}$$

$$\mathscr{P} = \mathscr{P}_0 + \frac{\mu v_0}{l_0} \breve{\mathscr{P}} \tag{5.1-15}$$

These expressions are then substituted directly into the equations of change (Eqs. 5.1-1 and 5.1-2). Equation 5.1-1 is thus written

$$\left([l_0 \breve{\nabla}] \cdot [v_0 \breve{\mathbf{v}}] \right) = 0 \tag{5.1-16}$$

---

[3] See fn. 1 in §2.2.

[4] **William Froude** (1810–1879) (rhymes with "food") studied at Oxford and worked as a civil engineer concerned with railways and steamships. The Froude number is sometimes defined as the square root of the group given in Eq. (5.1-10).

[5] **Moritz Weber** (1871–1951) (pronounced "Vayber") was a professor of naval architecture in Berlin; another dimensionless group involving the surface tension is the capillary number, defined as $\text{Ca} = [\![\mu v_0 / \sigma]\!]$.

Dividing through by the constants l_0 and v_0 gives

$$(\breve{\nabla} \cdot \breve{\mathbf{v}}) = 0 \tag{5.1-17}$$

For Eq. 5.1-2, the first term becomes

$$\rho \frac{\partial \left[v_0 \breve{\mathbf{v}}\right]}{\partial \left[(l_0/v_0)\,\breve{t}\right]} = \frac{\rho v_0^2}{l_0} \left[\frac{\partial \breve{\mathbf{v}}}{\partial \breve{t}}\right] \tag{5.1-18}$$

The second term becomes

$$\left[\rho(v_0 \breve{\mathbf{v}}) \cdot \left(\frac{1}{l_0}\breve{\nabla}\right)(v_0 \breve{\mathbf{v}})\right] = \frac{\rho v_0^2}{l_0} \left[\breve{\mathbf{v}} \cdot \breve{\nabla} \breve{\mathbf{v}}\right] \tag{5.1-19}$$

The third term becomes

$$-\left[\frac{1}{l_0}\breve{\nabla}\right]\left[\mathscr{P}_0 + \frac{\mu v_0}{l_0}\breve{\mathscr{P}}\right] = -\frac{\mu v_0}{l_0^2}\breve{\nabla}\breve{\mathscr{P}} \tag{5.1-20}$$

(here we have used the fact that $\breve{\nabla}\mathscr{P}_0 = \mathbf{0}$ because $\mathscr{P}_0$ is a constant). The fourth term becomes

$$\mu\left[\frac{1}{l_0}\breve{\nabla}\right]^2 \left[v_0 \breve{\mathbf{v}}\right] = \frac{\mu v_0}{l_0^2}\left[\breve{\nabla}^2 \breve{\mathbf{v}}\right] \tag{5.1-21}$$

Substituting each term back into Eq. 5.1-2 gives

$$\frac{\rho v_0^2}{l_0}\left[\frac{\partial \breve{\mathbf{v}}}{\partial \breve{t}}\right] + \frac{\rho v_0^2}{l_0}\left[\breve{\mathbf{v}} \cdot \breve{\nabla}\breve{\mathbf{v}}\right] = -\frac{\mu v_0}{l_0^2}\left[\breve{\nabla}\breve{\mathscr{P}}\right] + \frac{\mu v_0}{l_0^2}\left[\breve{\nabla}^2 \breve{\mathbf{v}}\right] \tag{5.1-22}$$

Dividing this equation by $\mu v_0/l_0^2$ gives

$$\mathrm{Re}\left(\frac{\partial \breve{\mathbf{v}}}{\partial \breve{t}} + \left[\breve{\mathbf{v}} \cdot \breve{\nabla}\breve{\mathbf{v}}\right]\right) = -\breve{\nabla}\breve{\mathscr{P}} + \breve{\nabla}^2 \breve{\mathbf{v}} \tag{5.1-23}$$

where $\mathrm{Re} = l_0 v_0 \rho/\mu$ as before.

There are several benefits of nondimensionalizing the equations for fluid flow:

1. The dimensionless equations typically involve fewer parameters than their dimensional versions. For example, the Navier-Stokes equation (Eq. 5.1-2) contains the two parameters ρ and μ, whereas the dimensionless versions (Eqs. 5.1-8 or 5.1-23) contain only the single parameter Re. Such a reduction in the number of parameters decreases the number of experiments necessary to characterize a particular fluid flow.
2. The process of nondimensionalization reveals the relevant dimensionless groups. For example, Eq. 5.1-2 implies that both ρ and μ are important quantities in fluid flow, but the dimensionless versions (Eqs. 5.1-8 or 5.1-23) show that only their ratio as expressed in the Reynolds number dictates behavior.
3. The dimensionless equations can help to illustrate physical interpretations of dimensionless groups. For example, when $\mathrm{Re} \gg 1$, both Eqs. 5.1-8 and 5.1-23 suggest that the viscous term $\breve{\nabla}^2\breve{\mathbf{v}}$ is not as important as the inertia term $\left[\breve{\mathbf{v}} \cdot \breve{\nabla}\breve{\mathbf{v}}\right]$. This implies that the Reynolds number can be interpreted as a measure of the relative importance of inertial and viscous forces. In fact, the Reynolds number can be written as the ratio of the magnitudes of the inertial force term $[\rho\mathbf{v} \cdot \nabla\mathbf{v}]$ (whose magnitude is approximately $\rho v_0^2/l_0$) and the viscous force term $\mu\nabla^2\mathbf{v}$ (whose magnitude is approximately $\mu v_0/l_0^2$)

$$\mathrm{Re} = \frac{\text{inertial force}}{\text{viscous force}} = \frac{\rho v_0^2/l_0}{\mu v_0/l_0^2} = \frac{l_0 v_0 \rho}{\mu} \tag{5.1-24}$$

4. A solution for the dimensionless velocity for one value of Re can be the solution for a wide variety of combinations of values of l_0, v_0, ρ and μ, as long as $\text{Re} = l_0 v_0 \rho/\mu$ is the same. This is often exploited in scaling up equipment from laboratory models to full-size industrial operations. More specifically, if the laboratory model and industrial equipment have the same shapes and ratios of dimensions (geometric similarity), as well as the same dimensionless initial conditions, boundary conditions, and values of dimensionless parameters (dynamic similarity), then the dimensionless velocity profiles for the two systems are identical. Examples of scale-up analyses are illustrated in the following sections.
5. A final reason for nondimensionalizing the governing equations is that simplifications can often be made. For example, if the scales for length, velocity, and pressure are chosen appropriately, then the terms $[\breve{\mathbf{v}} \cdot \breve{\nabla}\breve{\mathbf{v}}]$, $\breve{\nabla}\breve{\mathscr{P}}$, $\breve{\nabla}^2\breve{\mathbf{v}}$, etc. will all be of order of magnitude of 1. Thus, Eq. 5.1-8 shows that if $\text{Re} \gg 1$, then the term containing $\breve{\nabla}^2\breve{\mathbf{v}}$ is smaller than all of the other terms and can therefore be omitted. Equation 5.1-8 then reduces to a dimensionless version of the Euler equation for inviscid flow (Eq. 3.6-4). This analysis is valid when the pressure does indeed scale with ρv_0^2, which we learn from experience to be true for relatively large Reynolds numbers. For low Reynolds numbers, we learn from experience that the pressure scales with $\mu v_0/l_0$. In this case, the nondimensionalization performed in Example 5.1-1 is more appropriate, and the resulting dimensionless equation of motion is Eq. 5.1-23. For $\text{Re} \ll 1$, this equation tells us that the entire left side of the equation can be omitted. This produces a dimensionless version of the Stokes equation (Eq. 3.6-3).

Choosing appropriate scales can be quite challenging[6] and requires more experience than a student typically has prior to a first course on transport phenomena. However, given scales, beginning students should be able to produce dimensionless equations as illustrated in Example 5.1-1. Problems at the end of this chapter offer more opportunities for practicing this skill.

§5.2 TRANSVERSE FLOW AROUND A CIRCULAR CYLINDER[1]

The flow of an incompressible Newtonian fluid past a long circular cylinder is to be studied experimentally. We want to know how the flow patterns and pressure distribution depend on the cylinder diameter, the approach velocity, and the fluid density and viscosity. Here we will use dimensional analysis to determine how to organize the work so that the number of needed experiments will be minimized.

For the analysis we consider an idealized flow system: a cylinder of diameter D and infinite length, submerged in an unbounded fluid of constant density and viscosity. Initially the fluid and the cylinder are both at rest. At time $t = 0$, the cylinder is abruptly made to move with velocity v_∞ in the negative x direction. The subsequent fluid motion is analyzed by using coordinates fixed in the cylinder axis as shown in Fig. 5.2-1.

The differential equations describing the flow are the equation of continuity (Eq. 5.1-1) and the equation of motion (Eq. 5.1-2). The initial condition for $t = 0$ and all z is:

I. C. $$\text{for } x^2 + y^2 > \frac{1}{4}D^2, \qquad \mathbf{v} = \boldsymbol{\delta}_x v_\infty \tag{5.2-1}$$

[6] Approaches for selecting appropriate scales are discussed in W. M. Deen, *Analysis of Transport Phenomena*, 2nd edition, Oxford University Press, New York (2012), Chapter 3.

[1] This discussion is adapted from R. P. Feynman, R. B. Leighton, and M. Sands, *The Feynman Lectures on Physics*, Vol. **II**, Addison-Wesley, Reading MA (1964), §41-4.

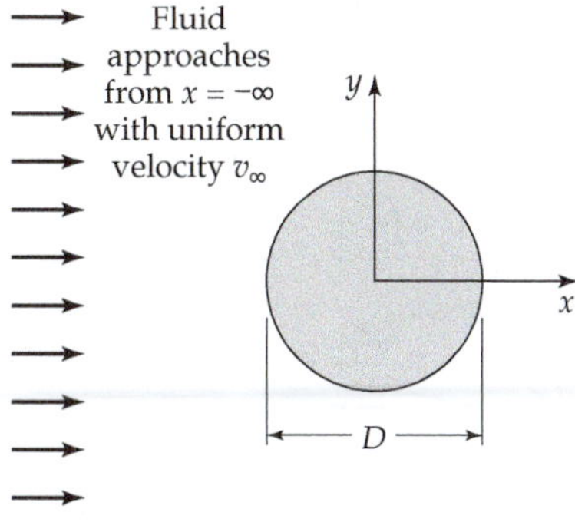

Fig. 5.2-1. Transverse flow around a cylinder.

The boundary conditions for $t \geq 0$ and all z are:

B. C. 1 $\quad$ as $x^2 + y^2 \to \infty,\quad \mathbf{v} \to \boldsymbol{\delta}_x v_\infty$ $\quad$ (5.2-2)

B. C. 2 $\quad$ at $x^2 + y^2 = \frac{1}{4}D^2,\quad \mathbf{v} = \mathbf{0}$ $\quad$ (5.2-3)

B. C. 3 $\quad$ as $x \to -\infty$ at $y = 0,\quad \mathscr{P} \to \mathscr{P}_\infty$ $\quad$ (5.2-4)

Now we rewrite the problem in terms of variables made dimensionless with the characteristic length D, velocity v_∞, and modified pressure $\mathscr{P}_\infty$. The resulting dimensionless equations of change are

$$(\breve{\nabla} \cdot \breve{\mathbf{v}}) = 0, \quad \text{and} \quad \frac{\partial \breve{\mathbf{v}}}{\partial \breve{t}} + [\breve{\mathbf{v}} \cdot \breve{\nabla}\breve{\mathbf{v}}] = -\breve{\nabla}\breve{\mathscr{P}} + \frac{1}{\text{Re}}\breve{\nabla}^2\breve{\mathbf{v}} \qquad (5.2\text{-}5, 6)$$

in which $\text{Re} = Dv_\infty\rho/\mu$. The corresponding initial and boundary conditions are, for all $\breve{z}$:

I. C. $\quad$ for $\breve{x}^2 + \breve{y}^2 > \frac{1}{4},\quad \breve{\mathbf{v}} \to \boldsymbol{\delta}_x$ $\quad$ (5.2-7)

B. C. 1 $\quad$ as $\breve{x}^2 + \breve{y}^2 \to \infty,\quad \breve{\mathbf{v}} \to \boldsymbol{\delta}_x$ $\quad$ (5.2-8)

B. C. 2 $\quad$ at $\breve{x}^2 + \breve{y}^2 = \frac{1}{4},\quad \breve{\mathbf{v}} = \mathbf{0}$ $\quad$ (5.2-9)

B. C. 3 $\quad$ as $\breve{x} \to -\infty$ at $\breve{y} = 0,\quad \breve{\mathscr{P}} \to 0$ $\quad$ (5.2-10)

If we were bright enough to be able to solve the dimensionless equations of change along with the dimensionless initial and boundary conditions, the solutions would *have* to be of the following form:

$$\breve{\mathbf{v}} = \breve{\mathbf{v}}\,(\breve{x},\breve{y},\breve{t},\text{Re}) \quad \text{and} \quad \breve{\mathscr{P}} = \breve{\mathscr{P}}\,(\breve{x},\breve{y},\breve{t},\text{Re}) \qquad (5.2\text{-}11, 12)$$

That is, the dimensionless velocity and dimensionless modified pressure can depend only on the dimensionless parameter Re and the dimensionless independent variables $\breve{x}$, $\breve{y}$, and $\breve{t}$ (the solution is independent of $\breve{z}$ for an infinite cylinder). For cylinders of finite length L, the solution will depend additionally on $\breve{z}$ and also on L/D.

This completes the dimensional analysis of the problem. We have not solved the flow problem, but have decided on a convenient set of dimensionless variables to restate the problem and suggest the form of the solution. The analysis shows that if we wish to catalog the flow patterns for flow past a long cylinder, it will suffice to record them (e.g., by capturing video images) for a series of Reynolds numbers $\text{Re} = Dv_\infty\rho/\mu$; thus, separate investigations into the roles of D, v_∞, ρ, and μ are unnecessary. Such a simplification saves

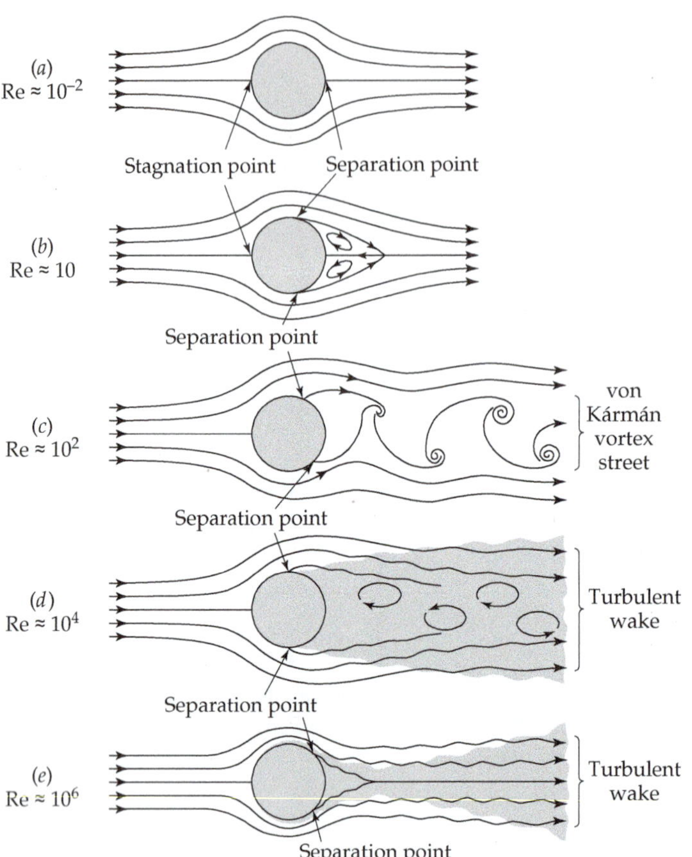

Fig. 5.2-2. The types of behavior for the flow around a cylinder, illustrating the various flow regimes that are observed as the Reynolds number increases. Regions of turbulent flow are shaded in gray. [R. P. Feynman, R. B. Leighton, and M. Sands, *The Feynman Lectures in Physics, Vol.* **II**, Addison-Wesley, Reading MA (1964), Fig. 41-6.]

a lot of time and expense. Similar comments apply to the tabulation of numerical results, in the event that one decides to make a numerical assault on the problem.[2,3]

Experiments involve some necessary departures from the above analysis: the cylinder has ends, the stream is finite in size, and small fluctuations of velocity are inevitably present at the initial state. Nevertheless, for Reynolds numbers below about 40, and locations away from the ends of the cylinder, Eqs. 5.2-11 and 5.2-12 describe the available observations quite well. For Re approaching 40, the damping of disturbances is slower, and beyond this approximate limit unsteady flow is always observed.

The observed flow patterns at large $\breve{t}$ vary strongly with the Reynolds number as shown in Fig. 5.2-2. For Re = 1, the flow is orderly, as shown in (*a*). At Re of about 10, a pair of vortices appears behind the cylinder, as may be seen in (*b*). This type of flow persists up to about Re = 40, when there appear two "separation points," at which the streamlines separate from the solid surface. Furthermore, the flow becomes permanently unsteady; vortices begin to "peel off" from the cylinder and move downstream. With further increase

---

[2]Analytical solutions of this problem at very small Re and infinite L/D are reviewed in L. Rosenhead (ed.), *Laminar Boundary Layers*, Oxford University Press (1963), Chapter IV. An important feature of this two-dimensional problem is the absence of a "creeping flow" solution. Thus, the $[\mathbf{v} \cdot \nabla \mathbf{v}]$ term in the equation of motion has to be included, even in the limit as Re → 0 (see Problem 3B.9). This is in sharp contrast to the situation for slow flow around a sphere (see §2.7) and around other finite, three-dimensional objects.

[3]For a numerical study of the flow around a long cylinder, see F. H. Harlow and J. E. From, *Scientific American*, **212**, 104–110 (1965), and J. E. From and F. H. Harlow, *Phys. Fluids*, **6**, 975–982 (1963).

in Re, the vortices separate regularly from alternate sides of the cylinder, as shown in (*c*); such a regular array of vortices is known as a "von Kármán vortex street." At still higher Re, there is a disorderly fluctuating motion (turbulence) in the wake of the cylinder, as shown in (*d*). Finally, at Re near 10^6, turbulence appears upstream of the separation point, and the wake abruptly narrows down as shown in (*e*). Clearly, the unsteady flows shown in the last three sketches would be very difficult to compute from the equations of change. It is much easier to observe them experimentally and correlate the results in terms of Eqs. 5.2-11 and 5.2-12.

Equations 5.2-11 and 5.2-12 can also be used for scale-up from a single experiment. Suppose that we wanted to predict the flow patterns around a cylinder of diameter $D_{\mathrm{I}} =$ 5 ft, around which air is to flow with an approach velocity $(v_\infty)_{\mathrm{I}} = 30$ ft/s, by means of an experiment on a scale model of diameter $D_{\mathrm{II}} = 1$ ft. To have dynamic similarity, we must choose conditions such that $\mathrm{Re}_{\mathrm{I}} = \mathrm{Re}_{\mathrm{II}}$. Then, if we use the same fluid in the small-scale experiment as in the large system, so that $\mu_{\mathrm{II}}/\rho_{\mathrm{II}} = \mu_{\mathrm{I}}/\rho_{\mathrm{I}}$, we find $(v_\infty)_{\mathrm{II}} = 150$ ft/s as the required air velocity in the small-scale model. With the Reynolds numbers thus equalized, the flow patterns in the model and the full-scale system will look alike—that is, they are geometrically and dynamically similar.

Furthermore, if Re is in the range of periodic vortex formation, the dimensionless time interval $t_v v_\infty/D$ between vortices will be the same in the two systems. Thus, the vortices will shed 25 times as fast in the model as in the full-scale system. The regularity of the vortex shedding at Reynolds numbers from about 10^2 to 10^4 has been utilized commercially for precise flow metering in large pipelines.

§5.3 STEADY FLOW IN AN AGITATED TANK

We next consider the problem of predicting the flow behavior in a large, unbaffled tank of oil, shown in Fig. 5.3-1, as a function of the impeller rotation speed. We propose to do this by means of model experiments in a smaller, geometrically similar system. We want to determine the conditions necessary for the model studies to provide a direct means of prediction.

We consider a tank of radius R, with a centered impeller of overall diameter D. At time $t = 0$, the system is stationary and contains liquid to a height H above the tank bottom. Immediately after time $t = 0$, the impeller begins rotating at a constant speed of N

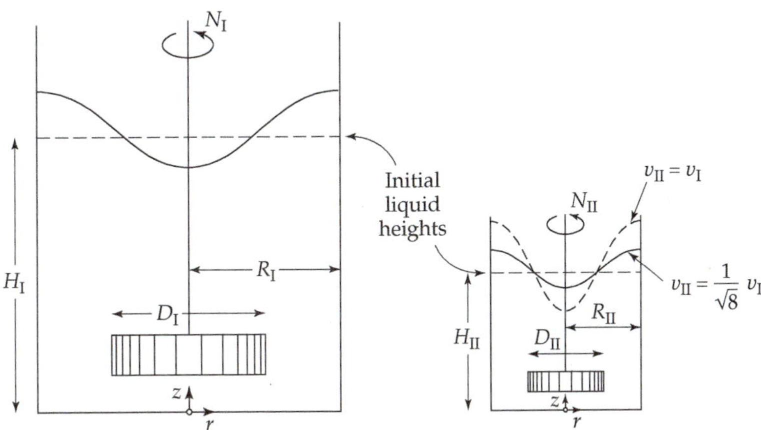

Fig. 5.3-1. Long-time average free-surface shapes, with $\mathrm{Re}_{\mathrm{I}} = \mathrm{Re}_{\mathrm{II}}$.

revolutions per minute. The drag of the atmosphere on the liquid surface is neglected. The impeller shape and initial position are described by the function $S_{imp}(r,\theta,z) = 0$.

The flow is governed by Eqs. 5.1-1 and 5.1-2, along with the initial condition

$$\text{I.C.} \qquad \text{at } t = 0, \text{ for } 0 \le r < R \text{ and } 0 < z < H, \qquad \mathbf{v} = \mathbf{0} \tag{5.3-1}$$

and the following boundary conditions for the liquid region:

$$\text{tank bottom} \qquad \text{at } z = 0 \text{ and } 0 \le r < R, \qquad \mathbf{v} = \mathbf{0} \tag{5.3-2}$$

$$\text{tank wall} \qquad \text{at } r = R, \qquad \mathbf{v} = \mathbf{0} \tag{5.3-3}$$

$$\text{impeller surface} \qquad \text{at } S_{imp}(r,\theta - 2\pi Nt,z) = 0, \qquad \mathbf{v} = 2\pi Nr\boldsymbol{\delta}_\theta \tag{5.3-4}$$

$$\text{gas-liquid interface} \qquad \text{at } S_{int}(r,\theta,z,t) = 0, \qquad (\mathbf{n} \cdot \mathbf{v}) = 0 \tag{5.3-5}$$

$$\text{and} \qquad \mathbf{n}p + [\mathbf{n} \cdot \boldsymbol{\tau}] = \mathbf{n}p_{atm} \tag{5.3-6}$$

Equations 5.3-2 and 5.3-4 are the no-slip conditions at solid-liquid interfaces; the surface $S_{imp}(r,\theta - 2\pi Nt,z) = 0$ describes the location of the impeller after Nt rotations. Equation 5.3-5 is the condition of no mass flow through the gas-liquid interface, described by $S_{int}(r,\theta,z,t) = 0$, which has a local unit normal vector $\mathbf{n}$. Equation 5.3-6 is a force balance on an element of this interface (or a statement of the continuity of the normal component of the momentum flux tensor $\boldsymbol{\pi}$, in which the viscous contributions from the gas side are neglected). This interface is initially stationary in the plane $z = H$, and its motion thereafter is best obtained by measurement, though it is also predictable in principle by numerical solution of this equation system, which describes the initial conditions and subsequent acceleration $D\mathbf{v}/Dt$ of every fluid element.

Next we nondimensionalize the equations using the characteristic quantities $v_0 = ND$, $l_0 = D$, and $\mathscr{P}_0 = \rho N^2D^2$ along with dimensionless cylindrical coordinates $\breve{r} = r/D$, θ, and $\breve{z} = z/D$. Then the equations of continuity and motion appear as in Eqs. 5.1-7 and 5.1-8, with $\text{Re} = D^2N\rho/\mu$. The initial condition takes the form

$$\text{I.C.} \qquad \text{at } \breve{t} = 0, \quad \text{for } \breve{r} = \left[\!\left[\frac{R}{D}\right]\!\right] \quad \text{and} \quad 0 < \breve{z} < \left[\!\left[\frac{H}{D}\right]\!\right], \qquad \mathbf{v} = \mathbf{0} \tag{5.3-7}$$

and the boundary conditions become:

$$\text{tank bottom} \qquad \text{at } \breve{z} = 0 \quad \text{and} \quad 0 < \breve{r} < \left[\!\left[\frac{R}{D}\right]\!\right], \qquad \breve{\mathbf{v}} = \mathbf{0} \tag{5.3-8}$$

$$\text{tank wall} \qquad \text{at } \breve{r} = \left[\!\left[\frac{R}{D}\right]\!\right], \qquad \breve{\mathbf{v}} = \mathbf{0} \tag{5.3-9}$$

$$\text{impeller surface} \qquad \text{at } \breve{S}_{imp}\left(\breve{r},\theta - 2\pi\breve{t},\breve{z}\right) = 0, \qquad \breve{\mathbf{v}} = 2\pi\breve{r}\boldsymbol{\delta}_\theta \tag{5.3-10}$$

$$\text{gas-liquid interface} \qquad \text{at } \breve{S}_{int}\left(\breve{r}, \theta, \breve{z}, \breve{t}\right) = 0 \qquad (\mathbf{n} \cdot \breve{\mathbf{v}}) = 0 \tag{5.3-11}$$

$$\text{and } \mathbf{n}\breve{\mathscr{P}} - \mathbf{n}\left[\!\left[\frac{g}{DN^2}\right]\!\right]\breve{z} - \left[\!\left[\frac{\mu}{D^2N\rho}\right]\!\right]\left[\mathbf{n} \cdot \breve{\dot{\boldsymbol{\gamma}}}\right] = \mathbf{0} \tag{5.3-12}$$

In going from Eqs. 5.3-6 to 5.3-12 we have used Newton's law of viscosity in the form of Eq. 1.2-13 (but with the last term omitted, as is appropriate for incompressible liquids). We have also used the abbreviation $\breve{\dot{\boldsymbol{\gamma}}} = \breve{\nabla}\breve{\mathbf{v}} + \left(\breve{\nabla}\breve{\mathbf{v}}\right)^\dagger$ for the dimensionless rate-of-deformation tensor, whose dimensionless Cartesian components are $\breve{\dot{\gamma}}_{ij} = (\partial\breve{v}_j/\partial\breve{x}_i) + (\partial\breve{v}_i/\partial\breve{x}_j)$.

The quantities in Eqs. 5.3-7 through 5.3-12 in double brackets are known dimensionless quantities. The function $\breve{S}_{imp}\left(\breve{r},\theta - 2\pi\breve{t},\breve{z}\right)$ is known for a given impeller design. The unknown function $\breve{S}_{int}\left(\breve{r},\theta,\breve{z},\breve{t}\right)$ is measurable by video image analysis, or in principle computable from the problem statement.

By inspection of the dimensionless equations, we find that the velocity and pressure profiles must have the form

$$\breve{\mathbf{v}} = \breve{\mathbf{v}}\left(\breve{r},\theta,\breve{z},\breve{t};\frac{R}{D},\frac{H}{D},\text{Re},\text{Fr}\right) \tag{5.3-13}$$

$$\breve{\mathscr{P}} = \breve{\mathscr{P}}\left(\breve{r},\theta,\breve{z},\breve{t};\frac{R}{D},\frac{H}{D},\text{Re},\text{Fr}\right) \tag{5.3-14}$$

for a given impeller shape and location. The corresponding locus of the free surface is given by

$$\breve{S}_{\text{int}} = \breve{S}_{\text{int}}\left(\breve{r},\theta,\breve{z},\breve{t};\frac{R}{D},\frac{H}{D},\text{Re},\text{Fr}\right) = 0 \tag{5.3-15}$$

in which $\text{Re} = D^2N\rho/\mu$ and $\text{Fr} = DN^2/g$. For time-smoothed observations at large $\breve{t}$, the dependence on $\breve{t}$ will disappear, as will the dependence on θ for this axisymmetric tank geometry.

These results provide the necessary conditions for the proposed model experiment: the two systems must be (i) geometrically similar (same values of R/D and H/D, same impeller geometry and location), and (ii) operated at the same values of the Reynolds and Froude numbers. Condition (ii) requires that

$$\frac{D_{\text{I}}^2N_{\text{I}}}{\nu_{\text{I}}} = \frac{D_{\text{II}}^2N_{\text{II}}}{\nu_{\text{II}}} \tag{5.3-16}$$

$$\frac{D_{\text{I}}N_{\text{I}}^2}{g_{\text{I}}} = \frac{D_{\text{II}}N_{\text{II}}^2}{g_{\text{II}}} \tag{5.3-17}$$

in which the kinematic viscosity $\nu = \mu/\rho$ is used. Normally both tanks will operate in the same gravitational field $g_{\text{I}} = g_{\text{II}}$, so that Eq. 5.3-17 requires

$$\frac{N_{\text{II}}}{N_{\text{I}}} = \left(\frac{D_{\text{I}}}{D_{\text{II}}}\right)^{1/2} \tag{5.3-18}$$

Substitution of this into Eq. 5.3-16 gives the requirement

$$\frac{\nu_{\text{II}}}{\nu_{\text{I}}} = \left(\frac{D_{\text{II}}}{D_{\text{I}}}\right)^{3/2} \tag{5.3-19}$$

This is an important result, namely, that the smaller tank (II) requires a fluid of smaller kinematic viscosity to maintain dynamic similarity. For example, if we use a scale model with $D_{\text{II}} = \frac{1}{2}D_{\text{I}}$, then we need to use a fluid with kinematic viscosity $\nu_{\text{II}} = \nu_{\text{I}}/\sqrt{8}$ in the scaled-down experiment. Evidently the requirements for dynamic similarity are more stringent here than in the previous example, because of the additional dimensionless group Fr.

In many practical cases, Eq. 5.3-19 calls for unattainably low values of ν_{II}. Exact scale-up from a single-model experiment is then not possible. In some circumstances, however, the effect of one or more dimensionless groups may be known to be small, or may be predictable from experience with similar systems; in such situations, approximate scale-up from a single experiment is still feasible.[1]

This example shows the importance of including the boundary conditions in a dimensional analysis. Here the Froude number appeared only in the free-surface boundary condition in Eq. 5.3-12. Failure to use this condition would result in the omission of the restriction in Eq. 5.3-18, and one might improperly choose $\nu_{\text{II}} = \nu_{\text{I}}$. If one did this, with $\text{Re}_{\text{II}} = \text{Re}_{\text{I}}$, the Froude number in the smaller tank would be too large, and the vortex would be too deep, as shown by the dotted line in Fig. 5.3-1.

---

[1] For an introduction to methods for scale-up with incomplete dynamic similarity, see R. W. Powell, *An Elementary Text in Hydraulics and Fluid Mechanics*, Macmillan, New York (1951).

§5.4 PRESSURE DROP FOR CREEPING FLOW IN A PACKED TUBE

Here we show that the mean axial gradient of the modified pressure $\mathscr{P}$ for flow of a fluid of constant ρ and μ through a tube of radius R, uniformly packed for a length $L \gg D_p$ with solid particles of characteristic size $D_p \ll R$, is

$$\frac{\Delta \langle \mathscr{P} \rangle}{L} = \frac{\mu \langle v_z \rangle}{D_p^2} K\,(\text{geom}) \tag{5.4-1}$$

Here $\langle \cdots \rangle$ denotes an average over a tube cross section within the packed length L, and the function K(geom) is a constant for a given bed geometry (i.e., a given shape and arrangement of the particles).

We choose D_p as the characteristic length, and $\langle v_z \rangle$ as the characteristic velocity. Since the flow is slow, we use the pressure scale $\mu \langle v_z \rangle / D_p$ as in Example 5.1-1. Then the interstitial fluid motion is determined by Eqs. 5.1-17 and 5.1-23, with $\check{\mathbf{v}} = \mathbf{v}/\langle v_z \rangle$ and $\check{\mathscr{P}} = (\mathscr{P} - \mathscr{P}_0)\, D_p/\mu \langle v_z \rangle$, along with no-slip conditions on the solid surfaces and the modified pressure difference $\Delta \langle \mathscr{P} \rangle = \langle \mathscr{P}_0 \rangle - \langle \mathscr{P}_L \rangle$. The solutions for $\check{\mathbf{v}}$ and $\check{\mathscr{P}}$ in creeping flow ($\text{Re} = D_p \langle v_z \rangle \rho/\mu \to 0$) accordingly depend only on $\check{r}$, θ, and $\check{z}$ for a given particle arrangement and shape. Then the mean axial dimensionless pressure gradient

$$\frac{D_p}{L} \int_0^{L/D_p} \left(-\frac{d\check{\mathscr{P}}}{d\check{z}} \right) d\check{z} = \frac{D_p}{L} \left(\check{\mathscr{P}}_0 - \check{\mathscr{P}}_L \right) \tag{5.4-2}$$

depends only on the bed geometry as long as R and L are large relative to D_p. That is,

$$\frac{D_p}{L} \left(\check{\mathscr{P}}_0 - \check{\mathscr{P}}_L \right) = \text{function of geometry} \tag{5.4-3}$$

Inserting the definition of the dimensionless pressure difference $\Delta \left\langle \check{\mathscr{P}} \right\rangle = \left(D_p/\mu \langle \overline{v}_z \rangle \right) \cdot \Delta \langle \mathscr{P} \rangle$ into Eq. 5.4-3 and rearranging then gives Eq. 5.4-1.

§5.5 THE BUCKINGHAM PI THEOREM[1]

It is sometimes possible to obtain the dimensionless groups necessary for scale-up calculations without nondimensionalizing the governing equations. Consider a flow problem in which it is known ahead of time that there are q relevant quantities, $x_1, x_2, \ldots, x_q$. The dimensions of these q quantities can be constructed from d primary dimensions (e.g., when the primary dimensions are mass, length, and time, then $d = 3$). To obtain a dimensionless group Π_i from these quantities, we write the dimensionless group as a product of each variable raised to a power

$$\Pi_i = \left(x_1^{n_1} \right) \left(x_2^{n_2} \right) \cdots \left(x_q^{n_q} \right) \tag{5.5-1}$$

The exponents $n_1, n_2, \ldots, n_q$ are then selected so that the right side of the equation becomes dimensionless. For typical problems, this procedure can yield $q - d$ different dimensionless groups for different choices of the exponents. The dimensionless groups are not unique, so some experience with common dimensionless groups can be helpful. More information about this method, as well as discussions about exceptions to the typical type of problem, are available elsewhere.[2,3]

According to the Buckingham pi theorem, the functional relationship among q quantities, whose dimensions can be given in terms of the d primary dimensions, may be written

---

[1] E. Buckingham, *Phys. Rev.*, **4**, 345–376 (1914).
[2] *The Handbook of Fluid Dynamics*, R. W. Johnson, ed., CRC Press, Boca Raton (1998), Chapter 5.
[3] A. A. Sonin, *PNAS*, **101**, 8525–8526 (2004).

as a relation among the $q - d$ dimensionless groups (the Π's). Then one can write a relation of the form

$$f\left(\Pi_1,\Pi_2,\Pi_3,\ldots,\Pi_{q-d}\right) = 0 \tag{5.5-2}$$

Any one of these Π's may be replaced by itself raised to a power, multiplied by any other Π, raised to any power. One can also solve Eq. 5.5-2 for any one of the Π's to get a relation of the form

$$\Pi_1 = g\left(\Pi_2,\Pi_3,\ldots,\Pi_{q-d}\right) \tag{5.5-3}$$

A difficulty in using the pi theorem is that one has to know all of the relevant variables, and sometimes this can be a major difficulty. For problems involving many variables, obtaining the most useful dimensionless groups can also be a challenge.

Several examples are presented below in which the Buckingham pi theorem is employed to obtain relationships between dimensionless groups.

EXAMPLE 5.5-1

Incompressible Flow in a Circular Tube

We seek a method of correlating the modified pressure difference for tube flow in terms of relevant variables. We start by listing possible quantities of interest along with their dimensions

$$\begin{array}{lll} \mathscr{P}_0 - \mathscr{P}_L\,[=]\,M/Lt^2 & \langle v_z\rangle\,[=]\,L/t & \rho\,[=]\,M/L^3 \\ \mu\,[=]\,M/Lt & R\,[=]\,L & L\,[=]\,L \end{array} \tag{5.5-4}$$

There are $q = 6$ quantities with $d = 3$ primary dimensions, and therefore $q - d = 6 - 3 = 3$ dimensionless groups. Each dimensionless group will have the form

$$\Pi = (\mathscr{P}_0 - \mathscr{P}_L)^a \langle v_z\rangle^b \rho^c \mu^d R^e L^f \tag{5.5-5}$$

By substituting the dimensions of the 6 variables into Eq. 5.5-5, we find that the dimensions of Π are

$$\Pi = M^{a+c+d} L^{-a+b-3c-d+e+f} t^{-2a-b-d} \tag{5.5-6}$$

The exponents are determined by requiring that Π be dimensionless. This gives the system of equations

$$a + c + d = 0 \tag{5.5-7}$$

$$-a + b - 3c - d + e + f = 0 \tag{5.5-8}$$

$$-2a - b - d = 0 \tag{5.5-9}$$

This is an underspecified system of 3 equations and 6 unknowns, and thus, we can select 3 of the unknowns $a - f$ to obtain dimensionless groups (as long as the 3 selected values do not violate Eqs. 5.5-7 through 5.5-9). Because we seek a relationship between $\mathscr{P}_0 - \mathscr{P}_L$ and all of the other variables, we select $a = 1$ for the first dimensionless group, and $a = 0$ for the remaining dimensionless groups.

For the dimensionless group with $a = 1$, we also select $d = 0$. Equation 5.5-7 then gives $c = -1$. Substituting these values into Eq. 5.5-9 gives $b = -2$. Substituting all of these values into Eq. 5.5-8 gives $e + f = 0$. For simplicity, we choose $e = 0$ which gives $f = 0$. The first dimensionless group can thus be written $(P_0 - P_L)/\rho\langle v_z\rangle^2$. We choose to add a factor of $\frac{1}{2}$ in the denominator because the factor $\frac{1}{2}\rho\langle v\rangle^2$ commonly appears in fluid mechanics. The first dimensionless group is thus

$$\Pi_1 = \frac{\mathscr{P}_0 - \mathscr{P}_L}{\frac{1}{2}\rho\langle v_z\rangle^2} = \text{a dimensionless pressure difference} \tag{5.5-10}$$

To get a second dimensionless group, we enforce $a = 0$, and select $b = 1$, because we expect a dimensionless average velocity will be a useful group. Equation 5.5-9 then gives $d = -1$. Substituting these values into Eq. 5.5-7 then gives $c = 1$. Substituting all of these values into Eq. 5.5-8 then gives $e + f = 0$. We arbitrarily set $f = 0$, which gives $e = 1$. The second dimensionless group then can be written $R\langle v_z\rangle \rho/\mu$. We add a factor of 2 to the numerator so that the second dimensionless group becomes the conventional Reynolds number for tube flow,

$$\Pi_2 = \frac{2R\langle v_z\rangle \rho}{\mu} = \text{Reynolds number} \tag{5.5-11}$$

For the third dimensionless group, we choose both $a = 0$ and $b = 0$, so that the modified pressure difference and average velocity only appear in one dimensionless group each. Equation 5.5-9 then gives $d = 0$, and then Eq. 5.5-7 gives $c = 0$. Substituting all of these values into Eq. 5.5-8 gives $e + f = 0$. We arbitrarily choose $f = 1$, which gives $e = -1$. The third dimensionless group is thus

$$\Pi_3 = \frac{L}{R} = \text{a dimensionless tube length} \tag{5.5-12}$$

We note that this analysis illustrates that these dimensionless groups are not unique, and others could have been obtained.

This analysis suggests that we might plot experimental data of Π_1 (the dimensionless modified pressure difference) versus Π_2 (the Reynolds number) for various values of Π_3 (L/R). This is essentially what we do in §6.2 for turbulent flow. For laminar flow in very long tubes, Eq. 2.3-22 divided by $\langle v_z\rangle$ is the same as $1 = \Pi_1\Pi_2/\Pi_3$, aside from a numerical factor. Thus, the theoretical result in §2.3 is consistent with this analysis of the Buckingham pi theorem.

EXAMPLE 5.5-2

Correlating Viscosity Data for Low-Density Gases

Develop the form of the expression for the temperature dependence of gas viscosity in terms of the molecular parameters of the Lennard-Jones potential and the molecular mass.

SOLUTION

We start by listing the $q = 6$ quantities involved in the relation along with their corresponding dimensions, whence we see that $d = 4$ (the primary dimensions are mass M, length L, time t and temperature T)

$$\begin{array}{lll} \mu\,[=]\,M/Lt & T\,[=]\,T & m\,[=]\,M \\ \sigma\,[=]\,L & \varepsilon\,[=]\,ML^2/t^2 & \kappa\,[=]\,ML^2/t^2T \end{array} \tag{5.5-13}$$

We find that the Boltzmann constant κ has to be included in order to construct a dimensionless group for the viscosity, as well as for the parameter ε. Then the $q - d = 6 - 4 = 2$ dimensionless groups may be found to be

$$\Pi_1 = \frac{\mu\sigma^2}{\sqrt{m\kappa T}} \qquad \Pi_2 = \frac{\kappa T}{\varepsilon} \tag{5.5-14}$$

When we write $\Pi_1 = F(\Pi_2)$, we see that this is the form of the kinetic theory expression in Eq. 1.6-14. The kinetic theory gives the numerical factor as well as the exact function of the dimensionless temperature $\kappa T/\varepsilon$ (see Table D.2).

At extremely low temperatures, one expects quantum effects to be important, and therefore that Planck's constant h should play a role in the theory of transport properties. Therefore, to the list of quantities in Eq. 5.5-13, we have to add $h\,[=]\,ML^2/t$, so that $q = 7$. Since d is still equal to 4, we have now $q - d = 7 - 4 = 3$ dimensionless groups. Then in addition to Π_1 and Π_2 in Eq. 5.5-14, we have a third dimensional group, $\Pi_3 = h/\sigma\sqrt{m\varepsilon}$, often called the *de Boer parameter*.[4] This parameter has the following values for the light gases:

Gas	He^3	He^4	H_2	D_2	Ne
$h/\sigma\sqrt{m\varepsilon}$	3.08	2.67	1.729	1.223	0.593

The larger the value of the de Boer parameter, the larger the deviation from classical behavior.

[4]J. de Boer, *Reports on Progress in Physics*, **12**, 305–374 (1949). **Jan de Boer** (1911–2010) was the founding director of the Institute for Theoretical Physics of the University of Amsterdam; he was a superb classroom teacher, who knew how to present difficult concepts in a neat and well-organized way.

§5.6 CONCLUDING COMMENTS

In this chapter it was shown how dimensional analysis of the equations of change can lead to better understanding of complex flow systems, even when the governing equations are not solved. By going through the analysis of three systems in some detail, we have demonstrated the utility of the approach.

The Buckingham pi theorem, which is about a century old, was then presented. It involves a certain amount of intuition about how a system behaves. But when applied skillfully, it can be helpful in determining the dimensionless representations that can be useful in correlating data on complex systems.

In the next chapter we exploit dimensional analysis to develop friction factor correlations.

QUESTIONS FOR DISCUSSION

1. What are the advantages of writing equations in dimensionless form?
2. Compare and contrast the use of the nondimensionalized equations of change and the Buckingham pi theorem.
3. Write Newton's law of viscosity in dimensionless form, using v_0 and l_0 as the scale factors for velocity and length, respectively.
4. Write the velocity distribution for laminar flow in a circular tube in dimensionless form. Use $\langle v_z \rangle$ and R as the scale factors for velocity and length, respectively. What other choices could be made?
5. Discuss the use of the Reynolds number as a means for classifying the types of flow patterns for the transverse flow of a fluid around a cylinder.
6. What special form does the Navier-Stokes equation take in the limit as Re approaches zero? As it approaches infinity?
7. What factors would need to be taken into account in designing a mixing tank for use on the moon, by using data from a similar tank on earth?
8. Do the Reynolds number, the Froude number, and the Weber number contain both the physical properties of the fluid and information about the geometry of the flow system and the strength of the flow?
9. The Weber number contains the interfacial tension. How is this fluid property measured?

PROBLEMS

5A.1 Calculating Reynolds numbers. Calculate values for the Reynolds number, $\text{Re} = l_0 v_0 \rho/\mu$, for the following cases:

(a) $l_0 = 0.1$ m, $v_0 = 0.2$ m/s, $\rho = 10^3$ kg/m^3, $\mu = 10^{-3}$ Pa · s

(b) $l_0 = 100$ μm, $v_0 = 10$ μm/s, $\rho = 10^3$ kg/m^3, $\mu = 10^{-3}$ Pa · s

(c) $l_0 = 3$ in., $v_0 = 2$ ft/s, $\rho = 63$ lb$_m$/ft^3, $\mu = 1.2$ cp

5A.2 Verifying dimensionless groups. Verify that the dimensionless groups identified in Examples 5.5-1 and 5.5-2 are actually dimensionless.

5B.1 Nondimensionalizing the substantial derivative. Show that the equation for the substantial derivative of the velocity

$$\frac{D\mathbf{v}}{Dt} = \frac{\partial \mathbf{v}}{\partial t} + [\mathbf{v} \cdot \nabla \mathbf{v}] \tag{5B.1-1}$$

becomes

$$\frac{D\breve{\mathbf{v}}}{D\breve{t}} = \frac{\partial \breve{\mathbf{v}}}{\partial \breve{t}} + [\breve{\mathbf{v}} \cdot \breve{\nabla} \breve{\mathbf{v}}] \tag{5B.1-2}$$

when nondimensionalized using the length and velocity scales, l_0 and v_0, respectively.

5B.2 Nondimensionalizing a different form of the Navier-Stokes equation. The Navier-Stokes equation can be written in terms of the actual pressure as

$$\rho\left(\frac{\partial \mathbf{v}}{\partial t} + [\mathbf{v}\cdot\nabla\mathbf{v}]\right) = -\nabla p + \mu\nabla^2\mathbf{v} + \rho\mathbf{g} \tag{5B.2-1}$$

Using the scales l_0 for length, v_0 for velocity, and ρv_0^2 for pressure, nondimensionalize this equation to obtain

$$\frac{\partial \breve{\mathbf{v}}}{\partial \breve{t}} + \left[\breve{\mathbf{v}}\cdot\breve{\nabla}\breve{\mathbf{v}}\right] = -\breve{\nabla}\breve{p} + \frac{1}{\text{Re}}\breve{\nabla}^2\breve{\mathbf{v}} + \frac{1}{\text{Fr}}\frac{\mathbf{g}}{g} \tag{5B.2-2}$$

How are the Reynolds number, Re, and Froude number, Fr, defined in this case?

5B.3 Dimensional analysis using the Buckingham pi theorem.

(a) Use the Buckingham pi theorem to determine the dependence of the torque required to rotate the inner cylinder of a Couette viscometer (see Example 3.7.3) on the other relevant variables. Show that the result is consistent with Eq. 3.7-33.

(b) Use the Buckingham pi theorem to determine the relationship between the force and other relevant variables in the parallel-disk compression viscometer in Problem 3C.3. Show that the result is consistent with Eq. 3C.3-14.

5C.1 Two-phase interfacial boundary conditions. In §2.1, boundary conditions for solving viscous flow problems were given. At that point no mention was made of the role of interfacial tension. At the interface between two immiscible fluids, I and II, the following boundary condition should be used:[7]

$$\mathbf{n}^{\text{I}}\left(p^{\text{I}} - p^{\text{II}}\right) + \left[\mathbf{n}^{\text{I}}\cdot\left(\boldsymbol{\tau}^{\text{I}} - \boldsymbol{\tau}^{\text{II}}\right)\right] = \mathbf{n}^{\text{I}}\left(\frac{1}{R_1} + \frac{1}{R_2}\right)\sigma \tag{5C.1-1}$$

This is essentially a momentum balance written for an interfacial element dS with no matter passing through it, and with no interfacial mass or viscosity. Here $\mathbf{n}^{\text{I}}$ is the unit vector normal to dS and pointing into phase I. The quantities R_1 and R_2 are the principal radii of curvature at dS, and each of these is positive if its center lies in phase I. The sum $(1/R_1) + (1/R_2)$ can also be expressed as $(\nabla\cdot\mathbf{n}^{\text{I}})$. The quantity σ is the interfacial tension, assumed constant.

(a) Show that, for a spherical droplet of I at rest in a second medium II, *Laplace's equation*

$$\left(p^{\text{I}} - p^{\text{II}}\right) = \left(\frac{1}{R_1} + \frac{1}{R_2}\right)\sigma \tag{5C.1-2}$$

relates the pressures inside and outside the droplet. Is the pressure in phase I greater than that in phase II, or the reverse? What is the relation between the pressures at a planar interface?

(b) Show that Eq. 5C.1-1 leads to the following dimensionless boundary condition

$$\begin{aligned}\mathbf{n}^{\text{I}}\left(\breve{\mathscr{P}}^{\text{I}} - \breve{\mathscr{P}}^{\text{II}}\right) &+ \mathbf{n}^{\text{I}}\left[\!\left[\frac{\rho^{\text{II}} - \rho^{\text{I}}}{\rho^{\text{I}}}\right]\!\right]\left[\!\left[\frac{g l_0}{v_0^2}\right]\!\right]\breve{h} \\ &- \left[\!\left[\frac{\mu^{\text{I}}}{l_0 v_0 \rho^{\text{I}}}\right]\!\right]\left[\mathbf{n}^{\text{I}}\cdot\breve{\dot{\boldsymbol{\gamma}}}^{\text{I}}\right] + \left[\!\left[\frac{\mu^{\text{II}}}{l_0 v_0 \rho^{\text{II}}}\right]\!\right]\left[\mathbf{n}^{\text{I}}\cdot\breve{\dot{\boldsymbol{\gamma}}}^{\text{II}}\right]\left[\!\left[\frac{\rho^{\text{II}}}{\rho^{\text{I}}}\right]\!\right] \\ &= \mathbf{n}^{\text{I}}\left(\frac{1}{\breve{R}_1} + \frac{1}{\breve{R}_2}\right)\left[\!\left[\frac{\sigma}{l_0 v_0^2 \rho^{\text{I}}}\right]\!\right]\end{aligned} \tag{5C.1-3}$$

in which $\breve{h} = (h - h_0)/l_0$ is the dimensionless elevation of dS, $\breve{\dot{\boldsymbol{\gamma}}}^{\text{I}}$ and $\breve{\dot{\boldsymbol{\gamma}}}^{\text{II}}$ are dimensionless rate-of-deformation tensors, and $\breve{R}_1 = R_1/l_0$ and $\breve{R}_2 = R_2/l_0$ are dimensionless radii of curvature. Furthermore,

$$\breve{\mathscr{P}}^{\text{I}} = \frac{p^{\text{I}} - p_0 + \rho^{\text{I}} g (h - h_0)}{\rho^{\text{I}} v_0^2};$$

$$\breve{\mathscr{P}}^{\text{II}} = \frac{p^{\text{II}} - p_0 + \rho^{\text{II}} g (h - h_0)}{\rho^{\text{I}} v_0^2} \tag{5C.1-4, 5}$$

In the above, the zero-subscripted quantities are the scale factors, valid in both phases. Identify the dimensionless groups that appear in Eq. 5C.1-3.

(c) Show how the result in (b) simplifies to Eq. 5.3-12 under the assumptions made in §5.3.

5C.2 Power input to an agitated tank. Show by dimensional analysis that the power, P, imparted by a rotating impeller to an incompressible fluid in an agitated tank may be correlated, for any specific tank and impeller shape, by the expression

$$\frac{P}{\rho N^3 D^5} = \Phi\left(\frac{D^2 N \rho}{\mu}, \frac{DN^2}{g}, Nt\right) \tag{5C.2-1}$$

Here N is the rate of rotation of the impeller, D is the impeller diameter, t is the time since the start of the operation, and Φ is a function whose form has to be determined experimentally.

For the commonly used geometry shown in Fig. 5C.2, the power is given by the sum of two integrals representing the contributions of friction drag of the cylindrical tank body and bottom and the form drag of the radial baffles, respectively:

$$P = NT_z = N\left[\int_S R\left(\frac{\partial v_\theta}{\partial n}\right)_{\text{surf}} dS + \int_A Rp_{\text{surf}}\, dA\right] \tag{5C.2-2}$$

Here T_z is the torque required to turn the impeller, S is the total surface area of the tank, A is the surface area of the baffles, (considered positive on the "upstream" side and negative on the "downstream side"), R is the radial distance to any surface element dS or dA from the impeller axis of rotation, and n is the distance measured normally into the fluid from any element of tank surface dS.

The desired solution may now be obtained by dimensional analysis of the equations of motion and continuity by rewriting the integrals above in dimensionless form. Here it is convenient to use D, DN, and $\rho N^2 D^2$ for the characteristic length, velocity, and pressure, respectively.

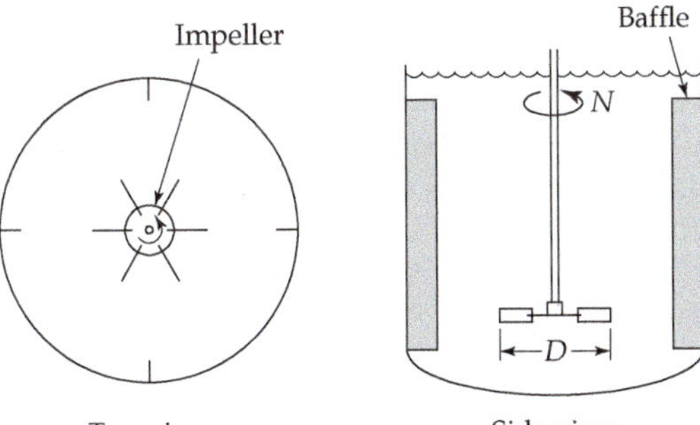

Fig. 5C.2 Agitated tank with a six-bladed impeller and four vertical baffles.

5C.3 An alternate analysis of the power input to an agitated tank. Repeat Problem 5C.2 using the Buckingham pi theorem as opposed to the dimensional analysis of the equations of continuity and motion.

Chapter 6

Interphase Transport in Isothermal Systems

In Chapters 2 and 3, we showed how laminar-flow problems may be formulated and solved. In Chapter 4, we presented some methods for solving turbulent-flow problems by dimensional arguments or by semi-empirical relations between the turbulent momentum flux and the gradient of the time-smoothed velocity. In Chapter 5, we introduced dimensional analysis, and showed how this can help organize data and analyze flow problems. In this chapter, we show how flow problems can be solved by a combination of dimensional analysis and experimental data. The technique presented here has been widely used in chemical, mechanical, aeronautical, and civil engineering, and it is useful for solving many practical problems. It is a topic worth learning well.

Many engineering flow problems fall into one of two broad categories: flow in channels and flow around submerged objects. Examples of channel flow are the pumping of oil through pipes, the flow of blood through arteries and veins, the flow of water in open channels, and extrusion of plastics through dies. Examples of flow around submerged objects are the motion of air around an airplane wing, motion of fluid around particles undergoing sedimentation, and flow across tube banks in heat exchangers.

For channel flow, the main object is usually to get a relationship between the volume rate of flow and the pressure drop and/or the elevation change. For problems involving flow around submerged objects, the desired information is generally the relation between the velocity of the approaching fluid and the drag force on the object. We have seen in the preceding chapters that, if one knows the velocity and pressure distributions in the system, then the desired relationships for these two cases may be obtained. The derivation of the Hagen-Poiseuille equation in §2.3 and the derivation of the Stokes drag equation in §2.7 illustrate the two categories we are discussing here.

For many systems, the velocity and pressure profiles cannot be easily calculated, particularly if the flow is turbulent or the geometry is complicated. One such system is the flow through a packed column; another is the flow in a tube in the shape of a helical coil. For such systems, we can take carefully chosen experimental data and then construct "correlations" of dimensionless variables that can be used to estimate the flow behavior in geometrically similar systems. This method is based on the discussion of dimensional analysis of Chapter 5.

We start in §6.1 by defining the "friction factor," and then we show in §6.2 and §6.3 how to construct friction factor charts for flow in circular tubes and flow around spheres. These are both systems we have already studied, and, in fact, several results from earlier chapters are included in these charts. Finally, in §6.4 we examine the flow in packed columns to illustrate the treatment of a geometrically complicated system. The more complex problem of fluidized beds is not included in this chapter.[1]

§6.1 DEFINITION OF FRICTION FACTORS

We consider the steadily driven flow of a fluid of constant density in one of two systems: (*a*) the fluid flows in a straight conduit of uniform cross section; (*b*) the fluid flows around a submerged object that has an axis of symmetry (or two planes of symmetry) parallel to the direction of the approaching fluid. There will be a force $\mathbf{F}_{f\to s}$ exerted by the fluid on the solid surfaces. It is convenient to split this force into two parts: $\mathbf{F}_s$, the force that would be exerted by the fluid even if it were stationary; and $\mathbf{F}_k$, the additional force associated with the motion of the fluid (see §2.7 for the discussion of $\mathbf{F}_s$ and $\mathbf{F}_k$ for flow around spheres). In systems of type (*a*), $\mathbf{F}_k$ points in the same direction as the average velocity $\langle\overline{\mathbf{v}}\rangle$ in the conduit, and in systems of type (*b*), $\mathbf{F}_k$ points in the same direction as the approach velocity $\mathbf{v}_\infty$.

For both systems we state that the magnitude of the force $\mathbf{F}_k$ is proportional to a characteristic area A and a characteristic kinetic energy K per unit volume

$$F_k = AKf \tag{6.1-1}$$[1]

in which the proportionality constant f is called the *friction factor.* Note that Eq. 6.1-1 is *not* a law of fluid dynamics, but just a definition for f. This is a useful definition, because the dimensionless quantity f can be given as a relatively simple function of the Reynolds number and the system shape.

Clearly, for any given flow system, f is not completely defined until A and K are specified. Let us now see what the customary definitions are:

a. Flow in conduits

Here, A is usually taken to be the wetted surface, and K is taken to be $\frac{1}{2}\rho\langle\bar{v}_z\rangle^2$, where $\langle\bar{v}_z\rangle$ is the time and spatially averaged axial velocity. Specifically, for circular tubes of radius R and length L we define f by

$$F_k = (2\pi RL)\left(\tfrac{1}{2}\rho\langle\bar{v}_z\rangle^2\right)f \tag{6.1-2}$$

Generally, the quantity measured is not F_k, but rather the pressure difference $p_0 - p_L$ and the elevation difference $h_0 - h_L$. An axial force balance on the fluid between 0 and L in the direction of flow gives for fully developed flow

$$\begin{aligned} F_k &= \left[(p_0 - p_L) + \rho g\left(h_0 - h_L\right)\right]\pi R^2 \\ &= \left(\mathscr{P}_0 - \mathscr{P}_L\right)\pi R^2 \end{aligned} \tag{6.1-3}$$

---

[1] R. Jackson, *The Dynamics of Fluidized Particles,* Cambridge University Press (2000).

[1] For systems lacking symmetry, the fluid exerts both a force and a torque on the solid. For a discussion of such systems, see J. Happel and H. Brenner, *Low Reynolds Number Hydrodynamics,* Martinus Nijhoff, The Hague (1983), Chapter 5; H. Brenner, in *Adv. Chem. Engr.,* **6**, 287–438 (1966); S. Kim and S. J. Karrila, *Microhydrodynamics: Principles and Selected Applications,* Butterworth-Heinemann, Boston (1991); Dover, Mineola, NY (2005), Chapter 5.

Elimination of F_k between the last two equations then gives

$$f = \frac{1}{4}\left(\frac{D}{L}\right)\left(\frac{\mathcal{P}_0 - \mathcal{P}_L}{\frac{1}{2}\rho\langle \bar{v}_z\rangle^2}\right) \tag{6.1-4}$$

in which $D = 2R$ is the tube diameter. Equation 6.1-4 shows how to calculate f from experimental data. The quantity f is sometimes called the *Fanning friction factor*.[2]

b. Flow around submerged objects

The characteristic area A is usually taken to be the area obtained by projecting the solid onto a plane perpendicular to the velocity of the approaching fluid; the quantity K is taken to be $\frac{1}{2}\rho v_\infty^2$, where v_∞ is the approach velocity of the fluid at a large distance from the object. For example, for flow around a sphere of radius R, we define f by the equation

$$F_k = \left(\pi R^2\right)\left(\tfrac{1}{2}\rho v_\infty^2\right) f \qquad (6.1\text{-}5)^3$$

If it is not possible to measure F_k, then we can measure the terminal velocity of the sphere when it falls through the fluid (in that case, v_∞ has to be interpreted as the terminal velocity of the sphere). For the steady-state fall of a sphere in a fluid, the force F_k is just counterbalanced by the gravitational force on the sphere less the buoyant force (cf. Eq. 2.7-16):

$$F_k = \tfrac{4}{3}\pi R^3 \rho_{\text{sph}} g - \tfrac{4}{3}\pi R^3 \rho g \tag{6.1-6}$$

Elimination of F_k between Eqs. 6.1-5 and 6.1-6 then gives

$$f = \frac{4}{3}\frac{gD}{v_\infty^2}\left(\frac{\rho_{\text{sph}} - \rho}{\rho}\right) \tag{6.1-7}$$

This expression can be used to obtain f from terminal-velocity data. The friction factor used in Eqs. 6.1-5 and 6.1-7 is sometimes called the *drag coefficient* and given the symbol c_D.

We have seen that the "drag coefficient" for submerged objects and the "friction factor" for channel flow are defined in the same general way. It is for this reason that we prefer to use the same symbol and name for both of them.

§6.2 FRICTION FACTORS FOR FLOW IN TUBES

We now combine the definition of f in Eq. 6.1-2 with the dimensional analysis of Chapter 5 to show what f must depend on in this kind of system. We consider a "test section" of

---

[2] This friction factor definition is due to J. T. Fanning, *A Practical Treatise on the Hydraulic and Water Supply Engineering*, Van Nostrand, New York, 1st edition (1877), 16th edition (1906); the name "Fanning" is used to avoid confusion with the "Moody friction factor," which is larger by a factor of 4 than the f used here [L. F. Moody, *Trans. ASME*, **66**, 671–684 (1944)].

If we use the "friction velocity" $v_* = \sqrt{\tau_0/\rho} = \sqrt{(\mathcal{P}_0 - \mathcal{P}_L)R/2L\rho}$, introduced in §4.3, then Eq. 6.1-4 assumes the form

$$f = 2\left(v_*/\langle \bar{v}_z\rangle\right)^2 \tag{6.1-4a}$$

John Thomas Fanning (1837–1911) studied architectural and civil engineering, served as an officer in the Civil War (1861–1865), and after the war became prominent in hydraulic engineering. The 14th edition of his book *A Practical Treatise on Hydraulic and Water-Supply Engineering* appeared in 1899.

[3] For the arbitrary, time-dependent motion of a sphere in three dimensions, one can write the kinetic force *approximately* in the quasi-steady form

$$\mathbf{F}_k(t; \text{Re}) = \left(\pi R^2\right)\left(\tfrac{1}{2}\rho v_\infty^2\right) f(\text{Re})\,\mathbf{n}(t) \tag{6.1-5a}$$

where $\mathbf{n}$ is a unit vector in the direction of $\mathbf{v}_\infty$, which may depend on time; see Problem 6C.1. An exact description of this force would include the time dependence of f as well.

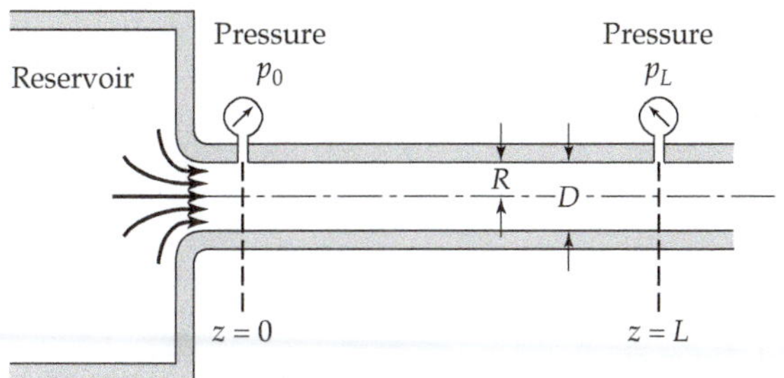

Fig. 6.2-1. Section of a circular pipe from $z = 0$ to $z = L$ for the discussion of dimensional analysis.

inner radius R and length L, shown in Fig. 6.2-1, carrying a fluid of constant density and viscosity at a steady mass flow rate. The pressures $\mathscr{P}_0$ and $\mathscr{P}_L$ at the ends of the test section are known.

The system is either in steady laminar flow or steadily driven turbulent flow (i.e., turbulent flow with a steady total throughput). In either case, the force in the z direction of the fluid on the inner wall of the test section is

$$F_k(t) = \int_0^L \int_0^{2\pi} \left(-\mu \frac{\partial v_z}{\partial r}\right)\bigg|_{r=R} R\,d\theta\,dz \tag{6.2-1}$$

In turbulent flow the force may be a function of time, not only because of the turbulent fluctuations, but also because of occasional ripping off of the boundary layer from the wall, which results in some disturbances with long time scales. In laminar flow it is understood that the force will be independent of time.

Equating Eqs. 6.2-1 and 6.1-2, we get the following expression for the friction factor:

$$f(t) = \frac{\displaystyle\int_0^L \int_0^{2\pi} \left(-\mu \frac{\partial v_z}{\partial r}\right)\bigg|_{r=R} R\,d\theta\,dz}{(2\pi RL)\left(\frac{1}{2}\rho\langle\overline{v}_z\rangle^2\right)} \tag{6.2-2}$$

Next we introduce the dimensionless quantities from Chapter 5: $\breve{r} = r/D$, $\breve{z} = z/D$, $\breve{v}_z = v_z/\langle\overline{v}_z\rangle$, $\breve{t} = \langle\overline{v}_z\rangle t/D$, $\breve{\mathscr{P}} = (\mathscr{P} - \mathscr{P}_0)/\rho\langle\overline{v}_z\rangle^2$, and $\text{Re} = D\langle\overline{v}_z\rangle\rho/\mu$. Then Eq. 6.2-2 may be rewritten as

$$f(\breve{t}) = \frac{1}{\pi}\frac{D}{L}\frac{1}{\text{Re}} \int_0^{L/D} \int_0^{2\pi} \left(-\frac{\partial \breve{v}_z}{\partial \breve{r}}\right)\bigg|_{\breve{r}=\frac{1}{2}} d\theta\,d\breve{z} \tag{6.2-3}$$

This relation is valid for laminar or turbulent flow in circular tubes. We see that for flow systems in which the drag depends on viscous forces alone (i.e., no "form drag") the product $f\text{Re}$ is essentially a dimensionless velocity gradient averaged over the surface.

Recall now that, in principle, $\partial\breve{v}_z/\partial\breve{r}$ can be evaluated from the dimensionless equations of change (e.g., Eqs. 5.1-7 and 5.1-8, or 5.1-7 and 5.1-23) along with the boundary conditions[1]

B.C.1:	at $\breve{r} = \frac{1}{2}$,	$\breve{\mathbf{v}} = \mathbf{0}$	for $z > 0$	(6.2-4)
B.C.2:	at $\breve{z} = 0$,	$\breve{\mathbf{v}} = \boldsymbol{\delta}_z$		(6.2-5)
B.C.3:	at $\breve{r} = 0$ and $\breve{z} = 0$,	$\breve{\mathscr{P}} = 0$		(6.2-6)

and appropriate initial conditions. The uniform inlet velocity profile in Eq. 6.2-5 is accurate except very near the wall, for a well-designed nozzle and upstream system. If Eqs. 5.1-7

[1]Here we follow the customary practice of neglecting the $(\partial^2/\partial\breve{z}^2)\,\breve{\mathbf{v}}$ terms of Eq. 5.1-8 and 5.1-23 because the velocity is expected to be nominally independent of z for long tubes. With those terms suppressed, we do not need an outlet boundary condition on $\mathbf{v}$.

and 5.1-8 could be solved with these boundary and initial conditions to get $\breve{\mathbf{v}}$ and $\breve{\mathscr{P}}$, the solutions would necessarily be of the form

$$\breve{\mathbf{v}} = \breve{\mathbf{v}}(\breve{r},\theta,\breve{z},\breve{t};\text{Re}) \tag{6.2-7}$$

$$\breve{\mathscr{P}} = \breve{\mathscr{P}}(\breve{r},\theta,\breve{z},\breve{t};\text{Re}) \tag{6.2-8}$$

That is, the functional dependence of $\breve{\mathbf{v}}$ and $\breve{\mathscr{P}}$ must, in general, include all the dimensionless variables and the one dimensionless group appearing in the differential equations. No additional dimensionless groups enter via the foregoing boundary conditions. As a consequence, $\partial\breve{v}_z/\partial\breve{r}$ must likewise depend on $\breve{r},\theta,\breve{z},\breve{t}$, and Re. When $\partial\breve{v}_z/\partial\breve{r}$ is evaluated at $\breve{r} = \frac{1}{2}$ and then integrated over $\breve{z}$ and θ in Eq. 6.2-3, the result depends only on $\breve{t}$, Re, and L/D (the latter appearing in the upper limit in the integration over $\breve{z}$). Therefore, we are led to the conclusion that $f(\breve{t}) = f(\text{Re},L/D,\breve{t})$, which, when time averaged, becomes

$$f = f(\text{Re},L/D) \tag{6.2-9}$$

when the time average is performed over an interval long enough to include any long-time turbulent disturbances. The measured friction factor then depends only on the Reynolds number and the length-to-diameter ratio.

The dependence of f on L/D arises from the development of the time-average velocity distribution from its flat entry shape toward more rounded profiles at downstream z values. This development occurs within an entrance region, of length $L_e \approx 0.03\,D\text{Re}$ for laminar flow or $L_e \approx 60\,D$ for turbulent flow, beyond which the shape of the velocity distribution is "fully developed." In the transportation of fluids, the entrance length is usually a small fraction of the total; then Eq. 6.2-9 reduces to the long-tube form

$$f = f(\text{Re}) \tag{6.2-10}$$

and f can be evaluated experimentally from Eq. 6.1-4, which was written for fully developed flow at the inlet and outlet.

Equations 6.2-9 and 6.2-10 are useful results, since they provide a guide for the systematic presentation of data on flow rate versus pressure difference for laminar and turbulent flow in circular tubes. For *long* tubes we need only a single curve of f plotted versus the single combination $D\langle\bar{v}_z\rangle\rho/\mu$. Just think how much simpler this is than the plotting of the pressure drop versus the flow rate for separate values of D,L,ρ, and μ, which is what the uninitiated might do.

There is much experimental information for pressure drop versus flow rate in tubes, and hence, f can be calculated from the experimental data by Eq. 6.1-4. Then f can be plotted versus Re for smooth tubes to obtain the *solid* curves shown in Fig. 6.2-2. These solid curves describe the laminar and turbulent behavior for fluids flowing in *long, smooth, circular* tubes.

Note that the *laminar* curve on the friction factor chart is just a plot of the *Hagen-Poiseuille* equation in Eq. 2.3-21. This can be seen by substituting the expression for $(\mathscr{P}_0 - \mathscr{P}_L)$ from Eq. 2.3-21 into Eq. 6.1-4 and using the relation $w = \rho\langle\bar{v}_z\rangle\pi R^2$; this gives

$$f(\text{Re}) = \frac{16}{\text{Re}}\left\{\begin{array}{ll}\text{Re} < 2100 & \text{stable} \\ \text{Re} > 2100 & \text{usually unstable}\end{array}\right\} \tag{6.2-11}$$

in which $\text{Re} = D\langle\bar{v}_z\rangle\rho/\mu$; this is exactly the laminar line in Fig. 6.2-2.

Analogous *turbulent* curves have been constructed by using *experimental data.* Some analytical curve-fit expressions are also available. For example, Eq. 4.1-6 can be put into the form

$$f(\mathrm{Re}) = \frac{0.0791}{\mathrm{Re}^{1/4}} \qquad 2.1 \times 10^3 < \mathrm{Re} < 10^5 \tag{6.2-12}$$

which is known as the *Blasius formula*.[2] Equation 4.5-1 (with 2.5 replaced by 2.45 and 1.75 by 2.00) is equivalent to

$$\frac{1}{\sqrt{f}} = 4.0 \log_{10} \mathrm{Re}\sqrt{f} - 0.4 \qquad 2.3 \times 10^3 < \mathrm{Re} < 4 \times 10^6 \tag{6.2-13}$$

which is known as the *Prandtl formula*.[3]

A further relation, which includes the dashed curves for rough pipes in Fig. 6.2-2, is the empirical *Haaland equation*[4]

$$\frac{1}{\sqrt{f}} = -3.6 \log_{10} \left[\frac{6.9}{\mathrm{Re}} + \left(\frac{k/D}{3.7} \right)^{10/9} \right] \qquad \begin{cases} 4 \times 10^4 < \mathrm{Re} < 10^8 \\ 0 < k/D < 0.05 \end{cases} \tag{6.2-14}$$

where k is the tube roughness, a measure of the average height of protuberances on the tube wall. This equation is stated[4] to be accurate within 1.5%. As can be seen in Fig. 6.2-2,

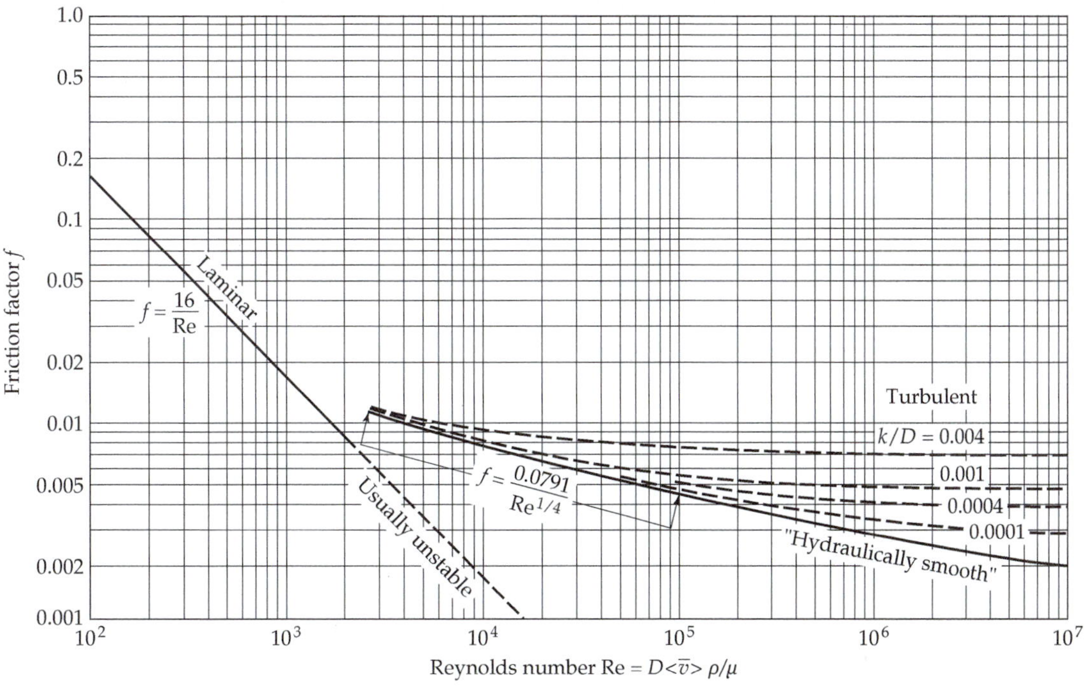

Fig. 6.2-2. Friction factor for tube flow (see definition of f in Eqs. 6.1-2 and 6.1-4). [Curves of L. F. Moody, *Trans. ASME*, **66**, 671–684 (1944) as presented in W. L. McCabe and J. C. Smith, *Unit Operations of Chemical Engineering*, McGraw-Hill, New York (1954).]

---

[2] H. Blasius, *Forschungsarbeiten des Ver. Deutsch. Ing.*, No. 131 (1913).

[3] L. Prandtl, *Essentials of Fluid Dynamics*, Hafner, New York (1952), p. 165.

[4] S. E. Haaland, *Trans. ASME, J. Fluids Engr.*, **105**, 89–90 (1983). For other empiricisms, see D. J. Zigrang and N. D. Sylvester, *AIChE Journal*, **28**, 514–515 (1982).

the frictional resistance to flow increases with the roughness. Of course, k has to enter into the correlation in a dimensionless fashion and hence appears via the ratio k/D (called the relative roughness).

For *turbulent flow* in *noncircular tubes*, it is common to use the following empiricism: First, we define a "mean hydraulic radius" R_h as follows:

$$R_h = S/Z \tag{6.2-15}$$

in which S is the cross-sectional area of the conduit and Z is the wetted perimeter. Then, we can use Eq. 6.1-4 and Fig. 6.2-2, with the diameter D of the circular pipe replaced by 4 R_h. That is, we calculate pressure differences by replacing Eq. 6.1-4 by

$$f = \left(\frac{R_h}{L}\right)\left(\frac{\mathscr{P}_0 - \mathscr{P}_L}{\frac{1}{2}\rho\langle\bar{v}_z\rangle^2}\right) \tag{6.2-16}$$

and getting f from Fig. 6.2-2 with a Reynolds number defined as

$$\text{Re}_h = \frac{4R_h\langle\bar{v}_z\rangle\rho}{\mu} \tag{6.2-17}$$

This estimation method of Eqs. 6.2-15 to 6.2-17 should *not* be used for *laminar flow*.

EXAMPLE 6.2-1

Pressure Drop Required for a Given Flow Rate

What pressure difference is required to cause N,N-diethylaniline, $C_6H_5N(C_2H_5)_2$, to flow in a horizontal smooth circular tube of inside diameter $D = 3\,\text{cm}$ at a mass rate of 1028 g/s at 20°C? At this temperature the density of diethylaniline is $\rho = 0.935\ \text{g/cm}^3$ and its viscosity is $\mu = 1.95$ cp.

SOLUTION

The Reynolds number for the flow is

$$\text{Re} = \frac{D\langle\bar{v}_z\rangle\rho}{\mu} = \frac{Dw}{(\pi D^2/4)\,\mu} = \frac{4w}{\pi D\mu}$$

$$= \frac{4\,(1028\ \text{g/s})}{\pi\,(3\ \text{cm})\,(1.95\times10^{-2}\ \text{g/cm}\cdot\text{s})} = 2.24\times10^4 \tag{6.2-18}$$

From Fig. 6.2-2, we find that for this Reynolds number the friction factor f has a value of 0.0063 for smooth tubes. Hence, the pressure difference required to maintain the flow is (according to Eq. 6.1-4)

$$\frac{p_0 - p_L}{L} = \left(\frac{4}{D}\right)\left(\frac{1}{2}\rho\langle\bar{v}_z\rangle^2\right)f = \frac{2}{D}\rho\left(\frac{4w}{\pi D^2\rho}\right)^2 f$$

$$= \frac{32w^2 f}{\pi^2 D^5\rho} = \frac{(32)\,(1028\ \text{g/s})^2\,(0.0063)}{\pi^2(3\ \text{cm})^5\,(0.935\ \text{g/cm}^3)}$$

$$= 95\frac{\text{g}}{\text{cm}^2\,\text{s}^2}\left(\frac{1\ \text{dyne}}{1\ \text{g}\cdot\text{cm/s}^2}\right)\left(\frac{760\ \text{mm Hg}}{1.0133\times10^6\ \text{dyne/cm}^2}\right)$$

$$= 0.071\frac{\text{mm Hg}}{\text{cm}} \tag{6.2-19}$$

EXAMPLE 6.2-2

Flow Rate for a Given Pressure Drop

Determine the mass flow rate, in lb_m/hr, of water at 68°F through a 1000 ft length of horizontal 8-in. schedule 40 steel pipe (internal diameter 7.981 in.) under a pressure difference of 3.00 psi. For such a pipe, use Fig. 6.2-2 and assume that $k/D = 2.3 \times 10^{-4}$.

SOLUTION

We want to use Eq. 6.1-4 and Fig. 6.2-2 to solve for $\langle \bar{v}_z \rangle$ when $p_0 - p_L$ is known. But the quantity $\langle \bar{v}_z \rangle$ appears explicitly on the right side of Eq. 6.1-4 and implicitly on the left side in f, which depends on $\text{Re} = D\langle \bar{v}_z \rangle \rho/\mu$. Because neither side can be evaluated, we cannot solve *explicitly* for $\langle \bar{v}_z \rangle$. One way to solve such problems is by using "trial and error." That is, guess a value for $\langle \bar{v}_z \rangle$, evaluate the right side of Eq. 6.1-4, then calculate $\text{Re} = D\langle \bar{v}_z \rangle \rho/\mu$ so that a value of f can be read from Fig. 6.2-2. This process can be repeated for various values of $\langle \bar{v}_z \rangle$ until the values determined for the right and left sides of Eq. 6.1-4 agree (plotting the difference between the two sides as a function of $\langle \bar{v}_z \rangle$ can facilitate this process, as one can obtain improved guesses for $\langle \bar{v}_z \rangle$ by interpolating or extrapolating results from previous guesses). Another approach is not to use Fig. 6.2-2 at all, but rather to use a correlation for $f(\text{Re})$ (e.g., Eqs. 6.2-11, 6.2-12, 6.2-13, or 6.2-14). If one substitutes the appropriate correlation for $f(\text{Re})$ in the left side of Eq. 6.1-4, the average velocity may be obtained directly, either analytically or numerically. This, however, is also ultimately a trial-and-error method, since one does not know ahead of time if the range of Re for the correlation chosen is consistent with the solution obtained.

Below we show two graphical solution methods. These methods have the advantage of producing solutions without trial and error. Understanding these methods can also help to improve our understanding of the relationship between friction factor definitions and their dependence on other parameters.

Method A. First, Fig. 6.2-2 is used to construct a plot[5] of Re versus the group $\text{Re}\sqrt{f}$. A variety of values of Re are selected, and corresponding values of f are read off of Fig. 6.2-2. For each (Re,f) pair, the quantity $\text{Re}\sqrt{f}$ can be calculated. One can then plot Re vs. $\text{Re}\sqrt{f}$. Such a plot is presented in Fig. 6.2-3(a) using the data from Fig. 6.2-2 for smooth tubes, as well as for tubes with $k/D = 0.00023$.

Next, we recognize that the quantity $\text{Re}\sqrt{f}$ is independent of $\langle \bar{v}_z \rangle$, as follows

$$\text{Re}\sqrt{f} = \frac{D\langle \bar{v}_z \rangle \rho}{\mu}\sqrt{\frac{(p_0 - p_L)D}{2L\rho\langle \bar{v}_z \rangle^2}} = \frac{D\rho}{\mu}\sqrt{\frac{(p_0 - p_L)D}{2L\rho}} \tag{6.2-20}$$

and thus the quantity $\text{Re}\sqrt{f}$ can be computed for this problem because all of the parameters in the right-most term are known. The corresponding value of the Reynolds number can then be read from the Re versus $\text{Re}\sqrt{f}$ plot. From Re the average velocity and mass flow rate can then be calculated.

For this problem, we have

$$\begin{aligned}
p_0 - p_L &= \left(\frac{3.00\ \text{lb}_f}{\text{in.}^2}\right)\left(\frac{32.174\ \text{lb}_m \cdot \text{ft}}{\text{lb}_f \cdot \text{s}^2}\right)\left(\frac{144\ \text{in.}^2}{\text{ft}^2}\right) \\
&= 1.39 \times 10^4 \frac{\text{lb}_m}{\text{ft} \cdot \text{s}^2} \\
D &= (7.981\ \text{in.})\left(\frac{1\ \text{ft}}{12\ \text{in.}}\right) = 0.665\ \text{ft} \\
L &= 1000\ \text{ft} \\
\rho &= 62.3 \frac{\text{lb}_m}{\text{ft}^3} \\
\mu &= (1.03\ \text{cp})\left(\frac{6.72 \times 10^{-4}\ \text{lb}_m/\text{ft} \cdot \text{s}}{\text{cp}}\right) = 6.92 \times 10^{-4} \frac{\text{lb}_m}{\text{ft} \cdot \text{s}}
\end{aligned}$$

---

[5]A related plot was proposed by T. von Kármán, *Nachr. Ges. Wiss. Göttingen, Fachgruppen*, Series **5**, 58–76 (1930).

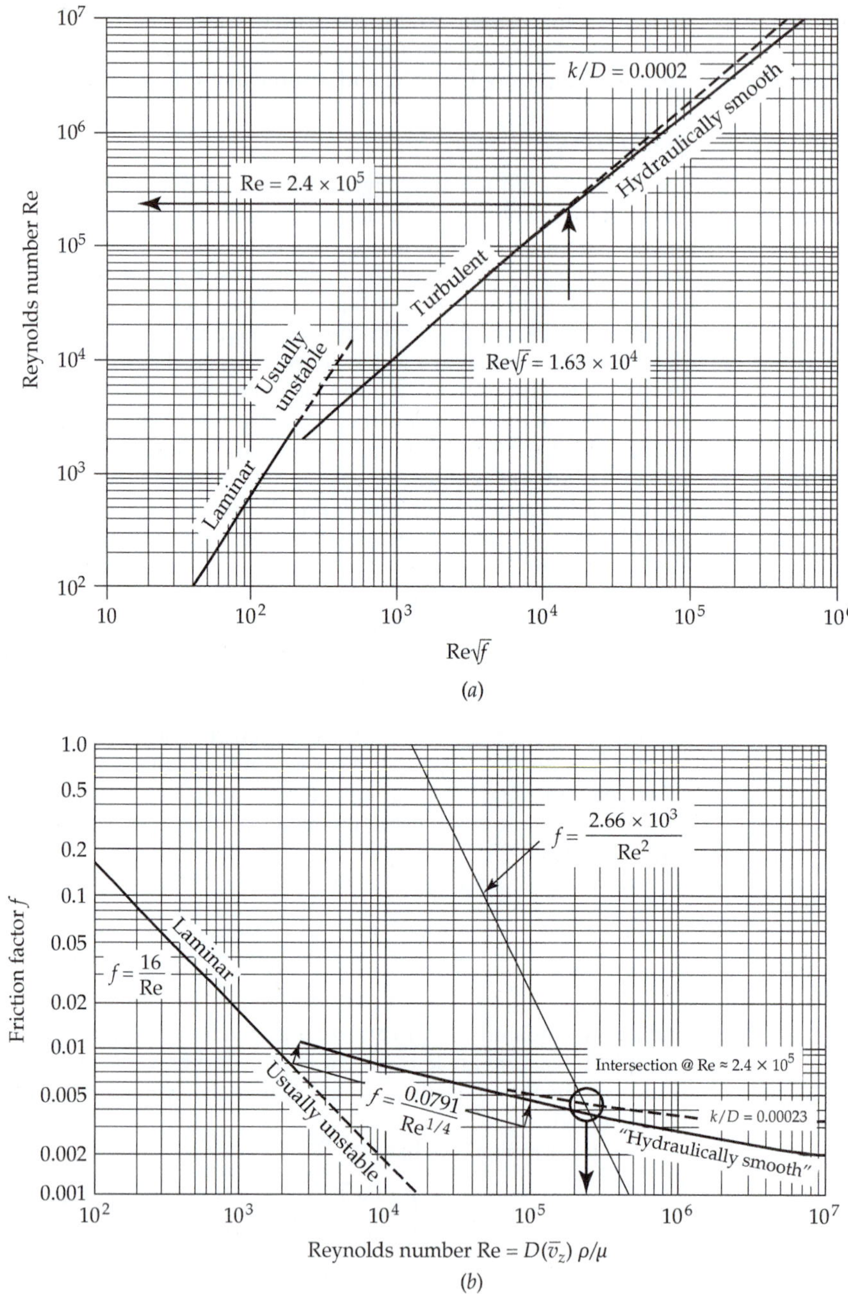

Fig. 6.2-3. (*a*) Reynolds number as a function of $\text{Re}\sqrt{f}$ used for "Method A" in Example 6.2-2. (*b*) Graphical procedure for "Method B" used in Example 6.2-2.

Then according to Eq. 6.2-20,

$$\begin{aligned}\text{Re}\sqrt{f} &= \frac{D\rho}{\mu}\sqrt{\frac{(p_0 - p_L)D}{2L\rho}} \\ &= \frac{(0.665\text{ ft})(62.3\text{ lb}_m/\text{ft}^3)}{(6.92\times10^{-4}\text{ lb}_m/\text{ft}\cdot\text{s})}\sqrt{\frac{(1.39\times10^{-4}\text{ lb}_m/\text{ft}\cdot\text{s}^2)(0.665\text{ ft})}{2(1000\text{ ft})(62.3\text{ lb}_m/\text{ft}^3)}} \\ &= 1.63\times10^4 \quad \text{(dimensionless)}\end{aligned} \tag{6.2-21}$$

This value for $\mathrm{Re}\sqrt{f}$ is then located on the horizontal axis of Fig. 6.2-3(a), and the value of $\mathrm{Re} = 2.4 \times 10^5$ is read from the vertical axis using the curve for $k/D = 0.00023$. With this value of Re, average velocity and mass flow rate can be computed:

$$\langle \bar{v}_z \rangle = \frac{\mu \mathrm{Re}}{D\rho} = \frac{(6.92 \times 10^{-4}\ \mathrm{lb}_m/\mathrm{ft}\cdot\mathrm{s})\,(2.4 \times 10^5)}{\left(0.665\ \mathrm{ft}\right)\left(62.3\ \mathrm{lb}_m/\mathrm{ft}^3\right)} = 4.01\frac{\mathrm{ft}}{\mathrm{s}} \tag{6.2-22}$$

$$\begin{aligned} w = Q\rho &= \left(\frac{\pi}{4}D^2 \langle \bar{v}_z \rangle\right)\rho \\ &= \frac{\pi}{4}(0.665\ \mathrm{ft})^2 \left(4.01\frac{\mathrm{ft}}{\mathrm{s}}\right)\left(62.3\frac{\mathrm{lb}_m}{\mathrm{ft}^3}\right)\left(\frac{3600\ \mathrm{s}}{\mathrm{hr}}\right) = 3.12 \times 10^5 \frac{\mathrm{lb}_m}{\mathrm{hr}} \end{aligned} \tag{6.2-23}$$

Method B. Figure 6.2-2 may also be used directly without any replotting, by devising a scheme that is equivalent to the graphical solution of two simultaneous equations. The two equations are:

$$f = f(\mathrm{Re}, k/D) \qquad \text{the curve given in Fig. 6.2-2} \tag{6.2-24}$$

$$f = \frac{\left(\mathrm{Re}\sqrt{f}\right)^2}{\mathrm{Re}^2} \qquad \text{a straight line of slope} - 2 \text{ on a log-log plot of } f \text{ vs Re} \tag{6.2-25}$$

The procedure is then to compute $\mathrm{Re}\sqrt{f}$ according to Eq. 6.2-20 using the parameters for the current problem as in Eq. 6.2-22, and then to plot Eqs. 6.2-24 and 6.2-25 on a log-log plot of f versus Re. The intersection of the two curves gives the Reynolds number of the flow, from which $\langle \bar{v}_z \rangle$ and w can then be computed.

From Method A, we have $\left(\mathrm{Re}\sqrt{f}\right)^2 = 2.66 \times 10^8$. Using this value, Eq. 6.2-25 is plotted along with Eq. 6.2-24 in Fig. 6.2-3(b). For the latter, we have included only the curves for smooth tubes and tubes with $k/D = 0.00023$. The intersection of these curves again gives $\mathrm{Re} = 2.4 \times 10^5$, from which $\langle \bar{v}_z \rangle$ and w can then be computed as in Method A.

§6.3 FRICTION FACTORS FOR FLOW AROUND SPHERES

In this section we use the definition of the friction factor in Eq. 6.1-5 along with the dimensional analysis of §5.1 to determine the behavior of f for a stationary sphere in an infinite stream of fluid approaching with a uniform, steady velocity v_∞. We have already studied the flow around a sphere in §2.7 for $\mathrm{Re} < 0.1$ (the "creeping flow" region). At Reynolds numbers above about 1 there is a significant unsteady eddy motion in the wake of the sphere. Therefore, it will be necessary to do a time average over a time interval long with respect to this eddy motion.

Recall from §2.7 that the total force acting in the z direction on the sphere can be written as the sum of a contribution from the normal stresses (F_n) and one from the tangential stresses (F_t). One part of the normal-stress contribution is the force that would be present even if the fluid were stationary, F_s. Thus, the "kinetic force," associated with the fluid motion, is

$$F_k = (F_n - F_s) + F_t = F_{\mathrm{form}} + F_{\mathrm{friction}} \tag{6.3-1}$$

The forces associated with the form drag and the friction drag are then obtained from

$$F_{\mathrm{form}}(t) = \int_0^{2\pi}\int_0^{\pi} \left(-\left(p + \rho g z - p_0\right)\Big|_{r=R} \cos\theta\right) R^2 \sin\theta \, d\theta \, d\phi \tag{6.3-2}$$

$$F_{\mathrm{friction}}(t) = \int_0^{2\pi}\int_0^{\pi} \left(-\mu \left[r\frac{\partial}{\partial r}\left(\frac{v_\theta}{r}\right) + \frac{1}{r}\frac{\partial v_r}{\partial \theta}\right]\Bigg|_{r=R} \sin\theta \right) R^2 \sin\theta \, d\theta \, d\phi \tag{6.3-3}$$

Since v_r is zero everywhere on the sphere surface, the term containing $\partial v_r/\partial\theta$ is zero.

If now we split f into two parts as follows

$$f = f_{\text{form}} + f_{\text{friction}} \tag{6.3-4}$$

then, from the definition in Eq. 6.1-5, we get

$$f_{\text{form}}(\breve{t}) = \frac{2}{\pi} \int_0^{2\pi} \int_0^{\pi} \left(-\breve{\mathscr{P}} \Big|_{\breve{r}=1} \cos\theta \right) \sin\theta \, d\theta \, d\phi \tag{6.3-5}$$

$$f_{\text{friction}}(\breve{t}) = -\frac{4}{\pi} \frac{1}{\text{Re}} \int_0^{2\pi} \int_0^{\pi} \left[\breve{r} \frac{\partial}{\partial \breve{r}} \left(\frac{\breve{v}_\theta}{\breve{r}} \right) \right] \Bigg|_{\breve{r}=1} \sin^2\theta \, d\theta \, d\phi \tag{6.3-6}$$

The friction factor has been expressed in terms of dimensionless variables

$$\breve{\mathscr{P}} \equiv \frac{\mathscr{P} - \mathscr{P}_0}{\rho v_\infty^2} = \frac{(p + \rho g z) - (p_0 + \rho g 0)}{\rho v_\infty^2}; \quad \breve{v}_\theta \equiv \frac{v_\theta}{v_\infty}; \quad \breve{r} \equiv \frac{r}{R}; \quad \breve{t} \equiv \frac{v_\infty t}{R} \tag{6.3-7}$$

and a Reynolds number defined as

$$\text{Re} \equiv \frac{D v_\infty \rho}{\mu} = \frac{2 R v_\infty \rho}{\mu} \tag{6.3-8}$$

In order to evaluate f one would have to know $\breve{\mathscr{P}}$ and $\breve{v}_\theta$ as functions of $\breve{r}$, θ, ϕ, and $\breve{t}$.

We know that for incompressible flow these distributions can *in principle* be obtained from the solution of Eqs. 5.1-7 and 5.1-8 along with the boundary conditions

B.C.1: at $\breve{r} = 1$, $\breve{v}_r = 0$ and $\breve{v}_\theta = 0$ (6.3-9)

B.C.2: as $\breve{r} \to \infty$, $\breve{v}_z \to 1$ (6.3-10)

B.C.3: as $\breve{r} \to \infty$, $\breve{\mathscr{P}} \to 0$ (6.3-11)

and some appropriate initial condition on $\breve{\mathbf{v}}$. Because no additional dimensionless groups enter via the boundary and initial conditions, we know that the dimensionless pressure and velocity profiles will have the following form:

$$\breve{\mathscr{P}} = \breve{\mathscr{P}}(\breve{r},\theta,\phi,\breve{t};\text{Re}) \qquad \breve{\mathbf{v}} = \breve{\mathbf{v}}(\breve{r},\theta,\phi,\breve{t};\text{Re}) \tag{6.3-12}$$

When these expressions are substituted into Eqs. 6.3-5 and 6.3-6, it is then evident that the friction factor in Eq. 6.3-4 must have the form $f(\breve{t}) = f(\text{Re},\breve{t})$, which, when time averaged over the turbulent fluctuations, simplifies to

$$f = f(\text{Re}) \tag{6.3-13}$$

by using arguments similar to those in §6.2. Hence, from the expression for the friction factor and the dimensionless form of the equations of change, we find that f must be a function of Re alone.

Many experimental measurements of the drag force on spheres are available, and when these are plotted in dimensionless form, Fig. 6.3-1 results. For this system there is no sharp transition from an unstable laminar-flow curve to a stable turbulent-flow curve as for long tubes at a Reynolds number of about 2100 (see Fig. 6.2-2). Instead, as the approach velocity increases, f varies smoothly and moderately up to Reynolds number of the order of 10^5. The kink in the curve at about $\text{Re} = 2 \times 10^5$ is associated with the shift of the boundary-layer separation zone from in front of the equator to in back of the equator of the sphere.[1]

[1] R. K. Adair, *The Physics of Baseball*, Harper and Row, New York (1990).

We have juxtaposed the discussions of tube flow and flow around a sphere in order to emphasize the fact that various flow systems behave quite differently. Several points of difference between the two systems are:

Flow in long tubes	**Flow around spheres**
• Rather well-defined laminar-turbulent transition at about Re = 2100	• No well-defined laminar-turbulent transition
• The only contribution to f is the friction drag (if the tubes are smooth)	• Contributions to f from both friction and form drag
• No boundary-layer separation	• There is a kink in the f vs. Re curve associated with a shift in the separation zone

The general shape of the curves in Figs. 6.2-2 and 6.3-1 should be carefully remembered.

For the *creeping flow region*, we already know that the drag force is given by *Stokes' law* (Eq. 2.7-15), which is a consequence of solving the continuity equation and the Navier-Stokes equation of motion without the $\rho D\mathbf{v}/Dt$ term. Stokes' law can be rearranged into the form of Eq. 6.1-5 to get

$$F_k = (\pi R^2)\left(\tfrac{1}{2}\rho v_\infty^2\right)\left(\frac{24}{Dv_\infty\rho/\mu}\right) \tag{6.3-14}$$

Hence, for *creeping flow* around a sphere

$$f = \frac{24}{\text{Re}} \quad \text{for} \quad \text{Re} < 0.1 \tag{6.3-15}$$

and this is the straight-line asymptote as Re → 0 of the friction factor curve in Fig. 6.3-1.

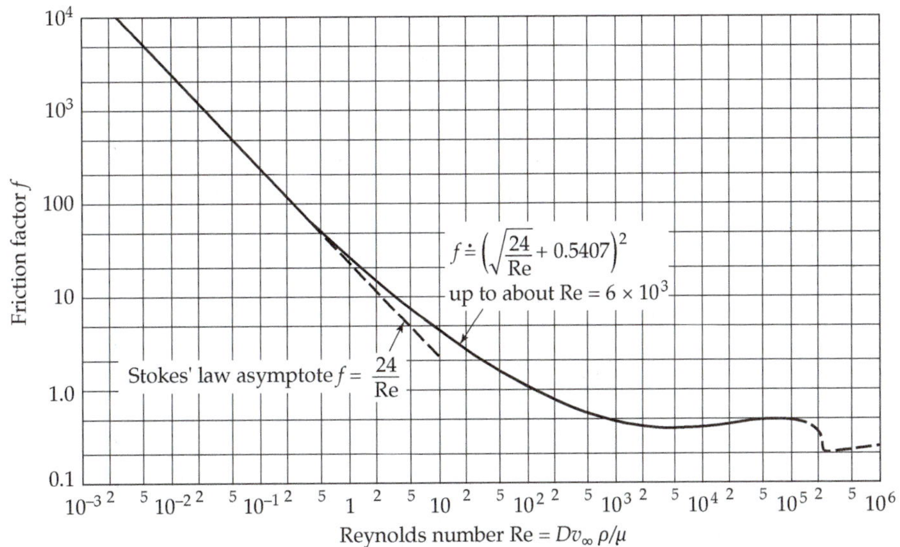

Fig. 6.3-1. Friction factor (or drag coefficient) for spheres moving relative to a fluid with a velocity v_∞. The definition of f is given in Eq. 6.1-5. [Curve taken from C. E. Lapple, "Dust and Mist Collection," in *Chemical Engineers' Handbook*, (J. H. Perry, ed.), McGraw-Hill, New York, 3rd edition (1950), p. 1018.]

For higher values of the Reynolds number, the empirical expression[2]

$$f = \left(\sqrt{\frac{24}{\mathrm{Re}}} + 0.5407\right)^2 \quad \text{for} \quad \mathrm{Re} < 6000 \tag{6.3-16}$$

is both simple and useful. It is important to remember that

$$f \approx 0.44 \quad \text{for} \quad 5 \times 10^2 < \mathrm{Re} < 1 \times 10^5 \tag{6.3-17}$$

which covers a remarkable range of Reynolds numbers. Equation 6.3-17 is sometimes called *Newton's resistance law*; it is handy for a seat-of-the-pants calculation. According to this, the drag force is proportional to the square of the approach velocity of the fluid.

Many extensions of Fig. 6.3-1 have been made, but a systematic study is beyond the scope of this text. Among the effects that have been investigated are wall effects[3] (see Problem 6C.2), fall of droplets with internal circulation,[4] hindered settling (i.e., fall of clusters of particles[5] that interfere with one another), unsteady flow,[6] and the fall of nonspherical particles.[7]

EXAMPLE 6.3-1

Determination of the Diameter of a Falling Sphere

Glass spheres of density $\rho_{\mathrm{sph}} = 2.62\ \mathrm{g/cm^3}$ fall through liquid CCl_4 at 20°C in an experiment for studying human reaction times in making time observations with stopwatches and more elaborate devices. At this temperature, the relevant properties of CCl_4 are: $\rho = 1.59\ \mathrm{g/cm^3}$ and $\mu = 9.58$ millipoise. What diameter should the spheres be in order to have a terminal velocity of about 65 cm/s?

SOLUTION

To find the sphere diameter, we have to solve Eq. 6.1-7 for D. However, in this equation one has to know D in order to get f; and f is given by the solid curve in Fig. 6.3-1. A trial-and-error procedure can be used, taking $f = 0.44$ as a first guess.

Alternatively, we can solve Eq. 6.1-7 for f and then note that f/Re is a quantity independent of D:

$$\frac{f}{\mathrm{Re}} = \frac{4}{3}\frac{g\mu}{\rho v_\infty^3}\left(\frac{\rho_{\mathrm{sph}} - \rho}{\rho}\right) \tag{6.3-18}$$

The quantity on the right side can be calculated with the information above, and we call it C. Hence, we have two simultaneous equations to solve:

$$f = C\mathrm{Re} \quad \text{from Eq. 6.3-18} \tag{6.3-19}$$

$$f = f(\mathrm{Re}) \quad \text{from Fig. 6.3-1} \tag{6.3-20}$$

Equation 6.3-19 is a straight line with slope of unity on the log-log plot of f versus Re. A graphical solution of these simultaneous equations is analogous to "Method B" introduced in Example 6.2-2.

---

[2] F. F. Abraham, *Physics of Fluids*, **13**, 2194 (1970); M. Van Dyke, *Physics of Fluids*, **14**, 1038–1039 (1971).

[3] J. R. Strom and R. C. Kintner, *AIChE Journal*, **4**, 153–156 (1958).

[4] L. Landau and E. M. Lifshitz, *Fluid Mechanics*, Pergamon, Oxford, 2nd edition (1987), pp. 65–66; S. Hu and R. C. Kintner, *AIChE Journal*, **1**, 42–48 (1955).

[5] C. E. Lapple, *Fluid and Particle Mechanics*, University of Delaware Press, Newark DE (1951), Chapter 13; R. F. Probstein, *Physicochemical Hydrodynamics*, Wiley, New York, 2nd edition (1994), §5.4.

[6] R. R. Hughes and E. R. Gilliland, *Chem. Eng. Prog.*, **48**, 497–504 (1952); L. Landau and E. M. Lifshitz, *Fluid Mechanics*, Pergamon, Oxford, 2nd edition (1987), pp. 90–91.

[7] E. S. Pettyjohn and E. B. Christiansen, *Chem. Eng. Prog.*, **44**, 147 (1948); H. A. Becker, *Can. J. Chem. Eng.*, **37**, 885–891 (1959); S. Kim and S. J. Karrila, *Microhydrodynamics: Principles and Selected Applications*, Butterworth-Heinemann, Boston (1991); Dover, Mineola, NY (2005), Chapter 5.

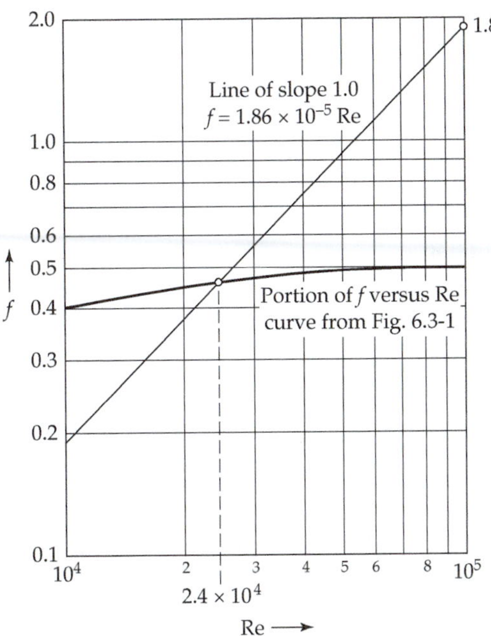

Fig. 6.3-2. Graphical procedure used in Example 6.3-1.

For the problem at hand we have

$$\frac{f}{\text{Re}} = \frac{4}{3}\frac{g\mu}{\rho v_\infty^3}\left(\frac{\rho_{\text{sph}} - \rho}{\rho}\right) \tag{6.3-21}$$

$$\begin{aligned} C &= \frac{4}{3}\frac{(980\,\text{cm/s}^2)(9.58 \times 10^{-3}\,\text{g/cm}\cdot\text{s})}{(1.59\,\text{g/cm}^3)(65\,\text{cm/s})^3}\left(\frac{(2.62 - 1.59)\,\text{g/cm}^3}{1.59\,\text{g/cm}^3}\right) \\ &= 1.86 \times 10^{-5} \end{aligned} \tag{6.3-22}$$

Hence, at Re $= 10^5$, according to Eq. 6.3-18, $f = 1.86$. The line of slope 1 passing through $f = 1.86$ at Re $= 10^5$ is shown in Fig. 6.3-2. This line intersects the curve of Eq. 6.3-20 (i.e., the curve of Fig. 6.3-1) at Re $= Dv_\infty\rho/\mu = 2.4 \times 10^4$. The sphere diameter is then found to be

$$D = \frac{\text{Re}\mu}{\rho v_\infty} = \frac{(2.4 \times 10^4)(9.58 \times 10^{-3}\,\text{g/cm}\cdot\text{s})}{(1.59\,\text{g/cm}^3)(65\,\text{cm/s})} = 2.2\,\text{cm} \tag{6.3-23}$$

---

§6.4 FRICTION FACTORS FOR PACKED COLUMNS

In the preceding two sections we have discussed the friction factor correlations for two simple flow systems of rather wide interest. Friction factor charts are available for a number of other systems, such as transverse flow past a cylinder, flow across tube banks, flow around baffles, and flow around rotating disks. These and many more are summarized in various reference works.[1] One complex system of considerable interest in chemical engineering is the packed column, widely used for catalytic reactors and for separation processes.

---

[1] P. C. Carman, *Flow of Gases through Porous Media*, Butterworths, London (1956); J. G. Richardson, Section 16 in *Handbook of Fluid Dynamics* (V. L. Streeter, ed.), McGraw-Hill, New York (1961); M. Kaviany, Chapter 21 in *The Handbook of Fluid Dynamics* (R. W. Johnson, ed.), CRC Press, Boca Raton, FL (1998).

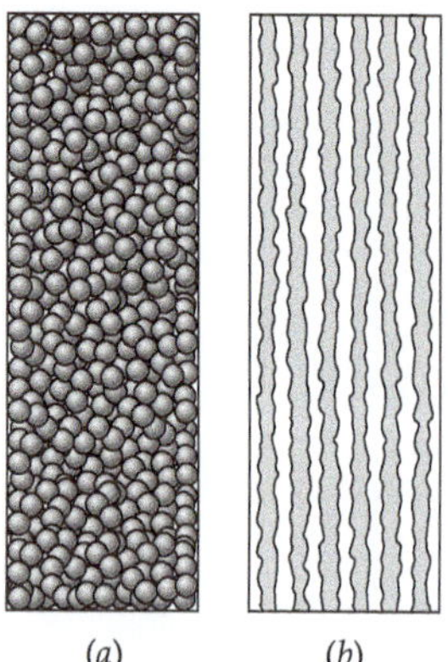

Fig. 6.4-1. (*a*) A cylindrical tube packed with spheres; (*b*) a "tube bundle" model for the packed column in (*a*).

There have been two main approaches for developing friction factor expressions for packed columns. In one method the packed column is visualized as a bundle of tangled tubes of weird cross section; the theory is then developed by applying the previous results for single straight tubes to the collection of crooked tubes. In the second method, the packed column is regarded as a collection of submerged objects, and the pressure drop is obtained by summing up the resistances of the submerged particles.[2] The tube bundle theories have been somewhat more successful, and we discuss them here. Figure 6.4-1(*a*) depicts a packed column, and Fig. 6.4-1(*b*) illustrates the tube bundle model.

A variety of materials may be used for the packing in columns: spheres, cylinders, Berl saddles, etc. It is assumed throughout the following discussion that the packing is statistically uniform, so that there is no "channeling" (in actual practice, channeling frequently occurs, and then the development given here does not apply). It is further assumed that the diameter of the packing particles is small in comparison to the diameter of the column in which the packing is contained, and that the column diameter is uniform.

We define the friction factor for the packed column analogously to Eq. 6.1-4:

$$f = \frac{1}{4}\left(\frac{D_p}{L}\right)\left(\frac{\breve{\mathscr{P}}_0 - \breve{\mathscr{P}}_L}{\frac{1}{2}\rho v_0^2}\right) \tag{6.4-1}$$

in which L is the length of the packed column, D_p is the particle diameter (defined presently), and v_0 is the *superficial velocity*; this is the volume flow rate divided by the empty column cross-sectional area, $v_0 = Q/S = w/\rho S$.

The pressure drop through one of the tubes in the tube bundle model is given by Eq. 6.2-16

$$\mathscr{P}_0 - \mathscr{P}_L = \tfrac{1}{2}\rho\langle\overline{v}_z\rangle^2\left(\frac{L}{R_h}\right)f_{\text{tube}} \tag{6.4-2}$$

in which $\langle\overline{v}_z\rangle$ is the average axial velocity in a model tube of mean hydraulic radius R_h (see §6.2 for the definition of R_h), and the friction factor for a single tube, f_{tube}, is a function of the Reynolds number $\text{Re}_h = 4R_h\langle\overline{v}_z\rangle\rho/\mu$. When this pressure difference is substituted into Eq. 6.4-1, we get

$$f = \frac{1}{4}\frac{D_p}{R_h}\frac{\langle\overline{v}_z\rangle^2}{v_0^2}f_{\text{tube}} = \frac{1}{4\varepsilon^2}\frac{D_p}{R_h}f_{\text{tube}} \tag{6.4-3}$$

---

[2] W. E. Ranz, *Chem. Eng. Prog.*, **48**, 274–253 (1952); H. C. Brinkman, *Appl. Sci. Research.*, **A1**, 27–34, 81–86, 333–346 (1949). **Henri Coenraad Brinkman** (1908–1961) did research on viscous dissipation heating, flow in porous media, and plasma physics; he taught at the University of Bandung, Indonesia, from 1949–1954, where he wrote *The Application of Spinor Invariants to Atomic Physics.*

In the second expression, we have introduced the *void fraction*, ε, the fraction of space in the column not occupied by the packing. Then, $v_0 = \langle \bar{v}_z \rangle \varepsilon$, which results from the definition of the superficial velocity. We now need an expression for R_h.

The hydraulic radius defined by Eq. 6.2-15 can be expressed in terms of the void fraction ε and the wetted surface a per unit volume of column as follows:

$$R_h = \left(\frac{\text{cross section available for flow}}{\text{wetted perimeter}}\right) = \left(\frac{\text{volume available for flow}}{\text{total wetted surface}}\right) = \frac{\left(\dfrac{\text{volume of voids}}{\text{volume of column}}\right)}{\left(\dfrac{\text{wetted surface}}{\text{volume of column}}\right)} = \frac{\varepsilon}{a} \tag{6.4-4}$$

The quantity a is related to the "specific surface" a_v (total particle surface per volume of particles) by

$$a_v = \frac{a}{1-\varepsilon} \tag{6.4-5}$$

The quantity a_v is in turn used to define the mean particle diameter D_p as follows:

$$D_p = \frac{6}{a_v} \tag{6.4-6}$$

This definition is chosen because, for spheres of uniform diameter, D_p is just the diameter of a sphere. From the last three expressions we find that the hydraulic radius is $R_h = D_p\varepsilon/6(1-\varepsilon)$. When this is substituted into Eq. 6.4-3, we get

$$f = \frac{3}{2}\left(\frac{1-\varepsilon}{\varepsilon^3}\right) f_{\text{tube}} \tag{6.4-7}$$

We now adapt this result to laminar and turbulent flows by inserting appropriate expressions for f_{tube}.

a. Laminar flow

For *laminar flow* in tubes, $f_{\text{tube}} = 16/\text{Re}_h$, where Re_h is defined by Eq. 6.2-17. This is exact for circular tubes only. To account for the fact that the fluid is flowing through tubes that are noncircular and that its path is quite tortuous, it has been found that replacing 16 by 100/3 allows the tube bundle model to describe the packed-column data. When this modified expression for the tube friction factor is used, Eq. 6.4-7 becomes

$$f = \frac{(1-\varepsilon)^2}{\varepsilon^3} \frac{75}{\left(D_p G_0/\mu\right)} \tag{6.4-8}$$

in which $G_0 = \rho v_0$ is the convective mass flux through the system. When this expression for f is substituted into Eq. 6.4-1, we get

$$\frac{\mathscr{P}_0 - \mathscr{P}_L}{L} = 150\left(\frac{\mu v_0}{D_p^2}\right)\frac{(1-\varepsilon)^2}{\varepsilon^3} \tag{6.4-9}$$

which is the *Blake-Kozeny equation*.[3] Equations 6.4-8 and 6.4-9 are generally good for $\left(D_p G_0/\mu(1-\varepsilon)\right) < 10$ and for void fractions less than $\varepsilon = 0.5$.

b. Highly turbulent flow

A treatment similar to the above can be given for turbulent flow. We begin again with the expression for the friction factor definition for flow in a circular tube. This time,

---

[3]F. C. Blake, *Trans. Amer. Inst. Chem. Engrs.*, **14**, 415–421 (1922); J. Kozeny, *Sitzungsber. Akad. Wiss. Wien, Abt. IIa*, **136**, 271–306 (1927).

however, we note that, for highly turbulent flow in tubes with any appreciable roughness, the friction factor is a function of the roughness only, and is independent of the Reynolds number. If we assume that the tubes in all packed columns have similar roughness characteristics, then the value of f_{tube} may be taken to be the same constant for all systems. Taking $f_{\text{tube}} = 7/12$ proves to be an acceptable choice. When this is inserted into Eq. 6.4-7, we get

$$f = \frac{7}{8}\left(\frac{1-\varepsilon}{\varepsilon^3}\right) \tag{6.4-10}$$

When this is substituted into Eq. 6.4-1, we get

$$\frac{\mathscr{P}_0 - \mathscr{P}_L}{L} = \frac{7}{4}\left(\frac{\rho v_0^2}{D_p}\right)\frac{1-\varepsilon}{\varepsilon^3} \tag{6.4-11}$$

which is the *Burke-Plummer*[4] equation valid for $\left(D_p G_0/\mu(1-\varepsilon)\right) > 1000$. Note that the dependence on the void fraction is different from that for laminar flow.

c. *Laminar-turbulent transition region*

Here, we may superpose the pressure-drop expressions for (a) and (b) above to get

$$\frac{\mathscr{P}_0 - \mathscr{P}_L}{L} = 150\left(\frac{\mu v_0}{D_p^2}\right)\frac{(1-\varepsilon)^2}{\varepsilon^3} + \frac{7}{4}\left(\frac{\rho v_0^2}{D_p}\right)\frac{1-\varepsilon}{\varepsilon^3} \tag{6.4-12}$$

For very small v_0, this simplifies to the Blake-Kozeny equation, and for very large v_0, to the Burke-Plummer equation. Such empirical superpositions of asymptotes often lead to satisfactory results. Equation 6.4-12 may be rearranged to form dimensionless groups:

$$\left(\frac{(\mathscr{P}_0 - \mathscr{P}_L)\rho}{G_0^2}\right)\left(\frac{D_p}{L}\right)\left(\frac{\varepsilon^3}{1-\varepsilon}\right) = 150\left(\frac{1-\varepsilon}{D_p G_0/\mu}\right) + \frac{7}{4} \tag{6.4-13}$$

This is the *Ergun equation*,[5] which is shown in Fig. 6.4-2 along with the Blake-Kozeny and Burke-Plummer equations and experimental data. The Ergun equation is valid for the entire range of Reynolds numbers in Fig. 6.4-2, and has been applied with success to gas flow through packed columns by using the density of the gas at the arithmetic average of the end pressures. Note that G_0 is constant through the column, whereas v_0 changes through the column for a compressible fluid. For large pressure drops, however, it seems more appropriate to apply Eq. 6.4-12 locally by expressing the pressure gradient in differential form.

The Ergun equation is but one of many[6] that have been proposed for describing packed columns. For example, the *Tallmadge equation*[7]

$$\left(\frac{(\mathscr{P}_0 - \mathscr{P}_L)\rho}{G_0^2}\right)\left(\frac{D_p}{L}\right)\left(\frac{\varepsilon^3}{1-\varepsilon}\right) = 150\left(\frac{1-\varepsilon}{D_p G_0/\mu}\right) + 4.2\left(\frac{1-\varepsilon}{D_p G_0/\mu}\right)^{1/6} \tag{6.4-14}$$

is reported to give good agreement with experimental data over the range $0.1 < \left(D_p G_0/\mu(1-\varepsilon)\right) < 10^5$.

The above discussion of packed beds illustrates how one can often combine solutions of elementary problems to create simple models for complex systems. The constants appearing in the models are then determined from experimental data. As better data become available, the modeling can be improved.

---

[4] S. P. Burke and W. B. Plummer, *Ind. Eng. Chem.*, **20**, 1196–1200 (1928).

[5] S. Ergun, *Chem. Engr. Prog.*, **48**, 89–94 (1952).

[6] I. F. Macdonald, M. S. El-Sayed, K. Mow, and F. A. Dullien, *Ind. Eng. Chem. Fundam.*, **18**, 199–208 (1979).

[7] J. A. Tallmadge, *AIChE Journal*, **16**, 1092–1093 (1970).

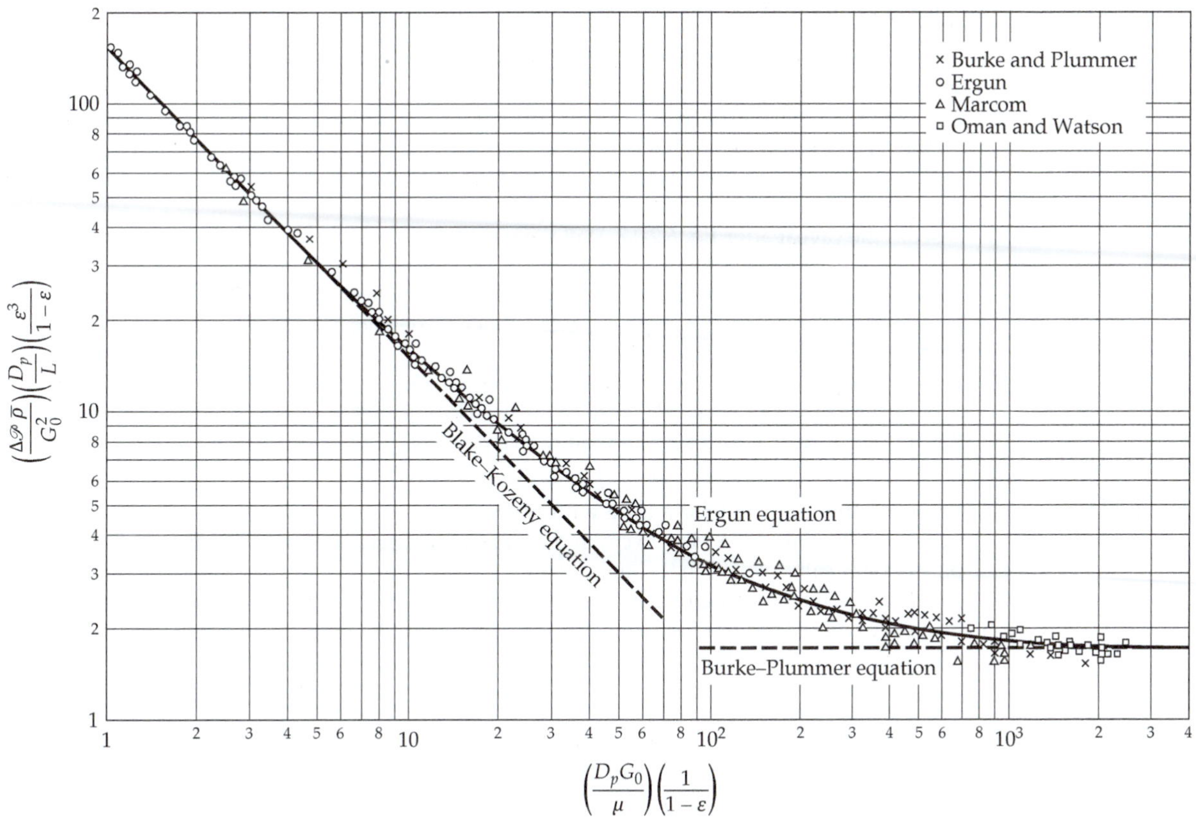

Fig. 6.4-2. The Ergun equation for flow in packed beds, and the two related asymptotes, the Blake-Kozeny equation and the Burke-Plummer equation. [S. Ergun, *Chem. Eng. Prog.*, **48**, 89–94 (1952).]

§6.5 CONCLUDING COMMENTS

In this chapter, we learned how to solve certain flow problems using correlations of experimental data. The correlations are developed by using dimensional analysis, thereby reducing the number of parameters in the relations.

In §6.2, we related the flow rate in tubes to the pressure drop using the Fanning friction factor. This friction factor depends only on the Reynolds number for the flow in the tube for large L/D values. Knowledge of this dependence allows one to solve for a variety of quantities given values for the other relevant flow parameters.

In §6.3, we related the force on a sphere to the ambient velocity using a similarly defined friction factor. Again, this friction factor, which depends only on the Reynolds number, can be used to interrelate a variety of flow parameters.

In §6.4, a similar analysis was performed to relate pressure drop to flow rate within packed beds. The resulting Ergun equation is valid for a wide range of Reynolds numbers.

Using the correlations presented in this chapter to solve flow problems is often much simpler than the methods employed in Chapters 2 and 3. However, the types of problems and information that can be determined using the correlations are much more limited. Thus, these approaches augment rather than replace the methods developed in Chapters 2, 3, and 4.

QUESTIONS FOR DISCUSSION

1. How are graphs of friction factors versus Reynolds numbers generated from experimental data, and why are they useful?
2. Compare and contrast the friction factor curves for flow in tubes and flow around spheres. Why do they not have the same shapes?
3. In Fig. 6.2-2, why does the f versus Re curve for turbulent flow lie above the curve for laminar flow rather than below?
4. Discuss the caveat after Eq. 6.2-17. Will the use of the mean hydraulic radius for laminar flow predict a pressure drop that is too high or too low for a given flow rate?
5. Can the friction factor correlations be used for unsteady flows?
6. Discuss the flow of water through a $\frac{1}{2}$-inch rubber garden hose that is attached to a house faucet with a pressure of 70 psig available.
7. Why was Eq. 6.4-12 rewritten in the form of Eq. 6.4-13?
8. A baseball announcer says: "Because of the high humidity today, the baseball cannot go so far through the heavy humid air as it would on a dry day." Comment critically on this statement.

PROBLEMS

6A.1 Pressure drop required for a pipe with fittings. What pressure drop is needed for pumping water at 20°C through a pipe 25 cm in diameter and length 1234 m at a rate of 1.97 m^3/s? The pipe is at the same elevation throughout and contains four standard-radius 90° elbows and two 45° elbows. The resistance of a standard-radius 90° elbow is roughly equivalent to that offered by a pipe whose length is 32 diameters; a 45° elbow, 15 diameters. (An alternative method for calculating losses in fittings is given in §7.5.)

Answer: 4.7×10^3psi = 33 MPa

6A.2 Pressure difference required for pipe with elevation change. Water at 68°F is to be pumped through 95 ft of standard 3-in. pipe (internal diameter 3.068 in.) into an overhead reservoir as illustrated in Fig. 6A.2.

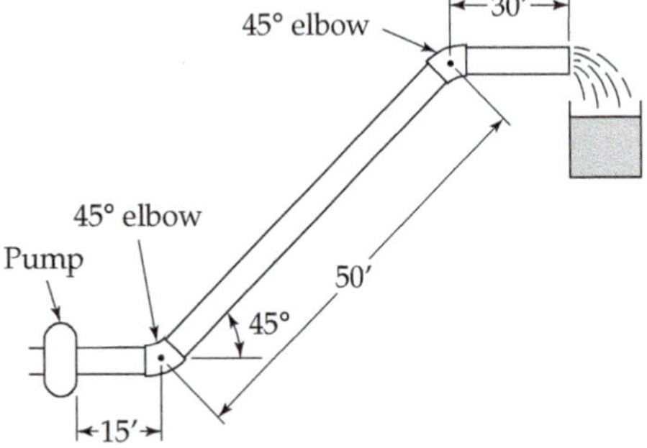

Fig. 6A.2 Pipe flow system.

(a) What pressure is required at the outlet of the pump to supply water to the overhead reservoir at a rate of 18 gal/min? At 68°F the viscosity of water is 1.002 cp and the density is 0.9982 g/ml.

(b) What percentage of the pressure drop (pump outlet to reservoir) is needed for overcoming the pipe friction?

Answer: **(a)** 15.3 psig

6A.3 Flow rate for a given pressure drop. How many gal/hr of water at 68°F can be delivered through a 1320-ft length of smooth 6.00-in. i.d. pipe under a pressure difference of 0.25 psi? Assume that the pipe is "hydraulically smooth."

(a) Solve by Method *A* of Example 6.2-2.

(b) Solve by Method *B* of Example 6.2-2.

Answer: 4.1×10^3gal/hr

6A.4 Motion of a sphere in a liquid. A hollow sphere, 5.00 mm in diameter, with a mass of 0.0500 g, is released in a column of liquid and attains a terminal velocity of 0.500 cm/s. The liquid density is 0.900 g/cm^3. The local gravitational acceleration is 980.7 cm/s^2. The sphere is far enough from the containing walls so that their effect can be neglected.

(a) Compute the drag force on the sphere in dynes.

(b) Compute the friction factor.

(c) Determine the viscosity of the liquid.

Answers: **(a)** 8.74 dynes; **(b)** $f = 396$; **(c)** 3.7 g/cm· s

6A.5 Sphere diameter for a given terminal velocity.

(a) Explain how to find the sphere diameter D corresponding to given values of v_∞, ρ, ρ_{sph}, μ, and g by making a direct construction on Fig. 6.3-1.

(b) Rework Problem 2A.4 by using Fig. 6.3-1.

(c) Rework (b) when the gas velocity is 10 ft/s.

6A.6 Estimation of void fraction of a packed column. A tube of 146 sq. in. cross section and 73 in. height is packed with spherical particles of diameter 2 mm. When a pressure difference of 158 psi is maintained across the column, a 60% aqueous sucrose solution at 20°C flows through the bed at a rate of 244 lb_m/min. At this temperature, the viscosity of the solution is 56.5 cp and its density is 1.2865 g/cm^3. What is the void fraction of the bed? Discuss the usefulness of this method of obtaining the void fraction.

Answer: 0.30

6A.7 Estimation of pressure drops in annular flow. For flow in an annulus formed by cylindrical surfaces of diameters D and κD (with $\kappa < 1$) the friction factors for laminar and turbulent flow are

Laminar
$$f = \frac{16}{\text{Re}_\kappa} \tag{6A.7-1}$$

Turbulent
$$\frac{1}{\sqrt{f}} = G\log_{10}\left(\text{Re}_\kappa\sqrt{f}\right) - H \tag{6A.7-2}$$

in which the Reynolds number is defined by

$$\text{Re}_\kappa = K\frac{D(1-\kappa)\langle\bar{v}_z\rangle\rho}{\mu} \tag{6A.7-3}$$

The values of G, H, and K are given as:[1]

κ	G	H	K
0.00	4.000	0.400	1.000
0.05	3.747	0.293	0.7419
0.10	3.736	0.239	0.7161
0.15	3.738	0.208	0.7021
0.20	3.746	0.186	0.6930
0.30	3.771	0.154	0.6820
0.40	3.801	0.131	0.6757
0.50	3.833	0.111	0.6719
0.60	3.866	0.093	0.6695
0.70	3.900	0.076	0.6681
0.80	3.933	0.060	0.6672
0.90	3.967	0.046	0.6668
1.00	4.000	0.031	0.6667

[1] D. M. Meter and R. B. Bird, *AIChE Journal*, 7, 41–45 (1961).

Equation 6A.7-2 is believed to be capable of reproducing the experimental data within about 3% up to Reynolds numbers of 20,000.

(a) Verify that, for developed laminar flow, Eqs. 6A.7-1 and 6A.7-3 with the tabulated K values are consistent with Eq. 2.4-17.

(b) An annular duct is formed from cylindrical surfaces of diameters 6 in. and 15 in. It is desired to pump water at 60°F at a rate of 1500 cu ft per second. How much pressure drop is required per unit length of conduit, if the annulus is horizontal? Use Eq. 6A.7-2.

(c) Repeat (b) using the "mean hydraulic radius" empiricism.

6A.8 Force on a water tower in a gale. A water tower has a spherical storage tank 40 ft in diameter. In a 100-mph gale, what is the force of the wind on the spherical tank at 0°C? Take the density of air to be 1.29 g/liter or 0.08 lb_m/ft^3 and the viscosity to be 0.017 cp.

Answer: $1.7 \times 10^4 \text{lb}_f = 5.4 \times 10^5$ poundals

6A.9 Flow of gas through a packed column. A horizontal tube with diameter 4 in. and length 5.5 ft is packed with glass spheres of diameter 1/16 in., and the void fraction is 0.41. Carbon dioxide is to be pumped through the tube at 300K, at which temperature its viscosity is known to be 1.495×10^{-4} g/cm · s. What will be the mass flow rate through the column when the inlet and outlet pressures are 25 atm and 3 atm, respectively?

Answer: 480 g/s

6A.10 Determination of pipe diameter. What size of circular pipe is needed to produce a flow rate of 250 firkins per fortnight when there is a pressure drop of 3×10^5 scruples per square barleycorn? The pipe is horizontal. (The authors are indebted to the late Professor R. S. Kirk of the University of Massachusetts who introduced them to these units.)

6B.1 Effect of error in friction factor calculations. In a calculation using the Blasius formula for turbulent flow in pipes, the Reynolds number used was too low by 4%. Calculate the resulting error in the friction factor.

Answer: Too high by 1%

6B.2 Friction factor for flow along a flat plate.[2]

(a) An expression for the drag force on a flat plate, wetted on both sides, is (see Problem 3C.2)

$$F_k = 1.328\sqrt{\rho\mu L W^2 v_\infty^3} \tag{6B.2-1}$$

This equation was derived by using *laminar* boundary-layer theory and is known to be in good agreement with experimental data. Define a friction factor and Reynolds number, and obtain the f versus Re relation.

(b) For *turbulent* flow, an approximate boundary-layer treatment based on the 1/7 power velocity distribution gives

$$F_k = 0.072\rho v_\infty^2 WL\left(Lv_\infty\rho/\mu\right)^{-1/5} \tag{6B.2-2}$$

When 0.072 is replaced by 0.074, this relation describes the drag force within experimental error for $5 \times 10^5 < Lv_\infty\rho/\mu < 2 \times 10^7$. Express the corresponding friction factor as a function of the Reynolds number.

6B.3 Friction factor for laminar flow in a slit. Use the results of Problem 2B.4 to show that for the laminar flow in a thin slit of width $2B$ the friction factor is $f = 12/\text{Re}$, if the Reynolds number is defined as $\text{Re} = 2B\langle \bar{v}_z \rangle \rho/\mu$. Compare this result for f with what one would get from the mean hydraulic radius empiricism.

6B.4 Friction factor for a rotating disk.[3] A thin circular disk of radius R is immersed in a large body of fluid with density ρ and viscosity μ. If a torque T_z is required to make the disk rotate

---

[2] H. Schlichting, *Boundary-Layer Theory*, McGraw-Hill, New York, 7th edition (1979), Chapter XXI.
[3] T. von Kármán, *Zeits. für angew. Math. u. Mech.*, **1**, 233–252 (1921).

at an angular velocity Ω, then a friction factor f may be defined analogously to Eq. 6.1-1 as follows,

$$T_z/R = AKf \tag{6B.4-1}$$

where reasonable definitions for K and A are $K = \frac{1}{2}\rho(\Omega R)^2$ and $A = 2(\pi R^2)$. An appropriate choice for the Reynolds number for the system is $\text{Re} = R^2\Omega\rho/\mu$.

For *laminar* flow, an exact boundary-layer development gives

$$T_z = 0.616\pi\rho R^4\sqrt{\mu\Omega^3/\rho} \tag{6B.4-2}$$

For *turbulent* flow, an approximate boundary-layer treatment based on the 1/7 power velocity distribution leads to

$$T_z = 0.073\rho\Omega^2R^5\sqrt[5]{\mu/R^2\Omega\rho} \tag{6B.4-3}$$

Express these results as relations between f and Re.

6B.5 Turbulent flow in horizontal pipes. A fluid is flowing with a mass flow rate w in a smooth horizontal pipe of length L and diameter D as the result of a pressure difference $p_0 - p_L$. The flow is known to be turbulent.

The pipe is to be replaced by one of diameter $D/2$ but with the same length. The same fluid is to be pumped at the same mass flow rate w. What pressure difference will be needed?

(a) Use Eq. 6.2-12 as a suitable equation for the friction factor.

(b) How can this problem be solved using Fig. 6.2-2 if Eq. 6.2-12 is not appropriate?

Answer: **(a)** A pressure difference 27 times greater will be needed.

6B.6 Inadequacy of mean hydraulic radius for laminar flow.

(a) For laminar flow in an annulus with radii κR and R, use Eqs. 6.2-16 and 6.2-17 to get an expression for the average velocity in terms of the pressure difference analogous to the exact expression given in Eq. 2.4-17.

(b) What is the percent error in the result in (a) for $\kappa = \frac{1}{2}$?

(c) Use Eqs. 6.2-16 and 6.2-17 to obtain expressions for the average velocity in laminar flow in terms of pressure difference for tubes of triangular (equilateral) and square cross sections. Compare these results to the exact results presented for laminar flow in Problems 3B.2 and 3B.3.

Answer: **(b)** 49%

6B.7 Falling sphere in Newton's drag-law region. A sphere initially at rest at $z = 0$ falls under the influence of gravity. Conditions are such that, after a negligible interval, the sphere falls with a resisting force proportional to the square of the velocity.

(a) Find the distance z that the sphere falls as a function of t.

(b) What is the terminal velocity of the sphere? Assume that the density of the fluid is much less than the density of the sphere.

Answer: **(a)** The distance is $z(t) = (1/c^2 g)\ln(\cosh cgt)$ where $c^2 = \frac{3}{8}(0.44)\left(\rho/\rho_{\text{sph}}\right)(1/gR)$; **(b)** $1/c$

6B.8 Design of an experiment to verify the f vs. Re chart for spheres. It is desired to design an experiment to test the friction factor chart in Fig. 6.3-1 for flow around a sphere. Specifically, we want to demonstrate that $f = 1$ at $\text{Re} = 100$. This is to be done by dropping bronze spheres $(\rho_{\text{sph}} = 8\ \text{g/cm}^3)$ in water $(\rho = 1\ \text{g/cm}^3,\ \mu = 10^{-2}\ \text{g/cm}\cdot\text{s})$. What sphere diameter must be used?

(a) Derive a formula that gives the sphere diameter as a function of f, Re, g, μ, ρ, and ρ_{sph} for terminal velocity conditions.

(b) Insert numerical values and find the value of the sphere diameter.

Answers: **(a)** $D = \sqrt[3]{\dfrac{3f\text{Re}^2\mu^2}{4\left(\rho_{\text{sph}} - \rho\right)\rho g}}$; **(b)** $D = 0.048$ cm.

6B.9 Friction factor for flow past an infinite cylinder.[4] The flow past a long cylinder (diameter D, length L) is very different from the flow past a sphere. It is found that, when the fluid approaches with a velocity v_∞, the kinetic force acting on a length L of the cylinder oriented perpendicular to the flow is

$$F_k = \frac{4\pi\mu v_\infty L}{\ln(7.4/\mathrm{Re})} \tag{6B.9-1}$$

The Reynolds number is defined here as $\mathrm{Re} = Dv_\infty\rho/\mu$. Equation 6B.9-1 is valid only up to about Re = 1. In this range of Re, what is the formula for the friction factor as a function of the Reynolds number?

6C.1 Two-dimensional particle trajectories. A sphere of radius R is fired horizontally (in the x direction) at high velocity in still air above level ground. As it leaves the propelling device, an identical sphere is dropped from the same height above the ground (in the y direction).

(a) Develop differential equations from which the particle trajectories can be computed, and which will permit comparison of the behavior of the two spheres. Include the effects of fluid friction, and make the assumption that steady-state friction factors may be used (this is a "quasi-steady-state assumption"; see Eq. 6.1-5(a)).

(b) Which sphere will reach the ground first?

(c) Would the answer to (b) have been the same if the sphere Reynolds numbers had been in the Stokes' law region?

Answers: **(a)**

$$\frac{dv_x}{dt} = -\frac{3}{8}\frac{v_x}{R}\sqrt{v_x^2 + v_y^2}\, f\frac{\rho_{\mathrm{air}}}{\rho_{\mathrm{sph}}},$$

$$\frac{dv_y}{dt} = -\frac{3}{8}\frac{v_y}{R}\sqrt{v_x^2 + v_y^2}\, f\frac{\rho_{\mathrm{air}}}{\rho_{\mathrm{sph}}} + \left(1 - \frac{\rho_{\mathrm{air}}}{\rho_{\mathrm{sph}}}\right) g$$

in which $f = f(\mathrm{Re})$ is given by Fig. 6.3-1, with

$$\mathrm{Re} = \frac{2R\sqrt{v_x^2 + v_y^2}\,\rho_{\mathrm{air}}}{\mu_{\mathrm{air}}}$$

[4] G. K. Batchelor, *An Introduction to Fluid Dynamics*, Cambridge University Press (1967), pp. 244–246, 257–261. For flow past finite cylinders, see J. Happel and H. Brenner, *Low-Reynolds Number Hydrodynamics*, Martinus Nijhoff, The Hague (1983), pp. 227–230.

Chapter 7

Macroscopic Balances for Isothermal Flow Systems

In the first three sections of Chapter 3, *equations of change* for isothermal systems were presented. The equations for mass, momentum, and angular momentum (where there is no interchange between internal and external angular momentum) were based on *conservation laws,* applied to a "microscopic system," namely, a tiny fixed element of volume through which the fluid is moving. The microscopic system has no solid bounding surfaces, and the interactions of the fluid with solid surfaces in specific flow systems are

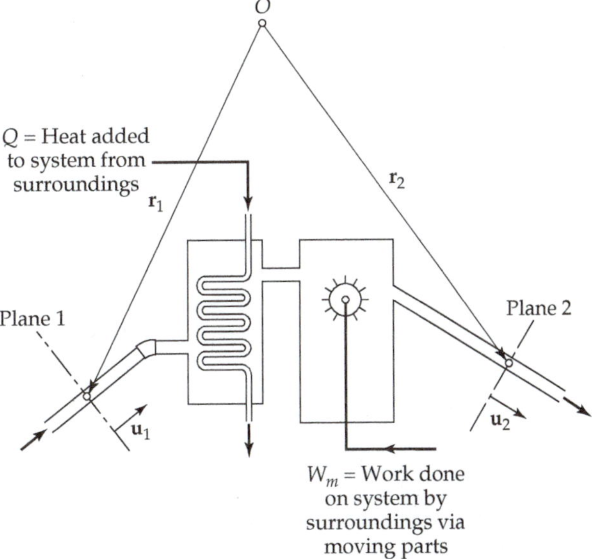

Fig. 7.0-1. Macroscopic flow system with fluid entering at plane 1 and leaving at plane 2. It may be necessary to add heat at a rate Q to maintain the system temperature constant. The rate of doing work *on* the system *by* the surroundings by means of moving surfaces is W_m. The symbols $\mathbf{u}_1$ and $\mathbf{u}_2$ denote *unit vectors* in the direction of flow at planes 1 and 2. The quantities $\mathbf{r}_1$ and $\mathbf{r}_2$ are position vectors giving the location of the centers of the inlet and outlet planes with respect to some designated origin of coordinates.

accounted for by boundary conditions on the differential equations. In the fourth section of Chapter 3, an additional equation of change, namely that for mechanical energy (which is not conserved), was obtained by taking the dot product of the velocity vector with the equation of motion, describing the conservation of momentum.

In this chapter we write similar conservation laws for mass, momentum, and angular momentum for "macroscopic systems," that is, the fluid within large pieces of equipment or parts thereof (see Fig. 7.0-1). However, we may not do this for mechanical energy, since it is not conserved. For unsteady-state systems, these macroscopic balances are ordinary differential equations, and for steady-state systems, they are algebraic equations. The macroscopic balances contain terms that account for the interactions of the fluid with the solid surfaces. The fluid can exert forces and torques on the surfaces of the system, and the surroundings can do work at a rate W_m on the system by means of moving surfaces.

The macroscopic balances can also be obtained from the equations of change by integrating the latter over the entire volume of the flow system:[1,2]

$$\int_{V(t)} (\text{eq. of continuity})\, dV = \text{macroscopic mass balance}$$

$$\int_{V(t)} (\text{eq. of motion})\, dV = \text{macroscopic momentum balance}$$

$$\int_{V(t)} (\text{eq. of angular momentum})\, dV = \text{macroscopic angular momentum balance}$$

$$\int_{V(t)} (\text{eq. of mechanical energy})\, dV = \text{macroscopic mechanical energy balance}$$

It is important to note that the first three of these macroscopic balances can be obtained either by writing the conservation laws directly for the macroscopic system or by performing the indicated integrations. However, to get the macroscopic mechanical energy balance, the corresponding equation of change must be integrated over the macroscopic system.

In §7.1 to §7.3 we set up the macroscopic mass, momentum, and angular momentum balances by writing the conservation laws. In §7.4 we state the macroscopic mechanical energy balance, postponing a detailed derivation until §7.7. In the macroscopic mechanical energy balance, there is a term called the "friction loss," and we devote §7.5 to methods for estimating this quantity. Then in §7.6 we show how the set of macroscopic balances can be used to solve flow problems.

The macroscopic balances have been widely used in many branches of engineering and applied science. They provide global descriptions of large systems without much regard for the details of the fluid dynamics inside the systems. Often they are useful for making an initial appraisal of an engineering problem and for making order-of-magnitude estimates of various quantities. Sometimes they are used to derive approximate relations, which can then be modified with the help of experimental data to compensate for terms that have been omitted or about which there is insufficient information.

In using the macroscopic balances one often has to decide which terms can be omitted, or one has to estimate some of the terms. This requires (i) intuition, based on experience with similar systems, (ii) some experimental data on the system, (iii) flow visualization studies, or (iv) order-of-magnitude estimates. This will be clear when we come to specific examples.

---

[1] R. B. Bird, *Chem. Eng. Sci.*, **6**, 123–131 (1957); *Chem. Eng. Educ.*, **27** (2), 102–109 (Spring 1993); *NPT Procestechnologie*, Sept. 1994. pp. 6–10; *Korean J. Chem. Eng.*, **15**, 105–123 (1998); R. B. Bird, W. E. Stewart, and E. N. Lightfoot, *Transport Phenomena*, Wiley, New York, Revised Second Edition (2007), §7.8.

[2] J. C. Slattery and R. A. Gaggioli, *Chem. Eng. Sci.*, **17**, 893–895 (1962).

The macroscopic balances make use of nearly all the topics covered thus far, and therefore, Chapter 7 provides a good opportunity for reviewing the preceding chapters.

§7.1 THE MACROSCOPIC MASS BALANCE

In the system shown in Fig. 7.0-1, the fluid enters the system at plane 1 with cross section S_1, and leaves at plane 2 with cross section S_2. The average velocity in the flow direction is $\langle v_1 \rangle$ at the entry plane and $\langle v_2 \rangle$ at the exit plane. In this and the following sections, we introduce two assumptions that are not very restrictive: (i) at the planes 1 and 2 the time-smoothed velocity is perpendicular to the relevant cross section, and (ii) at planes 1 and 2 the density and other physical properties are uniform over the cross section.

The law of conservation of mass for this system is then

$$\underbrace{\frac{d}{dt}m_{\text{tot}}}_{\substack{\text{rate of}\\\text{increase}\\\text{of mass}}} = \underbrace{\rho_1 \langle v_1 \rangle S_1}_{\substack{\text{rate of}\\\text{mass in}}} - \underbrace{\rho_2 \langle v_2 \rangle S_2}_{\substack{\text{rate of}\\\text{mass out}}} \tag{7.1-1}$$

Here $m_{\text{tot}} = \int \rho \, dV$ is the total mass of fluid contained in the system between planes 1 and 2. Also, in this chapter, we use the symbol $\langle v_z \rangle$ for the spatially averaged flow velocity, whether the flow is laminar or turbulent.

The average velocity can be related to the mass flow rate by

$$w = \rho Q = \rho \langle v \rangle S \tag{7.1-2}$$

where Q is the volumetric rate. We also introduce the notation $\Delta w = w_2 - w_1$ (exit value minus entrance value). Then the *unsteady-state macroscopic mass balance* becomes

Unsteady-state:
$$\boxed{\frac{d}{dt}m_{\text{tot}} = -\Delta w} \tag{7.1-3}$$

If the total mass of fluid does not change with time, then we get the *steady-state macroscopic mass balance.*

Steady-state:
$$\boxed{\Delta w = 0} \tag{7.1-4}$$

which is just the statement that the rate of mass entering equals the rate of mass leaving.

For the macroscopic mass balances we use the term "steady state" to mean that the time derivative on the left side (e.g., in Eq. 7.1-3) is zero. Within the system, because of the possibility for moving parts, flow instabilities, and turbulence, there may well be regions of unsteady flow.

EXAMPLE 7.1-1

Draining of a Spherical Tank

A spherical tank of radius R and its drainpipe of length L and diameter D are completely filled with a heavy oil. At time $t = 0$, the valve at the bottom of the drainpipe is opened. How long will it take to drain the tank? There is an air vent at the very top of the spherical tank. Ignore the amount of oil that clings to the inner surface of the tank, and assume that the flow in the drainpipe is laminar.

SOLUTION

We label three planes as in Fig. 7.1-1, and we let the instantaneous liquid level above plane 2 be $h(t)$. We define the system as the space within the tank between planes 1 and 2. Our strategy for solving this problem is to obtain expressions for each term in Eq. 7.1-3, and then use Eq. 7.1-3 to relate $h(t)$ to other quantities. With this information, we can determine the time required for h to decrease to zero.

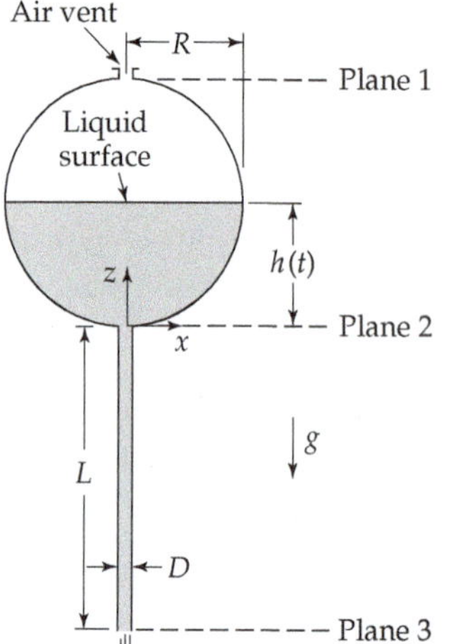

Fig. 7.1-1. Spherical tank with drainpipe.

We first determine the total mass of liquid in the system (between planes 1 and 2), which is equivalent to the total mass of liquid between plane 2 and the top of the liquid surface. We imagine that the sphere is generated by a circle in the xz plane, with its center at $z = R$ and $x = 0$. The tank is draining in the negative z direction, and the exit from the sphere into the attached tube is located at $z = 0$. The sphere is created by rotating the generating circle around the z axis.

The generating circle is described by the equation:

$$x^2 + (z - R)^2 = R^2 \quad \text{or} \quad x^2 = 2Rz - z^2 \tag{7.1-5}$$

We then visualize the liquid volume as being made up of a stack of thin circular disks of thickness dz, each with radius x given by Eq. 7.1-5, and a volume of

$$dV = \pi x^2 dz = \pi\left(2Rz - z^2\right) dz \tag{7.1-6}$$

The total volume of the liquid in the spherical tank at time t is then

$$V(t) = \int_0^{h(t)} \pi\left(2Rz - z^2\right) dz = \pi\left(Rz^2 - \frac{1}{3}z^3\right)\Bigg|_0^{h(t)}$$

$$= \pi R\left[h(t)\right]^2\left[1 - \frac{1}{3}\frac{h(t)}{R}\right] \tag{7.1-7}$$

and the total mass of liquid in the tank at time t is

$$m_{\text{tot}}(t) = \pi R\left[h(t)\right]^2\left[1 - \frac{1}{3}\frac{h(t)}{R}\right]\rho \tag{7.1-8}$$

This result may be checked by applying Eq. 7.1-8 to the cases when the tank is full, half full, and empty.

Since no fluid crosses plane 1, we know that $w_1 = 0$. The outlet mass flow rate w_2, as determined from the Hagen-Poiseuille formula (Eq. 2.3-21), is

$$w_2 = \frac{\pi\left(\mathscr{P}_2 - \mathscr{P}_3\right) D^4 \rho}{128\mu L} = \frac{\pi\left[\rho g h(t) + \rho g L\right] D^4 \rho}{128\mu L} \tag{7.1-9}$$

The Hagen-Poiseuille formula was derived for steady-state flow, but we use it here since the volume of liquid in the tank is changing slowly with time; this is an example of the "quasi-steady-state" approximation (where a slowly varying process—momentum transport in this case—is approximated as occurring at steady state).

When these expressions for m_{tot} and w_2 are substituted into Eq. 7.1-3, we get, after some rearrangement

$$-\frac{(2R - h)h}{h + L}\frac{dh}{dt} = \frac{\rho g D^4}{128\mu L} \tag{7.1-10}$$

We now abbreviate the constant on the right side of the equation as A. The equation is easier to integrate if we make the change of variable $H = h + L$ so that

$$\frac{[H-(2R+L)]\,(H-L)}{H}\frac{dH}{dt} = A \tag{7.1-11}$$

We now integrate this equation between $t = 0$ (when $h = 2R$ or $H = 2R + L$), and $t = t_{\text{efflux}}$ (when $h = 0$ or $H = L$). This gives for the efflux time

$$t_{\text{efflux}} = \frac{L^2}{A}\left[2\frac{R}{L}\left(1+\frac{R}{L}\right) - \left(1+2\frac{R}{L}\right)\ln\left(1+2\frac{R}{L}\right)\right] \tag{7.1-12}$$

in which A is given by the right side of Eq. 7.1-10. Note that we have obtained this result without any detailed analysis of the fluid motion within the sphere.

---

§7.2 THE MACROSCOPIC MOMENTUM BALANCE

We now apply the law of conservation of momentum to the system in Fig. 7.0-1, using the same two assumptions ((i) and (ii)) mentioned in the preceding section, plus two additional assumptions: (iii) the forces associated with the stress tensor $\boldsymbol{\tau}$ are neglected at planes 1 and 2, since they are generally small compared to the pressure forces at the entry and exit planes, and (iv) the pressure does not vary over the cross section at the entry and exit planes.

Since momentum is a vector quantity, each term in the balance must be a vector. We use unit vectors $\mathbf{u}_1$ and $\mathbf{u}_2$ to represent the direction of flow at planes 1 and 2. The law of conservation of momentum then reads

$$\frac{d}{dt}\mathbf{P}_{\text{tot}} = \rho_1\langle v_1^2\rangle S_1\mathbf{u}_1 - \rho_2\langle v_2^2\rangle S_2\mathbf{u}_2 + p_1S_1\mathbf{u}_1 - p_2S_2\mathbf{u}_2 + \mathbf{F}_{s\to f} + m_{\text{tot}}\mathbf{g} \tag{7.2-1}$$

rate of increase of momentum	rate of momentum in at plane 1	rate of momentum out at plane 2	pressure force on fluid at plane 1	pressure force on fluid at plane 2	force of solid surface on fluid	force of gravity on fluid

Here $\mathbf{P}_{\text{tot}} = \int \rho\mathbf{v}\,dV$ is the total momentum in the system. The equation states that the total momentum within the system changes because of convection of momentum into and out of the system, and because of the various forces acting on the system: the pressure forces at the ends of the system, the force of the solid surfaces acting on the fluid in the system, and the force of gravity acting on the fluid within the walls of the system. The subscript "$s \to f$" serves as a reminder of the direction of the force.

By introducing the mass rate of flow and the Δ symbol, we get finally for the *unsteady-state macroscopic momentum balance*

Unsteady-state:
$$\frac{d}{dt}\mathbf{P}_{\text{tot}} = -\Delta\left(\frac{\langle v^2\rangle}{\langle v\rangle}w + pS\right)\mathbf{u} + \mathbf{F}_{s\to f} + m_{\text{tot}}\mathbf{g} \tag{7.2-2}$$

If the total amount of momentum in the system does not change with time, then we get the *steady-state macroscopic momentum balance*

Steady-state:
$$\mathbf{F}_{f\to s} = -\Delta\left(\frac{\langle v^2\rangle}{\langle v\rangle}w + pS\right)\mathbf{u} + m_{\text{tot}}\mathbf{g} \tag{7.2-3}$$

Once again we emphasize that this is a vector equation. It is useful for computing the force of the fluid on the solid surfaces, $\mathbf{F}_{f\to s}$, such as the force on a pipe bend or a turbine blade. Actually we have already used a simplified version of the above equation in Eq. 6.1-3.

Notes Regarding Turbulent Flow: (i) For turbulent flow, it is customary to replace $\langle v \rangle$ by $\langle \overline{v} \rangle$ and $\langle v^2 \rangle$ by $\langle \overline{v}^2 \rangle$; in the latter we are neglecting the term $\langle \overline{v'^2} \rangle$, which is generally small with respect to $\langle \overline{v}^2 \rangle$. (ii) Then we further replace $\langle \overline{v}^2 \rangle / \langle \overline{v} \rangle$ by $\langle \overline{v} \rangle$. The error in doing this is quite small; for the empirical $\frac{1}{7}$-power law velocity profile given in Eq. 4.1-4, $\langle \overline{v}^2 \rangle / \langle \overline{v} \rangle = \frac{50}{49} \langle \overline{v} \rangle$, so that the error is about 2%. (iii) When we make this assumption, we will normally drop the angular brackets and overbars in order to simplify the notation. That is, we will let $\langle \overline{v}_1 \rangle \equiv v_1$ and $\langle \overline{v}_1^2 \rangle \equiv v_1^2$, with similar simplifications for quantities at plane 2.

EXAMPLE 7.2-1

Force Exerted by a Jet (Part (a))

A turbulent jet of water emerges from a tube of radius $R_1 = 2.5$ cm with a speed $v_1 = 6$ m/s, as shown in Fig. 7.2-1. The jet impinges on a disk-and-rod assembly of mass $m = 5.5$ kg, which is free to move vertically. The friction between the rod and the sleeve will be neglected. Find the height h at which the disk will "float" as a result of the jet.[1] Assume that the water is incompressible.

SOLUTION

To solve this problem, one has to imagine how the jet behaves. In Fig. 7.2-1(a) we assume that the jet has a constant radius, R_1, between the tube exit and the disk, whereas in Fig. 7.2-1(b) we assume that the jet spreads slightly. In this example, we make the first assumption, and in Example 7.4-1, we account for the jet spreading.

We apply the z component of the steady-state momentum balance between planes 1 and 2. The pressure terms can be omitted, since the pressure is atmospheric at both planes. The fluid velocity at plane 2 is zero. The momentum balance then becomes

$$mg = v_1 (\rho v_1 \pi R_1^2) - (\pi R_1^2 h) \rho g \tag{7.2-4}$$

where mg is the force exerted by the solid disk on the fluid in the $-z$ direction. When this is solved for h, we get (in SI units)

$$h = \frac{v_1^2}{g} - \frac{m}{\rho \pi R_1^2} = \frac{(6 \text{ m/s})^2}{(9.807 \text{ m/s}^2)} - \frac{(5.5 \text{ kg})}{(1000 \text{ kg/m}^3)\, \pi (0.025 \text{ m})^2} = 0.87 \text{ m} \tag{7.2-5}$$

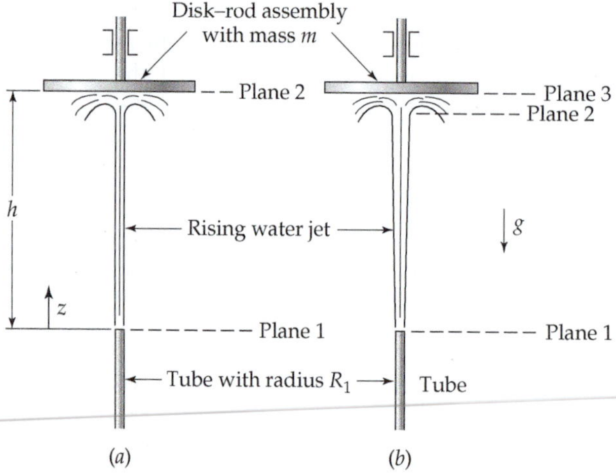

Fig. 7.2-1. Sketches corresponding to the two solutions to the jet-and-disk problem. In (a) the water jet is assumed to have a uniform radius R_1. In (b) allowance is made for the spreading of the liquid jet.

[1] K. Federhofer, *Aufgaben aus der Hydromechanik*, Springer-Verlag, Vienna (1954), p. 36 and p. 172.

§7.3 THE MACROSCOPIC ANGULAR MOMENTUM BALANCE

The development of the macroscopic angular momentum balance parallels that for the (linear) momentum balance in the preceding section. All we have to do is to replace "momentum" by "angular momentum," and "force" by "torque."

To describe the angular momentum and torque, we have to select an origin of coordinates with respect to which these quantities are evaluated. The origin is designated by "O" in Fig. 7.0-1, and the locations of the midpoints of planes 1 and 2 with respect to this origin are given by the position vectors $\mathbf{r}_1$ and $\mathbf{r}_2$.

Once again we make assumptions (i)–(iv) introduced in §7.1 and §7.2. With these assumptions the rate of entry of angular momentum at plane 1, which is $\int [\mathbf{r} \times \rho\mathbf{v}](\mathbf{v} \cdot \mathbf{u})\, dS$ evaluated at that plane, becomes $\rho_1 \langle v_1^2 \rangle S_1 [\mathbf{r}_1 \times \mathbf{u}_1]$, with a similar expression for the rate at which angular momentum leaves the system at 2.

The *unsteady-state macroscopic angular momentum balance* may now be written as

$$\underset{\substack{\text{rate of}\\\text{increase of}\\\text{angular}\\\text{momentum}}}{\frac{d}{dt}\mathbf{L}_{\text{tot}}} = \underset{\substack{\text{rate of angular}\\\text{momentum}\\\text{in at plane 1}}}{\rho_1 \langle v_1^2 \rangle S_1 [\mathbf{r}_1 \times \mathbf{u}_1]} - \underset{\substack{\text{rate of angular}\\\text{momentum}\\\text{out at plane 2}}}{\rho_2 \langle v_2^2 \rangle S_2 [\mathbf{r}_2 \times \mathbf{u}_2]}$$

$$+ \underset{\substack{\text{torque due to}\\\text{pressure on}\\\text{fluid at}\\\text{plane 1}}}{p_1 S_1 [\mathbf{r}_1 \times \mathbf{u}_1]} - \underset{\substack{\text{torque due to}\\\text{pressure on}\\\text{fluid at}\\\text{plane 2}}}{p_2 S_2 [\mathbf{r}_2 \times \mathbf{u}_2]} + \underset{\substack{\text{torque}\\\text{of solid}\\\text{surface}\\\text{on fluid}}}{\mathbf{T}_{s\to f}} + \underset{\substack{\text{external}\\\text{torque}\\\text{on fluid}}}{\mathbf{T}_{\text{ext}}} \tag{7.3-1}$$

Here $\mathbf{L}_{\text{tot}} = \int \rho [\mathbf{r} \times \mathbf{v}]\, dV$ is the total angular momentum within the system, and $\mathbf{T}_{\text{ext}} = \int [\mathbf{r} \times \rho\mathbf{g}]\, dV$ is the torque on the fluid in the system resulting from the gravitational force. This equation can also be written as

Unsteady-state:

$$\frac{d}{dt}\mathbf{L}_{\text{tot}} = -\Delta\left(\frac{\langle v^2 \rangle}{\langle v \rangle} w + pS\right)[\mathbf{r} \times \mathbf{u}] + \mathbf{T}_{s\to f} + \mathbf{T}_{\text{ext}} \tag{7.3-2}$$

Finally, the *steady-state macroscopic angular momentum balance* is:

Steady-state:

$$\mathbf{T}_{f\to s} = -\Delta\left(\frac{\langle v^2 \rangle}{\langle v \rangle} w + pS\right)[\mathbf{r} \times \mathbf{u}] + \mathbf{T}_{\text{ext}} \tag{7.3-3}$$

This gives the torque exerted by the fluid on the solid surfaces.

EXAMPLE 7.3-1

Torque on a Mixing Vessel[1]

A mixing vessel, shown in Fig. 7.3-1, is being operated at steady state. The fluid enters tangentially at plane 1 in turbulent flow with a velocity v_1 and leaves through the vertical pipe with a velocity v_2. Since the tank is baffled, there is no swirling motion of the fluid in the vertical exit pipe. Find the torque exerted on the mixing vessel.

SOLUTION

The origin of the coordinate system is taken to be on the tank axis in a plane passing through the axis of the entrance pipe and parallel to the tank top. Then the vector $[\mathbf{r}_1 \times \mathbf{u}_1]$ is a vector

[1] R.B. Bird, in *Selected Topics in Transport Phenomena*, CEP Symposium Series #58, **61**, 1–15 (1965).

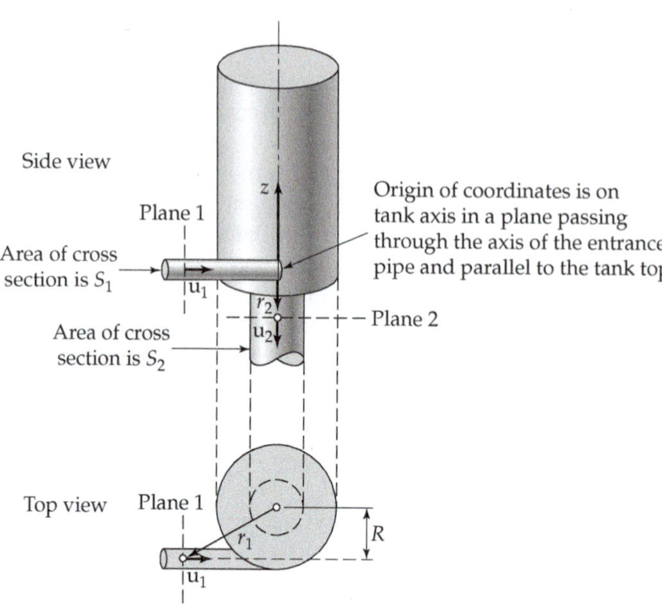

Fig. 7.3-1. Torque on a tank, showing side view and top view. [Adapted from R. B. Bird, in *Selected Topics in Transport Phenomena*, CEP Symposium Series #58, **61**, 1–15 (1965).]

pointing in the z direction with magnitude R. Furthermore, $[\mathbf{r}_2 \times \mathbf{u}_2] = 0$, since the two vectors are collinear. For this problem Eq. 7.3-3 gives

$$\mathbf{T}_{f\to s} = (\rho v_1^2 S_1 + p_1 S_1) R\boldsymbol{\delta}_z \tag{7.3-4}$$

Thus, the torque is just "force × lever arm," as would be expected. If the torque is sufficiently large, the equipment must be suitably braced to withstand the torque produced by the fluid motion and the inlet pressure.

EXAMPLE 7.3-2

Angular Velocity of a Lawn Sprinkler

A lawn sprinkler has four arms of length L and cross-sectional area S as illustrated in Fig. 7.3-2.[2] Water enters the sprinkler at the center at a mass flow rate w and splits into four streams. It is desired to find the angular velocity Ω of the sprinkler, when there is a frictional torque T_f per arm. (Note that the angular velocity *vector* of the sprinkler illustrated in Fig. 7.3-2 is $-\Omega\boldsymbol{\delta}_z$.)

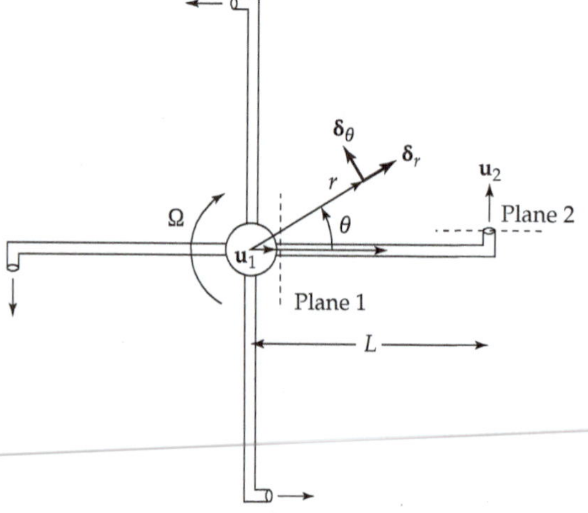

Fig. 7.3-2. The lawn sprinkler.

[2] F. M. White, *Fluid Mechanics*, McGraw-Hill, New York (1979), p. 174 describes the angular momentum balance for a one-armed sprinkler.

SOLUTION

The flow velocity in each arm is $v = w/4\rho S$, so that the velocity of an outlet stream, relative to the sprinkler arm, in the θ direction is (at plane 2)

$$\mathbf{v}_2 = \boldsymbol{\delta}_\theta \left(\frac{w}{4\rho S} - L\Omega \right) \tag{7.3-5}$$

Therefore, Eq. 7.3-3 gives for the angular momentum balance (z component) at steady state

$$0 = -w \left[L\boldsymbol{\delta}_r \times \left(\frac{w}{4\rho S} - L\Omega \right) \boldsymbol{\delta}_\theta \right] - T_f \boldsymbol{\delta}_z \tag{7.3-6}$$

The entering stream contributes nothing to the angular momentum balance because the flow is perpendicular to the z axis. The unit vectors may now be removed, since $[\boldsymbol{\delta}_r \times \boldsymbol{\delta}_\theta] = \boldsymbol{\delta}_z$, to give

$$0 = -wL \left(\frac{w}{4\rho S} - L\Omega \right) - T_f \tag{7.3-7}$$

Then, solving for Ω, we get

$$\Omega = \frac{w}{4\rho SL} - \frac{T_f}{wL^2} \tag{7.3-8}$$

for the angular velocity of the sprinkler.

§7.4 THE MACROSCOPIC MECHANICAL ENERGY BALANCE

Equations 7.1-3, 7.2-2, and 7.3.2 have been set up by applying the laws of conservation of mass, (linear) momentum, and angular momentum over the macroscopic system in Fig. 7.0-1. The three macroscopic balances thus obtained correspond to the equations of change in Eqs. 3.1-4, 3.2-9, and 3.4-1, and, in fact, they are very similar in structure. These three macroscopic balances can also be obtained by integrating the three equations of change over the volume of the flow system.

Next we want to set up the macroscopic mechanical energy balance, which corresponds to the equation of mechanical energy in Eq. 3.3-2. There is no way to do this directly as we have done in the preceding three sections, since there is no conservation law for mechanical energy. In this instance we *must* integrate the equation of change of mechanical energy over the volume of the flow system. The result, which has made use of the same assumptions (i–iv) used above, is the *unsteady-state macroscopic mechanical energy balance* (sometimes called the *engineering Bernoulli equation*). The equation is derived in §7.7; here we state the result and discuss its meaning:

$$\frac{d}{dt}\left(K_{\text{tot}} + \Phi_{\text{tot}}\right) = \left(\tfrac{1}{2}\rho_1 \langle v_1^3 \rangle + \rho_1 \hat{\Phi}_1 \langle v_1 \rangle\right) S_1 - \left(\tfrac{1}{2}\rho_2 \langle v_2^3 \rangle + \rho_2 \hat{\Phi}_2 \langle v_2 \rangle\right) S_2$$

rate of increase of kinetic and potential energy in system	rate at which kinetic and potential energy enter system at plane 1	rate at which kinetic and potential energy leave system at plane 2

$$+ \left(p_1 \langle v_1 \rangle S_1 - p_2 \langle v_2 \rangle S_2\right) + W_m + \int_{V(t)} p(\nabla \cdot \mathbf{v})\, dV + \int_{V(t)} (\boldsymbol{\tau} : \nabla \mathbf{v})\, dV \tag{7.4-1}$$

net rate at which the surroundings do work on the fluid at planes 1 and 2 by the pressure	rate of doing work on fluid by moving surfaces	rate at which mechanical energy increases or decreases because of expansion or compression of fluid	rate at which mechanical energy decreases because of viscous dissipation[1]

[1]This interpretation of the term is valid only for Newtonian fluids; polymeric liquids have elasticity and the interpretation given above no longer holds.

Here $K_{\text{tot}} = \int \frac{1}{2}\rho v^2 \, dV$ and $\Phi_{\text{tot}} = \int \frac{1}{2}\rho \hat{\Phi} \, dV$ are the total kinetic and potential energies within the system. According to Eq. 7.4-1, the total mechanical energy (i.e., kinetic plus potential) changes because of a difference in the rates of addition and removal of mechanical energy, because of work done on the fluid by the surroundings, and because of compressibility effects and viscous dissipation. Note that, at the system entrance (plane 1), the force $p_1 S_1$ multiplied by the velocity $\langle v_1 \rangle$ gives the rate at which the surroundings does work on the fluid. Furthermore, W_m is the rate of work done by the surroundings on the fluid by means of moving surfaces.

The macroscopic mechanical energy balance may now be written more compactly as

Unsteady state:
$$\frac{d}{dt}\left(K_{\text{tot}} + \Phi_{\text{tot}}\right) = -\Delta\left(\frac{1}{2}\frac{\langle v^3 \rangle}{\langle v \rangle} + \hat{\Phi} + \frac{p}{\rho}\right) w + W_m - E_c - E_v \tag{7.4-2}$$

in which the terms E_c and E_v are defined as follows:

$$E_c = -\int_{V(t)} p(\nabla \cdot \mathbf{v}) \, dV \quad \text{and} \quad E_v = -\int_{V(t)} (\boldsymbol{\tau} : \nabla \mathbf{v}) \, dV \tag{7.4-3,4}$$

The *compression term* E_c is positive in compression and negative in expansion; it is zero when the fluid is assumed to be incompressible. The term E_v is the *viscous dissipation* (or *friction-loss*) *term*, which is always positive for Newtonian liquids, as can be seen from Eq. 3.3-3. (For polymeric fluids, which are viscoelastic, E_v is not necessarily positive; these fluids are discussed in the next chapter.)

If the total kinetic-plus-potential energy in the system is not changing with time, we get

Steady state:
$$\Delta\left(\frac{1}{2}\frac{\langle v^3 \rangle}{\langle v \rangle} + gh + \frac{p}{\rho}\right) w = W_m - E_c - E_v \tag{7.4-5}$$

which is the *steady-state macroscopic mechanical energy balance*. Here h is the height above some arbitrarily chosen datum plane.

Next, if we assume that it is possible to draw a representative streamline through the system, we may combine the $\Delta\,(p/\rho)$ and E_c terms to get the following *approximate* relation (see §7.7)

$$\Delta\left(\frac{p}{\rho}\right) + E_c \approx w \int_1^2 \frac{1}{\rho} \, dp \tag{7.4-6}$$

Then, after dividing Eq. 7.4-5 by $w_1 = w_2 = w$, we get

Steady state (approx):
$$\Delta\left(\frac{1}{2}\frac{\langle v^3 \rangle}{\langle v \rangle}\right) + g\Delta h + \int_1^2 \frac{1}{\rho} \, dp = \hat{W}_m - \hat{E}_v \tag{7.4-7}$$

Here $\hat{W}_m = W_m/w$ and $\hat{E}_v = E_v/w$. Equation 7.4-7 is the version of the steady-state mechanical energy balance that is most often used. Each term in this equation has dimensions of energy/mass. For isothermal systems, the integral term can be calculated as long as an expression for density as a function of pressure is available.

Equation 7.4-7 should now be compared with Eq. 3.3-9, which is the "classical" Bernoulli equation for an inviscid fluid. If, to the right side of Eq. 3.3-9, we simply add the work $\hat{W}_m$ done by the surroundings and subtract the viscous-dissipation term $\hat{E}_v$, and reinterpret the velocities as appropriate averages over the cross sections, then we get Eq. 7.4-7. This provides a "plausibility argument" for Eq. 7.4-7 and still preserves the fundamental idea that the macroscopic mechanical energy balance is derived from the equation of motion (i.e., from the law of conservation of momentum). The full derivation of the macroscopic mechanical energy balance is given in §7.7 for those who are interested.

Notes for Turbulent Flow: (i) For turbulent flows we replace $\langle v^3 \rangle$ by $\langle \overline{v}^3 \rangle$, and ignore the contribution from the turbulent fluctuations. (ii) It is common practice to replace the quotient $\langle \overline{v}^3 \rangle / \langle \overline{v} \rangle$ by $\langle \overline{v} \rangle^2$. For the empirical $\frac{1}{7}$-power law velocity profile given in Eq. 4.1-4, it can be shown that $\langle \overline{v}^3 \rangle / \langle \overline{v} \rangle = \frac{43200}{40817} \langle \overline{v} \rangle^2$, so that the error amounts to about 6%. (iii) We further omit the brackets and overbars in order to simplify the notation in turbulent flow.

EXAMPLE 7.4-1

Force Exerted by a Jet (Part (b))

Continue the problem in Example 7.2-1 by accounting for the spreading of the jet as it moves upward.

SOLUTION

We now assume that the jet diameter increases with increasing z as shown in Fig. 7.2-1(*b*). It is convenient to work with three planes and to make balances between pairs of planes. The separation between planes 2 and 3 is taken to be quite small.

A mass balance between planes 1 and 2 gives

$$w_1 = w_2 \tag{7.4-8}$$

Next we apply the mechanical energy balance of Eq. 7.4-5 or 7.4-7 between the same two planes. The pressures at 1 and 2 are both atmospheric, and there is no work done by moving parts W_m. We assume that the viscous dissipation term E_v can be neglected. If z is measured upward from the tube exit, then $g\Delta h = g(h_2 - h_1) \approx g(h - 0)$, since planes 2 and 3 are so close together. Thus, the mechanical energy balance gives

$$\frac{1}{2}\left(v_2^2 - v_1^2\right) + gh = 0 \tag{7.4-9}$$

We now apply the z momentum balance between planes 2 and 3. Since the region is very small, we neglect the last term in Eq. 7.2-3. Both planes are at atmospheric pressure, so that the pressure terms do not contribute. The z component of the fluid velocity is zero at plane 3, so that there are only two terms left in the momentum balance

$$mg = v_2 w_2 \tag{7.4-10}$$

The above three equations can be rearranged to obtain h in terms of known quantities. Equation 7.4-9 can be rearranged to give

$$h = \frac{v_1^2}{2g}\left(1 - \frac{v_2^2}{v_1^2}\right) \tag{7.4-11}$$

Solving Eq. 7.4-10 for v_2 and inserting the result into Eq. 7.4-11 gives

$$h = \frac{v_1^2}{2g}\left(1 - \frac{\left(mg/w_2\right)^2}{v_1^2}\right) \tag{7.4-12}$$

Using Equation 7.4-8, this can be rewritten

$$h = \frac{v_1^2}{2g}\left(1 - \left(\frac{mg}{v_1 w_1}\right)^2\right) \tag{7.4-13}$$

in which mg and $v_1 w_1 = \pi R_1^2 \rho v_1^2$ are known. When the numerical values are substituted into Eq. 7.4-13, we get $h = 0.77$ m. This is probably a better result than the value of 0.87 m obtained in Example 7.2-1, since it accounts for the spreading of the jet. We have not, however, considered the clinging of the water to the disk, which gives the disk-rod assembly a somewhat greater effective mass. In addition, the frictional resistance of the rod in the sleeve has been neglected. It is necessary to run an experiment in order to assess the validity of Eq. 7.4-13.

§7.5 ESTIMATION OF THE VISCOUS LOSS

This section is devoted to methods for estimating the viscous loss (or friction loss), E_v, which appears in the macroscopic mechanical energy balance. The general expression for E_v is given in Eq. 7.4-4. For incompressible Newtonian fluids, Eq. 3.3-3 may be used to rewrite E_v as

$$E_v = \int \mu\Phi_v \, dV \tag{7.5-1}$$

which shows that it is the integral of the local rate of viscous dissipation over the volume of the entire flow system.

We now want to examine E_v from the point of view of dimensional analysis. The quantity Φ_v is a sum of squares of velocity gradients; hence, it has dimensions of $(v_0/l_0)^2$, where v_0 and l_0 are a characteristic velocity and length, respectively. We can therefore write

$$E_v = (\rho v_0^3 l_0^2)\left(\frac{\mu}{l_0 v_0 \rho}\right)\int \breve{\Phi}_v \, d\breve{V} \tag{7.5-2}$$

where $\breve{\Phi}_v = (l_0/v_0)^2\Phi_v$ and $d\breve{V} = l_0^{-3}\, dV$ are dimensionless quantities. If we make use of the dimensional arguments of §5.1 and §6.2, we see that the integral in Eq. 7.5-2 depends only on the various dimensionless groups in the equations of change and on various geometrical factors that enter into the boundary conditions. Hence, if the only significant dimensionless group is a Reynolds number, $\text{Re} = l_0 v_0 \rho/\mu$, then Eq. 7.5-2 must have the general form

$$E_v = (\rho v_0^3 l_0^2) \times \begin{pmatrix}\text{a dimensionless function of Re} \\ \text{and various geometrical ratios}\end{pmatrix} \tag{7.5-3}$$

In *steady-state flow* we prefer to work with the quantity $\hat{E}_v = E_v/w$, in which $w = \rho\langle v\rangle S$ is the mass rate of flow passing through *any* cross section of the flow system. If we select the reference velocity v_0 to be $\langle v\rangle$ and the reference length l_0 to be $\sqrt{S}$, then we may write

$$\hat{E}_v = \frac{1}{2}\langle v\rangle^2 e_v \tag{7.5-4}$$

in which e_v, the *friction-loss factor*, is a function of a Reynolds number and relevant dimensionless geometrical ratios. The factor $\frac{1}{2}$ has been introduced in keeping with the form of several related equations. We now want to summarize what is known about the friction-loss factor for the various parts of a piping system.

For a straight conduit the friction-loss factor is closely related to the friction factor. We consider only the steady flow of a fluid of constant density in a straight conduit of arbitrary, but constant, cross section S and length L. If the fluid is flowing in the z direction under the influence of a pressure gradient and gravity, then Eqs. 7.2-2 and 7.4-7 become

(*z momentum*)
$$F_{f\to s} = (p_1 - p_2)S + (\rho SL)g_z \tag{7.5-5}$$

(*mechanical energy*)
$$\hat{E}_v = \frac{1}{\rho}(p_1 - p_2) + Lg_z \tag{7.5-6}$$

Multiplication of the second of these by ρS and subtracting gives

$$\hat{E}_v = \frac{F_{f\to s}}{\rho S} \tag{7.5-7}$$

If, in addition, the flow is *turbulent*, then the expression for $F_{f\to s}$ in terms of the mean hydraulic radius R_h may be used (see Eqs. 6.2-15 to 6.2-17). Specifically, Eq. 6.2-16 can be rearranged to give

$$F_{f\to s} = S\Delta\mathcal{P} = S\left(\frac{L}{R_h}\right)\left(\frac{1}{2}\rho\langle v_z\rangle^2\right)f(\text{Re}_h) \tag{7.5-8}$$

Table 7.5-1. Brief Summary of Friction-Loss Factors for Use with Eqs. 7.5-11 and 7.5-12 (Approximate Values for Turbulent Flow)[a]

Disturbances	e_v
Sudden changes in cross-sectional area[b]	
Rounded entrance to pipe	0.05
Sudden contraction	$0.45\,(1-\beta)$
Sudden expansion[c]	$\left(\frac{1}{\beta}-1\right)^2$
Orifice (sharp-edged)	$2.7\,(1-\beta)\left(1-\beta^2\right)\frac{1}{\beta^2}$
Fittings and Valves	
90° elbows (rounded)	0.4–0.9
90° elbows (square)	1.3–1.9
45° elbows	0.3–0.4
Globe valve (open)	6–10
Gate valve (open)	0.2

[a]Taken from H. Kramers, *Physische Transportverschijnselen*, Technische Hogeschool Delft, Holland (1958), pp. 53–54.

[b]Here $\beta = $ (smaller cross-sectional area) / (larger cross-sectional area).

[c]See derivation from the macroscopic balances in Example 7.6-1. If $\beta = 0$, then $\hat{E}_v = \frac{1}{2}\langle v\rangle^2$, where $\langle v\rangle$ is the velocity *upstream* from the enlargement.

in which f is the friction factor discussed in Chapter 6. Inserting this into Eq. 7.5-7 gives

$$\hat{E}_v = \frac{1}{2}\langle v\rangle^2 \frac{L}{R_h} f(\mathrm{Re}_h) \tag{7.5-9}$$

Since this equation is of the form of Eq. 7.5-4, we get a simple relation between the friction-loss factor and the friction factor

$$e_v = \frac{L}{R_h} f(\mathrm{Re}_h) \tag{7.5-10}$$

for turbulent flow in sections of straight pipe with uniform cross section. For a similar treatment for conduits of variable cross section, see Problem 7B.2.

Most flow systems contain various "obstacles," such as fittings, sudden changes in diameter, valves, or flow meters. These also contribute to the friction loss $\hat{E}_v$. Such additional resistances may be written in the form of Eq. 7.5-4, with e_v determined by one of two methods: (*a*) simultaneous solution of the macroscopic balances, or (*b*) experimental measurement. Some rough values of e_v are tabulated in Table 7.5-1 for the convention that $\langle v\rangle$ is the average velocity *downstream* from the disturbance. These e_v values are for *turbulent flow* for which the Reynolds number dependence is not too important.

Now we are in a position to rewrite Eq. 7.4-7 in the *approximate* form frequently used for *turbulent flow* calculations in a system composed of various kinds of piping and additional resistances:

$$\frac{1}{2}\left(v_2^2 - v_1^2\right) + g\left(h_2 - h_1\right) + \int_{p_1}^{p_2} \frac{1}{\rho}\, dp = \hat{W}_m - \underbrace{\sum_i \left(\frac{1}{2}v^2 \frac{L}{R_h} f(\mathrm{Re}_h)\right)_i}_{\substack{\text{sum over all}\\ \text{sections of}\\ \text{straight conduits}}} - \underbrace{\sum_i \left(\frac{1}{2}v^2 e_v\right)_i}_{\substack{\text{sum over all}\\ \text{fittings, valves,}\\ \text{meters, etc.}}} \tag{7.5-11}$$

Here R_h is the mean hydraulic radius defined in Eq. 6.2-15, f is the friction factor defined in Eq. 6.1-4, and e_v is the friction-loss factor given in Table 7.5-1. Note that the v_1 and v_2 in the first term refer to the velocities at planes 1 and 2; the v in the first sum is the average velocity in the ith pipe segment; and the v in the second sum is the average velocity *downstream* from the ith fitting, valve, or other obstacle.

For the special case of turbulent flow in *tubes of circular cross section* (diameter D), the mean hydraulic radius is not needed, and Eq. 7.5-11 may be rewritten

$$\frac{1}{2}\left(v_2^2 - v_1^2\right) + g\left(h_2 - h_1\right) + \int_{p_1}^{p_2} \frac{1}{\rho}\,dp = \hat{W}_m - \underbrace{\sum_i \left(2v^2 \frac{L}{D} f(\text{Re})\right)_i}_{\substack{\text{sum over all}\\ \text{sections of}\\ \text{straight conduits}}} - \underbrace{\sum_i \left(\frac{1}{2}v^2 e_v\right)_i}_{\substack{\text{sum over all}\\ \text{fittings, valves,}\\ \text{meters, etc.}}} \tag{7.5-12}$$

EXAMPLE 7.5-1

Power Requirement for Pipeline Flow

What is the required power output from the pump at steady state in the system shown in Fig. 7.5-1? The water at 68°F ($\rho = 62.4\ \text{lb}_m/\text{ft}^3$; $\mu = 1$ cp) is to be delivered to the upper tank at a rate of 12 ft^3/min. All of the piping is 4-in. internal diameter smooth circular pipe.

SOLUTION

As indicated in Fig. 7.5-1, we select planes 1 and 2 as the liquid surfaces within the tanks. With these selections, the velocities at the entrance and exit are negligible (assuming that the tank surface area is much larger than the pipe cross-sectional area), and the pressures at the entrance and exit are atmospheric pressure. Thus, the first and third terms in Eq. 7.5-12 will be zero. The remaining terms are calculated as follows.

The average velocity *in the pipe* is (see Eq. 7.1-2)

$$\langle v \rangle = \frac{w}{\rho S} = \frac{Q}{\pi R^2} = \frac{\left(12\ \text{ft}^3/\text{min}\right)}{\pi (2\ \text{in.})^2} \left(\frac{1\ \text{min}}{60\ \text{s}}\right) \left(\frac{12\ \text{in.}}{\text{ft}}\right)^2 = 2.29\ \frac{\text{ft}}{\text{s}} \tag{7.5-13}$$

and the Reynolds number is

$$\begin{aligned} \text{Re} &= \frac{D\langle v \rangle \rho}{\mu} \\ &= \frac{(4\ \text{in.})\,(2.29\ \text{ft/s})\,(62.4\ \text{lb}_m/\text{ft}^3)}{(1\ \text{cp})} \left(\frac{1\ \text{ft}}{12\ \text{in.}}\right) \left(\frac{1\ \text{cp}}{6.72 \times 10^{-4}\ \text{lb}_m/\text{ft}\cdot\text{s}}\right) \\ &= 7.09 \times 10^4 \end{aligned} \tag{7.5-14}$$

Hence, the flow is *turbulent*. For this value of Re, the Blasius formula for the friction factor (Eq. 6.2-12) gives $f = 0.0048$.

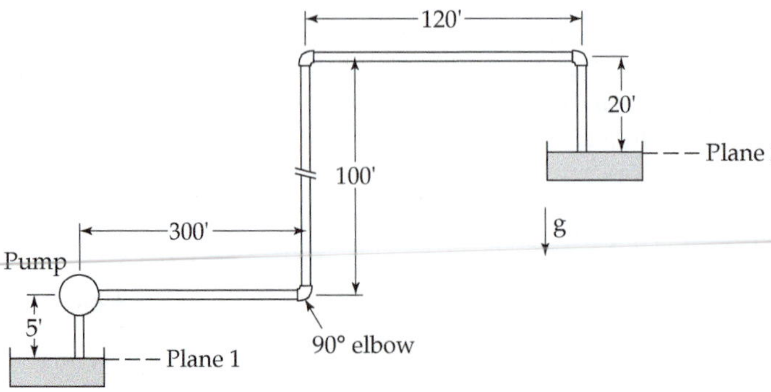

Fig. 7.5-1. Pipeline flow with friction losses because of fittings. Planes 1 and 2 are just under the surface of the liquid.

The contribution to $\hat{E}_v$ from the various lengths of pipe will be (see Eq. 7.5-12)

$$\sum_i \left(2v^2 \frac{L}{D} f(\mathrm{Re})\right)_i = \frac{2v^2 f}{D} \sum_i L_i$$
$$= \frac{2(2.29\ \mathrm{ft/s})^2 (0.0048)}{(4\ \mathrm{in.})} [(5 + 300 + 100 + 120 + 20)\ \mathrm{ft}] \left(\frac{12\ \mathrm{in.}}{1\ \mathrm{ft}}\right)$$
$$= 82\ \frac{\mathrm{ft}^2}{\mathrm{s}^2} \tag{7.5-15}$$

The contribution to $\hat{E}_v$ from the sudden contraction ($e_v = 0.45$), the three 90° elbows ($e_v = 0.5$ for each), and the sudden expansion ($e_v = 1$; see Table 7.5-1) will be

$$\sum_i \left(\frac{1}{2} v^2 e_v\right)_i = \frac{1}{2}\left(2.29 \frac{\mathrm{ft}}{\mathrm{s}}\right)^2 [0.45 + 3(0.5) + 1] = 8\ \frac{\mathrm{ft}^2}{\mathrm{s}^2} \tag{7.5-16}$$

Then from Eq. 7.5-12 we get

$$0 + \left(32.2\ \frac{\mathrm{ft}}{\mathrm{s}^2}\right) [(85 - 0)\ \mathrm{ft}] + 0 = \hat{W}_m - 82 \frac{\mathrm{ft}^2}{\mathrm{s}^2} - 8 \frac{\mathrm{ft}^2}{\mathrm{s}^2} \tag{7.5-17}$$

Solving for $\hat{W}_m$ we get

$$\hat{W}_m = (2740 + 82 + 8)\ \frac{\mathrm{ft}^2}{\mathrm{s}^2} = 2830\ \frac{\mathrm{ft}^2}{\mathrm{s}^2} \tag{7.5-18}$$

This is the work per unit mass of fluid done *on* the fluid *in* the pump. It may not be obvious that a quantity with units $\mathrm{ft}^2/\mathrm{s}^2$ has the dimensions of energy/mass. To convert this quantity to more familiar units, we can use the definition of the lb_f, $1\ \mathrm{lb}_f \equiv 32.2\ \mathrm{lb}_m \cdot \mathrm{ft/s}^2$ (e.g., consider Newton's law of motion, $F = ma$). Hence, $\hat{W}_m$ can be rewritten

$$\hat{W}_m = 2830\ \frac{\mathrm{ft}^2}{\mathrm{s}^2} \left(\frac{1\ \mathrm{lb}_f}{32.2\ \mathrm{lb}_m \cdot \mathrm{ft/s}^2}\right) = 88\ \frac{\mathrm{ft} \cdot \mathrm{lb}_f}{\mathrm{lb}_m} \tag{7.5-19}$$

Hence, the pump does 88 $\mathrm{ft} \cdot \mathrm{lb}_f/\mathrm{lb}_m$ of work on the fluid passing through the system. The mass rate of flow is

$$w = \rho Q = \left(62.4\ \frac{\mathrm{lb}_m}{\mathrm{ft}^3}\right)\left(12\ \frac{\mathrm{ft}^3}{\mathrm{min}}\right)\left(\frac{1\ \mathrm{min}}{60\ \mathrm{s}}\right) = 12.5\ \frac{\mathrm{lb}_m}{\mathrm{s}} \tag{7.5-20}$$

Consequently, the rate of work on the fluid is, in various units,

$$W_m = w\hat{W} = \left(12.5\ \frac{\mathrm{lb}_m}{\mathrm{s}}\right)\left(88\ \frac{\mathrm{ft} \cdot \mathrm{lb}_f}{\mathrm{lb}_m}\right) = 1100\ \frac{\mathrm{ft} \cdot \mathrm{lb}_f}{\mathrm{s}}$$
$$= 1100\ \frac{\mathrm{ft} \cdot \mathrm{lb}_f}{\mathrm{s}} \left(\frac{1\ \mathrm{hp} \cdot \mathrm{hr}}{1.98 \times 10^6\ \mathrm{ft} \cdot \mathrm{lb}_f}\right)\left(\frac{3600\ \mathrm{s}}{\mathrm{hr}}\right) = 2.0\ \mathrm{hp}$$
$$= 2.0\ \mathrm{hp} \left(\frac{2.6845 \times 10^6\ \mathrm{J}}{1\ \mathrm{hp} \cdot \mathrm{hr}}\right)\left(\frac{1\ \mathrm{hr}}{3600\ \mathrm{s}}\right)\left(\frac{10^{-3}\ \mathrm{kW}}{1\ \mathrm{J/s}}\right) = 1.5\ \mathrm{kW} \tag{7.5-21}$$

which is the power delivered by the pump.

§7.6 USE OF THE MACROSCOPIC BALANCES FOR SOLVING PROBLEMS

In §3.7 it was shown how to set up the differential equations to calculate the velocity and pressure profiles for isothermal flow systems by simplifying the equations of change. In this section, in Examples 7.6-1 to 7.6-5, we show how to use the set of steady-state macroscopic balances to obtain the algebraic equations for describing large systems.

Table 7.6-1. Steady-State Macroscopic Balances for Turbulent Flow in Isothermal Systems

Mass:	$\sum w_1 - \sum w_2 = 0$	(A)
Momentum:	$\sum (v_1 w_1 + p_1 S_1)\,\mathbf{u}_1 - \sum (v_2 w_2 + p_2 S_2)\,\mathbf{u}_2 + m_{\text{tot}}\mathbf{g} = \mathbf{F}_{f\to s}$	(B)
Angular momentum:	$\sum (v_1 w_1 + p_1 S_1)\,[\mathbf{r}_1 \times \mathbf{u}_1] - \sum (v_2 w_2 + p_2 S_2)\,[\mathbf{r}_2 \times \mathbf{u}_2] + \mathbf{T}_{\text{ext}} = \mathbf{T}_{f\to s}$	(C)
Mechanical energy:	$\sum \left(\frac{1}{2}v_1^2 + gh_1 + \frac{p_1}{\rho_1}\right) w_1 - \sum \left(\frac{1}{2}v_2^2 + gh_2 + \frac{p_2}{\rho_2}\right) w_2 = -W_m + E_c + E_v$	(D)

Notes:
[a] All formulas here assume flat velocity profiles.
[b] $\sum w_1 = w_{1a} + w_{1b} + w_{1c} + \cdots$, where $w_{1a} = \rho_{1a} v_{1a} S_{1a}$, etc.
[c] h_1 and h_2 are elevations above an arbitrary datum plane.
[d] All equations are written for compressible flow; for incompressible flow, $E_c = 0$.

For each problem, we start with the four macroscopic balances. By keeping track of the discarded or approximated terms, we automatically have a complete listing of the assumptions inherent in the final result. All of the examples given here are for isothermal, incompressible flow. The incompressibility assumption means that the velocity of the fluid must be less than the velocity of sound in the fluid and that the pressure changes must be small enough that the resulting density changes can be neglected.

The steady-state macroscopic balances may be easily generalized for systems with multiple inlet streams (called 1a, 1b, 1c, ...) and multiple outlet streams (called 2a, 2b, 2c, ...). These balances are summarized in Table 7.6-1 for turbulent flow (where the velocity profiles are regarded as flat).

Finally, in Example 7.6-6, we illustrate the solution of an unsteady-state problem using the time-dependent macroscopic balances.

EXAMPLE 7.6-1

Pressure Rise and Friction Loss in a Sudden Enlargement

An incompressible fluid flows from a small circular tube into a large tube in turbulent flow, as shown in Fig. 7.6-1. The cross-sectional areas of the tubes are S_1 and S_2. Obtain an expression for the pressure change between 1 and 2 and for the friction loss associated with the sudden enlargement in cross section. Let $\beta = S_1/S_2$, which is less than unity.

SOLUTION

(a) ***Mass balance.*** For steady flow the mass balance gives

$$w_1 = w_2 \qquad \text{or} \qquad \rho_1 v_1 S_1 = \rho_2 v_2 S_2 \tag{7.6-1}$$

For a fluid of constant density, this gives

$$\frac{v_1}{v_2} = \frac{1}{\beta} \tag{7.6-2}$$

(b) ***Momentum balance.*** We define the z direction as the tube axes in the direction of flow. We assume that the system volume is small so that the force of gravity can be neglected, so that the z component of the momentum balance is

$$F_{f\to s} = (v_1 w_1 - v_2 w_2) + (p_1 S_1 - p_2 S_2) \tag{7.6-3}$$

The axial force $F_{f\to s}$ can also be calculated directly from the molecular stresses. This force is composed of two parts: the viscous force on the cylindrical surfaces parallel to the direction

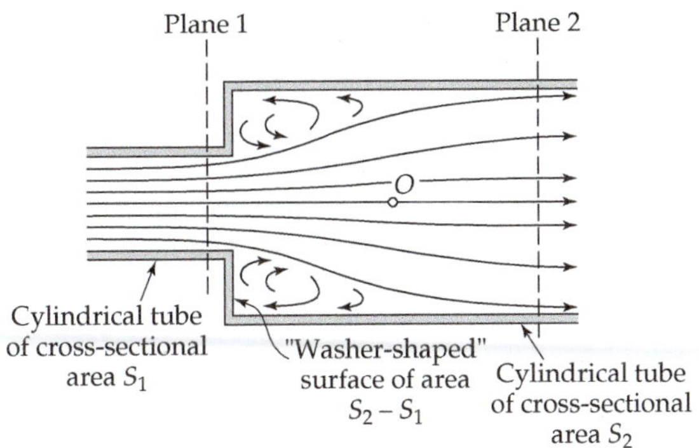

Fig. 7.6-1. Flow through a sudden enlargement.

of flow, and the pressure force on the "washer-shaped" surface just to the right of 1 and perpendicular to the flow axis. We neglect the viscous force contribution because (1) the surface area of the cylindrical surface is small, and (2) the recirculation flow in the corners illustrated in Fig. 7.6-1 implies that the wall shear stress changes sign along the axial direction and thus the shear stresses on the left half partly cancel the shear stresses on the right half. We take the pressure force to be $p_1(S_2 - S_1)$ by assuming that the pressure on the washer-shaped surface is the same as that at plane 1.

We then get, by using Eq. 7.6-1,

$$-p_1(S_2 - S_1) = \rho v_2 S_2 (v_1 - v_2) + (p_1 S_1 - p_2 S_2) \tag{7.6-4}$$

Solving for the pressure difference gives

$$p_2 - p_1 = \rho v_2 (v_1 - v_2) \tag{7.6-5}$$

This equation can be written in terms of the downstream velocity by using Eq. 7.6-2

$$p_2 - p_1 = \rho v_2^2 \left(\frac{1}{\beta} - 1\right) \tag{7.6-6}$$

Note that the momentum balance predicts (correctly) a *rise* in pressure. One might expect a rise in pressure because fast-moving fluid at plane 1 "runs into" slow-moving fluid; this creates an increase in pressure and thus a net force in the flow direction much like a water jet impinging on a surface generates a force in the flow direction.

(c) ***Angular momentum balance.*** This balance is not needed. If we take the origin of coordinates on the axis of the system at the center of gravity of the fluid located between planes 1 and 2, then $[\mathbf{r}_1 \times \mathbf{u}_1]$ and $[\mathbf{r}_2 \times \mathbf{u}_2]$ are both zero, and there are no torques on the fluid system.

(d) ***Mechanical energy balance.*** There is no compressive loss, no work done via moving parts, and negligible elevation change, so that the mechanical energy balances simplifies to

$$\hat{E}_v = \frac{1}{2}(v_1^2 - v_2^2) + \frac{1}{\rho}(p_1 - p_2) \tag{7.6-7}$$

Insertion of Eq. 7.6-6 for the pressure rise then gives, after some rearrangement,

$$\hat{E}_v = \frac{1}{2} v_2^2 \left(\frac{1}{\beta} - 1\right)^2 \tag{7.6-8}$$

which produces the entry for e_v for a sudden enlargement in Table 7.5-1.

This example has shown how to use the macroscopic balances to estimate the friction-loss factor for a simple resistance in a flow system. Because of the assumptions mentioned after Eq. 7.6-3, the results in Eqs. 7.6-6 and 7.6-8 are approximate. If great accuracy is needed, a correction factor based on experimental data should be introduced.

EXAMPLE 7.6-2

Performance of a Liquid-Liquid Ejector

A diagram of a liquid-liquid ejector is shown in Fig. 7.6-2. It is desired to analyze the mixing of the two streams, both of the same fluid, by means of the macroscopic balances. At plane 1 the two fluid streams merge. Stream 1a has a velocity v_0 and a cross-sectional area $\frac{1}{3}S_1$, and stream 1b has a velocity $\frac{1}{2}v_0$ and a cross-sectional area $\frac{2}{3}S_1$. Plane 2 is chosen far enough downstream so that the two streams have mixed and the velocity is almost uniform at v_2. The flow is turbulent and the velocity profiles at 1 and 2 are assumed to be flat. In the following analysis $\mathbf{F}_{f\to s}$ is neglected, since it is felt that it is less important than the other terms in the momentum balance.

SOLUTION

(a) ***Mass balance.*** At steady state, Eq. (A) of Table 7.6-1 gives

$$w_{1a} + w_{1b} = w_2 \tag{7.6-9}$$

or

$$\rho v_0 \left(\frac{1}{3}S_1\right) + \rho\left(\frac{1}{2}v_0\right)\left(\frac{2}{3}S_1\right) = \rho v_2 S_2 \tag{7.6-10}$$

Hence, since $S_1 = S_2$, this equation gives

$$v_2 = \frac{2}{3}v_0 \tag{7.6-11}$$

for the velocity of the exit stream. We also note, for later use, that $w_{1a} = w_{1b} = \frac{1}{2}w_2$.

(b) ***Momentum balance.*** From Eq. (B) of Table 7.6-1, the component of the momentum balance in the flow direction is

$$(v_{1a}w_{1a} + v_{1b}w_{1b} + p_1S_1) - (v_2w_2 + p_2S_2) = 0 \tag{7.6-12}$$

or using the relation at the end of (a)

$$\begin{aligned}(p_2 - p_1)S_2 &= \left(\frac{1}{2}(v_{1a} + v_{1b}) - v_2\right)w_2 \\ &= \left(\frac{1}{2}\left(v_0 + \frac{1}{2}v_0\right) - \frac{2}{3}v_0\right)\left(\rho\left(\frac{2}{3}v_0\right)S_2\right)\end{aligned} \tag{7.6-13}$$

which can be simplified to give

$$p_2 - p_1 = \frac{1}{18}\rho v_0^2 \tag{7.6-14}$$

This is the expression for the pressure rise resulting from the mixing of the two streams.

(c) ***Angular momentum balance.*** This balance is not needed.

(d) ***Mechanical energy balance.*** Equation (D) of Table 7.6-1 gives:

$$\left(\frac{1}{2}v_{1a}^2w_{1a} + \frac{1}{2}v_{1b}^2w_{1b}\right) - \left(\frac{1}{2}v_2^2 + \frac{p_2 - p_1}{\rho}\right)w_2 = E_v \tag{7.6-15}$$

or, using the relation at the end of (a), we get

$$\left(\frac{1}{2}v_0^2\left(\frac{1}{2}w_2\right) + \frac{1}{2}\left(\frac{1}{2}v_0\right)^2\left(\frac{1}{2}w_2\right)\right) - \left(\frac{1}{2}\left(\frac{2}{3}v_0\right)^2 + \frac{1}{18}v_0^2\right)w_2 = E_v \tag{7.6-16}$$

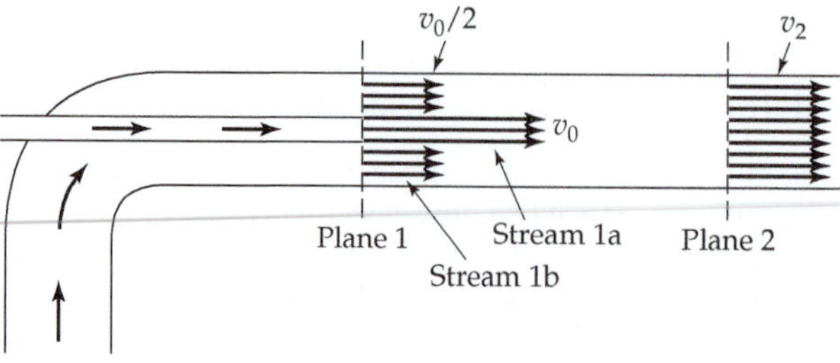

Fig. 7.6-2. Flow in a liquid-liquid ejector pump.

Hence,

$$\hat{E}_v = \frac{E_v}{w_2} = \frac{5}{144} v_0^2 \tag{7.6-17}$$

is the energy dissipation per unit mass. The foregoing analysis gives fairly good results for liquid-liquid ejector pumps. In gas-gas ejectors, however, the density varies significantly and it is necessary to include the macroscopic total-energy balance as well as an equation of state in the analysis. This is discussed in Example 15.3-2.

EXAMPLE 7.6-3

Thrust on a Pipe Bend

Water at 95°C is flowing at a rate of 2.0 ft³/s through a 60° bend, in which there is a contraction from 4-in. to 3-in. internal diameter (see Fig. 7.6-3). Compute the force exerted on the bend if the pressure at the downstream end is 1.1 atm. The density and viscosity of water at the temperature of the system are 0.962 g/cm³ and 0.299 cp, respectively.

SOLUTION

The Reynolds number for the flow in the 4-in. pipe is

$$\begin{aligned} \text{Re} &= \frac{D\langle v\rangle \rho}{\mu} = \frac{4w}{\pi D\mu} = \frac{4Q\rho}{\pi D\mu} \\ &= \frac{4\left(2\ \text{ft}^3/\text{s}\right)(0.962\ \text{g/cm}^3)}{\pi(4\ \text{in.})(0.299\ \text{cp})}\left(\frac{12\ \text{in.}}{\text{ft}}\right)^3\left(\frac{2.54\ \text{cm}}{\text{in.}}\right)^2\left(\frac{1\ \text{cp}}{10^{-2}\ \text{g/cm}\cdot\text{s}}\right) \\ &= 2.3\times 10^6 \end{aligned} \tag{7.6-18}$$

At this Reynolds number the flow is highly turbulent, and the assumption of flat velocity profiles is reasonable.

(a) ***Mass balance.*** For steady-state flow $w_1 = w_2$. If the density is constant throughout,

$$\frac{v_1}{v_2} = \frac{S_2}{S_1} = \beta \tag{7.6-19}$$

in which β is the ratio of the smaller to the larger cross section.

(b) ***Mechanical energy balance.*** For steady, incompressible flow, Eq. (D) of Table 7.6-1 becomes, for this problem,

$$\frac{1}{2}\left(v_2^2 - v_1^2\right) + g\left(h_2 - h_1\right) + \frac{1}{\rho}\left(p_2 - p_1\right) + \hat{E}_v = 0 \tag{7.6-20}$$

According to Table 7.5-1, the friction-loss factors for rounded elbows from 45° to 90° range from 0.3 to 0.9. Since a 60° elbow is closer to the lower end of this range, we estimate $e_v = 0.4$. According to Eq. 7.5-4, the friction loss $\hat{E}_v$ is thus approximately $\frac{1}{2}v_2^2(0.4) = 0.2v_2^2$. Inserting this

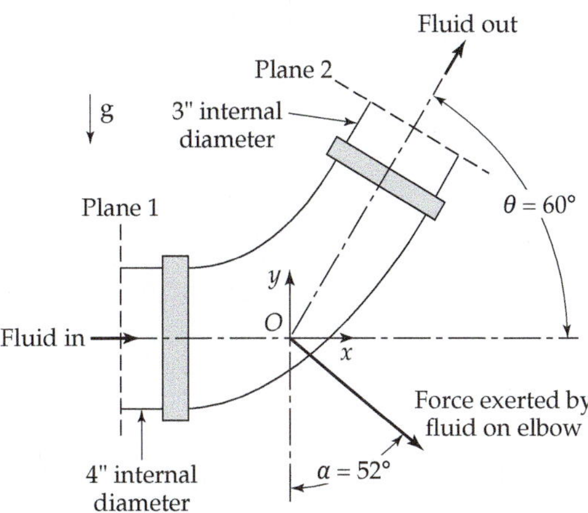

Fig. 7.6-3. Reaction force at a reducing bend in a pipe.

into Eq. 7.6-20 and using the mass balance we get

$$p_1 - p_2 = \rho v_2^2 \left(\frac{1}{2} - \frac{1}{2}\beta^2 + 0.2 \right) + \rho g \left(h_2 - h_1 \right) \tag{7.6-21}$$

This is the pressure drop through the bend in terms of the known velocity v_2 and the known geometrical factor β.

(c) ***Momentum balance.*** We now have to consider both the x and y components of the momentum balance. The inlet and outlet unit vectors will have x and y components given by: $u_{1x} = 1$, $u_{1y} = 0$, $u_{2x} = \cos\theta$, and $u_{2y} = \sin\theta$.

The x component of the momentum balance then gives

$$F_x = \left(v_1 w_1 + p_1 S_1 \right) - \left(v_2 w_2 + p_2 S_2 \right) \cos\theta \tag{7.6-22}$$

where F_x is the x component of $\mathbf{F}_{f\rightarrow s}$. Using $w = \rho v S$ for each plane, we get

$$\begin{aligned} F_x &= v_1 \left(\rho v_1 S_1 \right) - v_2 \left(\rho v_2 S_2 \right) \cos\theta + p_1 S_1 - p_2 S_2 \cos\theta \\ &= \rho v_2^2 S_2 \left(\beta - \cos\theta \right) + \left(p_1 - p_2 \right) S_1 + p_2 \left(S_1 - S_2 \cos\theta \right) \end{aligned} \tag{7.6-23}$$

Substituting into this the expression for $p_1 - p_2$ from Eq. 7.6-21 gives

$$\begin{aligned} F_x &= \rho v_2^2 S_2 \left(\beta - \cos\theta \right) + \rho v_2^2 S_2 \beta^{-1} \left(0.7 - \frac{1}{2}\beta^2 \right) \\ &\quad + \rho g \left(h_2 - h_1 \right) S_2 \beta^{-1} + p_2 S_2 \left(\beta^{-1} - \cos\theta \right) \\ &= w^2 \left(\rho S_2 \right)^{-1} \left(0.7\beta^{-1} - \cos\theta + \frac{1}{2}\beta \right) \\ &\quad + \rho g \left(h_2 - h_1 \right) S_2 \beta^{-1} + p_2 S_2 \left(\beta^{-1} - \cos\theta \right) \end{aligned} \tag{7.6-24}$$

The y component of the momentum balance is

$$F_y = -\left(v_2 w_2 + p_2 S_2 \right) \sin\theta - m_{\text{tot}} g \tag{7.6-25}$$

or

$$F_y = -w^2 \left(\rho S_2 \right)^{-1} \sin\theta - p_2 S_2 \sin\theta - \pi R^2 L \rho g \tag{7.6-26}$$

in which R and L are the radius and length of a cylinder of equivalent volume.

We now have the components of the reaction force in terms of known quantities. The numerical values needed are:

$$\rho = 60\ \frac{\text{lb}_m}{\text{ft}^3} \qquad \cos\theta = \frac{1}{2}$$

$$w = \left(2.0\ \frac{\text{ft}^3}{\text{s}} \right) \left(60 \frac{\text{lb}_m}{\text{ft}^3} \right) = 120\ \frac{\text{lb}_m}{\text{s}} \qquad \sin\theta = \frac{\sqrt{3}}{2}$$

$$S_2 = \pi \left(1.5\ \text{in.} \times \left(\frac{1\ \text{ft}}{12\ \text{in.}} \right) \right)^2 = 0.049\ \text{ft}^2 \qquad R \approx \frac{1}{8}\ \text{ft}$$

$$\beta = \frac{S_2}{S_1} = \left(\frac{3\ \text{in.}}{4\ \text{in.}} \right)^2 = 0.562 \qquad L \approx \frac{5}{6}\ \text{ft}$$

$$p_2 = 1.1\ \text{atm} \left(\frac{14.7\ \text{lb}_f/\text{in.}^2}{1\ \text{atm}} \right) = 16.2\ \frac{\text{lb}_f}{\text{in.}^2} \qquad h_2 - h_1 \approx \frac{1}{2}\ \text{ft}$$

(Values for R, L and $h_2 - h_1$ are only estimates.) With these values we then get

$$\begin{aligned} F_x = &\frac{\left(120\ \text{lb}_m/\text{s} \right)^2}{\left(60\ \text{lb}_m/\text{ft}^3 \right) \left(0.049\ \text{ft}^2 \right)} \left(\frac{1\ \text{lb}_f}{32.2\ \text{lb}_m \cdot \text{ft}/\text{s}^2} \right) \left(0.7 \times \frac{1}{0.562} - \frac{1}{2} + \frac{1}{2} \times 0.562 \right) \\ &+ \left(60\ \frac{\text{lb}_m}{\text{ft}^3} \right) \left(32.2\ \frac{\text{ft}}{\text{s}^2} \right) \left(\frac{1}{2}\ \text{ft} \right) \left(0.049\ \text{ft}^2 \right) \left(\frac{1}{0.562} \right) \left(\frac{1\ \text{lb}_f}{32.2\ \text{lb}_m \cdot \text{ft}/\text{s}^2} \right) \end{aligned}$$

$$+\left(16.2\,\frac{\text{lb}_f}{\text{in.}^2}\right)\left(0.049\text{ ft}^2\right)\left(\frac{1}{0.562}-\frac{1}{2}\right)\left(\frac{12\text{ in.}}{\text{ft}}\right)^2$$

$$= (155 + 2.6 + 146)\text{ lb}_f = 304\text{ lb}_f\left(\frac{1\text{ N}}{0.224881\text{ lb}_f}\right) = 1352\text{ N} \qquad (7.6\text{-}27)$$

$$F_y = -\frac{\left(120\text{ lb}_m/\text{s}\right)^2}{\left(60\text{ lb}_m/\text{ft}^3\right)\left(0.049\text{ ft}^2\right)}\left(\frac{1\text{ lb}_f}{32.2\text{ lb}_m\cdot\text{ft/s}^2}\right)\left(\frac{\sqrt{3}}{2}\right)$$

$$-\left(16.2\,\frac{\text{lb}_f}{\text{in.}^2}\right)\left(0.049\text{ ft}^2\right)\left(\frac{\sqrt{3}}{2}\right)\left(\frac{12\text{ in.}}{\text{ft}}\right)^2$$

$$-\pi\left(\frac{1}{8}\text{ ft}\right)^2\left(\frac{5}{6}\text{ ft}\right)\left(60\,\frac{\text{lb}_m}{\text{ft}^3}\right)\left(32.2\,\frac{\text{ft}}{\text{s}^2}\right)\left(\frac{1\text{ lb}_f}{32.2\text{ lb}_m\cdot\text{ft/s}^2}\right)$$

$$= (-132 - 99 - 2.5)\text{ lb}_f = -234\text{ lb}_f\left(\frac{1\text{ N}}{0.224881\text{ lb}_f}\right) = -1041\text{ N} \qquad (7.6\text{-}28)$$

Hence, the magnitude of the force is

$$|\mathbf{F}| = \sqrt{F_x^2 + F_y^2} = \sqrt{\left(304\text{ lb}_f\right)^2 + \left(234\text{ lb}_f\right)^2} = 384\text{ lb}_f\left(\frac{1\text{ N}}{0.224881\text{ lb}_f}\right) = 1708\text{ N} \qquad (7.6\text{-}29)$$

The angle that this force makes with the vertical is

$$\alpha = \tan^{-1}\left(\left|\frac{F_x}{F_y}\right|\right) = \tan^{-1}1.30 = 52° \qquad (7.6\text{-}30)$$

In looking back over the calculation, we see that all the effects we have included are important, with the possible exception of the gravity terms of 2.6 lb_f in F_x and 2.5 lb_f in F_y.

EXAMPLE 7.6-4

The Impinging Jet

A rectangular incompressible fluid jet of thickness b_1 emerges from a slot of width c, hits a flat plate, and splits into two streams of thicknesses b_{2a} and b_{2b} as shown in Fig. 7.6-4. The emerging turbulent jet stream has a velocity v_1 and a mass flow rate w_1. Find the velocities and mass rates of flow in the two streams on the plate.[1]

SOLUTION

We neglect viscous dissipation and gravity, and assume that the velocity profiles of all three streams are flat and that their pressures are equal. The macroscopic balances then give

(a) ***Mass balance.***

$$w_1 = w_{2a} + w_{2b} \qquad (7.6\text{-}31)$$

(b) ***Momentum balance*** (in the direction parallel to the plate).

$$v_1 w_1 \cos\theta = v_{2a}w_{2a} - v_{2b}w_{2b} \qquad (7.6\text{-}32)$$

(c) ***Mechanical energy balance.***

$$\frac{1}{2}v_1^2 w_1 = \frac{1}{2}v_{2a}^2 w_{2a} + \frac{1}{2}v_{2b}^2 w_{2b} \qquad (7.6\text{-}33)$$

---

[1] For alternative solutions to this problem, see G. K. Batchelor, *An Introduction to Fluid Dynamics*, Cambridge University Press (1967), pp. 392–394, and S. Whitaker, *Introduction to Fluid Dynamics*, Prentice-Hall, Englewood Cliffs, NJ (1968), p. 260. An application of the *compressible* impinging jet problem has been given by J. V. Foa, U.S. Patent 3,361,336 (Jan. 2, 1968). There, use is made of the fact that if the slot-shaped nozzle moves to the left in Fig. 7.6-4 (i.e., left with respect to the plate), then, for a compressible fluid, the right stream will be cooler than the jet and the left stream will be warmer.

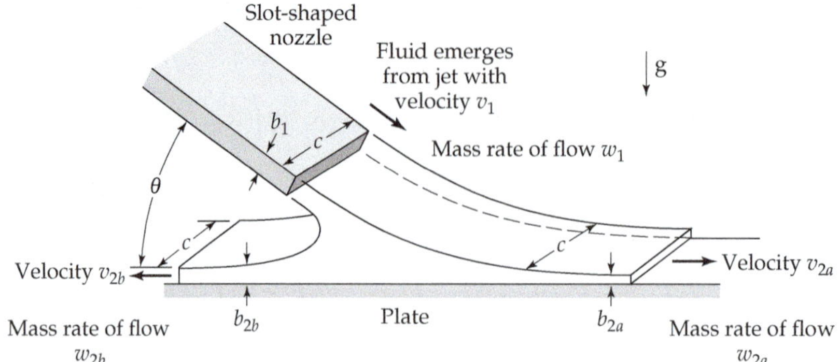

Fig. 7.6-4. Jet impinging on a wall and splitting into two streams. The point O, which is the origin of coordinates for the angular momentum balance, is taken to be the intersection of the center line of the incoming jet and a plane, which is at an elevation $\frac{1}{2}b_1$ above the flat plate.

(d) ***Angular momentum balance.*** (put the origin of coordinates on the center line of the jet and at an elevation of $\frac{1}{2}b_1$ above the flat plate; this is done so that there will be no angular momentum of the incoming jet)

$$0 = (v_{2a}w_{2a}) \cdot \frac{1}{2}(b_1 - b_{2a}) - (v_{2b}w_{2b}) \cdot \frac{1}{2}(b_1 - b_{2b}) \tag{7.6-34}$$

This last equation can be rewritten to eliminate the b's in favor of the w's. Since $w_1 = \rho v_1 b_1 c$ and $w_{2a} = \rho v_{2a} b_{2a} c$, we can replace $b_1 - b_{2a}$ by $(w_1/\rho v_1 c) - (w_{2a}/\rho v_{2a} c)$, and replace $b_1 - b_{2b}$ with the corresponding expression. Then the angular momentum balance becomes, after multiplying by ρc,

$$(v_{2a}w_{2a})\left(\frac{w_1}{v_1} - \frac{w_{2a}}{v_{2a}}\right) = (v_{2b}w_{2b})\left(\frac{w_1}{v_1} - \frac{w_{2b}}{v_{2b}}\right) \tag{7.6-35}$$

or

$$w_{2a}^2 - w_{2b}^2 = \frac{w_1}{v_1}(v_{2a}w_{2a} - v_{2b}w_{2b}) \tag{7.6-36}$$

Now Eqs. 7.6-31 to 7.6-33, and 7.6-36 are four equations with four unknowns. When these are solved we find that

$$v_{2a} = v_1; \qquad w_{2a} = \tfrac{1}{2}w_1(1 + \cos\theta) \tag{7.6-37,38}$$

$$v_{2b} = v_1; \qquad w_{2b} = \tfrac{1}{2}w_1(1 - \cos\theta) \tag{7.6-39,40}$$

Hence, the velocities of all three streams are equal. The same result is obtained by applying the classical Bernoulli equation for the flow of an inviscid fluid (see Example 3.3-1).

EXAMPLE 7.6-5

Isothermal Flow of a Liquid through an Orifice

A common method for determining the mass rate of flow through a pipe is to measure the pressure drop across some "obstacle" in the pipe. An example of this is the orifice, which is a thin plate with a hole in the middle (the cross-sectional area of the hole is S_0). There are pressure taps at planes 1 and 2, upstream and downstream of the orifice plate. Figure 7.6-5(a) shows the orifice meter, the pressure taps, and the general behavior of the velocity profiles as observed experimentally. The velocity profile at plane 1 will be assumed to be flat. In Fig. 7.6-5(b) we show an approximate velocity profile at plane 2, which we use in the application of the macroscopic balances. The standard orifice-meter equation is obtained by applying the steady-state macroscopic mass and mechanical energy balances.

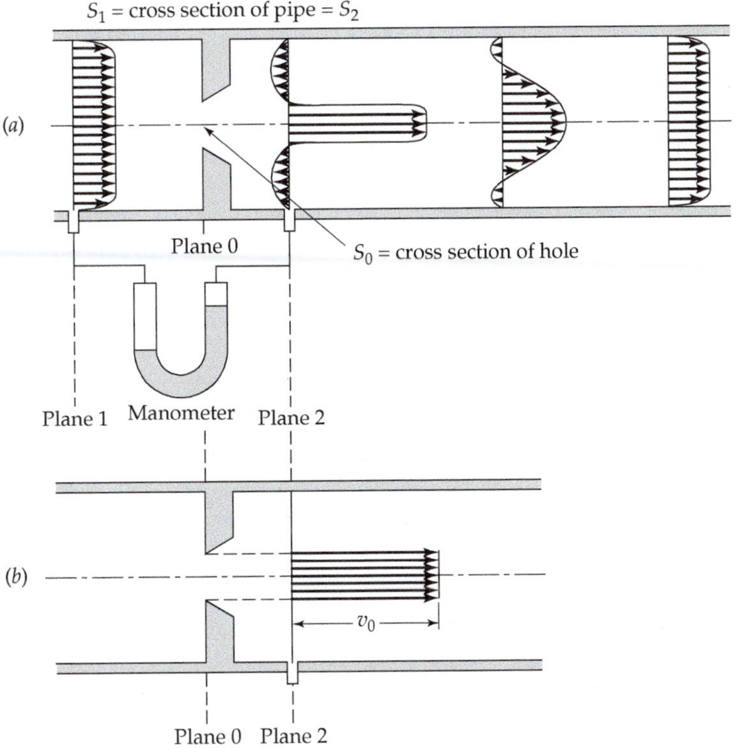

Fig. 7.6-5. (*a*) A sharp-edged orifice, showing the approximate velocity profiles at several planes near the orifice plate. The fluid jet emerging from the hole is somewhat smaller than the hole itself. In highly turbulent flow, this jet necks down to a minimum cross section at the *vena contracta*. The extent of this necking down can be given by the *contraction coefficient*, $C_c = (S_{\text{vena contracta}}/S_0)$. According to inviscid-flow theory, $C_c = \pi/(\pi + 2) = 0.611$ if $S_0/S_1 = 0$ (H. Lamb, *Hydrodynamics*, Dover, New York (1945), p. 99). Note that there is some back flow near the wall. (*b*) Approximate velocity profile at 2 used to estimate $\langle v_2^3 \rangle / \langle v_2 \rangle$.

SOLUTION

(a) ***Mass balance.*** For a fluid of constant density with a system for which $S_1 = S_2 = S$, the steady-state mass balance in Eq. 7.1-4 gives

$$\langle v_1 \rangle = \langle v_2 \rangle \tag{7.6-41}$$

With the assumed flat velocity profiles this becomes

$$v_1 = \frac{S_0}{S} v_0 \tag{7.6-42}$$

where v_0 is the velocity in the orifice. The mass rate of flow is $w = \rho v_1 S$.

(b) ***Mechanical energy balance.*** For a constant-density fluid in a flow system with no elevation change and no moving parts, Eq. 7.4-7 gives

$$\frac{1}{2}\frac{\langle v_2^3 \rangle}{\langle v_2 \rangle} - \frac{1}{2}\frac{\langle v_1^3 \rangle}{\langle v_1 \rangle} + \frac{p_2 - p_1}{\rho} + \hat{E}_v = 0 \tag{7.6-43}$$

The viscous loss $\hat{E}_v$ is neglected, even though it is certainly not equal to zero. With the assumed flat velocity profiles, Eq. 7.6-43 then becomes

$$\frac{1}{2}\left(v_0^2 - v_1^2\right) + \frac{p_2 - p_1}{\rho} = 0 \tag{7.6-44}$$

When Eqs. 7.6-42 and 7.6-44 are combined to eliminate v_0, we can solve for v_1 to get

$$v_1 = \sqrt{\frac{2(p_1 - p_2)}{\rho} \frac{1}{(S/S_0)^2 - 1}} \tag{7.6-45}$$

We can now multiply by ρS to get the mass rate of flow. Then to account for the errors introduced by neglecting $\hat{E}_v$ and by the assumptions regarding the velocity profiles, we include a *discharge coefficient*, C_d, and obtain

$$w = C_d S_0 \sqrt{\frac{2\rho(p_1 - p_2)}{1 - (S_0/S)^2}} \tag{7.6-46}$$

Experimental discharge coefficients have been correlated as a function of S_0/S and the Reynolds number.[2] For Reynolds numbers greater than 10^4, C_d approaches about 0.61 for all practical values of S_0/S.

This example has illustrated the use of the macroscopic balances to get the general form of the result, which is then modified by introducing a multiplicative function of dimensionless groups to correct for errors introduced by unwarranted assumptions. This combination of macroscopic balances and dimensional considerations is often used and can be quite useful.

EXAMPLE 7.6-6

Acceleration Effects in Unsteady Flow from a Cylindrical Tank

An open cylinder of height H and radius R is initially entirely filled with a liquid. For all times $t > 0$, the liquid is allowed to drain out through a small hole of radius R_0 at the bottom of the tank (see Fig. 7.6-6).

(a) Find the efflux time by using the unsteady-state mass balance and by assuming Torricelli's equation (see Example 3.3-2) to describe the relation between efflux velocity and the instantaneous height of the liquid.

(b) Find the efflux time using the unsteady-state mass and mechanical energy balances.

SOLUTION

(a) We apply Eq. 7.1-3 to the system in Fig. 7.6-6, taking plane 1 to be at the top of the tank (so that $w_1 = 0$). If the instantaneous liquid height is $h(t)$, then

$$\frac{d}{dt}\left(\pi R^2 h \rho\right) = -\rho v_2 \left(\pi R_0^2\right) \tag{7.6-47}$$

Here we have assumed that the velocity profile at plane 2 is flat. According to Torricelli's equation $v_2 = \sqrt{2gh}$, so that Eq. 7.6-47 becomes

$$\frac{dh}{dt} = -\left(\frac{R_0}{R}\right)^2 \sqrt{2gh} \tag{7.6-48}$$

---

[2] G. L. Tuve and R. E. Sprenkle, *Instruments*, **6**, 202–205, 225, 232–234 (1935); see also R. H. Perry and T. H. Chilton, *Chemical Engineers' Handbook*, McGraw-Hill, New York, 5th edition (1973), Fig. 5–18; *Fluid Meters: Their Theory and Applications*, 6th edition, American Society of Mechanical Engineers, New York (1971), pp. 58–65; Measurement of Fluid Flow Using Small Bore Precision Orifice Meters, American Society of Mechanical Engineers, MFC-14-M, New York (1995). **Thomas Hamilton Chilton** (1899–1972) had his entire professional career at the E. I. du Pont de Nemours Company, Inc., in Wilmington, Delaware; he was president of AIChE in 1951. After "retiring" he was a guest professor at a dozen or so universities.

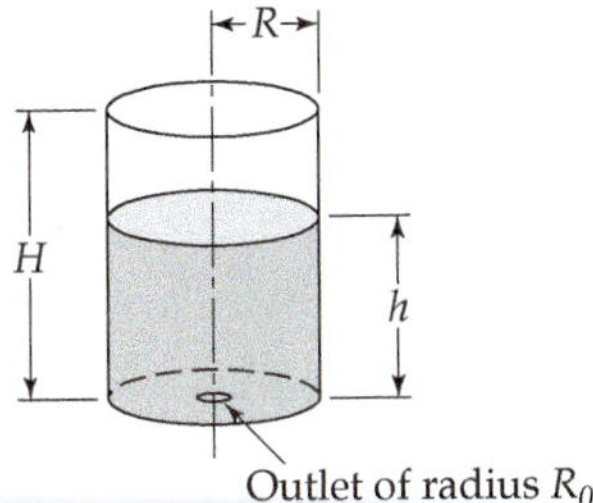

Fig. 7.6-6. Flow out of a cylindrical tank.

When this is integrated from $t = 0$ to $t = t_{\text{efflux}}$, we get

$$t_{\text{efflux}} = \sqrt{\frac{2NH}{g}} \tag{7.6-49}$$

in which $N = (R/R_0)^4 \gg 1$ and H is the initial height of the liquid in the tank. This is effectively a quasi-steady-state solution, since we have used the unsteady-state mass balance along with Torricelli's equation, which was derived for a steady-state flow.

(b) We now use Eq. 7.6-47 and the mechanical energy balance in Eq. 7.4-2. In the latter, the terms W_m and E_c are identically zero, and we assume that E_v is negligibly small, since the velocity gradients in the system will be small. We take the datum plane for the potential energy to be at the bottom of the tank, so that $\hat{\Phi}_2 = gz_2 = 0$. Because $w_1 = 0$, the "entrance" terms do not contribute, i.e., $\left(\langle v^3\rangle/2\langle v\rangle + \hat{\Phi} + p/\rho\right)_1 w_1 = 0$.

To get the total kinetic energy in the system at any time t, we have to know the velocity of every fluid element in the tank. At every point in the tank, we assume that the fluid is moving downward at the same velocity, namely, $v_z = dh/dt = -v_2(R_0/R)^2$ so that the kinetic energy per unit volume is everywhere $\frac{1}{2}\rho v_2^2(R_0/R)^4$.

To get the total potential energy in the system at any time t, we have to integrate the potential energy per unit volume ρgz over the volume of fluid from 0 to h. This gives $\pi R^2\rho g\left(\frac{1}{2}h^2\right)$.

Therefore, the mechanical energy balance in Eq. 7.4-2 becomes

$$\frac{d}{dt}\left[(\pi R^2 h)\left(\frac{1}{2}\rho v_2^2\right)(R_0/R)^4 + \pi R^2\rho g\left(\frac{1}{2}h^2\right)\right] = -\frac{1}{2}v_2^2\left(\rho v_2\pi R_0^2\right) \tag{7.6-50}$$

The unsteady-state mass balance is the same as in part (a), and thus, $v_2 = -(R/R_0)^2(dh/dt)$. When this is inserted into Eq. 7.6-50, we get (after dividing by $\left(\frac{1}{2}\pi R^2\rho g\right)dh/dt$)

$$2h\frac{d^2h}{dt^2} - (N-1)\left(\frac{dh}{dt}\right)^2 + 2gh = 0 \tag{7.6-51}$$

This is to be solved with the two initial conditions:

$$\text{I.C. 1:} \quad \text{at } t = 0, \quad h = H \tag{7.6-52}$$

$$\text{I.C. 2:} \quad \text{at } t = 0, \quad \frac{dh}{dt} = -(R_0/R)^2\sqrt{2gH} \tag{7.6-53}$$

where in the second of these we have assumed that Torricelli's equation can be used for the initial value of v_2.

The second-order differential equation for h can be converted to a first-order equation for the function $u(h)$ by making the change of variable $(dh/dt)^2 = u$. The first term of Eq. 7.6-51 becomes

$$2h\frac{d^2h}{dt^2} = 2h\frac{d}{dt}\left(\sqrt{u}\right) = 2h\left(\frac{1}{2}u^{-1/2}\frac{du}{dt}\right) = hu^{-1/2}\frac{du}{dh}\frac{dh}{dt} = h\frac{du}{dh} \tag{7.6-54}$$

so that Eq. 7.6-51 becomes

$$h\frac{du}{dh} - (N-1)u + 2gh = 0 \tag{7.6-55}$$

The solution to this first-order ordinary differential equation is[3]

$$u = Ch^{N-1} + 2gh/(N-2) \tag{7.6-56}$$

where C is a constant of integration. To verify that this is the solution to the differential equation, we substitute Eq. 7.6-56 into Eq. 7.6-55 to get

$$h\left[C(N-1)h^{N-2} + \frac{2g}{N-2}\right] - (N-1)\left[Ch^{N-1} + \frac{2gh}{N-2}\right] + 2gh = 0 \tag{7.6-57}$$

We then see that the terms containing C and those containing g separately sum to zero

$$g\text{-terms}: \quad \frac{2gh}{N-2} - (N-1)\frac{2gh}{N-2} + \frac{N-2}{N-2}2gh = 0 \tag{7.6-58}$$

$$C\text{-terms}: \quad C(N-1)h^{N-1} - (N-1)Ch^{N-1} = 0 \tag{7.6-59}$$

We now apply the second initial condition (Eq. 7.6-53), to evaluate the constant of integration

$$C = \frac{2g}{H^{N-2}}\left(\frac{1}{N} - \frac{1}{N-2}\right) = -\frac{4g}{H^{N-2}}\left(\frac{1}{N(N-2)}\right) \tag{7.6-60}$$

Even though the factor containing $(N-2)$ would cause C to become infinite for $N = 2$, this need cause no alarm, since $N = (R/R_0)^4$ is going to be a large number, i.e., when the outlet hole radius is small compared to the radius of the tank.

Now we take the square root of Eq. 7.6-56. Using $\sqrt{u} = dh/dt$ and C inserted, and introducing the dimensionless liquid height $\eta = h/H$, we get

$$\frac{d\eta}{dt} = \pm\sqrt{\frac{2g}{(N-2)H}}\sqrt{\left(\eta - \frac{2}{N}\eta^{N-1}\right)} \tag{7.6-61}$$

We then choose the minus sign, because we know that the height of the fluid will be decreasing with time, and therefore, $d\eta/dt$ must be negative. To get the efflux time, we integrate Eq. 7.6-61 from $t = 0$ when $\eta = 1$ (full tank) to $t = t_{\text{efflux}}$ (empty tank)

$$t_{\text{efflux}} = \int_0^{t_{\text{efflux}}} dt = \sqrt{\frac{(N-2)H}{2g}}\int_0^1 \frac{1}{\sqrt{\eta - (2/N)\eta^{N-1}}}d\eta \equiv \sqrt{\frac{NH}{2g}}\phi(N) \tag{7.6-62}$$

Keep in mind that $t_{\text{efflux}} = \sqrt{NH/2g}$ is the quasi-steady-state solution in Eq. 7.6-49—the solution that we would expect to be valid when N is extremely large, i.e., for the case that the outlet hole is so small that the system is never very far from steady state. The function $\phi(N)$ represents the deviation from the quasi-steady-state solution, and is given by

$$\phi(N) = \frac{1}{2}\sqrt{\frac{N-2}{N}}\int_0^1 \frac{1}{\sqrt{\eta - (2/N)\eta^{N-1}}}d\eta \tag{7.6-63}$$

This integral can probably not be performed analytically. However, in the expression under the square-root sign, for large N, the first term will predominate over the range of integration (from $\eta = 0$ to $\eta = 1$). Therefore, we take the first term outside of the radical and write

$$\phi(N) = \frac{1}{2}\sqrt{\frac{N-2}{N}}\int_0^1 \frac{1}{\sqrt{\eta}}\left(1 - \frac{2}{N}\eta^{N-2}\right)^{-1/2} d\eta \tag{7.6-64}$$

---

[3]See E. Kamke, *Differentialgleichungen: Lösungsmethoden und Lösungen*, Chelsea Publishing Company, New York (1948), p. 311, #1.94; G. M. Murphy, *Ordinary Differential Equations and Their Solutions*, Van Nostrand, Princeton, NJ (1960) p. 236, #157.

The quantity to the $-\frac{1}{2}$ power can then be expanded in a Taylor series (see Eq. C.2-1) about $(2/N)\,\eta^{N-2} = 0$ to get, after integrating term by term,

$$\phi(N) = \frac{1}{2}\sqrt{\frac{N-2}{N}} \int_0^1 \frac{1}{\sqrt{\eta}} \left[1 + \frac{1}{1!}\frac{1}{2}\left(\frac{2}{N}\eta^{N-2}\right) + \frac{1}{2!}\frac{1}{2}\frac{3}{2}\left(\frac{2}{N}\eta^{N-2}\right)^2 + \cdots \right] d\eta$$
$$= \frac{1}{2}\sqrt{\frac{N-2}{N}} \left(2 + \frac{1}{N\left(N-\frac{3}{2}\right)} + \frac{3}{4N^2\left(N-\frac{7}{4}\right)} + \cdots \right) \tag{7.6-65}$$

When this expression is expanded in a Taylor series around $1/N = 0$, we get

$$\phi(N) = \left[1 - \frac{1}{N} - \frac{1}{2}\frac{1}{N^2} + \mathrm{O}\left(\frac{1}{N^3}\right)\right] + \frac{1}{2}\left[\frac{1}{N^2} + \mathrm{O}\left(\frac{1}{N^3}\right)\right] + \frac{3}{8}\left[\mathrm{O}\left(\frac{1}{N^3}\right)\right] + \cdots$$
$$= 1 - \frac{1}{N} + \mathrm{O}\left(\frac{1}{N^3}\right) \tag{7.6-66}$$

in which O() means "term of the order of ( )." Since $N = \left(R/R_0\right)^4 \gg 1$, it is evident that the factor $\phi(N)$ differs only very slightly from unity.

It is instructive now to return to Eq. 7.6-50 and omit the term describing the change in total kinetic energy with time. If this is done, one obtains exactly the expression for efflux time in Eq. 7.6-49 (or Eq. 7.6-62, with $\phi(N) = 1$). We can therefore conclude that in this type of problem, the change in kinetic energy with time can safely be neglected.

---

§7.7 DERIVATION OF THE MACROSCOPIC MECHANICAL ENERGY BALANCE[1]

In Eq. 7.4-2 the macroscopic mechanical energy balance was presented without proof. In this section we show how the equation is obtained by integrating the equation of change for mechanical energy (Eq. 3.3-2) over the entire volume of the flow system of Fig. 7.0-1. We begin by doing the formal integration:

$$\int_{V(t)} \frac{\partial}{\partial t}\left(\frac{1}{2}\rho v^2 + \rho\hat{\Phi}\right) dV = -\int_{V(t)} \left(\nabla \cdot \left(\frac{1}{2}\rho v^2 + \rho\hat{\Phi}\right)\mathbf{v}\right) dV$$
$$-\int_{V(t)} (\nabla \cdot p\mathbf{v})\, dV - \int_{V(t)} (\nabla \cdot [\boldsymbol{\tau} \cdot \mathbf{v}])\, dV$$
$$+\int_{V(t)} p(\nabla \cdot \mathbf{v})\, dV + \int_{V(t)} (\boldsymbol{\tau} : \nabla\mathbf{v})\, dV \tag{7.7-1}$$

Next we apply the three-dimensional Leibniz formula (Eq. A.5-4) to the left side and the Gauss divergence theorem (Eq. A.5-1) to terms 1, 2, and 3 on the right side.

$$\frac{d}{dt}\int_{V(t)} \left(\frac{1}{2}\rho v^2 + \rho\hat{\Phi}\right) dV = -\int_{S(t)} \left(\mathbf{n} \cdot \left(\frac{1}{2}\rho v^2 + \rho\hat{\Phi}\right)(\mathbf{v} - \mathbf{v}_S)\right) dS$$
$$-\int_{S(t)} (\mathbf{n} \cdot p\mathbf{v})\, dS - \int_{S(t)} (\mathbf{n} \cdot [\boldsymbol{\tau} \cdot \mathbf{v}])\, dS$$
$$+\int_{V(t)} p(\nabla \cdot \mathbf{v})\, dV + \int_{V(t)} (\boldsymbol{\tau} : \nabla\mathbf{v})\, dV \tag{7.7-2}$$

---

[1] R. B. Bird, *Korean J. Chem. Eng.*, **15**, 105–123 (1998), §3.

The term containing $\mathbf{v}_S$, the velocity of the surface of the system, arises from the application of the Leibniz formula. The surface $S(t)$ consists of four parts:

- the fixed surfaces S_f, on which both $\mathbf{v}$ and $\mathbf{v}_S$ are zero
- the moving surfaces S_m, on which $\mathbf{v} = \mathbf{v}_S$, both being nonzero
- the cross section of the entry port S_1, where $\mathbf{v}_S = 0$
- the cross section of the exit port S_2, where $\mathbf{v}_S = 0$

Presently each of the surface integrals will be split into four parts corresponding to these four surfaces.

We now interpret the terms in Eq. 7.7-2 and, in the process, introduce several assumptions; these assumptions have already been mentioned in §7.1 to §7.4, but now the reasons for them will be made clear.

The term on the left side can be interpreted as the time rate of change of the total kinetic and potential energy (see Fig. 7.0-1) within the entire flow system, whose shape and volume may be changing with time.

We next examine one by one the five terms on the right side:

Term 1 (including the minus sign) contributes only at the entry and exit ports and gives the rates of influx and efflux of kinetic and potential energy:

$$\text{Term 1} = \left(\frac{1}{2}\rho_1 \langle v_1^3 \rangle S_1 + \rho_1 \hat{\Phi}_1 \langle v_1 \rangle S_1\right) - \left(\frac{1}{2}\rho_2 \langle v_2^3 \rangle S_2 + \rho_2 \hat{\Phi}_2 \langle v_2 \rangle S_2\right) \tag{7.7-3}$$

The angular brackets indicate an average over the cross section. To get this result, we have to assume that the fluid density and potential energy per unit mass are constant over the cross section, and that the fluid is flowing parallel to the tube walls at the entry and exit ports. The first term in Eq. 7.7-3 is positive, since at plane 1, $(-\mathbf{n} \cdot \mathbf{v}) = \left(\mathbf{u}_1 \cdot (\mathbf{u}_1 v_1)\right) = v_1$, and the second term is negative, since at plane 2, $(-\mathbf{n} \cdot \mathbf{v}) = \left(-\mathbf{u}_2 \cdot (\mathbf{u}_2 v_2)\right) = -v_2$.

Term 2 (including the minus sign) gives no contribution on S_f since $\mathbf{v}$ is zero there. On each surface element dS of S_m, there is a force $-\mathbf{n}p dS$ acting on a surface moving with a velocity $\mathbf{v}$, and the dot product of these quantities gives the rate at which the surroundings do work on the fluid through the moving surface element dS. We use the symbol $W_m^{(p)}$ to indicate the sum of all these surface terms. Furthermore, the integrals over the stationary surfaces S_1 and S_2 give the work required to push the fluid into the system at plane 1 minus the work required to push the fluid out of the system at plane 2. Therefore, Term 2 gives finally:

$$\text{Term 2} = p_1 \langle v_1 \rangle S_1 - p_2 \langle v_2 \rangle S_2 + W_m^{(p)} \tag{7.7-4}$$

Here we have assumed that the pressure does not vary over the cross section at the entry and exit ports.

Term 3 (including the minus sign) gives no contribution on S_f since $\mathbf{v}$ is zero there. The integral over S_m can be interpreted as the rate at which the surroundings do work on the fluid by means of the viscous forces, and this integral is designated as $W_m^{(\tau)}$. At the entry and exit ports it is conventional to neglect the work terms associated with the viscous forces, since they are generally quite small compared with the pressure contributions. Therefore, we get

$$\text{Term 3} = W_m^{(\tau)} \tag{7.7-5}$$

We now introduce the symbol $W_m = W_m^{(p)} + W_m^{(\tau)}$ to represent the total rate at which the surroundings do work on the fluid within the system through the agency of the moving surfaces.

Terms 4 and 5 cannot be further simplified, and hence, we define

$$\text{Term 4} = +\int_{V(t)} p\,(\nabla \cdot \mathbf{v})\,dV = -E_c \tag{7.7-6}$$

$$\text{Term 5} = +\int_{V(t)} (\boldsymbol{\tau} : \nabla\mathbf{v})\,dV = -E_v \tag{7.7-7}$$

For Newtonian fluids the viscous loss E_v is the rate at which mechanical energy is *irreversibly* degraded into thermal energy because of the viscosity of the fluid and is always a positive quantity (see Eq. 3.3-3). We have already discussed methods for estimating E_v in §7.5. (For viscoelastic fluids, which we discuss in Chapter 8, E_v has to be interpreted differently and may even be negative.) The compression term E_c is the rate at which mechanical energy is *reversibly* changed into thermal energy because of the compressibility of the fluid; it may be either positive or negative. If the fluid is being regarded as incompressible, E_c is zero.

When all the contributions are inserted into Eq. 7.7-2 we finally obtain the macroscopic mechanical energy balance:

$$\frac{d}{dt}\left(K_{\text{tot}} + \Phi_{\text{tot}}\right) = \left(\frac{1}{2}\rho_1 \left\langle v_1^3 \right\rangle S_1 + \rho_1\hat{\Phi}_1 \left\langle v_1 \right\rangle S_1 + p_1 \left\langle v_1 \right\rangle S_1\right) - \left(\frac{1}{2}\rho_2 \left\langle v_2^3 \right\rangle S_2 + \rho_2\hat{\Phi}_2 \left\langle v_2 \right\rangle S_2 + p_2 \left\langle v_2 \right\rangle S_2\right) + W_m - E_c - E_v \tag{7.7-8}$$

If, now, we use $w_1 = \rho_1 \left\langle v_1 \right\rangle S_1$ and $w_2 = \rho_2 \left\langle v_2 \right\rangle S_2$ for the mass rates of flow in and out, then Eq. 7.7-8 can be rewritten in the form of Eq. 7.4-2. Several assumptions have been made in this development, but normally they are not serious. If the situation warrants, one can go back and include the neglected effects.

It should be noted that the above derivation of the mechanical energy balance does not require that the system be isothermal. Therefore, the results in Eqs. 7.4-2 and 7.7-8 are valid for nonisothermal systems.

To get the mechanical energy balance in the form of Eq. 7.4-7, we have to develop an *approximate* expression for E_c. We imagine that there is a representative streamline running through the system, and we introduce the distance s along the streamline. We assume that pressure, density, and velocity do not vary over the cross section. We further imagine that at each position along the streamline, there is a cross section $S(s)$ perpendicular to the s coordinate, so that we can write $dV = S(s)\,ds$. If there are moving parts in the system and if the system geometry is complicated, it may not be possible to do this.

We start by using the fact that $(\nabla \cdot \rho\mathbf{v}) = 0$ *at steady state* so that

$$E_c = -\int_V p\,(\nabla \cdot \mathbf{v})\,dV = +\int_V \frac{p}{\rho}(\mathbf{v} \cdot \nabla\rho)\,dV \tag{7.7-9}$$

Then we use the assumption that the pressure and density are constant over the cross section to write approximately

$$E_c \approx \int_1^2 \frac{p}{\rho}\left(v\frac{d\rho}{ds}\right) S(s)\;ds \tag{7.7-10}$$

Even though ρ, v, and S are functions of the streamline coordinate s, their product, $w = \rho v S$, is a constant for steady-state operation and hence, may be taken outside the integral. This gives

$$E_c \approx w\int_1^2 \frac{p}{\rho^2}\frac{d\rho}{ds}\,ds = -w\int_1^2 p\frac{d}{ds}\left(\frac{1}{\rho}\right)ds \tag{7.7-11}$$

Then an integration by parts can be performed:

$$E_c \approx -w\left[\frac{p}{\rho}\bigg|_1^2 - \int_1^2 \frac{1}{p}\frac{dp}{ds}\,ds\right] = -w\Delta\left(\frac{p}{\rho}\right) + w\int_1^2 \frac{1}{\rho}\,dp \qquad (7.7\text{-}12)$$

When this result is put into Eq. 7.4-5, the approximate relation in Eq. 7.4-7 is obtained. Because of the questionable nature of the assumptions made (the existence of a representative streamline and the constancy of p and ρ over a cross section), it seems preferable to use Eq. 7.4-5 rather than Eq. 7.4-7. Also Eq. 7.4-5 is easily generalized to systems with multiple inlet and outlet ports, whereas Eq. 7.4-7 is not; the generalization is given in Eq. (D) of Table 7.6-1.

§7.8 CONCLUDING COMMENTS

The *steady-state macroscopic balances* have been part of the engineers' and applied scientists' tool kit for many decades. Armed only with limited mathematical skills and some intuition, one can solve many practical problems. Sometimes the macroscopic balances have to be supplemented with experimental data or visual observations. Sometimes adjustable constants (or "fudge factors") have to be introduced in order to get suitable agreement with the experimental facts.

The *unsteady-state macroscopic balances* are of more recent vintage and have not received much attention. They usually do not appear in engineering handbooks. When they are used, they have to be supplemented with intuition, good judgment, experimental data, and flow visualizations. Only experience will enable one to develop skill in their use.

Sometimes the macroscopic balances can be used for making "back-of-the-envelope" estimates before embarking on an extensive experimental program or a detailed numerical analysis of a process.

QUESTIONS

1. Discuss the origin, meaning, and use of the macroscopic balances, and explain what assumptions have been made in deriving them.
2. How does one decide which macroscopic balances to use for a given problem? What auxiliary information might one need in order to solve problems with the macroscopic balances?
3. Are friction factors and friction-loss factors related? If so, how?
4. Discuss the viscous loss E_v and the compression term E_c, with regard to physical interpretation, sign, and methods of estimation.
5. How is the macroscopic mechanical energy balance related to the Bernoulli equation for inviscid fluids? How is it derived?
6. What happens in Example 7.3-1 if one makes a different choice for the origin of the coordinate system?
7. In Example 7.5-1, what would be the error in the final result if the estimation of the viscous loss E_v were off by a factor of 2? Under what circumstances would such an error be more serious?
8. In Example 7.5-1, what would happen if 5 ft were replaced by 50 ft?
9. In Example 7.6-3, how would the results be affected if the outlet pressure were 11 atm instead of 1.1 atm?
10. List all the assumptions that are inherent in the equations given in Table 7.6-1.

PROBLEMS

7A.1 Pressure rise in a sudden enlargement. An aqueous salt solution is flowing through a sudden enlargement (see Fig. 7.6-1) at a rate of 450 U.S. gal/min = 0.0284 m^3/s. The inside diameter of the smaller pipe is 5 in. and that of the large pipe is 9 in. What is the pressure rise in pounds per square inch if the density of the solution is 63 lb_m/ft^3? Is the flow in the smaller pipe laminar or turbulent?

Answer: 0.157 psi = 1.08×10^3 N/m^2

7A.2 Pumping a hydrochloric acid solution. A dilute HCl solution of constant density and viscosity ($\rho = 64$ lb_m/ft^3, $\mu = 1$ cp), is to be pumped from tank 1 to tank 2 with no overall change in elevation (see Fig. 7A.2). The pressures in the gas spaces of the two tanks are $p_1 = 1$ atm and $p_2 = 4$ atm. The pipe radius is 2 in. and the Reynolds number is 7.11×10^4. The average velocity in the pipe is to be 2.30 ft/s. What power must be delivered by the pump?

Answer: 2.3 hp = 1.7 kW

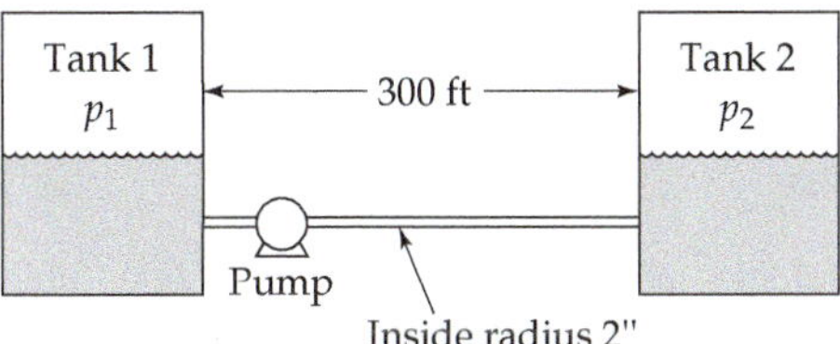

Fig. 7A.2 Pumping of a hydrochloric acid solution.

7A.3 Compressible gas flow in a cylindrical pipe. Gaseous nitrogen is in isothermal turbulent flow at 25°C through a straight length of horizontal pipe with 3-in. inside diameter at a rate of 0.28 lb_m/s. The absolute pressures at the inlet and outlet are 2 atm and 1 atm, respectively. Evaluate $\hat{E}_v$, assuming ideal gas behavior and radially uniform velocity distribution.

Answer: 26.3 Btu/lb_m = 6.11×10^4 J/kg

7A.4 Incompressible flow in an annulus. Water at 60°F is being delivered from a pump through a coaxial annular conduit 20.3 ft long at a rate of 241 U.S. gal/min. The inner and outer radii of the annular space are 3 in. and 7 in. The inlet is 5 ft lower than the outlet. Determine the power output required from the pump. Use the mean hydraulic radius empiricism to solve the problem. Assume that the pressures at the pump inlet and the annular outlet are the same.

Answer: 0.31 hp = 0.23 kW

7A.5 Force on a U-bend. Water at 68°F ($\rho = 62.4$ lb_m/ft^3, $\mu = 1$ cp), is flowing in turbulent flow in a U-shaped pipe bend at 3 ft^3/s (see Fig. 7A.5). What is the horizontal force exerted by the water on the U-bend?

Answer: 903 lb_f to the right

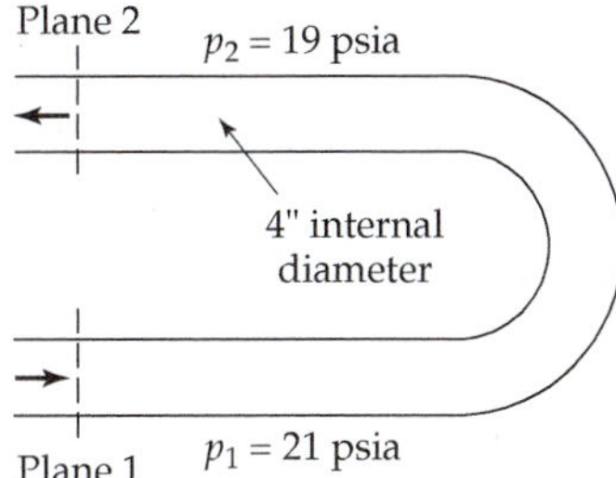

Fig. 7A.5 Flow in a U-bend; both arms of the bend are at the same elevation.

7A.6 Flow-rate calculation. For the system shown in Fig. 7A.6, calculate the volume flow rate of water at 68°F.

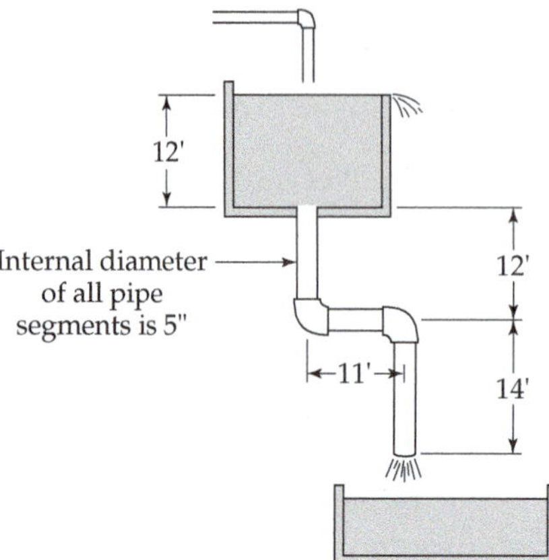

Fig. 7A.6 Flow from a constant-head tank.

7A.7 Evaluation of various velocity averages from Pitot tube data. Following are some experimental data[1] for a Pitot tube traverse for the flow of water in a pipe of internal radius 3.06 in.:

Position	Distance from tube center (in.)	Local velocity (ft/s)	Position	Distance from tube center (in.)	Local velocity (ft/s)
1	2.80	7.85	6	0.72	11.70
2	2.17	10.39	7	1.43	11.47
3	1.43	11.31	8	2.17	11.10
4	0.72	11.66	9	2.80	9.26
5	0.00	11.79			

Plot these data and find out whether the flow is laminar or turbulent. Then use Simpson's rule for numerical integration to compute $\langle v\rangle / v_{\max}$, $\langle v^2\rangle / v_{\max}^2$, and $\langle v^3\rangle / v_{\max}^3$. Are these results consistent with the values of $\langle v^2\rangle / \langle v\rangle^2 = \frac{50}{49}$ (given just before Example 7.2-1) and $\langle v^3\rangle / \langle v\rangle^3 = \frac{43200}{40817}$ (given just before Example 7.4-1)?

7B.1 Velocity averages from the $\frac{1}{7}$ power law. Evaluate the velocity ratios in Problem 7A.7 according to the velocity distribution in Eq. 4.1-4.

7B.2 Relation between force and viscous loss for flow in conduits of variable cross section. Equation 7.5-7 gives the relation $F_{f\to s} = \rho S \hat{E}_v$ between the drag force and viscous loss for straight conduits of arbitrary, but constant, cross section. Here we consider a straight channel whose cross section varies gradually with the downstream distance. We restrict ourselves to axisymmetrical channels, so that the force $F_{f\to s}$ is axially directed.

If the cross section and pressure at the entrance are S_1 and p_1, and those at the exit are S_2 and p_2, then prove that the relation analogous to Eq. 7.5-7 is

$$F_{f\to s} = \rho S_m \hat{E}_v + p_m (S_1 - S_2) \tag{7B.2-1}$$

where

$$\frac{1}{S_m} = \frac{1}{2}\left(\frac{1}{S_1} + \frac{1}{S_2}\right) \tag{7B.2-2}$$

[1] B. Bird, C. E. Thesis, University of Wisconsin (1915).

$$p_m = \frac{p_1 S_1 + p_2 S_2}{S_1 + S_2} \tag{7B.2-3}$$

Interpret the results.[2]

7B.3 Flow through a sudden enlargement. A fluid is flowing through a sudden enlargement, in which the initial and final diameters are D_1 and D_2, respectively (see Fig. 7.6-1). At what ratio D_2/D_1 will the pressure rise $p_2 - p_1$ be a maximum for a given value of v_1?

Answer: $D_2/D_1 = \sqrt{2}$

7B.4 Flow between two tanks. *Case I*: A fluid flows between two tanks A and B because $p_A > p_B$. The tanks are at the same elevation and there is no pump in the line. The connecting line has a cross-sectional area S_I and the mass rate of flow is w for a pressure drop of $(p_A - p_B)_\text{I}$ (see Fig. 7B.4).

Case II: It is desired to replace the connecting line by two lines, each with cross section $S_\text{II} = \frac{1}{2}S_\text{I}$. What pressure difference $(p_A - p_B)_\text{II}$ is needed to give the same total mass flow rate as in Case I? Assume turbulent flow and use the Blasius formula (Eq. 6.2-12) for the friction factor. Neglect entrance and exit losses.

Answer: $(p_A - p_B)_\text{II}/(p_A - p_B)_\text{I} = 2^{5/8}$

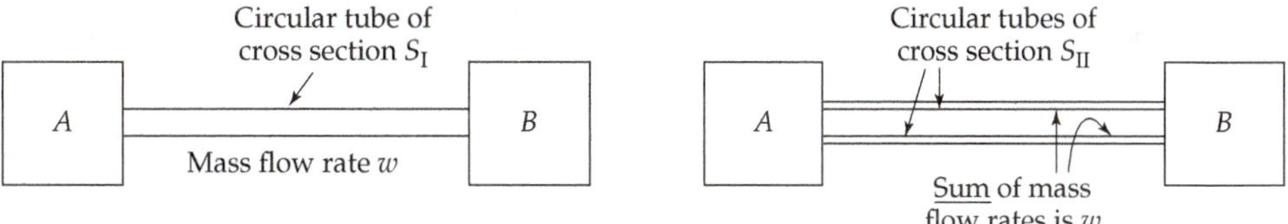

Fig. 7B.4 Flow between two tanks.

7B.5 Revised design of an air duct. A straight, horizontal air duct was to be installed in a factory. The duct was supposed to be 4 ft × 4 ft in cross section. Because of an obstruction, the duct may be only 2 ft high, but it may have any width (see Fig. 7B.5). How wide should the duct be to have the same terminal pressures and same volume rate of flow? Assume that the flow is turbulent and that the Blasius formula (Eq. 6.2-12) is satisfactory for this calculation. Air can be regarded as incompressible in this situation.

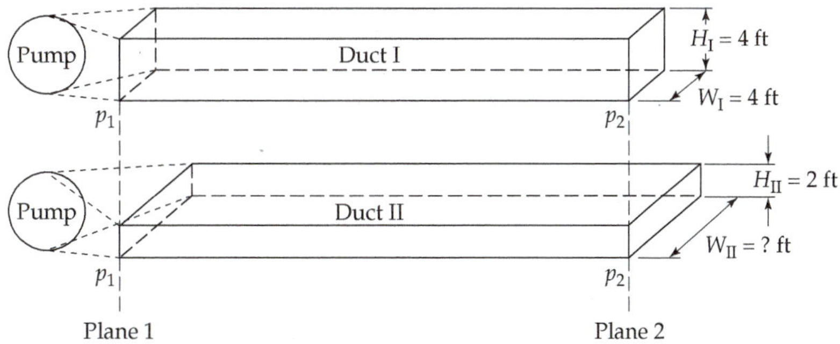

Fig. 7B.5 Installation of an air duct.

[2] R. B. Bird, in *Selected Topics in Transport Phenomena*, CEP Symposium Series #58, **61**, 1–15 (1965), pp. 14 & 15.

(a) Write the simplified versions of the mechanical energy balance for ducts I and II.

(b) Equate the pressure drops for the two ducts and obtain an equation relating the widths and heights of the two ducts.

(c) Solve the equation in (b) numerically to find the width that should be used for duct II.

Answer: **(c)** 9.2 ft

7B.6 Multiple discharge into a common conduit.[3] Extend Example 7.6-1 to an incompressible fluid discharging from several tubes into a larger tube with a net increase in cross section, as shown in Fig. 7B.6. Such systems are important in heat exchangers of certain types, for which the expansion and contraction losses account for an appreciable fraction of the overall pressure drop. The flows in the small tubes and the large tube may be laminar or turbulent. Analyze this system by means of the macroscopic mass, momentum, and mechanical energy balances.

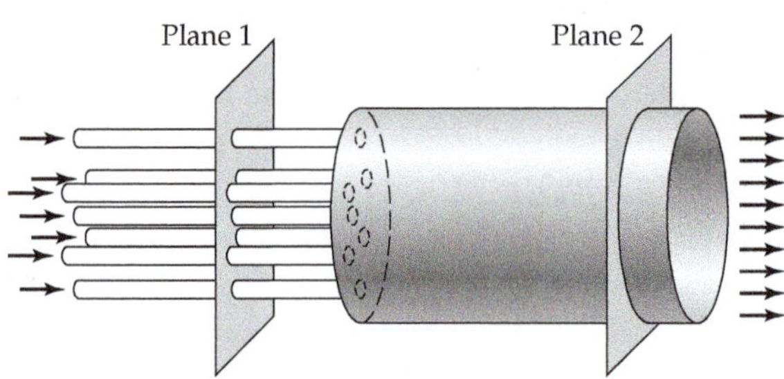

Fig. 7B.6 Multiple discharge into a common conduit. The total cross-sectional area at plane 1 available for flow is S_1, and that at plane 2 is S_2.

7B.7 Inventory variations in a gas reservoir. A natural gas reservoir is to be supplied from a pipeline at a steady-state rate of w_1 lb$_m$/hr. During a 24-hour period, the fuel demand from the reservoir, w_2, varies approximately as follows:

$$w_2 = A + B\cos\omega t \tag{7B.7-1}$$

where ωt is a dimensionless time measured from the time of peak demand (approximately 6 a.m.).

(a) Determine the maximum, minimum, and average values of w_2 for a 24-hour period in terms of A and B.

(b) Determine the required value of w_1 in terms of A and B.

(c) Let $m_{\text{tot}} = m^0_{\text{tot}}$ at $t = 0$, and integrate the unsteady mass balance with this initial condition to obtain m_{tot} as a function of time.

(d) If $A = 5000$ lb$_m$/hr, $B = 2000$ lb$_m$/hr, and $\rho = 0.044$ lb$_m$/ft^3 in the reservoir, determine the absolute minimum reservoir capacity in cubic feet to meet the demand without interruption. At what time of day must the reservoir be full to permit such operation?

(e) Determine the minimum reservoir capacity in cubic feet required to permit maintaining at least a three-day reserve at all times.

Answer: **(d)** 3.47×10^5 ft^3; **(e)** 8.53×10^6 ft^3

7B.8 Change in liquid height with time.

(a) In Example 7.1-1, obtain the expression for the liquid height h as a function of time t (see Fig. 7.1-1).

(b) Make a graph of Eq. 7.1-12 using dimensionless quantities. Is this useful?

---

[3] W. M. Kays, *Trans. ASME*, **72**, 1067–1074 (1950).

7B.9 Draining of a cylindrical tank with exit pipe.

(a) Rework Example 7.1-1, but with a cylindrical tank instead of a spherical tank (see Fig. 7B.9). Use the quasi-steady-state approach; that is, use the unsteady-state mass balance along with the Hagen-Poiseuille equation for the laminar flow in the pipe.

(b) Rework the problem for turbulent flow in the pipe.

Answer: **(a)** $t_{\text{efflux}} = \dfrac{128\mu LR^2}{\rho g D^4} \ln\left(1 + \dfrac{H}{L}\right)$

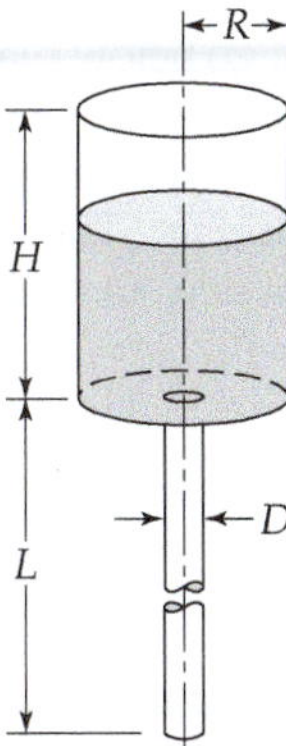

Fig. 7B.9 A cylindrical tank with a long pipe attached. The fluid surfaces and pipe exits are open to the atmosphere.

7B.10 Efflux time for draining a conical tank. A conical tank, with dimensions given in Fig. 7B.10, is initially filled with a liquid. The liquid is allowed to drain out by gravity. Determine the efflux time. In parts (a)–(c) take the liquid in the cone to be the "system."

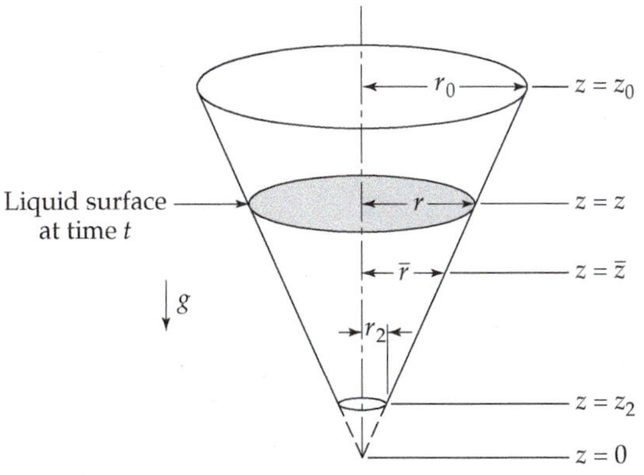

Fig. 7B.10 A conical container from which a fluid is allowed to drain. The quantity r is the radius of the liquid surface at height z, and $\bar{r}$ is the radius of the cone at some arbitrary height $\bar{z}$.

(a) First, use an unsteady macroscopic mass balance to show that the exit velocity is related to the liquid height z by

$$v_2 = -\frac{z^2}{z_2^2}\frac{dz}{dt} \tag{7B.10-1}$$

(b) Write the unsteady-state mechanical energy balance for the system. Discard the viscous loss term and the term containing the time derivative of the kinetic energy, and give reasons for doing so. Show that Eq. 7B.10-1 then leads to

$$v_2 = \sqrt{2g(z - z_2)} \tag{7B.10-2}$$

(c) Combine the results of (a) and (b). Solve the resulting differential equation with an appropriate initial condition to get the liquid level z as a function of t. From this, get the efflux time

$$t_{\text{efflux}} = \frac{1}{5}\left(\frac{z_0}{z_2}\right)^2 \sqrt{\frac{2z_0}{g}} \tag{7B.10-3}$$

List all the assumptions that have been made and discuss how serious they are. How could these assumptions be avoided?

(d) Rework part (b) by choosing plane 1 to be stationary and slightly below the liquid surface at time t. It is understood that the liquid surface does not go below plane 1 during the differential time interval dt over which the unsteady mechanical energy balance is made. With this choice of plane 1, the derivative $d\Phi_{\text{tot}}/dt$ is zero and there is no work term W_m. Furthermore, the conditions at plane 1 are very nearly those at the liquid surface. Then with the quasi-steady-state approximation that the derivative dK_{tot}/dt is approximately zero and the neglect of the viscous loss term, show that the mechanical energy balance, with $w_1 = w_2$, takes the form

$$0 = \frac{1}{2}\left(v_1^2 - v_2^2\right) + g\left(h_1 - h_2\right) \tag{7B.10-4}$$

Is this result reasonable? Could you get this result from the classical Bernoulli equation in Ch. 3?

7B.11 Disintegration of wood chips. In the manufacture of paper pulp, the cellulose fibers of wood chips are freed from the lignin binder by heating in alkaline solutions under pressure in large cylindrical tanks called digesters (see Fig. 7B.11). At the end of the "cooking" period, a small port in one end of the digester is opened, and the slurry of softened wood chips is allowed to blow against an impact plate to complete the breakup of the chips and the separation of the fibers. Estimate the velocity of the discharging stream and the additional force on the impact plate shortly after the discharge begins. Frictional effects inside the digester, and the small kinetic energy of the fluid inside the tank, may be neglected. (*Note*: See Problem 7B.10 for two different methods for selecting the entrance and exit planes.)

Answer: 2810 lb_m/s (or 1275 kg/s); 10,900 lb_f (or 48,500 N)

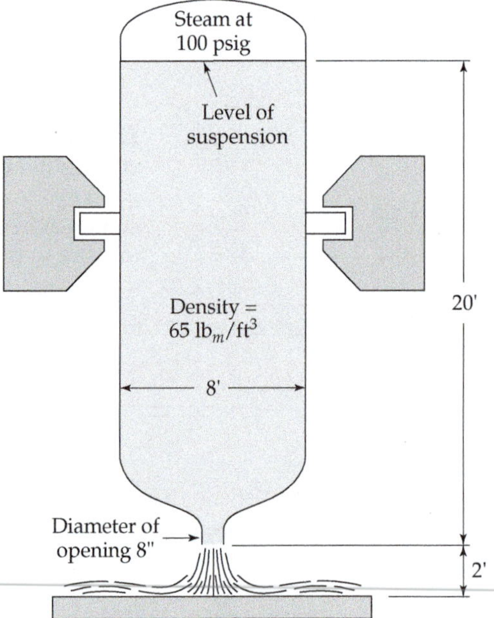

Fig. 7B.11 Pulp digester.

7B.12 Criterion for vapor-free flow in a pipeline. To ensure that a pipeline is completely filled with liquid, it is necessary that $p > p_{\text{vap}}$ at every point. Apply this criterion to the system in Fig. 7.5-1 by using a mechanical energy balance over appropriate portions of the system. Where in the system is the pressure a minimum?

7C.1 End corrections in tube viscometers.[4] In analyzing tube-flow viscometric data to determine viscosity, one compares pressure drop vs. flow rate data with the theoretical expression (the Hagen-Poiseuille equation of Eq. 2.3-21). The latter assumes that the flow is fully developed in the region between the two planes at which the pressure is measured. In an apparatus such as that shown in Fig. 7C.1, the pressure is known at the tube exit (2) and also above the fluid in the reservoir (1). However, in the entrance region of the tube, the velocity profiles are not yet fully developed. Hence, the theoretical expression relating the pressure drop to the flow rate is not valid.

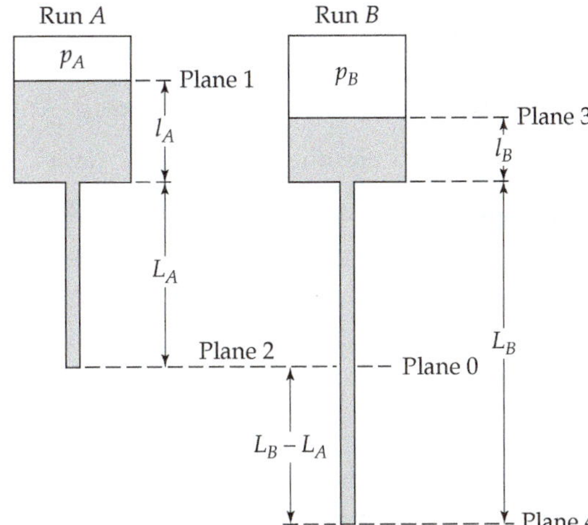

Fig. 7C.1 Two tube viscometers with the same flow rate and the same exit pressure. The pressures p_A and p_B are maintained by an inert gas.

There is, however, a method in which the Hagen-Poiseuille equation can be used, by making flow measurements in two tubes of different lengths, L_A and L_B; the shorter of the two tubes must be long enough so that the velocity profiles are fully developed at the exit. Then the end section of the long tube, of length $L_B - L_A$, will be a region of fully developed flow. If we knew the value of $\mathscr{P}_0 - \mathscr{P}_4$ for this region, then we could apply the Hagen-Poiseuille equation.

Show that proper combination of the mechanical energy balances, written for the systems 1–2, 3–4, and 0–4 gives the following expression for $\mathscr{P}_0 - \mathscr{P}_4$ when each viscometer has the *same flow rate*

$$\frac{\mathscr{P}_0 - \mathscr{P}_4}{L_B - L_A} = \frac{p_B - p_A}{L_B - L_A} + \rho g\left(1 + \frac{l_B - l_A}{L_B - L_A}\right) \tag{7C.1-1}$$

where $\mathscr{P}_0 = p_0 + \rho g z_0$. Explain carefully how you would use Eq. 7C.1-1 to analyze experimental measurements. Is Eq. 7C.1-1 valid for ducts with noncircular, uniform cross section?

[4] A. G. Fredrickson, PhD Thesis, University of Wisconsin (1959); *Principles and Applications of Rheology*, Prentice-Hall, Englewood Cliffs, NJ (1964), §9.2.

Chapter 8

Non-Newtonian Liquids

In the first seven chapters only *Newtonian fluids* have been considered. The relations between stresses and velocity gradients are described by Eq. 1.2-2 for simple shear flow and by Eq. 1.2-13 (or Eqs. 1.2-7 to 1.2-12) for arbitrary time-dependent flows. For the Newtonian fluid, two material parameters are needed—the viscosity μ and the dilatational viscosity κ—which depend on temperature, pressure, and composition, but not on the velocity gradients or on time. In most situations κ is not needed. All gases and all liquids composed of "small" molecules (up to molecular weights of about 5000 g/g-mol) are accurately described by the Newtonian fluid model.

There are many fluids that are not described by Eq. 1.2-13, and these are called *non-Newtonian fluids.* These structurally complex fluids include polymer solutions, polymer melts, soap solutions, suspensions, colloidal dispersions, pastes, foodstuffs, foams, and some biological fluids.

Because of their complexity, these fluids have behavior qualitatively different from that of Newtonian fluids. For example, they may have viscosities that depend strongly on the velocity gradients and time, and, in addition, may display pronounced "elastic effects." Also in the steady simple shear flow between two parallel plates, unequal normal stresses (τ_{xx}, τ_{yy}, and τ_{zz}) arise. In §8.1 we describe some experiments that emphasize the "funny fluid flow fenomena" of polymeric liquids and other non-Newtonian fluids.

In dealing with Newtonian fluids, the science of the measurement of viscosity is called *viscometry*, and in earlier chapters we have seen examples of simple flow systems that can be used as *viscometers* (the circular tube, the cone-plate system, and coaxial cylinders). To characterize non-Newtonian fluids, we must measure not only the viscosity, but the normal stresses and the viscoelastic responses as well. The science of measurement of these properties is called *rheometry*, and the instruments are called *rheometers*; the rotational cone-and-plate rheometer is the most famous of these. We treat this subject briefly in §8.2. The science of *rheology* includes all aspects of the study of deformation and flow of both non-Hookean solids and non-Newtonian liquids.

After the first two sections, which deal with experimental facts, we turn to the presentation of various non-Newtonian "models" (that is, empirical expressions for the stress tensor) that are commonly used for describing non-Newtonian liquids. In §8.3 we start with the *generalized Newtonian models*, which are relatively simple, but which can describe only the non-Newtonian viscosity (and not the normal stresses and the viscoelastic effects). Then in §8.4 we give examples of *linear viscoelastic models*, which can describe the viscoelastic responses, but only in flows with exceedingly small displacement gradients. In §8.5 we give examples of *nonlinear viscoelastic models*, which are capable of describing most of the phenomena observed in polymeric fluids, at least qualitatively. Finally, in §8.6, we give a simple molecular theory that describes nonlinear effects.

Polymeric liquids are encountered in the fabrication of plastic objects, and as additives to lubricants, foodstuffs, and inks. They represent a vast and important class of liquids, and many scientists and engineers must deal with them. Polymer fluid dynamics, heat transfer, and diffusion form a rapidly growing part of the subject of transport phenomena, and there are many textbooks,[1] treatises,[2] and journals devoted to the subject. The subject has also been approached from the kinetic theory standpoint, and molecular theories of the subject have contributed much to our understanding of the mechanical, thermal, and diffusional behavior of these fluids.[3] Finally, for those interested in the history of the subject, the reader is referred to the book by Tanner and Walters.[4]

§8.1 "PHUNNY PHLUID PHLOW PHENOMENA"

In this section we discuss several experiments that contrast the flow behavior of Newtonian and non-Newtonian fluids.[1] Most of the emphasis is placed on polymeric fluids, but the two-phase fluids are discussed at the very end of the section. The reason for this lopsided treatment is that the flow phenomena of polymers have been much more extensively studied than those of suspensions and emulsions, which are every bit as important as polymers are.

a. Steady axial flow in tubes

Even for the steady-state, axial, laminar flow in circular tubes, there is an important difference between the behavior of Newtonian liquids and that of polymeric liquids. For Newtonian liquids, the velocity distribution, average velocity, and mass flow rate are given by Eqs. 2.3-19, 2.3-22, and 2.3-23, respectively.

---

[1] A. S. Lodge, *Elastic Liquids*, Academic Press, New York (1964). **Arthur Scott Lodge** (1922–2005) was an eminent rheologist and a professor at the University of Wisconsin-Madison from 1968 to 2005. His two books, *Elastic Liquids* and *Body Tensor Fields in Continuum Mechanics*, provided guidance to many generations of rheologists. He was the inventor of the "Lodge Stressmeter." He was a recipient of the Bingham Medal of The Society of Rheology, and the Gold Medal of the British Society of Rheology; he was elected to the National Academy of Engineering in 1992. R. B. Bird, R. C. Armstrong, and O. Hassager, *Dynamics of Polymeric Liquids, Vol 1, Fluid Mechanics*, Wiley-Interscience, New York, 2nd edition (1987); R. I. Tanner, *Engineering Rheology*, Clarendon Press, Oxford (1985); 2nd edition (2000).

[2] H. A. Barnes, J. F. Hutton, and K. Walters, *An Introduction to Rheology*, Elsevier, Amsterdam (1989); H. Giesekus, *Phänomenologische Rheologie: Eine Einführung*, Springer Verlag, Berlin (1994). Books emphasizing the engineering aspects of the subject include Z. Tadmor and C. G. Gogos, *Principles of Polymer Processing*, Wiley, New York, 2nd edition (2006); D. G. Baird and D. I. Collias, *Polymer Processing: Principles and Design*, Butterworth-Heinemann, Boston (1995); J. Dealy and K. Wissbrun, *Melt Rheology and its Role in Plastics Processing*, Van Nostrand Reinhold, New York (1990).

[3] R. B. Bird, C. F. Curtiss, R. C. Armstrong, and O. Hassager, *Dynamics of Polymeric Liquids, Vol. 2, Kinetic Theory*, Wiley-Interscience, New York, 2nd edition (1987); C. F. Curtiss and R. B. Bird, *Adv. Polymer Sci*, **125**, 1–101 (1996) and *J. Chem. Phys.* **111**, 10362–10370 (1999).

[4] R. I. Tanner and K. Walters, *Rheology: An Historical Perspective*, Elsevier, Amsterdam (1998).

[1] More details about these and other experiments can be found in R. B. Bird, R. C. Armstrong, and O. Hassager, *Dynamics of Polymeric Liquids, Vol. 1, Fluid Dynamics*, Wiley-Interscience, New York (1977), Chapter 3; 2nd edition (1987), Chapter 2. See also A. S. Lodge, *Elastic Liquids*, Academic Press, New York (1964), Chapter 10.

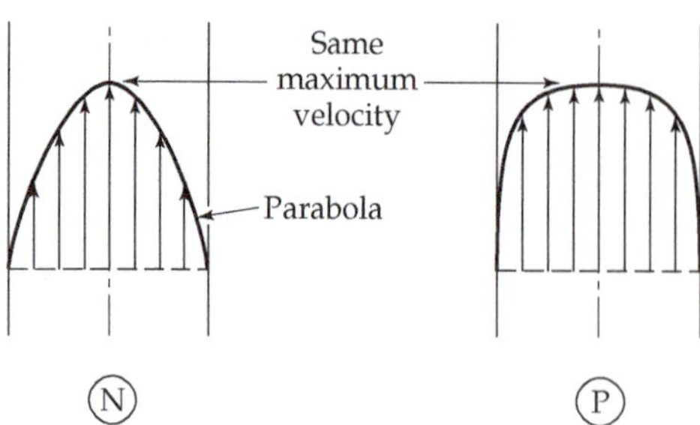

Fig. 8.1-1. Laminar flow in a circular tube. The symbols Ⓝ(Newtonian liquid) and Ⓟ (Polymeric liquid) are used in this and the next six figures.

For *polymeric fluids*, experimental data suggest that the following equations are reasonable:

$$\frac{v_z(r)}{v_{z,\max}} \approx 1 - \left(\frac{r}{R}\right)^{(1/n)+1} \quad \text{and} \quad \frac{\langle v_z \rangle}{v_{z,\max}} \approx \frac{(1/n)+1}{(1/n)+3} \tag{8.1-1,2}$$

where n is a positive parameter characterizing the fluid, usually with a value less than unity. That is, the velocity profile is more blunt than that for the Newtonian fluid, for which $n = 1$. It is further found experimentally that

$$w \approx (\mathscr{P}_0 - \mathscr{P}_L)^{1/n} \tag{8.1-3}$$

The mass flow rate for polymers increases much more rapidly with the pressure drop than for Newtonian fluids, for which the relation is linear.

In Fig. 8.1-1 we show typical velocity profiles for laminar flow of Newtonian and polymeric fluids for the same maximum velocity. The blunted velocity profiles for polymeric liquids suggest that they have a viscosity that depends on the velocity gradient. This point will be elaborated on in §8.3.

For *laminar flow in tubes of noncircular cross section*, polymeric liquids exhibit secondary flows superposed on the axial motion. Recall that for turbulent Newtonian flows secondary flows are also observed—in Fig. 4.1-3 it is shown that the fluid moves from the center directly toward the corners of the conduit and then back along the walls. For laminar flow of polymeric fluids, the secondary flows go in just the opposite direction—from the corners of the conduit directly toward the center and then back toward the walls.[2] In turbulent Newtonian flows, the secondary flows result from inertial effects, whereas in the flow of polymers the secondary flows are associated with the "normal stress differences."

b. Recoil after cessation of steady-state flow in a circular tube

We start with a fluid at rest in a circular tube and, with a syringe, we "draw" a dye line radially in the fluid as shown in Fig. 8.1-2. Then we pump the fluid and watch the dye deform.[3]

For a Newtonian fluid, the dye line deforms into a continuously stretching parabola. If the pump is turned off, the dye-parabola stops moving. After some time diffusion occurs and the parabola begins to get fuzzy, of course.

For a *polymeric liquid*, the dye line deforms into a curve that is more blunt than a parabola (cf. Eq. 8.1-1). If the pump is stopped and the fluid is not axially constrained, the fluid will begin to "recoil" and will retreat from this maximum stretched shape. That is, the fluid snaps back somewhat like a rubber band. But, whereas a rubber band returns to its original shape, the fluid retreats only part way towards its original configuration.

---

[2]B. Gervang and P. S. Larsen, *J. Non-Newtonian Fluid Mech.*, **39**, 217–237 (1991).

[3]For the details of this experiment see N. N. Kapoor, M.S. Thesis, University of Minnesota, Minneapolis (1964), as well as A. G. Fredrickson, *Principles and Applications of Rheology*, Prentice-Hall, Englewood Cliffs, NJ (1964), p. 120.

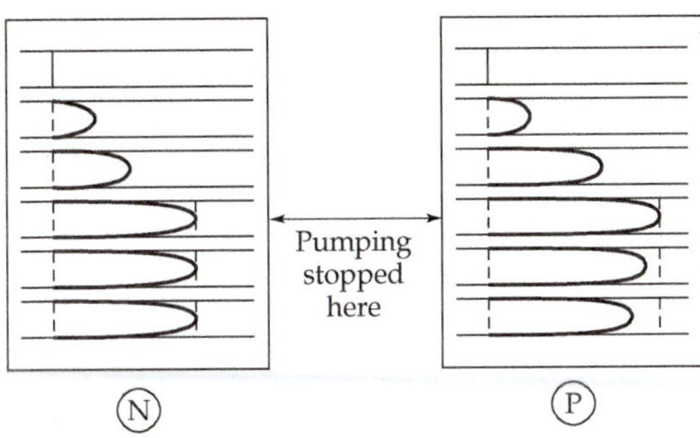

Fig. 8.1-2. Constrained recoil after cessation of flow in a circular tube, observed in polymeric liquids, but not in Newtonian liquids.

If we permit ourselves an anthropomorphism, we can say that a rubber band has "perfect memory," since it returns to its initial unstressed state. The polymeric fluid, on the other hand, has a "fading memory," since it gradually "forgets" its original state. That is, as it recoils, its memory becomes weaker and weaker.

Fluid recoil is a manifestation of *elasticity*, and any complete description of polymeric fluids must be able to incorporate the idea of elasticity into the expression for the stress tensor. The theory must also include the notion of fading memory.

c. "Normal stress" effects

Other striking differences in the behavior of Newtonian and polymeric liquids appear in the "normal stress" effects. The reason for this nomenclature will be given in the next section.

A rotating rod in a beaker of a Newtonian fluid causes the fluid to undergo a tangential motion. At steady state, the fluid surface is lower near the rotating rod. Intuitively we know that this comes about because the centrifugal force causes the fluid to move radially toward the beaker wall. For a *polymeric liquid*, on the other hand, the fluid moves toward the rotating rod, and, at steady state, the fluid surface is as shown in Fig. 8.1-3. This phenomenon is called the *Weissenberg rod-climbing effect*.[4] Evidently some kinds of forces are induced that cause the polymeric liquid to behave in a way that is qualitatively different from that of a Newtonian liquid.

In a closely related experiment, we can put a rotating disk on the surface of a fluid in a cylindrical container as shown in Fig. 8.1-4. If the fluid is Newtonian, the rotating disk causes the fluid to move in a tangential direction (the "primary flow"), but, in addition, the fluid moves slowly outward toward the cylinder wall because of the centrifugal force,

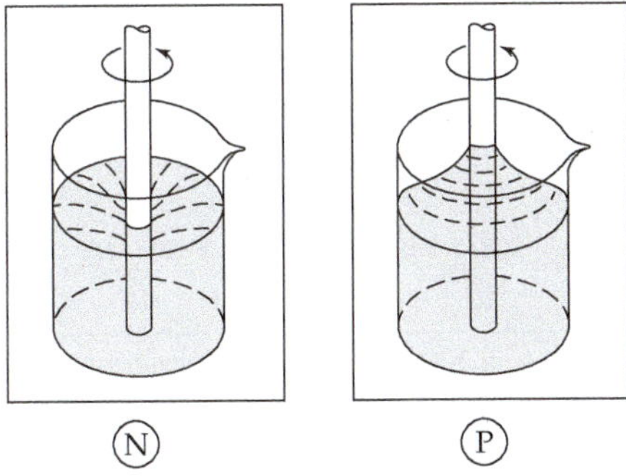

Fig. 8.1-3. The free surface of a liquid near a rotating rod. The polymeric liquid shows the Weissenberg rod-climbing effect.

[4]This phenomenon was first described by F. H. Garner and A. H. Nissan, *Nature*, **158**, 634–635 (1946) and by R. J. Russel, Ph.D. Thesis, Imperial College, University of London (1946), p. 58. The experiment was then analyzed by K. Weissenberg, *Nature*, **159**, 310–311 (1947).

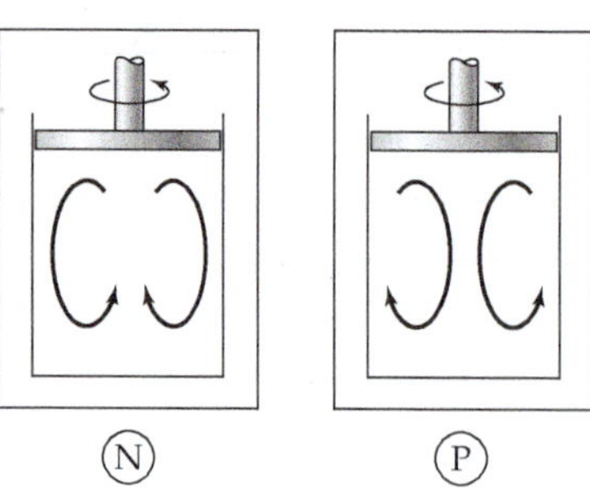

Fig. 8.1-4. The secondary flows in a cylindrical container with a rotating disk at the liquid surface have the opposite directions for Newtonian and polymeric fluids. For polymeric fluids, photographs suggest that the secondary flow may not be symmetric as illustrated in the figure.[5]

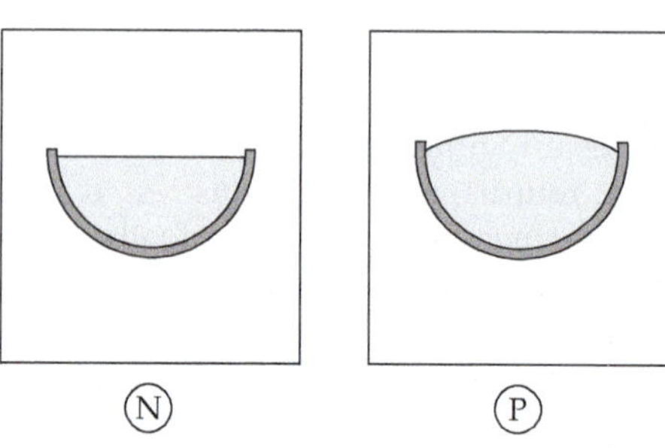

Fig. 8.1-5. Flow down a tilted semicylindrical trough. The convexity of the polymeric liquid surface is somewhat exaggerated here.

then moves downward, and then back up along the cylinder axis. This superposed radial and axial flow is weaker than the primary flow and is termed a "secondary flow." For a *polymeric liquid*, the fluid also develops a primary tangential flow with a weak radial and axial secondary flow, but the latter goes in a direction opposite to that seen in the Newtonian fluid.[5]

In another experiment we can let a liquid flow down a tilted, semi-cylindrical trough as shown in Fig. 8.1-5. If the fluid is Newtonian, the liquid surface is flat—except for the meniscus effects at the outer edges. But for most *polymeric liquids*, the liquid surface is found to be slightly convex. The effect is small but reproducible.[6]

d. Some other experiments

The operation of a simple siphon is familiar to everyone. We know from experience that, if the fluid is Newtonian, the removal of the siphon tube from the liquid causes the siphoning action to stop. However, as may be seen in Fig. 8.1-6, for *polymeric liquids* the siphoning can continue even when the siphon is lifted several centimeters above the liquid surface. This is called the *tubeless siphon* effect. One can also just lift some of the fluid up over the edge of the beaker with one's finger and then the fluid will flow upwards along the inside of the beaker and then down the outside until the beaker is nearly empty.[7]

In another experiment a long cylindrical rod, with its axis in the z direction, is made to oscillate back and forth in the x direction with the axis parallel to the z axis (see Fig. 8.1-7). In a Newtonian fluid, a secondary flow is induced, whereby the fluid moves toward the cylinder from above and below (i.e., from the $+y$ and $-y$ directions), and moves away to the left and right (i.e., toward the $-x$ and $+x$ directions). For the *polymeric liquid*, however, the induced secondary motion is in the opposite direction: the fluid moves inward

---

[5]C. T. Hill, J. D. Huppler, and R. B. Bird, *Chem. Eng. Sci.*, **21**, 815–817 (1966); C. T. Hill, *Trans. Soc. Rheol.*, **16**, 213–245 (1972). Theoretical analyses have been given by J. M. Kramer and M. W. Johnson, Jr., *Trans. Soc. Rheol.*, **16**, 197–212 (1972), and by J. P. Nirschl and W. E. Stewart, *J. Non-Newtonian Fluid Mech.*, **16**, 233–250 (1984).

[6]This experiment was first done by R. I. Tanner, *Trans. Soc. Rheol.*, **14**, 483–507 (1970), prompted by a suggestion by A. S. Wineman and A. C. Pipkin, *Acta Mech.*, **2**, 104–115 (1966). See also R. I. Tanner, *Engineering Rheology*, Oxford University Press (1985), 102–105; 2nd edition, 107–109 (2000).

[7]D. F. James, *Nature*, **212**, 754–756 (1966).

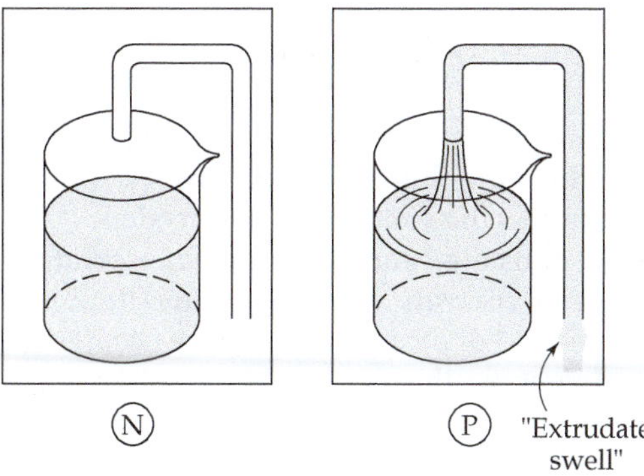

Fig. 8.1-6. Siphoning continues to occur when the tube is raised above the surface of a polymeric liquid, but not so for a Newtonian liquid. Note the swelling of the polymeric liquid as it leaves the siphon tube.

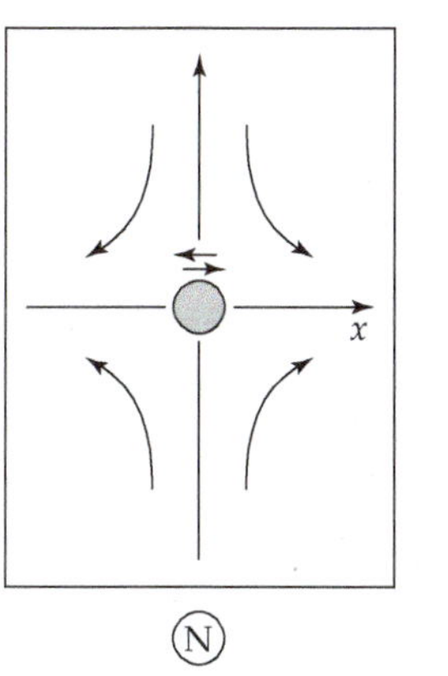

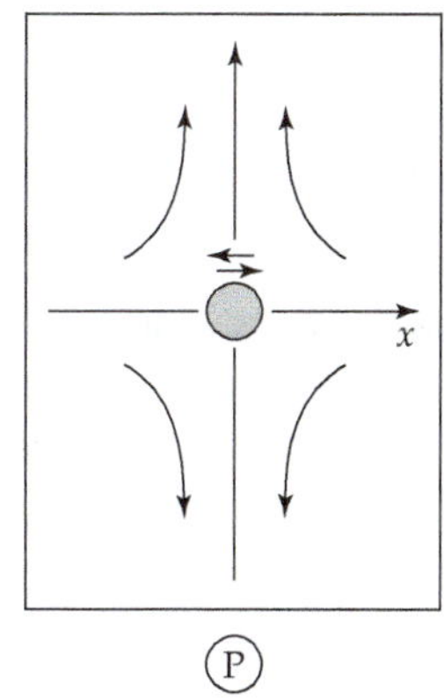

Fig. 8.1-7. The "acoustical streaming" near a laterally oscillating rod, showing that the induced secondary flow goes in the opposite directions for Newtonian and polymeric fluids.

from the left and right along the x axis and outward in the up and down directions along the y axis.[8]

The above examples are just a few of many interesting experiments that have been performed.[9] These phenomena can be illustrated easily and inexpensively with a 0.5% aqueous solution of polyethylene oxide.

There are also some fascinating effects that occur when only tiny quantities of polymers are present. The most striking of these is the phenomenon of *drag reduction*.[10] With just parts per million of some polymers ("drag-reducing agents"), the friction loss in turbulent pipe flow may be lowered dramatically—by 30% to 50%. Such polymeric drag-reducing agents are used by fire departments to increase the flow of water, and by oil companies to lower the costs for pumping crude oil over long distances.

e. Two-phase systems

Particulate suspensions, emulsions, and other two-phase systems can exhibit many of the non-Newtonian features exhibited by polymer solutions. For example, two-phase systems often exhibit blunted velocity profiles in tube flow that arise from a velocity

---

[8] C. F. Chang and W. R. Schowalter, *J. Non-Newtonian Fluid Mech.*, **6**, 47–67 (1979).

[9] The book by D. V. Boger and K. Walters, *Rheological Phenomena in Focus*, Elsevier, Amsterdam (1993), contains many photographs of fluid behavior in a variety of non-Newtonian flow systems.

[10] This is sometimes called the *Toms phenomenon*, since it was perhaps first reported in B. A. Toms, *Proc. Int. Congress on Rheology*, North-Holland, Amsterdam (1949). The phenomenon has also been studied in connection with the drag-reducing nature of fish slime [T. L. Daniel, *Biol. Bull.*, **160**, 376–382 (1981)], which is thought to explain, at least in part, "Gray's paradox"—the fact that fish seem to be able to swim faster than energy considerations permit.

gradient-dependent viscosity, just as for polymers. The recoil phenomenon is observed for emulsions and colloidal gels. The rod-climbing effect occurs for suspensions of elongated fibers.[11,12]

For concentrated suspensions and pastes, one often encounters a *yield stress*, that is, a shear stress below which the fluid does not flow. For example, toothpaste and tomato ketchup will not flow unless the yield stress is exceeded.

For suspensions, emulsions, and some polymer solutions, one can also encounter *thixotropy*, the decrease of viscosity with time as the fluid is being sheared. This change occurs because of some kind of change of structure occurring during flow. Less common is the phenomenon of *rheopexy*, the increase of viscosity with time.

More information about the non-Newtonian character of suspensions and other two-phase systems can be found elsewhere.[13,14,15]

§8.2 RHEOMETRY AND MATERIAL FUNCTIONS

The experiments described in §8.1 make it abundantly clear that many complex fluids do not obey Newton's law of viscosity. In this section we discuss several simple, controllable flows in which the stress components can be measured. From these experiments one can measure a number of *material functions* that describe the mechanical response of complex fluids. Whereas incompressible Newtonian fluids are described by just one material constant (the viscosity), one can measure many different material functions for non-Newtonian liquids. Here we show how a few of the more commonly used material functions are defined and measured. Information about the actual measurement equipment and other material functions can be found elsewhere.[1,2] It is assumed throughout that the polymeric liquids or suspensions can be regarded as incompressible.

a. Steady simple shear flow

We consider now the steady shear flow between a pair of parallel plates, where the velocity profile is given by $v_x = \dot{\gamma}y$, the other velocity components being zero (see Fig. 8.2-1). The quantity $\dot{\gamma}$, here taken to be positive, is called the "shear rate" (i.e., $\dot{\gamma} \equiv |dv_x/dy|$). For an incompressible Newtonian fluid, the shear stress τ_{yx} is given by Eq. 1.2-2, and the normal stresses (τ_{xx}, τ_{yy}, and τ_{zz}) are all zero.

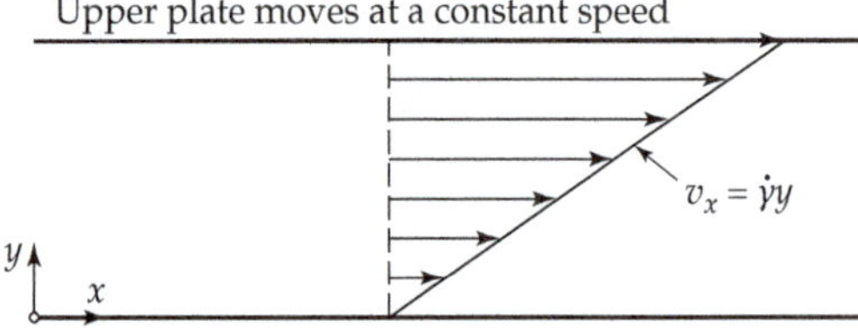

Fig. 8.2-1. Steady simple shear flow between parallel plates, with shear rate $\dot{\gamma}$. For Newtonian fluids in this flow, $\tau_{xx} = \tau_{yy} = \tau_{zz} = 0$, but for polymeric fluids the normal stresses are in general nonzero and unequal.

[11] M. A. Nawab and S. G. Mason, *J. Phys. Chem.*, **62**, 1248–1253 (1958).

[12] J. Mewis and A. B. Metzner, *J. Fluid Mech.*, **62**, 593–600 (1974).

[13] R. J. Hunter, *Foundations of Colloid Science*, Vol. **2**, Oxford (1989).

[14] W. B. Russel, D. A. Saville, and W. R. Schowalter, *Colloidal Dispersions*, Cambridge University Press (1989).

[15] J. Mewis and N. J. Wagner, *Colloidal Suspension Rheology*, Cambridge University Press (2012).

[1] J. R. Van Wazer, J. W. Lyons, K. Y. Kim, and R. E. Colwell, *Viscosity and Flow Measurement*, Interscience (Wiley), New York (1963).

[2] K. Walters, *Rheometry*, Wiley, New York (1975).

For incompressible non-Newtonian fluids, the normal stresses are nonzero and unequal. For these fluids, it is conventional to define three material functions as follows:

$$\tau_{yx} = -\eta \frac{dv_x}{dy} \tag{8.2-1}$$

$$\tau_{xx} - \tau_{yy} = -\Psi_1 \left(\frac{dv_x}{dy} \right)^2 \tag{8.2-2}$$

$$\tau_{yy} - \tau_{zz} = -\Psi_2 \left(\frac{dv_x}{dy} \right)^2 \tag{8.2-3}$$

in which η is the non-Newtonian viscosity, Ψ_1 the first normal stress coefficient, and Ψ_2 the second normal stress coefficient. These three quantities—η, Ψ_1, Ψ_2—are all functions of the shear rate $\dot{\gamma}$. For many polymeric liquids, η may decrease by a factor of as much as 10^4 as the shear rate increases. Similarly, the normal stress coefficients may decrease by a factor of as much as 10^7 over a typical range of shear rates. For polymeric fluids made up of flexible macromolecules, the functions $\eta(\dot{\gamma})$ and $\Psi_1(\dot{\gamma})$ have been found experimentally to be positive, whereas $\Psi_2(\dot{\gamma})$ is almost always negative and perhaps only 1/10 to 1/3 of $\Psi_1(\dot{\gamma})$ in magnitude. It can be shown that for positive $\Psi_1(\dot{\gamma})$ the fluid behaves as though it were under tension in the flow (or x) direction, and that the negative $\Psi_2(\dot{\gamma})$ means that the fluid is under compression in the transverse (or z) direction. For suspensions of rigid spherical particles,[3] both $\Psi_1(\dot{\gamma})$ and $\Psi_2(\dot{\gamma})$ have been measured and found to be negative, with $|\Psi_1| > |\Psi_2|$. For the Newtonian fluid $\eta = \mu$, $\Psi_1 = 0$, and $\Psi_2 = 0$.

The strongly shear-rate-dependent non-Newtonian viscosity is connected with the behavior given in Eqs. 8.1-1 to 8.1-3, as is shown in the next section. The positive Ψ_1 is primarily responsible for the Weissenberg rod-climbing effect. Because of the tangential flow, there is a tension in the tangential direction, and this tension pulls the fluid toward the rotating rod, overcoming the centrifugal force. The secondary flows in the disk-and-cylinder experiment (Fig. 8.1-4) can also be explained qualitatively in terms of the positive Ψ_1. Also, the negative Ψ_2 can be shown to explain the convex surface shape in the tilted-trough experiment (Fig. 8.1-5).

Many ingenious devices have been developed to measure the three material functions for steady shearing flow, and the theories needed for the use of the instruments are explained in detail elsewhere.[2] See Problem 8C.2 for deducing $\eta(\dot{\gamma})$ from tube-flow data and Problem 8C.1 for the use of the cone-and-plate instrument for measuring all three of the material functions.

b. Small-amplitude oscillatory motion

A standard method for measuring the elastic response of a fluid is the small-amplitude oscillatory shear flow experiment, depicted in Fig. 8.2-2. Here the top plate moves back and forth in sinusoidal fashion, and with a tiny amplitude. If the plate spacing is extremely small and the fluid has a very high viscosity, then the velocity profile will be nearly linear, so that $v_x(y,t) = \dot{\gamma}^0 y \cos \omega t$, in which $\dot{\gamma}^0$, a real quantity, gives the amplitude of the shear rate excursion.

The shear stress required to maintain the oscillatory motion will also be periodic in time and, in general, of the form

$$\tau_{yx}(t) = -\eta' \dot{\gamma}^0 \cos \omega t - \eta'' \dot{\gamma}^0 \sin \omega t \tag{8.2-4}$$

in which $\eta'(\omega)$ and $-\eta''(\omega)$ are the real and imaginary parts of the *complex viscosity*, $\eta^* = \eta' - i\eta''$, which is a function of the frequency (where $i = \sqrt{-1}$). The first (in-phase with the shear rate) term is the viscous response, and the second (out-of-phase with the shear

[3] I. E. Zarraga, D. A. Hill, and D. T. Leighton, Jr., *J. Rheol.*, **44**, 185–220 (2000); A. Singh and P. R. Nott, *J. Non-Newtonian Fluid Mech.*, **49**, 293–330 (2003).

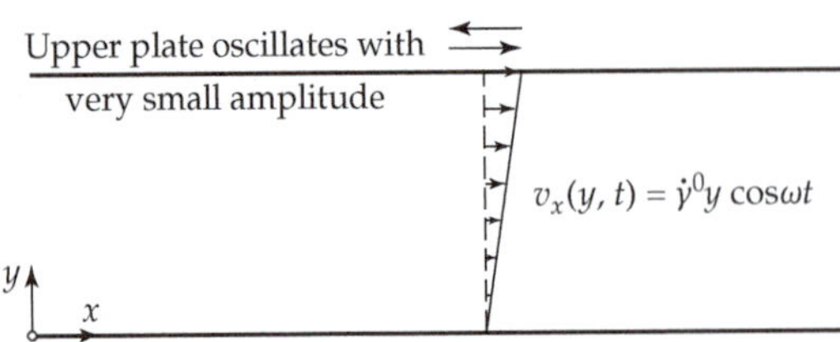

Fig. 8.2-2. Small-amplitude oscillatory motion. For small plate spacing and highly viscous fluids, the velocity profile may be *assumed* to be linear.

rate) term is the elastic response.[4] Polymer scientists and engineers use the curves of $\eta'(\omega)$ and $\eta''(\omega)$ (or the storage and loss moduli, $G' = \eta''\omega$ and $G'' = \eta'\omega$) for "characterizing" polymers and suspensions, since much is known about the connection between the shapes of these curves and the chemical structure of the polymers or particle interactions in the suspensions.[5] For the Newtonian fluid, $\eta' = \mu$ and $\eta'' = 0$.

c. Steady-state elongational flow

A third experiment that can be performed involves the stretching of the fluid, in which the velocity distribution is given by $v_z = \dot{\varepsilon}z$, $v_x = -\frac{1}{2}\dot{\varepsilon}x$, and $v_y = -\frac{1}{2}\dot{\varepsilon}y$ (see Fig. 8.2-3), where the constant quantity $\dot{\varepsilon}$ is called the "elongation rate." Then the relation

$$\tau_{zz} - \tau_{xx} = -\bar{\eta}\frac{dv_z}{dz} \tag{8.2-5}$$

defines the *elongational viscosity* $\bar{\eta}$, which depends on $\dot{\varepsilon}$. When $\dot{\varepsilon}$ is negative, the flow is referred to as *biaxial stretching* because the fluid is being stretched in the x and y directions. For the Newtonian fluid, it can be shown that $\bar{\eta} = 3\mu$, and this is sometimes called the "Trouton viscosity."

The elongational viscosity $\bar{\eta}$ cannot be measured for all fluids, since a steady-state elongational flow cannot always be attained.[6]

The three experiments described above are just a few of the many rheometric tests that can be performed. Other tests include stress relaxation after cessation of flow, stress growth at the inception of flow, recoil, and creep—each of which can be performed in shear, elongation, or in other types of flow. Each experiment results in the definition of one or more material functions. These can be used for fluid characterization and also for determining the constants in the models described in §8.3 to §8.6.

Some sample material functions are displayed in Figs. 8.2.4 to 8.2.6. Since the chemical structure and constitution of complex fluids vary widely, there are many types of

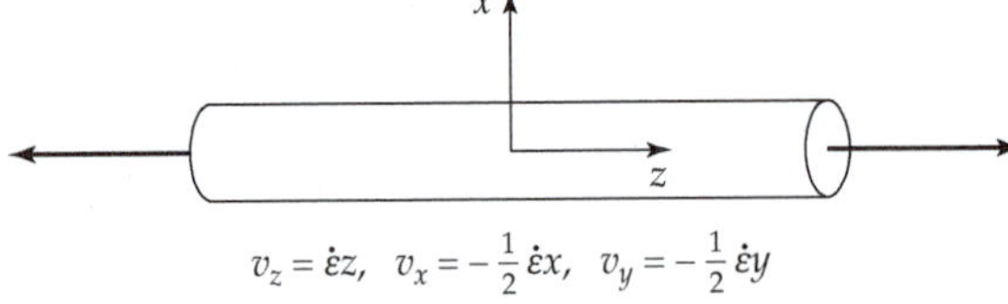

Fig. 8.2-3. Steady elongational flow with elongation rate $\dot{\varepsilon} = \partial v_z/\partial z$.

---

[4]For a discussion of the sign associated with $\eta''(\omega)$, see A. J. Giacomin and R. B. Bird, "Erratum: Official Nomenclature of The Society of Rheology," *J. Rheol.*, **55**, 921–923 (2011).

[5]J. D. Ferry, *Viscoelastic Properties of Polymers*, Wiley, New York, 3rd edition (1980). **John Douglass Ferry** (1912–2002) was a professor of chemistry at the University of Wisconsin-Madison from 1946 to 2002. He was a leader in experimental aspects of linear viscoelasticity. The three editions of his book, *Viscoelastic Properties of Polymers*, are highly respected, standard references for viscoelasticity. He was a recipient of the Bingham Medal of The Society of Rheology, and was a member of both the National Academy of Engineering and the National Academy of Sciences. The term *complex viscosity* was first proposed by A. Gemant, *Trans. Faraday Soc.*, **31**, 1582–1590 (1935); see also R. B. Bird and A. J. Giacomin, *Rheologica Acta*, **51**, 481–486 (2012).

[6]C. J. S. Petrie, *Elongational Flows*, Pitman, London (1979); J. Meissner, *Chem. Engr. Commun.*, **33**, 159–180 (1985).

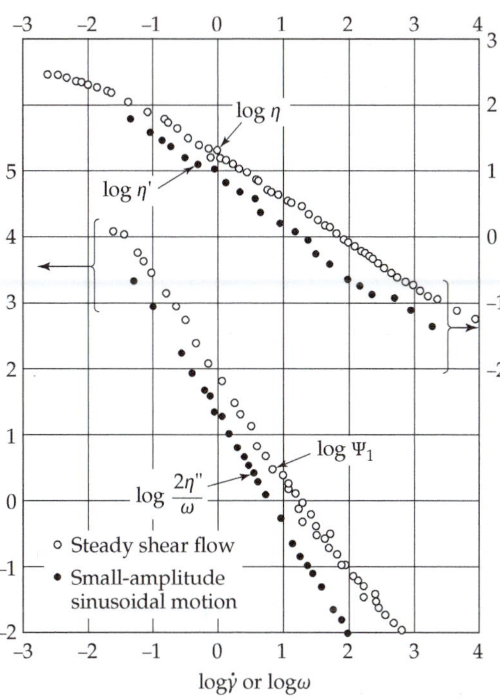

Fig. 8.2-4. The material functions $\eta(\dot{\gamma})$, $\Psi_1(\dot{\gamma})$, $\eta'(\omega)$, and $\eta''(\omega)$ for a 1.5% polyacrylamide solution in a 50/50 mixture of water and glycerin. The quantities η, η', and η'' are given in Pa · s, and Ψ_1 in Pa · s^2. Both $\dot{\gamma}$ and ω are given in s^{-1}. The data are from J. D. Huppler, E. Ashare, and L. Holmes, *Trans. Soc. Rheol.*, **11**, 159–179 (1967), as replotted by J. M. Wiest. The oscillatory normal stresses have also been studied experimentally and theoretically (see M. C. Williams and R. B. Bird, *Ind. Eng. Chem. Fundam.*, **3**, 42–48 (1964); M. C. Williams, *J. Chem. Phys.*, **42**, 2988–2989 (1965); E. B. Christiansen and W. R. Leppard, *Trans. Soc. Rheol.*, **18**, 65–86 (1974), in which the ordinate of Fig. 15 should be multiplied by 39.27).

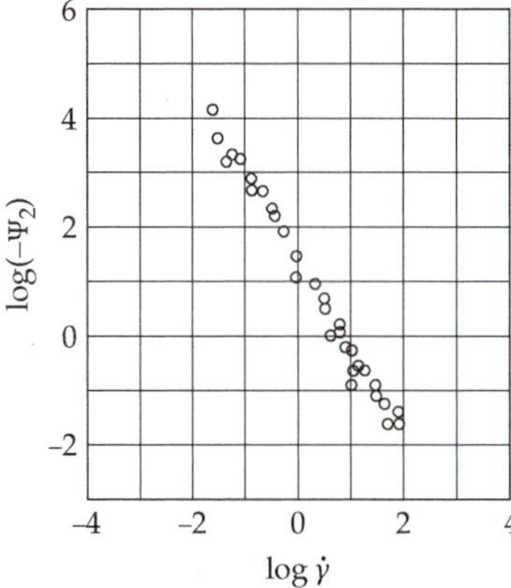

Fig. 8.2-5. Dependence of the second normal stress coefficient on shear rate for a 2.5% solution of polyacrylamide in a 50/50 mixture of water and glycerin. The quantity Ψ_2 is given in Pa · s^2, and $\dot{\gamma}$ in s^{-1}. The data of E. B. Christiansen and W. R. Leppard, *Trans. Soc. Rheol.*, **18**, 65–86 (1974), have been replotted by J. M. Wiest.

mechanical responses in these various experiments. More complete discussions of the data obtained in rheometric experiments are given elsewhere.[7]

The next three sections are devoted to stress tensor expressions for non-Newtonian fluids. One might say, very roughly, that these three sections satisfy the following three different groups of people:

§8.3 The *generalized Newtonian models* are empirical expressions, primarily for describing steady-state shear flows. They have been widely used by *engineers* for designing time-independent flow systems.

§8.4 The *linear viscoelastic models* are primarily used to describe unsteady-state flows in systems with very small displacement gradients and have been used mainly by *scientists and engineers* interested in characterizing polymer and suspension structure.

[7] R. B. Bird, R. C. Armstrong, and O. Hassager, *Dynamics of Polymeric Liquids, Vol. 1, Fluid Mechanics*, Wiley-Interscience, 2nd Edition (1987).

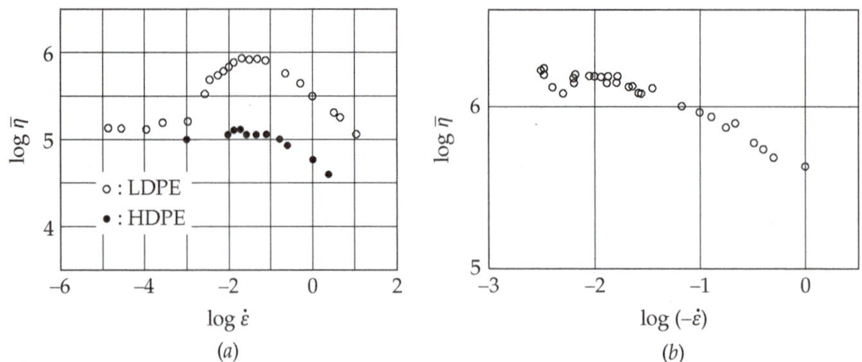

Fig. 8.2-6. (*a*) Elongational viscosity for uniaxial stretching of low- and high-density polyethylene. [From H. Münstedt and H. M. Laun, *Rheol. Acta*, **20**, 211–221 (1981).] (*b*) Elongational viscosity for squeeze flow of polydimethyl siloxane. [From Sh. Chatraei, C. W. Macosko, and H. H. Winter, *J. Rheol.*, **25**, 433 (1981).] In both graphs the quantity $\bar{\eta}$ is given in Pa · s and $\dot{\varepsilon}$ in s^{-1}.

§8.5 The *nonlinear viscoelastic models* represent an attempt to describe all types of flow (including the two listed above) and have been developed largely by *physicists, applied mathematicians,* and, increasingly, *engineers* interested in finding an all-inclusive theory. Most of these models are empirical, but molecular theories are being used more and more to suggest the form of the equations.

Actually the three classes of models are interrelated, and each is important for understanding the subject of non-Newtonian flow. In the following discussion of non-Newtonian models, we assume throughout that the fluids are incompressible. Finally, after discussing these phenomenological models, in §8.6 we turn our attention to a simple molecular theory.

§8.3 NON-NEWTONIAN VISCOSITY AND THE GENERALIZED NEWTONIAN MODELS

The *generalized Newtonian models*[1] discussed here are the simplest of the three types of models to be discussed. However, they can describe only the non-Newtonian viscosity, and none of the normal stress effects, time-dependent effects, or elastic effects. Nonetheless, in many processes in the polymer industry, such as pipe flow with heat transfer, distributor design, extrusion, and injection molding, the non-Newtonian viscosity and its enormous variation with shear rate are central to describing the flows of interest.

For incompressible Newtonian fluids the expression for the stress tensor is given by Eq. 1.2-13 with the last term omitted

$$\boldsymbol{\tau} = -\mu(\nabla \mathbf{v} + (\nabla \mathbf{v})^{\dagger}) \equiv -\mu\, \dot{\boldsymbol{\gamma}} \tag{8.3-1}$$

Here we have introduced the symbol $\dot{\boldsymbol{\gamma}} = \nabla \mathbf{v} + (\nabla \mathbf{v})^{\dagger}$, the *rate-of-strain tensor* (or *rate-of-deformation tensor*). The generalized Newtonian fluid model is obtained by simply replacing the constant viscosity μ by the shear-rate-dependent non-Newtonian viscosity $\eta(\dot{\gamma})$. The shear rate in general can be written as the "magnitude of the rate-of-strain

[1]K. Hohenemser and W. Prager, *Zeits. f. Math. u. Mech.*, **12**, 216–226 (1932); J. G. Oldroyd, *Proc. Camb. Phil. Soc.*, **45**, 595–611 (1949), and **47**, 410–418 (1950). **James Gardner Oldroyd** (1921–1982), a professor at the University of Liverpool, made many contributions to the theory of non-Newtonian fluids, and in particular his ideas on the construction of constitutive equations and the principles of continuum mechanics.

tensor" $\dot{\gamma} = \sqrt{\frac{1}{2}(\dot{\boldsymbol{\gamma}} : \dot{\boldsymbol{\gamma}})}$; it is understood that when the square root is taken, the sign must be so chosen that $\dot{\gamma}$ is a positive quantity. Then the generalized Newtonian fluid model is:

$$\boxed{\boldsymbol{\tau} = -\eta(\nabla \mathbf{v} + (\nabla \mathbf{v})^\dagger) \equiv -\eta\, \dot{\boldsymbol{\gamma}}} \qquad \text{with } \eta = \eta(\dot{\gamma}) \tag{8.3-2}$$

The components of the rate-of-strain tensor $\dot{\boldsymbol{\gamma}}$ can be obtained in Cartesian, cylindrical, and spherical coordinates from the right sides of the equations in §B.1 by omitting the $(\nabla \cdot \mathbf{v})$ terms as well as the factor $(-\mu)$ in the remaining terms.

We now have to give an empiricism for the non-Newtonian viscosity function $\eta(\dot{\gamma})$. Dozens of such expressions have been proposed, but we mention only five here, the last three being expressions that contain a yield stress.

(a) The simplest empiricism for $\eta(\dot{\gamma})$ is the two-parameter *power-law* expression[2]

$$\boxed{\eta(\dot{\gamma}) = m\dot{\gamma}^{n-1}} \tag{8.3-3}$$

in which m and n are constants characterizing the fluid. This simple relation describes the non-Newtonian viscosity curve over the linear portion of the log-log plot of the viscosity vs. shear rate for many materials (see, for example, the viscosity data in Fig. 8.2-4). The parameter m has units of $\text{Pa} \cdot \text{s}^n$, and $n - 1$ is the slope of the $\log \eta$ vs. $\log \dot{\gamma}$ plot. Some sample values of power-law parameters are given in Table 8.3-1.

Although the power-law model was proposed as an empirical expression, the simple rigid dumbbell molecular theory for dilute polymer solutions[3] leads to a power-law expression for high shear rates, with $n = \frac{1}{3}$ (see §8.6).

(b) A better curve fit for most data can be obtained by using the four-parameter *Carreau equation*,[4] which is

$$\boxed{\frac{\eta(\dot{\gamma}) - \eta_\infty}{\eta_0 - \eta_\infty} = [1 + (\lambda\dot{\gamma})^2]^{(n-1)/2}} \tag{8.3-4}$$

Table 8.3-1. Power Law Parameters for Aqueous Solutions[a]

Solution	Temperature (K)	$m(\text{Pa} \cdot \text{s}^n)$	n(-)
2.0% hydroxyethylcellulose	293	93.5	0.189
	313	59.7	0.223
	333	38.5	0.254
0.5% hydroxyethylcellulose	293	0.84	0.509
	313	0.30	0.595
	333	0.136	0.645
1.0% polyethylene oxide	293	0.994	0.532
	313	0.706	0.544
	333	0.486	0.599

[a] R. M. Turian, Ph.D. Thesis, University of Wisconsin, Madison (1964), pp. 142–148.

[2] W. Ostwald, *Kolloid-Zeitschrift*, **36**, 99–117 (1925); A. de Waele, *Oil Color Chem. Assoc. J.*, **6**, 33–88 (1923).

[3] R. B. Bird, W. E. Stewart, and E. N. Lightfoot, *Transport Phenomena*, Wiley, New York, 2nd edition (2002), p. 255.

[4] P. J. Carreau, Ph.D. Thesis, University of Wisconsin, Madison (1968). See also K. Yasuda, R. C. Armstrong, and R. E. Cohen, *Rheol. Acta*, **20**, 163–178 (1981) for a similar five-parameter equation.

Table 8.3-2. Parameters in the Carreau Model for Some Solutions of Linear Polystyrene in 1-Chloronaphthalene[a]

Properties of solution		Parameters in Eq. 8.3-4 (η_∞ is taken to be zero)		
$\overline{M}_w$(g/mol)	c(g/ml)	η_0(Pa · s)	λ(s)	n(---)
3.9×10^5	0.45	8080	1.109	0.304
3.9×10^5	0.30	135	3.61×10^{-2}	0.305
1.1×10^5	0.52	1180	9.24×10^{-2}	0.441
1.1×10^5	0.45	166	1.73×10^{-2}	0.538
3.7×10^4	0.62	3930	1×10^{-1}	0.217

[a]Values of the parameters are taken from K. Yasuda, R. C. Armstrong, and R. E. Cohen, *Rheol. Acta*, **20**, 163–178 (1981).

in which η_0 is the zero-shear-rate viscosity, η_∞ is the infinite-shear-rate viscosity, λ is a fluid time constant, and n is a dimensionless parameter. Some sample parameters for the Carreau model are given in Table 8.3-2.

(c) For materials with a yield stress (primarily thick suspensions), the simplest model is the two-constant *Bingham fluid*[5]

$$\boxed{\dot{\gamma} = 0} \qquad \text{when } \tau \leq \tau_0 \tag{8.3-5a}$$

$$\boxed{\eta(\dot{\gamma}) = \mu_0 + \frac{\tau_0}{\dot{\gamma}}} \qquad \text{when } \tau \geq \tau_0 \tag{8.3-5b}$$

in which τ_0 is the yield stress, the stress below which no flow occurs, and μ_0 is a parameter sometimes called the "plastic viscosity." The quantity $\tau = \sqrt{\frac{1}{2}(\boldsymbol{\tau} : \boldsymbol{\tau})}$ is the magnitude of the stress tensor.

(d) Closely related to the Bingham equation is the two-constant *Casson equation*, proposed for pigment-oil suspensions, but widely used for blood[6]

$$\boxed{\dot{\gamma} = 0} \qquad \text{when } \tau \leq \tau_0 \tag{8.3-6a}$$

$$\boxed{\sqrt{\eta(\dot{\gamma})} = \sqrt{\mu_0} + \sqrt{\frac{\tau_0}{\dot{\gamma}}}} \qquad \text{when } \tau \geq \tau_0 \tag{8.3-6b}$$

(e) A somewhat more flexible equation, containing a yield stress τ_0, is the three-constant *Herschel-Bulkley equation*[7]

$$\boxed{\dot{\gamma} = 0} \qquad \text{when } \tau \leq \tau_0 \tag{8.3-7a}$$

$$\boxed{\eta = m\dot{\gamma}^{n-1} + \frac{\tau_0}{\dot{\gamma}}} \qquad \text{when } \tau \geq \tau_0 \tag{8.3-7b}$$

where m and n are constants, as in the power-law model.

[5]E. C. Bingham, *Fluidity and Plasticity*, McGraw-Hill, New York (1922), pp. 215–218. See R. B. Bird, G. C. Dai, and B. J. Yarusso, *Reviews in Chemical Engineering*, **1**, 1–70 (1982) for a review of models with a yield stress.

[6]N. Casson, *Rheology of Disperse Systems*, C. C. Mill, ed., Pergamon, London (1959), pp. 84–104.

[7]W. H. Herschel and R. Bulkley, *Kolloid Z.*, **39**, 291–300 (1926); *ASTM*, Part II, **26**, 621–629 (1926).

We now give some examples of how to use the power-law and Casson models. These are extensions of problems discussed in Chapters 2 and 3 for Newtonian fluids.[8,9]

EXAMPLE 8.3-1

Laminar Flow of an Incompressible Power-Law Fluid in a Circular Tube[8,9]

Derive the expression for the mass flow rate of a polymer that can be described by the power-law fluid flowing in a circular tube of radius R and length L, as a result of a pressure difference, gravity, or both.

SOLUTION

In §2.3, when beginning the analysis of tube flow, we started out by making a momentum balance on a cylindrical shell of thickness Δr and length L. There we were going to study the flow of a Newtonian fluid. However, Eq. 2.3-14 for the shear-stress distribution is valid for *any* fluid, Newtonian or non-Newtonian, flowing in a circular tube. Therefore, into Eq. 2.3-14 we can insert the shear stress for the power-law fluid (instead of using Eq. 2.3-15). The shear-stress expression may be obtained from Eqs. 8.3-2 and 8.3-3 above,

$$\tau_{rz}(r) = -m\dot{\gamma}^{n-1}\frac{dv_z}{dr} \tag{8.3-8}$$

Since v_z is postulated to be a function of r alone, from Eq. B.1-13 we find that $\dot{\gamma} = \sqrt{\frac{1}{2}(\dot{\boldsymbol{\gamma}}:\dot{\boldsymbol{\gamma}})} = \sqrt{(dv_z/dr)^2} = |dv_z/dr|$. Since dv_z/dr is negative in tube flow, $\dot{\gamma} = |dv_z/dr| = -dv_z/dr$, and thus Eq. 8.3-8 can be rewritten

$$\tau_{rz}(r) = -m\left(-\frac{dv_z}{dr}\right)^{n-1}\frac{dv_z}{dr} = m\left(-\frac{dv_z}{dr}\right)^{n} \tag{8.3-9}$$

Combining Eqs. 8.3-9 and 2.3-14 then gives the following differential equation for the velocity:

$$m\left(-\frac{dv_z}{dr}\right)^{n} = \left(\frac{\mathscr{P}_0 - \mathscr{P}_L}{2L}\right)r \tag{8.3-10}$$

After taking the n th root, the equation may be integrated, and, when the no-slip boundary condition at $r = R$ is used, we get

$$v_z(r) = \left(\frac{(\mathscr{P}_0 - \mathscr{P}_L)R}{2mL}\right)^{1/n}\frac{R}{(1/n)+1}\left[1 - \left(\frac{r}{R}\right)^{(1/n)+1}\right] \tag{8.3-11}$$

for the velocity-distribution (see Eq. 8.1-1). When this is multiplied by the density ρ and integrated over the cross section of the circular tube, we get

$$\begin{aligned} w &= \int_0^{2\pi}\int_0^R \rho v_z(r) r\,dr\,d\theta = 2\pi\rho\int_0^R v_z(r) r\,dr \\ &= \frac{\pi R^3\rho}{(1/n)+3}\left(\frac{(\mathscr{P}_0 - \mathscr{P}_L)R}{2mL}\right)^{1/n} \end{aligned} \tag{8.3-12}$$

This simplifies to the mass flow rate form of the Hagen-Poiseuille law for Newtonian fluids (Eq. 2.3-23) when $n = 1$ and $m = \mu$. Equation 8.3-12 can be used along with data on pressure drop vs. flow rate to determine the power-law parameters m and n.

---

[8]For additional examples, including nonisothermal flows, see R. B. Bird, R. C. Armstrong, and O. Hassager, *Dynamics of Polymeric Liquids, Vol. 1, Fluid Mechanics*, Wiley-Interscience, New York, 2nd edition (1987), Chapter 4.

[9]M. Reiner, *Deformation, Strain and Flow*, Interscience, New York, 2nd edition (1960), pp. 243–245.

EXAMPLE 8.3-2

Flow of a Power-Law Fluid in a Narrow Slit[4]

The flow of a Newtonian fluid in a narrow slit is solved in Problem 2B.4. Find the velocity distribution and the mass flow rate for a power-law fluid flowing in the slit.

SOLUTION

The expression for the shear stress τ_{xz} as a function of position x in Eq. 2B.4-1 can be taken over here, since it does not depend on the type of fluid. The power-law formula for τ_{xz} from Eqs. 8.3-2 and 8.3-3 is

$$\tau_{xz}(x) = m\left(-\frac{dv_z}{dx}\right)^n \quad \text{for } 0 \le x \le B \tag{8.3-13}$$

$$\tau_{xz}(x) = -m\left(\frac{dv_z}{dx}\right)^n \quad \text{for } -B \le x \le 0 \tag{8.3-14}$$

To get the velocity distribution for $0 \le x \le B$, we substitute τ_{xz} from Eq. 8.3-13 into Eq. 2B.4-1 to get:

$$m\left(-\frac{dv_z}{dx}\right)^n = \frac{(\mathscr{P}_0 - \mathscr{P}_L)x}{L} \qquad 0 \le x \le B \tag{8.3-15}$$

Integrating and using the no-slip boundary condition at $x = B$ gives

$$v_z(x) = \left(\frac{(\mathscr{P}_0 - \mathscr{P}_L)B}{mL}\right)^{1/n} \frac{B}{(1/n)+1}\left[1 - \left(\frac{x}{B}\right)^{(1/n)+1}\right] \qquad 0 \le x \le B \tag{8.3-16}$$

Since we expect the velocity profile to be symmetric about the midplane $x = 0$, we can get the mass rate of flow as follows:

$$\begin{aligned} w &= \int_0^W \int_{-B}^B \rho v_z(x)\,dx\,dy = 2W\rho\int_0^B v_z(x)\,dx \\ &= 2\left(\frac{(\mathscr{P}_0 - \mathscr{P}_L)B}{mL}\right)^{1/n} \frac{WB^2\rho}{(1/n)+1}\int_0^1 \left[1 - \left(\frac{x}{B}\right)^{(1/n)+1}\right] d\left(\frac{x}{B}\right) \\ &= \frac{2WB^2\rho}{(1/n)+2}\left(\frac{(\mathscr{P}_0 - \mathscr{P}_L)B}{mL}\right)^{1/n} \end{aligned} \tag{8.3-17}$$

When $n = 1$ and $m = \mu$, the Newtonian result in Problem 2B.4 is recovered. Experimental data on pressure drop and mass flow rate through a narrow slit can be used with Eq. 8.3-17 to determine the power-law parameters.

EXAMPLE 8.3-3

Tangential Annular Flow of a Power-Law Fluid[8,9]

Rework Example 3.7-3 for a power-law fluid.

SOLUTION

Equations 3.7-22 and 3.7-24 remain unchanged for a non-Newtonian fluid, but in lieu of Eq. 3.7-23 we write the θ component of the equation of motion in terms of the shear stress by using §B.5:

$$0 = -\frac{1}{r^2}\frac{d}{dr}(r^2\tau_{r\theta}) \tag{8.3-18}$$

For the postulated velocity profile, we get for the power-law model (with the help of §B.1)

$$\begin{aligned} \tau_{r\theta} &= -\eta r\frac{d}{dr}\left(\frac{v_\theta}{r}\right) \\ &= m\left(-r\frac{d}{dr}\left(\frac{v_\theta}{r}\right)\right)^{n-1}\left(-r\frac{d}{dr}\left(\frac{v_\theta}{r}\right)\right) \\ &= m\left(-r\frac{d}{dr}\left(\frac{v_\theta}{r}\right)\right)^n \end{aligned} \tag{8.3-19}$$

Combining Eqs. 8.3-18 and 8.3-19 we get

$$\frac{d}{dr}\left(r^2 m\left(-r\frac{d}{dr}\left(\frac{v_\theta}{r}\right)\right)^n\right) = 0 \tag{8.3-20}$$

Integration gives

$$r^2\left(-r\frac{d}{dr}\left(\frac{v_\theta}{r}\right)\right)^n = C_1 \tag{8.3-21}$$

Dividing by r^2 and taking the nth root gives a first-order differential equation for the angular velocity

$$\frac{d}{dr}\left(\frac{v_\theta}{r}\right) = -\frac{1}{r}\left(\frac{C_1}{r^2}\right)^{1/n} \tag{8.3-22}$$

This may be integrated with the boundary conditions in Eqs. 3.7-29 and 3.7-30 to give

$$\frac{v_\theta(r)}{\Omega_i r} = \frac{(R/r)^{2/n} - 1}{(1/\kappa)^{2/n} - 1} \tag{8.3-23}$$

The (z component of the) torque needed on the inner cylinder to maintain the motion is then:

$$\begin{aligned} T_z &= \tau_{r\theta}\big|_{r=\kappa R} \cdot 2\pi\kappa RL \cdot \kappa R \\ &= m\left(-r\frac{d}{dr}\left(\frac{v_\theta}{r}\right)\right)^n\Bigg|_{r=\kappa R} \cdot 2\pi\kappa RL \cdot \kappa R \end{aligned} \tag{8.3-24}$$

Combining Eqs. 8.3-23 and 8.3-24 then gives

$$T_z = 2\pi m(\kappa R)^2 L\left(\frac{2\Omega_i/n}{1-\kappa^{2/n}}\right)^n \tag{8.3-25}$$

The Newtonian result can be recovered by setting $n = 1$ and $m = \mu$. Equation 8.3-25 can be used along with torque vs. angular velocity data to determine the power-law parameters m and n.

EXAMPLE 8.3-4

Flow of a Casson Fluid in a Circular Tube[10,11]

As mentioned in Example 8.3-1, the shear-stress expression in Eq. 2.3-14 is valid for any kind of non-Newtonian fluid. According to this equation, the shear stress goes from zero at the tube center to some value $\tau_R = (\mathscr{P}_0 - \mathscr{P}_L)R/2L$ at the tube wall, where R and L are the radius and length of the tube. If τ_R is less than or equal to the yield stress τ_0, there will be no flow at all. If τ_R is greater than τ_0, then, instead of the situation shown in Fig. 2.3-2, we will have a velocity distribution such as that shown in Fig. 8.3-1. That is, for $\tau_{rz} \le \tau_0$, there will be a plug-flow region (i.e., a region in which $dv_z/dr = 0$), and for $\tau_{rz} \ge \tau_0$, there will be a nonzero velocity gradient. The plug-flow region is bounded by the radial coordinate, $r = r_0$, where r_0 is defined by $\tau_0 = ((\mathscr{P}_0 - \mathscr{P}_L)/2L)r_0$.

We note that $\dot\gamma = -dv_z/dr$ for flow in circular tubes, according to the definition given just before Eq. 8.3-2. For the region $r_0 \le r \le R$, we then get from Eq. 8.3-6(b)

$$\sqrt{\eta\dot\gamma} = \sqrt{\mu_0\dot\gamma} + \sqrt{\tau_0} \qquad \text{or} \qquad \sqrt{\tau_{rz}} = \sqrt{\mu_0\dot\gamma} + \sqrt{\tau_0} \tag{8.3-26a,b}$$

Combining Eqs. 2.3-14 and 8.3-26b and rearranging gives

$$\sqrt{\frac{(\mathscr{P}_0 - \mathscr{P}_L)r}{2L}} = \sqrt{\mu_0\dot\gamma} + \sqrt{\tau_0} \tag{8.3-27}$$

---

[10] M. M. Lih, *Transport Phenomena in Medicine and Biology*, Wiley, New York (1975), 378–380.
[11] E. N. Lightfoot, *Transport in Living Systems*, Wiley, New York (1974), pp. 35, 430, 438, 440.

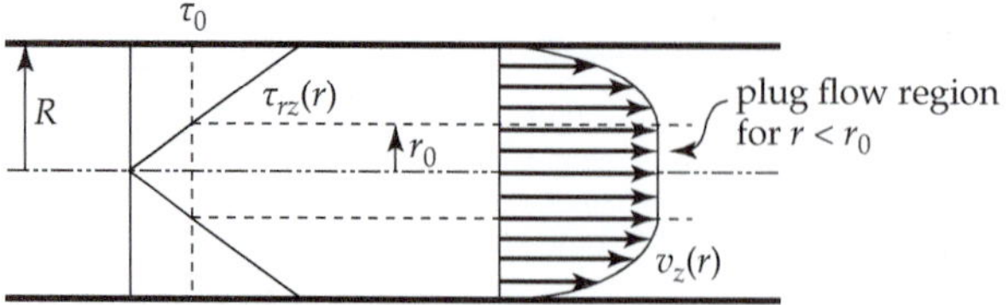

Fig. 8.3-1. The momentum flux distribution $\tau_{rz}(r)$ and velocity distribution $v_z(r)$ for the flow of a Casson fluid in a circular tube. For $r < r_0$, the velocity is independent of r (plug flow region).

Solving Eq. 8.3-27 for $\dot{\gamma}$ we get

$$\dot{\gamma} = \frac{1}{\mu_0}\left[\frac{(\mathscr{P}_0 - \mathscr{P}_L)r}{2L} - 2\sqrt{\frac{(\mathscr{P}_0 - \mathscr{P}_L)r\tau_0}{2L}} + \tau_0\right] \equiv \frac{\tau_0}{\mu_0}\left(\frac{r}{r_0} - 2\sqrt{\frac{r}{r_0}} + 1\right) \tag{8.3-28}$$

We define $v_z^>(r)$ to be the velocity in the region $r_0 \le r \le R$, and $v_z^<(r)$ in the region $0 \le r \le r_0$. Inserting $\dot{\gamma} = -dv_z^>/dr$, we obtain a differential equation for the dimensionless velocity profile $\phi^> = (\mu_0/r_0\tau_0)v_z^>$ as a function of the dimensionless radial coordinate $\xi = r/r_0$

$$-\frac{d\phi^>}{d\xi} = \xi - 2\sqrt{\xi} + 1 \tag{8.3-29}$$

Integration of this equation, and using the boundary condition that $\phi^> = 0$ at $\xi = R/r_0$, we get for $r_0 \le r \le R$

$$\phi^>(r) = \frac{1}{2}\left(\frac{R}{r_0}\right)^2\left[1 - \left(\frac{r}{R}\right)^2\right] - \frac{4}{3}\left(\frac{R}{r_0}\right)^{3/2}\left[1 - \left(\frac{r}{R}\right)^{3/2}\right] + \frac{R}{r_0}\left(1 - \frac{r}{R}\right) \tag{8.3-30}$$

and hence for $0 \le r \le r_0$

$$\phi^<(r) = \phi^>(r_0) = \frac{1}{2}\left(\frac{R}{r_0}\right)^2 - \frac{4}{3}\left(\frac{R}{r_0}\right)^{3/2} + \frac{R}{r_0} - \frac{1}{6} \tag{8.3-31}$$

The mass rate of flow can then be obtained by using the integral in Eq. 2.3-21 after multiplying by ρ and doing an integration by parts to give (the dashed underlined terms are zero)

$$\begin{aligned}
w &= 2\pi\rho\left(\underline{\frac{1}{2}v_z r^2\Big|_{r=0}^{r=R}} - \int_0^R \frac{1}{2}r^2\frac{dv_z}{dr}dr\right) \\
&= -\pi\rho\left(\underline{\int_0^{r_0} r^2\frac{dv_z}{dr}dr} + \int_{r_0}^R r^2\frac{dv_z}{dr}dr\right) \\
&= -\pi\rho\left(\frac{r_0^3\tau_0}{\mu_0}\right)\int_1^{R/r_0}\xi^2\frac{d\phi^>}{d\xi}d\xi = +\pi\rho\left(\frac{r_0^3\tau_0}{\mu_0}\right)\int_1^{R/r_0}\xi^2(\xi - 2\sqrt{\xi} + 1)d\xi \\
&= \pi\rho\left(\frac{r_0^3\tau_0}{\mu_0}\right)\left\{\frac{1}{4}\left[\left(\frac{R}{r_0}\right)^4 - 1\right] - \frac{4}{7}\left[\left(\frac{R}{r_0}\right)^{7/2} - 1\right] + \frac{1}{3}\left[\left(\frac{R}{r_0}\right)^3 - 1\right]\right\} \\
&= \pi\rho\left(\frac{\tau_0 R^4}{4\mu_0 r_0}\right)\left[1 - \frac{16}{7}\left(\frac{r_0}{R}\right)^{1/2} + \frac{4}{3}\left(\frac{r_0}{R}\right) - \frac{1}{21}\left(\frac{r_0}{R}\right)^4\right] \\
&= \frac{\pi(\mathscr{P}_0 - \mathscr{P}_L)R^4\rho}{8\mu_0 L}\left[\begin{array}{l} 1 - \dfrac{16}{7}\sqrt{\dfrac{2L\tau_0}{(\mathscr{P}_0 - \mathscr{P}_L)R}} + \dfrac{4}{3}\dfrac{2L\tau_0}{(\mathscr{P}_0 - \mathscr{P}_L)R} \\ \qquad - \dfrac{1}{21}\left(\dfrac{2L\tau_0}{(\mathscr{P}_0 - \mathscr{P}_L)R}\right)^4 \end{array}\right]
\end{aligned} \tag{8.3-32}$$

When $\tau_0 \to 0$ and $\mu_0 \to \mu$, the Newtonian result (Eq. 2.3-23) is recovered.

§8.4 ELASTICITY AND THE LINEAR VISCOELASTIC MODELS

In §1.2, in the discussion about generalizing Newton's "law of viscosity" to arbitrary flows, we assumed that the stress does not depend explicitly on time. In this section, we allow for the inclusion of a time derivative, but still require a linear relation between $\boldsymbol{\tau}$ and $\dot{\boldsymbol{\gamma}}$. This leads to a *linear viscoelastic* model.

We start by writing down Newton's expression for the stress tensor for an incompressible viscous liquid along with Hooke's analogous expression for the stress tensor for an incompressible elastic solid:[1]

Newton: $$\boldsymbol{\tau} = -\mu\left(\nabla\mathbf{v} + (\nabla\mathbf{v})^{\dagger}\right) \equiv -\mu\dot{\boldsymbol{\gamma}} \quad (8.4\text{-}1)$$

Hooke: $$\boldsymbol{\tau} = -G\left(\nabla\mathbf{u} + (\nabla\mathbf{u})^{\dagger}\right) \equiv -G\boldsymbol{\gamma} \quad (8.4\text{-}2)$$

In the second of these expressions, G is the *elastic shear modulus*, and $\mathbf{u}$ is the *displacement vector*, which gives the distance and direction that a point in the solid has moved from some initial position as a result of the applied stresses. The quantity $\boldsymbol{\gamma}$ is called the *infinitesimal strain tensor*. The rate-of-strain tensor and the infinitesimal strain tensor are related by $\dot{\boldsymbol{\gamma}} = \partial\boldsymbol{\gamma}/\partial t$. The Hookean solid has a perfect memory; when imposed stresses are removed, the solid returns to its initial configuration. Hooke's law is valid only for very small displacement gradients, $\nabla\mathbf{u}$. Now we want to combine the ideas embodied in Eqs. 8.4-1 and 8.4-2 to describe viscoelastic fluids.

A simple way to describe a fluid that is both viscous and elastic is the *Maxwell model*[2]

$$\boldsymbol{\tau} + \lambda\frac{\partial}{\partial t}\boldsymbol{\tau} = -\eta_0\dot{\boldsymbol{\gamma}} \quad (8.4\text{-}3)$$

Here $\lambda = \eta_0/G$ is a time constant (the *relaxation time*) and η_0 is the *zero-shear-rate viscosity*. When the stress tensor changes imperceptibly with time, then Eq. 8.4-3 has the form of Eq. 8.4-1 for a Newtonian liquid. When there are very rapid changes in the stress tensor with time, then the first term on the left side of Eq. 8.4-3 can be omitted, and when the equation is integrated with respect to time, we get an equation of the form of Eq. 8.4-2 for the Hookean solid. In that sense, Eq. 8.4-3 incorporates both viscosity and elasticity, but is restricted to small displacement gradients, where Hooke's law is valid.

A simple experiment that illustrates the behavior of a viscoelastic liquid involves "silly putty." This material flows easily when squeezed slowly between the palms of the hands, and this indicates that it behaves as a viscous fluid. However, when it is rolled into a ball, the ball will bounce when dropped onto a hard surface. During the impact, the stresses change rapidly, and the material behaves as an elastic solid.

When Eq. 8.4-3 is integrated, with the condition that the fluid is at rest at $t = -\infty$, we get the integral form of the Maxwell model

$$\boldsymbol{\tau}(t) = -\int_{-\infty}^{t}\left\{\frac{\eta_0}{\lambda}e^{-(t-t')/\lambda}\right\}\dot{\boldsymbol{\gamma}}(t')\,dt' \quad (8.4\text{-}4)$$

In this form, the idea of "fading memory" is clearly present: the stress at time t depends on the velocity gradient at all past times t', but, because of the exponential, greatest weight is given to the velocity gradient at times t' that are near t. That is, the fluid's "memory" is better for recent times than for times more remote in the past. That part of the integrand in Eq. 8.4-4 enclosed in braces is called the *relaxation modulus* of the fluid and is generally denoted by $G(t - t')$. The main problem with the Maxwell model is that it fails to predict

[1]R. Hooke, *Lectures de Potentia Restitutiva* (1678).

[2]This relation was proposed by J. C. Maxwell, *Phil. Trans. Roy. Soc.*, **A157**, 49–88 (1867), to demonstrate that gases are *not* viscoelastic.

the shear-rate dependence of viscosity that all polymeric liquids exhibit, and is captured by the generalized Newtonian models in the previous section.

EXAMPLE 8.4-1

Small-Amplitude Oscillatory Motion

Obtain an expression for the components of the complex viscosity by using the Maxwell model.

SOLUTION

We use the yx component of Eq. 8.4-4, and for this problem the yx component of the rate-of-strain tensor is:

$$\dot{\gamma}_{yx}(t) = \frac{\partial v_x}{\partial y} = \dot{\gamma}_0 \cos \omega t \tag{8.4-5}$$

where ω is the angular frequency. When this is substituted into Eq. 8.4-4, with the relaxation modulus (in braces) expressed as $G(t - t')$, we get

$$\begin{aligned} \tau_{yx}(t) &= -\int_{-\infty}^{t} G(t-t')\,\dot{\gamma}^0 \cos \omega t'\, dt' \\ &= -\dot{\gamma}^0 \int_0^{\infty} G(s)\, \cos \omega(t-s)\, ds \\ &= -\dot{\gamma}^0 \left[\int_0^{\infty} G(s)\, \cos \omega s\, ds\right] \cos \omega t - \dot{\gamma}^0 \left[\int_0^{\infty} G(s)\, \sin \omega s\, ds\right] \sin \omega t \end{aligned} \tag{8.4-6}$$

in which $s = t - t'$. When this equation is compared with Eq. 8.2-4, we obtain

$$\eta'(\omega) = \int_0^{\infty} G(s) \cos \omega s\, ds \tag{8.4-7}$$

$$\eta''(\omega) = \int_0^{\infty} G(s) \sin \omega s\, ds \tag{8.4-8}$$

for the components of the complex viscosity $\eta^* = \eta' - i\eta''$. When the Maxwell expression for the relaxation modulus is introduced and the integrals are evaluated, we find that

$$\frac{\eta'(\omega)}{\eta_0} = \frac{1}{1 + (\lambda\omega)^2} \quad \text{and} \quad \frac{\eta''(\omega)}{\eta_0} = \frac{\lambda\omega}{1 + (\lambda\omega)^2} \tag{8.4-9,10}$$

Equations 8.4-9 and 8.4-10 show that η' and η'' both decrease with increasing frequency ω. The shapes of the curves agree only qualitatively with the shapes of the experimental curves.

However, if one uses a superposition of Maxwell models

$$\tau = \sum_{k=1}^{\infty} \tau_k \quad \text{where} \quad \tau_k + \lambda_k \frac{\partial}{\partial t} \tau_k = -\eta_k \dot{\gamma}_k \tag{8.4-11,12}$$

then the experimental data may be described with great accuracy. In this formulation, the λ_k and η_k are constants, and $\Sigma_k \eta_k = \eta_0$. Alternatively, the stress tensor may be written in integral form as

$$\tau(t) = -\int_{-\infty}^{t} G(t-t')\,\dot{\gamma}(t')dt' \tag{8.4-13}$$

with

$$G(t-t') = \sum_{k=1}^{\infty} (\eta_k/\lambda_k) \exp[-(t-t')/\lambda_k] \tag{8.4-14}$$

Instead of using an expression containing the infinite sets of constants λ_k and η_k, one can make use of the Spriggs empiricisms[3]

$$\eta_k = \frac{\eta_0 \lambda_k}{\Sigma_k \lambda_k} \quad \text{and} \quad \lambda_k = \frac{\lambda}{k^a} \tag{8.4-15,16}$$

---

[3]T. W. Spriggs, *Chem. Eng. Sci.*, **20**, 931–940 (1965).

This enables one to fit the experimental data quite well with just three parameters: the zero-shear-rate viscosity η_0, a time constant λ (the longest relaxation time), and a dimensionless parameter α.

§8.5 OBJECTIVITY AND THE NONLINEAR VISCOELASTIC MODELS

We have seen how to get (non-Newtonian) viscosity and elasticity into the expressions for the stress tensor, but something is still missing. The expression for the stress tensor should be independent of the instantaneous orientation of a fluid element in space, and this leads to the requirement of *rheological invariance* or *objectivity*.[1] In order for a stress-tensor expression to be objective, it is necessary to replace $\partial/\partial t$ in differential expressions by a different kind of time derivative, one that will satisfy the principle of objectivity. There are several ways to do this. Here we will use the simplest way, and replace $\partial\boldsymbol{\tau}/\partial t$ by the "corotational time derivative" or "Jaumann derivative"[2,3]

$$\frac{\mathscr{D}}{\mathscr{D}t}\boldsymbol{\tau} = \frac{D}{Dt}\boldsymbol{\tau} + \frac{1}{2}\{\boldsymbol{\omega}\cdot\boldsymbol{\tau} - \boldsymbol{\tau}\cdot\boldsymbol{\omega}\} \tag{8.5-1}$$

in which $\boldsymbol{\omega} = \nabla\mathbf{v} - (\nabla\mathbf{v})^\dagger$ is the *vorticity tensor*. This derivative tells how the stress tensor changes with time as one goes along with a fluid element and rotates with it. If this time derivative replaces $\partial\boldsymbol{\tau}/\partial t$ in Eq. 8.4-3, then we get the "corotational Maxwell model"

$$\boldsymbol{\tau} + \lambda\frac{\mathscr{D}}{\mathscr{D}t}\boldsymbol{\tau} = -\eta_0\dot{\boldsymbol{\gamma}} \tag{8.5-2}$$

which gives a non-Newtonian viscosity (which drops off too steeply with increasing shear rate, as shown in the example below) a positive first normal stress coefficient, and a negative second normal stress coefficient (which is too large in magnitude). Nonetheless, the corotational Maxwell model can be useful for getting trends in the physical properties and learning how to manipulate nonlinear viscoelastic models.

EXAMPLE 8.5-1

Obtaining the Steady-State Material Functions for the Corotational Maxwell Model

Show how to get the viscosity, the first normal-stress coefficient, and the second normal-stress coefficient for the steady-state shear flow of a fluid described by the corotational Maxwell model. The imposed flow is $v_x = \dot{\gamma}y$ with $v_y = 0$ and $v_z = 0$.

SOLUTION

In learning how to discuss nonlinear viscoelastic models, it is useful to display the various parts of the models in matrix form. We start by getting $\nabla\mathbf{v}$, $(\nabla\mathbf{v})^\dagger$, $\dot{\boldsymbol{\gamma}}$, and $\boldsymbol{\omega}$ for the flow field being considered.

$$\nabla\mathbf{v} = \begin{pmatrix} (\nabla\mathbf{v})_{xx} & (\nabla\mathbf{v})_{xy} & (\nabla\mathbf{v})_{xz} \\ (\nabla\mathbf{v})_{yx} & (\nabla\mathbf{v})_{yy} & (\nabla\mathbf{v})_{yz} \\ (\nabla\mathbf{v})_{zx} & (\nabla\mathbf{v})_{zy} & (\nabla\mathbf{v})_{zz} \end{pmatrix} = \begin{pmatrix} 0 & 0 & 0 \\ 1 & 0 & 0 \\ 0 & 0 & 0 \end{pmatrix}\dot{\gamma} \qquad (\nabla\mathbf{v})^\dagger = \begin{pmatrix} 0 & 1 & 0 \\ 0 & 0 & 0 \\ 0 & 0 & 0 \end{pmatrix}\dot{\gamma} \tag{8.5-3,4}$$

---

[1] For a discussion of rheological invariance, see R. B. Bird, R. C. Armstrong, and O. Hassager, *Dynamcs of Polymeric Liquids, Vol. 1*, 2nd Edition, Wiley, New York (1987), pp. 482 and 483.

[2] G. Jaumann, *Grundlagen der Bewegungslehre*, Leipzig (1905). S. Zaremba, Bull. Int. Academie Sci., Cracovie, 594–614, 614–621 (1903). **Gustav Andreas Johannes Jaumann** (1863–1924) (pronounced "Yow-mahn") taught at the German university in Brünn (now Brno in the Czech Republic). He was an important contributor to the field of continuum mechanics at the beginning of the twentieth century.

[3] J. G. Oldroyd, *Proc. Roy. Soc.*, **A245**, 278–297 (1958); L. E. Wedgewood, *Rheol. Acta*, **38**, 91–99 (1999).

$$\dot{\boldsymbol{\gamma}} = \begin{pmatrix} \dot{\gamma}_{xx} & \dot{\gamma}_{xy} & \dot{\gamma}_{xz} \\ \dot{\gamma}_{yx} & \dot{\gamma}_{yy} & \dot{\gamma}_{yz} \\ \dot{\gamma}_{zx} & \dot{\gamma}_{zy} & \dot{\gamma}_{zz} \end{pmatrix} = \begin{pmatrix} 0 & 1 & 0 \\ 1 & 0 & 0 \\ 0 & 0 & 0 \end{pmatrix} \dot{\gamma} \qquad \boldsymbol{\omega} = \begin{pmatrix} 0 & -1 & 0 \\ 1 & 0 & 0 \\ 0 & 0 & 0 \end{pmatrix} \dot{\gamma} \tag{8.5-5,6}$$

Since the stress tensor $\boldsymbol{\tau}$ does not depend on position and time, in the Jaumann derivative of $\boldsymbol{\tau}$, the substantial derivative will not contribute and we are left with only $\frac{1}{2}\{\boldsymbol{\omega} \cdot \boldsymbol{\tau} - \boldsymbol{\tau} \cdot \boldsymbol{\omega}\}$.

$$\{\boldsymbol{\omega} \cdot \boldsymbol{\tau}\} = \begin{pmatrix} 0 & -1 & 0 \\ 1 & 0 & 0 \\ 0 & 0 & 0 \end{pmatrix} \begin{pmatrix} \tau_{xx} & \tau_{xy} & \tau_{xz} \\ \tau_{yx} & \tau_{yy} & \tau_{yz} \\ \tau_{zx} & \tau_{zy} & \tau_{zz} \end{pmatrix} \dot{\gamma} = \begin{pmatrix} -\tau_{yx} & -\tau_{yy} & -\tau_{yz} \\ \tau_{xx} & \tau_{xy} & \tau_{xz} \\ 0 & 0 & 0 \end{pmatrix} \dot{\gamma} \tag{8.5-7}$$

$$\{\boldsymbol{\tau} \cdot \boldsymbol{\omega}\} = \begin{pmatrix} \tau_{xx} & \tau_{xy} & \tau_{xz} \\ \tau_{yx} & \tau_{yy} & \tau_{yz} \\ \tau_{zx} & \tau_{zy} & \tau_{zz} \end{pmatrix} \begin{pmatrix} 0 & -1 & 0 \\ 1 & 0 & 0 \\ 0 & 0 & 0 \end{pmatrix} \dot{\gamma} = \begin{pmatrix} \tau_{xy} & -\tau_{xx} & 0 \\ \tau_{yy} & -\tau_{xy} & 0 \\ \tau_{zy} & -\tau_{xz} & 0 \end{pmatrix} \dot{\gamma} \tag{8.5-8}$$

Hence, the Jaumann derivative is (keep in mind that the stress tensor is symmetric)

$$\frac{\mathscr{D}\boldsymbol{\tau}}{\mathscr{D}t} = \begin{pmatrix} -\tau_{yx} & \frac{1}{2}\left(\tau_{xx} - \tau_{yy}\right) & -\frac{1}{2}\tau_{yz} \\ \frac{1}{2}(\tau_{xx} - \tau_{yy}) & \tau_{yx} & \frac{1}{2}\tau_{xz} \\ -\frac{1}{2}\tau_{yz} & \frac{1}{2}\tau_{xz} & 0 \end{pmatrix} \dot{\gamma} \tag{8.5-9}$$

Thus, for simple shear flow, the corotational Maxwell model (Eq. 8.5-2) in matrix form is

$$\begin{pmatrix} \tau_{xx} & \tau_{xy} & \tau_{xz} \\ \tau_{yx} & \tau_{yy} & \tau_{yz} \\ \tau_{zx} & \tau_{zy} & \tau_{zz} \end{pmatrix} + \lambda \begin{pmatrix} -\tau_{yx} & \frac{1}{2}\left(\tau_{xx} - \tau_{yy}\right) & -\frac{1}{2}\tau_{yz} \\ \frac{1}{2}(\tau_{xx} - \tau_{yy}) & \tau_{yx} & \frac{1}{2}\tau_{xz} \\ -\frac{1}{2}\tau_{yz} & \frac{1}{2}\tau_{xz} & 0 \end{pmatrix} \dot{\gamma}$$
$$= -\eta_0 \begin{pmatrix} 0 & 1 & 0 \\ 1 & 0 & 0 \\ 0 & 0 & 0 \end{pmatrix} \dot{\gamma} \tag{8.5-10}$$

From this matrix equation, we can write down the following algebraic equations

xx component minus yy component: $\quad (\tau_{xx} - \tau_{yy}) - 2\tau_{yx}\lambda\dot{\gamma} = 0 \quad$ (8.5-11)

yy component minus zz component: $\quad (\tau_{yy} - \tau_{zz}) + \tau_{yx}\lambda\dot{\gamma} = 0 \quad$ (8.5-12)

yx component: $\quad \tau_{yx} + \dfrac{1}{2}(\tau_{xx} - \tau_{yy})\lambda\dot{\gamma} = -\eta_0\dot{\gamma} \quad$ (8.5-13)

xz component: $\quad \tau_{xz} - \dfrac{1}{2}\tau_{yz}\lambda\dot{\gamma} = 0 \quad$ (8.5-14)

yz component: $\quad \tau_{yz} + \dfrac{1}{2}\tau_{xz}\lambda\dot{\gamma} = 0 \quad$ (8.5-15)

From the last two equations we see that $\tau_{xz} = \tau_{yz} = 0$. From solving the first three simultaneous equations for τ_{yx}, $\tau_{xx} - \tau_{yy}$, and $\tau_{yy} - \tau_{zz}$ we find

$$\tau_{yx} = -\eta_0\dot{\gamma}\frac{1}{1 + (\lambda\dot{\gamma})^2} \quad \text{or} \quad \frac{\eta}{\eta_0} = \frac{1}{1 + (\lambda\dot{\gamma})^2} \tag{8.5-16,17}$$

$$\tau_{xx} - \tau_{yy} = -\eta_0\dot{\gamma}\frac{2\lambda\dot{\gamma}}{1 + (\lambda\dot{\gamma})^2} \quad \text{or} \quad \frac{\Psi_1}{\eta_0\lambda} = \frac{2}{1 + (\lambda\dot{\gamma})^2} \tag{8.5-18,19}$$

$$\tau_{yy} - \tau_{zz} = +\eta_0\dot{\gamma}\frac{\lambda\dot{\gamma}}{1 + (\lambda\dot{\gamma})^2} \quad \text{or} \quad \frac{\Psi_2}{\eta_0\lambda} = -\frac{1}{1 + (\lambda\dot{\gamma})^2} \tag{8.5-20,21}$$

We know that the shear stress keeps increasing as the velocity gradient increases, but Eq. 8.5-16 indicates otherwise. In fact, the shear stress goes through a maximum at $\lambda\dot{\gamma} = 1$, and therefore, we see that Eqs. 8.5-16 and 8.5-17 are limited to $\lambda\dot{\gamma} < 1$. The second normal stress is negative, and that is in agreement with measurements for polymeric liquids. However, generally one expects Ψ_2/Ψ_1 to be in the range from −0.1 to −0.4. Hence, from examination of this one particular flow, it is evident that caution must be used when drawing any conclusions from this model. However, for a model with just two parameters, it seems to be promising for predicting qualitative behavior for low shear rates.

Considerably better results can be obtained by superposing a large number of corotational Maxwell models[4] with different time constants λ_k and viscosity constants η_k

$$\boldsymbol{\tau} = \sum_{k=1}^{\infty} \boldsymbol{\tau}_k \quad \text{with} \quad \boldsymbol{\tau}_k + \lambda_k \frac{\mathscr{D}}{\mathscr{D}t}\boldsymbol{\tau}_k = -\eta_k \dot{\boldsymbol{\gamma}} \tag{8.5-22,23}$$

This leads to

$$\eta = \sum_{k=1}^{\infty} \frac{\eta_k}{1+(\lambda_k\dot{\gamma})^2} \quad \text{and} \quad \frac{1}{2}\Psi_1 = \sum_{k=1}^{\infty} \frac{\eta_k\lambda_k}{1+(\lambda_k\dot{\gamma})^2} = -\Psi_2 \tag{8.5-24,25}$$

or, using the Spriggs empiricisms of Eqs. 8.4-15 and 8.4-16

Low-shear-rate region:

$$\frac{\eta}{\eta_0} = 1 - \frac{(\lambda\dot{\gamma})^2}{\zeta(\alpha)} \sum_{k=1}^{\infty} \frac{1}{k^{\alpha}\left(k^{2\alpha} + (\lambda\dot{\gamma})^2\right)} \tag{8.5-26}$$

$$\frac{\Psi_1}{2\eta_0\lambda} = -\frac{\Psi_2}{\eta_0\lambda} = \frac{\zeta(2\alpha)}{\zeta(\alpha)} - \frac{(\lambda\dot{\gamma})^2}{\zeta(\alpha)} \sum_{k=1}^{\infty} \frac{1}{k^{2\alpha}\left(k^{2\alpha} + (\lambda\dot{\gamma})^2\right)} \tag{8.5-27}$$

High-shear-rate (power-law) region:

$$\frac{\eta}{\eta_0} = \frac{1}{\zeta(\alpha)} \left[\frac{\pi(\lambda\dot{\gamma})^{(1/\alpha)-1}}{2\alpha \sin\left((\alpha+1)\pi/2\alpha\right)} \right] \tag{8.5-28}$$

$$\frac{\Psi_1}{2\eta_0\lambda} = -\frac{\Psi_2}{\eta_0\lambda} = \frac{1}{\zeta(\alpha)} \left[\frac{\pi(\lambda\dot{\gamma})^{(1/\alpha)-2}}{2\alpha \sin(\pi/2\alpha)} - \frac{(\lambda\dot{\gamma})^{-2}}{2} \right] \tag{8.5-29}$$

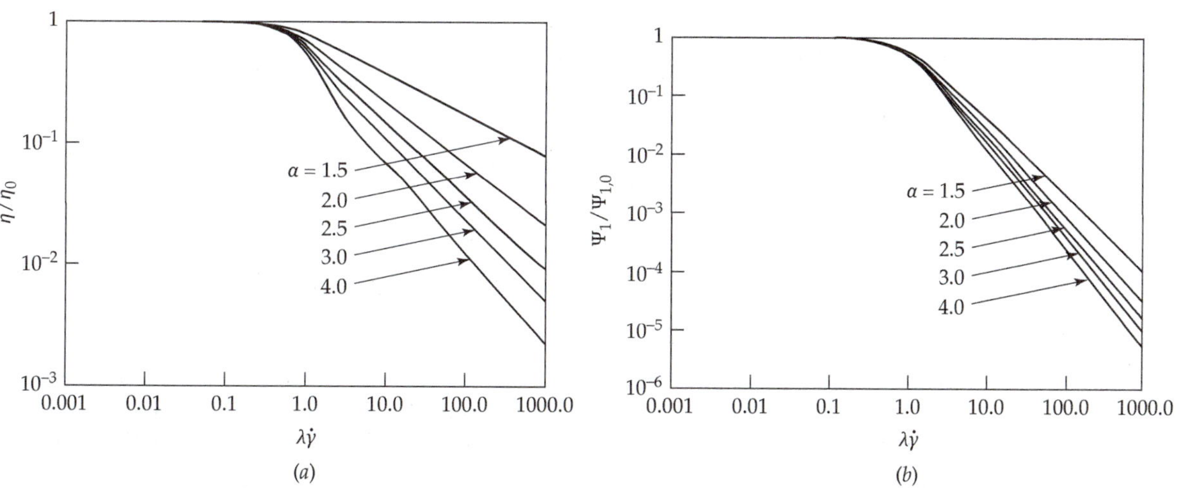

Fig. 8.5-1. Simple shear flow material functions for the corotational Maxwell model. (*a*) Viscosity curves. It is not clear if the "wiggles" evident for larger values of α are in agreement with experimental observations; (*b*) Normal-stress curves. Here $\Psi_{1,0}$ is the value of Ψ_1 in the limit as $\dot{\gamma} \rightarrow 0$.

---

[4] R. B. Bird, R. C. Armstrong, and O. Hassager, *Dynamics of Polymeric Liquids*, 1st Edition, Vol. **1**, Wiley, New York (1977), pp. 335–338; A. J. Giacomin, R. B. Bird, L. M. Johnson, and A. W. Mix, *J. Non-Newtonian Fluid Mechanics*, **166**, 1081–1099 (2011); Errata: In Eq. (66), "20De^2" and "10De^2 – 50De^4" should be "20De" and "10De – 50De^3"; after Eq. (119), "$(\zeta\alpha)$" should be "$\zeta(\alpha)$"; in Eq. (147), "$n - 1$" should be "$n = 1$." See also A. J. Giacomin and R. B. Bird, *Rheologica Acta*, **50**, 741–752 (2011).

where $\zeta(\alpha) = \sum_{k=1}^{\infty} k^{-\alpha}$ is the Riemann zeta function[5]: $\zeta(2) = \pi^2/6$, $\zeta(4) = \pi^4/90$, $\zeta(6) = \pi^6/495$, etc. It should be noted that from the slope of the power-law part of the viscosity curve that n of the power law model (Eq. 8.3-3) is the same as $1/\alpha$. Equations 8.5-26 to 8.5-29 were used to plot the curves shown in Figures 8.5-1(a) and (b).

Many rheological properties have been well described by the generalized corotational Maxwell model of Eqs. 8.5-22 and 8.5-23.

§8.6 A MOLECULAR THEORY AND A NONLINEAR VISCOELASTIC MODEL

The simplest molecular model that can describe the nonlinear rheological behavior is the dilute suspension of *rigid dumbbells*. This is the suspension in a Newtonian liquid of dumbbells that each consist of two identical "beads" connected by a massless rigid "rod." The dumbbells are presumed to be far enough apart that they do not in any way interact with one another. Although this bead-rod model may appear to be quite unrealistic, it has been found to describe rather well the responses of dilute, and even concentrated, solutions of flexible polymers.

We let the length of the dumbbell be L and the friction coefficient for the bead be ζ. The latter is defined as the coefficient of proportionality between the force $\mathbf{F}$ acting on the bead, and the velocity $\mathbf{v}$ with which the bead moves; if Stokes' law is used, then $\zeta = 6\pi\eta_s r_0$, where r_0 is the bead radius and η_s is the viscosity of the suspending medium. Then one can form a fluid "time constant" λ as follows:

$$\lambda = \frac{\zeta L^2}{12\kappa T} \tag{8.6-1}$$

We do not attempt to give a discussion of the details of the kinetic theory here, as it can be found elsewhere,[1] but simply present the results.

In steady-state shear flow, it is found that the non-Newtonian viscosity and normal stress functions are given by the following power series expansions for small values of the shear rate:

$$\eta(\dot{\gamma}) - \eta_s = n\kappa T\lambda\left[1 - \frac{18}{35}(\lambda\dot{\gamma})^2 + \frac{1326}{1825}(\lambda\dot{\gamma})^4 - \cdots\right] \tag{8.6-2}$$

$$\Psi_1(\dot{\gamma}) = n\kappa T\lambda^2\left[1 - \frac{38}{35}(\lambda\dot{\gamma})^2 + \frac{38728}{25025}(\lambda\dot{\gamma})^4 - \cdots\right] \tag{8.6-3}$$

$$\Psi_2(\dot{\gamma}) = 0 \tag{8.6-4}$$

where n is the number of dumbbells per unit volume. By using a technique known as "orthogonal collocation," Stewart and Sørensen[2] showed that, for very large values of the shear rate, the viscosity and first normal stresses have the following asymptotic power-law expressions:

$$\eta(\dot{\gamma}) - \eta_s \doteq 0.678 n\kappa T\lambda(\lambda\dot{\gamma})^{-1/3} \tag{8.6-5}$$

$$\Psi_1(\dot{\gamma}) \doteq 1.20 n\kappa T\lambda^2(\lambda\dot{\gamma})^{-4/3} \tag{8.6-6}$$

[5]M. Abramowitz and I. A. Stegun, Eds., *Handbook of Mathematical Functions*, Dover, New York, 9th Printing (1973).

[1]R. B. Bird, H. R. Warner, Jr., and D. C. Evans, *Adv. in Polymer Sci.*, **8**, 1–70 (1971); R. B. Bird, C. F. Curtiss, R. C. Armstrong, and O. Hassager, *Dynamics of Polymeric Liquids*, Vol. **2**, 2nd edition, Wiley, New York (1987), Chapter 14; R. B. Bird, O. Hassager, R. C. Armstrong, and C. F. Curtiss, *Dynamics of Polymeric Liquids*, Vol. **2**, 1st edition, Wiley, New York (1977), Chapter 11.

[2]W.E. Stewart and J.P. Sørensen, *Trans. Soc. Rheol.*, **16**, 1–13 (1972).

The limiting slopes provided by these equations are fairly realistic for most polymer solutions and melts.

For oscillatory shearing flow with very small shear rate amplitude, the rigid dumbbell suspension model gives

$$\eta'(\omega) - \eta_s = n\kappa T\lambda \left[\frac{1 + \frac{2}{5}(\lambda\omega)^2}{1 + (\lambda\omega)^2} \right] \tag{8.6-7}$$

$$\eta''(\omega) - \eta_s = n\kappa T\lambda \left[\frac{\frac{3}{5}(\lambda\omega)}{1 + (\lambda\omega)^2} \right] \tag{8.6-8}$$

These expressions capture the main features of the behavior of the linear viscoelastic behavior of polymeric fluids. In addition, the normal stress difference responses in small amplitude oscillatory shear flow have been predicted.[1]

For steady elongational flow, the rigid dumbbell model gives the following:

$$\frac{\bar{\eta}(\dot{\varepsilon}) - 3\eta_s}{3n\kappa T\lambda} = \frac{1}{2} \mp \frac{3}{4X} \pm \frac{3}{4\sqrt{X}} \frac{\exp(\pm X)}{\int_0^{\sqrt{X}} \exp(\pm Y^2) dY} \tag{8.6-9}$$

in which $X = (9/2)\lambda|\dot{\varepsilon}|$. This predicts that for positive elongation rates, $(\bar{\eta} - 3\eta_s)/3n\kappa T\lambda$ increases monotonically from zero to 2.00 as the elongation rate goes from $\dot{\varepsilon} = 0$ to $\dot{\varepsilon} = \infty$; it also predicts that for negative elongation rates (biaxial extension), $(\bar{\eta} - 3\eta_s)/3n\kappa T\lambda$ decreases monotonically from zero to $-1/2$ as the elongation rate goes from $\dot{\varepsilon} = 0$ to $\dot{\varepsilon} = -\infty$. These predictions seem to agree, at least qualitatively, with the behavior of some polymers.

The rheological responses in even more complex flows have been studied. For example, the behavior of the dumbbell suspension model in large-amplitude oscillatory shear flow has been worked out.[3] In addition, by using a superposition of dumbbells with different time constants to model a mixture of dumbbells of different lengths, one can get refinements in the predictions of the various rheological properties.[4] The subject has been pursued well beyond the discussion of this section, including the development of theories for freely jointed bead-spring and bead-rod chains.[4]

§8.7 CONCLUDING COMMENTS

It has been demonstrated that many experiments show that polymeric fluids (as well as suspensions) do not behave according to Newton's law of viscosity. We have also seen that for such fluids, measuring the viscosity is not sufficient to characterize the fluids. One can measure the normal stress differences and the viscosity as functions of the shear rate. In addition, many other quantities can be measured: the elongational viscosity, the stress relaxation after cessation of steady shear flow, the stress growth after the start-up of steady shear flow, and many others. Suspensions exhibit some, but not all, phenomena observed for polymers.

For describing steady shearing motions, the generalized fluid models, with two or three constants, are quite sufficient and these have served the polymer industry well. For

---

[3] R. B. Bird, A. J. Giacomin, A. M. Schmalzer, and C. Aumnate, *J. Chem. Phys.*, **140**, 074904 (2014); Errata: in Eq. 91, η' should be η''; in caption to Fig. 3, $\psi[P_2^2 s_2]$ should be cos $3\omega t$, and $\psi[P_2^0 c_0, P_2^2 c_2]$ should be sin $3\omega t$.

[4] R. B. Bird, C. F. Curtiss, R. C. Armstrong, and O. Hassager, *Dynamics of Polymeric Liquids*, Vol. **2**, 2nd edition, Wiley, New York (1987); Problem 14C.2 (for dumbbell mixtures) and Chapters 15 through 19 (for chainlike molecules and other more complex models).

small amplitude oscillatory flow, the Maxwell model can be used to get general trends, but is unsuitable for quantitative calculations; however, the three-constant Maxwell model (using the Spriggs empiricism) can give quantitative agreement with experiments. For nonlinear viscoelastic flows, recourse has to be taken to expressions for the stress tensor that take into account the rotation of fluid elements. The simplest model of this type is the two-constant corotational Maxwell model; this model is inadequate for describing the nonlinear behavior of rheological phenomena, but the corresponding three-constant generalized corotational model does rather well. The two-constant dumbbell molecular theory gives the material functions as well as information regarding the molecular orientation.

This chapter is but an elementary introduction to a huge subject with a vast and rapidly growing literature, particularly in the field of nonlinear viscoelastic models. The molecular theories for polymers and structural models for suspensions will likely, in the future, provide a more reliable basis for stress tensor expressions for these complex fluids.

QUESTIONS FOR DISCUSSION

1. Compare the behavior of Newtonian liquids and polymeric liquids in the various experiments discussed in §8.1 and §8.2.
2. Why do we deal only with differences in normal stresses for incompressible liquids (see Eqs. 8.2-2 and 8.2-3)?
3. In §8.2*b* the postulated velocity profile is linear in y. What would you expect the velocity distribution to look like if the gap between the plates were not small and the fluid had a very low viscosity?
4. How is the parameter n in Eq. 8.3-3 related to the parameter n in Eq. 8.3-4?
5. What limitations have to be placed on use of the generalized Newtonian models and the two-constant corotational Maxwell model?
6. For what kinds of industrial problems would you use the various kinds of models described in this chapter?
7. Why may the power-law model be unsatisfactory for describing the axial flow in an annulus?
8. In Eq. 8.3-32, why do we integrate by parts?

PROBLEMS

8A.1 Flow of a polyisoprene solution in a pipe. A 13.5% (by weight) solution of polyisoprene in isopentane has the following power-law parameters at 323 K: $n = 0.2$ and $m = 5 \times 10^3$ Pa · s^n. It is being pumped (in laminar flow) through a horizontal pipe that has a length of 10.2 m and an internal diameter of 1.3 cm. It is desired to use another pipe with a length of 30.6 m with the same mass flow rate and the same pressure drop. What should the pipe radius be?

8A.2 Pumping of a polyethylene oxide solution. A 1% aqueous solution of polyethylene oxide at 333 K has power-law parameters $n = 0.6$ and $m = 0.50$ Pa · s^n. The solution is being pumped between two tanks, with the first tank at pressure p_1, and the second at pressure p_2. The pipe carrying the solution has a length of 14.7 m and an internal diameter of 0.27 m.

It has been decided to replace the single pipe by a pair of pipes of the same length, but with smaller diameter. What diameter should these pipes have so that the mass flow rate will be the same as in the single pipe?

8B.1 Flow of a polymeric film. Work the problem in §2.2 for the power-law fluid. Show that the result simplifies properly to the Newtonian result.

8B.2 Power-law flow in a narrow slit. In Example 8.3-2, show how to derive the velocity distribution for the region $-B \leq x \leq 0$. Is it possible to combine this result with that in Eq. 8.3-16 into one equation?

8B.3 Non-Newtonian flow in an annulus. Rework Problem 2B.8 for the annular flow of a power-law fluid with the flow being driven by the axial motion of the inner cylinder.

(a) Show that the velocity distribution for the fluid is

$$\frac{v_z(r)}{v_0} = \frac{(r/R)^{1-(1/n)} - 1}{\kappa^{1-(1/n)} - 1} \tag{8B.3-1}$$

(b) Verify that the result in (a) simplifies to the Newtonian result when n goes to 1.

(c) Show that the mass flow rate in the annular region is given by

$$w = \frac{2\pi R^2 v_0 \rho}{\kappa^{1-(1/n)} - 1}\left(\frac{1-\kappa^{3-(1/n)}}{3-(1/n)} - \frac{1-\kappa^2}{2}\right) \qquad \left(\text{for } n \neq \frac{1}{3}\right) \tag{8B.3-2}$$

(d) What is the mass flow rate for fluids with $n = \frac{1}{3}$?

(e) Simplify Eq. 8B.3-2 for the Newtonian fluid.

8B.4 Flow of a polymeric liquid in a tapered tube. Work Problem 2B.11 for a power-law fluid, using the lubrication approximation.

8B.5 Derivation of the Buckingham-Reiner equation.[1] Rework Example 8.3-1 (or Example 8.3-4) for the Bingham model. First show that the velocity distribution is

$$v_z^{>}(r) = \frac{(\mathscr{P}_0 - \mathscr{P}_L)R^2}{4\mu_0 L}\left[1 - \left(\frac{r}{R}\right)^2\right] - \frac{\tau_0 R}{\mu_0}\left[1 - \left(\frac{r}{R}\right)\right] \qquad r \geq r_0 \tag{8B.5-1}$$

$$v_z^{<}(r) = \frac{(\mathscr{P}_0 - \mathscr{P}_L)R^2}{4\mu_0 L}\left(1 - \frac{r_0}{R}\right) \qquad r \leq r_0 \tag{8B.5-2}$$

where r_0 is the radius of the plug-flow region. Then show that the mass rate of flow is given by

$$w = \frac{\pi(\mathscr{P}_0 - \mathscr{P}_L)R^4\rho}{8\mu_0 L}\left[1 - \frac{4}{3}\left(\frac{\tau_0}{\tau_R}\right) + \frac{1}{3}\left(\frac{\tau_0}{\tau_R}\right)^4\right] \tag{8B.5-3}$$

in which $\tau_R = (\mathscr{P}_0 - \mathscr{P}_L)R/2L$ is the shear stress at the tube wall. This expression is valid only when $\tau_R \geq \tau_0$.

8B.6 Draining of a tank with an exit pipe. Rework Problem 7B.9(a) for the system depicted in Fig. 7B.9, for the power-law fluid.

8B.7 Radial flow between parallel disks for a power-law fluid. First review the solution to the problem of a power-law fluid through a thin slit in Example 8.3-2.

Then solve the problem of slow, steady radial flow between two fixed parallel circular disks separated by a distance $2B$. Let the inner and outer radii of the disks be R_1 and R_2 (see Fig. 8B.7). Obtain the mass rate of flow by adapting the results of Example 8.3-2 locally in the region between the two disks. The result is

$$w = \frac{4\pi B^2 \rho}{(1/n)+2}\left(\frac{(p_1 - p_2)\,B(1-n)}{m(R_2^{1-n} - R_1^{1-n})}\right)^{1/n} \tag{8B.7-1}$$

This equation has been used by several groups[2,3,4] to describe the radial flow data for moderately viscoelastic polymer solutions.

---

[1] E. Buckingham, *Proc. ASTM*, **21**, 1154–1161 (1921); M. Reiner, *Deformation and Flow*, Lewis, London (1949).

[2] T. Y. Na and A. G. Hansen, *Int. J. Nonlinear Mech.*, **2**, 261–273 (1967).

[3] B. R. Laurencena and M. C. Williams, *Trans. Soc. Rheol.*, **18**, 331–355 (1974).

[4] H. Amadou, P. M. Adler, and J.-M. Piau, *J. Non-Newtonian Fluid Mech.*, **4**, 229–237 (1978).

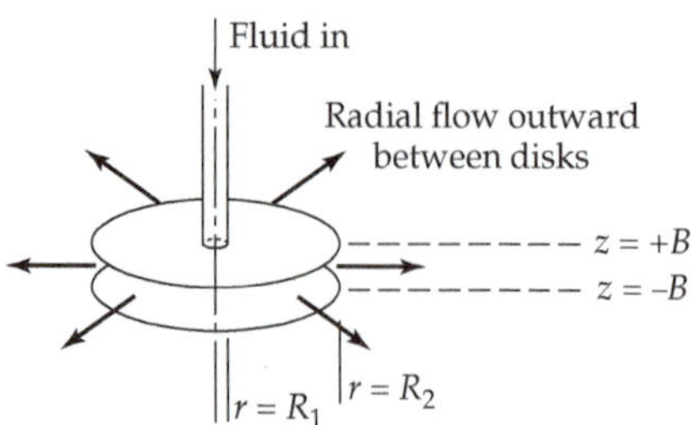

Fig. 8B.7 Outward radial flow in the space between two parallel, circular disks.

8B.8 Axial annular flow of a Bingham fluid. Repeat the problem discussed in §2.4 for a Bingham fluid. The flow of a Newtonian fluid is pictured in Fig. 2.4-1, and the flow of a Bingham fluid is shown in Fig. 8B.8.

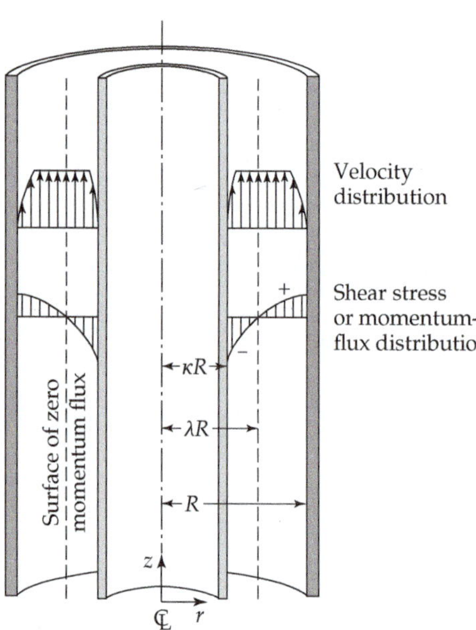

Fig. 8B.8 The momentum-flux distribution and velocity distribution for the upward flow of a Bingham fluid in a cylindrical annulus.

8C.1 The cone-and-plate viscometer.[5] Review the Newtonian analysis of the cone-and-plate instrument shown in Fig. 2.6-2 and discussed in §2.6. Then do the following:

(a) Show that the shear rate $\dot{\gamma}$ is uniform throughout the gap (for small ψ_0) and equal to $\dot{\gamma} = -\dot{\gamma}_{\theta\phi} = \Omega/\psi_0$. Because of the uniformity of $\dot{\gamma}$, the components of the stress tensor are also constant throughout the gap.

(b) Show that the non-Newtonian viscosity is then obtained from measurements of the torque T_z and rotation speed Ω by using

$$\eta(\dot{\gamma}) = \frac{3T_z\psi_0}{2\pi R^3\Omega} \tag{8C.1-1}$$

(c) Show that for the cone-and-plate system the radial component of the equation of motion is

$$0 = -\frac{\partial p}{\partial r} - \frac{1}{r^2}\frac{\partial}{\partial r}(r^2\tau_{rr}) + \frac{\tau_{\theta\theta} + \tau_{\phi\phi}}{r} \tag{8C.1-2}$$

[5] R. B. Bird, R. C. Armstrong, and O. Hassager, *Dynamics of Polymeric Liquids, Vol. 1, Fluid Mechanics*, Wiley-Interscience, New York, 2nd Edition (1987), pp. 521–524.

if the centrifugal force term $-\rho v_\phi^2/r$ can be neglected. Rearrange this to get

$$0 = -\frac{\partial \pi_{rr}}{\partial \ln r} + (\tau_{\phi\phi} - \tau_{\theta\theta}) + 2(\tau_{\theta\theta} - \tau_{rr}) \tag{8C.1-3}$$

Then introduce the normal stress coefficients, and use the result of (a) to replace $\partial\pi_{rr}/\partial \ln r$ by $\partial\pi_{\theta\theta}/\partial \ln r$, to get

$$\frac{\partial \pi_{\theta\theta}}{\partial \ln r} = -(\Psi_1 + 2\Psi_2)\dot{\gamma}^2 \tag{8C.1-4}$$

Integrate this from r to R and use the boundary condition $\pi_{rr}(R) = p_a$ to get

$$\begin{aligned} \pi_{\theta\theta}(r) &= \pi_{\theta\theta}(R) - (\Psi_1 + 2\Psi_2)\dot{\gamma}^2 \ln(r/R) \\ &= p_a - \Psi_2\dot{\gamma}^2 - (\Psi_1 + 2\Psi_2)\dot{\gamma}^2 \ln(r/R) \end{aligned} \tag{8C.1-5}$$

in which p_a is the atmospheric pressure acting on the fluid at the rim of the cone-and-plate instrument.

(d) Show that the total thrust in the z direction exerted by the fluid on the cone is

$$F_z = \int_0^{2\pi}\int_0^R [\pi_{\theta\theta}(r) - p_a]\, r\, dr\, d\theta = \frac{1}{2}\pi R^2 \Psi_1 \dot{\gamma}^2 \tag{8C.1-6}$$

From this one can obtain the first normal-stress coefficient by measuring the force that the fluid exerts.

(e) Suggest a method for measuring the second normal-stress coefficient using results in part (c) if small pressure transducers are flush-mounted in the plate at several different radial locations.

8C.2 **Deduction of $\eta(\dot{\gamma})$ from tube-flow data.** In Problem 8C.1 it is shown how to get the function $\eta(\dot{\gamma})$ from data obtained with a cone-and-plate viscometer. In the absence of this kind of device, one may have to extract the function from tube-flow data. Show how to get the relation for the non-Newtonian viscosity $\eta(\dot{\gamma})$ from experimental data on mass rate of flow w vs. pressure drop $\mathscr{P}_0 - \mathscr{P}_L$ for flow through circular tubes.

We know from §2.3 that, for any kind of fluid $\tau_{rz} = \tau_R r/R$, where $\tau_R = (\mathscr{P}_0 - \mathscr{P}_L)R/2L$ is the shear stress at the tube wall $r = R$. The mass rate of flow through the tube is

$$w = 2\pi\rho\int_0^R v_z r dr = -2\pi\rho\int_0^R \left(\frac{dv_z}{dr}\right)\left(\frac{r^2}{2}\right) dr = +\pi\rho\int_0^R \dot{\gamma} r^2 dr \tag{8C.2-1}$$

the second form being obtained by an integration by parts. In the third form we have made the replacement $\dot{\gamma} = -dv_z/dr$.

Next change the variable of integration from r to τ_{rz}

$$\frac{w}{\pi R^3 \rho} = \frac{1}{\tau_R^3}\int_0^{\tau_R} \dot{\gamma}\tau_{rz}^2\, d\tau_{rz} \tag{8C.2-2}$$

In this equation, the shear rate $\dot{\gamma}$ is to be regarded as a function of the shear stress. Equation 8C.2-2 says that data taken in tubes of different lengths and radii should collapse onto a single curve when plotted as $w/\pi R^3\rho$ vs. τ_R. Now multiply Eq. 8C.2-2 by τ_R^3 and then differentiate with respect to τ_R to get

$$\frac{d}{d\tau_R}\left(\tau_R^3 \frac{w}{\pi R^3 \rho}\right) = \dot{\gamma}_R \tau_R^2 \tag{8C.2-3}$$

where the Leibniz formula for differentiating an integral has been used (see §C.3). This is the *Weissenberg-Rabinowitsch equation.* It tells how the wall shear rate $\dot{\gamma}_R$ can be obtained by differentiating the flow-rate vs pressure-drop data.

Now put Eq. 8C.2-3 into a different form. Differentiate the product with respect to τ_R to get

$$\frac{w}{\pi R^3 \rho}\cdot 3\tau_R^2 + \tau_R^3 \frac{d}{d\tau_R}\left(\frac{w}{\pi R^3 \rho}\right) = \dot{\gamma}_R \tau_R^2 \tag{8C.2-4}$$

and then divide by $\tau_R^2(w/\pi R^3\rho)$ to obtain

$$3 + \frac{d\ln(w/\pi R^3\rho)}{d\ln\tau_R} = \dot{\gamma}_R\frac{1}{w/\pi R^3\rho} \tag{8C.2-5}$$

Then finally

$$\eta(\dot{\gamma}_R) = \frac{\tau_R}{\dot{\gamma}_R} = \frac{\tau_R}{w/\pi R^3\rho}\left[3 + \frac{d\ln\left(w/\pi R^3\rho\right)}{d\ln\tau_R}\right]^{-1} \tag{8C.2-6}$$

Equation 8C.2-6 gives the viscosity as a function of shear rate at the wall from experimental measurements of w and $\mathscr{P}_0 - \mathscr{P}_L$. If it is assumed that $\eta(\tau_R)$ at the wall is the same as $\eta(\dot{\gamma})$ throughout the tube, then Eq. 8C.2-6 gives $\eta(\dot{\gamma})$. This assumption seems to be valid for typical polymeric fluids. It would not, however, be expected to hold for suspensions of fibers, because of the change in the fluid microstructure near the wall. The above analysis is applicable only when there is no appreciable viscous heating.

8C.3 Squeezing flow between parallel disks.[6] Rework Problem 3C.3(e) for the power-law fluid (see Fig. 3C.3). This device can be useful for determining the power-law parameters for materials that are highly viscous. Show that the power-law analog of Eq. 3C.3-15 is

$$\frac{1}{H^{(n+1)/n}} = \frac{1}{H_0^{(n+1)/n}} + \frac{n+1}{2n+1}\left(\frac{n+3}{2\pi m R^{n+3}}\right)^{1/n} F_0^{1/n} t \tag{8C.3-1}$$

This equation reduces to that for a Newtonian fluid for $m \to \mu$ and $n \to 1$.

[6] R. B. Bird, R. C. Armstrong, and O. Hassager, *Dynamics of Polymeric Liquids, Vol. 1, Fluid Mechanics*, Wiley-Interscience, New York, 2nd Edition (1987), pp. 189–192.

Part Two

Energy Transport

Chapter 9

Thermal Conductivity and the Mechanisms of Energy Transport

In Chapter 1, when we discussed the mechanisms of momentum transport, we discussed first the convective momentum flux and then the molecular momentum flux. In this chapter we follow the same approach for energy transport.

In §9.1, we discuss the convective transport of kinetic energy and internal energy; in Eqs. 0.4-10 and 0.4-11, we have already obtained expressions for describing how these forms of energy are swept along by the fluid motion. This leads to the definition of the *convective energy-flux vector* $\mathbf{q}^{(c)}$.

Then we turn to the molecular transport of energy, but here we must include two different types of transport. In §9.2, we discuss the conductive heat flux, which involves the thermal conductivity k of the material, and in §9.3, we explain the work flux, which involves the viscosity μ of the material. In these sections we will define the *conductive heat-flux vector* $\mathbf{q}$, and the *work-flux vector* $\mathbf{w} = [\boldsymbol{\pi} \cdot \mathbf{v}] = p\mathbf{v} + [\boldsymbol{\tau} \cdot \mathbf{v}]$. These two molecular fluxes will appear as $\mathbf{q} + \mathbf{w}$, and this will be reminiscent of the combination $Q + W$ that appears in classical thermodynamics.

Having defined the three fluxes mentioned above, in §9.4, we combine them and define the *total energy-flux vector* $\mathbf{e} = \mathbf{q}^{(c)} + \mathbf{q} + \mathbf{w}$. This quantity will be used in Chapter 10 in setting up shell energy balances, and again in Chapter 11 in setting up the equation of change for energy. This is a conservation equation for energy; it is actually a generalization of the first law of thermodynamics, in that it can describe flow systems and time-dependent problems, whereas classical thermodynamics is restricted to equilibrium systems that are time-independent.

The remainder of the chapter concerns the thermal conductivity of various materials—gases, liquids, and solids. In §9.5, we give a brief overview of the experimental data, in order to give an idea of the range of numerical values of this physical property. Then in §9.6, we show how the principle of corresponding states can give a qualitative picture of the behavior of the thermal conductivity of gases and liquids as a function of temperature and pressure. Then we present a brief discussion of the kinetic theory of gases, because this gives a clear picture of the mechanism for energy transport for gaseous systems. We then conclude the chapter with brief discussions of the thermal conductivity of liquids, solids, and solid composites.

However, this chapter does not tell the whole story about energy transport. A very important mechanism is that of radiant energy transport, which can even take place in a vacuum. This is so important that it deserves a chapter of its own, namely Chapter 16. Furthermore, energy can also be transported by mass diffusion, and this mechanism is postponed until Chapter 24.

§9.1 CONVECTIVE ENERGY-FLUX VECTOR

Energy may be transported by the bulk motion of the fluid. This convective energy transport mechanism was introduced in Chapter 0 (§0.4) for the special case of unidirectional, steady flow (students may find it helpful to review that short section before proceeding). Here we consider the convective transport of energy for arbitrary flows. In Fig. 9.1-1 we show three mutually perpendicular elements of area dS at the point P, where the fluid velocity is $\mathbf{v}$. The volume rate of flow across the surface element dS perpendicular to the x axis is $v_x\,dS$. The rate at which kinetic plus internal energy is being swept across the same surface element in the $+x$ direction is then

$$\left(\frac{1}{2}\rho v^2 + \rho\hat{U}\right) v_x dS \tag{9.1-1}$$

in which $\frac{1}{2}\rho v^2 = \frac{1}{2}\rho(v_x^2 + v_y^2 + v_z^2)$ is the kinetic energy per unit volume, and $\rho\hat{U}$ is the internal energy per unit volume. When we divide Eq. 9.1-1 by dS, we get the *flux* of kinetic plus

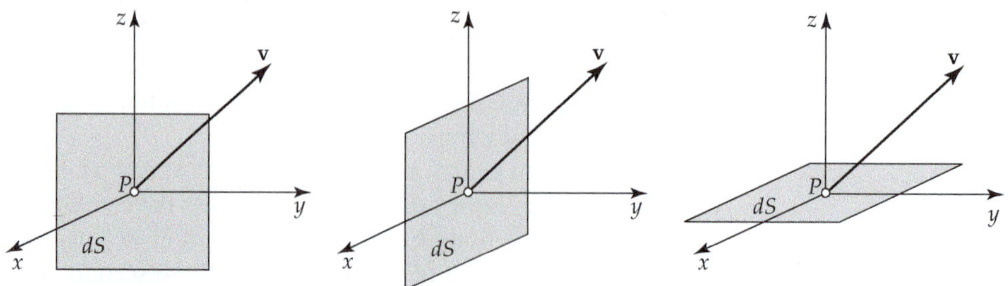

Fig. 9.1-1. Three mutually perpendicular surface elements of area dS across which energy is being transported by convection by the fluid moving with velocity $\mathbf{v}$. The volume rate of flow across the face perpendicular to the x axis is $v_x\,dS$, and the rate of flow of energy across dS is then $\left(\frac{1}{2}\rho v^2 + \rho\hat{U}\right) v_x\,dS$. Similar expressions can be written for the surface elements perpendicular to the y and z axes.

internal energy in the $+x$ direction, $\left(\frac{1}{2}\rho v^2 + \rho\hat{U}\right) v_x$, which has dimensions of energy/time · area.

We can write expressions similar to Eq. 9.1-1 for the rate at which energy is being swept through the surface elements perpendicular to the y and z axes to give the rates of energy transport in the $+y$ and $+z$ directions, respectively. If we now multiply each of the three expressions by the corresponding unit vector and add vectorially, we then get, after division by dS, the *convective energy-flux vector* $\mathbf{q}^{(c)}$,

$$\mathbf{q}^{(c)} = \left(\frac{1}{2}\rho v^2 + \rho\hat{U}\right) v_x \boldsymbol{\delta}_x + \left(\frac{1}{2}\rho v^2 + \rho\hat{U}\right) v_y \boldsymbol{\delta}_y + \left(\frac{1}{2}\rho v^2 + \rho\hat{U}\right) v_z \boldsymbol{\delta}_z \tag{9.1-2}$$

or

$$\boxed{\mathbf{q}^{(c)} = \left(\frac{1}{2}\rho v^2 + \rho\hat{U}\right) \mathbf{v}} \tag{9.1-3}$$

It is understood that this is the flux from the negative side of the surface to the positive side (e.g., $q_x^{(c)}$ is the flux in the positive x direction). This flux should be compared to the convective momentum flux illustrated in Fig. 1.1-2.

The definition of the internal energy in a nonequilibrium situation requires some care. From the *continuum* point of view, the internal energy at position $\mathbf{r}$ and time t is assumed to be the same function of the local, instantaneous density and temperature that one would have at equilibrium. From the *molecular* point of view, the internal energy consists of the sum of the kinetic energies of all the constituent atoms (relative to the flow velocity $\mathbf{v}$), the intramolecular potential energies, and the intermolecular energies, within a tiny region about the point $\mathbf{r}$ at time t.

Recall that, in the discussion of molecular collisions in §0.3, we found it convenient to regard the energy of a colliding pair of molecules to be the sum of the kinetic energies referred to the center of mass of the molecule plus the intramolecular potential energy of the molecule. Here also we split the energy of the fluid (regarded as a continuum) into kinetic energy associated with the bulk fluid motion and the internal energy associated with the kinetic energy of the molecules with respect to the flow velocity and the intra- and intermolecular potential energies.

§9.2 CONDUCTIVE HEAT-FLUX VECTOR—FOURIER'S LAW

Energy can also be transported even when there is no flow. Consider a slab of solid material of area A located between two large parallel plates a distance Y apart. We imagine that initially (for time $t < 0$) the solid material is at a temperature T_0 throughout. At $t = 0$, the lower plate is suddenly brought to a slightly higher temperature T_1 and maintained at that temperature. As time proceeds, the temperature profile in the slab changes, and ultimately a linear steady-state temperature distribution is attained (as shown in Fig. 9.2-1). When this steady-state condition has been reached, a constant rate of heat flow Q through the slab is required to maintain the temperature difference $\Delta T = T_1 - T_0$. It is found then that for sufficiently small values of ΔT the following relation holds:

$$\frac{Q}{A} = k\frac{\Delta T}{Y} \tag{9.2-1}$$

That is, the rate of heat flow per unit area is proportional to the temperature decrease over the distance Y. The constant of proportionality k is the *thermal conductivity* of the slab. Equation 9.2-1 is also valid if a liquid or gas is placed between the two plates, provided that suitable precautions are taken to eliminate convection and radiation.

In subsequent chapters it is better to work with the above equation in differential form. That is, we use the limiting form of Eq. 9.2-1 as the slab thickness approaches zero.

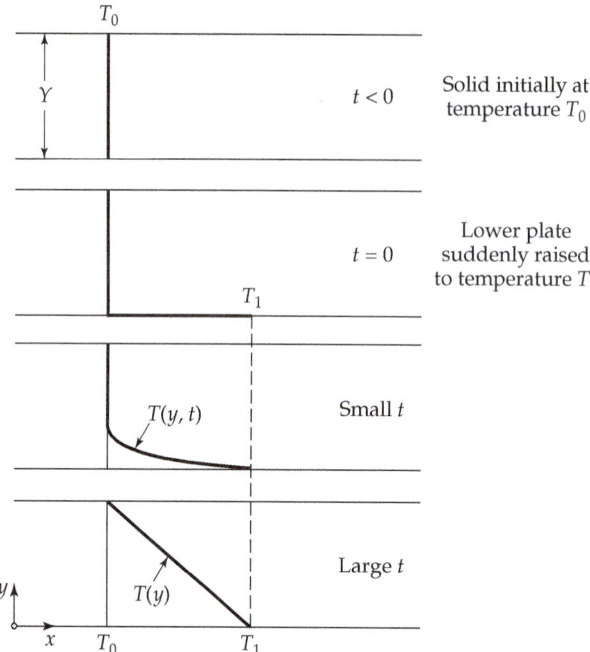

Fig. 9.2-1. Development of the steady-state temperature profile for a solid slab between two parallel plates. See Fig. 1.1-1 for the analogous situation for momentum transport.

The local rate of heat flow per unit area (heat flux) in the positive y direction is designated by q_y. In this notation Eq. 9.2-1 becomes

$$q_y = -k\frac{dT}{dy} \tag{9.2-2}$$

This equation, which serves to define k, is the one-dimensional form of *Fourier's law of heat conduction.*[1,2] It states that the heat flux by conduction is proportional to the temperature gradient, or, to put it pictorially, "heat slides downhill on the temperature-versus-distance graph." Actually Eq. 9.2-2 is not really a "law" of nature, but rather a suggestion, which has proven to be a very useful empiricism. However, it does have a theoretical basis, as discussed in §9.7.

If the temperature varies in all three directions, then we can write an equation like Eq. 9.2-2 for each of the coordinate directions

$$q_x = -k\frac{\partial T}{\partial x} \qquad q_y = -k\frac{\partial T}{\partial y} \qquad q_z = -k\frac{\partial T}{\partial z} \tag{9.2-3,4,5}$$

---

[1] J. B. Fourier, *Théorie analytique de la chaleur, Oeuvres de Fourier,* Gauthier-Villars et Fils, Paris (1822). **(Baron) Jean-Baptiste-Joseph Fourier** (pronounced "Foo-ree-ay")(1768–1830) was not only a brilliant mathematician and the originator of the Fourier series and the Fourier transform, but also famous as an Egyptologist and a political figure (he was Prefect of the Province of Isère).

[2] Some authors prefer to write Eq. 9.2-2 in the form

$$q_y = -J_e k\frac{dT}{dy} \tag{9.2.2a}$$

in which J_e is the "mechanical equivalent of heat," which displays explicitly the conversion of thermal units into mechanical units. For example, in the c.g.s. system one would use the following units: q_y [=] erg/cm^2 · s, k [=] cal/cm · s · °C, T [=] °C, y [=] cm, and J_e [=] erg/cal. We will not use Eq. 9.2-2a in this book.

If each of these equations is multiplied by the appropriate unit vector and the equations are then added vectorially, we get

$$\boxed{\mathbf{q} = -k\nabla T} \tag{9.2-6}$$

which is the three-dimensional form of Fourier's law; we call $\mathbf{q}$ the *conductive heat-flux vector*. This equation describes the molecular transport of heat in isotropic media. By "isotropic" we mean that the material has no preferred direction, so that heat is conducted with the same thermal conductivity k in all directions.[3]

The reader will have noticed that Eq. 9.2-2 for heat conduction and Eq. 1.2-2 for viscous flow are quite similar. In both equations the flux is proportional to the negative of the gradient of a macroscopic variable, and the coefficient of proportionality is a physical property characteristic of the material and dependent on the temperature and pressure. For the situations in which there is three-dimensional transport, we find that Eq. 9.2-6 for heat conduction and Eq. 1.2-13 for viscous flow differ in appearance. This difference arises because energy is a scalar, whereas momentum is a vector, and the conductive heat flux $\mathbf{q}$ is a vector with three components, whereas the momentum flux $\boldsymbol{\tau}$ is a second-order tensor with nine components. We can anticipate that the transport of energy and momentum will in general not be mathematically analogous except in certain geometrically simple situations.

In addition to the thermal conductivity k, defined by Eq. 9.2-2, a quantity known as the *thermal diffusivity* α is widely used. It is defined as

$$\alpha = \frac{k}{\rho\hat{C}_p} \tag{9.2-7}$$

Here $\hat{C}_p$ is the heat capacity at constant pressure; the circumflex (^) over the symbol indicates a quantity "per unit mass." Occasionally we will need to use the symbol $\tilde{C}_p$ in which the tilde (~) over the symbol stands for a quantity "per mole."

The thermal diffusivity α has the same dimensions as the kinematic viscosity ν—namely, (length)^2/time. When the assumption of constant physical properties is made, the quantities ν and α occur in similar ways in the equations of change for momentum and energy transport. Their ratio ν/α indicates the relative ease of momentum and energy transport in flow systems. This dimensionless ratio

$$\mathrm{Pr} = \frac{\nu}{\alpha} = \frac{\hat{C}_p\mu}{k} \tag{9.2-8}$$

is called the *Prandtl number*.[4] Another dimensionless group that we will encounter in subsequent chapters is the *Péclet number*,[5] Pé = RePr.

The units that are commonly used for thermal conductivity and related quantities are given in Table 9.2-1. Other units, as well as the interrelations among the various systems, may be found in Appendix E. Values for thermal conductivities and related quantities are tabulated below in §9.5.

---

[3]Conduction in anisotropic media is described in R. B. Bird, E. N. Lightfoot, and W. E. Stewart, *Transport Phenomena*, Revised Second Edition, Wiley, New York (2007).

[4]This dimensionless group, named for Ludwig Prandtl, involves only the physical properties of the fluid.

[5]**Jean-Claude-Eugène Péclet** (pronounced "Pay-clay" with the second syllable accented) (1793–1857) authored several books including one on heat conduction.

Table 9.2-1. Summary of Units for Quantities in Eqs. 9.2-2 and 9.2-8

	SI	c.g.s.	EE
q_y	$\mathrm{W/m^2}$	$\mathrm{cal/cm^2 \cdot s}$	$\mathrm{Btu/ft^2 \cdot hr}$
T	K	°C	°F
y	m	cm	ft
k	$\mathrm{W/m \cdot K}$	$\mathrm{cal/cm \cdot s \cdot {}^\circ C}$	$\mathrm{Btu/ft \cdot hr \cdot {}^\circ F}$
$\hat{C}_p$	$\mathrm{J/kg \cdot K}$	$\mathrm{cal/g \cdot {}^\circ C}$	$\mathrm{Btu/lb_m \cdot {}^\circ F}$
α	$\mathrm{m^2/s}$	$\mathrm{cm^2/s}$	$\mathrm{ft^2/s}$
μ	$\mathrm{Pa \cdot s}$	$\mathrm{g/cm \cdot s}$	$\mathrm{lb_m/ft \cdot s}$
Pr	—	—	—

Note: The watt (W) is the same as J/s, the joule (J) is the same as $\mathrm{N \cdot m}$, the newton (N) is $\mathrm{kg \cdot m/s^2}$, and the Pascal (Pa) is $\mathrm{N/m^2}$. For more information on interconversion of units, see Appendix E.

EXAMPLE 9.2-1

Measurement of Thermal Conductivity

A plastic panel of area $A = 1\ \mathrm{ft}^2$ and thickness $Y = 0.252$ in. was found to conduct heat at a rate of 3.0 W at steady state with temperatures $T_0 = 24.00\,°\mathrm{C}$ and $T_1 = 26.00\,°\mathrm{C}$ imposed on the two main surfaces. What is the thermal conductivity of the plastic in $\mathrm{cal/cm \cdot s \cdot K}$ at 25 °C?

SOLUTION

First, convert units with the aid of Appendix E:

$$A = 144\ \mathrm{in.}^2 \times \left(\frac{2.54\ \mathrm{cm}}{\mathrm{in.}}\right)^2 = 929\ \mathrm{cm}^2$$

$$Y = 0.252\ \mathrm{in.} \times \left(\frac{2.54\ \mathrm{cm}}{\mathrm{in.}}\right) = 0.640\ \mathrm{cm}$$

$$Q = 3.0\ \mathrm{W} \times \left(\frac{0.23901\ \mathrm{cal/s}}{\mathrm{W}}\right) = 0.717\ \mathrm{cal/s}$$

$$\Delta T = (26.00 - 24.00)\mathrm{K} = 2.00\ \mathrm{K}$$

Substitution into Eq. 9.2-1 then gives

$$k = \frac{QY}{A\Delta T} = \frac{(0.717\ \mathrm{cal/s})(0.640\ \mathrm{cm})}{(929\ \mathrm{cm}^2)(2.00\ \mathrm{K})} = 2.47 \times 10^{-4} \frac{\mathrm{cal}}{\mathrm{cm \cdot s \cdot K}} \tag{9.2-9}$$

For ΔT as small as 2°C, it is reasonable to assume that the value of k applies at the average temperature, which in this case is 25 °C.

§9.3 WORK-FLUX VECTOR

The last energy transport mechanism that we will consider in this chapter is the energy transported by doing work on a material. First we recall that, when a force **F** acts on a body and causes it to move through a distance $d\mathbf{r}$, the work done on the body is $dW = (\mathbf{F} \cdot d\mathbf{r})$. Then the rate of doing work is $dW/dt = (\mathbf{F} \cdot d\mathbf{r}/dt) = (\mathbf{F} \cdot \mathbf{v})$, that is, the dot product of the force and the velocity. We now apply this formula to the three perpendicular planes at a point P in space shown in Fig. 9.3-1.

First we consider the surface element perpendicular to the x axis. The fluid on the minus side (lesser values of x) of the surface exerts a force $\boldsymbol{\pi}_x dS$ on the fluid that is on the plus side (greater values of x; see Table 1.2-2). Since the fluid is moving with a velocity **v**,

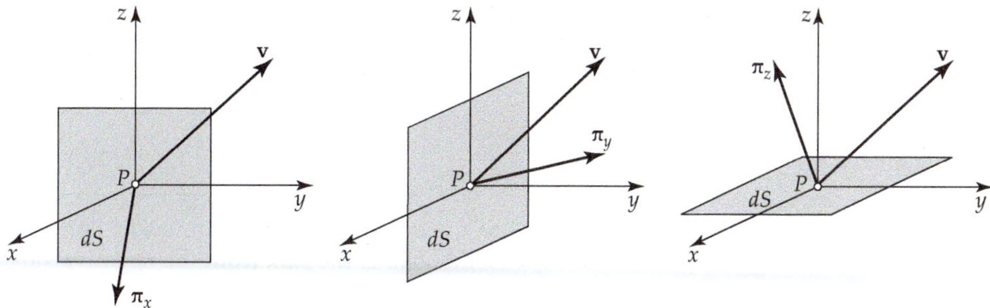

Fig. 9.3-1. Three mutually perpendicular surface elements of area dS at point P along with the stress vectors $\boldsymbol{\pi}_x$, $\boldsymbol{\pi}_y$, $\boldsymbol{\pi}_z$ acting on these surfaces. In the first figure, the rate at which work is done by the fluid on the minus side of dS on the fluid on the plus side of dS is then $(\boldsymbol{\pi}_x \cdot \mathbf{v})dS = [\boldsymbol{\pi} \cdot \mathbf{v}]_x dS$. Similar expressions hold for the surface elements perpendicular to the other two coordinate axes.

the rate at which work is done by the fluid on the minus side of the surface on the fluid on the plus side of the surface is $(\boldsymbol{\pi}_x \cdot \mathbf{v})dS$. Similar expressions may be written for the work done across the other two surface elements. When written out in component form, these rate-of-work expressions, per unit area, become

$$(\boldsymbol{\pi}_x \cdot \mathbf{v}) = \pi_{xx}v_x + \pi_{xy}v_y + \pi_{xz}v_z \equiv [\boldsymbol{\pi} \cdot \mathbf{v}]_x \tag{9.3-1}$$

$$(\boldsymbol{\pi}_y \cdot \mathbf{v}) = \pi_{yx}v_x + \pi_{yy}v_y + \pi_{yz}v_z \equiv [\boldsymbol{\pi} \cdot \mathbf{v}]_y \tag{9.3-2}$$

$$(\boldsymbol{\pi}_z \cdot \mathbf{v}) = \pi_{zx}v_x + \pi_{zy}v_y + \pi_{zz}v_z \equiv [\boldsymbol{\pi} \cdot \mathbf{v}]_z \tag{9.3-3}$$

When these scalar components are multiplied by the unit vectors and added, we get the "rate-of-doing-work (vector) per unit area," and we can call this, for short, the *work-flux vector*

$$\mathbf{w} = \boldsymbol{\delta}_x(\boldsymbol{\pi}_x \cdot \mathbf{v}) + \boldsymbol{\delta}_y(\boldsymbol{\pi}_y \cdot \mathbf{v}) + \boldsymbol{\delta}_z(\boldsymbol{\pi}_z \cdot \mathbf{v}) \tag{9.3-4}$$

or

$$\boxed{\mathbf{w} = [\boldsymbol{\pi} \cdot \mathbf{v}]} \tag{9.3-5}$$

Furthermore, the rate of doing work across a unit area of surface with orientation given by the unit vector $\mathbf{n}$ is $(\mathbf{n} \cdot [\boldsymbol{\pi} \cdot \mathbf{v}])$.

In curvilinear coordinates, Eqs. 9.3-1 to 9.3-5 are easily written down: for cylindrical coordinates, replace x,y,z by r,θ,z, and for spherical coordinates, replace x,y,z by r,θ,ϕ.

§9.4 TOTAL ENERGY-FLUX VECTOR

We now define, for later use, the *total energy-flux vector* $\mathbf{e}$ as follows:

$$\boxed{\mathbf{e} = \mathbf{q}^{(c)} + \mathbf{q} + \mathbf{w} = \left(\frac{1}{2}\rho v^2 + \rho\hat{U}\right)\mathbf{v} + \mathbf{q} + [\boldsymbol{\pi} \cdot \mathbf{v}]} \tag{9.4-1}$$

The $\mathbf{e}$ vector is the sum of (a) the convective energy-flux vector, (b) the conductive heat-flux vector, and (c) the work-flux vector. All the terms in Eq. 9.4-1 have the same sign convention, so that e_x is the energy transport in the positive x direction per unit area per unit time.

The total molecular stress tensor $\boldsymbol{\pi}$ can now be split up into two parts $\boldsymbol{\pi} = p\boldsymbol{\delta} + \boldsymbol{\tau}$ so that $[\boldsymbol{\pi}\cdot\mathbf{v}] = p[\boldsymbol{\delta}\cdot\mathbf{v}] + [\boldsymbol{\tau}\cdot\mathbf{v}] = p\mathbf{v} + [\boldsymbol{\tau}\cdot\mathbf{v}]$. The term $p\mathbf{v}$ can then be combined with the internal energy term $\rho\hat{U}\mathbf{v}$ to give an enthalpy term $\rho\hat{U}\mathbf{v} + p\mathbf{v} = \rho(\hat{U} + (p/\rho))\mathbf{v} = \rho(\hat{U} + p\hat{V})\mathbf{v} = \rho\hat{H}\mathbf{v}$, so that

$$\boxed{\mathbf{e} = \left(\frac{1}{2}\rho v^2 + \rho\hat{H}\right)\mathbf{v} + [\boldsymbol{\tau}\cdot\mathbf{v}] + \mathbf{q}} \tag{9.4-2}$$

We shall usually use the $\mathbf{e}$ vector in this form, understanding that the total energy flux across the surface element dS is the flux from the negative side to the positive side of the surface.

In Table 9.4-1 we summarize the notation for the various energy-flux vectors introduced in this section. All of them have the same sign convention.

To evaluate the enthalpy in Eq. 9.4-2, we make use of the standard equilibrium thermodynamics formula[1]

$$d\hat{H} = \left(\frac{\partial\hat{H}}{\partial T}\right)_p dT + \left(\frac{\partial\hat{H}}{\partial p}\right)_T dp = \hat{C}_p dT + \left[\hat{V} - T\left(\frac{\partial\hat{V}}{\partial T}\right)_p\right] dp \tag{9.4-3}$$

When this is integrated from some reference state p^o, T^o to the state p,T, we then get[1]

$$\hat{H}(p,T) - \hat{H}^o = \int_{T^o}^{T} \hat{C}_p dT + \int_{p^o}^{p} \left[\hat{V} - T\left(\frac{\partial\hat{V}}{\partial T}\right)_p\right] dp \tag{9.4-4}$$

in which $\hat{H}^o$ is the enthalpy per unit mass at the reference state. The integral over p is zero for an ideal gas and $(p - p^o)/\rho$ for fluids of constant density. The integral over T becomes $\hat{C}_p(T - T^o)$ if the heat capacity can be regarded as constant over the relevant temperature range, which is often a reasonable assumption for incompressible liquids. Thus, we have the following simplifications for Eq. 9.4-4 for two special cases:

ideal gases:
$$\hat{H}(p,T) - \hat{H}^o = \hat{C}_p(\overline{T})(T - T^o) \tag{9.4-5}$$

incompressible liquids:
$$\hat{H}(p,T) - \hat{H}^o = \hat{C}_p(T - T^o) + \frac{p - p^o}{\rho} \tag{9.4-6}$$

In Eq. 9.4-5, $\hat{C}_p(\overline{T}) \equiv \left(\int_{T^o}^{T} \hat{C}_p\, dT\right)/(T - T^o)$ is the mean heat capacity for the relevant temperature range. In Eq. 9.4-6, the pressure term is often negligible as illustrated in

Table 9.4-1. Summary of Notation for Energy Fluxes

Symbol	Meaning	Reference
$\mathbf{q}^{(c)} = \left(\frac{1}{2}\rho v^2 + \rho\hat{U}\right)\mathbf{v}$	convective energy-flux vector	Eq. 9.1-3
$\mathbf{q} = -k\nabla T$	conductive heat-flux vector	Eq. 9.2-6
$\mathbf{w} = [\boldsymbol{\pi}\cdot\mathbf{v}] = p\mathbf{v} + [\boldsymbol{\tau}\cdot\mathbf{v}]$	work-flux vector	Eq. 9.3-5
$\mathbf{e} = \left(\frac{1}{2}\rho v^2 + \rho\hat{U}\right)\mathbf{v} + \mathbf{q} + [\boldsymbol{\pi}\cdot\mathbf{v}]$ $= \left(\frac{1}{2}\rho v^2 + \rho\hat{H}\right)\mathbf{v} + \mathbf{q} + [\boldsymbol{\tau}\cdot\mathbf{v}]$	total energy-flux vector	Eq. 9.4-1 Eq. 9.4-2

[1]See, for example, R. J. Silbey, R. A. Alberty, and M. G. Bawendi, *Physical Chemistry*, Wiley, 4th edition (2005), §2.7 and §2.8.

Example 9.4-1 below. It is assumed that Eqs. 9.4-4 through 9.4-6 are valid in nonequilibrium systems, where p and T are the *local* values of the pressure and temperature.

EXAMPLE 9.4-1

Comparison of Heat Capacity and Pressure Terms in Eq. 9.4-6

Compare the magnitudes of the two terms on the right side of Eq. 9.4-6 for reasonable values of the variables.

SOLUTION

We choose the following values for the variables in Eq. 9.4-6: $\hat{C}_p = 4184\ \text{J/kg}\cdot{}^\circ\text{C}$ and $\rho = 10^3\ \text{kg/m}^3$ (for water near room temperature), $T - T^o = 10^\circ\text{C}$, and $p - p^o = 1\ \text{atm} = 1.01325 \times 10^5\ \text{N/m}^2$. For these values, the ratio of the pressure and heat capacity terms in Eq. 9.4-6 is

$$\frac{p - p^o}{\rho\hat{C}_p(T - T^o)} = \frac{1.01325 \times 10^5\ \text{N/m}^2}{(10^3\ \text{kg/m}^3)(4184\ \text{J/kg}\cdot{}^\circ\text{C})(10^\circ\text{C})}\left(\frac{1\ \text{J}}{\text{N}\cdot\text{m}}\right) = 0.0024 \qquad (9.4\text{-}7)$$

Thus, for these variable values, the pressure term is negligible.

§9.5 THERMAL CONDUCTIVITY DATA FROM EXPERIMENTS

The thermal conductivity can vary all the way from about 0.01 W/m · K for gases up to about 1000 W/m · K for pure metals. Some experimental values of the thermal conductivity of gases, liquids, liquid metals, and solids are given in Tables 9.5-1, 9.5-2, 9.5-3, and 9.5-4. In making calculations, experimental values should be used when

Table 9.5-1. Thermal Conductivities, Heat Capacities, and Prandtl Numbers of Some Common Gases at 1 atm Pressure[a]

Gas	Temperature T (K)	Thermal conductivity k (W/m · K)	Heat Capacity $\hat{C}_p$ (J/kg · K)	Prandtl Number Pr (---)
H_2	100	0.06799	11,192	0.682
	200	0.1282	13,665	0.724
	300	0.1779	14,316	0.720
O_2	100	0.00904	910	0.764
	200	0.01833	911	0.734
	300	0.02657	920	0.716
NO	200	0.01778	1015	0.781
	300	0.02590	997	0.742
CO_2	200	0.00950	734	0.783
	300	0.01665	846	0.758
CH_4	100	0.01063	2073	0.741
	200	0.02184	2087	0.721
	300	0.03427	2227	0.701

[a] Taken from J. O. Hirschfelder, C. F. Curtiss, and R. B. Bird, *Molecular Theory of Gases and Liquids*, Wiley, New York, 2nd corrected printing (1964), Table 8.4-10. The k values are measured, the $\hat{C}_p$ values are calculated from spectroscopic data, and μ is calculated from Eq. 1.6-14. The values of $\hat{C}_p$ for H_2 represent a 3:1 ortho-para mixture.

Table 9.5-2. Thermal Conductivities, Heat Capacities, and Prandtl Numbers for Some Nonmetallic Liquids at Their Saturation Pressures[a]

Liquid	Temperature T (K)	Thermal conductivity k (W/m · K)	Viscosity $\mu \times 10^4$ (Pa · s)	Heat Capacity $\hat{C}_p \times 10^{-3}$ (J/kg · K)	Prandtl Number Pr (—)
1-Pentene	200	0.1461	6.193	1.948	8.26
	250	0.1307	3.074	2.070	4.87
	300	0.1153	1.907	2.251	3.72
CCl_4	250	0.1092	20.32	0.8617	16.0
	300	0.09929	8.828	0.8967	7.97
	350	0.08935	4.813	0.9518	5.13
$(C_2H_5)_2O$	250	0.1478	3.819	2.197	5.68
	300	0.1274	2.213	2.379	4.13
	350	0.1071	1.387	2.721	3.53
C_2H_5OH	250	0.1808	30.51	2.120	35.8
	300	0.1676	10.40	2.454	15.2
	350	0.1544	4.486	2.984	8.67
Glycerol	300	0.2920	7949	2.418	6580
	350	0.2977	365.7	2.679	329
	400	0.3034	64.13	2.940	62.2
H_2O	300	0.6089	8.768	4.183	6.02
	350	0.6622	3.712	4.193	2.35
	400	0.6848	2.165	4.262	1.35

[a]The entries in this table were prepared from functions provided by T. E. Daubert, R. P. Danner, H. M. Sibul, C. C. Stebbins, J. L. Oscarson, R. L. Rowley, W. V. Wilding, M. E. Adams, T. L. Marshall, and N. A. Zundel, *DIPPR® Data Compilation of Pure Compound Properties*, Design Institute for Physical Property Data®, AIChE, New York, NY (2000).

possible. In the absence of experimental data, one can make estimates by using the methods outlined in the next several sections, or by consulting various engineering handbooks.[1]

§9.6 THERMAL CONDUCTIVITY AND THE PRINCIPLE OF CORRESPONDING STATES

When thermal conductivity data for a particular compound cannot be found, one can make an estimate by using the corresponding-states chart in Fig. 9.6-1, which is based on experimental thermal conductivity data for several monatomic substances. This chart, which is similar to that for viscosity shown in Fig. 1.5-1, is a plot of the reduced thermal conductivity $k_r = k/k_c$, which is the thermal conductivity at pressure p and temperature T divided by the thermal conductivity at the critical point. This quantity is plotted as a function of the reduced temperature $T_r = T/T_c$ for various values of the reduced pressure

[1]For example, W. M. Rohsenow, J. P. Hartnett, and Y. I. Cho, Eds., *Handbook of Heat Transfer*, McGraw-Hill, New York (1998); Landolt-Börnstein, *Zahlenwerte und Funktionen*, Vol. **II**, 5, Springer (1968–1969).

Table 9.5-3. Thermal Conductivities, Heat Capacities, and Prandtl Numbers of Some Liquid Metals at Atmospheric Pressure[a]

Metal	Temperature T (K)	Thermal conductivity k (W/m · K)	Heat capacity $\hat{C}_p$ (J/kg · K)	Prandtl number Pr (—)
Hg	273.2	8.20	140.2	0.0288
	373.2	10.50	137.2	0.0162
	473.2	12.34	156.9	0.0116
Pb	644.2	15.9	15.9	0.024
	755.2	15.5	15.5	0.017
	977.2	15.1	14.6[b]	0.013[b]
Bi	589.2	16.3	14.4	0.0142
	811.2	15.5	15.4	0.0110
	1033.2	15.5	16.4	0.0083
Na	366.2	86.2	13.8	0.011
	644.2	72.8	13.0	0.0051
	977.2	59.8	12.6	0.0037
K	422.2	45.2	795	0.0066
	700.2	39.3	753	0.0034
	977.2	33.1	753	0.0029
Na-K alloy[c]	366.2	25.5	1130	0.026
	644.2	27.6	1054	0.0091
	977.2	28.9	1042	0.0058

[a]Data taken from *Liquid Metals Handbook*, 2nd edition, U.S. Government Printing Office, Washington, DC (1952), and from E. R. G. Eckert and R. M. Drake, Jr., *Heat and Mass Transfer*, McGraw-Hill, New York, 2nd edition (1959), Appendix A.

[b]Based on an extrapolated heat capacity.

[c]56% Na by weight, 44% K by weight.

$p_r = p/p_c$. Figure 9.6-1 is based on a limited amount of experimental data for monatomic substances, but may be used for rough estimates for polyatomic materials. It should *not* be used in the neighborhood of the critical point.[1]

It can be seen that the thermal conductivity of a gas approaches a limiting function of T at low pressures; for most gases, this limit is reached at about 1 atm pressure. The thermal conductivities of *gases* at low density *increase* with increasing temperature, whereas the thermal conductivities of most *liquids decrease* with increasing temperature. The correlation is less reliable in the liquid region; polar or associated liquids, such as water, may exhibit a maximum in the curve of k versus T. The main virtue of the corresponding-states chart is that one gets a global view of the behavior of the thermal conductivity of gases and liquids.

[1]In the vicinity of the critical point, where the thermal conductivity diverges, it is customary to write $k = k^b + \Delta k$, where k^b is the "background" contribution and Δk is the "critical enhancement" contribution. The k_c being used in the corresponding states correlation is the background contribution. For the behavior of transport properties near the critical point, see J. V. Sengers and J. Luettmer Strathmann, in *Transport Properties of Fluids* (J. H. Dymond, J. Millat, and C. A. Nieto de Castro, eds.), Cambridge University Press (1995); E. P. Sakonidou, H. R. van den Berg, C. A. ten Seldam, and J. V. Sengers, *J. Chem. Phys.*, **105**, 10535–10555 (1996) and **109**, 717–736 (1998).

Table 9.5-4. Experimental Values of Thermal Conductivities of Some Solids[a]

Substance	Temperature T (K)	Thermal conductivity k (W/m · K)
Aluminum	373.2	205.9
	573.2	268
	873.2	423
Cadmium	273.2	93.0
	373.2	90.4
Copper	291.2	384.1
	373.2	379.9
Steel	291.2	46.9
	373.2	44.8
Tin	273.2	63.93
	373.2	59.8
Brick (common red)	—	0.63
Concrete (stone)	—	0.92
Earth's crust (average)	—	1.7
Glass (soda)	473.2	0.71
Graphite	—	5.0
Sand (dry)	—	0.389
Wood (fir)		
parallel to axis	—	0.126
normal to axis	—	0.038

[a]Data taken from the *Reactor Handbook*, Vol. **2**, Atomic Energy Commission, AECD-3646, U. S. Government Printing Office, Washington, DC (May 1955), pp. 1766 *et seq*.

The quantity k_c may be estimated in one of two ways: (i) given k at a known temperature and pressure, preferably close to the conditions at which k is to be estimated, one can read k_r from the chart and compute $k_c = k/k_r$; or (ii) one can estimate a value of k in the low-density region by the methods given in §9.7 and then proceed as in (i). Values of k_c obtained by method (i) are given in Appendix D.

For mixtures, one might estimate the thermal conductivity by methods analogous to those described in §1.5. Very little is known about the accuracy of pseudocritical procedures as applied to thermal conductivity, largely because there are so few data on mixtures at elevated pressures.

EXAMPLE 9.6-1

Effect of Pressure on Thermal Conductivity

Estimate the thermal conductivity of ethane at 153°F and 191.9 atm from the experimental value[2] $k = 0.0159$ Btu/hr · ft · °F at 1 atm and 153°F.

SOLUTION

Since a measured value of k is known, we use method (i). First we calculate p_r and T_r at the condition of the measured value, using the critical values from Table D.1:

$$T_r = \frac{(153 + 460)^\circ\text{R}}{(305.4\text{ K})(1.8\,^\circ\text{R/K})} = 1.115; \quad p_r = \frac{1\text{ atm}}{48.2\text{ atm}} = 0.021 \tag{9.6-1}$$

[2]J. M. Lenoir, W. A. Junk, and E. W. Comings, *Chem. Eng. Progr.*, **49**, 539–542 (1949).

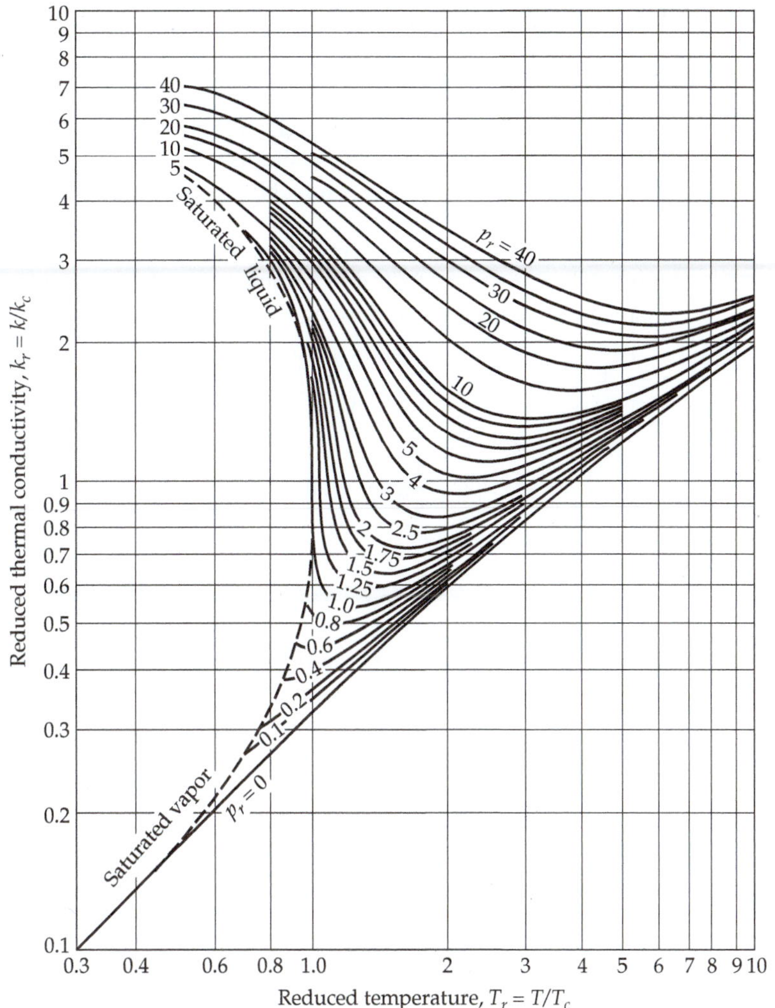

Fig. 9.6-1. Reduced thermal conductivity for monatomic substances as a function of the reduced temperature and pressure [E. J. Owens and G. Thodos, *AIChE Journal*, **3**, 454–461 (1957)]. A large-scale version of this chart may be found in O. A. Hougen, K. M. Watson, and R. A. Ragatz, *Chemical Process Principles Charts*, 2nd edition, Wiley, New York (1960).

From Fig. 9.6-1 we read $k_r = 0.36$. Hence, k_c is

$$k_c = \frac{k}{k_r} = \frac{0.0159 \text{ Btu/hr} \cdot \text{ft} \cdot {}^\circ\text{F}}{0.36} = 0.0442 \text{ Btu/hr} \cdot \text{ft} \cdot {}^\circ\text{F} \tag{9.6-2}$$

At 153°F ($T_r = 1.115$) and 191.9 atm ($p_r = 191.9$ atm/48.2 atm $= 3.98$), we read from the chart $k_r = 2.07$. The predicted thermal conductivity is then

$$k = k_r k_c = (2.07)(0.0442 \text{ Btu/hr} \cdot \text{ft} \cdot {}^\circ\text{F}) = 0.0914 \text{ Btu/hr} \cdot \text{ft} \cdot {}^\circ\text{F} \tag{9.6-3}$$

An observed value of 0.0453 Btu/hr · ft · °F has been reported.[2] This poor agreement shows that one should not rely heavily on this correlation for polyatomic substances or for conditions near the critical point.

§9.7 THERMAL CONDUCTIVITY OF GASES AND KINETIC THEORY

The thermal conductivities of dilute *monatomic* gases are well understood and can be described by the kinetic theory of gases at low density. Although detailed theories for *polyatomic* gases have been developed,[1] it is customary to use some simple approximate theories. Here, as in §1.6, we give a simplified mean-free-path derivation for monatomic gases, and then summarize the result of the more realistic Chapman-Enskog kinetic theory of gases.

We use the model of rigid, nonattracting spheres of mass m and diameter d. The gas as a whole is at rest ($\mathbf{v} = 0$), but the molecular motions must be accounted for.

As in §1.6, we use the following results for a rigid-sphere gas:

$$\bar{u} = \sqrt{\frac{8\kappa T}{\pi m}} = \text{mean molecular speed} \tag{9.7-1}$$

$$Z = \frac{1}{4}n\bar{u} = \text{wall collision frequency per unit area} \tag{9.7-2}$$

$$\lambda = \frac{1}{\sqrt{2}\pi d^2 n} = \text{mean-free path} \tag{9.7-3}$$

The molecules reaching any plane in the gas have, on an average, had their last collision at a distance a from the plane, where

$$a = \frac{2}{3}\lambda \tag{9.7-4}$$

In these equations κ is the Boltzmann constant, n is the number of molecules per unit volume, and m is the mass of a molecule.

The only form of energy that can be exchanged in a collision between two smooth rigid spheres is translational kinetic energy. The mean translational energy per molecule under equilibrium conditions is

$$\frac{1}{2}m\overline{u^2} = \frac{3}{2}\kappa T \tag{9.7-5}$$

For such a gas, the molar heat capacity at constant volume is

$$\tilde{C}_V = \left(\frac{\partial \tilde{U}}{\partial T}\right)_V = \tilde{N}\frac{d}{dT}\left(\frac{1}{2}m\overline{u^2}\right) = \frac{3}{2}R \tag{9.7-6}$$

in which $\tilde{N}$ is Avogadro's number and R is the gas constant. Equation 9.7-6 is satisfactory for monatomic gases up to temperatures of several thousand degrees.

To determine the thermal conductivity, we examine the behavior of the gas under a temperature gradient dT/dy (see Fig. 9.7-1). We assume that Eqs. 9.7-1 to 9.7-6 remain valid in this nonequilibrium situation, except that $\frac{1}{2}m\overline{u^2}$ in Eq. 9.7-5 is taken as the average kinetic energy for molecules that had their last collision in a region of temperature T. The heat flux q_y across any plane of constant y is found by summing the kinetic energies of the

---

[1] C. S. Wang Chang, G. E. Uhlenbeck, and J. de Boer, *Studies in Statistical Mechanics*, Wiley-Interscience, New York, Vol. **II** (1964), pp. 241–265; E. A. Mason and L. Monchick, *J. Chem. Phys.*, **35**, 1676–1697 (1961); **36**, 1622–1639, 2746–2757 (1962); L. Monchick, A. N. G. Pereira, and E. A. Mason, *J. Chem. Phys.*, **42**, 3241–3256 (1965). For an introduction to the kinetic theory of the transport properties, see R. S. Berry, S. A. Rice, and J. Ross, *Physical Chemistry*, 2nd edition (2000), Chapter 28.

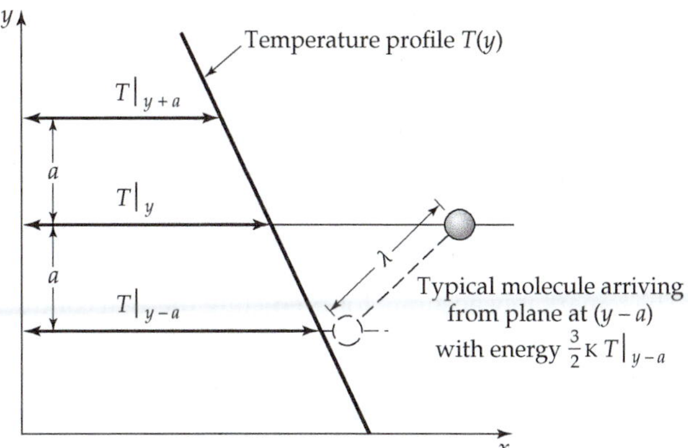

Fig. 9.7-1. Molecular transport of (kinetic) energy from plane at $(y - a)$ to plane at y.

molecules that cross the plane per unit time in the positive y direction and subtracting the kinetic energies of the equal number that cross in the negative y direction:

$$
\begin{aligned}
q_y &= Z\left(\frac{1}{2}m\overline{u^2}\Big|_{y-a} - \frac{1}{2}m\overline{u^2}\Big|_{y+a}\right) \\
&= \frac{3}{2}\kappa Z(T|_{y-a} - T|_{y+a})
\end{aligned}
\tag{9.7-7}
$$

Equation 9.7-7 is based on the assumption that all molecules have velocities representative of the region of their last collision and that the temperature profile $T(y)$ is linear for a distance of several mean-free paths. In view of the latter assumption we may write

$$T|_{y-a} = T|_y - \frac{2}{3}\lambda\frac{dT}{dy} \tag{9.7-8}$$

$$T|_{y+a} = T|_y + \frac{2}{3}\lambda\frac{dT}{dy} \tag{9.7-9}$$

By combining the last three equations we get

$$q_y = -\frac{1}{2}n\kappa\bar{u}\lambda\frac{dT}{dy} \tag{9.7-10}$$

This corresponds to Fourier's law of heat conduction (Eq. 9.2-2) with the thermal conductivity given by

$$k = \frac{1}{2}n\kappa\bar{u}\lambda = \frac{1}{2}\rho\hat{C}_V\bar{u}\lambda \qquad \text{(monatomic gas)} \tag{9.7-11}$$

in which $\rho = nm$ is the gas density, and $\hat{C}_V = \frac{3}{2}\kappa/m$ (from Eq. 9.7-6).

Substitution of the expressions for $\bar{u}$ and λ from Eqs. 9.7-1 and 9.7-3 then gives

$$k = \frac{\sqrt{m\kappa T/\pi}}{\pi d^2}\frac{\kappa}{m} = \frac{2}{3\pi}\frac{\sqrt{\pi m\kappa T}}{\pi d^2}\hat{C}_V \qquad \text{(monatomic gas)} \tag{9.7-12}$$

which is the thermal conductivity of a dilute gas composed of rigid spheres of diameter d. This equation predicts that k is independent of pressure. Figure 9.6-1 indicates that this prediction is in good agreement with experimental data up to about 10 atm for most gases. The predicted temperature dependence is too weak, just as was the case for viscosity.

For a more accurate treatment of the monatomic gas, we turn again to the rigorous Chapman-Enskog treatment discussed in §1.6. The Chapman-Enskog formula[2] for the thermal conductivity of a monatomic gas at low density at temperature T is

$$k = \frac{25}{32}\frac{\sqrt{\pi m \kappa T}}{\pi \sigma^2 \Omega_k}\hat{C}_V \quad \text{or} \quad k = 1.9891 \times 10^{-4}\frac{\sqrt{T/M}}{\sigma^2 \Omega_k} \quad \text{(monatomic gas)} \tag{9.7-13}$$

In the second form of this equation, k [=] cal/cm · s · K, T [=] K, σ [=] Å, and the "collision integral" for thermal conductivity, Ω_k, is identical to that for viscosity, Ω_μ in §1.6. Values of $\Omega_k = \Omega_\mu$ are given for the Lennard-Jones intermolecular potential in Table D.2 as a function of the dimensionless temperature $\kappa T/\varepsilon$. Equation 9.7-13, together with Table D.2, has been found to be remarkably accurate for predicting thermal conductivities of *monatomic* gases when the parameters σ and ε deduced from viscosity measurements are used (that is, the values given in Table D.1).

Equation 9.7-13 is very similar to the corresponding viscosity formula, Eq. 1.6-14. From these two equations we can then get

$$k = \frac{15}{4}\frac{R}{M}\mu = \frac{5}{2}\hat{C}_V\mu \quad \text{(monatomic gas)} \tag{9.7-14}$$

The simplified rigid-sphere theory (see Eqs. 1.6-8 and 9.7-11) gives $k = \hat{C}_V\mu$ and is thus in error by a factor 2.5. This is not surprising in view of the many approximations that were made in the simple treatment.

So far we have discussed only *monatomic* gases. We know from the discussion in §0.3, that, in binary collisions between diatomic molecules, there may be interchanges between kinetic and internal (i.e., vibrational and rotational) energy. Such interchanges are not taken into account in the Chapman-Enskog theory for monatomic gases. It can therefore be anticipated that the Chapman-Enskog theory will not be adequate for describing the thermal conductivity of polyatomic molecules.

A simple semiempirical method of accounting for the energy exchange in polyatomic gases was developed by Eucken.[3] His equation for thermal conductivity of a polyatomic gas at low density is

$$k = \left(\hat{C}_p + \frac{5}{4}\frac{R}{M}\right)\mu \quad \text{(polyatomic gas)} \tag{9.7-15}$$

This *Eucken formula* includes the monatomic formula (Eq. 9.7-14) as a special case, because $\hat{C}_p = \frac{5}{2}(R/M)$ for monatomic gases. Hirschfelder[4] has obtained a formula similar to that of Eucken by using multicomponent-mixture theory. Other theories, correlations, and empirical formulas are also available.[5,6]

Equation 9.7-15 provides a simple method for estimating the Prandtl number, defined in Eq. 9.2-8:

$$\text{Pr} = \frac{\hat{C}_p\mu}{k} = \frac{\tilde{C}_p}{\tilde{C}_p + \frac{5}{4}R} \quad \text{(polyatomic gas)} \tag{9.7-16}$$

This equation is fairly satisfactory for nonpolar polyatomic gases at low density, as can be seen in Table 9.7-1; it is less accurate for polar molecules.

---

[2]J. O. Hirschfelder, C. F. Curtiss, and R. B. Bird, *Molecular Theory of Gases and Liquids*, Wiley, New York, 2nd corrected printing (1964), p. 534.

[3]A. Eucken, *Physik. Z.*, **14**, 324–333 (1913); "Eucken" is pronounced "Oy-ken."

[4]J. O. Hirschfelder, *J. Chem. Phys.*, **26**, 274–281, 282–285 (1957).

[5]J. H. Ferziger and H. G. Kaper, *Mathematical Theory of Transport Processes in Gases*, North-Holland, Amsterdam (1972).

[6]B. E. Poling, J. M. Prausnitz, and J. P. O'Connell, *The Properties of Gases and Liquids*, McGraw-Hill, New York, 5th edition (2001).

Table 9.7-1. Predicted and Observed Values of the Prandtl Number for Gases at Atmospheric Pressure[a]

Gas	T(K)	$\hat{C}_p\mu/k$ from Eq. 9.7-16	$\hat{C}_p\mu/k$ from observed values of $\hat{C}_p$, μ, and k
Ne[b]	273.2	0.667	0.66
Ar[b]	273.2	0.667	0.67
H_2	90.6	0.68	0.68
	273.2	0.73	0.70
	673.2	0.74	0.65
N_2	273.2	0.74	0.73
O_2	273.2	0.74	0.74
Air	273.2	0.74	0.73
CO	273.2	0.74	0.76
NO	273.2	0.74	0.77
Cl_2	273.2	0.76	0.76
H_2O	373.2	0.77	0.94
	673.2	0.78	0.90
CO_2	273.2	0.78	0.78
SO_2	273.2	0.79	0.86
NH_3	273.2	0.77	0.85
C_2H_4	273.2	0.80	0.80
C_2H_6	273.2	0.83	0.77
$CHCl_3$	273.2	0.86	0.78
CCl_4	273.2	0.89	0.81

[a]Calculated from values given by M. Jakob, *Heat Transfer*, Wiley, New York (1949), pp. 75–76.
[b]J. O. Hirschfelder, C. F. Curtiss, and R. B. Bird, *Molecular Theory of Gases and Liquids*, Wiley, New York, corrected printing (1964), p. 16.

Thermal conductivities of gas mixtures can be calculated using an extension of the Chapman-Enskog theory.[2,7] A variety of empirical approaches can also be used to give satisfactory results for gas mixtures.[6,8,9]

EXAMPLE 9.7-1

Computation of the Thermal Conductivity of a Monatomic Gas at Low Density

Compute the thermal conductivity of Ne at 1 atm and 373.2 K.

SOLUTION

From Table D.1 the Lennard-Jones constants for neon are $\sigma = 2.789$ Å and $\varepsilon/\kappa = 35.7$ K, and its molecular weight M is 20.180 g/g-mol. Then, at 373.2 K, we have $\kappa T/\varepsilon = (373.2 \text{ K})/(35.7 \text{ K}) =$

[7]C. F. Curtiss and J. O. Hirschfelder, *J. Chem. Phys.*, **17**, 550–555 (1949).

[8]R. B. Bird, W. E. Stewart, and E. N. Lightfoot, *Transport Phenomena*, Wiley, New York, Revised Second Edition (2007).

[9]E. A. Mason and S. C. Saxena, *Physics of Fluids*, **1**, 361–369 (1958). Their method is an approximation to a more accurate method given by J. O. Hirschfelder, *J. Chem. Phys.*, **26**, 274–281, 282–285 (1957).

10.45. From Table D.2 we find that $\Omega_k = \Omega_\mu = 0.821$. Substitution into Eq. 9.7-13 gives

$$\begin{aligned} k &= 1.9891 \times 10^{-4} \frac{\text{cal} \cdot \text{Å}^2 \cdot \text{g}^{1/2}}{\text{cm} \cdot \text{s} \cdot \text{K}^{3/2} \cdot \text{g-mol}^{1/2}} \frac{\sqrt{T/M}}{\sigma^2 \Omega_k} \\ &= 1.9891 \times 10^{-4} \frac{\text{cal} \cdot \text{Å}^2 \cdot \text{g}^{1/2}}{\text{cm} \cdot \text{s} \cdot \text{K}^{3/2} \cdot \text{g-mol}^{1/2}} \frac{\sqrt{(373.2 \text{ K})/(20.180 \text{ g/g-mol})}}{(2.789 \text{ Å})^2(0.821)} \\ &= 1.338 \times 10^{-4} \frac{\text{cal}}{\text{cm} \cdot \text{s} \cdot \text{K}} \end{aligned} \tag{9.7-17}$$

A measured value of = 1.35×10^{-4} cal/cm · s · K has been reported[10] at 1 atm and 373.2 K.

EXAMPLE 9.7-2

Estimation of the Thermal Conductivity of a Polyatomic Gas at Low Density

Estimate the thermal conductivity of molecular oxygen at 300 K and low pressure.

SOLUTION

The molecular weight of O_2 is 32.0000 g/g-mol; its molar heat capacity $\tilde{C}_p$ at 300°C and low pressure is 7.019 cal/g-mol · K. From Table D.1 we find the Lennard-Jones parameters for molecular oxygen to be $\sigma = 3.433$ Å and $\varepsilon/\kappa = 113$ K. At 300 K, then, $\kappa T/\varepsilon = (300 \text{ K})/(113\text{K}) = 2.655$. From Table D.2, we find $\Omega_\mu = 1.074$. The viscosity, from Eq. 1.6-15, is

$$\begin{aligned} \mu &= 2.6693 \times 10^{-5} \frac{\text{g}^{1/2} \cdot \text{Å}^2 \cdot \text{g-mol}^{1/2}}{\text{cm} \cdot \text{s} \cdot \text{K}^{1/2}} \frac{\sqrt{MT}}{\sigma^2 \Omega_\mu} \\ &= 2.6693 \times 10^{-5} \frac{\text{g}^{1/2} \cdot \text{Å}^2 \cdot \text{g-mol}^{1/2}}{\text{cm} \cdot \text{s} \cdot \text{K}^{1/2}} \frac{\sqrt{(32.0000 \text{ g/g-mol})(300 \text{ K})}}{(3.433 \text{ Å})^2(1.074)} \\ &= 2.065 \times 10^{-4} \frac{\text{g}}{\text{cm} \cdot \text{s}} \end{aligned} \tag{9.7-18}$$

Then, from Eq. 9.7-15, the Eucken approximation to the thermal conductivity is

$$\begin{aligned} k &= \left(\hat{C}_p + \frac{5}{4}\frac{R}{M}\right)\mu = \left(\tilde{C}_p + \frac{5}{4}R\right)\frac{\mu}{M} \\ &= \left(\left(7.019 + \frac{5}{4} \cdot 1.987\right) \text{cal/g-mol} \cdot \text{K}\right) \frac{(2.065 \times 10^{-4} \text{ g/cm} \cdot \text{s})}{(32.0000 \text{ g/g-mol})} \\ &= 6.14 \times 10^{-5} \frac{\text{cal}}{\text{cm} \cdot \text{s} \cdot \text{K}} \left(\frac{\text{W/m} \cdot \text{K}}{2.3901 \times 10^{-3} \text{ cal/cm} \cdot \text{s} \cdot \text{K}}\right) \\ &= 0.0257 \frac{\text{W}}{\text{m} \cdot \text{K}} \end{aligned} \tag{9.7-19}$$

This compares favorably with the experimental value of = 0.02657 W/m · K in Table 9.5-1.

§9.8 THERMAL CONDUCTIVITY OF LIQUIDS

A very detailed kinetic theory for the thermal conductivity of monatomic liquids was developed over a half-century ago,[1] but it has not yet been possible to implement it for practical calculations. As a result we have to use rough theories or empirical estimation methods.[2]

[10] W. G. Kannuluik and E. H. Carman, *Proc. Phys. Soc.* (London), **65B**, 701–704 (1952).

[1] J. H. Irving and J. G. Kirkwood, *J. Chem. Phys.*, **18**, 817–829 (1950). This theory has been extended to polymeric liquids by C. F. Curtiss and R. B. Bird, *J. Chem. Phys.*, **107**, 5254–5267 (1997).

[2] B. E. Poling, J. M. Prausnitz, and J. P. O'Connell, *The Properties of Gases and Liquids*, McGraw-Hill, New York, 5th edition (2001); L. Riedel, *Chemie-Ing.-Techn.*, **27**, 209–213 (1955).

We choose to discuss here Bridgman's simple theory[3] of energy transport in pure liquids. He assumed that the molecules are arranged in a cubic lattice, with a center-to-center spacing given by $(\tilde{V}/\tilde{N})^{1/3}$, in which $\tilde{V}/\tilde{N}$ is the volume per molecule. He further assumed energy to be transferred from one lattice plane to the next at the sonic velocity v_s for the given fluid. The development is based on a reinterpretation of Eq. 9.7-11 of the rigid-sphere gas theory:

$$k = \frac{1}{3}\rho\hat{C}_V\bar{u}\lambda = \rho\hat{C}_V\overline{|u_y|}a \tag{9.8-1}$$

The heat capacity at constant volume of a monatomic liquid is about the same as for a solid at high temperature, which is given by the Dulong and Petit formula[4] $\hat{C}_V = 3(\kappa/m)$. The mean molecular speed in the y direction, $\overline{|u_y|}$, is replaced by the sonic velocity v_s. The distance a that the energy travels between two successive collisions is taken to be the lattice spacing $(\tilde{V}/\tilde{N})^{1/3}$. Making these substitutions in Eq. 9.8-1 gives

$$k = 3(\tilde{N}/\tilde{V})^{2/3}\kappa v_s \tag{9.8-2}$$

which is *Bridgman's equation*. Experimental data show good agreement with Eq. 9.8-2, even for polyatomic liquids, but the numerical coefficient is somewhat too high. Better agreement is obtained if the coefficient is changed to 2.80:

$$k = 2.80(\tilde{N}/\tilde{V})^{2/3}\kappa v_s \tag{9.8-3}$$ [5]

This equation is limited to densities well above the critical density, because of the tacit assumption that each molecule oscillates in a "cage" formed by its nearest neighbors. The success of this equation for polyatomic fluids seems to imply that the energy transfer in collisions of polyatomic molecules is incomplete, since the heat capacity used here, $\hat{C}_V = 3(\kappa/m)$, is less than the heat capacities of polyatomic liquids.

The velocity of low-frequency sound is given by

$$v_s = \sqrt{\frac{C_p}{C_V}\left(\frac{\partial p}{\partial \rho}\right)_T} \tag{9.8-4}$$

The quantity $(\partial p/\partial \rho)_T$ may be obtained from isothermal compressibility measurements, or from an equation of state, and (C_p/C_V) is very nearly unity for liquids, except near the critical point.

EXAMPLE 9.8-1

Prediction of the Thermal Conductivity of a Liquid

The density of liquid CCl_4 at 20°C and 1 atm is 1.595 g/cm³, and its isothermal compressibility $(1/\rho)(\partial\rho/\partial p)_T$ is 90.7×10^{-6} atm⁻¹. What is its thermal conductivity?

SOLUTION

First compute

$$\begin{aligned}\left(\frac{\partial p}{\partial \rho}\right)_T &= \frac{1}{\rho(1/\rho)(\partial\rho/\partial p)_T} = \frac{1}{(1.595\ \mathrm{g/cm^3})(90.7\times10^{-6}\ \mathrm{atm^{-1}})}\\ &= 6.91\times10^{3}\frac{\mathrm{atm\cdot cm^3}}{\mathrm{g}}\left(\frac{1.0133\times10^{6}\ \mathrm{g/cm\cdot s^2}}{1\ \mathrm{atm}}\right)\\ &= 7.00\times10^{9}\frac{\mathrm{cm^2}}{\mathrm{s^2}}\end{aligned} \tag{9.8-5}$$

---

[3]P. W. Bridgman, *Proc. Am. Acad. Arts and Sci.*, **59**, 141–169 (1923). Bridgman's equation is often misquoted, because he gave it in terms of a little-known gas constant equal to $\frac{3}{2}\kappa$.

[4]This empirical equation has been justified, and extended, by A. Einstein [*Ann. Phys.* [4], **22**, 180–190 (1907)] and P. Debye [*Ann. Phys.*, [4] **39**, 789–839 (1912)].

[5]Equation (9.8-3) is in approximate agreement with a formula derived by R. E. Powell, W. E. Roseveare, and H. Eyring, *Ind. Eng. Chem.*, **33**, 430–435 (1941).

Assuming that $C_p/C_V = 1$, we get from Eq. 9.8-4

$$v_s = \sqrt{(1.0)\left(7.00 \times 10^9 \frac{\text{cm}^2}{\text{s}^2}\right)} = 8.37 \times 10^4 \frac{\text{cm}}{\text{s}} \tag{9.8-6}$$

The molar volume is $\tilde{V} = M/\rho = (153.84 \text{ g/g-mol})/(1.595 \text{ g/cm}^3) = 96.5 \text{ cm}^3/\text{g-mol}$. Substitution of these values in Eq. 9.8-3 gives

$$\begin{aligned} k &= 2.80(\tilde{N}/\tilde{V})^{2/3}\kappa v_s \\ &= 2.80\left(\frac{6.023 \times 10^{23} \text{ 1/g-mol}}{96.5 \text{ cm}^3/\text{g-mol}}\right)^{2/3}\left(1.3805 \times 10^{-16}\frac{\text{erg}}{\text{K}}\right)\left(8.37 \times 10^4 \frac{\text{cm}}{\text{s}}\right) \\ &= 1.10 \times 10^4 \frac{\text{erg}}{\text{cm}\cdot\text{s}\cdot\text{K}}\left(\frac{10^{-7}\text{J}}{1 \text{ erg}}\right)\left(\frac{1 \text{ W}}{1 \text{ J/s}}\right)\left(\frac{100 \text{ cm}}{1 \text{ m}}\right) \\ &= 0.110 \frac{\text{W}}{\text{m}\cdot\text{K}} \end{aligned} \tag{9.8-7}$$

The experimental value given in Table 9.5-2 is 0.101 W/m · K.

§9.9 THERMAL CONDUCTIVITY OF SOLIDS

Thermal conductivities of solids have to be measured experimentally, since they depend on many factors that are difficult to measure or predict.[1] In crystalline materials, the phase and crystallite size are important; in amorphous solids the degree of molecular orientation has a considerable effect. In porous solids, the thermal conductivity is strongly dependent on the void fraction, the pore size, and the fluid contained in the pores. A detailed discussion of thermal conductivity of solids has been given by Jakob.[2]

In general, metals are better heat conductors than nonmetals, and crystalline materials conduct heat more readily than amorphous materials. Dry porous solids are very poor heat conductors and are therefore excellent for thermal insulation. The conductivities of most pure metals decrease with increasing temperature, whereas the conductivities of nonmetals increase; alloys show intermediate behavior. Perhaps the most useful of the rules of thumb is that thermal and electrical conductivity go hand in hand.

For pure metals, the thermal conductivity k and the electrical conductivity k_e are related approximately[3] as follows:

$$\frac{k}{k_e T} = L = \text{constant} \tag{9.9-1}$$

This is the *Wiedemann-Franz-Lorenz equation*; this equation can also be explained theoretically (see Problem 9A.6). The "Lorenz number" L is about 22 to 29×10^{-9} volt^2/K^2 for pure metals at 0°C and changes but little with temperature above 0°C, increases of 10% to 20% per 1000°C being typical. At very low temperatures (−269.4°C for mercury) metals become superconductors of electricity, but not of heat, and L thus varies strongly with temperature near the superconducting region. Equation 9.9-1 is of limited use for alloys, since L varies strongly with composition and, in some cases, with temperature.

---

[1]A. Goldsmith, T. E. Waterman, and H. J. Hirschhorn, eds., *Handbook of Thermophysical Properties of Solids*, Macmillan, New York (1961).

[2]M. Jakob, *Heat Transfer*, Wiley, New York (1949), Vol. **I**, Chapter 6. See also W. H. Rohsenow, J. P. Hartnett, and Y. I. Cho, eds., *Handbook of Heat Transfer*, McGraw-Hill, New York (1998).

[3]G. Wiedemann and R. Franz, *Ann. Phys. u. Chemie*, **89**, 497–531 (1853); L. Lorenz, *Poggendorff's Annalen*, **147**, 429–452 (1872).

The success of Eq. 9.9-1 for pure metals is due to the fact that free electrons are the major heat carriers in pure metals. The equation is not suitable for nonmetals, in which the concentration of free electrons is so low that energy transport by molecular motion predominates.

§9.10 EFFECTIVE THERMAL CONDUCTIVITY OF COMPOSITE SOLIDS

Up to this point we have discussed homogeneous materials. Now we turn our attention briefly to the thermal conductivity of two-phase solids—one solid phase dispersed in a second solid phase, or solids containing pores, such as granular materials, sintered metals, and plastic foams. A complete description of the heat transport through such materials is extremely complicated. However, for steady conduction these materials can be regarded as homogeneous materials with an *effective thermal conductivity* k_{eff}, and the temperature and heat-flux components are reinterpreted as the analogous quantities averaged over a volume that is large with respect to the scale of the heterogeneity, but small with respect to the overall dimensions of the heat conduction system.

The first major contribution to the estimation of the conductivity of heterogeneous solids was by Maxwell.[1] He considered a material made of spheres of thermal conductivity k_1 embedded in a continuous solid phase with thermal conductivity k_0. The volume fraction ϕ of embedded spheres is taken to be sufficiently small that the spheres do not "interact" thermally; that is, one needs to consider only the thermal conduction in a large medium containing only one embedded sphere. Then by means of a surprisingly simple derivation Maxwell showed that for *small volume fraction* ϕ

$$\frac{k_{\text{eff}}}{k_0} = 1 + \frac{3\phi}{\left(\dfrac{k_1 + 2k_0}{k_1 - k_0}\right) - \phi} \tag{9.10-1}$$

For *complex nonspherical inclusions*, often encountered in practice, no exact treatment is possible, but some approximate relations are available.[2,3,4] For simple unconsolidated granular beds the following expression has proven successful:

$$\frac{k_{\text{eff}}}{k_0} = \frac{(1-\phi) + \alpha\phi(k_1/k_0)}{(1-\phi) + \alpha\phi} \tag{9.10-2}$$

in which

$$\alpha = \frac{1}{3}\sum_{k=1}^{3}\left[1 + \left(\frac{k_1}{k_0} - 1\right)g_k\right]^{-1} \tag{9.10-3}$$

The g_k are "shape factors" for the granules of the medium,[5] and they must satisfy $g_1 + g_2 + g_3 = 1$. For spheres $g_1 = g_2 = g_3 = \frac{1}{3}$, and Eq. 9.10-2 reduces to Eq. 9.10-1. For unconsolidated soils[3] $g_1 = g_2 = \frac{1}{8}$ and $g_3 = \frac{3}{4}$. The structure of consolidated porous beds—for

---

[1]Maxwell's derivation was for electrical conductivity, but the same arguments apply for thermal conductivity. See J. C. Maxwell, *A Treatise on Electricity and Magnetism*, Oxford University Press, 3rd edition (1891, reprinted 1998), Vol. **1**, §314; H. S. Carslaw and J. C. Jaeger, *Conduction of Heat in Solids*, Clarendon Press, Oxford, 2nd edition (1959), p. 428.

[2]V. I. Odelevskii, *J. Tech. Phys.* (USSR), **24**, 667 and 697 (1954); F. Euler, *J. Appl. Phys.*, **28**, 1342–1346 (1957).

[3]D. A. de Vries, Mededelingen van de Landbouwhogeschool te Wageningen, (1952); D. A. de Vries, Chapter 7 in *Physics of Plant Environment*, W. R. van Wijk, ed., Wiley, New York (1963).

[4]W. Woodside and J. H. Messmer, *J. Appl. Phys.*, **32**, 1688–1699, 1699–1706 (1961).

[5]A. L. Loeb, *J. Amer. Ceramic Soc.*, **37**, 96–99 (1954).

example, sandstones—is considerably more complex. Some success is claimed for predicting the effective conductivity of such substances,[2,4,6] but the generality of the methods is not yet known.

§9.11 CONCLUDING COMMENTS

This chapter is concerned with the various mechanisms for energy transport and the corresponding definitions of the energy fluxes: the convective energy-flux vector $\mathbf{q}^{(c)}$, the conductive heat-flux vector $\mathbf{q}$, the work-flux vector $\mathbf{w}$, and the sum of these fluxes, which is the total energy-flux vector $\mathbf{e} = \mathbf{q}^{(c)} + \mathbf{q} + \mathbf{w}$. These quantities are summarized in Table 9.4-1, and mastery of the definitions and meanings of all symbols in that table is essential for understanding Part II of this book.

It is also necessary to be acquainted with the dimensions and various units used for the thermal conductivity and the thermal diffusivity. In this connection, the tables of unit conversions in Appendix E will prove to be very helpful. Also, in this chapter several new dimensionless groups have been introduced, namely, the Prandtl number and the Péclet number.

For solving heat conduction and heat-transfer problems, values of the thermal conductivity are needed. It is always preferable to use experimental values for the thermal conductivity, but in the absence of these, familiarity with available theories and empiricisms is necessary. Therefore, one has to be familiar with the published literature on the physical properties encountered in practical problems. The last several sections of this chapter are intended to give you a slight acquaintance with what is available.

QUESTIONS FOR DISCUSSION

1. Define and give the dimensions of: thermal conductivity k, thermal diffusivity α, heat capacity $\hat{C}_p$, conductive heat-flux vector $\mathbf{q}$, the work-flux vector $\mathbf{w}$, and the total energy-flux vector $\mathbf{e}$. Use for the dimensions M = mass, L = length, T = temperature, t = time.
2. Compare the orders of magnitude of the thermal conductivities of gases, liquids, and solids.
3. In what ways are Newton's law of viscosity and Fourier's law of heat conduction similar? Dissimilar?
4. Are gas viscosities and thermal conductivities related? If so, how?
5. Compare the temperature dependence of the thermal conductivities of gases, liquids, and solids.
6. Compare the orders of magnitudes of Prandtl numbers for gases and liquids.
7. Are the thermal conductivities of gaseous Ne^{20} and Ne^{22} the same?
8. Is the relation $\tilde{C}_p - \tilde{C}_V = R$ true only for ideal gases, or is it also true for liquids? If it is not true for liquids, what formula should be used?
9. What is the kinetic energy flux in the axial direction for the laminar flow of a Newtonian liquid in a circular tube (Poiseuille flow)?
10. What is $\mathbf{w} = [\boldsymbol{\pi} \cdot \mathbf{v}] = p\mathbf{v} + [\boldsymbol{\tau} \cdot \mathbf{v}]$ for Poiseuille flow?

PROBLEMS

9A.1 Prediction of thermal conductivities of gases at low density.

(a) Compute the thermal conductivity of argon at 100°C and atmospheric pressure, using the Chapman-Enskog theory and the Lennard-Jones constants derived from viscosity data. Compare your result with the observed value[1] of 506×10^{-7} cal/cm · s · K.

---

[6] Sh. N. Plyat, *Soviet Physics JETP*, **2**, 2588–2589 (1957).

[1] W. G. Kannuluik and E. H. Carman, *Proc. Phys. Soc. (London)*, **65B**, 701–704 (1952).

(b) Compute the thermal conductivities of NO and CH_4 at 300 K and atmospheric pressure from the following data for these conditions:

	$\mu \times 10^7$ (g/cm · s)	$\tilde{C}_p$ (cal/g-mol · K)
NO	1929	7.15
CH_4	1116	8.55

Compare your results with the experimental values gives in Table 9.5-1.

9A.2 Computation of the Prandtl numbers for gases at low density.

(a) By using the Eucken formula and experimental heat capacity data, estimate the Prandtl number at 1 atm and 300 K for each of the gases listed in the table.

(b) For the same gases, compute the Prandtl number directly by substituting the following values of the physical properties into the defining formula $\text{Pr} = \hat{C}_p\mu/k$, and compare the values with the results obtained in (a). All properties are given at low pressure and 300 K.

Gas[a]	$\hat{C}_p \times 10^{-3}$ (J/kg · K)	$\mu \times 10^5$ (Pa · s)	k (W/m · K)
He	5.193	1.995	0.1546
Ar	0.5204	2.278	0.01784
H_2	14.28	0.8944	0.1789
Air	1.001	1.854	0.02614
CO_2	0.8484	1.506	0.01661
H_2O	1.864	1.041	0.02250

[a] The entries in this table were prepared from functions provided by T. E. Daubert, R. P. Danner, H. M. Sibul, C. C. Stebbins, J. L. Oscarson, R. L. Rowley, W. V. Wilding, M. E. Adams, T. L. Marshall, and N. A. Zundel, *DIPPR® Data Compilation of Pure Compound Properties*, Design Institute for Physical Property Data®, AIChE, New York, NY (2000).

9A.3 Estimation of the thermal conductivity of a dense gas. Predict the thermal conductivity of methane at 110.4 atm and 127°F by the following methods:

(a) Use Fig. 9.6-1. Obtain the necessary critical properties from Appendix D.

(b) Use the Eucken formula to get the thermal conductivity at 127°F and low pressure. Then apply a pressure correction by using Fig. 9.6-1. The experimental value[2] is 0.0282 Btu/hr · ft · °F.

Answer: **(a)** 0.0294 Btu/hr · ft · °F.

9A.4 Dimensions of thermal quantities. Verify that the following equations are dimensionally correct: Eq. 9.7-12, Eq. 9.4-2, and Eq. 9.4-4. The table of notation, just before the author index, will be helpful.

9A.5 Estimation of the thermal conductivity of a pure liquid. Predict the thermal conductivity of liquid H_2O at 40°C and 40 bars pressure (1 bar = 10^6 dyn/cm^2). The isothermal compressibility, $(1/\rho)(\partial\rho/\partial p)_T$, is 38×10^{-6} bar^{-1} and the density is 0.9938 g/cm^3. Assume that $\tilde{C}_p = \tilde{C}_V$.

Answer: 0.375 Btu/hr · ft · °F

[2] J. M. Lenoir, W. A. Junk, and E. W. Comings, *Chem. Engr. Prog.* **49**, 539–542 (1953).

9A.6 Calculation of the Lorenz number.

(a) Application of kinetic theory to the "electron gas" in a metal[3] gives for the Lorenz number

$$L = \frac{\pi^2}{3}\left(\frac{\kappa}{e}\right)^2 \tag{9A.6-1}$$

in which κ is the Boltzmann constant and e is the charge on the electron. Compute L in the units given under Eq. 9.9-1.

(b) The electrical resistivity, $1/k_e$, of copper at 20°C is 1.72×10^{-6} ohm · cm. Estimate its thermal conductivity in W/m · K using Eq. 9A.6-1, and compare your result with the experimental value given in Table 9.5-4.

Answers: **(a)** 2.44×10^{-8} volt^2/K^2; **(b)** 416 W/m · K

9A.7 Corroboration of the Wiedemann-Franz-Lorenz law. Given the following experimental data at 20°C for pure metals, compute the corresponding values of the Lorenz number, L, defined in Eq. 9.9-1.

Metal	$(1/k_e)$(ohm · cm)	k (cal/cm · s · K)
Na	4.6×10^{-6}	0.317
Ni	6.9×10^{-6}	0.140
Cu	1.69×10^{-6}	0.92
Al	2.62×10^{-6}	0.50

9A.8 Thermal conductivity and Prandtl number of a polyatomic gas.

(a) Estimate the thermal conductivity of CH_4 at 1500 K and 1.37 atm. The molar heat capacity at constant pressure[4] at 1500 K is 20.71 cal/g-mol · K.

(b) What is the Prandtl number at the same pressure and temperature?

Answers: **(a)** 5.06×10^{-4} cal/cm · s · K; **(b)** 0.89

9A.9 Thermal conductivity of gaseous chlorine. Use Eq. 9.7-15 to calculate the thermal conductivity of gaseous chlorine. To do this you will need to use Eq. 1.6-14 to estimate the viscosity, and also the following values of the heat capacity:

T (K)	200	300	400	500	600
$\tilde{C}_p$ (cal/g-mol · K)	(8.06)	8.12	8.44	8.62	8.74

Check to see how well the calculated values agree with the following experimental thermal conductivity data:[5]

[3] P. Drude, *Ann. Phys.*, **1**, 566–613 (1900); J. E. Mayer and M. G. Mayer, *Statistical Mechanics*, Wiley, New York (1946), p. 412; **Maria Goeppert Mayer** (1906–1972) received a Ph.D. in Physics from the University of Göttingen. She won the Nobel Prize for Physics in 1963, for the shell model of the atomic nucleus.

[4] O. A. Hougen, K. M. Watson, and R. A. Ragatz, *Chemical Process Principles*, Vol. **1**, Wiley, New York (1954), p. 253.

[5] Interpolated from data of E. U. Frank, *Z. Elektrochem.*, **55,** 636 (1951), as reported in *Nouveau Traité de Chimie Minerale*, P. Pascal, ed., Masson et Cie, Paris (1960), pp. 158–159.

T (K)	p (mm Hg)	$k \times 10^5$ (cal/cm · s · K)
198	50	1.31 ± 0.03
275	220	1.90 ± 0.02
276	120	1.93 ± 0.01
	220	1.92 ± 0.01
363	100	2.62 ± 0.02
	200	2.61 ± 0.02
395	210	3.04 ± 0.02
453	150	3.53 ± 0.03
	250	3.42 ± 0.02
495	250	3.72 ± 0.07
553	100	4.14 ± 0.04
583	170	4.43 ± 0.04
	210	4.45 ± 0.08
676	150	5.07 ± 0.10
	250	4.90 ± 0.03

9A.10 Thermal conductivity of quartz sand. A typical sample of quartz sand has the following properties at 20°C:

Component	Volume Fraction	k_i (cal/cm · s · K)
$i = 1$: Silica	0.510	20.4×10^{-3}
$i = 2$: Feldspar	0.063	7.0×10^{-3}
Continuous phase ($i = 0$) is one of the following:		
(i) Water	0.427	1.42×10^{-3}
(ii) Air	0.427	0.0615×10^{-3}

Estimate the effective thermal conductivity of the sand (i) when it is water-saturated, and (ii) when it is completely dry.

(a) Use the following generalization of Eqs. 9.10-2 and 9.10-3:

$$\frac{k_{\text{eff}}}{k_0} = \frac{\sum_{i=0}^{N} \alpha_i (k_i/k_0)\phi_i}{\sum_{i=0}^{N} \alpha_i \phi_i} \qquad \text{(9A.10-1)}$$

$$\alpha_i = \frac{1}{3}\sum_{j=1}^{3}\left[1 + \left(\frac{k_i}{k_0} - 1\right) g_j\right]^{-1} \qquad \text{(9A.10-2)}$$

Here N is the number of solid phases. Compare the prediction for spheres $\left(g_1 = g_2 = g_3 = \frac{1}{3}\right)$ with the recommendation of de Vries $\left(g_1 = g_2 = \frac{1}{8}; g_3 = \frac{3}{4}\right)$. The latter g_i values closely approximate the fitted ones[6] for the present sample. The right-hand member of Eq. 9A.10-1 is to be multiplied by 1.25 for completely dry sand.[6]

(b) Use Eq. 9.10-1 with $k_1 = 18.9 \times 10^{-3}$ cal/cm · s · K, which is the volume-average thermal conductivity of the two solids. Observed values, accurate within about 3%, are 6.2 and 0.58 × 10^{-3} cal/cm · s · K for wet and dry sand, respectively.[6]

Answers in cal/cm · s · K for wet and dry sand, respectively:

(a) Eq. 9A.10-1 gives $k_{\text{eff}} = 4.9 \times 10^{-3}$ and 0.38×10^{-3} with $g_1 = g_2 = g_3 = \frac{1}{3}$ vs. 6.2 × 10^{-3} and 0.64×10^{-3} with $g_1 = g_2 = \frac{1}{8}; g_3 = \frac{3}{4}$.

(b) Eq. 9.10-1 gives $k_{\text{eff}} = 5.1 \times 10^{-3}$ and 0.30×10^{-3}.

9A.11 Calculation of molecular diameters from transport properties.

(a) Determine the molecular diameter d for argon from Eq. 1.6-9 and the experimental viscosity given in Problem 9A.2.

(b) Repeat part (a), but using Eq. 9.7-12 and the measured thermal conductivity in Problem 9A.2. Compare this result with the value obtained in (a).

(c) Calculate and compare the values of the Lennard-Jones collision diameter σ from the same experimental data used in (a) and (b), using ε/κ from Table D.1.

(d) What can be concluded from the above calculations?

Answer: **(a)** 2.95 Å; **(b)** 1.88 Å; **(c)** 3.415 Å from Eq. 1.6-14, 3.409 Å from Eq. 9.7-13

9B.1 Enthalpy for an ideal monatomic gas. Show that for an ideal monatomic gas Eq. 9.4-4 becomes

$$\hat{H} - \hat{H}^{\circ} = \frac{5}{2}\frac{R}{M}(T - T^{\circ}) \tag{9B.1-1}$$

[6]The behavior of partially wetted soil has been treated by D. A. de Vries, Chapter 7 in *Physics and Plant Environment*, W. R. van Wijk, ed., Wiley, New York (1963).

Chapter 10

Shell Energy Balances and Temperature Distributions in Solids and Laminar Flow

In Chapter 2 it was shown how certain simple viscous flow problems can be solved by applying conservation of momentum to a shell that is thin in the direction in which the velocity varies. The main steps are: (i) a momentum balance made over the shell leads to a first-order differential equation for the momentum-flux distribution; (ii) insertion of an expression for the momentum flux in terms of the fluid velocity (which contains Newton's law of viscosity) gives a differential equation for the fluid velocity as a function of position. These differential equations can be solved using boundary conditions, which specify the velocity or momentum flux at the bounding surfaces.

In this chapter we show how a number of energy transport problems are solved by an analogous procedure: (i) an energy balance made over a shell that is thin in the direction in which the temperature varies leads to a first-order differential equation for the energy-flux distribution; (ii) substitution of an expression for the total energy flux in terms of the temperature (which contains Fourier's law of heat conduction) gives a differential equation for the temperature as a function of position. These differential equations can be solved using boundary conditions for the temperature or heat flux at the bounding surfaces.

It should be clear from the similar wording of the foregoing two paragraphs that the mathematical methods used in this chapter are the same as those introduced in Chapter 2—only the notation and terminology are different. However, we will encounter here a number of physical phenomena that have no counterpart in Chapter 2.

After a brief introduction to the shell energy balance in §10.1, we analyze heat conduction in a series of uncomplicated systems. Although these examples are somewhat idealized, the results find application in numerous standard engineering calculations. The problems were chosen to introduce the beginner to a number of important physical concepts associated with the heat-transfer field. In addition, they serve to show how to use a variety of boundary conditions and to illustrate problem solving in different coordinate systems. In §10.2 to §10.5 four problems on heat conduction in solids are given. In §10.6 to §10.8, three problems with heat sources are presented. Finally, in §10.9 and §10.10, we analyze two limiting cases of heat transfer in moving fluids: forced convection and free convection. The study of these topics paves the way for the general equations of change in Chapter 11.

§10.1 SHELL ENERGY BALANCES; BOUNDARY CONDITIONS

The problems discussed in this chapter are set up by means of shell energy balances. We select a shell whose surfaces are parallel or perpendicular to the direction of energy flow, and that is thin in the direction that the temperature varies. We then write for this system a statement of the law of conservation of energy. For *steady-state* (i.e., time-independent) systems, we write:

$$\left\{\begin{matrix}\text{Total rate}\\ \text{of energy}\\ \text{transported}\\ \textit{in}\end{matrix}\right\} - \left\{\begin{matrix}\text{Total rate}\\ \text{of energy}\\ \text{transported}\\ \textit{out}\end{matrix}\right\} + \left\{\begin{matrix}\text{Rate of work}\\ \text{done } \textit{on}\\ \text{system by}\\ \text{external forces}\end{matrix}\right\} + \left\{\begin{matrix}\text{Rate of}\\ \text{energy}\\ \text{production}\end{matrix}\right\} = 0 \quad (10.1\text{-}1)$$

The rates of energy transported in and out in the first two terms will be expressed by the appropriate components of the total energy-flux vector $\mathbf{e}$ ([=] energy/time · area) multiplied by the relevant areas. The total energy-flux vector $\mathbf{e}$ given in §9.4 includes all transport mechanisms: the *convective energy-flux vector* $\mathbf{q}^{(c)}$ and the two contributions to the *molecular energy flux*, the *conductive heat-flux vector* $\mathbf{q}$ and the *work-flux vector* $\mathbf{w}$. In setting up problems here and in the next chapter, we will use the $\mathbf{e}$ vector along with the expression for the enthalpy in Eq. 9.4-4, often employing the simplifications of Eqs. 9.4-5 or 9.4-6. Note that in nonflowing systems (for which $\mathbf{v}$ is zero) the $\mathbf{e}$ vector reduces to the conductive heat-flux vector $\mathbf{q}$, which is given by Fourier's law.

The *energy production* terms in Eq. 10.1-1 include (i) the degradation of electrical energy into heat, (ii) the heat produced by slowing down of neutrons and nuclear fragments liberated in the fission process, (iii) the heat produced in chemical reactions, and (iv) the heat produced by viscous dissipation. The chemical-reaction heat source will be discussed further in Chapter 19.

Equation 10.1-1 is a statement of the first law of thermodynamics, written for an "open" system at steady-state conditions. In Chapter 11 this same statement—extended to unsteady-state systems—will be written down as an equation of change for energy.

After Eq. 10.1-1 has been written for a thin shell of material, the thickness of the slab or shell is allowed to approach zero. This procedure leads ultimately to an expression for the temperature distribution containing constants of integration, which we evaluate by use of boundary conditions. The most common types of boundary conditions are:

a. The temperature may be specified at a surface.

b. The heat flux normal to a surface may be given (this is equivalent to specifying the temperature gradient).

c. At interfaces the continuity of temperature and the heat flux normal to the interface are required.

d. At a solid-fluid interface, the normal heat-flux component may be related to the difference between the solid-surface temperature $T|_s$ and the "bulk" fluid temperature T_b:

$$q_n = h(T|_s - T_b) \tag{10.1-2}$$

where the normal vector is directed into the fluid. This relation is referred to as *Newton's law of cooling*. It is not really a "law" but rather the defining equation for h, which is called the *heat-transfer coefficient*. Chapter 14 deals with methods for estimating heat-transfer coefficients.

All four types of boundary conditions are encountered in this chapter. Still other kinds of boundary conditions are possible, and they will be introduced as needed.

§10.2 HEAT CONDUCTION IN A STEAM PIPE

In this section we will obtain the radial temperature distribution in the wall of a pipe through which steam is flowing. The temperature at the inner pipe surface at $r = \kappa R$ is fixed at the steam temperature T_s. For the condition at the outer pipe surface at $r = R$, we will consider three different types of boundary conditions. These different cases, illustrated in Fig. 10.2-1(*a*)–(*c*), are considered separately below. For each situation, the goals are to obtain the temperature distribution $T(r)$ for $\kappa R \leq r \leq R$, as well as the rate of heat loss from the pipe.

a. Temperature at the outer surface of the pipe is specified.

Here the temperature at the outer surface of the pipe $r = R$, is maintained at T_o, as illustrated in Fig. 10.2-1(*a*). To find the temperature distribution within the pipe wall, $T(r)$, we start by writing a steady-state energy balance over a cylindrical shell of thickness Δr and length L within the pipe wall. The shell is thin in only the r direction because we

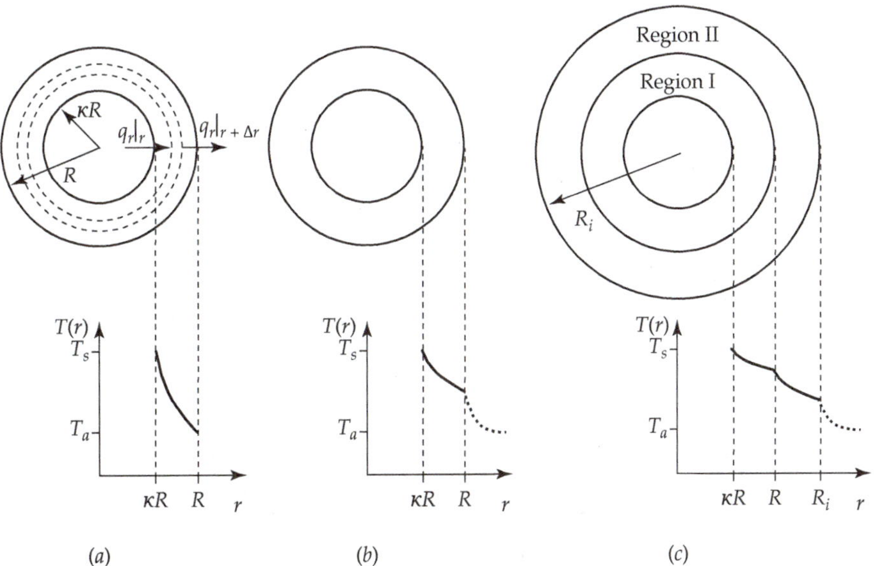

Fig. 10.2-1. Radial temperature distribution in the wall of a steam pipe, (*a*) when the temperature of the outer surface is specified; (*b*) when heat is lost to the ambient air (described by Newton's law of cooling); (*c*) when the pipe is surrounded by a layer of insulating material and heat is lost to the ambient air (described by Newton's law of cooling).

assume that the temperature depends only on r. Since energy is being transported only by conduction ($\mathbf{v} = 0$), the total energy flux $\mathbf{e}$ reduces to the conductive flux $\mathbf{q}$. The rates of energy transport in and out of the shell will thus consist of the appropriate components of the conductive heat-flux vector $\mathbf{q}$ ([=] energy/time · area), multiplied by the relevant areas. Because there are no sources of energy production, and no work done on the system by external forces, the only contributions to the energy balance are

Rate of energy transported in across the cylindrical surface at r:

$$q_r|_r \cdot 2\pi rL = (2\pi rLq_r)|_r \tag{10.2-1}$$

Rate of energy transported out across the cylindrical surface at $r + \Delta r$:

$$q_r|_{r+\Delta r} \cdot 2\pi(r + \Delta r)L = (2\pi rLq_r)|_{r+\Delta r} \tag{10.2-2}$$

The notation q_r means "heat flux in the r direction," and $(\cdots)|_{r+\Delta r}$ means "evaluated at $r + \Delta r$." Note that we take "in" and "out" to be in the positive r direction.

The steady-state energy balance can then be written as

$$(2\pi rLq_r)|_r - (2\pi rLq_r)|_{r+\Delta r} = 0 \tag{10.2-3}$$

If we now divide the equation by $2\pi L\Delta r$ and then take the limit as $\Delta r \to 0$, we get

$$\lim_{\Delta r \to 0} \frac{(rq_r)|_{r+\Delta r} - (rq_r)|_r}{\Delta r} = \frac{d}{dr}(rq_r) = 0 \tag{10.2-4}$$

where use has been made of the definition of the first derivative.

Integration with respect to r gives

$$rq_r = C_1' \tag{10.2-5}$$

where C_1' is the constant of integration. Insertion of Fourier's law of heat conduction $q_r = -k\partial T/\partial r$ gives us then

$$-kr\frac{dT}{dr} = C_1' = kC_1 \tag{10.2-6}$$

where we have chosen to replace C_1' by kC_1 so that the k will drop out. Another integration with respect to r then gives

$$T(r) = C_1 \ln r + C_2 \tag{10.2-7}$$

To determine the constants of integration, we must specify the boundary conditions. These have been discussed above, and are as follows:

B. C. 1: $$\text{at } r = \kappa R, \quad T = T_s \tag{10.2-8}$$

B. C. 2: $$\text{at } r = R, \quad T = T_o \tag{10.2-9}$$

Application of these boundary conditions to Eq. 10.2-7 gives the two equations

B. C. 1: $$T_s = C_1 \ln \kappa R + C_2 \tag{10.2-10}$$

B. C. 2: $$T_o = C_1 \ln R + C_2 \tag{10.2-11}$$

From these relations we get $C_1 = (T_s - T_o)/\ln \kappa$ and $C_2 = T_o - (T_s - T_o) \ln R/\ln \kappa$, and thus, the temperature profile is

$$\boxed{\frac{T(r) - T_o}{T_s - T_o} = \frac{\ln(r/R)}{\ln \kappa}} \tag{10.2-12}$$

Note that both sides of this equation are dimensionless, which suggests some natural dimensionless variables. In fact, this problem can be solved in terms of dimensionless variables from the very beginning (see Problem 10B.21).

Finally we get the rate of heat loss from a length L of pipe

$$Q = (2\pi RL)q_r|_{r=R} = (2\pi RL)\left(-k\frac{dT}{dr}\bigg|_{r=R}\right)$$
$$= (2\pi RL)(-k)\frac{T_s - T_o}{\ln\kappa}\frac{1}{r}\bigg|_{r=R} = \frac{2\pi kL(T_s - T_o)}{\ln(1/\kappa)} \tag{10.2-13}$$

Here Eq. 10.2-12 has been used to evaluate the temperature gradient at the outside surface of the pipe.

b. Heat is lost to the surroundings from the outer surface of the pipe according to Newton's law of cooling

Here we consider a different boundary condition at the outer surface, namely, Newton's law of cooling,

B. C. 2: $$\text{at } r = R, \quad q_r = -k\frac{\partial T}{\partial r} = h(T - T_a) \tag{10.2-14}$$

in place of the known, fixed temperature T_o. Here, T_a is the air temperature far from the pipe surface (the "ambient" temperature).

Because this case differs from the previous case only by a boundary condition, the shell energy balance analysis of part (a) up to and including Eq. 10.2-7 still applies here. Thus, the temperature profile will still have the form

$$T(r) = C_1 \ln r + C_2 \tag{10.2-15}$$

but the values of the constants C_1 and C_2 will differ from that in part (a). The boundary conditions are now

B.C. 1: $$\text{at } r = \kappa R, \quad T = T_s \tag{10.2-16}$$

B.C. 2: $$\text{at } r = R, \quad -k\frac{dT}{dr} = h(T - T_a) \tag{10.2-17}$$

When we apply the boundary conditions, we get two equations for the constants of integration C_1 and C_2

B.C. 1: $$T_s = C_1 \ln\kappa R + C_2 \tag{10.2-18}$$

B.C. 2: $$-\frac{kC_1}{R} = h(C_1 \ln R + C_2 - T_a) \tag{10.2-19}$$

Note that the profile $T(r)$ from Eq. 10.2-15 was substituted into both sides of Eq. 10.2-17 to arrive at Eq. 10.2-19. Solving these equations for C_1 and C_2 gives the temperature distribution

$$\boxed{\frac{T(r) - T_a}{T_s - T_a} = \frac{1 - \text{Bi}\cdot\ln(r/R)}{1 - \text{Bi}\cdot\ln\kappa}} \quad \text{for } \kappa R \le r \le R \tag{10.2-20}$$

where $\text{Bi} \equiv hR/k$ is the Biot number.[1]

The heat loss from a length L of pipe can be obtained from either of the following expressions:

$$Q = h\cdot(2\pi RL)\cdot(T|_{r=R} - T_a) \quad \text{or} \quad Q = 2\pi RL\left(-k\frac{\partial T}{\partial r}\bigg|_{r=R}\right) \tag{10.2-21}$$

---

[1]The Biot number is named after **Jean Baptiste Biot** (1774–1862) (pronounced "Bee-oh"). A professor of physics at the Collège de France, he received the Rumford Medal for his development of a simple, nondestructive test to determine sugar concentration. Note that the Biot number contains the thermal conductivity of the *solid*, whereas the Nusselt number (see §10.8) contains the thermal conductivity of the *fluid*. Both dimensionless groups have the same structure: (heat transfer coefficient)(length)/(thermal conductivity).

When the profile $T(r)$ is substituted into either expression, we obtain

$$Q = 2\pi kL(T_s - T_a)\frac{\text{Bi}}{1 + \text{Bi} \cdot \ln(1/\kappa)} \tag{10.2-22}$$

It is instructive to examine the temperature at the outer pipe surface,

$$T|_{r=R} = T_a + (T_s - T_a)\frac{1}{1 + \text{Bi} \cdot \ln(1/\kappa)} \tag{10.2-23}$$

In the limit $\text{Bi} = hR/k \to \infty$, we see that $T|_{r=R} \to T_a$. In this limit, heat transfer in the surrounding air is so fast relative to conduction within the pipe wall (i.e., h is so large relative to k/R) that the temperature of the pipe surface is equal to the ambient temperature. One can verify that the temperature profile Eq. 10.2-20 reduces to Eq. 10.2-12 (with T_o replaced by T_a) in this limit. In the opposite limit $\text{Bi} = hR/k \to 0$, we see that $T|_{r=R} \to T_s$. Here the heat transfer in the surrounding air is so slow relative to conduction within the pipe wall (i.e., h is so small relative to k/R) that the entire pipe wall is at the steam temperature T_s. From the analysis of these two limits, it is thus apparent that the Biot number may be interpreted qualitatively as the following ratios:

$$\begin{aligned}\text{Bi} = \frac{hR}{k} &= \frac{\text{heat-transfer rate in the surrounding fluid}}{\text{heat-transfer rate in the solid}} \\ &= \frac{\text{heat-transfer resistance in the solid}}{\text{heat-transfer resistance in the surrounding fluid}}\end{aligned} \tag{10.2-24}$$

c. *Steam pipe with an insulating layer, and heat loss according to Newton's law of cooling*

Next we add a layer of insulating material on the outside of the steam pipe as illustrated in Fig. 10.2-1(c). The thermal conductivity of the insulating layer is k_i (the thermal conductivity of the steam pipe still being k), and the outside of the insulation has a radius R_i. The heat transfer to the surrounding air is still modeled using Newton's law of cooling. We will obtain the temperature distribution over the entire region $\kappa R \le r \le R_i$, that is, in the steam pipe wall (region I, $\kappa R \le r \le R$) and in the insulation (region II, $R \le r \le R_i$). We *assume* that the insulation fits tightly around the pipe, so that there are no air spaces between the pipe and the insulation. This enables us to specify as boundary conditions at the pipe/insulation interface that the temperature and radial heat flux are continuous.

Shell energy balances over regions I and II proceed as in cases (a) and (b), and give the forms for the temperature profiles

$$T^{\text{I}}(r) = C_1^{\text{I}} \ln r + C_2^{\text{I}} \tag{10.2-25}$$

$$T^{\text{II}}(r) = C_1^{\text{II}} \ln r + C_2^{\text{II}} \tag{10.2-26}$$

The boundary conditions are

B. C. 1: at $r = \kappa R$, $\quad T^{\text{I}} = T_s \quad$ (10.2-27)

B. C. 2: at $r = R$, $\quad T^{\text{I}} = T^{\text{II}} \quad$ (10.2-28)

B. C. 3: at $r = R$, $\quad -k\dfrac{\partial T^{\text{I}}}{\partial r} = -k_i\dfrac{\partial T^{\text{II}}}{\partial r} \quad$ (10.2-29)

B. C. 4: at $r = R_i$, $\quad -k_i\dfrac{\partial T^{\text{II}}}{\partial r} = h(T^{\text{II}} - T_a) \quad$ (10.2-30)

When the profiles in Eqs. 10.2-25 and 10.2-26 are substituted into Eqs. 10.2-27 to 10.2-30, we obtain the following four equations that can be solved simultaneously to obtain the four constants of integration:

B.C. 1: $$T_s = C_1^{\mathrm{I}} \ln \kappa R + C_2^{\mathrm{I}} \tag{10.2-31}$$

B.C. 2: $$C_1^{\mathrm{I}} \ln R + C_2^{\mathrm{I}} = C_1^{\mathrm{II}} \ln R + C_2^{\mathrm{II}} \tag{10.2-32}$$

B.C. 3: $$kC_1^{\mathrm{I}} = k_i C_1^{\mathrm{II}} \tag{10.2-33}$$

B.C. 4: $$-k_i \frac{C_1^{\mathrm{II}}}{R_i} = h(C_1^{\mathrm{II}} \ln R_i + C_2^{\mathrm{II}} - T_a) \tag{10.2-34}$$

Solving for the constants of integration, we obtain the temperature profiles

$$\boxed{\frac{T^{\mathrm{I}}(r) - T_a}{T_s - T_a} = \frac{1 - \mathrm{Bi}_i\left[(k_i/k)\ln(r/R) - \ln(R_i/R)\right]}{1 - \mathrm{Bi}_i\left[(k_i/k)\ln(\kappa) - \ln(R_i/R)\right]}} \quad \text{for } \kappa R \leq r \leq R \tag{10.2-35}$$

$$\boxed{\frac{T^{\mathrm{II}}(r) - T_a}{T_s - T_a} = \frac{1 - \mathrm{Bi}_i\left[\ln(r/R) - \ln(R_i/R)\right]}{1 - \mathrm{Bi}_i\left[(k_i/k)\ln(\kappa) - \ln(R_i/R)\right]}} \quad \text{for } R \leq r \leq R_i \tag{10.2-36}$$

where $\mathrm{Bi}_i \equiv hR_i/k_i$. In addition, we can get the heat loss from a length L of the pipe-plus-insulation

$$\begin{aligned} Q &= h \cdot (2\pi R_i L) \cdot \left(T^{\mathrm{II}}\Big|_{r=R_i} - T_a \right) \\ &= h(2\pi R_i L)(T_s - T_a)\frac{1}{1 - \mathrm{Bi}_i\left[(k_i/k)\ln \kappa - \ln(R_i/R)\right]} \end{aligned} \tag{10.2-37}$$

One can verify that this expression reduces to Eq. 10.2-22 as $R_i \to R$.

§10.3 HEAT CONDUCTION THROUGH COMPOSITE WALLS

In industrial heat-transfer problems one is often concerned with conduction through walls made up of layers of various materials, each with its own characteristic thermal conductivity. In this section we show how the various resistances to heat transfer are combined into a total resistance.

In Fig. 10.3-1 we show a composite wall made up of three materials of different thicknesses, $x_1 - x_0$, $x_2 - x_1$, and $x_3 - x_2$, and different thermal conductivities k_{01}, k_{12}, and k_{23}. At $x = x_0$, substance 01 is in contact with a fluid with ambient temperature T_a, and at $x = x_3$, substance 23 is in contact with a fluid at temperature T_b. The heat transfer at the boundaries $x = x_0$ and $x = x_3$ is given by Newton's law of cooling with heat-transfer coefficients h_0 and h_3, respectively. The anticipated temperature profile is sketched in Fig. 10.3-1.

First, we set up the energy balance for the problem. Since we are dealing with heat conduction in a solid, the terms containing velocity in the $\mathbf{e}$ vector can be discarded, and the only relevant contribution is the $\mathbf{q}$ vector, describing heat conduction. For the shell of thickness Δx (width W and height H) in region 01 illustrated in Fig. 10.3-1, the rate of energy transported into the shell is $q_x|_x WH$, and the rate of energy transported out of the shell is $q_x|_{x+\Delta x} WH$. The energy balance for the shell in region 01 is thus

Region 01: $$q_x|_x WH - q_x|_{x+\Delta x} WH = 0 \tag{10.3-1}$$

which just states that the energy entering at x must be equal to the energy leaving at $x + \Delta x$, since no energy is produced within the region. After division by $WH\Delta x$ and taking the limit as $\Delta x \to 0$, we get

Region 01: $$\frac{dq_x}{dx} = 0 \tag{10.3-2}$$

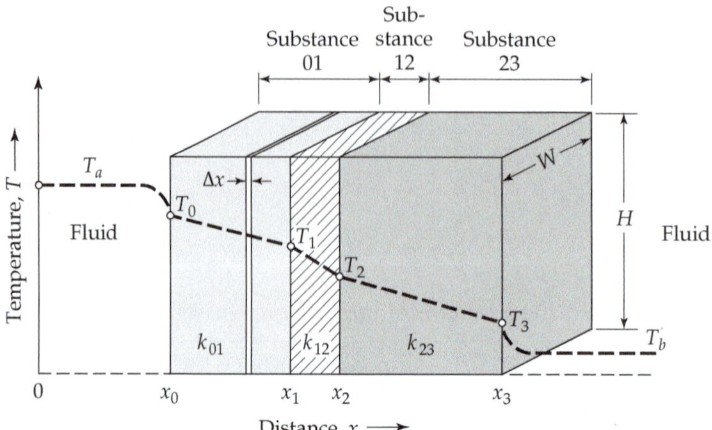

Fig. 10.3-1. Heat conduction through a composite wall, located between two fluid streams at temperatures T_a and T_b.

Integration of this equation gives

Region 01: $$q_x = q_0 \quad \text{(a constant)} \tag{10.3-3}$$

The constant of integration, q_0, is the heat flux at the plane $x = x_0$. The development in Eqs. 10.3-1, 10.3-2, and 10.3-3 can be repeated for regions 12 and 23, so that the heat flux is constant and the same for all three regions:

Regions 01, 12, 23: $$q_x = q_0 \tag{10.3-4}$$

with the same constant for each of the regions (i.e., by requiring continuity of the heat flux at the boundaries). We may now introduce Fourier's law for each of the three regions and get

Region 01: $$-k_{01}\frac{dT}{dx} = q_0 \tag{10.3-5}$$

Region 12: $$-k_{12}\frac{dT}{dx} = q_0 \tag{10.3-6}$$

Region 23: $$-k_{23}\frac{dT}{dx} = q_0 \tag{10.3-7}$$

We now assume that k_{01}, k_{12}, and k_{23} are constants. Then we integrate each equation over the entire thickness of the relevant slab of material to get

Region 01: $$T_0 - T_1 = q_0\left(\frac{x_1 - x_0}{k_{01}}\right) \tag{10.3-8}$$

Region 12: $$T_1 - T_2 = q_0\left(\frac{x_2 - x_1}{k_{12}}\right) \tag{10.3-9}$$

Region 23: $$T_2 - T_3 = q_0\left(\frac{x_3 - x_2}{k_{23}}\right) \tag{10.3-10}$$

In addition we have the two statements regarding the heat transfer at the surfaces according to Newton's law of cooling:

At surface 0: $$T_a - T_0 = \frac{q_0}{h_0} \tag{10.3-11}$$

At surface 3: $$T_3 - T_b = \frac{q_0}{h_0} \tag{10.3-12}$$

Addition of these last five equations then gives:

$$T_a - T_b = q_0 \left(\frac{1}{h_0} + \frac{x_1 - x_0}{k_{01}} + \frac{x_2 - x_1}{k_{12}} + \frac{x_3 - x_2}{k_{23}} + \frac{1}{h_3} \right) \tag{10.3-13}$$

or

$$q_0 = \frac{T_a - T_b}{\left(\frac{1}{h_0} + \sum_{j=1}^{3} \frac{x_j - x_{j-1}}{k_{j-1,j}} + \frac{1}{h_3} \right)} \tag{10.3-14}$$

Sometimes this result is rewritten in a form reminiscent of Newton's law of cooling, either in terms of the heat flux q_0 ([=] J/m$^2 \cdot$ s) or the heat flow Q_0 ([=] J/s)

$$q_0 = U(T_a - T_b) \quad \text{or} \quad Q_0 = U(WH)(T_a - T_b) \tag{10.3-15}$$

The quantity U, called the "overall heat-transfer coefficient," is given then by the following famous formula for the "additivity of resistances":

$$\frac{1}{U} = \frac{1}{h_0} + \sum_{j=1}^{n} \frac{x_j - x_{j-1}}{k_{j-1,j}} + \frac{1}{h_n} \tag{10.3-16}$$

Here we have generalized the formula to a system with n slabs of material. Equations 10.3-15 and 10.3-16 are useful for calculating the heat-transfer rate through a composite wall separating two fluid streams, when the heat-transfer coefficients and thermal conductivities are known. The estimation of heat-transfer coefficients is discussed in Chapter 14.

In the above development it has been tacitly *assumed* that the solid slabs are contiguous with no intervening "air spaces." If the solid surfaces touch each other only at several points, the resistance to heat transfer will be appreciably increased.

EXAMPLE 10.3-1

Composite Cylindrical Walls

Develop a formula for the overall heat-transfer coefficient for the composite cylindrical pipe wall shown in Fig. 10.3-2.

SOLUTION

An energy balance on a shell of volume $2\pi rL\Delta r$ for region 01 is

Region 01:
$$q_r|_r 2\pi rL - q_r|_{r+\Delta r} 2\pi(r + \Delta r)L = 0 \tag{10.3-17}$$

which can also be written as:

Region 01:
$$(2\pi rLq_r)|_r - (2\pi rLq_r)|_{r+\Delta r} = 0 \tag{10.3-18}$$

Division by $2\pi L\Delta r$ and taking the limit as Δr goes to zero gives

Region 01:
$$\frac{d}{dr}(rq_r) = 0 \tag{10.3-19}$$

Integration of this equation gives:

$$rq_r = \text{constant} = r_0 q_0 \tag{10.3-20}$$

in which r_0 is the inner radius of region 01, and q_0 is the heat flux there. In regions 12 and 23, rq_r is equal to the same constant (i.e., the total radial heat-transfer rate $2\pi rLq_r$ is the same everywhere at steady state). Application of Fourier's law to the three regions gives:

Region 01:
$$-k_{01} r \frac{dT}{dr} = r_0 q_0 \tag{10.3-21}$$

Region 12:
$$-k_{12} r \frac{dT}{dr} = r_0 q_0 \tag{10.3-22}$$

Region 23:
$$-k_{23} r \frac{dT}{dr} = r_0 q_0 \tag{10.3-23}$$

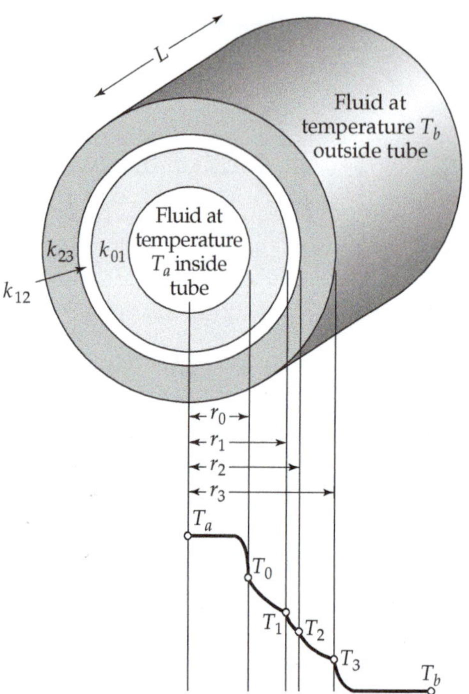

Fig. 10.3-2. Heat conduction through a laminated tube with a fluid at temperature T_a inside the tube and temperature T_b outside.

If we assume that the thermal conductivities in the three annular regions are constants, then each of the above three equations can be integrated across its region to give.

Region 10:
$$T_0 - T_1 = r_0 q_0 \frac{\ln(r_1/r_0)}{k_{01}} \tag{10.3-24}$$

Region 12:
$$T_1 - T_2 = r_0 q_0 \frac{\ln(r_2/r_1)}{k_{12}} \tag{10.3-25}$$

Region 23:
$$T_2 - T_3 = r_0 q_0 \frac{\ln(r_3/r_2)}{k_{23}} \tag{10.3-26}$$

At the two fluid-solid interfaces we can write Newton's law of cooling:

Surface 0:
$$T_a - T_0 = \frac{q_0}{h_0} \tag{10.3-27}$$

Surface 3:
$$T_3 - T_b = \frac{q_3}{h_3} = \frac{q_0}{h_3}\frac{r_0}{r_3} \tag{10.3-28}$$

Addition of the preceding five equations gives an equation for $T_a - T_b$, which can be solved for the heat-transfer rate to give

$$Q_0 = 2\pi L r_0 q_0 = \frac{2\pi L(T_a - T_b)}{\left(\dfrac{1}{r_0 h_0} + \dfrac{\ln(r_1/r_0)}{k_{01}} + \dfrac{\ln(r_2/r_1)}{k_{12}} + \dfrac{\ln(r_3/r_2)}{k_{23}} + \dfrac{1}{r_3 h_3}\right)} \tag{10.3-29}$$

We now define an "overall heat-transfer coefficient based on the inner surface" U_0 by

$$Q_0 = 2\pi L r_0 q_0 = U_0(2\pi L r_0)(T_a - T_b) \tag{10.3-30}$$

Combination of the last two equations gives, on generalizing to a system with n annular layers,

$$\frac{1}{r_0 U_0} = \left(\frac{1}{r_0 h_0} + \sum_{j=1}^{n} \frac{\ln\left(r_j/r_{j-1}\right)}{k_{j-1,j}} + \frac{1}{r_n h_n}\right) \tag{10.3-31}$$

The subscript "0" on U_0 indicates that the overall heat-transfer coefficient is referred to the radius r_0. Equation 10.3-31 can be used to determine the heat loss rates directly for the systems considered in §10.2.

§10.4 HEAT CONDUCTION WITH TEMPERATURE-DEPENDENT THERMAL CONDUCTIVITY

In this section we study the heat conduction in the θ direction in the annular region shown in Fig. 10.4-1 of inner radius r_1, outer radius r_2, and length L. The curved surfaces and the end surfaces (both shaded in the figure) are insulated. The surface at $\theta = 0$ of area $(r_2 - r_1)L$ is maintained at temperature T_0, and the surface at $\theta = \pi$, also of area $(r_2 - r_1)L$, is kept at temperature T_π.

In the preceding sections we have taken the thermal conductivities to be constants. In this problem we show how to take into account the dependence of the thermal conductivity on the temperature. Specifically, here we assume that the thermal conductivity of the solid varies linearly with the temperature from k_0 at $T = T_0$, to k_π at $T = T_\pi$. That is,

$$k(T) = k_0 + (k_\pi - k_0)\left(\frac{T(\theta) - T_0}{T_\pi - T_0}\right) = k_0 + (k_\pi - k_0)\Theta \tag{10.4-1}$$

in which $\Theta(\theta)$ is a dimensionless temperature difference. We want to find the dimensionless temperature difference Θ as a function of the angle θ.

Since T (or Θ) is a function of θ alone, we make a shell energy balance over a shell of thickness $\Delta\theta$. Because the velocity is zero, the energy flux **e** reduces to the conductive energy flux **q**. Thus, for conduction across the surfaces of area $(r_2 - r_1)L$ bounding the shell between θ and $\theta + \Delta\theta$, the shell energy balance is

$$q_\theta|_\theta (r_2 - r_1)L - q_\theta|_{\theta+\Delta\theta}(r_2 - r_1)L = 0 \tag{10.4-2}$$

Dividing by $(r_2 - r_1)L\Delta\theta$ and taking the limit as $\Delta\theta \to 0$ gives

$$\frac{dq_\theta}{d\theta} = 0 \tag{10.4-3}$$

We then insert Fourier's law with the temperature-dependent thermal conductivity $k(\Theta)$ to get

$$\frac{d}{d\theta}\left[-\left(k_0 + (k_\pi - k_0)\,\Theta\right)\frac{1}{r}\frac{d\Theta}{d\theta}\right] = 0 \tag{10.4-4}$$

where the dimensionless temperature difference Θ is defined in Eq. 10.4-1. To get Eq. 10.4-4, we have made use of the expression for q_θ given in Eq. B.2-5.

Integrating once we get

$$\left(k_0 + (k_\pi - k_0)\,\Theta\right)\frac{d\Theta}{d\theta} = C_1 \tag{10.4-5}$$

and a second integration gives

$$k_0\Theta + \frac{1}{2}(k_\pi - k_0)\Theta^2 = C_1\theta + C_2 \tag{10.4-6}$$

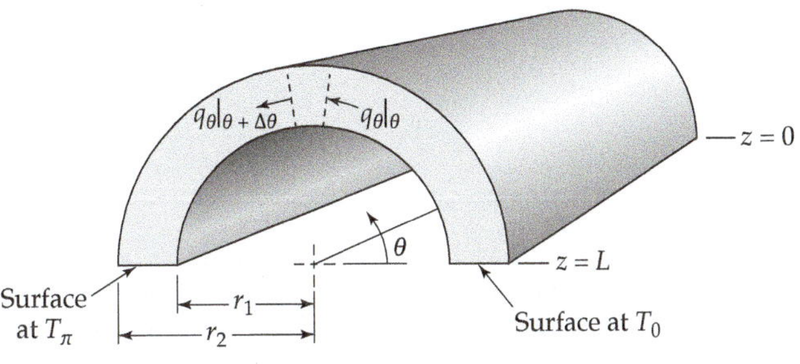

Fig. 10.4-1. Tangential heat conduction in an annular region.

When the constants are determined from the boundary conditions, $\Theta(0) = 0$ and $\Theta(\pi) = 1$, we get $C_2 = 0$ and $C_1 = (k_0 + k_\pi)/2\pi$. Therefore, the temperature distribution in the solid is given by the quadratic equation

$$k_0\Theta + \frac{1}{2}(k_\pi - k_0)\Theta^2 = \frac{(k_0 + k_\pi)\theta}{2\pi} \tag{10.4-7}$$

or

$$\frac{1}{2}(k_0 - k_\pi)\Theta^2 - k_0\Theta + \frac{(k_0 + k_\pi)\theta}{2\pi} = 0 \tag{10.4-8}$$

This quadratic equation has the solution

$$\boxed{\Theta(\theta) = \frac{k_0 \pm \sqrt{[1 - (\theta/\pi)]k_0^2 + (\theta/\pi)k_\pi^2}}{k_0 - k_\pi}} \tag{10.4-9}$$

The minus sign in Eq. 10.4-9 has to be chosen in order that the boundary conditions be satisfied.

Next, let us get the rate of heat flow through the surface at $\theta = 0$. This is

$$\begin{aligned} Q &= \int_0^L \int_{r_1}^{r_2} \left(-k\frac{1}{r}\frac{dT}{d\theta}\right)\bigg|_{\theta=0} dr\,dz = (T_0 - T_\pi)\int_0^L \int_{r_1}^{r_2} \left(k(\Theta)\frac{1}{r}\frac{d\Theta}{d\theta}\right)\bigg|_{\theta=0} dr\,dz \\ &= (T_0 - T_\pi)\int_0^L \int_{r_1}^{r_2} \frac{1}{r}\left(\frac{(k_0 + k_\pi)}{2\pi}\right) dr\,dz \\ &= \left(\frac{L}{\pi}\ln\frac{r_2}{r_1}\right)\left(\frac{k_0 + k_\pi}{2}\right)(T_0 - T_\pi) \end{aligned} \tag{10.4-10}$$

where in the second line we have used the fact that $k(\Theta)d\Theta/d\theta$ is equal to the constant C_1. One can show that Eq. 10.4-10 is the heat-transfer rate one would obtain if the thermal conductivity were constant and equal to $(k_0 + k_\pi)/2$.

§10.5 HEAT CONDUCTION IN A COOLING FIN[1]

Another simple, but practical application of heat conduction is the calculation of the efficiency of a cooling fin. Fins are used to increase the area available for heat transfer between metal walls and poorly conducting fluids such as gases. A simple rectangular fin is shown in Fig. 10.5-1. The wall temperature is T_w and the ambient air temperature is T_a.

A reasonably good description of the system may be obtained by approximating the true physical situation by a simplified model:

True situation	Model
1. T is a function of x, y, and z, the dependence on z is most important.	1. T is a function of z alone.
2. A small quantity of heat is lost from the fin at the end (area $2BW$) and at the edges (area $2 \times 2BL$).	2. No heat is lost from the end or from the edges.
3. The heat-transfer coefficient is a function of position.	3. The heat flux normal to the surface, from the solid to the fluid, is given by $q_n = h(T - T_a)$, where h is a constant and T depends on z.

[1]For further information on fins, see M. Jakob, *Heat Transfer*, Vol. I, Wiley, New York (1949), Chapter 11; and H. D. Baehr and K. Stephan, *Heat and Mass Transfer*, Springer, Berlin (1998), §2.2.3.

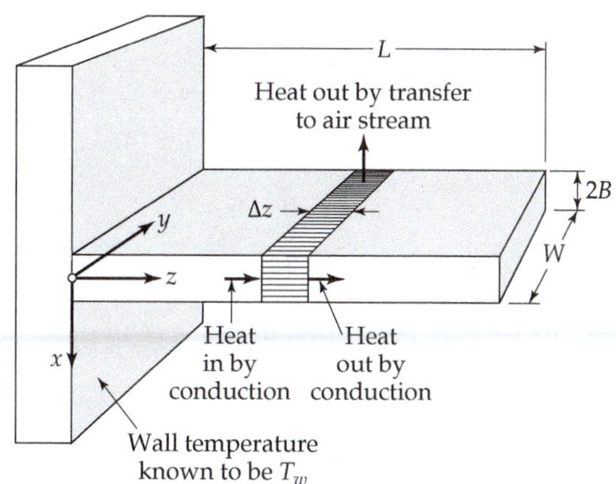

Fig. 10.5-1. A simple cooling fin with $B << L$ and $B << W$.

The energy balance is made over a segment Δz of the fin; the shell is thin in only the z direction because it has been assumed that the temperature depends only on z. Since the fin is stationary, the terms containing $\mathbf{v}$ in the total energy-flux vector $\mathbf{e}$ may be discarded, and the only contribution to the energy flux is the conductive heat flux $\mathbf{q}$. Therefore, the energy balance is

$$2BWq_z|_z - 2BWq_z|_{z+\Delta z} - h(2W\Delta z)(T(z) - T_a) = 0 \tag{10.5-1}$$

The last term in the above equation represents the heat transferred to the surrounding fluid from the top and bottom surfaces. In contrast to previous problems, Newton's law of cooling appears here in the energy balance (as opposed to a boundary condition) because the assumption that T depends only on z requires that the shell system includes the surfaces in contact with the surrounding fluid.

Division of Eq. 10.5-1 by $2BW\Delta z$ and taking the limit as Δz approaches zero gives

$$-\frac{dq_z}{dz} - \frac{h}{B}(T(z) - T_a) = 0 \tag{10.5-2}$$

We now insert Fourier's law ($q_z = -kdT/dz$), in which k is the thermal conductivity of the metal. If we assume that k is constant, we then get

$$\frac{d^2T}{dz^2} - \frac{h}{kB}(T(z) - T_a) = 0 \tag{10.5-3}$$

This equation is to be solved with the boundary conditions

B. C. 1: $$\text{at } z = 0, \quad T = T_w \tag{10.5-4}$$

B. C. 2: $$\text{at } z = L, \quad \frac{dT}{dz} = 0 \tag{10.5-5}$$

The second boundary condition is consistent with the assumption that the energy lost from the end of the fin is negligible. We now introduce the following dimensionless quantities:

$$\Theta(\zeta) = \frac{T(z) - T_a}{T_w - T_a} = \text{dimensionless temperature difference} \tag{10.5-6}$$

$$\zeta = \frac{z}{L} = \text{dimensionless distance} \tag{10.5-7}$$

$$N^2 = \frac{hL^2}{kB} = \text{dimensionless heat-transfer coefficient}^2 \tag{10.5-8}$$

---

[2]The quantity N^2 may be rewritten as $N^2 = (hL/k)(L/B) = \text{Bi}(L/B)$, where Bi is the *Biot number* introduced in §10.2.

The problem then takes the form:

$$\frac{d^2\Theta}{d\zeta^2} - N^2\Theta = 0 \tag{10.5-9}$$

B. C. 1: at $\zeta = 0$, $\Theta = 1$ (10.5-10)

B. C. 2: at $\zeta = 1$, $\dfrac{d\Theta}{d\zeta} = 0$ (10.5-11)

The solution to Eq. 10.5-9 may be written in terms of hyperbolic functions (see Eq. C.1-4 and §C.5)

$$\Theta(\zeta) = C_1 \cosh N\zeta + C_2 \sinh N\zeta \tag{10.5-12}$$

Application of the boundary conditions gives

B. C. 1: $1 = C_1 \cdot 1 + C_2 \cdot 0$ (10.5-13)

B. C. 2: $0 = C_1 N \sinh N + C_2 N \cosh N$ (10.5-14)

From these last two equations the constants of integration may be determined ($C_1 = 1$ and $C_2 = -\tanh N$) and we get

$$\begin{aligned}\Theta(\zeta) &= \cosh N\zeta - (\sinh N/\cosh N)\sinh N\zeta \\ &= \frac{(\cosh N\zeta)\cosh N - (\sinh N\zeta)\sinh N}{\cosh N}\end{aligned} \tag{10.5-15}$$

This may be rewritten with the help of Eq. C.5-6 to give finally

$$\boxed{\Theta(\zeta) = \frac{\cosh N(1-\zeta)}{\cosh N}} \tag{10.5-16}$$

This result is reasonable only if the heat lost at the end and at the edges is negligible.

The "effectiveness" of the fin surface is defined[3] by

$$\eta = \frac{\text{actual rate of heat loss from the fin}}{\text{rate of heat loss from an isothermal fin at } T_w} \tag{10.5-17}$$

For the problem being considered here η is then

$$\eta = \frac{\int_0^W \int_0^L h(T(z) - T_a)dzdy}{\int_0^W \int_0^L h(T_w - T_a)dzdy} = \frac{\int_0^1 \Theta(\zeta)d\zeta}{\int_0^1 d\zeta} \tag{10.5-18}$$

or

$$\eta = \frac{1}{\cosh N}\left(-\frac{1}{N}\sinh N(1-\zeta)\right)\Bigg|_0^1 = \frac{\tanh N}{N} \tag{10.5-19}$$

in which N is the dimensionless quantity defined in Eq. 10.5-8.

EXAMPLE 10.5-1

Error in Thermocouple Measurement

In Fig. 10.5-2 a thermocouple is shown in a cylindrical well inserted into a gas stream. Estimate the true temperature of the gas stream if

[3] M. Jakob, *Heat Transfer*, Wiley, New York (1949), Vol. **I**, p. 235.

$T_1 = 500°\text{F}$ = temperature indicated by thermocouple
$T_w = 350°\text{F}$ = wall temperature
$h = 120\ \text{Btu/hr} \cdot \text{ft}^2 \cdot °\text{F}$ = heat-transfer coefficient
$k = 60\ \text{Btu/hr} \cdot \text{ft} \cdot °\text{F}$ = thermal conductivity of well wall
$B = 0.08$ in. = thickness of well wall
$L = 0.2$ ft = length of well

SOLUTION

The thermocouple well wall of thickness B is in contact with the gas stream on one side only, and the tube thickness is small compared with the diameter. Hence, the temperature distribution along this wall will be about the same as that along a fin of thickness $2B$, in contact with the gas stream on both sides. According to Eq. 10.5-16, the temperature at the end of the well (that registered by the thermocouple) satisfies

$$\frac{T_1 - T_a}{T_w - T_a} = \frac{\cosh 0}{\cosh N} = \frac{1}{\cosh \sqrt{hL^2/kB}}$$

$$= \frac{1}{\cosh \sqrt{(120\ \text{Btu/hr} \cdot \text{ft}^2 \cdot °\text{F})(0.2\ \text{ft})^2 \Big/ \left[(60\ \text{Btu/hr} \cdot \text{ft} \cdot °\text{F})\,(0.08\ \text{in.})\left(\dfrac{1\ \text{ft}}{12\ \text{in.}}\right)\right]}}$$

$$= \frac{1}{\cosh(3.464)} = 0.0625 \tag{10.5-20}$$

Hence, the actual ambient gas temperature is obtained by solving this equation for T_a:

$$\frac{500°\text{F} - T_a}{350°\text{F} - T_a} = 0.0625 \tag{10.5-21}$$

and the result is

$$T_a = 510°\text{F} \tag{10.5-22}$$

Therefore, the reading is 10°F too low.

This example has focused on one kind of error that can occur in thermometry. Frequently a simple analysis, such as the foregoing, can be used to estimate the measurement errors.[4]

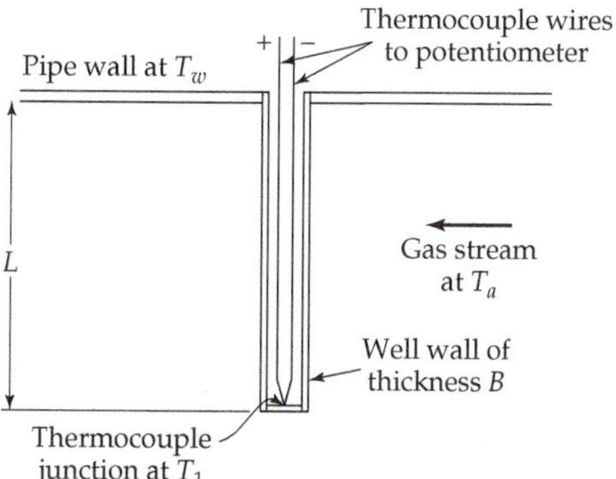

Fig. 10.5-2. A thermocouple in a cylindrical well.

[4]For further discussion, see M. Jakob, *Heat Transfer*, Vol. **II**, Wiley, New York (1949), Chapter 33, pp. 147–201.

§10.6 ENERGY TRANSPORT WITH ENERGY PRODUCTION: ELECTRICAL ENERGY CONVERSION IN A WIRE

In discussing the shell energy balances in §10.1, a "production" term was included. In this section we discuss a heat source in a homogeneous medium, specifically a heat source resulting from electrical heating in a wire. In §10.7 we take up a heat source in a heterogeneous medium, taking as an example the heat production by chemical reaction in a granular bed. And in §10.8 we discuss heat production in a flowing viscous fluid, where the heat is produced by viscous dissipation.

In this section the system being studied involves the production of heat by electrical conduction. But other heat sources can be treated similarly. In Problems 10B.2 and 10B.3 we get the temperature profile in spherical and cylindrical nuclear fuel elements resulting from the heating associated with the nuclear reaction.

The first system we consider is an electric wire of circular cross section with radius R, and electrical conductivity k_e ohm^{-1}cm^{-1}. Through this wire there is an electric current with current density I amp/cm^2. The transmission of an electric current is an irreversible process, and some electrical energy is converted into thermal energy. The rate of thermal energy production per unit volume is given by the expression

$$S_e = \frac{I^2}{k_e} \tag{10.6-1}$$

The quantity S_e is the heat source owing to electrical energy dissipation. We assume here that the temperature rise in the wire is not so large that the temperature dependence of either the thermal or electrical conductivity need be considered. As a result, the rate of energy production per unit volume S_e is a constant. The surface of the wire is maintained at temperature T_0. We now show how to find the radial temperature distribution within the wire.

For the energy balance, we take the system to be a cylindrical shell of thickness Δr and length L (see Fig. 10.6-1). Since $\mathbf{v} = 0$ in this system, the only contributions to the energy balance are

Rate of heat in across cylindrical surface at r:

$$q_r|_r \cdot 2\pi rL = (2\pi rLq_r)|_r \tag{10.6-2}$$

Rate of heat out across cylindrical surface at r + Δr:

$$q_r|_{r+\Delta r} \cdot 2\pi(r+\Delta r)L = (2\pi rLq_r)|_{r+\Delta r} \tag{10.6-3}$$

Rate of thermal energy production by electrical energy dissipation:

$$(2\pi r\Delta rLS_e) \tag{10.6-4}$$

We now substitute these quantities into the energy balance of Eq. 10.1-1. Division by $2\pi L\Delta r$ and taking the limit as Δr goes to zero gives

$$\lim_{\Delta r \to 0} \frac{(rq_r)|_{r+\Delta r} - (rq_r)|_r}{\Delta r} = S_e r \tag{10.6-5}$$

The expression on the left side is just the first derivative of rq_r with respect to r, so that Eq. 10.6-5 becomes

$$\frac{d}{dr}(rq_r) = S_e r \tag{10.6-6}$$

This is a first-order differential equation for the energy flux, and it may be integrated to give

$$q_r = \frac{S_e r}{2} + \frac{C_1}{r} \tag{10.6-7}$$

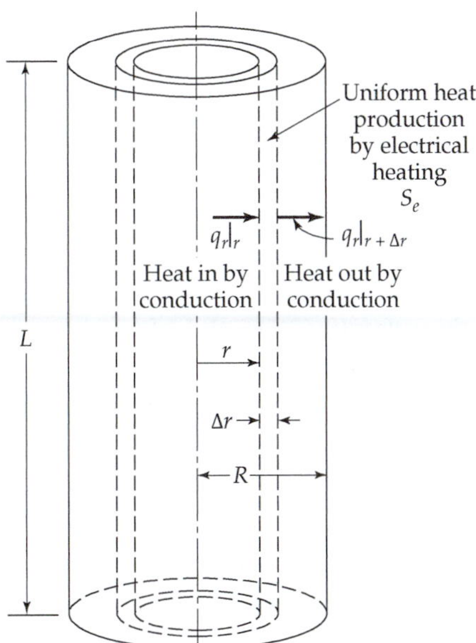

Fig. 10.6-1. Electrically heated wire, showing the cylindrical shell over which the energy balance is made.

The integration constant C_1 must be zero because of the boundary condition that

B. C. 1: at $r = 0$, q_r is not infinite (10.6-8)

Hence, the final expression for the heat-flux distribution is

$$\boxed{q_r = \frac{S_e r}{2}} \tag{10.6-9}$$

This states that the heat flux increases linearly with r.

We now substitute Fourier's law in the form $q_r = -k(dT/dr)$ (see Eq. B.2-4) into Eq. 10.6-9 to obtain

$$-k\frac{dT}{dr} = \frac{S_e r}{2} \tag{10.6-10}$$

When k is assumed to be constant, this first-order differential equation can be integrated to give

$$T(r) = -\frac{S_e r^2}{4k} + C_2 \tag{10.6-11}$$

The integration constant is determined from

B. C. 2: at $r = R$, $T = T_0$ (10.6-12)

Hence, $C_2 = (S_e R^2/4k) + T_0$ and Eq. 10.6-11 becomes

$$\boxed{T(r) - T_0 = \frac{S_e R^2}{4k}\left[1 - \left(\frac{r}{R}\right)^2\right]} \tag{10.6-13}$$

Equation 10.6-13 gives the temperature rise as a parabolic function of the distance r from the wire axis.

Once the temperature and heat-flux distributions are known, various information about the system may be obtained:

(i) *Maximum temperature rise* (at $r = 0$)

$$T_{\max} - T_0 = \frac{S_e R^2}{4k} \tag{10.6-14}$$

(ii) *Average temperature rise*

$$\langle T \rangle - T_0 = \frac{\int_0^{2\pi} \int_0^R [T(r) - T_0] r \, dr \, d\theta}{\int_0^{2\pi} \int_0^R r \, dr \, d\theta} = \frac{S_e R^2}{8k} \tag{10.6-15}$$

Thus, the temperature rise, averaged over the cross section, is just half the maximum temperature rise.

(iii) *Heat outflow at the surface* (for a length L of wire)

$$Q|_{r=R} = 2\pi RL \cdot q_r|_{r=R} = 2\pi RL \cdot \frac{S_e R}{2} = \pi R^2 L \cdot S_e \tag{10.6-16}$$

This result is not surprising, since, at steady state, all the thermal energy produced by electrical dissipation in the volume $\pi R^2 L$ must leave through the surface $r = R$.

The reader, while going through this development, may well have had the feeling of *déja vu*. There is, after all, a pronounced similarity between the heated wire problem and the viscous flow in a circular tube. Only the notation is different:

	Tube flow	Heated wire
First integration gives	$\tau_{rz}(r)$	$q_r(r)$
Second integration gives	$v_z(r)$	$T(r) - T_0$
Boundary condition at $r = 0$	τ_{rz} = finite	q_r = finite
Boundary condition at $r = R$	$v_z = 0$	$T - T_0 = 0$
Transport property	μ	k
Source term	$(\mathscr{P}_0 - \mathscr{P}_L)/L$	S_e
Assumptions	μ = constant	k, k_e = constant

That is, when proper quantities are chosen, the differential equations *and* the boundary conditions for the two problems are identical, and the physical processes are said to be "analogous." Not all problems in momentum transfer have analogs in energy and mass transport. However, when such analogies can be found, they may be useful in taking over known results from one field and applying them in another. For example, the reader should have no trouble in finding a heat-conduction analog for the viscous flow in a liquid film on an inclined plane.

There are many examples of heat-conduction problems in the electrical industry.[1] The minimizing of temperature rises inside electrical machinery prolongs insulation life. One example is the use of internally liquid-cooled stator conductors in very large (500,000 kW) AC generators.

To illustrate further problems in electrical heating, we give two examples concerning the temperature rise in wires: the first indicates the order of magnitude of the heating effect, and the second shows how to handle different boundary conditions. In addition, in Problem 10C.1 it is shown how to take into account the temperature dependence of the thermal and electrical conductivities.

[1] M. Jakob, *Heat Transfer*, Vol. I, Wiley, New York (1949), Chapter 10, pp. 167–199.

EXAMPLE 10.6-1

Voltage Required for a Given Temperature Rise in a Wire Heated by an Electric Current

A copper wire has a radius of 2 mm and a length of 5 m. For what voltage drop would the temperature rise at the wire axis be 10°C, if the surface temperature of the wire is 20°C?

SOLUTION

Combination of Eqs. 10.6-14 and 10.6-1 gives

$$T_{\max} - T_0 = \frac{I^2 R^2}{4kk_e} \tag{10.6-17}$$

The current density is related to the voltage drop E over a length L by

$$I = k_e \frac{E}{L} \tag{10.6-18}$$

Hence,

$$T_{\max} - T_0 = \frac{E^2 R^2}{4L^2}\left(\frac{k_e}{k}\right) \tag{10.6-19}$$

from which

$$E = 2\frac{L}{R}\sqrt{\frac{k}{k_e T_0}}\sqrt{T_0(T_{\max} - T_0)} \tag{10.6-20}$$

For copper, the Lorenz number of §9.9 is $k/k_e T_0 = 2.23 \times 10^{-8}$ volt^2/K^2. Therefore, the voltage drop needed to cause a 10°C temperature rise is

$$\begin{aligned} E &= 2\left(\frac{5000 \text{ mm}}{2 \text{ mm}}\right)\sqrt{2.23 \times 10^{-8}\frac{\text{volt}}{\text{K}}}\sqrt{(293 \text{ K})(10 \text{ K})} \\ &= 40 \text{ volts} \end{aligned} \tag{10.6-21}$$

EXAMPLE 10.6-2

Heated Wire with Specified Heat-Transfer Coefficient and Ambient Air Temperature

Repeat the analysis in §10.6, assuming that T_0 is not known, but that instead the heat flux at the wall is given by Newton's "law of cooling" (Eq. 10.1-2). Assume that the heat-transfer coefficient h and the ambient air temperature T_{air} are known.

SOLUTION

The solution proceeds as before through Eq. 10.6-11, but the second integration constant is determined from the new boundary condition

$$\text{B. C. 2}': \qquad \text{at } r = R, \quad -k\frac{dT}{dr} = h(T - T_{\text{air}}) \tag{10.6-22}$$

Substituting the temperature profile $T(r)$ given by Eq. 10.6-11 into both sides of Eq. 10.6-22 gives $C_2 = (S_e R/2h) + (S_e R^2/4k) + T_{\text{air}}$, and the temperature profile is then

$$T(r) - T_{\text{air}} = \frac{S_e R^2}{4k}\left[1 - \left(\frac{r}{R}\right)^2\right] + \frac{S_e R}{2h} \tag{10.6-23}$$

From this, the surface temperature of the wire is found to be $T_{\text{air}} + S_e R/2h$.

§10.7 ENERGY TRANSPORT WITH ENERGY PRODUCTION: CHEMICAL ENERGY CONVERSION IN A REACTOR

In this section, we discuss the heat production in a granular system, in which a chemical reaction is occurring as a fluid flows through a "packed bed." We ignore the granular structure and calculate the temperature profiles as though the system were homogeneous.

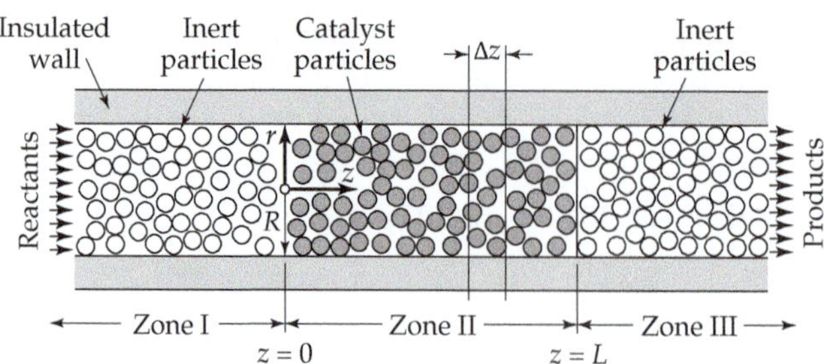

Fig. 10.7-1. Fixed-bed axial-flow reactor. Reactants enter at $z = -\infty$ and leave at $z = +\infty$. The reaction zone extends from $z = 0$ to $z = L$.

A chemical reaction is being carried out in a tubular, fixed-bed flow reactor with inner radius R as shown in Fig. 10.7-1. The reactor extends from $z = -\infty$ to $z = +\infty$ and is divided into three zones:

Zone I: Entrance zone packed with noncatalytic spheres

Zone II: Reaction zone packed with catalyst spheres, extending from $z = 0$ to $z = L$

Zone III: Exit zone packed with noncatalytic spheres

It is assumed that the fluid proceeds through the reactor tube in "plug flow"—that is, with axial velocity uniform at a superficial value $v_0 = w/\pi R^2 \rho$ (see just below Eq. 6.4-1 for the definition of "superficial velocity"). The density, mass flow rate, and superficial velocity are all treated as independent of r and z. In addition, the reactor wall is assumed to be well insulated, so that the temperature can be considered essentially independent of r. It is desired to find the steady-state axial temperature distribution $T(z)$ when the fluid enters at $z = -\infty$ with a uniform temperature T_1.

When a chemical reaction occurs, thermal energy is produced or consumed when the reactant molecules rearrange to form the products. The rate of thermal energy production per unit volume by chemical reaction, S_c, is in general a complicated function of pressure, temperature, composition, and catalyst activity. For simplicity, we represent S_c here as a function of temperature only: $S_c = S_{c1}F(\Theta)$, where $\Theta = (T(z) - T_0)/(T_1 - T_0)$ is the dimensionless temperature difference. Here $T(z)$ is the local temperature in the catalyst bed (assumed equal for catalyst and fluid), and S_{c1} and T_0 are empirical constants for the given reactor inlet conditions.

For the shell balance we select a disk of radius R and thickness Δz in the catalyst zone (see Fig. 10.7-1), and we choose Δz to be much larger than the catalyst particle dimensions. In setting up the energy balance, we use the total energy-flux vector **e** inasmuch as we are dealing with a flow system (i.e., $\mathbf{v} \neq 0$). Then, at steady state, the energy balance is

$$\pi R^2 e_z|_z - \pi R^2 e_z|_{z+\Delta z} + (\pi R^2 \Delta z)S_c = 0 \tag{10.7-1}$$

Next we divide by $\pi R^2 \Delta z$ and take the limit as Δz goes to zero. Strictly speaking, this operation is not "legal," since we are not dealing with a continuum, but rather with a granular structure. Nevertheless, we perform this limiting process with the understanding that the resulting equation describes, not point values, but rather average values of e_z and S_c for reactor cross sections at a particular value of z. This gives

$$\frac{de_z}{dz} = S_c \tag{10.7-2}$$

Now we substitute the z component of Eq. 9.4-2 into this equation to get

$$\frac{d}{dz}\left(\left(\frac{1}{2}\rho v^2 + \rho\hat{H}\right)v_z + \tau_{zz}v_z + q_z\right) = S_c \tag{10.7-3}$$

We now use Fourier's law for q_z, Eq. 1.2-9 for τ_{zz}, and the enthalpy expression in Eq. 9.4-6 (with the assumption that the heat capacity is constant) to get

$$\frac{d}{dz}\left(\frac{1}{2}\rho v_z^2 v_z + \rho\hat{C}_p(T - T^\circ)v_z + (p - p^\circ)v_z + \rho\hat{H}^\circ v_z - 2\mu v_z\frac{dv_z}{dz} - k_{\text{eff}}\frac{dT}{dz}\right) = S_c \qquad (10.7\text{-}4)$$

in which an effective thermal conductivity k_{eff} has been used to describe heat conduction in the z direction. The first, fourth, and fifth terms on the left side may be discarded, since the velocity is not changing with z. The third term may be discarded if the pressure does not change significantly in the axial direction. Then in the second term we replace v_z by the superficial velocity v_0, because the latter is the fluid velocity in the reactor, averaged over any cross section. Then Eq. 10.7-4 becomes

$$\rho\hat{C}_p v_0\frac{dT}{dz} = k_{\text{eff}}\frac{d^2T}{dz^2} + S_c \qquad (10.7\text{-}5)$$

This is the differential equation for the temperature in Zone II. The same equation applies in Zones I and III with the source term set equal to zero. The differential equations for the temperature are then

Zone I $\quad (z < 0) \qquad \rho\hat{C}_p v_0\frac{dT^{\text{I}}}{dz} = k_{\text{eff}}\frac{d^2T^{\text{I}}}{dz^2} \qquad (10.7\text{-}6)$

Zone II $\quad (0 > z > L) \qquad \rho\hat{C}_p v_0\frac{dT^{\text{II}}}{dz} = k_{\text{eff}}\frac{d^2T^{\text{II}}}{dz^2} + S_{c1}F(\Theta) \qquad (10.7\text{-}7)$

Zone III $\quad (z > L) \qquad \rho\hat{C}_p v_0\frac{dT^{\text{III}}}{dz} = k_{\text{eff}}\frac{d^2T^{\text{III}}}{dz^2} \qquad (10.7\text{-}8)$

Here we have assumed that we can use the same value of the effective thermal conductivity in all three zones. These three second-order differential equations are subject to the following six boundary conditions:

B. C. 1: as $z \to -\infty$, $\quad T^{\text{I}} \to T_1 \qquad (10.7\text{-}9)$

B. C. 2: at $z = 0$, $\quad T^{\text{I}} = T^{\text{II}} \qquad (10.7\text{-}10)$

B. C. 3: at $z = 0$, $\quad k_{\text{eff}}\frac{dT^{\text{I}}}{dz} = k_{\text{eff}}\frac{dT^{\text{II}}}{dz} \qquad (10.7\text{-}11)$

B. C. 4: at $z = L$, $\quad T^{\text{II}} = T^{\text{III}} \qquad (10.7\text{-}12)$

B. C. 5: at $z = L$, $\quad k_{\text{eff}}\frac{dT^{\text{II}}}{dz} = k_{\text{eff}}\frac{dT^{\text{III}}}{dz} \qquad (10.7\text{-}13)$

B. C. 6: as $z \to \infty$, $\quad T^{\text{III}} = \text{finite} \qquad (10.7\text{-}14)$

Equations 10.7-10 to 10.7-13 express the continuity of temperature and heat flux at the boundaries between the zones. Equations 10.7-9 and 10.7-14 specify requirements at the two ends of the system.

The solution of Eqs. 10.7-6 to 10.7-14 is considered here for arbitrary $F(\Theta)$. In many cases of practical interest, the convective heat transport is far more important than the axial conductive heat transport. Therefore, here we drop the conductive terms entirely (those containing k_{eff}). The relative importance of conductive and convective energy transport is discussed in more detail in Chapter 13.

If we introduce a dimensionless axial coordinate $Z = z/L$ and a dimensionless chemical heat source $N = S_{c1}L/\rho\hat{C}_p v_0(T_1 - T_0)$, then Eqs. 10.7-6 to 10.7-8 become

Zone I $\quad (Z < 0) \qquad \frac{d\Theta^{\text{I}}}{dZ} = 0 \qquad (10.7\text{-}15)$

Zone II $\quad (0 > Z > 1) \qquad \frac{d\Theta^{\text{II}}}{dZ} = NF(\Theta) \qquad (10.7\text{-}16)$

Zone III $\quad (Z > 1) \qquad \frac{d\Theta^{\text{III}}}{dZ} = 0 \qquad (10.7\text{-}17)$

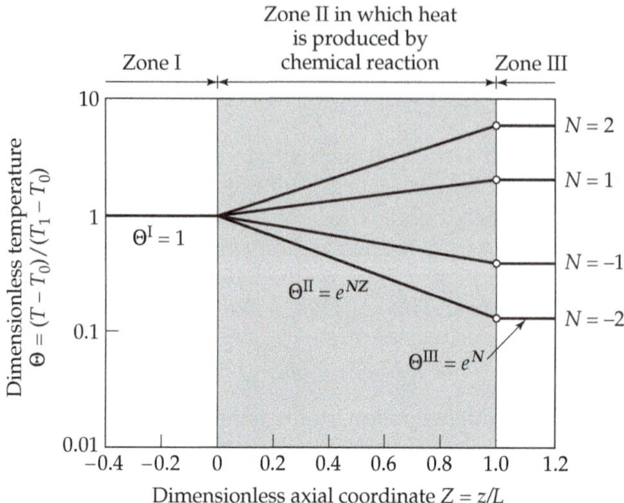

Fig. 10.7-2. Predicted temperature profiles in a fixed-bed axial-flow reactor when the heat production varies linearly with the temperature and when there is negligible axial conduction.

for which we now need only three boundary conditions:

B. C. 1: as $Z \to -\infty$, $\Theta^{\mathrm{I}} = 1$ (10.7-18)

B. C. 2: at $Z = 0$, $\Theta^{\mathrm{I}} = \Theta^{\mathrm{II}}$ (10.7-19)

B. C. 3: at $Z = 1$, $\Theta^{\mathrm{II}} = \Theta^{\mathrm{III}}$ (10.7-20)

The above first-order, separable differential equations, with boundary conditions, are integrated directly to get

$$\boxed{\Theta^{\mathrm{I}} = 1} \tag{10.7-21}$$

$$\boxed{\int_{\Theta^{\mathrm{I}}}^{\Theta^{\mathrm{II}}} \frac{1}{F(\Theta)} d\Theta = NZ} \tag{10.7-22}$$

$$\boxed{\Theta^{\mathrm{III}} = \Theta^{\mathrm{II}}|_{Z=1}} \tag{10.7-23}$$

These results are shown in Fig. 10.7-2 for a simple choice for the source function—namely, $F(\Theta) = \Theta$—which is reasonable for small changes in temperature, if the reaction rate is insensitive to concentration.

Here in this section we ended up discarding the axial conduction terms. In Problem 10B.18, these terms are not discarded, and then the solution shows that there may be some preheating (or precooling) in Region I.

§10.8 ENERGY TRANSPORT WITH ENERGY PRODUCTION: MECHANICAL ENERGY CONVERSION BY VISCOUS DISSIPATION

Next we consider the flow of an incompressible Newtonian fluid between two coaxial cylinders as shown in Fig. 10.8-1. The surfaces of the inner and outer cylinders are maintained at $T = T_0$ and $T = T_b$, respectively. We can expect that T will be a function of r alone.

As the outer cylinder rotates, each cylindrical shell of fluid "rubs" against an adjacent shell of fluid. This friction between adjacent layers of the fluid produces heat; that is,

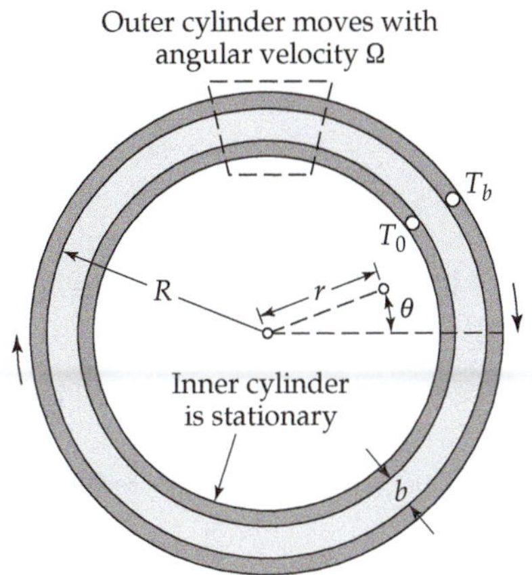

Fig. 10.8-1. Flow between cylinders with energy production by viscous dissipation. That part of the system enclosed within the dotted lines is shown in modified form in Fig. 10.8-2.

the mechanical energy is degraded into thermal energy. The volume heat source resulting from this "viscous dissipation," which can be designated by S_v, appears automatically in the shell balance, when we use the total energy-flux vector **e** defined at the end of Chapter 9, as we shall see presently. Thus, in this section we actually *derive* an expression for S_v, whereas the previous two sections we did not arrive at expressions for S_e or S_c from first principles. We simply stated that there would be heat production, but we gave no "recipe" for creating specific formulas for these quantities. The reason that we did not derive expressions for S_e and S_c via shell balances is that the electrical and chemical bond contributions to the energy flux were not explicitly accounted for; because we did account for viscous work in the energy flux, an expression for S_v will arise naturally. We note that the need to identify energy production expressions is not necessarily a problem, as the energy production phenomena are well understood in general.

If the slit width b is small with respect to the radius R of the outer cylinder, then the problem can be solved approximately by using the somewhat simplified system depicted in Fig 10.8-2. That is, we ignore curvature effects and solve the problem in Cartesian coordinates. The velocity distribution is then $v_z(x) = v_b(x/b)$, where $v_b = \Omega R$.

We now make an energy balance over a shell of thickness Δx, width W, and length L. Since the fluid is in motion, we use the total energy-flux vector **e** as written in Eq. 9.4-1. The balance then reads

$$WLe_x|_x - WLe_x|_x = 0 \tag{10.8-1}$$

Dividing by $WL\Delta x$ and letting the shell thickness Δx go to zero then gives

$$\frac{de_x}{dx} = 0 \tag{10.8-2}$$

This equation may be integrated to give

$$e_x = C_1 \tag{10.8-3}$$

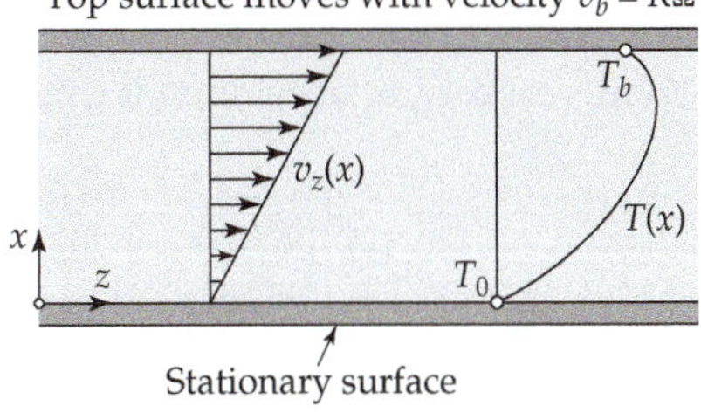

Fig. 10.8-2. Modification of a portion of the flow system in Fig. 10.8-1, in which the curvature of the bounding surfaces is neglected.

Since we do not know any boundary conditions for e_x, we cannot evaluate the integration constant at this point.

We now insert the expression for e_x from Eq. 9.4-1. Since the velocity component in the x direction is zero, the term $\left(\frac{1}{2}\rho v^2 + \rho\hat{U}\right)\mathbf{v}$ can be discarded. The x component of $\mathbf{q}$ is $-k(dT/dx)$ according to Fourier's law. The x component of $[\boldsymbol{\tau}\cdot\mathbf{v}]$ is, by analogy with Eq. 9.3-1, $\tau_{xx}v_x + \tau_{xy}v_y + \tau_{xz}v_z$. Since the only nonzero component of the velocity is v_z, and since $\tau_{xz} = -\mu(dv_z/dx)$ according to Newton's law of viscosity, the x component of $[\boldsymbol{\tau}\cdot\mathbf{v}]$ is $-\mu v_z(dv_z/dx)$. We conclude, then, that Eq. 10.8-3 becomes

$$-k\frac{dT}{dx} - \mu v_z\frac{dv_z}{dx} = C_1 \tag{10.8-4}$$

When the linear velocity profile $v_z(x) = v_b(x/b)$ is inserted, we get

$$-k\frac{dT}{dx} - \mu x\left(\frac{v_b}{b}\right)^2 = C_1 \tag{10.8-5}$$

in which $\mu(v_b/b)^2$ [=] energy/vol · time can be identified as the rate of thermal energy production by viscous dissipation per unit volume S_v.

When Eq. 10.8-5 is integrated, we get

$$T(x) = -\left(\frac{\mu}{k}\right)\left(\frac{v_b}{b}\right)^2\frac{x^2}{2} - \frac{C_1}{k}x + C_2 \tag{10.8-6}$$

The two integration constants are determined from the boundary conditions:

B. C. 1: at $x = 0$, $T = T_0$ (10.8-7)

B. C. 2: at $x = b$, $T = T_b$ (10.8-8)

This yields finally, for $T_b \neq T_0$

$$\boxed{\frac{T(x) - T_0}{T_b - T_0} = \frac{1}{2}\mathrm{Br}\frac{x}{b}\left(1 - \frac{x}{b}\right) + \frac{x}{b}} \tag{10.8-9}$$

in which $\mathrm{Br} = \mu v_b^2/k(T_b - T_0)$ is the dimensionless *Brinkman number*,[1] which is a measure of the importance of the viscous dissipation term (i.e., the importance of rate of viscous dissipation relative to the rate of conduction). If $T_b = T_0$, then Eq. 10.8-9 can be written as

$$\frac{T(x) - T_0}{T_0} = \frac{1}{2}\frac{\mu v_b^2}{kT_0}\frac{x}{b}\left(1 - \frac{x}{b}\right) \tag{10.8-10}$$

and the maximum temperature is at $x/b = \frac{1}{2}$.

If the temperature rise is appreciable, the temperature dependence of the viscosity has to be taken into account. This is discussed in Problem 10C.2.

The viscous heating term $S_v = \mu(v_b/b)^2$ may be understood by the following arguments. For the system in Fig. 10.8-2, the rate at which work is done is the force acting on the upper plate times the velocity with which it moves, or $(-\tau_{xz}WL)(v_b)$. The rate of energy addition per unit volume is then obtained by dividing this quantity by WLb, which gives $(-\tau_{xz}/b)(v_b) = \mu(v_b/b)^2$. This energy all appears as heat and is hence S_v.

In most flow problems viscous heating is not important. However, if there are large velocity gradients or if the fluid viscosity is large, then it cannot be neglected. Examples

---

[1] H. C. Brinkman, *Appl. Sci. Research*, **A2**, 120–124 (1951), solved the viscous dissipation heating problem for the Poiseuille flow in a circular tube. Other dimensionless groups that may be used for characterizing viscous heating have been summarized by R. B. Bird, R. C. Armstrong, and O. Hassager, *Dynamics of Polymeric Liquids*, Vol. **1**, 2nd edition, Wiley, New York, (1987), pp. 207–208.

of situations where viscous heating must be accounted for include: (i) flow of a lubricant between rapidly moving parts, (ii) flow of molten polymers through dies in high-speed extrusion, (iii) flow of highly viscous fluids in high-speed viscometers, and (iv) flow of air in the boundary layer near an earth satellite or rocket during reentry into the earth's atmosphere. The first two of these are further complicated because many lubricants and molten plastics are non-Newtonian fluids. Viscous heating for non-Newtonian fluids is illustrated in Problem 10B.6.

§10.9 FORCED CONVECTION

In the preceding sections the emphasis has been placed predominately on heat conduction in solids. In this and the following section we study two limiting types of heat transport in fluids: *forced convection* and *free convection* (also called *natural convection*). The main differences between these two modes of convection are shown in Fig. 10.9-1. Most industrial heat-transfer problems are usually put into either one or the other of these two limiting categories. In some problems, however, both effects must be taken into account, and then we speak of *mixed convection* (see §14.6 for some empiricisms for handling this situation).

In this section we consider forced convection in a circular tube, a limiting case of which is simple enough to be solved analytically.[1,2]

A viscous fluid with physical properties (μ, k, ρ, and $\hat{C}_p$) assumed constant is in laminar flow in a circular tube of radius R. For $z < 0$, the fluid temperature is uniform at the inlet

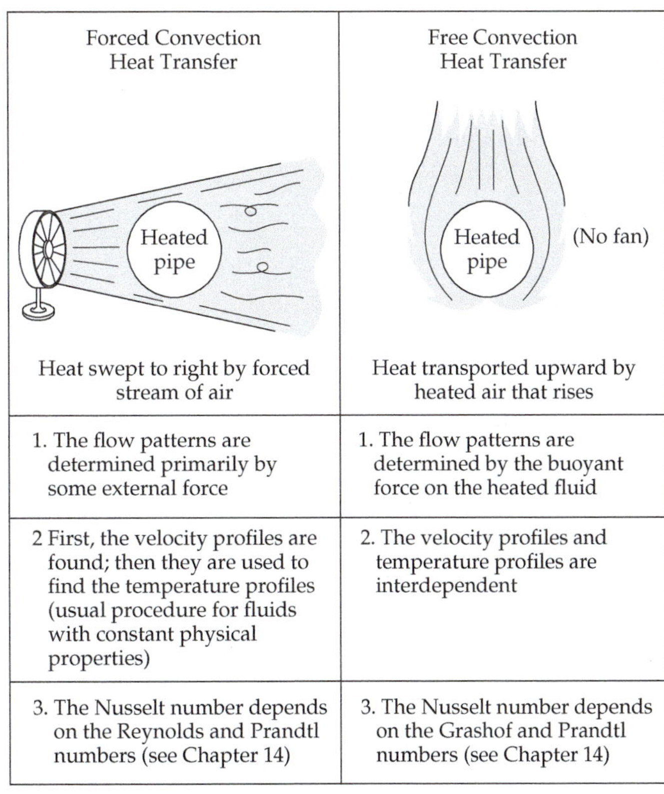
Forced Convection
Heat Transfer
Free Convection
Heat Transfer
Heated pipe
Heated pipe
(No fan)
Heat swept to right by forced stream of air
Heat transported upward by heated air that rises
1. The flow patterns are determined primarily by some external force
1. The flow patterns are determined by the buoyant force on the heated fluid
2 First, the velocity profiles are found; then they are used to find the temperature profiles (usual procedure for fluids with constant physical properties)
2. The velocity profiles and temperature profiles are interdependent
3. The Nusselt number depends on the Reynolds and Prandtl numbers (see Chapter 14)
3. The Nusselt number depends on the Grashof and Prandtl numbers (see Chapter 14)

Fig. 10.9-1. A comparison of forced and free convection in nonisothermal systems.

[1] A. Eagle and R. M. Ferguson, *Proc. Roy. Soc. (London)*, **A127**, 540–566 (1930).

[2] S. Goldstein, *Modern Developments in Fluid Dynamics*, Oxford University Press (1938), Dover Edition (1965), Vol. II, p. 622.

temperature T_1. For $z > 0$, there is a constant radial heat flux $q_r = -q_0$ at the wall. Such a situation exists, for example, when a pipe is wrapped uniformly with an electrical heating coil, in which case q_0 is positive. If the pipe is being chilled, then q_0 is negative.

As indicated in Fig. 10.9-1, the first step in solving a forced convection heat-transfer problem is the calculation of the velocity profiles in the system. We have seen in §2.3 how this may be done for tube flow by using the shell balance method. We know that the velocity distribution so obtained is $v_r = 0$, $v_\theta = 0$, and

$$v_z(r) = \frac{(\mathscr{P}_0 - \mathscr{P}_L)R^2}{4\mu L}\left[1 - \left(\frac{r}{R}\right)^2\right] = v_{z,\max}\left[1 - \left(\frac{r}{R}\right)^2\right] \tag{10.9-1}$$

This parabolic distribution is valid sufficiently far downstream from the inlet that the entrance length has been exceeded.

In this problem, heat is being transported in both the r and the z directions. Therefore, for the energy balance we use a "washer-shaped" system, which is formed by the intersection of an annular region of thickness Δr with a slab of thickness Δz (see Fig. 10.9-2). In this problem, we are dealing with a flowing fluid, and therefore all terms in the **e** vector will be retained. The various contributions to the energy balance (Eq. 10.1-1) are

Total energy in at r:	$e_r\|_r \cdot 2\pi r\Delta z = (2\pi r e_r)\|_r \Delta z$	(10.9-2)
Total energy out at r + Δr:	$e_r\|_{r+\Delta r} \cdot 2\pi(r+\Delta r)\Delta z = (2\pi r e_r)\|_{r+\Delta r}\Delta z$	(10.9-3)
Total energy in at z:	$e_z\|_z \cdot 2\pi r\Delta r$	(10.9-4)
Total energy out at z + Δz:	$e_z\|_{z+\Delta z} \cdot 2\pi r\Delta r$	(10.9-5)
Work done on fluid by gravity:	$\rho v_z g_z \cdot 2\pi r\Delta r\Delta z$	(10.9-6)

The last contribution is the rate at which work is done on the fluid within the ring by gravity—that is, the force per unit volume ρg_z times the volume $2\pi r\Delta r\Delta z$ multiplied by the velocity of the fluid.

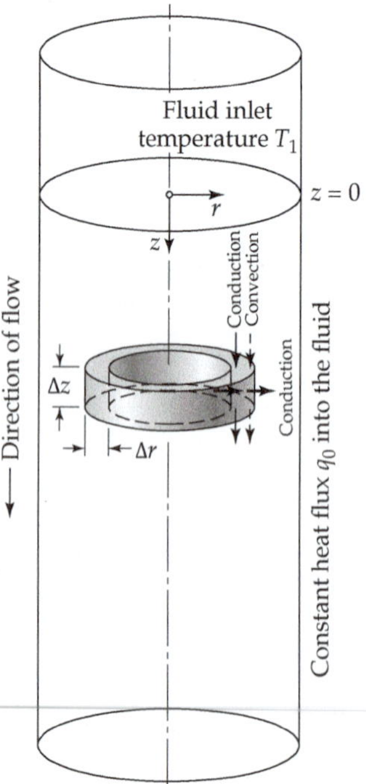

Fig. 10.9-2. Heating of a fluid in laminar flow through a circular tube, showing the annular ring over which the energy balance is made.

The energy balance is obtained by summing these contributions and equating the sum to zero. Then we divide by $2\pi\Delta r\Delta z$ to get

$$-\frac{(re_r)|_{r+\Delta r} - (re_r)|_r}{\Delta r} - r\frac{e_z|_{z+\Delta z} - e_z|_z}{\Delta z} + \rho v_z g_z r = 0 \tag{10.9-7}$$

In the limits as Δr and Δz go to zero, we find

$$-\frac{1}{r}\frac{\partial}{\partial r}(re_r) - \frac{\partial e_z}{\partial z} + \rho v_z g_z = 0 \tag{10.9-8}$$

Next we use Eqs. 9.4-2 and 9.4-6 to write out the expressions for the r and z components of the total energy-flux vector, using the fact that the only nonzero component of $\mathbf{v}$ is the axial component v_z:

$$e_r = \tau_{rz}v_z + q_r = \left(-\mu\frac{\partial v_z}{\partial r}\right)v_z - k\frac{\partial T}{\partial r} \tag{10.9-9}$$

$$\begin{aligned} e_z &= \left(\frac{1}{2}\rho v_z^2\right)v_z + \rho\hat{H}v_z + \tau_{zz}v_z + q_z \\ &= \left(\frac{1}{2}\rho v_z^2\right)v_z + \rho\hat{H}^{\mathrm{o}}v_z + \rho\hat{C}_p(T - T^{\mathrm{o}})v_z + (p - p^{\mathrm{o}})v_z \\ &\quad + \left(-2\mu\frac{\partial v_z}{\partial z}\right)v_z - k\frac{\partial T}{\partial z} \end{aligned} \tag{10.9-10}$$

These expressions may be simplified using a few approximations. The first term in Eq. 10.9-9 contributes only to energy production by viscous dissipation, which we assume to be negligible (i.e., this is not a problem of lubrication, polymer extrusion, high-speed viscometry, or spacecraft reentry; see §10.8). The first and second terms of Eq. 10.9-10 may be ignored because, in the next step, we will be taking the derivative of this expression with respect to z, and v_z and $\hat{H}^{\mathrm{o}}$ do not depend on z. We ignore the fourth term because the pressure contributions to the enthalpy are often negligible compared to the thermal contribution (the third term) (see Example 9.4-1). The fifth term can be omitted because v_z depends only on r. Thus, the approximate energy-flux expressions employed in this problem are

$$e_r \approx -k\frac{\partial T}{\partial r} \tag{10.9-11}$$

$$e_z \approx \rho\hat{C}_p(T - T^0)v_z - k\frac{\partial T}{\partial z} \tag{10.9-12}$$

Substitution of these flux expressions into Eq. 10.9-8 and using the fact that v_z depends only on r gives, after some rearrangement,

$$\rho\hat{C}_p v_z(r)\frac{\partial T}{\partial z} = k\left[\frac{1}{r}\frac{\partial}{\partial r}\left(r\frac{\partial T}{\partial r}\right) + \frac{\partial^2 T}{\partial z^2}\right] \tag{10.9-13}$$

where we have also neglected the gravity term in Eq. 10.9-8, because this term is typically small (see Example 10.9-1) (an alternative, more rigorous approach for obtaining this equation is illustrated in Problem 10B.20). The last term in the brackets corresponds to heat conduction in the axial direction. We omit this term since we know from experience that it is usually small in comparison with the convection of energy in the axial direction (the term on the left side of the equation). Therefore, after substituting the expression for $v_z(r)$, the equation that we want to solve here is

$$\rho\hat{C}_p v_{z,\max}\left[1 - \left(\frac{r}{R}\right)^2\right]\frac{\partial T}{\partial z} = k\left[\frac{1}{r}\frac{\partial}{\partial r}\left(r\frac{\partial T}{\partial r}\right)\right] \tag{10.9-14}$$

This partial differential equation, when solved, describes the temperature in the fluid as a function of r and z. The boundary conditions are

B. C. 1: at $r = 0$, $T = \text{finite}$ (10.9-15)

B. C. 2: at $r = R$, $k\dfrac{\partial T}{\partial r} = q_0$ (constant) (10.9-16)

B. C. 3: at $z = 0$, $T = T_1$ (10.9-17)

We now put the problem statement into dimensionless form. The choice of the dimensionless quantities is arbitrary. We choose:

$$\Theta = \frac{T - T_1}{q_0 R/k} \qquad \xi = \frac{r}{R} \qquad \zeta = \frac{z}{\rho \hat{C}_p v_{z,\max} R^2/k} \tag{10.9-18,19,20}$$

Generally one tries to select dimensionless quantities so as to minimize the number of parameters in the final problem formulation. In this problem, the choice of $\xi = r/R$ is a natural one, because of the appearance of r/R in the differential equation. The choice for the dimensionless temperature difference Θ is suggested by the second and third boundary conditions. Having specified these two dimensionless variables, the choice of dimensionless axial coordinate follows naturally.

The resulting problem statement, in dimensionless form, is now

$$(1 - \xi^2)\frac{\partial \Theta}{\partial \zeta} = \frac{1}{\xi}\frac{\partial}{\partial \xi}\left(\xi \frac{\partial \Theta}{\partial \xi}\right) \tag{10.9-21}$$

with boundary conditions

B. C. 1: at $\xi = 0$, $\Theta = \text{finite}$ (10.9-22)

B. C. 2: at $\xi = 1$, $\dfrac{\partial \Theta}{\partial \xi} = 1$ (10.9-23)

B. C. 3: at $\zeta = 0$, $\Theta = 0$ (10.9-24)

The partial differential equation in Eq. 10.9-21 has been solved for these boundary conditions,[3] but in this section we do not give the complete solution.

It is, however, instructive to obtain the *asymptotic solution* to Eq. 10.9-21 for large ζ. After the fluid is sufficiently far downstream from the beginning of the heated section, one expects that the constant heat flux through the wall will result in an increase of the fluid temperature that is linear in ζ. One further expects that the shape of the temperature profiles as a function of ξ will ultimately not undergo further change with increasing ζ (see Fig. 10.9-3). Hence, a solution of the following form seems reasonable for large ζ

$$\Theta(\xi,\zeta) = C_0\zeta + \Psi(\xi) \tag{10.9-25}$$

in which C_0 is a constant to be determined presently.

The function in Eq. 10.9-25 is clearly not the complete solution to the problem; it does allow the partial differential equation and boundary conditions 1 and 2 to be satisfied, but clearly does not satisfy boundary condition 3. Hence, we replace the latter by a statement that the energy entering through the wall from 0 to z is the same as the difference between the energy leaving by convection through the cross section at z and that entering at $z = 0$ (see Fig. 10.9-4),

Condition 4: $$2\pi R z q_0 = \int_0^{2\pi}\int_0^R \rho \hat{C}_p (T(r,z) - T_1) v_z(r) r\, dr\, d\theta \tag{10.9-26}$$

---

[3] R. Siegel, E. M. Sparrow, and T. M. Hallman, *Appl. Sci. Research*, **A7**, 386–392 (1958). See Example 11.5-3 for the asymptotic solution for small ζ.

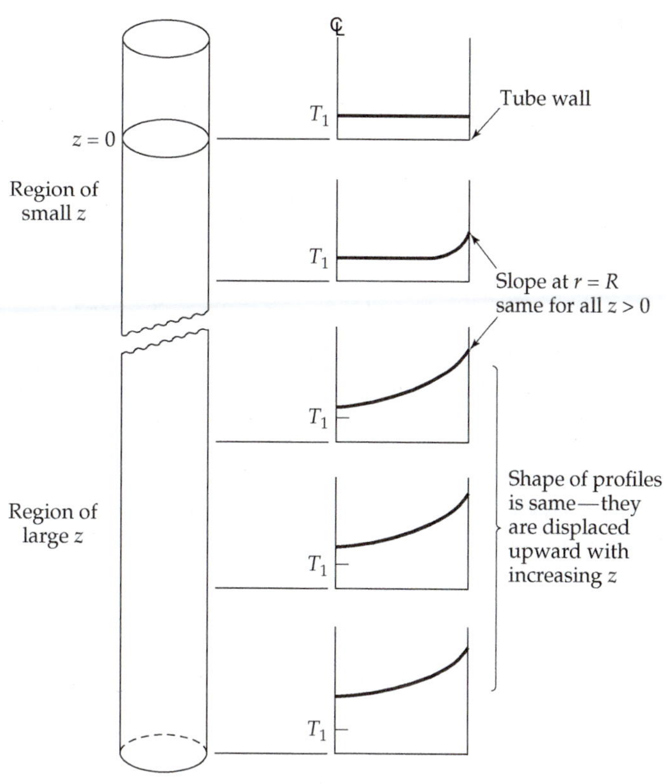

Fig. 10.9-3. Sketch showing how one expects the temperature $T(r,z)$ to look for the system shown in Fig. 10.9-2 when the fluid is heated by means of a heating coil wrapped uniformly around the tube (corresponding to q_0 positive).

We next integrate over θ, and then put Eq. 10.9-26 into dimensionless form to get

Condition 4:
$$\zeta = \int_0^1 \Theta(\xi,\zeta)(1-\xi^2)\xi\, d\xi \tag{10.9-27}$$

Using this condition guarantees that we satisfy an energy balance over the entire tube.

Substitution of the postulated function of Eq. 10.9-25 into Eq. 10.9-21 leads to the following ordinary differential equation for Ψ (see Eq. C.1-11):

$$\frac{1}{\xi}\frac{d}{d\xi}\left(\xi\frac{d\Psi}{d\xi}\right) = C_0(1-\xi^2) \tag{10.9-28}$$

This equation may be integrated twice with respect to ξ, and the result substituted into Eq. 10.9-25 to give

$$\Theta(\xi,\zeta) = C_0\zeta + C_0\left(\frac{\xi^2}{4} - \frac{\xi^4}{16}\right) + C_1 \ln \xi + C_2 \tag{10.9-29}$$

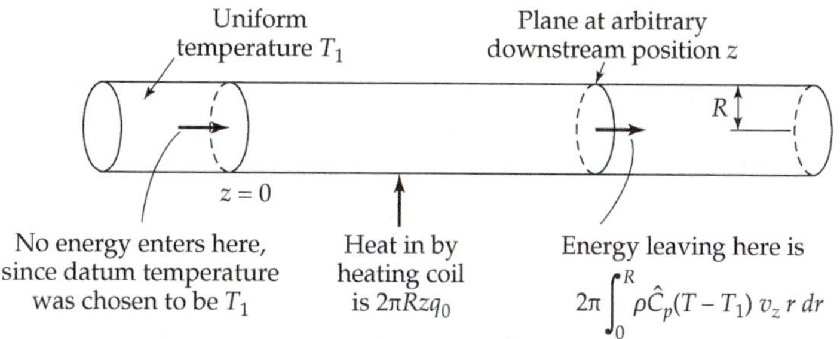

Fig. 10.9-4. Energy balance used for boundary condition 4 given in Eq. 10.9-26.

The three constants are determined from the conditions 1, 2, and 4 above:

B. C. 1: $$C_1 = 0 \tag{10.9-30}$$

B. C. 2: $$C_0 = 4 \tag{10.9-31}$$

Condition 4: $$C_2 = -\frac{7}{24} \tag{10.9-32}$$

Substitution of these values into Eq. 10.9-29 gives finally

$$\boxed{\Theta(\xi,\zeta) = 4\zeta + \xi^2 - \frac{1}{4}\xi^4 - \frac{7}{24}} \tag{10.9-33}$$

This result gives the dimensionless temperature difference as a function of the dimensionless radial and axial coordinates. It is exact in the limit as $\zeta \to \infty$; for $\zeta > 0.1$, it predicts the local value of Θ to within about 2%. An asymptotic solution to this problem in the limit of small ζ is given in §11.5 (see Example 11.5-3).

Once the temperature distribution is known, one can get various derived quantities. There are two kinds of average temperatures commonly used in connection with the flow of fluids with constant ρ and $\hat{C}_p$:

$$\langle T \rangle = \frac{\int_0^{2\pi}\int_0^R T(r,z) r\,dr\,d\theta}{\int_0^{2\pi}\int_0^R r\,dr\,d\theta} = T_1 + \left(4\zeta + \frac{1}{8}\right)\frac{q_0 R}{k} \tag{10.9-34}$$

$$T_b(z) = \frac{\langle v_z T \rangle}{\langle v_z \rangle} = \frac{\int_0^{2\pi}\int_0^R v_z(r)T(r,z) r\,dr\,d\theta}{\int_0^{2\pi}\int_0^R v_z(r) r\,dr\,d\theta} = T_1 + (4\zeta)\frac{q_0 R}{k} \tag{10.9-35}$$

Both averages are functions of z. The quantity $\langle T \rangle$ is just the arithmetic average of the temperatures over the cross section at z. The "bulk temperature" T_b is the temperature one would measure if the tube were chopped off at z and if the fluid exiting there were collected in a container and thoroughly mixed. This average temperature is sometimes referred to as the "cup-mixing temperature" or the "flow-average temperature."

Now let us evaluate the local heat-transfer driving force, $T_0 - T_b$, which is just the difference between the wall temperature $T_0(z) = T(R,z)$ (from Eq. 10.9-33) and bulk temperature (from Eq. 10.9-35) at a distance z down the tube:

$$T_0 - T_b = \frac{11}{24}\frac{q_0 R}{k} = \frac{11}{48}\frac{q_0 D}{k} \tag{10.9-36}$$

where D is the tube diameter. We may now rearrange this result in the form of a dimensionless wall heat flux

$$\text{Nu} = \frac{q_0 D}{k(T_0 - T_b)} = \frac{48}{11} \tag{10.9-37}$$

The quantity Nu, in Chapter 14, will be identified as the *Nusselt number*.

Before leaving this section, we point out that the dimensionless axial coordinate ζ introduced above may be rewritten in the following way:

$$\zeta = \left[\!\left[\frac{\mu}{D\langle v_z \rangle \rho}\right]\!\right]\left[\!\left[\frac{k}{\hat{C}_p \mu}\right]\!\right]\left[\!\left[\frac{z}{R}\right]\!\right] = \frac{1}{\text{Re}\,\text{Pr}}\left[\!\left[\frac{z}{R}\right]\!\right] = \frac{1}{\text{Pé}}\left[\!\left[\frac{z}{R}\right]\!\right] \tag{10.9-38}$$

Here D is the tube diameter, Re is the Reynolds number used in Part I, and Pr and Pé are the Prandtl and Péclet numbers introduced in Chapter 9. We shall find in Chapter 13 that the Reynolds and Prandtl numbers can be expected to appear in forced convection problems. This point will be reinforced in Chapter 14 in connection with correlations for heat-transfer coefficients.

EXAMPLE 10.9-1

Comparison of the Magnitudes of the Gravity and Convection Terms

Show that the gravity term omitted from Eq. 10.9-8 is indeed negligible for reasonable values of the variables.

SOLUTION

Equation 10.9-8 shows that the omitted gravity term is $\rho v_z g_z$. We wish to compare this to the magnitude of the convection term in Eq. 10.9-13, $\rho \hat{C}_p v_z(\partial T/\partial z)$. The ratio of the gravity term to the convection term is $g_z/\hat{C}_p(\partial T/\partial z)$. Choosing the values for the variables $g_z = 9.8$ m/s^2, $\hat{C}_p = 4184$ J/kg · °C (for water near room temperature) and $\partial T/\partial z = 10$ °C/m, the ratio of the terms is

$$\frac{g_z}{\hat{C}_p(\partial T/\partial z)} = \left(\frac{9.8\ \text{m}}{\text{s}^2}\right)\left(\frac{\text{kg}\cdot{}^\circ\text{C}}{4184\ \text{J}}\right)\left(\frac{\text{m}}{10\ {}^\circ\text{C}}\right)\left(\frac{1\ \text{J}}{1\ \text{kg}\cdot\text{m}^2/\text{s}^2}\right)$$
$$= 2.3 \times 10^{-4} \qquad (10.9\text{-}39)$$

Thus, the gravity term is negligible, even for magnitudes of the temperature gradient smaller than 10 °C/m.

§10.10 FREE CONVECTION

In §10.9 we gave an example of forced convection. In this section we turn our attention to an elementary free convection problem, namely, the flow between two parallel walls maintained at different temperatures (see Fig. 10.10-1).

A fluid with density ρ and viscosity μ is located between two vertical walls a distance $2B$ apart. The heated wall at $y = -B$ is maintained at temperature T_2, and the cooled wall at $y = +B$ is maintained at temperature T_1. It is assumed that the temperature difference is sufficiently small that the fluid density varies linearly with the temperature.

Because of the temperature gradient in the system, the fluid near the hot wall rises and that near the cold wall descends. The system is closed at the top and bottom, so that the fluid is continuously circulating between the plates. The mass rate of flow of the fluid in the upward-moving stream is the same as that in the downward-moving stream. The

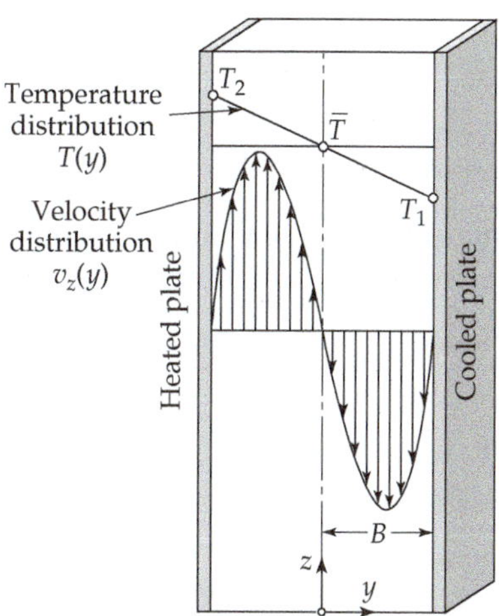

Fig. 10.10-1. Laminar, free-convection flow between two vertical plates at two different temperatures. The velocity is a cubic function of the coordinate y.

plates are presumed to be very tall, so that end effects near the top and bottom can be disregarded. Then, for all practical purposes, the temperature is a function of y alone.

An energy balance can now be made over a thin slab of fluid of thickness Δy, using the y component of the total energy flux vector $\mathbf{e}$ as given in Eq. 9.4-2. The term containing the kinetic and internal energy can be disregarded, since the y component of the $\mathbf{v}$ vector is zero. The y component of the term $[\boldsymbol{\tau} \cdot \mathbf{v}]$ is $\tau_{yz}v_z = -\mu(dv_z/dy)v_z$, which would lead to the viscous heating contribution discussed in §10.8. However, in the very slow flows encountered in free convection, this term will be extremely small and can be neglected. The energy balance then leads to the equation

$$-\frac{dq_y}{dy} = 0 \quad \text{or} \quad k\frac{d^2T}{dy^2} = 0 \tag{10.10-1}$$

for constant k. The temperature equation is to be solved with the boundary conditions:

B. C. 1: at $y = -B$, $T = T_2$ (10.10-2)

B. C. 2: at $y = +B$, $T = T_1$ (10.10-3)

The solution to this problem is

$$\boxed{T(y) = \overline{T} - \frac{1}{2}\Delta T\frac{y}{B}} \tag{10.10-4}$$

in which $\Delta T = T_2 - T_1$ is the difference of the wall temperatures, and $\overline{T} = \frac{1}{2}(T_1 + T_2)$ is their arithmetic mean.

By making a momentum balance over the same slab of thickness Δy, one arrives at a differential equation for the velocity distribution

$$\mu\frac{d^2v_z}{dy^2} = \frac{dp}{dz} + \rho g \tag{10.10-5}$$

Here the viscosity has been assumed constant (see Problem 10B.12 for a solution with temperature-dependent viscosity).

The phenomenon of free convection results from the fact that when the fluid is heated, the density (usually) decreases, and the fluid rises. The mathematical description of the phenomenon must take this essential feature of the phenomenon into account. Because the temperature difference $\Delta T = T_2 - T_1$ is taken to be small in this problem, it can be expected that the density changes in the system will be small. This suggests that we should expand ρ in a Taylor series (§C.2) about the temperature $\overline{T} = \frac{1}{2}(T_1 + T_2)$ thus:

$$\begin{aligned} \rho(T) &= \rho|_{T=\overline{T}} + \left.\frac{d\rho}{dT}\right|_{T=\overline{T}}(T - \overline{T}) + \cdots \\ &= \overline{\rho} - \overline{\rho}\overline{\beta}(T - \overline{T}) + \cdots \end{aligned} \tag{10.10-6}$$

Here $\overline{\rho}$ and $\overline{\beta}$ are the density and coefficient of volume expansion evaluated at the temperature $\overline{T}$. The coefficient of volume expansion is defined as

$$\beta = \frac{1}{V}\left(\frac{\partial V}{\partial T}\right)_p = \frac{1}{(1/\rho)}\left(\frac{\partial(1/\rho)}{\partial T}\right)_p = -\frac{1}{\rho}\left(\frac{\partial \rho}{\partial T}\right)_p \tag{10.10-7}$$

We now introduce the "Taylor-made" equation of state of Eq. 10.10-6 (keeping only the first two terms) into the equation of motion in Eq. 10.10-5 to get

$$\mu\frac{d^2v_z}{dy^2} = \frac{dp}{dz} + \overline{\rho}g - \overline{\rho}g\overline{\beta}(T(y) - \overline{T}) \tag{10.10-8}$$

This equation describes the balance among the viscous force, the pressure force, the gravity force, and the buoyant force $-\bar{\rho}g\bar{\beta}(T(y) - \bar{T})$ (all per unit volume). Into this we now substitute the temperature distribution given in Eq. 10.10-4 to get the differential equation

$$\mu \frac{d^2 v_z}{dy^2} = \left(\frac{dp}{dz} + \bar{\rho} g \right) + \frac{1}{2} \bar{\rho} g \bar{\beta} \Delta T \frac{y}{B} \tag{10.10-9}$$

which is to be solved with the boundary conditions

B. C. 1: at $y = -B, \quad v_z = 0$ (10.10-10)

B. C. 2: at $y = +B, \quad v_z = 0$ (10.10-11)

The solution is

$$v_z(y) = \frac{(\bar{\rho} g \bar{\beta} \Delta T) B^2}{12\mu} \left[\left(\frac{y}{B} \right)^3 - \left(\frac{y}{B} \right) \right] + \frac{B^2}{12\mu} \left(\frac{dp}{dz} + \bar{\rho} g \right) \left[\left(\frac{y}{B} \right)^2 - 1 \right] \tag{10.10-12}$$

We now require that the net mass flow in the z direction be zero, that is,

$$\int_{-B}^{+B} \rho v_z dy = 0 \tag{10.10-13}$$

Substitution of v_z from Eq. 10.10-12 and ρ from Eqs. 10.10-6 and 10.10-4 into this integral leads to the conclusion that

$$\frac{dp}{dz} = -\bar{\rho} g \tag{10.10-14}$$

when terms containing the square of the small quantity ΔT are neglected (which has been employed in Eq. 10.10-9). Equation 10.10-14 states that the pressure gradient in the system is due solely to the weight of the fluid, and the usual hydrostatic pressure distribution prevails. Therefore, the second term on the right side of Eq. 10.10-12 drops out, and the final expression for the velocity distribution is

$$\boxed{v_z(y) = \frac{(\bar{\rho} g \bar{\beta} \Delta T) B^2}{12\mu} \left[\left(\frac{y}{B} \right)^3 - \left(\frac{y}{B} \right) \right]} \tag{10.10-15}$$

The average velocity in the upward-moving stream is

$$\langle v_z \rangle = \frac{(\bar{\rho} g \bar{\beta} \Delta T) B^2}{48\mu} \tag{10.10-16}$$

The motion of the fluid is thus a direct result of the buoyant force term in Eq. 10.10-8, associated with the temperature gradient in the system. The velocity distribution of Eq. 10.10-15 is shown in Fig. 10.10-1. It is this sort of velocity distribution that occurs in the air space in a double-pane window or in double-wall panels in buildings. It is also this kind of flow that occurs in a Clusius-Dickel column used for separating isotopes or organic liquids mixtures by the combined effects of thermal diffusion and free convection.[1]

The velocity distribution in Eq. 10.10-15 may be rewritten using a dimensionless velocity $\breve{v}_z = Bv_z\bar{\rho}/\mu$ and a dimensionless coordinate $\breve{y} = y/B$ thus:

$$\breve{v}_z(\breve{y}) = \frac{1}{12} \mathrm{Gr} (\breve{y}^3 - \breve{y}) \tag{10.10-17}$$

[1] Thermal diffusion is the diffusion resulting from a temperature gradient (see Chapter 24). For a lucid discussion of the Clusius-Dickel column, see K. E. Grew and T. L. Ibbs, *Thermal Diffusion in Gases*, Cambridge University Press (1952), pp. 94–106.

Here Gr is the dimensionless *Grashof* number,[2] defined by

$$\text{Gr} = \left[\!\left[\frac{(\bar{\rho}^2 g \bar{\beta} \Delta T)\, B^3}{\mu^2} \right]\!\right] = \left[\!\left[\frac{\bar{\rho} g B^3 \Delta\rho}{\mu^2} \right]\!\right] \tag{10.10-18}$$

where $\Delta\rho = \rho_1 - \rho_2$. The second form of the Grashof number is obtained from the first form by using Eq. 10.10-6. The Grashof number is the characteristic group occurring in analyses of free convection, as is shown by dimensional analysis in Chapter 13. It arises in heat-transfer coefficient correlations in Chapter 14.

§10.11 CONCLUDING COMMENTS

In this chapter we have presented a wide variety of problems that can be solved by shell balance methods. However, these have been restricted to rather simple problems in which heat flows in just one coordinate direction (an exception to this is the forced convection problem of §10.9 where two coordinate directions are involved). We have given illustrations of problems involving Cartesian and cylindrical coordinates. In the problems at the end of the chapter, spherical coordinates are also used.

In this chapter we have given three examples of problems in which there is a heat source term. The heating of a current-carrying wire was an example of the "electrical heat source" S_e. The example of the flow with chemical reactions introduced the "chemical heat source" S_c, and the example of viscous heating in the flow between two moving surfaces illustrated the "viscous heat source" S_v. These examples point out the differences between heat sources that have to be added artificially (such as S_e and S_c) and those that are automatically built into the equations (such as S_v, which arose naturally from the expression for the energy flux). In the problems at the end of the chapter, we introduce another artificial source term, namely, the heat source associated with nuclear reaction heating, S_n. The source terms S_c, S_e, and S_n have to be added on an ad hoc basis, because the energy-flux expression used here is not actually sufficiently general to account for all the chemical, electrical, or nuclear processes occurring in the material. More will be said about this in the next chapter.

In Chapter 11, we will set up the general energy equation based on the law of conservation of energy. This equation of change will enable us to set up problems more quickly and efficiently, in that the shell balances do not have to be written down for each problem. The equation of change for energy is not completely general in that the chemical, electrical, and nuclear processes are not fully accounted for. Nonetheless, within this limitation it will be found to be most useful. It will be found to extend the notion of the first law of thermodynamics, which applies only to time-independent equilibrium systems, so that we can then handle time-dependent nonequilibrium systems.

QUESTIONS FOR DISCUSSION

1. Verify that the Brinkman, Biot, Prandtl, Nusselt, and Grashof numbers are indeed dimensionless.
2. Why does **e** simplify to **q** for problems in heat conduction in solids?
3. To what problem in electrical circuits is the addition of thermal resistances analogous?
4. What is the coefficient of volume expansion for an ideal gas? What is the corresponding expression for the Grashof number?

---

[2]Named for **Franz Grashof** (1826–1893) (pronounced "Grahss-hoff"). He was professor of applied mechanics in Karlsruhe and one of the founders of the Verein Deutscher Ingenieure in 1856.

5. What might be some consequences of large temperature gradients produced by viscous heating in viscometry, lubrication, and plastics extrusion?
6. In §10.9, would there be any advantage to choosing the dimensionless temperature and dimensionless axial coordinate to be $\Theta = (T - T_1)/T_1$ and $\zeta = z/R$?
7. What would happen in §10.10 if the fluid were water and $\bar{T}$ were 4°C?
8. Is there any advantage to solving Eq. 10.5-9 in terms of exponential functions rather than hyperbolic functions (see Eq. C.1-4a and b)?
9. In going from Eq. 10.9-11 to Eq. 10.9-12 the axial conduction term was neglected with respect to the axial convection term. To justify this, put in some reasonable numerical values to estimate the relative sizes of the terms.
10. How serious is it to neglect the dependence of viscosity on temperature in solving forced convection problems? Viscous dissipation heating problems?
11. At steady state the temperature profiles in a laminated system appear thus:

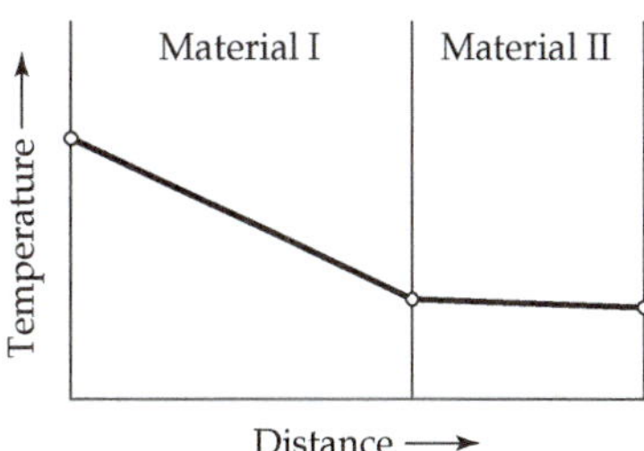

Which material has the higher thermal conductivity?

12. Show that Eq. 10.3-4 can be obtained directly by rewriting Eq. 10.3-1 with $x + \Delta x$ replaced by x_0. Similarly, one gets Eq. 10.3-20 from Eq. 10.3-17, with $r + \Delta r$ replaced by r_0.

PROBLEMS

10A.1 Heat loss from an insulated pipe. A standard Schedule 40, 2-in. steel pipe (inside diameter 2.067 in., and wall thickness 0.154 in.) carrying steam is insulated with 2 in. of 85% magnesia covered in turn with 2 in. of cork. Estimate the heat loss per hour per foot of pipe if the inner surface of the pipe is at 250°F and the outer surface of the cork is at 90°F. The thermal conductivities (in Btu/hr · ft · °F) of the substances concerned are: steel, 26.1; 85% magnesia, 0.04; cork, 0.03.

Answer: 24 Btu/hr · ft

10A.2 Heat loss from a rectangular fin. Calculate the heat loss from a rectangular fin (see Fig. 10.5-1) for the following conditions:

Air temperature	350°F
Wall temperature	500°F
Thermal conductivity of fin	60 Btu/hr · ft · °F
Thermal conductivity of air	0.0022 Btu/hr · ft · °F
Heat-transfer coefficient	120 Btu/hr · ft^2 · °F
Length of fin	0.2 ft
Width of fin	1.0 ft
Thickness of fin	0.16 in.

Answer: 2074 Btu/hr

10A.3 Maximum temperature in a lubricant. An oil is acting as a lubricant for a pair of cylindrical surfaces such as those shown in Fig. 10.8-1. The angular velocity of the outer cylinder is 7908 rpm. The outer cylinder has a radius of 5.06 cm, and the clearance between the cylinders is 0.027 cm. What is the maximum temperature in the oil if both wall temperatures are known to be 158°F? The physical properties of the oil are assumed constant at the following values:

Viscosity	92.3 cp
Density	1.22 g/cm^3
Thermal conductivity	0.055 cal/s · cm · °C

Answer: 174°F

10A.4 Current-carrying capacity of wire. A copper wire of 0.040 in. in diameter is insulated uniformly with plastic to an outer diameter of 0.12 in. and is exposed to surroundings at 100°F. The heat-transfer coefficient from the outer surface of the plastic to the surroundings is 1.5 Btu/hr · ft^2 · °F. What is the maximum steady current, in amperes, that this wire can carry without heating any part of the plastic above its operating limit of 200°F? The thermal and electrical conductivities may be assumed constant at the values given here:

	k(Btu/hr · ft · °F)	k_e(ohm^{-1}cm^{-1})
Copper	220	5.1×10^5
Plastic	0.20	0.0

Answer: 13.4 amp

10A.5 Free-convection velocity.

(a) Verify the expression for the average velocity in the upward-moving stream in Eq. 10.10-16.

(b) Evaluate $\bar{\beta}$ for the conditions given below.

(c) What is the average velocity in the upward-moving stream in the system described in Fig. 10.10-1 for air flowing under these conditions?

Pressure	1 atm
Temperature of the heated wall	100°C
Temperature of the cooled wall	20°C
Spacing between the walls	0.6 cm

Answer: **(c)** 2.3 cm/s

10A.6 Insulating power of a wall. The "insulating power" of a wall can be measured by means of the arrangement shown in Fig. 10A.6. One places a plastic panel against the wall. In the panel two thermocouples are mounted flush with the panel surfaces. The thermal conductivity and thickness of the plastic panel are known. From the measured steady-state temperatures shown in the figure, calculate:

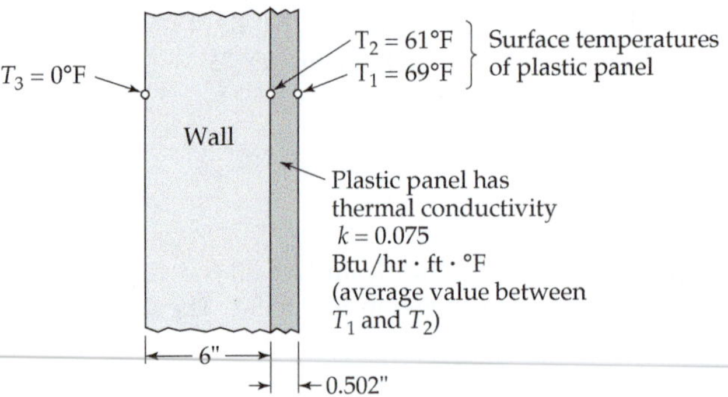

Fig. 10A.6 Determination of the thermal resistance of a wall.

(a) The steady-state heat flux through the wall (and panel).

(b) The "thermal resistance" (wall thickness divided by thermal conductivity).

Answers: **(a)** 14.3 Btu/hr · ft^2; **(b)** 4.2 ft^2 · hr · °F/Btu

10A.7 Viscous heating in a ballpoint pen. You are asked to decide whether the apparent decrease in viscosity in ballpoint pen inks during writing results from "shear thinning" (decrease in viscosity because of non-Newtonian effects) or "temperature thinning" (decrease in viscosity because of temperature rise caused by viscous heating). If the temperature rise is less than 1 K, then "temperature thinning" will not be important. Estimate the temperature rise using Eq. 10.8-9 and the following estimated data:

Clearance between ball and holding cavity	5×10^{-5} in.
Diameter of ball	1 mm
Viscosity of ink	10^4 cp
Speed of writing	100 in./min
Thermal conductivity of ink (rough guess)	5×10^{-4} cal/s · cm · °C

10A.8 Temperature rise in an electrical wire.

(a) A copper wire, 5 mm in diameter and 15 ft long, has a voltage drop of 0.6 volts. Find the maximum temperature in the wire if the ambient air temperature is 25°C and the heat-transfer coefficient h is 5.7 Btu/hr · ft^2 · °F.

(b) Compare the temperature drops across the wire and the surrounding air.

10B.1 Heat conduction from a sphere to a stagnant fluid. A heated sphere of radius R is suspended in a large, motionless body of fluid. It is desired to study the heat conduction in the fluid surrounding the sphere in the absence of convection.

(a) Set up the differential equation describing the temperature T in the surrounding fluid as a function of r, the distance from the center of the sphere. The thermal conductivity k of the fluid is considered constant.

(b) Integrate the differential equation and use these boundary conditions to determine the integration constants: at $r = R$, $T = T_R$; and as $r \to \infty$, $T \to T_\infty$.

(c) From the temperature profile, obtain an expression for the heat flux at the surface. Equate this result to the heat flux given by "Newton's law of cooling" and show that a dimensionless heat-transfer coefficient (known as the *Nusselt number*) is given by

$$\text{Nu} = \frac{hD}{k} = 2 \tag{10B.1-1}$$

in which D is the sphere diameter. This well-known result provides the limiting value of Nu for heat transfer from spheres at low Reynolds and Grashof numbers (see §14.4).

(d) In what respect are the Biot number and the Nusselt number different?

10B.2 Viscous heating in slit flow. Find the temperature profile for the viscous heating problem shown in Fig. 10.8-2, when given the following boundary conditions: at $x = 0$, $T = T_0$; at $x = b$, $q_x = 0$.

Answer: $\dfrac{T(x) - T_0}{\mu v_b^2/k} = \left(\dfrac{x}{b}\right) - \dfrac{1}{2}\left(\dfrac{x}{b}\right)^2$

10B.3 Heat conduction in a spherical nuclear fuel element. Consider a spherical nuclear fuel element as shown in Fig. 10B.3. It consists of a sphere of fissionable material with radius R_F, surrounded by a spherical shell of aluminum "cladding" with outer radius R_C. Inside the fuel element, fission fragments are produced that have very high kinetic energies. Collisions between these fragments and the atoms of the fissionable material provide the major source of thermal energy in the reactor. Such a volume source of thermal energy resulting from nuclear fission we call S_n ([=] energy/volume · time). This source will not be uniform throughout the sphere of fissionable material; it will be the smallest at the center of the sphere. For the purpose of this problem, we assume that the source can be approximated by a simple parabolic function

$$S_n = S_{n0}\left[1 + b\left(\frac{r}{R_F}\right)^2\right] \tag{10B.3-1}$$

Here S_{n0} is the volume rate of heat production at the center of the sphere, b is a dimensionless positive constant, and r is the radial distance from the center of the sphere. The known temperature at the outside cladding surface (at $r = R_C$) is T_0. The thermal conductivities of the fuel element and cladding are k_F and k_C. Determine the temperature profiles in the fuel element and the cladding.

Answers:

$$T_F(r) = \frac{S_{n0}R_F^2}{6k_F}\left\{\left[1-\left(\frac{r}{R_F}\right)^2\right] + \frac{3}{10}b\left[1-\left(\frac{r}{R_F}\right)^4\right]\right\} + \frac{S_{n0}R_F^2}{3k_C}\left(1+\frac{3}{5}b\right)\left(1-\frac{R_F}{R_C}\right) + T_0$$

$$T_C(r) = \frac{S_{n0}R_F^2}{3k_C}\left(1+\frac{3}{5}b\right)\left(\frac{R_F}{r}-\frac{R_F}{R_C}\right) + T_0$$

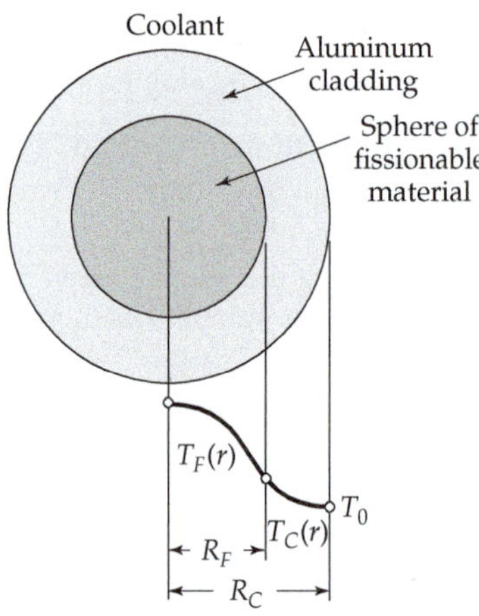

Fig. 10B.3 A spherical nuclear fuel assembly, showing the temperature distribution within the system.

10B.4 Heat conduction in a nuclear fuel rod assembly. Consider a long cylindrical nuclear fuel rod, surrounded by an annular layer of aluminum cladding (**see** Fig. 10B.4). Within the fuel rod heat is produced by fission; the rate energy production per unit volume depends on position approximately as

$$S_n = S_{n0}\left[1 + b\left(\frac{r}{R_F}\right)^2\right] \tag{10B.4-1}$$

Here S_{n0} and b are known constants, and r is the radial coordinate measured from the axis of the cylindrical fuel rod. Calculate the maximum temperature in the fuel rod if the outer surface of the cladding is in contact with a liquid coolant at temperature T_L. The heat-transfer coefficient at the cladding-coolant interface is h_L, and the thermal conductivities of the fuel rod and cladding are k_F and k_C.

Answer: $T_{F,\max} - T_L = \dfrac{S_{n0}R_F^2}{4k_F}\left(1+\dfrac{b}{4}\right) + \dfrac{S_{n0}R_F^2}{2k_C}\left(1+\dfrac{b}{2}\right)\left(\dfrac{k_C}{R_C h_L} + \ln\dfrac{R_C}{R_F}\right)$

Fig. 10B.4 Temperature distribution in a cylindrical fuel-rod assembly.

10B.5 Heat conduction in an annulus.

(a) Heat is flowing through an annular wall of inside radius r_0 and outside radius r_1 as in Fig. 10B.5. The thermal conductivity varies linearly with temperature from k_0 at T_0 to k_1 at T_1. Develop an expression for the heat flow through the wall.

(b) Show how the expression in (a) can be simplified when $r_1 - r_0$ is very small. Interpret the result physically.

Answers: **(a)** $Q = 2\pi L(T_0 - T_1)\left(\dfrac{k_0 + k_1}{2}\right)\left(\ln \dfrac{r_1}{r_0}\right)^{-1}$;

(b) $Q = 2\pi r_0 L \left(\dfrac{k_0 + k_1}{2}\right)\left(\dfrac{T_0 - T_1}{r_1 - r_0}\right)$

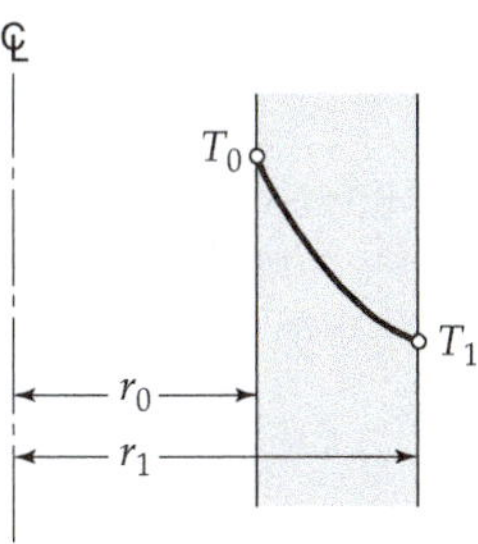

Fig. 10B.5 Temperature profile in an annular wall.

10B.6 Viscous heat generation in a polymer melt. Rework the problem discussed in §10.8 for a molten polymer, whose viscosity can be adequately described by the power-law model (see Chapter 8). Show that the temperature distribution is the same as that in Eq. 10.8-9 but with the Brinkman number replaced by

$$\text{Br}_n = \left[\!\left[\frac{m v_b^{n+1}}{b^{n-1} k (T_b - T_0)} \right]\!\right] \tag{10B.6-1}$$

10B.7 Insulation thickness for a furnace wall. A furnace wall consists of three layers: (i) a layer of heat-resistant or refractory brick, (ii) a layer of insulating brick, and (iii) a steel plate, 0.25 in. thick, for mechanical protection, as shown in Fig. 10B.7. Calculate the thickness of each layer of brick to give minimum total wall thickness if the heat loss through the wall is to be

5000 Btu/ft^2 · hr, assuming that the layers are in excellent thermal contact. The following information is available:

Material	Maximum allowable temperature	Thermal conductivity (Btu/hr · ft · °F) at 100°F	at 2000°F
Refractory brick	2600°F	1.8	3.6
Insulating brick	2000°F	0.9	1.8
Steel	—	26.1	—

Answer: Refractory brick, 0.39 ft; insulating brick, 0.51 ft.

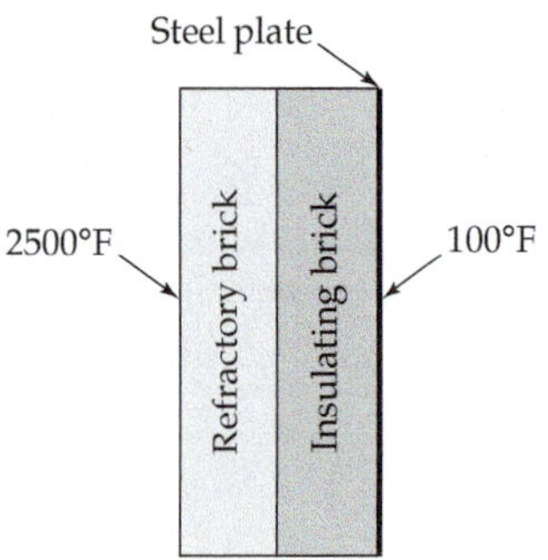

Fig. 10B.7 A composite furnace wall.

10B.8 Forced-convection heat transfer in flow between parallel plates. A viscous fluid with temperature-independent physical properties is in fully developed laminar flow between two flat surfaces placed a distance $2B$ apart as shown in Fig. 10B.8. For $z < 0$, the fluid temperature is uniform at $T = T_1$. For $z \geq 0$, heat is added at a constant, uniform flux q_0 at both walls. Find the temperature distribution $T(x,z)$ for large z.

(a) Make a shell energy balance to obtain the differential equation for $T(x,z)$. Then discard the viscous dissipation term and the axial heat-conduction term.

(b) Recast the problem in terms of the dimensionless quantities

$$\Theta = \frac{T - T_1}{q_0 B/k} \qquad \sigma = \frac{x}{B} \qquad \zeta = \frac{kz}{\rho \hat{C}_p v_{z,\max} B^2} \tag{10B.8-1,2,3}$$

(c) Obtain the asymptotic solution for large z

$$\Theta(\sigma,\zeta) = \frac{3}{2}\zeta + \frac{3}{4}\sigma^2 - \frac{1}{8}\sigma^4 - \frac{39}{280} \tag{10B.8-4}$$

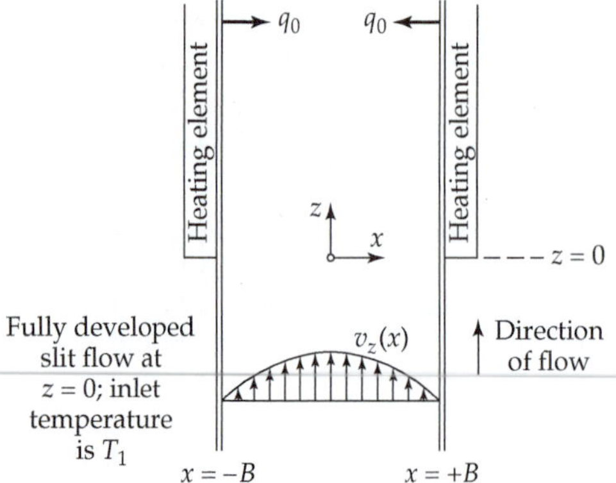

Fig. 10B.8 Laminar, incompressible flow between parallel plates, both of which are being heated by a uniform wall heat flux q_0 starting at $z = 0$.

10B.9 Electrical heating of a pipe. In the manufacture of glass-coated steel pipes, it is common practice first to heat the pipe to the melting range of glass and then to contact the hot pipe surface with glass granules (Fig 10B.9). These granules melt and wet the pipe surface to form a tightly adhering nonporous coat. In one method of preheating the pipe, an electric current is passed along the pipe, with the result that the pipe is heated (as in §10.6). For the purpose of this problem make the following assumptions:

(i) The electrical conductivity of the pipe k_e is constant over the temperature range of interest. The local rate of electrical heat production per unit volume S_e is then uniform throughout the pipe wall.

(ii) The top and bottom of the pipe are capped in such a way that heat losses through them are negligible.

(iii) The heat flux from the outer surface of the pipe to the surroundings is given by Newton's law of cooling: $q_r = h(T_1 - T_a)$. Here h is a suitable heat-transfer coefficient.

How much electrical power is needed to maintain the inner pipe surface at some desired value of the temperature, T_κ, for known k, T_a, h, and pipe dimensions?

Answer: $$P = \frac{\pi R^2 L(1-\kappa^2)(T_\kappa - T_a)}{\dfrac{(1-\kappa^2)R}{2h} - \dfrac{(\kappa R)^2}{4k}\left[\left(1-\dfrac{1}{\kappa^2}\right) - 2\ln\kappa\right]}$$

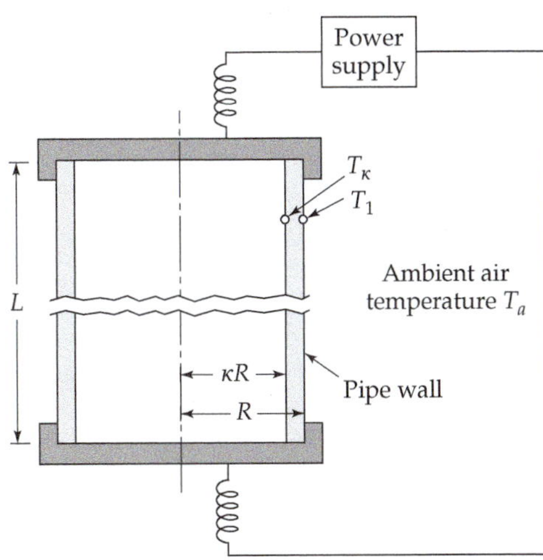

Fig. 10B.9 Electrical heating of a pipe.

10B.10 Plug flow with forced-convection heat transfer. Very thick slurries and pastes sometimes move in channels almost as a solid plug. Thus, one can approximate the velocity profile by a constant value v_0 over the conduit cross section.

(a) Rework the problem of §10.9 for plug flow in a *circular tube* of radius R. Show that the temperature distribution analogous to Eq. 10.9-33 is

$$\Theta(\xi,\zeta) = 2\zeta + \frac{1}{2}\xi^2 - \frac{1}{4} \tag{10B.10-1}$$

in which $\zeta = kz/\rho\hat{C}_p v_0 R^2$, and Θ and ξ are defined as in §10.9.

(b) Show that for plug flow in a *plane slit* of width $2B$ the temperature distribution analogous to Eq. 10B.8-4 is

$$\Theta(\sigma,\zeta) = \zeta + \frac{1}{2}\sigma^2 - \frac{1}{6} \tag{10B.10-2}$$

in which $\zeta = kz/\rho\hat{C}_p v_0 B^2$, and Θ and σ are defined as in Problem 10B.8.

10B.11 Free convection in an annulus of finite height. A fluid is contained in a vertical annulus closed at the top and bottom as shown in Fig. 10B.11. The inner wall of radius κR is maintained at the temperature T_κ, and the outer wall of radius R is kept at temperature T_1. Using the assumptions and approach of §10.10, obtain the velocity distribution produced by free convection.

(a) First, derive the temperature distribution

$$\frac{T_1 - T(\xi)}{T_1 - T_\kappa} = \frac{\ln \xi}{\ln \kappa} \tag{10B.11-1}$$

in which $\xi = r/R$.

(b) Then show that the equation of motion is

$$\frac{1}{\xi}\frac{d}{d\xi}\left(\xi \frac{dv_z}{d\xi}\right) = A + B \ln \xi \tag{10B.11-2}$$

in which $A = (R^2/\mu)(dp/dz + \rho_1 g)$ and $B = (\rho_1 g \beta_1 \Delta T)R^2/\mu \ln \kappa$ where $\Delta T = T_1 - T_\kappa$.

(c) Integrate the equation of motion (see Eq. C.1-11) and apply the boundary conditions to evaluate the constants of integration. Then show that A can be evaluated by the requirement of no net mass flow through any plane z = constant, with the final result that

$$v_z(\xi) = \frac{\rho_1 g \beta_1 \Delta T R^2}{16\mu}\left[\frac{(1-\kappa^2)(1-3\kappa^2) - 4\kappa^4 \ln \kappa}{(1-\kappa^2)^2 + (1-\kappa^4)\ln \kappa}\left((1-\xi^2) - (1-\kappa^2)\frac{\ln \xi}{\ln \kappa}\right) + 4(\xi^2 - \kappa^2)\frac{\ln \xi}{\ln \kappa}\right] \tag{10B.11-3}$$

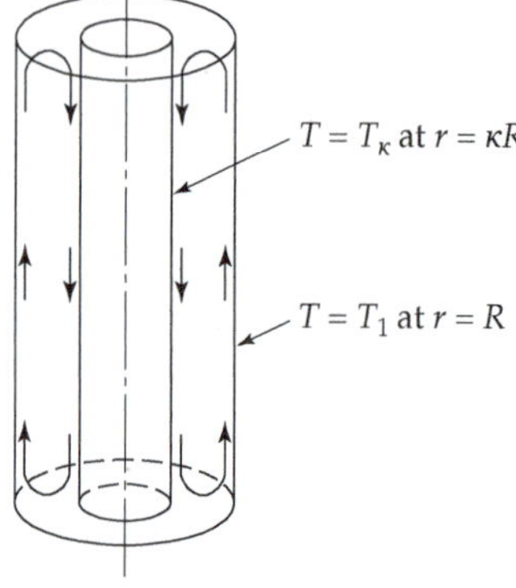

Fig. 10B.11 Free convection pattern in an annular space with $T_1 > T_\kappa$.

10B.12 Free convection with temperature-dependent viscosity. Rework the problem in §10.10 taking into account the variation of viscosity with temperature. Assume that the "fluidity" (reciprocal of viscosity) is the following linear function of the temperature

$$\frac{1}{\mu(T)} = \frac{1}{\bar{\mu}}[1 + \bar{\beta}_\mu (T - \bar{T})] \tag{10B.12-1}$$

Use the dimensionless quantities $\breve{y}$, $\breve{v}_z$, and Gr defined in §10.10 and in addition

$$b_T = \frac{1}{2}\bar{\beta}\Delta T, \quad b_\mu = \frac{1}{2}\bar{\beta}_\mu \Delta T \quad \text{and} \quad P = \frac{\bar{\rho}B^3}{\bar{\mu}^2}\left(\frac{dp}{dz} + \bar{\rho}g\right) \tag{10B.12-2,3}$$

and show that the differential equation for the velocity distribution is

$$\frac{d}{d\breve{y}}\left(\frac{1}{1 - b_\mu \breve{y}}\frac{d\breve{v}_z}{d\breve{y}}\right) = P + \frac{1}{2}\text{Gr}\breve{y} \tag{10B.12-4}$$

Follow the procedure in §10.10, discarding terms containing the third and higher powers of ΔT. Show that this leads to $P = -\frac{1}{30}\text{Gr}b_T + \frac{1}{15}\text{Gr}b_\mu$, and finally

$$\breve{v}_z(\breve{y}) = \frac{1}{12}\text{Gr}(\breve{y}^3 - \breve{y}) - \frac{1}{60}\text{Gr}b_T(\breve{y}^2 - 1) - \frac{1}{80}\text{Gr}b_\mu(\breve{y}^2 - 1)(5\breve{y}^2 - 1) \tag{10B.12-5}$$

Sketch the result to show how the velocity profile becomes skewed because of the temperature-dependent viscosity.

10B.13 Flow reactor with exponentially temperature-dependent source. Formulate the function $F(\Theta)$ of Eq. 10.7-7 for a zero-order reaction with the temperature dependence

$$S_c = Ke^{-E/RT} \tag{10B.13-1}$$

in which K and E are constants, and R is the gas constant. Then insert $F(\Theta)$ into Eqs. 10.7-15 through 10.7-20 and solve for the dimensionless temperature profile with k_{eff} neglected.

10B.14 Evaporation loss from an oxygen tank.

(a) Liquefied gases are sometimes stored in well-insulated spherical containers vented to the atmosphere. Develop an expression for the steady-state heat-transfer rate through the walls of such a container, with the radii of the inner and outer walls being r_0 and r_1, respectively, and the temperatures at the inner and outer walls being T_0 and T_1. The thermal conductivity of the insulation varies linearly with temperature from k_0 at T_0 to k_1 at T_1.

(b) Estimate the rate of evaporation of liquid oxygen from a spherical container of 6 ft inside diameter covered with a 1-ft-thick annular evacuated jacket filled with particulate insulation. The following information is available:

Temperature at inner surface of insulation	−183°C
Temperature at outer surface of insulation	0°C
Boiling point of O_2	−183°C
Heat of vaporization of O_2	1636 cal/g-mol
Thermal conductivity of insulation at 0°C	9.0×10^{-4} Btu/hr · ft · °F
Thermal conductivity of insulation at −183°C	7.2×10^{-4} Btu/hr · ft · °F

Answers: **(a)** $Q_0 = 4\pi r_0 r_1 \left(\frac{k_0 + k_1}{2}\right)\left(\frac{T_0 - T_1}{r_1 - r_0}\right)$; **(b)** 0.198 kg/hr

10B.15 Radial temperature gradients in an annular chemical reactor. A catalytic reaction is being carried out at constant pressure in a packed bed between coaxial cylindrical walls with inner radius r_0 and outer radius r_1. Such a configuration occurs when temperatures are measured with a centered thermowell, and is in addition useful for controlling temperature gradients if a thin annulus is used. The entire inner wall is at uniform temperature T_0, and it can be assumed that there is no heat transfer through this surface. The reaction releases heat at a uniform volumetric rate S_c throughout the reactor. The effective thermal conductivity of the reactor contents is to be treated as a constant throughout.

(a) By a shell energy balance, derive a second-order differential equation that describes the temperature profile, assuming that the temperature gradient in the axial direction can be neglected. What boundary conditions must be used?

(b) Rewrite the differential equation and boundary conditions in terms of the dimensionless radial coordinate and dimensionless temperature difference defined as

$$\xi = \frac{r}{r_0}; \qquad \Theta = \frac{T - T_0}{S_c r_0^2 / 4k_{\text{eff}}} \tag{10B.15-1,2}$$

Explain why these are logical choices.

(c) Integrate the dimensionless differential equation to get the radial temperature profile. To which viscous flow problem is this conduction problem analogous?

(d) Develop expressions for the temperature at the outer wall and for the volumetric average temperature of the catalyst bed.

(e) Calculate the outer wall temperature when $r_0 = 0.45$ in., $r_1 = 0.50$ in., $k_{\text{eff}} = 0.3$ Btu/hr · ft · °F, $T_0 = 900°\text{F}$, and $S_c = 4800$ cal/hr · cm³.

(f) How would the results of part (e) be affected if the inner and outer radii were doubled?

Answer: (e) 885°F

10B.16 Temperature distribution in a hot-wire anemometer. A hot-wire anemometer is essentially a fine wire, usually made of platinum, which is heated electrically and exposed to a flowing fluid. Its temperature, which is a function of the fluid temperature, fluid velocity, and the rate of heating, may be determined by measuring its electrical resistance. It is used for measuring velocities and velocity fluctuations in flow systems. In this problem we analyze the temperature distribution in the wire element.

We consider a wire of diameter D and length $2L$ supported at its ends ($z = -L$ and $z = +L$) and mounted perpendicular to an air stream. An electric current of density I amp/cm² flows through the wire, and the heat thus generated is partially lost by convection to the air stream (see Eq. 10.1-2) and partially by conduction toward the ends of the wire. Because of their size and their high electrical and thermal conductivity, the supports are not appreciably heated by the current, but remain at the temperature T_L, which is the same as that of the approaching air stream. Heat loss by radiation is to be neglected.

(a) Derive an equation for the steady-state temperature distribution in the wire, assuming that T depends on z alone; that is, the radial temperature variation in the wire is neglected. Further, assume uniform thermal and electrical conductivities in the wire, and a uniform heat transfer coefficient from the wire to the air stream.

(b) Sketch the temperature profile obtained in (a).

(c) Compute the current, in amperes, required to heat a platinum wire to a midpoint temperature of 50°C under the following conditions:

$$T_L = 20°\text{C} \qquad h = 100 \text{ Btu/hr} \cdot \text{ft}^2 \cdot °\text{F}$$
$$D = 0.127 \text{ mm} \qquad k = 40.2 \text{ Btu/hr} \cdot \text{ft} \cdot °\text{F}$$
$$L = 0.5 \text{ cm} \qquad k_e = 1.00 \times 10^5 \text{ ohm}^{-1}\text{cm}^{-1}$$

Answers: **(a)** $T(z) - T_L = \dfrac{DI^2}{4hk_e}\left(1 - \dfrac{\cosh\sqrt{4h/kD}z}{\cosh\sqrt{4h/kD}L}\right)$; **(c)** 1.01 amp

10B.17 Non-Newtonian flow with forced-convection heat transfer.[1] For estimating the effect of non-Newtonian viscosity on heat transfer in ducts, the power-law model of Chapter 8 gives velocity profiles that show rather well the deviation from parabolic shape.

(a) Rework the problem of §10.9 (heat transfer in a *circular tube*) for the power-law model given in Eqs. 8.3-2 and 8.3-3. Show that the final temperature profile is

$$\Theta(\xi,\zeta) = \frac{2(s+3)}{(s+1)}\zeta + \frac{(s+3)}{2(s+1)}\xi^2 - \frac{2}{(s+1)(s+3)}\xi^{s+3} - \frac{(s+3)^3 - 8}{4(s+1)(s+3)(s+5)} \tag{10B.17-1}$$

in which $s = 1/n$.

(b) Rework Problem 10B.8 (heat transfer in a *plane slit*) for the power-law model. Obtain the dimensionless temperature profile:

$$\Theta(\sigma,\zeta) = \frac{(s+2)}{(s+1)}\left[\zeta + \frac{1}{2}\sigma^2 - \frac{1}{(s+2)(s+3)}|\sigma|^{s+3} - \frac{(s+2)(s+3)(2s+5) - 6}{6(s+3)(s+4)(2s+5)}\right] \tag{10B.17-2}$$

Note that these results contain the Newtonian results ($s = 1$) and the plug flow results ($s = \infty$).

[1] R. B. Bird, *Chem.-Ing. Technik*, **31**, 569–572 (1959).

10B.18 Reactor temperature profiles with axial heat flux.[2]

(a) Show that for a heat source that depends linearly on the temperature, Eqs. 10.7-6 to 10.7-14 have the solutions[1] (for $m_+ \neq m_-$)

$$\Theta^{\text{I}} = 1 + \frac{m_+m_-(\exp m_+ - \exp m_-)}{m_+^2 \exp m_+ - m_-^2 \exp m_-} \exp[(m_+ + m_-)Z] \tag{10B.18-1}$$

$$\Theta^{\text{II}} = \frac{m_+(\exp m_+)(\exp m_- Z) - m_-(\exp m_-)(\exp m_+ Z)}{m_+^2 \exp m_+ - m_-^2 \exp m_-}(m_+ + m_-) \tag{10B.18-2}$$

$$\Theta^{\text{III}} = \frac{m_+^2 - m_-^2}{m_+^2 \exp m_+ - m_-^2 \exp m_-} \exp(m_+ + m_-) \tag{10B.18-3}$$

Here $m_\pm = \frac{1}{2}B\left[1 \pm \sqrt{1-(4N/B)}\right]$, in which $B = \rho v_0 \hat{C}_p L/k_{\text{eff}}$. Some profiles calculated from these equations are shown in Fig. 10B.18.

(b) Show that, in the limit as B goes to infinity, the above solution agrees with that in Eqs. 10.7-21, 10.7-22, and 10.7-23.

(c) Make numerical comparisons of the results in Eq. 10.7-22 and Fig. 10B.18 for $N = 2$ at $Z =$ 0.0, 0.5, 0.9, and 1.0.

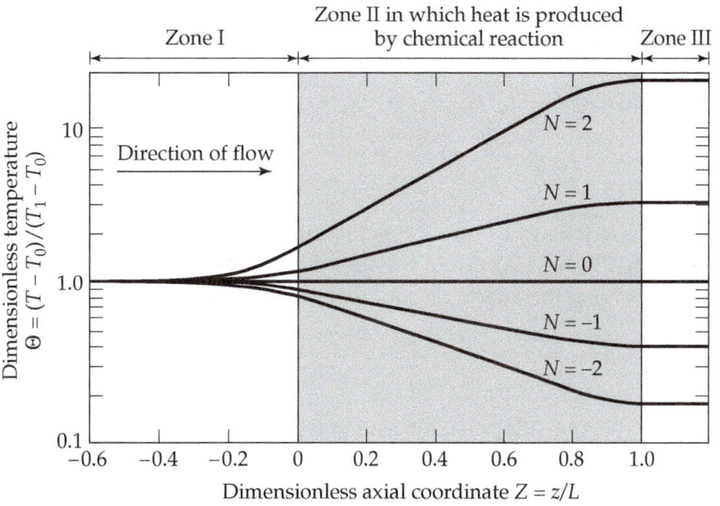

Fig. 10B.18 Predicted temperature profiles in a fixed-bed axial-flow reactor for $B = 8$ and various values of N.

10B.19 Heat conduction in a conical region. A solid is bounded by the surface $\theta = \theta_0$, which is insulated, and surfaces $r = r_1$ and $r = r_2$ in spherical coordinates, as illustrated in Fig. 10B.19. Surface $r = r_1$ is maintained at temperature $T = T_1$, and surface $r = r_2$ at $T = T_2$. We take the thermal conductivity to be a linear function of the temperature

$$k(T) = k_1 + (k_2 - k_1)\left(\frac{T - T_1}{T_2 - T_1}\right) \equiv k_1 + (k_2 - k_1)\Theta \tag{10B.19-1}$$

[2]Taken from the corresponding results of G. Damköhler, *Z. Elektrochem.*, **43**, 1–8, 9–13 (1937), and J. F. Wehner and R. H. Wilhelm, *Chem. Engr. Sci.*, **6**, 89–93 (1956); **8**, 309 (1958), for isothermal flow reactors with longitudinal diffusion and first-order reaction. **Gerhard Damköhler** (1908–1944) achieved fame for his work on chemical reactions in flowing, diffusing systems; a key publication was in *Der Chemie-Ingenieur*, Leipzig (1937), pp. 359–485. **Richard Herman Wilhelm** (1909–1968), chairman of the Chemical Engineering Department at Princeton University, was well known for his work on fixed-bed catalytic reactors, fluidized transport, and the "parametric pumping" separation process.

It is desired to find the temperature in this solid as a function of position. Because of the symmetry of this problem, we assume that T depends only on the radial position r.

(a) Show that a lens-shaped region defined by a surface of constant r within the cone has the area $2\pi r^2(1 - \cos\theta_0)$.

(b) Make an energy balance over a lens-shaped shell of thickness Δr to get

$$2\pi(1-\cos\theta_0)(r^2 q_r)|_r - 2\pi(1-\cos\theta_0)(r^2 q_r)|_{r+\Delta r} = 0 \tag{10B.19-2}$$

which leads to the differential equation

$$\frac{d}{dr}(r^2 q_r) = 0 \tag{10B.19-3}$$

Integrate this, and then use Fourier's law with the temperature-dependent thermal conductivity given above to get finally

$$k_1\Theta + \frac{1}{2}(k_2 - k_1)\Theta^2 = \frac{C}{r} + D \tag{10B.19-4}$$

where C and D are constants of integration.

(c) Determine the constants of integration from the boundary conditions at $r = r_1$ and $r = r_2$ to get

$$K\Theta^2 + L\Theta - f(r) = 0 \tag{10B.19-5}$$

where $K = (k_2 - k_1)/k_{av}$, $k_{av} \equiv \frac{1}{2}(k_1 + k_2)$, $L = k_1/k_{av}$, and

$$f(r) = \frac{(1/r) - (1/r_1)}{(1/r_2) - (1/r_1)} \tag{10B.19-6}$$

(d) Verify that Eq. 10B.19-6 is the dimensionless temperature distribution when the thermal conductivity k is constant throughout (i.e., take $k_2 = k_1$).

(e) Solving the quadratic equation in Eq. 10B.19-5 then gives

$$\Theta(r) = \frac{-L \pm \sqrt{L^2 + 4Kf(r)}}{2K} \tag{10B.19-7}$$

How do you determine the sign in front of the square root?

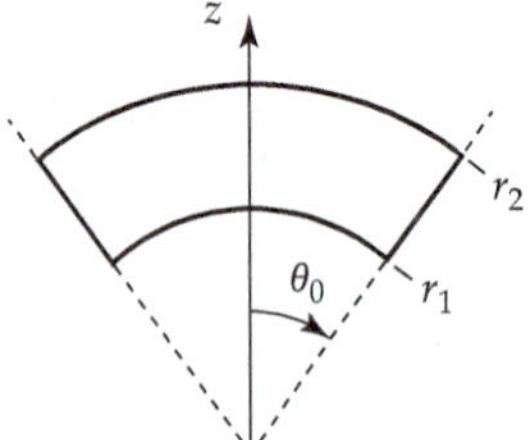

Fig. 10B.19 Heat conduction in a conical region.

10B.20 Alternate derivation for forced convection in a tube. The goal of this problem is re-derive Eq. 10.9-13 using a more rigorous approach than that employed in §10.9.

(a) Insert the flux expressions given by Eqs. 10.9-9 and 10.9-10 into the energy balance given by Eq. 10.9-8. Show that this gives

$$\begin{aligned}\rho\hat{C}_p v_z(r)\frac{\partial T}{\partial z} &= k\left[\frac{1}{r}\frac{\partial}{\partial r}\left(r\frac{\partial T}{\partial r}\right) + \frac{\partial^2 T}{\partial z^2}\right] + \mu\left(\frac{\partial v_z}{\partial r}\right)^2 \\ &\quad + v_z(r)\left[-\frac{\partial p}{\partial z} + \mu\frac{1}{r}\frac{\partial}{\partial r}\left(r\frac{\partial v_z}{\partial r}\right) + \rho g_z\right]\end{aligned} \tag{10B.20-1}$$

(b) Show that at steady state, the last term in square brackets is zero for an incompressible, Newtonian fluid in laminar flow in a tube. (Hint: Consider the Navier-Stokes equations from Chapter 3.)

(c) Show that Eq. 10B.20-1 reduces to Eq. 10.9-13 when energy production by viscous dissipation is neglected.

10B.21 Dimensionless solutions to heat conduction in a steam pipe. Repeat the analysis of §10.2 using dimensionless variables.

(a) For the system depicted in Fig. 10.2-1(*a*), use the dimensionless variables $\Theta = (T - T_o)/(T_s - T_o)$ and $\xi = r/R$. Show that the Eq. 10.2-7 becomes

$$\Theta(\xi) = D_1 \ln \xi + D_2 \tag{10B.21-1}$$

where the new constants D_1 and D_2 are related to C_1 and C_2.

(b) Nondimensionalize the boundary conditions in Eqs. 10.2-8 and 10.2-9 to obtain boundary conditions for $\Theta(\xi)$.

(c) Apply the new boundary conditions and obtain the integration constants D_1 and D_2, as well as the dimensionless profile $\Theta(\xi)$.

(d) For the system depicted in Fig. 10.2-1(*b*), use the dimensionless variables $\Theta = (T - T_a)/(T_s - T_a)$ and $\xi = r/R$. Show that the Eq. 10.2-15 becomes

$$\Theta(\xi) = D_1 \ln \xi + D_2 \tag{10B.21-2}$$

where the new constants D_1 and D_2 are related to C_1 and C_2.

(e) Nondimensionalize the boundary conditions in Eqs. 10.2-16 and 10.2-17 to obtain boundary conditions for $\Theta(\xi)$.

(f) Apply the new boundary conditions and obtain the integration constants D_1 and D_2, as well as the dimensionless profile $\Theta(\xi)$.

(g) For the system depicted in Fig. 10.2-1(*c*), use the dimensionless variables $\Theta = (T - T_a)/(T_s - T_a)$ and $\xi = r/R$. Show that the Eqs. 10.2-25 and 26 become

$$\Theta^{\rm I}(\xi) = D_1^{\rm I} \ln \xi + D_2^{\rm I} \tag{10B.21-3}$$

$$\Theta^{\rm II}(\xi) = D_1^{\rm II} \ln \xi + D_2^{\rm II} \tag{10B.21-4}$$

where the new constants $D_1^{\rm I}$, $D_2^{\rm I}$, $D_1^{\rm II}$, and $D_2^{\rm II}$, are related to $C_1^{\rm I}$, $C_2^{\rm I}$, $C_1^{\rm II}$, and $C_2^{\rm II}$.

(h) Nondimensionalize the boundary conditions in Eqs. 10.2-27 through 10.2-30 to obtain boundary conditions for $\Theta^{\rm I}(\xi)$ and $\Theta^{\rm II}(\xi)$.

(i) Apply the new boundary conditions and obtain the integration constants $D_1^{\rm I}$, $D_2^{\rm I}$, $D_1^{\rm II}$, and $D_2^{\rm II}$, as well as the dimensionless profiles $\Theta^{\rm I}(\xi)$ and $\Theta^{\rm II}(\xi)$.

10C.1 Heating of an electric wire with temperature-dependent electrical and thermal conductivity.[3] Find the temperature distribution in an electrically heated wire when the thermal and electrical conductivities vary with temperature as follows:

$$\frac{k}{k_0} = 1 - \alpha_1 \Theta - \alpha_2 \Theta^2 + \cdots \tag{10C.1-1}$$

$$\frac{k_e}{k_{e0}} = 1 - \beta_1 \Theta - \beta_2 \Theta^2 + \cdots \tag{10C.1-2}$$

Here k_0 and k_{e0} are the values of the conductivities at temperature T_0, and $\Theta = (T - T_0)/T_0$ is a dimensionless temperature rise. The coefficients α_i and β_i are constants. Such series expansions are useful over moderate temperature ranges.

---

[3] The solution given here was suggested by Professor L. J. F. Broer of the Technical University of Delft (personal communication, 20 August 1958).

(a) Because the temperature depends on the radial position r in the wire, the electrical conductivity is a function of position, $k_e(r)$. Therefore, the current density is also a function of r: $I(r) = k_e(r) \cdot (E/L)$, and the electrical heat source also is position dependent: $S_e(r) = k_e(r) \cdot (E/L)^2$. The equation for the temperature distribution is then

$$-\frac{1}{r}\frac{d}{dr}\left(rk(r)\frac{dT}{dr}\right) = k_e(r)\left(\frac{E}{L}\right)^2 \tag{10C.1-3}$$

Now introduce the dimensionless quantities $\xi = r/R$ and $B = k_{e0}R^2E^2/k_0L^2T_0$ and show that Eq. 10C.1-3 then becomes

$$-\frac{1}{\xi}\frac{d}{d\xi}\left(\frac{k}{k_0}\xi\frac{d\Theta}{d\xi}\right) = B\frac{k_e}{k_{e0}} \tag{10C.1-4}$$

When the power-series expressions for the conductivities are inserted into this equation we get

$$-\frac{1}{\xi}\frac{d}{d\xi}\left((1 - \alpha_1\Theta - \alpha_2\Theta^2 + \cdots)\,\xi\frac{d\Theta}{d\xi}\right) = B(1 - \beta_1\Theta - \beta_2\Theta^2 + \cdots) \tag{10C.1-5}$$

This is the equation that is to be solved for the dimensionless temperature distribution.

(b) Begin by noting that if all the α_i and β_i were zero (that is, both conductivities are constant), then Eq. 10C.1-5 would simplify to

$$-\frac{1}{\xi}\frac{d}{d\xi}\left(\xi\frac{d\Theta}{d\xi}\right) = B \tag{10C.1-6}$$

When this is solved with the boundary conditions that $\Theta =$ finite at $\xi = 0$, and $\Theta = 0$ at $\xi = 1$, we get:

$$\Theta = \frac{1}{4}B(1 - \xi^2) \tag{10C.1-7}$$

This is just Eq. 10.6-13 in dimensionless notation.

Note that Eq. 10C.1-5 will have the solution in Eq. 10C.1-7 for small values of B—that is, for weak heat sources. For stronger heat sources, postulate that the temperature distribution can be expressed as a power series in the dimensionless heat source strength B:

$$\Theta = \frac{1}{4}B(1 - \xi^2)(1 + B\Theta_1 + B^2\Theta_2 + \cdots) \tag{10C.1-8}$$

Here the Θ_n are functions of ξ but not of B. Substitute Eq. 10C.1-8 into Eq. 10C.1-5, and equate the coefficients of like powers of B to get a set of ordinary differential equations for the Θ_n, with $n = 1, 2, 3,\ldots$ These may be solved with the boundary conditions that $\Theta_n =$ finite at $\xi = 0$ and $\Theta_n = 0$ at $\xi = 1$. In this way obtain

$$\Theta = \frac{1}{4}B(1 - \xi^2)\left[1 + B\left(\frac{1}{8}\alpha_1(1 - \xi^2) - \frac{1}{16}\beta_1(3 - \xi^2)\right) + O(B^2)\right] \tag{10C.1-9}$$

where $O(B^2)$ means "terms of the order of B^2 and higher."

(c) For materials that are described by the Wiedemann-Franz-Lorenz law (see §9.9), the ratio k/k_eT is a constant (independent of temperature). Hence,

$$\frac{k}{k_eT} = \frac{k_0}{k_{e0}T_0} \tag{10C.1-10}$$

Combine this with Eqs. 10C.1-1 and 10C.1-2 to get

$$1 - \alpha_1\Theta - \alpha_2\Theta^2 + \cdots = (1 - \beta_1\Theta - \beta_2\Theta^2 + \cdots)(1 + \Theta) \tag{10C.1-11}$$

Equate coefficients of equal powers of the dimensionless temperature to get relations among the α_i and the β_i: $\alpha_1 = \beta_1 - 1$, $\alpha_2 = \beta_1 + \beta_2$, and so on. Use these relations to get

$$\Theta = \frac{1}{4}B(1 - \xi^2)\left[1 - \frac{1}{16}B\left((\beta_1 + 2) + (\beta_1 - 2)\xi^2\right) + O(B^2)\right] \tag{10C.1-12}$$

10C.2 Viscous heating with temperature-dependent viscosity and thermal conductivity. Consider the flow situation shown in Fig. 10.8-2. Both the stationary surface and the moving surface are maintained at a constant temperature T_0. The temperature dependences of k and μ are given by

$$\frac{k}{k_0} = 1 + \alpha_1\Theta + \alpha_2\Theta^2 + \cdots \tag{10C.2-1}$$

$$\frac{\mu_0}{\mu} = \frac{\varphi}{\varphi_0} = 1 + \beta_1\Theta + \beta_2\Theta^2 + \cdots \tag{10C.2-2}$$

in which the α_i and β_i are constants, $\varphi = 1/\mu$ is the fluidity, and the subscript "0" means "evaluated at $T = T_0$." The dimensionless temperature is defined as $\Theta = (T - T_0)/T_0$.

(a) Show that the differential equations describing the viscous flow and heat conduction may be written in the forms

$$\frac{d}{d\xi}\left(\frac{\mu}{\mu_0}\frac{d\phi}{d\xi}\right) = 0 \tag{10C.2-3}$$

$$\frac{d}{d\xi}\left(\frac{k}{k_0}\frac{d\Theta}{d\xi}\right) + \text{Br}\frac{\mu}{\mu_0}\left(\frac{d\phi}{d\xi}\right)^2 = 0 \tag{10C.2-4}$$

in which $\phi = v_z/v_b$, $\xi = x/b$, and $\text{Br} = \mu_0 v_b^2/k_0 T_0$ (the Brinkman number).

(b) The equation for the dimensionless velocity distribution may be integrated once to give $d\phi/d\xi = C_1 \cdot (\varphi/\varphi_0)$, in which C_1 is an integration constant. This expression is then substituted into the energy equation to get

$$\frac{d}{d\xi}\left((1 + \alpha_1\Theta + \alpha_2\Theta^2 + \cdots)\frac{d\Theta}{d\xi}\right) + \text{Br}C_1^2(1 + \beta_1\Theta + \beta_2\Theta^2 + \cdots)^2 = 0 \tag{10C.2-5}$$

Obtain the first two terms of a solution in the form

$$\Theta(\xi; \text{Br}) = \text{Br}\Theta_1(\xi) + \text{Br}^2\Theta_2(\xi) + \cdots \tag{10C.2-6}$$

$$\phi(\xi; \text{Br}) = \phi_0 + \text{Br}\phi_1(\xi) + \text{Br}^2\phi_2(\xi) + \cdots \tag{10C.2-7}$$

It is further suggested that the constant of integration C_1 also be expanded as a power series in the Brinkman number, thus

$$C_1(\text{Br}) = C_{10} + \text{Br}C_{11} + \text{Br}^2C_{12} + \cdots \tag{10C.2-8}$$

(c) Repeat the problem changing the boundary condition at $y = b$ to $q_x = 0$ (instead of specifying the temperature).[4]

Answers: **(b)** $\phi = \xi - \frac{1}{12}\text{Br}\beta_1(\xi - 3\xi^2 + 2\xi^3) + \cdots$

$$\Theta = \frac{1}{2}\text{Br}(\xi - \xi^2) - \frac{1}{8}\text{Br}^2\alpha_1(\xi^2 - 2\xi^3 + \xi^4) - \frac{1}{24}\text{Br}^2\beta_1(\xi - 2\xi^2 + 2\xi^3 - \xi^4) + \cdots$$

(c) $\phi = \xi - \frac{1}{6}\text{Br}\beta_1(2\xi - 3\xi^2 + \xi^3) + \cdots$

$$\Theta = \text{Br}\left(\xi - \frac{1}{2}\xi^2\right) - \frac{1}{8}\text{Br}^2\alpha_1(4\xi^2 - 4\xi^3 + \xi^4) + \frac{1}{24}\text{Br}^2\beta_1(-8\xi + 8\xi^2 - 4\xi^3 + \xi^4) + \cdots$$

[4] R. M. Turian and R. B. Bird, *Chem. Eng. Sci.*, **18**, 689–696 (1963).

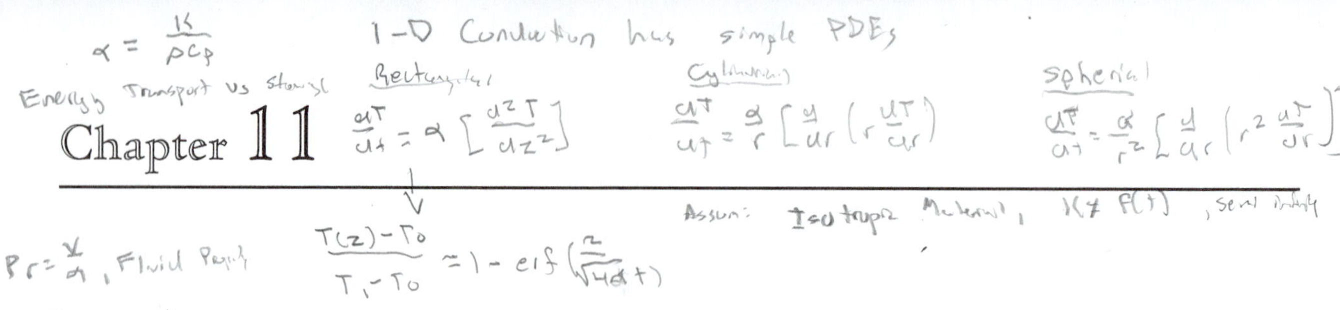

Chapter 11

The Equations of Change for Nonisothermal Systems

In Chapter 10 we introduced the shell energy balance method for solving relatively simple, time-independent heat-flow problems. We obtained the temperature profiles, as well as some derived properties, such as the average temperature and heat-transfer rate. In this chapter, we generalize the shell energy balance and obtain the *equation of energy*, a partial differential equation that describes the transport of energy in a homogeneous fluid or solid.

This chapter is also closely related to Chapter 3, where we introduced the equation of continuity (conservation of mass) and the equation of motion (conservation of momentum). The addition of the equation of energy (conservation of energy) allows us to extend our problem-solving ability to include nonisothermál systems.

We begin in §11.1 by deriving the equation of change for the *total energy*. Just as in Chapter 10, we use the total energy-flux vector **e** in applying the law of conservation of energy. In §11.2 we subtract the *mechanical energy* equation (given in §3.3) from the total energy equation to get an equation of change for the *internal energy*. From the latter we can get an equation of change for the *temperature*, and it is this form of the energy equation that is most commonly used.

Although our main concern in this chapter will be with the various energy equations just mentioned, we find it useful to discuss in §11.3 an approximate equation of motion that is convenient for solving problems involving free convection.

In §11.4 we summarize the equations of change encountered up to this point. Then we proceed to illustrate the use of these equations in a series of examples, in which we begin with the general equations and discard terms that are not needed. In this way we have a standard procedure for setting up and solving problems.

In §11.5 we show how to solve more complicated problems, and, in particular, ones that involve time dependence.

§11.1 THE ENERGY EQUATION

The equation of change for energy is obtained by applying the law of conservation of energy to a tiny element of volume $\Delta x \Delta y \Delta z$ (see Fig. 3.1-1) and then allowing the dimensions of the volume element to become vanishingly small. The law of conservation of energy is an extension of the first law of classical thermodynamics, the latter concerning itself with the difference in internal energies of two equilibrium states of a closed system because of the heat added to the system and the work done on the system (i.e., the familiar $\Delta U = Q + W$).[1]

Here we focus on a stationary volume element, fixed in space, through which a fluid is flowing. The kinetic plus internal energy within the element increases with time because kinetic and internal energy may be entering and leaving the element by convection, because energy may be entering and leaving the element by conduction, and because work is being done on the element by molecular processes. These effects are all accounted for in the **e** vector—the total energy flux. In addition, external forces (such as gravity) are acting on the fluid within the element of volume.

We can summarize the preceding paragraph by writing the conservation energy in words as follows:

$$\left\{\begin{matrix}\text{rate of}\\ \text{increase of}\\ \text{kinetic and}\\ \text{internal}\\ \text{energy}\end{matrix}\right\} = \left\{\begin{matrix}\text{net rate of}\\ \text{energy addition}\\ \text{by all mechanisms}\end{matrix}\right\} + \left\{\begin{matrix}\text{rate of work}\\ \text{done on system}\\ \text{by external}\\ \text{forces}\\ \text{(e.g., by gravity)}\end{matrix}\right\} \qquad (11.1\text{-}1)$$

Several comments need to be made before proceeding:

i. By *kinetic energy* (per unit volume) we mean the energy associated with the observable motion of the fluid, which is, $\frac{1}{2}\rho v^2 \equiv \frac{1}{2}\rho(\mathbf{v}\cdot\mathbf{v})$, per-unit-volume. Here $\mathbf{v}$ is the fluid velocity vector that appears in the equation of motion.

ii. By *internal energy* we mean the kinetic energies of the constituent molecules calculated in a frame moving with the velocity $\mathbf{v}$, plus the energies associated with the vibrational and rotational motions of the molecules and also the energies of interaction among all the molecules (see §0.3). It is *assumed* that the internal energy U for a flowing fluid is the same function of temperature and density as that for a fluid at equilibrium. Keep in mind that a similar assumption is made for the thermodynamic pressure $p(\rho, T)$ for a flowing fluid.

iii. The *potential energy* does not appear in Eq. 11.1-1, since we prefer instead to consider the work done on the system by gravity. At the end of this section, however, we show how to express this work in terms of the potential energy.

iv. In Eq. 10.1-1 various *source terms* were included in the shell energy balance. In §10.8 the viscous heat source S_v appeared automatically, because the mechanical energy terms in **e** were properly accounted for; the same situation prevails here, and the viscous heating term $-(\boldsymbol{\tau} : \nabla\mathbf{v})$ will appear automatically in Eq. 11.2-1. The chemical, electrical, and nuclear source terms (S_c, S_e, and S_n) do not appear automatically, since chemical reactions, electrical effects, and nuclear disintegrations have *not* been included in the energy balance, or in the flux expressions. In Chapter 19, where the energy equation for mixtures with chemical reactions is considered, the chemical heat source S_c appears naturally, as does a "diffusive source term," $\Sigma_\alpha(\mathbf{j}_\alpha \cdot \mathbf{g}_\alpha)$.

---

[1]R. J. Silbey, R. A. Alberty, and M. G. Bawendi, *Physical Chemistry*, Wiley, New York, 4th edition (2005), §2.2.

We now translate Eq. 11.1-1 into mathematical terms. The rate of increase of kinetic and internal energy within the volume element $\Delta x \Delta y \Delta z$ is

$$\Delta x \Delta y \Delta z \frac{\partial}{\partial t}\left(\frac{1}{2}\rho v^2 + \rho \hat{U}\right) \tag{11.1-2}$$

Here $\hat{U}$ is the internal energy per unit mass (sometimes called the "specific internal energy"). The product $\rho\hat{U}$ is the internal energy per unit volume, and $\frac{1}{2}\rho v^2 = \frac{1}{2}\rho(v_x^2 + v_y^2 + v_z^2)$ is the kinetic energy per unit volume.

Next we have to state how much energy enters and leaves across the six faces of the volume element $\Delta x \Delta y \Delta z$. This is

$$\Delta y \Delta z(e_x|_x - e_x|_{x+\Delta x}) + \Delta x \Delta z(e_y|_y - e_y|_{y+\Delta y}) + \Delta x \Delta y(e_z|_z - e_z|_{z+\Delta z}) \tag{11.1-3}$$

Keep in mind that the **e** vector includes the convective transport of kinetic and internal energy, as well as the heat conduction and the work associated with molecular processes.

The rate at which work is done on the fluid by the external force is the dot product of the fluid velocity **v** and the force acting on the fluid $(\rho \Delta x \Delta y \Delta z)\mathbf{g}$, or

$$\rho \Delta x \Delta y \Delta z(v_x g_x + v_y g_y + v_z g_z) \tag{11.1-4}$$

We now insert these various contributions into Eq. 11.1-1 and then divide by $\Delta x \Delta y \Delta z$. When Δx, Δy, and Δz are allowed to go to zero, we get

$$\frac{\partial}{\partial t}\left(\frac{1}{2}\rho v^2 + \rho \hat{U}\right) = -\left(\frac{\partial e_x}{\partial x} + \frac{\partial e_y}{\partial y} + \frac{\partial e_z}{\partial z}\right) + \rho(v_x g_x + v_y g_y + v_z g_z) \tag{11.1-5}$$

This equation may be written more compactly in vector notation as

$$\frac{\partial}{\partial t}\left(\frac{1}{2}\rho v^2 + \rho \hat{U}\right) = -(\nabla \cdot \mathbf{e}) + \rho(\mathbf{v} \cdot \mathbf{g}) \tag{11.1-6}$$

Then, if we insert the expression for the **e** vector from Eq. 9.4-1, we get

$$\frac{\partial}{\partial t}\left(\frac{1}{2}\rho v^2 + \rho \hat{U}\right) = -\left(\nabla \cdot \left(\frac{1}{2}\rho v^2 + \rho \hat{U}\right)\mathbf{v}\right) - (\nabla \cdot \mathbf{q}) - (\nabla \cdot \mathbf{w}) + \rho(\mathbf{v} \cdot \mathbf{g}) \tag{11.1-7}$$

The similarity with the first law of classical thermodynamics now becomes apparent. For *flow systems*, Eq. 11.1-7 becomes $\rho(D/Dt)\left(\frac{1}{2}v^2 + \hat{U}\right)\mathbf{v} = -(\nabla \cdot (\mathbf{q} + \mathbf{w}))$, in the absence of external forces. This states that, in a small region of fluid going along with the flow, the sum of its internal energy and kinetic energy changes because of the heat being added to the fluid and work being done on it. For *non-flow* systems, there is a similar relationship, $\Delta U = Q + W$, between the change in internal energy ΔU and the heat added to the system Q and work W done on it. Whereas Eq. 11.1-7 describes how the various quantities change with time, the simple classical thermodyamics expression for ΔU relates properties at two successive equilibrium states.

Next we insert the expression for the **w** vector from Eq. 9.3-5 and use Eq. A.4-26 to get the *equation of energy* (a statement of the law of conservation of energy)

$$\underbrace{\frac{\partial}{\partial t}\left(\frac{1}{2}\rho v^2 + \rho \hat{U}\right)}_{\substack{\text{rate of increase of} \\ \text{energy per unit volume}}} = \underbrace{-\left(\nabla \cdot \left(\frac{1}{2}\rho v^2 + \rho \hat{U}\right)\mathbf{v}\right)}_{\substack{\text{rate of energy addition per} \\ \text{unit volume by convective} \\ \text{transport}}} \underbrace{- (\nabla \cdot \mathbf{q})}_{\substack{\text{rate of energy addition} \\ \text{per unit volume by} \\ \text{heat conduction}}}$$
$$\underbrace{- (\nabla \cdot p\mathbf{v})}_{\substack{\text{rate of work done} \\ \text{on fluid per unit} \\ \text{volume by} \\ \text{pressure forces}}} \underbrace{-(\nabla \cdot [\boldsymbol{\tau} \cdot \mathbf{v}])}_{\substack{\text{rate of work done} \\ \text{on fluid per unit} \\ \text{volume by viscous} \\ \text{forces}}} \underbrace{+ \rho(\mathbf{v} \cdot \mathbf{g})}_{\substack{\text{rate of work done} \\ \text{on fluid per unit} \\ \text{volume by external} \\ \text{forces}}} \tag{11.1-8}$$

This equation does not include nuclear, radiative, electromagnetic, or chemical forms of energy. For viscoelastic fluids, the next-to-last term has to be reinterpreted by replacing "viscous" by "viscoelastic."

Equation 11.1-8 is the main result of this section, and it provides the basis for the remainder of the chapter. The equation may be written in another form to include the potential energy per unit mass, $\hat{\Phi}$, which has been defined earlier by $\mathbf{g} = -\nabla\hat{\Phi}$ (see §3.3). For moderate elevation changes, this gives $\hat{\Phi} = gh$, where h is a coordinate in the direction opposed to the gravitational field. For terrestrial problems, where the gravitational field is independent of time, we can write

$$\begin{aligned} \rho(\mathbf{v}\cdot\mathbf{g}) &= -(\rho\mathbf{v}\cdot\nabla\hat{\Phi}) \\ &= -(\nabla\cdot\rho\mathbf{v}\hat{\Phi}) + \hat{\Phi}(\nabla\cdot\rho\mathbf{v}) && \text{Use vector identityin Eq. A.4−19} \\ &= -(\nabla\cdot\rho\mathbf{v}\hat{\Phi}) - \hat{\Phi}\frac{\partial\rho}{\partial t} && \text{Use Eq. 3.1−4} \\ &= -(\nabla\cdot\rho\mathbf{v}\hat{\Phi}) - \frac{\partial}{\partial t}(\rho\hat{\Phi}) && \text{Use } \hat{\Phi} \text{ independent of } t \end{aligned} \tag{11.1-9}$$

When this result is inserted into Eq. 11.1-8, we get

$$\begin{aligned} \frac{\partial}{\partial t}\left(\frac{1}{2}\rho v^2 + \rho\hat{U} + \rho\hat{\Phi}\right) = -\left(\nabla\cdot\left(\frac{1}{2}\rho v^2 + \rho\hat{U} + \rho\hat{\Phi}\right)\mathbf{v}\right) \\ -(\nabla\cdot\mathbf{q}) - (\nabla\cdot p\mathbf{v}) - (\nabla\cdot[\boldsymbol{\tau}\cdot\mathbf{v}]) \end{aligned} \tag{11.1-10}$$

Sometimes it is convenient to use the energy equation in this form, as we will see in Chapter 15.

§11.2 SPECIAL FORMS OF THE ENERGY EQUATION

The most useful form of the energy equation is one in which the temperature appears. The object of this section is to arrive at such an equation, which can be used for prediction of temperature profiles.

First, we subtract the mechanical energy equation in Eq. 3.3-1 from the energy equation in Eq. 11.1-8. This leads to the following *equation of change for internal energy*:

$$\frac{\partial}{\partial t}\rho\hat{U} = -(\nabla\cdot\rho\hat{U}\mathbf{v}) - (\nabla\cdot\mathbf{q}) - p(\nabla\cdot\mathbf{v}) - (\boldsymbol{\tau}:\nabla\mathbf{v}) \tag{11.2-1}$$

$\frac{\partial}{\partial t}\rho\hat{U}$	rate of increase in internal energy per unit volume
$-(\nabla\cdot\rho\hat{U}\mathbf{v})$	net rate of addition of internal energy by convective transport, per unit volume
$-(\nabla\cdot\mathbf{q})$	rate of internal energy addition by heat conduction, per unit volume
$-p(\nabla\cdot\mathbf{v})$	*reversible* rate of internal energy increase per unit volume by compression
$-(\boldsymbol{\tau}:\nabla\mathbf{v})$	*irreversible* rate of internal energy increase per unit volume by viscous dissipation

It is now of interest to compare the mechanical energy equation of Eq. 3.3-1 and the internal energy equation of Eq. 11.2-1. Note that the terms $p(\nabla\cdot\mathbf{v})$ and $(\boldsymbol{\tau}:\nabla\mathbf{v})$ appear in both equations—but with opposite signs. Therefore, these terms describe the interconversion of mechanical and thermal energy. The term $p(\nabla\cdot\mathbf{v})$ can be either positive or negative, depending on whether the fluid is expanding or contracting; therefore, it represents a *reversible* mode of interchange. On the other hand, for Newtonian fluids, the quantity $-(\boldsymbol{\tau}:\nabla\mathbf{v})$ is always positive (see Eq. 3.3-3) and therefore represents an *irreversible* degradation of mechanical into internal energy. For viscoelastic fluids, discussed in Chapter 8, the quantity $-(\boldsymbol{\tau}:\nabla\mathbf{v})$ does not have to be positive, since some energy may be stored as elastic energy.

It was pointed out in §3.5 that the equations of change can be written somewhat more compactly by using the substantial derivative (see Table 3.5-1). Equation 11.2-1 can be put in the substantial derivative form by using Eq. 3.5-6. This gives, with no further assumptions,

$$\rho\frac{D\hat{U}}{Dt} = -(\nabla\cdot\mathbf{q}) - p(\nabla\cdot\mathbf{v}) - (\boldsymbol{\tau}:\nabla\mathbf{v}) \tag{11.2-2}$$

Next it is convenient to switch from internal energy to enthalpy, as we did at the very end of §9.4. That is, in Eq. 11.2-2 we set $\hat{U} = \hat{H} - p\hat{V} = \hat{H} - (p/\rho)$, making the standard assumption that thermodynamic formulas derived from equilibrium thermodynamics may be applied locally for nonequilibrium systems. When this formula is substituted into Eq. 11.2-2 and use is made of the equation of continuity (Eq. A of Table 3.5-1), we get

$$\begin{aligned}\rho\frac{D\hat{H}}{Dt} &= -(\nabla\cdot\mathbf{q}) - p(\nabla\cdot\mathbf{v}) + \rho\frac{D}{Dt}\left(\frac{p}{\rho}\right) - (\boldsymbol{\tau}:\nabla\mathbf{v})\\ &= -(\nabla\cdot\mathbf{q}) - p(\nabla\cdot\mathbf{v}) + \frac{Dp}{Dt} - p\frac{1}{\rho}\frac{D\rho}{Dt} - (\boldsymbol{\tau}:\nabla\mathbf{v})\\ &= -(\nabla\cdot\mathbf{q}) - (\boldsymbol{\tau}:\nabla\mathbf{v}) + \frac{Dp}{Dt}\end{aligned} \tag{11.2-3}$$

Next we may use Eq. 9.4-3, which presumes that the enthalpy is a function of p and T (this restricts the subsequent development to *Newtonian fluids*). Then we may get an expression for the change in the enthalpy in an element of fluid moving with the fluid velocity,

$$\begin{aligned}\rho\frac{D\hat{H}}{Dt} &= \rho\hat{C}_p\frac{DT}{Dt} + \rho\left[\hat{V} - T\left(\frac{\partial\hat{V}}{\partial T}\right)_p\right]\frac{Dp}{Dt}\\ &= \rho\hat{C}_p\frac{DT}{Dt} + \rho\left[\frac{1}{\rho} - T\left(\frac{\partial(1/\rho)}{\partial T}\right)_p\right]\frac{Dp}{Dt}\\ &= \rho\hat{C}_p\frac{DT}{Dt} + \left[1 + \left(\frac{\partial\ln\rho}{\partial\ln T}\right)_p\right]\frac{Dp}{Dt}\end{aligned} \tag{11.2-4}$$

Equating the right sides of Eqs. 11.2-3 and 11.2-4 gives

$$\boxed{\rho\hat{C}_p\frac{DT}{Dt} = -(\nabla\cdot\mathbf{q}) - (\boldsymbol{\tau}:\nabla\mathbf{v}) - \left(\frac{\partial\ln\rho}{\partial\ln T}\right)_p\frac{Dp}{Dt}} \tag{11.2-5}$$

This is the *equation of change for temperature*, in terms of the heat-flux vector $\mathbf{q}$ and the viscous momentum-flux tensor $\boldsymbol{\tau}$.[1] This equation is tabulated in Cartesian, cylindrical, and spherical coordinates in §B.8. To use this equation, we need expressions for these fluxes:

(i) When Fourier's law of Eq. 9.2-6 is used, the term $-(\nabla\cdot\mathbf{q})$ becomes $+(\nabla\cdot k\nabla T)$, or, if the thermal conductivity is assumed constant, $+k\nabla^2 T$.

(ii) When Newton's law of Eq. 1.2-13 is used, the term $-(\boldsymbol{\tau}:\nabla\mathbf{v})$ becomes $\mu\Phi_v + \kappa\Psi_v$, the quantity given explicitly in Eq. 3.3-3.

We do not perform the substitutions here, because the equation of change for temperature is almost never used in its complete generality.

We now discuss several special *restricted* versions of the equation of change for temperature. In all of these we use Fourier's law with constant k, and we omit the viscous dissipation term, since it is important only in flows with enormous velocity gradients or for fluids of very large viscosity:

---

[1] An equation of change for temperature in polymeric liquid mixtures has been derived by C. F. Curtiss and R. B. Bird, *Adv. Polymer Sci.*, **125**, 1–101 (1996); this derivation assumes that polymers may be represented by a bead-spring-chain model.

(i) For an *ideal gas*, $(\partial \ln \rho/\partial \ln T)_p = -1$, and

$$\rho \hat{C}_p \frac{DT}{Dt} = k\nabla^2 T + \frac{Dp}{Dt} \tag{11.2-6}$$

Or, if use is made of the relation $\tilde{C}_p - \tilde{C}_V = R$, the equation of state in the form $pM = \rho RT$, and the equation of continuity as written in Eq. A of Table 3.5-1, we get

$$\rho \hat{C}_V \frac{DT}{Dt} = k\nabla^2 T - p(\nabla \cdot \mathbf{v}) \tag{11.2-7}$$

(ii) For a *fluid flowing in a constant pressure system*, $Dp/Dt = 0$, and

$$\rho \hat{C}_p \frac{DT}{Dt} = k\nabla^2 T \tag{11.2-8}$$

(iii) For a *fluid with constant density*,[2] $(\partial \ln \rho/\partial \ln T)_p = 0$, and

$$\rho \hat{C}_p \frac{DT}{Dt} = k\nabla^2 T \tag{11.2-9}$$

(iv) For a *stationary solid*, $\mathbf{v}$ is zero and

$$\rho \hat{C}_p \frac{\partial T}{\partial t} = k\nabla^2 T \tag{11.2-10}$$

These last five equations are the ones most frequently encountered in textbooks and research publications. Equation 11.2-8 (or equivalently, Eq. 11.2-9) is tabulated in Cartesian, cylindrical, and spherical coordinates in §B.9, with the viscous dissipation term $-(\boldsymbol{\tau} : \nabla \mathbf{v}) = +\mu\Phi_v$ added for completeness. One can also use the table in §B.9 to obtain Eqs. 11.2-6, 11.2-7, or 11.2-10 in component form by adding or subtracting the relevant terms. Of course, one can always go back to Eq. 11.2-5 and develop less restrictive equations when needed. Also, one can add chemical, electrical, and nuclear source terms on an ad hoc basis, just as was done in Chapter 10.

Equation 11.2-10 is the heat conduction equation for stationary solids, and much has been written about this famous equation developed first by Fourier.[3] The famous reference work by Carslaw and Jaeger deserves special mention. It contains hundreds of solutions of this equation for a wide variety of boundary and initial conditions.[4]

§11.3 THE BOUSSINESQ EQUATION OF MOTION FOR FORCED AND FREE CONVECTION

The equation of motion given in Eq. 3.2-9 (or Eq. B of Table 3.5-1) is valid for both isothermal and nonisothermal flow. In nonisothermal flow, the fluid density and viscosity depend in general on temperature as well as on pressure. The variation in the density is

---

[2]The assumption of constant density is made here, instead of the less stringent assumption that $(\partial \ln \rho/\partial \ln T)_p = 0$, since Eq. 11.2-9 is customarily used along with Eq. 3.1-5 (equation of continuity for constant density) and Eq. 3.6-1 (equation of motion for constant density and viscosity). It should be noted that the hypothetical equation of state $\rho =$ constant has to be supplemented by the statement that $(\partial p/\partial T)_\rho =$ finite, in order to permit the evaluation of certain thermodynamic derivatives. For example, the relation

$$\hat{C}_p - \hat{C}_V = -\frac{1}{\rho}\left(\frac{\partial \ln \rho}{\partial \ln T}\right)_p \left(\frac{\partial p}{\partial T}\right)_\rho \tag{11.2-9a}$$

leads to the result that $\hat{C}_p = \hat{C}_V$ for the "incompressible fluid" thus defined.

[3]J. B. Fourier, *Théorie analytique de la chaleur, Œuvres de Fourier*, Gauthier-Villars et Fils, Paris (1822).

[4]H. S. Carslaw and J. C. Jaeger, *Conduction of Heat in Solids*, Oxford University Press, 2nd edition (1959).

particularly important because it gives rise to buoyant forces, and thus to free convection, as we have already seen in §10.10.

The buoyant force appears automatically when an equation of state is inserted into the equation of motion. For example, we can use the simplified equation of state introduced in Eq. 10.10-6 (this is called the *Boussinesq approximation*)[1]

$$\rho(T) = \bar{\rho} - \bar{\rho}\bar{\beta}(T - \bar{T}) \tag{11.3-1}$$

in which $\bar{\beta}$ is $-(1/\rho)(\partial\rho/\partial T)_p$ evaluated at $T = \bar{T}$. This equation is obtained by writing the Taylor series for ρ as a function of T, considering the pressure p to be constant, and keeping only the first two terms of the series. When Eq. 11.3-1 is substituted into the $\rho\mathbf{g}$ term (but not into the $\rho(D\mathbf{v}/Dt)$ term) of Eq. B of Table 3.5-1, we get the *Boussinesq equation*:

$$\boxed{\rho\frac{D\mathbf{v}}{Dt} = (-\nabla p + \bar{\rho}\mathbf{g}) - [\nabla \cdot \boldsymbol{\tau}] - \bar{\rho}\mathbf{g}\bar{\beta}(T - \bar{T})} \tag{11.3-2}$$

This form of the equation of motion is very useful for heat-transfer analyses. It describes the limiting cases of forced convection and free convection (see Fig. 10.9-1), and the region between these extremes as well. In *forced convection* the buoyancy term $-\bar{\rho}\mathbf{g}\bar{\beta}(T - \bar{T})$ is neglected. In *free convection* (or *natural convection*) the term $(-\nabla p + \bar{\rho}\mathbf{g})$ is small, and omitting it is usually appropriate, particularly for vertical, rectilinear flow and for the flow near submerged objects in large bodies of fluid. Setting $(-\nabla p + \bar{\rho}\mathbf{g})$ equal to zero is equivalent to assuming that the pressure distribution is just that for a fluid at rest.

It is also customary to replace ρ on the left side of Eq. 11.3-2 by $\bar{\rho}$. This substitution has been successful for free convection at moderate temperature differences. Under these conditions the fluid motion is slow, and the acceleration term $D\mathbf{v}/Dt$ is small compared to $\mathbf{g}$.

However, in systems where the acceleration term is large with respect to $\mathbf{g}$, one must also use Eq. 11.3-1 for the density on the left side of the equation of motion. This is particularly true, for example, in gas turbines and near hypersonic missiles, where the term $(\rho - \bar{\rho})D\mathbf{v}/Dt$ may be at least as important as $\bar{\rho}\mathbf{g}$.

§11.4 THE EQUATIONS OF CHANGE AND SOLVING STEADY-STATE PROBLEMS WITH ONE INDEPENDENT VARIABLE

In §3.1 to §3.4 and in §11.1 to §11.3 we have derived various equations of change for a pure fluid or solid. It seems appropriate here to present a summary of these equations for future reference. Such a summary is given in Table 11.4-1, with each equation given in the D/Dt form. Reference is also made to the first place where each equation has been presented.

Although Table 11.4-1 is a useful summary, for problem solving we use the equations written out explicitly in the several commonly used coordinate systems. This has been done in Appendix B, and the readers should thoroughly familiarize themselves with the tables there.

In general, to describe the nonisothermal flow of a Newtonian fluid one needs

- the equation of continuity
- the equation of motion (containing μ and κ)
- the equation of energy (containing μ, κ, and k)
- the thermal equation of state ($p = p(\rho,T)$)
- the caloric equation of state ($\hat{C}_p = \hat{C}_p(\rho,T)$)

[1] J. Boussinesq, *Théorie Analytique de Chaleur*, Vol. **2**, Gauthier-Villars, Paris (1903).

Table 11.4-1. Equations of Change for Pure Fluids in Terms of the Fluxes

Eq.	Special Form			Comments
Mass	General	$\frac{D\rho}{Dt} = -\rho(\nabla \cdot \mathbf{v})$	Table 3.5-1 (A)	For $\rho =$ constant, simplifies to $(\nabla \cdot \mathbf{v}) = 0$
Momentum	General	$\rho \frac{D\mathbf{v}}{Dt} = -\nabla p - [\nabla \cdot \boldsymbol{\tau}] + \rho \mathbf{g}$	Table 3.5-1 (B)	For $\boldsymbol{\tau} = 0$, this becomes Euler's equation
	Approximate	$\rho \frac{D\mathbf{v}}{Dt} = -\nabla p - [\nabla \cdot \boldsymbol{\tau}] + \bar{\rho}\mathbf{g} - \bar{\rho}\mathbf{g}\bar{\beta}(T - \bar{T})$	11.3-2 (C)	Displays buoyancy term
Energy	In terms of $\hat{U}$ and $\frac{1}{2}v^2$	$\rho \frac{D}{Dt}\left(\frac{1}{2}v^2 + \hat{U}\right) = -(\nabla \cdot \mathbf{q}) - (\nabla \cdot p\mathbf{v}) - (\nabla \cdot [\boldsymbol{\tau} \cdot \mathbf{v}]) + (\rho \mathbf{v} \cdot \mathbf{g})$	– (D)	Eq. 11.1-8 with the continuity equation
	In terms of $\hat{H}$	$\rho \frac{D}{Dt}\hat{H} = -(\nabla \cdot \mathbf{q}) - (\boldsymbol{\tau} : \nabla \mathbf{v}) + \frac{Dp}{Dt}$	Eq. 11.2-3 (E)	
	In terms of $\hat{C}_V$ and T	$\rho \hat{C}_V \frac{DT}{Dt} = -(\nabla \cdot \mathrm{q}) - T\left(\frac{\partial p}{\partial T}\right)_\rho (\nabla \cdot \mathbf{v}) - (\boldsymbol{\tau} : \nabla \mathbf{v})$	– (F)	For an ideal gas $T(\partial p/\partial T)_\rho = p$
	In terms of $\hat{C}_p$ and T	$\rho \hat{C}_p \frac{DT}{Dt} = -(\nabla \cdot \mathbf{q}) - \left(\frac{\partial \ln \rho}{\partial \ln T}\right)_p \frac{Dp}{Dt} - (\boldsymbol{\tau} : \nabla \mathbf{v})$	Eq. 11.2-5 (G)	For an ideal gas $(\partial \ln \rho / \partial \ln T)_p = -1$

Notes:

[a] To convert these equations from the *D/Dt* form to the form, see Eqs. 3.5-6 and 3.5-7.

[b] Equations A, B and G are written out in various coordinate systems in Appendix B.

as well as expressions for the density and temperature dependence of the viscosity, dilatational viscosity, and thermal conductivity. In addition one needs the boundary and initial conditions. In the most general case, Eq. 11.2-5 for the equation energy is used. This equation is tabulated in Cartesian, cylindrical, and spherical coordinates in §B.8. The entire set of equations listed above can then—in principle—be solved to get the pressure, density, velocity, and temperature as functions of position and time. If one wishes to solve such a detailed problem, numerical methods generally have to be used. Commercial software packages are available for such approaches.

Often one may be content with a restricted solution, just for making an order-of-magnitude analysis of a problem, or for investigating limiting cases prior to doing a complete numerical solution. This is done by making some standard assumptions:

(i) *Assumption of constant physical properties.* If it can be assumed that all physical properties are constant, then the equations become considerably simpler, and in some cases analytical solutions can be found. In this case, the forms of the energy equation in Eqs. 11.2-6 through 11.2-10 are typically used. Equation 11.2-8 (or 11.2-9) is tabulated in Cartesian, cylindrical, and spherical coordinates in §B.9 (with the viscous dissipation term included). Other forms may be obtained by adding or removing relevant terms.

(ii) *Assumption of zero fluxes.* Setting $\boldsymbol{\tau}$ and $\mathbf{q}$ equal to zero may be useful for (a) adiabatic flow processes in systems designed to minimize frictional effects (such as Venturi meters and turbines), and (b) high-speed flows around streamlined objects. The solutions obtained would be of no use for describing the situation near fluid–solid boundaries, but may be adequate for analysis of phenomena far from the solid boundaries.

To illustrate the solution of problems in which the energy equation plays a significant role, we solve a series of (idealized) problems. We restrict ourselves here to steady-state flow problems and consider unsteady-state problems in §11.5. In each problem we start by listing the postulates that lead us to simplified versions of the equations of change. We then write the simplified equation in component form with the help of §B.8 or §B.9, employ assumptions in the problem at hand to omit appropriate terms, then solve using the boundary conditions to obtain constants of integration.

Of course, one can also practice solving problems with the equation of energy by reworking the examples in Chapter 10, using the appropriate form of the equation of energy instead of the shell energy balance.

EXAMPLE 11.4-1

Steady-State Forced-Convection Heat Transfer in Laminar Flow in a Circular Tube

Show how to set up the equations for the problem considered in §10.9, namely, that of finding the fluid temperature profiles for the fully developed laminar flow in a tube.

SOLUTION

We assume constant physical properties, and we postulate a solution of the following form: $\mathbf{v} = \boldsymbol{\delta}_z v_z(r)$, $\mathscr{P} = \mathscr{P}(z)$, and $T = T(r,z)$. Then the equations of change, as given in Appendix B, may be simplified to

Equation of continuity:
$$0 = 0 \tag{11.4-1}$$

z Equation of motion:
$$0 = -\frac{d\mathscr{P}}{dz} + \mu\left[\frac{1}{r}\frac{d}{dr}\left(r\frac{dv_z}{dr}\right)\right] \tag{11.4-2}$$

Equation of energy:
$$\rho\hat{C}_p v_z\frac{\partial T}{\partial z} = k\left[\frac{1}{r}\frac{\partial}{\partial r}\left(r\frac{\partial T}{\partial r}\right) + \frac{\partial^2 T}{\partial z^2}\right] + \mu\left(\frac{dv_z}{dr}\right)^2 \tag{11.4-3}$$

The r and θ components of the equation of motion do not add helpful information. The equation of continuity is automatically satisfied as a result of the postulates. The z component of the equation of motion, when solved as in Example 3.6-1, gives the velocity distribution (the parabolic velocity profile). This expression is then substituted into the convective heat transport term on the left side of Eq. 11.4-3 and into the viscous dissipation heating term on the right side.

Next, as in §10.9, we make two assumptions: (i) in the z direction, heat conduction is much smaller than heat convection, so that the term $\partial^2 T/\partial z^2$ can be neglected, and (ii) the flow is not sufficiently fast so that viscous heating is significant, and hence the term $\mu(dv_z/dr)^2$ can be omitted. When these assumptions are made, Eq. 11.4-3 becomes the same as Eq. 10.9-14. From that point on, the asymptotic solution, valid for large z only, proceeds as in §10.9. Note that we have gone through three types of restrictive processes: (i) *postulates*, in which a tentative guess is made as to the form of the solution; (ii) *assumptions*, in which we eliminate some physical phenomena or effects by discarding terms or assuming physical properties to be constant; and (iii) an *asymptotic solution*, in which we obtain only a portion of the entire mathematical solution. It is important to distinguish among these various kinds of restrictions.

EXAMPLE 11.4-2

Tangential Flow in an Annulus with Viscous Heat Generation

Determine the temperature distribution in an incompressible liquid confined between two coaxial cylinders, the outer one of which is rotating at a steady angular velocity Ω_o as illustrated in Fig. 11.4-1. The inner cylinder has radius κR and the outer cylinder has radius R. Consider the radius ratio κ to be fairly small so that the curvature of the fluid streamlines must be taken into account. Gravity acts in the $-z$ direction.

The temperatures of the inner and outer surfaces of the annular region are maintained at T_κ and T_1, respectively, with $T_\kappa \neq T_1$. Assume steady laminar flow, and neglect the temperature dependence of the physical properties.

This is an example of a forced convection problem: the equations of continuity and motion are solved to get the velocity distribution, and then the energy equation is solved to get the temperature distribution. This problem is of interest in connection with heat effects in coaxial cylinder viscometers[1] and in lubrication systems.

SOLUTION

We begin by postulating that $\mathbf{v} = \boldsymbol{\delta}_\theta v_\theta(r)$, that $\mathscr{P} = \mathscr{P}(r,z)$, and that $T = T(r)$. Because the physical properties are assumed to be constant, Eqs. B.6-4 through B.6-6 can be used for the equation of motion, and Eq. B.9-2 can be used for the equation of energy (all in cylindrical coordinates). Then the postulates stated above lead to the following simplifications of the equations of change

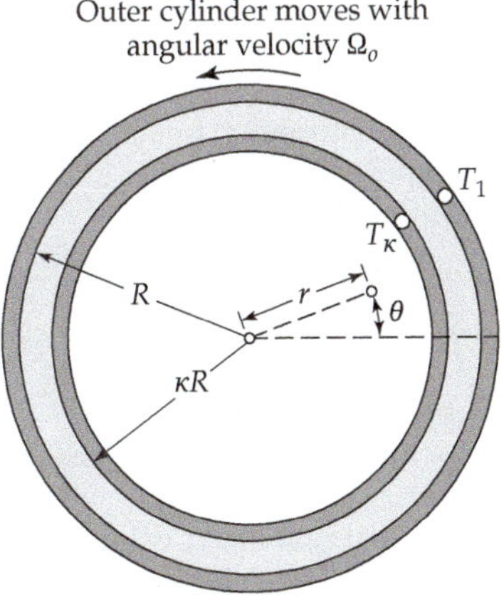

Fig. 11.4-1. Tangential annular flow between concentric cylinders. The outer cylinder at $r = R$ rotates with angular velocity Ω_o, and the inner cylinder at $r = \kappa R$ is stationary.

[1] J. R. Van Wazer, J. W. Lyons, K. Y. Kim, and R. E. Colwell, *Viscosity and Flow Measurement*, Wiley, New York (1963), pp. 82–85.

for motion and energy:

r Equation of motion: $$-\rho\frac{v_\theta^2}{r} = -\frac{\partial p}{\partial r} \tag{11.4-4}$$

θ Equation of motion: $$0 = \frac{d}{dr}\left(\frac{1}{r}\frac{d}{dr}(rv_\theta)\right) \tag{11.4-5}$$

z Equation of motion: $$0 = -\frac{\partial p}{\partial z} - \rho g \tag{11.4-6}$$

Equation of energy: $$0 = k\frac{1}{r}\frac{d}{dr}\left(r\frac{dT}{dr}\right) + \mu\left[r\frac{d}{dr}\left(\frac{v_\theta}{r}\right)\right]^2 \tag{11.4-7}$$

When the solution to the θ component of the equation of motion, given in Eq. 3.7-34, is substituted into the energy equation, we get

$$0 = k\frac{1}{r}\frac{d}{dr}\left(r\frac{dT}{dr}\right) + \frac{4\mu\Omega_o^2\kappa^4R^4}{(1-\kappa^2)^2}\frac{1}{r^4} \tag{11.4-8}$$

This is the differential equation for the temperature distribution. It may be rewritten in terms of dimensionless quantities using

$$\xi = \frac{r}{R} \qquad \Theta = \frac{T - T_\kappa}{T_1 - T_\kappa} \qquad N = \frac{\mu\Omega_o^2R^2}{k(T_1 - T_\kappa)}\cdot\frac{\kappa^4}{(1-\kappa^2)^2} \tag{11.4-9,10,11}$$

The parameter N is closely related to the Brinkman number of §10.8. Equation 11.4-8 now becomes

$$\frac{1}{\xi}\frac{d}{d\xi}\left(\xi\frac{d\Theta}{d\xi}\right) = -4N\frac{1}{\xi^4} \tag{11.4-12}$$

This ordinary differential equation is of the form of Eq. C.1-11 and has the solution

$$\Theta(\xi) = -N\frac{1}{\xi^2} + C_1\ln\xi + C_2 \tag{11.4-13}$$

The integration constants are found from the boundary conditions

B. C. 1: $$\text{at } \xi = \kappa, \quad \Theta = 0 \tag{11.4-14}$$

B. C. 2: $$\text{at } \xi = 1, \quad \Theta = 1 \tag{11.4-15}$$

Determination of the constants then leads to

$$\Theta(\xi) = \left(1 - \frac{\ln\xi}{\ln\kappa}\right) + N\left[\left(1 - \frac{1}{\xi^2}\right) - \left(1 - \frac{1}{\kappa^2}\right)\frac{\ln\xi}{\ln\kappa}\right] \tag{11.4-16}$$

When $N = 0$, we obtain the temperature distribution for a motionless cylindrical shell of thickness $R(1-\kappa)$ with inner and outer temperatures T_κ and T_1. If N is large enough, there will be a maximum in the temperature distribution, located at

$$\xi_{max} = \sqrt{\frac{2\ln(1/\kappa)}{(1/\kappa^2) - 1 - (1/N)}} \tag{11.4-17}$$

with the temperature at this point greater than either T_κ or T_1.

Although this example provides an illustration of the use of the tabulated equations of change in cylindrical coordinates, in most viscometric and lubrication applications, the clearance between the cylinders is so small that numerical values computed from Eq. 11.4-16 will not differ substantially from those computed from Eq. 10.8-9.

EXAMPLE 11.4-3

Steady Flow in a Nonisothermal Film

A liquid is flowing downward in steady laminar flow along an inclined plane surface, as shown in Figs. 2.2-1 to 2.2-3. The free liquid surface is maintained at temperature T_0, and the solid surface at $x = \delta$ is maintained at T_δ. At these temperatures the liquid viscosity has values μ_0 and μ_δ, respectively, and the liquid density and thermal conductivity may be assumed constant. Find the velocity distribution in this nonisothermal flow system, neglecting end effects and recognizing that viscous heating is unimportant in this flow. Assume that the temperature dependence of viscosity may be expressed by an equation of the form $\mu = Ae^{B/T}$, with A and B being empirical constants; this is suggested by the Eyring theory given in §1.7.

We first solve the energy equation to get the temperature profile, and then use the latter to find the dependence of viscosity on position. Then the equation of motion can be solved to get the velocity profile.

SOLUTION

We postulate that $T = T(x)$ and that $\mathbf{v} = \boldsymbol{\delta}_z v_z(x)$. Because the density and thermal conductivity are assumed to be constant, we can use the form of the energy equation expressed by Eq. B.9-1 (i.e., Eq. 11.2-9 in Cartesian coordinates). Using the assumptions in this problem, the energy equation simplifies to

$$\frac{d^2T}{dx^2} = 0 \tag{11.4-18}$$

This can be integrated, and the boundary conditions $T(0) = T_0$ and $T(\delta) = T_\delta$ used to determine the constants of integration. The resulting temperature profile is

$$\frac{T(x) - T_0}{T_\delta - T_0} = \frac{x}{\delta} \tag{11.4-19}$$

The dependence of viscosity on temperature may be written as

$$\frac{\mu(T)}{\mu_0} = \exp\left[B\left(\frac{1}{T} - \frac{1}{T_0}\right)\right] \tag{11.4-20}$$

in which B is a constant, which can be determined from experimental data for viscosity versus temperature. To get the dependence of viscosity on position, we combine the last two equations to get

$$\frac{\mu(x)}{\mu_0} = \exp\left[B\frac{T_0 - T_\delta}{T_0 T}\left(\frac{x}{\delta}\right)\right] \approx \exp\left[B\frac{T_0 - T_\delta}{T_0 T_\delta}\left(\frac{x}{\delta}\right)\right] \tag{11.4-21}$$

The second expression is a good approximation if the temperature does not change greatly through the film. When this equation is combined with Eq. 11.4-20, written for $T = T_\delta$, we then get

$$\frac{\mu(x)}{\mu_0} = \exp\left[\left(\ln\frac{\mu_\delta}{\mu_0}\right)\left(\frac{x}{\delta}\right)\right] = \left(\frac{\mu_\delta}{\mu_0}\right)^{x/\delta} \tag{11.4-22}$$

This is the same as the expression used in Example 2.2-2, if we set α equal to $-\ln(\mu_\delta/\mu_0)$. Therefore, we may take over the result from Example 2.2-2 and write the velocity profile as:

$$v_z(x) = \left(\frac{\rho g \cos\beta}{\mu_0}\right)\left(\frac{\delta}{\ln(\mu_\delta/\mu_0)}\right)^2\left[\frac{1 + (x/\delta)\ln(\mu_\delta/\mu_0)}{(\mu_\delta/\mu_0)^{x/\delta}} - \frac{1 + \ln(\mu_\delta/\mu_0)}{(\mu_\delta/\mu_0)}\right] \tag{11.4-23}$$

This completes the analysis of the problem begun in Example 2.2-2, by providing the appropriate value of the constant α.

EXAMPLE 11.4-4

Transpiration Cooling[2]

A system with two concentric porous spherical shells of radii κR and R is shown in Fig. 11.4-2. The inner surface of the outer shell is at temperature T_1, and the outer surface of the inner one is at a lower temperature T_κ. Dry air at T_κ is blown outward radially from the inner shell into the intervening space and then through the outer shell. Develop an expression for the required rate of heat removal from the inner sphere as a function of the mass rate of flow of the gas. Assume steady laminar flow and low gas velocity. We may also assume that the density, viscosity and thermal conductivity are constants.

[2]M. Jakob, *Heat Transfer*, Vol. **II**, Wiley, New York (1957), pp. 394–415.

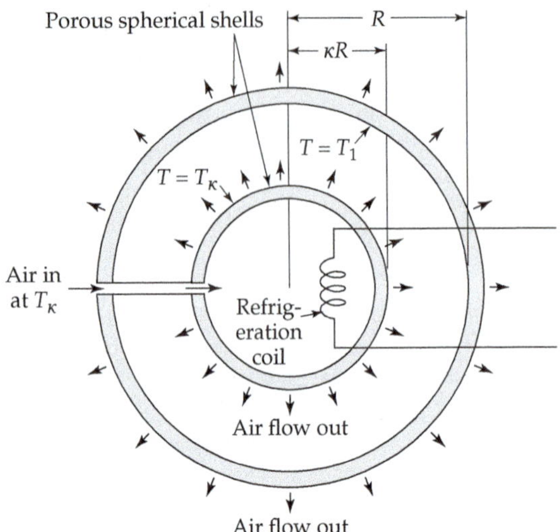

Fig. 11.4-2. Transpiration cooling. The inner sphere is being cooled by means of the refrigeration coil to maintain its temperature at T_κ. When air is blown outward, as shown, less refrigeration is required.

In this example the equations of continuity and energy are solved to get the temperature distribution. The equation of motion gives information about the pressure distribution in the system.

SOLUTION

We postulate that for this system $\mathbf{v} = \boldsymbol{\delta}_r v_r(r)$, $T = T(r)$, and $\mathscr{P} = \mathscr{P}(r)$. The *equation of continuity* in spherical coordinates (Eq. B.4-3) then becomes

$$\frac{1}{r^2}\frac{d}{dr}(r^2 \rho v_r) = 0 \tag{11.4-24}$$

This equation can be integrated to give

$$r^2 \rho v_r = \text{const.} = \frac{w_r}{4\pi} \tag{11.4-25}$$

Here w_r is the radial mass flow rate of the gas.

The r component of the *equation of motion* in spherical coordinates is, from Eq. B.6-7

$$\rho v_r \frac{dv_r}{dr} = -\frac{d\mathscr{P}}{dr} + \mu\left(\frac{1}{r^2}\frac{d^2}{dr^2}(r^2 v_r)\right) \tag{11.4-26}$$

The viscosity term drops out because of Eq. 11.4-24, along with the assumption of constant density. The left side can be written in terms of w_r using Eq. 11.4-25. Integration of Eq. 11.4-26 from $r = R$ to an arbitrary radial position r then gives

$$\mathscr{P}(r) - \mathscr{P}(R) = \frac{w_r}{32\pi^2 \rho R^4}\left[1 - \left(\frac{R}{r}\right)^4\right] \tag{11.4-27}$$

Hence, the modified pressure $\mathscr{P}$ increases with r, but only very slightly for the low gas velocity assumed here.

Thus far, we have assumed constant density and thermal conductivity. If we also assume that energy production by viscous dissipation can be neglected—a reasonable assumption for the slow flow here—then the appropriate form of the *energy equation* in terms of the temperature for this problem is Eq. 11.2-8, which is expressed in spherical coordinates, by Eq. B.9-3 (when the viscous dissipation term is neglected). Using $T = T(r)$, this equation becomes

$$\rho \hat{C}_p v_r \frac{dT}{dr} = k\frac{1}{r^2}\frac{d}{dr}\left(r^2\frac{dT}{dr}\right) \tag{11.4-28}$$

When Eq. 11.4-25 for the velocity distribution is used for v_r in Eq. 11.4-28, we obtain the following differential equation for the temperature distribution $T(r)$ of the gas between the two shells:

$$\frac{dT}{dr} = \frac{4\pi k}{w_r \hat{C}_p} \frac{d}{dr}\left(r^2 \frac{dT}{dr}\right) \tag{11.4-29}$$

We make the change of variable $u = r^2(dT/dr)$ and obtain a first-order, separable differential equation for $u(r)$, $du/dr = (w_r\hat{C}_p/4\pi k)u(r)/r^2$. This may be integrated to obtain $u(r)$. After substituting $u = r^2(dT/dr)$, the resulting first-order differential equation for $T(r)$ can be integrated. The constants of integration are evaluated using the boundary conditions $T(\kappa R) = T_\kappa$ and $T(R) = T_1$, to obtain finally the temperature profile

$$\frac{T(r) - T_1}{T_\kappa - T_1} = \frac{\exp(-R_0/r) - \exp(-R_0/R)}{\exp(-R_0/\kappa R) - \exp(-R_0/R)} \tag{11.4-30}$$

in which $R_0 = w_r\hat{C}_p/4\pi k$ is a constant with units of length.

The rate of heat flow towards the inner sphere is

$$Q = -4\pi\kappa^2 R^2 q_r|_{r=\kappa R} \tag{11.4-31}$$

and this is the required rate of heat removal by the refrigerant (note that the convective energy flow in the inlet stream is equal to the convective energy flow radially outward at $r = \kappa R$, and hence, Q is due entirely to conduction at $r = \kappa R$). Insertion of Fourier's law for the r component of the heat flux gives

$$Q = +4\pi\kappa^2 R^2 k \left.\frac{dT}{dr}\right|_{r=\kappa R} \tag{11.4-32}$$

Next we evaluate the temperature gradient at the surface with the aid of Eq. 11.4-30 to obtain the expression for the heat-removal rate.

$$Q = \frac{4\pi R_0 k(T_1 - T_\kappa)}{\exp[(R_0/\kappa R)(1-\kappa)] - 1} \tag{11.4-33}$$

In the limit that the mass flow rate of the gas is zero, so that $R_0 = 0$, the heat-removal rate becomes

$$Q_0 = \frac{4\pi\kappa R k(T_1 - T_\kappa)}{1-\kappa} \tag{11.4-34}$$

The fractional reduction in heat removal as a result of the transpiration of the gas is then

$$\frac{Q_0 - Q}{Q_0} = 1 - \frac{\phi}{e^\phi - 1} \tag{11.4-35}$$

Here $\phi = R_0(1-\kappa)/\kappa R = w_r\hat{C}_p(1-\kappa)/4\pi\kappa R k$ is the "dimensionless transpiration rate." Equation 11.4-35 is shown graphically in Fig. 11.4-3. For small values of ϕ, the quantity $(Q_0 - Q)/Q_0$ approaches the asymptote $\frac{1}{2}\phi$.

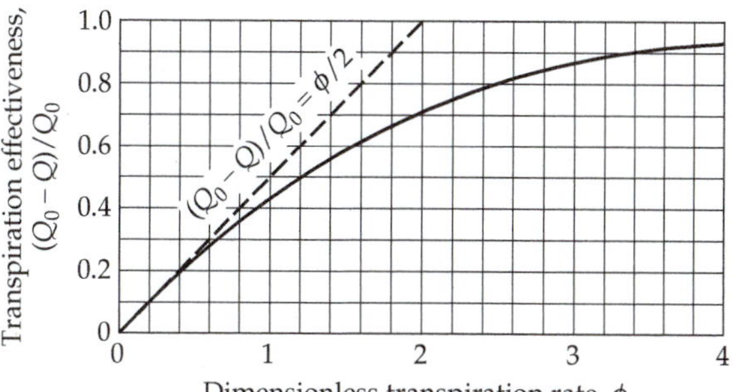

Fig. 11.4-3. The effect of transpiration cooling.

EXAMPLE 11.4-5

Adiabatic Frictionless Processes in an Ideal Gas

Develop equations for the relationship of local pressure to density or temperature in a stream of ideal gas in which the momentum flux $\boldsymbol{\tau}$ and the heat flux $\mathbf{q}$ are negligible.

SOLUTION

With $\boldsymbol{\tau}$ and $\mathbf{q}$ neglected, the equation of energy (Eq. 11.2-5 or Eq. (G) in Table 11.4-1 with $\rho = 1/\hat{V}$) may be rewritten as

$$\rho \hat{C}_p \frac{DT}{Dt} = \left(\frac{\partial \ln \hat{V}}{\partial \ln T} \right)_p \frac{Dp}{Dt} \tag{11.4-36}$$

For an ideal gas, $p\hat{V} = RT/M$, where M is the molecular weight of the gas, so that $(\partial \ln \hat{V}/\partial \ln T)_p = 1$. Hence, Eq. 11.4-36 becomes

$$\rho \hat{C}_p \frac{DT}{Dt} = \frac{Dp}{Dt} \tag{11.4-37}$$

Dividing this equation by p and assuming the molar heat capacity $\tilde{C}_p = M\hat{C}_p$ to be constant, we can again use the ideal-gas law to get

$$\frac{D}{Dt}\left(\frac{\tilde{C}_p}{R} \ln T - \ln p \right) = 0 \tag{11.4-38}$$

Hence, the quantity in parentheses is a constant following an element of the fluid, as is its antilogarithm, so that we have

$$T^{\tilde{C}_p/R} p^{-1} = \text{constant} \tag{11.4-39}$$

This relation among the thermodynamic properties applies to all thermodynamic states p, T that the fluid element encounters as it moves along with the fluid.

When we introduce the definition $\gamma = \hat{C}_p/\hat{C}_V$ and the ideal-gas relations $\tilde{C}_p - \tilde{C}_V = R$ and $p = \rho RT/M$, Eq. 11.4-39 can be rearranged to give the related expressions

$$p^{(\gamma-1)/\gamma} T^{-1} = \text{constant} \tag{11.4-40}$$

and

$$p\rho^{-\gamma} = \text{constant} \tag{11.4-41}$$

These last three equations find frequent use in the study of frictionless adiabatic processes in ideal-gas dynamics. Equation 11.4-41 is a famous relation well worth remembering.

When the viscous momentum flux $\boldsymbol{\tau}$ and the heat flux $\mathbf{q}$ are zero, there is no change in entropy following an element of fluid. Hence, the derivative $d \ln p / d \ln T = \gamma/(\gamma - 1)$ following the fluid motion has to be understood to mean $(\partial \ln p/\partial \ln T)_S = \gamma/(\gamma - 1)$. This equation is then just a standard formula from equilibrium thermodynamics.

EXAMPLE 11.4-6

One-Dimensional Compressible Flow: Velocity, Temperature, and Pressure Profiles in a Stationary Shock Wave

We consider here the adiabatic expansion[3–8] of an ideal gas through a convergent-divergent nozzle under such conditions that a stationary shock wave is formed. The gas enters the nozzle from a reservoir, where the pressure is p_0, and discharges to the atmosphere, where the pressure is p_a. In the absence of a shock wave, the flow through a well-designed nozzle is virtually frictionless (hence, *isentropic* for the adiabatic situation being considered). If, in addition, p_a/p_0 is sufficiently small, it is known that the flow is essentially sonic at the throat (the region of minimum cross section) and is supersonic in the divergent portion of the nozzle. Under these conditions, the pressure will continually *decrease*, and the velocity will *increase* in the direction of the flow, as indicated by the curves in Fig. 11.4-4.

---

[3]H. W. Liepmann and A. Roshko, *Elements of Gas Dynamics*, Wiley, New York (1957), §5.4 and §13.12.

[4]J. O. Hirschfelder, C. F. Curtiss, and R. B. Bird, *Molecular Theory of Gases and Liquids*, Wiley, New York, 2nd corrected printing (1964), pp. 791–797.

[5]M. Morduchow and P. A. Libby, *J. Aeronautical Sci.*, **16**, 674–684 (1948).

[6]R. von Mises, *J. Aeronautical Sci.*, **17**, 551–554 (1950).

[7]G. S. S. Ludford, *J. Aeronautical Sci.*, **18**, 830–834 (1951).

[8]C. R. Illingworth, "Shock waves," in *Modern Developments in Fluid Dynamics. High Speed Flow*, vol **1.**, L. Howarth, ed., Clarendon Press, Oxford (1953), pp. 122–130.

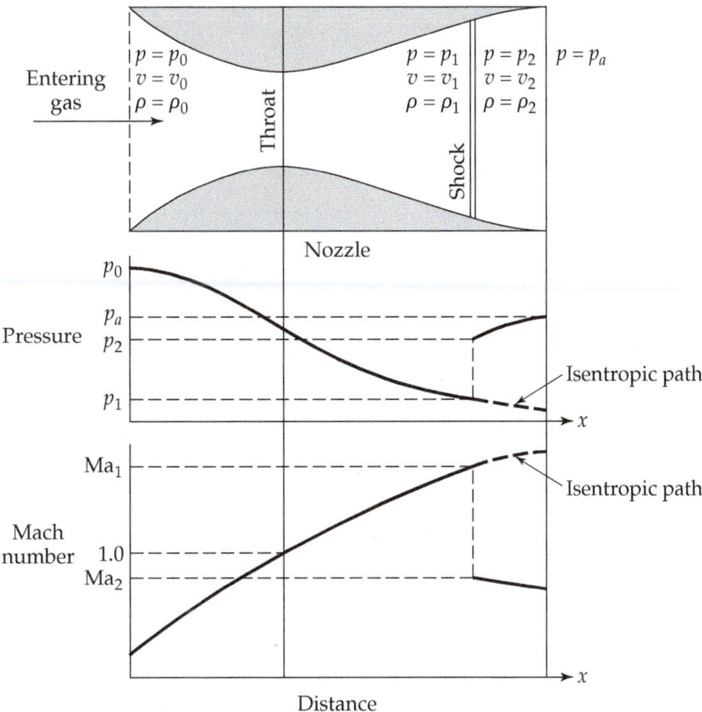

Fig. 11.4-4. Formation of a shock wave in a nozzle.

However, for any nozzle design there is a range of p_a/p_0 for which such an isentropic flow produces a pressure less than p_a at the exit. Then the isentropic flow becomes unstable. The simplest of many possibilities is a stationary normal shock wave, shown schematically in Fig. 11.4-4 as a pair of closely spaced parallel lines. Here the velocity falls off very rapidly to a subsonic value, while both the pressure and the density rise. These changes take place in an extremely thin region, which may therefore be considered locally one-dimensional and laminar, and they are accompanied by a very substantial dissipation of mechanical energy. Viscous dissipation and heat conduction effects are thus concentrated in an extremely small region of the nozzle, and it is the purpose of this example to explore the fluid behavior there. For simplicity the shock wave will be considered normal to the fluid streamlines; in practice, much more complicated shapes are often observed. The velocity, pressure, and temperature just upstream of the shock can be calculated and will be considered as known for the purposes of this example.

Use the three equations of change to determine the conditions under which a shock wave is possible and to find the velocity, temperature, and pressure distributions in such a shock wave. Assume steady, one-dimensional flow of an ideal gas, neglect the dilatational viscosity κ, and ignore changes of μ, k, and $\hat{C}_p$ with temperature and pressure.

SOLUTION

The assumption of one-dimensional flow implies that the density, x component of the velocity, temperature, and pressure depend only on the position x along the flow direction, and that the y and z components of the velocity are zero. The equations of change in the neighborhood of the stationary (steady-state) shock wave may be simplified to

Equation of continuity:
$$\frac{d}{dx}\rho v_x = 0 \qquad (11.4\text{-}42)$$

x Equation of motion:

$$\rho v_x \frac{dv_x}{dx} = -\frac{dp}{dx} + \frac{4}{3}\frac{d}{dx}\left(\mu \frac{dv_x}{dx}\right) \tag{11.4-43}$$[9]

Equation of energy:

$$\rho \hat{C}_p v_x \frac{dT}{dx} = \frac{d}{dx}\left(k\frac{dT}{dx}\right) + v_x\frac{dp}{dx} + \frac{4}{3}\mu\left(\frac{dv_x}{dx}\right)^2 \tag{11.4-44}$$

The energy equation is in the form of Eq. 11.2-5 or Eq. (G) of Table 11.4-1, written for an ideal gas at steady state.

The equation of continuity may be integrated to give

$$\rho v_x = \rho_1 v_1 \tag{11.4-45}$$

in which ρ_1 and v_1 are quantities evaluated a short distance upstream from the shock.

In the energy equation we eliminate ρv_x by use of Eq. 11.4-45 and dp/dx by using the equation of motion to get (after combining the two terms containing the viscosity)

$$\begin{aligned}\rho_1 \hat{C}_p v_1 \frac{dT}{dx} &= \frac{d}{dx}\left(k\frac{dT}{dx}\right) + v_x\left(-\rho_1 v_1 \frac{dv_x}{dx} + \frac{4}{3}\frac{d}{dx}\left(\mu\frac{dv_x}{dx}\right)\right) + \frac{4}{3}\mu\left(\frac{dv_x}{dx}\right)^2 \\ &= \frac{d}{dx}\left(k\frac{dT}{dx}\right) - \rho_1 v_1 \frac{d}{dx}\left(\frac{1}{2}v_x^2\right) + \frac{4}{3}\mu\frac{d}{dx}\left(v_x\frac{dv_x}{dx}\right)\end{aligned} \tag{11.4-46}$$

We next move the second term on the right side over to the left side and divide the entire equation by $\rho_1 v_1$. Then each term is integrated with respect to x to give

$$\hat{C}_p T + \frac{1}{2}v_x^2 = \frac{k}{\rho_1 \hat{C}_p v_1}\frac{d}{dx}\left(\hat{C}_p T + \left(\frac{4}{3}\text{Pr}\right)\frac{1}{2}v_x^2\right) + C_{\text{I}} \tag{11.4-47}$$

in which C_{I} is a constant of integration and $\text{Pr} = \hat{C}_p\mu/k$. For most gases Pr is between 0.65 and 0.85, with an average value close to 0.75. Therefore, in order to simplify the problem, we set Pr equal to $\frac{3}{4}$. Then Eq. 11.4-47 becomes a first-order, linear ordinary differential equation, for which the solution is

$$\hat{C}_p T + \frac{1}{2}v_x^2 = C_{\text{I}} + C_{\text{II}}\exp[(\rho_1 \hat{C}_p v_1/k)x] \tag{11.4-48}$$

Since $\hat{C}_p T + \frac{1}{2}v_x^2$ cannot increase without limit in the positive x direction, the second integration constant C_{II} must be zero. The first integration constant is evaluated just upstream from the shock, so that

$$\hat{C}_p T + \frac{1}{2}v_x^2 = \hat{C}_p T_1 + \frac{1}{2}v_1^2 \equiv C_{\text{I}} \tag{11.4-49}$$

Of course, if we had not chosen Pr to be $\frac{3}{4}$, a numerical integration of Eq. 11.4-47 would have been required.

Next we substitute the integrated continuity equation into the equation of motion and integrate once to obtain

$$\rho_1 v_1 v_x = -p + \frac{4}{3}\mu\frac{dv_x}{dx} + C_{\text{III}} \tag{11.4-50}$$

Evaluation of the constant C_{III} from conditions just upstream from the shock, where $v_x = v_1$ and $dv_x/dx \approx 0$ (i.e., the velocity upstream from the shock varies slowly with x compared to the changes occurring within the shock wave), gives $C_{\text{III}} = \rho_1 v_1^2 + p_1 = \rho_1[v_1^2 + (RT_1/M)]$. We now multiply both sides by v_x and divide by $\rho_1 v_1$. Then, with the help of the ideal-gas law, $p = \rho RT/M$, and Eqs. 11.4-45 and 11.4-49, we may eliminate p from Eq. 11.4-44 to obtain a relation

[9]The term containing (4/3) comes from the $-[\nabla \cdot \boldsymbol{\tau}]$ term in the equation of motion, whose xx component is, according to Eq. 1.2-7 (with $\kappa = 0$)

$$-\frac{d}{dx}\tau_{xx} = -\frac{d}{dx}\left(-2\mu\frac{dv_x}{dx} + \frac{2}{3}\mu\frac{dv_x}{dx}\right) = +\frac{d}{dx}\left(\frac{4}{3}\mu\frac{dv_x}{dx}\right) \tag{11.4-43a}$$

containing only v_x and x as variables:

$$\frac{4}{3}\frac{\mu}{\rho_1 v_1} v_x \frac{dv_x}{dx} = \frac{\gamma+1}{2\gamma} v_x^2 + \frac{\gamma-1}{\gamma} C_{\mathrm{I}} - \frac{C_{\mathrm{III}}}{\rho_1 v_1} v_x \tag{11.4-51}$$

where $\gamma = C_p/C_V$. This equation may now be rewritten dimensionless form:

$$\phi \frac{d\phi}{d\xi} = \beta \mathrm{Ma}_1 (\phi - 1)(\phi - \alpha) \tag{11.4-52}$$

The relevant dimensionless quantities are:

$$\phi = \frac{v_x}{v_1} = \text{dimensionless velocity} \tag{11.4-53}$$

$$\xi = \frac{x}{\lambda} = \text{dimensionless coordinate} \tag{11.4-54}$$

$$\mathrm{Ma}_1 = \frac{v_1}{\sqrt{\gamma R T_1/M}} = \text{Mach number at the upstream condition} \tag{11.4-55}$$

$$\alpha = \frac{\gamma-1}{\gamma+1} + \frac{2}{\gamma+1}\frac{1}{\mathrm{Ma}_1^2} \tag{11.4-56}$$

$$\beta = \frac{9}{8}(\gamma+1)\sqrt{\pi/8\gamma} \tag{11.4-57}$$

The reference length λ is the mean-free path defined in Eq. 1.6-3 (with d^2 eliminated by use of Eq. 1.6-9):

$$\lambda = \frac{3\mu_1}{\rho_1}\sqrt{\frac{\pi M}{8RT_1}} \tag{11.4-58}$$

This choice of dimensionless quantities is far from obvious. Equation 11.4-52 may be verified by working backward from Eq. 11.4-52 to Eq. 11.4-51, and even that is tedious. The solution to Eq. 11.4-52 is

$$\frac{1-\phi(\xi)}{(\phi(\xi)-\alpha)^{\alpha}} = C\exp[\beta \mathrm{Ma}_1(1-\alpha)\xi] \tag{11.4-59}$$

where C is an arbitrary constant of integration. Note that Eq. 11.4-59 satisfies the boundary condition that $\phi \to 1$ as $\xi \to -\infty$. Setting the constant of integration to 1, Eq. 11.4-59 can be rewritten

$$\frac{1-\phi(\xi)}{(\phi(\xi)-\alpha)^{\alpha}} = \exp[\beta \mathrm{Ma}_1(1-\alpha)(\xi-\xi_0)] \qquad (\alpha < \phi < 1) \tag{11.4-60}$$

where $\xi_0 = x_0/\lambda$ and x_0 is the location of the shock wave. This equation describes the dimensionless velocity distribution $\phi(\xi)$ where ξ_0 is considered to be known. It can be seen from the plot of Eq. 11.4-60 in Fig. 11.4-5 that shock waves are indeed very thin. The temperature and pressure distributions may be determined from Eq. 11.4-60 and Eqs. 11.4-49 and 11.4-50. Since $\phi(\xi)$ must approach unity as $\xi \to -\infty$, the constant α is less than 1. This can be true only if $\mathrm{Ma}_1 > 1$, that is, if the upstream flow is supersonic. It can also be seen that for very large positive ξ, the dimensionless velocity ϕ approaches α. The Mach number Ma_1 is defined as the ratio of v_1 to the velocity of sound at T_1 (see Problem 11C.1).

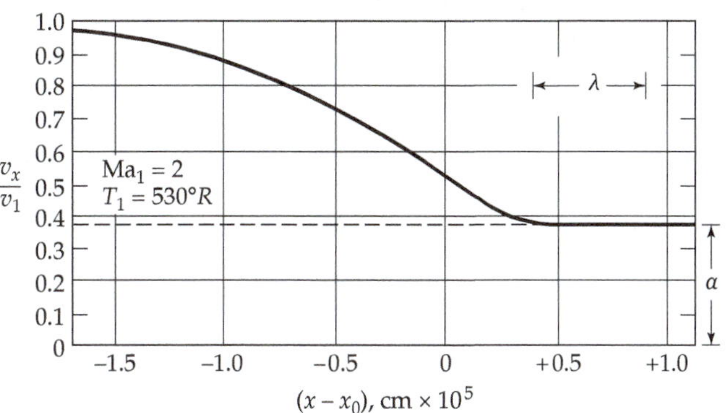

Fig. 11.4-5. Velocity distribution in a stationary shock wave.

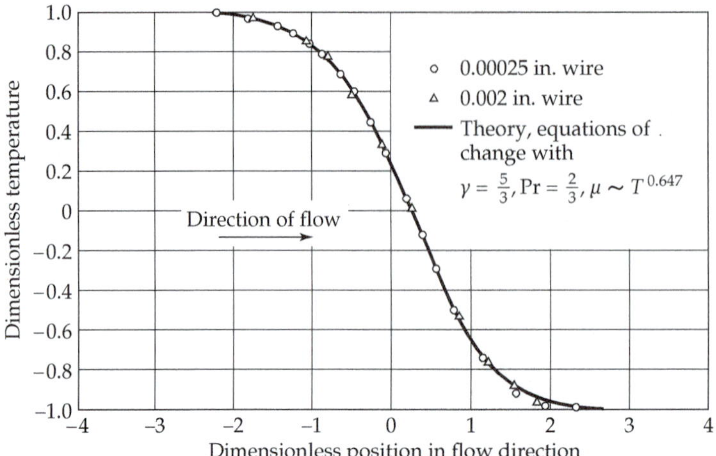

Fig. 11.4-6. Temperature profile through a shock wave for helium with $\text{Ma}_1 = 1.82$. The experimental values were measured with a resistance-wire thermometer. [Adapted from H. W. Liepmann and A. Roshko, *Elements of Gas Dynamics*, Wiley, New York (1957), p. 333.]

In the above development we chose the Prandtl number Pr to be $\frac{3}{4}$, but the solution has been extended[5] to include other values of Pr, as well as the temperature variation of the viscosity.

The tendency of a gas in supersonic flow to revert spontaneously to subsonic flow is important in wind tunnels and in the design of high-velocity systems, e.g., in turbines and rocket engines. Note that the changes taking place in shock waves are irreversible and that, since the velocity gradients are so very steep, a considerable amount of mechanical energy is dissipated.

In view of the thinness of the predicted shock wave, one may question the applicability of the analysis given here, based on the continuum equations of change. Therefore, it is desirable to compare the theory with experiment. In Fig. 11.4-6 experimental temperature measurements for a shock wave in helium are compared with the theory for $\gamma = \frac{5}{3}$, $\text{Pr} = \frac{2}{3}$, and $\mu \sim T^{0.647}$. It can be seen that the agreement is excellent. Nevertheless we should recognize that this is a simple system, inasmuch as helium is monatomic, and therefore internal degrees of freedom are not involved. The corresponding analysis for a diatomic or polyatomic gas would need to consider the exchange of energy between translational and internal degrees of freedom, which typically requires hundreds of collisions, broadening the shock wave considerably. Further discussion of this matter can be found in Chapter 11 of Ref. 4.

§11.5 THE EQUATIONS OF CHANGE AND SOLVING PROBLEMS WITH TWO INDEPENDENT VARIABLES

The illustrative examples in the preceding section involved only one independent variable. In this section we discuss several problems involving two independent variables. For problems with two independent variables, the equations of change produce partial differential equations that can be quite challenging to solve. In the examples below, the problems are solved by using a few well-known methods. There are numerous other methods that can be exploited to solve flow problems, which are beyond the scope of this text. The reasons for including some examples here are (1) to provide additional examples where the equations of change are simplified using postulates, and boundary (and initial) conditions are used to evaluate unknown constants; (2) to illustrate some useful and interesting transport phenomena; and (3) to show the reader that with a little more experience with mathematics, many more complicated problems in transport phenomena can be solved.

Example 11.5-1 addresses the heating of a semi-infinite slab. The similarity solution method employed draws upon an analogy with a mathematically equivalent momentum transport problem solved previously in Chapter 3. Example 11.5-2 considers the heating of a finite slab. Here we introduce the powerful method of separation of variables. Example 11.5-3 focuses on forced convection of a fluid heated in a tube in laminar flow. An approximate solution valid for large z was presented in §10.9. Here we exploit the similarity method to develop an approximate solution valid for small z.

EXAMPLE 11.5-1

Heating of a Semi-Infinite Solid Slab

A solid material occupying the space from $y = 0$ to $y = \infty$ is initially at temperature T_0. At time $t = 0$, the surface at $y = 0$ is suddenly raised to temperature T_1 and maintained at that temperature for $t > 0$. Find the time-dependent temperature profile $T(y,t)$. Assume that the density, heat capacity, and thermal conductivity are all constant.

SOLUTION

For a solid with $\mathbf{v} = \mathbf{0}$, with constant physical properties, the appropriate form of the equation of energy in §11.2 is Eq. 11.2-10. Assuming $T = T(y,t)$, this equation can be further simplified to (with the help of Eq. B.9-1)

$$\rho\hat{C}_p\frac{\partial T}{\partial t} = k\frac{\partial^2 T}{\partial y^2} \tag{11.5-1}$$

The initial and boundary conditions are

I. C.:	at $t \leq 0$,	$T = T_0$ for all y	(11.5-2)
B. C. 1:	at $y = 0$,	$T = T_1$ for all $t > 0$	(11.5-3)
B. C. 2:	as $y \to \infty$,	$T \to T_0$ for all $t > 0$	(11.5-4)

We introduce the dimensionless temperature difference $\Theta = (T - T_0)/(T_1 - T_0)$. The equation of energy can then be written as

$$\frac{\partial \Theta}{\partial t} = \alpha\frac{\partial^2 \Theta}{\partial y^2} \tag{11.5-5}$$

where $\alpha = k/\rho\hat{C}_p$ is the thermal diffusivity. The initial and boundary conditions are then rewritten as

I. C.:	at $t \leq 0$,	$\Theta = 0$ for all y	(11.5-6)
B. C. 1:	at $y = 0$,	$\Theta = 1$ for all $t > 0$	(11.5-7)
B. C. 2:	as $y \to \infty$,	$\Theta \to 0$ for all $t > 0$	(11.5-8)

This problem is mathematically analogous to that given by Eq. 3.8-5 along with the subsequently stated boundary conditions (for flow near a wall suddenly set in motion). Hence, the solution in Eq. 3.8-16 can be taken over directly by appropriate changes in notation:

$$\Theta(y/\sqrt{4\alpha t}) = 1 - \frac{2}{\sqrt{\pi}}\int_0^{y/\sqrt{4\alpha t}} e^{-\eta^2}d\eta \tag{11.5-9}$$

or

$$\frac{T(y,t) - T_0}{T_1 - T_0} = 1 - \text{erf}\frac{y}{\sqrt{4\alpha t}} \tag{11.5-10}$$

The solution shown in Fig. 3.8-2(*b*) describes the temperature profiles when the ordinate is labeled $(T - T_0)/(T_1 - T_0)$ and the abscissa is labeled $y/\sqrt{4\alpha t}$.

Since the error function reaches a value of 0.99 when the argument is about 2, the *thermal penetration thickness* δ_T is typically defined to be

$$\delta_T = 4\sqrt{\alpha t} \tag{11.5-11}$$

That is, for distances $y > \delta_T$, the temperature has changed by less than 1% of the difference $T_1 - T_0$. If it is necessary to calculate the temperature in a slab of finite thickness, the solution in Eq. 11.5-10 will be a good approximation when δ_T is small with respect to the slab thickness. However, when δ_T is of the order of magnitude of the slab thickness or greater, then the series solution of Example 11.5-2 has to be used.

The wall heat flux can be calculated from Eq. 11.5-10 as follows:

$$q_y|_{y=0} = -k\left.\frac{\partial T}{\partial y}\right|_{y=0} = \frac{k}{\sqrt{\pi \alpha t}}(T_1 - T_0) \tag{11.5-12}$$

Hence, the wall heat flux varies as $t^{-1/2}$, whereas the penetration thickness varies as $t^{1/2}$.

EXAMPLE 11.5-2

Heating of a Solid Slab of Finite Thickness

A solid slab occupying the space between $y = -b$ and $y = +b$ is initially at temperature T_0. At time $t = 0$, the surfaces at $y = \pm b$ are suddenly raised to T_1 and kept at that temperature. Find $T(y,t)$.

SOLUTION

As in the previous example, we have $\mathbf{v} = \mathbf{0}$ and assume that the physical properties are constant. With the assumption that $T = T(y,t)$, the equation of energy expressed by Eq. 11.2-10 becomes

$$\rho \hat{C}_p \frac{\partial T}{\partial t} = k\frac{\partial^2 T}{\partial y^2} \tag{11.5-13}$$

The initial and boundary conditions are

I. C.: at $t \leq 0$, $T = T_0$ for $-b \leq y \leq b$ (11.5-14)

B. C. 1: at $y = -b$, $T = T_1$ for all $t > 0$ (11.5-15)

B. C. 2: at $y = +b$, $T = T_1$ for all $t > 0$ (11.5-16)

For this problem we define the following dimensionless variables:

Dimensionless temperature: $$\Theta(\eta,\tau) = \frac{T_1 - T(y,t)}{T_1 - T_0} \tag{11.5-17}$$

Dimensionless coordinate: $$\eta = \frac{y}{b} \tag{11.5-18}$$

Dimensionless time: $$\tau = \frac{\alpha t}{b^2} \tag{11.5-19}$$

With these dimensionless variables, the differential equation and boundary conditions are

$$\frac{\partial \Theta}{\partial \tau} = \frac{\partial^2 \Theta}{\partial \eta^2} \tag{11.5-20}$$

I. C.: at $\tau = 0$, $\Theta = 1$ for $-1 \leq \eta \leq 1$ (11.5-21)

B. C. 1 and 2: at $\eta = \pm 1$, $\Theta = 0$ for $\tau > 0$ (11.5-22)

Note that no parameters appear when the problem is restated thus.

We can solve this problem by the method of separation of variables. We start by postulating that a solution of the following product form can be obtained:

$$\Theta(\eta,\tau) = f(\eta)g(\tau) \tag{11.5-23}$$

Substitution of this trial function into Eq. 11.5-20 and subsequent division by the product $f(\eta)g(\tau)$ gives

$$\frac{1}{g}\frac{dg}{d\tau} = \frac{1}{f}\frac{d^2 f}{d\eta^2} \tag{11.5-24}$$

The left side is a function of τ alone, and the right side a function of η alone. This can be true only if both sides equal a constant, which we call $-\lambda$. Then we get two ordinary differential equations

$$\frac{dg}{d\tau} = -\lambda g \tag{11.5-25}$$

$$\frac{d^2 f}{d\eta^2} = -\lambda f \tag{11.5-26}$$

Equation 11.5-25 is of the form of Eq. C.1-1 and may be integrated to give

$$g(\tau) = A\exp(-\lambda\tau) \tag{11.5-27}$$

in which A is a constant of integration. We do not expect the temperature to grow exponentially in time, and thus Eq. 11.5-27 implies that λ must be positive. It is then convenient to let $\lambda = c^2$ to enforce this requirement (where c is a real number). Equation 11.5-27 is thus rewritten

$$g(\tau) = A\exp(-c^2\tau) \tag{11.5-28}$$

Equation 11.5-26 is of the form of Eq. C.1-3, and the general solution for $f(\eta)$ is

$$f(\eta) = B\sin c\eta + C\cos c\eta \tag{11.5-29}$$

where B and C are constants of integration.

Because of the symmetry about the xz plane, we must have $\Theta(\eta,\tau) = \Theta(-\eta,\tau)$, and thus $f(\eta) = f(-\eta)$. Since the sine function does not have this kind of behavior, we have to require that B be zero. Use of either of the two boundary conditions then gives

$$C\cos c = 0 \tag{11.5-30}$$

Clearly C cannot be zero, because that choice leads to an inadmissible solution (i.e., we would not be able to satisfy the initial condition). However, the equality can be satisfied by many different choices of c, which we call c_n:

$$c_n = \left(n + \frac{1}{2}\right)\pi, \qquad n = 0, \pm 1, \pm 2, \pm 3 \ldots \pm \infty \tag{11.5-31}$$

Hence, Eq. 11.5-20 can be satisfied by

$$\Theta_n(\eta,\tau) = A_n C_n \exp\left[-\left(n+\frac{1}{2}\right)^2 \pi^2\tau\right] \cos\left(n+\frac{1}{2}\right)\pi\eta \tag{11.5-32}$$

The subscripts n remind us that A and C may be different for each value of n. Because of the linearity of the differential equation, we may now superpose all the solutions of the form of Eq. 11.5-32. This superposition is necessary because Eq. 11.5-32 alone cannot satisfy the initial condition. When we add all of the solutions of the form of Eq. 11.5-32, we note that the exponentials and cosines for n have the same values as those for $-(n+1)$, so that the terms with negative indices combine with those with positive indices. The superposition then gives

$$\Theta(\eta,\tau) = \sum_{n=0}^{\infty} D_n \exp\left[-\left(n+\frac{1}{2}\right)^2 \pi^2\tau\right] \cos\left(n+\frac{1}{2}\right)\pi\eta \tag{11.5-33}$$

in which $D_n = A_n C_n + A_{-(n+1)}C_{-(n+1)}$.

The D_n are now determined by applying the initial condition, which gives

$$1 = \sum_{n=0}^{\infty} D_n \cos\left(n+\frac{1}{2}\right)\pi\eta \tag{11.5-34}$$

Multiplication by $\cos\left(m+\frac{1}{2}\right)\pi\eta$ and integration from $\eta = -1$ to $\eta = +1$ gives

$$\int_{-1}^{+1} \cos\left(m+\frac{1}{2}\right)\pi\eta\, d\eta = \sum_{n=0}^{\infty} D_n \int_{-1}^{+1} \cos\left(m+\frac{1}{2}\right)\pi\eta \cos\left(n+\frac{1}{2}\right)\pi\eta\, d\eta \tag{11.5-35}$$

When the integrations are performed, all integrals on the right side are identically zero, except for the term in which $n = m$. Hence, we get

$$\left.\frac{\sin\left(m+\frac{1}{2}\right)\pi\eta}{\left(m+\frac{1}{2}\right)\pi}\right|_{\eta=-1}^{\eta=+1} = D_m \left.\frac{\frac{1}{2}\left(m+\frac{1}{2}\right)\pi\eta + \frac{1}{4}\sin 2\left(m+\frac{1}{2}\right)\pi\eta}{\left(m+\frac{1}{2}\right)\pi}\right|_{\eta=-1}^{\eta=+1} \tag{11.5-36}$$

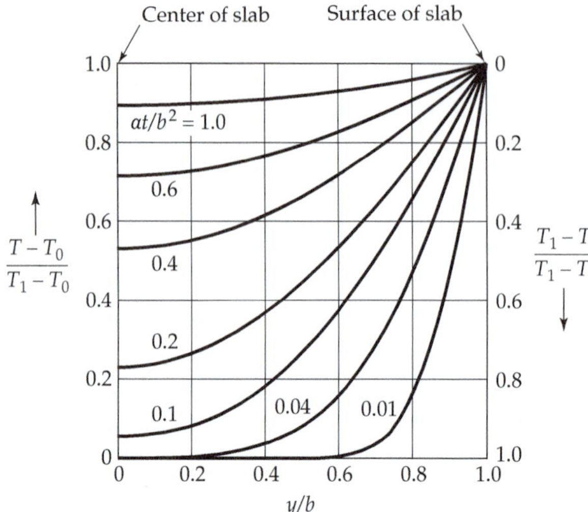

Fig. 11.5-1. Temperature profiles for unsteady-state heat conduction in a slab of finite thickness $2b$. The initial temperature of the slab is T_0, and T_1 is the temperature imposed at the slab surfaces for time $t > 0$. [H. S. Carslaw and J. C. Jaeger, *Conduction of Heat in Solids*, 2nd edition, Oxford University Press (1959), p. 101.]

After inserting the limits, we may solve for D_m to get

$$D_m = \frac{2(-1)^m}{\left(m + \frac{1}{2}\right)\pi} \tag{11.5-37}$$

Substitution of this expression into Eq. 11.5-33 gives the temperature profile, which we now rewrite in terms of the original variables[1]

$$\frac{T_1 - T(y,t)}{T_1 - T_0} = 2\sum_{n=0}^{\infty} \frac{(-1)^n}{\left(n + \frac{1}{2}\right)\pi} \exp\left[-\left(n + \frac{1}{2}\right)^2 \pi^2 at/b^2\right] \cos\left(n + \frac{1}{2}\right)\frac{\pi y}{b} \tag{11.5-38}$$

The solutions to many unsteady-state heat-conduction problems come out as infinite series, such as that just obtained here. These series converge rapidly for large values[1] of the dimensionless time, at/b^2. For very short times the convergence is very slow, and in the limit as at/b^2 approaches zero, the solution in Eq. 11.5-38 may be shown to approach that given in Eq. 11.5-10 (with the origin properly translated). The dimensionless temperature profile is plotted as a function of y/b for several values of $\tau = at/b^2$ in Fig. 11.5-1. From the figure it is clear that when the dimensionless time $\tau = at/b^2$ is 0.1, the energy has "penetrated" measurably to the center plane of the slab, and that at $\tau = 1.0$ the heating is 90% complete at the center plane.

Results analogous to Fig. 11.5-1 are given for infinite cylinders and for spheres in Figs. 11.5-2 and 11.5-3.

EXAMPLE 11.5-3

Laminar Tube Flow with Constant Heat Flux at the Wall: Asymptotic Solution for the Entrance Region

Recall that in §10.9 we studied the laminar tube flow with constant heat flux at the wall, but we obtained only the asymptotic solution for large distances z down the tube. Here we develop an expression for $T(r,z)$ that is useful for small values of z.

SOLUTION

The fluid enters the heated region with an initially uniform temperature of $T = T_1$. For small z, only the fluid very close to the wall, near $r = R$, will increase in temperature, as illustrated in Fig. 10.9-3. Limiting our attention to this thin region near the wall motivates the following three approximations that lead to results that are accurate in the limit as $z \to 0$:

---

[1]H. S. Carslaw and J. C. Jaeger, *Conduction of Heat in Solids*, Oxford University Press, 2nd edition (1959), p. 97, Eq. (8); the alternate solution in Eq. (9) converges rapidly for small times.

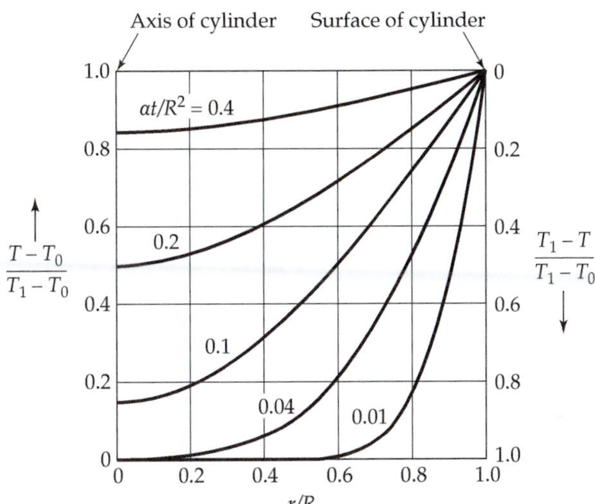

Fig. 11.5-2. Temperature profiles for unsteady-state heat conduction in a cylinder of radius R. The initial temperature of the cylinder is T_0, and T_1 is the temperature imposed at the cylinder surface for time $t > 0$. [H. S. Carslaw and J. C. Jaeger, *Conduction of Heat in Solids*, 2nd edition, Oxford University Press (1959), p. 200.]

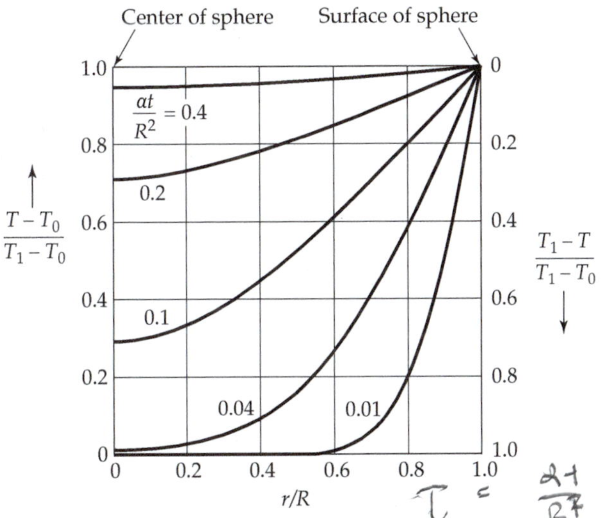

Fig. 11.5-3. Temperature profiles for unsteady-state heat conduction in a sphere of radius R. The initial temperature of the sphere is T_0, and T_1 is the temperature imposed at the sphere surface for time $t > 0$. [H. S. Carslaw and J. C. Jaeger, *Conduction of Heat in Solids*, 2nd edition, Oxford University Press (1959), p. 234.]

a. Curvature effects may be neglected and the problem treated as though the wall were flat; call the distance from the wall $y = R - r$.

b. The fluid may be regarded as extending from the (flat) heat-transfer surface ($y = 0$) to $y = \infty$.

c. The velocity profile may be regarded as linear, with a slope given by the slope of the parabolic velocity profile at the wall: $v_z(y) = v_0 y/R$, in which $v_0 = (\mathscr{P}_0 - \mathscr{P}_L)R^2/2\mu L$.

This is the way the system would appear to a tiny "observer" who is located within the very thin shell of heated fluid. To this observer, the wall would seem flat, the fluid would appear to be of infinite extent, and the velocity profile would seem to be linear.

As in §10.9, we assume that the physical properties are constant, that energy production by viscous dissipation is negligible, that the steady-state temperature profile depends on only y (or r) and z, and that conduction in the z direction is negligible compared to convection in the z direction. The energy equation then becomes, in the region just slightly beyond $z = 0$,

$$v_0 \frac{y}{R} \frac{\partial T}{\partial z} = \alpha \frac{\partial^2 T}{\partial y^2} \tag{11.5-39}$$

The boundary conditions for this differential equation are

B.C. 1: at $z = 0$, $T = T_1$ (11.5-40)

B.C. 2: at $y = 0$, $q_y = -k\partial T/\partial y = q_0$ (11.5-41)

B.C. 3: as $y \to \infty$, $T \to T_1$ (11.5-42)

Actually it is easier to work with the corresponding equation for the heat flux in the y direction ($q_y = -k\partial T/\partial y$). This equation is obtained by dividing Eq. 11.5-39 by y and differentiating with respect to y:

$$v_0 \frac{1}{R}\frac{\partial q_y}{\partial z} = \alpha \frac{\partial}{\partial y}\left(\frac{1}{y}\frac{\partial q_y}{\partial y}\right) \tag{11.5-43}$$

The boundary conditions for this differential equation are

B.C. 1: at $z = 0$, $q_y = 0$ (11.5-44)

B.C. 2: at $y = 0$, $q_y = -k\partial T/\partial y = q_0$ (11.5-45)

B.C. 3: as $y \to \infty$, $q_y \to 0$ (11.5-46)

The first and third boundary conditions arise because the derivative of a constant temperature is zero. It is more convenient to work with dimensionless variables defined as

$$\psi = \frac{q_y}{q_0} \qquad \eta = \frac{y}{R} \qquad \lambda = \frac{\alpha z}{v_0 R^2} \tag{11.5-47}$$

Then Eq. 11.5-43 becomes

$$\frac{\partial \psi}{\partial \lambda} = \frac{\partial}{\partial \eta}\left(\frac{1}{\eta}\frac{\partial \psi}{\partial \eta}\right) \tag{11.5-48}$$

with the boundary conditions

B. C. 1: at $\lambda = 0$, $\psi = 0$ (11.5-49)

B. C. 2: at $\eta = 0$, $\psi = 1$ (11.5-50)

B. C. 3: as $\eta \to \infty$, $\psi \to 0$ (11.5-51)

Boundary conditions 1 and 3 are similar and motivate us to seek a similarity solution, similar to that employed in Example 3.8-1. We assume that the dimensionless temperature ψ is a function of only a single combined variable, $\chi = \eta/\lambda^b$, where the exponent b is yet to be determined. Substituting $\psi = \psi(\chi)$ into Eq. 11.5-48 and performing the required derivatives, one can show that the resulting ordinary differential equation does not contain η or λ explicitly for $b = 1/3$. Thus, we define the new combined independent variable $\chi = \eta/\sqrt[3]{9\lambda}$, to a obtain a differential equation for ψ with χ as the only independent variable (the "9" inside the cube root is for convenience later). Then Eq. 11.5-48 becomes

$$\chi\frac{d^2\psi}{d\chi^2} + (3\chi^3 - 1)\frac{d\psi}{d\chi} = 0 \tag{11.5-52}$$

The boundary conditions are:

B.C. 1′: at $\chi = 0$, $\psi = 1$ (11.5-53)

B.C. 2′: as $\chi \to \infty$, $\psi \to 0$ (11.5-54)

The solution of Eq. 11.5-52 is found by first letting $d\psi/d\chi = p$, and getting the first-order equation for p

$$\chi\frac{dp}{d\chi} + (3\chi^3 - 1)p = 0 \tag{11.5-55}$$

Multiplication by $(1/p\chi)d\chi$ allows the variables to be separated so that the equation can be integrated to give

$$p = \frac{d\psi}{d\chi} = C_1\chi e^{-\chi^3} \tag{11.5-56}$$

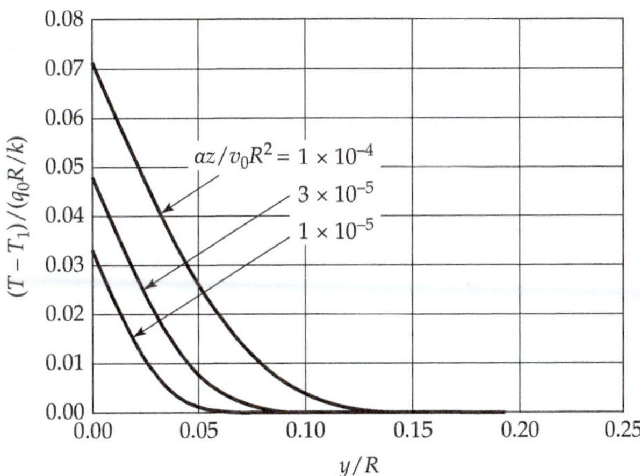

Fig. 11.5-4. The dimensionless temperature $(T - T_0)/(q_0R/k)$ as a function of y/R for several values of $\alpha z/v_0R^2$ (from Eq. 11.5-62).

This first-order separable differential equation can in turn be integrated to give

$$\psi(\chi) = C_1 \int \chi e^{-\chi^3} d\chi + C_2 \tag{11.5-57}$$

The constants of integration may be obtained from the boundary conditions to give the dimensionless heat flux as

$$\psi(\chi) = \frac{\int_\chi^\infty \bar{\chi} e^{-\bar{\chi}^3} d\bar{\chi}}{\int_0^\infty \chi e^{-\chi^3} d\chi} = \frac{3}{\Gamma\left(\frac{2}{3}\right)} \int_\chi^\infty \bar{\chi} e^{-\bar{\chi}^3} d\bar{\chi} \tag{11.5-58}$$

where $\Gamma(n)$ is the gamma function defined in §C.4 $\left(\Gamma\left(\frac{2}{3}\right) \approx 1.35412\right)$. The temperature is then obtained by integrating the heat flux

$$\int_T^{T_1} d\bar{T} = -\frac{1}{k} \int_y^\infty \bar{q}_y d\bar{y} \tag{11.5-59}$$

or, in dimensionless form

$$\Theta(\eta,\lambda) = \frac{T - T_1}{q_0R/k} = \sqrt[3]{9\lambda} \int_\chi^\infty \psi d\bar{\chi} \tag{11.5-60}$$

Then the expression for ψ is inserted into the integral, to get

$$\Theta(\eta,\lambda) = \sqrt[3]{9\lambda} \frac{3}{\Gamma\left(\frac{2}{3}\right)} \int_\chi^\infty \left[\int_{\bar{\chi}}^\infty \bar{\bar{\chi}} e^{-\bar{\bar{\chi}}^3} d\bar{\bar{\chi}} \right] d\bar{\chi} \tag{11.5-61}$$

The order of integration is now interchanged (§C.7), so that

$$\begin{aligned}
\Theta(\eta,\lambda) &= \sqrt[3]{9\lambda} \frac{3}{\Gamma\left(\frac{2}{3}\right)} \int_\chi^\infty \left[\int_\chi^{\bar{\bar{\chi}}} \bar{\bar{\chi}} e^{-\bar{\bar{\chi}}^3} d\bar{\chi} \right] d\bar{\bar{\chi}} \\
&= \sqrt[3]{9\lambda} \frac{3}{\Gamma\left(\frac{2}{3}\right)} \int_\chi^\infty [\bar{\bar{\chi}} e^{-\bar{\bar{\chi}}^3} (\bar{\bar{\chi}} - \chi)] d\bar{\bar{\chi}} \\
&= \sqrt[3]{9\lambda} \left[\frac{e^{-\chi^3}}{\Gamma\left(\frac{2}{3}\right)} - \chi \left(\frac{\Gamma\left(\frac{2}{3}, \chi^3\right)}{\Gamma\left(\frac{2}{3}\right)} \right) \right]
\end{aligned} \tag{11.5-62}$$

where $\Gamma(n,x)$ is an incomplete gamma function,[2] also described in App. C. The dimensionless temperature predicted by Eq. 11.5-62 is plotted as a function of $\eta = y/R$ for several values of $\lambda = \alpha z/v_0 R^2$ in Figure 11.5-4 (the dimensionless temperature $\Theta = (T - T_1)/(q_0 R/k)$ is defined identically in §10.9 and here). Each of the curves in Fig. 11.5-4 has the same slope for small y/R, which is required by the boundary condition of an imposed heat flux at the wall. The temperature profile can be characterized as a thermal boundary layer, which penetrates further into the fluid with increasing z. Keep in mind that this approximate solution is only valid when the boundary layer is of small extent compared to R.

§11.6 CONCLUDING COMMENTS

We began this chapter by deriving an equation for the conservation of kinetic plus internal energy. We then subtracted from this the equation of change for the kinetic energy (from Chapter 3), and this resulted in an equation of change for the internal energy. Next, using some formulas from equilibrium thermodynamics, the internal energy equation was converted into an equation for the temperature. It is this latter equation that forms the basis for most of the remainder of the chapter. With it one can formulate many important problems in heat transfer. Many problems so formulated may be solved by standard mathematical methods, but many more have to be solved by numerical methods.

The energy conservation equation set forth here is really incomplete. For example, we have not accounted for electromagnetic effects. To do that, we would have to add to the equations of continuity, motion, and energy (appropriately generalized to include electric and magnetic energies), and the Maxwell equations for electrodynamics. The fields of electrohydrodynamics and magneto-hydrodynamics are well–established fields. Rather than go into these fields, we have chosen to "splice in" an electrical source term S_e and then solve some simple problems, as we did in Chapter 10. The same comment applies to the nuclear source terms and chemical source terms that have been treated there.

In the remainder of the chapter we have shown how a number of relatively uncomplicated problems can be solved with the equations of change. In §11.4, we explained how to set up and solve problems involving one independent variable, and in §11.5 we did several problems involving two independent variables. There are many treatises on heat transfer, and also handbooks, in which the solutions to many more problems may be found.

QUESTIONS FOR DISCUSSION

1. Define energy, potential energy, kinetic energy, and internal energy. What common units are used for these?
2. How does one assign the physical meaning to the individual terms in Eqs. 11.1-8 and 11.2-1?
3. In getting Eq. 11.2-7 we used the relation $\tilde{C}_p - \tilde{C}_V = R$, which is valid for ideal gases. What is the analogous equation for nonideal gases and liquids?
4. Summarize all the assumptions made in obtaining the equation of change for the temperature.
5. Compare and contrast forced convection and free convection, with regard to methods of problem solving, and occurrence in industrial and meteorological problems.
6. If a rocket nose cone were made of a porous material and a volatile liquid were forced slowly through the pores during reentry into the atmosphere, how would the cone surface temperature be affected and why?
7. What is Archimedes's principle, and how is it related to the term $\bar{\rho}g\bar{\beta}(T - \bar{T})$ in Eq. 11.3-2?

[2] M. Abramowitz and I. A. Stegun, Eds., *Handbook of Mathematical Functions*, Dover, New York, 9th Printing (1973), pp. 255 et seq.

8. When, if ever, can the equation of energy be completely and exactly solved without detailed knowledge of the velocity profiles of the system?
9. When, if ever, can the equation of motion be completely solved for a nonisothermal system, without detailed knowledge of the temperature profiles of the system?

PROBLEMS

11A.1 Temperature in a friction bearing. Calculate the maximum temperature in the friction bearing of Problem 3A.1, assuming the thermal conductivity of the lubricant to be 4.0×10^{-4} cal/s · cm · °C, the metal temperature 200°C, and the rate of rotation 4000 rpm.

Answer: About 217°C (from both Eq. 11.4-16 and Eq. 10.8-9)

11A.2 Viscosity variation and velocity gradients in a nonisothermal film. Water is falling over a vertical wall in a film 0.1 mm thick. The water temperature is 100°C at the free liquid surface and 80°C at the wall surface.

(a) Show that the maximum fractional deviation between viscosities predicted by Eqs. 11.4-20 and 11.4-21 occurs when $T = \sqrt{T_0 T_\delta}$.

(b) Calculate the maximum fractional deviation for the conditions given.

Answer: **(b)** 0.6%

11A.3 Transpiration cooling

(a) Calculate the temperature distribution between the two shells of Example 11.4-4 for radial mass flow rates of zero and 10^{-5} g/s for the following conditions

$$R = 500 \text{ microns} \qquad T_R = 300°\text{C}$$
$$\kappa R = 100 \text{ microns} \qquad T_\kappa = 100°\text{C}$$
$$k = 6.13 \times 10^{-5} \text{ cal/cm} \cdot \text{s} \cdot °\text{C}$$
$$\hat{C}_p = 0.25 \text{ cal/g} \cdot °\text{C}$$

(b) Compare the rates of heat conduction to the surface at κR in the presence and absence of convection.

11A.4 Velocity, temperature, and pressure changes in a shock wave. Air at 1 atm and 70°F is flowing at an upstream Mach number of 2 across a stationary shock wave. Calculate the following quantities, assuming that γ is constant at 1.4 and that $\hat{C}_p = 0.24$ Btu/lb$_m$ · °F:

(a) The initial velocity of the air.

(b) The velocity, temperature, and pressure downstream from the shock wave.

(c) The changes of internal and kinetic energy across the shock wave.

Answers: **(a)** 2250 ft/s;
(b) 844 ft/s; 888°R; 4.48 atm;
(c) $\Delta\hat{U} = +61.4$ Btu/lb$_m$; $\Delta\hat{K} = -86.9$ Btu/lb$_m$

11A.5 Adiabatic frictionless compression of an ideal gas. Calculate the temperature attained by compressing air, initially at 100°F and 1 atm, to 0.1 of its initial volume. It is assumed that $\gamma = 1.40$, and that the compression is frictionless and adiabatic. Discuss the result in relation to the operation of an internal combustion engine.

Answer: 950°F

11B.1 Adiabatic frictionless processes in an ideal gas

(a) Note that a gas that obeys the ideal-gas law may deviate appreciably from $\tilde{C}_p$ = constant. Hence, rework Example 11.4-5 using a molar heat capacity expression of the form

$$\tilde{C}_p = a + bT + cT^2 \tag{11B.1-1}$$

(b) Determine the final pressure, p_2, required if methane (CH_4) is to be heated from 300K and 1 atm to 800K by adiabatic frictionless compression. The recommended empirical constants[1] for methane are: $a = 2.322$ cal/g-mol · K, $b = 38.04 \times 10^{-3}$ cal/g-mol · K^2, and $c = -10.97 \times 10^{-6}$ cal/g-mol · K^3.

Answers: **(a)** $pT^{-a/R} \exp[-(b/R)T - (c/2R)T^2] = \text{constant}$;
(b) 270 atm

11B.2 Viscous heating in laminar tube flow (asymptotic solutions).
(a) Show that for fully developed laminar Newtonian flow in a circular tube of radius R, the energy equation becomes

$$\rho \hat{C}_p v_{z,\max} \left[1 - \left(\frac{r}{R}\right)^2\right] \frac{\partial T}{\partial z} = k \frac{1}{r} \frac{\partial}{\partial r}\left(r \frac{\partial T}{\partial r}\right) + \frac{4\mu v_{z,\max}^2}{R^2}\left(\frac{r}{R}\right)^2 \tag{11B.2-1}$$

if the viscous dissipation terms are not neglected. Here $v_{z,\max}$ is the maximum velocity in the tube. What restrictions have to be placed on any solutions of Eq. 11B.2-1?

(b) For the *isothermal wall* problem ($T = T_0$ at $r = R$ for $z > 0$ and at $z = 0$ for all r), find the asymptotic expression for $T(r)$ at large z. Do this by recognizing that $\partial T/\partial z$ will be zero at large z. Solve Eq. 11B.2-1 and obtain

$$T(r) - T_0 = \frac{\mu v_{z,\max}^2}{4k}\left[1 - \left(\frac{r}{R}\right)^4\right] \tag{11B.2-2}$$

(c) For the *adiabatic wall* problem ($q_r = 0$ at $r = R$ for all $z > 0$) an asymptotic expression for large z may be found as follows: Multiply Eq. 11B.2-1 by $r dr$ and then integrate from $r = 0$ to $r = R$. Then integrate the resulting equation over z to get

$$T_b - T_1 = (4\mu v_{z,\max}/\rho \hat{C}_p R^2)z \tag{11B.2-3}$$

in which T_1 is the inlet temperature at $z = 0$. Postulate now that the asymptotic temperature profile at large z is of the form

$$T(r,z) - T_1 = (4\mu v_{z,\max}/\rho \hat{C}_p R^2)z + f(r) \tag{11B.2-4}$$

Substitute this into Eq. 11B.2-1 and integrate the resulting equation for $f(r)$ to obtain

$$T(r,z) - T_1 = \frac{4\mu v_{z,\max}}{\rho \hat{C}_p R^2} z + \frac{\mu v_{z,\max}^2}{k}\left[\left(\frac{r}{R}\right)^2 - \frac{1}{2}\left(\frac{r}{R}\right)^4 - \frac{1}{4}\right] \tag{11B.2-5}$$

after determining the integration constant by an energy balance of the tube from 0 to z. Keep in mind that the solutions in Eqs. 11B.2-2 and 11B.2-5 are valid solutions only for large z.

11B.3 Velocity distribution in a nonisothermal film. Show that Eq. 11.4-23 meets the following requirements:
(a) At $x = \delta$, $v_z = 0$.
(b) At $x = 0$, $\partial v_z/\partial x = 0$.
(c) $\lim_{\mu_\delta \to \mu_0} v_z(x) = (\rho g \delta^2 \cos\beta / 2\mu_0)[1 - (x/\delta)^2]$

11B.4 Heat conduction in a spherical shell. The solid spherical shell shown in Fig. 11B.4 has inner and outer radii R_1 and R_2. A hole is made in the shell at the north pole by cutting out the conical segment in the region $0 \le \theta \le \theta_1$. A similar hole is made at the south pole by removing the portion $(\pi - \theta_1) \le \theta \le \pi$. The surface $\theta = \theta_1$ is kept at temperature $T = T_1$, and the surface at $\theta = \pi - \theta_1$ is held at $T = T_2$. Find the steady-state temperature distribution.

---

[1] O. A. Hougen, K. M. Watson, and R. A. Ragatz, *Chemical Process Principles*, Part I, 2nd edition, Wiley, New York (1958), p. 255. See also Part II, pp. 646–653, for a fuller discussion of isentropic process calculations.

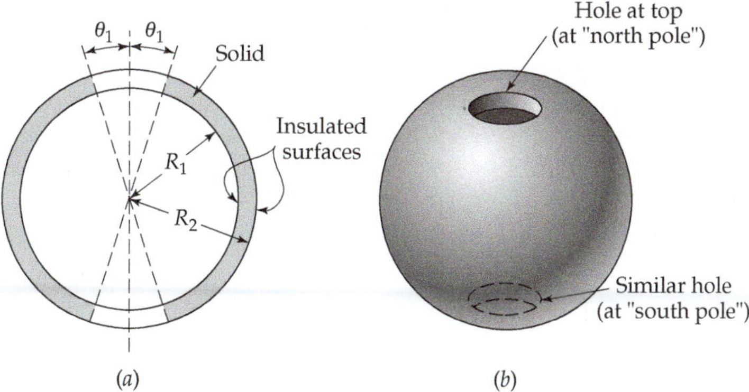

Fig. 11B.4 Heat conduction in a spherical shell: (*a*) cross section containing the z axis; (*b*) view of the sphere from above.

11B.5 Axial heat conduction in a wire.[2] A solid wire of constant density ρ moves downward with uniform speed v into a liquid metal bath at temperature T_0 as shown in Fig. 11B.5. It is desired to find the temperature profile $T(z)$. Assume that $T = T_\infty$ at $z = \infty$, and that resistance to radial heat conduction is negligible. Assume further that the wire temperature is $T = T_0$ at $z = 0$.

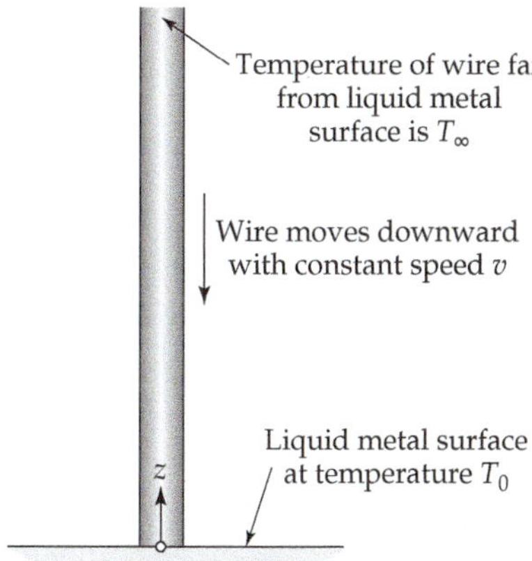

Fig. 11B.5 Wire moving into a liquid metal bath.

(a) First solve the problem for constant physical properties $\hat{C}_p$ and k. Obtain

$$\Theta(z) = \frac{T(z) - T_\infty}{T_0 - T_\infty} = \exp\left(-\frac{\rho \hat{C}_p v z}{k}\right) \tag{11B.5-1}$$

(b) Next solve the problem when $\hat{C}_p$ and k are known functions of the dimensionless temperature $\Theta(z)$: $k = k_\infty K(\Theta)$ and $\hat{C}_p = \hat{C}_{p\infty} L(\Theta)$. Obtain the temperature profile:

$$-\left(\frac{\rho \hat{C}_{p\infty} v z}{k_\infty}\right) = \int_1^\Theta \frac{K(\overline{\Theta})d\overline{\Theta}}{\int_0^{\overline{\Theta}} L(\overline{\overline{\Theta}})d\overline{\overline{\Theta}}} \tag{11B.5-2}$$

(c) Verify that the solution in (b) satisfies the differential equation from which it was derived.

[2]Suggested by Prof. G. L. Borman, Mechanical Engineering Department, University of Wisconsin.

11B.6 Transpiration cooling in a planar system. Two large flat porous horizontal plates are separated by a relatively small distance L. The upper plate at $y = L$ is at temperature T_L, and the lower one at $y = 0$ is to be maintained at a lower temperature T_0. To reduce the amount of heat that must be removed from the lower plate, an ideal gas at T_0 is blown upward through both plates at a steady rate. Develop expressions for the temperature distribution and the amount of heat q_0 that must be removed from the cold plate per unit area as functions of the fluid properties and gas flow rate. Use the abbreviation $\phi = \rho\hat{C}_p v_y L/k$.

Answer: $\dfrac{T(y) - T_L}{T_0 - T_L} = \dfrac{e^{\phi y/L} - e^{\phi}}{1 - e^{\phi}}$; $q_0 = \dfrac{k(T_L - T_0)}{L}\left(\dfrac{\phi}{e^{\phi} - 1}\right)$

11B.7 Reduction of evaporation losses by transpiration. It is proposed to reduce the rate of evaporation of liquefied oxygen in small containers by taking advantage of transpiration. To do this, the liquid is to be stored in a spherical container surrounded by a spherical shell of a porous insulating material as shown in Fig. 11B.7. A thin space is to be left between the container and insulation, and the opening in the insulation is to be stoppered. In operation, the evaporating oxygen is to leave the container proper, move through the gas space, and then flow uniformly out through the porous insulation.

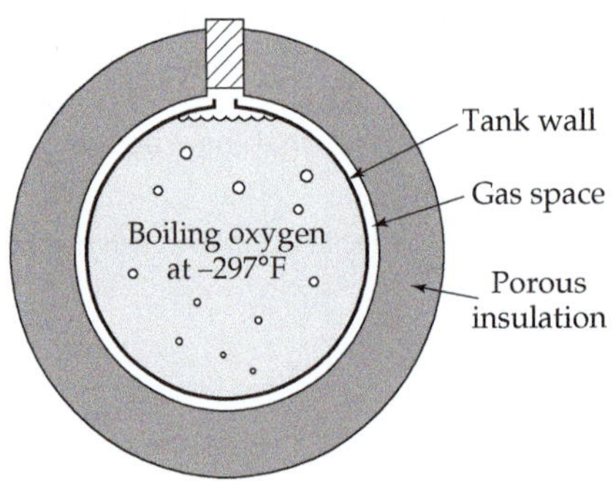

Fig. 11B.7 Use of transpiration to reduce the evaporation rate.

Calculate the rate of heat gain and evaporation loss from a tank 1 ft in diameter covered with a shell of insulation 6 in. thick under the following conditions with and without transpiration.

Temperature of liquid oxygen	−297°F
Temperature of outer surface of insulation	30°F
Effective thermal conductivity of insulation	0.02 Btu/hr · ft · °F
Heat of evaporation of oxygen	91.7 Btu/lb$_m$
Average $\hat{C}_p$ of O_2 flowing through insulation	0.22 Btu/lb$_m$ · °F

Neglect the thermal resistance of the liquid oxygen, container wall, and gas space, and neglect heat losses through the stopper. Assume the particles of insulation to be in local thermal equilibrium with the gas.

Answers: 82 Btu/hr without transpiration; 61 Btu/hr with transpiration

11B.8 Temperature distribution in an embedded sphere. A solid sphere of radius R and thermal conductivity k_1 is embedded in an infinite solid of thermal conductivity k_0. The center of the sphere is located at the origin of coordinates, and there is a constant temperature gradient A in the positive z direction far from the sphere. The temperature at the center of the sphere is T°.

The steady-state temperature distributions in the sphere T_1 and in the surrounding medium T_0 have been shown to be:[3]

$$T_1(r,\theta) - T^{\circ} = \left[\frac{3k_0}{k_1 + 2k_0}\right] Ar\cos\theta \quad r \leq R \tag{11B.8-1}$$

[3] L. D. Landau and E. M. Lifshitz, *Fluid Mechanics*, 2nd edition, Pergamon Press, Oxford (1987), p. 199.

$$T_0(r,\theta) - T^\circ = \left[1 - \frac{k_1 - k_0}{k_1 + 2k_0}\left(\frac{R}{r}\right)^3\right] Ar\cos\theta \quad r \geq R \tag{11B.8-2}$$

(a) What are the partial differential equations that must be satisfied by Eqs. 11B.8-1 and 11B.8-2?

(b) Write down the boundary conditions that apply at $r = R$.

(c) Show that T_1 and T_0 satisfy their respective partial differential equations in (a).

(d) Show that Eqs. 11B.8-1 and 11B.8-2 satisfy the boundary conditions in (b).

11B.9 Heat flow in a solid bounded by two conical surfaces. A solid object has the shape depicted in Fig. 11B.9. The conical surfaces θ_1 = constant and θ_2 = constant are held at temperatures T_1 and T_2, respectively. The spherical surface at $r = R$ is insulated. For steady-state heat conduction, find

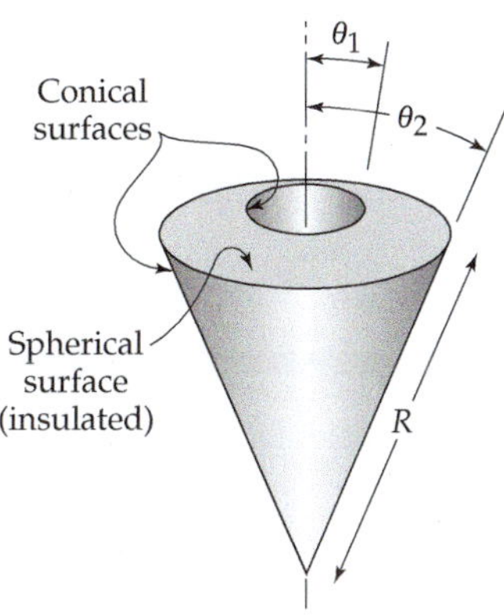

Fig. 11B.9 Body formed from the intersection of two cones and a sphere.

(a) The partial differential equation that $T(\theta)$ must satisfy.

(b) The solution to the differential equation in (a) containing two constants of integration.

(c) Expressions for the constants of integration.

(d) The expression for the θ component of the heat-flux vector.

(e) The total heat flow (cal/s) across the conical surface at $\theta = \theta_1$.

Answer: **(e)** $Q = \dfrac{2\pi Rk(T_1 - T_2)}{\ln\left(\dfrac{\tan\frac{1}{2}\theta_2}{\tan\frac{1}{2}\theta_1}\right)}$

11B.10 Freezing of a spherical drop. To evaluate the performance of an atomizing nozzle, it is proposed to atomize a nonvolatile liquid wax into a stream of cool air. The atomized wax particles are expected to solidify in the air, from which they may later be collected and examined. The wax droplets leave the atomizer only slightly above their melting point (see Fig. 11B.10).

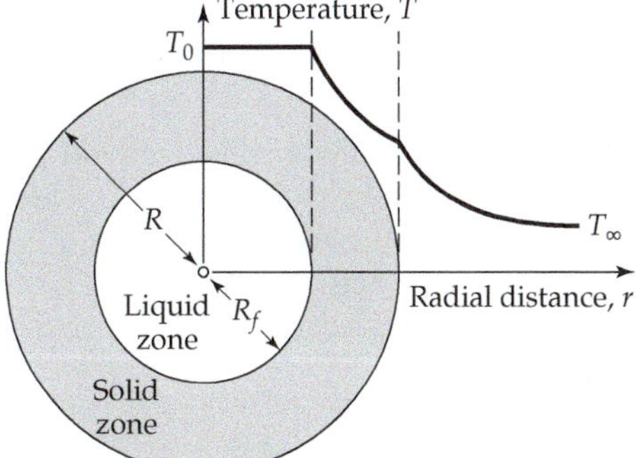

Fig. 11B.10 Temperature profile in the freezing of a spherical drop.

Estimate the time t_f required for a drop of radius R to freeze completely, if the drop is initially at its melting point T_0 and the surrounding air is at T_∞. Heat is lost from the drop to the surrounding air according to Newton's law of cooling, with a constant heat-transfer coefficient h. Assume that there is no volume change in the solidification process. Solve the problem by using a quasi-steady-state method.

(a) First solve the steady-state heat-conduction problem in the solid phase in the region between $r = R_f$ (the liquid–solid interface) and $r = R$ (the solid-air interface). Let k be the thermal conductivity of the solid phase. Then find the radial heat flow Q across the spherical surface at $r = R$.

(b) Then write an unsteady-steady energy balance, by equating the heat liberation at $r = R_f(t)$ resulting from the freezing of the solid to the heat flow Q across the spherical surface at $r = R$. Integrate the resulting separable, first-order differential equation between the limits 0 and R, to obtain the time that it takes for the drop to solidify. Let $\Delta\hat{H}_f$ be the latent heat of freezing (per unit mass).

Answers: **(a)** $Q = \dfrac{h \cdot 4\pi R^2(T_0 - T_\infty)}{[1 - (hR/k)] + (hR^2/kR_f)}$; **(b)** $t_f = \dfrac{\rho\Delta\hat{H}_f R}{h(T_0 - T_\infty)}\left[\dfrac{1}{3} + \dfrac{1}{6}\dfrac{hR}{k}\right]$

11B.11 Temperature rise in a spherical catalyst pellet. A catalyst pellet, shown in Fig. 11B.11, has a radius R and a thermal conductivity k (which may be assumed constant). Because of the chemical reaction occurring within the porous pellet, heat is generated at a rate S_c cal/cm$^3 \cdot$ s. Heat is lost at the outer surface of the pellet to a gas stream at constant temperature T_g by convective heat transfer with heat-transfer coefficient h. Find the steady-state temperature profile, assuming that S_c is constant throughout.

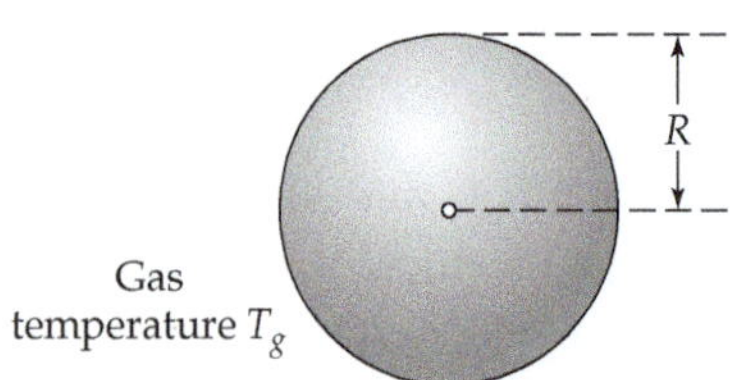

Fig. 11B.11 Sphere with heat generation.

(a) Set up the differential equation by making a shell energy balance.

(b) Set up the differential equation by simplifying the appropriate form of the energy equation.

(c) Integrate the differential equation to get the temperature profile. Sketch the function $T(r)$.

(d) What is the limiting form of $T(r)$ when $h \to \infty$.

(e) What is the maximum temperature in the system?

(f) Where in the derivation would one modify the procedure to account for variable k and variable S_c?

11B.12 Stability of an exothermic reaction system.[3] Consider a porous slab of thickness $2B$, width W, and length L, with $B << W$ and $B << L$. Within the slab an exothermic reaction occurs, with a temperature-dependent rate of heat production $S_c(T) = S_{c0} \exp A(T - T_0)$. Let the coordinate x vary across the thin dimension with $-B \leq x \leq B$. The temperatures at surfaces at $x = \pm B$ are both fixed at $T = T_0$.

(a) Use the energy equation to obtain a differential equation for the temperature in the slab. Assume constant physical properties, and postulate a steady-state solution $T(x)$,

(b) Write the differential equation and boundary conditions in terms of these dimensionless quantities: $\xi = x/B$, $\Theta = A(T - T_0)$, and $\lambda = S_{c0}AB^2/k$.

(c) Integrate the differential equation. *Hint*: First multiply by $2d\Theta/d\xi$ to obtain

$$\left(\frac{d\Theta}{d\xi}\right)^2 = 2\lambda(\exp\Theta_0 - \exp\Theta) \tag{11B.12-1}$$

in which Θ_0 is an auxiliary constant representing the value of Θ at $\xi = 0$.

(d) Integrate the result of (c) and make use of the boundary conditions to obtain the relation between the slab thickness and midplane temperature

$$\exp\left(-\frac{1}{2}\Theta_0\right) \operatorname{arc\,cosh}\left(\exp\left(\frac{1}{2}\Theta_0\right)\right) = \sqrt{\frac{1}{2}\lambda} \tag{11B.12-2}$$

(e) Calculate λ at $\Theta_0 = 0.5, 1.0, 1.2, 1.4$, and 2.0; graph these results to find the maximum value of λ for steady-state conditions. If this value of λ is exceeded, the system will explode.

11B.13 Laminar annular flow with constant wall heat flux. Repeat the development of §10.9 for flow in an annulus of inner and outer radii κR and R respectively, starting with the equations of change. Heat is added to the fluid through the inner cylinder wall at a rate q_0 (energy per unit area per unit time), and the outer cylinder wall is thermally insulated.

11B.14 Unsteady-state heating of a sphere. A sphere of radius R and thermal diffusivity α is initially at a uniform temperature T_0. For $t > 0$ the sphere is immersed in a well-stirred water bath maintained at a temperature $T_1 > T_0$. The temperature within the sphere is then a function of the radial coordinate r and the time t. The solution to the heat conduction equation is given by:[4]

$$\frac{T(r,t) - T_0}{T_1 - T_0} = 1 + 2\sum_{n=1}^{\infty} (-1)^n \left(\frac{R}{n\pi r}\right) \sin\left(\frac{n\pi r}{R}\right) \exp(-\alpha n^2\pi^2 t/R^2) \tag{11B.14-1}$$

It is desired to verify that this equation satisfies the differential equation, the boundary conditions, and the initial condition.

(a) Write down the differential equation describing the problem.

(b) Show that Eq. 11B.14-1 for $T(r,t)$ satisfies the differential equation in (a).

(c) Show that the boundary condition at $r = R$ is satisfied.

(d) Show that T is finite at $r = 0$.

(e) To show that Eq. 11B.14-1 satisfies the initial condition, set $t = 0$ and $T = T_0$ and obtain the following:

$$-1 = 2\sum_{n=1}^{\infty} (-1)^n \left(\frac{R}{n\pi r}\right) \sin\left(\frac{n\pi r}{R}\right) \tag{11B.14-2}$$

To show that this is true, multiply both sides by $(r/R)\sin(m\pi r/R)$, where m is any integer from 1 to ∞, and integrate from $r = 0$ to $r = R$. In the integration all terms with $m \neq n$ vanish on the right side. The term with $m = n$, when integrated, just equals the integral on the left side.

11C.1 The speed of propagation of sound waves. Sound waves are harmonic compression waves of very small amplitude traveling through a compressible fluid. The velocity of propagation of such waves may be estimated by assuming that the momentum-flux tensor $\boldsymbol{\tau}$ and the heat-flux vector $\mathbf{q}$ are zero and that the velocity of the fluid $\mathbf{v}$ is small.[5] The neglect of $\boldsymbol{\tau}$ and $\mathbf{q}$ is equivalent to assuming that the entropy is constant following the motion of a given fluid element.

(a) Use equilibrium thermodynamics to show that

$$\left(\frac{\partial p}{\partial \rho}\right)_S = \gamma\left(\frac{\partial p}{\partial \rho}\right)_T \tag{11C.1-1}$$

in which $\gamma = C_p/C_V$.

---

[4] H. S. Carslaw and J. C. Jaeger, *Conduction of Heat in Solids*, 2nd edition, Oxford University Press (1959), p. 233, Eq. (4).

[5] See L. Landau and E. M. Lifshitz, *Fluid Mechanics*, 2nd edition, Pergamon, Oxford (1987), Chapter VIII; R. J. Silbey, R. A. Alberty, and M. G. Bawendi, *Physical Chemistry*, Wiley, New York, 4th edition (2005), §17.4.

(b) When sound is being propagated through a fluid, there are slight perturbations in the pressure, density, and velocity from the rest state: $p = p_0 + p'$, $\rho = \rho_0 + \rho'$, and $\mathbf{v} = \mathbf{v}_0 + \mathbf{v}'$, the subscript-0 quantities being constants associated with the rest state (with $\mathbf{v}_0$ being zero), and the primed quantities being very small. Show that when these quantities are substituted into the equation of continuity and the equation of motion (with the $\boldsymbol{\tau}$ and $\mathbf{g}$ terms omitted) and products of the small primed quantities are omitted, we get

Equation of continuity
$$\frac{\partial \rho}{\partial t} = -\rho_0(\nabla \cdot \mathbf{v}) \tag{11C.1-2}$$

Equation of motion
$$\rho_0 \frac{\partial \mathbf{v}}{\partial t} = -\nabla p \tag{11C.1-3}$$

(c) Next use the result in (a) to rewrite the equation of motion as

$$\rho_0 \frac{\partial \mathbf{v}}{\partial t} = -v_s^2 \nabla \rho \tag{11C.1-4}$$

in which $v_s^2 = \gamma(\partial p/\partial \rho)_T$.

(d) Show how Eqs. 11C.1-2 and 11C.1-4 can be combined to give

$$\frac{\partial^2 \rho}{\partial t^2} = v_s^2 \nabla^2 \rho \tag{11C.1-5}$$

(e) Show that a solution of Eq. 11C.1-5 is

$$\rho(z,t) = \rho_0 \left[1 + A \sin\left(\frac{2\pi}{\lambda}(z - v_s t)\right)\right] \tag{11C.1-6}$$

This solution represents a harmonic wave of wavelength λ and amplitude $\rho_0 A$ traveling in the z direction at a speed v_s. More general solutions may be constructed by a superposition of waves of different wavelengths and directions.

11C.2 Free convection in a slot. A fluid of constant viscosity, with density given by Eq. 11.3-1, is confined in a rectangular slot. The slot has vertical walls at $x = \pm B$, $y = \pm W$, and a top and bottom at $z = \pm H$, with $H >> W >> B$. Gravity acts in the $-z$ direction. The walls are nonisothermal, with temperature distribution $T_w = \overline{T} + Ay$, so that the fluid circulates by free convection. The velocity profiles are to be predicted, for steady laminar-flow conditions and small deviations from the mean density $\overline{\rho}$.

(a) Simplify the equations of continuity, motion, and energy according to the postulates: $\mathbf{v} = \boldsymbol{\delta}_z v_z(x,y)$, $\partial^2 v_z/\partial y^2 << \partial^2 v_z/\partial x^2$, and $T = T(y)$. These postulates are reasonable for slow flow, except near the edges $y = \pm W$ and $z = \pm H$.

(b) List the boundary conditions to be used with the problem as simplified in (a).

(c) Solve for the temperature, pressure, and velocity profiles.

(d) When making diffusion measurements in closed chambers, free convection can be a serious source of error, and temperature gradients must be avoided. By way of illustration, compute the maximum tolerable temperature gradient, A, for an experiment with water at 20°C in a chamber with $B = 0.1$ mm, $W = 2.0$ mm, and $H = 2$ cm, if the maximum permissible convective movement is 0.1% of H in a one-hour experiment.

Answers: **(c)** $v_z(x,y) = \dfrac{\overline{\rho} g \overline{\beta} A}{2\mu}(x^2 - B^2)y$; **(d)** 2.7×10^{-3} K/cm

11C.3 Heating of a semi-infinite slab (approximate boundary-layer treatment). Rework Example 11.5-1 by the method used in Problem 3C.1.

11C.4 Boundary-layer development for the flow along a heated flat plate (approximate method).[6-8] In Problem 3C.2 the use of boundary-layer approximations for steady, laminar flow of incompressible fluids at constant temperature was discussed. In this problem we extend the previous "von Kármán integral" development by including the boundary-layer equation for energy transport, so that the temperature profiles near solid surfaces can be obtained.

As in Problem 3C.2 we consider the steady two-dimensional flow around a flat plate (see Fig. 11C.4). In the vicinity of the solid surface, the equations of change may be simplified to:

Equation of continuity
$$\frac{\partial v_x}{\partial x} + \frac{\partial v_y}{\partial y} = 0 \tag{11C.4-1}$$

x Equation of motion
$$\rho\left(v_x \frac{\partial v_x}{\partial x} + v_y \frac{\partial v_x}{\partial y}\right) = \mu \frac{\partial^2 v_x}{\partial y^2} \tag{11C.4-2}$$

Equation of energy
$$\rho \hat{C}_p\left(v_x \frac{\partial T}{\partial x} + v_y \frac{\partial T}{\partial y}\right) = k \frac{\partial^2 T}{\partial y^2} \tag{11C.4-3}$$

Here ρ, μ, k, and $\hat{C}_p$ are regarded as constants. The viscous heating term in the energy equation, $\mu(\partial v_x/\partial y)^2$, has been omitted in this problem. Solutions of these equations are asymptotically accurate for small momentum diffusivity $\nu = \mu/\rho$ in Eq. 11C.4-2, and for small thermal diffusivity $\alpha = k/\rho\hat{C}_p$ in Eq. 11C.4-3.

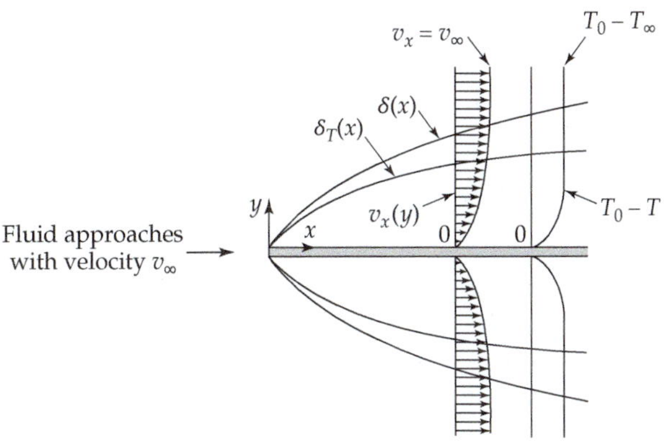

Fig. 11C.4 Boundary-layer development for the flow along a heated flat plate, showing the thermal boundary layer for $\Delta = \delta_T(x)/\delta(x) < 1$. The surface of the plate is at temperature T_0, and the approaching fluid is at T_∞.

Equation 11C.4-1 is the same as Eq. 3C.2-1, and Eq. 11C.4-2 is the same as Eq. 3C.2-2. Equation 11C.4-3 is obtained from Eq. 11.2-9 by neglecting the heat conduction in the x direction. More complete forms of the boundary-layer equations may be found elsewhere.[7,8]

The usual boundary conditions for Eqs. 11C.4-1 and 11C.4-2 are that $v_x = v_y = 0$ at the solid surface, and that the velocity merges into the flow at the outer edge of the *velocity boundary layer*, so that $v_x \to v_\infty$. For Eq. 11C.4-3 the temperature T is specified to be T_0 at the solid surface and T_∞ at the outer edge of the *thermal boundary layer*. That is, the velocity and temperature are different from v_∞ and T_∞ only in thin layers near the solid surface. However, the velocity and temperature boundary layers will be of different thicknesses corresponding to the relative ease of the diffusion of momentum and heat. Since $\text{Pr} = \nu/\alpha$, for $\text{Pr} > 1$ the temperature boundary layer usually lies inside the velocity boundary layer, whereas for $\text{Pr} < 1$ the relative thicknesses

---

[6] H. Schlichting, *Boundary-Layer Theory*, 7th edition, McGraw-Hill, New York (1979), Chapter 12.

[7] K. Stewartson, *The Theory of Laminar Boundary Layers in Compressible Fluids*, Oxford University Press (1964).

[8] E. R. G. Eckert and R. M. Drake, Jr., *Analysis of Heat and Mass Transfer*, McGraw-Hill, New York, (1972), Chapters 6 and 7.

are just reversed (keep in mind that for gases Pr is about $\frac{3}{4}$, whereas for ordinary liquids Pr > 1 and for liquid metals Pr << 1).

In Problem 3C.2 it was shown that the boundary-layer equation of motion could be integrated formally from $y = 0$ to $y = \infty$, if use is made of the equation of continuity. In a similar fashion, integrate Eqs. 11C.4-1 to 11.C.4-3 to get for the flat-plate problem

$$\textit{Momentum} \qquad \mu \left.\frac{\partial v_x}{\partial y}\right|_{y=0} = \frac{d}{dx}\int_0^\infty \rho v_x(v_\infty - v_x)dy \tag{11C.4-4}$$

$$\textit{Energy} \qquad k \left.\frac{\partial T}{\partial y}\right|_{y=0} = \frac{d}{dx}\int_0^\infty \rho \hat{C}_p v_x(T_\infty - T)dy \tag{11C.4-5}$$

Equations 11C.4-4 and 11C.4-5 are the *von Kármán momentum and energy balances*. The no-slip condition, $v_y = 0$ at $y = 0$, has been used here.

Obtain the temperature profiles near a flat plate, along which a Newtonian is flowing, as shown in Fig. 11C.4. The wetted surface of the plate is maintained at temperature T_0 and the temperature of the approaching fluid is T_∞.

(a) In order to use the von Kármán balances, postulate the following reasonable forms for the velocity and temperature profiles. The following polynomial form gives 0 at the wall and 1 at the outer limit of the boundary layer, with a slope of zero at the outer limit:

$$\begin{cases} \dfrac{v_x(y)}{v_\infty} = 2\left(\dfrac{y}{\delta}\right) - 2\left(\dfrac{y}{\delta}\right)^3 + \left(\dfrac{y}{\delta}\right)^4 & y \le \delta(x) \\ \dfrac{v_x(y)}{v_\infty} = 1 & y \ge \delta(x) \end{cases} \tag{11C.4-6,7}$$

$$\begin{cases} \dfrac{T(y) - T_0}{T_\infty - T_0} = 2\left(\dfrac{y}{\delta_T}\right) - 2\left(\dfrac{y}{\delta_T}\right)^3 + \left(\dfrac{y}{\delta_T}\right)^4 & y \le \delta_T(x) \\ \dfrac{T(y) - T_0}{T_\infty - T_0} = 1 & y \ge \delta_T(x) \end{cases} \tag{11C.4-8.9}$$

That is, assume that the dimensionless velocity and temperature profiles have the same form within their respective boundary layers. We further *assume* that the boundary-layer thicknesses $\delta(x)$ and $\delta_T(x)$ have a constant ratio, so that $\Delta = \delta_T(x)/\delta(x)$ is independent of x. Two possibilities have to be considered: $\Delta \le 1$ and $\Delta \ge 1$. We consider here $\Delta \le 1$, and leave the other case as an optional assignment.

(b) The use of Eqs. 11C.4-4 and 11C.4-5 is now straightforward but tedious. Show that substitution of Eqs. 11C.4-6 through 11C.4-9 into the integrals gives

$$\begin{aligned} \textstyle\int_0^\infty \rho v_x(v_\infty - v_x)\,dy &= \rho v_\infty^2 \delta(x) \textstyle\int_0^1 (2\eta - 2\eta^3 + \eta^4)(1 - 2\eta + 2\eta^3 - \eta^4)\,d\eta \\ &= \frac{37}{315}\rho v_\infty^2 \delta(x) \end{aligned} \tag{11C.4-10}$$

and

$$\begin{aligned} \textstyle\int_0^\infty \rho \hat{C}_p v_x(T_\infty - T)\,dy &= \rho \hat{C}_p v_\infty (T_\infty - T_0)\delta_T(x) \textstyle\int_0^\infty (2\eta_T\Delta - 2\eta_T^3\Delta^3 + \eta_T^4\Delta^4) \\ &\quad \times (1 - 2\eta_T + 2\eta_T^3 - \eta_T^4)\,d\eta_T \\ &= \left(\frac{2}{15}\Delta - \frac{3}{140}\Delta^3 + \frac{1}{180}\Delta^4\right)\rho \hat{C}_p v_\infty (T_\infty - T_0)\delta_T(x) \end{aligned} \tag{11C.4-11}$$

In these integrals $\eta = y/\delta(x)$ and $\eta_T = y/\delta_T(x) = y/\Delta\delta(x)$.

(c) Next substitute these integrals into Eqs. 11C.4-4 and 11C.4-5 to get differential equations for the boundary-layer thicknesses.

(d) Integrate the first-order differential equations obtained in (c) to get

$$\delta(x) = \sqrt{\frac{1260}{37}\left(\frac{\nu x}{v_\infty}\right)} \tag{11C.4-12}$$

$$\delta_T(x) = \sqrt{\frac{4}{\frac{2}{15}\Delta - \frac{3}{140}\Delta^3 + \frac{1}{180}\Delta^4}\left(\frac{\alpha x}{v_\infty}\right)} \tag{11C.4-13}$$

(e) Thus, the boundary-layer thicknesses are now determined, except for the evaluation of Δ in Eq. 11C.4-13. Take the ratio of Eq. 11C.4-12 to Eq. 11C.4-13 to get an equation for Δ as a function of the Prandtl number:

$$\frac{2}{15}\Delta^3 - \frac{3}{140}\Delta^5 + \frac{1}{180}\Delta^6 = \frac{37}{315}\text{Pr}^{-1} \qquad \Delta \leq 1 \tag{11C.4-14}$$

When this sixth-order algebraic equation is solved for Δ as a function of Pr, it is found that the solution may be curve-fitted by the simple relation[9]

$$\Delta = \text{Pr}^{-1/3} \qquad \Delta < 1 \tag{11C.4-15}$$

within about 5%.

(f) Show that the temperature profile is then (for $\Delta \leq 1$)

$$\frac{T(y) - T_0}{T_\infty - T_0} = 2\left(\frac{y}{\Delta\delta}\right) - 2\left(\frac{y}{\Delta\delta}\right)^3 + \left(\frac{y}{\Delta\delta}\right)^4 \tag{11C.4-16}$$

in which $\Delta \approx \text{Pr}^{-1/3}$ and $\delta(x) = \sqrt{(1260/37)(\nu x/v_\infty)}$. The assumption of laminar flow made here is valid for $x < x_{\text{crit}}$, where $x_{\text{crit}} v_\infty \rho/\mu$ is usually greater than 10^5.

(g) Finally, obtain the rate of heat loss from both sides of a heated plate of width W and length L from Eqs. 11C.4-5, 11C.4-11, 11C.4-12, 11C.4-15, and 11C.4-16:

$$\begin{aligned} Q &= 2\int_0^W \int_0^L q_y|_{y=0} dx\, dz \\ &= 2W(+k)(T_0 - T_\infty)\int_0^L \frac{\partial}{\partial y}\left[2\left(\frac{y}{\Delta\delta}\right) - 2\left(\frac{y}{\Delta\delta}\right)^3 + \left(\frac{y}{\Delta\delta}\right)^4\right]\bigg|_{y=0} dx \\ &= 2W\frac{(k)(T_0 - T_\infty)}{\Delta}\int_0^L 2\sqrt{\frac{37}{1260}\left(\frac{v_\infty}{\nu x}\right)}dx \\ &\approx \sqrt{\frac{148}{315}}(2WL)(T_0 - T_\infty)\left(\frac{k}{L}\right)\text{Pr}^{1/3}\text{Re}_L^{1/2} \end{aligned} \tag{11C.4-17}$$

in which $\text{Re}_L = Lv_\infty\rho/\mu$. Thus, the boundary-layer approach allows one to obtain the dependence of the rate of heat loss Q on the dimensions of the plate, the flow conditions, and the thermal properties of the fluid. Equation 11C.4-17 is in good agreement with more detailed solutions of Eqs. 11C.4-1 to 11C.4-3.

11C.5 Heat conduction with phase change (the *Neumann–Stefan problem*).[10] A liquid, contained in a long cylinder, is initially at temperature T_1, as illustrated in Fig. 11C.5. For all times $t > 0$,

---

[9]H. Schlichting, *Boundary-Layer Theory*, 7th edition, McGraw-Hill, New York (1979), p. 292–308.

[10]For literature references and related problems, see H. S. Carslaw and J. C. Jaeger, *Conduction of Heat in Solids*, 2nd edition, Oxford University Press (1959), Chapter XI; on pp. 283–286 the problem considered here is worked out for the situation that the physical properties of the liquid and solid phases are different. See also S. G. Bankoff, *Advances in Chemical Engineering*, Vol. **5**, Academic Press, New York (1964), pp. 75–150; J. Crank, *Free and Moving Boundary Problems*, Oxford University Press (1984); J. M. Hill, *One-Dimensional Stefan Problems*, Longmans (1987); A. M. Schmalzer, A. M. Mertz, D. N. Githuku, and A. J. Giacomin, *J. Adv. Eng.*, **7**, 135–143 (July, 2012), Appendix A.

the bottom of the container is maintained at a temperature T_0, which is below the melting point T_m. We want to estimate the movement of the solid–liquid interface, $Z(t)$, during the freezing process.

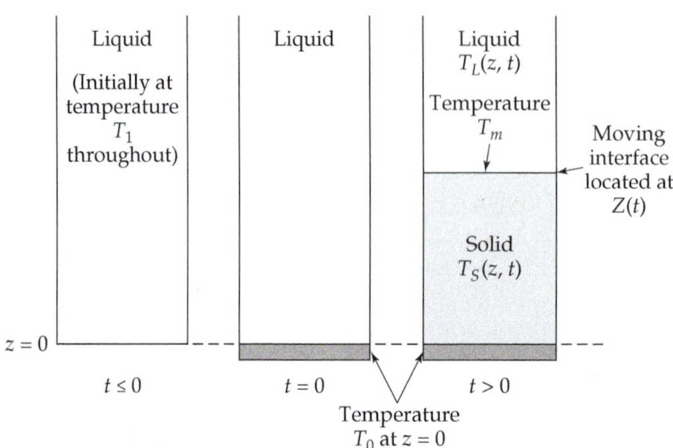

Fig. 11C.5 Heat conduction with solidification.

For the sake of simplicity, we assume here that the physical properties ρ, k, and $\hat{C}_p$ are constants and the same in both the solid and liquid phases. Let $\Delta\hat{H}_f$ be the heat of fusion per gram, and use the abbreviation $\Lambda = \Delta\hat{H}_f/\hat{C}_p(T_1 - T_0)$.

(a) Write the equation for heat conduction for the liquid (L) and solid (S) regions; state the boundary and initial conditions.

(b) Assume solutions of the form:

$$\Theta_S \equiv \frac{T_S - T_0}{T_1 - T_0} = C_1 + C_2 \operatorname{erf}\frac{z}{\sqrt{4\alpha t}} \tag{11C.5-1}$$

$$\Theta_L \equiv \frac{T_L - T_0}{T_1 - T_0} = C_3 + C_4 \operatorname{erf}\frac{z}{\sqrt{4\alpha t}} \tag{11C.5-2}$$

(c) Use the boundary condition at $z = 0$ to show that $C_1 = 0$, and the condition as $z \to \infty$ to show that $C_3 = 1 - C_4$. Then use the fact that $T_S = T_L = T_m$ at $z = Z(t)$ to conclude that $Z(t) = \lambda\sqrt{4\alpha t}$ where λ is some (as yet undetermined) constant. Then get C_3 and C_4 in terms of λ. Use the remaining boundary condition to get λ in terms of Λ and $\Theta_m = (T_m - T_0)/(T_1 - T_0)$:

$$\sqrt{\pi}\Lambda\lambda \exp \lambda^2 = \frac{\Theta_m}{\operatorname{erf}\lambda} - \frac{1 - \Theta_m}{1 - \operatorname{erf}\lambda} \tag{11C.5-3}$$

What is the final expression for $Z(t)$? (*Note:* In this problem it has been assumed that a phase change occurs instantaneously and that no supercooling of the liquid phase occurs. It turns out that in the freezing of many liquids, this assumption is untenable. That is, to describe the solidification process correctly, one has to take into account the kinetics of the crystallization process.[11])

---

[11]H. Janeschitz-Kriegl, *Plastics and Rubber Processing and Applications*, **4**, 145–158 (1984); H. Janeschitz-Kriegl, in *One-Hundred Years of Chemical Engineering* (N. A. Peppas, ed.), Kluwer Academic Publishers, Dordrecht (The Netherlands) (1989), pp. 111–124; H. Janeschitz-Kriegl, E. Ratajski, and G. Eder, *Ind. Eng. Chem. Res.*, **34**, 3481–3487 (1995); G. Astarita and J. M. Kenny, *Chem. Eng. Comm.*, **53**, 69–84 (1987).

Chapter 12

Temperature Distributions in Turbulent Flow

In Chapters 10 and 11 we have shown how to obtain temperature distributions in solids and in fluids in laminar motion. The procedure has involved solving the equations of change with appropriate boundary and initial conditions. Once the temperature profiles are known, practical problems can be solved, such as determining overall heat transfer rates.

We now turn to the problem of analyzing energy transport and finding temperature profiles in turbulent flow. This discussion is quite similar to that given in Chapter 4. We begin by time-smoothing the equations of change. In the time-smoothed energy equation there appears a turbulent heat flux $\overline{\mathbf{q}}^{(t)}$, which is expressed in terms of the correlation of velocity and temperature fluctuations. There are several rather useful empiricisms for $\overline{\mathbf{q}}^{(t)}$, which enable one to predict time-smoothed temperature distributions in wall turbulence and in free turbulence. We use heat transfer in tube flow and in jets to illustrate the method.

The most apparent influence of turbulence on heat transport is the enhanced transport perpendicular to the main flow. Consider the flow of a fluid in an externally heated circular tube (e.g., as in §10.9). If the flow is laminar with the only nonzero velocity component in the axial direction, then energy transport perpendicular to the tube wall can only occur by conduction. This mode of energy transport can be very slow. On the other hand, if the flow is turbulent, there is a nonzero, fluctuating component of the velocity perpendicular to the tube walls, which results in the convective transport of energy from the tube wall to the bulk fluid. This mode of transport is typically much faster. This enhanced transport can also be interpreted as mixing—the turbulent motion mixes the fluid within the tube, creating a more uniform temperature profile than observed in laminar flow. This mixing process is worked out in some detail here for flow in tubes and in circular jets.

§12.1 TIME-SMOOTHED EQUATIONS OF CHANGE FOR INCOMPRESSIBLE NONISOTHERMAL FLOW

In §4.2 the notions of time-smoothed quantities and turbulent fluctuations were introduced. In this chapter we shall be primarily concerned with the temperature profiles.

We introduce the time-smoothed temperature $\overline{T}$ and temperature fluctuation T', and write analogously to Eq. 4.2-1 for an arbitrary point in the fluid

$$T(t) = \overline{T} + T'(t) \tag{12.1-1}$$

Clearly the time average of T' is zero ($\overline{T'} = 0$), but quantities like $\overline{v_x'T'}$, $\overline{v_y'T'}$, $\overline{v_z'T'}$ will not be zero because of the "correlation" between the velocity and temperature fluctuations at any point.

For a nonisothermal pure fluid we need three equations of change, and we want here to discuss their time-smoothed forms. The time-smoothed equations of continuity and motion for a fluid with constant density and viscosity were given in Eqs. 4.2-10 and 4.2-12, and need not be repeated here. For a fluid with constant μ, ρ, $\hat{C}_p$, and k, Eq. 11.2-5, when put in the $\partial/\partial t$-form by using Eq. 3.5-6, and with Newton's and Fourier's laws included, becomes

$$\begin{aligned}\frac{\partial}{\partial t}\rho\hat{C}_pT = &-\left(\frac{\partial}{\partial x}\rho\hat{C}_pv_xT + \frac{\partial}{\partial y}\rho\hat{C}_pv_yT + \frac{\partial}{\partial z}\rho\hat{C}_pv_zT\right)\\ &+k\left(\frac{\partial^2 T}{\partial x^2} + \frac{\partial^2 T}{\partial y^2} + \frac{\partial^2 T}{\partial z^2}\right)\\ &+\mu\left[2\left(\frac{\partial v_x}{\partial x}\right)^2 + \left(\frac{\partial v_x}{\partial y}\right)^2 + 2\left(\frac{\partial v_x}{\partial y}\right)\left(\frac{\partial v_y}{\partial x}\right) + \cdots\right]\end{aligned} \tag{12.1-2}$$

in which only a few sample terms in the viscous dissipation term $-(\boldsymbol{\tau} : \nabla\mathbf{v}) = \mu\Phi_v$ have been written down (see Eq. B.7-1 for the complete expression).

In Eq. 12.1-2 we replace T by $T = \overline{T} + T'$, v_x by $\overline{v}_x + v_x'$, etc. Then the entire equation is time-smoothed (using the time average defined by Eq. 4.2-2) to give

$$\begin{aligned}\frac{\partial}{\partial t}\rho\hat{C}_p\overline{T} = &-\left(\frac{\partial}{\partial x}\rho\hat{C}_p\overline{v}_x\overline{T} + \frac{\partial}{\partial y}\rho\hat{C}_p\overline{v}_y\overline{T} + \frac{\partial}{\partial z}\rho\hat{C}_p\overline{v}_z\overline{T}\right)\\ &-\underline{\left(\frac{\partial}{\partial x}\rho\hat{C}_p\overline{v_x'T'} + \frac{\partial}{\partial y}\rho\hat{C}_p\overline{v_y'T'} + \frac{\partial}{\partial z}\rho\hat{C}_p\overline{v_z'T'}\right)}\\ &+k\left(\frac{\partial^2 \overline{T}}{\partial x^2} + \frac{\partial^2 \overline{T}}{\partial y^2} + \frac{\partial^2 \overline{T}}{\partial z^2}\right)\\ &+\mu\left[2\left(\frac{\partial \overline{v}_x}{\partial x}\right)^2 + \left(\frac{\partial \overline{v}_x}{\partial y}\right)^2 + 2\left(\frac{\partial \overline{v}_x}{\partial y}\right)\left(\frac{\partial \overline{v}_y}{\partial x}\right) + \cdots\right]\\ &+\underline{\mu\left[2\overline{\left(\frac{\partial v_x'}{\partial x}\right)\left(\frac{\partial v_x'}{\partial x}\right)} + \overline{\left(\frac{\partial v_x'}{\partial y}\right)\left(\frac{\partial v_x'}{\partial y}\right)} + 2\overline{\left(\frac{\partial v_x'}{\partial y}\right)\left(\frac{\partial v_y'}{\partial x}\right)} + \cdots\right]}\end{aligned} \tag{12.1-3}$$

Comparison of this equation with the preceding one shows that the time-smoothed equation has the same form as the original equation, except for the appearance of the terms indicated by dashed underlines, which are concerned with the turbulent fluctuations. We are thus led to the definition of the turbulent heat-flux vector $\overline{\mathbf{q}}^{(t)}$ with components

$$\overline{q}_x^{(t)} = \rho\hat{C}_p\overline{v_x'T'}; \qquad \overline{q}_y^{(t)} = \rho\hat{C}_p\overline{v_y'T'}; \qquad \overline{q}_z^{(t)} = \rho\hat{C}_p\overline{v_z'T'} \tag{12.1-4}$$

and the turbulent energy dissipation function $\overline{\Phi}_v^{(t)}$:

$$\overline{\Phi}_v^{(t)} = \sum_{i=1}^{3}\sum_{j=1}^{3}\overline{\left(\left(\frac{\partial v_i'}{\partial x_j}\right)\left(\frac{\partial v_i'}{\partial x_j}\right) + \left(\frac{\partial v_i'}{\partial x_j}\right)\left(\frac{\partial v_j'}{\partial x_i}\right)\right)} \tag{12.1-5}$$

The similarity between the components of $\overline{\mathbf{q}}^{(t)}$ in Eq. 12.1-4 and those of $\overline{\boldsymbol{\tau}}^{(t)}$ in Eq. 4.2-8 should be noted. In Eq. 12.1-5, v_1', v_2', and v_3' are synonymous with v_x', v_y', and v_z', and x_1, x_2, and x_3 have the same meaning as x, y, and z.

To summarize, we list all three time-smoothed equations of change for turbulent flows of pure fluids with constant μ, ρ, $\hat{C}_p$, and k in their D/Dt form (the first two were given in Eqs. 4.2-10 and 4.2-12):

Continuity:
$$(\nabla \cdot \overline{\mathbf{v}}) = 0 \tag{12.1-6}$$

Motion:
$$\rho \frac{D\overline{\mathbf{v}}}{Dt} = -\nabla \overline{p} - \left[\nabla \cdot \left(\overline{\boldsymbol{\tau}}^{(v)} + \underline{\overline{\boldsymbol{\tau}}^{(t)}}\right)\right] + \rho \mathbf{g} \tag{12.1-7}$$

Energy:
$$\rho \hat{C}_p \frac{D\overline{T}}{Dt} = -\left(\nabla \cdot \left(\overline{\mathbf{q}}^{(v)} + \underline{\overline{\mathbf{q}}^{(t)}}\right)\right) + \mu \left(\overline{\Phi}_v^{(v)} + \underline{\overline{\Phi}_v^{(t)}}\right) \tag{12.1-8}$$

Here it is understood that $D/Dt = \partial/\partial t + \overline{\mathbf{v}} \cdot \nabla$. Also $\overline{\mathbf{q}}^{(v)} = -k\nabla\overline{T}$, and $\overline{\Phi}_v^{(v)}$ is the viscous dissipation function of Eq. B.7-1, but with all the v_i replaced by $\overline{v}_i$.

In discussing turbulent energy transport problems, it has been customary to drop the viscous dissipation terms. Then, one sets up the problem as for laminar flow, except that $\boldsymbol{\tau}$ and $\mathbf{q}$ are replaced by $\overline{\boldsymbol{\tau}}^{(v)} + \overline{\boldsymbol{\tau}}^{(t)}$ and $\overline{\mathbf{q}}^{(v)} + \overline{\mathbf{q}}^{(t)}$, respectively, and that time-smoothed $\overline{p}$, $\overline{\mathbf{v}}$, and $\overline{T}$ are used in the remaining terms.

§12.2 THE TIME-SMOOTHED TEMPERATURE PROFILE NEAR A WALL[1]

Before giving empiricisms for $\overline{\mathbf{q}}^{(t)}$ in the next section, we present a short discussion of some results that do not depend on any empiricism.

We consider the turbulent flow along a flat wall as shown in Fig. 12.2-1, and we inquire as to the temperature in the inertial sublayer. We pattern the development after that in

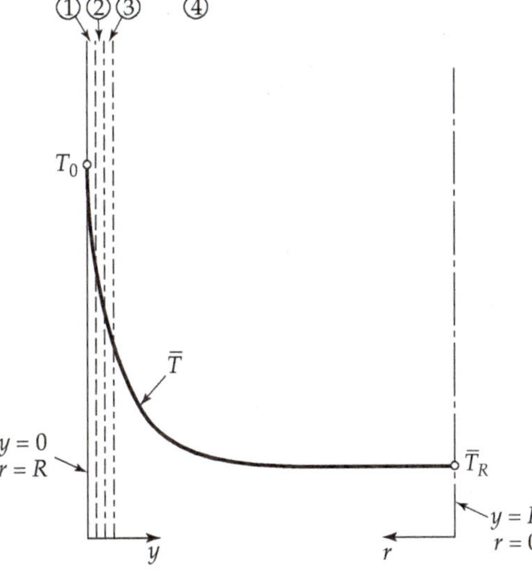

Fig. 12.2-1. Temperature profile in a tube with turbulent flow. The regions are (1) viscous sublayer, (2) buffer layer, (3) inertial sublayer, and (4) main turbulent stream.

[1] L. Landau and E. M. Lifshitz, *Fluid Mechanics*, 2nd edition, Pergamon Press, New York, (1987), §54.

§4.3(a). We let the heat flux into the fluid at $y = 0$ be $q_0 = \bar{q}_y|_{y=0}$, and we postulate that the heat flux in the inertial sublayer will not be very different from that at the wall.

We seek to relate q_0 to the time-smoothed temperature gradient in the inertial sublayer. Because transport in this region is dominated by turbulent convection, the viscosity μ and the thermal conductivity k will not play an important role. Therefore, the only parameters on which $d\bar{T}/dy$ can depend are q_0, $v_* = \sqrt{\tau_0/\rho}$, ρ, $\hat{C}_p$, and y. We must further use the fact that the linearity of the energy equation implies that $d\bar{T}/dy$ must be proportional to q_0. The only combination that satisfies these requirements is

$$-\frac{d\bar{T}}{dy} = \frac{\beta q_0}{\kappa \rho \hat{C}_p v_* y} \tag{12.2-1}$$

in which κ is the dimensionless constant in Eq. 4.3-8, and β is an additional constant (which will turn out to be the turbulent Prandtl number $\text{Pr}^{(t)} = \nu^{(t)}/\alpha^{(t)}$).

When Eq. 12.2-1 is integrated, we get

$$T_0 - \bar{T}(y) = \frac{\beta q_0}{\kappa \rho \hat{C}_p v_*} \ln y + C \tag{12.2-2}$$

where T_0 is the wall temperature and C is a constant of integration. The constant is to be determined by matching the logarithmic expression with the expression for $\bar{T}(y)$ that holds at the outer edge of the viscous sublayer. The latter expression will involve both μ and k; hence, C will necessarily contain μ and k, and will therefore include the dimensionless group $\text{Pr} = \hat{C}_p\mu/k$. If, in addition, we introduce the dimensionless coordinate yv_*/ν, then Eq. 12.2-2 can be rewritten as

$$T_0 - \bar{T}(y) = \frac{\beta q_0}{\kappa \rho \hat{C}_p v_*} \left[\ln\left(\frac{yv_*}{\nu}\right) + f(\text{Pr})\right] \qquad \text{for } \frac{yv_*}{\nu} > 1 \tag{12.2-3}$$

in which $f(\text{Pr})$ is a function representing the thermal resistance between the wall and the inertial sublayer. Landau and Lifshitz (see Ref. 1 on p. 409) estimate, from a mixing-length argument (see Eq. 12.3-3), that, for large Prandtl numbers, $f(\text{Pr}) = \text{constant} \cdot \text{Pr}^{3/4}$; however, Example 12.3-1 implies that the function $f(\text{Pr}) = \text{constant} \cdot \text{Pr}^{2/3}$ is better. Keep in mind that Eq. 12.2-3 can be expected to be valid only in the inertial sublayer and that it should not be used in the immediate neighborhood of the wall.

§12.3 EMPIRICAL EXPRESSIONS FOR THE TURBULENT HEAT FLUX

In §12.1 it was shown that the time-smoothing of the energy equation gives rise to a turbulent heat-flux vector $\bar{\mathbf{q}}^{(t)}$. In order to solve the energy equation for the time-smoothed temperature profiles, it is customary to postulate a relation between $\bar{\mathbf{q}}^{(t)}$ and the time-smoothed temperature gradient. We summarize here two of the most popular empirical expressions; more empiricisms can be found in the heat-transfer literature.

a. Eddy thermal conductivity

By analogy with the Fourier law of heat conduction we may write

$$\bar{q}_y^{(t)} = -k^{(t)} \frac{d\bar{T}}{dy} \tag{12.3-1}$$

in which the quantity $k^{(t)}$ is called the *turbulent thermal conductivity* or the *eddy thermal conductivity*. This quantity is not a physical property of the fluid, but depends on position, direction, and the nature of the turbulent flow.

The eddy kinematic viscosity $\nu^{(t)} = \mu^{(t)}/\rho$ and the eddy thermal diffusivity $\alpha^{(t)} = \mu^{(t)}/\rho\hat{C}_p$ have the same dimensions. Their ratio is a dimensionless group

$$\text{Pr}^{(t)} = \frac{\nu^{(t)}}{\alpha^{(t)}} \tag{12.3-2}$$

called the *turbulent Prandtl number*. This dimensionless quantity is of the order of unity, values in the literature varying from 0.5 to 1.0. For gas flow in conduits, $\text{Pr}^{(t)}$ ranges from 0.7 to 0.9 (for circular tubes the value 0.85 has been recommended[1]), whereas, for flow in jets and wakes, the value is more nearly 0.5. The assumption that $\text{Pr}^{(t)} = 1$ is called the *Reynolds analogy*.

b. The mixing-length expression of Prandtl and Taylor

According to Prandtl's theory, momentum and energy are transferred in turbulent flow by the same mechanism. Hence, by analogy with Eq. 4.4-4, one obtains

$$\bar{q}_y^{(t)} = -\rho\hat{C}_p l^2 \left|\frac{d\bar{v}_x}{dy}\right| \frac{d\bar{T}}{dy} \tag{12.3-3}$$

where l is the Prandtl mixing length introduced in §4.4. Note that this expression predicts that $\text{Pr}^{(t)} = \nu^{(t)}/\alpha^{(t)} = 1$. The Taylor vorticity transport theory[2] gives $\text{Pr}^{(t)} = \nu^{(t)}/\alpha^{(t)} = \frac{1}{2}$.

EXAMPLE 12.3-1

An Approximate Relation for the Wall Heat Flux for Turbulent Flow in a Tube

Use the Reynolds analogy ($\nu^{(t)} = \alpha^{(t)}$), along with Eq. 4.4-2 for the eddy viscosity, to estimate the wall heat flux q_0 for the turbulent flow in a tube of diameter $D = 2R$. Express the result in terms of the temperature-difference driving force $T_0 - \bar{T}_R$, where T_0 is the temperature at the wall ($y = 0$) and $\bar{T}_R$ is the time-smoothed temperature at the tube axis ($y = R$).

SOLUTION

The time-smoothed radial heat flux in a tube is given by the sum of $\bar{q}_r^{(v)}$ and $\bar{q}_r^{(t)}$:

$$\begin{aligned} \bar{q}_r &= -(k + k^{(t)})\frac{d\bar{T}}{dr} = -\left(1 + \frac{\alpha^{(t)}}{\alpha}\right)k\frac{d\bar{T}}{dr} \\ &= +\left(1 + \frac{\nu^{(t)}}{\alpha}\right)k\frac{d\bar{T}}{dy} \end{aligned} \tag{12.3-4}$$

Here we have used Eq. 12.3-1 and the Reynolds analogy, and we have switched to the coordinate y, which is the distance from the wall. We now use the empirical expression of Eq. 4.4-2, which applies across the viscous sublayer next to the wall

$$\bar{q}_y = -\left[1 + \text{Pr}\left(\frac{yv_*}{14.5\nu}\right)^3\right]k\frac{d\bar{T}}{dy} \qquad \text{for } \frac{yv_*}{\nu} < 5 \tag{12.3-5}$$

where $\bar{q}_r = -\bar{q}_y$ has been used.

If now we approximate the heat flux $\bar{q}_y$ in Eq. 12.3-5 by its wall value q_0, then integration from $y = 0$ to $y = R$ gives:

$$q_0 \int_0^R \frac{dy}{1 + \text{Pr}(yv_*/14.5\nu)^3} = k(T_0 - \bar{T}_R) \tag{12.3-6}$$

[1] W. M. Kays and M. E. Crawford, *Convective Heat and Mass Transfer*, 3rd edition, McGraw-Hill, New York (1993), pp. 259–266.

[2] G. I. Taylor, *Proc. Roy. Soc. (London)*, **A135**, 685–702 (1932); *Phil. Trans.*, **A215**, 1–26 (1915).

For very large Prandtl numbers, the upper limit R in the integral can be replaced by ∞, since the integrand is decreasing rapidly with increasing y. Then, when the integration on the left side is performed and the result is put into dimensionless form, we get

$$\frac{q_0 D}{k(T_0 - \overline{T}_R)} = \frac{3\sqrt{3}}{2\pi(14.5)}\left(\frac{v_*}{\langle \overline{v}_z \rangle}\right) \text{Re Pr}^{1/3} = \frac{1}{17.5}\sqrt{\frac{f}{2}} \text{Re Pr}^{1/3} \tag{12.3-7}$$

in which Eq. 6.1-4a has been used to eliminate v_* in favor of the friction factor.

The above development is only approximate. We have not taken into account the change of the bulk temperature as the fluid moves axially through the tube, nor have we taken into account the change in the heat flux throughout the tube. Furthermore, the result is restricted to very high Pr, because of the extension of the integration to $y = \infty$. Another derivation is given in the next section, which is free from these assumptions. However, it will be seen that at large Prandtl numbers the result in Eq. 12.4-21 simplifies to that in Eq. 12.3-7 but with a different numerical constant.

§12.4 TEMPERATURE DISTRIBUTION FOR TURBULENT FLOW IN TUBES

In §10.9 we showed how to get the asymptotic behavior of the temperature profiles for large z in a fluid in laminar flow in a circular tube. We repeat that problem here, but for a fluid in fully developed turbulent flow. The fluid enters the tube of radius R at an inlet temperature T_1. For $z > 0$, the fluid is heated because of a radial heat flux at the wall $\overline{q}_r|_{r=R} = -q_0$ ($q_0 > 0$ for a heated fluid; see Fig. 12.4-1).

We start from the energy equation, Eq. 12.1-8, written in cylindrical coordinates

$$\rho \hat{C}_p \overline{v}_z \frac{\partial \overline{T}}{\partial z} = -\frac{1}{r}\frac{\partial}{\partial r}\left(r\left(\overline{q}_r^{(v)} + \overline{q}_r^{(t)}\right)\right) \tag{12.4-1}$$

Then insertion of the expression for the radial heat flux from Eq. 12.3-4 gives

$$\overline{v}_z \frac{\partial \overline{T}}{\partial z} = \frac{1}{r}\frac{\partial}{\partial r}\left(r\left(\alpha + \alpha^{(t)}\right)\frac{\partial \overline{T}}{\partial r}\right) \tag{12.4-2}$$

This is to be solved with the boundary conditions

B. C. 1:	at $r = 0$,	$\overline{T} = \text{finite}$	(12.4-3)
B. C. 2:	at $r = R$,	$+k\dfrac{\partial \overline{T}}{\partial r} = q_0$ (a constant)	(12.4-4)
B. C. 3:	at $z = 0$,	$\overline{T} = T_1$	(12.4-5)

We now use the same dimensionless variables as already given in Eqs. 10.9-18 to 10.9-20 (with $\overline{T}$ in place of T in the definition of the dimensionless temperature). Then Eq. 12.4-2

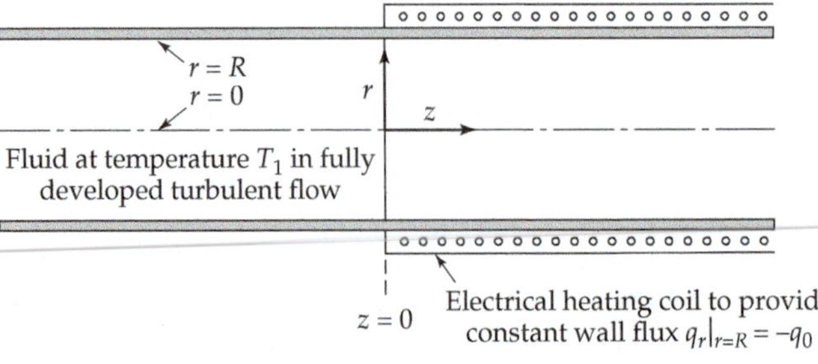

Fig. 12.4-1. System used for heating a liquid in fully developed turbulent flow with constant heat flux for $z > 0$.

in dimensionless form is

$$\phi\frac{\partial \Theta}{\partial \zeta} = \frac{1}{\xi}\frac{\partial}{\partial \xi}\left(\xi\left(1+\frac{\alpha^{(t)}}{\alpha}\right)\frac{\partial \Theta}{\partial \xi}\right) \tag{12.4-6}$$

in which $\phi(\xi) = \bar{v}_z/\bar{v}_{z,\max}$ is the dimensionless turbulent velocity profile. This equation is to be solved with the dimensionless boundary conditions

B. C. 1: at $\xi = 0$, $\Theta = \text{finite}$ (12.4-7)

B. C. 2: at $\xi = 1$, $+\dfrac{\partial \Theta}{\partial \xi} = 1$ (12.4-8)

B. C. 3: at $\zeta = 0$, $\Theta = 0$ (12.4-9)

The complete solution to this problem has been given,[1] but we content ourselves here with the solution for large z.

We begin by assuming an asymptotic solution of the form of Eq. 10.9-25

$$\Theta(\xi,\zeta) = C_0\zeta + \Psi(\xi) \tag{12.4-10}$$

which must satisfy the differential equation, together with B. C. 1 and 2 and Condition 4 in Eq. 10.9-27 (with T and $v_z = v_{\max}(1-\xi^2)$ replaced by $\bar{T}$ and $\bar{v}_z = \bar{v}_{\max}\phi(\xi)$). The resulting equation for Ψ is

$$\frac{1}{\xi}\frac{d}{d\xi}\left(\xi\left(1+\frac{\alpha^{(t)}}{\alpha}\right)\frac{d\Psi}{d\xi}\right) = C_0\phi \tag{12.4-11}$$

Integrating this equation twice and then constructing the function Θ using Eq. 12.4-10, we get

$$\Theta(\xi,\zeta) = C_0\zeta + C_0\int_0^{\xi}\frac{I(\bar{\xi})}{\bar{\xi}\left[1+\left(\alpha^{(t)}/\alpha\right)\right]}d\bar{\xi} + C_1\int_0^{\xi}\frac{1}{\bar{\xi}\left[1+\left(\alpha^{(t)}/\alpha\right)\right]}d\bar{\xi} + C_2 \tag{12.4-12}$$

in which it is understood that $\alpha^{(t)}$ is a function of $\bar{\xi}$, and $I(\bar{\xi})$ is shorthand for the integral

$$I(\bar{\xi}) = \int_0^{\bar{\xi}}\phi(\bar{\bar{\xi}})\bar{\bar{\xi}}\,d\bar{\bar{\xi}} \tag{12.4-13}$$

The constant of integration C_1 is set equal to zero in order to satisfy B. C. 1. The constants C_0 and C_2 are found by applying B. C. 2 and Condition 4 (see Eq. 10.9-27), respectively; we thus get

$$C_0 = \left(\int_0^1 \phi\xi\,d\xi\right)^{-1} = [I(1)]^{-1} \tag{12.4-14}$$

$$C_2 = \int_0^1 \frac{[I(\xi)/I(1)]^2 - [I(\xi)/I(1)]}{\xi\left[1+\left(\alpha^{(t)}/\alpha\right)\right]}d\xi \tag{12.4-15}$$

We next get an expression for the dimensionless temperature difference $\Theta_0 - \Theta_b$, the "driving force" for the heat transfer at the tube wall:

$$\begin{aligned}\Theta_0 - \Theta_b &= C_0\int_0^1\frac{I(\xi)}{\xi\left[1+\left(\alpha^{(t)}/\alpha\right)\right]}d\xi - \frac{C_0}{I(1)}\int_0^1\phi\xi\left[\int_0^{\xi}\frac{I(\bar{\xi})}{\bar{\xi}\left[1+\left(\alpha^{(t)}/\alpha\right)\right]}d\bar{\xi}\right]d\xi \\ &= C_0\int_0^1\frac{I(\xi)}{\xi\left[1+\left(\alpha^{(t)}/\alpha\right)\right]}d\xi - \frac{C_0}{I(1)}\int_0^1\frac{I(\bar{\xi})}{\bar{\xi}\left[1+\left(\alpha^{(t)}/\alpha\right)\right]}\left[\int_{\bar{\xi}}^1\phi\xi\,d\xi\right]d\bar{\xi}\end{aligned} \tag{12.4-16}$$

---

[1] R. H. Notter and C. A. Sleicher, *Chem. Eng. Sci.*, **27**, 2073–2093 (1972).

In the second line, the order of integration of the double integral has been reversed. The inner integral in the second term on the right is just $I(1) - I(\bar{\xi})$, and the portion containing $I(1)$ exactly cancels the first term in Eq. 12.4-16. Hence, we get, when Eq. 12.4-14 is used,

$$\Theta_0 - \Theta_b = \int_0^1 \frac{[I(\xi)/I(1)]^2}{\xi\left[1 + \left(\alpha^{(t)}/\alpha\right)\right]} d\xi \tag{12.4-17}$$

But the quantity $I(1)$ appearing in Eq. 12.4-17 has a simple interpretation:

$$I(1) = \int_0^1 \phi\xi \, d\xi = \left(\int_0^R \bar{v}_z r \, dr\right) \frac{1}{\bar{v}_{z,\max} R^2} = \frac{1}{2} \frac{\langle \bar{v}_z \rangle}{\bar{v}_{z,\max}} \tag{12.4-18}$$

Finally we want to get the dimensionless wall heat flux,

$$\frac{q_0 D}{k(\bar{T}_0 - \bar{T}_b)} = \frac{2}{\Theta_0 - \Theta_b} \tag{12.4-19}$$

the reciprocal of which is[2]

$$\frac{k(\bar{T}_0 - \bar{T}_b)}{q_0 D} = 2\left(\frac{\bar{v}_{z,\max}}{\langle \bar{v}_z \rangle}\right)^2 \int_0^1 \frac{[I(\xi)]^2}{\xi\left[1 + \left(\nu^{(t)}/\nu\right)\left(\text{Pr}/\text{Pr}^{(t)}\right)\right]} d\xi \tag{12.4-20}$$

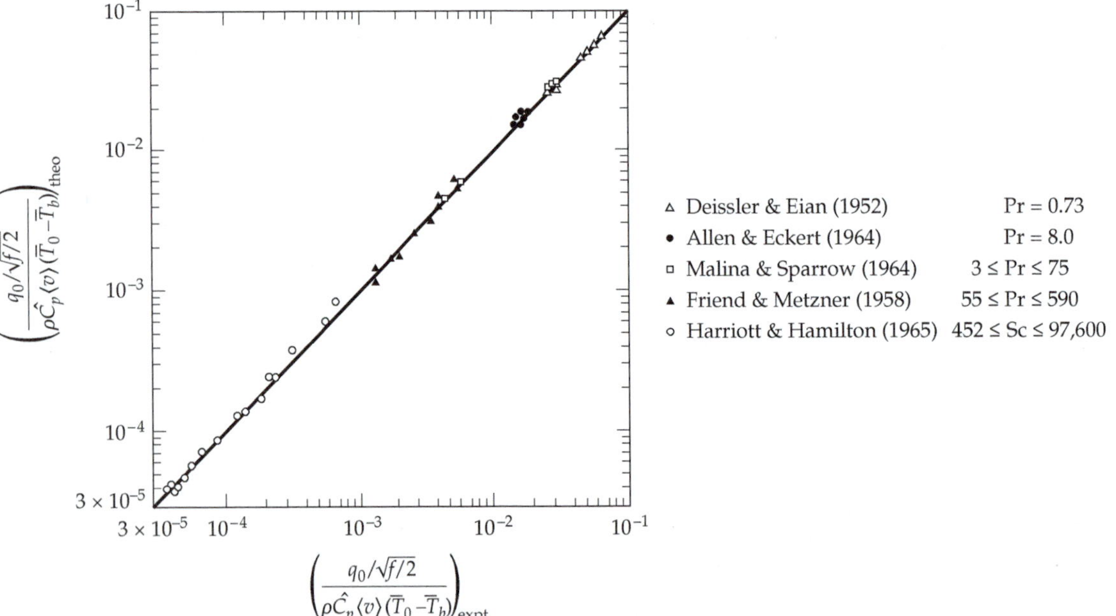

Fig. 12.4-2. Comparison of the expression in Eq. 12.4-21 for the wall heat flux in fully developed turbulent flow with the experimental data of R. G. Deissler and C. S. Eian, *NACA Tech. Note #2629* (1952); R. W. Allen and E. R. G. Eckert, *J. Heat Transfer, Trans. ASME, Ser. C.*, **86**, 301–310 (1964); J. A. Malina and E. M. Sparrow, *Chem. Eng. Sci*, **19**, 953–962 (1964); W. L. Friend and A. B. Metzner, *AIChE Journal*, **4**, 393–402 (1958); P. Harriott and R. M. Hamilton, *Chem. Eng. Sci.*, **20**, 1073–1078 (1965). The data of Harriott and Hamilton are for the analogous mass-transfer experiment, for which Eq. 12.4-21 also applies. Adapted from O. C. Sandall, O. T. Hanna, and P. R. Mazet, *Can. J. Chem. Eng.*, **58**, 443–447 (1980).

[2]Equation 12.4-20 was first developed by R. N. Lyon, *Chem. Eng. Prog.*, **47**, 75–79 (1950) in a paper on liquid-metal heat transfer. The left side of Eq. 12.4-20 is the reciprocal of the Nusselt number, $\text{Nu} = hD/k$, which is a dimensionless heat-transfer coefficient. This nomenclature is discussed in Chapter 14.

To use this result, we need an expression for the time-smoothed velocity distribution $\bar{v}_z$ (which appears in $I(\xi)$), the turbulent kinematic viscosity $\nu^{(t)}$ as a function of position, and a postulate for the turbulent Prandtl number $\text{Pr}^{(t)}$.

Extensive calculations based on Eq. 12.4-20 were performed by Sandall, Hanna, and Mazet.[3] These authors took the turbulent Prandtl number to be unity. They divided the region of integration into two parts, one near the wall and the other for the turbulent core. In the "wall region" they used a modified van Driest equation for the turbulent kinematic viscosity, and in the "core region" they used a logarithmic velocity distribution. Their final result[3] is given as

$$\frac{q_0 D}{k(\bar{T}_0 - \bar{T}_b)} = \frac{\text{RePr}\sqrt{f/2}}{12.48\,\text{Pr}^{2/3} - 7.853\,\text{Pr}^{1/3} + 3.613\ln \text{Pr} + 5.8 + 2.78\ln\left(\frac{1}{45}\text{Re}\sqrt{f/8}\right)} \tag{12.4-21}$$

In obtaining this result, Eq. 6.1-4a has been used.

Equation 12.4-21 agrees with the available data on heat transfer within 3.6% and 8.1% over the range $0.73 < \text{Pr} < 590$, depending on the sets of data studied. The analogous mass-transfer expression, containing $\text{Sc} = \mu/\rho\mathscr{D}_{AB}$ instead of Pr, was reported[3] to agree with the mass-transfer data within 8% over the range $452 < \text{Sc} < 97600$. The agreement of the theory with the heat-transfer and mass-transfer data, shown in Fig. 12.4-2 is quite convincing.

§12.5 TEMPERATURE DISTRIBUTION FOR TURBULENT FLOW IN JETS[1]

In §4.6 we derived an expression for the velocity distribution in a circular fluid jet discharging into an infinite expanse of the same fluid (see Fig. 4.6-1). Here we wish to extend this problem by considering an incoming jet with temperature T_0 higher than that of the surrounding fluid T_1. The problem then is to find the time-smoothed temperature distribution $\bar{T}(r,z)$ in a steadily driven jet. We expect that this distribution will be monotone decreasing in both the r and z directions.

We start by assuming that viscous dissipation is negligible, and we neglect the contribution $\bar{\mathbf{q}}^{(v)}$ to the heat-flux vector as well as the axial contribution to $\bar{\mathbf{q}}^{(t)}$. Then Eq. 12.1-8 takes the time-averaged form

$$\rho\hat{C}_p\left(\bar{v}_r\frac{\partial\bar{T}}{\partial r} + \bar{v}_z\frac{\partial\bar{T}}{\partial z}\right) = -\frac{1}{r}\frac{\partial}{\partial r}(r\bar{q}_r^{(t)}) \tag{12.5-1}$$

Then we express the turbulent heat flux in terms of the turbulent thermal conductivity introduced in Eq. 12.3-1:

$$\bar{q}_r^{(t)} = -k^{(t)}\frac{\partial\bar{T}}{\partial r} = -\rho\hat{C}_p\alpha^{(t)}\frac{\partial\bar{T}}{\partial r} = -\rho\hat{C}_p\frac{\nu^{(t)}}{\text{Pr}^{(t)}}\frac{\partial\bar{T}}{\partial r} \tag{12.5-2}$$

When Eq. 12.5-1 is written in terms of $\bar{\Theta}(\xi,\zeta) = (\bar{T} - T_1)/(T_0 - T_1)$, a dimensionless temperature function, it becomes

$$\left(\bar{v}_r\frac{\partial\bar{\Theta}}{\partial r} + \bar{v}_z\frac{\partial\bar{\Theta}}{\partial z}\right) = \frac{\nu^{(t)}}{\text{Pr}^{(t)}}\frac{1}{r}\frac{\partial}{\partial r}\left(r\frac{\partial\bar{\Theta}}{\partial r}\right) \tag{12.5-3}$$

---

[3] O. C. Sandall, O. T. Hanna, and P. R. Mazet, *Canad. J. Chem. Eng.*, **58**, 443–447 (1980). See also O. T. Hanna and O. C. Sandall, *AIChE Journal*, **18**, 527–533 (1972).

[1] J. O. Hinze, *Turbulence*, 2nd edition, McGraw-Hill, New York (1975), pp. 531–546.

Here it has been assumed that the turbulent Prandtl number and the turbulent kinematic viscosity are constants (see the discussion after Eq. 4.6-3). This equation is to be solved with the boundary conditions:

B. C. 1:	at $z = 0$,	$\overline{\Theta} = 1$	(12.5-4)
B. C. 2:	at $r = 0$,	$\overline{\Theta}$ is finite	(12.5-5)
B. C. 3:	at $r \to \infty$,	$\overline{\Theta} \to 0$	(12.5-6)

Equation 12.5-3 along with the boundary conditions can be solved to give[2]

$$\frac{\overline{\Theta}(\xi,\zeta)}{\overline{\Theta}(0,\zeta)} = \frac{\overline{\Theta}(\xi,\zeta)}{\overline{\Theta}_{\max}(\zeta)} = \left(1 + \frac{1}{4}\left(C_3\xi\right)^2\right)^{-2\mathrm{Pr}^{(t)}} \tag{12.5-7}$$

where $\xi = r/z$, $\zeta = (\rho v^{(t)}/w)z$, w is the total mass flow rate in the jet, and

$$\overline{\Theta}_{\max}(\zeta) = \frac{1}{\zeta}\frac{1 + 2\mathrm{Pr}^{(t)}}{8\pi} \tag{12.5-8}$$

Finally, comparison of Eq. 12.5-7 with Eq. 4.6-9 shows that the shapes of the time-smoothed temperature and axial velocity profiles are closely related,

$$\frac{\Theta(\xi,\zeta)}{\Theta_{\max}(\zeta)} = \left(\frac{\overline{v}_z(\xi,\zeta)}{\overline{v}_{z,\max}(\zeta)}\right)^{\mathrm{Pr}^{(t)}} \tag{12.5-9}$$

where $\overline{v}_z(\xi,\zeta)$ is given by Eq. 4.6-9, with z replaced by $\zeta/(\rho v^{(t)}/w)$. Equation 12.5-9 is attributed to Reichardt.[3] This theory provides a moderately satisfactory explanation for the shapes of the temperature profiles.[1] The turbulent Prandtl (or Schmidt) number deduced from temperature (or concentration) measurements in circular jets is about 0.7.

§12.6 CONCLUDING COMMENTS

This chapter on the turbulent energy flux closely parallels Chapter 4, which deals with turbulent momentum flux. In both chapters, we started by writing the variables as time-smoothed values plus fluctuations. Then we time-smoothed the relevant equations of change to get equations that have to be solved in order to get the time-smoothed velocity and temperature distributions. Solutions cannot be obtained until suitable expressions for the turbulent momentum and energy fluxes have been stated.

For many years, empirical expressions have been used to describe the turbulent fluxes, such as the mixing-length expressions, the expressions containing the eddy viscosity and eddy thermal conductivity, and others. These have led to some moderately useful results, but they should be carefully tested against experimental data.

In this chapter, we have presented a few fairly reliable results: the expression for the wall heat flux for turbulent flow in circular tubes (Eq. 12.3-7); the time-smoothed temperature distribution for flow in circular tubes with constant wall heat flux (Eqs. 12.4-12 through 12.4-15); and the time-smoothed temperature distribution for flow in circular jets (Eq. 12.5-7). By working through these three illustrations, the reader will get "the flavor" of the semi-empirical approach to the subject presented here.

---

[2]See R. B. Bird, W. E. Stewart, and E. N. Lightfoot, *Transport Phenomena*, Revised Second Edition, Wiley, New York (2007), §5.6 and §13.5.

[3]H. Reichardt, *Zeits. f. angew. Math. u. Mech.*, **24**, 268–272 (1944).

QUESTIONS FOR DISCUSSION

1. Compare turbulent thermal conductivity and turbulent viscosity as to definition, order of magnitude, and dependence upon physical properties and the nature of the flow.
2. What is the "Reynolds analogy," and what is its significance?
3. Is there any connection between Eq. 12.2-3 and Eq. 12.4-12, after the integration constants in the latter have been evaluated?
4. Is the analogy between Fourier's law of heat conduction and Eq. 12.3-1 a valid one?
5. What is the physical significance of the fact that the turbulent Prandtl number is of the order of unity?

PROBLEMS

12B.1 Wall heat flux for turbulent flow in tubes (approximate). Work through Example 12.3-1, and fill in the missing steps. In particular, verify the integration in going from Eq. 12.3-6 to Eq. 12.3-7.

12B.2 Wall heat flux for turbulent flow in tubes.

(a) Summarize the assumptions in §12.4.

(b) Work through the mathematical details of that section, taking particular care with the steps connecting Eq. 12.4-12 and Eq. 12.4-17.

(c) When is it not necessary to find the constant C_2 in Eq. 12.4-12?

12C.1 Wall heat flux for turbulent flow between two parallel plates.

(a) Work through the development in §12.4, and then perform a similar derivation for turbulent flow in a thin slit shown in Fig. 2B.4. Show that the analog of Eq. 12.4-20 is:

$$\frac{k(\overline{T}_0 - \overline{T}_b)}{q_0 B} = \left(\frac{\overline{v}_{z,\max}}{\langle \overline{v}_z \rangle}\right)^2 \int_0^1 \frac{[J(\xi)]^2}{[1 + (\nu^{(t)}/\nu)(\Pr/\Pr^{(t)})]} d\xi \tag{12C.1-1}$$

in which $\xi = x/B$, and $J(\xi) = \int_0^\xi \phi(\overline{\xi})\, d\overline{\xi}$.

(b) Show how the above result simplifies for laminar flow of Newtonian fluids, and for "plug flow" (flat velocity profiles).

Answer: **(b)** $k(\overline{T}_0 - \overline{T}_b)/q_0 B = \frac{17}{35}, \frac{1}{3}$

Chapter 13

Dimensional Analysis in Nonisothermal Systems

Many problems in nonisothermal flow are extremely difficult to solve analytically, either because of complicated geometrical arrangements, because of awkward boundary conditions, or because the differential equations are not easily soluble. In such situations, one can often get a partial solution by using the methods of dimensional analysis—either by using the Buckingham pi theorem or by using the equations of change, with their boundary and initial conditions. The purpose of this chapter is to introduce the dimensional analysis approach by giving several examples.

§13.1 DIMENSIONAL ANALYSIS OF THE EQUATIONS OF CHANGE FOR NONISOTHERMAL SYSTEMS

In Chapter 11, we have shown how to use the equations of change for nonisothermal systems to solve some representative heat-transport problems. Then in Chapter 12 we extended the application of the equations of change to turbulent flow. Now we discuss the dimensional analysis of these equations.

Just as the dimensional analysis discussion in Chapter 5 provided an introduction for the discussion of friction factors in Chapter 6, the material in this section provides the background needed for the discussion of heat-transfer coefficient correlations in Chapter 14. As in Chapter 5, we write the equations of change and boundary conditions in dimensionless form. In this way we find some dimensionless parameters that can be used to characterize nonisothermal flow systems.

We shall see, however, that the analysis of nonisothermal systems leads us to a larger number of dimensionless groups than we had in Chapter 5. As a result, greater reliance has to be placed on judicious simplifications of the equations of change and on carefully chosen physical models. An example of the latter is the Boussinesq equation of motion for free convection (§11.3).

As in Chapter 5, for the sake of simplicity, we restrict ourselves to a fluid with constant physical properties, μ, k, and $\hat{C}_p$. The density is taken to be $\rho = \bar{\rho} - \bar{\rho}\bar{\beta}(T - \bar{T})$ in the $\rho\mathbf{g}$ term in the equation of motion, and $\rho = \bar{\rho}$ everywhere else (the "Boussinesq approximation"). The equations of change then become

equation of continuity:
$$(\nabla \cdot \mathbf{v}) = 0 \tag{13.1-1}$$

equation of motion:
$$\bar{\rho}\frac{D\mathbf{v}}{Dt} = -\nabla\mathscr{P} + \mu\nabla^2\mathbf{v} + \bar{\rho}\mathbf{g}\bar{\beta}(T - \bar{T}) \tag{13.1-2}$$

equation of energy:
$$\bar{\rho}\hat{C}_p\frac{DT}{Dt} = k\nabla^2 T + \mu\Phi_v \tag{13.1-3}$$

where $\mathscr{P} = p + \bar{\rho}gh$. We now introduce quantities made dimensionless, indicated by a breve over each symbol, with characteristic quantities (subscript 0 or 1) as follows:

$$\breve{x} = \frac{x}{l_0}; \qquad \breve{y} = \frac{y}{l_0}; \qquad \breve{z} = \frac{z}{l_0}; \qquad \breve{t} = \frac{v_0 t}{l_0}; \tag{13.1-4}$$

$$\breve{\mathbf{v}} = \frac{\mathbf{v}}{v_0}; \qquad \breve{\mathscr{P}} = \frac{\mathscr{P} - \mathscr{P}_0}{\bar{\rho}v_0^2}; \qquad \breve{T} = \frac{T - T_0}{T_1 - T_0}; \tag{13.1-5}$$

$$\breve{\Phi}_v = \left(\frac{l_0}{v_0}\right)^2 \Phi_v; \qquad \breve{\nabla} = l_0\nabla; \qquad \frac{D}{D\breve{t}} = \left(\frac{l_0}{v_0}\right)\frac{D}{Dt} \tag{13.1-6}$$

Here l_0, v_0, and $\mathscr{P}_0$ are the reference quantities introduced in Chapter 5, T_0 is a reference temperature, and $T_1 - T_0$ is a reference temperature difference. In Eq. 13.1-2 the value $\bar{T}$ is the temperature around which the density ρ was expanded.

In terms of these dimensionless variables, the equations of change in Eqs. 13.1-1 to 13.1-3 take the forms

Equation of continuity:
$$(\breve{\nabla} \cdot \breve{\mathbf{v}}) = 0 \tag{13.1-7}$$

Equation of motion:
$$\frac{D\breve{\mathbf{v}}}{D\breve{t}} = -\breve{\nabla}\breve{\mathscr{P}} + \left[\!\left[\frac{\mu}{l_0 v_0 \bar{\rho}}\right]\!\right]\breve{\nabla}^2\breve{\mathbf{v}} - \left[\!\left[\frac{g l_0 \bar{\beta}(T_1 - T_0)}{v_0^2}\right]\!\right]\left(\frac{\mathbf{g}}{g}\right)(\breve{T} - \bar{\breve{T}}) \tag{13.1-8}$$

Equation of energy:
$$\frac{D\breve{T}}{D\breve{t}} = \left[\!\left[\frac{k}{l_0 v_0 \bar{\rho}\hat{C}_p}\right]\!\right]\breve{\nabla}^2\breve{T} + \left[\!\left[\frac{\mu v_0}{l_0 \bar{\rho}\hat{C}_p(T_1 - T_0)}\right]\!\right]\breve{\Phi}_v \tag{13.1-9}$$

These equations are typically simplified further for specific classes of problems. For these different cases, different choices for the characteristic velocity arise naturally. Below we consider two of these different classes in detail; the characteristic velocities and the resulting simplifications of Eqs. 13.1-7 through 13.1-9 are also summarized in Table 13.1-1.

a. Forced convection.

In forced-convection problems, the flow is imposed externally. The characteristic velocity is generally taken to be the approach velocity (for flow past submerged objects) or the average velocity (for flow in conduits). The buoyant forces are also typically ignored leading to the following simplified equations of change *for forced convection problems:*

Equation of continuity:
$$(\breve{\nabla} \cdot \breve{\mathbf{v}}) = 0 \tag{13.1-10}$$

Equation of motion:
$$\frac{D\breve{\mathbf{v}}}{D\breve{t}} = -\breve{\nabla}\breve{\mathscr{P}} + \frac{1}{\text{Re}}\breve{\nabla}^2\breve{\mathbf{v}} \tag{13.1-11}$$

Equation of energy:
$$\frac{D\breve{T}}{D\breve{t}} = \frac{1}{\text{Re Pr}}\breve{\nabla}^2\breve{T} + \frac{\text{Br}}{\text{Re Pr}}\breve{\Phi}_v \tag{13.1-12}$$

where the dimensionless Reynolds (Re), Prandtl (Pr), and Brinkmann (Br) numbers have been defined in previous chapters, and are summarized again in Table 13.1-2. These

Table 13.1-1. Dimensionless Groups in Equations 13.1-7, 13.1-8, and 13.1-9

Special cases →	Forced convection	Intermediate	Free convection (A)	Free convection (B)
Choice for v_0 →	v_0	v_0	ν/l_0	α/l_0
$\left[\!\left[\dfrac{\mu}{l_0 v_0 \bar{\rho}}\right]\!\right]$	$\dfrac{1}{\text{Re}}$	$\dfrac{1}{\text{Re}}$	1	Pr
$\left[\!\left[\dfrac{g l_0 \bar{\beta}(T_1 - T_0)}{v_0^2}\right]\!\right]$	Neglect	$\dfrac{\text{Gr}}{\text{Re}^2}$	Gr	GrPr^2
$\left[\!\left[\dfrac{k}{l_0 v_0 \bar{\rho}\hat{C}_p}\right]\!\right]$	$\dfrac{1}{\text{Re Pr}}$	$\dfrac{1}{\text{Re Pr}}$	$\dfrac{1}{\text{Pr}}$	1
$\left[\!\left[\dfrac{\mu v_0}{l_0 \bar{\rho}\hat{C}_p (T_1 - T_0)}\right]\!\right]$	$\dfrac{\text{Br}}{\text{Re Pr}}$	$\dfrac{\text{Br}}{\text{Re Pr}}$	Neglect	Neglect

Notes:

[a]For forced convection and forced-plus-free ("intermediate") convection, v_0 is generally taken to be the approach velocity (for flow around submerged objects) or an average velocity in the system (for flow in conduits).

[b]For free convection there are two standard choices for v_0, labeled as A and B. In §10.10 Case A arises naturally. Case B proves convenient if the assumption of creeping flow is appropriate, so that $D\breve{\mathbf{v}}/D\breve{t}$ can be neglected (see §13.3). Then a new dimensionless pressure difference $\widehat{\mathcal{P}} = \text{Pr}\breve{\mathcal{P}}$, different from $\breve{\mathcal{P}}$ in Eq. 5.1-12, can be introduced, so that when the equation of motion is divided by Pr, the only dimensionless group appearing in the equation is GrPr. Note that in Case B, no dimensionless groups appear in the equation of energy.

Table 13.1-2. Dimensionless Groups Used in Nonisothermal Systems

$\text{Re} = [\![l_0 v_0 \rho/\mu]\!] = [\![l_0 v_0/\nu]\!]$	= *Reynolds number*
$\text{Pr} = [\![\hat{C}_p \mu/k]\!] = [\![\nu/\alpha]\!]$	= *Prandtl number*
$\text{Gr} = [\![g\beta(T_1 - T_0)l_0^3/\nu^2]\!]$	= *Grashof number*
$\text{Br} = [\![\mu v_0^2/k(T_1 - T_0)]\!]$	= *Brinkman number*
$\text{Fr} = [\![v_0^2/l_0 g]\!]$	= *Froude number*
$\text{We} = [\![l_0 v_0^2 \rho/\sigma]\!]$	= *Weber number*
$\text{Ma} = [\![v_0/v_s]\!]$	= *Mach number*
Pé = Re Pr	= *Péclet number*
Ra = Gr Pr	= *Rayleigh number*
Ec = Br/Pr	= *Eckert number*
Ca = We/Re	= *Capillary number*

equations tell us that the dimensionless temperature profile depends only on the parameters Re, Pr, and Br. For the intermediate case listed in Table 13.1-1, the dimensionless temperature profile would also depend on the Grashof number Gr.

As in §10.9, we see that the product RePr appears in forced-convection problems. Equation 13.1-12 suggests that this product represents the relative importance of energy transport by convection and conduction. In fact, the product RePr can be written as the ratio of magnitudes of the convective energy transport term $\rho\hat{C}_p(\mathbf{v}\cdot\nabla T)$ (whose magnitude is $\rho\hat{C}_p v_0(T - T_0)/l_0$) and the magnitude of the conductive energy transport term $k\nabla^2 T$ (whose magnitude is $k(T - T_0)/l_0^2$),

$$\text{RePr} = \frac{\text{energy transport by convection}}{\text{energy transport by conduction}} = \frac{\rho\hat{C}_p v_0(T - T_0)/l_0}{k(T - T_0)/l_0^2} = \left(\frac{l_0 v_0 \rho}{\mu}\right)\left(\frac{\hat{C}_p \mu}{k}\right) \tag{13.1-13}$$

Similarly, the Brinkman number represents the relative importance of energy production by viscous dissipation and energy transport by conduction (see §10.8), and the quantity Br/RePr represents the relative importance of energy production by viscous dissipation and energy transport by convection.

b. Free convection.

In free-convection problems, the flow arises because of buoyant forces that are caused by temperature gradients. Unlike forced-convection problems, there are no imposed velocities—all velocities must be determined by solving the equations of change. The characteristic velocity is thus selected from combinations of parameters that are relevant to the transport problem. One such choice for the characteristic velocity is $v_0 = \mu/\rho l_0 = \nu/l_0$. Using this quantity, the simplified equations of change for free-convection problems become

Equation of continuity:
$$(\breve{\nabla} \cdot \breve{\mathbf{v}}) = 0 \tag{13.1-14}$$

Equation of motion:
$$\frac{D\breve{\mathbf{v}}}{D\breve{t}} = -\breve{\nabla}\breve{\mathscr{P}} + \breve{\nabla}^2\breve{\mathbf{v}} - \text{Gr}\frac{\mathbf{g}}{g}(\breve{T} - \bar{\breve{T}}) \tag{13.1-15}$$

Equation of energy:
$$\frac{D\breve{T}}{D\breve{t}} = \frac{1}{\text{Pr}}\breve{\nabla}^2\breve{T} \tag{13.1-16}$$

Here, Gr is the Grashof number, which was introduced in §10.10 and listed again in Table 13.1-2. The Grashof number quantifies the relative importance of buoyant and viscous forces.

EXAMPLE 13.1-1

Dimensional Analysis of Free Convection with a Different Velocity Scale

Verify the results in the last column of Table 13.1-1. More specifically, using the alternate velocity scale $v_0 = \alpha/l_0$, show that the dimensionless groups in Eqs. 13.1-8 and 13.1-9 simplify to the corresponding dimensionless groups listed in Table 13.1-1.

SOLUTION

Using $v_0 = \alpha/l_0$, the dimensionless groups in Eqs. 13.1-8 and 13.1-9 simplify as follows:

$$\left[\!\left[\frac{\mu}{l_0 v_0 \bar{\rho}}\right]\!\right] = \left[\!\left[\frac{\mu}{l_0(\alpha/l_0)\bar{\rho}}\right]\!\right] = \left[\!\left[\frac{\mu}{\alpha\bar{\rho}}\right]\!\right] = \left[\!\left[\frac{\mu}{(k/\bar{\rho}\hat{C}_p)\bar{\rho}}\right]\!\right] = \left[\!\left[\frac{\hat{C}_p\mu}{k}\right]\!\right] = \text{Pr} \tag{13.1-17}$$

$$\left[\!\left[\frac{g l_0 \bar{\beta}(T_1 - T_0)}{v_0^2}\right]\!\right] = \left[\!\left[\frac{g l_0 \bar{\beta}(T_1 - T_0)}{(\alpha/l_0)^2}\right]\!\right] = \left[\!\left[\frac{g l_0^3 \bar{\beta}(T_1 - T_0)}{(k/\bar{\rho}\hat{C}_p)^2}\right]\!\right]$$
$$= \left[\!\left[\frac{\bar{\rho}^2 g l_0^3 \bar{\beta}(T_1 - T_0)}{\mu^2}\right]\!\right]\left[\!\left[\left(\frac{\hat{C}_p\mu}{k}\right)^2\right]\!\right] = \text{Gr Pr}^2 \tag{13.1-18}$$

$$\left[\!\left[\frac{k}{l_0 v_0 \bar{\rho}\hat{C}_p}\right]\!\right] = \left[\!\left[\frac{k}{l_0(\alpha/l_0)\bar{\rho}\hat{C}_p}\right]\!\right] = \left[\!\left[\frac{k}{(k/\bar{\rho}\hat{C}_p)\bar{\rho}\hat{C}_p}\right]\!\right] = 1 \tag{13.1-19}$$

$$\left[\!\left[\frac{\mu v_0}{l_0\bar{\rho}\hat{C}_p(T_1-T_0)}\right]\!\right]=\left[\!\left[\frac{\mu v_0^2}{l_0\bar{\rho}\hat{C}_p v_0(T_1-T_0)}\right]\!\right]=\left[\!\left[\frac{\mu v_0^2}{l_0\bar{\rho}\hat{C}_p(\alpha/l_0)(T_1-T_0)}\right]\!\right]$$
$$=\left[\!\left[\frac{\mu v_0^2}{\bar{\rho}\hat{C}_p\left(k/\bar{\rho}\hat{C}_p\right)(T_1-T_0)}\right]\!\right]$$
$$=\left[\!\left[\frac{\mu v_0^2}{k(T_1-T_0)}\right]\!\right]=\text{Br} \tag{13.1-20}$$

In the last equation, we retained a factor of v_0^2 so that the Brinkman number could be more easily identified. Of course, for slow flow problems, we do not expect energy production by viscous dissipation to be important, and thus Br should be small enough to neglect.

We already saw in Chapter 10 how several dimensionless groups appeared in the solution of nonisothermal problems. Here we have seen that the same groupings appear naturally when the equations of change are made dimensionless. These dimensionless groups are used widely in correlations of heat-transfer coefficients. Further dimensionless groups, listed in Table 13.1-2, may arise in the boundary conditions or in the equation of state. The Froude and Weber numbers have already been introduced in Chapter 5, and the Mach number in Example 11.4-6.

We have also seen that dimensionless groups can be interpreted as ratios of forces or phenomena. Such ratios are listed for various dimensionless groups in Table 13.1-3.

The values of the dimensionless groups can be helpful understanding flow problems and their solutions. A low value for the Reynolds number means that viscous forces are large in comparison with inertial forces. A low value of the Brinkman number indicates that the heat produced by viscous dissipation can be transported away quickly by heat conduction. When Gr/Re^2 is large, the buoyant force is important in determining the flow pattern.

Exploiting dimensional analysis to analyze problems and extract useful information is somewhat of an art, requiring judgment and experience. In the next three sections, we give three illustrative examples that demonstrate the power of dimensional analysis. In the first two we analyze forced and free convection in simple geometries. In the third we discuss scale-up problems in a relatively complex piece of equipment.

Table 13.1-3. Physical Interpretation of Dimensionless Groups

$$\text{Re}=\frac{\rho v_0^2/l_0}{\mu v_0/l_0^2}=\frac{\text{inertial force}}{\text{viscous force}}$$

$$\text{Fr}=\frac{\rho v_0^2/l_0}{\rho g}=\frac{\text{inertial force}}{\text{gravity force}}$$

$$\frac{\text{Gr}}{\text{Re}^2}=\frac{\rho g\bar{\beta}(T_1-T_0)}{\rho v_0^2/l_0}=\frac{\text{buoyant force}}{\text{inertial force}}$$

$$\text{Re Pr}=\frac{\rho\hat{C}_p v_0(T_1-T_0)/l_0}{k(T_1-T_0)/l_0^2}=\frac{\text{energy transport by convection}}{\text{energy transport by conduction}}$$

$$\text{Br}=\frac{\mu(v_0/l_0)^2}{k(T_1-T_0)/l_0^2}=\frac{\text{energy production by viscous dissipation}}{\text{energy transport by conduction}}$$

§13.2 TEMPERATURE DISTRIBUTION ABOUT A LONG CYLINDER

Suppose it is desired to predict the temperature distribution in a gas flowing about a long, internally cooled cylinder (System I) from experimental measurements on a one-quarter scale model (System II). If possible the same fluid should be used in the model as in the full-scale system. The system, shown in Fig. 13.2-1, is the same as that in §5.2 except that it is now nonisothermal. The fluid approaching the cylinder has a speed v_∞ and a temperature T_∞ , and the cylinder surface is maintained at T_0, for example, by the boiling of a refrigerant contained within it.

We want to show by means of dimensional analysis how suitable experimental conditions can be chosen for the model studies. We will perform the dimensional analysis for the "intermediate case" in Table 13.1-1.

The two systems, I and II, are geometrically similar. To ensure dynamical similarity, as pointed out in Chapter 5, the dimensionless differential equations and boundary conditions must be the same, and the dimensionless groups appearing in them must have the same numerical values.

Here we choose the characteristic length to be the diameter D of the cylinder, the characteristic velocity to be the approach velocity v_∞ of the fluid, the characteristic pressure to be the pressure at $x = -\infty$ and $y = 0$, and the characteristic temperatures to be the temperature T_∞ of the approaching fluid and the temperature T_0 of the cylinder wall. These characteristic quantities will carry a label I or II depending on which system is being described.

Both systems are described by the dimensionless differential equations given in Eqs. 13.1-7 to 13.1-9, and by boundary conditions

B. C. 1 $\quad$ as $\breve{x}^2 + \breve{y}^2 \to \infty,\quad \breve{\mathbf{v}} \to \boldsymbol{\delta}_x,\quad \breve{T} \to 1$ (13.2-1)

B. C. 2 $\quad$ at $\breve{x}^2 + \breve{y}^2 = \frac{1}{4},\quad \breve{\mathbf{v}} = 0,\quad \breve{T} = 0$ (13.2-2)

B. C. 3 $\quad$ at $\breve{x} \to -\infty$ and $\breve{y} = 0,\quad \breve{\mathscr{P}} = 0$ (13.2-3)

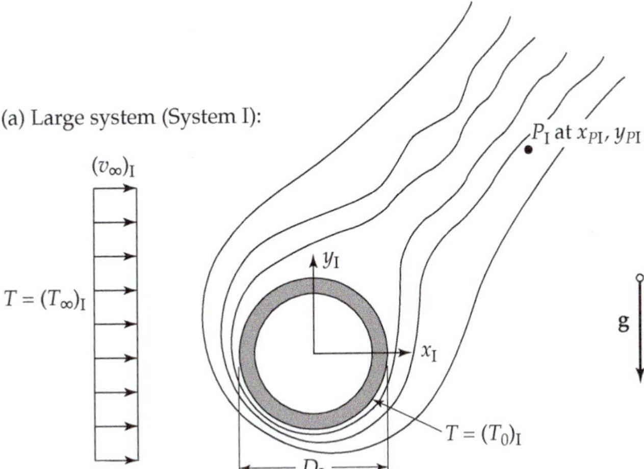

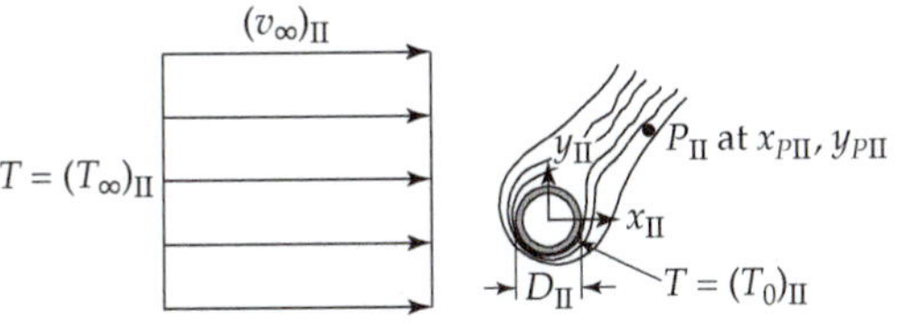

Fig. 13.2-1. Temperature profiles about long heated cylinders. The contour lines in the two figures represent surfaces of constant temperature.

in which $\check{T} = (T - T_0)/(T_\infty - T_0)$. For this simple geometry, the boundary conditions contain no dimensionless groups. Therefore, the requirement that the differential equations and boundary conditions in dimensionless form be identical is that the following dimensionless groups be equal in the two systems: $\text{Re} = Dv_\infty\rho/\mu$, $\text{Pr} = \hat{C}_p\mu/k$, $\text{Br} = \mu v_\infty^2/k(T_\infty - T_0)$, and $\text{Gr} = \rho^2 g\bar{\beta}(T_\infty - T_0)D^3/\mu^2$. In the latter group we use the ideal-gas expression $\beta = 1/T$.

To obtain the necessary equality of the four governing dimensionless groups, we may use different values of the four parameters at our disposal in the two systems: the approach velocity v_∞, the fluid temperature T_∞, the approach pressure $\mathscr{P}_\infty$, and the cylinder temperature T_0.

The similarity requirements are then (for $D_\text{I} = 4D_\text{II}$):

Equality of Pr:
$$\frac{\nu_\text{I}}{\nu_\text{II}} = \frac{\alpha_\text{I}}{\alpha_\text{II}} \tag{13.2-4}$$

Equality of Re:
$$\frac{\nu_\text{I}}{\nu_\text{II}} = 4\frac{v_{\infty\text{I}}}{v_{\infty\text{II}}} \tag{13.2-5}$$

Equality of Gr:
$$\left(\frac{\nu_\text{I}}{\nu_\text{II}}\right)^2 = 64\frac{T_{\infty\text{II}}}{T_{\infty\text{I}}}\frac{(T_\infty - T_0)_\text{I}}{(T_\infty - T_0)_\text{II}} \tag{13.2-6}$$

Equality of Br:
$$\left(\frac{\text{Pr}_\text{I}}{\text{Pr}_\text{II}}\right)\left(\frac{v_{\infty\text{I}}}{v_{\infty\text{II}}}\right)^2 = \frac{\hat{C}_{p\text{I}}}{\hat{C}_{p\text{II}}}\frac{(T_\infty - T_0)_\text{I}}{(T_\infty - T_0)_\text{II}} \tag{13.2-7}$$

Here $\nu = \mu/\rho$ is the kinematic viscosity and $\alpha = k/\rho\hat{C}_p$ is the thermal diffusivity.

The simplest way to satisfy Eq. 13.2-4 is to use the same fluid at the same approach pressure $\mathscr{P}_\infty$ and temperature T_∞ in the two systems. If that is done, Eq. 13.2-5 requires that the approach velocity in the small model (II) be four times that used in the full-scale system (I). If the fluid velocity is moderately large and the temperature differences small, the equality of Pr and Re in the two systems provides a sufficient approximation to thermal similarity. This is the limiting case of forced convection with negligible viscous dissipation.

If, however, the temperature differences $T_\infty - T_0$ are large, free-convection effects may be appreciable. Under these conditions, according to Eq. 13.2-6, temperature differences in the model must be 64 times those in the large system to ensure similarity.

From Eq. 13.2-7 it may be seen that such a ratio of temperature differences will not permit equality of the Brinkman number. For the latter a ratio of 16 would be needed. This conflict will not normally arise, however, as free-convection and viscous heating effects are seldom important simultaneously. Free-convection effects arise in low-velocity systems, whereas viscous heating occurs to a significant degree only when velocity gradients are very large.

§13.3 FREE CONVECTION IN A HORIZONTAL FLUID LAYER; FORMATION OF BÉNARD CELLS

In this section we are going to investigate the free-convection motion in the system shown in Fig. 13.3-1. It consists of a thin layer of fluid between two horizontal parallel plates, the lower one at temperature T_0, and the upper one at T_1, with $T_1 < T_0$. In the absence of fluid motion, the conductive heat flux will be the same for all z, and a nearly uniform temperature gradient will be established at steady state. This temperature gradient will in turn cause a density gradient. If the density decreases with increasing z, the system will be stable, but if it increases, a potentially unstable situation occurs. It appears possible in this latter case that any chance disturbance may cause the more dense fluid to move downward and displace the lighter fluid beneath it. If the temperatures of the top and bottom surfaces are maintained constant, the result may be a continuing free-convection

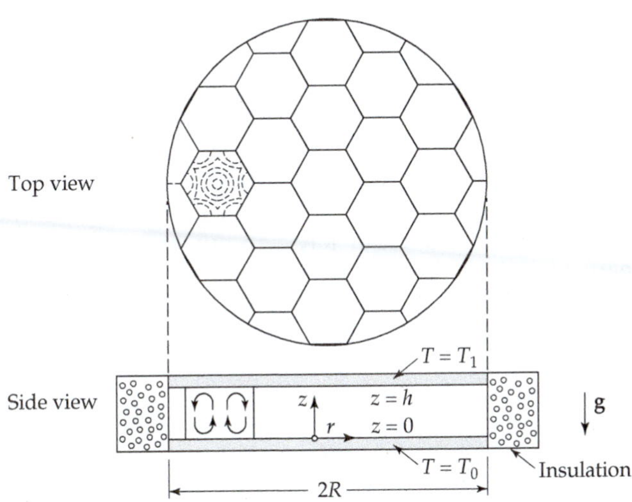

Fig. 13.3-1. Bénard cells formed in the region between two horizontal parallel plates, with the bottom plate at a higher temperature than the upper one. If the Rayleigh number exceeds a certain critical value, the system becomes unstable and hexagonal Bénard cells are produced.

motion. This motion will, however, be opposed by viscous forces and may, therefore, occur only if the temperature difference tending to cause it is greater than some critical minimum value.

We want to determine by means of dimensional analysis the functional dependence of this fluid motion and the conditions under which it may be expected to arise. The system is described by Eqs. 13.1-1 to 13.1-3 along with the following boundary conditions:

B.C. 1	at $z = 0$,	$\mathbf{v} = 0$	$T = T_0$	(13.3-1)
B.C. 2	at $z = h$,	$\mathbf{v} = 0$	$T = T_1$	(13.3-2)
B.C. 3	at $r = R$,	$\mathbf{v} = 0$	$\partial T/\partial r = 0$	(13.3-3)

We now restate the problem in dimensionless form, using $l_0 = h$. We use the dimensionless quantities listed under Case "B" in Table 13.1-1, and we select the reference temperature $\overline{T}$ to be $\frac{1}{2}(T_0 + T_1)$, so that

Equation of continuity: $$(\breve{\nabla} \cdot \breve{\mathbf{v}}) = 0 \tag{13.3-4}$$

Equation of motion: $$\frac{D\breve{\mathbf{v}}}{D\breve{t}} = -\breve{\nabla}\breve{\mathscr{P}} + \text{Pr}\breve{\nabla}^2\breve{\mathbf{v}} - \text{GrPr}^2\left(\frac{\mathbf{g}}{g}\right)\left(\breve{T} - \frac{1}{2}\right) \tag{13.3-5}$$

Equation of energy: $$\frac{D\breve{T}}{D\breve{t}} = \breve{\nabla}^2\breve{T} \tag{13.3-6}$$

with dimensionless boundary conditions:

B. C. 1:	at $\breve{z} = 0$,	$\breve{\mathbf{v}} = 0$	$\breve{T} = 0$	(13.3-7)
B. C. 2:	at $\breve{z} = 1$,	$\breve{\mathbf{v}} = 0$	$\breve{T} = 1$	(13.3-8)
B. C. 3:	at $\breve{r} = R/h$,	$\breve{\mathbf{v}} = 0$,	$\partial\breve{T}/\partial\breve{r} = 0$	(13.3-9)

If the above dimensionless equations could be solved along with the dimensionless boundary conditions, we would find that the velocity and temperature profiles would depend only on Gr, Pr, and R/h. Furthermore, the larger the ratio R/h is, the less prominent its effect will be, and in the limit of extremely large horizontal plates, the system behavior will depend solely on Gr and Pr.

If we consider only steady creeping flows, then the term $D\breve{\mathbf{v}}/D\breve{t}$ may be set equal to zero. Then we define a new dimensionless pressure difference as $\widehat{\mathscr{P}} = \text{Pr}\breve{\mathscr{P}}$. With the left side of Eq. 13.3-5 equal to zero, we may now divide by Pr and the resulting equation contains only one dimensionless group, namely the Rayleigh number[1] $\text{Ra} = \text{GrPr} = \rho^2 g\bar{\beta}(T_1 - T_0)h^3\hat{C}_p/\mu k$, whose value will determine the behavior of the system. This illustrates how one may reduce the number of dimensionless groups that are needed to describe a nonisothermal flow system.

The above analysis suggests that there may be a critical value of the Rayleigh number, and when this critical value is exceeded, fluid motion will occur. This suggestion has been amply confirmed experimentally[2,3] and the critical Rayleigh number has been found to be 1700 $\pm$ 51 for $R/h \gg 1$. For Rayleigh numbers below the critical value, the fluid is stationary, as evidenced by the observation that the heat flux across the liquid layer is the same as that predicted for conduction through a static fluid: $q_z = k(T_0 - T_1)/h$. As soon as the critical Rayleigh number is exceeded, however, the heat flux rises sharply, because of convective energy transport. An increase of the thermal conductivity reduces the Rayleigh number, thus moving Ra towards its stable range.

The assumption of creeping flow is a reasonable one for this system and is asymptotically correct when $\text{Pr} \to \infty$. It is also very convenient, inasmuch as it allows analytic solutions of the relevant equations of change.[4] One such solution, which agrees well with experiment, is sketched qualitatively in Fig. 13.3-1. This flow pattern is cellular and hexagonal, with upflow at the center of each hexagon and downflow at the periphery. The units of this fascinating pattern are called *Bénard cells*.[5] The analytic solution also confirms the existence of a critical Rayleigh number. For the boundary conditions of this problem and very large R/h it has been calculated[4] to be 1708, which is in excellent agreement with the experimental result cited above.

Similar behavior is observed for other boundary conditions. If the upper plate of Fig. 13.3-1 is replaced by a liquid-gas interface, so that the surface shear stress in the liquid is negligible, cellular convection is predicted theoretically[3] for Rayleigh numbers above about 1101. A spectacular example of this type of instability occurs in the occasional spring "turnover" of water in northern lakes. If the lake water is cooled to near freezing during the winter, an adverse density gradient will occur as the surface waters warm toward 4°C, the temperature of maximum density for water.

In shallow liquid layers with free surfaces, instabilities can also arise from surface-tension gradients. The resulting surface stresses produce cellular convection superficially similar to that resulting from temperature gradients, and the two effects may be easily confused. Indeed, it appears that the steady flows first seen by Bénard, and ascribed to buoyancy effects, may actually have been produced by surface-tension gradients.[6]

§13.4 SURFACE TEMPERATURE OF AN ELECTRICAL HEATING COIL

An electrical heating coil of diameter D is being designed to keep a large tank of liquid above its freezing point. It is desired to predict the temperature that will be reached at the coil surface as a function of the heating rate Q and the bulk liquid temperature T_0. This

---

[1] The Rayleigh number is named after Lord Rayleigh (J. W. Strutt), *Phil. Mag.*, (6) **32**, 529–546 (1916).

[2] P. L. Silveston, *Forsch. Ingenieur-Wesen*, **24**, 29–32, 59–69 (1958).

[3] S. Chandrasekhar, *Hydrodynamic and Hydromagnetic Instability*, Oxford University Press (1961); T. E. Faber, *Fluid Dynamics for Physicists*, Cambridge University Press (1995), §8.7.

[4] A. Pellew and R. V. Southwell, *Proc. Roy. Soc.*, **A176**, 312–343 (1940).

[5] H. Bénard, *Revue générale des sciences pures et appliquées*, **11**, 1261–1271, 1309–1328 (1900); *Annales de Chimie et de Physique*, **23**, 62–144 (1901).

[6] C. V. Sternling and L. E. Scriven, *AIChE Journal*, **5**, 514–523 (1959); L. E. Scriven and C. V. Sternling, *J. Fluid Mech.*, **19**, 321–340 (1964).

prediction is to be made on the basis of experiments with a smaller, geometrically similar apparatus filled with the same liquid.

We need to outline a suitable experimental procedure for making the desired prediction. The temperature dependence of the physical properties, other than the density, may be neglected. The entire heating coil surface may be assumed to be at a uniform temperature T_1.

This is a free-convection problem, and we use the column labeled A in Table 13.1-1 for the dimensionless groups. If we could solve the equations of change for this complicated system, we know that the dimensionless temperature $\breve{T} = (T - T_0)/(T_1 - T_0)$ would be a function of the dimensionless coordinates, and would depend on the dimensionless groups Pr and Gr.

The total energy input rate through the coil surface is

$$Q = -k \int_S \left. \frac{\partial T}{\partial r} \right|_S dS \tag{13.4-1}$$

Here r is the coordinate measured outward from and normal to the coil surface, S is the surface area of the coil, and the temperature gradient is that of the fluid immediately adjacent to the coil surface. In dimensionless form this relation is

$$\frac{Q}{k(T_1 - T_0)D} = -\int_{\breve{S}} \left. \frac{\partial \breve{T}}{\partial \breve{r}} \right|_{\breve{S}} d\breve{S} = \psi(\text{Pr},\text{Gr}) \tag{13.4-2}$$

in which ψ is a function of $\text{Pr} = \hat{C}_p\mu/k$ and $\text{Gr} = \rho^2 g\bar{\beta}(T_1 - T_0)D^3/\mu^2$. Since the large-scale and small-scale systems are geometrically similar, the dimensionless function $\breve{S}$ describing the surface of integration will be the same for both systems and hence does not need to be included in the function ψ. Similarly, if we write the boundary conditions for temperature, velocity, and pressure at the coil and tank surfaces, we will obtain only size ratios that will be identical in the two systems.

We now note that the desired quantity $(T_1 - T_0)$ appears on both sides of Eq. 13.4-2. If we multiply both sides of the equation by the Grashof number, then $(T_1 - T_0)$ appears only on the right side:

$$\frac{Q\rho^2 g\bar{\beta}D^2}{k\mu^2} = \text{Gr} \cdot \psi(\text{Pr},\text{Gr}) \tag{13.4-3}$$

In principle, we may solve Eq. 13.4-3 for Gr and obtain an expression for $(T_1 - T_0)$. Since we are neglecting the temperature dependence of physical properties, we may consider the Prandtl number constant for the given fluid and rearrange Eq. 13.4-3 to obtain

$$T_1 - T_0 = \frac{\mu^2}{\rho^2 g\beta D^3} \cdot \phi\left(\frac{Q\rho^2 g\bar{\beta}D^2}{k\mu^2}\right) \tag{13.4-4}$$

Here ϕ is some function of the dimensionless group $Q\rho^2 g\bar{\beta}D^2/k\mu^2$ to be determined experimentally. We may then construct a plot of Eq. 13.4-4 from the experimental measurements of T_1, T_0, and Q for the small-scale system, and the known physical properties of the fluid. This plot may then be used to predict the behavior of the large-scale system.

Since we have neglected the temperature dependence of the fluid properties, we may go even further. If we maintain the ratio of the Q values in the two systems inversely proportional to the square of the ratio of the diameters, then the corresponding ratio of the values of $(T_1 - T_0)$ will be inversely proportional to the cube of the ratio of the diameters.

§13.5 THE BUCKINGHAM PI THEOREM

It was pointed out in §5.5 that the Buckingham pi theorem can often be useful in suggesting relations among dimensionless groups of the quantities occurring in complicated problems. According to this theorem, if we are dealing with q quantities, $x_1, x_2, \ldots, x_q$, involving d dimensions, then one can obtain a relation among $q - d$ different dimensionless groups.

Here we illustrate the use of this theorem for nonisothermal flow problems.

EXAMPLE 13.5-1

Heating of a Fluid in a Pipe

Suppose we seek a relation for the heat flux q that should be supplied through the wall of a tube of diameter D and length L, through which a fluid with constant physical properties (density ρ, viscosity μ, thermal conductivity k, heat capacity $\hat{C}_p$) is flowing. Let the average velocity in the tube be v and the difference between the average wall temperature and the average bulk fluid temperature be ΔT, which is the driving force for heat transfer (other definitions of ΔT are possible that do not change the conclusions of this analysis; see Ch. 14).

What dimensionless groups should we expect to encounter in the correlation of experimental data?

SOLUTION

In this problem we have 9 physical quantities that involve 4 dimensions: q [=] M/t^3, D [=] L, L [=] L, ρ [=] M/L^3, μ [=] M/Lt, k [=] ML/t^3T, $\hat{C}_p$ [=] L^2/t^2T, v [=] L/t, ΔT [=] T. Therefore, we can expect to have 5 dimensionless groups, each of the form

$$\Pi = q^a D^b L^c \rho^d \mu^e k^f \hat{C}_p^g v^h \Delta T^i \tag{13.5-1}$$

By substituting the dimensions of the 9 quantities into Eq. 13.5-1, we find that dimensions of Π are

$$\Pi\ [=]\ M^{a+d+e+f} L^{b+c-3d-e+f+2g+h} t^{-3a-e-3f-2g-h} T^{-f-g+i} \tag{13.5-2}$$

The exponents can be determined by requiring that Π be dimensionless. This gives the following system of equations:

$$a + d + e + f = 0 \tag{13.5-3}$$

$$b + c - 3d - e + f + 2g + h = 0 \tag{13.5-4}$$

$$-3a - e - 3f - 2g - h = 0 \tag{13.5-5}$$

$$-f - g + i = 0 \tag{13.5-6}$$

This is an underspecified system of 4 equations and 9 unknowns, and thus we are free to select values for 5 of the unknowns a-i to obtain a dimensionless group (as long as the 5 selected values do not violate Eqs. 13.5-3 through 13.5-6). Because we seek a relationship between q and the other quantities, we will select $a = 1$ for one of the dimensionless groups, and $a = 0$ for the other 4 dimensionless groups.

For the dimensionless group that contains q, we use our knowledge of energy transport to select $b = 1, f = -1$, and $i = -1$ (i.e., we know that q has the same dimensions as $k\nabla T$). Setting all of the remaining unknowns to zero satisfies Eqs. 13.5-3 through 13.5-6, and thus our first dimensionless group is

$$\frac{qD}{k\Delta T} = \text{dimensionless heat flux} \tag{13.5-7}$$

For the dimensionless groups with $a = 0$, we again use our experience. The parameters L and D have the same dimensions, and thus their ratio is dimensionless. We therefore choose another dimensionless group to be

$$\frac{L}{D} = \text{a geometrical ratio} \tag{13.5-8}$$

and one can verify that the corresponding choices $b = -1, c = 1$, and $a = d = e = f = g = h = i = 0$ satisfy Eqs. 13.5-3 through 13.5-6. Again using our knowledge of transport phenomena, we expect two additional groups to be

$$\frac{Dv\rho}{\mu} = \text{Reynolds number} \tag{13.5-9}$$

$$\frac{\hat{C}_p\mu}{k} = \text{Prandtl number} \tag{13.5-10}$$

And again, we can verify that the corresponding choices for the unknowns a-i satisfy Eqs. 13.5-3 through 13.5-6.

For the last dimensionless group, we choose to obtain a ratio that contains k and ΔT as the only two parameters that include the dimension of temperature. Since this requires that $g = 0$, Eq. 15.5-6 then gives $f = i$. We choose $f = i = -1$ for convenience (see below). We also select $b = c = 0$, hoping that the dependence on system size is already captured in the existing dimensionless groups. The remaining 3 unknowns, d, e, and h, are determined from Eqs. 13.5-3, 13.5-4, and 13.5-5 to be $d = 0$, $e = 1$, and $h = 2$. Thus, our final dimensionless group is

$$\frac{\mu v^2}{k\Delta T} = \text{Brinkman number} \tag{13.5-11}$$

Had we chosen a different value for $f = i$, we would have obtained the Brinkman number raised to a power.

The above selections for the dimensionless groups are not unique. However, any choices for dimensionless groups that are consistent with Eqs. 13.5-3 through 13.5-6 can be written as combinations of the dimensionless groups obtained here.

In developing a correlation of experimental data, these would be dimensionless groups that we could use. We know, however, from the discussion in §10.8, that the Brinkman number will not be important in flows involving low viscosity fluids and/or small velocity gradients. Thus, we expect to be able to correlate data using the form

$$\frac{qD}{k\Delta T} = \phi\left(\text{Re,Pr,}\frac{L}{D}\right) \tag{13.5-12}$$

where ϕ is a function yet to be determined.

In the next chapter, several data correlations based on the above analysis are given, including the geometric ratio L/D. It will also be shown how the temperature dependence of the viscosity can be incorporated into the correlations of experimental data. We will also introduce the heat-transfer coefficient, h, by the definition $q = h\Delta T$, and then the dimensionless heat flux in Eq. 13.5-12 becomes $\text{Nu} = hD/k$, which is called the Nusselt number. The Buckingham pi theorem does not tell us anything about the functional relation $\text{Nu} = \text{Nu}(\text{Re, Pr}, D/L)$; that has to be found from experimental data or from the solution of the governing equations.

EXAMPLE 13.5-2

Heat Loss from a Heated Sphere

A sphere of diameter D with surface temperature T_0 is located in a stream of fluid approaching with a velocity v_∞ and temperature T_∞ (less than T_0). The density, viscosity, thermal conductivity, and heat capacity of the fluid are known and assumed here to be constant. The mean heat flux from the surface of the sphere is q_m. What functional dependence would one expect to find for q_m?

SOLUTION

Here we have the 8 physical quantities q_m, D, v_∞, ρ, μ, k, $\hat{C}_p$, and $\Delta T = (T_0 - T_\infty)$ and 4 dimensions. Therefore, we expect a relationship among 4 dimensionless groups. This example is similar to the previous, except that here we do not have a parameter L. Thus, we can expect to arrive at the same dimensionless groups, excluding L/D. In addition, for systems with negligible heat generation, we would not expect the heat flux to depend on the Brinkman number. We would therefore expect that $q_m D/k\Delta T$ would be a function of the Reynolds number $Dv_\infty\rho/\mu$ and Prandtl number $\hat{C}_p\mu/k$.

However, we know from Problem 10B.1 that, if there is no flow past the sphere, $q_m = 2k\Delta T/D$. This suggests that a suitable expression for $q_m D/k\Delta T$ might be

$$q_m D/k\Delta T = 2 + f(\text{Re,Pr}) \tag{13.5-13}$$

Often, investigators have attempted to find simple power-law relations of the form

$$q_m D/k\Delta T = 2 + C\text{Re}^m\text{Pr}^n \tag{13.5-14}$$

where C is a numerical constant; this is, indeed, the function proposed in 1938 by Frössling (see §14.4). In later refinements of the correlation, it was found that such a simple expression was not adequate, and that furthermore, one additional dimensionless group, namely μ_∞/μ_0, made it possible to get a better correlation.

Thus, although the pi theorem may in some instances prove helpful, it does require some intuition as well as auxiliary information. It is far better to make use of the equations of change if possible.

§13.6 CONCLUDING COMMENTS

In the opening section of this chapter, we showed how a variety of dimensionless groups can arise by putting the equations of change into dimensionless form. We discussed the physical meaning of these dimensionless groups, and classified the types of flows in which these groups would appear. These are by no means all of the groups that one can expect to encounter in nonisothermal flow systems. After all we encountered the Biot and Nusselt numbers in §10.2 and the Mach number in §11.4. In flow systems with free surfaces, we would expect dimensionless groups containing the interfacial tension to arise.

The three examples we have given in this chapter—the transverse flow around a cylinder, the formation of Bénard cells, and the electrical heating coil—provide a few illustrations of the power of the dimensional analysis method. From these examples it should be clear that some intuition and inventiveness is required. Heat–transfer treatises and handbooks can provide additional examples. We have also illustrated the use of the Buckingham pi theorem to develop relationships among dimensionless group. This approach requires even more insight, and is thus less desirable than dimensional analysis of the equations of change.

QUESTIONS FOR DISCUSSION

1. What are some of the advantages of writing the equations of change in dimensionless form?
2. Why is nondimensionalizing the equations of change preferred over the method illustrated in §13.5 for obtaining dimensionless groups? Under what conditions would the method in §13.5 be preferred?
3. Which dimensionless groups contain only physical properties? Which contain physical properties as well as information about the geometry, the strength of flow, or imposed temperatures?
4. How does one decide if a problem is one of forced convection, of free convection, or of an intermediate case?
5. Can the Prandtl number be interpreted as a ratio of phenomena? Explain.
6. When listing the dimensions of the quantities that are important in a problem, what could be possible implications of a fundamental dimension appearing in only one of the quantities?
7. Given a set of q quantities with d dimensions, describe how the $q - d$ dimensionless groups may be obtained from a system of d linear equations.

PROBLEMS

13A.1 Calculating Prandtl numbers. Calculate values for the Prandtl number, $\text{Pr} = \hat{C}_p\mu/k$, for the following cases:

(a) $\hat{C}_p = 4184\ \text{J/kg} \cdot {}^\circ\text{C}$, $\mu = 10^{-3}\ \text{Pa} \cdot \text{s}$, $k = 0.61\ \text{W/m} \cdot \text{K}$

(b) $\hat{C}_p = 1.0\ \text{Btu/lb}_m \cdot {}^\circ\text{F}$, $\mu = 6.7 \times 10^{-4}\ \text{lb}_m/\text{ft} \cdot \text{s}$, $k = 0.35\ \text{Btu/hr} \cdot \text{ft} \cdot {}^\circ\text{F}$

(c) $\tilde{C}_p = 29.44\ \text{J/g-mol} \cdot \text{K}$, $\mu = 2.07 \times 10^{-5}\ \text{Pa} \cdot \text{s}$, $k = 0.027\ \text{W/m} \cdot \text{K}$, $M = 32\ \text{g/g-mol}$

13B.1 Verifying dimensionless groups. Verify that the Grashof and Brinkman numbers, listed in Table 13.1-2, are dimensionless.

13B.2 Dimensional analysis with source terms. Repeat the nondimensionalization of the energy equation that leads to Eq. 13.1-9, this time including a term accounting for energy production by conversion of electrical energy. The rate of energy production per unit volume is S_e.

13B.3 Nondimensionalization of boundary conditions. Using the same dimensional scales employed in §13.1, nondimensionalize the boundary conditions below. What dimensionless groups appear?

(a) $-k\left.\dfrac{\partial T}{\partial x}\right|_s = q_0$, where q_0 is a prescribed constant heat flux

(b) $-k\left.\dfrac{\partial T}{\partial x}\right|_s = h(T|_s - T_0)$

13C.1 Free-convection heat transfer from a vertical flat plate (Fig. 13C.1). A flat plate of height H and width W (with $W \gg H$) heated to a temperature T_0 is suspended in a large body of fluid, which is at ambient temperature T_1. In the neighborhood of the heated plate, the fluid rises because of the buoyant force. From the equations of change, deduce the dependence of the heat loss on the system variables. The physical properties of the fluid are considered constant, except that the change in density with temperature will be accounted for by the Boussinesq approximation.

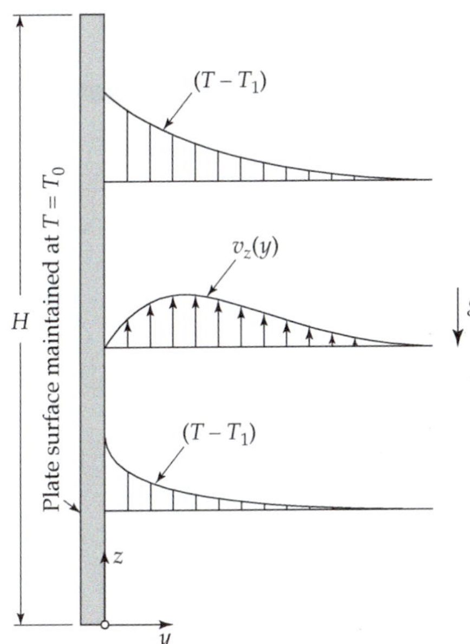

Fig. 13C.1 The temperature and velocity profiles on one side of a thin heated plate suspended in a large body of fluid at temperature T_1.

Begin by postulating that $\mathbf{v} = \boldsymbol{\delta}_y v_y(y,z) + \boldsymbol{\delta}_z v_z(y,z)$ and that $T = T(y,z)$. We assume that the fluid moves almost directly upward so that $v_y \ll v_z$. The x and y components of Eq. 11.3-2 give $p = p(z)$, so that the pressure distribution is given to a very good approximation by $-dp/dz - \bar{\rho}g = 0$, which is the hydrostatic pressure distribution.

(a) Show that the remaining equations of change simplify to

continuity
$$\frac{\partial v_y}{\partial y} + \frac{\partial v_z}{\partial z} = 0 \tag{13C.1-1}$$

z motion
$$\bar{\rho}\left(v_y\frac{\partial}{\partial y} + v_z\frac{\partial}{\partial z}\right)v_z = \mu\left(\frac{\partial^2}{\partial y^2} + \underline{\frac{\partial^2}{\partial z^2}}\right)v_z + \bar{\rho}g\bar{\beta}(T - T_1) \tag{13C.1-2}$$

energy
$$\bar{\rho}\hat{C}_p\left(v_y\frac{\partial}{\partial y} + v_z\frac{\partial}{\partial z}\right)(T - T_1) = k\left(\frac{\partial^2}{\partial y^2} + \underline{\frac{\partial^2}{\partial z^2}}\right)(T - T_1) \tag{13C.1-3}$$

in which $\bar{\rho}$ and $\bar{\beta}$ are evaluated at temperature T_1. The dashed-underlined terms will be omitted on the grounds that momentum and energy transport by molecular processes in the z direction are small compared with the corresponding convective terms on the left sides of the equations. These omissions give a satisfactory description of the system except for a small region around the bottom of the plate. With this simplification, the following boundary conditions suffice to analyze the system up to $z = H$:

B. C. 1 at $y = 0$, $v_y = v_z = 0$ and $T = T_0$ (13C.1-4)

B. C. 2 as $y \to \infty$, $v_z \to 0$ and $T \to T_1$ (13C.1-5)

B. C. 3 at $z = 0$, $v_z = 0$ (13C.1-6)

Note that the temperature rise appears in the equation of motion and the velocity distribution appears in the energy equation. Thus, these equations are coupled. Analytic solutions of such coupled, nonlinear differential equations are very difficult to obtain, and we content ourselves here with a dimensional analysis approach.

(b) Using the following dimensionless variables

$$\Theta = \frac{T - T_1}{T_0 - T_1} = \text{dimensionless temperature} \tag{13C.1-7}$$

$$\zeta = \frac{z}{H} = \text{dimensionless vertical coordinate} \tag{13C.1-8}$$

$$\eta = \left(\frac{B}{\mu \alpha H}\right)^{1/4} y = \text{dimensionless horizontal coordinate} \tag{13C.1-9}$$

$$\phi_z = \left(\frac{\mu}{\alpha B H}\right)^{1/2} v_z = \text{dimensionless vertical velocity} \tag{13C.1-10}$$

$$\phi_y = \left(\frac{\mu H}{\alpha^3 B}\right)^{1/4} v_y = \text{dimensionless horizontal velocity} \tag{13C.1-11}$$

in which $\alpha = k/\bar{\rho}\hat{C}_p$ and $B = \bar{\rho} g \bar{\beta}(T_0 - T_1)$, show that the dimensionless equations of change become (with the dashed-underlined terms omitted)

continuity
$$\frac{\partial \phi_y}{\partial \eta} + \frac{\partial \phi_z}{\partial \zeta} = 0 \tag{13C.1-12}$$

z motion
$$\frac{1}{\text{Pr}}\left(\phi_y \frac{\partial}{\partial \eta} + \phi_z \frac{\partial}{\partial \zeta}\right)\phi_z = \frac{\partial^2 \phi_z}{\partial \eta^2} + \Theta \tag{13C.1-13}$$

energy
$$\left(\phi_y \frac{\partial}{\partial \eta} + \phi_z \frac{\partial}{\partial \zeta}\right)\Theta = \frac{\partial^2 \Theta}{\partial \eta^2} \tag{13C.1-14}$$

(c) Show that the boundary conditions become

B. C. 1 at $\eta = 0$, $\phi_y = \phi_z = 0$ and $\Theta = 1$ (13C.1-15)

B. C. 2 as $\eta \to \infty$, $\phi_z \to 0$ and $\Theta \to 1$ (13C.1-16)

B. C. 3 at $\zeta = 0$, $\phi_z = 0$ (13C.1-17)

From these equations and boundary conditions, it is apparent that the dimensionless velocity components ϕ_y and ϕ_z and the dimensionless temperature Θ will depend on η and ζ and also on the Prandtl number, Pr. Since the flow is usually very slow in free convection, the term in which Pr appears will generally by small; setting this term equal to zero corresponds to the creeping flow assumption. Hence, we expect the dependence of the solution on the Prandtl number to be weak.

(d) The average heat flux from the plate is given by

$$q_{\text{avg}} = \frac{1}{H}\int_0^H \left(-k\frac{\partial T}{\partial y}\right)\bigg|_{y=0} dz \tag{13C.1-18}$$

Write this integral in dimensionless form, and use our knowledge of the dependence of Θ on system variables to show that

$$q_{\text{avg}} = C \cdot \frac{k}{H}(T_0 - T_1)(\text{GrPr})^{1/4} \tag{13C.1-19}$$

where the dimensionless parameter C is given by

$$C = \int_0^1 \left(-\frac{\partial \Theta}{\partial \eta} \right) \bigg|_{\eta=0} d\zeta \tag{13C.1-20}$$

Thus, C depends only weakly on Pr. Note that if we use the definition of the heat-transfer coefficient in Eq. 10.1-2, namely, $h \equiv q/\Delta T$, then Eq. 13C.1-19 can be written

$$\frac{hH}{k} \equiv \text{Nu} = C \cdot (\text{GrPr})^{1/4} \tag{13C.1-21}$$

We will find in the next chapter that this equation is a starting point for determining heat-transfer coefficients for free convection past submerged objects.

This analysis shows that, even without solving the partial differential equations, we can predict that the average heat flux is proportional to the $\frac{5}{4}$-power of the temperature difference $(T_0 - T_1)$ and inversely proportional to the $\frac{1}{4}$-power of H. Both predictions are confirmed by experiment. The only thing we could not do was to determine C as a function of Pr. To determine that function, we have to make experimental measurements or solve Eqs. 13C.1-12 to 13C.1-17. In 1881, Lorenz[1] obtained an approximate solution to these equations and found $C = 0.548$. More refined calculations[2] gave the following dependence of C on Pr:

Pr:	0.73	1	10	100	1000	∞
C:	0.518	0.535	0.620	0.653	0.665	0.67

These values of C are nearly in exact agreement with the best experiments in the laminar flow range (i.e., for $\text{GrPr} < 10^9$).[3]

[1]L. Lorenz, *Wiedemann's Ann. der Physik u. Chemie,* **13**, 422–447, 582–606 (1881). See also U. Grigull, *Die Grundgesetze der Wärmeübertragung*, Springer-Verlag, Berlin, 3rd edition (1955), pp. 263–269.

[2]See S. Whitaker, *Fundamental Principles of Heat Transfer*, Krieger, Malabar, FL (1977), §5.11. The limiting case of $\text{Pr} \to \infty$ has been worked out numerically by E. J. LeFevre [Heat Div. Paper 113, Dept. Sci. and Ind. Res., Mech. Engr. Lab. (Great Britain), Aug. 1956] and it was found that

$$\frac{\partial \Theta}{\partial \eta}\bigg|_{\eta=0} = \frac{0.5028}{\zeta^{1/4}} \qquad \frac{\partial \phi_z}{\partial \eta}\bigg|_{\eta=0} = \frac{1.16}{\zeta^{1/4}} \tag{13C.1-19a,b}$$

Equation 13C.1-19a corresponds to the value $C = 0.670$ above. This result has been verified experimentally by C. R. Wilke, C. W. Tobias, and M. Eisenberg, *J. Electrochem. Soc.*, **100**, 513–523 (1953), for the analogous mass-transfer problem.

[3]For an analysis of free convection in three-dimensional creeping flow, see W. E. Stewart, *Int. J. Heat and Mass Transfer*, **14**, 1013–1031 (1971).

Chapter 14

Interphase Transport in Nonisothermal Systems

In Chapter 10 we saw how shell energy balances may be set up for various simple problems and how these balances lead to differential equations from which the temperature profiles may be calculated. We have also seen in Chapter 11 that the energy balance over an arbitrary differential fluid element leads to a partial differential equation—the energy equation—which may be used to set up more complex problems. It was further seen in Chapter 12 that the time-smoothed energy equation, together with empirical expressions for the turbulent heat flux, provides a useful basis for summarizing and extrapolating temperature-profile measurements in turbulent systems. Hence, at this point the reader should have a fairly good appreciation for the meaning of the equations of change for nonisothermal flow and their range of applicability.

It should be apparent that all of the problems discussed have pertained to systems of rather simple geometry and furthermore that most of these problems have contained assumptions, such as temperature-independent viscosity and constant fluid density. For some purposes, these solutions may be adequate, especially for order-of-magnitude estimates. Furthermore, the study of simple systems provides the stepping stone to the discussion of more complex problems.

In this chapter we turn to some of the problems in which it is convenient or necessary to use a less detailed analysis. In such problems the usual engineering approach is to formulate energy balances over pieces of equipment, or parts thereof, as described in Chapter 15. In the macroscopic energy balance thus obtained, there are usually terms that require estimating the heat that is transferred through the system boundaries. This requires knowing the *heat-transfer coefficient* for describing the interphase transport. Usually the heat-transfer coefficient is given, for the flow system of interest, as an empirical correlation of the *Nusselt number*[1] (a dimensionless wall heat flux or heat transfer coefficient) as a function of the relevant dimensionless quantities, such as the Reynolds and Prandtl numbers.

This situation is not unlike that in Chapter 6, where we learned how to use dimensionless correlations of the friction factor to solve momentum-transfer problems. However, for nonisothermal problems the number of dimensionless groups is larger, the types of boundary conditions are more numerous, and the temperature dependence of the physical properties is often important. In addition, the phenomena of free convection, condensation, and boiling are encountered in nonisothermal systems.

We have purposely limited ourselves here to a small number of heat-transfer formulas and correlations—just enough to introduce the reader to the subject without attempting to be encyclopedic. Many treatises and handbooks treat the subject in much greater depth.[2,3,4,5,6]

§14.1 DEFINITIONS OF HEAT-TRANSFER COEFFICIENTS

Let us consider a flow system with the fluid flowing either in a conduit or around a solid object. Suppose that the solid surface is warmer than the fluid, so that heat is being transferred from the solid to the fluid. Then the rate of heat flow across the solid-fluid interface would be expected to depend on the area of the interface and on the temperature drop between the fluid and the solid. It is customary to define a proportionality factor h (the *heat-transfer coefficient*) by

$$Q = hA\Delta T \tag{14.1-1}$$

in which Q is the rate of heat flow into the fluid ([=] J/hr or Btu/hr), A is a characteristic area, and ΔT is a characteristic temperature difference. Equation 14.1-1 can also be used when the fluid is cooled. Equation 14.1-1, in slightly different form, has been encountered in Eq. 10.1-2. Note that h is not completely defined until the area A and the temperature difference ΔT have been specified. We now consider the usual definitions for h for two types of flow geometry.

As an example of *flow in conduits*, we consider a fluid flowing through a circular tube of diameter D (see Fig. 14.1-1), in which there is a heated wall section of length L and varying inside surface temperature $T_0(z)$, going from T_{01} to T_{02}. Suppose that the bulk temperature T_b of the fluid (defined in Eq. 10.9-35 for fluids with constant ρ and $\hat{C}_p$) increases from T_{b1} to T_{b2} in the heated section. Then there are three conventional definitions of heat-transfer coefficients for the fluid in the heated section:

$$Q = h_1(\pi DL)(T_{01} - T_{b1}) \equiv h_1(\pi DL)\Delta T_1 \tag{14.1-2}$$

---

[1]This dimensionless group is named for **Ernst Kraft Wilhelm Nusselt** (1882–1957), the German engineer who was the first major figure in the field of convective heat and mass transfer. See, for example, W. Nusselt, *Zeits. d. Ver. deutsch. Ing.*, **53**, 1750–1755 (1909), *Forschungsarb. a. d. Geb. d. Ingenieurwesen*, No. 80, 1–38, Berlin (1910), and *Gesundheits-Ing.*, **38**, 477–482, 490–496 (1915).

[2]M. Jakob, *Heat Transfer*, Vol. I (1949) and Vol. II (1957), Wiley, New York.

[3]W. M. Kays and M. E. Crawford, *Convective Heat and Mass Transfer*, 3rd edition, McGraw-Hill, New York (1993).

[4]H. D. Baehr and K. Stephan, *Heat and Mass Transfer*, Springer, Berlin (1998).

[5]W. M. Rohsenow, J. P. Hartnett, and Y. I. Cho (eds.), *Handbook of Heat Transfer*, McGraw-Hill, New York (1998).

[6]H. Gröber, S. Erk, and U. Grigull, *Die Grundgesetze der Wärmeübertragung*, Springer, Berlin, 3rd edition (1961).

$$Q = h_a(\pi DL)\left(\frac{(T_{01} - T_{b1}) + (T_{02} - T_{b2})}{2}\right) \equiv h_a(\pi DL)\Delta T_a \tag{14.1-3}$$

$$Q = h_{\ln}(\pi DL)\left(\frac{(T_{01} - T_{b1}) - (T_{02} - T_{b2})}{\ln(T_{01} - T_{b1}) - \ln(T_{02} - T_{b2})}\right) \equiv h_{\ln}(\pi DL)\Delta T_{\ln} \tag{14.1-4}$$

That is, h_1 is based on the temperature difference ΔT_1 at the inlet, h_a is based on the *arithmetic mean* ΔT_a of the terminal temperature differences, and $h_{\ln}$ is based on the corresponding *logarithmic mean* temperature difference $\Delta T_{\ln}$. For most calculations $h_{\ln}$ is preferable, because it is less dependent on L/D than the other two, although it is not always used.[1] In using heat-transfer correlations from treatises and handbooks, one must be careful to note the definitions of the heat-transfer coefficients.

If the wall temperature distribution is initially unknown, or if the fluid properties change appreciably along the pipe, it is difficult to predict the heat-transfer coefficients defined above. Under these conditions, it is customary to rewrite Eq. 14.1-2 in the differential form:

$$dQ = h_{\text{loc}}(\pi D dz)(T_0 - T_b)_{\text{loc}} \equiv h_{\text{loc}}(\pi D dz)\Delta T_{\text{loc}} \tag{14.1-5}$$

Here dQ is the rate of heat added to the fluid over a distance dz along the pipe, ΔT_{loc} is the local temperature difference (at position z), and h_{loc} is the *local heat-transfer coefficient*. This equation is widely used in engineering design. Actually, the definition of h_{loc} and ΔT_{loc} is not complete without specifying the shape of the element of area. In Eq. 14.1-5 we have set $dA = \pi D dz$, which means that h_{loc} and ΔT_{loc} are the mean values for the darker shaded area dA in Fig. 14.1-1.

As an example of *flow around submerged objects*, consider a fluid flowing around a sphere of radius R, whose surface temperature is maintained at a uniform value T_0. Suppose that the fluid approaches the sphere with a uniform temperature T_∞. Then we may define a *mean heat-transfer coefficient*, h_m, for the entire surface of the sphere by the relation

$$Q = h_m(4\pi R^2)(T_0 - T_\infty) \tag{14.1-6}$$

The characteristic area is here taken to be the heat-transfer surface (as in Eqs. 14.1-2 to 14.1-5), whereas in Eq. 6.1-5 we used the sphere cross section.

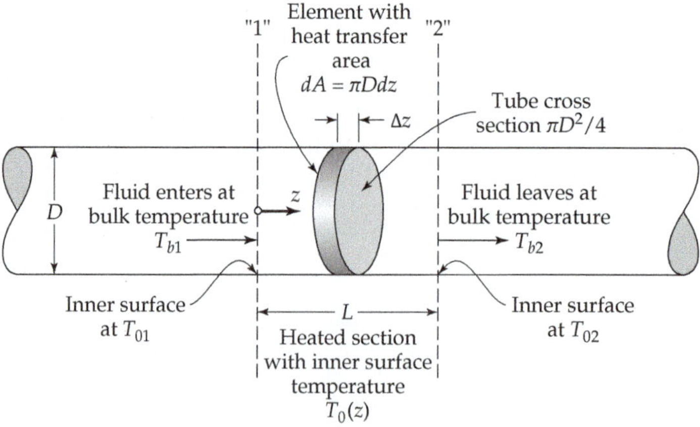

Fig. 14.1-1. Heat transfer in a circular tube.

[1] If $\Delta T_2/\Delta T_1$ is between 0.5 and 2.0, then ΔT_a may be substituted for $\Delta T_{\ln}$, and h_a for $h_{\ln}$, with a maximum error of 4%. This degree of accuracy is acceptable in most heat-transfer calculations.

A local coefficient can also be defined for submerged objects by analogy with Eq. 14.1-5:

$$dQ = h_{loc}(dA)(T_0 - T_\infty) \qquad (14.1\text{-}7)$$

This coefficient is more informative than h_m because it predicts how the heat flux is distributed over the surface. However, most experimentalists report only h_m, which is easier to measure.

Let us emphasize that the definitions of A and ΔT must be made clear before h is defined. Keep in mind, also, that h is not a constant characteristic of the fluid medium. On the contrary, the heat-transfer coefficient depends in a complicated way on many variables, including the fluid properties ($k, \mu, \rho, \hat{C}_p$), the system geometry, and the flow velocity. The remainder of this chapter is devoted to describing the dependence of h on these quantities. Usually this is done by using experimental data and dimensional analysis to develop correlations. It is also possible, for some very simple systems, to calculate the heat-transfer coefficient directly from the equations of change. Some sample values of h are given in Table 14.1-1.

We saw in §10.3 that, in the calculation of heat-transfer rates between two fluid streams separated by one or more solid layers, it is convenient to use an *overall heat-transfer coefficient*, U_0, which expresses the combined effect of the series of resistances through which the heat flows. We give here a definition of U_0 and show how to calculate it in the special case of heat exchange between two coaxial streams with bulk temperatures T_h ("hot") and T_c ("cold"), separated by a cylindrical tube of inside diameter D_0 and outside diameter D_1:

$$dQ = U_0(\pi D_0 dz)(T_h - T_c) \qquad (14.1\text{-}8)$$

$$\frac{1}{D_0 U_0} = \left(\frac{1}{D_0 h_0} + \frac{\ln(D_1/D_0)}{2k_{01}} + \frac{1}{D_1 h_1}\right)_{loc} \qquad (14.1\text{-}9)$$

Note that U_0 is defined as a local coefficient. This is the definition implied in most design procedures (see Example 15.4-1).

Equations 14.1-8 and 14.1-9 are, of course, restricted to thermal resistances connected in *series*. In some situations there may be an appreciable *parallel* heat flux at one or both surfaces by radiation, and Eqs. 14.1-8 and 14.1-9 will require special modification (see Example 16.5-2).

Table 14.1-1. Typical Orders of Magnitude for Heat-Transfer Coefficients.[a]

System	h (W/m$^2 \cdot$ K) or (kcal/m$^2 \cdot$ hr $\cdot$ °C)	h (Btu/ft$^2 \cdot$ hr $\cdot$ °F)
Free convection		
Gases	3–20	1–4
Liquids	100–600	20–120
Boiling water	1000–20,000	200–4000
Forced convection		
Gases	10–100	2–20
Liquids	50–500	10–100
Water	500–10,000	100–2000
Condensing vapors	1000–100,000	200–20,000

[a] Taken from H. Gröber, S. Erk, and U. Grigull, *Wärmeübertragung*, Springer, Berlin, 3rd edition (1955), p. 158. When given h in kcal/m$^2 \cdot$ hr $\cdot$ °C, multiply by 0.204 to get h in Btu/ft$^2 \cdot$ hr $\cdot$ °F, and by 1.162 to get h in W/m$^2 \cdot$ K. For additional conversion factors, see Appendix E.

To illustrate the physical significance of heat-transfer coefficients and illustrate one method of measuring them, we conclude this section with an analysis of a hypothetical set of heat-transfer data.

EXAMPLE 14.1-1

Calculation of Heat-Transfer Coefficients from Experimental Data

A series of simulated steady-state experiments on the heating of air in tubes is shown in Fig. 14.1-2. In the first experiment, air at $T_{b1} = 200.0°F$ is flowing in a 0.5-in. i.d. tube with fully developed laminar velocity profile in the isothermal pipe section for $z < 0$. At $z = 0$ the wall temperature is suddenly increased to $T_0 = 212.0°F$ and maintained at that value for the remaining tube length L_A. At $z = L_A$ the fluid flows into a mixing chamber in which the cup-mixing (or "bulk") temperature T_{b2} is measured. Similar experiments are done with tubes of different lengths, L_B, L_C, etc., with the following results:

Experiment	A	B	C	D	E	F	G
L (in.)	1.5	3.0	6.0	12.0	24.0	48.0	96.0
T_{b2}(°F)	201.4	202.2	203.1	204.6	206.6	209.0	211.0

In all experiments, the air flow rate w is 3.0 lb$_m$/hr. Calculate $h_1, h_a, h_{\ln}$, and the exit value of h_{loc} as functions of the L/D ratio.

SOLUTION

Our strategy is to use an energy balance to relate the heat-transfer rate Q to the inlet and outlet conditions. Once Q is known, then the various heat-transfer coefficients can be obtained via Eqs. 14.1-2 through 14.1-5. First we make a steady-state energy balance over a length L of the tube, by stating that the heat in through the walls plus the energy entering at $z = 0$ by convection equals the energy leaving the tube at $z = L$. The axial energy flux at the tube entry and exit may be calculated from Eq. 9.4-2. If it is assumed that the flow is fully developed throughout, then the kinetic energy flux $\frac{1}{2}\rho v^2\mathbf{v}$ and the work term $[\boldsymbol{\tau} \cdot \mathbf{v}]$ are the same at the inlet and outlet. We also assume that $q_z << \rho\hat{H}v_z$, so that the axial heat-conduction term may be neglected. Hence, the only contribution to the energy flux entering and leaving with the flow will be the term containing the enthalpy, which can be computed with the help of Eq. 9.4-6 and the assumptions that the heat capacity and density of the fluid are constant throughout. Therefore, the steady-state energy balance becomes simply "rate of energy flow in = rate of energy flow out," or

$$Q + w\hat{C}_pT_{b1} = w\hat{C}_pT_{b2} \tag{14.1-10}$$

The heat-transfer rate Q can be expressed in terms of the heat-transfer coefficient h_1 using Eq. 14.1-2. After rearranging, this gives

$$w\hat{C}_p(T_{b2} - T_{b1}) = h_1(\pi DL)(T_0 - T_{b1}) \tag{14.1-11}$$

from which

$$h_1 = \frac{w\hat{C}_p}{\pi D^2}\frac{(T_{b2} - T_{b1})}{(T_0 - T_{b1})}\left(\frac{D}{L}\right) \tag{14.1-12}$$

This gives us the formula for calculating h_1 from the data given above.

Analogously, use of Eqs. 14.1-3 and 14.1-4 gives

$$h_a = \frac{w\hat{C}_p}{\pi D^2}\frac{(T_{b2} - T_{b1})}{(T_0 - T_b)_a}\left(\frac{D}{L}\right) \tag{14.1-13}$$

$$h_{\ln} = \frac{w\hat{C}_p}{\pi D^2}\frac{(T_{b2} - T_{b1})}{(T_0 - T_b)_{\ln}}\left(\frac{D}{L}\right) \tag{14.1-14}$$

for obtaining h_a and $h_{\ln}$ from the data.

To evaluate h_{loc}, we have to use the preceding data to construct a continuous curve $T_b(z)$, as in Fig. 14.1-2, to represent the change in bulk temperature with z in the longest (96-in.) tube. Then Eq. 14.1-10 becomes

$$Q(z) + w\hat{C}_p T_{b1} = w\hat{C}_p T_b(z) \tag{14.1-15}$$

By differentiating this expression with respect to z and combining the result with Eq. 14.1-5, we get

$$w\hat{C}_p \frac{dT_b}{dz} = h_{loc}\pi D(T_0 - T_b) \tag{14.1-16}$$

or

$$h_{loc} = \frac{w\hat{C}_p}{\pi D} \frac{1}{(T_0 - T_b)} \frac{dT_b}{dz} \tag{14.1-17}$$

Since T_0 is constant, this becomes

$$h_{loc} = -\frac{w\hat{C}_p}{\pi D^2} \frac{d\ln(T_0 - T_b)}{d(z/L)} \left(\frac{D}{L}\right) \tag{14.1-18}$$

The derivative in this equation is conveniently determined as a function of z/L from the slope of a plot of $\ln(T_0 - T_b)$ versus z/L at various positions z/L. Because numerical differentiation of data is involved, it is difficult to determine h_{loc} precisely.

The calculated results are shown in Fig. 14.1-3. Note that all of the coefficients decrease with increasing L/D, but that h_{loc} and h_{ln} vary less than the others and approach a common asymptote (see Problem 14B.5 and Fig. 14.1-3). Somewhat similar behavior is observed in turbulent flow with constant wall temperature, except that h_{loc} approaches the asymptote much more rapidly (see Fig. 14.3-2).

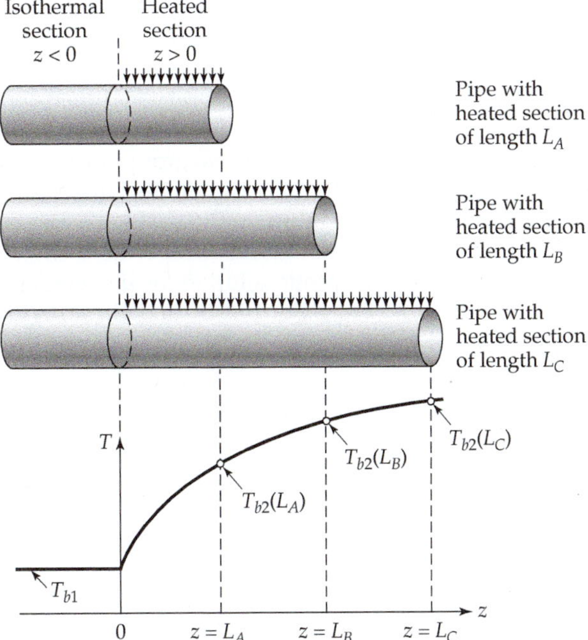

Fig. 14.1-2. Series of experiments for measuring heat-transfer coefficients.

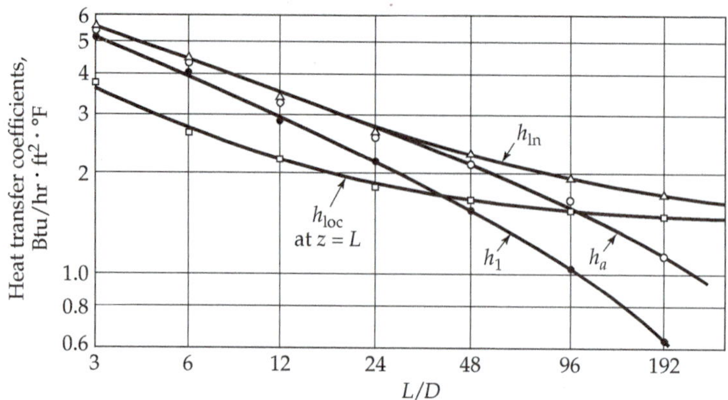

Fig. 14.1-3. Heat-transfer coefficients calculated in Example 14.1-1.

---

§14.2 HEAT-TRANSFER COEFFICIENTS FOR FORCED CONVECTION THROUGH TUBES AND SLITS OBTAINED FROM SOLUTIONS OF THE EQUATIONS OF CHANGE

The main value of heat-transfer coefficients is that they relate heat-transfer rates to temperature differences (as in Eqs. 14.1-2 through 14.1-7), without requiring knowledge of the entire temperature profile. Thus, the challenges of solving for temperature profiles can often be avoided. However, it is instructive to use the knowledge of temperature profiles that have been determined for idealized problems to predict heat-transfer coefficients, and to begin to understand the dependence of heat-transfer coefficients on other variables.

Recall from Chapter 6, where we defined and discussed friction factors, that for some very simple laminar flow systems we could obtain analytical formulas for the (dimensionless) friction factor as a function of the (dimensionless) Reynolds number. We would like to do the same for the heat-transfer coefficient, h, which, however, is not dimensionless. Nonetheless we can construct with it a dimensionless quantity, $\text{Nu} = hD/k$, the *Nusselt number*, using the fluid thermal conductivity k and a characteristic length D, which must be specified for each flow system. Two other related dimensionless groups are commonly used: the *Stanton number*, $\text{St} = \text{Nu}/\text{RePr}$, and the *Chilton-Colburn j-factor* for heat transfer, $j_H = \text{Nu}/\text{RePr}^{1/3}$. Each of these dimensionless groups may be "decorated" with a subscript 1, a, ln, or m, corresponding to the subscript on the Nusselt number.

By way of illustration, let us return to §10.9 where we discussed the heating of a fluid in laminar flow in a tube, with all the fluid properties being considered constant. Specifically considered was the asymptotic solution for the temperature profile valid for large distances z down the tube. From Eq. 10.9-35 and Eq. 10.9-33 we can get the difference between the wall temperature $T_0(z) = T(r,z)|_{r=R}$ and the bulk temperature:

$$T_0(z) - T_b(z) = \left(4\zeta + \frac{11}{24}\right)\left(\frac{q_0 R}{k}\right) - 4\zeta\left(\frac{q_0 R}{k}\right)$$

$$= \frac{11}{24}\left(\frac{q_0 R}{k}\right) = \frac{11}{48}\left(\frac{q_0 D}{k}\right) \tag{14.2-1}$$

in which R and D are the radius and diameter of the tube. Note that even though T_0 and T_b are both functions of z, their difference is independent of z. Solving for the wall heat flux we get

$$q_0 = \frac{48}{11}\left(\frac{k}{D}\right)(T_0 - T_b) \tag{14.2-2}$$

Then making use of the definition of the local heat-transfer coefficient h_{loc} in Eq. 14.1-5—namely, that $q_0 = h_{loc}(T_0 - T_b)$—we find that

$$h_{loc} = \frac{48}{11}\left(\frac{k}{D}\right) \quad \text{or} \quad \text{Nu}_{loc} = \frac{h_{loc}D}{k} = \frac{48}{11} \tag{14.2-3}$$

This result illustrates that for these conditions—namely, laminar tube flow with constant physical properties and a constant heat flux at the wall—the local Nusselt number and thus the local heat-transfer coefficient approach constants at large distances. One can show that the log-mean Nusselt number $\text{Nu}_{ln} = h_{ln}D/k$ also approaches a constant at large distances for these conditions (see Problem 14B.5).

Next consider laminar flow with constant physical properties and a constant wall heat flux for small values of z (the "entrance region"). This situation was addressed in Example 11.5-3. Near the entry of the tube, the bulk fluid temperature is approximately the entrance temperature T_1, and the wall temperature is given by $T_0(z) = T(y,z)|_{y=0}$. In terms of dimensionless variables used in Example 11.5-3, the wall temperature is expressed as $T_0(z) = T_1 + (q_0R/k)\Theta(\eta = 0, \lambda = \alpha z/v_0R^2) = T_1 + (q_0R/k)(9\alpha z/v_0R^2)^{1/3}/\Gamma\left(\frac{2}{3}\right)$, where $\Theta(\eta,\lambda) \equiv (T - T_1)/(q_0R/k)$ is the dimensionless temperature profile given by Eq. 11.5-62. The local Nusselt number is thus

$$\begin{aligned}\text{Nu}_{loc} = \frac{h_{loc}D}{k} &= \frac{q_0D/k}{(T_0(z) - T_1)} \\ &= \frac{2\Gamma\left(\frac{2}{3}\right)}{9^{1/3}}\left(\frac{\langle v_z\rangle D^2}{\alpha z}\right)^{1/3} \\ &= C\cdot\left(\text{RePr}\frac{D}{z}\right)^{1/3}\end{aligned} \tag{14.2-4}$$

where we have used $\alpha = k/\rho\hat{C}_p$, T_0, and $C = 2\Gamma\left(\frac{2}{3}\right)/9^{1/3} \approx 1.302$ (the gamma function $\Gamma(n)$ is given in App. C). Here we see that the Nusselt number is quite large for small values of z, and then decreases as z increases. This occurs because the wall flux is constant, and the temperature difference $T_0(z) - T_1$ is small near the entrance and increases with increasing z. Equation 14.2-4 is valid for small values of z, given by $\alpha z/\langle v_z\rangle D^2 \ll 1$.

The above results are illustrated in Fig. 14.2-1 where Nu_{loc} is plotted as a function of $\alpha z/\langle v_z\rangle D^2 = (\text{RePr}D/z)^{-1}$. Also shown are analogous results for laminar flow in a tube with constant wall temperature, as well as the constant wall flux and constant wall temperature results for flow in a slit of thickness $2B$. The asymptotic results for small and large z for both flow in a tube and flow in a slit are also summarized in Table 14.2-1.[1] Keep in mind that these results are valid for laminar flow of Newtonian fluids with constant physical properties.

For *turbulent flow* in a circular tube with constant heat flux, the Nusselt number can be obtained from Eq. 12.4-21:[2]

$$\text{Nu}_{loc} = \frac{\text{RePr}\sqrt{f/2}}{12.48\text{Pr}^{2/3} - 7.853\text{Pr}^{1/3} + 3.613\ \ln\ \text{Pr} + 5.8 + 2.78\ \ln\left(\frac{1}{45}\text{Re}\sqrt{f/8}\right)} \tag{14.2-5}$$

[1] This table is adapted from R. B. Bird, R. C. Armstrong, and O. Hassager, *Dynamics of Polymeric Liquids, Vol. 1, Fluid Mechanics*, 2nd edition, Wiley, New York (1987), pp. 212–213. They are based, in turn, on W. J. Beek and R. Eggink, *De Ingenieur*, **74**, (35) Ch. 81–Ch. 89 (1962) and J. M. Valstar and W. J. Beek, *De Ingenieur*, **75**, (1), Ch. 1–Ch. 7 (1963).

[2] O. C. Sandall, O. T. Hanna, and P. R. Mazet, *Canad. J. Chem. Eng.*, **58**, 443–447 (1980).

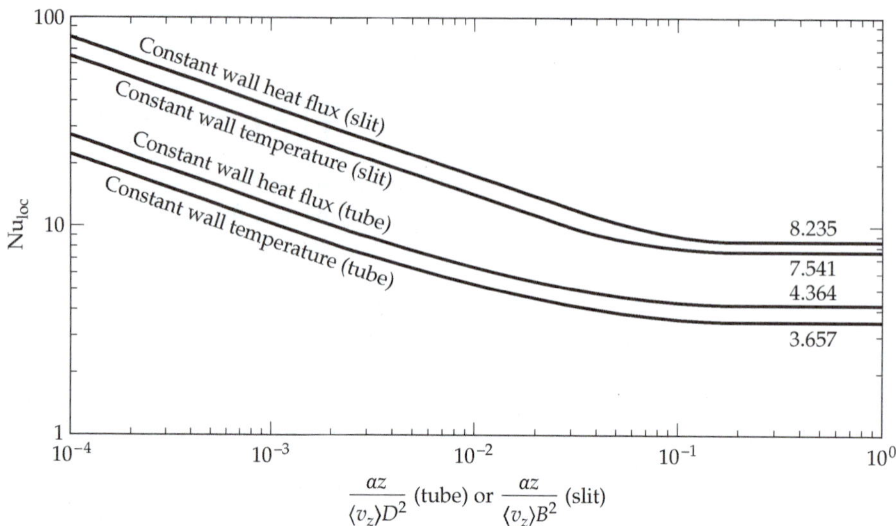

Fig. 14.2-1. The Nusselt number for fully developed, laminar flow of Newtonian fluids with constant physical properties: $\mathrm{Nu_{loc}} = h_{loc}D/k$ for circular tubes of diameter D, and $\mathrm{Nu_{loc}} = 4h_{loc}B/k$ for slits of half-width B. The limiting expressions are given in Table 14.2-1.

This is valid only for $az/\langle v_z\rangle D^2 \gg 1$, for fluids with constant physical properties, and for tubes with no roughness. It has been applied successfully over the Prandtl-number range $0.7 < \mathrm{Pr} < 590$. Note that, for very large Prandtl numbers, Eq. 14.2-5 gives

$$\mathrm{Nu_{loc}} \approx \frac{\mathrm{RePr}\sqrt{f/2}}{12.48\,\mathrm{Pr}^{2/3}} = 0.0566\,\mathrm{Re}\,\mathrm{Pr}^{1/3}\sqrt{f} \tag{14.2-6}$$

The $\mathrm{Pr}^{1/3}$ dependence is in exact agreement with Eq. 12.3-7. For turbulent flow, there is little difference between Nu for constant wall temperature and for constant wall heat flux.

For the turbulent flow of *liquid metals,* for which the Prandtl numbers are generally much less than unity, there are two results of importance. Notter and Sleicher[3] solved the energy equation numerically, using a realistic turbulent velocity profile, and obtained the rates of heat transfer through the wall. The final results were curve-fitted to simple analytical expressions for two cases:

Constant wall temperature: $$\mathrm{Nu_{loc}} = 4.8 + 0.0156\,\mathrm{Re}^{0.85}\mathrm{Pr}^{0.93} \tag{14.2-7}$$

Constant wall heat flux: $$\mathrm{Nu_{loc}} = 6.3 + 0.0167\,\mathrm{Re}^{0.85}\mathrm{Pr}^{0.93} \tag{14.2-8}$$

These equations are limited to $L/D > 60$ and constant physical properties. Equation 14.2-7 is displayed in Fig. 14.2-2.

It has been emphasized that all of the results of this section are limited to fluids with constant physical properties. When there are large temperature differences in the system, it is necessary to take into account the temperature dependence of the viscosity, density, heat capacity, and thermal conductivity. Usually this is done by means of an empiricism, namely by evaluating the physical properties at some appropriate average temperature.

---

[3] R. H. Notter and C. A. Sleicher, *Chem. Eng. Sci.*, **27**, 2073–2093 (1972).

Table 14.2-1. Asymptotic Results for Local Nusselt Numbers[a] in Tube Flow[b] ($\text{Nu}_{\text{loc}} = h_{\text{loc}}D/k$) and in Slit Flow[c] ($\text{Nu}_{\text{loc}} = 4h_{\text{loc}}B/k$)

Tube flow	Constant wall temperature 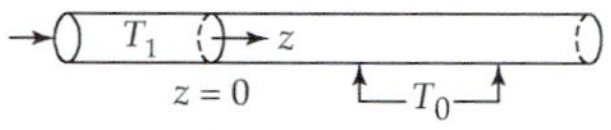			Constant wall heat flux 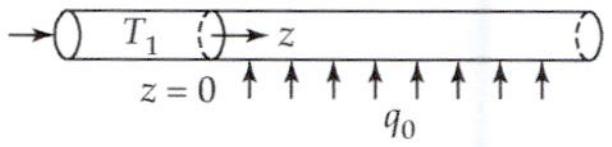		
Thermal entrance region[d] $\dfrac{\langle v_z\rangle D^2}{\alpha z} \gg 1$	Plug flow	$\text{Nu}_{\text{loc}} = \dfrac{1}{\sqrt{\pi}}\left(\dfrac{\langle v_z\rangle D^2}{\alpha z}\right)^{1/2}$	(A)	Plug flow	$\text{Nu}_{\text{loc}} = \dfrac{\sqrt{\pi}}{2}\left(\dfrac{\langle v_z\rangle D^2}{\alpha z}\right)^{1/2}$	(E)
	Laminar Newtonian flow	$\text{Nu}_{\text{loc}} = \dfrac{2}{9^{1/3}\Gamma(\frac{4}{3})}\left(\dfrac{\langle v_z\rangle D^2}{\alpha z}\right)^{1/3}$	(B)	Laminar Newtonian flow	$\text{Nu}_{\text{loc}} = \dfrac{2\Gamma(\frac{2}{3})}{9^{1/3}}\left(\dfrac{\langle v_z\rangle D^2}{\alpha z}\right)^{1/3}$	(F)
Thermally fully developed flow $\dfrac{\langle v_z\rangle D^2}{\alpha z} \ll 1$	Plug flow	$\text{Nu}_{\text{loc}} = 5.772$	(C)	Plug flow	$\text{Nu}_{\text{loc}} = 8$	(G)
	Laminar Newtonian flow	$\text{Nu}_{\text{loc}} = 3.657$	(D)	Laminar Newtonian flow	$\text{Nu}_{\text{loc}} = \dfrac{48}{11} = 4.364$	(H)
Slit flow	Constant wall temperature 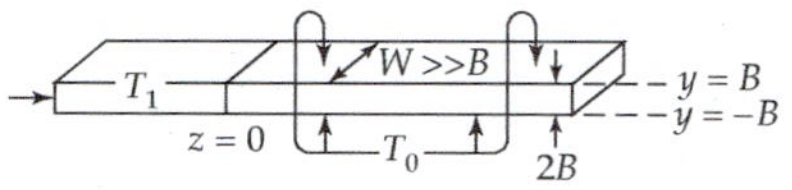			Constant wall heat flux 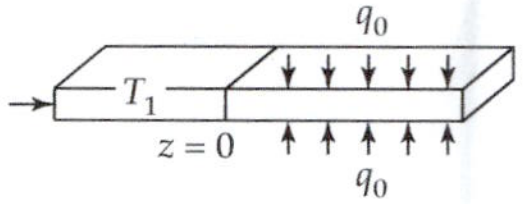		
Thermal entrance region[d] $\dfrac{\langle v_z\rangle B^2}{\alpha z} \gg 1$	Plug flow	$\text{Nu}_{\text{loc}} = \dfrac{4}{\sqrt{\pi}}\left(\dfrac{\langle v_z\rangle B^2}{\alpha z}\right)^{1/2}$	(I)	Plug flow	$\text{Nu}_{\text{loc}} = 2\sqrt{\pi}\left(\dfrac{\langle v_z\rangle B^2}{\alpha z}\right)^{1/2}$	(M)
	Laminar Newtonian flow	$\text{Nu}_{\text{loc}} = \dfrac{4}{3^{1/3}\Gamma(\frac{4}{3})}\left(\dfrac{\langle v_z\rangle B^2}{\alpha z}\right)^{1/3}$	(J)	Laminar Newtonian flow	$\text{Nu}_{\text{loc}} = \dfrac{4\Gamma(\frac{2}{3})}{3^{1/3}}\left(\dfrac{\langle v_z\rangle B^2}{\alpha z}\right)^{1/3}$	(N)
Thermally fully developed flow $\dfrac{\langle v_z\rangle B^2}{\alpha z} \ll 1$	Plug flow	$\text{Nu}_{\text{loc}} = \pi^2 = 9.870$	(K)	Plug flow	$\text{Nu}_{\text{loc}} = 12$	(O)
	Laminar Newtonian flow	$\text{Nu}_{\text{loc}} = 7.541$	(L)	Laminar Newtonian flow	$\text{Nu}_{\text{loc}} = \dfrac{140}{17} = 8.235$	(P)

[a]Note: for tube flow $\langle v_z\rangle D^2/\alpha z = \text{RePr}(D/z)$, with $\text{Re} = D\langle v_z\rangle\rho/\mu$; for slit flow $\langle v_z\rangle B^2/\alpha z = 4\text{RePr}(B/z)$, with $\text{Re} = 4B\langle v_z\rangle\rho/\mu$. Here $\alpha = k/\rho\hat{C}_p$.

[b]W.J. Beek and R. Eggink, *De Ingenieur*, **74**, No. 35, Ch. 81-89 (1962); erratum, 75, No. 1, Ch. 7 (1963).

[c]J.M. Valstar and W.J. Beek, *De Ingenieur*, **75**, No. 1, Ch. 1-7 (1963).

[d]The groupings $\langle v_z\rangle D^2/\alpha z$ and $\langle v_z\rangle B^2/\alpha z$ are sometimes written as $\text{Gz}\cdot(L/z)$, where Gz is the Graetz number. For tube flow, $\text{Gz} = \langle v_z\rangle D^2/\alpha L$, and for slit flow, $\text{Gz} = \langle v_z\rangle B^2/\alpha L$.

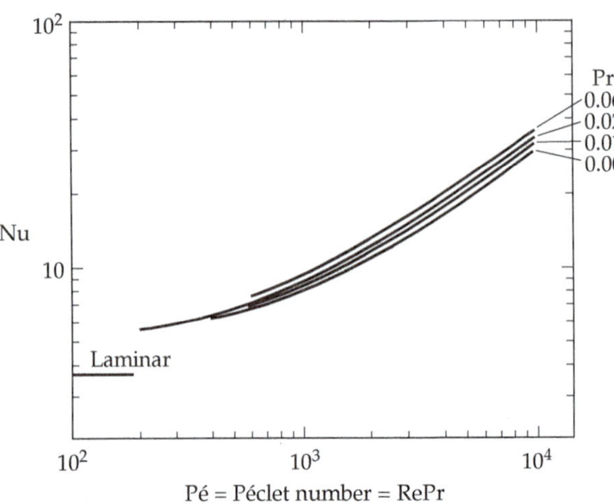

Fig. 14.2-2. Nusselt number for turbulent flow of liquid metals in circular tubes, based on the theoretical calculations of R. H. Notter and C. A. Sleicher, *Chem. Eng. Sci.*, **27**, 2073–2093 (1972).

Throughout this chapter, unless explicitly stated otherwise, it is understood that all physical properties are to be calculated at the film temperature T_f defined as follows:[4]

a. For *tubes, slits, and other ducts*

$$T_f = \frac{1}{2}(T_{0a} + T_{ba}) \tag{14.2-9}$$

in which T_{0a} is the arithmetic average of the surface temperatures at the two ends, $T_{0a} = \frac{1}{2}(T_{01} + T_{02})$, and T_{ba} is the arithmetic average of the inlet and outlet bulk temperatures, $T_{ba} = \frac{1}{2}(T_{b1} + T_{b2})$. It is also recommended that the Reynolds number be written as $\text{Re} = D\langle \rho v \rangle/\mu = Dw/S\mu$, to account for viscosity, velocity, and density changes over the cross section of area S.

b. For *submerged objects* with uniform surface temperature T_0 in a stream of liquid approaching with uniform temperature T_∞

$$T_f = \frac{1}{2}(T_0 + T_\infty) \tag{14.2-10}$$

For flow systems involving more complicated geometries, it is preferable to use experimental correlations of the heat-transfer coefficients. In the following sections we show how such correlations can be established by a combination of dimensional analysis and experimental data.

§14.3 EMPIRICAL CORRELATIONS FOR HEAT-TRANSFER COEFFICIENTS FOR FORCED CONVECTION IN TUBES

In the previous section we have shown that the Nusselt numbers for circular tubes can in some special cases be computed from first principles. In this section we show how dimensional analysis leads us to a general form for the dependence of the Nusselt number on various dimensionless groups, and that this form includes not only laminar flows, but turbulent flows as well. Then we present a dimensionless plot of Nusselt numbers that was

[4]W. J. M. Douglas and S. W. Churchill, *Chem. Eng. Prog. Symposium Series*, No. 18, **52**, 23–28 (1956); E. R. G. Eckert, *Recent Advances in Heat and Mass Transfer*, McGraw-Hill, New York (1961), pp. 51–81, Eq. (20); more detailed reference states have been proposed by W. E. Stewart, R. Kilgour, and K.-T. Liu, University of Wisconsin-Madison Mathematics Research Center Report #1310 (June 1973).

obtained by correlating experimental data, as well as various empirical equations valid for specific ranges of parameters.

First we extend the dimensional analysis given in §13.1 to obtain a general form for correlations of heat-transfer coefficients in forced convection. Consider the steadily driven laminar or turbulent flow of a Newtonian fluid through a straight tube of inner radius R, as shown in Fig. 14.3-1. The fluid enters the tube at $z = 0$ with velocity uniform out to very near the wall, and with a uniform inlet temperature $T_1(= T_{b1})$. The tube wall is insulated except in the region $0 < z \leq L$, where a uniform inner-surface temperature T_0 is maintained, for example, by heat from vapor condensing on the outer surface. For the moment, we assume constant physical properties ρ, μ, k, and $\hat{C}_p$. Later we will extend the empiricism given in §14.2 to provide a fuller allowance for the temperature dependence of these properties.

We follow the same procedure used in §6.2 for friction factors. We start by writing the expression for the instantaneous heat flow from the tube wall into the fluid in the system described above

$$Q(t) = \int_0^L \int_0^{2\pi} \left(+k\frac{\partial T}{\partial r}\right)\bigg|_{r=R} R\, d\theta\, dz \tag{14.3-1}$$

which is valid for laminar or turbulent flow (in laminar flow, Q would, of course, be independent of time). The + sign appears here because the heat is added to the system in the negative r direction.

Equating the expressions for Q given in Eqs. 14.1-2 and 14.3-1 and solving for h_1, we get

$$h_1(t) = \frac{1}{\pi DL(T_0 - T_{b1})} \int_0^L \int_0^{2\pi} \left(+k\frac{\partial T}{\partial r}\right)\bigg|_{r=R} R\, d\theta\, dz \tag{14.3-2}$$

Introducing dimensionless quantities $\breve{r} = r/D$, $\breve{z} = z/D$, and $\breve{T} = (T - T_0)/(T_{b1} - T_0)$, and multiplying by D/k, we get an expression for the Nusselt number $\text{Nu}_1 = h_1 D/k$:

$$\text{Nu}_1(t) = \frac{1}{2\pi L/D} \int_0^{L/D} \int_0^{2\pi} \left(-\frac{\partial \breve{T}}{\partial \breve{r}}\right)\bigg|_{\breve{r}=1/2} d\theta\, d\breve{z} \tag{14.3-3}$$

Thus, the (instantaneous) Nusselt number is basically a *dimensionless temperature gradient averaged over the heat-transfer surface.*

The dimensionless temperature gradient appearing in Eq. 14.3-3 could, in principle, be evaluated by differentiating the expression for $\breve{T}$ obtained by solving Eqs. 13.1-10, 13.1-11,

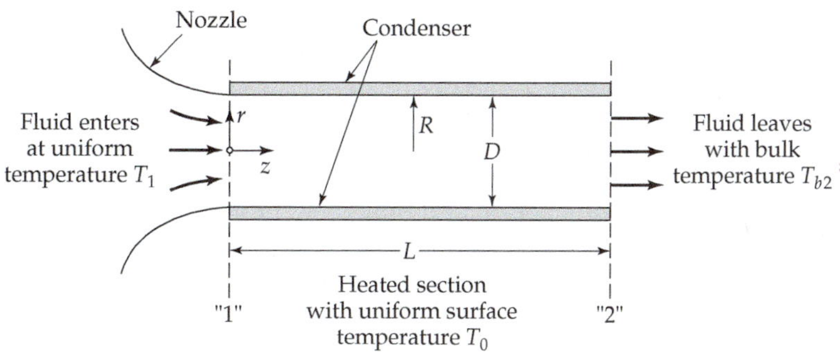

Fig. 14.3-1. Heat transfer in the entrance region of a tube.

and 13.1-12 with the boundary conditions

$$\text{at } \breve{z} = 0, \qquad \breve{\mathbf{v}} = \boldsymbol{\delta}_z \qquad \text{for } 0 \leq \breve{r} \leq \frac{1}{2} \tag{14.3-4}$$

$$\text{at } \breve{r} = \frac{1}{2}, \qquad \breve{\mathbf{v}} = \mathbf{0} \qquad \text{for } \breve{z} \geq 0 \tag{14.3-5}$$

$$\text{at } \breve{r} = 0 \text{ and } \breve{z} = 0, \qquad \breve{\mathscr{P}} = 0 \tag{14.3-6}$$

$$\text{at } \breve{z} = 0, \qquad \breve{T} = 1 \qquad \text{for } 0 \leq \breve{r} \leq \frac{1}{2} \tag{14.3-7}$$

$$\text{at } \breve{r} = \frac{1}{2} \qquad \breve{T} = 0 \qquad \text{for } 0 \leq \breve{z} \leq L/D \tag{14.3-8}$$

where $\breve{\mathbf{v}} = \mathbf{v}/\langle v_z \rangle_1$ and $\breve{\mathscr{P}} = (\mathscr{P} - \mathscr{P}_1)/\rho\langle v_z \rangle_1^2$. As in §6.2, we have neglected the $\partial^2/\partial\breve{z}^2$ terms (i.e., because of (1) incompressibility, and (2) negligible axial heat conduction). With those terms suppressed, upstream transport of heat and momentum are excluded, so that the solutions upstream of plane 2 in Fig. 14.3-1 do not depend on L/D.

From Eqs. 13.1-10, 13.1-11, and 13.1-12 and these boundary conditions, we conclude that the dimensionless instantaneous temperature distribution must be of the following form:

$$\breve{T}(t) \quad = \quad \breve{T}(\breve{r},\theta,\breve{z},\breve{t};\,\text{Re},\text{Pr},\text{Br}) \quad \text{for} \quad 0 \leq \breve{z} \leq L/D \tag{14.3-9}$$

Substitution of this relation into Eq. 14.3-3 leads to the conclusion that $\text{Nu}_1(\breve{t}) = \text{Nu}_1(\text{Re},\text{Pr},\text{Br},L/D,\breve{t})$. When time averaged over an interval long enough to include all the turbulent disturbances, this becomes

$$\text{Nu}_1 = \text{Nu}_1(\text{Re},\text{Pr},\text{Br},L/D) \tag{14.3-10}$$

If, as is often the case, the viscous dissipation heating is small, then the Brinkman number can be omitted. Then Eq. 14.3-10 simplifies to

$$\text{Nu}_1 = \text{Nu}_1(\text{Re},\text{Pr},L/D) \tag{14.3-11}$$

Therefore, dimensional analysis tells us that, for forced-convection heat transfer in circular tubes with constant wall temperature, experimental values of the heat-transfer coefficient h_1 can be correlated by giving Nu_1 as a function of the Reynolds number, the Prandtl number, and the geometric ratio L/D. This should be compared with the similar, but simpler, situation with the friction factor (Eqs. 6.2-9 and 6.2-10).

The same reasoning leads us to similar expressions for the other heat-transfer coefficients we have defined. It can be shown (see Problem 14.B-4) that:

$$\text{Nu}_a = \text{Nu}_a(\text{Re},\text{Pr},L/D) \tag{14.3-12}$$

$$\text{Nu}_{\text{ln}} = \text{Nu}_{\text{ln}}(\text{Re},\text{Pr},L/D) \tag{14.3-13}$$

$$\text{Nu}_{\text{loc}} = \text{Nu}_{\text{loc}}(\text{Re},\text{Pr},L/D) \tag{14.3-14}$$

in which $\text{Nu}_a = h_a D/k$, $\text{Nu}_{\text{ln}} = h_{\text{ln}} D/k$, and $\text{Nu}_{\text{loc}} = h_{\text{loc}} D/k$. That is, to each of the heat-transfer coefficients, there is a corresponding Nusselt number. These Nusselt numbers are, of course, interrelated (see Problem 14.B-5). These general functional forms for the Nusselt numbers have a firm scientific basis, since they involve only the dimensional analysis of the equations of change and boundary conditions.

Thus far we have assumed that the physical properties are constants over the temperature range encountered in the flow system. At the end of §14.2 we indicated that evaluating the physical properties at the film temperature is a suitable empiricism. However, for very large temperature differences, the viscosity variations may result in such a large distortion of the velocity profiles that it is necessary to account for this by introducing an additional dimensionless group, μ_b/μ_0, where μ_b is the viscosity at the arithmetic average bulk temperature $\frac{1}{2}(T_{b1} + T_{b2})$ and μ_0 is the viscosity at the arithmetic average wall temperature $\frac{1}{2}(T_{01} + T_{02})$.[1] Then we may express the dependence of the Nusselt number as

$$\text{Nu} = \text{Nu}(\text{Re},\text{Pr},L/D,\mu_b/\mu_0) \tag{14.3-15}$$

for each of the different Nusselt numbers Nu_1, Nu_a, Nu_{ln}, and Nu_{loc}. This type of correlation seems to have first been presented by Sieder and Tate.[2] If, in addition, the density varies significantly, then some free convection may occur. This effect can be accounted for in correlations by including the Grashof number along with the other dimensionless groups. Correlations for free-convection heat transfer are discussed in §14.6.

Let us now pause to reflect on the significance of the above discussion for constructing heat-transfer correlations. The heat-transfer coefficient h depends on *eight* physical quantities $(D,\langle v\rangle,\rho,\mu_0,\mu_b,\hat{C}_p,k,L)$. But Eq. 14.3-15 tells us that this dependence can be expressed more concisely by giving Nu as a function of only *four* dimensionless groups $(\text{Re},\text{Pr},L/D,\mu_b/\mu_0)$. Thus, instead of taking data on h for 5 values of each of the eight individual physical quantities ($5^8 \approx 3.9 \times 10^5$ tests), we can measure h for 5 values of the dimensionless groups ($5^4 = 625$ tests)—a rather dramatic saving of time and effort.

A good global view of heat transfer in circular tubes with nearly constant wall temperature can be obtained from the Sieder and Tate[2] correlation given in Fig. 14.3-2. This is of the form of Eq. 14.3-15. It has been found empirically[2,3] that transition to turbulence usually begins at about Re = 2100, even when the viscosity varies appreciably in the radial direction.

For *highly turbulent flow*, the curves for $L/D > 10$ converge to a single curve. For Re > 20,000 this curve is described by the equation

$$\text{Nu}_{\text{ln}} = 0.026\,\text{Re}^{0.8}\text{Pr}^{1/3}\left(\frac{\mu_b}{\mu_0}\right)^{0.14} \tag{14.3-16}$$

This equation reproduces available experimental data within about ±20% in the ranges $10^4 < \text{Re} < 10^5$ and $0.6 < \text{Pr} < 100$.

For *laminar flow*, the descending lines at the left are given by the equation

$$\text{Nu}_{\text{ln}} = 1.86\left(\text{RePr}\frac{D}{L}\right)^{1/3}\left(\frac{\mu_b}{\mu_0}\right)^{0.14} \tag{14.3-17}$$

---

[1] One can arrive at the viscosity ratio by inserting into the equations of change a temperature-dependent viscosity, described, for example, by a Taylor expansion about the wall temperature:

$$\mu = \mu_0 + \left.\frac{\partial \mu}{\partial T}\right|_{T=T_0}(T - T_0) + \cdots \tag{14.3-15a}$$

When the series is truncated and the differential quotient is approximated by a difference quotient, we get

$$\mu \approx \mu_0 + \left(\frac{\mu_b - \mu_0}{T_b - T_0}\right)(T - T_0) \tag{14.3-15b}$$

or, with some rearrangement

$$\frac{\mu}{\mu_0} \approx 1 + \left(\frac{\mu_b}{\mu_0} - 1\right)\left(\frac{T - T_0}{T_b - T_0}\right) \tag{14.3-15c}$$

Thus, the viscosity ratio appears in the equation of motion and hence in the dimensionless correlation.

[2] E. N. Sieder and G. E. Tate, *Ind. Eng. Chem.*, **28**, 1429–1435 (1936).

[3] A. P. Colburn, *Trans. AIChE*, **29**, 174–210 (1933). **Allan Philip Colburn** (1904–1955), Provost at the University of Delaware (1950–1955), made important contributions to the fields of heat and mass transfer, including the "Chilton-Colburn relations."

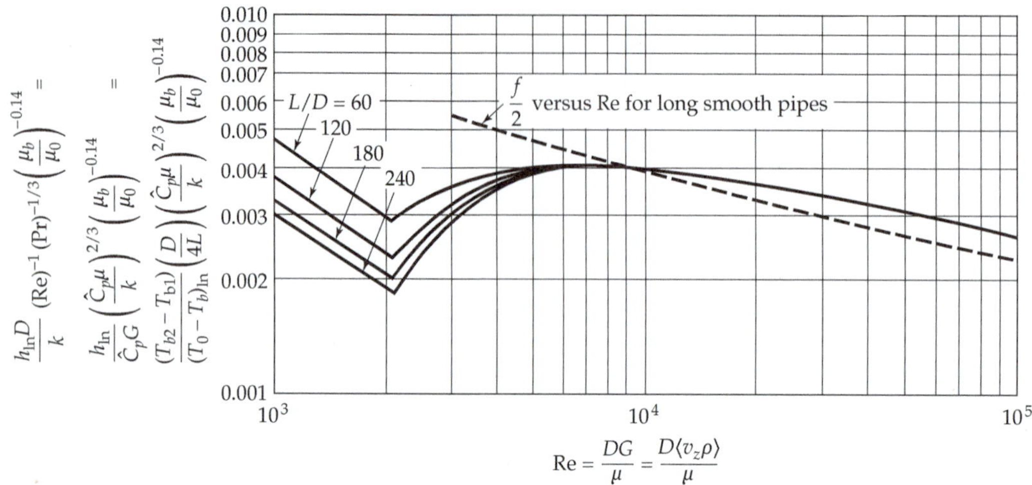

Fig. 14.3-2. Heat-transfer coefficients for fully developed flow in smooth tubes. The lines for laminar flow should not be used in the range $\text{RePr}D/L < 10$, which corresponds to $(T_0 - T_b)_2/(T_0 - T_b)_1 < 0.2$. The laminar curves are based on data for $\text{RePr}D/L \gg 10$ and nearly constant wall temperature; under these conditions h_a and $h_{\ln}$ are indistinguishable. We recommend using $h_{\ln}$, as opposed to the h_a, suggested by Sieder and Tate, because this choice is conservative in the usual heat-exchanger design calculations. [E. N. Sieder and G. E. Tate, *Ind. Eng. Chem.*, **28**, 1429–2435 (1936).]

which is based on Eq. (B) of Table 14.2-1.[4] The numerical coefficient in Eq. (B) has been multiplied by a factor of $\frac{3}{2}$ to convert from h_{loc} to $h_{\ln}$, and then further modified empirically to account for the deviations due to variable physical properties. This illustrates how a satisfactory empirical correlation can be obtained by modifying the result of an analytical derivation. Equation 14.3-17 is good within about 20% for $\text{RePr}D/L > 10$, but at lower values of $\text{RePr}D/L$, it underestimates $h_{\ln}$ considerably.

The *transition region*, roughly $2100 < \text{Re} < 8000$ in Fig. 14.3-2, is not well understood and is usually avoided in design if possible. The curves in this region are supported by experimental measurements[2] but are less reliable than the rest of the plot.

The general characteristics of the curves in Fig. 14.3-2 deserve careful study. Note that for a heated section of given L and D, and a fluid of given physical properties, the ordinate is proportional to the dimensionless temperature rise of the fluid passing through, that is, $(T_{b2} - T_{b1})/(T_0 - T_b)_{\ln}$. Under these conditions, as the flow rate (or Reynolds number) is increased, the exit fluid temperature will first decrease, until Re reaches about 2100, then increase, until Re reaches about 8000, and then finally decrease again. Note also that the influence of L/D on $h_{\ln}$ is marked in laminar flow but becomes insignificant for $\text{Re} > 8000$ with $L/D > 60$.

Note also that Fig. 14.3-2 somewhat resembles the friction-factor plot in Fig. 6.2-2, although the physical situation is quite different. In the highly turbulent range ($\text{Re} > 10{,}000$) the heat-transfer ordinate agrees approximately with $f/2$ for the long smooth pipes under consideration. This was first pointed out by Colburn,[3] who proposed

[4]Equation (B) of Table 14.2-1 is an asymptotic solution of the *Graetz problem*, one of the classic problems of heat convection: L. Graetz, *Ann. d. Physik*, **13**, 79–94 (1883), **25**, 337–357 (1885); see J. Lévêque, *Ann. mines* (Series 12), **13**, 201–299, 305–362, 381–415 (1928) for the asymptote in Eq. (B). An extensive summary can be found in M. A. Ebadian and Z. F. Dong, Chapter 5 of Handbook of Heat Transfer, 3rd edition, (W. M. Rohsenow, J. P. Hartnett, and Y. I. Cho, eds.), McGraw-Hill, New York, (1998).

the following empirical analogy for long, smooth tubes

$$j_{H,\ln} \approx \frac{1}{2}f \qquad (\text{Re} > 10{,}000) \tag{14.3-18}$$

in which

$$j_{H,\ln} = \frac{\text{Nu}_{\ln}}{\text{Re}\,\text{Pr}^{1/3}} = \frac{h_{\ln}}{\langle \rho v\rangle \hat{C}_p}\left(\frac{\hat{C}_p\mu}{k}\right)^{2/3} = \frac{h_{\ln}S}{w\hat{C}_p}\left(\frac{\hat{C}_p\mu}{k}\right)^{2/3} \tag{14.3-19}$$

where S is the area of the tube cross section, w is the mass rate of flow through the tube, and $f/2$ is obtainable from Fig. 6.2-2 using $\text{Re} = Dw/S\mu = 4w/\pi D\mu$. Clearly the analogy of Eq. 14.3-18 is not valid below Re = 10,000. For rough tubes with fully developed turbulent flow the analogy breaks down completely, because f is affected more by roughness than j_H is.

One additional remark about the use of Fig. 14.3-2 has to do with the application to conduits of noncircular cross section. For *highly turbulent* flow, one may use the mean hydraulic radius of Eq. 6.2-15. To apply that empiricism, D is replaced by $4R_h$ everywhere in the Reynolds and Nusselt numbers.

EXAMPLE 14.3-1

Design of a Tubular Heater

Air at 70°F and 1 atm is to be pumped through a straight 2-in. i.d. pipe at a rate of 70 lb_m/hr. A section of the pipe is to be heated to an inside wall temperature of 250°F to raise the air temperature to 230°F. What heated length is required?

SOLUTION

The arithmetic average bulk temperature is $T_{ba} = 150°\text{F}$, and the film temperature is $T_f = \frac{1}{2}(150°\text{F} + 250°\text{F}) = 200°\text{F}$. At this temperature the properties of air are: $\mu = 0.052\ \text{lb}_m/\text{ft}\cdot\text{hr}$, $\hat{C}_p = 0.242\ \text{Btu/lb}_m\cdot°\text{F}$, $k = 0.0180\ \text{Btu/hr}\cdot\text{ft}\cdot°\text{F}$, and $\text{Pr} = \hat{C}_p\mu/k = 0.70$. The viscosities of air at 150°F and 250°F are 0.049 and 0.055 $\text{lb}_m/\text{ft}\cdot\text{hr}$, respectively, so that the viscosity ratio is $\mu_b/\mu_0 = 0.049/0.055 = 0.89$.

The Reynolds number, evaluated at the film temperature, 200°F, is then

$$\text{Re} = \frac{Dw}{S\mu} = \frac{4w}{\pi D\mu} = \frac{4(70\ \text{lb}_m/\text{hr})}{\pi(2\ \text{in.})(0.052\ \text{lb}_m/\text{ft}\cdot\text{hr})}\left(\frac{12\ \text{in.}}{1\ \text{ft}}\right) = 1.03\times 10^4 \tag{14.3-20}$$

From Fig. 14.3-2 we obtain

$$\frac{(T_{b2}-T_{b1})}{(T_0-T_b)_{\ln}}\frac{D}{4L}\text{Pr}^{2/3}\left(\frac{\mu_b}{\mu_0}\right)^{-0.14} = 0.0039 \tag{14.3-21}$$

When this is solved for L/D, we get

$$\begin{aligned}\frac{L}{D} &= \frac{1}{4(0.0039)}\frac{(T_{b2}-T_{b1})}{(T_0-T_b)_{\ln}}\text{Pr}^{2/3}\left(\frac{\mu_b}{\mu_0}\right)^{-0.14}\\ &= \frac{1}{4(0.0039)}\frac{(230°\text{F}-70°\text{F})}{72.8°\text{F}}(0.70)^{2/3}(0.89)^{-0.14}\\ &= 113\end{aligned} \tag{14.3-22}$$

Hence, the required length is

$$L = 113D = (113)(2\ \text{in.})\left(\frac{1\ \text{ft}}{12\ \text{in.}}\right) = 19\ \text{ft} \tag{14.3-23}$$

If Re had been much smaller, it would have been necessary to estimate L/D before reading Fig. 14.3-2, thus initiating a trial-and-error process.

Note that in this problem we did not have to calculate h. Numerical evaluation of h is necessary, however, in more complicated problems such as heat exchange between two fluids with an intervening wall.

§14.4 HEAT-TRANSFER COEFFICIENTS FOR FORCED CONVECTION AROUND SUBMERGED OBJECTS

Another topic of industrial and natural importance is the transfer of heat to or from an object around which a fluid is flowing. The object may be relatively simple, such as a single cylinder or sphere, or it may be more complex, such as a "tube bundle" made up of a set of cylindrical tubes with a stream of gas or liquid flowing between them. We examine here only a few selected correlations for simple systems: the flat plate, the sphere, and the cylinder. Many additional correlations may be found in the references cited in the introduction to the chapter.

a. Flow along a flat plate

We first examine the flow along a flat plate, oriented parallel to the flow, with its surface maintained at T_0 and the approaching stream having a uniform temperature T_∞ and a uniform velocity v_∞. The local heat-transfer coefficient $h_{\text{loc}} = q_0/(T_0 - T_\infty)$ and the local friction factor $f_{\text{loc}} = \tau_0/\frac{1}{2}\rho v_\infty^2$ are shown in Fig. 14.4-1. For the laminar region, which normally exists near the leading edge of the plate, the following expressions can be obtained from an approximate boundary analysis (see Problems 3C.2 and 11C.4):

$$\frac{1}{2}f_{\text{loc}} = +\frac{\mu}{\rho v_\infty^2}\left.\frac{\partial v_x}{\partial y}\right|_{y=0} = \frac{3}{2}\sqrt{\frac{13}{280}}\text{Re}_x^{-1/2} = 0.323\text{Re}_x^{-1/2} \tag{14.4-1}$$

$$\text{Nu}_{\text{loc}} = \frac{h_{\text{loc}}x}{k} = \frac{x}{(T_\infty - T_0)}\left.\frac{\partial T}{\partial y}\right|_{y=0} = 2\sqrt{\frac{37}{1260}}\text{Re}_x^{1/2}\text{Pr}^{1/3} = 0.343\text{Re}_x^{1/2}\text{Pr}^{1/3} \tag{14.4-2}$$

More accurate analyses[1] of these laminar-flow problems yield exactly the same numerical coefficient in both expressions above, namely 0.332. If we use this value, then Eq. 14.4-2 gives

$$j_{H,\text{loc}} = \frac{\text{Nu}_{\text{loc}}}{\text{RePr}^{1/3}} = \frac{h_{\text{loc}}}{\rho\hat{C}_p v_\infty}\left(\frac{\hat{C}_p\mu}{k}\right)^{2/3} = 0.332\ \text{Re}_x^{-1/2} \tag{14.4-3}$$

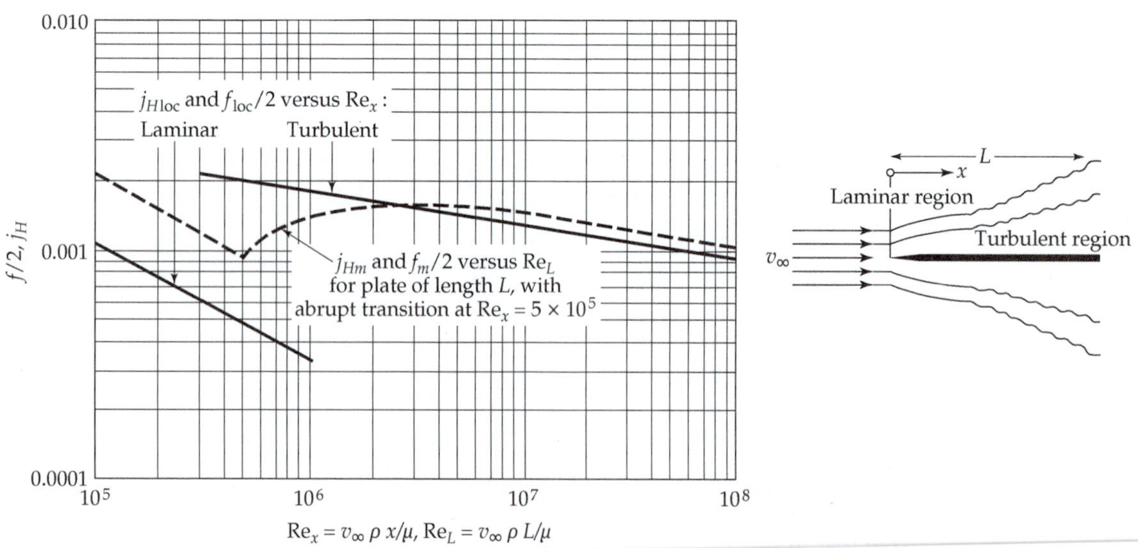

Fig. 14.4-1. Transfer coefficients for a smooth flat plate in tangential flow. [Adapted from H. Schlichting, *Boundary-Layer Theory*, McGraw-Hill, New York (1955), pp. 438–439.]

Since the numerical coefficient in Eq. 14.4-3 is the same as that in the more accurate version of Eq. 14.4-1, we then get

$$j_{H,\mathrm{loc}} = \frac{1}{2}f_{\mathrm{loc}} = 0.332\ \mathrm{Re}_x^{-1/2} \tag{14.4-4}$$

for the Colburn analogy between heat transfer and fluid friction. This was to be expected, because there is no "form drag" in this flow geometry.

Equation 14.4-4 was derived for fluids with constant physical properties.[1] When the physical properties are evaluated at the film temperature $T_f = \frac{1}{2}(T_0 + T_\infty)$, Eq. 14.4-3 is known to work well for gases.[2] The analogy of Eq. 14.4-4 is accurate within 2% for Pr > 0.6, but becomes inaccurate at lower Prandtl numbers.

For highly turbulent flows, the Colburn analogy still holds with fair accuracy, with f_{loc} given by the empirical curve in Fig. 14.4-1. The transition between laminar and turbulent flow resembles that for pipes in Fig. 14.3-1, but the limits of the transition region are harder to predict. For smooth, sharp-edged flat plates in an isothermal flow the transition usually begins at a Reynolds number $\mathrm{Re}_x = xv_\infty\rho/\mu$ of 100,000 to 300,000 and is almost complete at a 50% higher Reynolds number.

b. Flow around a sphere

In Problem 10B.1 it is shown that the Nusselt number for a sphere in a stationary fluid is 2. For the sphere with constant surface temperature T_0 in a flowing fluid approaching with a uniform velocity v_∞, the mean Nusselt number is given by the following empiricism[3]

$$\mathrm{Nu}_m = 2 + 0.60\ \mathrm{Re}^{1/2}\mathrm{Pr}^{1/3} \tag{14.4-5}$$

This result is useful for predicting the heat loss from droplets or bubbles.

Another correlation that has proven successful[4] is

$$\mathrm{Nu}_m = 2 + (0.4\ \mathrm{Re}^{1/2}+0.06\ \mathrm{Re}^{2/3})\mathrm{Pr}^{0.4}\left(\frac{\mu_\infty}{\mu_0}\right)^{1/4} \tag{14.4-6}$$

in which the physical properties appearing in Nu_m, Re, and Pr are evaluated at the approaching stream temperature. This correlation is recommended for $3.5 < \mathrm{Re} < 7.6 \times 10^4$, $0.71 < \mathrm{Pr} < 380$, and $1.0 < \mu_\infty/\mu_0 < 3.2$. In contrast to Eq. 14.4-5, it is not valid in the limit that $\mathrm{Pr} \to \infty$.

c. Flow around a cylinder

A cylinder in a stationary fluid of infinite extent does not admit a steady-state solution. Therefore, the Nusselt number for a cylinder does not have the same form as that for a sphere. Whitaker recommends for the mean Nusselt number[4] for flow transverse to a cylinder

$$\mathrm{Nu}_m = (0.4\ \mathrm{Re}^{1/2}+0.06\ \mathrm{Re}^{2/3})\mathrm{Pr}^{0.4}\left(\frac{\mu_\infty}{\mu_0}\right)^{1/4} \tag{14.4-7}$$

in the range $1.0 < \mathrm{Re} < 1.0 \times 10^5$, $0.67 < \mathrm{Pr} < 300$, and $0.25 < \mu_\infty/\mu_0 < 5.2$. Here, as in Eq. 14.4-6, the values of viscosity and thermal conductivity in Re and Pr are those at the

[1] Eq. 14.4-1 with a numerical coefficient of 0.332 was first obtained by H. Blasius, *Z. Math. Phys.*, **56**, 1–37 (1908), and Eq. 14.4-2 with a numerical coefficient of 0.332 was obtained by E. Pohlhausen, *Z. angew. Math. Mech.*, **1**, 115–121 (1921).

[2] E. R. G. Eckert, *Trans. ASME*, **56**, 1273–1283 (1956). This article also includes high-velocity flows, for which compressibility and viscous dissipation become important.

[3] W. E. Ranz and W. R. Marshall, Jr., *Chem. Eng. Prog.*, **48**, 141–146, 173–180 (1952). N. Frössling, *Gerlands Beitr. Geophys.*, **52**, 170–216 (1938), first gave a correlation of this form, with a coefficient of 0.552 in lieu of 0.60 in the last term.

[4] S. Whitaker, *Fundamental Principles of Heat Transfer*, Krieger Publishing Co., Malabar, FL (1977), pp. 340–342; *AIChE Journal*, **18**, 361–371 (1972).

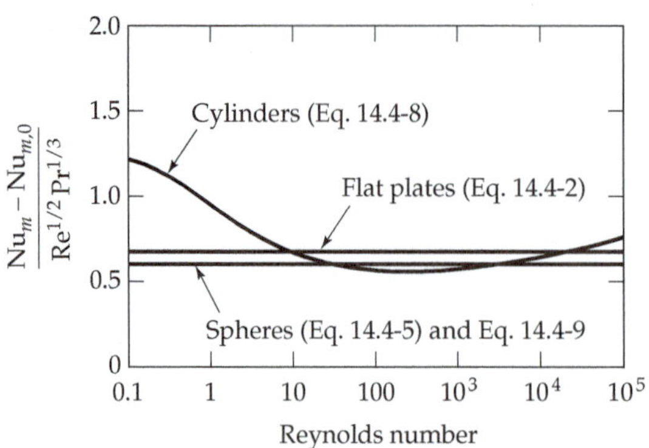

Fig. 14.4-2. Graph comparing the Nusselt numbers for flow around flat plates, spheres, and cylinders with Eq. 14.4-9.

approaching stream temperature. Similar results are available for banks of cylinders, which may be encountered in certain types of heat exchangers.[4]

Another correlation,[5] based on a curve-fit of McAdams' compilation of heat-transfer coefficient data,[6] is

$$\begin{aligned} Nu_m &= (0.376\,Re^{1/2} + 0.057\,Re^{2/3})Pr^{1/3} \\ &\quad + 0.92\left[\ln\left(\frac{7.4055}{Re}\right) + 4.18\,Re\right]^{-1/3} Re^{1/3}Pr^{1/3} \end{aligned} \tag{14.4-8}$$

This correlation has the proper behavior in the limit that $Pr \to \infty$, and also behaves properly for small values of the Reynolds number. This result can be used for analyzing the steady-state performance of hot-wire anemometers, which typically operate at low Reynolds numbers.

d. Flow around other objects

We learn from the above three discussions that, for the flow around objects of shapes other than those described above, a fairly good guess for the heat-transfer coefficients can be obtained by using the relation

$$Nu_m - Nu_{m,0} = 0.6\,Re^{1/2}Pr^{1/3} \tag{14.4-9}$$

in which $Nu_{m,0}$ is the mean Nusselt number at zero Reynolds number. This generalization, which is shown in Fig. 14.4-2, is often useful in estimating the heat transfer from irregularly shaped objects.

§14.5 HEAT-TRANSFER COEFFICIENTS FOR FORCED CONVECTION THROUGH PACKED BEDS

Heat-transfer coefficients between particles and fluid in packed beds are important in the design of fixed-bed catalytic reactors, absorbers, driers, and pebble-bed heat exchangers. The velocity profiles in packed beds exhibit a strong maximum near the wall, attributable partly to the higher void fraction there and partly to the more ordered interstitial passages along this smooth boundary. The resulting segregation of the flow into a fast outer stream and a slower interior one, which mix at the exit of the bed, leads to complicated behavior of mean Nusselt numbers in deep packed beds,[1] unless the tube-to-particle diameter ratio

---

[5]W. E. Stewart (unpublished).

[6]W. H. McAdams, *Heat Transmission*, 3rd edition, McGraw-Hill, New York (1954), p. 259.

[1]H. Martin, *Chem. Eng. Sci.*, **33**, 913–919 (1978).

D_t/D_p is very large or close to unity. Experiments with wide, shallow beds show simpler behavior and are used in the following discussion.

We define h_{loc} for a representative volume Sdz of particles and fluid by the following modification of Eq. 14.1-5:

$$dQ = h_{\text{loc}}(aSdz)(T_0 - T_b) \tag{14.5-1}$$

Here a is the outer surface area of particles per unit bed volume, as in §6.4. Equations 6.4-5 and 6.4-6 give the effective particle size D_p as $6/a_v = 6(1-\varepsilon)/a$ for a packed bed with void fraction ε.

Extensive data on forced convection for the flow of gases[2] and liquids[3] through shallow packed beds have been critically analyzed[4] to obtain the following local heat-transfer correlation,

$$j_H = 2.19\,\text{Re}^{-2/3} + 0.78\,\text{Re}^{-0.381} \tag{14.5-2}$$

and an identical formula for the mass-transfer function j_D defined in §22.3. Here the Chilton-Colburn j_H factor and the Reynolds number are defined by

$$j_H = \frac{h_{\text{loc}}}{\hat{C}_p G_0}\left(\frac{\hat{C}_p \mu}{k}\right)^{2/3} \tag{14.5-3}$$

$$\text{Re} = \frac{D_p G_0}{(1-\varepsilon)\mu\psi} = \frac{6G_0}{a\mu\psi} \tag{14.5-4}$$

In this equation the physical properties are all evaluated at the film temperature $T_f = \frac{1}{2}(T_0 + T_b)$, and $G_0 = w/S$ is the superficial mass velocity introduced in §6.4. The quantity ψ is a particle-shape factor, with a defined value of 1 for spheres and a fitted value[4] of 0.92 for cylindrical pellets. A related shape factor was used by Gamson[5] in Re and j_H; the present factor ψ is used in Re only.

For small Re, Eq. 14.5-2 yields the asymptote

$$j_H = 2.19\,\text{Re}^{-2/3} \tag{14.5-5}$$

or

$$\text{Nu}_{\text{loc}} = \frac{h_{\text{loc}} D_p}{k(1-\varepsilon)\psi} = 2.19(\text{RePr})^{1/3} \tag{14.5-6}$$

consistent with boundary-layer theory[6] for creeping flow with $\text{RePr} >> 1$. The latter restriction gives $\text{Nu} >> 1$ corresponding to a thin thermal boundary layer relative to $D_p/(1-\varepsilon)\psi$. This asymptote represents the creeping-flow mass-transfer data for liquids[3] very well.

The exponent $\frac{2}{3}$ in Eq. 14.5-3 is a high-Pr asymptote given by boundary-layer theory for steady laminar flows[6] and for steadily driven turbulent flows.[7] This dependence is consistent with the cited data over the full range $\text{Pr} > 0.6$ and the corresponding range of the dimensionless group Sc for mass transfer.

---

[2]B. W. Gamson, G. Thodos, and O. A. Hougen, *Trans. AIChE*, **39**, 1–35 (1943); C. R. Wilke and O. A. Hougen, *Trans. AIChE*, **41**, 445–451 (1945).

[3]L. K. McCune and R. H. Wilhelm, *Ind. Eng. Chem.*, **41**, 1124–1134 (1949); J. E. Williamson, K. E. Bazaire, and C. J. Geankoplis, *Ind. Eng. Chem. Fund.*, **2**, 126–129 (1963); E. J. Wilson and C. J. Geankoplis, *Ind. Eng. Chem. Fund.*, **5**, 9–14 (1966).

[4]W. E. Stewart, unpublished.

[5]B. W. Gamson, *Chem. Eng. Prog.*, **47**, 19–28 (1951).

[6]W. E. Stewart, *AIChE Journal*, **9**, 528–535 (1963); R. Pfeffer, *Ind. Eng. Chem. Fund.*, **3**, 380–383 (1964); J. P. Sørensen and W. E. Stewart, *Chem. Eng. Sci.*, **29**, 833–837 (1974).

[7]W. E. Stewart, *AIChE Journal*, **33**, 2008–2016 (1987); corrigenda **34**, 1030 (1988).

§14.6 HEAT-TRANSFER COEFFICIENTS FOR FREE AND MIXED CONVECTION FOR SUBMERGED OBJECTS[1]

In this section we consider free-convection heat transfer from objects submerged in large volumes of otherwise stagnant fluids. We will consider only the situation in which the surface temperature of the object is uniform at temperature T_0, and the fluid temperature far from the object is also uniform at temperature T_∞.

A set of simplified equations of continuity, motion, and energy were presented in §13.1 for such free-convection problems. If one could solve these equations subject to appropriate boundary conditions to obtain the temperature profile, one could determine the rate of heat transfer from the object to the fluid using

$$Q = \int_S \left(-k\frac{\partial T}{\partial n}\right) dS \tag{14.6-1}$$

where the integral is over the surface of the object, and n is the local coordinate normal to the surface of the object, directed into the fluid. The mean heat-transfer coefficient h_m is defined by

$$Q = h_m A(T_0 - T_\infty) \tag{14.6-2}$$

These last two equations can be combined to eliminate Q and express h_m in terms of the temperature profile. In dimensionless form this results in

$$\text{Nu}_m = \frac{h_m l_0}{k} = \frac{1}{\breve{S}}\int_{\breve{S}} \left(-\frac{\partial \breve{T}}{\partial \breve{n}}\right) d\breve{S} \tag{14.6-3}$$

where Nu_m is the mean Nusselt number, l_0 is an appropriate length scale, $\breve{S} = S/l_0^2$, $\breve{n} = n/l_0$, and $\breve{T} = (T - T_\infty)/(T_0 - T_\infty)$.

In §13.1, the simplified equations of change for free-convection problems were nondimensionalized, with the resulting dimensionless equations of continuity, motion, and energy presented in Eqs. 13.1-14, 13.1-15, and 13.1-16. The only dimensionless groups that appear in these equations are the Grashof number $\text{Gr} = \rho^2 g\bar{\beta}(T_0 - T_\infty)l_0^3/\mu^2$ and Prandtl number $\text{Pr} = \hat{C}_p\mu/k$. As a result, the dimensionless temperature profile can be expressed as

$$\breve{T} = \breve{T}(\breve{x}, \breve{y}, \breve{z}, \breve{t}; \text{Gr,Pr,shape}) \tag{14.6-4}$$

Substitution of this relation into Eq. 14.6-3 leads to the conclusion that $\text{Nu}_m(\breve{t}) = \text{Nu}_m(\breve{t}; \text{Gr,Pr,shape})$. When time averaged, this becomes

$$\text{Nu}_m = \text{Nu}_m(\text{Gr,Pr,shape}) \tag{14.6-5}$$

Thus, the Nusselt number depends only on Gr and Pr, with perhaps different functionalities for different object shapes.

Determining the specific dependence of Nu_m on Gr and Pr can be quite challenging. Some progress can be gained by approximate solutions to idealized problems. Consider the natural convection near a vertical heated plate where the cooler adjacent fluid flows upward as a result of the buoyant force. If the flow is assumed to be slow, laminar, and confined to a thin layer near the plate, an approximate boundary-layer analysis gives (see Problem 13C.1)

$$\text{Nu}_m = C(\text{GrPr})^{1/4} \tag{14.6-6}$$

where C is a weak function of Pr. Equation 14.6-6 does a reasonable job of reproducing experimental trends for intermediate values of GrPr, but fails to describe behavior for small and large values of GrPr; for small values of GrPr, the thin boundary-layer assumption is not valid, while for large values of GrPr, the buoyancy driven flows are turbulent instead of laminar.

Despite the shortcomings of Eq. 14.6-6, development of correlations for mean Nusselt numbers for free convection from submerged objects often begins with a form similar to Eq. 14.6-6, with various empirical "corrections" added to better describe experimental data. This is the approach taken here, which is based on the development described by Raithby and Hollands.[1] The method for calculating mean Nusselt numbers thus proceeds as follows:

a. A Nusselt number based on the thin, laminar boundary analysis is first calculated. This result is labeled $\mathrm{Nu}_m^{\mathrm{thin}}$.

b. The thin boundary layer result is then modified to account for thick boundary-layer effects. This result is labeled $\mathrm{Nu}_m^{\mathrm{lam}}$.

c. A Nusselt number for turbulent flow is then calculated. This result is labeled $\mathrm{Nu}_m^{\mathrm{turb}}$.

d. The laminar- and turbulent-flow results are combined in a single function that permits the calculation of Nu_m for essentially arbitrary values of GrPr.

e. For objects of small-aspect ratio such as spheres, one must also correct for the effects of heat transfer via conduction. The value of the Nusselt number for conduction only, $\mathrm{Nu}_m^{\mathrm{cond}}$, can simply be added to the thin boundary-layer result to obtain an improved estimate of the mean Nusselt number. Such a correction does not appear for large-aspect ratio objects; the solution to the corresponding two-dimensional conduction problem does not admit a steady-state solution.

These steps are described in detail for various objects below. Discussions for other geometries, as well for free convection in other situations (such as prescribed heat fluxes, stratified media, and transient conditions) are presented elsewhere.[1] All physical properties appearing in the equations below should be evaluated at the film temperature $T_f = (T_0 + T_\infty)/2$. A method for estimating Nusselt numbers when both free and forced convection are significant is briefly discussed following the discussion of free-convection results below.

a. Vertical Plates

The thin boundary-layer estimate of the Nusselt number for vertical flat plates of height L and width W ($W \gg L$) is

$$\mathrm{Nu}_m^{\mathrm{thin}} = \bar{C}_\ell(\mathrm{Pr})(\mathrm{GrPr})^{1/4} \tag{14.6-7}$$

where Gr is the Grashof number based on the height of the plate ($l_0 = L$) and

$$\bar{C}_\ell(\mathrm{Pr}) = \frac{0.671}{[1 + (0.492/\mathrm{Pr})^{9/16}]^{4/9}} \tag{14.6-8}$$

The correction for thick boundary layers is then expressed

$$\mathrm{Nu}_m^{\mathrm{lam}} = \frac{2.0}{\ln(1 + 2.0/\mathrm{Nu}_m^{\mathrm{thin}})} \tag{14.6-9}$$

The estimate of the mean Nusselt number for turbulent flow is

$$\mathrm{Nu}_m^{\mathrm{turb}} = \frac{0.13\mathrm{Pr}^{0.22}}{\left(1 + 0.61\mathrm{Pr}^{0.81}\right)^{0.42}} \frac{(\mathrm{GrPr})^{1/3}}{\left(1 + 1.4 \times 10^9/\mathrm{Gr}\right)} \tag{14.6-10}$$

[1] G. D. Raithby and K. G. T. Hollands, Chapter 4 in W. M. Rohsenow, J. P. Hartnett, and Y. I. Cho, eds., *Handbook of Heat Transfer*, 3rd edition, McGraw-Hill, New York (1998).

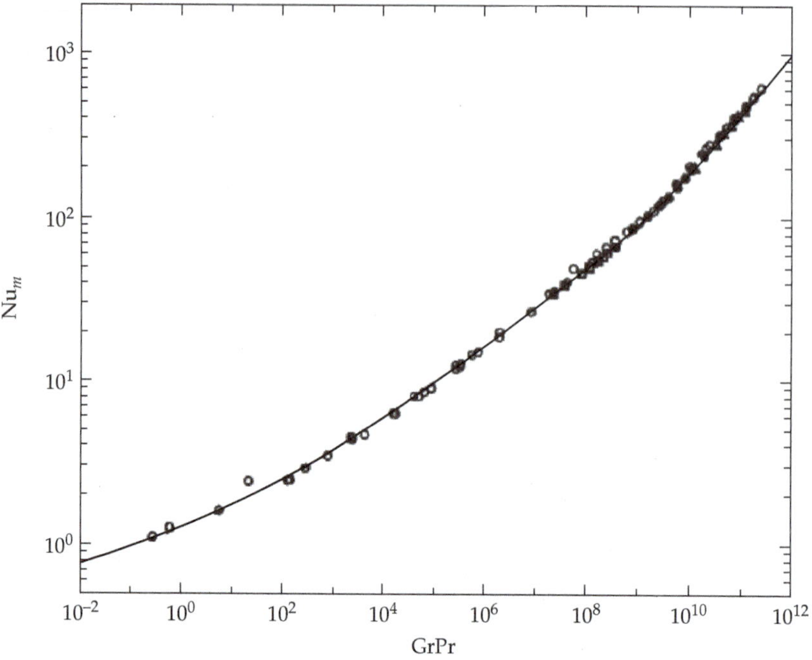

Fig. 14.6-1. Comparison of predicted (Eq. 14.6-11 along with Eqs. 14.6-7 through 14.6-10) and measured values of Nusselt numbers for free convection from vertical flat plates (for air, Pr = 0.71). Adapted from G. D. Raithby and K. G. T. Hollands, Chapter 4 in W. M. Rohsenow, J. P. Hartnett, and Y. I. Cho, *Handbook of Heat Transfer*, 3rd edition, McGraw-Hill, New York (1998). The data are those of O. A. Saunders (∘) [*Proc. Royal Soc., Ser. A*, **157**, 278–291 (1936)], C. Y. Warner and V. S. Arpaci (•) [*Int. J. Heat Mass Transfer*, **11**, 397–406 (1968)], A. Pirovano, S. Viannay, and M. Jannot (△) [Rep. *EUR 4489f*, Commission of the European Communities (1970); Proc. 4th Int. Heat Transfer Conf., Elsevier, Amsterdam, paper NC 1.8 (1970)], and A. M. Clausing and S. N. Kempka (□) [*J. Heat Transfer*, **103**, 609–612 (1981)].

The last two equations are combined to give a final estimate of the mean Nusselt number[2]

$$\mathrm{Nu}_m = \frac{h_m L}{k} = [(\mathrm{Nu}_m^{\mathrm{lam}})^6 + (\mathrm{Nu}_m^{\mathrm{turb}})^6]^{1/6} \tag{14.6-11}$$

Nusselt numbers predicted by Eq. 14.6-11 are compared with measured values in Fig. 14.6-1. The data agree well with predictions over the entire range $10^{-2} \le \mathrm{GrPr} \le 10^{12}$. Effects of turbulent flow are apparent for $\mathrm{GrPr} > 10^{10}$, where Nu_m increases more rapidly with GrPr. Deviations from the thin boundary-layer behavior (i.e., $\mathrm{Nu}_m^{\mathrm{thin}} \propto (\mathrm{GrPr})^{1/4}$) are apparent for $\mathrm{GrPr} < 10^4$.

b. Horizontal Plates

For heated horizontal plates facing upward, or cooled horizontal plates facing downward, the equations for the various contributions to the mean Nusselt number are

$$\mathrm{Nu}_m^{\mathrm{thin}} = 0.835\, \overline{C}_\ell(\mathrm{Pr})(\mathrm{GrPr})^{1/4} \tag{14.6-12}$$

$$\mathrm{Nu}_m^{\mathrm{lam}} = \frac{1.4}{\ln(1 + 1.4/\mathrm{Nu}_m^{\mathrm{thin}})} \tag{14.6-13}$$

[2]The form $\mathrm{Nu}_m = [(\mathrm{Nu}_m^{\mathrm{lam}})^m + (\mathrm{Nu}_m^{\mathrm{turb}})^m]^{1/m}$, where m depends on the shape of the object, was suggested by S. W. Churchill and R. Usagi, *AIChE Journal*, **23**, 1121–1128 (1972).

$$\mathrm{Nu}_m^{\mathrm{turb}} = 0.14\frac{1+0.0107\mathrm{Pr}}{1+0.01\mathrm{Pr}}(\mathrm{GrPr})^{1/3} \tag{14.6-14}$$

$$\mathrm{Nu}_m = \frac{h_m L^*}{k} = [(\mathrm{Nu}_m^{\mathrm{lam}})^{10} + (\mathrm{Nu}_m^{\mathrm{turb}})^{10}]^{1/10} \tag{14.6-15}$$

In these equations, the Grashof and Nusselt numbers are based on the length scale $l_0 = L^* = A/P$, where A is the area and P is the perimeter of the horizontal plate. The term $\overline{C}_\ell(\mathrm{Pr})$ is given by Eq. 14.6-8.

For heated horizontal surfaces facing downward and cooled horizontal surfaces facing upward, the following correlation[3] is recommended:

$$\mathrm{Nu}_m^{\mathrm{thin}} = \frac{0.527}{[1+(1.9/\mathrm{Pr})^{9/10}]^{2/9}}(\mathrm{GrPr})^{1/5} \tag{14.6-16}$$

$$\mathrm{Nu}_m = \mathrm{Nu}_m^{\mathrm{lam}} = \frac{2.5}{\ln(1+2.5/\mathrm{Nu}_m^{\mathrm{thin}})} \tag{14.6-17}$$

Since the buoyancy-driven flow here is toward the surface, the flow is typically laminar up to large values of GrPr and thus the turbulent correction is not needed.

c. Vertical Cylinders

For vertical cylinders of diameter D and length L, one uses Eqs. 14.6-7 and 14.6-8 to obtain the thin and thick boundary-layer values for the mean Nusselt numbers for a vertical plate of height L (termed "$\mathrm{Nu}_{m,\,\mathrm{plate}}^{\mathrm{thin}}$" and "$\mathrm{Nu}_{m,\,\mathrm{plate}}^{\mathrm{lam}}$"). The thick boundary-layer estimate for the vertical cylinder is then modified using

$$\mathrm{Nu}_m^{\mathrm{lam}} = \frac{\zeta}{\ln(1+\zeta)}\mathrm{Nu}_{m,\,\mathrm{plate}}^{\mathrm{lam}} \tag{14.6-18}$$

where

$$\zeta = \frac{1.8L/D}{\mathrm{Nu}_{m,\,\mathrm{plate}}^{\mathrm{thin}}} \tag{14.6-19}$$

The correction for turbulent flow then proceeds as in Eq. 14.6-10 and 14.6-11. The Grashof and Nusselt numbers in this case are based on the length of the cylinder ($l_0 = L$).

d. Horizontal Cylinders

The equations for long, horizontal cylinders of diameter D are

$$\mathrm{Nu}_m^{\mathrm{thin}} = 0.772\,\overline{C}_\ell(\mathrm{Pr})(\mathrm{GrPr})^{1/4} \tag{14.6-20}$$

$$\mathrm{Nu}_m^{\mathrm{lam}} = \frac{2f}{\ln(1+2f/\mathrm{Nu}_m^{\mathrm{thin}})} \tag{14.6-21}$$

$$\mathrm{Nu}_m^{\mathrm{turb}} = \overline{C}_t(\mathrm{GrPr})^{1/3} \tag{14.6-22}$$

$$\mathrm{Nu}_m = \frac{h_m D}{k} = [(\mathrm{Nu}_m^{\mathrm{lam}})^{10} + (\mathrm{Nu}_m^{\mathrm{turb}})^{10}]^{1/10} \tag{14.6-23}$$

where the Grashof and Nusselt numbers are based on the cylinder diameter ($l_0 = D$), $\overline{C}_\ell(\mathrm{Pr})$ is given by Eq. 14.6-8, $f = 1 - 0.13/\mathrm{Nu}_m^T$ for $\mathrm{GrPr} < 10^{-4}$ and $f = 0.8$ for $\mathrm{GrPr} \geq 10^{-4}$. Values of $\overline{C}_t$ are tabulated for different values of Pr in Table 14.6-1.

Nusselt numbers predicted by Eq. 14.6-23 are compared with measured values in Fig. 14.6-2. The agreement is excellent over the entire range $10^{-10} \leq \mathrm{GrPr} \leq 10^7$.

[3] T. Fujii, M. Honda, and I. Morioka, *Int. J. Heat and Mass Transfer*, **15**, 755–767 (1972).

Table 14.6-1. Values of the Parameter $\overline{C}_t$ for Horizontal Cylinders and Spheres[1]

Pr	$\overline{C}_t$ Horizontal Cylinders	$\overline{C}_t$ Spheres
0.71	0.103	0.104
6.0	0.109	0.111
100	0.097	0.098
2000	0.088	0.086

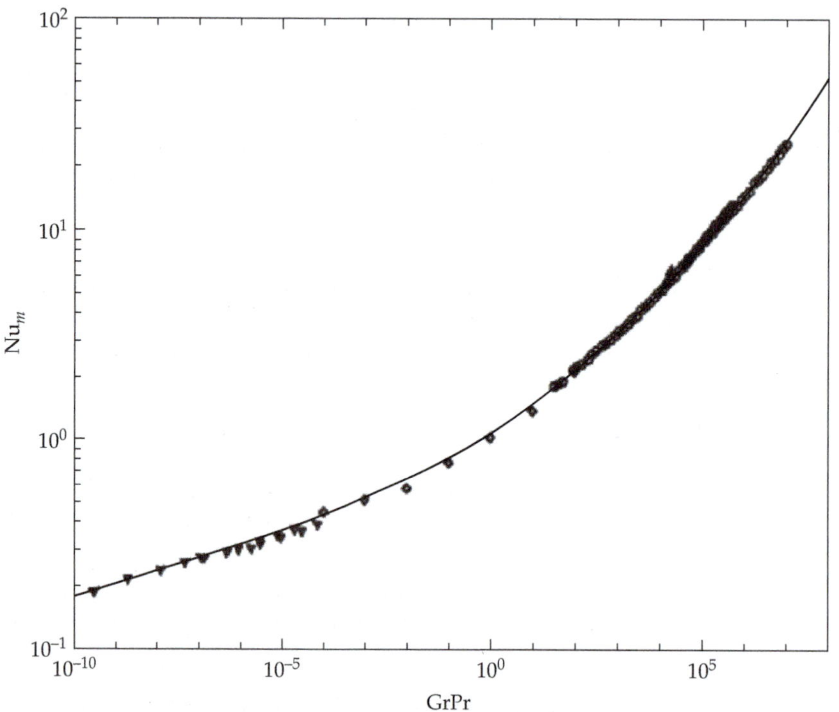

Fig. 14.6-2. Comparison of predicted (Eq. 14.6-23 along with Eqs. 14.6-20 through 22) and measured values of Nusselt numbers for free convection from long horizontal cylinders (for air, Pr = 0.71). Adapted from G. D. Raithby and K. G. T. Hollands, Chapter 4 in W. M. Rohsenow, J. P. Hartnett, and Y. I. Cho, *Handbook of Heat Transfer*, 3rd edition, McGraw-Hill, New York (1998). The data are those of S. B. Clemes, K. G. T. Hollands, and A. P. Brunger (∘) [*J. Heat Transfer*, **116,** 96–104 (1994)], L. M. De Socio (•) [*Int. J. Heat Mass Transfer*, **26**, 1669–1677 (1983)], G. Hesse and E. M. Sparrow (◇) [*J. Heat Transfer*, **17,** 796–798 (1974)], J. Li and J. D. Tarasuk (△) [*Experimental Thermal and Fluid Science*, **5**, 235–242 (1992)], and D. C. Collis and M. J. Williams (∇) [Aeronautical Research Laboratory, Note 140, Melbourne, Australia (1954)].

e. Spheres

For spheres of diameter D, the equations are

$$\mathrm{Nu}_m^{\mathrm{thin}} = 0.878\,\overline{C}_\ell(\mathrm{Pr})(\mathrm{GrPr})^{1/4} \tag{14.6-24}$$

$$\mathrm{Nu}_m^{\mathrm{lam}} = 2 + \mathrm{Nu}_m^{\mathrm{thin}} \tag{14.6-25}$$

$$\mathrm{Nu}_m^{\mathrm{turb}} = \overline{C}_t(\mathrm{GrPr})^{1/3} \tag{14.6-26}$$

$$\mathrm{Nu}_m = \frac{h_m D}{k} = [(\mathrm{Nu}_m^{\mathrm{lam}})^{15} + (\mathrm{Nu}_m^{\mathrm{turb}})^{15}]^{1/15} \tag{14.6-27}$$

where the Grashof and Nusselt numbers are based on the sphere diameter ($l_0 = D$), $\overline{C}_\ell(\mathrm{Pr})$ is given by Eq. 14.6-8, and values of $\overline{C}_t$ are tabulated for different values of Pr in Table 14.6-1 along with the values for horizontal cylinders. Raithby and Hollands[1] report that Nusselt numbers predicted by Eq. 14.6-27 are within 20% of measured values for $1 \leq \mathrm{GrPr} \leq 10^{12}$.

f. Mixed Free and Forced Convection

Finally, when one must deal with the problem of simultaneous free and forced convection, this is again done through the use of an empirical combining rule:[4]

$$\mathrm{Nu}_m^{\mathrm{total}} = [(\mathrm{Nu}_m^{\mathrm{free}})^3 + (\mathrm{Nu}_m^{\mathrm{forced}})^3]^{1/3} \tag{14.6-28}$$

where $\mathrm{Nu}_m^{\mathrm{free}}$ is the estimate for the mean Nusselt number for free convection obtained above, and $\mathrm{Nu}_m^{\mathrm{forced}}$ is the value estimated for forced convection using the appropriate equations in §14.4. This rule appears to hold reasonably well for all geometries and situations, provided only that the forced and free convection have the *same* primary flow direction.

EXAMPLE 14.6-1

Heat Loss by Free Convection from a Horizontal Pipe

Estimate the rate of heat loss by free convection from a unit length of a long horizontal pipe, 6 in. in outside diameter, if the outer surface temperature is 100°F and the surrounding air is at 1 atm and 80°F.

SOLUTION

The properties of air at 1 atm and a film temperature $T_f = 90°\mathrm{F}$ (or 550°R) are

$$\begin{aligned}
\mu &= 0.0190 \text{ cp} = 0.0460 \text{ lb}_m/\text{ft}\cdot\text{hr} \\
\overline{\rho} &= 0.0723 \text{ lb}_m/\text{ft}^3 \\
\hat{C}_p &= 0.241 \text{ Btu/lb}_m\cdot{}^\circ\text{R} \\
k &= 0.0152 \text{ Btu/hr}\cdot\text{ft}\cdot{}^\circ\text{R} \\
\overline{\beta} &= 1/T_f = (1/550)\ {}^\circ\text{R}^{-1}
\end{aligned}$$

Other relevant values are $D = 0.5$ ft, $\Delta T = 20°\mathrm{R}$, and $g = 4.17 \times 10^8$ ft/hr². From these data we obtain (using Table 13.1-1, with length scale $l_0 = D$)

$$\begin{aligned}
\mathrm{GrPr} &= \left(\frac{\overline{\rho}^2 g\overline{\beta}(T_1 - T_0)D^3}{\mu^2}\right)\left(\frac{\hat{C}_p\mu}{k}\right) \\
&= \left(\frac{(0.0723 \text{ lb}_m/\text{ft}^3)^2(4.17\times 10^8 \text{ ft/hr}^2)\left(\frac{1}{550}\ {}^\circ\text{R}^{-1}\right)(20\ {}^\circ\text{R})(0.5 \text{ ft})^3}{(0.0460 \text{ lb}_m/\text{ft}\cdot\text{hr})^2}\right) \\
&\quad\times\left(\frac{(0.241 \text{ Btu/lb}_m\cdot{}^\circ\text{R})(0.0460 \text{ lb}_m/\text{ft}\cdot\text{hr})}{(0.0152 \text{ Btu/hr}\cdot\text{ft}\cdot{}^\circ\text{R})}\right) \\
&= (4.68\times 10^6)(0.729) = 3.4\times 10^6
\end{aligned} \tag{14.6-29}$$

[4] E. Ruckenstein, *Adv. in Chem. Eng.*, **13**, 111–112 (1987); E. Ruckenstein and R. Rajagopalan, *Chem. Eng. Communications*, **4**, 15–29 (1980).

Then from Eqs. 14.6-8 and 14.6-20 we get the thin boundary-layer estimate

$$\mathrm{Nu}_m^{\text{thin}} = 0.772 \left(\frac{0.671}{[1 + (0.492/0.729)^{9/16}]^{4/9}} \right) (3.4 \times 10^6)^{1/4}$$

$$= 0.772 \left(\frac{0.671}{1.30} \right) (42.9) = 17.1 \tag{14.6-30}$$

The thick laminar boundary-layer correction for the Nusselt number is obtained from Eq. 14.6-21,

$$\mathrm{Nu}_m^{\text{lam}} = \frac{2(0.8)}{\ln(1 + 2(0.8)/17.1)} = 17.9 \tag{14.6-31}$$

The Nusselt number for turbulent free convection is obtained from Eq. 14.6-22,

$$\mathrm{Nu}_m^{\text{turb}} = (0.103)(3.4 \times 10^6)^{1/3} = 15.5 \tag{14.6-32}$$

The final estimate of the mean Nusselt number is then obtained from Eq. 14.6-23,

$$\mathrm{Nu}_m = [(17.9)^{10} + (15.5)^{10}]^{1/10} = 18.3 \tag{14.6-33}$$

The heat-transfer coefficient is then

$$h_m = \mathrm{Nu}_m \frac{k}{D} = 18.3 \left(\frac{0.0152 \text{ Btu/hr} \cdot \text{ft} \cdot {}^\circ\text{R}}{0.5 \text{ ft}} \right) = 0.56 \text{ Btu/hr} \cdot \text{ft}^2 \cdot {}^\circ\text{F} \tag{14.6-34}$$

The rate of heat loss per unit length of the pipe is

$$\frac{Q}{L} = \frac{h_m A \Delta T}{L} = h_m \pi D \Delta T$$

$$= \left(0.56 \frac{\text{Btu}}{\text{hr} \cdot \text{ft}^2 \cdot {}^\circ\text{F}} \right) (\pi)(0.5 \text{ ft})(20^\circ\text{F}) = 18 \text{ Btu/hr} \cdot \text{ft} \tag{14.6-35}$$

This is the heat loss by convection only. The radiation loss for the same problem is obtained in Example 16.5-2.

---

§14.7 HEAT-TRANSFER COEFFICIENTS FOR CONDENSATION OF PURE VAPORS ON SOLID SURFACES

The condensation of a pure vapor on a solid surface is a particularly complicated heat-transfer process, because it involves two flowing fluid phases: the vapor and the condensate. Condensation occurs industrially in many types of equipment; for simplicity, we consider here only the common cases of condensation of a slowly moving vapor on the outside of horizontal tubes, vertical tubes, and vertical flat walls.

The condensation process on a vertical wall is illustrated schematically in Fig. 14.7-1. Vapor flows over the condensing surface and is moved toward it by the small pressure gradient near the liquid surface.[1] Some of the molecules from the vapor phase strike the liquid surface and bounce off; others penetrate the surface and give up their latent heat of condensation. The energy thus released must then move through the condensate to the wall, thence to the coolant on the other side of the wall. At the same time, the condensate must drain from the surface by gravity flow.

The condensate on the wall is normally the sole important resistance to heat transfer on the condensing side of the wall. If the solid surface is clean, the condensate will usually form a continuous film over the surface, but if traces of certain impurities are present

---

[1]Note that there occur small but abrupt changes in pressure and temperature at an interface. These discontinuities are essential to the condensation process, but are generally of negligible magnitude in engineering calculations for pure fluids. For mixtures, they may be important. See R. W. Schrage, *Interphase Mass Transfer*, Columbia University Press (1953).

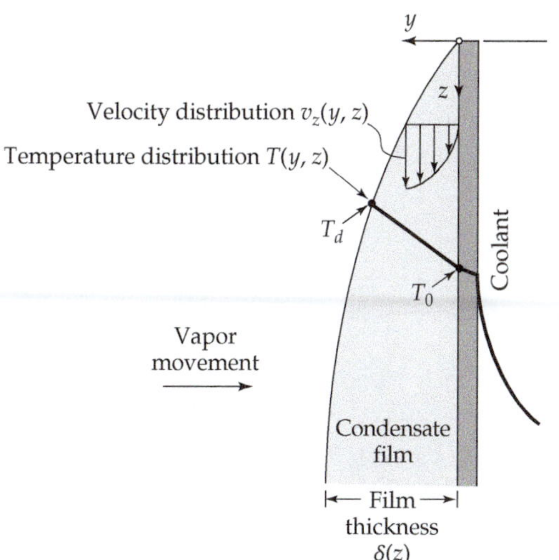

Fig. 14.7-1. Film condensation on a vertical surface (interfacial temperature discontinuity exaggerated).

(such as fatty acids in a steam condenser), the condensate will form in droplets. "Dropwise condensation"[2] gives much higher rates of heat transfer than "film condensation," but is difficult to maintain, so that it is common practice to assume film condensation in condenser design. The correlations that follow apply only to film condensation.

The usual definition of h_m for condensation of a pure vapor on a solid surface of area A and uniform temperature T_0 is

$$Q = h_m A(T_d - T_0) = w\Delta\hat{H}_{\text{vap}} \tag{14.7-1}$$

in which Q is the rate of heat flow into the solid surface, and T_d is the *dew point* of the vapor approaching the wall surface—that is, the temperature at which the vapor would condense if cooled slowly at the prevailing pressure. This temperature is very nearly that of the liquid at the liquid-gas interface. Therefore, h_m may be regarded as a heat-transfer coefficient for the liquid film. In Eq. (14.7-1), w is the total mass rate of condensation.

Expressions for h_m have been derived[3] for *laminar nonrippling* condensate flow by approximate solution of the equations of energy and motion for a falling liquid film (see Problem 14C.1). For film condensation on a horizontal tube of diameter D, length L, and constant surface temperature T_0, the result of Nusselt[3] may be written as

$$h_m = 0.954\left(\frac{k^3\rho^2 gL}{\mu w}\right)^{1/3} \tag{14.7-2}$$

Here w/L is the mass rate of condensation per unit length of tube, and it is understood that all the physical properties of the condensate are to be calculated at the film temperature, $T_f = \frac{1}{2}(T_d + T_0)$.

For moderate temperature differences, Eq. 14.7-2 may be rewritten with the aid of an energy balance on the condensate to give

$$h_m = 0.725\left(\frac{k^3\rho^2 g\Delta\hat{H}_{\text{vap}}}{\mu D(T_d - T_0)}\right)^{1/4} \tag{14.7-3}$$

---

[2]Dropwise condensation and boiling is discussed at length by J. G. Collier and J. R. Thome, *Convective Boiling and Condensation*, 3rd edition, Oxford University Press (1996).

[3]W. Nusselt, *Z. Ver. deutsch. Ing.*, **60**, 541–546, 596–575 (1916).

Equations 14.7-2 and 14.7-3 have been confirmed experimentally within ±10% for single horizontal tubes. They also seem to give satisfactory results for bundles of horizontal tubes,[4] in spite of the complications introduced by condensate dripping from tube to tube.

For film condensation on *vertical tubes* or *vertical walls* of height L, the theoretical results corresponding to Eqs. 14.7-2 and 14.7-3 are

$$h_m = \frac{4}{3}\left(\frac{k^3\rho^2 g}{3\mu\Gamma}\right)^{1/3} \tag{14.7-4}$$

and

$$h_m = \frac{2\sqrt{2}}{3}\left(\frac{k^3\rho^2 g\Delta\hat{H}_{\text{vap}}}{\mu L(T_d - T_0)}\right)^{1/4} \tag{14.7-5}$$

respectively. The quantity Γ in Eq. 14.7-4 is the total rate of condensate flow from the bottom of the condensing surface per unit width of that surface. For a vertical tube, $\Gamma = w/\pi D$, where w is the total mass rate of condensation on the tube. For *short vertical tubes* ($L <$ 0.5 ft), the experimental values of h_m confirm the theory well, but the measured values for *long vertical tubes* ($L >$ 8 ft) may exceed the theory for a given $T_d - T_0$ by as much as 70%. This discrepancy is attributed to ripples that attain greatest amplitude on long vertical tubes.[5]

We now turn to the empirical expressions for *turbulent* condensate flow. Turbulent flow begins, on *vertical tubes or walls*, at a Reynolds number $\text{Re} = \Gamma/\mu$ of about 350. For higher Reynolds numbers, the following empirical formula has been proposed:[6]

$$h_m = 0.003\left(\frac{k^3\rho^2 g(T_d - T_0)L}{\mu^3\Delta\hat{H}_{\text{vap}}}\right)^{1/2} \tag{14.7-6}$$

This equation is equivalent, for small $T_d - T_0$, to the formula

$$h_m = 0.021\left(\frac{k^3\rho^2 g\Gamma}{\mu^3}\right)^{1/3} \tag{14.7-7}$$

Equations 14.7-4 to 14.7-7 are summarized in Fig. 14.7-2, for convenience of making calculations, and to show the extent of agreement with the experimental data. Somewhat better agreement could have been obtained by using a family of lines in the turbulent range to represent the effect of Prandtl number. However, in view of the scatter of the data, a single line is adequate.

Turbulent condensate flow is very difficult to obtain on horizontal tubes, unless the tube diameters are very large or high temperature differences are encountered. Equations 14.7-2 and 14.7-3 are believed to be satisfactory up to the estimated transition Reynolds number, $\text{Re} = w_T/L\mu$, of about 1000, where w_T is the *total condensate flow* leaving a given tube, including the condensate from the tubes above.[7]

The inverse process of vaporization of a pure fluid is considerably more complicated than condensation. We do not attempt to discuss heat transfer to boiling liquids here, but refer the reader to some reviews.[2,8]

---

[4] B. E. Short and H. E. Brown, *Proc. General Disc. Heat Transfer*, London (1951), pp. 27–31. See also D. Butterworth, in *Handbook of Heat Exchanger Design* (G. F. Hewitt, ed.), Oxford University Press, London (1977), pp. 426–462.

[5] W. H. McAdams, *Heat Transmission*, 3rd edition, McGraw-Hill, New York (1954) p. 333.

[6] U. Grigull, *Forsch. Ingenieurswesen*, **13**, 49–57 (1942); *Z. Ver. dtsch. Ing.*, **86**, 444–445 (1942).

[7] W. H. McAdams, *Heat Transmission*, 3rd edition, McGraw-Hill, New York (1954), pp. 338–339.

[8] H. D. Baehr and K. Stephan, *Heat and Mass Transfer*, Springer, Berlin (1998), Chapter 4.

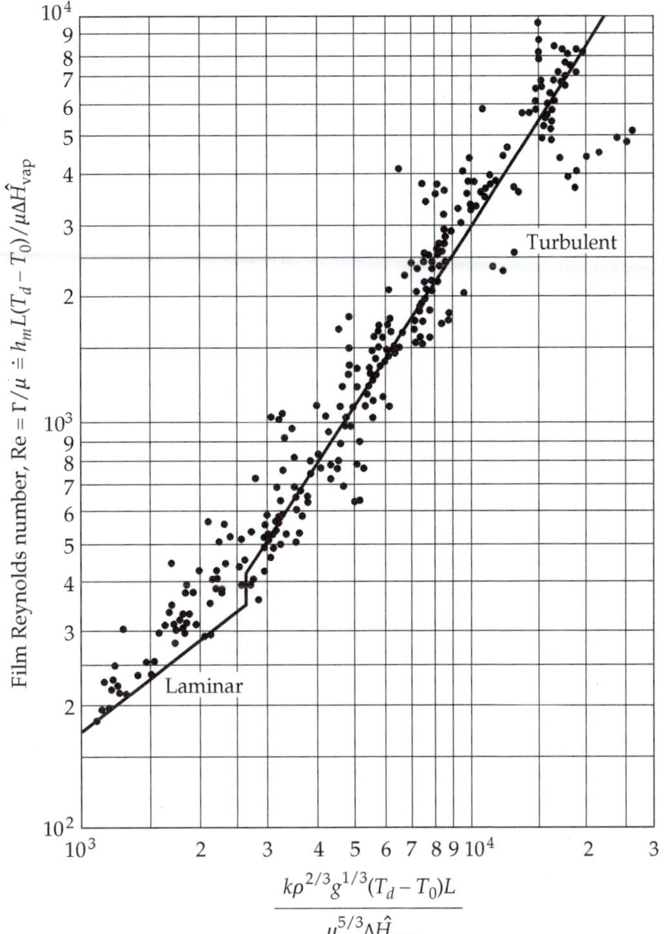

Fig. 14.7-2. Correlation of heat-transfer data for film condensation of pure vapors on vertical surfaces. [H. Gröber, S. Erk, and U. Grigull, *Die Grundgesetze der Wärmeübertragung*, 3rd edition, Springer-Verlag, Berlin (1955), p. 296.]

EXAMPLE 14.7-1

Condensation of Steam on a Vertical Surface

A boiling liquid flowing in a vertical tube is being heated by condensation of steam on the outside of the tube. The steam-heated tube section is 10 ft high and 2 in. in outside diameter. If saturated steam is used, what steam temperature is required to supply 92,000 Btu/hr of heat to the tube at a tube-surface temperature of 200°F? Assume film condensation.

SOLUTION

The fluid properties depend on the unknown temperature T_d. We make a *guess* of $T_d = T_0 = 200°\text{F}$. Then the physical properties at the film temperature (also 200°F) are

$$\Delta\hat{H}_{vap} = 978 \text{ Btu/lb}_m$$

$$k = 0.393 \text{ Btu/hr} \cdot \text{ft} \cdot °\text{F}$$

$$\rho = 60.1 \text{ lb}_m/\text{ft}^3$$

$$\mu = 0.738 \text{ lb}_m/\text{ft} \cdot \text{hr}$$

If it is assumed that the steam gives up only latent heat (the assumption $T_d = T_0 = 200°\text{F}$ implies this), an energy balance around the tube gives

$$Q = w\Delta\hat{H}_{vap} = \pi D\Gamma\Delta\hat{H}_{vap} \tag{14.7-8}$$

in which Q is the heat flow into the tube wall. The film Reynolds number is

$$\frac{\Gamma}{\mu} = \frac{Q}{\pi D \mu \Delta \hat{H}_{\text{vap}}} = \frac{92{,}000 \text{ Btu/hr}}{\pi\left(\frac{2}{12}\text{ ft}\right)(0.738 \text{ lb}_m/\text{hr}\cdot\text{ft})(978 \text{ Btu/lb}_m)} = 244 \quad (14.7\text{-}9)$$

Reading Fig. 14.7-2 at this value of the ordinate, we find that the flow is laminar. Equation 14.7-4 is applicable, but it is more convenient to use the line based on this equation in Fig. 14.7-2, which gives

$$\frac{k\rho^{2/3}g^{1/3}(T_d - T_0)L}{\mu^{5/3}\Delta\hat{H}_{\text{vap}}} = 1700 \quad (14.7\text{-}10)$$

from which

$$\begin{aligned} T_d - T_0 &= 1700\frac{\mu^{5/3}\Delta\hat{H}_{\text{vap}}}{k\rho^{2/3}g^{1/3}L} \\ &= 1700\frac{(0.738 \text{ lb}_m/\text{hr}\cdot\text{ft})^{5/3}(978 \text{ Btu/lb}_m)}{(0.393 \text{ Btu/hr}\cdot\text{ft}\cdot{}^\circ\text{F})(60.1 \text{ lb}_m/\text{ft}^3)^{2/3}(4.17\times 10^8 \text{ ft/hr}^2)^{1/3}(10 \text{ ft})} \\ &= 22^\circ\text{F} \end{aligned} \quad (14.7\text{-}11)$$

Therefore, the first approximation to the steam temperature is $T_d = 222°$F. This result is close enough; evaluation of the physical properties in accordance with this result gives $T_d = 220°$F as a second approximation. It is apparent from Fig. 14.7-2 that this result represents an upper limit. If rippling occurs, the temperature drop through the condensate film may be as little as half that predicted here.

§14.8 CONCLUDING COMMENTS

In this chapter, we have discussed a variety of methods for estimating heat-transfer coefficients. In §14.2 we showed that in some cases it is possible to obtain heat-transfer coefficients from the solution of the equations of change.

Many times, however, the physical systems of interest are too complicated to permit analytical solutions, and thus we are forced to use either numerical solutions of the equations or resort to experiments to obtain heat-transfer coefficients. Most of the rest of the chapter is devoted to presenting correlations for heat-transfer coefficients obtained largely from experimental measurements.

We have also made extensive use of dimensional analysis in this chapter. This greatly simplifies the tasks of determining and presenting correlations, as the number dimensionless variables is significantly less than the number of relevant dimensional variables.

QUESTIONS FOR DISCUSSION

1. Define the heat-transfer coefficient, the Nusselt number, the Stanton number, and the Colburn j_H factor. How can each of these be "decorated" to indicate the type of temperature-difference driving force that is being used?
2. What are the characteristic dimensionless groups that arise in the correlations for Nusselt numbers for forced convection? For free convection? For mixed convection?
3. To what extent can the Nusselt numbers be calculated a priori from analytical solutions?
4. Explain how one develops an experimental correlation for Nusselt numbers as a function of the relevant dimensionless groups.
5. To what extent can empirical correlations be developed in which the Nusselt number is given as the product of the relevant dimensionless groups, each raised to a characteristic power?
6. In addition to the Nusselt number, we have met up with the Reynolds number Re, the Prandtl number Pr, the Grashof number Gr, the Péclet number Pé, and the Rayleigh number Ra. Define each of these and explain their meaning and usefulness.
7. Discuss the concept of wind-chill temperature.

PROBLEMS

14A.1 Average heat-transfer coefficients. Ten thousand pounds per hour of an oil with a heat capacity of 0.6 Btu/lb_m · °F are being heated from 100°F to 200°F in the simple heat exchanger, shown Fig. 14A.1. The oil is flowing through the tubes, which are copper, 1 in. in outside diameter, with 0.065-in. walls. The combined length of the tubes is 300 ft. The required heat is supplied by condensation of saturated steam at 15.0 psia on the outside of the tubes. Calculate h_1, h_a, and h_{ln} for the oil, assuming that the inside surfaces of the tubes are at the saturation temperature of the steam, 213°F.

Answers: 78, 139, 190 Btu/hr · ft^2 · °F

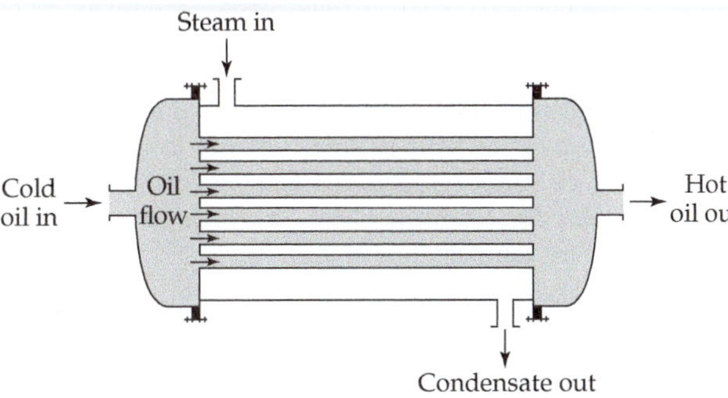

Fig. 14A.1 A single-pass "shell-and-tube" heat exchanger.

14A.2 Heat transfer in laminar tube flow. One hundred pounds per hour of oil at 100°F are flowing through a 1-in. i.d. copper tube, 20 ft long. The inside surface of the tube is maintained at 215°F by condensing steam on the outside surface. Fully developed flow may be assumed through the length of the tube, and the physical properties of the oil may be considered constant at the following values: $\rho = 55$ lb_m/ft^3, $\hat{C}_p = 0.49$ Btu/lb_m · °F, $\mu = 1.42$ lb_m/hr · ft, $k = 0.0825$ Btu/hr · ft · °F.

(a) Calculate Pr.

(b) Calculate Re.

(c) Calculate the exit temperature of the oil.

Answers: **(a)** 8.43; **(b)** 1076; **(c)** 155°F

14A.3 Effect of flow rate on exit temperature from a heat exchanger.

(a) Repeat parts (b) and (c) of Problem 14A.2 for oil flow rates of 200, 400, 800, 1600, and 3200 lb_m/hr.

(b) Calculate the total heat flow through the tube wall for each of the oil flow rates in (a).

14A.4 Local heat-transfer coefficient for turbulent forced convection in a tube. Water is flowing in a 2-in. i.d. tube at a mass flow rate $w = 15{,}000$ lb_m/hr. The inner wall temperature at some point along the tube is 160°F, and the bulk fluid temperature at that point is 60°F. What is the local heat flux q_r at the pipe wall? Assume that h_{loc} has attained a constant asymptotic value (see Fig. 14.3-2).

Answer: -8.8×10^4 Btu/hr · ft^2

14A.5 Heat transfer from condensing vapors.

(a) The outer surface of a vertical tube 1 in. in outside diameter and 1 ft long is maintained at 190°F. If this tube is surrounded by saturated steam at 1 atm, what will be the total rate of heat transfer through the tube wall?

(b) What would the rate of heat transfer be if the tube were horizontal?

Answers: **(a)** 8500 Btu/hr; **(b)** 12,000 Btu/hr

14A.6 Forced-convection heat transfer from an isolated sphere.

(a) A solid sphere 1 in. in diameter is placed in an otherwise undisturbed air stream, which approaches at a velocity of 100 ft/s, a pressure of 1 atm, and a temperature of 100°F. The sphere surface is maintained at 200°F by means of an imbedded electric heating coil. What must be the rate of electrical heating in cal/s to maintain the stated conditions? Neglect radiation, and use Eq. 14.4-5.

(b) Repeat the problem in (a), but use Eq. 14.4-6.

Answer: **(a)** 13.2 W = 3.2 cal/s; **(b)** 16.8 W = 4.0 cal/s

14A.7 Free convection heat transfer from an isolated sphere. If the sphere of Problem 14A.6 is suspended in still air at 1 atm pressure and 100°F ambient air temperature, and if the sphere surface is again maintained at 200°F, what rate of electrical heating would be needed? Neglect radiation.

Answer: 1.06 W = 0.25 cal/s

14A.8 Heat loss by free convection from a horizontal pipe immersed in a liquid. Estimate the rate of heat loss by free convection from a unit length of a long horizontal pipe, 6 in. in outside diameter, if the outer surface temperature is 100°F and the surrounding water is at 80°F. Compare the result with that obtained in Example 14.6-1, in which air is the surrounding medium. The properties of water at a film temperature of 90°F (or 32.3°C) are: $\mu = 0.7632$ cp, $\hat{C}_p = 0.9986$ cal/g · °C, $k = 0.363$ Btu/hr · ft · °F. Also the density of water in the neighborhood of 90°F is:

T (°C)	30.3	31.3	32.3	33.3	34.3
ρ (g/cm^3)	0.99558	0.99528	0.99496	0.99463	0.99430

Answer: $Q/L =$ 2550 Btu/hr · ft

14A.9 The ice fisherman on Lake Mendota. Compare the rates of heat loss of an ice fisherman, when he is fishing in calm weather (wind velocity zero) and when the wind velocity is 20 mph out of the north. The ambient air temperature is −10°F. Assume that a bundled-up ice fisherman can be approximated as a sphere 3 ft in diameter.

14B.1 Limiting local Nusselt number for plug flow with constant heat flux.

(a) Equation 10B.10-1 gives the asymptotic temperature distribution for cooling a fluid of constant physical properties in plug flow in a long tube with constant heat flux at the wall. Use this temperature profile to show that the limiting Nusselt number for these conditions is Nu = 8.

(b) The asymptotic temperature distribution for the analogous problem for plug flow in a plane slit is given in Eq. 10B.10-2. Use this to show that the limiting Nusselt number is Nu = 12.

14B.2 Local overall heat-transfer coefficient. In Problem 14A.1 the thermal resistances of the condensed steam film and wall were neglected. Justify this neglect by calculating the actual inner-surface temperature of the tubes at that cross section in the exchanger at which the oil temperature is 150°F. You may assume that for the oil h_{loc} is constant throughout the exchanger at 190 Btu/hr · ft^2 · °F. The tubes are horizontal.

14B.3 The hot-wire anemometer.[1] A hot-wire anemometer is essentially a fine wire, usually made of platinum, which is heated electrically and inserted into a flowing fluid. The wire temperature, which is a function of the fluid temperature, fluid velocity, and the rate of heating, may be determined by measuring its electrical resistance.

(a) A straight cylindrical wire 0.5 in. long and 0.01 in. in diameter is exposed to a stream of air at 70°F flowing past the wire at 100 ft/s. What must the rate of energy input be in watts to maintain the wire surface at 600°F? Neglect radiation as well as heat conduction along the wire.

(b) It has been reported[2] that for a given fluid and wire at given fluid and wire temperatures (hence a given wire resistance)

$$I^2 = B\sqrt{v_\infty} + C \tag{14B.3-1}$$

[1]See, for example, G. Comte-Bellot, Chapter 34 in *The Handbook of Fluid Dynamics* (R. W. Johnson, ed.), CRC Press, Boca Raton, FL (1999).

[2]L. V. King, *Phil. Trans. Roy. Soc.* (London), **A214**, 373–432 (1914).

in which I is the current required to maintain the desired temperature, v_∞ is the velocity of the approaching fluid, and B and C are constants. How well does this equation agree with the predictions of Eq. 14.4-7 or Eq. 14.4-8 for the fluid and wire of (a) over a fluid velocity range of 100 to 300 ft/s? What is the significance of the constant C in Eq. 14B.3-1?

14B.4 Dimensional analysis. Consider the flow system described in the first paragraph of §14.3, for which dimensional analysis has already given the dimensionless parameters on which the dimensionless velocity profile (Eq. 6.2-7) and temperature profile (Eq. 14.3-9) depend.

(a) Use Eqs. 6.2-7 and 14.3-9 and the definition of cup-mixing temperature to get the time-averaged expression

$$\frac{T_{b2} - T_{b1}}{T_0 - T_{b1}} = \text{a function of Re, Pr, } L/D \tag{14B.4-1}$$

(b) Use the result just obtained and the definitions of the heat-transfer coefficients to derive Eqs. 14.3-12, 14.3-13, and 14.3-14.

14B.5 Relation between h_{loc} and h_{ln}. In many industrial tubular heat exchangers (see Example 15.4-1) the tube-surface temperature T_0 varies linearly with the bulk fluid temperature T_b. For this common situation h_{loc} and h_{ln} may be simply interrelated.

(a) Starting with Eq. 14.1-5, show that

$$h_{\text{loc}}(\pi D dz)(T_b - T_0) = -\left(\frac{1}{4}\pi D^2\right)(\rho \hat{C}_p \langle v \rangle dT_b) \tag{14B.5-1}$$

and therefore that

$$\int_0^L h_{\text{loc}}\,dz = \frac{1}{4}\rho \hat{C}_p D \langle v \rangle \frac{T_b(L) - T_b(0)}{(T_0 - T_b)_{\text{ln}}} \tag{14B.5-2}$$

(b) Combine the result in (a) with Eq. 14.1-4 to show that

$$h_{\text{ln}} = \frac{1}{L}\int_0^L h_{\text{loc}}\,dz \tag{14B.5-3}$$

in which L is the total tube length, and therefore that (if $(\partial h_{\text{loc}}/\partial L)_z = 0$, which is equivalent to the statement that axial heat conduction is neglected)

$$h_{\text{loc}}|_{z=L} = h_{\text{ln}} + L\frac{dh_{\text{ln}}}{dL} \tag{14B.5-4}$$

14B.6 Heat loss by free convection from a pipe. In Example 14.6-1, would the heat loss be higher or lower if the pipe-surface temperature were 200°F and the air temperature were 180°F?

14C.1 The Nusselt expression for condensing vapor heat-transfer coefficients (Fig. 14.7-1). Consider a laminar film of condensate flowing down a vertical wall, and assume that this liquid film constitutes the sole heat-transfer resistance on the vapor side of the wall. Further assume that (i) the shear stress between liquid and vapor may be neglected; (ii) the physical properties in the film may be evaluated at the arithmetic mean of vapor and cooling-surface temperatures and that the cooling-surface temperature may be assumed constant; (iii) acceleration of fluid elements in the film may be neglected compared to the gravitational and viscous forces; (iv) sensible heat changes, $\hat{C}_p dT$, in the condensate film are unimportant compared to the latent heat transferred through it; and (v) the heat flux is very nearly normal to the wall surface.

(a) Recall from §2.2 that the average velocity of a film of constant thickness δ is $\langle v_z \rangle = \rho g \delta^2 / 3\mu$. Assume that this relation is valid for any value of z.

(b) Write the energy equation for the film, neglecting film curvature and convection. Show that the heat flux through the film toward the cold surface is

$$-q_y = k\left(\frac{T_d - T_0}{\delta}\right) \tag{14C.1-1}$$

(c) As the film proceeds down the wall, it picks up additional material by the condensation process. In this process, heat is liberated to the extent of $\Delta\hat{H}_{\text{vap}}$ per unit mass of material that undergoes the change in state. Show that equating the heat liberation by condensation with the heat flowing through the film in a segment dz of the film leads to

$$\rho\Delta\hat{H}_{\text{vap}}d(\langle v_z\rangle\delta) = k\left(\frac{T_d - T_0}{\delta}\right)dz \tag{14C.1-2}$$

(d) Insert the expression for the average velocity from (a) into Eq. 14C.1-2 and integrate from $z = 0$ to $z = L$ to obtain

$$\delta(L) = \left(\frac{4k(T_d - T_0)\mu L}{\rho^2 g\Delta\hat{H}_{\text{vap}}}\right)^{1/4} \tag{14C.1-3}$$

(e) Use the definition of the heat-transfer coefficient and the result in (d) to obtain Eq. 14.7-5.

(f) Show that Eqs. 14.7-4 and 14.7-5 are equivalent for the conditions of this problem.

14C.2 Heat-transfer correlations for agitated tanks. A liquid of essentially constant physical properties is being continuously heated by passage through an agitated tank, as shown in Fig. 14C.2. Heat is supplied by condensation of steam on the outer wall of the tank. The thermal resistance of the condensate film and the tank wall may be considered small compared to that of the fluid in the tank, and the unjacketed portion of the tank may be assumed to be well insulated. The rate of liquid flow through the tank has a negligible effect on the flow pattern in the tank.

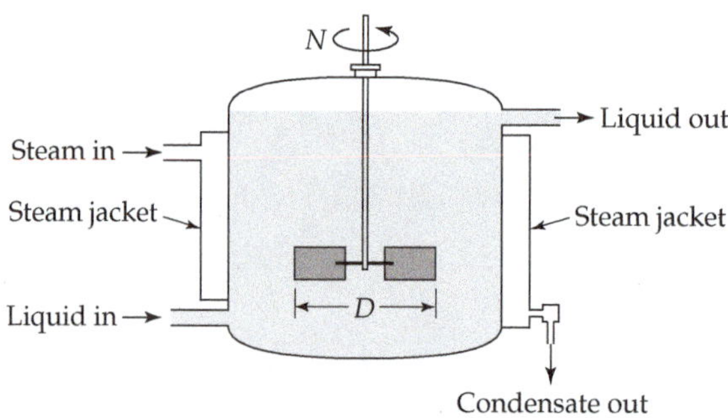

Fig. 14C.2 Continuous heating of a liquid in an agitated tank.

Develop a general form of dimensionless heat-transfer correlation for the tank corresponding to the correlation for tube flow in §14.3. Choose the following reference quantities: reference length, D, the impeller diameter; reference velocity, ND, where N is the rate of shaft rotation in revolutions per unit time; reference pressure, ρN^2D^2, where ρ is the fluid density.

Corrugated Tubes disrupt laminar sublayer →

Chapter 15

Macroscopic Balances for Nonisothermal Systems

In Chapter 7 we discussed the macroscopic mass, momentum, angular momentum, and mechanical energy balances. The treatment there was restricted to systems at constant temperature. Actually that restriction is somewhat artificial, because in real flow systems mechanical energy is always being converted into thermal energy by viscous dissipation. What we really assumed in Chapter 7 is that any heat so produced is either too small to change the fluid properties or is immediately conducted away through the walls of the system containing the fluid. In this chapter we extend the previous results to describe the overall behavior of nonisothermal macroscopic flow systems.

For a nonisothermal system there are five macroscopic balances that describe the relations between the inlet and outlet conditions of the stream. They may be derived by integrating the equations of change over the macroscopic system:

$$\int_{V(t)} (\text{eq. of continuity})\, dV = \text{macroscopic mass balance}$$

$$\int_{V(t)} (\text{eq. of motion})\, dV = \text{macroscopic momentum balance}$$

$$\int_{V(t)} (\text{eq. of angular momentum})\, dV = \text{macroscopic angular momentum balance}$$

$$\int_{V(t)} (\text{eq. of mechanical energy})\, dV = \text{macroscopic mechanical energy balance}$$

$$\int_{V(t)} (\text{eq. of (total) energy})\, dV = \text{macroscopic (total) energy balance}$$

The first four of these were discussed in Chapter 7, and their derivations suggest that they can be applied to nonisothermal systems just as well as to isothermal systems. In this chapter we add the fifth balance, namely that for the total energy. This is derived in §15.1, not by performing the integration indicated above, but rather by applying the law of conservation of total energy directly to the system shown in Fig. 7.0-1. Then in §15.2 we revisit the mechanical energy balance and examine it in light of the discussion of the (total) energy balance. Next in §15.3 we give the simplified versions of the macroscopic balances for steady-state systems and illustrate their use.

In §15.4 we give the differential forms (*d*-forms) of the steady-state balances. In these forms, the entry and exit planes 1 and 2 are taken to be only a differential distance apart. The "*d*-forms" are frequently useful for problems involving flow in conduits in which the velocity, temperature, and pressure are continually changing in the flow direction.

Finally in §15.5 we present several illustrations of unsteady-state problems that can be solved by the macroscopic balances.

This chapter will make use of nearly all the topics we have covered so far and provides an excellent opportunity to review the preceding chapters. Once again we remind the reader that in using the macroscopic balances, it may be necessary to omit some terms and to estimate the values of others. This requires good intuition or some extra experimental data.

§15.1 THE MACROSCOPIC ENERGY BALANCE

We consider the system sketched in Fig. 7.0-1 and make the same assumptions that were made in Chapter 7 with regard to quantities at the entrance and exit planes:

(i) The time-smoothed velocity is perpendicular to the entry and exit cross sections.

(ii) The density and other physical properties are uniform over the relevant cross sections.

(iii) The forces associated with the stress tensor $\boldsymbol{\tau}$ are neglected.

(iv) The pressure does not vary over the cross section.

To these we add (likewise at the entry and exit planes):

(v) The energy transport by conduction $\mathbf{q}$ is small compared to the convective energy transport and can be neglected.

(vi) The work associated with $[\boldsymbol{\tau} \cdot \mathbf{v}]$ is negligible when compared with $p\mathbf{v}$.

We now apply the statement of conservation of energy to the fluid in the macroscopic flow system. In doing this, we make use of the concept of potential energy to account for the work done against the external forces (this corresponds to using Eq. 11.1-10, rather than Eq. 11.1-8, as the equation of change for energy).

The statement of the law of conservation of energy thus takes the form:

$$\underbrace{\frac{d}{dt}(U_{\text{tot}} + K_{\text{tot}} + \Phi_{\text{tot}})}_{\substack{\text{rate of increase of}\\ \text{internal, kinetic, and}\\ \text{potential energy in}\\ \text{the system}}} = \underbrace{(\rho_1 \hat{U}_1 \langle v_1 \rangle + \tfrac{1}{2}\rho_1 \langle v_1^3 \rangle + \rho_1 \hat{\Phi}_1 \langle v_1 \rangle) S_1}_{\substack{\text{rate at which internal, kinetic, and}\\ \text{potential energy enter the system}\\ \text{at plane 1 by flow}}}$$

$$\underbrace{- (\rho_2 \hat{U}_2 \langle v_2 \rangle + \tfrac{1}{2}\rho_2 \langle v_2^3 \rangle + \rho_2 \hat{\Phi}_2 \langle v_2 \rangle) S_2}_{\substack{\text{rate at which internal, kinetic, and}\\ \text{potential energy leave the system}\\ \text{at plane 2 by flow}}} \qquad (15.1\text{-}1)$$

$$\underbrace{+Q}_{\substack{\text{rate at which}\\ \text{heat is added}\\ \text{to the system}\\ \text{across boundary}}} \quad \underbrace{+W_m}_{\substack{\text{rate at which work is done on}\\ \text{the system by the surroundings}\\ \text{by means of the moving}\\ \text{surfaces}}} \quad \underbrace{+(p_1 \langle v_1 \rangle S_1 - p_2 \langle v_2 \rangle S_2)}_{\substack{\text{rate at which work is}\\ \text{done on the system by the}\\ \text{surroundings at planes 1}\\ \text{and 2}}}$$

Here $U_{\text{tot}} = \int \rho \hat{U} dV$, $K_{\text{tot}} = \int \frac{1}{2}\rho v^2 dV$, and $\Phi_{\text{tot}} = \int \rho \hat{\Phi} dV$ are the total internal, kinetic, and potential energy within the system, the integrations being performed over the entire volume of the system. The last two terms represent the rate of work done by fluid at planes 1 and 2. The rate at which work is done on the system at plane 1 is the force $p_1 S_1$, exerted on the surface S_1, times the velocity $\langle v_1 \rangle$; the rate at which work is done by the system at plane 2 is the force $p_2 S_2$, exerted on the surface S_2, times the velocity $\langle v_2 \rangle$.

Equation 15.1-1 may be written in a more compact form by introducing the mass rates of flow $w_1 = \rho_1 \langle v_1 \rangle S_1$ and $w_2 = \rho_2 \langle v_2 \rangle S_2$, the total energy $E_{\text{tot}} = U_{\text{tot}} + K_{\text{tot}} + \Phi_{\text{tot}}$. We thus get for the *unsteady-state macroscopic energy balance*

$$\frac{d}{dt} E_{\text{tot}} = -\Delta \left[\left(\hat{U} + p\hat{V} + \frac{1}{2} \frac{\langle v^3 \rangle}{\langle v \rangle} + \hat{\Phi} \right) w \right] + Q + W_m \tag{15.1-2}$$

It is clear, from the derivation of Eq. 15.1-1, that the "work done on the system by the surroundings" consists of two parts: (1) the work done by the moving surfaces W_m, and (2) the work done at the ends of the system (planes 1 and 2), which appears as $-\Delta(p\hat{V}w)$ in Eq. 15.1-2. Although we have combined the pV-terms with the internal, kinetic, and potential energy terms on the right side of Eq. 15.1-2, it is inappropriate to say that "pV-energy enters and leaves the system" at the inlet and outlet. The pV-terms originate as work terms, along with W_m, and should be thought of as such.

We now consider the situation where the system is operating at steady state so that the total energy E_{tot} is constant, and the mass rates of flow in and out are equal ($w_1 = w_2 = w$). Then it is convenient to introduce the symbols $\hat{Q} = Q/w$ (the heat addition per unit mass of flowing fluid) and $\hat{W}_m = W_m/w$ (the work done on a unit mass of flowing fluid). Then the *steady-state macroscopic energy balance* is

$$\Delta \left(\hat{H} + \frac{1}{2} \frac{\langle v^3 \rangle}{\langle v \rangle} + gh \right) = \hat{Q} + \hat{W}_m \tag{15.1-3}$$

Here we have written $\hat{\Phi}_1 = gh_1$ and $\hat{\Phi}_2 = gh_2$, where h_1 and h_2 are heights above an arbitrarily chosen datum plane (see the discussion just before Eq. 3.3-2). Similarly, $\hat{H}_1 = \hat{U}_1 + p_1\hat{V}_1$ and $\hat{H}_2 = \hat{U}_2 + p_2\hat{V}_2$ are enthalpies per unit mass measured with respect to an arbitrarily specified reference state. The explicit formula for the enthalpy is given in Eq. 9.4-4.

For many problems in the chemical industry, the kinetic energy, potential energy, and work terms are negligible compared with the thermal terms in Eq. 15.1-3, and the steady-state energy balance becomes simply $\hat{H}_2 - \hat{H}_1 = \hat{Q}$, often incorrectly called an "enthalpy balance." However, this relation should not be construed as a statement of the conservation of enthalpy.

§15.2 THE MACROSCOPIC MECHANICAL ENERGY BALANCE

The macroscopic mechanical energy balance, given in §7.4 and derived in §7.7, is repeated here for comparison with Eqs. 15.1-2 and 15.1-3. The *unsteady-state macroscopic mechanical energy balance*, as given in Eq. 7.4-2, is:

$$\frac{d}{dt}(K_{\text{tot}} + \Phi_{\text{tot}}) = -\Delta \left(\frac{1}{2} \frac{\langle v^3 \rangle}{\langle v \rangle} + \hat{\Phi} + \frac{p}{\rho} \right) w + W_m - E_c - E_v \tag{15.2-1}$$

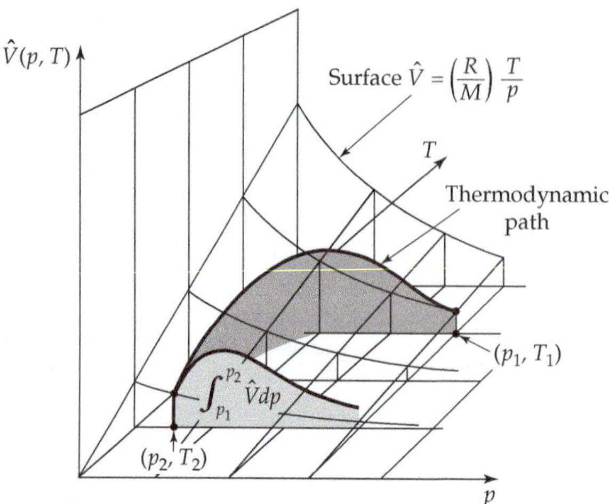

Fig. 15.2-1. Graphical representation of the integral in Eq. 15.2-2. The lightly shaded area is $\int_{p_1}^{p_2} \hat{V}dp = \int_{p_1}^{p_2}(1/\rho)\,dp$. Note that the value of this integral is negative here, because we are integrating from right to left.

where E_v and E_c are defined in Eqs. 7.4-3 and 7.4-4. An approximate form of the *steady-state macroscopic mechanical balance,* as given in Eq. 7.4-7, is:

$$\Delta\left(\frac{1}{2}\frac{\langle v^3\rangle}{\langle v\rangle}\right) + g\Delta h + \int_1^2 \frac{1}{\rho}dp = \hat{W}_m - \hat{E}_v \tag{15.2-2}$$

The details of the approximation introduced here are explained in Eqs. 7.7-9 to 7.7-12.

The integral in Eq. 15.2-2 must be evaluated along a "representative streamline" in the system. To do this, one must know the equation of state $\rho = \rho(p,T)$ and also how T changes with p along the streamline. In Fig. 15.2-1 the surface $\hat{V} = \hat{V}(p,T)$ for an ideal gas is shown. In the pT-plane there is shown a curve beginning at p_1,T_1 (the inlet stream conditions) and ending at p_2,T_2 (the outlet stream conditions). The curve in the pT-plane indicates the succession of states through which the gas passes in going from the initial state to the final state. The integral $\int_1^2(1/\rho)\,dp$ is then the projection of the shaded area in Fig. 15.2-1 onto the $p\hat{V}$-plane. It is evident that the value of this integral changes as the "thermodynamic path" of the process from 1 to 2 is altered. If one knows the path and the equation of state then one can compute $\int_1^2(1/\rho)\,dp$.

In several special situations, it is not difficult to evaluate the integral:

a. For *isothermal systems,* the integral is evaluated by prescribing the isothermal equation of state—that is, by giving a relation for ρ as a function of p. For example, for ideal gases, $\rho = pM/RT$ and

$$\int_1^2 \frac{1}{\rho}\,dp = \frac{RT}{M}\int_{p_1}^{p_2}\frac{1}{p}\,dp = \frac{RT}{M}\ln\frac{p_2}{p_1} \quad \text{(ideal gases)} \tag{15.2-3}$$

b. For *incompressible liquids,* ρ is constant so that

$$\int_1^2 \frac{1}{\rho}\,dp = \frac{1}{\rho}(p_2 - p_1) \quad \text{(incompressible liquids)} \tag{15.2-4}$$

c. For frictionless *adiabatic flow of ideal gases* with constant heat capacity, p and ρ are related by the expression $p\rho^{-\gamma} = \text{constant}$, in which $\gamma = \hat{C}_p/\hat{C}_V$, as shown in Example 11.4-5. Then the integral becomes

$$\int_1^2 \frac{1}{\rho}\,dp = \frac{p_1^{1/\gamma}}{\rho_1}\int_{p_1}^{p_2} \frac{1}{p^{1/\gamma}}\,dp = \frac{p_1}{\rho_1}\frac{\gamma}{\gamma-1}\left[\left(\frac{p_2}{p_1}\right)^{(\gamma-1)/\gamma} - 1\right]$$
$$= \frac{p_1}{\rho_1}\frac{\gamma}{\gamma-1}\left[\left(\frac{\rho_2}{\rho_1}\right)^{\gamma-1} - 1\right] \qquad (15.2\text{-}5)$$

Hence, for this special case of nonisothermal flow, the integration can be performed analytically.

We now conclude with several comments involving both the mechanical energy balance and the total energy balance. It was emphasized in §7.7 that Eq. 7.4-2 (same as Eq. 15.2-1) is derived by taking the dot product of **v** with the equation of motion and then integrating the result over the volume of the flow system. Since we start with the equation of motion—which is a statement of the law of conservation of linear momentum—the mechanical energy balance contains information different from that of the (total) energy balance, which is a statement of the law of conservation of energy. Therefore, in general, both balances are needed for problem solving. The mechanical energy balance is *not* "an alternative form" of the energy balance.

Furthermore, if we subtract the mechanical energy balance in Eq. 15.2-1 from the total energy balance in Eq. 15.1-2, we get the *macroscopic balance for the internal energy*

$$\boxed{\frac{dU_{\text{tot}}}{dt} = -\Delta\hat{U}w + Q + E_c + E_v} \qquad (15.2\text{-}6)$$

This states that the total internal energy in the system changes because of the difference in the amount of internal energy entering and leaving the system by fluid flow ($-\Delta\hat{U}w$), because of the net heat entering (or leaving) the system through walls of the system (Q), because of the net heat produced (or consumed) within the fluid by compression (or expansion) (E_c), and because of the heat produced in the system because of viscous dissipation (E_v). Equation 15.2-6 cannot be written down a priori, because there is no conservation law for internal energy. It can, however, be obtained by integrating the equation of change for internal energy, Eq. 11.2-1, over the entire flow system.

§15.3 USE OF THE MACROSOPIC BALANCES TO SOLVE STEADY-STATE PROBLEMS WITH FLAT VELOCITY PROFILES

The most important applications of the macroscopic balances are to steady-state problems. Furthermore, it is usually assumed that the flow is turbulent so that the variation of the velocity over the entry and exit cross sections can be safely neglected (see "Notes" after Eqs. 7.2-3 and 7.4-7). The five macroscopic balances, with these additional restrictions, are summarized in Table 15.3-1. They have also been generalized to multiple inlet and outlet ports so as to accommodate a larger set of problems.

EXAMPLE 15.3-1

The Cooling of an Ideal Gas

Two hundred pounds per hour of dry air enter the inner tube of the heat exchanger shown in Fig. 15.3-1 at 300°F and 30 psia, with a velocity of 100 ft/sec. The air leaves the exchanger at 0°F and 15 psia, at 10 ft above the exchanger entrance (at plane 1).

Calculate the rate of energy removal across the tube wall. Assume turbulent flow and ideal gas behavior, and use the following expression for the heat capacity of air:

$$\tilde{C}_p = \left(6.39\frac{\text{Btu}}{\text{lb-mol}\cdot{}^\circ\text{R}}\right) + \left(9.8\times10^{-4}\frac{\text{Btu}}{\text{lb-mol}\cdot{}^\circ\text{R}^2}\right)T - \left(8.18\times10^{-8}\frac{\text{Btu}}{\text{lb-mol}\cdot{}^\circ\text{R}^3}\right)T^2 \qquad (15.3\text{-}1)$$

where $\tilde{C}_p$ is in Btu/lb-mol · °R and T is in °R.

Table 15.3-1. Steady-State Macroscopic Balances for Turbulent Flow in Nonisothermal Systems

Mass:	$\sum w_1 - \sum w_2 = 0$	(A)
Momentum:	$\sum (v_1 w_1 + p_1 S_1)\mathbf{u}_1 - \sum (v_2 w_2 + p_2 S_2)\mathbf{u}_2 + m_{\text{tot}}\mathbf{g} = \mathbf{F}_{f\to s}$	(B)
Angular momentum:	$\sum (v_1 w_1 + p_1 S_1)[\mathbf{r}_1 \times \mathbf{u}_1] - \sum (v_2 w_2 + p_2 S_2)[\mathbf{r}_2 \times \mathbf{u}_2] + \mathbf{T}_{\text{ext}} = \mathbf{T}_{f\to s}$	(C)
Mechanical energy:	$\sum \left(\frac{1}{2}v_1^2 + gh_1 + \frac{p_1}{\rho_1}\right) w_1 - \sum \left(\frac{1}{2}v_2^2 + gh_2 + \frac{p_2}{\rho_2}\right) w_2 = -W_m + E_c + E_v$	(D)
(Total) energy:	$\sum \left(\frac{1}{2}v_1^2 + gh_1 + \hat{H}_1\right) w_1 - \sum \left(\frac{1}{2}v_2^2 + gh_2 + \hat{H}_2\right) w_2 = -W_m - Q$	(E)

Notes:

[a] All formulas here imply flat velocity profiles.

[b] $\sum w_1 = w_{1a} + w_{1b} + w_{1c} + \cdots$, where $w_{1a} = \rho_{1a} v_{1a} S_{1a}$, etc.

[c] h_1 and h_2 are elevations above an arbitrary datum plane.

[d] $\hat{H}_1$ and $\hat{H}_2$ are enthalpies per unit mass relative to some arbitrarily chosen reference state (see Eq. 9.4-4).

[e] The quantities E_c and E_v are defined in Eqs. 7.4-3 and 7.4-4. All equations are written for compressible flow; for incompressible flow, $E_c = 0$.

[f] $\mathbf{u}_1$ and $\mathbf{u}_2$ are unit vectors in the direction of flow.

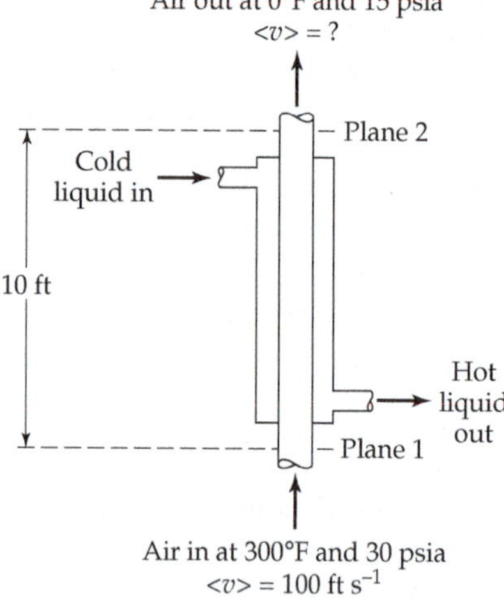

Fig. 15.3-1. The cooling of air in a countercurrent flow heat exchanger.

SOLUTION

For this system, the (steady-state) macroscopic energy balance, Eq. 15.1-3, becomes

$$(\hat{H}_2 - \hat{H}_1) + \tfrac{1}{2}(v_2^2 - v_1^2) + g(h_2 - h_1) = \hat{Q} \tag{15.3-2}$$

The enthalpy difference may be obtained from Eq. 9.4-4, and the velocity may be obtained as a function of temperature and pressure with the aid of the macroscopic mass balance, $\rho_1 v_1 = \rho_2 v_2$, and the ideal gas law $p = \rho RT/M$. Hence, Eq. 15.3-2 becomes

$$\frac{1}{M}\int_{T_1}^{T_2} \tilde{C}_p\, dT + \frac{1}{2}v_1^2 \left[\left(\frac{p_1 T_2}{p_2 T_1}\right)^2 - 1\right] + g(h_2 - h_1) = \hat{Q} \tag{15.3-3}$$

The explicit expression for $\tilde{C}_p$ in Eq. 15.3-1 may then be inserted into Eq. 15.3-3 and the integration performed. Then substitution of the numerical values gives the heat removal per pound of fluid passing through the heat exchanger:

$$\begin{aligned}
-\hat{Q} &= \left(\frac{1}{29}\frac{\text{lb-mol}}{\text{lb}_m}\right)\left[\begin{array}{l}\left(6.39\dfrac{\text{Btu}}{\text{lb-mol}\cdot{}^\circ\text{R}}\right)(300^\circ\text{R}) \\ +\dfrac{1}{2}\left(9.8\times10^{-4}\dfrac{\text{Btu}}{\text{lb-mol}\cdot{}^\circ\text{R}^2}\right)(5.78-2.12)(10^5)({}^\circ\text{R}^2) \\ -\dfrac{1}{3}\left(8.18\times10^{-8}\dfrac{\text{Btu}}{\text{lb-mol}\cdot{}^\circ\text{R}^3}\right)(4.39-0.97)(10^8)({}^\circ\text{R}^3)\end{array}\right] \\
&\quad -\frac{1}{2}\left(\frac{(100\ \text{ft/s})^2}{(32.2\ \text{ft}\cdot\text{lb}_m/\text{lb}_f\cdot\text{s}^2)(778\ \text{ft-lb}_f/\text{Btu})}\right)[(1.21)^2-1] \\
&\quad -\left(\frac{(32.2\ \text{ft/s}^2)(10\ \text{ft})}{(32.2\ \text{ft}\cdot\text{lb}_m/\text{lb}_f\cdot\text{s}^2)(778\ \text{ft}\cdot\text{lb}_f/\text{Btu})}\right) \\
&= (72.0-0.093-0.0128)\ \text{Btu/lb}_m \\
&= 71.9\ \text{Btu/lb}_m
\end{aligned} \tag{15.3-4}$$

The rate of heat removal is then

$$-\hat{Q}w = 14{,}380\ \text{Btu/hr} \tag{15.3-5}$$

Note that, in Eq. 15.3-4, the kinetic and potential energy contributions (–0.093 and –0.0128, repectively) are negligible in comparison with the enthalpy change (72.0).

EXAMPLE 15.3-2

Mixing of Two Ideal-Gas Streams

Two steady, turbulent streams of the same ideal gas flowing at different velocities, temperatures, and pressures are mixed as shown in Fig. 15.3-2. Calculate the velocity, temperature, and pressure of the resulting stream.

SOLUTION

The fluid behavior in this example is more complex than that for the incompressible, isothermal situation discussed in Example 7.6-2, because here changes in density and temperature may be important. We need to use the steady-state macroscopic energy balance, Eq. 15.1-3, and the ideal-gas equation of state, in addition to the mass and momentum balances. With these exceptions, we proceed as in Example 7.6-2.

We choose the inlet planes (1a and 1b) to be cross sections at which the fluids first begin to mix. The outlet plane (2) is taken far enough downstream that complete mixing has occurred. As in Example 7.6-2 we assume flat velocity profiles, negligible shear stresses on the pipe wall, and no changes in the potential energy. In addition, we neglect the changes in the heat capacity

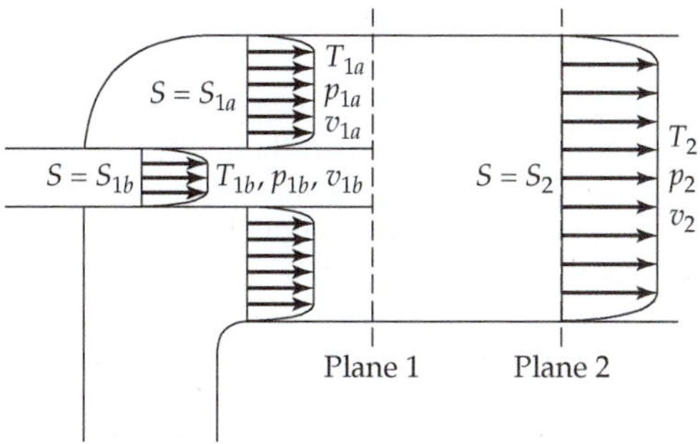

Fig. 15.3-2. The mixing of two ideal-gas streams.

of the fluid and assume adiabatic operation. We now write the following equations for this system with two entry ports and one exit port:

Mass: $$w_2 = w_{1a} + w_{1b} \tag{15.3-6}$$

Momentum: $$v_2 w_2 + p_2 S_2 = v_{1a} w_{1a} + p_{1a} S_{1a} + v_{1b} w_{1b} + p_{1b} S_{1b} \tag{15.3-7}$$

Energy: $$w_2[\hat{C}_p(T_2 - T_{\text{ref}}) + \tfrac{1}{2}v_2^2] = w_{1a}[\hat{C}_p(T_{1a} - T_{\text{ref}}) + \tfrac{1}{2}v_{1a}^2] + w_{1b}[\hat{C}_p(T_{1b} - T_{\text{ref}}) + \tfrac{1}{2}v_{1b}^2] \tag{15.3-8}$$

Equation of state: $$p_2 = \rho_2 R T_2 / M \tag{15.3-9}$$

In this set of equations we know all the quantities at 1a and 1b, and the four unknowns are p_2, T_2, ρ_2, and v_2. T_{ref} is the reference temperature for the enthalpy. By multiplying Eq. 15.3-6 by $\hat{C}_p T_{\text{ref}}$ and adding the result to Eq. 15.3-8, we get

$$w_2[\hat{C}_p T_2 + \tfrac{1}{2}v_2^2] = w_{1a}[\hat{C}_p T_{1a} + \tfrac{1}{2}v_{1a}^2] + w_{1b}[\hat{C}_p T_{1b} + \tfrac{1}{2}v_{1b}^2] \tag{15.3-10}$$

The right sides of Eqs. 15.3-6, 15.3-7, and 15.3-10 contain known quantities and we designate them by w, P, and E, respectively. Note that w, P, and E are not independent, because the pressure, temperature, and density of each inlet stream must be related by the equation of state.

We now solve Eq. 15.3-7 for v_2 and eliminate p_2 by using the ideal-gas law. In addition, we write w_2 as $\rho_2 v_2 S_2$. This gives

$$v_2 + \frac{RT_2}{Mv_2} = \frac{P}{w} \tag{15.3-11}$$

Equation 15.3-11 can be solved for T_2, which is inserted into Eq. 15.3-10 to give

$$w\left[\hat{C}_p\left(\frac{Mv_2}{R}\right)\left(\frac{P}{w} - v_2\right) + \frac{1}{2}v_2^2\right] = E \tag{15.3-12}$$

Then rearranging and introducing the molar heat capacity gives

$$\frac{1}{2}v_2^2 + \frac{\tilde{C}_p P v_2}{Rw} - \frac{\tilde{C}_p}{R}v_2^2 = \frac{E}{w} \tag{15.3-13}$$

Next we make use of the ideal gas relation $\tilde{C}_p/R = \gamma/(\gamma - 1)$ to rewrite Eq. 15.3-13 as

$$\left(1 - 2\frac{\gamma}{\gamma - 1}\right)v_2^2 + 2\left(\frac{\gamma}{\gamma - 1}\frac{P}{w}\right)v_2 - \frac{2E}{w} = 0 \tag{15.3-14}$$

or

$$\left(-\frac{\gamma + 1}{\gamma - 1}\right)v_2^2 + 2\left(\frac{\gamma}{\gamma - 1}\frac{P}{w}\right)v_2 - \frac{2E}{w} = 0 \tag{15.3-15}$$

Finally, we multiply by $-(\gamma - 1)/(\gamma + 1)$ to obtain

$$v_2^2 - \left[2\left(\frac{\gamma}{\gamma + 1}\right)\frac{P}{w}\right]v_2 + 2\left(\frac{\gamma - 1}{\gamma + 1}\right)\frac{E}{w} = 0 \tag{15.3-16}$$

in which $\gamma = C_p/C_V$, a quantity that varies from about 1.1 to 1.667 for gases. Here we have used the fact that $\tilde{C}_p/R = \gamma/(\gamma - 1)$ for an ideal gas. When Eq. 15.3-16 is solved for v_2, we get

$$v_2 = \left(\frac{\gamma}{\gamma + 1}\right)\frac{P}{w}\left[1 \pm \sqrt{1 - 2\left(\frac{\gamma^2 - 1}{\gamma^2}\right)\frac{wE}{P^2}}\right] \tag{15.3-17}$$

On physical grounds, the radicand cannot be negative (i.e., we expect the velocity to be a real number). It can be shown (see Problem 15B.4) that, when the radicand is zero, the velocity of the final stream is sonic. Therefore, in general, one of the solutions for v_2 is supersonic and one is subsonic. Only the lower (subsonic) solution can be obtained in the turbulent mixing process under consideration, since supersonic duct flow is unstable. The transition from supersonic to subsonic duct flow is illustrated in Example 11.4-6.

Once the velocity v_2 is known, the pressure and temperature may be calculated from Eqs. 15.3-7 and 15.3-11. The mechanical energy balance can be used to get $(E_c + E_v)$.

§15.4 THE *d*-FORMS OF THE MACROSCOPIC BALANCES

The estimation of E_v in the mechanical energy balance and Q in the total energy balance often presents some difficulties in nonisothermal systems.

For example, for E_v, consider the following two nonisothermal situations:

a. For liquids, the average flow velocity in a tube of constant cross section is nearly constant. However, the viscosity may change markedly in the direction of the flow because of the temperature changes, so that f in Eqs. 7.5-11 and 7.5-12 changes with distance. Hence Eqs. 7.5-11 and 7.5-12 cannot be applied to the entire pipe.

b. For gases, the viscosity does not change much with pressure, so that the local Reynolds number and local friction factor are nearly constant for ducts of constant cross section. However, the average velocity may change considerably along the duct as a result of the change in density with temperature. Hence, Eqs. 7.5-11 and 7.5-12 cannot be applied to the entire duct.

Similarly for pipe flow with the wall temperature changing with distance, it may be necessary to use local heat-transfer coefficients. For such a situation, we can write Eq. 15.1-3 on an incremental basis and generate a differential equation. Or the cross-sectional area of the conduit may be changing with downstream distance, and this situation also results in a need for handling the problem on an incremental basis.

It is therefore useful to rewrite the steady-state macroscopic mechanical energy balance and the total energy balance by taking planes 1 and 2 to be a differential distance dl apart. We then obtain what we call the "d-forms" of the balances:

i. The d-form of the mechanical energy balance

If we take planes 1 and 2 to be a differential distance apart, then we may write Eq. 15.2-2 in the following differential form:

$$d\left(\tfrac{1}{2}v^2\right) + g\,dh + \frac{1}{\rho}dp = d\hat{W} - d\hat{E}_v \tag{15.4-1}$$

Then using Eq. 7.5-9 for a differential length dl, we write

$$v\,dv + g\,dh + \frac{1}{\rho}dp = d\hat{W} - \tfrac{1}{2}v^2\frac{f}{R_h}dl \tag{15.4-2}$$

in which f is the local friction factor, and R_h is the local value of the mean hydraulic radius. In most applications we omit the $d\hat{W}$ term, since work is usually done at isolated points along the flow path (e.g., by pumps). The term $d\hat{W}$ would be needed, however, in tubes with extensible walls, magnetically driven flows, or systems with transport by rotating screws.

ii. The d-form of the total energy balance

If we write Eq. 15.1-3 in differential form, we have

$$d\left(\tfrac{1}{2}v^2\right) + g\,dh + d\hat{H} = d\hat{Q} + d\hat{W} \tag{15.4-3}$$

Then, using Eq. 9.4-4 for $\hat{H}$ and Eq. 14.1-8 for $d\hat{Q}$, we get

$$v\,dv + g\,dh + \hat{C}_p dT + \left[\hat{V} - T\left(\frac{\partial \hat{V}}{\partial T}\right)_p\right]dp = \frac{U_{\text{loc}}Z\Delta T}{w}dl + d\hat{W} \tag{15.4-4}$$

in which U_{loc} is the local overall heat-transfer coefficient, Z is the corresponding local conduit perimeter, and ΔT is the local temperature difference between the fluids inside and outside of the conduit.

Applications of Eqs. 15.4-2 and 15.4-4 are illustrated in the examples that follow.

EXAMPLE 15.4-1

Cocurrent or Countercurrent Heat Exchangers

It is desired to describe the performance of the simple double-pipe heat exchanger, shown in Fig. 15.4-1, in terms of the heat-transfer coefficients of the two streams and the thermal resistance of the pipe wall. The exchanger consists of two coaxial pipes with one fluid stream flowing through the inner pipe and another in the annular space; heat is transferred across the wall of the inner pipe. Both streams may flow in the same direction (cocurrent flow), as indicated in the figure, but normally it is more efficient to reverse the direction of one stream so that either w_h or w_c is negative (countercurrent flow). Steady-state turbulent flow may be assumed, and the heat losses to the surroundings may be neglected. Assume further that the local overall heat-transfer coefficient is constant along the exchanger.

SOLUTION

(a) ***Macroscopic energy balance for each stream as a whole.*** We designate quantities referring to the hot stream with a subscript h and the cold stream with subscript c. The steady-state energy balance in Eq. 15.1-3 becomes, for negligible changes in kinetic and potential energy,

$$w_h(\hat{H}_{h2} - \hat{H}_{h1}) = Q_h \tag{15.4-5}$$

$$w_c(\hat{H}_{c2} - \hat{H}_{c1}) = Q_c \tag{15.4-6}$$

Because there is no heat loss to the surroundings, $Q_h = -Q_c$. For incompressible liquids with a pressure drop that is not too large, or for ideal gases, Eq. 9.4-4 gives for constant $\hat{C}_p$ the relation $\Delta\hat{H} = \hat{C}_p\Delta T$. Hence, Eqs. 15.4-5 and 15.4-6 can be rewritten as

$$w_h\hat{C}_{ph}(T_{h2} - T_{h1}) = Q_h \tag{15.4-7}$$

$$w_c\hat{C}_{pc}(T_{c2} - T_{c1}) = Q_c = -Q_h \tag{15.4-8}$$

(b) ***d-form of the macroscopic energy balance.*** Application of Eq. 15.4-4 to the hot stream gives

$$\hat{C}_{ph}dT_h = \frac{U_0(2\pi r_0)(T_c - T_h)}{w_h}dl \tag{15.4-9}$$

where r_0 is the outside radius of the inner tube, and U_0 is the overall heat-transfer coefficient based on the radius r_0 (see Eq. 14.1-8).

Rearrangement of Eq. 15.4-9 gives

$$\frac{dT_h}{T_c - T_h} = U_0\frac{(2\pi r_0)dl}{w_h\hat{C}_{ph}} \tag{15.4-10}$$

The corresponding equation for the cold stream is

$$-\frac{dT_c}{T_c - T_h} = U_0\frac{(2\pi r_0)dl}{w_c\hat{C}_{pc}} \tag{15.4-11}$$

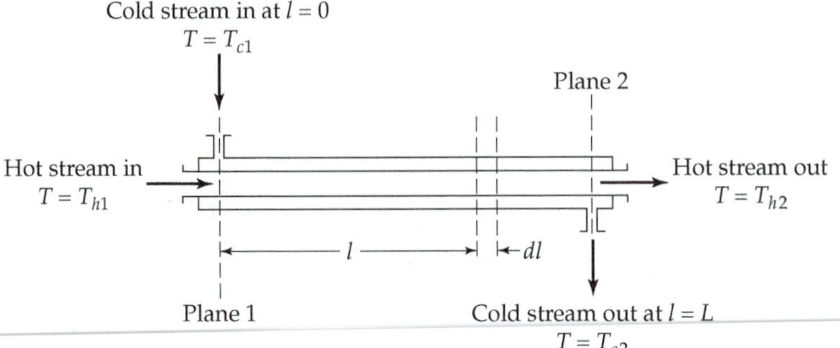

Fig. 15.4-1. A double-pipe heat exchanger.

Adding Eqs. 15.4-10 and 15.4-11 gives a differential equation for the temperature difference of the two fluids as a function of the coordinate l:

$$-\frac{d(T_h - T_c)}{T_h - T_c} = U_0\left(\frac{1}{w_h\hat{C}_{ph}} + \frac{1}{w_c\hat{C}_{pc}}\right)(2\pi r_0)dl \tag{15.4-12}$$

By assuming that U_0 is independent of l and integrating from plane 1 to plane 2, we get

$$\ln\left(\frac{T_{h1} - T_{c1}}{T_{h2} - T_{c2}}\right) = U_0\left(\frac{1}{w_h\hat{C}_{ph}} + \frac{1}{w_c\hat{C}_{pc}}\right)(2\pi r_0)L \tag{15.4-13}$$

where L is the total length of either tube. This expression relates the terminal temperatures to the stream rates and exchanger dimensions, and it can thus be used to describe the performance of the exchanger. However, it is conventional to rearrange Eq. 15.4-13 by taking advantage of the steady-state energy balances in Eq. 15.4-7 and 15.4-8. We solve each of these latter equations for $w\hat{C}_p$ and substitute the results into Eq. 15.4-13 to obtain

$$Q_c = U_0(2\pi r_0 L)\left(\frac{(T_{h2} - T_{c2}) - (T_{h1} - T_{c1})}{\ln[(T_{h2} - T_{c2})/(T_{h1} - T_{c1})]}\right) \tag{15.4-14}$$

or

$$Q_c = U_0 A_0 (T_h - T_c)_{\ln} \tag{15.4-15}$$

Here $A_0 = 2\pi r_0 L$ is the total outer surface of the inner tube, and $(T_h - T_c)_{\ln}$ is the "logarithmic mean temperature difference" between the two streams. Equations 15.4-14 and 15.4-15 describe the rate of heat exchange between the two streams and find wide application in engineering practice. Note that the stream mass flow rates do not appear explicitly in these equations, which are valid for both cocurrent and countercurrent exchangers (see Problem 15A.1).

From Eqs. 15.4-10 and 15.4-11, we can also get the stream temperatures as functions of l if desired. Considerable care must be used in applying the results of this example to laminar flow, for which the variation of the overall heat-transfer coefficient may be quite large. An example of a problem with variable U_0 is given in Problem 15B.1.

EXAMPLE 15.4-2

Power Requirement for Pumping a Compressible Fluid through a Long Pipe

A natural gas, which may be considered to be pure methane, is to be pumped through a long, smooth pipeline with an inside diameter of two feet. The gas enters the line at 100 psia with a velocity of 40 ft/s and at the ambient temperature of 70°F. Pumping stations are provided every 10 miles along the line, and at each of these stations the gas is recompressed and cooled to its original temperature and pressure (see Fig. 15.4-2). Estimate the power that must be expended on the gas at each pumping station, assuming ideal-gas behavior, flat velocity profiles, and negligible changes in elevation.

SOLUTION

We consider the pipe and compressor separately. First we apply Eq. 15.4-2 to a length dl of the pipe. We then integrate this equation between planes 1 and 2 to obtain the unknown pressure p_2. Once this is known, we may apply Eq. 15.2-2 to the system between planes 2 and 3 to obtain the work done by the pump.

(a) ***Flow through the pipe.*** For this portion of the system, Eq. 15.4-2 becomes

$$v\,dv + \frac{1}{\rho}dp + \frac{2v^2 f}{D}dl = 0 \tag{15.4-16}$$

where D is the pipe diameter. Since the pipe is quite long, we assume that the fluid is isothermal at 70°F. We may then eliminate both v and ρ from Eq. 15.4-16 by use of the assumed equation

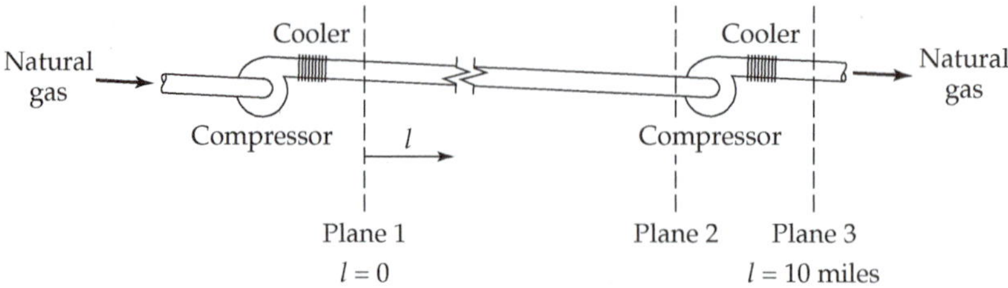

Fig. 15.4-2. Pumping a compressible fluid through a pipeline.

of state, $p = \rho RT/M$, and the macroscopic mass balance, which may be written $\rho v = \rho_1 v_1$. With ρ and v written in terms of the pressure, Eq. 15.4-16 then becomes

$$-\frac{1}{p}dp + \frac{RT_1}{M(p_1 v_1)^2}p dp + \frac{2f}{D}dl = 0 \tag{15.4-17}$$

It was pointed out in §1.6 that the viscosity of ideal gases is independent of the pressure. From this it follows that the Reynolds number of the gas, $\text{Re} = Dw/S\mu$, and hence the friction factor f, must be constants. We may then integrate Eq. 15.4-17 from plane 1 to plane 2 to obtain

$$-\ln\frac{p_2}{p_1} + \frac{1}{2}\left[\left(\frac{p_2}{p_1}\right)^2 - 1\right]\frac{RT_1}{Mv_1^2} + \frac{2fL}{D} = 0 \tag{15.4-18}$$

This equation gives p_2 in terms of quantities that are already known, except for f which is easily calculated: the kinematic viscosity of methane at 100 psi and 70°F is about 2.61×10^{-5} ft^2/s, and therefore $\text{Re} = Dv/\nu = (2\text{ ft})(40\text{ ft/s})/(2.61 \times 10^{-5}\text{ ft}^2/\text{s}) = 3.07 \times 10^6$. The friction factor can then be estimated to be 0.0025, from Fig. 6.2-2.

Substituting numerical values into Eq. 15.4-18 gives

$$-\ln\frac{p_2}{p_1} + \frac{1}{2}\left[\left(\frac{p_2}{p_1}\right)^2 - 1\right]\frac{(1545\text{ ft}\cdot\text{lb}_f/\text{lb-mol}\cdot{}^\circ\text{R})(530^\circ\text{R})(32.2\text{ ft}\cdot\text{lb}_m/\text{lb}_f\cdot\text{s}^2)}{(16.04\text{ lb}_m/\text{lb-mol})(40\text{ ft/s})^2}$$

$$+\frac{(2)(0.0025)(52{,}800\text{ ft})}{(2\text{ ft})} = 0 \tag{15.4-19}$$

or

$$-\ln\frac{p_2}{p_1} + 513\left[\left(\frac{p_2}{p_1}\right)^2 - 1\right] + 132 = 0 \tag{15.4-20}$$

By solving this equation numerically, with $p_1 = 100$ psia, we obtain $p_2 = 86$ psia.

(b) ***Flow through the compressor.*** We are now ready to apply the mechanical energy balance to the compressor. We start by putting Eq. 15.2-1 into the form

$$\hat{W}_m = \tfrac{1}{2}(v_3^2 - v_2^2) + \int_{p_2}^{p_3}\frac{1}{\rho}dp + \hat{E}_v \tag{15.4-21}$$

To evaluate the integral in this equation, we assume that the compression is adiabatic (i.e., the compressor surface area available for heat transfer is small and thus the actual rate of heat of transfer Q should also be small) and further that $\hat{E}_v$ between planes 2 and 3 can be neglected. We may use Eq. 15.2-5 to rewrite Eq. 15.4-21 as

$$\begin{aligned}\hat{W}_m &= \tfrac{1}{2}(v_3^2 - v_2^2) + \frac{p_2^{1/\gamma}}{\rho_2}\int_{p_2}^{p_3} p^{-1/\gamma}dp \\ &= \frac{v_1^2}{2}\left[1 - \left(\frac{p_1}{p_2}\right)^2\right] + \frac{RT_2}{M}\frac{\gamma}{\gamma - 1}\left[\left(\frac{p_1}{p_2}\right)^{(\gamma-1)/\gamma} - 1\right]\end{aligned} \tag{15.4-22}$$

in which $\hat{W}_m$ is the energy required of the compressor. By substituting numerical values into Eq. 15.4-22, we get

$$\hat{W}_m = \frac{(40\text{ ft/s})^2}{2(32.2\text{ ft}\cdot\text{lb}_m/\text{lb}_f\cdot\text{s}^2)}\left[1-\left(\frac{100\text{ psia}}{86\text{ psia}}\right)^2\right]$$
$$+\frac{(1545\text{ ft}\cdot\text{lb}_f/\text{lb-mol}\cdot{}^\circ\text{R})(530^\circ\text{R})}{16.04\text{ lb}_m/\text{lb-mol}}\frac{1.3}{0.3}\left[\left(\frac{100\text{ psia}}{86\text{ psia}}\right)^{0.3/1.3}-1\right]$$
$$= -9 + 7834 = 7825\text{ ft}\cdot\text{lb}_f/\text{lb}_m \tag{15.4-23}$$

The power required to compress the fluid is

$$w\hat{W}_m = \left(\frac{\pi D^2}{4}\right)\left(\frac{p_1 M}{RT_1}\right)v_1\hat{W}_m$$
$$= \pi(1\text{ ft}^2)\frac{(100\text{ psia})(16.04\text{ lb}_m/\text{lb-mol})}{(10.73\text{ ft}^3\cdot\text{psia/lb-mol}\cdot{}^\circ\text{R})(530^\circ\text{R})}(40\text{ ft/s})(7825\text{ ft}\cdot\text{lb}_f/\text{lb}_m)$$
$$= 277{,}000\text{ ft}\cdot\text{lb}_f/\text{s} = 504\text{ hp} \tag{15.4-24}$$

The power required would be virtually the same if the flow in the pipeline were adiabatic (see Problem 15A.2).

The assumptions used here—assuming the compression to be adiabatic and neglecting the viscous dissipation—are conventional in the design of compressor-cooler combinations. Note that the energy required to run the compressor is greater than the calculated work, $\hat{W}_m$, by (i) $\hat{E}_v$ between planes 2 and 3, (ii) mechanical losses in the compressor itself, and (iii) errors in the assumed p-ρ path. Normally the energy required at the pump shaft is at least 15 to 20% greater than $\hat{W}_m$.

---

§15.5 USE OF THE MACROSCOPIC BALANCES TO SOLVE UNSTEADY-STATE PROBLEMS AND PROBLEMS WITH NON-FLAT VELOCITY PROFILES

In Table 15.5-1 we summarize all five macroscopic balances for unsteady state and nonflat velocity profiles, and for systems with multiple entry and exit ports. One practically never needs to use these balances in this degree of completeness, but it is convenient to have the entire set of equations collected in one place. We illustrate their use in the examples that follow.

EXAMPLE 15.5-1

Heating of a Liquid in an Agitated Tank[1]

A cylindrical tank capable of holding 1000 ft^3 of liquid is equipped with an agitator having sufficient power to keep the liquid contents at a uniform temperature (see Fig. 15.5-1). Heat is transferred to the contents by means of a coil arranged in such a way that the area available for heat transfer is proportional to the quantity of liquid in the tank. This heating coil consists of 10 turns, 4 ft in diameter, of 1-in. o. d. tubing. Water at 20°C is fed into this tank at a rate of 20 lb$_m$/min, starting with no water in the tank at time $t = 0$. Steam at 105°C flows through the heating coil, and the overall heat-transfer coefficient is 100 Btu/hr · ft^2 · °F. What is the temperature of the water when the tank is filled?

---

[1]This problem is taken in modified form from W. R. Marshall, Jr., and R. L. Pigford, *Applications of Differential Equations to Chemical Engineering Problems*, University of Delaware Press, Newark DE (1947), pp. 16–18.

Table 15.5-1. Unsteady-State Macroscopic Balances for Flow in Nonisothermal Systems

Mass:	$\frac{d}{dt}m_{tot} = \sum w_1 - \sum w_2 = \sum \rho_1 \langle v_1 \rangle S_1 - \sum \rho_2 \langle v_2 \rangle S_2$	(A)
Momentum:	$\frac{d}{dt}\mathbf{P}_{tot} = \sum \left(\frac{\langle v_1^2 \rangle}{\langle v_1 \rangle} w_1 + p_1 S_1 \right) \mathbf{u}_1 - \sum \left(\frac{\langle v_2^2 \rangle}{\langle v_2 \rangle} w_2 + p_2 S_2 \right) \mathbf{u}_2 + m_{tot}\mathbf{g} - \mathbf{F}_{f \to s}$	(B)
Angular momentum:	$\frac{d}{dt}\mathbf{L}_{tot} = \sum \left(\frac{\langle v_1^2 \rangle}{\langle v_1 \rangle} w_1 + p_1 S_1 \right) [\mathbf{r}_1 \times \mathbf{u}_1] - \sum \left(\frac{\langle v_2^2 \rangle}{\langle v_2 \rangle} w_2 + p_2 S_2 \right) [\mathbf{r}_2 \times \mathbf{u}_2] + \mathbf{T}_{ext} - \mathbf{T}_{f \to s}$	(C)
Mechanical energy:	$\frac{d}{dt}(K_{tot} + \Phi_{tot}) = \sum \left(\frac{1}{2}\frac{\langle v_1^3 \rangle}{\langle v_1 \rangle} + gh_1 + \frac{p_1}{\rho_1} \right) w_1 - \sum \left(\frac{1}{2}\frac{\langle v_2^3 \rangle}{\langle v_2 \rangle} + gh_2 + \frac{p_2}{\rho_2} \right) w_2 + W_m - E_c - E_v$	(D)
(Total) energy:	$\frac{d}{dt}(K_{tot} + \Phi_{tot} + U_{tot}) = \sum \left(\frac{1}{2}\frac{\langle v_1^3 \rangle}{\langle v_1 \rangle} + gh_1 + \hat{H}_1 \right) w_1 - \sum \left(\frac{1}{2}\frac{\langle v_2^3 \rangle}{\langle v_2 \rangle} + gh_2 + \hat{H}_2 \right) w_2 + W_m + Q$	(E)

Notes:

[a] $\sum w_1 = w_{1a} + w_{1b} + w_{1c} + \cdots$, where $w_{1a} = \rho_{1a} v_{1a} S_{1a}$, etc.

[b] h_1 and h_2 are elevations above an arbitrary datum plane.

[c] $\hat{H}_1$ and $\hat{H}_2$ are enthalpies per unit mass relative to some arbitrarily chosen reference state; the formula for $\hat{H}$ is given in Eq. 9.4-4.

[d] The quantities E_c and E_v are defined in Eqs. 7.4-3 and 7.4-4. All equations are written for compressible flow; for incompressible flow, $E_c = 0$.

[e] $\mathbf{u}_1$ and $\mathbf{u}_2$ are unit vectors in the direction of flow.

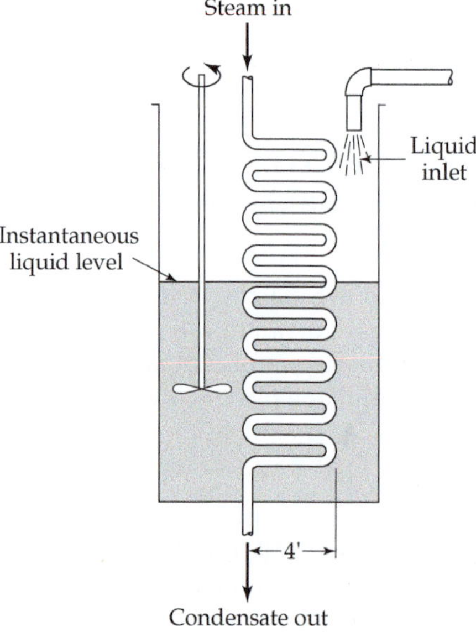

Fig. 15.5-1. Heating of a liquid in a tank with a variable liquid level.

SOLUTION

We make the following assumptions:

a. The steam temperature is uniform throughout the coil.

b. The density and heat capacity do not change very much with temperature.

c. The fluid is approximately incompressible so that $\hat{C}_p \approx \hat{C}_V$.

d. The agitator maintains uniform temperature throughout the liquid.

e. The heat-transfer coefficient is independent of position and time.

f. The walls of the tank are perfectly insulated so that no heat loss occurs.

We select the fluid within the tank as the system to be considered, and we make a time-dependent energy balance over this system. Such a balance is provided by Eq. (E) of Table 15.5-1. On the left side of the equation, the time rates of change of kinetic and potential energies can be neglected, since they are probably small compared with that of the internal energy. On the right side, we can omit the work term, and the kinetic and potential energy terms can be discarded, since they will be small compared with the other terms. Inasmuch as there is no outlet stream, we can set w_2 equal to zero. Hence, for this system, the total energy balance simplifies to

$$\frac{d}{dt}U_{\text{tot}} = w_1\hat{H}_1 + Q \tag{15.5-1}$$

This states that the internal energy of the system increases because of the enthalpy added by the incoming fluid, and because of the addition of heat through the steam coil.

Because U_{tot} and $\hat{H}_1$ cannot be given absolutely, we now select the inlet temperature T_1 as the thermal datum plane. Then $\hat{H}_1 = 0$ and $U_{\text{tot}} = \rho\hat{C}_V V(T - T_1) \approx \rho\hat{C}_p V(T - T_1)$, where T and V are the instantaneous temperature and volume of the liquid. Furthermore, the rate of heat addition to the liquid Q is given by $Q = U_0 A(T_s - T)$, in which T_s is the steam temperature, and A is the instantaneous heat-transfer area. Hence, Eq. 15.5-1 becomes

$$\rho\hat{C}_p\frac{d}{dt}V(T - T_1) = U_0 A(T_s - T) \tag{15.5-2}$$

The expressions for $V(t)$ and $A(t)$ are

$$V(t) = \frac{w_1}{\rho}t \qquad A(t) = \frac{V(t)}{V_0}A_0 = \frac{w_1 t}{\rho V_0}A_0 \tag{15.5-3}$$

in which V_0 and A_0 are the volume and heat-transfer area when the tank is full. Hence, the energy balance equation becomes

$$w_1\hat{C}_p t\frac{d}{dt}(T - T_1) + w_1\hat{C}_p(T - T_1) = \frac{w_1 t}{\rho V_0}U_0 A_0(T_s - T) \tag{15.5-4}$$

which is to be solved with the initial condition that $T = T_1$ at $t = 0$.

The equation is somewhat easier to solve in dimensionless form. We divide both sides by $w_1\hat{C}_p(T_s - T_1)$ to get

$$t\frac{d}{dt}\left(\frac{T - T_1}{T_s - T_1}\right) + \left(\frac{T - T_1}{T_s - T_1}\right) = \frac{U_0 A_0 t}{\rho\hat{C}_p V_0}\left(\frac{T_s - T}{T_s - T_1}\right) \tag{15.5-5}$$

This equation suggests that suitable definitions of dimensionless temperature and time are

$$\Theta(\tau) = \left(\frac{T(t) - T_1}{T_s - T_1}\right) \quad \text{and} \quad \tau = \frac{U_0 A_0 t}{\rho\hat{C}_p V_0} \tag{15.5-6,7}$$

Then Eq. 15.5-5 becomes after some rearranging

$$\frac{d\Theta}{d\tau} + \left(1 + \frac{1}{\tau}\right)\Theta = 1 \tag{15.5-8}$$

and the initial condition requires that $\Theta = 0$ at $\tau = 0$.

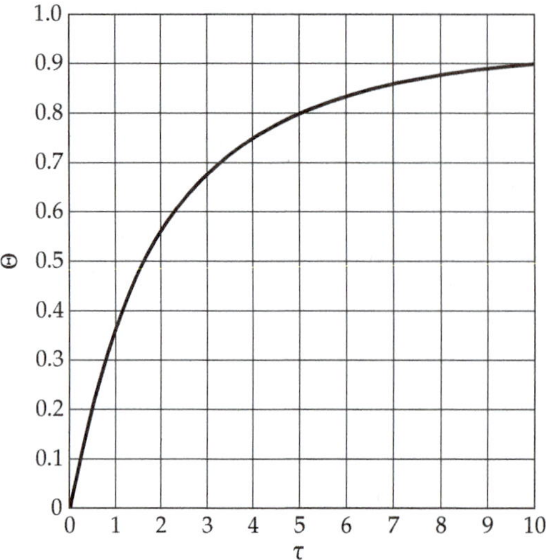

Fig. 15.5-2. Plot of dimensionless temperature, $\Theta = (T - T_1)/(T_s - T_1)$, versus dimensionless time, $\tau = (U_0 A_0/\rho \hat{C}_p V_0)t$, according to Eq. 15.5-10. [W. R. Marshall and R. L. Pigford, *Application of Differential Equations to Chemical Engineering Problems*, University of Delaware Press, Newark, DE (1947), p. 18.]

This is a first-order linear differential equation of the form of Eq. C.1-2. It may be solved as follows:

$$\begin{aligned}\Theta(\tau) &= e^{-\int\left(1+\frac{1}{\tau}\right)d\tau}\left[\int e^{\int\left(1+\frac{1}{\tau}\right)d\tau}d\tau + C\right]\\ &= e^{-(\tau+\ln\tau)}\left[\int e^{(\tau+\ln\tau)}d\tau + C\right]\\ &= \frac{e^{-\tau}}{\tau}\left[\int \tau e^{\tau}d\tau + C\right]\\ &= \frac{e^{-\tau}}{\tau}[e^{\tau}(\tau-1)+C]\\ &= 1 - \frac{1 - Ce^{-\tau}}{\tau}\end{aligned} \tag{15.5-9}$$

The constant of integration, C, can be obtained from the initial condition after first multiplying Eq. 15.5-9 by τ. In that way it is found that $C = 1$, so that the final solution is

$$\Theta(\tau) = 1 - \frac{1 - e^{-\tau}}{\tau} \tag{15.5-10}$$

This function is shown in Fig. 15.5-2.

Finally, the temperature T_0 of the liquid in the tank, when it has been filled, is given by Eq. 15.5-10 when $t = \rho V_0/w_1$ (from Eq. 15.5-3) or $\tau = U_0 A_0/w_1 \hat{C}_p$ (from Eq. 15.5-7). Therefore, in terms of the original variables

$$\frac{T_0 - T_1}{T_s - T_1} = 1 - \frac{1 - \exp(-U_0 A_0/w_1 \hat{C}_p)}{U_0 A_0/w_1 \hat{C}_p} \tag{15.5-11}$$

Thus, it can be seen that the final liquid temperature is determined entirely by the dimensionless group $U_0 A_0/w_1 \hat{C}_p$, which for this problem is

$$\frac{U_0 A_0}{w_1 \hat{C}_p} = \frac{(100\ \text{Btu/hr}\cdot\text{ft}^2\cdot{}^\circ\text{F})(10)(\pi\cdot 4\ \text{ft})\left(\pi\cdot\frac{1}{12}\ \text{ft}\right)}{(20\ \text{lb}_m/\text{min})(60\ \text{min/hr})(0.998\ \text{Btu/lb}_m\cdot{}^\circ\text{F})} = 2.74 \tag{15.5-12}$$

Knowing this we can find from Eq. 15.5-11 that

$$\frac{T_0 - T_1}{T_s - T_1} = 1 - \frac{1 - \exp(-2.74)}{2.74} = 0.659 \tag{15.5-13}$$

whence $T_0 = 76°\text{C}$.

EXAMPLE 15.5-2

Operation of a Simple Temperature Controller

A well-insulated agitated tank is shown in Fig. 15.5-3. Liquid enters at a temperature $T_1(t)$, which may vary with time. It is desired to control the temperature, $T_2(t)$, of the fluid leaving the tank. It is presumed that the stirring device is sufficiently thorough that the temperature in the tank is uniform and equal to the exit temperature. The volume of the liquid in the tank, V, and the mass rate of liquid flow, w, are both constant.

To accomplish the desired control, a metallic electric heating coil of surface area A is placed in the tank, and a temperature-sensing element is placed in the exit stream to measure $T_2(t)$. These devices are connected to a temperature controller that supplies energy to the heating coil at a rate $Q_e = b(T_{max} - T_2(t))$, in which T_{max} is the maximum temperature for which the controller is designed to operate, and b is a known parameter. It may be assumed that the liquid temperature $T_2(t)$ is always less than T_{max} in normal operation. The heating coil supplies energy to the liquid in the tank at a rate $Q = UA\left[T_c(t) - T_2(t)\right]$, where U is the overall heat-transfer coefficient between the coil and the liquid, and $T_c(t)$ is the instantaneous coil temperature, considered uniform at any instant.

Up to time $t = 0$, the system has been operating at steady state with liquid inlet temperature $T_1(t) = T_{10}$ and exit temperature $T_2(t) = T_{20}$. At time $t = 0$, the inlet stream temperature is suddenly increased to $T_1(t) = T_{1\infty}$ and held there. As a consequence of this disturbance, the tank temperature will begin to rise, and the temperature indicator in the outlet stream will signal the controller to decrease the power supplied to the heating coil. Ultimately, the liquid temperature in the tank will attain a new steady-state value $T_{2\infty}$. It is desired to describe the behavior of the liquid temperature $T_2(t)$. A qualitative sketch showing the various temperatures is given in Fig. 15.5-4.

SOLUTION

We first write the unsteady-state macroscopic energy balances (Eq. (E) of Table 15.5-1) for the liquid in the tank and for the heating coil:

(liquid) $$\rho\hat{C}_p V\frac{dT_2}{dt} = w\hat{C}_p(T_1 - T_2) + UA(T_c - T_2) \tag{15.5-14}$$

(coil) $$\rho_c\hat{C}_{pc} V_c\frac{dT_c}{dt} = b(T_{max} - T_2) - UA(T_c - T_2) \tag{15.5-15}$$

Note that in applying the macroscopic energy balance to the liquid, we have neglected kinetic and potential energy changes as well as the power input to the agitator.

(a) ***Steady-state behavior for*** $t \leq 0$. When the time derivatives in Eqs. 15.5-14 and 15.5-15 are set equal to zero and the equations added, we get for $t \leq 0$, where $T_1 = T_{10}$:

$$T_{20} = \frac{w\hat{C}_p T_{10} + bT_{max}}{w\hat{C}_p + b} \tag{15.5-16}$$

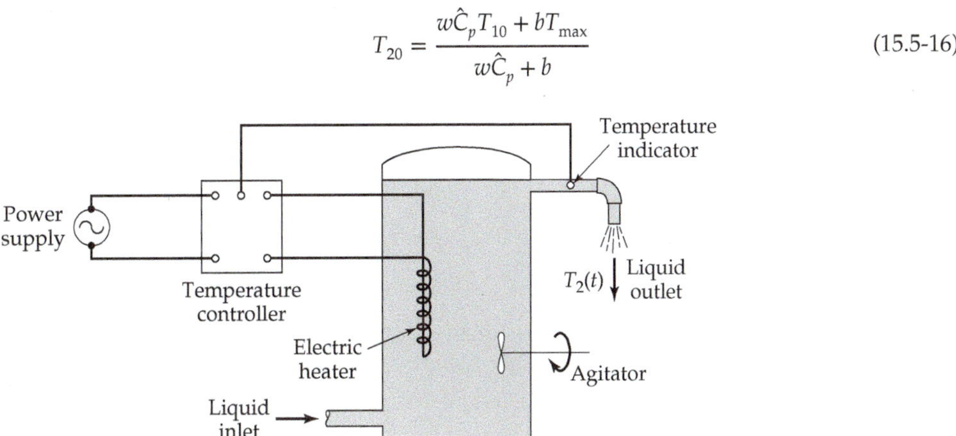

Fig. 15.5-3. An agitated tank with a temperature controller.

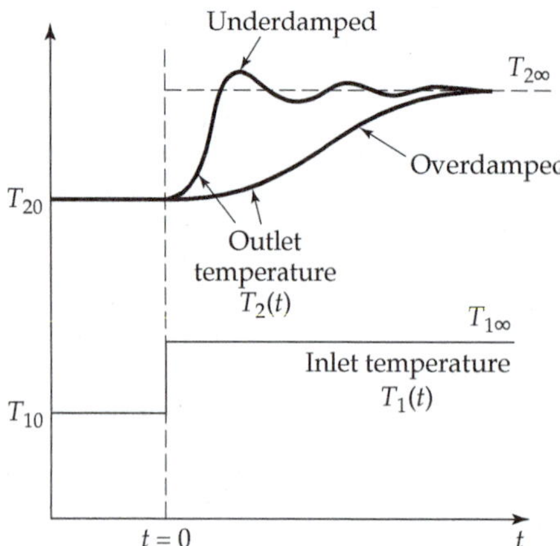

Fig. 15.5-4. Inlet and outlet temperatures as functions of time.

Then from Eq. 15.5-15 we can get the initial temperature of the coil

$$T_{c0} = T_{20}\left(1 - \frac{b}{UA}\right) + \frac{bT_{\max}}{UA} \tag{15.5-17}$$

(b) ***Steady-state behavior for*** $t \to \infty$. When similar operations are performed with $T_1 = T_{1\infty}$, we get

$$T_{2\infty} = \frac{w\hat{C}_p T_{1\infty} + bT_{\max}}{w\hat{C}_p + b} \tag{15.5-18}$$

and

$$T_{c\infty} = T_{2\infty}\left(1 - \frac{b}{UA}\right) + \frac{bT_{\max}}{UA} \tag{15.5-19}$$

for the final temperature of the coil.

(c) ***Unsteady-state behavior for*** $t > 0$. It is convenient to define dimensionless variables using the steady-state quantities for $t \leq 0$ and $t \to \infty$:

$$\Theta_2(\tau) = \frac{T_2(t) - T_{2\infty}}{T_{20} - T_{2\infty}} = \text{dimensionless liquid temperature} \tag{15.5-20}$$

$$\Theta_c(\tau) = \frac{T_c(t) - T_{c\infty}}{T_{c0} - T_{c\infty}} = \text{dimensionless coil temperature} \tag{15.5-21}$$

$$\tau = \frac{UAt}{\rho\hat{C}_p V} = \text{dimensionless time} \tag{15.5-22}$$

In addition we define three dimensionless parameters:

$$R = \rho\hat{C}_p V/\rho_c\hat{C}_{pc}V_c = \text{ratio of thermal capacities} \tag{15.5-23}$$

$$F = w\hat{C}_p/UA = \text{flow-rate parameter} \tag{15.5-24}$$

$$B = b/UA = \text{controller parameter} \tag{15.5-25}$$

In terms of these quantities, the unsteady-state balances in Eqs. 15.5-14 and 15.5-15 become (after considerable manipulation)

$$\frac{d\Theta_2}{d\tau} = -(1+F)\Theta_2 + (1-B)\Theta_c \tag{15.5-26}$$

$$\frac{d\Theta_c}{d\tau} = R(\Theta_2 - \Theta_c) \tag{15.5-27}$$

We may now eliminate Θ_c between these equations as follows: Take the derivative with respect τ of Eq. 15.5-26 and then insert the expression for $d\Theta_c/d\tau$ from Eq. 15.5-27 to get

$$\frac{d^2\Theta_2}{d\tau^2} + (1+F)\frac{d\Theta_2}{d\tau} - (1-B)R(\Theta_2 - \Theta_c) = 0 \tag{15.5-28}$$

Then multiply Eq.15.5-26 by R to get

$$R\frac{d\Theta_2}{d\tau} + (1+F)R\Theta_2 - (1-B)R\Theta_c = 0 \tag{15.5-29}$$

Addition of Eqs. 15.5-28 and 15.5-29 gives a single second-order linear ordinary differential equation for the exit liquid temperature as a function of time

$$\frac{d^2\Theta_2}{d\tau^2} + (1+R+F)\frac{d\Theta_2}{d\tau} + R(B+F)\Theta_2 = 0 \tag{15.5-30}$$

This equation is in the form of Eq. C.1-7. The general solution is then

$$\Theta_2(\tau) = C_+e^{m_+\tau} + C_-e^{m_-\tau} \qquad (m_+ \neq m_-) \tag{15.5-31}$$

or

$$\Theta_2(\tau) = C_1e^{m\tau} + C_2\tau e^{m\tau} \qquad (m_+ = m_- = m) \tag{15.5-32}$$

where

$$m_\pm = \tfrac{1}{2}\left[-(1+R+F) \pm \sqrt{(1+R+F)^2 - 4R(B+F)}\right] \tag{15.5-33}$$

The behavior of the tank temperature thus depends on the value of the discriminant $(1+R+F)^2 - 4R(B+F)$ in Eq. 15.5-33:

(a) If $(1+R+F)^2 > 4R(B+F)$, the system is *overdamped*, and the liquid temperature changes slowly and monotonically to its asymptotic value $T_{2\infty}$.

(b) If $(1+R+F)^2 < 4R(B+F)$, the system is *underdamped*, and liquid temperature oscillates about its asymptotic value $T_{2\infty}$, with the oscillations becoming smaller and smaller (see Eq. C.1-7c).

(c) If $(1+R+F)^2 = 4R(B+F)$, the system is *critically damped*, and the liquid temperature approaches its asymptotic value in most rapid, monotonic fashion.

The system parameters appear in the dimensionless time variable, as well as in the parameters B,F, and R. Therefore, numerical calculations are needed to decide whether in a particular system the temperature will oscillate or not.

EXAMPLE 15.5-3

Flow of Compressible Fluids through Head Meters

Extend the development of Example 7.6-5 to the steady-state flow of compressible fluids through orifice meters and Venturi tubes.

SOLUTION

We begin, as in Example 7.6-5, by writing down the steady-state mass and mechanical energy balances between reference planes 1 and 2 of the two flow meters shown in Fig. 15.5-5. For compressible fluids, these may be expressed as

$$w = \rho_1\langle v_1\rangle S_1 = \rho_2\langle v_2\rangle S_2 \tag{15.5-34}$$

$$\frac{\langle v_2\rangle^2}{2\alpha_2} - \frac{\langle v_1\rangle^2}{2\alpha_1} + \int_1^2 \frac{1}{\rho}\,dp + \frac{1}{2}\langle v_2\rangle^2 e_v = 0 \tag{15.5-35}$$

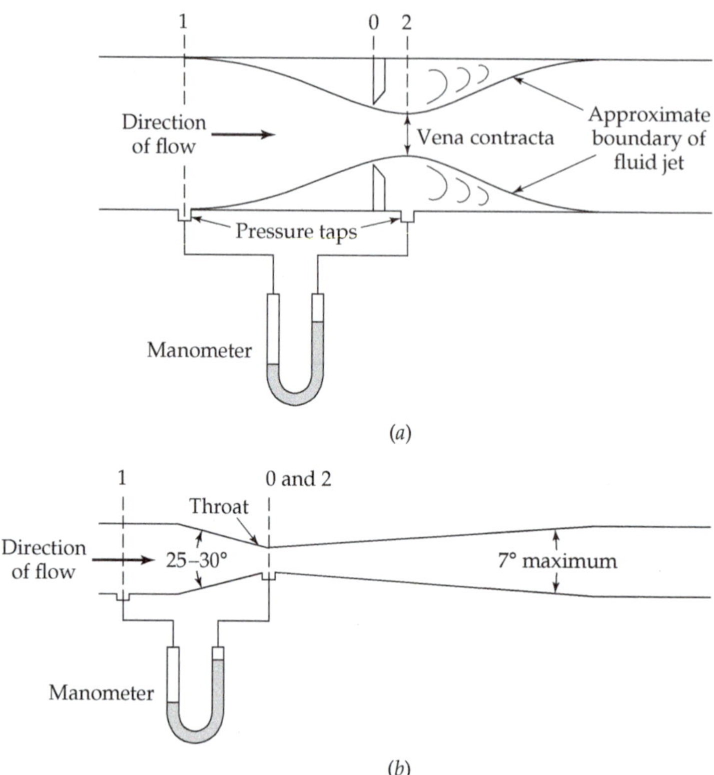

Fig. 15.5-5. Measurement of mass flow rate by use of (*a*) an orifice meter, and (*b*) a Venturi tube.

in which the quantities $\alpha_i = \langle v_i \rangle^3 / \langle v_i^3 \rangle$ are included to allow for the replacement of the average of the cube by the cube of the average.

We next eliminate $\langle v_1 \rangle$ and $\langle v_2 \rangle$ from the above two equations to get an expression for the mass flow rate:

$$w = \rho_2 S_2 \sqrt{\frac{-2\alpha_2 \int_1^2 (1/\rho) dp}{1 - (\alpha_2/\alpha_1)(\rho_2 S_2 / \rho_1 S_1)^2 + \alpha_2 e_v}} \tag{15.5-36}$$

We now repeat the assumptions of Example 7.6-5: (i) $e_v = 0$, (ii) $\alpha_1 = 1$, and (iii) $\alpha_2 = (S_0/S_2)^2$. Then Eq. 15.5-36 becomes

$$w = C_d \rho_2 S_0 \sqrt{\frac{-2\alpha_2 \int_1^2 (1/\rho) dp}{1 - (\rho_2 S_0 / \rho_1 S_1)^2}} \tag{15.5-37}$$

The empirical "discharge coefficient," C_d, is included in this equation to permit correction of this expression for errors introduced by the three assumptions and must be determined experimentally.

For *Venturi meters,* it is convenient to put plane 2 at the point of minimum cross section of the meter so that $S_2 = S_0$. Then α_2 is very nearly unity, and it has been found experimentally that C_d is almost the same for compressible and incompressible fluids, that is, about 0.98 for well-designed Venturi meters. For *orifice meters,* the degree of contraction of a compressible fluid stream at plane 2 is somewhat less than for incompressible fluids, especially at high flow rates, and a different discharge coefficient[2] is required.

In order to use Eq. 15.5-37, the fluid density must be known as a function of pressure. That is, one must know both the path of the expansion and the equation of state of the fluid. In most cases, the assumption of frictionless adiabatic behavior appears to be acceptable. For ideal gases, one may write $p\rho^{-\gamma}$ = constant, where $\gamma = C_p/C_V$ (see Eq. 15.2-5). Then Eq. 15.5-37 becomes

$$w = C_d \rho_2 S_0 \sqrt{\frac{2(p_1/\rho_1)[\gamma/(\gamma-1)][1-(p_2/p_1)^{(\gamma-1)/\gamma}]}{1-(S_0/S_1)^2(p_2/p_1)^{2/\gamma}}} \tag{15.5-38}$$

This formula expresses the mass flow rate as a function of measurable quantities and the discharge coefficient C_d. Values of the latter may be found in engineering handbooks.[2]

EXAMPLE 15.5-4

Free Batch Expansion of a Compressible Fluid

A compressible gas, initially at $T = T_0$, $p = p_0$, and $\rho = \rho_0$, is discharged from a large stationary insulated tank through a small convergent nozzle, as shown in Fig. 15.5-6. Show how the fractional remaining mass of fluid in the tank, ρ/ρ_0, may be determined as a function of time. Develop working equations, assuming that the gas is ideal.

SOLUTION

For convenience, we divide the tank into two parts, separated by the surface 1 as shown in the figure. We assume that surface 1 is near enough to the tank exit that essentially all of the fluid mass is to left of it, but far enough from the exit that the fluid velocity through the surface 1 is negligible. We further assume that the average fluid properties to the left of 1 are identical to those at surface 1. We now consider the behavior of these two parts of the system separately.

(a) ***The bulk of the fluid in the tank.*** For the region V to the left of 1, the unsteady-state mass balance in Eq. (A) of Table 15.5-1 is

$$\frac{d}{dt}(\rho_1 V) = -w_1 \tag{15.5-39}$$

For the same region, the energy balance of Eq. (E) of Table 15.5-1 becomes

$$\frac{d}{dt}(\rho_1 V(\hat{U}_1 + \hat{\Phi}_1)) = -w_1\left(\hat{U}_1 + \frac{p_1}{\rho_1} + \hat{\Phi}_1\right) \tag{15.5-40}$$

in which V is the total volume in the system being considered, and w_1 is the mass rate of flow of gas leaving the system. In writing this equation, we have neglected the kinetic energy of the fluid.

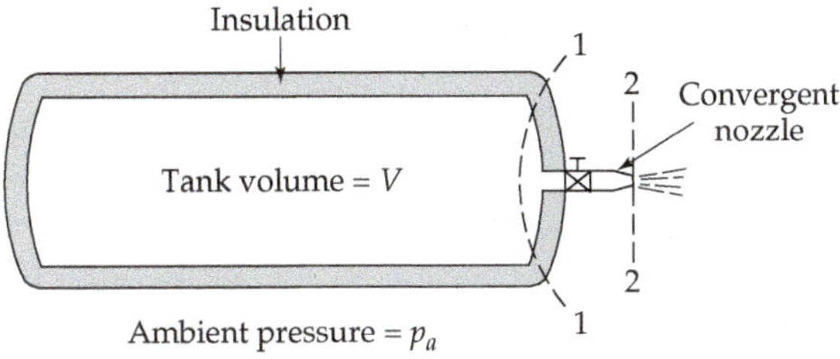

Fig. 15.5-6. Free batch expansion of a compressible fluid. The sketch shows the locations of surfaces 1 and 2.

[2] R. H. Perry, *Chemical Engineers' Handbook*, 7th edition, McGraw-Hill, New York, (1997); see also, Chapter 15 of *Handbook of Fluid Dynamics and Fluid Machinery* (J. A. Schertz and A. E. Fuhs, eds.), Wiley, New York (1996).

Substituting the mass balance into both sides of the energy equation (with $V =$ constant) gives

$$\rho_1 \left(\frac{d\hat{U}_1}{dt} + \frac{d\hat{\Phi}_1}{dt} \right) = \frac{p_1}{\rho_1} \frac{d\rho_1}{dt} \tag{15.5-41}$$

For a stationary system under the influence of no external forces other than gravity, $d\hat{\Phi}_1/dt = 0$, so that Eq. 15.5-41 becomes

$$\frac{d\hat{U}_1}{d\rho_1} = \frac{p_1}{\rho_1^2} \tag{15.5-42}$$

This equation may be combined with the thermal and caloric equations of state for the fluid in order to obtain $p_1(\rho_1)$ and $T_1(\rho_1)$. We find, thus, that the condition of the fluid in the tank depends only upon the degree to which the tank has been emptied and not upon the rate of discharge. For the special case of an ideal gas with constant $\hat{C}_V$, for which $d\hat{U} = \hat{C}_V dT$ and $p = \rho RT/M$, we may integrate Eq. 15.5-42 to obtain

$$p_1 \rho_1^{-\gamma} = p_0 \rho_0^{-\gamma} \tag{15.5-43}$$

in which $\gamma = C_p/C_V$. This result also follows from Eq. 11.4-41.

(b) *Discharge of the gas through the nozzle.* For the sake of simplicity we assume here that the flow between surfaces 1 and 2 is both frictionless and adiabatic. Also, since w_1 is not far different from w_2, it is also appropriate to consider at any one instant that the flow is quasi-steady-state. Then we can use the macroscopic mechanical energy balance in the form of Eq. 15.2-2 with the second, fourth, and fifth terms omitted. That is

$$\tfrac{1}{2} v_2^2 + \int_1^2 \frac{1}{\rho} dp = 0 \tag{15.5-44}$$

Since we are dealing with an ideal gas, we may use the result in Eq. 15.5-38 to get the instantaneous discharge rate. Since in this problem the ratio S_2/S_1 is very small and its square is even smaller, we can replace the denominator under the square root sign in Eq. 15.5-38 by unity. Then the ρ_2 outside the square root sign is moved inside and use is made of Eq. 15.5-43. This gives

$$w = -V\frac{d\rho_1}{dt} = S_2 \sqrt{2 p_1 \rho_1 \left[\frac{\gamma}{(\gamma - 1)} \right] \left[\left(\frac{p_2}{p_0} \right)^{2/\gamma} - \left(\frac{p_2}{p_0} \right)^{(\gamma+1)/\gamma} \right]} \tag{15.5-45}$$

in which S_2 is the cross-sectional area of the nozzle opening.

Now we use Eq. 15.5-43 to eliminate p_1 from Eq. 15.5-45. Then we have a first-order differential equation for ρ_1, which may be integrated to give

$$t = \frac{V/S_2}{\sqrt{2(p_0/\rho_0)[\gamma/(\gamma-1)]}} \int_{\rho_1/\rho_0}^{1} \frac{d(\overline{\rho}_1/\rho_0)}{\sqrt{(p_2/p_0)^{2/\gamma}(\overline{\rho}_1/\rho_0)^{\gamma-1} - (p_2/p_0)^{(\gamma+1)/\gamma}}} \tag{15.5-46}$$

From this equation we can obtain the time required to discharge any given fraction of the original gas.

At low flow rates the pressure p_2 at the nozzle opening is equal to the ambient pressure. However, examination of Eq. 15.5-45 shows that, as the ambient pressure is reduced, the calculated mass rate of flow reaches a maximum at a critical pressure ratio

$$r \equiv \left(\frac{p_2}{p_1} \right)_{\text{crit}} = \left(\frac{2}{\gamma + 1} \right)^{\gamma/(\gamma-1)} \tag{15.5-47}$$

For air ($\gamma = 1.4$), this critical pressure ratio is 0.53. If the ambient pressure is further reduced, the pressure just inside the nozzle will remain at the value of p_2 calculated from Eq. 15.5-47, and the mass rate of flow will become independent of ambient pressure p_a. Under these conditions, the discharge rate is

$$w_{\text{max}} = S_2 \sqrt{p_1 \rho_1 \gamma \left(\frac{2}{\gamma + 1} \right)^{(\gamma+1)/(\gamma-1)}} \tag{15.5-48}$$

Then, for $p_a/p_1 < r$, we may write Eq. 15.5-46 more simply

$$t = \frac{V/S_2}{\sqrt{(p_0/\rho_0)\gamma(2/(\gamma+1))^{(\gamma+1)/(\gamma-1)}}} \int_{\rho_1/\rho_0}^{1} \frac{dx}{x^{(\gamma+1)/2}} \quad (15.5\text{-}49)$$

or

$$t = \frac{V/S_2}{\sqrt{(\gamma RT_0/M)(2/(\gamma+1))^{(\gamma+1)/(\gamma-1)}}} \left(\frac{2}{\gamma-1}\right)\left[\left(\frac{\rho_1}{\rho_0}\right)^{(1-\gamma)/2} - 1\right] \quad (p_a/p_1 < r) \quad (15.5\text{-}50)$$

If p_a/p_1 is initially less than r, both Eqs. 15.5-50 and 15.5-47 will be useful for calculating the total discharge time.

---

§15.6 CONCLUDING COMMENTS

At this point, the reader should be impressed by the variety of problems that can be tackled by means of the macroscopic balances for nonisothermal systems, both for time-independent and time-dependent systems. We have even included systems with multiple entry and exit points. Most of the problems discussed here have not involved much difficult mathematics, but some experience is needed for deciding what assumptions are permitted, which balance equations are needed, and which terms in the equations may be neglected.

The problems at the end of the chapter should provide an opportunity for further exploring the macroscopic balances. In addition, there are plenty of opportunities for getting more experience in conversions of units. By now the readers should have a good appreciation for the usefulness of the material in Appendix E on constants and conversion factors.

QUESTIONS FOR DISCUSSION

1. Give the physical significance of each term in the five macroscopic balances.
2. How are the macroscopic balances related to the equations of change?
3. Explain how one goes from Eq. 15.1-1 to Eq. 15.1-2.
4. Does each of the four terms within the parentheses in Eq. 15.1-2 represent a form of energy? Explain.
5. How is the macroscopic (total) energy balance related to the first law of thermodynamics, $\Delta U = Q + W$?
6. Explain how the averages $\langle v \rangle$ and $\langle v^3 \rangle$ arise in Eq. 15.1-1.
7. What is the physical significance of the quantities E_c and E_v? What sign do they have? How are they related to the velocity distribution? How can they be estimated?
8. How is the macroscopic balance for internal energy derived?
9. What information can be obtained from Eq. 15.2-2 about a fluid at rest?
10. Explain how the factor 32.2 ft · $\text{lb}_m/\text{lb}_f \cdot \text{s}^2$ arises in Eq. 15.3-4.
11. What is the usefulness of the *d*-forms of the macroscopic balances?
12. What is meant by the "logarithmic mean temperature difference"?
13. Define the terms "underdamped" and "overdamped" in the discussion of the temperature controller.

PROBLEMS

15A.1 Rates of heat transfer in a double-pipe heat exchanger.

(a) Hot oil entering the heat exchanger in Example 15.4-1 at surface 2 is to be cooled by water entering at surface 1. That is, the exchanger is being operated in *countercurrent* flow. Compute the required exchanger area A, if the heat-transfer coefficient U is 200 Btu/hr · ft^2 · °F and the fluid streams have the following properties:

	Mass flow rate (lb_m/hr)	Heat capacity (Btu/$lb_m \cdot$ °F)	Temperature entering (°F)	Temperature leaving (°F)
Oil	10,000	0.60	200	100
Water	5,000	1.00	60	—

(b) Repeat the calculation of part (a) if $U_1 = 50$ and $U_2 = 350$ Btu/hr · ft² · °F. Assume that U varies linearly with the water temperature, and use the results of Problem 15B.1.

(c) What is the minimum amount of water that can be used in parts (a) and (b) to obtain the desired temperature change for the oil? What is the minimum amount of water that can be used in cocurrent flow?

(d) Calculate the required heat exchanger area for cocurrent flow operation, if the mass rate of flow of water is 15,500 lb_m/hr, and U is constant at 200 Btu/hr · ft² · °F.

Answers: **(a)** 104 ft²; **(b)** 122 ft²; **(c)** 4290 lb_m/hr, 15,000 lb_m/hr; **(d)** about 101 ft²

15A.2 Adiabatic flow of natural gas in a pipeline. Recalculate the power requirement $w\hat{W}$ in Example 15.4-2 if the flow in the pipeline were adiabatic rather than isothermal.

(a) Use the result of Problem 15B.3(d) to determine the density of the gas at plane 2.

(b) Use your answer to part (a), along with the result of Problem 15B.3(e), to obtain p_2.

(c) Calculate the power requirement, as in Example 15.4-2.

Answers: **(a)** 0.243 lb_m/ft^3; **(b)** 86 psia; **(c)** 504 hp

15A.3 Mixing of two ideal-gas streams.

(a) Calculate the resulting velocity, temperature, and pressure when the following two air streams are mixed in an apparatus such as that described in Example 15.3-2. The molar heat capacity $\tilde{C}_p$ of air may be considered constant at 6.97 Btu/lb-mol · °F. The properties of the two streams are:

	w (lb_m/hr)	v (ft/s)	T (°F)	p (atm)
Stream 1a:	1000	1000	80	1.00
Stream 2a:	10,000	100	80	1.00

(b) What would the calculated velocity be, if the fluid density were treated as constant?

(c) Estimate $\hat{E}_v$ for this operation, basing your calculation on the results of part (b).

Answers: **(a)** 11,000 lb_m/hr; about 110 ft/s; 86.5°F; 1.00 atm; **(b)** 109 ft/s; **(c)** 1.4×10^3 ft · lb_f/lb_m

15A.4 Flow through a Venturi tube. A Venturi tube, with a throat 3 in. in diameter, is placed in a circular pipe 1 ft in diameter carrying dry air. The discharge coefficient C_d of the meter is 0.98. Calculate the mass flow rate of air in the pipe if the air enters the Venturi at 70°F and 1 atm and the throat pressure is 0.75 atm.

(a) Assume adiabatic frictionless flow and $\gamma = 1.4$.

(b) Assume isothermal flow.

(c) Assume incompressible flow at the entering density.

Answers: **(a)** 2.08 lb_m/s; **(b)** 1.96 lb_m/s; **(c)** 2.44 lb_m/s

15A.5 Free batch expansion of a compressible fluid. A tank with volume $V = 10$ ft³ (see Fig. 15.5-6) is filled with air ($\gamma = 1.4$) at $T_0 = 300$K and $p_0 = 100$ atm. At time $t = 0$ the valve is opened, allowing the air to expand to the ambient pressure of 1 atm through a convergent nozzle, with a throat cross section $S_2 = 0.1\ \text{ft}^2$.

(a) Calculate the pressure and temperature at the throat of the nozzle, just after the start of the discharge.

(b) Calculate the pressure and temperature within the tank when p_2 attains its final value of 1 atm.

(c) How long will it take for the system to attain the state described in (b)?

15A.6 Heating of air in a tube. A horizontal tube of 20 ft length is heated by means of an electrical heating element wrapped uniformly around it. Dry air enters at 5°F and 40 psia at a velocity 75 ft/s and 185 lb_m/hr. The heating element provides heat at a rate of 800 Btu/hr per foot of tube. At what temperature will the air leave the tube, if the exit pressure is 15 psia? Assume turbulent flow and ideal-gas behavior. For air in the range of interest, the heat capacity at constant pressure in Btu/lb-mol · °F is

$$\tilde{C}_p = 6.39 + (9.8 \times 10^{-4})T - (8.18 \times 10^{-8})T^2 \tag{15A.6-1}$$

where T is expressed in degrees Rankine.

Answer: $T_2 = 354°\text{F}$

15A.7 Operation of a simple double-pipe heat exchanger. A cold-water stream, 5400 lb_m/hr at 70°F, is to be heated by 8100 lb_m/hr of hot water at 200°F in a simple double-pipe heat exchanger. The cold water is to flow through the inner pipe, and the hot water through the annular space between the pipes. Two 20-ft lengths of heat exchanger are available, and also all the necessary fittings.

(a) By means of a sketch, show the way in which the two double-pipe heat exchangers should be hooked up in order to get the most effective heat transfer.

(b) Calculate the exit temperature of the cold stream for the arrangement decided on in (a) for the following situation:

(i) The heat-transfer coefficient for the annulus, based on the heat-transfer area of the inner surface of the inner pipe, is 2000 Btu/hr · ft² · °F.

(ii) The inner pipe has the following properties: total length, 40 ft; inside diameter 0.0875 ft; heat-transfer surface per foot, 0.2745 ft²; capacity at average velocity of 1 ft/s is 1345 lb_m/hr

(iii) The average properties of the water in the inner pipe are:

$$\begin{aligned}\mu &= 0.45\ \text{cp} = 1.09\ \text{lb}_m/\text{hr}\cdot\text{ft}\\ \hat{C}_p &= 1.00\ \text{Btu/lb}_m\cdot{}^\circ\text{F}\\ k &= 0.376\ \text{Btu/hr}\cdot\text{ft}\cdot{}^\circ\text{F}\\ \rho &= 61.5\ \text{lb}_m/\text{ft}^3\end{aligned}$$

(iv) The combined resistance of the pipe wall and encrustations combined is 0.001 hr · ft² · °F/Btu based on the inner pipe surface area.

(c) Sketch the temperature profile in the exchanger.

Answer: **(b)** 136°F

15B.1 Performance of a double-pipe heat exchanger with variable overall heat-transfer coefficient. Develop an expression for the amount of heat transferred in an exchanger of the type discussed in Example 15.4-1, if the overall heat-transfer coefficient U varies linearly with the temperature of either stream.

(a) Since $T_h - T_c$ is a linear function of both T_h and T_c, show that

$$\frac{U - U_1}{U_2 - U_1} = \frac{\Delta T - \Delta T_1}{\Delta T_2 - \Delta T_1} \tag{15B.1-1}$$

in which $\Delta T = T_h - T_c$, and the subscripts 1 and 2 refer to the conditions at control surfaces 1 and 2.

(b) Substitute the result in (a) for $T_h - T_c$ into Eq. 15.4-12, and integrate the equation thus obtained over the length of the exchanger. Use this result to show that[1]

$$Q = A\frac{U_1\Delta T_2 - U_2\Delta T_1}{\ln(U_1\Delta T_2/U_2\Delta T_1)} \tag{15B.1-2}$$

15B.2 Pressure drop in turbulent flow in a slightly converging tube. Consider the turbulent flow of an incompressible fluid in a circular tube with a diameter that varies linearly with distance according to the relation

$$D = D_1 + (D_2 - D_1)\frac{z}{L} \tag{15B.2-1}$$

as shown in Fig. 15B.2. At $z = 0$, the velocity is v_1 and may be assumed constant over the cross section. The Reynolds number for the flow is such that f is given approximately by the Blasius formula of Eq. 6.2-12,

$$f = \frac{0.0791}{\text{Re}^{1/4}} \tag{15B.2-2}$$

Obtain the pressure drop $p_1 - p_2$ in terms of v_1, D_1, D_2, ρ, L, and $\nu = \mu/\rho$.

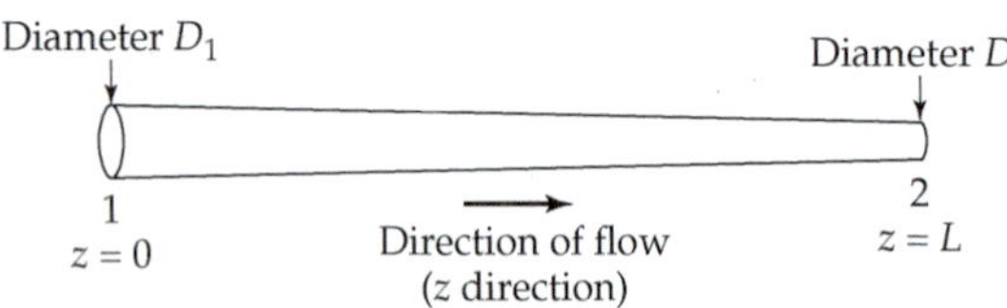

Fig. 15B.2 Turbulent flow in a horizontal, slightly tapered tube (D_1 is slightly greater than D_2).

(a) Integrate the *d*-form of the mechanical energy balance to get

$$\frac{1}{\rho}(p_1 - p_2) = \frac{1}{2}(v_2^2 - v_1^2) + 2\int_0^L \frac{v^2 f}{D}dz \tag{15B.2-3}$$

and then eliminate v_2 from the equation.

(b) Show that both v and f are functions of D:

$$v = v_1\left(\frac{D_1}{D}\right)^2 \qquad f = \frac{0.0791}{(D_1 v_1/\nu)^{1/4}}\left(\frac{D}{D_1}\right)^{1/4} \tag{15B.2-4}$$

Of course, D is a function of z according to Eq. 15B.2-1.

(c) Make a change of variable in the integral in Eq. 15B.2-3 and show that

$$\int_0^L \frac{v^2 f}{D}dz = \frac{L}{D_2 - D_1}\int_{D_1}^{D_2} \frac{v^2 f}{D}dD \tag{15B.2-5}$$

(d) Combine the results of (b) and (c) to get finally

$$\frac{1}{\rho}(p_1 - p_2) = \frac{1}{2}v_1^2\left[\left(\frac{D_1}{D_2}\right)^4 - 1\right] + \frac{2Lv_1^2}{D_1 - D_2}\frac{\frac{4}{15}(0.0791)}{(D_1 v_1/\nu)^{1/4}}\left[\left(\frac{D_1}{D_2}\right)^{15/4} - 1\right] \tag{15B.2-6}$$

(e) Show that this result simplifies properly for $D_1 = D_2$.

15B.3 Steady flow of ideal gases in ducts of constant cross section.

(a) Show that, for the horizontal flow of any fluid in a circular duct of uniform diameter D, the *d*-form of the mechanical energy balance, Eq. 15.4-1, may be written as

$$vdv + \frac{1}{\rho}dp + \tfrac{1}{2}v^2 de_v = 0 \tag{15B.3-1}$$

in which $de_v = (4f/D)dL$. Assume flat velocity profiles.

[1] A. P. Colburn, *Ind. Eng. Chem.*, **25**, 873 (1933).

(b) Show that Eq. 15B.3-1 may be rewritten as

$$vdv + d\left(\frac{p}{\rho}\right) + \left(\frac{p}{\rho^2}\right)d\rho + \tfrac{1}{2}v^2 de_v = 0 \tag{15B.3-2}$$

Show further that, when use is made of the d-form of the mass balance, Eq. 15B.3-2 becomes for *isothermal flow* of an ideal gas

$$de_v = \frac{2RT}{M}\frac{dv}{v^3} - 2\frac{dv}{v} \tag{15B.3-3}$$

(c) Integrate Eq. 15B.3-3 between any two pipe cross sections 1 and 2 enclosing a total pipe length L. Make use of the ideal-gas equation of state and the macroscopic mass balance to show that $v_2/v_1 = \rho_1/\rho_2 = p_1/p_2$, so that the "mass velocity" G can be put in the form

$$G \equiv \rho_1 v_1 = \sqrt{\frac{\rho_1 p_1(1-r)}{e_v - \ln r}} \quad \text{(\textit{isothermal} flow of ideal gases)} \tag{15B.3-4}$$

in which $r = (p_2/p_1)^2$. Show that, for any given value of e_v and conditions at 1, the quantity G reaches its maximum possible value at a critical value of r defined by $\ln r_c + (1 - r_c)/r_c = e_v$. See also Problem 15B.4.

(d) Show that, for the *adiabatic flow* of an ideal gas with constant $\hat{C}_p$ in a horizontal duct of constant cross section, the d-form of the total energy balance (Eq. 15.4-4) yields

$$p\hat{V} + \left(\frac{\gamma-1}{\gamma}\right)\tfrac{1}{2}v^2 = \text{constant} \tag{15B.3-5}$$

where $\gamma = C_p/C_V$. Combine this result with Eq. 15B.3-2 to get

$$\frac{\gamma+1}{\gamma}\frac{dv}{v} - 2\left(\frac{p_1}{\rho_1} + \left(\frac{\gamma-1}{\gamma}\right)\tfrac{1}{2}v_1^2\right)\frac{dv}{v^3} = -de_v \tag{15B.3-6}$$

Integrate this equation between sections 1 and 2 enclosing the resistance e_v, assuming γ constant. Rearrange the result with the aid of the macroscopic mass balance to obtain the following relation for the mass velocity G:

$$G \equiv \rho_1 v_1 = \sqrt{\frac{\rho_1 p_1}{\dfrac{e_v - [(\gamma+1)/2\gamma]\ln s}{1-s} - \dfrac{\gamma-1}{2\gamma}}} \quad \text{(\textit{adiabatic} flow of ideal gases)} \tag{15B.3-7}$$

in which $s = (\rho_2/\rho_1)^2$.

(e) Show by use of the macroscopic energy and mass balances that for horizontal adiabatic flow of ideal gases with constant γ,

$$\frac{p_2}{p_1} = \frac{\rho_2}{\rho_1}\left[1 + \frac{\left[1 - (\rho_1/\rho_2)^2\right]G^2}{\rho_1 p_1}\left(\frac{\gamma-1}{2\gamma}\right)\right] \tag{15B.3-8}$$

This equation can be combined with Eq. 15B.3-7 to show that, as for isothermal flow, there is a critical pressure ratio p_2/p_1 corresponding to the maximum possible mass flow rate.

15B.4 The Mach number in the mixing of two fluid streams.

(a) Show that when the radicand in Eq. 15.3-17 is zero, the Mach number of the final stream is unity. Note that the Mach number, Ma, which is the ratio of the local fluid velocity to the velocity of sound at the local conditions, may be written for an ideal gas as $v/v_s = v/\sqrt{\gamma RT/M}$ (see Problem 11C.1).

(b) Show how the results of Example 15.3-2 may be used to predict the behavior of a gas passing through a sudden enlargement of duct cross section.

15B.5 Limiting discharge rates for Venturi meters.

(a) Starting with Eq. 15.5-38 (for *adiabatic flow of an ideal gas*), show that as the throat pressure in a Venturi meter is reduced, the mass rate of flow reaches a maximum when the ratio $r = p_2/p_1$ of throat pressure to entrance pressure is defined by the expression

$$\frac{\gamma+1}{r^{2/\gamma}} - \frac{2}{r^{(\gamma+1)/\gamma}} - \frac{\gamma-1}{(S_1/S_0)^2} = 0 \tag{15B.5-1}$$

(b) Show that for $S_1 >> S_0$ the mass flow rate under these limiting conditions is

$$w = C_d p_1 S_0 \sqrt{\frac{\gamma M}{RT_1}\left(\frac{2}{\gamma+1}\right)^{(\gamma+1)/(\gamma-1)}} \tag{15B.5-2}$$

(c) Obtain results analogous to Eqs. 15B.5-1 and 15B.5-2 for *isothermal flow*.

15B.6 Flow of a compressible fluid through a convergent-divergent nozzle. In many applications, such as steam turbines or rockets, hot compressed gases are expanded through nozzles of the kind shown in Fig. 15B.6 in order to convert the gas enthalpy into kinetic energy. This operation is in many ways similar to the flow of gases through orifices. Here, however, the purpose of the expansion is to produce power, for example, by the impingement of the fast-moving fluid on a turbine blade, or by direct thrust, as in a rocket engine.

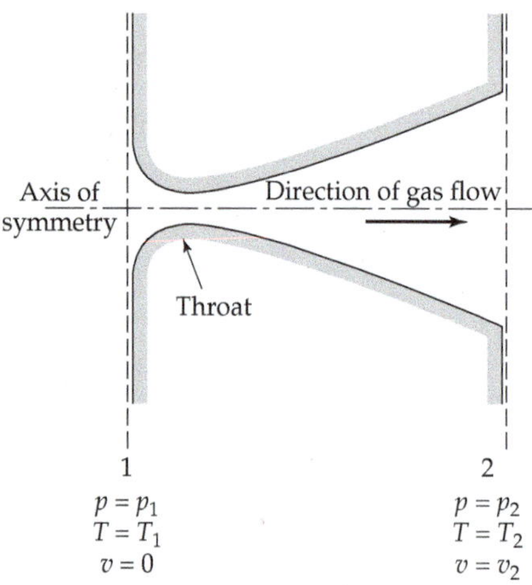

Fig. 15B.6 Schematic cross section of a convergent-divergent nozzle.

To explain the behavior of such a system and to justify the general shape of the nozzle described, follow the path of expansion of an ideal gas. Assume that the gas is initially in a very large reservoir at essentially zero velocity and that it expands through an adiabatic frictionless nozzle to zero pressure. Further assume flat velocity profiles, and neglect changes in elevation.

(a) Show, by writing the macroscopic mechanical energy balance or the total energy balance between planes 1 and 2, that

$$\tfrac{1}{2}v_2^2 = \frac{RT_1}{M}\frac{\gamma}{\gamma-1}\left[1 - \left(\frac{p_2}{p_1}\right)^{(\gamma-1)/\gamma}\right] \tag{15B.6-1}$$

(b) Show, by use of the ideal-gas law, the steady-state macroscopic mass balance, and Eq. 15B.6-1, that the cross section S of the expanding stream goes through a minimum at a critical pressure

$$p_{2,\text{crit}} = p_1\left(\frac{2}{\gamma+1}\right)^{\gamma/\gamma-1} \tag{15B.6-2}$$

(c) Show that the Mach number, $\text{Ma} = v_2/v_s(T_2)$, of the fluid at this minimum cross section is unity (v_s for low-frequency sound waves is derived in Problem 11C.1). How do the results of part (a) above compare with those of Problem 15B.5?

(d) Calculate fluid velocity v, fluid temperature T, and stream cross section S as a function of the local pressure p for the discharge of 10 lb-moles of air per second from 560°R and 10 atm to zero pressure. Discuss the significance of your results.

Answer:

p, atm	10	9	8	7	6	5.28	5	4	3	2	1	0
v, ft s^{-1}	0	449	645	807	956	1058	1099	1245	1398	1574	1798	2591
T,°R	560	543	525	506	484	466	459	431	397	353	290	0
S, ft^2	∞	0.977	0.739	0.650	0.613	0.606	0.607	0.628	0.688	0.816	1.171	∞

15B.7 Transient thermal behavior of a chromatographic device. You are a consultant to an industrial concern that is experimenting, among other things, with transient thermal phenomena in gas chromatography. One of the employees first shows you some reprints of a well-known researcher and says that he is trying to apply some of the researcher's new approaches, but that he is currently stuck on a heat-transfer problem. Although the problem is only ancillary to the main study, it must nonetheless be understood in connection with his interpretation of the data and the application of the new theories.

A very tiny chromatographic column is contained within a coil, which is in turn inserted into a pipe through which a gas is blown to control the temperature (see Fig. 15B.7(*a*)). The gas temperature will be called $T_g(t)$. The temperature at the ends of the coil (outside the pipe) is T_0, which is not very much different from the initial value of T_g. The actual temperature within the chromatographic column (i.e., within the coil) will be called $T(t)$. Initially the gas and the coil are both at the temperature T_{g0}. Then beginning at time, $t = 0$, the gas temperature is increased linearly according to the equation

$$T_g(t) = T_{g0}\left(1 + \frac{t}{t_0}\right) \tag{15B.7-1}$$

where t_0 is a known constant with dimensions of time.

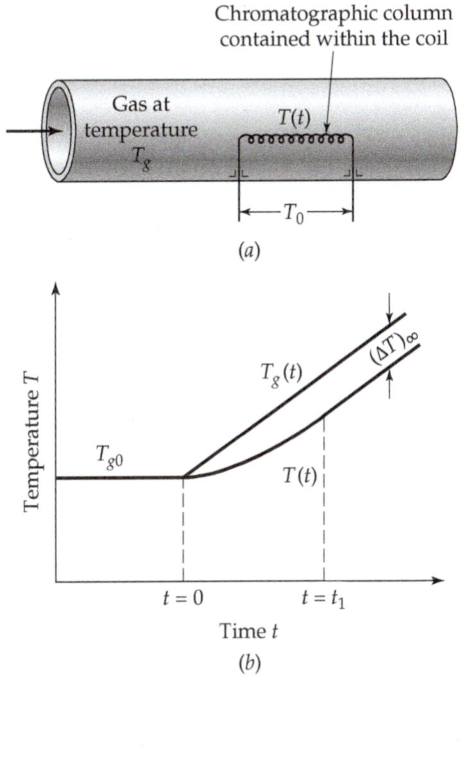

Fig. 15B.7 (*a*) Chromatographic device; (*b*) temperature response $T(t)$ of the chromatographic system.

You are told that, by inserting thermocouples into the column itself, the people in the lab have obtained temperature curves that look like those in Fig. 15B.7(*b*). The $T(t)$ curve seems to become parallel to the $T_g(t)$ curves for large t. You are asked to explain the above pair of curves by means of some kind of theory. Specifically you are asked to find out the following:

(a) At any time t, what will $T_g - T$ be?

(b) What will the limiting value of $T_g - T$ be when $t \to \infty$? Call this quantity $(\Delta T)_\infty$.

(c) What time interval t_1 is required for $T_g - T$ to come within, say, 1% of $(\Delta T)_\infty$?

(d) What assumptions had to be made to model the system?

(e) What physical constants, physical properties, etc., have to be known in order to make a comparison between the measured and theoretical values of $(\Delta T)_\infty$?

Devise the simplest possible theory to account for the temperature curves and to answer the above five questions.

15B.8 Continuous heating of slurry in an agitated tank. A slurry is being heated by pumping it through a well-stirred heating tank, as shown in Fig. 15B.8. The inlet temperature of the slurry is T_i and the temperature of the outer surface of the steam coil is T_s. Use the following symbols:

V = volume of the slurry in the tank
$\rho, \hat{C}_p$ = density and heat capacity of the slurry
w = mass rate of flow of slurry through the tank
U = overall heat-transfer coefficient of heating coil
A = total heat-transfer area of the coil

Assume that the stirring is sufficiently thorough that the fluid temperature in the tank is uniform and the same as the outlet fluid temperature.

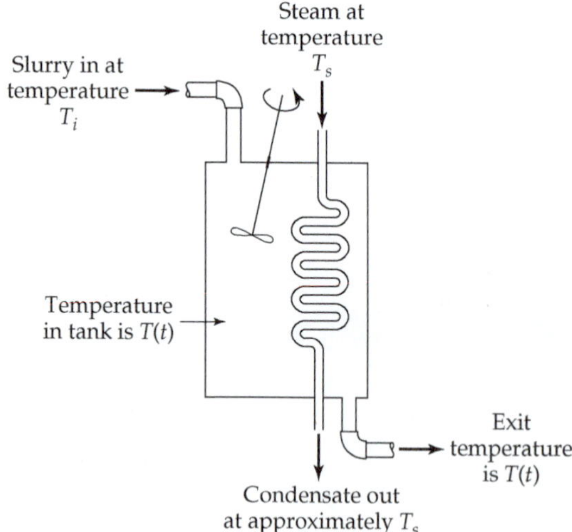

Fig. 15B.8 Heating of a slurry in an agitated tank.

(a) By means of an energy balance, show that the slurry temperature $T(t)$ is described by the differential equation

$$\frac{dT}{dt} = \left(\frac{UA}{\rho \hat{C}_p V}\right)(T_s - T) - \left(\frac{w}{\rho V}\right)(T - T_i) \qquad (15B.8\text{-}1)$$

The variable t is the time since the start of heating.

(b) Rewrite this differential equation in terms of the dimensionless variables

$$\tau = \frac{wt}{\rho V} \qquad \Theta = \frac{T - T_\infty}{T_i - T_\infty} \qquad (15B.8\text{-}2,3)$$

where

$$T_\infty = \frac{(UA/w\hat{C}_p)T_s + T_i}{(UA/w\hat{C}_p) + 1} \tag{15B.8-4}$$

What is the physical significance of τ, Θ, and T_∞?

(c) Solve the dimensionless equation obtained in (b) for the initial condition that $T = T_i$ at $t = 0$.

(d) Check the solution to see that the differential equation and initial condition are satisfied. How does the system behave at large time? Is this limiting behavior in agreement with your intuition?

(e) How is the temperature at infinite time affected by the flow rate? Is this reasonable?

Answer: **(c)** $\dfrac{T(t) - T_\infty}{T_i - T_\infty} = \exp\left[-\left(\dfrac{UA}{\rho\hat{C}_pV} + \dfrac{w}{\rho V}\right)t\right]$

15C.1 Cocurrent/countercurrent heat exchangers. In the heat exchanger shown in Fig. 15C.1, the "tube fluid" (fluid A) enters and leaves at the same end of the heat exchanger, whereas the "shell fluid" (fluid B) always moves in the same direction. Thus, there is both cocurrent flow and countercurrent flow in the same apparatus. This flow arrangement is one of the simplest examples of "mixed flow," often used in practice to reduce exchanger length.[2] The behavior of this kind of equipment may be analyzed by making the following assumptions:

(i) Steady-state conditions exist.

(ii) The overall heat-transfer coefficient U and the heat capacities of the two fluids are constants.

(iii) The shell-fluid temperature T_B is constant over any cross section perpendicular to the flow direction

(iv) There is an equal amount of heating area in each "pass," that is, for tube fluid streams I and II in the figure.

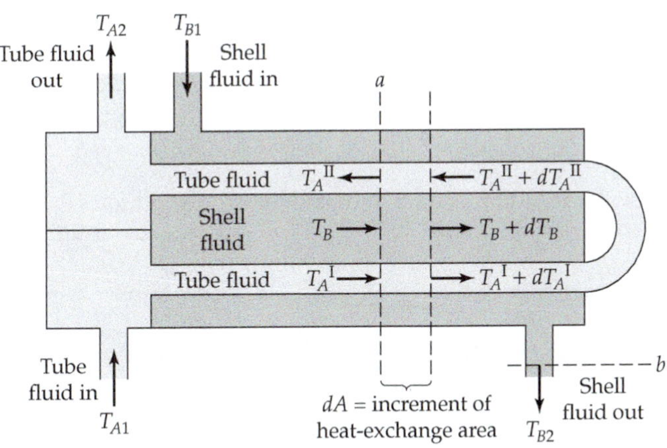

Fig. 15C.1 A cocurrent/countercurrent heat exchanger.

(a) Show by an energy balance (Eq. 15.1-3, with only the enthalpy terms contributing) over the portion of the system between planes a and b that

$$T_B - T_{B2} = R(T_A^{\rm II} - T_A^{\rm I}) \qquad \text{where } R = |w_A\hat{C}_{pA}/w_B\hat{C}_{pB}| \tag{15C.1-1}$$

[2]See D. Q. Kern, *Process Heat Transfer*, McGraw-Hill, New York (1950), pp. 127–189; J. H. Perry, *Chemical Engineers' Handbook*, 3rd edition, McGraw-Hill, New York (1950), pp. 464–465; W. M. Rohsenow, J. P. Hartnett, and Y. I. Cho, *Handbook of Heat Transfer*, 3rd edition, McGraw-Hill, New York (1998), Chapter 17; S. Whitaker, *Fundamentals of Heat Transfer*, corrected edition, Krieger Publishing Company, Malabar, FL (1983), Chapter 11.

(b) Show that application of Eq. 15.4-4 over a differential section of the exchanger, including a *total* heat exchange surface dA gives

$$\frac{dT_A^{\mathrm{I}}}{da} = \frac{1}{2}(T_B - T_A^{\mathrm{I}}) \tag{15C.1-2}$$

$$\frac{dT_A^{\mathrm{II}}}{da} = \frac{1}{2}(T_A^{\mathrm{II}} - T_B) \tag{15C.1-3}$$

$$\frac{1}{R}\frac{dT_B}{da} = -\left[T_B - \frac{1}{2}\left(T_A^{\mathrm{I}} + T_A^{\mathrm{II}}\right)\right] \tag{15C.1-4}$$

in which $da = (U/w_A\hat{C}_{pA})dA$, and w_A and $\hat{C}_{pA}$ are defined as in Example 15.4-1.

(c) Show that when T_A^{I} and T_A^{II} are eliminated between these three equations, a differential equation for the shell fluid can be obtained:

$$\frac{d^2\Theta}{da^2} + R\frac{d\Theta}{da} - \frac{1}{4}\Theta = 0 \tag{15C.1-5}$$

in which $\Theta(a) = (T_B - T_{B2})/(T_{B1} - T_{B2})$. [Hint: Start by differentiating Eq. 15C.1-4 with respect to a and then eliminating the first derivatives in the parentheses by means of Eqs. 15C.1-2 and 15C.1-3, and then using Eq. 15C.1-1.] Solve Eq. 15C.1-5 (see Eq. C.1-7a) with the boundary conditions

B.C.1: at $a = 0$, $\Theta = 1$ (15C.1-6)

B.C.2: at $a = (UA_T/w_A\hat{C}_{pA})$, $\Theta = 0$ (15C.1-7)

in which A_T is the total heat-transfer surface area of the exchanger. Obtain

$$\begin{aligned}\Theta(a) &= C_1 \exp\left[-\tfrac{1}{2}\left(R - \sqrt{R^2+1}\right)a\right] + C_2 \exp\left[-\tfrac{1}{2}\left(R + \sqrt{R^2+1}\right)a\right] \\ &\equiv C_1 e^{m_- a} + C_2 e^{m_+ a}\end{aligned} \tag{15C.1-8}$$

in which

$$C_1 = \frac{1}{1 - \exp(\sqrt{R^2+1}\,a_T)} \qquad C_2 = \frac{1}{1 - \exp(-\sqrt{R^2+1}\,a_T)} \tag{15C.1-9a,b}$$

where $a_T = UA_T/w_A\hat{C}_{pA}$.

(d) Use the result of part (c) to obtain an expression for dT_B/da. Eliminate dT_B/da from this expression with the aid of Eq. 15C.1-4 and evaluate the resulting equation at $a = 0$ to obtain

$$\frac{R}{T_{B1} - T_{B2}}\left[-T_{B1} + \tfrac{1}{2}\left(T_{A1} + T_{A2}\right)\right] = \frac{m_-}{1 - e^{\sqrt{R^2+1}\,a_T}} + \frac{m_+}{1 - e^{-\sqrt{R^2+1}\,a_T}} \tag{15C.1-10}$$

Then manipulate the left side by using Eq. 15C.1-1 in such a way as to obtain the ratio $\Psi = (T_{A2} - T_{A1})/(T_{B1} - T_{A1})$. Finally obtain

$$\begin{aligned}e^{\sqrt{R^2+1}\,a_T} &= \frac{\left(-(1/\Psi) + \frac{1}{2}\right) - m_-}{\left(-(1/\Psi) + \frac{1}{2}\right) - m_+} = \frac{2 - \Psi(1 - 2m_-)}{2 - \Psi(1 - 2m_+)} \\ &= \frac{2 - \Psi(R + 1 - \sqrt{R^2+1})}{2 - \Psi(R + 1 + \sqrt{R^2+1})}\end{aligned} \tag{15C.1-11}$$

From this one can then get the following relation for the performance of the exchanger:

$$a_T = \frac{UA_T}{w_A\hat{C}_{pA}} = \frac{1}{\sqrt{R^2+1}}\ln\left[\frac{2 - \Psi\left(R + 1 - \sqrt{R^2+1}\right)}{2 - \Psi\left(R + 1 + \sqrt{R^2+1}\right)}\right] \tag{15C.1-12}$$

(e) Use Eq. 15C.1-12 to obtain the following expression for the rate of heat transfer in the exchanger:

$$Q = UA(\Delta T)_{\ln} \cdot Y \tag{15C.1-13}$$

in which

$$(\Delta T)_{\ln} = \frac{(T_{B1} - T_{A2}) - (T_{B2} - T_{A1})}{\ln[(T_{B1} - T_{A2})/(T_{B2} - T_{A1})]} \tag{15C.1-14}$$

$$Y = \frac{\sqrt{R^2+1}\,\ln[(1-\Psi)/(1-R\Psi)]}{(R-1)\ln\left[\dfrac{2-\Psi\left(R+1-\sqrt{R^2+1}\right)}{2-\Psi\left(R+1+\sqrt{R^2+1}\right)}\right]} \tag{15C.1-15}$$

The quantity Y represents the ratio of the heat transferred in the "1-2 cocurrent-countercurrent exchanger" shown to that transferred in a true countercurrent exchanger of the same area and terminal fluid temperatures. Values of $Y(R,\Psi)$ are given graphically in Perry's handbook.[2] It may be seen that $Y(R,\Psi)$ is always less than unity.

Chapter 16

Energy Transport by Radiation

We concluded Part I of this book with a chapter about fluids that cannot be described by Newton's law of viscosity, but that require various kinds of nonlinear and time-dependent expressions. We now end Part II with a brief discussion of radiative energy transport, which cannot be described by Fourier's law.

In Chapters 9 to 15 the transport of energy by conduction and by convection has been discussed. Both modes of transport rely on the presence of a material medium. For *heat conduction* to occur, there must be temperature inequalities between neighboring points. For *heat convection* to occur, there must be a fluid that is free to move and transport energy with it. In this chapter, we turn our attention to a third mechanism for energy transport, namely, *radiation*. This is basically an electromagnetic mechanism, which allows energy to be transported with the speed of light through a region of space that is devoid of matter. The rate of energy transport between two "black" bodies in a vacuum is proportional to the difference of the fourth powers of their absolute temperatures. This mechanism is qualitatively very different from the three transport mechanisms considered elsewhere in this book: momentum transport in Newtonian fluids, proportional to the velocity gradient; energy transport by heat conduction, proportional to a temperature gradient; and mass transport by diffusion, proportional to a concentration gradient. Because of the uniqueness of radiation as a means of transport, and because of the importance of radiant heat transfer in industrial calculations, we have devoted a separate chapter to this subject.

A thorough understanding of the physics of radiative transport requires the use of several different disciplines:[1,2] electromagnetic theory is needed to describe the essentially

[1]M. Planck, *Theory of Heat*, Macmillan, London (1932), Parts III and IV. Nobel Laureate **Max Karl Ernst Ludwig Planck** (1858–1947) was the first to hypothesize the quantization of energy and thereby introduce a new fundamental constant h (Planck's constant); his name is also associated with the "Fokker-Planck" equation of stochastic dynamics.

[2]W. Heitler, *Quantum Theory of Radiation*, 2nd edition, Oxford University Press (1944).

wavelike nature of radiation, in particular the energy and pressure associated with electromagnetic waves; thermodynamics is useful for obtaining some relations among the "bulk properties" of an enclosure containing radiation; quantum mechanics is necessary in order to describe in detail the atomic and molecular processes that occur when radiation is produced within matter and when it is absorbed by matter; and statistical mechanics is needed to describe the way in which the energy of radiation is distributed over the wavelength spectrum. All we can do in this elementary discussion is define the key quantities and set forth the results of theory and experiment. We then show how some of these results can be used to compute the rate of heat transfer by radiant processes in simple systems.

In §16.1 and §16.2 we introduce the basic concepts and definitions. Then in §16.3 some of the principal physical results concerning black-body radiation are given. In the following section, §16.4, the rate of heat exchange between two black bodies is discussed. This section introduces no new physical principles, the basic problems being those of geometry. Next, §16.5 is devoted to an extension of the preceding section to nonblack surfaces. Finally, in §16.6, there is a brief discussion of radiation processes in absorbing media.[3]

§16.1 THE SPECTRUM OF ELECTROMAGNETIC RADIATION

When a solid body is heated, for example by an electric coil, the surface of the solid emits radiation of wavelength primarily in the range 0.1 to 10 microns. Such radiation is usually referred to as *thermal radiation.* A quantitative description of the atomic and molecular mechanisms by which the radiation is produced is given by quantum mechanics and is outside the scope of this discussion. A qualitative description, however, is possible: When energy is supplied to a solid body, some of the constituent molecules and atoms are raised to "excited states." There is a tendency for the atoms or molecules to return spontaneously to lower energy states. When this occurs, energy is emitted in the form of electromagnetic radiation. Because the emitted radiation results from changes in the electronic, vibrational, and rotational states of the atoms and molecules, the radiation will be distributed over a range of wavelengths.

Actually, thermal radiation represents just a small part of the total spectrum of electromagnetic radiation. Figure 16.1-1 shows roughly the kinds of mechanisms that are responsible for the various parts of the radiation spectrum. The various kinds of radiation are distinguished from one another only by the range of wavelengths they include. In a vacuum, all these forms of radiant energy travel with the speed of light c. The wavelength λ, characterizing an electromagnetic wave, is then related to its frequency ν by the equation

$$\lambda = \frac{c}{\nu} \tag{16.1-1}$$

in which $c = 2.998 \times 10^8$ m/s. Within the visible part of the spectrum, the various wavelengths are associated with the "color" of the light.

For some purposes, it is convenient to think of electromagnetic radiation from a corpuscular point of view. Then we associate with an electromagnetic wave of frequency ν a *photon*, which is a particle with charge zero and mass zero with an energy given by

$$\varepsilon = h\nu \tag{16.1-2}$$

[3]For additional information on radiative heat transfer and engineering applications, see the comprehensive textbook by R. Siegel and J. R. Howell, *Thermal Radiation Heat Transfer*, 3rd edition, Hemisphere Publishing Co., New York (1992). See also J. R. Howell and M. P. Mengöç, in *Handbook of Heat Transfer*, 3rd edition (W. M. Rohsenow, J. P. Hartnett, and Y. I. Cho, eds.), McGraw-Hill, New York (1998), Chapter 7.

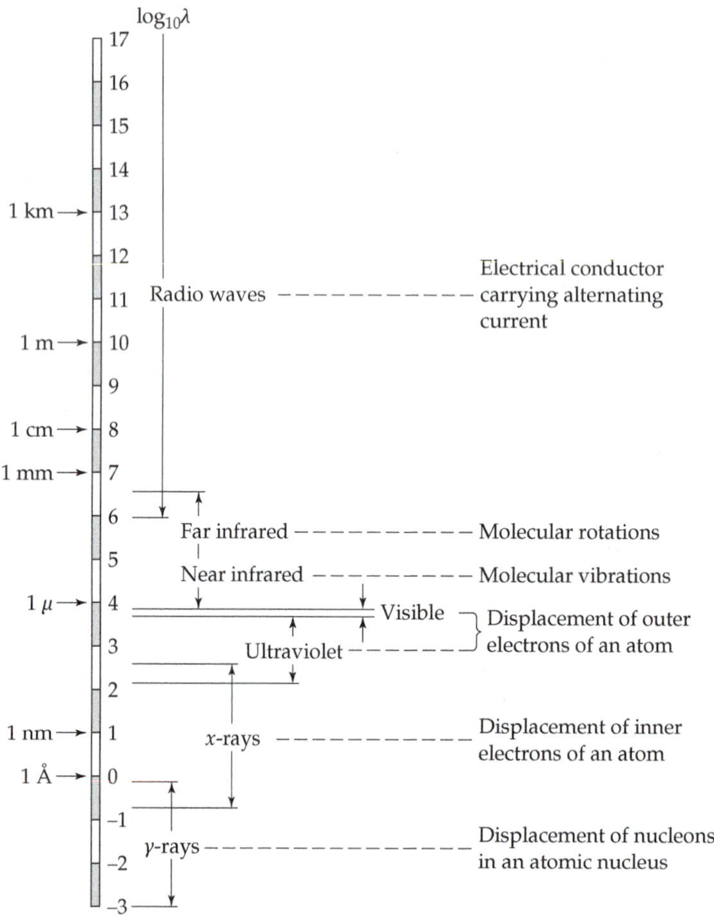

Fig. 16.1-1. The spectrum of electromagnetic radiation, showing roughly the mechanisms by which various types of radiation are produced (1Å = Ångström = 10^{-8} cm = 0.1 nm; 1μ = 1 micron = 10^{-6} m).

Here $h = 6.626 \times 10^{-34}$ J · s is Planck's constant. From these two equations and the information from Fig. 16.1-1, we see that decreasing the wavelength of electromagnetic radiation corresponds to increasing the energy of the corresponding photons. This fact ties in with the various mechanisms that produce the radiation. For example, relatively small energies are released when a molecule decreases its speed of rotation, and the associated radiation is in the infrared. On the other hand, relatively large energies are released when an atomic nucleus goes from a high-energy state to a lower one, and the associated radiation is either gamma- or x-radiation. The foregoing statements also make it seem reasonable that the radiant energy emitted from heated objects will tend towards shorter wavelengths (higher-energy photons) as the temperature of the body is raised.

Thus far we have sketched the phenomenon of the *emission* of radiant energy or photons when a molecular or atomic system goes from a high- to a low-energy state. The reverse process, known as *absorption*, occurs when the addition of radiant energy to a molecular or atomic system causes the system to go from a low- to a high-energy state. The latter process is then what occurs when radiant energy impinges on a solid surface and causes its temperature to rise.

§16.2 ABSORPTION AND EMISSION AT SOLID SURFACES

Having introduced the concepts of absorption and emission in terms of the atomic picture, we now proceed to the discussion of the same processes from a macroscopic viewpoint. We restrict the discussion here to opaque solids.

Radiation impinging on the surface of an opaque solid is either absorbed or reflected. The fraction of the incident radiation that is absorbed is called the *absorptivity* and is given the symbol a. Also the fraction of the incident radiation with frequency ν that is absorbed is designated by a_ν. That is, a and a_ν are defined as

$$a = \frac{q^{(a)}}{q^{(i)}}; \qquad a_\nu = \frac{q_\nu^{(a)}}{q_\nu^{(i)}} \tag{16.2-1,2}$$

in which $q_\nu^{(a)}d\nu$ and $q_\nu^{(i)}d\nu$ are the absorbed and incident radiation per unit area per unit time ([=] energy/area · time) in the frequency range ν to $\nu + d\nu$. For any *real body*, a_ν will be less than unity and will vary considerably with the frequency. A hypothetical body for which a_ν is a constant, less than unity, over the entire frequency range and at all temperatures is called a *gray body*. That is, a gray body always absorbs the same fraction of the incident radiation, regardless of the frequency. A limiting case of the gray body is that for which $a_\nu = 1$ for all frequencies and all temperatures. This limiting behavior defines a *black body*.

All solid surfaces emit radiant energy. The total radiant energy emitted per unit area per unit time is designated by $q^{(e)}$, and that emitted in the frequency range ν to $\nu + d\nu$ is called $q_\nu^{(e)}d\nu$. The corresponding rates of energy emission from a black body are given the symbols $q_b^{(e)}$ and $q_{b\nu}^{(e)}d\nu$. In terms of these quantities, the *emissivity* for the total radiant-energy emission as well as that for a given frequency are defined as

$$e = \frac{q^{(e)}}{q_b^{(e)}}; \qquad e_\nu = \frac{q_\nu^{(e)}}{q_{b\nu}^{(e)}} \tag{16.2-3,4}$$

The emissivity is also a quantity less than unity for real, nonfluorescing surfaces and is equal to unity for black bodies. At any given temperature the radiant energy emitted by a black body represents an upper limit to the radiant energy emitted by real, nonfluorescing surfaces.

We now consider the radiation within an evacuated enclosure or "cavity" with isothermal walls. We imagine that the entire system is at equilibrium. Under this condition, there is no net flux of energy across the interfaces between the solid and the cavity. We now show that the radiation in such a cavity is independent of the nature of the walls and dependent solely on the temperature of the walls of the cavity. We connect two cavities, the walls of which are at the same temperature, but are made of two different materials, as shown in Fig. 16.2-1. If the radiation intensities in the two cavities were different, there would be a net transport of radiant energy from one cavity to the other. Because such a flux would violate the second law of thermodynamics, the radiation intensities in the two cavities must be equal, regardless of the compositions of the cavity surfaces. Furthermore, it can be shown that the radiation is uniform and unpolarized throughout the cavity. This *cavity radiation* plays an important role in the development of

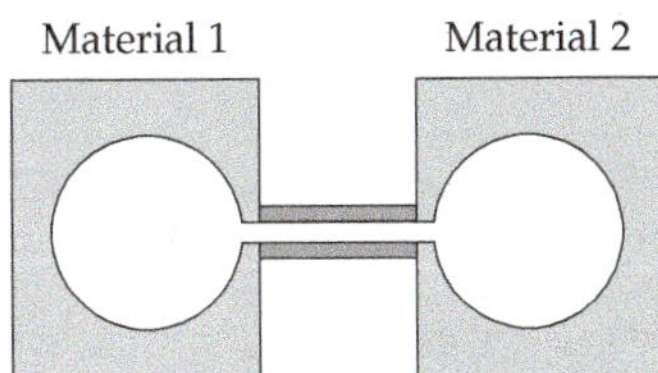

Fig. 16.2-1. Thought experiment for proof that cavity radiation is independent of the wall materials.

Planck's law. We designate the intensity of the radiation as $q^{(\mathrm{cav})}$. This is the radiant energy that would impinge on a solid surface of unit area placed anywhere within the cavity.

We now perform two additional thought experiments. In the first, we put into a cavity a small black body at the same temperature as the walls of the cavity. There will be no net interchange of energy between the black body and the cavity walls. Hence, the energy impinging on the black-body surface must equal the energy emitted by the black body:

$$q^{(\mathrm{cav})} = q_b^{(e)} \tag{16.2-5}$$

From this result, we draw the important conclusion that the radiation emitted by a black body is the same as the equilibrium radiation intensity within a cavity at the same temperature.

In the second thought experiment, we put a small nonblack body into the cavity, once again specifying that its temperature be the same as that of the cavity walls. There is no net heat exchange between the nonblack body and the cavity walls. Hence, we can state that the energy absorbed by the nonblack body will be the same as that radiating from it:

$$aq^{(\mathrm{cav})} = q^{(e)} \tag{16.2-6}$$

Comparison of Eqs. 16.2-5 and 16.2-6 leads to the result

$$a = \frac{q^{(e)}}{q_b^{(e)}} \tag{16.2-7}$$

The definition of the emissivity e in Eq. 16.2-3 allows us to conclude that

$$\boxed{e = a} \tag{16.2-8}$$

This is *Kirchhoff's law*,[1] which states that at a given temperature the emissivity and absorptivity of any solid surface are the same when the radiation is in equilibrium with the solid surface. It can be shown that Eq. 16.2-8 is also valid for each wavelength separately:

$$\boxed{e_\nu = a_\nu} \tag{16.2-9}$$

Values of the total emissivity e for some solids are given in Table 16.2-1. Actually, e depends also on the frequency and on the angle of emission, but the averaged values given there have found widespread use. The tabulated values are, with a few exceptions, for emission normal to the surface, but they may be used for hemispheric emissivity, particularly for rough surfaces. Unoxidized, clean, metallic surfaces have very low emissivities, whereas most nonmetals and metallic oxides have emissivities above 0.8 at room temperature or higher. Note that emissivity increases with increasing temperature for nearly all materials.

We have indicated that the radiant energy emitted by a black body is an upper limit to the radiant energy emitted by real surfaces and that this energy is a function of the temperature. It has been shown experimentally that the total emitted energy flux from a *black* surface is

$$\boxed{q_b^{(e)} = \sigma T^4} \tag{16.2-10}$$

in which T is the absolute temperature. This is known as the *Stefan-Boltzmann law*.[2] The Stefan-Boltzmann constant σ has been found to have the value 0.1712×10^{-8} Btu/hr · ft^2 · °R^4 or 1.355×10^{-12} cal/s · cm^2 · K^4. In the next section we indicate two routes by which

---

[1] G. Kirchhoff, *Monatsber. d. preuss. Akad. d. Wissenschaften*, p. 24, 783–787 (1859); *Poggendorffs Annalen*, **109**, 275–301 (1860). **Gustav Robert Kirchhoff** (1824–1887) published his famous laws for electrical circuits while still a graduate student; he taught at Breslau, Heidelberg, and Berlin.

[2] J. Stefan, *Sitzber. Akad. Wiss. Wien*, **79**, part 2, 391–428 (1879); L. Boltzmann, *Ann. Phys.* (*Wied. Ann.*), Ser. 2, **22**, 291–294 (1884). Slovenian-born **Josef Stefan** (1835–1893), rector of the University of Vienna

Table 16.2-1. The Total Emissivities of Various Surfaces for Perpendicular Emission[a]

	T (°R)	e	T (°R)	e
Aluminum				
Highly polished, 98.3% pure	900	0.039	1530	0.057
Oxidized at 1110°F	850	0.11	1570	0.19
Al-coated roofing	560	0.216		
Copper				
Highly polished, electrolytic	636	0.018		
Oxidized at 1110°F	850	0.57	1570	0.57
Iron				
Highly polished, electrolytic	810	0.052	900	0.064
Completely rusted	527	0.685		
Cast iron, polished	852	0.21		
Cast iron, oxidized at 1100°F	850	0.64	1570	0.78
Asbestos paper	560	0.93	1160	0.945
Brick				
Red, rough	530	0.93		
Silica, unglazed, rough	2292	0.80		
Silica, glazed, rough	2472	0.85		
Lampblack, 0.003 in. or thicker	560	0.945	1160	0.945
Paints				
Black shiny lacquer on iron	536	0.875		
White lacquer	560	0.80	660	0.95
Oil paints, 16 colors	672	0.92–0.96		
Aluminum paints, varying age and lacquer content	672	0.27–0.67		
Refractories, 40 different				
Poor radiators	1570	0.65–0.70	2290	0.75
Good radiators	1570	0.80–0.85	2290	0.85–0.90
Water, liquid, thick layer[b]	492	0.95	672	0.963

[a]Selected values from the table compiled by H. C. Hottel for W. H. McAdams, *Heat Transmission*, 3rd edition, McGraw-Hill, New York (1954), pp. 472–479.

[b]Calculated from spectroscopic data.

this important formula has been obtained theoretically. For *nonblack* surfaces at temperature T the emitted energy flux is

$$\boxed{q_b^{(e)} = e\sigma T^4} \tag{16.2-11}$$

in which e must be evaluated at temperature T. The use of Eqs. 16.2-10 and 16.2-11 to calculate radiant heat-transfer rates between heated surfaces is discussed in §16.4 and §16.5.

We have mentioned that the Stefan-Boltzmann constant has been experimentally determined. This implies that we have a true black body at our disposal. Solids with

(1876–1877), in addition to being known for the law of radiation that bears his name, also contributed to the theory of multicomponent diffusion and to the problem of heat conduction with phase change. **Ludwig Eduard Boltzmann** (1844–1906), who held professorships in Vienna, Graz, Munich, and Leipzig, developed the basic differential equation for gas kinetic theory and the fundamental entropy-probability relation, $S = \kappa \ln W$, which is engraved on his tombstone in Vienna; κ is called the Boltzmann constant.

perfectly black surfaces do not exist. However, we can get an excellent approximation to a black surface by piercing a very small hole in the wall of an isothermal cavity. The hole itself is then very nearly a black surface. The extent to which this is a good approximation may be seen from the following relation, which gives the effective emissivity of the hole, e_{hole}, in a rough-walled enclosure in terms of the actual emissivity e of the cavity walls and the fraction f of the total internal cavity area that is cut away by the hole:

$$e_{\text{hole}} \cong \frac{e}{e + f(1 - e)} \tag{16.2-12}$$

If $e = 0.8$ and $f = 0.001$, then $e_{\text{hole}} = 0.99975$. Therefore, 99.975% of the radiation that falls on the hole will be absorbed. The radiation that emerges from the hole will then be very nearly black-body radiation.

§16.3 PLANCK'S DISTRIBUTION LAW, WIEN'S DISPLACEMENT LAW, AND THE STEFAN-BOLTZMANN LAW[1,2,3]

The Stefan-Boltzmann law may be deduced from thermodynamics, provided that certain results of the theory of electromagnetic fields are known. Specifically, it can be shown that for cavity radiation the energy density (that is, the energy per unit volume) within the cavity of volume V is

$$u^{(r)} = \frac{4}{c} q_b^{(e)} \tag{16.3-1}$$

Since the radiant energy emitted by a black body depends on temperature alone, the energy density $u^{(r)}$ must also be a function of temperature only. It can be further shown that the electromagnetic radiation exerts a pressure $p^{(r)}$ on the walls of the cavity given by

$$p^{(r)} = \tfrac{1}{3} u^{(r)} \tag{16.3-2}$$

The preceding results for cavity radiation can also be obtained by considering the cavity to be filled with a gas made up of photons, each endowed with an energy $h\nu$ and momentum $h\nu/c$. We now apply the thermodynamic formula

$$\left(\frac{\partial U}{\partial V}\right)_T = T\left(\frac{\partial p}{\partial T}\right)_V - p \tag{16.3-3}$$

to the photon gas or radiation in the cavity. Insertion of $U^{(r)} = Vu^{(r)}$ and $p^{(r)} = \frac{1}{3}u^{(r)}$ into this relation gives the following ordinary differential equation for $u^{(r)}(T)$:

$$u^{(r)} = \tfrac{1}{3} T \frac{du^{(r)}}{dT} - \tfrac{1}{3} u^{(r)} \tag{16.3-4}$$

This equation can be integrated to give

$$u^{(r)} = bT^4 \tag{16.3-5}$$

in which b is a constant of integration. Combination of this result with Eq. 16.3-1 gives the radiant energy emitted from the surface of a black body per unit area per unit time:

$$q_b^{(e)} = \frac{c}{4} u^{(r)} = \frac{cb}{4} T^4 = \sigma T^4 \tag{16.3-6}$$

---

[1] J. de Boer, Chapter VII in *Leerboek der Natuurkunde*, 3rd edition (R. Kronig, ed.), Scheltema and Holkema, Amsterdam (1951).

[2] H. B. Callen, *Thermodynamics and an Introduction to Thermostatistics*, 2nd edition, Wiley, New York (1985), pp. 78–79.

[3] M. Planck, *Vorlesungen über die Theorie der Wärmestrahlung*, 5th edition, Barth, Leipzig (1923); *Ann. Phys.*, **4**, 553–563, 564–566 (1901).

This is just the Stefan-Boltzmann law. Note that the thermodynamic development does not predict the numerical value of σ.

The second way of deducing the Stefan-Boltzmann law is by integrating the *Planck distribution law*. This famous equation gives the radiated energy flux $q_{b\lambda}^{(e)}$ in the wavelength range λ to $\lambda + d\lambda$ from a black surface:

$$q_{b\lambda}^{(e)} = \frac{2\pi c^2 h}{\lambda^5} \frac{1}{e^{ch/\lambda\kappa T} - 1} \tag{16.3-7}$$

Here h is Planck's constant. The result can be derived by applying quantum statistics to a photon gas in a cavity, the photons obeying Bose-Einstein statistics.[4,5] The Planck distribution, which is shown in Fig. 16.3-1, correctly predicts the entire energy versus wavelength curve and the shift of the maximum towards shorter wavelengths at higher temperatures. When Eq. 16.3-7 is integrated over all wavelengths, we get

$$\begin{aligned} q_b^{(e)} &= \int_0^\infty q_{b\lambda}^{(e)}\, d\lambda \\ &= 2\pi c^2 h \int_0^\infty \frac{\lambda^{-5}}{e^{ch/\lambda\kappa T} - 1}\, d\lambda \\ &= \frac{2\pi\kappa^4 T^4}{c^2 h^3} \int_0^\infty \frac{x^3}{e^x - 1}\, dx \\ &= \frac{2\pi\kappa^4 T^4}{c^2 h^3} \left(6 \sum_{n=1}^{\infty} \frac{1}{n^4} \right) \\ &= \frac{2\pi\kappa^4 T^4}{c^2 h^3} \left(\frac{\pi^4}{15} \right) \end{aligned} \tag{16.3-8}$$

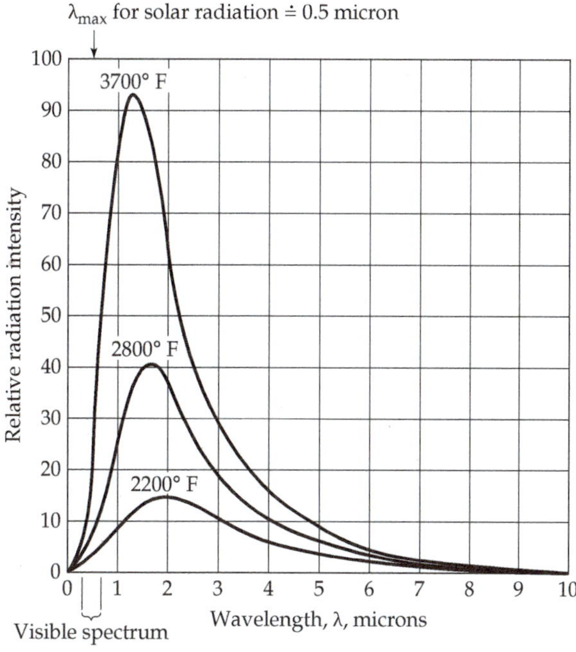

Fig. 16.3-1. The spectrum of equilibrium radiation as given by Planck's law, Eq. 16.3-7. [M. Planck, *Verh. der deutschen Physik. Gesell.*, **2**, 202, 237 (1900); *Ann. der Physik*, **4**, 553–563, 564–566 (1901).]

---

[4] J. E. Mayer and M. G. Mayer, *Statistical Mechanics*, Wiley, New York (1940), pp. 363–374.

[5] L. D. Landau and E. M. Lifshitz, *Statistical Physics*, 3rd edition, Part 1, Pergamon, Oxford (1980), §63.

In the above integration we changed the variable of integration from λ to $x = ch/\lambda\kappa T$. Then the integration over x was performed by expanding $1/(e^x - 1)$ in a Taylor series in e^x (see §C.2) and then integrating term by term. The quantum statistical approach thus gives the details of the spectral distribution of the radiation and also the expression for the Stefan-Boltzmann constant,

$$\boxed{\sigma = \frac{2}{15}\frac{\pi^5\kappa^4}{c^2h^3}} \tag{16.3-9}$$

which has the value 1.355×10^{-12} cal/s · cm^2 · K^4, which is confirmed within experimental uncertainty by direct radiation measurements. Equation 16.3-9 is an amazing formula, interrelating as it does the σ from radiation, the κ from statistical mechanics, the speed of light c from electromagnetism, and the h from quantum mechanics!

In addition to obtaining the Stefan-Boltzmann law from the Planck distribution, we can get an important relation pertaining to the maximum in the Planck distribution. First we rewrite Eq. 16.3-7 in terms of x and then set $dq_{b\lambda}^{(e)}/dx = 0$. This gives the following equation for x_{max}, which is the value of x for which the Planck distribution shows a maximum:

$$x_{max} = 5(1 - e^{-x_{max}}) \tag{16.3-10}$$

The solution to this equation is found numerically to be $x_{max} = 4.9651\ldots$. Hence, at a given temperature T

$$\lambda_{max}T = \frac{ch}{\kappa x_{max}} \tag{16.3-11}$$

Inserting the values of the universal constants and the value for x_{max}, we then get

$$\lambda_{max}T = 0.2884 \text{ cm K} \tag{16.3-12}$$

This result, originally found experimentally,[6] is known as *Wien's displacement law.* It is useful primarily for estimating the temperature of remote objects. The law predicts, in agreement with experience, that the apparent color of radiation shifts from red (long wavelengths) toward blue (short wavelengths) as the temperature increases.

Finally, we may reinterpret some of our previous remarks in terms of the Planck distribution law. In Fig. 16.3-2 we have sketched three curves: the Planck distribution law for a hypothetical black body, the distribution curve for a hypothetical gray body, and a distribution curve for some real body. It is thus clear that when we used the total emissivity values, such as those in Table 16.2-1, we are just accounting empirically for the deviations from Planck's law over the entire spectrum.

We should not leave the subject of the Planck distribution without pointing out that the "Planck distribution law" in Eq. 16.3-7 was presented at the October 1900 meeting of the German Physical Society as an *empiricism* that fitted the available data.[7] However, before the end of the year,[8] Planck succeeded in *deriving* the equation, but at the expense of introducing the radical notion of the quantization of energy, an idea that was met with little enthusiasm. Planck himself had misgivings, as clearly stated in his textbook.[9] In a letter in 1931, he wrote: "...what I did can be described as an act of desperation...I had been wrestling unsuccessfully for six years...with the problem of equilibrium between

---

[6] W. Wien, *Sitzungsber. d. kglch. preuss. Akad. d. Wissenschaften*, (VI), p. 55–62 (1893).

[7] O. Lummer and E. Pringsheim, *Wied. Ann.*, **63**, 396 (1897); *Ann. der Physik*, **3**, 159 (1900).

[8] M. Planck, *Verhandl. d. deutsch. Physik. Ges.*, **2**, 202 and 237 (1900); *Ann. Phys.*, **4**, 553–563, 564–566 (1901).

[9] M. Planck, *The Theory of Heat Radiation*, Dover, New York (1991), English translation of *Vorlesungen über die Theorie der Wärmestrahlung* (1913), p. 154.

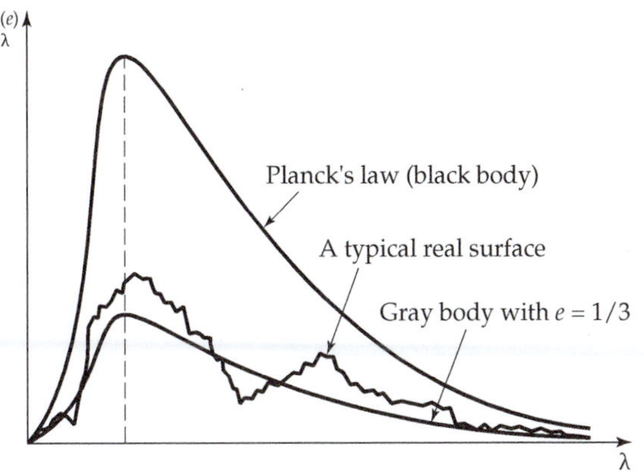

Fig. 16.3-2. Comparison of the emitted radiation from black, gray, and real surfaces.

radiation and matter, and I knew that the problem was of fundamental importance...." Then Planck went on to say that he was "ready to sacrifice every one of my previous convictions about physical laws," except for the first and second laws of thermodynamics.[10] Planck's radical proposal ushered in a new and exciting era of physics, and quantum mechanics penetrated into chemistry and other fields in the twentieth century.

EXAMPLE 16.3-1

Temperature and Radiant-Energy Emission of the Sun

For approximate calculations, the sun may be considered to be a black body, emitting radiation with a maximum intensity at $\lambda = 0.5$ microns (5000 Å). With this information, estimate **(a)** the surface temperature of the sun, and **(b)** the emitted heat flux at the sun's surface.

SOLUTION

(a) From Wien's displacement law, Eq. 16.3-12,

$$T = \frac{0.2884}{\lambda_{\text{max}}} = \frac{0.2884 \text{ cm K}}{0.5 \times 10^{-4} \text{ cm}} = 5760 \text{ K} = 10{,}400°\text{R} \tag{16.3-13}$$

(b) From the Stefan-Boltzmann law, Eq. 16.2-10,

$$\begin{aligned} q_b^{(e)} &= \sigma T^4 = (0.1712 \times 10^{-8} \text{ Btu/hr} \cdot \text{ft}^2 \cdot °\text{R}^4)(10{,}400°\text{R})^4 \\ &= 2.0 \times 10^7 \text{ Btu/hr} \cdot \text{ft}^2 \end{aligned} \tag{16.3-14}$$

§16.4 DIRECT RADIATION BETWEEN BLACK BODIES IN VACUO AT DIFFERENT TEMPERATURES

In the preceding sections we have given the Stefan-Boltzmann law, which describes the total radiant-energy emission from a perfectly black surface. In this section we discuss the radiant-energy transfer between two black bodies of arbitrary geometry and orientation. Hence, we need to know how the radiant energy emanating from a black body is distributed with respect to angle. Because black-body radiation is isotropic, the following relation, known as *Lambert's cosine law*,[1] can be deduced:

$$q_{b\theta}^{(e)} = \frac{q_b^{(e)}}{\pi} \cos\theta = \frac{\sigma T^4}{\pi} \cos\theta \tag{16.4-1}$$

[10] A. Hermann, *The Genesis of Quantum Theory*, MIT Press (1971), pp. 23–24.

[1] J. H. Lambert, *Photometria*, Augsburg (1760).

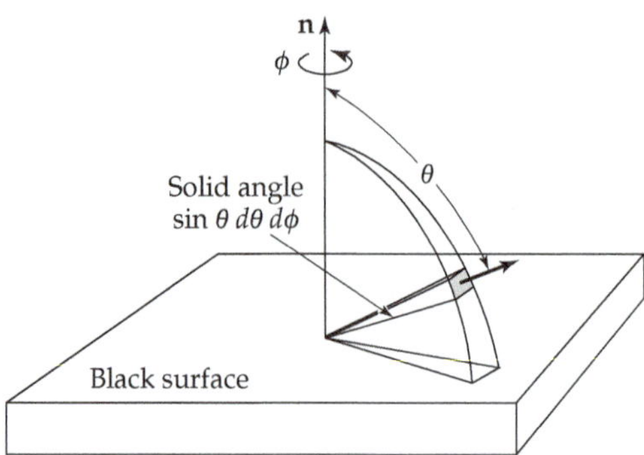

Fig. 16.4-1. Radiation at an angle θ from the normal to the surface into a solid angle $\sin\theta\, d\theta\, d\phi$.

in which $q_{b\theta}^{(e)}$ is the energy emitted per unit area per unit time per unit solid angle in a direction θ (see Fig. 16.4-1). The energy emitted through the shaded solid angle is then $q_{b\theta}^{(e)} \sin\theta\, d\theta\, d\phi$ per unit area of black solid surface. Integration of the foregoing expression for $q_{b\theta}^{(e)}$ over the entire hemisphere gives the known total energy emission:

$$\int_0^{2\pi}\int_0^{\pi/2} q_{b\theta}^{(e)} \sin\theta\, d\theta\, d\phi = \frac{\sigma T^4}{\pi}\int_0^{2\pi}\int_0^{\pi/2} \cos\theta \sin\theta\, d\theta\, d\phi$$
$$= \sigma T^4 = q_b^{(e)} \tag{16.4-2}$$

This justifies the inclusion of the factor of $1/\pi$ in Eq. 16.4-1.

We are now in a position to get the net heat-transfer rate from body 1 to body 2, where these are black bodies of any shape and orientation (see Fig. 16.4-2). We do this by getting the net heat-transfer rate between a pair of surface elements dA_1 and dA_2 that can "see" each other, and then integrating over all such possible pairs of areas. The elements dA_1 and dA_2 are joined by a straight line of length r_{12}, which makes an angle θ_1 with the normal to dA_1 and an angle θ_2 with the normal to dA_2.

We start by writing an expression for the energy radiated from dA_1 into a solid angle $\sin\theta_1\, d\theta_1\, d\phi_1$ about r_{12}. We choose this solid angle large enough that dA_2 will lie entirely within the "beam" (see Fig 16.4-2). According to Lambert's cosine law, the energy radiated in a unit time will be

$$\left(\frac{\sigma T_1^4}{\pi}\cos\theta_1\right) dA_1 \sin\theta_1\, d\theta_1\, d\phi_1 \tag{16.4-3}$$

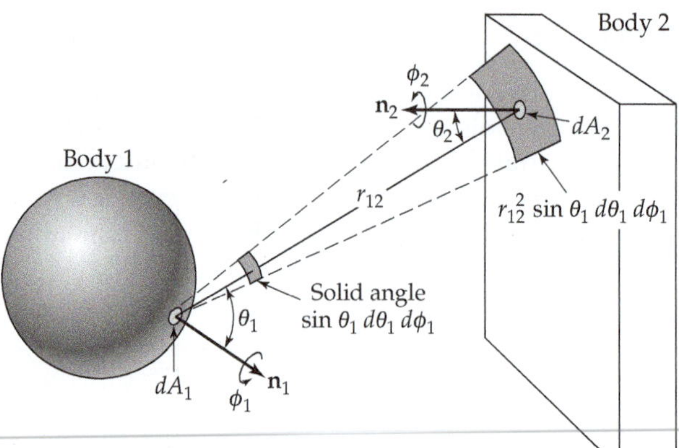

Fig. 16.4-2. Radiant interchange between two black bodies.

Of the energy leaving dA_1 at an angle θ_1, only the fraction given by the following ratio will be intercepted by dA_2

$$\frac{\begin{pmatrix}\text{area of } dA_2 \text{ projected onto a} \\ \text{plane perpendicular to } r_{12}\end{pmatrix}}{\begin{pmatrix}\text{area formed by the intersection} \\ \text{of the solid angle } \sin\theta_1\, d\theta_1\, d\phi_1 \\ \text{with a sphere of radius } r_{12} \text{ with} \\ \text{center at } dA_1\end{pmatrix}} = \frac{dA_2 \cos\theta_2}{r_{12}^2 \sin\theta_1\, d\theta_1\, d\phi_1} \tag{16.4-4}$$

Multiplication of these last two expressions then gives

$$dQ_{\vec{12}} = \frac{\sigma T_1^4}{\pi} \frac{\cos\theta_1 \cos\theta_2}{r_{12}^2} dA_1\, dA_2 \tag{16.4-5}$$

This is the radiant energy emitted by dA_1 and intercepted by dA_2 per unit time. In a similar way we can write

$$dQ_{\vec{21}} = \frac{\sigma T_2^4}{\pi} \frac{\cos\theta_1 \cos\theta_2}{r_{12}^2} dA_1\, dA_2 \tag{16.4-6}$$

which is the radiant energy emitted by dA_2 that is intercepted by dA_1 per unit time. The net rate of energy transport from dA_1 to dA_2 is then

$$\begin{aligned} dQ_{12} &= dQ_{\vec{12}} - dQ_{\vec{21}} \\ &= \frac{\sigma}{\pi}(T_1^4 - T_2^4) \frac{\cos\theta_1 \cos\theta_2}{r_{12}^2} dA_1\, dA_2 \end{aligned} \tag{16.4-7}$$

Therefore, the net rate of energy transfer from an isothermal black body 1 to another isothermal black body 2 is

$$Q_{12} = \frac{\sigma}{\pi}(T_1^4 - T_2^4) \iint \frac{\cos\theta_1 \cos\theta_2}{r_{12}^2} dA_1\, dA_2 \tag{16.4-8}$$

Here it is understood that the integration is restricted to those pairs of areas dA_1 and dA_2 that are in full view of each other. This result is conventionally written in the form

$$Q_{12} = A_1 F_{12} \sigma (T_1^4 - T_2^4) = A_2 F_{21} \sigma (T_1^4 - T_2^4) \tag{16.4-9}$$

where A_1 and A_2 are usually chosen to be the total areas of bodies 1 and 2. The dimensionless quantities F_{12} and F_{21}, called *view factors* (or *angle factors* or *configuration factors*), are given by

$$F_{12} = \frac{1}{\pi A_1} \iint \frac{\cos\theta_1 \cos\theta_2}{r_{12}^2} dA_1\, dA_2 \tag{16.4-10}$$

$$F_{21} = \frac{1}{\pi A_2} \iint \frac{\cos\theta_1 \cos\theta_2}{r_{12}^2} dA_1\, dA_2 \tag{16.4-11}$$

and the two view factors are related by $A_1 F_{12} = A_2 F_{21}$. The view factor F_{12} represents the fraction of radiation leaving body 1 that is directly intercepted by body 2.

The actual calculation of view factors is a difficult problem, except for some very simple situations. In Fig. 16.4-3 and Fig. 16.4-4 some view factors for direct radiation are shown.[2,3,4] When such charts are available, the calculations of energy interchanges by Eq. 16.4-9 are straightforward.

---

[2] H. C. Hottel and A. F. Sarofim, *Radiative Transfer*, McGraw-Hill, New York (1967).

[3] H. C. Hottel, Chapter 4 in W. H. McAdams, *Heat Transmission*, McGraw-Hill, New York (1954).

[4] R. Siegel and J. R. Howell, *Thermal Radiation Heat Transfer*, 3rd edition, Hemisphere Publishing Co., New York (1992).

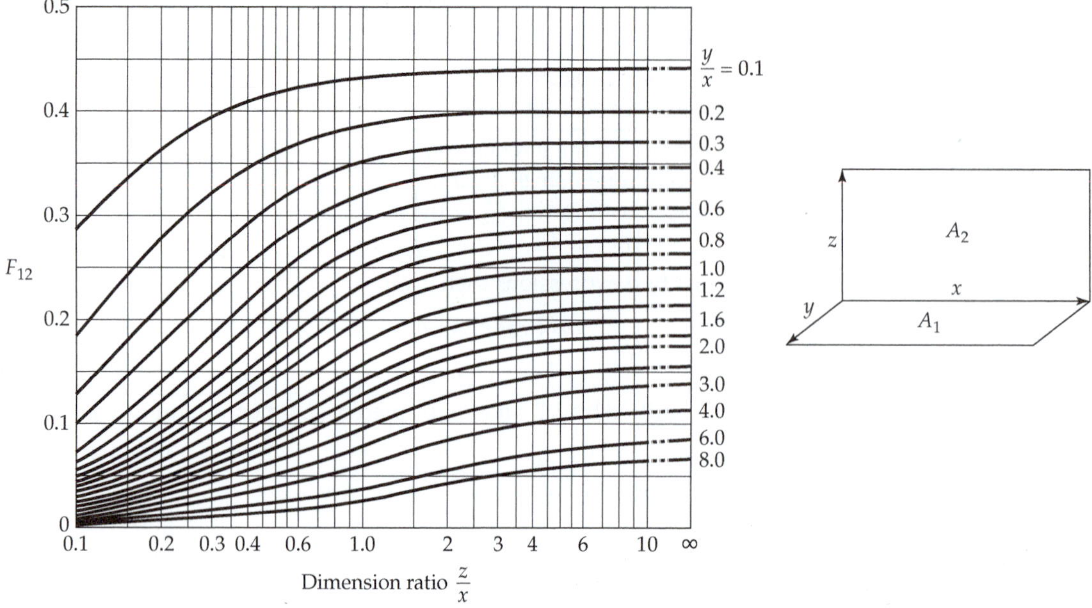

Fig. 16.4-3. View factors for direct radiation between adjacent rectangles in perpendicular planes. [H. C. Hottel, Chapter 3 in W. H. McAdams, *Heat Transmission*, McGraw-Hill, New York (1954), p. 68.]

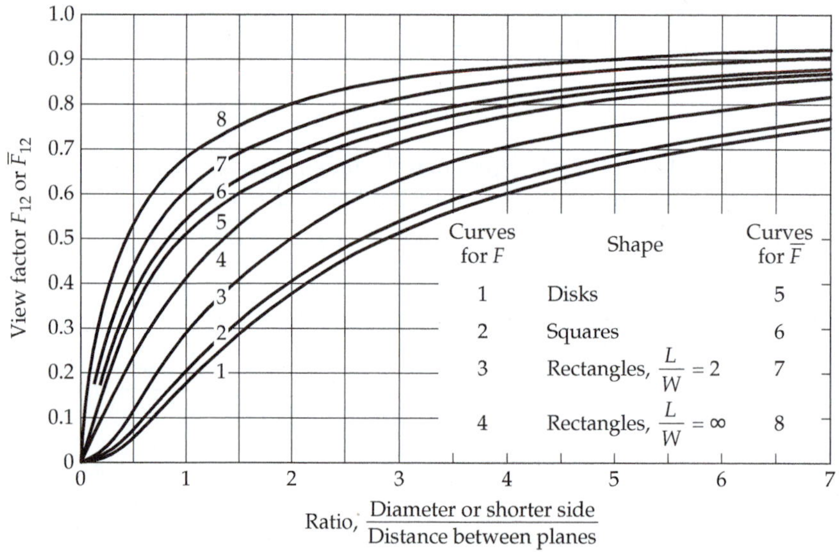

Fig. 16.4-4. View factors for direct radiation between opposed identical shapes in parallel planes. [H. C. Hottel, Chapter 3 in W. H. McAdams, *Heat Transmission*, McGraw-Hill, New York (1954), Third Edition, p. 69.]

In the above development, we have assumed that Lambert's law and the Stefan-Boltzmann law may be used to describe the nonequilibrium transport process, in spite of the fact that they are strictly valid only for radiative equilibrium. The errors thus introduced do not seem to have been studied thoroughly, but apparently the resulting formulas give a good quantitative description.

Thus far we have concerned ourselves with the radiative interactions between two black bodies. We now wish to consider a set of black surfaces 1, 2,..., n, which form the walls of a complete enclosure. The surfaces are maintained at temperatures $T_1, T_2, \ldots, T_n$, respectively. The net heat flow from any surface i to the enclosure surfaces is

$$Q_i = \sigma A_i \sum_{j=1}^{n} F_{ij}(T_i^4 - T_j^4) \qquad i = 1,2,\ldots, n \tag{16.4-12}$$

or

$$Q_i = \sigma A_i \left(T_i^4 - \sum_{j=1}^{n} F_{ij} T_j^4 \right) \qquad i = 1,2,\ldots, n \tag{16.4-13}$$

In writing the second form, we have used the relations

$$\sum_{j=1}^{n} F_{ij} = 1 \qquad i = 1,2,\ldots, n \tag{16.4-14}$$

The sums in Eqs. 16.4-13 and 16.4-14 include the term F_{ii}, which is zero for any object that intercepts none of its own rays. The set of n equations given in Eq. 16.4-12 (or Eq. 16.4-13) may be solved to get the temperatures or heat flows according to the data available.

A simultaneous solution of Eqs. 16.4-13 of special interest is that for which $Q_3 = Q_4 = \cdots = Q_n = 0$. Surfaces such as 3, 4,..., n are here called "adiabatic." In this situation one can eliminate the temperatures of all surfaces except 1 and 2 from the heat-flow calculation and obtain an exact solution for the net heat flow from surface 1 to surface 2:

$$Q_{12} = A_1 \bar{F}_{12} \sigma (T_1^4 - T_2^4) = A_2 \bar{F}_{21} \sigma (T_1^4 - T_2^4) \tag{16.4-15}$$

Values of $\bar{F}_{12}$ for use in this equation are given in Fig. 16.4-4. These values apply only when the adiabatic walls are formed from line elements perpendicular to 1 and 2.

The use of these view factors F and $\bar{F}$ greatly simplifies the calculations for black-body radiation, when the temperatures of surfaces 1 and 2 are known to be uniform. The reader wishing further information on radiative heat exchange in enclosures is referred to the literature.[4]

EXAMPLE 16.4-1

Estimation of the Solar Constant

The radiant heat flux entering the earth's atmosphere from the sun is called the "solar constant" and is important in solar-energy utilization as well as in meteorology. Designate the sun as body 1 and the earth as body 2, and use the following data to calculate the solar constant: $D_1 = 8.60 \times 10^5$ miles; $r_{12} = 9.29 \times 10^7$ miles; $q_{b1}^{(e)} = 2.0 \times 10^7$ Btu/hr · ft^2 (from Example 16.3-1).

SOLUTION

From Eq. 16.4-5, the radiant energy arriving at dA_2 from dA_1 (see Fig. 16.4-5) is $dQ_{\overrightarrow{12}} / \cos\theta_2\, dA_2 = (\sigma T_1^4 / \pi r_{12}^2) \cos\theta_1\, dA_1$. Then the total radiant energy arriving at dA_2 from the sun is obtained by integrating $dQ_{\overrightarrow{12}} / \cos\theta_2\, dA_2$ over that part of the sun's surface that is visible from the earth:

$$\begin{aligned}
\text{Solar constant} &= \frac{dQ_{\overrightarrow{12}}}{\cos\theta_2\, dA_2} = \frac{\sigma T_1^4}{\pi r_{12}^2} \int \cos\theta_1\, dA_1 \\
&= \frac{\sigma T_1^4}{\pi r_{12}^2} \left(\frac{\pi D_1^2}{4} \right) = \frac{q_{b1}^{(e)}}{4} \left(\frac{D_1}{r_{12}} \right)^2 \\
&= \frac{(2.0 \times 10^7 \text{ Btu/hr} \cdot \text{ft}^2)}{4} \left(\frac{8.60 \times 10^5 \text{ miles}}{9.29 \times 10^7 \text{ miles}} \right)^2 \\
&= 430 \text{ Btu/hr} \cdot \text{ft}^2
\end{aligned} \tag{16.4-16}$$

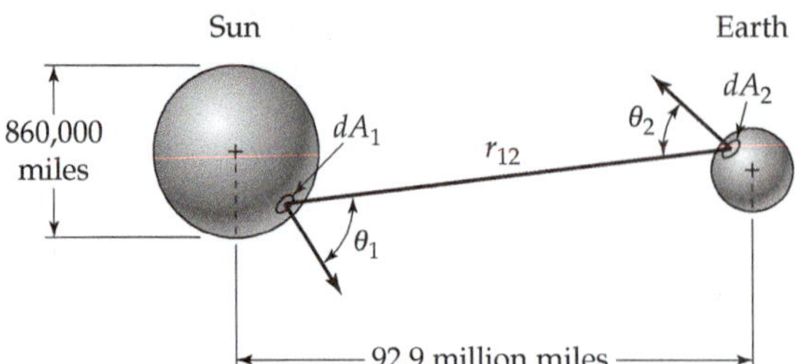

Fig. 16.4-5. Estimation of the solar constant.

This is in satisfactory agreement with other estimates that have been made. The treatment of r_{12}^2 as a constant in the integrand is permissible here because the distance r_{12} varies by less than 0.5% over the visible surface of the sun. The remaining integral, $\int \cos\theta_1 \, dA_1$, is the projected area of the sun as seen from the earth, or very nearly $\pi D_1^2/4$.

EXAMPLE 16.4-2

Radiant Heat Transfer between Disks

Two black disks of diameter 2 ft are placed directly opposite one another at a distance of 4 ft. Disk 1 is maintained at 2000°R, and disk 2 at 1000°R. Calculate the heat flow between the two disks **(a)** when no other surfaces are present, and **(b)** when the two disks are connected by an adiabatic right-cylindrical black surface.

SOLUTION

(a) From Eq. 16.4-9 and curve 1 of Fig. 16.4-4,

$$\begin{aligned} Q_{12} &= A_1 F_{12} \sigma (T_1^4 - T_2^4) \\ &= (\pi \cdot 1\ \text{ft}^2)(0.06)(0.1712 \times 10^{-8}\ \text{Btu/hr} \cdot \text{ft}^2 \cdot {}^\circ\text{R}^4)\left[(2000^\circ\text{R})^4 - (1000^\circ\text{R})^4\right] \\ &= 4.83 \times 10^3\ \text{Btu/hr} \end{aligned} \tag{16.4-17}$$

(b) From Eq. 16.4-15 and curve 5 of Fig. 16.4-4,

$$\begin{aligned} Q_{12} &= A_1 \overline{F}_{12} \sigma (T_1^4 - T_2^4) \\ &= (\pi \cdot 1\ \text{ft}^2)(0.34)(0.1712 \times 10^{-8}\ \text{Btu/hr} \cdot \text{ft}^2 \cdot {}^\circ\text{R}^4)\left[(2000^\circ\text{R})^4 - (1000^\circ\text{R})^4\right] \\ &= 27.4 \times 10^3\ \text{Btu/hr} \end{aligned} \tag{16.4-18}$$

§16.5 RADIATION BETWEEN NONBLACK BODIES AT DIFFERENT TEMPERATURES

In principle, radiation between nonblack surfaces can be treated by differential analysis of emitted rays and their successive reflected components. For nearly black surfaces this is feasible, as only one or two reflections need be considered. For highly reflecting surfaces, however, the analysis is complicated, and the distributions of emitted and reflected rays with respect to angle and wavelength are not usually known with enough accuracy to justify a detailed calculation.

A reasonably accurate treatment is possible for a small convex surface in a large, nearly isothermal enclosure (i.e., a "cavity"), such as a steam pipe in a room with walls at constant temperature. The rate of energy emission from a nonblack surface 1 to the surrounding enclosure 2 is given by

$$Q_{\vec{12}} = e_1 A_1 \sigma T_1^4 \tag{16.5-1}$$

and the rate of energy absorption from the surroundings by surface 1 is

$$Q_{\vec{21}} = a_1 A_1 \sigma T_2^4 \tag{16.5-2}$$

Here we have made use of the fact that the radiation impinging on surface 1 is very nearly cavity radiation or black-body radiation corresponding to temperature T_2. Since A_1 is convex, it intercepts none of its own rays; hence, F_{12} has been set equal to unity. The net radiation rate from A_1 to the surroundings is therefore

$$Q_{12} = \sigma A_1(e_1 T_1^4 - a_1 T_2^4) \tag{16.5-3}$$

In Eq. 16.5-3, e_1 is the value of the emissivity of surface 1 at T_1. The absorptivity a_1 is usually *estimated* as the value of e at T_2.

Next we consider an enclosure formed by n gray, opaque, diffuse-reflecting surfaces $A_1, A_2, A_3, \cdots, A_n$ at temperatures $T_1, T_2, T_3, \cdots, T_n$. Following Oppenheim[1] we define the *radiosity* J_i for each surface A_i as the sum of the fluxes of reflected and emitted radiant energy from A_i. Then the net radiant energy flow from A_i to A_k is expressed as

$$Q_{ik} = A_i F_{ik}(J_i - J_k) \qquad i,j = 1,2,3,\cdots,n \tag{16.5-4}$$

that is, by Eq. 16.4-9 with substitution of radiosities J_i in place of the black-body emissive fluxes σT_i^4.

The definition of J_i gives, for an opaque surface

$$J_i = (1 - e_i)I_i + e_i \sigma T_i^4 \tag{16.5-5}$$

in which I_i is the incident radiant flux on A_i. Elimination of I_i in favor of the net radiant flux Q_{ie}/A_i from A_i into the enclosure gives

$$\frac{Q_{ie}}{A_i} = J_i - I_i = J_i - \frac{J_i - e_i \sigma T_i^4}{1 - e_i} \tag{16.5-6}$$

whence

$$\frac{Q_{ie}}{A_i} = \frac{e_i}{1 - e_i} A_i(\sigma T_i^4 - J_i) \tag{16.5-7}$$

Finally, an energy balance on each surface gives

$$Q_i = Q_{ie} = \sum_{k=1}^{n} Q_{ik} \tag{16.5-8}$$

Here Q_i is the rate of heat addition to surface A_i by nonradiative means.

Equations analogous to Eqs. 16.5-4, 16.5-7, and 16.5-8 arise in the analysis of direct-current circuits, from Ohm's law of conduction and Kirchhoff's law of charge conservation. Hence, we have the following analogies:

Electrical	*Radiative*
Current	Q
Voltage	J or σT^4
Resistance	$(1 - e_i)/e_i A_i$ or $1/A_i F_{ij}$

This analogy allows easy diagramming of equivalent circuits for visualization of simple enclosure radiation problems. For example, the system in Fig. 16.5-1 gives the equivalent circuit shown in Fig. 16.5-2 so that the radiant heat-transfer rate is

$$Q_1 = \frac{\sigma(T_1^4 - T_2^4)}{\dfrac{1 - e_1}{e_1 A_1} + \dfrac{1}{A_1 F_{12}} + \dfrac{1 - e_2}{e_2 A_2}} \tag{16.5-9}$$

[1]A. K. Oppenheim, *Trans. ASME*, **78**, 725–735 (1956); for earlier work, see G. Poljak, *Tech. Phys. USSR*, **1**, 555–590 (1935).

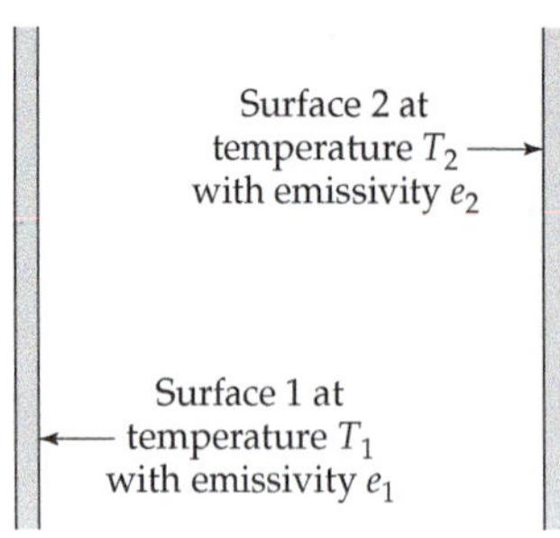

Fig. 16.5-1. Radiation between two infinite, parallel gray surfaces.

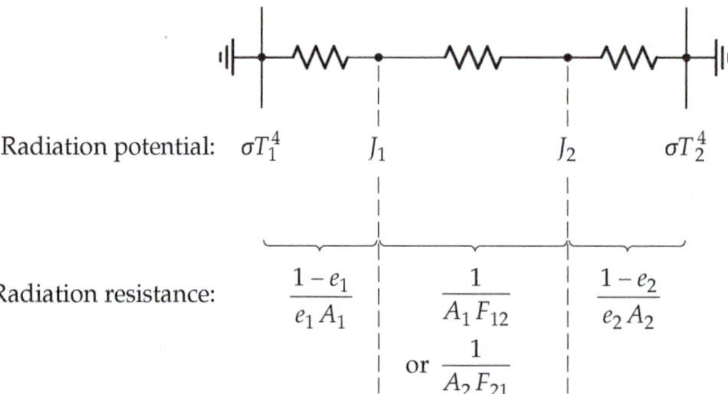

Fig. 16.5-2. Equivalent circuit for system shown in Fig. 16.5-1.

The short-cut solution summarized in Eq. 16.4-15 has been similarly generalized to nonblack-walled enclosures giving

$$Q_{12} = A_1 \overline{F}_{12}(J_1 - J_2) \tag{16.5-10}$$

in place of Eq. 16.5-8, for an enclosure with $Q_i = 0$ for $i = 2,3,\cdots,n$. The result is like that in Eq. 16.5-9, except that $\overline{F}_{12}$ must be used instead of F_{12} to include indirect paths from A_1 to A_2, thus giving a larger heat-transfer rate.

EXAMPLE 16.5-1

Radiation Shields

Develop an expression for the reduction in radiant heat transfer between two infinite parallel gray planes having the same area, A, when a thin parallel gray sheet of very high thermal conductivity is placed between them as shown in Fig. 16.5-3.

SOLUTION

The radiation balance between planes 1 and 2 is given by

$$Q_{12} = \frac{A\sigma(T_1^4 - T_2^4)}{\dfrac{1-e_1}{e_1} + 1 + \dfrac{1-e_2}{e_2}} = \frac{A\sigma(T_1^4 - T_2^4)}{\dfrac{1}{e_1} + \dfrac{1}{e_2} - 1} \tag{16.5-11}$$

since both planes have the same area A and the view factor is unity. Similarly the heat transfer between planes 2 and 3 is

$$Q_{23} = \frac{A\sigma(T_2^4 - T_3^4)}{\dfrac{1-e_2}{e_2} + 1 + \dfrac{1-e_3}{e_3}} = \frac{A\sigma(T_2^4 - T_3^4)}{\dfrac{1}{e_2} + \dfrac{1}{e_3} - 1} \tag{16.5-12}$$

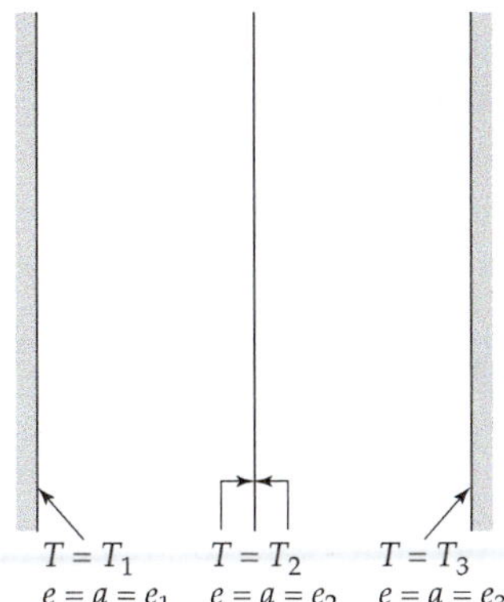

Fig. 16.5-3. Radiation shield.

These last two equations may be combined to eliminate the temperature of the radiation shield, T_2, giving

$$Q_{12}\left(\frac{1}{e_1}+\frac{1}{e_2}-1\right)+Q_{23}\left(\frac{1}{e_2}+\frac{1}{e_3}-1\right)=A\sigma(T_1^4-T_3^4) \tag{16.5-13}$$

Then, since $Q_{12} = Q_{23} = Q_{13}$, we get

$$Q_{13}=\frac{A\sigma(T_1^4-T_3^4)}{\left(\frac{1}{e_1}+\frac{1}{e_2}-1\right)+\left(\frac{1}{e_2}+\frac{1}{e_3}-1\right)} \tag{16.5-14}$$

Finally the ratio of radiant-energy transfer with a shield to that without one is

$$\frac{(Q_{13})_{\text{with}}}{(Q_{13})_{\text{without}}}=\frac{\left(\frac{1}{e_1}+\frac{1}{e_3}-1\right)}{\left(\frac{1}{e_1}+\frac{1}{e_2}-1\right)+\left(\frac{1}{e_2}+\frac{1}{e_3}-1\right)} \tag{16.5-15}$$

EXAMPLE 16.5-2

Radiation and Free-Convection Heat Losses from a Horizontal Pipe

Predict the total rate of heat loss, by radiation and free convection, from a unit length of horizontal pipe covered with asbestos. The outside diameter of the insulation is 6 in. The outer surface of the insulation is at 100°F (560°R), and the surrounding walls and air in the room are at 80°F (540°R).

SOLUTION

Let the outer surface of the insulation be surface 1 and the walls of the room be surface 2. Then Eq. 16.5-3 gives

$$Q_{12} = \sigma A_1 F_{12}(e_1 T_1^4 - a_1 T_2^4) \tag{16.5-16}$$

Since the pipe surface is convex and completely enclosed by surface 2, F_{12} is unity. From Table 16.2-1, we find $e_1 = 0.93$ at 560°R and $a_1 = 0.93$ at 540°R. Substitution of numerical values into Eq. 16.5-16 then gives for 1 ft of pipe

$$\begin{aligned}Q_{12} &= (0.1712\times10^{-8}\ \text{Btu/hr}\cdot\text{ft}^2\cdot{}^\circ\text{R}^4)\left(\pi\cdot\left(\tfrac{1}{2}\ \text{ft}\right)(1\ \text{ft})\right)(1.00)\times\left[0.93(560^\circ\text{R})^4-0.93(540^\circ\text{R})^4\right]\\ &= 33\ \text{Btu/hr}\end{aligned} \tag{16.5-17}$$

By adding the convection heat loss from Example 14.6-1, we obtain the total heat loss from the pipe:

$$Q = Q^{(\text{conv})} + Q^{(\text{rad})} = 18 + 33 = 51\ \text{Btu/hr} \tag{16.5-18}$$

Note that in this situation radiation accounts for more than half of the heat loss. If the fluid were *not* transparent, the convection and radiation processes would not be independent, and the convective and radiative contributions could not be added directly.

EXAMPLE 16.5-3

Combined Radiation and Convection

A body directly exposed to the night sky will be cooled below ambient temperature because of radiation to outer space. This effect can be used to freeze water in shallow trays well insulated from the ground. Estimate the maximum air temperature for which freezing is possible, neglecting evaporation.

SOLUTION

As a first approximation, the following assumptions may be made:

- **a.** All heat received by the water is by free convection from the surrounding air, which is assumed to be quiescent.
- **b.** The heat effect of evaporation or condensation of water is not significant.
- **c.** Steady state has been achieved.
- **d.** The pan of water is square in cross section.
- **e.** Back radiation from the atmosphere may be neglected.

The maximum permitted air temperature at the water surface is $T_1 = 492°\text{R}$. The rate of *heat loss by radiation* is

$$\begin{aligned} Q^{(\text{rad})} &= \sigma A_1 e_1 T_1^4 \\ &= (1.712 \times 10^{-9}\ \text{Btu/hr} \cdot \text{ft}^2 \cdot °\text{R}^4)(L\ \text{ft})^2(0.95)(492°\text{R})^4 \\ &= 95L^2\ \text{Btu/hr} \end{aligned} \tag{16.5-19}$$

in which L is the length in feet of one edge of the pan, and 0.95 is the value of the emissivity of a thick layer of water, from Table 16.2-1.

To get the *heat gain by convection*, we use the relation

$$Q^{(\text{conv})} = hL^2(T_{\text{air}} - T_{\text{water}}) \tag{16.5-20}$$

in which h is the heat-transfer coefficient for free convection. For cooling atmospheric air by a horizontal square facing upward, the heat-transfer coefficient is given by[2]

$$h = (0.2\ \text{Btu/hr} \cdot \text{ft}^2 \cdot °\text{F} \cdot (°\text{R})^{1/4})(T_{\text{air}} - T_{\text{water}})^{1/4} \tag{16.5-21}$$

in which h is expressed in $\text{Btu/hr} \cdot \text{ft}^2 \cdot °\text{F}$ and the temperature is given in degrees Rankine.

When the foregoing expressions for heat loss by radiation and heat gain by free convection are equated, we get

$$(95\ \text{Btu/hr} \cdot \text{ft}^2)L^2 = (0.2\ \text{Btu/hr} \cdot \text{ft}^2 \cdot (°\text{R})^{5/4})L^2(T_{\text{air}} - 492\ °\text{R})^{5/4} \tag{16.5-22}$$

From this we find that the maximum ambient air temperature is 630°R or 170°F. Except under desert conditions, back radiation and moisture condensation from the surrounding air greatly lower the required air temperature.

---

§16.6 RADIANT-ENERGY TRANSPORT IN ABSORBING MEDIA[1]

The methods given in the preceding sections are applicable only to materials that are completely transparent or completely opaque. To describe energy transport in nontransparent media, we write differential equations for the local rate of change of energy as viewed from both from the material and radiation standpoint. That is, we regard a material medium

---

[2]W. H. McAdams, in *Chemical Engineers' Handbook* (J. H. Perry, Ed.), McGraw-Hill, New York (1950), 3rd edition, p. 474.

[1]G. C. Pomraning, *Radiation Hydrodynamics*, Pergamon Press, New York (1973); R. Siegel and J. R. Howell, *Thermal Radiation and Heat Transfer*, 3rd edition, Hemisphere Publishing Co., New York (1992).

traversed by electromagnetic radiation as two coexisting "phases": a "material phase," consisting of all the mass in the system, and a "photon phase," consisting of the electromagnetic radiation.

In Chapter 11 we have already given an energy balance equation for a system containing no radiation. Here we extend Eq. 11.2-1 for the material phase to take into account the energy that is being interchanged with the photon phase by emission and absorption processes:

$$\frac{\partial}{\partial t}\rho\hat{U} = -(\nabla \cdot \rho\hat{U}\mathbf{v}) - (\nabla \cdot \mathbf{q}) - p(\nabla \cdot \mathbf{v}) - (\boldsymbol{\tau} : \nabla\mathbf{v}) - (\mathscr{E} - \mathscr{A}) \qquad (16.6\text{-}1)$$

Here we have introduced $\mathscr{E}$ and $\mathscr{A}$, which are the local rates of photon emission and absorption per unit volume, respectively. That is, $\mathscr{E}$ represents the energy lost by the material phase resulting from the emission of photons by molecules, and $\mathscr{A}$ represents the local gain of energy by the material phase resulting from photon absorption by the molecules (see Fig. 16.6-1). The $\mathbf{q}$ in Eq. 16.6-1 is the conduction heat flux given by Fourier's law.

For the "photon phase," we may write an equation describing the local rate of change of radiant-energy density $u^{(r)}$:

$$\frac{\partial}{\partial t}u^{(r)} = -(\nabla \cdot \mathbf{q}^{(r)}) + (\mathscr{E} - \mathscr{A}) \qquad (16.6\text{-}2)$$

in which $\mathbf{q}^{(r)}$ is the radiant energy flux. This equation may be obtained by writing a radiant-energy balance on an element of volume fixed in space. Note that there is no convective term in Eq. 16.6-2, since the photons move independently of the local material velocity. Note further that the term $(\mathscr{E} - \mathscr{A})$ appears with opposite signs in Eqs. 16.6-1 and 16.6-2, indicating that a net gain of radiant energy occurs at the expense of molecular energy. Equation 16.6-2 can also be written for the radiant energy within a frequency range ν to $\nu + d\nu$:

$$\frac{\partial}{\partial t}u_\nu^{(r)} = -(\nabla \cdot \mathbf{q}_\nu^{(r)}) + (\mathscr{E}_\nu - \mathscr{A}_\nu) \qquad (16.6\text{-}3)$$

This expression is obtained by differentiating Eq. 16.6-2 with respect to ν.

For the purpose of simplifying the discussion, we consider a steady-state nonflow system in which the radiation travels only in the $+z$ direction. Such a system can be closely approximated by passing a collimated light beam through a solution at temperatures sufficiently low that the emission by the solution is unimportant. (If emission were important, it would be necessary to consider radiation in all directions.) These are the conditions

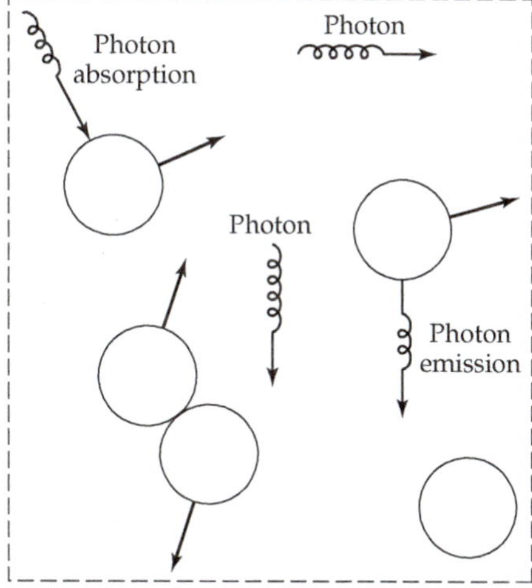

Fig. 16.6-1. Volume element over which energy balances are made. The circles represent molecules.

commonly encountered in spectrophotometry. For such a system, Eqs. 16.6-1 and 16.6-2 become

$$0 = -\frac{d}{dz}q_z + \mathscr{A} \tag{16.6-4}$$

$$0 = -\frac{d}{dz}q_z^{(r)} - \mathscr{A} \tag{16.6-5}$$

In order to use these equations, we need information about the volumetric absorption rate $\mathscr{A}$. For a unidirectional beam a conventional expression is

$$\mathscr{A} = m_a q^{(r)} \tag{16.6-6}$$

in which m_a is known as the *extinction coefficient*. Basically, this just states that the probability for photon absorption is proportional to the concentration of photons.

EXAMPLE 16.6-1

Absorption of a Monochromatic Radiant Beam

A monochromatic radiant beam of frequency ν, focused parallel to the z axis, passes through an absorbing fluid. The local rate of energy absorption is given by $m_{a\nu}q_\nu^{(r)}$, in which $m_{a\nu}$ is the extinction coefficient for radiation of frequency ν. Determine the distribution of the radiant flux $q_\nu^{(r)}(z)$ in the system.

SOLUTION

We neglect refraction and scattering of the incident beam. Also, we assume that the liquid is cooled so that re-radiation can be neglected. Then Eq. 16.6-5 becomes for steady state

$$0 = -\frac{d}{dz}q_\nu^{(r)} - m_{a\nu}q_\nu^{(r)} \tag{16.6-7}$$

Integration with respect to z gives

$$q_\nu^{(r)}(z) = q_\nu^{(r)}(0)\exp(-m_{a\nu}z) \tag{16.6-8}$$

This is *Lambert's law of absorption*,[2] widely used in spectrometry. For any given pure material, $m_{a\nu}$ depends in a characteristic way on ν. The shape of the absorption spectrum is therefore a useful tool for qualitative analysis.

§16.7 CONCLUDING COMMENTS

The theory of radiant-heat transfer is a very large topic, one about which entire volumes have been written. All we have done in this chapter is present some of the key ideas and show how to make some fairly simple calculations.

Keep in mind that it was Planck's persistence in trying to understand the distribution of black-body radiation over a wide range of frequencies that led him to the radical notion that energy is quantized. This resulted in the development of an entire new body of physics, namely quantum mechanics, one of the two gigantic basic developments in the twentieth century—the other being Einstein's development of the theory of relativity.

QUESTIONS FOR DISCUSSION

1. The "named laws" in this chapter are important. What is the physical content of the laws associated with the following scientists' names: Stefan and Boltzmann, Planck, Kirchhoff, Lambert, Wien?
2. How are the Stefan-Boltzmann law and the Wien displacement law related to the Planck black-body distribution law?
3. Do black bodies exist? Why is the concept of a black body useful?

[2] J. H. Lambert, *Photometria*, Augsburg (1760).

4. In specular (mirrorlike) reflection, the angle of incidence equals the angle of reflection. How are these angles related for diffuse reflection?
5. What is the physical significance of the view factor, and how can it be calculated?
6. What are the units of $q^{(e)}$, $q_\nu^{(e)}$, and $q_\lambda^{(e)}$?
7. Under what conditions is the effect of geometry on radiant-heat interchange completely expressible in terms of view factors?
8. Which of the equations in this chapter show that the apparent brightness of a black body with a uniform surface temperature is independent of the position (distance and direction) from which it is viewed through a transparent medium?
9. What relation is analogous to Eq. 16.3-2 for an ideal monatomic gas?
10. Check the dimensional consistency of Eq. 16.3-9.

PROBLEMS

16A.1 Approximation of a black body by a hole in a sphere. A thin sphere of copper, with its internal surface highly oxidized, has a diameter of 6 in. How small a hole must be made in the sphere to make an opening that will have an absorptivity of 0.99?

Answer: Radius = 0.70 in.

16A.2 Efficiency of a solar engine. A device for utilizing solar energy, developed by Abbot,[1] consists of a parabolic mirror that focuses the impinging sunlight onto a Pyrex tube containing a high-boiling, nearly black liquid. This liquid is circulated to a heat exchanger in which the heat energy is transferred to superheated water at 25 atm pressure. Steam may be withdrawn and used to run an engine. The most efficient design requires a mirror 10 ft in diameter to generate 2 hp, when the axis of the mirror is pointed directly toward the sun. What is the overall efficiency of the device?

Answer: 15%

16A.3 Radiant heating requirement. A shed is rectangular in shape, with the floor 15 ft by 30 ft and the roof 7.5 ft above the floor. The floor is heated by hot water running through coils. On cold winter days the exterior walls and roof are about −10°F. At what rate must heat be supplied through the floor in order to maintain the floor temperature at 75°F? (Assume that all the surfaces of the system are black, and that convective heat transfer within the shed can be neglected.)

16A.4 Steady-state temperature of a roof. Estimate the maximum temperature attained by a level roof at 45° north latitude on June 21 in clear weather. Radiation from sources other than the sun may be neglected, and a convection-heat-transfer coefficient of 2.0 Btu/hr · ft^2 · °F may be assumed. A maximum temperature of 100°F may be assumed for the surrounding air. The solar constant of Example 16.4-1 may be used, and the absorption and scattering of the sun's rays by the atmosphere may be neglected.

(a) Solve for a perfectly black roof.

(b) Solve for an aluminum-coated roof, with an absorptivity of 0.3 for solar radiation and an emissivity of 0.07 at the temperature of the roof.

16A.5 Radiation errors in temperature measurements. The temperature of an air stream in a duct is being measured by means of a thermocouple. The thermocouple wires and junction are cylindrical, 0.05 in. in diameter, and extend across the duct perpendicular to the flow with the junction in the center of the duct. Assuming a junction emissivity of $e = 0.8$, estimate the temperature of the gas stream from the following data obtained under steady conditions:

Thermocouple junction temperature = 500°F
Duct wall temperature = 300°F
Convection heat-transfer coefficient from wire to air = 50 Btu/hr · ft^2 · °F

---

[1] C. G. Abbot, in *Solar Energy Research* (F. Daniels and J. A. Duffie, eds.), University of Wisconsin Press, Madison (1955), pp. 91–95; see also U.S. Patent No. 2,460,482 (Feb. 1, 1945).

The wall temperature is constant at the value given for 20 duct diameters upstream and downstream of the thermocouple installation. The thermocouple leads are positioned so that the effect of heat conduction along them on the junction temperature may be neglected.

16A.6 Surface temperatures on the earth's moon.

(a) Estimate the surface temperature of our moon at the point nearest the sun by a quasi-steady-state radiant energy balance, regarding the lunar surface as a gray sphere. Neglect radiation and reflection from the planets. The solar constant at earth is given in Example 16.4-1.

(b) Extend part (a) to give the lunar surface temperature as a function of angular displacement from the hottest point.

16B.1 Reference temperature for effective emissivity. Show that, if the emissivity increases linearly with the temperature, Eq. 16.5-3 may be written as

$$Q_{12} = e_1^{\circ}\sigma A_1(T_1^4 - T_2^4) \tag{16B.1-1}$$

in which e_1° is the emissivity of surface 1 evaluated at a reference temperature T° given by

$$T^{\circ} = \frac{T_1^5 - T_2^5}{T_1^4 - T_2^4} \tag{16B.1-2}$$

16B.2 Radiation across an annular gap. Develop an expression for the radiant heat transfer between two long, gray coaxial cylinders 1 and 2. Show that

$$Q_{12} = \frac{\sigma(T_1^4 - T_2^4)}{\dfrac{1}{A_1 e_1} + \dfrac{1}{A_2}\left(\dfrac{1}{e_2} - 1\right)} \tag{16B.2-1}$$

where A_1 is the surface area of the inner cylinder.

16B.3 Multiple radiation shields.

(a) Develop an equation for the rate of radiant heat transfer through a series of n very thin, flat, parallel metal sheets, each having a different emissivity e, when the first sheet is at temperature T_1 and the nth sheet is at temperature T_n. Give your result in terms of the radiation resistances

$$R_{i,i+1} = \frac{\sigma(T_i^4 - T_{i+1}^4)}{Q_{i,i+1}} \tag{16B.3-1}$$

for the successive pairs of planes. Edge effects and conduction across the air gaps between the sheets are to be neglected.

(b) Determine the ratio of the radiant heat-transfer rate for n identical sheets to that for two identical sheets.

(c) Compare your results for three sheets with that obtained in Example 16.5-1.

The marked reduction in heat-transfer rates produced by a number of radiation shields in series has led to the use of multiple layers of metal foils for high-temperature insulation.

16B.4 Radiation and conduction through absorbing media. A glass slab, bounded by planes $z = 0$ to $z = \delta$, is of infinite extent in the x and y directions. The temperatures of the surfaces at $z = 0$ and $z = \delta$ are maintained at T_0 and T_δ, respectively. A uniform monochromatic radiant beam of intensity $q_0^{(r)}$ in the z direction impinges on the face at $z = 0$. Emission within the slab, and incident radiation in the $-z$ direction can be neglected.

(a) Determine the temperature distribution in the slab, assuming m_a and k to be constants (see §16.6).

(b) How does the distribution of the conductive heat flux q_z depend on m_a?

16B.5 Cooling of a black body in vacuo. A thin black body of very high thermal conductivity has a volume V, surface area A, density ρ, and heat capacity $\hat{C}_p$. At time $t = 0$, this body at temperature T_1 is placed in a black enclosure, the walls of which are maintained permanently at temperature T_2 (with $T_2 < T_1$). Derive an expression for the temperature T of the black body as a function of time.

16B.6 Heat loss from an insulated pipe. A standard Schedule 40 two-inch horizontal steel pipe (inside diameter 2.067 in., wall thickness 0.154 in.; $k = 26$ Btu/hr · ft · °F) carrying steam is lagged (i.e., insulated) with 2 in. of 85% magnesia ($k = 0.04$ Btu/hr · ft · °F) and tightly wrapped with a single outer layer of clean aluminum foil ($e = 0.05$). The inner surface of the pipe is at 250°F, and the pipe is surrounded by air at 1 atm and 80°F.

(a) Compute the conductive heat flow per unit length, $Q^{(\text{cond})}/L$, through the pipe wall and insulation for assumed temperatures, T_0, of 100°F and 250°F at the outer surface of the aluminum foil.

(b) Compute the radiative and free-convective heat losses, $Q^{(\text{rad})}/L$ and $Q^{(\text{conv})}/L$, for the same assumed outer surface temperatures T_0.

(c) Plot or interpolate the foregoing results to obtain the steady-state values of T_0 and $Q^{(\text{cond})}/L = Q^{(\text{rad})}/L + Q^{(\text{conv})}/L$.

16C.1 Integration of the view-factor integral for a pair of disks. Two identical, perfectly black disks of radius R are placed a distance H apart as shown in Fig. 16C.1. Integrate the view-factor integrals for this case and show that

$$F_{12} = F_{21} = \frac{1 + 2B^2 - \sqrt{1 + 4B^2}}{2B^2} \qquad (16C.1\text{-}1)^2$$

in which $B = R/H$.

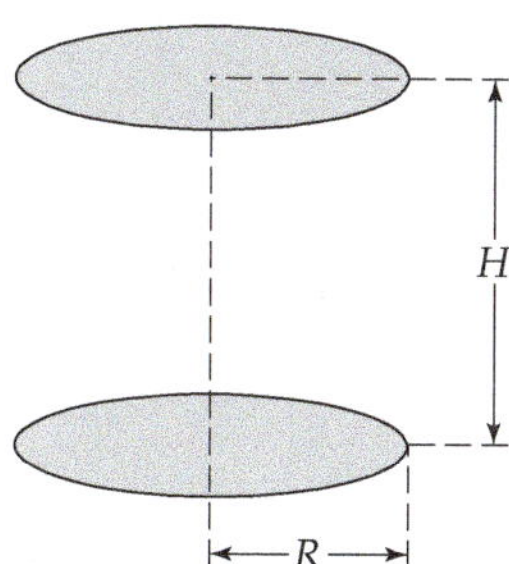

Fig. 16C.1 Two parallel, perfectly black disks.

---

[2] C. Christiansen, *Wiedemann's Ann. d. Physik*, **19**, 267–283 (1883); see also M. Jakob, *Heat Transfer*, Vol. **II**, Wiley, New York (1957), p. 14.

Part Three

Mass Transport

Chapter 17

Diffusivity and the Mechanisms of Mass Transport

In Chapter 1, we began by describing convective momentum transport, and then we discussed molecular momentum transport (Newton's law of viscosity); in Chapter 9, we began by describing convective energy transport, and then we discussed molecular energy transport (Fourier's law of heat conduction). In this chapter we start by describing convective mass transport, and then discuss the molecular mechanism of mass transport. The latter is described mathematically by Fick's law of diffusion, which relates the movement of chemical species A through a binary mixture of A and B to the concentration gradient of A. Fick's law of diffusion is the mass transfer analog of Newton's law of viscosity for momentum transport, and Fourier's law of heat conduction for energy transport.

In §17.1, we preface the discussion of mass transport by defining several ways of specifying the concentration of species A in a mixture of A and B. Both mass concentrations and molar concentrations are discussed. It is helpful to know how these two ways of specifying concentration are related, since both of them will be used. When dealing with chemical reactions, molar units are more useful, but when solving problems in conjunction with the equation of motion, mass units may be more useful. Often the molar units are preferred for gaseous mixtures, because usually the total molar density c can be assumed to be constant. Similarly, for liquid mixtures the mass units are sometimes preferred, because the total mass density ρ is often very nearly constant.

Then in §17.2, we begin by defining the *mass average velocity* $\mathbf{v}$ and the *molar average velocity* $\mathbf{v}^*$ of a mixture of A and B. It is the mass average velocity that is measured by means of a Pitot tube or by laser-Doppler velocimetry, and it is identical to the $\mathbf{v}$ in the equation of motion. The molar average velocity is not amenable to simple measurement.

After defining these average velocities, we then can describe how species A and B are swept along by the average rate of flow—i.e., we can formulate the *convective mass flux vector* and the *convective molar flux vector*.

In §17.3, we turn to the molecular mass transport and describe how species A and B "diffuse" in a mixture. The movement of a chemical species from a region of high concentration to a region of low concentration can be observed by dropping a small crystal of potassium permanganate ($KMnO_4$) into a beaker of water. The $KMnO_4$ begins to dissolve in the water, and very near the crystal there is a dark purple, concentrated solution of $KMnO_4$. Because of the concentration gradient that is established, the $KMnO_4$ diffuses away from the crystal—from a region of high concentration to a region of low concentration. The progress of this molecular diffusion can then be followed by observing the growth of the purple region. Molecular diffusion is described by Fick's law, which relates the *diffusive mass flux vector* or the *diffusive molar flux vector* to a concentration gradient. The relevant physical property is the *diffusivity* $\mathscr{D}_{AB}$ for the pair A-B.

The diffusive flux owes its existence to the seemingly random thermal motion of the molecules. Through this motion, individual molecules can travel large distances with time. This process is also known as diffusion, and in fact can be shown to be related to Fick's law. While we will not dwell on this perspective, it is important to note that the molecular motion that gives rise to diffusion and Fick's law is also responsible for viscous stress and Newton's law of viscosity, as well as thermal conduction and Fourier's law of heat conduction.

In §17.4, the convective mass (molar) flux vector and the diffusive mass (molar) flux vector are combined to give the *total mass and molar flux vectors*. This is the flux that will be used for setting up problems by shell mass (or molar) balances in Chapter 18 and establishing the equations of change in Chapter 19.

The remainder of the chapter is devoted to various ways of estimating the diffusivity $\mathscr{D}_{AB}$. As in Chapters 1 and 9, we discuss the experimental data (§17.5), the use of corresponding states (§17.6), the kinetic theory for gas mixtures (§17.7), and empiricisms for liquid mixtures (§17.8).

This entire chapter, as well as subsequent chapters, are restricted to two-component (binary) mixtures. The multicomponent equivalent of Fick's law (the *Maxwell-Stefan equations*) are discussed briefly in the last section of Chapter 24. However, the reader should be aware of the fact that there exists a sizeable volume of literature devoted to multicomponent mixtures.[1]

§17.1 SPECIES CONCENTRATIONS

As mentioned in the introduction to this chapter, we shall find it necessary to use both mass units and molar units for describing concentrations. In *mass* units, the concentration of A in a binary mixture of A and B will be given as mass of A per unit volume of the mixture and represented by the symbol ρ_A. It is sometimes convenient to express the composition in terms of the mass fraction of A, defined as the mass of A divided by the total mass and represented by the symbol ω_A. In *molar units*, the concentration of A in a binary mixture will be given as the number of moles of A per unit volume of the mixture and represented

---

[1] R. Taylor and R. Krishna, *Multicomponent Mass Transfer*, Wiley, New York (1993). R. B. Bird, W. E. Stewart, and E. N. Lightfoot, *Transport Phenomena*, Wiley, New York, 2nd Revised Edition (2007), §17.9 and §§24.1, 24.2, 24.4. C. F. Curtiss and R. B. Bird, *Ind. Eng. Chem. Research*, **38**, 2515–2522 (1999), erratum 40, 1791 (2001). R. B. Bird and D. J. Klingenberg, *Advances in Water Resources*, **62**, 238–242 (2013); *errata*: in Eq. 3.15, $\omega_\alpha \omega_\beta$ on the right side should be in the denominator instead of the numerator; also, in the first sentence of Sec. 4, the reference should be to Eq. 3.15 rather than Eq. 3.13.

Table 17.1-1. Species Concentrations in a Binary Mixture of A and B

ρ_A = mass of A per unit volume	(A)[a]	c_A = moles of A per unit volume	(F)
$\rho = \rho_A + \rho_B$ = mass density of mixture	(B)	$c = c_A + c_B$ = molar density of mixture	(G)
$\omega_A = \rho_A/\rho$ = mass fraction of A	(C)	$x_A = c_A/c$ = mole fraction of A	(H)
$\omega_A + \omega_B = 1$	(D)	$x_A + x_B = 1$	(I)
$\nabla\omega_A = -\nabla\omega_B$	(E)	$\nabla x_A = -\nabla x_B$	(J)

[a]We emphasize that ρ_A is the mass concentration of species A in a mixture. The notation $\rho^{(A)}$ is reserved for the density of pure species A when the need arises. Similar comments apply to the symbols c_A and $c^{(A)}$.

Table 17.1-2. Relations among Concentrations and Molecular Weights

$\rho_A = c_A M_A$	(A)	$c_A = \rho_A/M_A$	(F)
$M = x_A M_A + x_B M_B$	(B)	$1/M = \omega_A/M_A + \omega_B/M_B$	(G)
$\rho = cM$	(C)	$c = \rho/M$	(H)
$\omega_A = \dfrac{x_A M_A}{M}$	(D)	$x_A = \dfrac{\omega_A/M_A}{1/M}$	(I)
$\nabla\omega_A = \dfrac{M_A M_B \nabla x_A}{M^2}$	(E)	$\nabla x_A = \dfrac{M^2}{M_A M_B}\nabla\omega_A$	(J)
$= \dfrac{\omega_A\omega_B}{x_A x_B}\nabla x_A$	(E′)	$= \dfrac{x_A x_B}{\omega_A\omega_B}\nabla\omega_A$	(J′)

by the symbol c_A. The mole fraction of A is the number of moles of A divided by the total number of moles and is represented by the symbol x_A.

In Table 17.1-1, we list the basic definitions along with some important relations among the defined quantities. In Table 17.1-2, we summarize some of the relations that link the mass quantities with the molar quantities. These involve the species molecular weights M_A and M_B. Note that M, with no subscript, is the *molar mean molecular weight* of the mixture, defined in Eqs. (B) and (G) of Table 17.1-2. With the relations in Table 17.1-2, it is possible to transform relations in mass units into the analogous relations in molar units and vice versa.

In the first four sections of this chapter, the tables are so arranged that mass units are given at the left, and the corresponding molar units at the right.

§17.2 CONVECTIVE MASS AND MOLAR FLUX VECTORS

Before discussing the convective mass and molar flux vectors, we must take a small detour and define species velocities and average velocities.

In a mixture of A and B, the chemical species are generally moving at different velocities. By $\mathbf{v}_A$, the velocity of species A, we do *not* mean the velocity of an individual molecule of species A, but rather the average of all the velocities of molecules of species A within a small volume (this volume should be large enough to contain many A molecules, but much smaller than the system of interest). Then, for a mixture of two chemical species, the local *mass average velocity* $\mathbf{v}$ is defined as

$$\mathbf{v} = \frac{\rho_A \mathbf{v}_A + \rho_B \mathbf{v}_B}{\rho_A + \rho_B} = \frac{\rho_A \mathbf{v}_A + \rho_B \mathbf{v}_B}{\rho} = \omega_A \mathbf{v}_A + \omega_B \mathbf{v}_B \tag{17.2-1}$$

Note that $\rho\mathbf{v}$ is the local rate at which mass passes through a unit cross section perpendicular to the velocity $\mathbf{v}$ (see §0.4).

Similarly one may define the local *molar average velocity* $\mathbf{v}^*$ by

$$\mathbf{v}^* = \frac{c_A \mathbf{v}_A + c_B \mathbf{v}_B}{c_A + c_B} = \frac{c_A \mathbf{v}_A + c_B \mathbf{v}_B}{c} = x_A \mathbf{v}_A + x_B \mathbf{v}_B \tag{17.2-2}$$

The product $c\mathbf{v}^*$ is the local rate at which moles pass through a unit cross section perpendicular to the velocity $\mathbf{v}^*$. Other types of averages may be defined, but we shall not need them here.

a. Convective mass flux vector

Species A may be transported by the bulk motion of the fluid. This convective mass transport mechanism was introduced in Chapter 0 (§0.4) for the special case of unidirectional, steady-state flow (readers may find it helpful to review that short section before proceeding). Here we consider the convective transport of species A for arbitrary flows. In Fig. 17.2-1 we show three mutually perpendicular elements of area dS at the point P, where the fluid velocity is $\mathbf{v}$. The volume rate of flow across the surface element dS perpendicular to the x axis is $v_x dS$. The rate at which mass of species A is being swept across the same surface element in the $+x$ direction is then

$$\rho_A v_x dS = \omega_A \rho v_x dS \tag{17.2-3}$$

When we divide Eq. 17.2-3 by dS, we get the *convective mass flux* of species A in the $+x$ direction, $\rho_A v_x = \omega_A \rho v_x$, which has dimensions of mass/time · area.

We can write expressions similar to Eq. 17.2-3 for the rate at which mass of species A is being swept through the surface elements perpendicular to the y and z axes to give the rates of species A mass transport in the $+y$ and $+z$ directions, respectively. If we now multiply each of the three expressions by the corresponding unit vector and add vectorially, we then get, after division by dS, the *convective mass flux vector* of species A,

$$\mathbf{j}_A^{(c)} = \rho_A(v_x \boldsymbol{\delta}_x + v_y \boldsymbol{\delta}_y + v_z \boldsymbol{\delta}_z) = \omega_A \rho(v_x \boldsymbol{\delta}_x + v_y \boldsymbol{\delta}_y + v_z \boldsymbol{\delta}_z) \tag{17.2-4}$$

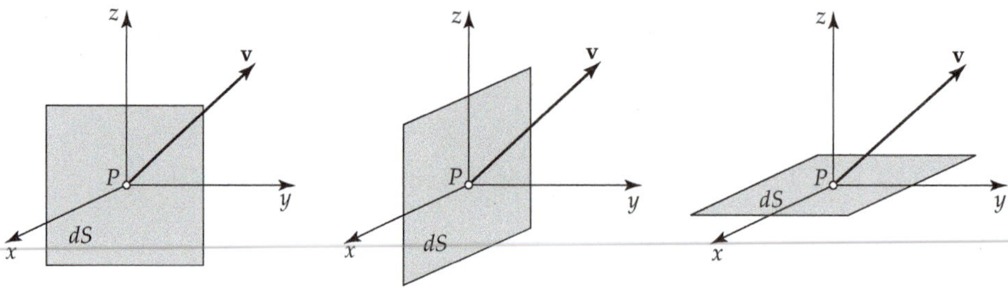

Fig. 17.2-1. Three mutually perpendicular surface elements of area dS across which species A is being transported by convection by the fluid moving with the velocity $\mathbf{v}$. The volume rate of flow across the face perpendicular to the x axis is $v_x\, dS$, and the rate of flow of mass of species A across dS is then $\rho_A v_x\, dS$. Similar expressions can be written for the surface elements perpendicular to the y and z axes.

Table 17.2-1. Average Velocities and Convective Mass and Molar Fluxes

$\mathbf{v}_A$ = velocity of species A with respect to axes fixed in space
$\mathbf{v}_B$ = velocity of species B with respect to axes fixed in space

$\mathbf{v} = \omega_A \mathbf{v}_A + \omega_B \mathbf{v}_B$ = mass average velocity (A)	$\mathbf{v}^* = x_A \mathbf{v}_A + x_B \mathbf{v}_B$ = molar average velocity (C)
$\mathbf{j}_A^{(c)} = \rho_A \mathbf{v}$ = convective mass flux vector of A (B)	$\mathbf{J}_A^{*(c)} = c_A \mathbf{v}^*$ = convective molar flux vector of A (D)

or

$$\boxed{\mathbf{j}_A^{(c)} = \rho_A \mathbf{v} = \omega_A (\rho \mathbf{v})} \tag{17.2-5}$$

It is understood that this is the flux from the negative side of the surface to the positive side (e.g., $j_{A,x}^{(c)}$ is the flux in the positive x direction). This flux should be compared to the convective momentum flux discussed in §1.1, and the convective energy flux discussed in §9.1.

b. Convective molar flux vector

The convective molar flux vector of species A can be developed similarly. The rate at which moles of species A are being swept in the $+x$ direction across the surface element dS oriented perpendicular to the x axis is

$$c_A v_x^* dS = x_A c v_x^* dS \tag{17.2-6}$$

When we divide Eq. 17.2-6 by dS, we get the *convective molar flux* of species A in the $+x$ direction, $c_A v_x^* = x_A c v_x^*$, which has dimensions of moles/time · area.

We can write expressions similar to Eq. 17.2-6 for the rate at which moles of species A are being swept through the surface elements perpendicular to the y and z axes to give the rates of species A molar transport in the $+y$ and $+z$ directions, respectively. If we now multiply each of the three expressions by the corresponding unit vector and add vectorially, we then get, after division by dS, the *convective molar flux vector* of species A,

$$\mathbf{J}_A^{*(c)} = c_A (v_x^* \boldsymbol{\delta}_x + v_y^* \boldsymbol{\delta}_y + v_z^* \boldsymbol{\delta}_z) = x_A c (v_x^* \boldsymbol{\delta}_x + v_y^* \boldsymbol{\delta}_y + v_z^* \boldsymbol{\delta}_z) \tag{17.2-7}$$

or

$$\boxed{\mathbf{J}_A^{*(c)} = c_A \mathbf{v}^* = x_A (c \mathbf{v}^*)} \tag{17.2-8}$$

It is understood that this is the molar flux from the negative side of the surface to the positive side (e.g., $J_{A,x}^{*(c)}$ is the molar flux in the positive x direction). Note that $\mathbf{j}_A^{(c)} = \rho_A \mathbf{v}$ is the local rate at which mass of A passes through a unit cross section placed perpendicular to the velocity $\mathbf{v}$, and that $\mathbf{J}_A^{*(c)} = c_A \mathbf{v}^*$ is the local rate at which moles of A pass through a unit cross section placed perpendicular to the velocity $\mathbf{v}^*$. The key ideas in this section are summarized in Table 17.2-1.

§17.3 DIFFUSIVE MASS AND MOLAR FLUX VECTORS—FICK'S LAW

Having discussed the convective mass and molar flux vectors, we now turn to the diffusive mass and molar flux vectors, that is, the fluxes associated with the random molecular motion.

Consider a thin, horizontal, fused-silica plate of area A and thickness Y. Suppose that initially (for time $t < 0$) both horizontal surfaces of the plate are in contact with air, which we regard as completely insoluble in silica. At time $t = 0$, the air below the plate is suddenly replaced by pure helium, which is appreciably soluble in silica. The helium slowly penetrates into the plate by virtue of its molecular motion and ultimately appears in the gas above. This molecular transport of one substance relative to another is known as *diffusion* (also known as *mass diffusion, concentration diffusion,* or *ordinary diffusion*). The air above the plate is being replaced rapidly with pure air, so that there is no appreciable buildup of helium there. We thus have the situation represented in Fig. 17.3-1; this process is analogous to those described in §1.2 and §9.2 where viscosity and thermal conductivity, respectively, were defined.

a. Diffusive mass flux vector

In this system, we will call helium "species A" and silica "species B." The concentrations will be given by the mass fractions ω_A and ω_B. The mass fraction ω_A is the mass of helium divided by the mass of helium-plus-silica in a given microscopic volume element. The mass fraction ω_B is defined analogously.

For time t less than zero, the mass fraction of helium, ω_A, is everywhere equal to zero. For time t greater than zero, at the lower surface, $y = 0$, the mass fraction of helium is equal to ω_{A0}. This latter quantity is the solubility of helium in silica, expressed as mass fraction, just inside the solid. As time proceeds, the mass fraction profile develops, with $\omega_A = \omega_{A0}$ at the bottom surface of the plate and $\omega_A = 0$ at the top surface of the plate. As indicated in Fig. 17.3-1, the profile tends toward a straight line with increasing t.

At steady state, it is found that the mass flow rate w_{Ay} (in grams per second, say) of helium in the positive y direction can be described to a very good approximation by

$$\frac{w_{Ay}}{A} = \rho \mathscr{D}_{AB} \frac{\omega_{A0} - 0}{Y} \tag{17.3-1}$$

That is, the diffusive mass flow rate of helium per unit area (or *diffusive mass flux*) is proportional to the mass fraction difference divided by the plate thickness. Here ρ is the

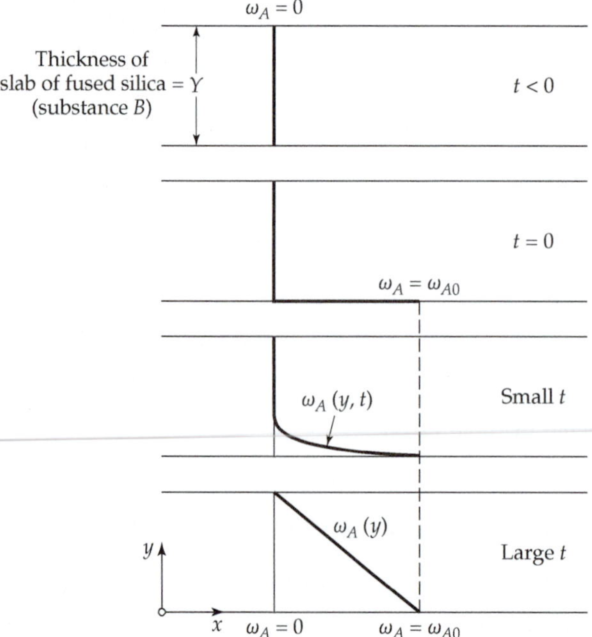

Fig. 17.3-1. Buildup to the steady-state concentration profile for the diffusion of helium (substance A) through fused silica (substance B). The symbol ω_A stands for the mass fraction of helium, and ω_{A0} is the solubility of helium in fused silica, expressed as the mass fraction. See Figs. 1.1-1 and 9.2-1 for analogous momentum and heat-transport situations.

density of the silica-helium system, and the proportionality factor $\mathscr{D}_{AB}$ is the *diffusivity* of the silica-helium system. We now rewrite Eq. 17.3-1 for a differential element within the slab:

$$j_{Ay} = -\rho \mathscr{D}_{AB} \frac{d\omega_A}{dy} \tag{17.3-2}$$

Here w_{Ay}/A as been replaced by j_{Ay}, the *diffusive mass flux* of helium in the positive y direction. Note that the first index, A, designates the chemical species (in this case, helium), and the second index indicates the direction in which diffusive transport is taking place (in this case, the y direction).

Equation 17.3-2 is the one-dimensional form of *Fick's first law of diffusion.*[1] It is valid for any binary solid, liquid, or gas solution, provided that j_{Ay} is defined as the mass flux relative to the *mass average velocity* of the mixture v_y,

$$v_y = \omega_A v_{Ay} + \omega_B v_{By} \tag{17.3-3}$$

as defined in §17.2. For the system examined in Fig. 17.3-1, the helium is moving rather slowly and its concentration is very small, so that v_y is negligibly different from zero during the diffusion process. As pointed out at the beginning of §17.2, the species velocity $\mathbf{v}_A$ is not the instantaneous molecular velocity of a molecule of A, but rather the arithmetic average of the velocities of all the molecules of A within a tiny volume element.

Equation 17.3-2 relates the diffusive mass flux to the concentration gradient (y component), and may be interpreted as the definition of the diffusivity $\mathscr{D}_{AB}$. The diffusive mass flux j_{Ay} itself can be defined more generally in terms of the velocities discussed above. Following the derivation of the convective mass flux in §17.2, the y component of the diffusive mass flux is

$$j_{Ay} = \rho\omega_A (v_{Ay} - v_y) \tag{17.3-4}$$

which clearly illustrates that this is the flux relative to the mass average velocity, or equivalently, the flux relative to a coordinate system moving with the mass average velocity $\mathbf{v}$. The diffusive mass flux of B is defined analogously.

As the two chemical species interdiffuse, there is, locally, a shifting of the center of mass in the y direction if the molecular weights of A and B differ. The mass fluxes j_{Ay} and j_{By} are so defined that $j_{Ay} + j_{By} = 0$ (this result can be obtained by using Eqs. 17.3-3 and 17.3-4; see Problem 17B.3). This again illustrates that the fluxes j_{Ay} and j_{By} are measured with respect to the motion of the center of mass.

When equations similar to Eq. 17.3-2 are written for diffusion in the x and z directions, and the three equations added vectorially, we then get the vector form of Fick's law for the *diffusive mass flux vector* of species A:

$$\boxed{\mathbf{j}_A = -\rho \mathscr{D}_{AB} \nabla \omega_A} \tag{17.3-5}$$

A similar relation can be written down for the *diffusive mass flux vector* of species B:

$$\boxed{\mathbf{j}_B = -\rho \mathscr{D}_{BA} \nabla \omega_B} \tag{17.3-6}$$

It is shown in Example 17.3-2 that $\mathscr{D}_{AB} = \mathscr{D}_{BA}$. Thus, for the pair A-B, there is just one diffusivity; in general it will be a function of pressure, temperature, and composition. This equation is written in component form in Cartesian, cylindrical, and spherical coordinates in Appendix B.3.

[1] A. Fick, *Ann. der Physik*, **94**, 59–86 (1855). Fick's second law, the diffusional analog of the heat-conduction equation in Eq. 11.2-10, is given in Eq. 19.1-22. **Adolf Eugen Fick** (1829–1901) was a medical doctor who taught in Zürich and Marburg, and then became the Rector of the University of Würzburg. He postulated the laws of diffusion by analogy with heat conduction, not by experiment.

A frequently asked question is: Why don't we write Eq. 17.3-5 as $\mathbf{j}_A = -\mathscr{D}_{AB}\nabla\rho_A$ so that Fick's law will have the same form as Fourier's law $\mathbf{q} = -k\nabla T$? If the total mass density ρ is constant, then the two forms are the same. However, if the total mass density is a function of position and if ω_A is constant throughout, then $\nabla\rho_A = \rho\nabla\omega_A + \omega_A\nabla\rho$ would imply that there is a diffusive flux of A because of the gradient of the mass density ρ. We want $\mathbf{j}_A$, the mass flux of A, to depend only on the nonuniformity in the mass fraction of A, that is on the gradient of ω_A. Furthermore, requiring that the flux be proportional to $\nabla\omega_A$ guarantees that $\mathbf{j}_A + \mathbf{j}_B = \mathbf{0}$ (i.e., $\mathbf{j}_A$ and $\mathbf{j}_B$ are fluxes relative to the mass average velocity), whereas making the flux proportional to $\nabla\rho_A$ only produces this result in special situations. Therefore, Eq. 17.3-5 is the only reasonable expression for the mass flux. We note that Eq. 17.3-5 is in complete agreement with the kinetic theory of gases[2] and with the thermodynamics of irreversible processes.[3]

b. Diffusive molar flux vector

The diffusive fluxes in molar units are written analogously. For the process illustrated in Fig. 17.3-1, the molar flow rate of species A at steady state is

$$\frac{W_{Ay}}{A} = c\mathscr{D}_{AB}\frac{x_{A0} - 0}{Y} \tag{17.3-7}$$

where W_{Ay} is the number of moles of species A transported in the y direction per unit time, c is the total molar concentration of the silica-helium system, and x_{A0} is the mole fraction of species A at $y = 0$, which is also the solubility of species A expressed as a mole fraction. Equation 17.3-7 can be rewritten for a differential element within the slab as

$$J^*_{Ay} = -c\mathscr{D}_{AB}\frac{dx_A}{dy} \tag{17.3-8}$$

Here W_{Ay}/A as been replaced by J^*_{Ay}, the *diffusive molar flux* of helium in the positive y direction. As before, the first index, A, designates the chemical species (in this case, helium), and the second index indicates the direction in which diffusive transport is taking place (in this case, the y direction). Note that the diffusivity that appears in Eqs. 17.3-7 and 17.3-8 is the same diffusivity that appears in Eqs. 17.3-1 and 17.3-2.

Equation 17.3-8 is another one-dimensional form of *Fick's first law of diffusion*.[1] It is valid for any binary solid, liquid, or gas solution, provided that J^*_{Ay} is defined as the molar flux relative to the *molar average velocity* of the mixture v^*_y,

$$v^*_y = x_A v_{Ay} + x_B v_{By} \tag{17.3-9}$$

as defined in §17.2. For the system examined in Fig. 17.3-1, the helium is moving rather slowly and its concentration is very small, so that v^*_y is negligibly small.

The diffusive molar flux J^*_{Ay} is defined in general as

$$J^*_{Ay} = cx_A(v_{Ay} - v^*_y) \tag{17.3-10}$$

which again illustrates that this is the flux relative to the molar average velocity. The diffusive molar flux of B is defined analogously. Equations 17.3-9 and 17.3-10 imply that J^*_{Ay} and J^*_{By} are defined such that $J^*_{Ay} + J^*_{By} = 0$.

---

[2]S. Chapman and T. G. Cowling, *The Mathematical Theory of Non-Uniform Gases*, Third Edition, Cambridge University Press (1970), p. 257, Eq. 14.1,1.

[3]L. Landau and E. M. Lifshitz, *Fluid Mechanics*, Pergamon Press, London (1959), p. 224, Eq. 58.11; see also R. B. Bird, W. E. Stewart, and E. N. Lightfoot, *Transport Phenomena*, Revised Second Edition, Wiley (2007), Eqs. 24.1-8 and 24.2-4.

For diffusion in three dimensions, the vector form of Fick's law for the *diffusive molar flux vectors* are written for species A and B as

$$\boxed{\mathbf{J}_A^* = -c\mathscr{D}_{AB}\nabla x_A} \tag{17.3-11}$$

$$\boxed{\mathbf{J}_B^* = -c\mathscr{D}_{BA}\nabla x_B} \tag{17.3-12}$$

where again, $\mathscr{D}_{AB} = \mathscr{D}_{BA}$. These equations can be written in component form in Cartesian, cylindrical, or spherical coordinates by using Appendix B.3 along with the appropriate variable substitutions.

In Table 17.3-1, we summarize the main results of the above discussion of the various expressions for the diffusive mass and molar fluxes.

The mass diffusivity $\mathscr{D}_{AB}$, the thermal diffusivity $\alpha = k/\rho\hat{C}_p$, and the momentum diffusivity (kinematic viscosity) $\nu = \mu/\rho$, all have dimensions of (length)^2/time. The ratios of these three quantities are therefore dimensionless groups:

The Prandtl number: $$\mathrm{Pr} = \frac{\nu}{\alpha} = \frac{\hat{C}_p\mu}{k} \tag{17.3-13}$$

The Schmidt number:[4] $$\mathrm{Sc} = \frac{\nu}{\mathscr{D}_{AB}} = \frac{\mu}{\rho\mathscr{D}_{AB}} \tag{17.3-14}$$

The Lewis number:[4] $$\mathrm{Le} = \frac{\alpha}{\mathscr{D}_{AB}} = \frac{k}{\rho\hat{C}_p\mathscr{D}_{AB}} \tag{17.3-15}$$

Table 17.3-1. Diffusion Velocities and Diffusive Mass and Molar Flux Vectors

$\mathbf{v}_A - \mathbf{v}$	= diffusion velocity of A relative to the mass average velocity $\mathbf{v}$	(A)	$\mathbf{v}_A - \mathbf{v}^*$	= diffusion velocity of A relative to the molar average velocity $\mathbf{v}^*$	(E)
$\mathbf{j}_A = \rho_A(\mathbf{v}_A - \mathbf{v})$	= diffusive mass flux vector of A	(B)	$\mathbf{J}_A^* = c_A(\mathbf{v}_A - \mathbf{v}^*)$	= diffusive molar flux vector of A	(F)
$\mathbf{j}_A = -\rho\mathscr{D}_{AB}\nabla\omega_A$	diffusive mass flux vector is proportional to mass fraction gradient (Fick's law)	(C)	$\mathbf{J}_A^* = -c\mathscr{D}_{AB}\nabla x_A$	diffusive molar flux vector is proportional to mole fraction gradient (Fick's law)	(G)
$\mathbf{j}_A + \mathbf{j}_B = 0$		(D)	$\mathbf{J}_A^* + \mathbf{J}_B^* = 0$		(H)

The connection between diffusive mass and molar flux vectors is:

$$\frac{\mathbf{j}_A}{\rho\omega_A\omega_B} = \frac{\mathbf{J}_A^*}{cx_Ax_B} \tag{I}$$

Note: Vector components in Cartesian, cylindrical, and spherical coordinates can be obtained using Appendix A. The components of Eq. C are written explicitly in Cartesian, cylindrical, and spherical coordinates in Appendix B.3.

[4]These groups were named for: **Ernst Heinrich Wilhelm Schmidt** (1892–1975), who taught at the universities in Gdansk, Braunschweig, and Munich (where he was the successor to Nusselt); **Warren Kendall Lewis** (1882–1975), who taught at MIT and was a coauthor of the pioneering textbook: W. H. Walker, W. K. Lewis, and W. H. McAdams, *Principles of Chemical Engineering*, McGraw-Hill, New York (1923). Sometimes the Lewis number is erroneously attributed to Bernard Lewis, famous for his work in flames, combustion, and explosions.

These dimensionless groups of fluid properties play a prominent role in dimensionless equations for systems in which competing transport processes are occurring. (*Note*: Sometimes the Lewis number is defined as the inverse of the expression above.)

A special case of binary diffusion is *self-diffusion*, namely, the interdiffusion of two chemically identical species A and A^*. Here A^* is a "tagged" species, which may differ physically from A by virtue of radioactivity or other nuclear properties such as the mass, magnetic moment, or spin.[5]

EXAMPLE 17.3-1

Diffusion of Helium through a Pyrex Glass Plate

Calculate the steady-state mass flux j_{Ay} of helium for the system of Fig. 17.3-1 at 500°C. The partial pressure of helium is 1 atm at $y = 0$ and zero at the upper surface of the plate. The thickness Y of the Pyrex plate is 10^{-2} mm, and its density $\rho^{(B)}$ is 2.6 g/cm^3. The solubility and diffusivity of helium in Pyrex are reported[6] as 0.0084 volumes of gaseous helium per volume of glass, and $\mathscr{D}_{AB} = 0.2 \times 10^{-7}$ cm^2/s, respectively. Show that the neglect of the mass average velocity implicit in Eq. 17.3-4 is reasonable.

SOLUTION

The mass concentration of helium in the glass at the lower surface is obtained from the solubility data and the ideal-gas law:

$$\begin{aligned}\rho_{A0} &= (0.0084)\frac{p_{A0}M_A}{RT} \\ &= (0.0084)\frac{(1.0 \text{ atm})(4.00 \text{ g/g-mol})}{(82.05 \text{ cm}^3\text{atm/g-mol K})(773 \text{ K})} \\ &= 5.3 \times 10^{-7} \text{ g/cm}^3\end{aligned} \tag{17.3-16}$$

The mass fraction of helium in the solid phase at the lower surface is then

$$\omega_{A0} = \frac{\rho_{A0}}{\rho_{A0} + \rho_{B0}} = \frac{5.3 \times 10^{-7} \text{ g/cm}^3}{5.3 \times 10^{-7} + 2.6 \text{ g/cm}^3} = 2.04 \times 10^{-7} \tag{17.3-17}$$

We may now calculate the flux of helium from Eq. 17.3-1 as

$$\begin{aligned}j_{Ay} &= (2.6 \text{ g/cm}^3)(2.0 \times 10^{-8} \text{ cm}^2\text{/s})\frac{2.04 \times 10^{-7}}{10^{-3} \text{ cm}} \\ &= 1.05 \times 10^{-11} \text{ g/cm}^2\text{s}\end{aligned} \tag{17.3-18}$$

Next, the velocity of the helium can be obtained from Eq. 17.3-4:

$$v_{Ay} = \frac{j_{Ay}}{\rho_A} + v_y \tag{17.3-19}$$

At the lower surface of the plate ($y = 0$), this velocity has the value

$$v_{Ay}|_{y=0} = \frac{1.05 \times 10^{-11} \text{ g/cm}^2\text{s}}{5.3 \times 10^{-7} \text{ g/cm}^3} + v_{y0} = 1.98 \times 10^{-5} \text{ cm/s } + v_{y0} \tag{17.3-20}$$

---

[5] E. O. Stejskal and J. E. Tanner, *J. Chem. Phys.*, **42**, 288–292 (1965); P. Stilbs, *Prog. NMR Spectros.*, **19**, 1–45 (1987); P. T. Callaghan and J. Stepišnik, *Adv. Magn. Opt. Reson.*, **19**, 325–388 (1996).

[6] C. C. Van Voorhis, *Phys. Rev.* 23, 557 (1924), as reported by R. M. Barrer, *Diffusion in and through Solids*, corrected printing, Cambridge University Press (1951).

The corresponding value v_{y0} of the mass average velocity of the glass-helium system at $y = 0$ is then obtained from Eq. 17.3-3

$$v_{y0} = (2.04 \times 10^{-7})(1.98 \times 10^{-5}\ \text{cm/s} + v_{y0}) + (1 - 2.04 \times 10^{-7})(0)$$

$$v_{y0} = \frac{(2.04 \times 10^{-7})(1.98 \times 10^{-5}\ \text{cm/s})}{1 - 2.04 \times 10^{-7}}$$
$$= 4.04 \times 10^{-12}\ \text{cm/s} \tag{17.3-21}$$

Thus, it is safe to neglect v_y in Eq. 17.3-19 (i.e., Eq. 17.3-4), and the analysis of the experiment in Fig. 17.3-1 at steady state is accurate.

EXAMPLE 17.3-2

The Equivalence of $\mathscr{D}_{AB}$ and $\mathscr{D}_{BA}$

Show that only one diffusivity is needed to describe the diffusional behavior of an isotropic binary mixture.

SOLUTION

We begin by writing Eq. 17.3-6 as follows:

$$\mathbf{j}_B = -\rho\mathscr{D}_{BA}\nabla\omega_B = +\rho\mathscr{D}_{BA}\nabla\omega_A \tag{17.3-22}$$

The second form of this equation follows from the fact that $\omega_A + \omega_B = 1$. We next use the vector equivalents of Eqs. 17.3-3 and 17.3-4 to write

$$\mathbf{j}_A = \rho\omega_A(\mathbf{v}_A - \mathbf{v})$$
$$= \rho\omega_A(\mathbf{v}_A - \omega_A\mathbf{v}_A - \omega_B\mathbf{v}_B)$$
$$= \rho\omega_A((1 - \omega_A)\mathbf{v}_A - \omega_B\mathbf{v}_B)$$
$$= \rho\omega_A\omega_B(\mathbf{v}_A - \mathbf{v}_B) \tag{17.3-23}$$

Interchanging A and B in this expression shows that $\mathbf{j}_A = -\mathbf{j}_B$. Combining this with the second form of Eq. 17.3-22 then gives

$$\mathbf{j}_A = -\rho\mathscr{D}_{BA}\nabla\omega_A \tag{17.3-24}$$

Comparing this with Eq. 17.3-5 gives $\mathscr{D}_{AB} = \mathscr{D}_{BA}$. We find that the order of subscripts is unimportant for a binary system and that only one diffusivity is required to describe the diffusional behavior.

Keep in mind that the diffusivities are concentration-dependent, and that $\mathscr{D}_{AB} = \mathscr{D}_{BA}$ is true at each composition. However, it may well be that the diffusivity for a dilute solution of A in B and that for a dilute solution of B in A are numerically different, because these two cases refer to two different compositions.

In this section we have discussed the diffusion that occurs as a result of a concentration gradient in the system, i.e., *mass diffusion, concentration diffusion,* or *ordinary diffusion.* There are, however, still more kinds of diffusion: *thermal diffusion,* which results from a temperature gradient; *pressure diffusion,* resulting from a pressure gradient; and *forced diffusion,* which is caused by unequal external forces acting on the chemical species. For the time being, we consider only concentration diffusion, and we postpone discussion of the other mechanisms until Chapter 24. Also, in that chapter we discuss the use of the activity gradient, rather than the concentration gradient, as the driving force for ordinary diffusion.

§17.4 TOTAL MASS AND MOLAR FLUX VECTORS

In Chapter 1 we combined the convective momentum flux $\boldsymbol{\pi}^{(c)} = \rho\mathbf{vv}$ and the molecular moment flux $\boldsymbol{\pi} = \boldsymbol{\tau} + p\boldsymbol{\delta}$ to create the total momentum flux $\boldsymbol{\phi}$. In Chapter 9 we combined the convective energy flux $\mathbf{q}^{(c)} = \rho\left(\hat{U} + \frac{1}{2}v^2\right)\mathbf{v}$ with the molecular energy flux $\mathbf{q} + \mathbf{w} = -k\nabla T + [\boldsymbol{\pi} \cdot \mathbf{v}]$ to create the total energy flux $\mathbf{e}$. These total fluxes were then used to solve momentum and energy balances. In this section we combine the convective and diffusive mass and molar flux vectors to create total mass and molar flux vectors. These total fluxes will then be used in Chapters 18 and 19 for setting up shell balances and deriving equations of change to solve mass-transfer problems.

a. Total mass flux vector

The total mass flux vector $\mathbf{n}_A$ is obtained by summing the convective mass flux vector $\mathbf{j}_A^{(c)} = \rho_A\mathbf{v}$ and the diffusive mass flux vector $\mathbf{j}_A$,

$$\mathbf{n}_A = \rho_A\mathbf{v} + \mathbf{j}_A \tag{17.4-1}$$

with an analogous equation for $\mathbf{n}_B$. The total flux can also be written in terms of the velocity of species A. Following the derivation of the convective mass flux in §17.2, the total flux relative to a fixed coordinate system can be expressed $\mathbf{n}_A = \rho_A\mathbf{v}_A$. Adding this to the analogous expression for species B, we obtain $\mathbf{n}_A + \mathbf{n}_B = (\rho_A\mathbf{v}_A + \rho_B\mathbf{v}_B) = \rho(\omega_A\mathbf{v}_A + \omega_B\mathbf{v}_B) = \rho\mathbf{v}$. Thus, $\mathbf{v}$ in Eq. 17.4-1 can be replaced by $(\mathbf{n}_A + \mathbf{n}_B)/\rho$. If we also replace $\mathbf{j}_A$ by $-\rho\mathscr{D}_{AB}\nabla\omega_A$ by using Fick's law of diffusion (Eq. 17.3-5), Eq. 17.4-1 can be rewritten as

$$\boxed{\mathbf{n}_A = \omega_A(\mathbf{n}_A + \mathbf{n}_B) - \rho\mathscr{D}_{AB}\nabla\omega_A} \tag{17.4-2}$$

This equation can be written in component form in Cartesian, cylindrical, and spherical coordinates with the help of Appendix B.3.

b. Total molar flux vector

The total molar flux vector $\mathbf{N}_A$ is defined analogously, by summing the convective molar flux vector $\mathbf{J}_A^{*(c)} = c_A\mathbf{v}^*$ and the diffusive molar flux vector $\mathbf{J}_A^*$,

$$\mathbf{N}_A = c_A\mathbf{v}^* + \mathbf{J}_A^* \tag{17.4-3}$$

This total flux can also be written in terms of just the velocity of species A. Following the derivation of the convective molar flux in §17.2, the total molar flux relative to a fixed coordinate system can be expressed $\mathbf{N}_A = c_A\mathbf{v}_A$. Adding this to the analogous expression for species B, we get $\mathbf{N}_A + \mathbf{N}_B = (c_A\mathbf{v}_A + c_B\mathbf{v}_B) = c(x_A\mathbf{v}_A + x_B\mathbf{v}_B) = c\mathbf{v}^*$. Thus, $\mathbf{v}^*$ in Eq. 17.4-3 can be replaced by $(\mathbf{N}_A + \mathbf{N}_B)/c$. If we also replace $\mathbf{J}_A^*$ by $-c\mathscr{D}_{AB}\nabla x_A$ by using Fick's law of diffusion (Eq. 17.3-11), Eq. 17.4-3 can be rewritten as

$$\boxed{\mathbf{N}_A = x_A(\mathbf{N}_A + \mathbf{N}_B) - c\mathscr{D}_{AB}\nabla x_A} \tag{17.4-4}$$

This equation can be written in component form in Cartesian, cylindrical, and spherical coordinates with the help of Appendix B.3 along with the appropriate variable substitutions. In Table 17.4-1, we give a summary of the notation used for the total mass and molar fluxes as well as several useful relations. In Table 17.4-2 we give several common expressions using Fick's law of diffusion, both in mass units and in molar units.

In Chapter 18 we will use the total molar flux $\mathbf{N}_A$ extensively in the form of Eq. 17.4-4. It is this form that has generally been used in chemical engineering. In many problems

Table 17.4-1. Total Mass and Molar Flux Vectors

Total mass flux of A:		Total molar flux of A:	
$\mathbf{n}_A = \mathbf{j}_A^{(c)} + \mathbf{j}_A = \rho_A \mathbf{v}_A$	(A)	$\mathbf{N}_A = \mathbf{J}_A^{*(c)} + \mathbf{J}_A^* = c_A \mathbf{v}_A$	(E)
$= \rho_A \mathbf{v} + \mathbf{j}_A$	(B)	$= c_A \mathbf{v}^* + \mathbf{J}_A^*$	(F)
$= \rho_A \mathbf{v} - \rho \mathscr{D}_{AB} \nabla \omega_A$	(C)	$= c_A \mathbf{v}^* - c \mathscr{D}_{AB} \nabla x_A$	(G)
Sum of total mass flux vectors:		Sum of total molar flux vectors:	
$\mathbf{n}_A + \mathbf{n}_B = \rho \mathbf{v}$	(D)	$\mathbf{N}_A + \mathbf{N}_B = c \mathbf{v}^*$	(H)

Interrelation:

$$\mathbf{n}_A = M_A \mathbf{N}_A \qquad \text{(I)}$$

Note: Vector components in Cartesian, cylindrical, and spherical coordinates can be obtained using Appendix A.

Table 17.4-2. Common Expressions Using Fick's Law for Binary Mixtures

$\mathbf{j}_A = -\rho \mathscr{D}_{AB} \nabla \omega_A$	(A)	$\mathbf{J}_A^* = -c \mathscr{D}_{AB} \nabla x_A$	(D)
$\mathbf{n}_A = \omega_A (\mathbf{n}_A + \mathbf{n}_B) - \rho \mathscr{D}_{AB} \nabla \omega_A$	(B)	$\mathbf{N}_A = x_A (\mathbf{N}_A + \mathbf{N}_B) - c \mathscr{D}_{AB} \nabla x_A$	(E)
$\rho(\mathbf{v}_A - \mathbf{v}_B) = -\dfrac{\rho \mathscr{D}_{AB}}{\omega_A \omega_B} \nabla \omega_A$	(C)	$c(\mathbf{v}_A - \mathbf{v}_B) = -\dfrac{c \mathscr{D}_{AB}}{x_A x_B} \nabla x_A$	(F)

Note: Vector components in Cartesian, cylindrical, and spherical coordinates can be obtained using Appendix A.

something is known about the relation between $\mathbf{N}_A$ and $\mathbf{N}_B$, for example, from reaction stoichiometry or from boundary conditions. Therefore, $\mathbf{N}_B$ can be eliminated from Eq. 17.4-4 giving a direct relation between $\mathbf{N}_A$ and x_A for the particular problem.

§17.5 DIFFUSIVITY DATA FROM EXPERIMENTS

In Tables 17.5-1 through 17.5-4 some values of $\mathscr{D}_{AB}$ in cm^2/s are given for a few gas, liquid, solid, and polymer systems. These values can be converted to m^2/s by multiplication by 10^{-4}. Diffusivities of gases at low density are almost independent of ω_A, increase with temperature, and vary inversely with pressure. Liquid and solid diffusivities are strongly concentration-dependent and generally increase with temperature. There are numerous ways of measuring diffusivities, and some of these are described in subsequent chapters.[1]

For *gas mixtures*, the Schmidt number can range from about 0.2 to 3, as can be seen in Table 17.5-1. For *liquid mixtures*, values up to 40,000 have been observed.[2]

[1]For an extensive discussion see W. E. Wakeham, A. Nagashima, and J. V. Sengers, *Measurement of the Transport Properties of Fluids: Experimental Thermodynamics, Vol. III*, CRC Press, Boca Raton, FL (1991).

[2]D. A. Shaw and T. J. Hanratty, *AIChE Journal*, **23**, 28–37, 160–169 (1977); P. Harriott and R. M. Hamilton, *Chem. Eng. Sci.*, **20**, 1073–1078 (1965).

Table 17.5-1. Experimental Diffusivities[a] and Limiting Schmidt Numbers[b] of Gas Pairs at 1 Atmosphere Pressure

Gas Pair A - B	*Temperature* (K)	$\mathscr{D}_{AB}$ (cm^2/s)	Sc = $\nu/\mathscr{D}_{AB}$ $x_A \to 1$	$x_B \to 1$
CO_2 - N_2O	273.2	0.096	0.73	0.72
CO_2 - CO	273.2	0.139	0.50	0.96
CO_2 - N_2	273.2	0.144	0.48	0.91
	288.2	0.158	0.49	0.92
	298.2	0.165	0.50	0.93
N_2 - C_2H_6	298.2	0.148	1.04	0.51
N_2 - nC_4H_{10}	298.2	0.0960	1.60	0.33
N_2 - O_2	273.2	0.181	0.72	0.74
H_2 - SF_6	298.2	0.420	3.37	0.055
H_2 - CH_4	298.2	0.726	1.95	0.23
H_2 - N_2	273.2	0.674	1.40	0.19
NH_3 - H_2[c]	263	0.58	0.19[e]	1.53
NH_3 - N_2[c]	298	0.233	0.62[e]	0.65
H_2O - N_2[c]	308	0.259	0.58[e]	0.62
H_2O - O_2[c]	352	0.357	0.56[e]	0.59
C_3H_8 - nC_4H_{10}[d]	378.2	0.0768	0.95	0.66
	437.7	0.107	0.91	0.63
C_3H_8 - iC_4H_{10}[d]	298.0	0.0439	1.04	0.73
	378.2	0.0823	0.89	0.63
	437.8	0.112	0.87	0.61
C_3H_8 - neo-C_5H_{12}[d]	298.1	0.0431	1.06	0.56
	378.2	0.0703	1.04	0.55
	437.7	0.0945	1.03	0.55
nC_4H_{10} - neo-C_5H_{12}[d]	298.0	0.0413	0.76	0.59
	378.2	0.0644	0.78	0.61
	437.8	0.0839	0.80	0.62
iC_4H_{10} - neo-C_5H_{12}[d]	298.1	0.0362	0.89	0.67
	378.2	0.0580	0.89	0.67
	437.7	0.0786	0.87	0.66

[a]Unless otherwise indicated, the values are taken from J. O. Hirschfelder, C. F. Curtiss, and R. B. Bird, *Molecular Theory of Gases and Liquids*, 2nd corrected printing, Wiley, New York (1964), p. 579. All values are given for 1 atmosphere pressure.

[b]Calculated using the Lennard-Jones parameters of Table D.1. The parameters for sulfur hexafluoride were obtained from second virial coefficient data.

[c]Values of $\mathscr{D}_{AB}$ for the water and ammonia mixtures are taken from the tabulation of B. E. Poling, J. M. Prausnitz, and J. P. O'Connell, *The Properties of Gases and Liquids*, 5th edition, McGraw-Hill, New York (2001).

[d]Values of $\mathscr{D}_{AB}$ for the hydrocarbon-hydrocarbon pairs are taken from S. Gotoh, M. Manner, J. P. Sørensen, and W. E. Stewart, *J. Chem. Eng. Data*, **19**, 169–171 (1974).

[e]Values of μ for water and ammonia were calculated from functions provided by T. E. Daubert, R. P. Danner, H. M. Sibul, C. C. Stebbins, J. L. Oscarson, R. L. Rowley, W. V. Wilding, M. E. Adams, T. L. Marshall, and N. A. Zundel, *DIPPR®, Data Compilation of Pure Compound Properties*, Design Institute for Physical Property Data®, AIChE, New York, NY (2000).

Table 17.5-2. Experimental Diffusivities in the Liquid State [a,b]

A	B	T (°C)	x_A	$\mathscr{D}_{AB} \times 10^5$ (cm^2/s)
Chlorobenzene	Bromobenzene	10.10	0.0332	1.007
			0.2642	1.069
			0.5122	1.146
			0.7617	1.226
			0.9652	1.291
		39.92	0.0332	1.584
			0.2642	1.691
			0.5122	1.806
			0.7617	1.902
			0.9652	1.996
Water	*n*-Butanol	30	0.131	1.24
			0.222	0.920
			0.358	0.560
			0.454	0.437
			0.524	0.267
Ethanol	Water	25	0.026	1.076
			0.266	0.368
			0.408	0.405
			0.680	0.743
			0.880	1.047
			0.944	1.181

[a]The data for the first two pairs above are taken from a review article by P. A. Johnson and A. L. Babb, *Chem. Revs.*, **56,** 387–453 (1956). Other summaries of experimental results may be found in: P. W. M. Rutten, *Diffusion in Liquids*, Delft University Press, Delft, The Netherlands (1992); L. J. Gosting, *Adv. in Protein Chem.*, Vol. XI, Academic Press, New York (1956); A. Vignes, *IEC Fundamentals*, **5**, 189–199 (1966).
[b]The ethanol-water data were taken from M. T. Tyn and W. F. Calus, *J. Chem. Eng. Data*, **20**, 310–316 (1975).

Table 17.5-3. Experimental Diffusivities in the Solid State [a]

A	B	T (°C)	$\mathscr{D}_{AB}$ (cm^2/s)
He	SiO_2	20	$2.4 - 5.5 \times 10^{-10}$
He	Pyrex	20	4.5×10^{-11}
		500	2×10^{-8}
H_2	SiO_2	500	$0.6 - 2.1 \times 10^{-8}$
H_2	Ni	85	1.16×10^{-8}
		165	10.5×10^{-8}
Bi	Pb	20	1.1×10^{-16}
Hg	Pb	20	2.5×10^{-15}
Sb	Ag	20	3.5×10^{-21}
Al	Cu	20	1.3×10^{-30}
Cd	Cu	20	2.7×10^{-15}

[a]It is presumed that in each of the above pairs, component A is present only in very small amounts. The data are taken from R. M. Barrer, *Diffusion in and through Solids*, Macmillan, New York (1941), pp. 141, 222, and 275.

Table 17.5-4. Experimental Diffusivities of Gases in Polymers.[a] Diffusivities, $\mathscr{D}_{AB}$, are given in units of 10^{-6} (cm²/s). The values for N_2 and O_2 are for 298K, and those for CO_2 and H_2 are for 198K

	N_2	O_2	CO_2	H_2
Polybutadiene	1.1	1.5	1.05	9.6
Silicone rubber	15	25	15	75
Trans-1,4-polyisoprene	0.50	0.70	0.47	5.0
Polystyrene	0.06	0.11	0.06	4.4

[a]Excerpted from D. W. van Krevelen, *Properties of Polymers*, 3rd edition, Elsevier, Amsterdam (1990), pp. 544–545. Another relevant reference is S. Pauly, in *Polymer Handbook*, 4th edition (J. Brandrup and E. H. Immergut, eds.), Wiley-Interscience, New York (1999), Chapter VI.

§17.6 DIFFUSIVITY AND THE PRINCIPLE OF CORRESPONDING STATES

In this section we discuss the prediction of the $\mathscr{D}_{AB}$ for binary systems by corresponding-states methods. These methods are also useful for extrapolating existing data. Comparisons of many alternative methods are available in the literature.[1,2]

For binary gas mixtures at low pressure, $\mathscr{D}_{AB}$ is inversely proportional to the pressure, increases with increasing temperature, and is almost independent of the composition for a given gas pair. The following equation for estimating $\mathscr{D}_{AB}$ at low pressures has been developed[3] from a combination of kinetic theory and corresponding-states arguments:

$$\frac{p\mathscr{D}_{AB}}{(p_{cA}p_{cB})^{1/3}(T_{cA}T_{cB})^{5/12}(1/M_A+1/M_B)^{1/2}}=a\left(\frac{T}{\sqrt{T_{cA}T_{cB}}}\right)^b \tag{17.6-1}$$

Here $\mathscr{D}_{AB}$ [=] cm²/s, p [=] atm, and T [=] K. Analysis of experimental data gives $a = 2.745\times10^{-4}$ $(\text{cm}^2/\text{s})(\text{atm})^{1/3}(\text{K})^{-5/6}(\text{g/g-mol})^{1/2}$ and $b = 1.823$ (dimensionless) for nonpolar gas pairs, excluding helium and hydrogen, and $a = 3.640\times10^{-4}$ and $b = 2.334$ for pairs consisting of H_2O and a nonpolar gas. Equation 17.6-1 fits the experimental data at atmospheric pressure within an average deviation of 6% to 8%. If the gases A and B are nonpolar and their Lennard-Jones parameters are known, the kinetic theory method described in the next section usually gives somewhat better accuracy.

At high pressures, and in the liquid state, the behavior of $\mathscr{D}_{AB}$ is more complicated. The simplest and best understood situation is that of self-diffusion (interdiffusion of labeled molecules of the same chemical species). We discuss this case first and then extend the results approximately to binary mixtures.

A corresponding-states plot of the self-diffusivity $\mathscr{D}_{AA^*}$ for nonpolar substances is given in Fig. 17.6-1.[4] This plot is based on self-diffusion measurements, supplemented

[1]B. E. Poling, J. M. Prausnitz, and J. P. O'Connell, *The Properties of Gases and Liquids*, 5th edition, McGraw-Hill, New York (2001), Chapter 11.

[2]E. N. Fuller, P. D. Shettler, and J. C. Giddings, *Ind. Eng. Chem.*, **58**, No. 5, 19–27 (1966); Erratum: *ibid.* **58**, No. 8, 81 (1966). This paper gives a useful method for predicting binary gas diffusivities from the molecular formulas of the two species.

[3]J. C. Slattery and R. B. Bird, *AIChE Journal*, **4**, 137–142 (1958).

[4]Other correlations for self-diffusivity at elevated pressures have appeared in Ref. 3 and in L. S. Tee, G. F. Kuether, R. C. Robinson, and W. E. Stewart, *API Proceedings, Division of Refining*, 235–243 (1966); R. C. Robinson and W. E. Stewart, *IEC Fundamentals*, **7**, 90–95 (1968); J. L. Bueno, J. Dizy, R. Alvarez, and J. Coca, *Trans. Inst. Chem. Eng.*, **68**, Part A, 392–397 (1990).

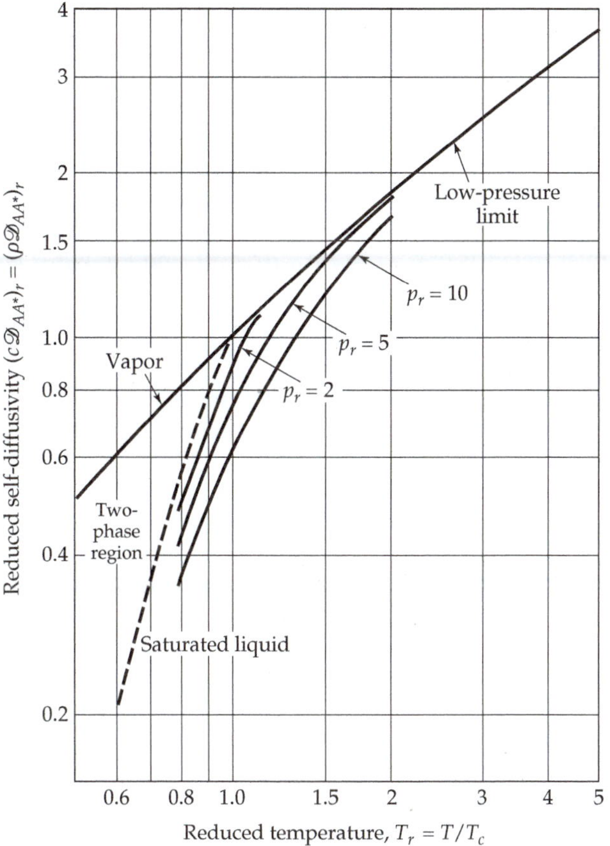

Fig. 17.6-1. A corresponding-states plot for the reduced self-diffusivity. Here $(c\mathscr{D}_{AA^*})_r = (\rho\mathscr{D}_{AA^*})_r$ for Ar, Kr, Xe, and CH_4 is plotted as a function of reduced temperature for several values of the reduced pressure. This chart is based on diffusivity data of J. J. van Loef and E. G. D. Cohen, *Physica A*, **156**, 522–533 (1989), the compressibility function of B. I. Lee and M. G. Kessler, *AIChE Journal*, **21**, 510–527 (1975), and Eq. 17.7-11 for the low-pressure limit.

by molecular dynamics simulations, and by kinetic theory for the low-pressure limit. The ordinate is $c\mathscr{D}_{AA^*}$ at pressure p and temperature T, divided by $c\mathscr{D}_{AA^*}$ at the critical point. This quantity is plotted as a function of the reduced pressure $p_r = p/p_c$ and the reduced temperature $T_r = T/T_c$. Because of the similarity of species A and the labeled species A^*, the critical properties are all taken as those of species A.

From Fig. 17.6-1 we see that $c\mathscr{D}_{AA^*}$ increases strongly with temperature, especially for liquids. At each temperature, $c\mathscr{D}_{AA^*}$ decreases toward zero with increasing pressure. With decreasing pressure, $c\mathscr{D}_{AA^*}$ increases toward a low-pressure limit, as predicted by kinetic theory (see §17.7). The reader is warned that this chart is tentative, and that the curves, except for the low-density limit, are based on data for a very few substances: Ar, Kr, Xe, and CH_4.

The quantity $(c\mathscr{D}_{AA^*})_c$ may be estimated by one of the following three methods:

(i) Given $c\mathscr{D}_{AA^*}$ at a known temperature and pressure, one can read $(c\mathscr{D}_{AA^*})_r$ from the chart and get $(c\mathscr{D}_{AA^*})_c = (c\mathscr{D}_{AA^*})/(c\mathscr{D}_{AA^*})_r$.

(ii) One can predict a value of $c\mathscr{D}_{AA^*}$ in the low-density region by the methods given in §17.7 and then proceed as in (i).

(iii) One can use the empirical formula (see Problem 17A.9)

$$(c\mathscr{D}_{AA^*})_c = 2.96 \times 10^{-6}\left(\frac{1}{M_A} + \frac{1}{M_{A^*}}\right)^{1/2} \frac{p_{cA}^{2/3}}{T_{cA}^{1/6}} \tag{17.6-2}$$

This equation, like Eq. 17.6-1, should not be used for helium or hydrogen isotopes. Here c [=] g-mol/cm^3, $\mathscr{D}_{AA^*}$ [=] cm^2/s, T_c [=] K, and p_c [=] atm; the constant 2.96×10^{-6} has units of (g-mol/cm · s) (g/g-mol)$^{1/2}$ K$^{1/6}$ atm$^{-2/3}$.

Thus far, the discussion of high-density behavior has been concerned with self-diffusion. We turn now to the binary diffusion of chemically dissimilar species. In the absence of other information, it is suggested that Fig. 17.6-1 may be used for crude estimation of $c\mathscr{D}_{AB}$, with p_{cA} and T_{cA} replaced everywhere by $\sqrt{p_{cA}p_{cB}}$ and $\sqrt{T_{cA}T_{cB}}$, respectively (see Problem 17A.9 for the basis for this empiricism). The ordinate of the plot is then interpreted as $(c\mathscr{D}_{AB})_r = c\mathscr{D}_{AB}/(c\mathscr{D}_{AB})_c$ and Eq. 17.6-2 is replaced by

$$(c\mathscr{D}_{AB})_c = 2.96 \times 10^{-6}\left(\frac{1}{M_A} + \frac{1}{M_B}\right)^{1/2} \frac{(p_{cA}p_{cB})^{1/3}}{(T_{cA}T_{cB})^{1/12}} \tag{17.6-3}$$

With these substitutions, accurate results are obtained in the low-pressure limit. At higher pressures, very few data are available for comparison, and the method must be regarded as provisional.

The results in Fig. 17.6-1, and their extensions to binary systems, are expressed in terms of $c\mathscr{D}_{AA^*}$ and $c\mathscr{D}_{AB}$ rather than $\mathscr{D}_{AA^*}$ and $\mathscr{D}_{AB}$. This is done because the products of concentration and diffusivity are more frequently required in mass-transfer calculations, and their dependence on pressure and temperature is simpler.

EXAMPLE 17.6-1

Estimation of Diffusivity at Low Density

Estimate $\mathscr{D}_{AB}$ for the system CO-CO_2 at 296.1K and 1 atm total pressure.

SOLUTION

The properties needed for Eq. 17.2-1 are (see Table D.1):

Label	Species	M (g/g-mol)	T_c (K)	p_c (atm)
A	CO	28.01	133	34.5
B	CO_2	44.01	304.2	72.9

Therefore,

$$(p_{cA}p_{cB})^{1/3} = (34.5 \text{ atm} \times 72.9 \text{ atm})^{1/3} = 13.60 \text{ atm}^{2/3}$$
$$(T_{cA}T_{cB})^{5/12} = (133 \text{ K} \times 304.2 \text{ K})^{5/12} = 83.1 \text{ K}^{5/6}$$

$$\left(\frac{1}{M_A} + \frac{1}{M_B}\right)^{1/2} = \left(\frac{1}{28.01}\frac{\text{g-mol}}{\text{g}} + \frac{1}{44.01}\frac{\text{g-mol}}{\text{g}}\right)^{1/2} = 0.2417\left(\frac{\text{g-mol}}{\text{g}}\right)^{1/2}$$

$$a\left(\frac{T}{\sqrt{T_{cA}T_{cB}}}\right)^b = 2.745 \times 10^{-4} \text{ (cm}^2\text{/s)(atm)}^{1/3}\text{(K)}^{-5/6}\text{(g/g-mol)}^{1/2}$$
$$\times\left(\frac{296.1 \text{ K}}{\sqrt{133 \text{ K} \times 304.2 \text{ K}}}\right)^{1.823} = 5.56 \times 10^{-4} \text{ (cm}^2\text{/s)(atm)}^{1/3}\text{(K)}^{-5/6}\text{(g/g-mol)}^{1/2}$$

Substitution of these values into Eq. 17.6-1 gives

$$(1.0 \text{ atm})\mathscr{D}_{AB} = 5.56 \times 10^{-4} \text{ (cm}^2\text{/s)(atm)}^{1/3}\text{(K)}^{-5/6}\text{(g/g-mol)}^{1/2}$$
$$\times\, 13.60 \text{ atm}^{2/3}(83.1 \text{ K}^{5/6})(0.2417 \text{ (g/g-mol)}^{-1/2}) \tag{17.6-4}$$

This gives $\mathscr{D}_{AB} = 0.152$ cm^2/s, in agreement with the experimental value.[5] This is unusually good agreement.

This problem can also be solved by means of Fig. 17.6-1 and Eq. 17.6-3, together with the ideal-gas law $p = cRT$. The result is $\mathscr{D}_{AB} = 0.140$ cm^2/s, in fair agreement with the data.

EXAMPLE 17.6-2

Estimation of Self-Diffusivity at High Density

Estimate $c\mathscr{D}_{AA^*}$ for $C^{14}O_2$ in ordinary CO_2 at 171.7 atm and 373K. It is known[6] that $\mathscr{D}_{AA^*} = 0.113$ cm^2/s at 1.00 atm and 298K, at which condition $c = p/RT = 4.12 \times 10^{-5}$ g-mol/cm^3.

SOLUTION

Since a measured value of $\mathscr{D}_{AA^*}$ is given, we use Method (i). The reduced conditions of the measurement are: $T_r = 298/304.2 = 0.980$ and $p_r = 1.00/72.9 = 0.014$. Then from Fig. 17.6-1, we get the value $(c\mathscr{D}_{AA^*})_r = 0.98$. Hence,

$$(c\mathscr{D}_{AA^*})_c = \frac{c\mathscr{D}_{AA^*}}{(c\mathscr{D}_{AA^*})_r} = \frac{(4.12 \times 10^{-5} \text{ g-mol/cm}^3)(0.113 \text{ cm}^2/\text{s})}{0.98}$$
$$= 4.75 \times 10^{-6} \text{ g-mol/cm} \cdot \text{s} \qquad (17.6\text{-}5)$$

At the conditions of prediction ($T_r = 373/304.2 = 1.23$ and $p_r = 171.7/72.9 = 2.36$), we read $(c\mathscr{D}_{AA^*})_r = 1.21$. The predicted value is then

$$c\mathscr{D}_{AA^*} = (c\mathscr{D}_{AA^*})_r(c\mathscr{D}_{AA^*})_c = (1.21)(4.75 \times 10^{-6} \text{ g-mol/cm} \cdot \text{s})$$
$$= 5.75 \times 10^{-6} \text{ g-mol/cm} \cdot \text{s} \qquad (17.6\text{-}6)$$

The data of O'Hern and Martin[7] give $c\mathscr{D}_{AA^*} = 5.89 \times 10^{-6}$ g-mol/cm · s at these conditions. This good agreement is not unexpected, inasmuch as their low-pressure data were used in the estimation of $(c\mathscr{D}_{AA^*})_c$.

This problem can also be solved by Method (iii) without an experimental value of $c\mathscr{D}_{AA^*}$. Equation 17.4-2 gives directly

$$(c\mathscr{D}_{AA^*})_c = 2.96 \times 10^{-6} \text{ (g-mol/cm s) (g/g-mol)}^{1/2} \text{ K}^{1/6} \text{ atm}^{-2/3}$$
$$\times \left(\frac{1}{44.01}\frac{\text{g-mol}}{\text{g}} + \frac{1}{46}\frac{\text{g-mol}}{\text{g}}\right)^{1/2} \frac{(72.9 \text{ atm})^{2/3}}{(304.2 \text{ K})^{1/6}}$$
$$= 4.20 \times 10^{-6} \text{ g-mol/cm} \cdot \text{s} \qquad (17.6\text{-}7)$$

The resulting predicted value of $c\mathscr{D}_{AA^*}$ is 5.1×10^{-6} g-mol/cm · s.

EXAMPLE 17.6-3

Estimation of Binary Diffusivity at High Density

Estimate $c\mathscr{D}_{AB}$ for a mixture of 80 mole% CH_4 and 20 mole% C_2H_6 at 136 atm and 313 K. It is known that, at 1 atm and 293 K, the molar density is $c = 4.17 \times 10^{-5}$ g-mol/cm^3 and $\mathscr{D}_{AB} = 0.163$ cm^2/s.

SOLUTION

Figure 17.6-1 is used, with Method (i). The reduced conditions for the known data are

$$T_r = \frac{T}{\sqrt{T_{cA}T_{cB}}} = \frac{293 \text{ K}}{\sqrt{(190.7 \text{ K})(305.4 \text{ K})}} = 1.22 \qquad (17.6\text{-}8)$$

$$p_r = \frac{p}{\sqrt{p_{cA}p_{cB}}} = \frac{1.0 \text{ atm}}{\sqrt{(45.8 \text{ atm})(48.2 \text{ atm})}} = 0.021 \qquad (17.6\text{-}9)$$

---

[5] B. A. Ivakin and P. E. Suetin, *Sov. Phys. Tech. Phys.* (English translation), **8**, 748–751 (1964).

[6] E. B. Wynn, *Phys. Rev.*, **80**, 1024–1027 (1950).

[7] H. A. O'Hern and J. J. Martin, *Ind. Eng. Chem.*, **47**, 2081–2086 (1955).

From Fig. 17.6-1 at these conditions, we obtain $(c\mathscr{D}_{AB})_r = 1.21$. The critical value $(c\mathscr{D}_{AB})_c$ is therefore

$$(c\mathscr{D}_{AB})_c = \frac{c\mathscr{D}_{AB}}{(c\mathscr{D}_{AB})_r} = \frac{(4.17 \times 10^{-5}\ \text{g-mol/cm}^3)(0.163\ \text{cm}^2/\text{s})}{1.21}$$
$$= 5.62 \times 10^{-6}\ \text{g-mol/cm} \cdot \text{s} \tag{17.6-10}$$

Next we calculate the reduced conditions for the prediction ($T_r = 1.30$, $p_r = 2.90$) and read the value $(c\mathscr{D}_{AB})_r = 1.31$ from Fig. 17.6-1. The predicted value of $c\mathscr{D}_{AB}$ is therefore

$$c\mathscr{D}_{AB} = (c\mathscr{D}_{AB})_r(c\mathscr{D}_{AB})_c = (1.31)(5.62 \times 10^{-6}\ \text{g-mol/cm} \cdot \text{s})$$
$$= 7.4 \times 10^{-6}\ \text{g-mol/cm} \cdot \text{s} \tag{17.6-11}$$

Experimental measurements[8] give $c\mathscr{D}_{AB} = 6.0 \times 10^{-6}$ g-mol/cm · s, so that the predicted value is 23% high. Deviations of this magnitude are not unusual in the estimation of $c\mathscr{D}_{AB}$ at high densities.

An alternative solution may be obtained by Method (iii). Substitution into Eq. 17.6-3 gives

$$(c\mathscr{D}_{AA^*})_c = 2.96 \times 10^{-6}\ (\text{g-mol/cm} \cdot \text{s})\ (\text{g/g-mol})^{1/2}\ \text{K}^{1/6}\ \text{atm}^{-2/3}$$
$$\times \left(\frac{1}{16.04}\frac{\text{g-mol}}{\text{g}} + \frac{1}{30.07}\frac{\text{g-mol}}{\text{g}} \right)^{1/2} \frac{(45.8\ \text{atm} \times 48.2\ \text{atm})^{1/3}}{(190.7\ \text{K} \times 305.4\ \text{K})^{1/12}}$$
$$= 4.78 \times 10^{-6}\ \text{g-mol/cm} \cdot \text{s} \tag{17.6-12}$$

Multiplication by $(c\mathscr{D}_{AB})_r$ at the desired condition gives

$$c\mathscr{D}_{AB} = (4.78 \times 10^{-6}\ \text{g-mol/cm} \cdot \text{s})(1.31)$$
$$= 6.26 \times 10^{-6}\ \text{g-mol/cm} \cdot \text{s} \tag{17.6-13}$$

This is in closer agreement with the measured value.[8]

§17.7 DIFFUSIVITY OF GASES AND KINETIC THEORY

The diffusivity $\mathscr{D}_{AB}$ for binary mixtures of nonpolar gases is predictable within about 5% by kinetic theory. As in the earlier kinetic theory discussions in §1.6 and §9.7, we start with a simplified derivation to illustrate the mechanisms involved and then present the more accurate results of the Chapman-Enskog theory.

Consider a large body of gas containing molecular species A and A^*, which are identical except for labeling. We wish to determine the self-diffusivity $\mathscr{D}_{AA^*}$ in terms of the molecular properties on the assumption that the molecules are rigid spheres of equal mass m_A and diameter d_A.

Since the properties of A and A^* are nearly the same, we can use the following results of the kinetic theory for a pure rigid-sphere gas at low density in which the gradients of temperature, pressure, and velocity are small:

$$\bar{u} = \sqrt{\frac{8\kappa T}{\pi m_A}} = \text{mean molecular speed relative to } \mathbf{v} \tag{17.7-1}$$

$$Z = \frac{1}{4}n\bar{u} = \text{wall collision frequency per unit area in a stationary gas} \tag{17.7-2}$$

$$\lambda = \frac{1}{\sqrt{2}\pi d_A^2 n} = \text{mean free path} \tag{17.7-3}$$

[8] V. J. Berry, Jr., and R. C. Koeller, *AIChE Journal*, **6**, 274–280 (1960).

The molecules reaching any plane in the gas have, on the average, had their last collision at a distance a from the plane, where

$$a = \frac{2}{3}\lambda \tag{17.7-4}$$

In these equations, n is the number density (total number of molecules per unit volume).

To predict the self-diffusivity $\mathscr{D}_{AA^*}$, we consider the motion of species A in the y direction under a mass fraction gradient $d\omega_A/dy$ (see Fig. 17.7-1), where the fluid mixture moves in the y direction at a finite mass average velocity v_y throughout. The temperature T and the total molar mass concentration ρ are considered constant. We assume that Eqs. 17.7-1 to 17.7-4 remain valid in this nonequilibrium situation. The net mass flux of species A crossing a unit area of any plane of constant y is found by writing an expression for the mass of A crossing the plane in the positive y direction and subtracting the mass of A crossing in the negative y direction:

$$(\rho\omega_A v_y)|_y + \left[\left(\frac{1}{4}\rho\omega_A\bar{u}\right)\bigg|_{y-a} - \left(\frac{1}{4}\rho\omega_A\bar{u}\right)\bigg|_{y+a}\right] \tag{17.7-5}$$

Here the first term is the mass transport in the y direction because of the mass motion of the fluid—that is, the convective transport—and the last two terms give the diffusive transport relative to v_y.

It is assumed that the concentration profile $\omega_A(y)$ is very nearly linear over distances of several mean free paths. Then we may write

$$\omega_A|_{y\pm a} = \omega_A|_y \pm \frac{2}{3}\lambda\frac{d\omega_A}{dy} \tag{17.7-6}$$

Combination of the last two equations then gives for the *total mass flux* at plane y:

$$\begin{aligned} n_{Ay} &= \rho\omega_A v_y - \frac{1}{3}\rho\bar{u}\lambda\frac{d\omega_A}{dy} \\ &\equiv \rho\omega_A v_y - \rho\mathscr{D}_{AA^*}\frac{d\omega_A}{dy} \end{aligned} \tag{17.7-7}$$

This is the *convective mass flux* plus the *diffusive mass flux*, the latter being given by Eq. 17.3-2. Therefore, we get the following expression for the self-diffusivity:

$$\mathscr{D}_{AA^*} = \frac{1}{3}\bar{u}\lambda \tag{17.7-8}$$

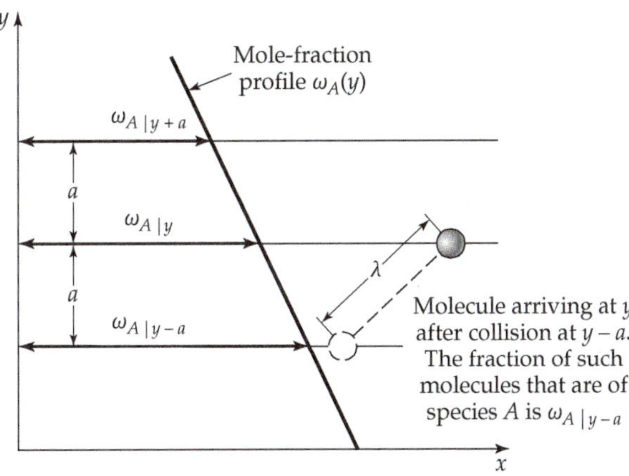

Fig. 17.7-1. Molecular transport of species A from the plane at $(y - a)$ to the plane at y.

Finally, making use of Eqs. 17.7-1 and 17.7-3, we get

$$\mathscr{D}_{AA^*} = \frac{2}{3}\frac{\sqrt{\kappa T/\pi m_A}}{\pi d_A^2}\frac{1}{n} = \frac{2}{3\pi}\frac{\sqrt{\pi m_A \kappa T}}{\pi d_A^2}\frac{1}{\rho} \tag{17.7-9}$$

which can be compared with Eq. 1.6-9 for the viscosity and Eq. 9.7-12 for the thermal conductivity.

The development of a formula for $\mathscr{D}_{AB}$ for rigid spheres of unequal masses and diameters is considerably more difficult. We simply quote the result[1] here:

$$\mathscr{D}_{AB} = \frac{2}{3}\sqrt{\frac{\kappa T}{\pi}}\sqrt{\frac{1}{2}\left(\frac{1}{m_A}+\frac{1}{m_B}\right)}\frac{1}{\pi\left(\frac{1}{2}(d_A+d_B)\right)^2}\frac{1}{n} \tag{17.7-10}$$

That is, $1/m_A$ is replaced by the arithmetic average of $1/m_A$ and $1/m_B$, and d_A by the arithmetic average of d_A and d_B.

The preceding discussion shows how the diffusivity can be obtained by mean-free path arguments. For accurate results, the Chapman-Enskog kinetic theory should be used. The Chapman-Enskog results for viscosity and thermal conductivity were given in §1.6 and §9.7, respectively. The corresponding formula for $c\mathscr{D}_{AB}$ is:[2,3]

$$\begin{aligned} c\mathscr{D}_{AB} &= \frac{3}{16}\sqrt{\frac{2RT}{\pi}\left(\frac{1}{M_A}+\frac{1}{M_B}\right)}\frac{1}{\tilde{N}\sigma_{AB}^2\Omega_{\mathscr{D},AB}} \\ &= 2.2646\times10^{-5}\sqrt{T\left(\frac{1}{M_A}+\frac{1}{M_B}\right)}\frac{1}{\sigma_{AB}^2\Omega_{\mathscr{D},AB}} \end{aligned} \tag{17.7-11}$$

Or, if we approximate c by the ideal-gas law $p = cRT$, we get for $\mathscr{D}_{AB}$

$$\begin{aligned} \mathscr{D}_{AB} &= \frac{3}{16}\sqrt{\frac{2(RT)^3}{\pi}\left(\frac{1}{M_A}+\frac{1}{M_B}\right)}\frac{1}{\tilde{N}p\sigma_{AB}^2\Omega_{\mathscr{D},AB}} \\ &= 0.0018583\sqrt{T^3\left(\frac{1}{M_A}+\frac{1}{M_B}\right)}\frac{1}{p\sigma_{AB}^2\Omega_{\mathscr{D},AB}} \end{aligned} \tag{17.7-12}$$

In the second lines of Eqs. 17.7-11 and 17.7-12, $\mathscr{D}_{AB}$ [=] cm^2/s, σ_{AB} [=] Å, T [=] K, and p [=] atm; the numerical constants in Eqs. 17.7-11 and 17.7-12 have units of (g-mol/cm s)(K$^{-1/2}$)(g/g-mol)$^{-1/2}$Å^2 and (cm^2/s)K$^{-3/2}$(g/g-mol)$^{1/2}$atm Å^2, respectively.

The dimensionless quantity $\Omega_{\mathscr{D},AB}$—the "collision integral" for diffusion—is a function of the dimensionless temperature $\kappa T/\varepsilon_{AB}$. The parameters σ_{AB} and ε_{AB} are those appearing in the Lennard-Jones potential between one molecule of A and one of B (cf. Eq. 1.6-10),

$$\varphi_{AB}(r) = 4\varepsilon_{AB}\left[\left(\frac{\sigma_{AB}}{r}\right)^{12} - \left(\frac{\sigma_{AB}}{r}\right)^{6}\right] \tag{17.7-13}$$

and are given for nonpolar gas pairs by[4]

$$\sigma_{AB} = \frac{1}{2}(\sigma_A+\sigma_B) \quad \text{and} \quad \varepsilon_{AB} = \sqrt{\varepsilon_A\varepsilon_B} \tag{17.7-14,15}$$

---

[1] A similar result is given by R. D. Present, *Kinetic Theory of Gases*, McGraw-Hill, New York (1958), p. 55.

[2] S. Chapman and T. G. Cowling, *The Mathematical Theory of Non-Uniform Gases*, Third Edition, Cambridge University Press (1970), Chapters 10 and 14.

[3] J. O. Hirschfelder, C. F. Curtiss, and R. B. Bird, *Molecular Theory of Gases and Liquids*, 2nd Corrected Printing, Wiley, New York (1964), p. 539.

[4] J. O. Hirschfelder, R. B. Bird, and E. L. Spotz, *Chem. Revs.*, **44**, 205–231 (1949); S. Gotoh, M. Manner, J. P. Sørensen, and W. E. Stewart, *J. Chem. Eng. Data*, **19**, 169–171 (1974).

The function $\Omega_{\mathscr{D},AB}$ is given in Table D.2 and Eq. D.2-2. From these results, one can compute that $\mathscr{D}_{AB}$ increases roughly as the 2.0 power of T at low temperatures and as the 1.65 power of T at very high temperatures; see the $p_r \to 0$ curve in Fig 17.6-1. For rigid spheres, $\Omega_{\mathscr{D},AB}$ would be unity at all temperatures and a result analogous to Eq. 17.7-10 would be obtained.

The parameters σ_{AB} and ε_{AB} could, in principle, be determined directly from accurate measurements of $\mathscr{D}_{AB}$ over a wide range of temperatures. Suitable data are not yet available for many gas pairs, and hence, Eqs. 17.7-14 and 17.7-15 are normally used.

For isotopic pairs, $\sigma_{AA^*} = \sigma_A = \sigma_{A^*}$ and $\varepsilon_{AA^*} = \varepsilon_A = \varepsilon_{A^*}$; that is, the intermolecular force fields for the various pairs A - A^*, A^* - A^*, and A - A are virtually identical, and the parameters σ_A and ε_A may be obtained from viscosity data on pure A. If, in addition, M_A is large, Eq. 17.7-11 simplifies to

$$c\mathscr{D}_{AA^*} = 3.2027 \times 10^{-5}\sqrt{\frac{T}{M_A}}\frac{1}{\sigma_A^2\Omega_{\mathscr{D},AA^*}} \tag{17.7-16}$$

the numerical constant having the units of (g-mol/cm · s)(K$^{-1/2}$)(g/g-mol)$^{-1/2}$Å^2. The corresponding equation for the rigid-sphere model is given in Eq. 17.7-9.

Comparison of Eq. 17.7-16 with Eq. 1.6-14 shows that the self-diffusivity $\mathscr{D}_{AA^*}$ and the viscosity μ (or kinematic viscosity ν) are related as follows for heavy isotopic gas pairs at low density:

$$\frac{\mu}{\rho\mathscr{D}_{AA^*}} = \frac{\nu}{\mathscr{D}_{AA^*}} = \frac{5}{6}\frac{\Omega_{\mathscr{D},AA^*}}{\Omega_\mu} \tag{17.7-17}$$

in which $\Omega_\mu \approx 1.1\Omega_{\mathscr{D},AA^*}$ over a wide of $\kappa T/\varepsilon_A$, as may be seen in Table D.2. Thus, $\mathscr{D}_{AA^*} \approx 1.32\nu$ for the *self-diffusivity.* The relation between ν and the *binary diffusivity* $\mathscr{D}_{AB}$ is not so simple, because ν may vary considerably with the composition. The Schmidt number $\mathrm{Sc} = \mu/\rho\mathscr{D}_{AB}$ is in the range from 0.2 to 5.0 for most gas pairs.

Equations 17.7-11, 17.7-12, 17.7-16, and 17.7-17 were derived for monatomic nonpolar gases but have been found useful for polyatomic nonpolar gases as well. In addition, these equations may be used to predict $\mathscr{D}_{AB}$ for interdiffusion of a polar gas and a nonpolar gas by using combining laws different[5] from those given in Eqs. 17.7-14 and 17.7-15.

EXAMPLE 17.7-1

Computation of Mass Diffusivity for Low-Density Gases

Predict the value of $\mathscr{D}_{AB}$ for the system CO - CO_2 at 296.1 K and 1.0 atm total pressure.

SOLUTION

From Table D.1 we obtain the following parameters:

CO:	$M_A = 28.01$ g/g-mol	$\sigma_A = 3.590$Å	$\varepsilon_A/\kappa = 110$K
CO_2:	$M_B = 44.01$ g/g-mol	$\sigma_B = 3.996$Å	$\varepsilon_B/\kappa = 190$K

The mixture parameters are then estimated from Eqs. 17.7-14 and 17.7-15:

$$\sigma_{AB} = \frac{1}{2}(3.590 + 3.996\ \text{Å}) = 3.793\ \text{Å} \tag{17.7-18}$$

$$\varepsilon_{AB}/\kappa = \sqrt{(110\ \text{K})(190\ \text{K})} = 144.6\ \text{K} \tag{17.7-19}$$

[5]J. O. Hirschfelder, C. F. Curtiss, and R. B. Bird, *Molecular Theory of Gases and Liquids*, 2nd corrected printing, Wiley, New York (1964), §8.6*b* and p. 1201. Polar gases and gas mixtures are discussed by E. A. Mason and L. Monchick, *J. Chem. Phys.*, **36**, 2746–2757 (1962).

The dimensionless temperature is then $\kappa T/\varepsilon_{AB} = (296.1\text{ K})/(144.6\text{ K}) = 2.048$. From Table D.2 we can find the collision integral for diffusion, $\Omega_{\mathscr{D},AB} = 1.067$. Substitution of the preceding values in Eq. 17.7-12 gives

$$\begin{aligned}\mathscr{D}_{AB} &= 0.0018583\ (\text{cm}^2/\text{s})\text{K}^{-3/2}(\text{g/g-mol})^{1/2}\text{atm Å}^2 \\ &\times\sqrt{(296.1\text{ K})^3\left(\frac{1}{28.01}\frac{\text{g-mol}}{\text{g}}+\frac{1}{44.01}\frac{\text{g-mol}}{\text{g}}\right)}\frac{1}{(1.0\text{ atm})(3.793\text{ Å})^2(1.067)} \\ &= 0.149\text{ cm}^2/\text{s}\end{aligned} \tag{17.7-20}$$

§17.8 DIFFUSIVITY OF LIQUIDS

The kinetic theory for diffusion in simple liquids is not as well developed as that for dilute gases, and it cannot presently give accurate analytical predictions of diffusivities.[1–3] As a result, our understanding of liquid diffusion depends primarily on the rather crude hydrodynamic and activated-state models. These in turn have spawned a number of empirical correlations, which provide the best available means for prediction. These correlations permit estimation of diffusivities in terms of more easily measured properties such as viscosity and molar volume.

The *hydrodynamic theory* takes as its starting point the Nernst-Einstein equation,[4] which states that the diffusivity of a single particle or solute molecule A through a stationary medium B is given by

$$\mathscr{D}_{AB} = \kappa T(u_A/F_A) \tag{17.8-1}$$

in which u_A/F_A is the "mobility" of a particle A (that is, the steady-state velocity attained by the particle under the action of a unit force). If the shape and size of A are known, the mobility can be calculated by the solution of the creeping-flow equation of motion[5] (Eq. 3.6-3). Thus, if A is spherical and if one takes into account the possibility of "slip" at the fluid–solid interface, one obtains[6]

$$\frac{u_A}{F_A} = \left(\frac{3\mu_B + R_A\beta_{AB}}{2\mu_B + R_A\beta_{AB}}\right)\frac{1}{6\pi\mu_B R_A} \tag{17.8-2}$$

in which μ_B is the viscosity of the pure solvent, R_A is the radius of the solute particle, and β_{AB} is the *coefficient of sliding friction* (formally the same as the μ/ζ of Problem 2B.10). The limiting cases of $\beta_{AB} \to \infty$ and $\beta_{AB} \to 0$ are of particular interest:

a. The limit $\beta_{AB} \to \infty$ (no slip condition)

In this case Eq. 17.8-2 becomes Stokes' law (Eq. 2.7-15) and Eq. 17.8-1 becomes:

$$\frac{\mathscr{D}_{AB}\mu_B}{\kappa T} = \frac{1}{6\pi R_A} \tag{17.8-3}$$

which is usually called the *Stokes-Einstein equation* (or the *Stokes-Einstein-Sutherland equation*[7]). This equation applies well to the diffusion of very large spherical molecules

---

[1]R. J. Bearman and J. G. Kirkwood, *J. Chem. Phys.*, **28**, 136–145 (1958).

[2]R. J. Bearman, *J. Phys. Chem.*, **65**, 1961–1968 (1961).

[3]C. F. Curtiss and R. B. Bird, *J. Chem. Phys.*, **111**, 10362–10370 (1999).

[4]See E. A. Moelwyn-Hughes, *Physical Chemistry*, 2nd edition, corrected printing, Macmillan, New York (1964), pp. 62–74. See also R. J. Silbey, R. A. Alberty, and M. G. Bawendi, *Physical Chemistry*, 4th edition, Wiley, New York, (2005), §20.2.

[5]S. Kim and S. J. Karrila, *Microhydrodynamics: Principles and Selected Applications*, Butterworth-Heinemann, Boston (1991); Dover, Mineola, NY (2005).

[6]H. Lamb, *Hydrodynamics*, 6th edition, Cambridge University Press (1932), Dover, New York (1945), §337.

[7]I. C. Carpen and J. F. Brady, *J. Rheol.*, **49**, 1483–1502 (2005)

in solvents of low molecular weight[8] and to suspended particles. Analogous expressions developed for nonspherical particles have been used for estimating the shapes of protein molecules.[9,10]

b. The limit $\beta_{AB} \to 0$ (complete slip condition)
In this case Eq. 17.8-2 leads to

$$\frac{\mathscr{D}_{AB}\mu_B}{\kappa T} = \frac{1}{4\pi R_A} \tag{17.8-4}$$

If the molecules A and B are identical (that is for *self-diffusion*) and if they can be assumed to form a cubic lattice with the adjacent molecules just touching, then $2R_A = (\tilde{V}_A/\tilde{N})^{1/3}$ and

$$\frac{\mathscr{D}_{AA}\mu_A}{\kappa T} = \frac{1}{2\pi}\left(\frac{\tilde{N}}{\tilde{V}_A}\right)^{1/3} \tag{17.8-5}$$

Equation 17.8-5 has been found[11] to agree with self-diffusion data for a number of liquids, including polar and associated substances, liquid metals, and molten sulfur, to within about 12%. The hydrodynamic model has proven less useful for binary diffusion (that is, for A not identical to B), although the predicted temperature and viscosity dependences are approximately correct.

It should be kept in mind that the above formulas apply only to dilute solutions of A in B. Some attempts have been made, however, to extend the hydrodynamic model to more concentrated solutions.[12]

The *Eyring activated-state theory* attempts to explain transport behavior via a quasi-crystalline model of the liquid state.[13] It is assumed in this theory that there is some unimolecular rate process in terms of which diffusion can be described, and it is further assumed that in this process there is some configuration that can be identified as the "activated state." The Eyring theory of reaction rates is applied to this elementary process in a manner analogous to that described in §1.7 for estimation of liquid viscosity. A modification of the original Eyring model by Ree, Eyring, and coworkers[14] yields an expression similar to Eq. 17.8-5 for traces of A in solvent B:

$$\frac{\mathscr{D}_{AB}\mu_B}{\kappa T} = \frac{1}{\xi}\left(\frac{\tilde{N}}{\tilde{V}_B}\right)^{1/3} \tag{17.8-6}$$

Here ξ is a "packing parameter," which in the theory represents the number of nearest neighbors of a given solvent molecule. For the special case of self-diffusion, ξ is found to be very close to 2π, so that Eqs. 17.8-5 and 17.8-6 are in good agreement despite the difference between the models from which they were developed.

The Eyring theory is based on an oversimplified model of the liquid state, and consequently, the conditions required for its validity are not clear. However, Bearman[2] has shown that the Eyring model gives results consistent with the statistical mechanics for

---

[8]A. Polson, *J. Phys. Colloid Chem.*, **54**, 649–652 (1950).

[9]H. J. V. Tyrrell, *Diffusion and Heat Flow in Liquids*, Butterworths, London (1961), Chapter 6.

[10]Creeping motion around finite bodies in a fluid of infinite extent has been reviewed by J. Happel and H. Brenner, *Low Reynolds Number Hydrodynamics*, Prentice-Hall, Englewood Cliffs, New Jersey (1965); Martinus Nijhoff, The Hague (1983); see also S. Kim and S. J. Karrila, *Microhydrodynamics: Principles and Selected Applications*, Butterworth-Heinemann, Boston (1991); Dover, Mineola, NY (2005). G. K. Youngren and A. Acrivos, *J. Chem. Phys.* **63**, 3846–3848 (1975) have calculated the rotational friction coefficient for benzene, thereby establishing the validity of the no-slip condition at molecular dimensions.

[11]J. C. M. Li and P. Chang, *J. Chem. Phys.*, **23**, 518–520 (1955).

[12]C. W. Pyun and M. Fixman, *J. Chem. Phys.*, **41**, 937–944 (1964).

[13]S. Glasstone, K. J. Laidler, and H. Eyring, *Theory of Rate Processes*, McGraw-Hill, New York (1941), Chapter IX.

[14]H. Eyring, D. Henderson, B. J. Stover, and E. M. Eyring, *Statistical Mechanics and Dynamics*, Wiley, New York (1964), §16.8.

"regular solutions," that is, for mixtures of molecules that have similar size, shape, and intermolecular forces. For this limiting situation, Bearman also obtains an expression for the concentration dependence of the diffusivity,

$$\frac{\mathscr{D}_{AB}\mu_B}{(\mathscr{D}_{AB}\mu_B)_{x_A\to 0}} = \left[1 + x_A\left(\frac{\overline{V}_A}{\overline{V}_B} - 1\right)\right]\left(\frac{\partial \ln a_A}{\partial \ln x_A}\right)_{T,p} \tag{17.8-7}$$

in which $\mathscr{D}_{AB}$ and μ_B are the diffusivity and viscosity of the mixture at the composition x_A, and a_A is the thermodynamic activity of species A. For regular solutions, the partial molar volumes, $\overline{V}_A$ and $\overline{V}_B$, are equal to the molar volumes of the pure components. Bearman suggests on the basis of his analysis that Eq. 17.8-7 should be limited to regular solutions, and it has in fact been found to apply well only to nearly ideal solutions.

Because of the unsatisfactory nature of the theory for diffusion in liquids, it is necessary to rely on *empirical* expressions. For example, the Wilke-Chang equation[15] gives the diffusivity in cm^2/s for small concentrations of A in B as:

$$\mathscr{D}_{AB} = 7.4\times 10^{-8}\frac{\sqrt{\psi_B M_B}\,T}{\mu \tilde{V}_A^{0.6}} \tag{17.8-8}$$

Here $\tilde{V}_A$ is the molar volume of the solute A in $cm^3/g\text{-mol}$ as liquid at its normal boiling point, μ is the viscosity of the solution in centipoises, ψ_B is a dimensionless "association parameter" for the solvent, and T is the absolute temperature in K; the numerical constant has dimensions $(cm^2/s)(g\text{-mol}/g)^{1/2}K^{-1}(cp)(cm^3/g\text{-mol})^{0.6}$. Recommended values of ψ_B are: 2.6 for water; 1.9 for methanol; and 1.0 for benzene, ether, heptane, and other unassociated solvents. Equation 17.8-8 is good only for dilute solutions of nondissociating solutes. For such solutions, it is usually good within $\pm 10\%$.

Other empiricisms, along with their relative merits, have been summarized by Poling, Prausnitz, and O'Connell.[16]

EXAMPLE 17.8-1

Estimation of Liquid Diffusivity

Estimate $\mathscr{D}_{AB}$ for a dilute solution of TNT (2,4,6-trinitro-toluene) in benzene at 15°C.

SOLUTION

Use the equation of Wilke and Chang, taking TNT as component A and benzene and component B. The required data are

$$\begin{aligned} \mu &= 0.705\text{ cp (the viscosity for pure benzene)}\\ \tilde{V}_A &= 140\text{ cm}^3/\text{g-mol (for TNT)}\\ \psi_B &= 1.0\text{ (for benzene)}\\ M_B &= 78.11\text{ g/g-mol (for benzene)} \end{aligned}$$

Substitution into Eq. 17.8-8 gives

$$\begin{aligned} \mathscr{D}_{AB} &= 7.4\times 10^{-8}\,(\text{cm}^2/\text{s})(\text{g-mol/g})^{1/2}(\text{K})^{-1}(\text{cp})(\text{cm}^3/\text{g-mol})^{0.6}\\ &\quad\times\frac{\sqrt{(1.0)(78.11\text{ g/g-mol})}\,(273+15\text{ K})}{(0.705\text{ cp})(140\text{ cm}^3/\text{g-mol})^{0.6}}\\ &= 1.38\times 10^{-5}\text{ cm}^2/\text{s} \end{aligned} \tag{17.8-9}$$

This result compares well with the measured value of 1.39×10^{-5} cm^2/s.

---

[15] C. R. Wilke, *Chem. Eng. Prog.*, **45**, 218–224 (1949); C. R. Wilke and P. Chang, *AIChE Journal*, **1**, 264–270 (1955).

[16] B. E. Poling, J. M. Prausnitz, and J. P. O'Connell, *The Properties of Gases and Liquids*, 5th edition, McGraw-Hill, New York (2001), Chapter 11.

§17.9 CONCLUDING COMMENTS

One of the biggest problems in the subject of diffusion is that one encounters, at the very outset, a veritable zoo of symbols. This cannot be avoided, however, and the person studying this for the first time has to bite the bullet and make their acquaintance. Both mass and molar units are in common use. In addition, each of the various reference frames—axes fixed in space (for $\mathbf{n}_A$ and $\mathbf{N}_A$), axes moving with the mass average velocity (for $\mathbf{j}_A$), and axes moving with the molar average velocity (for $\mathbf{J}_A^*$)—has its own advantages. Perhaps the following will help in remembering the notation:

a. The mass density and concentrations are written with Greek letters (ρ, ρ_A, ω_A), whereas the molar concentrations are written with Roman letters (c, c_A, x_A).

b. The mass flux vectors are in lowercase letters ($\mathbf{j}_A, \mathbf{n}_A$), and the molar flux vectors are in capital letters ($\mathbf{J}_A^*, \mathbf{N}_A$).

c. Quantities associated with the molar average velocity are decorated with asterisks ($\mathbf{v}^*, \mathbf{J}_A^*$), and quantities associated with the mass average velocity are not decorated ($\mathbf{v}, \mathbf{j}_A$).

Compared with the viscosity and thermal conductivity, the diffusivity has been neglected. Whereas experimental data for viscosity and thermal conductivity are plentiful, data on diffusivity are generally lacking. Part of the reason for this is that, for N compounds, there are N sets of data for μ and for k, but for $\mathscr{D}_{AB}$ one needs N^2 sets of data.

Many problems in the chemical industry and in biology are concerned with multicomponent mixtures. Here the situation is even worse regarding the accumulation of reliable experimental data. Furthermore, solution of multicomponent diffusion problems is no easy task. This topic is addressed briefly at the end of Chapter 24.

QUESTIONS FOR DISCUSSION

1. How is the binary diffusivity defined? How is self-diffusion defined? Give typical orders of magnitude of diffusivities for gases, liquids, and solids.
2. Summarize the notation for the molecular, convective, and total fluxes for the three transport processes. How does one calculate the flux of mass, momentum, and energy across a surface with orientation $\mathbf{n}$?
3. Define the Prandtl, Schmidt, and Lewis numbers. What ranges of Pr and Sc can one expect to encounter for gases and liquids?
4. How can you estimate the Lennard-Jones potential for a binary mixture, if you know the parameters for the two components of the mixture?
5. Of what value are the hydrodynamic theories of diffusion?
6. Compare and contrast the relation between binary diffusivity and viscosity for gases and for liquids.
7. In a binary mixture, does the vanishing of $\mathbf{N}_A$ imply the vanishing of ∇x_A?
8. Why do we use both mass and molar units in the description of diffusing systems? Write Fick's law in both systems of units.
9. Compare and contrast the mass average velocity and the molar average velocity.
10. When were Newton's law of viscosity, Fourier's law of heat conduction, and Fick's law of diffusion proposed?

PROBLEMS

17A.1 Prediction of a low-density binary diffusivity. Estimate $\mathscr{D}_{AB}$ for the methane-ethane system at 293 K and 1 atm by the following methods:

(a) Equation 17.6-1.

(b) The corresponding-states chart in Fig. 17.6-1 along with Eq. 17.6-3.

(c) The Chapman-Enskog relation (Eq. 17.7-12) with Lennard-Jones parameters from Appendix D.

(d) The Chapman-Enskog relation (Eq. 17.7-12) with the Lennard-Jones parameters estimated from critical properties.

Answers (all in cm^2/s): **(a)** 0.152; **(b)** 0.138; **(c)** 0.146; **(d)** 0.138

17A.2 Extrapolation of binary diffusivity to a very high temperature. A value of $\mathscr{D}_{AB} = 0.151\ cm^2/s$ has been reported[1] for the system CO_2-air at 293K and 1 atm. Extrapolate $\mathscr{D}_{AB}$ to 1500K by the following methods:

(a) Equation 17.6-1.

(b) Equation 17.7-10.

(c) Equations 17.7-12, 17.7-14, and 17.7-15, with Table D.2.

What do you conclude from comparing these results with the experimental value[1] of 2.45 cm^2/s?

Answers (all in cm^2/s): **(a)** 2.96; **(b)** 1.75; **(c)** 2.51

17A.3 Self-diffusion in liquid mercury. The diffusivity of Hg^{203} in normal liquid Hg has been measured,[2] along with its viscosity and volume per unit mass. Compare the experimentally measured $\mathscr{D}_{AA^*}$ with the values calculated with Eq. 17.8-5.

T (K)	$\mathscr{D}_{AA^*}$ (cm^2/s)	μ (cp)	$\hat{V}$ (cm^3/g)
275.7	1.52×10^{-5}	1.68	0.0736
289.6	1.68×10^{-5}	1.56	0.0737
364.2	2.57×10^{-5}	1.27	0.0748

17A.4 Schmidt numbers for binary gas mixtures at low density. Use Eq. 17.7-11 and the viscosity data given below[3] to compute $Sc = \mu/\rho\mathscr{D}_{AB}$ for binary mixtures of hydrogen (A) and Freon-12 (B, dichlorodifluoromethane) at $x_A = 0.00, 0.25, 0.50, 0.75$, and 1.00, and at 25°C and 1 atm.

Mole fraction of H_2:	0.00	0.25	0.50	0.75	1.00
$\mu \times 10^6$ (poise):	124.0	128.1	131.9	135.1	88.4

Sample answers: As $x_A \to 0.00$, Sc $\to$ 0.057; as $x_A \to 1.00$, Sc $\to$ 2.44

17A.5 Estimation of diffusivity for a binary mixture at high density. Predict $c\mathscr{D}_{AB}$ for an equimolar mixture of N_2 and C_2H_6 at 288.2 K and 40 atm.

(a) Use the value of $\mathscr{D}_{AB}$ at 1 atm from Table 17.5-1, along with Fig. 17.6-1.

(b) Use Eq. 17.6-3 and Fig. 17.6-1.

Answers: **(a)** 5.8×10^{-6} g-mol/cm · s; **(b)** 5.3×10^{-6} g-mol/cm · s

17A.6 Diffusivity and Schmidt number for chlorine-air mixtures.

(a) Predict $\mathscr{D}_{AB}$ for chlorine (A)-air (B) mixtures at 75°F and 1 atm. Treat air as a single substance with Lennard-Jones parameters as given in Appendix D. Use the Chapman-Enskog theory results in §17.7.

[1] Ts. M. Klibanova, V. V. Pomerantsev, and D. A. Frank-Kamenetskii, *J. Tech. Phys. (USSR)*, **12**, 14–30 (1942), as quoted by C. R. Wilke and C. Y. Lee, *Ind. Eng. Chem.*, **47**, 1253 (1955).

[2] R. E. Hoffman, *J. Chem. Phys.*, **20**, 1567–1570 (1952).

[3] J. W. Buddenberg and C. R. Wilke, *Ind. Eng. Chem.* **41**, 1345–1347 (1949).

(b) Repeat (a) using Eq. 17.6-1.

(c) Use the results of (a) to estimate Schmidt numbers for chlorine-air mixtures at 297K and 1 atm for the following mole fractions of chlorine: 0, 0.25, 0.50, 0.75, and 1.00. The mixture viscosities for these compositions have been predicted to be 0.0183, 0.0164, 0.0150, 0.0139, and 0.0130 cp, respectively.[4]

Answers: **(a)** 0.121 cm^2/s; **(b)** 0.124 cm^2/s; **(c)** Sc = 1.27, 0.832, 0.602, 0.463, 0.372

17A.7 The Schmidt number for self-diffusion.

(a) Use Eqs. 1.5-1b and 17.6-2 to predict the self-diffusion Schmidt number $\text{Sc} = \mu/\rho\mathscr{D}_{AA^*}$ at the critical point for a system with $M_A \approx M_{A^*}$.

(b) Use the above result, along with Fig. 1.5-1 and Fig. 17.6-1, to predict $\text{Sc} = \mu/\rho\mathscr{D}_{AA^*}$ at the following states:

Phase	Gas	Gas	Gas	Liquid	Gas	Gas
T_r	0.7	1.0	5.0	0.7	1.0	2.0
p_r	0.0	0.0	0.0	saturation	1.0	1.0

17A.8 Correction of high-density diffusivity for temperature. The measured value[5] of $c\mathscr{D}_{AB}$ for a mixture of 80 mole% CH_4 and 20 mole% C_2H_6 at 313K and 136 atm is 6.0×10^{-6} g-mol/cm · s (see Example 17.6-3). Predict $c\mathscr{D}_{AB}$ for the same mixture at 136 atm at 351K, using Fig. 17.6-1.

Answer: 6.6×10^{-6} g-mol/cm · s
Observed:[5] 6.33×10^{-6} g-mol/cm · s

17A.9 Prediction of the critical $c\mathscr{D}_{AB}$ values. Figure 17.6-1 gives the low-pressure limit $(c\mathscr{D}_{AA^*})_r = 1.01$ at $T_r = 1$ and $p_r \to 0$. At this limit, Eq. 17.7-11 gives

$$1.01(c\mathscr{D}_{AA^*})_c = 2.2646 \times 10^{-5}\sqrt{T_{cA}\left(\frac{1}{M_A} + \frac{1}{M_{A^*}}\right)}\frac{1}{\sigma^2_{AA^*}\Omega_{\mathscr{D},AA^*}} \tag{17A.9-1}$$

The argument $\kappa T_{cA}/\varepsilon_{AA^*}$ of $\Omega_{\mathscr{D},AA^*}$ is reported[6] to be about 1.225 for Ar, Kr, and Xe. Here we use the value 1/0.77 from Eq. 1.6-11a as representative average over many fluids.

(a) Combine Eq. 17A.9-1 with the relations

$$\sigma_{AA^*} = 2.44(T_{cA}/p_{cA})^{1/3} \qquad \varepsilon_{AA^*}/\kappa = 0.77T_{cA} \tag{17A.9-2,3}$$

and Table D.2 to obtain Eq. 17.6-2 for $(c\mathscr{D}_{AA^*})_c$.

(b) Show that the approximations

$$\sigma_{AB} = \sqrt{\sigma_A\sigma_B} \qquad \varepsilon_{AB} = \sqrt{\varepsilon_A\varepsilon_B} \tag{17A.9-4,5}$$

for the Lennard-Jones parameters for the *A-B* interaction give

$$\sigma_{AB} = 2.44\left(\frac{T_{cA}T_{cB}}{p_{cA}p_{cB}}\right)^{1/6} \qquad \frac{\varepsilon_{AB}}{\kappa} = 0.77\sqrt{T_{cA}T_{cB}} \tag{17A.9-6,7}$$

when the molecular parameters of each species are predicted according to Eqs. 1.6-11a, c. Combine these expressions with Eq. 17A.9-1 (with A^* replaced by B, and T_{cA} by $\sqrt{T_{cA}T_{cB}}$) to obtain Eq. 17.6-3 for $(c\mathscr{D}_{AB})_c$. The corresponding replacement of p_c and T_c in Fig. 17.6-1 by $\sqrt{p_{cA}p_{cB}}$ and

---

[4]R. B. Bird, W.E. Stewart, and E. N. Lightfoot, Transport Phenomena, revised 2nd edition, McGraw-Hill, 2007 (New York), Chapter 1.

[5]V. J. Berry and R. C. Koeller, *AIChE Journal*, **6**, 274–280 (1960).

[6]J. J. van Loef and E. G. D. Cohen, *Physica A*, **156**, 522–533 (1989).

$\sqrt{T_{cA}T_{cB}}$ amounts to regarding the A-B collisions as dominant over collisions of like molecules in the determine the value of $c\mathscr{D}_{AB}$.

17A.10 Estimation of liquid diffusivities.

(a) Estimate the diffusivity for a dilute aqueous solution of acetic acid at 12.5°C, using the Wilke-Chang equation. The density of pure acetic acid is 0.937 g/cm^3 at its boiling point.

(b) The diffusivity of a dilute aqueous solution of methanol at 15°C is about 1.28×10^{-5} cm^2/s. Estimate the diffusivity for the same solution at 100°C.

Answer: **(b)** 6.7×10^{-5} cm^2/s

17B.1 Dimensional checks of equations. Verify that Eq. 17.8-4, Eq. (F) of Table 17.3-1, and Eq. (C) of Table 17.4-2 are dimensionally consistent.

17B.2 Interrelation of composition variables in binary mixtures. Using the basic definitions in Table 17.1-1, derive the relations in Eqs. (C) through (E) and Eqs. (H) through (J) of Table 17.1-2.

17B.3 Relations between fluxes in binary systems.

(a) Verify Eqs. (D) and (H) of Table 17.3-1 using only the definitions of concentrations, velocities, and fluxes.

(b) Verify the relation in Eq. (I) of Table 17.3-1.

17B.4 Equivalence of various forms of Fick's law for binary mixtures.

(a) Starting with Eq. (A) of Table 17.4-2, derive the other equations in the table.

(b) Show that Eq. (F) of Table 17.4-2 can be rewritten as

$$\mathbf{v}_A - \mathbf{v}_B = -\mathscr{D}_{AB}\nabla \ln \frac{x_A}{x_B} \tag{17B.4-1}$$

17C.1 Mass flux with respect to volume-average velocity. Let the *volume-average velocity* in an N-component mixture be defined by

$$\mathbf{v}^{\square} = \sum_{\alpha=1}^{N} \rho_\alpha(\overline{V}_\alpha/M_\alpha)\mathbf{v}_\alpha = \sum_{\alpha=1}^{N} c_\alpha \overline{V}_\alpha \mathbf{v}_\alpha \tag{17C.1-1}$$

in which $\overline{V}_\alpha$ is the partial molar volume of species α (see Example 19.3-1 for the definition of a partial molar property). Then define

$$\mathbf{j}_\alpha^{\square} = \rho_\alpha(\mathbf{v}_\alpha - \mathbf{v}^{\square}) \tag{17C.1-2}$$

as the mass flux with respect to the volume-average velocity.

(a) Show that for a binary system of A and B,

$$\mathbf{j}_A^{\square} = \rho(\overline{V}_B/M_B)\mathbf{j}_A \tag{17C.1-3}$$

To do this you will need to use the identity $c_A\overline{V}_A + c_B\overline{V}_B = 1$. Where does this come from?

(b) Show that Fick's first law then assumes the form

$$\mathbf{j}_A^{\square} = -\mathscr{D}_{AB}\nabla\rho_A \tag{17C.1-4}$$

To verify this you will need the relation $\overline{V}_A\nabla c_A + \overline{V}_B\nabla c_B = 0$. What is the origin of this?

Chapter 18

Shell Mass Balances and Concentration Distributions in Solids and in Laminar Flow

In Chapter 2 it was shown how a number of steady-state viscous flow problems can be set up and solved by making a *shell momentum balance*. In Chapter 10 it was further shown how steady-state heat-conduction problems are handled by means of a *shell energy balance*. In this chapter we show how steady-state diffusion problems may be formulated by *shell mass balances*. The procedure used here is virtually the same as that used previously:

a. A mass balance is made over a thin shell perpendicular to the direction of mass transport, and this shell balance leads to a first-order differential equation, which may be solved to get the mass flux distribution.

b. Into this expression we insert the relation between mass flux and concentration gradient. The result is a second-order differential equation for the concentration profile. The integration constants that appear in the resulting expression are determined by the boundary conditions on the concentration and/or mass flux.

In Chapter 17 it was pointed out that several kinds of mass and molar fluxes are in common use to describe the transport of species. For simplicity, in this chapter we use the total molar flux $\mathbf{N}_A$, that is, the number of moles of A that go through a unit area in unit time, the unit area being fixed in space. We shall relate the molar flux to the concentration gradient by Eq. (E) of Table 17.4-2, written here for the z component

$$\underbrace{N_{Az}}_{\text{total flux of } A} = \underbrace{x_A(N_{Az} + N_{Bz})}_{\text{convective flux of } A} \underbrace{- c\mathscr{D}_{AB}\frac{\partial x_A}{\partial z}}_{\text{diffusive flux of } A} \tag{18.0-1}$$

This equation differs from Newton's law of viscosity, $\tau_{zx} = -\mu\partial v_x/\partial z$, and Fourier's law of heat conduction, $q_z = -k\partial T/\partial z$—both *molecular* fluxes—because here the *convective* flux has been included. That is, Eq. 18.0-1 is not analogous to Newton's law and Fourier's law.

However, the convection term can be omitted in several situations, illustrated here for transport in the z direction:

- Very small x_A (i.e., $x_A \ll 1$) *and* "stagnant B" (i.e., $N_{Bz} = 0$) or "nearly stagnant B" (i.e., $N_{Bz} \approx 0$). In these cases, the convective flux term is negligible compared to the diffusive flux term.
- "Equimolar counterdiffusion" (i.e., $N_{Bz} = -N_{Az}$). In this case, the convective flux vanishes.

Even if these assumptions aren't strictly valid, the assumption of negligible convection can sometimes be a good starting place for analyzing concentration profiles and mass transport.

When the convection terms cannot be ignored, there are several useful approaches that can sometimes be used to simplify the convection term and obtain an explicit expression for N_{Az}:

- Specification of N_{Bz}:
  - If species B is stagnant (i.e., $N_{Bz} = 0$), then Eq. 18.0-1 can be rearranged to the form

$$N_{Az} = -\frac{c\mathscr{D}_{AB}}{1 - x_A}\frac{\partial x_A}{\partial z} \tag{18.0-2}$$

  - If a heterogeneous reaction occurs, the ratio N_{Bz}/N_{Az} at steady state may be determined from stoichiometry. Letting $N_{Bz}/N_{Az} = -\lambda$, Eq. 18.0-1 can be rearranged to the form

$$N_{Az} = -\frac{c\mathscr{D}_{AB}}{1 - (1 - \lambda)x_A}\frac{\partial x_A}{\partial z} \tag{18.0-3}$$

- When the convective flux in the z direction dominates over the diffusive flux, Eq. 18.0-1 can be written

$$N_{Az} = x_A(N_{Az} + N_{Bz}) = c_A v_z^* \approx c_A v_z \tag{18.0-4}$$

  where the velocity component v_z is determined by the momentum balance.

The special cases given in this and the preceding paragraph are often very useful in simplifying and solving problems. Examples of all of these will be given in what follows.

In this chapter we study diffusion and convection in both *nonreacting* and *reacting* systems. When chemical reactions occur, we distinguish between two types: *homogeneous*, in which the chemical change occurs in the entire volume of the fluid, and *heterogeneous*, in which the chemical change takes place only in a restricted region, such as the surface of a catalyst. Not only is the physical picture different for homogeneous and heterogeneous reactions, but there is also a difference in the way the two types of reactions appear in the problem formulation. The rate of production of a chemical species by *homogeneous* reaction appears as a source term in the differential equation obtained from the shell balance, just as the thermal source term appears in the shell energy balance. The rate of production by a *heterogeneous* reaction, on the other hand, appears not in the differential equation, but rather in the boundary condition at the surface on which the reaction occurs.

In order to set up problems involving chemical reactions, some information has to be available about the rate at which the various chemical species appear or disappear by reaction. This brings us to the vast subject of *chemical kinetics*, that branch of physical chemistry that deals with the mechanisms of chemical reactions and the rates at which they occur.[1] In this chapter we assume that the reaction rates are described by means of simple functions of the concentrations of the reacting species.

Mention needs to be made at this point of the notation to be used for the chemical rate constants. For homogeneous reactions, the molar rate of production of species A may be given by an expression of the form

Homogeneous reaction:
$$R_A = k_n''' c_A^n \qquad (18.0\text{-}5)$$

in which R_A [=] moles/vol · time and c_A [=] moles/volume. The index n indicates the "order" of the reaction;[2] for a first-order reaction ($n = 1$), k_1''' [=] 1/time. For heterogeneous reactions, the molar rate of production at the reacting surface may often be specified at steady state by a relation of the form

Heterogeneous reaction:
$$N_{Az}|_{\text{surface}} = k_n'' c_A^n|_{\text{surface}} \qquad (18.0\text{-}6)$$

in which N_{Az} [=] moles/area · time and c_A [=] moles/volume. Here, for $n = 1$, k_1'' [=] length/time. Note that the triple prime on the rate constant indicates a volume source and the double prime a surface source.

We begin in §18.1 with a statement of the shell balance and the kinds of boundary conditions that may arise in solving diffusion problems. §18.2 to §18.4 focus on diffusion problems in which the convection term in the flux expression (Eq. 18.0-1) can be ignored. In §18.2 we discuss the diffusion of helium through a Pyrex wall, the helium being only slightly soluble. In §18.3 we talk about the diffusion of a very slightly soluble sphere in a surrounding liquid. In §18.4 we take up the diffusion accompanied by homogeneous chemical reaction, but restricting the problem to one in which the diffusing species is present only in very low concentration. In §18.5 we introduce the reader to diffusion with a heterogeneous chemical reaction occurring at a catalytic surface. The problem is first solved by omitting the convection term in the flux expression (Eq. 18.0-1), and then it is solved again with the convection term retained. Here the ratio N_{Bz}/N_{Az} can be specified from the stoichiometry of the chemical reaction. Then in §18.6 we discuss the evaporation of one gas, A, that then diffuses through another gas, B, which is motionless, so that N_{Bz} is exactly zero throughout the system; this provides one means for measuring the diffusivity of a gas pair. Here again the convection term in the flux expression is retained. Following that, in §18.7, we discuss the interdiffusion of two gases, in which $N_{Az} = -N_{Bz}$; that is, for every mole of A that moves, say, upwards, there will be a mole of B that moves downwards. This "two-bulb experiment" provides another method for measuring diffusivity. In the two sections that follow, §18.8 and §18.9, we study forced convection with diffusion; in the first section, the absorption of a gas into a liquid film falling along a vertical wall, and in the second, the dissolution of the wall into the falling film. Then in §18.10, we give an approximate treatment of the diffusion inside a catalyst pellet.

---

[1] R. J. Silbey, R. A. Alberty, and M. G. Bawendi, *Physical Chemistry*, Wiley, New York (2004), 4th edition, Chapter 18.

[2] Not all rate expressions are of the simple form of Eq. 18.0-5. The reaction rate may depend in a complicated way on the concentration of all species present. Similar remarks hold for Eq. 18.0-6. For detailed information on reaction rates, see *Table of Chemical Kinetics, Homogeneous Reactions*, National Bureau of Standards, Circular 510 (1951), Supplement No. 1 to Circular 510 (1956). See also C. G. Hill, *Introduction to Chemical Engineering Kinetics and Reactor Design*, Wiley, New York (1977); C. G. Hill and T. W. Root, *Introduction to Chemical Engineering Kinetics and Reactor Design*, 2nd edition, Wiley, New York (2013); J. B. Rawlings and J. G. Ekerdt, *Chemical Reactor Analysis and Design Fundamentals*, 2nd edition, Nob Hill Publishing, Madison, WI (2012).

§18.1 SHELL MASS BALANCES; BOUNDARY CONDITIONS

The diffusion problems in this chapter are solved by making mass balances for one of the chemical species over a thin shell of solid or fluid. Having selected an appropriate system, the law of conservation of mass for species A in a binary system is written for steady state over the volume of the shell in the form

$$\left\{\begin{matrix}\text{rate of}\\ \text{mass of}\\ A \text{ in}\end{matrix}\right\} - \left\{\begin{matrix}\text{rate of}\\ \text{mass of}\\ A \text{ out}\end{matrix}\right\} + \left\{\begin{matrix}\text{rate of production of}\\ \text{mass of } A \text{ by}\\ \text{homogeneous reaction}\end{matrix}\right\} = 0 \tag{18.1-1}$$

The conservation statement may, of course, be also expressed in terms of moles. The chemical species A may enter or leave the system by virtue of the overall motion of the fluid (i.e., by convection), or by diffusion (i.e., by molecular motion), both of these being included in the molar flux $\mathbf{N}_A$. In addition, species A may be produced or consumed by homogeneous chemical reactions.

After a balance is made on a shell of finite thickness by means of Eq. 18.1-1, we then let the thickness become infinitesimally small. As a result of this process, a differential equation for the mass (or molar) flux is generated. If, into this equation, we substitute the expression for the mass (or molar) flux in terms of a concentration gradient, we get a differential equation for the concentration.

When this differential equation has been integrated, constants of integration appear, and these have to be determined by the use of boundary conditions. The types of possible boundary conditions are very similar to those used in heat conduction (see §10.1):

a. The concentration at a surface can be specified; for example, at a solid–liquid interface, $x_A = x_{A0}$, where x_{A0} is the solubility of the solid in the liquid.

b. The molar flux at a surface can be specified, that is, $N_{Az} = N_{A0}$; for example, $N_{Az} = 0$ at a surface impermeable to A.

c. If diffusion is occurring in a solid or a liquid, it may happen that, at an interface, substance A is lost to a surrounding stream according to the relation

$$N_{A0} = k_c(c_{A0} - c_{Ab}) \quad \text{or} \quad N_{A0} = k_x(x_{A0} - x_{Ab}) \tag{18.1-2}$$

in which N_{A0} is the molar flux at the surface (into the surrounding stream), c_{A0} (or x_{A0}) is the surface concentration, c_{Ab} or (x_{Ab}) is the concentration in the bulk fluid stream, and the proportionality constant k_c (k_x) is a "mass-transfer coefficient." Methods of correlating mass-transfer coefficients are discussed in Chapter 22. Equation 18.1-2 is analogous to "Newton's law of cooling" given in Eq. 10.1-2.

d. The rate of chemical reaction at the surface can be specified. For example, if substance A disappears at a surface by a first-order chemical reaction, $N_{A0} = k_1''c_A$, where N_{A0} is the rate of transport to the surface. That is, at steady state, the rate of disappearance at a surface is proportional to the surface concentration, the proportionality constant k_1'' being a first-order chemical rate constant.

§18.2 DIFFUSION OF GASES THROUGH SOLIDS

Many gases are only very slightly soluble in solids, and in these cases the convection term in Eq. 18.0-1 may be neglected. Such is the situation with Pyrex glass, which is almost impermeable to all gases but helium. For example, the diffusivity of He through Pyrex is about 1,000 times the diffusivity of H_2 through Pyrex, the closest "competitor" in the

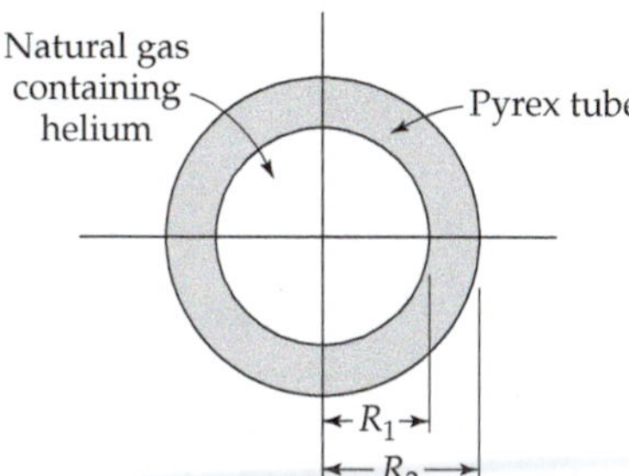

Fig. 18.2-1. Diffusion of helium through Pyrex tubing. The length of the tubing is L.

diffusion process. This fact suggests that a method for separating helium from natural gas could be based on the relative diffusion rates through Pyrex.[1]

Suppose that a natural gas mixture is contained in a Pyrex tube with dimensions shown in Fig. 18.2-1. We want to obtain an expression for the rate at which helium will leak out of the tube, in terms of the diffusivity of helium (A) through Pyrex (B), the interfacial concentrations of helium in the Pyrex, and the dimensions of the tube.

We begin by making a steady-state shell balance on helium over a cylindrical shell of thickness Δr and length L, located at an arbitrary radial position between R_1 and R_2,

$$N_{Ar}|_r \cdot 2\pi Lr - N_{Ar}|_{r+\Delta r} \cdot 2\pi L(r+\Delta r) = 0 \tag{18.2-1}$$

This balance may be rewritten as

$$2\pi L(rN_{Ar})|_r - 2\pi L(rN_{Ar})|_{r+\Delta r} = 0 \tag{18.2-2}$$

Division by $2\pi L \Delta r$ and taking the limit as $\Delta r \to 0$ gives

$$\lim_{\Delta r \to 0} \frac{(rN_{Ar})|_{r+\Delta r} - (rN_{Ar})|_r}{\Delta r} = 0 \tag{18.2-3}$$

Then according to the definition of the first derivative, this is

$$\frac{d}{dr} rN_{Ar} = 0 \tag{18.2-4}$$

Into this we insert Eq. 18.0-1, without the convective term (because $x_A \ll 1$ and $N_{Br} = 0$), to get

$$\frac{d}{dr}\left(rc\mathscr{D}_{AB} \frac{dx_A}{dr} \right) = 0 \tag{18.2-5}$$

We now make use of the fact that $\mathscr{D}_{AB}$ does not depend on the concentration of helium, and hence not on the coordinate r, so that $\mathscr{D}_{AB}$ may be moved to the left of the first derivative operator. Furthermore, because species A is very dilute, we assume that the total concentration c is a constant. Then Eq. 18.2-5 becomes

$$\frac{d}{dr}\left(r \frac{dc_A}{dr} \right) = 0 \tag{18.2-6}$$

This equation may be integrated, thus:

$$r\frac{dc_A}{dr} = C_1 \quad \text{and} \quad c_A(r) = C_1 \ln r + C_2 \tag{18.2-7,8}$$

The constants of integration have to be determined from the boundary conditions

B.C. 1: at $r = R_1$, $c_A = c_{A1}$ (18.2-9)

B.C. 2: at $r = R_2$, $c_A = c_{A2}$ (18.2-10)

[1] *Scientific American*, **199**, 46–54, (1958) (see p. 52) describes briefly the method developed by K. B. McAfee of Bell Telephone Laboratories.

Here c_{A1} and c_{A2} are the helium concentrations in the Pyrex, just inside the Pyrex phase at R_1 and R_2, respectively. It is found that the constants of integration are

$$C_1 = \frac{c_{A1} - c_{A2}}{\ln(R_1/R_2)} \qquad C_2 = c_{A2} - \frac{(c_{A1} - c_{A2})\ln R_2}{\ln(R_1/R_2)} \tag{18.2-11}$$

Insertion of these expressions into Eq. 18.2-8 then gives

$$\boxed{\frac{c_A(r) - c_{A2}}{c_{A1} - c_{A2}} = \frac{\ln(r/R_2)}{\ln(R_1/R_2)}} \tag{18.2-12}$$

Using this concentration profile we can get the flux of helium through the Pyrex as

$$N_{Ar}(r) = -\mathscr{D}_{AB}\frac{dc_A}{dr} = -\mathscr{D}_{AB}\frac{(c_{A1} - c_{A2})}{r\,\ln(R_1/R_2)} \tag{18.2-13}$$

and finally

$$W_A = 2\pi rLN_{Ar} = \frac{2\pi L\mathscr{D}_{AB}(c_{A1} - c_{A2})}{\ln(R_2/R_1)} \tag{18.2-14}$$

is the total molar flow rate, the moles of helium per unit time diffusing out of the tube.

§18.3 DIFFUSION AWAY FROM A SLIGHTLY SOLUBLE SPHERE

A sphere of solid material A ($KMnO_4$) is suspended in a stationary body of liquid B (water). We consider a system such that A is only very slightly soluble in liquid B, so that the molar flux of A may be well described by Eq. 18.0-1 without the convection term because x_A is so small. We want to describe the dissolution of A in B at steady state. That is, we know the solubility of A, so that the concentration x_{AR} at the sphere surface at $r = R$ is known, and it is presumed that the concentration far from the sphere $x_{A\infty}$ is also known. The situation is pictured in Fig. 18.3-1. This problem is exactly analogous to the heat conduction away from a sphere described in Problem 10B.1.

A shell balance over a spherical surface of thickness Δr within the liquid phase at a distance r from the sphere center gives

$$N_{Ar}|_r \cdot 4\pi r^2 - N_{Ar}|_{r+\Delta r} \cdot 4\pi(r + \Delta r)^2 = 0 \tag{18.3-1}$$

which can also be written in the form

$$4\pi(r^2N_{Ar})|_r - 4\pi(r^2N_{Ar})|_{r+\Delta r} = 0 \tag{18.3-2}$$

Division by $4\pi\Delta r$ and taking the limit as $\Delta r \to 0$ then gives

$$\lim_{\Delta r\to 0}\frac{(r^2N_{Ar})|_{r+\Delta r} - (r^2N_{Ar})|_r}{\Delta r} = 0 \tag{18.3-3}$$

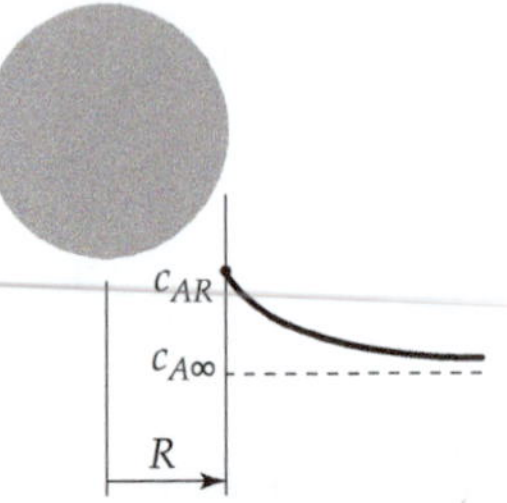

Fig. 18.3-1. A sphere of material A dissolving in a liquid B.

Using the definition of the first derivative, this may be written as

$$\frac{d}{dr}(r^2 N_{Ar}) = 0 \tag{18.3-4}$$

This equation may be integrated to give

$$r^2 N_{Ar} = C_1 \tag{18.3-5}$$

Insertion of N_{Ar} from Eq. 18.0-1, without the convection term, into Eq. 18.3-5 then gives

$$-c\mathscr{D}_{AB}\frac{dx_A}{dr} = \frac{C_1}{r^2} \tag{18.3-6}$$

A further integration then yields

$$-c\mathscr{D}_{AB}x_A(r) = -\frac{C_1}{r} + C_2 \tag{18.3-7}$$

The total molar concentration c will be virtually constant throughout the system, because of the extremely small concentration of A. Similarly, although $\mathscr{D}_{AB}$ may vary considerably with concentration in liquid systems, here we are dealing with only a slight amount of A and therefore $\mathscr{D}_{AB}$ may be treated as a constant. Therefore, we may define new constants of integration $C_1 \equiv -c\mathscr{D}_{AB}C_1'$ and $C_2 \equiv -c\mathscr{D}_{AB}C_2'$. Division of Eq. 18.3-7 by $-c\mathscr{D}_{AB}$ then gives

$$x_A(r) = -\frac{C_1'}{r} + C_2' \tag{18.3-8}$$

We next apply the boundary conditions

$$\text{B.C. 1:} \qquad \text{at } r = R, \qquad x_A = x_{AR} \tag{18.3-9}$$

$$\text{B.C. 2:} \qquad \text{as } r \to \infty, \qquad x_A \to x_{A\infty} \tag{18.3-10}$$

This enables us to obtain $C_1' = -R(x_{AR} - x_{A\infty})$ and $C_2' = x_{A\infty}$. Using these constants in Eq. 18.3-8, we get

$$\boxed{\frac{x_A(r) - x_{A\infty}}{x_{AR} - x_{A\infty}} = \frac{R}{r}} \tag{18.3-11}$$

which shows the concentration profile at steady state as A diffuses outward from the sphere. Furthermore,

$$N_{Ar}|_{r=R} = -c\mathscr{D}_{AB}\left.\frac{dx_A}{dr}\right|_{r=R} = +c\mathscr{D}_{AB}\frac{x_{AR} - x_{A\infty}}{R} \tag{18.3-12}$$

which is the molar flux at the sphere surface.

§18.4 DIFFUSION WITH A HOMOGENEOUS CHEMICAL REACTION

As the next illustration of setting up a mass balance, we consider the system shown in Fig. 18.4-1. Here gas A dissolves sparingly in liquid B in a beaker and diffuses isothermally into the liquid phase. As it diffuses, A also undergoes an irreversible first-order homogeneous reaction: $A + B \to AB$, with the moles of A consumed per unit time per unit volume given by $R_A = k_1'''c_A$, where k_1''' is a first-order rate constant. An example of such a system would be the absorption of CO_2 by a concentrated aqueous solution of NaOH.

We treat this as a binary solution of A and B, ignoring the small amount of AB that is present (the *pseudobinary assumption*). Then the steady-state mass balance on species A over a thickness Δz of the liquid phase becomes

$$N_{Az}|_z S - N_{Az}|_{z+\Delta z} S - k_1'''c_A S\Delta z = 0 \tag{18.4-1}$$

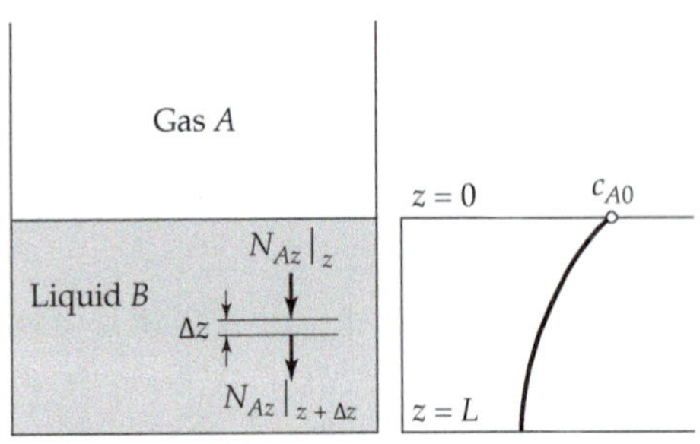

Fig. 18.4-1. Absorption of A by B with a homogeneous reaction in the liquid phase.

where S is the cross-sectional area of the liquid. Division of Eq. 18.4-1 by $S\Delta z$ and taking the limit as $\Delta z \to 0$ gives

$$\frac{dN_{Az}}{dz} + k_1'''c_A = 0 \tag{18.4-2}$$

If the concentration of A is very small, then we may to a good approximation neglect the convection term in Eq. 18.0-1 and write

$$N_{Az} = -\mathscr{D}_{AB}\frac{dc_A}{dz} \tag{18.4-3}$$

since the total molar concentration c is virtually uniform throughout the liquid. Combining the last two equations gives

$$\mathscr{D}_{AB}\frac{d^2c_A}{dz^2} - k_1'''c_A = 0 \tag{18.4-4}$$

This is to be solved with the following boundary conditions:

B.C. 1: at $z = 0$, $c_A = c_{A0}$ (18.4-5)

B.C. 2: at $z = L$, $N_{Az} = 0$ (or $dc_A/dz = 0$) (18.4-6)

The first boundary condition asserts that the concentration of A at the surface in the liquid is maintained at a fixed value c_{A0}. The second states that no A diffuses through the bottom of the container at $z = L$.

If Eq. 18.4-4 is multiplied by $L^2/c_{A0}\mathscr{D}_{AB}$, then it can be written in terms of dimensionless variables in the form of Eq. C.1-4

$$\frac{d^2\Gamma}{d\zeta^2} - \phi^2\Gamma = 0 \tag{18.4-7}$$

where $\Gamma = c_A/c_{A0}$ is a dimensionless concentration, $\zeta = z/L$ is a dimensionless length, and $\phi = \sqrt{k_1''' L^2/\mathscr{D}_{AB}}$ is a dimensionless group, known as the *Thiele modulus*.[1] This group represents the relative influence of the rates of chemical reaction $k_1'''c_A$ and diffusion $c_A\mathscr{D}_{AB}/L^2$. Equation 18.4-7 is to be solved with the dimensionless boundary conditions that at $\zeta = 0$, $\Gamma = 1$, and at $\zeta = 1$, $d\Gamma/d\zeta = 0$. The general solution is (see §C.1, as well as the discussion of hyperbolic functions in §C.5)

$$\Gamma(\zeta) = C_1 \cosh \phi\zeta + C_2 \sinh \phi\zeta \tag{18.4-8}$$

---

[1]E. W. Thiele, *Ind. Eng. Chem.*, **31**, 916–920 (1939). **Ernest William Thiele** (pronounced "tee-lee") (1895–1993) is noted for his work on catalyst effectiveness factors and his part in the development of the "McCabe-Thiele" diagram; after 35 years with Standard Oil of Indiana, he taught for a decade at Notre Dame University.

When the constants of integration are evaluated, we find that $C_1 = 1$ and $C_2 = -(\sinh\phi/\cosh\phi)$, so that

$$\Gamma(\zeta) = \frac{\cosh\phi\,\cosh\phi\zeta - \sinh\phi\,\sinh\phi\zeta}{\cosh\phi} = \frac{\cosh[\phi(1-\zeta)]}{\cosh\phi} \tag{18.4-9}$$

Then reverting to the original notation

$$\boxed{\frac{c_A(z)}{c_{A0}} = \frac{\cosh\left[\sqrt{k_1''' L^2/\mathscr{D}_{AB}}\,(1-(z/L))\right]}{\cosh\sqrt{k_1''' L^2/\mathscr{D}_{AB}}}} \tag{18.4-10}$$

The concentration profile thus obtained is plotted in Fig. 18.4-1.

Once we have the concentration profile, we may evaluate other quantities, such as the average concentration in the liquid phase

$$\begin{aligned}\frac{c_{A,\text{avg}}}{c_{A0}} &= \frac{\int_0^L (c_A/c_{A0})dz}{\int_0^L dz} \\ &= \frac{1}{\phi\cosh\phi}\int_0^\phi \cosh x\,dx = \frac{1}{\phi\cosh\phi}\sinh|_0^\phi = \frac{\tanh\phi}{\phi}\end{aligned} \tag{18.4-11}$$

in which we used the variable $x = \phi[1-(z/L)]$. Furthermore, the molar flux at plane $z = 0$ can be found to be (from Eq. 18.4-3)

$$N_{Az}|_{z=0} = -\mathscr{D}_{AB}\left.\frac{dc_A}{dz}\right|_{z=0} = \left(\frac{c_{A0}\mathscr{D}_{AB}}{L}\right)\phi\tanh\phi \tag{18.4-12}$$

This result shows how the chemical reaction influences the rate of absorption of a gas A by liquid B.

The reader may wonder how the solubility c_{A0} and the diffusivity $\mathscr{D}_{AB}$ can be determined experimentally if there is a chemical reaction occurring. First, k_1''' can be measured in a separate experiment in a well-stirred vessel. Then, in principle, c_{A0} and $\mathscr{D}_{AB}$ can be obtained from the measured absorption rates for various liquid depths L.

EXAMPLE 18.4-1

Gas Absorption with Chemical Reaction in an Agitated Tank[2]

Estimate the effect of chemical reaction rate on the rate of gas absorption in an agitated tank (see Fig. 18.4-2). Consider a system in which the dissolved gas A undergoes an irreversible

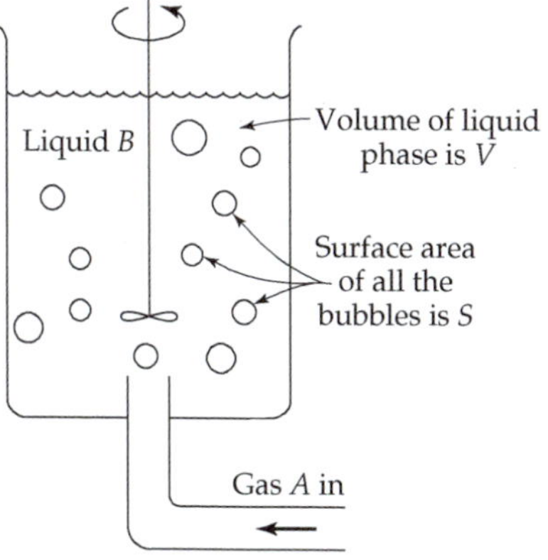

Fig. 18.4-2. Gas-absorption apparatus.

[2]E. N. Lightfoot, *AIChE Journal*, **4**, 499–500 (1958), **8**, 710–712 (1962).

first-order reaction with the liquid B—that is, A disappears within the liquid phase at a rate proportional to the local concentration of A. An example of such a system would be the absorption of SO_2 or H_2S in aqueous NaOH solutions.

SOLUTION

An exact analysis of this situation is not possible because of the complexity of the gas-absorption process. But a useful semiquantitative understanding can be obtained by the analysis of a relatively simple model, in which a spherical coordinate system is fixed on a bubble. The model we use involves the following *assumptions*:

- **a.** Each gas bubble is surrounded by a stagnant liquid film of thickness δ, which is small relative to the bubble diameter.
- **b.** A quasi-steady concentration profile is quickly established in the liquid film after the bubble is formed.
- **c.** The gas A is only sparingly soluble in the liquid, so that we can neglect the convection term in Eq. 18.0-1.
- **d.** The liquid outside the stagnant film is at concentration $c_{A\delta}$, which changes so slowly with respect to time that it can be considered constant.

The differential equation describing the diffusion with chemical reaction is the same as that in Eq. 18.4-4, but the boundary conditions are now

B.C. 1: $$\text{at } z = 0, \qquad c_A = c_{A0} \tag{18.4-13}$$

B.C. 2: $$\text{at } z = \delta, \qquad c_A = c_{A\delta} \tag{18.4-14}$$

The concentration c_{A0} is the interfacial concentration of A in the liquid phase, which is assumed to be at equilibrium with the gas phase at the interface, and $c_{A\delta}$ is the concentration of A in the main body of the liquid. The solution of Eq. 18.4-4 with these boundary conditions is

$$\frac{c_A(\zeta)}{c_{A0}} = \frac{\sinh\phi\cosh\phi\zeta + (B - \cosh\phi)\sinh\phi\zeta}{\sinh\phi} \tag{18.4-15}$$

in which $\zeta = z/\delta$, $B = c_{A\delta}/c_{A0}$, and $\phi = \sqrt{k_1''' \delta^2/\mathscr{D}_{AB}}$. This result is plotted in Fig. 18.4-3.

Next we use assumption (d) above and equate the amount of A entering the main body of liquid at $z = \delta$ over the total bubble surface S in the tank to the amount of A consumed in the bulk of the liquid by chemical reaction (that is, we consider this as a quasi-steady-state problem):

$$-S\mathscr{D}_{AB}\left.\frac{dc_A}{dz}\right|_{z=\delta} = Vk_1'''c_{A\delta} \tag{18.4-16}$$

Substitution of $c_A(z)$ from Eq. 18.4-15 into Eq. 18.4-16 gives an expression for B

$$B = \frac{1}{\cosh\phi + (V/S\delta)\phi\sinh\phi} \tag{18.4-17}$$

When this result is inserted into Eq. 18.4-15, we obtain an expression for c_A/c_{A0} in terms of ϕ and $V/S\delta$.

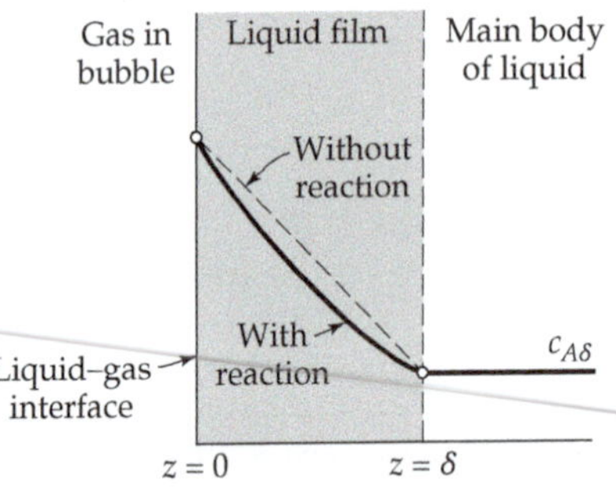

Fig. 18.4-3. Predicted concentration profile in the liquid film near the bubble.

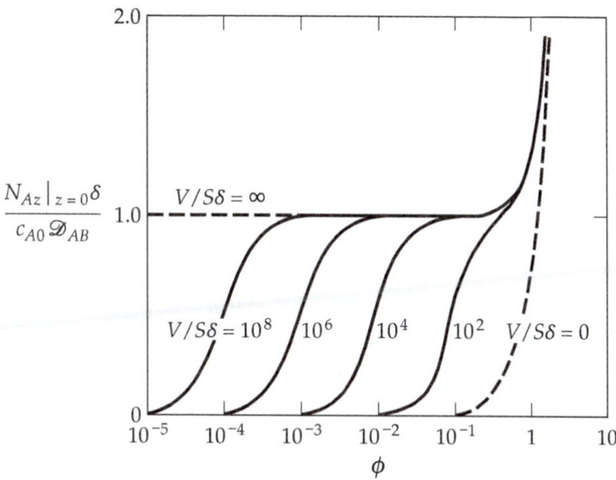

Fig. 18.4-4. Gas absorption accompanied by an irreversible first-order reaction. [Adapted from E. N. Lightfoot, *AIChE J.*, **4**, 499–500 (1958).]

From this expression for the concentration profile, we can then get the total rate of absorption with chemical reaction from $N_{Az} = -\mathscr{D}_{AB}(dc_A/dz)$ evaluated at $z = 0$, thus:

$$\check{N} = \frac{N_{Az}|_{z=0}\delta}{c_{A0}\mathscr{D}_{AB}} = \phi\left(\frac{\sinh\phi + (V/S\delta)\,\phi\cosh\phi}{\cosh\phi + (V/S\delta)\phi\sinh\phi}\right) \tag{18.4-18}$$

The result is plotted in Fig. 18.4-4.

It is seen here that the dimensionless absorption rate per unit area of interface, $\check{N}$, increases with ϕ for all finite values of $V/S\delta$. At very low values of ϕ—that is, for very slow reactions—$\check{N}$ approaches zero. For this limiting situation the liquid is nearly saturated with dissolved gas, and the "driving force" for absorption is very small. At large values of ϕ the dimensionless surface mass flux $\check{N}$ increases rapidly with ϕ and becomes very nearly independent of $V/S\delta$. Under the latter circumstances, the reaction is so rapid that almost all of the dissolving gas is consumed within the film. Then B is very nearly zero, and the bulk of the liquid plays no significant role. In the limit as ϕ becomes very large, $\check{N}$ approaches ϕ.

Somewhat more interesting behavior is observed for intermediate values of ϕ. It may be noted that, for moderately large $V/S\delta$, there is a considerable range of ϕ for which $\check{N}$ is very nearly unity. In this region, the chemical reaction is fast enough to keep the bulk of the solution almost solute free, but slow enough to have little effect on solute transport in the film. Such a situation will arise when the ratio $V/S\delta$ of bulk to film volume is sufficient to offset the higher volumetric reaction rate in the film. The absorption rate is then equal to the physical absorption rate (that is, the rate for $k_1''' = 0$) for a solute-free tank. This behavior is frequently observed in practice, and operation under such conditions has proven a useful means of characterizing mass-transfer behavior of a variety of gas absorbers.[2]

---

§18.5 DIFFUSION WITH A HETEROGENEOUS CHEMICAL REACTION

Let us now consider a simple model for a heterogeneous catalytic reactor, such as that shown in Fig. 18.5-1(*a*), in which a reaction $2A \rightarrow B$ is being carried out. An example of a reaction of this type would be the solid-catalyzed dimerization of $CH_3CH = CH_2$.

We imagine that each catalyst particle is surrounded by a stagnant gas film through which A has to diffuse to reach the catalyst surface, as shown in Fig. 18.5-1(*b*). At the catalyst surface we presume that the reaction $2A \rightarrow B$ occurs instantaneously, and that

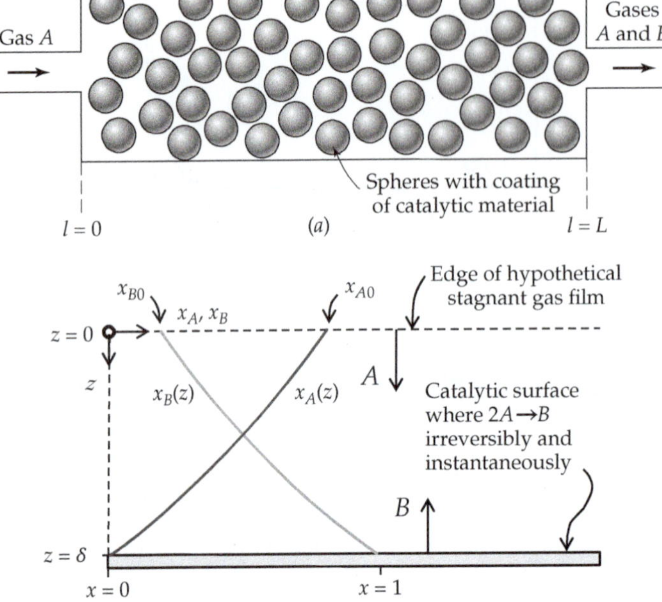

Fig. 18.5-1. (*a*) Schematic diagram of a catalytic reactor in which *A* is being converted to *B*; (*b*) Idealized picture (or "model") of the diffusion problem near a catalyst particle.

the product *B* then diffuses back out through the gas film to the main turbulent stream composed of *A* and *B*. We want to get an expression for the local rate of conversion from *A* to *B* when the effective gas-film thickness and the main stream concentrations x_{A0} and x_{B0} are known. We *assume* that the gas film is isothermal, although in many catalytic reactions the heat generated by the reaction cannot be neglected.

We solve this problem in two stages: (a) with the assumption that *A* is present in such a small concentration that the convection term in Eq. 18.0-1 may be safely omitted; and (b) without the assumption of a very small concentration so that the convection term in Eq. 18.0-1 must be included.

a. With the assumption of very small concentration of A

We start by making a mass balance on species *A* over a thin slab of thickness Δz in the gas film. This leads directly to the equation (see Eq. 18.6-2 below)

$$\frac{dN_{Az}}{dz} = 0 \tag{18.5-1}$$

When Eq. 18.0-1, without the convection term, is inserted into this mass balance, we get for constant *c* and $\mathscr{D}_{AB}$

$$\frac{d^2 x_A}{dz^2} = 0 \tag{18.5-2}$$

This may be integrated to give

$$x_A(z) = C_1 + C_2 z \tag{18.5-3}$$

We then apply the boundary conditions

B.C. 1: at $z = 0$, $x_A = x_{A0}$ (18.5-4)

B.C. 2: at $z = \delta$, $x_A = 0$ (18.5-5)

From these conditions we get the integration constants $C_1 = x_{A0}$ and $C_2 = -x_{A0}/\delta$, so that

$$x_A(z) = x_{A0}\left(1 - \frac{z}{\delta}\right) \tag{18.5-6}$$

From this we get the molar flux through the gas film

$$N_{Az} = -c\mathscr{D}_{AB}\frac{dx_A}{dz} = \frac{c\mathscr{D}_{AB}x_{A0}}{\delta} \tag{18.5-7}$$

This may be interpreted as the local rate of reaction per unit area of catalyst surface.

One point deserves to be emphasized. Although the chemical reaction occurs instantaneously at the catalytic surface, the conversion of A to B proceeds at a finite rate because of the diffusion process that is "in series" with the reaction process. Hence we speak of the conversion of A to B as being *diffusion controlled.*

b. Without the assumption of very small concentration of A

Once again we start with the molar balance over a thin region of thickness Δz, which leads to Eq. 18.5-1. Into this we have to substitute the expression for the molar flux of A.

Because x_A is not small, we must retain the convection term in the flux expression (Eq. 18.0-1). To obtain a molar balance for species A that does not contain N_{Bz}, we must get information about the ratio N_{Bz}/N_{Az}. We can get this from the stoichiometry of the reaction. We know that there is *one* mole of B moving in the *minus* z direction for every *two* moles of A moving in the *plus* z direction. Therefore, at steady state

$$N_{Bz} = -\tfrac{1}{2}N_{Az} \tag{18.5-8}$$

at any value of z. Therefore, Eq. 18.0-1 can be rewritten for this problem as

$$\begin{aligned} N_{Az} &= x_A(N_{Az} + N_{Bz}) - c\mathscr{D}_{AB}\frac{dx_A}{dz} = x_A\left(N_{Az} - \tfrac{1}{2}N_{Az}\right) - c\mathscr{D}_{AB}\frac{dx_A}{dz} \\ &= \tfrac{1}{2}x_A N_{Az} - c\mathscr{D}_{AB}\frac{dx_A}{dz} \end{aligned} \tag{18.5-9}$$

From this we may solve for N_{Az} in the form of Eq. 18.0-3,

$$N_{Az}(z) = -\frac{c\mathscr{D}_{AB}}{1 - \frac{1}{2}x_A}\frac{dx_A}{dz} \tag{18.5-10}$$

This is now inserted into the molar balance (Eq. 18.5-1), which gives (for constant c and $\mathscr{D}_{AB}$)

$$\frac{d}{dz}\left(\frac{1}{1 - \frac{1}{2}x_A}\frac{dx_A}{dz}\right) = 0 \tag{18.5-11}$$

Integration twice with respect to z gives

$$-2\ln\left(1 - \tfrac{1}{2}x_A(z)\right) = C_1 z + C_2 = -(2\ln K_1)z - (2\ln K_2) \tag{18.5-12}$$

or, in terms of the alternative integration constants K_1 and K_2,

$$1 - \tfrac{1}{2}x_A(z) = K_1^z K_2 \tag{18.5-13}$$

Then, from the boundary conditions

B.C. 1: at $z = 0$, $x_A = x_{A0}$ (18.5-14)

B.C. 2: at $z = \delta$, $x_A = 0$ (18.5-15)

we find $K_1 = \left(1 - \frac{1}{2}x_{A0}\right)^{-1/\delta}$ and $K_2 = \left(1 - \frac{1}{2}x_{A0}\right)$, so that the final result is then

$$\boxed{1 - \tfrac{1}{2}x_A(z) = \left(1 - \tfrac{1}{2}x_{A0}\right)^{1-(z/\delta)}} \tag{18.5-16}$$

for the concentration profile for species A in the gas film. Equation 18.5-16 may now be substituted into Eq. 18.5-10 to get the molar flux of reactant through the film as follows:

$$\begin{aligned} N_{Az} &= -\frac{c\mathscr{D}_{AB}}{\left(1-\frac{1}{2}x_{A0}\right)^{1-(z/\delta)}}(-2)\frac{d}{dz}\left(1-\frac{1}{2}x_{A0}\right)^{1-(z/\delta)} \\ &= +2c\mathscr{D}_{AB}\frac{d}{dz}\ln\left(1-\frac{1}{2}x_{A0}\right)^{1-(z/\delta)} \\ &= +2c\mathscr{D}_{AB}\ln\left(1-\frac{1}{2}x_{A0}\right)\cdot\frac{d}{dz}\left(1-\left(\frac{z}{\delta}\right)\right) \\ &= +2c\mathscr{D}_{AB}\ln\left(1-\frac{1}{2}x_{A0}\right)\cdot\left(-\frac{1}{\delta}\right) \\ &= \frac{2c\mathscr{D}_{AB}}{\delta}\ln\left(\frac{1}{1-\frac{1}{2}x_{A0}}\right) \end{aligned} \tag{18.5-17}$$

The quantity N_{Az} may also be interpreted as the local rate of reaction per unit area of catalytic surface. This information can be combined with other information about the catalytic reactor sketched in Fig. 18.5-1(a) to get the overall conversion rate in the entire reactor. (Note that Eq. 18.5-7, which is valid for small x_{A0}, may be obtained from Eq. 18.5-17 by expanding the logarithm using Eq. C.2-3 and retaining only the first term.)

Here again, although the chemical reaction occurs instantaneously at the catalytic surface, the conversion of A to B proceeds at a finite rate because this is still a diffusion-controlled reaction.

In the example above we have assumed that the reaction occurs instantaneously at the catalytic surface. In the following example, we show how to account for a reaction at the catalytic surface that does not occur infinitely fast.

EXAMPLE 18.5-1

Diffusion with a Slow Heterogeneous Reaction

Rework the problem just considered when the reaction $2A \rightarrow B$ is not instantaneous at the catalytic surface at $z = \delta$. Instead, assume that the rate at which A disappears at the catalyst-coated surface is proportional to the concentration of A in the fluid at the interface,

$$N_{Az} = k_1''c_A = k_1''cx_A \tag{18.5-18}$$

in which k_1'' is a rate constant for the pseudo-first-order surface reaction.

SOLUTION

We proceed exactly as before, except that B.C. 2 in Eq. 18.5-15 must be replaced by

$$\text{B.C. 2}': \qquad \text{at } z = \delta, \qquad x_A = \frac{N_{Az}}{k_1''c} \tag{18.5-19}$$

N_{Az} being, of course, a constant at steady state. The determination of the integration constants from B.C. 1 and B.C. 2′ gives

$$K_1 = \left[\frac{1-\frac{1}{2}\left(N_{Az}/k_1''c\right)}{1-\frac{1}{2}x_{A0}}\right]^{1/\delta}; \qquad K_2 = 1-\frac{1}{2}x_{A0} \tag{18.5-20}$$

so that

$$1-\frac{1}{2}x_A(z) = \left(1-\frac{1}{2}\frac{N_{Az}}{k_1''c}\right)^{z/\delta}\left(1-\frac{1}{2}x_{A0}\right)^{1-(z/\delta)} \tag{18.5-21}$$

From this we evaluate $(dx_A/dz)|_{z=0}$ and substitute it into Eq. 18.5-10, evaluated at $z = 0$, to get

$$N_{Az} = \frac{2c\mathscr{D}_{AB}}{\delta}\ln\left(\frac{1-\frac{1}{2}\left(N_{Az}/k_1''c\right)}{1-\frac{1}{2}x_{A0}}\right) \tag{18.5-22}$$

This is a transcendental equation for N_{Az} as a function of x_{A0}, k_1'', $c\mathscr{D}_{AB}$, and δ. When k_1'' is large, the logarithm of $1 - \frac{1}{2}(N_{Az}/k_1''c)$ may be expanded in a Taylor series (see Eq. C.2-3) and all terms discarded but the first. This gives

$$\begin{aligned} N_{Az} &= \frac{2c\mathscr{D}_{AB}}{\delta}\left[\ln\left(1 - \tfrac{1}{2}\left(N_{Az}/k_1''c\right)\right) - \ln\left(1 - \tfrac{1}{2}x_{A0}\right)\right] \\ &= \frac{2c\mathscr{D}_{AB}}{\delta}\left[-\tfrac{1}{2}\left(N_{Az}/k_1''c\right) - \tfrac{1}{4}\left(N_{Az}/k_1''c\right)^2 - \cdots - \ln\left(1 - \tfrac{1}{2}x_{A0}\right)\right] \\ &\approx \frac{2c\mathscr{D}_{AB}}{\delta}\left[-\left(\frac{1}{2}\frac{N_{Az}}{k_1''c}\right) + \ln\left(\frac{1}{1 - \frac{1}{2}x_{A0}}\right)\right] \end{aligned} \tag{18.5-23}$$

We can now solve for N_{Az} to get

$$N_{Az} = \frac{2c\mathscr{D}_{AB}/\delta}{1 + \mathscr{D}_{AB}/k_1''\delta}\ln\left(\frac{1}{1 - \frac{1}{2}x_{A0}}\right) \quad (k_1'' \text{ large}) \tag{18.5-24}$$

Note once again that we have obtained the rate of the *combined* reaction and diffusion process. We see that the dimensionless group $\mathscr{D}_{AB}/k_1''\delta$ describes the effect of the surface reaction kinetics on the overall diffusion–reaction process. The reciprocal of this group is known as the *second Damköhler number*[1] $\text{Da}^{\text{II}} = k_1''\delta/\mathscr{D}_{AB}$. Evidently we get the result in Eq. 18.5-17 in the limit as $\text{Da}^{\text{II}} \to \infty$.

§18.6 DIFFUSION THROUGH A STAGNANT GAS FILM

Let us now analyze the diffusion-convection system shown in Fig. 18.6-1 in which liquid A is evaporating into gas B. We imagine that there is some device that maintains the liquid level at $z = z_1$. Right at the liquid–gas interface, the gas-phase concentration of A, expressed as mole fraction, is x_{A1}. This is taken to be the gas-phase concentration of A

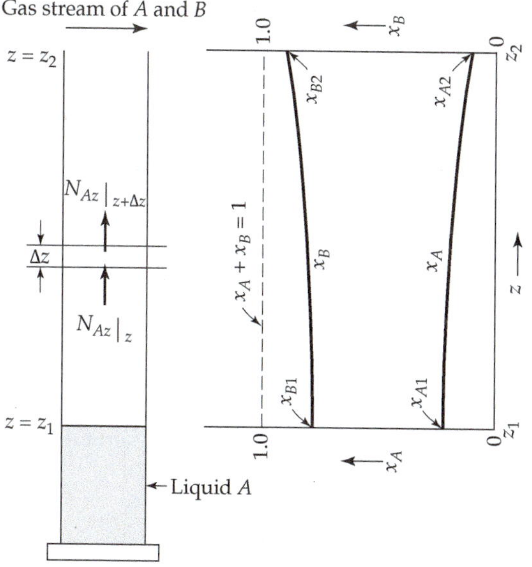

Fig. 18.6-1. Steady-state diffusion of A through stagnant B with the liquid–vapor interface maintained at a fixed position. The graph shows how the concentration profiles deviate from straight lines, because of the convective contribution to the mass flux.

[1] G. Damköhler, *Z. Elektrochem.*, **42**, 846–862 (1936). **Gerhard Damköhler (1908–1944)** was a key figure in the subject of diffusion with chemical reaction; an important publication was in Der Chemie-Ingenieur, Leipzig (1937), pp. 359–485.

corresponding to equilibrium[1] with the liquid at the interface. That is, x_{A1} is the vapor pressure of A divided by the total pressure, p_A^{vap}/p, provided that A and B form an ideal-gas mixture and that the solubility of gas B in liquid A is negligible.

A stream of gas mixture A-B of concentration x_{A2} flows slowly past the top of the tube, to maintain the mole fraction of A at x_{A2} at $z = z_2$. The entire system is kept at constant temperature and pressure. Gases A and B are assumed to be ideal.

We know that there will be a net flow of gas upward from the gas–liquid interface, and that the gas velocity near the cylinder wall will be smaller than that in the center of the tube. To simplify the problem, we ignore this effect and *assume* that there is no dependence of the z component of the velocity on the radial coordinate.

A steady-state mass balance (in molar units) over an increment Δz of the column states that the amount of A entering at plane z equals the amount of A leaving at plane $z + \Delta z$:

$$SN_{Az}|_z - SN_{Az}|_{z+\Delta z} = 0 \tag{18.6-1}$$

Here S is the cross-sectional area of the column. Division by $S\Delta z$ and taking the limit as $\Delta z \to 0$ gives

$$-\frac{dN_{Az}}{dz} = 0 \tag{18.6-2}$$

When this evaporating system attains a steady state, there is a net motion of A away from the interface and the species B is stationary. Hence, the molar flux of A is given by Eq. 18.0-1 with $N_{Bz} = 0$, which can therefore be written in the form of Eq. 18.0-2, since x_A is not small,

$$N_{Az} = x_A N_{Az} - c\mathscr{D}_{AB}\frac{dx_A}{dz} \quad \text{or} \quad N_{Az}(z) = -\frac{c\mathscr{D}_{AB}}{1 - x_A}\frac{dx_A}{dz} \tag{18.6-3}$$

Substitution of Eq. 18.6-3 into Eq. 18.6-2 then gives

$$\frac{d}{dz}\left(\frac{c\mathscr{D}_{AB}}{1 - x_A}\frac{dx_A}{dz}\right) = 0 \tag{18.6-4}$$

For an ideal-gas mixture, the equation of state is $p = cRT$, so that, at constant temperature and pressure, c must be a constant. Furthermore, for gases, $\mathscr{D}_{AB}$ is very nearly independent of the concentration, and hence is also position-independent. Therefore, $c\mathscr{D}_{AB}$ can be moved to the left of the derivative operator to get

$$\frac{d}{dz}\left(\frac{1}{1 - x_A}\frac{dx_A}{dz}\right) = 0 \tag{18.6-5}$$

This is a second-order differential equation for the concentration profile expressed as mole fraction of A. Integration with respect to z gives

$$\frac{1}{1 - x_A}\frac{dx_A}{dz} = C_1 \tag{18.6-6}$$

A second integration then gives

$$-\ln(1 - x_A(z)) = C_1 z + C_2 \tag{18.6-7}$$

If we replace C_1 by $-\ln K_1$ and C_2 by $-\ln K_2$, and then take the exponential of both sides, Eq. 18.6-7 becomes

$$1 - x_A(z) = K_1^z K_2 \tag{18.6-8}$$

---

[1] L. J. Delaney and L. C. Eagleton [*AIChE Journal*, **8**, 418–420 (1962)] conclude that, for evaporating systems, the interfacial equilibrium assumption is reasonable, with errors in the range 1.3% to 7.0% possible.

The two constants of integration, K_1 and K_2, may then be determined from the boundary conditions

$$\text{B.C. 1:} \qquad \text{at } z = z_1, \qquad x_A = x_{A1} \tag{18.6-9}$$

$$\text{B.C. 2:} \qquad \text{at } z = z_2, \qquad x_A = x_{A2} \tag{18.6-10}$$

From these boundary conditions, the constants are found to be

$$K_1 = \left(\frac{1 - x_{A2}}{1 - x_{A1}}\right)^{\frac{1}{z_2 - z_1}} \qquad K_2 = \left(\frac{1 - x_{A2}}{1 - x_{A1}}\right)^{-\frac{z_1}{z_2 - z_1}}(1 - x_{A1}) \tag{18.6-11, 12}$$

so that from Eqs. 18.6-8, 18.6-11, and 18.6-11, we get finally

$$\boxed{\left(\frac{1 - x_A(z)}{1 - x_{A1}}\right) = \left(\frac{1 - x_{A2}}{1 - x_{A1}}\right)^{\frac{z - z_1}{z_2 - z_1}}} \tag{18.6-13}$$

The profiles for gas B are obtained by using $x_B = 1 - x_A$. The concentration profiles are shown in Fig. 18.6-1. It can be seen there that the slope dx_A/dz is not constant, although N_{Az} is; this could have been anticipated from Eq. 18.6-3.

Once the concentration profiles are known, we can get average values and mass fluxes at surfaces. For example, the average concentration of B in the region between z_1 and z_2 is obtained as follows:

$$\frac{x_{B,\text{avg}}}{x_{B1}} = \frac{\int_{z_1}^{z_2} (x_B/x_{B1})dz}{\int_{z_1}^{z_2} dz} = \frac{\int_0^1 (x_{B2}/x_{B1})^{\zeta} d\zeta}{\int_0^1 d\zeta} = \left.\frac{(x_{B2}/x_{B1})^{\zeta}}{\ln(x_{B2}/x_{B1})}\right|_0^1 \tag{18.6-14}$$

in which $\zeta = (z - z_1)/(z_2 - z_1)$ is a dimensionless length variable. Equation 18.6-14 may be rewritten as

$$x_{B,\text{avg}} = \frac{x_{B2} - x_{B1}}{\ln(x_{B2}/x_{B1})} \tag{18.6-15}$$

That is, the average value of x_B is the logarithmic mean of the terminal concentrations, i.e., $(x_B)_{\ln} = (x_{B2} - x_{B1})/(\ln x_{B2} - \ln x_{B1})$.

The rate of mass transfer at the liquid–gas interface—that is, the rate of evaporation of A—may be obtained from Eq. 18.6-3 as follows:

$$N_{Az}|_{z=z_1} = -\frac{c\mathscr{D}_{AB}}{1 - x_{A1}} \left.\frac{dx_A}{dz}\right|_{z=z_1} = +\frac{c\mathscr{D}_{AB}}{x_{B1}} \left.\frac{dx_B}{dz}\right|_{z=z_1} = \frac{c\mathscr{D}_{AB}}{z_2 - z_1} \ln\left(\frac{x_{B2}}{x_{B1}}\right) \tag{18.6-16}$$

By combining Eqs. 18.6-15 and 18.6-16 we get finally

$$N_{Az}|_{z=z_1} = \frac{c\mathscr{D}_{AB}}{(z_2 - z_1)(x_B)_{\ln}}(x_{A1} - x_{A2}) \tag{18.6-17}$$

This expression gives the evaporation rate in terms of the characteristic driving force $x_{A1} - x_{A2}$.

By expanding the solution in Eq. 18.6-17 in a Taylor series, we can get (see §C.2 and Problem 18B.17)

$$N_{Az}|_{z=z_1} = \frac{c\mathscr{D}_{AB}(x_{A1} - x_{A2})}{(z_2 - z_1)} \left[1 + \tfrac{1}{2}(x_{A1} + x_{A2}) + \tfrac{1}{3}(x_{A1}^2 + x_{A1}x_{A2} + x_{A2}^2) + \cdots\right] \tag{18.6-18}$$

The quantity in front of the bracketed expression is the simple result that one would get if the convection term were entirely omitted in Eq. 18.0-1. The bracketed expression then gives the correction resulting from including the convection term. Another way of interpreting Eq. 18.6-18 is that the simple result corresponds to joining the end points of the

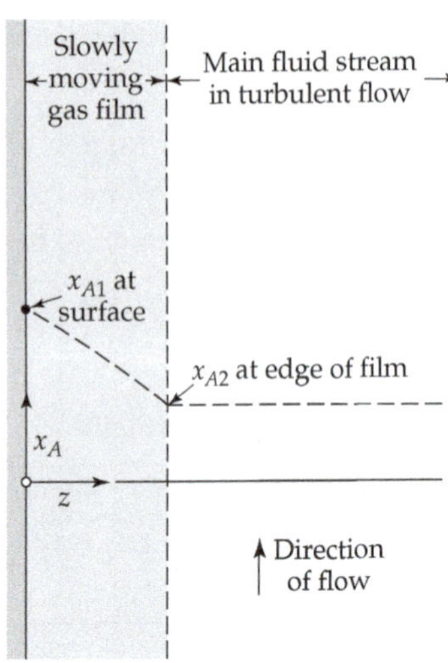

Fig. 18.6-2. Film model for mass transfer; component A is diffusing from the surface into the gas stream through a hypothetical stagnant gas film.

curves in Fig. 18.6-1 by a straight line, and the complete result corresponds to using the curve of x_A versus z. If the terminal mole fractions are small, the correction term in brackets in Eq. 18.6-18 is only slightly greater than unity.

The results of this section have been used for the experimental determinations of gas diffusivities.[2] Furthermore, these results are used in the "film models" of mass transfer. In Fig. 18.6-2 a solid or liquid surface is shown along which a gas is flowing. Near the surface is a slowly moving film through which A diffuses. This film is bounded by the surfaces $z = z_1$ and $z = z_2$. In this "model" it is assumed that there is a sharp transition from a stagnant film to a well-mixed fluid in which the concentration gradients are negligible. Although this model is physically unrealistic, it has nevertheless proven useful as a simplified picture for correlating mass-transfer coefficients.

EXAMPLE 18.6-1

Evaporation with a Moving Interface

We want now to examine a problem that is slightly different from the one just discussed. Instead of maintaining the liquid–gas interface at a constant height, we allow the liquid level to subside as the evaporation proceeds, as shown in Fig. 18.6-3. Since the liquid retreats very slowly, we can use a *quasi-steady-state* method with confidence.

SOLUTION

First we equate the rate at which moles of A enter the gas phase to the molar rate of evaporation of A from the liquid phase, the latter being given by Eq. 18.6-17 with z_1 replaced by $z_1(t)$

$$S\frac{\rho^{(A)}}{M_A}\left(-\frac{dz_1}{dt}\right) = \frac{c\mathscr{D}_{AB}}{(z_2 - z_1(t))(x_B)_{\ln}}(x_{A1} - x_{A2})S \tag{18.6-19}$$

Here $\rho^{(A)}$ is the density of pure liquid A, and M_A is the molecular weight. On the right side of Eq. 18.6-19, we have used the steady-state evaporation rate evaluated at the current liquid column height (this is the quasi-steady-state approximation). We first rewrite Eq. 18.6-19 in terms of $h(t) = z_1(0) - z_1(t)$, the distance that the interface has descended in time t, and $H = z_2 - z_1(0)$, the initial height of the gas column. Then the equation can be integrated to give

$$\int_0^h (H + \bar{h})d\bar{h} = \frac{c\mathscr{D}_{AB}(x_{A1} - x_{A2})}{(\rho^{(A)}/M_A)(x_B)_{\ln}}\int_0^t d\bar{t} \tag{18.6-20}$$

---

[2] C. Y. Lee and C. R. Wilke, *Ind. Eng. Chem.*, **46**, 2381–2387 (1954).

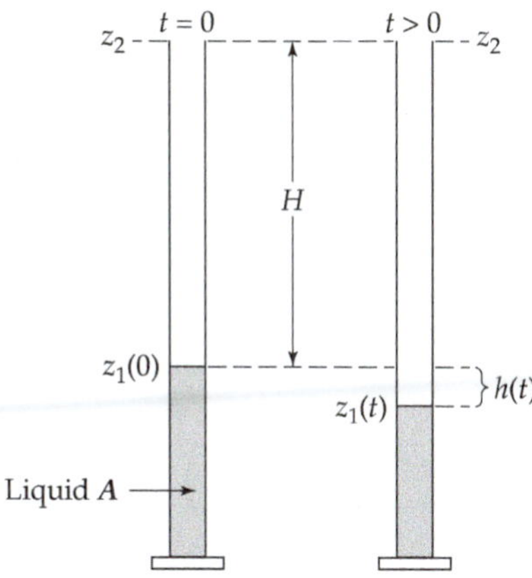

Fig. 18.6-3. Evaporation with quasi-steady-state diffusion. The liquid level goes down very slowly as the liquid evaporates. A gas mixture of composition x_{A2} flows across the top of the tube.

where the overbars on h and t indicate dummy variables of integration. When we abbreviate the entire right side of Eq. 18.6-20 by $\frac{1}{2}Ct$, the equation can be integrated and then solved for h to give

$$h(t) = H\left(\sqrt{1 + (Ct/H^2)} - 1\right) \tag{18.6-21}$$

In principle, one could use this experiment to get the diffusivity from measurements of the liquid level as a function of time.

EXAMPLE 18.6-2

Determination of Diffusivity

The diffusivity of the gas pair O_2 - CCl_4 is being determined by observing the steady-state evaporation of carbon tetrachloride into a tube containing oxygen, as shown in Fig. 18.6-1. The distance between the CCl_4 liquid level and the top of the tube is $z_2 - z_1 = 17.1$ cm. The total pressure on the system is 755 mm Hg, and the temperature is 0°C. The vapor pressure of CCl_4 at that temperature is 33.0 mm Hg and its liquid density is 1.629 g/cm^3. The cross-sectional area of the diffusion tube is 0.82 cm^2. It is found that 0.0208 cm^3 of liquid CCl_4 evaporates in a 10-hour period after steady state has been attained. It may be assumed that the partial pressure of CCl_4 at $z = z_2$ is zero. What is the diffusivity of the gas pair O_2 - CCl_4?

SOLUTION

Let A stand for CCl_4 and B for O_2. The molar flux of A is then

$$\begin{aligned} N_{Az} &= \frac{\text{moles of CCl}_4\text{ evaporated}}{\text{area} \cdot \text{time}} \\ &= \frac{(0.0208\text{ cm}^3)(1.629\text{ g/cm}^3)/(153.82\text{ g/g-mol})}{(0.82\text{ cm}^2)(3.6 \times 10^4\text{ s})} \\ &= 7.46 \times 10^{-9}\text{ g-mol/cm}^2 \cdot \text{s} \end{aligned} \tag{18.6-22}$$

Then from Eq. 18.6-16 we get

$$\begin{aligned} \mathscr{D}_{AB} &= \frac{(N_{Az}|_{z=z_1})(z_2 - z_1)}{c\ln(x_{B2}/x_{B1})} = \frac{(N_{Az}|_{z=z_1})(z_2 - z_1)RT}{p\ln(p_{B2}/p_{B1})} \\ &= \frac{(7.46 \times 10^{-9}\text{ g-mol/cm}^2 \cdot \text{s})(17.1\text{ cm})(82.06\text{ cm}^3\text{atm/g-mol} \cdot \text{K})(273\text{ K})}{((755/760)\text{ atm})(\ln(755/(755 - 33)))} \\ &= 0.0644\text{ cm}^2/\text{s} \end{aligned} \tag{18.6-23}$$

This method of determining gas-phase diffusivities suffers from several defects: the cooling of the liquid by evaporation, the concentration of nonvolatile impurities at the interface, the climbing of the liquid up the walls of the tube by surface tension, and the curvature of the meniscus.

EXAMPLE 18.6-3

Diffusion through a Nonisothermal Spherical Film

(a) Derive expressions for diffusion through a spherical shell that are analogous to Eq. 18.6-13 (concentration profile) and Eq. 18.6-16 (molar flux). The system under consideration is shown in Fig. 18.6-4.

(b) Extend these results to describe the diffusion in a nonisothermal film in which the temperature varies radially according to

$$\frac{T(r)}{T_1} = \left(\frac{r}{r_1}\right)^n \tag{18.6-24}$$

in which T_1 is the temperature at $r = r_1$. Assume as a *rough* approximation that $\mathscr{D}_{AB}$ varies as the $\frac{3}{2}$-power of the temperature:

$$\frac{\mathscr{D}_{AB}(T)}{\mathscr{D}_{AB,1}} = \left(\frac{T}{T_1}\right)^{3/2} \tag{18.6-25}$$

in which $\mathscr{D}_{AB,1}$ is the diffusivity at $T = T_1$. Problems of this kind arise in connection with drying of droplets and diffusion through gas films near spherical catalyst pellets.

The temperature distribution in Eq. 18.6-24 has been chosen solely for mathematical simplicity. This example is included to emphasize that, in nonisothermal systems, Eq. 18.0-1 is the correct starting point rather than $N_{Az} = -\mathscr{D}_{AB}(dc_A/dz) + x_A(N_{Az} + N_{Bz})$, as has been given in some textbooks. (See the discussion in the paragraph following Eq. 17.3-6.)

SOLUTION

(a) A steady-state mass balance on a spherical shell leads to

$$\frac{d}{dr}(r^2 N_{Ar}) = 0 \tag{18.6-26}$$

We now substitute into this equation the expression for the molar flux N_{Ar}, with N_{Br} set equal to zero, since B is insoluble in liquid A. The molar flux is thus the r component (spherical coordinates) analog of Eq. 18.0-2, which gives upon substitution in Eq. 18.6-26

$$\frac{d}{dr}\left(r^2 \frac{c\mathscr{D}_{AB}}{1 - x_A}\frac{dx_A}{dr}\right) = 0 \tag{18.6-27}$$

For *constant temperature* the product $c\mathscr{D}_{AB}$ is constant, and Eq. 18.6-27 may be integrated to give the concentration distribution

$$\left(\frac{1 - x_A(r)}{1 - x_{A1}}\right) = \left(\frac{1 - x_{A2}}{1 - x_{A1}}\right)^{\frac{(1/r_1) - (1/r)}{(1/r_1) - (1/r_2)}} \tag{18.6-28}$$

From Eq. 18.6-28 we can then get

$$W_A = 4\pi r_1^2 N_{Ar}|_{r=r_1} = \frac{4\pi c\mathscr{D}_{AB}}{(1/r_1) - (1/r_2)} \ln\left(\frac{1 - x_{A2}}{1 - x_{A1}}\right) \tag{18.6-29}$$

which is the molar flow of A across any spherical surface of radius r between r_1 and r_2.

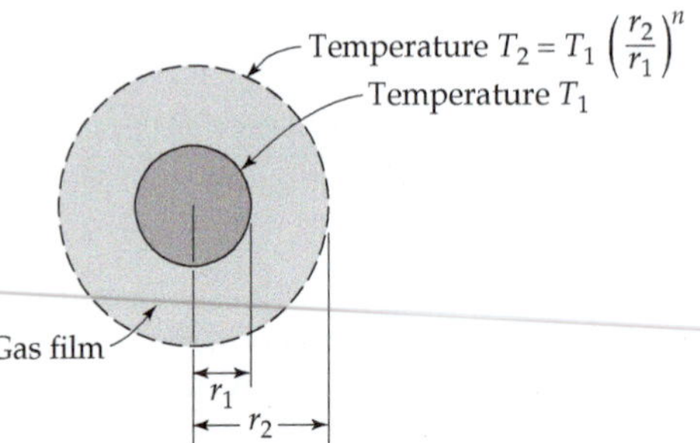

Fig. 18.6-4. Diffusion through a hypothetical spherical stagnant gas film surrounding a droplet of liquid A.

(b) For the nonisothermal problem, combination of Eqs. 18.6-24 and 18.6-25 gives the variation of diffusivity with position:

$$\frac{\mathscr{D}_{AB}(r)}{\mathscr{D}_{AB,1}} = \left(\frac{r}{r_1}\right)^{3n/2} \tag{18.6-30}$$

When this expression is inserted into Eq. 18.6-27 and c is set equal to p/RT, we get

$$\frac{d}{dr}\left(r^2 \frac{p\mathscr{D}_{AB,1}/RT_1}{1 - x_A}\left(\frac{r}{r_1}\right)^{n/2}\frac{dx_A}{dr}\right) = 0 \tag{18.6-31}$$

After integrating between r_1 and r_2, we obtain (for $n \neq -2$)

$$W_A = 4\pi r_1^2 N_{Ar}|_{r=r_1} = \frac{4\pi(p\mathscr{D}_{AB,1}/RT_1)[1 + (n/2)]}{[(1/r_1)^{1+(n/2)} - (1/r_2)^{1+(n/2)}]r_1^{n/2}} \ln\left(\frac{1 - x_{A2}}{1 - x_{A1}}\right) \tag{18.6-32}$$

For $n = 0$, this result becomes that in Eq. 18.6-29.

---

§18.7 DIFFUSION OF GASES IN A TWO-BULB EXPERIMENT

One way to measure the diffusivity of gas pairs is by means of the two-bulb experiment[1] pictured in Fig. 18.7-1. The lower bulb and the tube from $z = -L$ to $z = 0$ are filled with gas A. The upper bulb and the tube from $z = 0$ to $z = +L$ are filled with gas B ($\rho^{(B)} \leq \rho^{(A)}$, so that there is no gravity-driven flow). At time $t = 0$ the stopcock is opened, and diffusion begins. Because diffusion occurs, the concentration of A in the two well-stirred bulbs changes. One measures the mole fraction of A in the upper bulb, x_A^+, as a function of time, and from this one can deduce $\mathscr{D}_{AB}$, as shown below. We wish to derive the equations describing the diffusion. We assume that the system is maintained at constant temperature and pressure, so that the molar density c is constant throughout.

Since the volume of the bulbs is large compared with that of the tube, x_A^+ and x_A^- change *very slowly* with time. Hence, at any instant, the diffusion in the tube can be treated as a

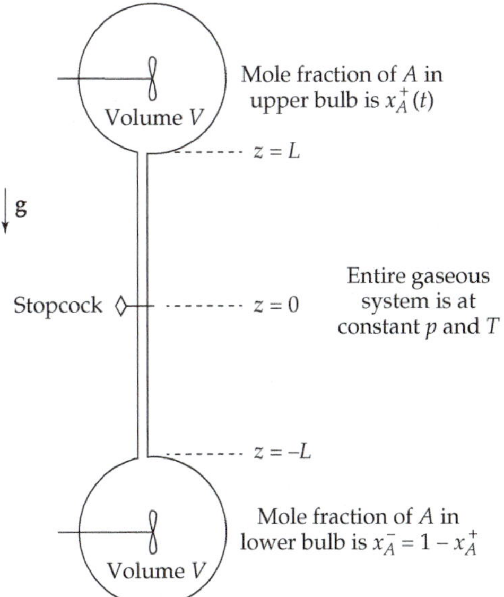

Fig. 18.7-1. Sketch of a two-bulb apparatus for measuring gas diffusivities. The stirrers maintain uniform concentration in the bulbs.

---

[1] S. P. S. Andrew, *Chem. Eng. Sci.*, **4**, 269–272 (1955).

quasi-steady-state problem, with the boundary conditions that $x_A = x_A^-$ and $z = -L$, and that $x_A = x_A^+$ at $z = +L$. That is, we temporarily treat x_A^- and x_A^+ as constants and solve a steady-state diffusion problem; then later, we allow x_A^- and x_A^+ to be functions of time.

A molar shell balance on a segment Δz of the tube with cross section S gives for species A at steady state

$$SN_{Az}|_z - SN_{Az}|_{z+\Delta z} = 0 \tag{18.7-1}$$

Division by $S\Delta z$ and taking the limit as Δz goes to zero gives

$$\frac{dN_{Az}}{dz} = 0 \tag{18.7-2}$$

or N_{Az} = constant. Equation 18.0-1, for this problem, may be simplified thus

$$N_{Az} = x_A(N_{Az} + N_{Bz}) - c\mathscr{D}_{AB}\frac{dx_A}{dz} = -c\mathscr{D}_{AB}\frac{dx_A}{dz} \tag{18.7-3}$$

since $N_{Az} = -N_{Bz}$; this is true, because, in a closed system at constant molar concentration c, for every mole of A that moves upward, a mole of B must move downward. This situation is referred to as *equimolar counterdiffusion.*

Equation 18.7-3, with N_{Az} = constant, is a differential equation for $x_A(z)$, which can be integrated to give

$$x_A(z) = -\frac{N_{Az}}{c\mathscr{D}_{AB}}z + C_1 \tag{18.7-4}$$

At $z = L$, we know that $x_A = x_A^+$, so that

$$x_A^+ = -\frac{N_{Az}}{c\mathscr{D}_{AB}}L + C_1 \tag{18.7-5}$$

and thus $C_1 = N_{Az}L/c\mathscr{D}_{AB} + x_A^+$. Substitution of this value of C_1 into Eq. 18.7-4 gives

$$x_A(z) - x_A^+ = \frac{N_{Az}}{c\mathscr{D}_{AB}}(L - z) \tag{18.7-6}$$

At $z = -L$, we know that $x_A = x_A^- = 1 - x_A^+$ (i.e., neglecting the small volume of the tubes, the total number of moles of A in the two bulbs is constant), so that

$$(1 - x_A^+) - x_A^+ = \frac{N_{Az}}{c\mathscr{D}_{AB}}(L - (-L)) \tag{18.7-7}$$

from which

$$N_{Az} = \left(\tfrac{1}{2} - x_A^+\right)\frac{c\mathscr{D}_{AB}}{L} \tag{18.7-8}$$

Thus, we have solved the *quasi-steady-state* diffusion problem. Now we have to take into account the fact that x_A^+ is changing with time, albeit rather slowly.

We make an unsteady-state molar balance on A over the upper bulb, stating that the increase in the moles of A in the upper bulb equals the number of moles of A that is diffusing upward through the connecting tube, to get

$$\frac{d}{dt}(Vcx_A^+) = SN_{Az} \tag{18.7-9}$$

Then by combining the last two equations we get

$$Vc\frac{dx_A^+}{dt} = S\left(\tfrac{1}{2} - x_A^+\right)\frac{c\mathscr{D}_{AB}}{L} \tag{18.7-10}$$

This differential equation for $x_A^+(t)$ can be integrated to give

$$-\ln\left(\tfrac{1}{2} - x_A^+(t)\right) = \frac{S\mathscr{D}_{AB}}{VL}t + C_2 \tag{18.7-11}$$

The integration constant may be obtained by using the initial condition that, at $t = 0$, the mole fraction of A in the upper bulb is zero, so that $C_2 = -\ln\left(\frac{1}{2} - 0\right)$. Therefore,

$$\ln\left(\tfrac{1}{2} - x_A^+(t)\right) - \ln\tfrac{1}{2} = -\frac{S\mathscr{D}_{AB}}{VL}t \tag{18.7-12}$$

The left side may be rewritten as

$$\ln\tfrac{1}{2}(1 - 2x_A^+) - \ln\tfrac{1}{2} = \ln\frac{\tfrac{1}{2}(1 - 2x_A^+)}{\tfrac{1}{2}} = \ln(1 - 2x_A^+) \tag{18.7-13}$$

so that, if we take the exponential of both sides, we may write Eq. 18.7-12 as

$$\boxed{1 - 2x_A^+(t) = \exp\left(-\frac{S\mathscr{D}_{AB}}{VL}t\right)} \tag{18.7-14}$$

This suggests that, if we plot $-(LV/S)\ln(1 - 2x_A^+)$ vs. t, the slope will be the diffusivity $\mathscr{D}_{AB}$. Therefore, the two-bulb experiment gives another method for measuring the diffusivity of a binary gas mixture.

§18.8 DIFFUSION INTO A FALLING LIQUID FILM (GAS ABSORPTION)[1]

In this section we present an illustration of *forced-convection* mass transfer, in which viscous flow and diffusion occur under such conditions that the velocity field can be considered virtually unaffected by the diffusion. Specifically, we consider the absorption of gas A by a laminar falling film of liquid B. The material A is only slightly soluble in B, so that the viscosity of the liquid is unaffected. We shall make the further restriction that the diffusion takes place so slowly in the liquid film that A will not "penetrate" very far into the film—that is, that the penetration distance will be small in comparison with the film thickness. The system is sketched in Fig. 18.8-1. An example of this kind of system occurs in the absorption of O_2 in H_2O.

Let us now set up the differential equations describing the diffusion process. First, we have to solve the momentum-transfer problem to obtain the velocity profile $v_z(x)$ for the film; this has already been worked out in §2.2 in the absence of mass transfer at the fluid surface, and we know that the result is

$$v_z(x) = v_{\max}\left[1 - \left(\frac{x}{\delta}\right)^2\right] \tag{18.8-1}$$

provided that "end effects" are ignored.

Next we have to establish a mass balance on component A. We note that c_A will be changing with both x and z, as illustrated in Fig. 18.8-1. Hence, as the element of volume for the mass balance, we select the volume formed by the intersection of a slab of thickness Δz with a slab of thickness Δx. Then the mass balance on A over this segment of a film of width W in the y direction becomes:

$$N_{Az}|_z W\Delta x - N_{Az}|_{z+\Delta z} W\Delta x + N_{Ax}|_x W\Delta z - N_{Ax}|_{x+\Delta x} W\Delta z = 0 \tag{18.8-2}$$

[1] S. Lynn, J. R. Straatemeier, and H. Kramers, *Chem. Engr. Sci.*, **4**, 49–67 (1955). **Professor Hendrik ("Hans") Kramers** (1917–2006) was the founder of the Laboratory of Physical Technology (now called the "Kramers Laboratory") at the Technical University of Delft; in this laboratory, many fascinating experiments were performed that involved the combination of diffusion, chemical reactions, and flow. He was the first person in Europe to give a university course in transport phenomena for engineers. He was an excellent teacher and mentor, much admired by his colleagues and his former research students.

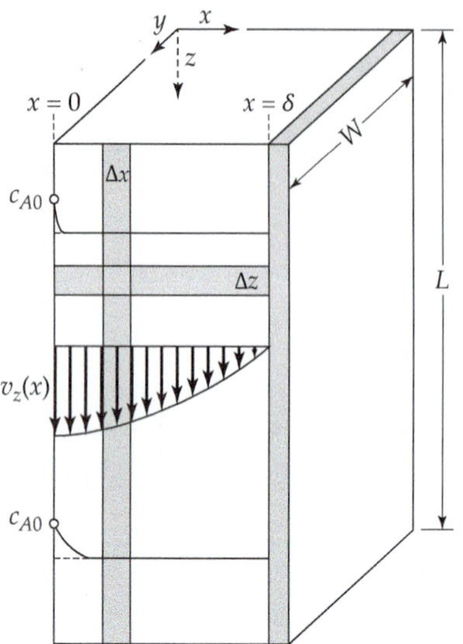

Fig. 18.8-1. Absorption of A into a falling film of liquid B.

Dividing by $W\Delta x\Delta z$ and performing the usual limiting processes as the volume element becomes infinitesimally small, we get

$$\frac{\partial N_{Az}}{\partial z} + \frac{\partial N_{Ax}}{\partial x} = 0 \tag{18.8-3}$$

Into this equation we now insert the expressions for N_{Az} and N_{Ax}, making appropriate simplifications of Eq. 18.0-1. For the molar flux in the z direction, we write, assuming constant c

$$N_{Az} = x_A(N_{Az} + N_{Bz}) - \underline{\mathscr{D}_{AB}\frac{\partial c_A}{\partial z}} \approx c_A v_z(x) \tag{18.8-4}$$

We discard the dashed-underlined term, since the transport of A in the z direction will be primarily by convection. We have made use of Eq. (F) in Table 17.4-1 and the fact that **v** is almost the same as **v*** in dilute solutions. The molar flux in the x direction is

$$N_{Ax} = \underline{x_A(N_{Ax} + N_{Bx})} - \mathscr{D}_{AB}\frac{\partial c_A}{\partial x} \approx -\mathscr{D}_{AB}\frac{\partial c_A}{\partial x} \tag{18.8-5}$$

Here we neglect the dashed-underlined term because in the x direction A moves predominantly by diffusion, there being almost no convective transport normal to the wall on account of the very slight solubility of A in B. Combining the last three equations, we then get for constant $\mathscr{D}_{AB}$

$$v_z(x)\frac{\partial c_A}{\partial z} = \mathscr{D}_{AB}\frac{\partial^2 c_A}{\partial x^2} \tag{18.8-6}$$

Finally, insertion of Eq. 18.8-1 for the velocity distribution gives

$$v_{\max}\left[1 - \left(\frac{x}{\delta}\right)^2\right]\frac{\partial c_A}{\partial z} = \mathscr{D}_{AB}\frac{\partial^2 c_A}{\partial x^2} \tag{18.8-7}$$

as the differential equation for $c_A(x,z)$.

Equation 18.8-7 is to be solved with the following boundary conditions:

B.C. 1: at $z = 0$, $c_A = 0$ (18.8-8)

B.C. 2: at $x = 0$, $c_A = c_{A0}$ (18.8-9)

B.C. 3: at $x = \delta$, $\dfrac{\partial c_A}{\partial x} = 0$ (18.8-10)

The first boundary condition corresponds to the fact that the film consists of pure B at the top ($z = 0$), and the second indicates that at the liquid–gas interface the concentration of A is determined by the solubility of A in B (that is, c_{A0}). The third boundary condition states that A cannot diffuse through the solid wall. This problem has been solved analytically in the form of an infinite series,[2] but we do not give that solution here. Instead, we seek only a limiting expression valid for "short contact times," that is, for small values of $L/v_{\max}$.

If, as indicated in Fig. 18.8-1, the substance A has penetrated only a short distance into the film, then the species A "has the impression" that the film is moving throughout with a velocity equal to $v_{\max}$. Furthermore, if A does not penetrate very far, it does not "sense" the presence of the solid wall at $x = \delta$. Hence, if the film were of infinite thickness moving with the velocity $v_{\max}$, the diffusing material "would not know the difference." This physical argument suggests (correctly) that we will get a very good asymptotic result (in the limit of short contact times) if we replace Eq. 18.8-7 and its boundary conditions by

$$v_{\max}\frac{\partial c_A}{\partial z} = \mathscr{D}_{AB}\frac{\partial^2 c_A}{\partial x^2} \tag{18.8-11}$$

B.C. 1: at $z = 0$, $c_A = 0$ (18.8-12)

B.C. 2: at $x = 0$, $c_A = c_{A0}$ (18.8-13)

B.C. 3: as $x \to \infty$, $c_A \to 0$ (18.8-14)

An exactly analogous problem occurred in Example 3.8-1, which was solved by the method of combination of variables. It is therefore possible to take over the solution to that problem just by changing the notation. The solution is[3]

$$\frac{c_A(x,z)}{c_{A0}} = 1 - \frac{2}{\sqrt{\pi}}\int_0^{x/\sqrt{4\mathscr{D}_{AB}z/v_{\max}}} \exp(-\xi^2)\,d\xi \tag{18.8-15}$$

or

$$\boxed{\frac{c_A(x,z)}{c_{A0}} = 1 - \operatorname{erf}\frac{x}{\sqrt{4\mathscr{D}_{AB}z/v_{\max}}} = \operatorname{erfc}\frac{x}{\sqrt{4\mathscr{D}_{AB}z/v_{\max}}}} \tag{18.8-16}$$

In these expressions "erf x" and "erfc x" are the "error function" and the "complementary error function" of x, respectively. They are discussed in §C.6 and tabulated in standard reference works.[4]

Once the concentration profiles are known, the local mass flux at the gas–liquid interface may be found from Eq. 18.8-5 as follows:

$$\begin{aligned} N_{Ax}|_{x=0} &= -\mathscr{D}_{AB}\frac{\partial c_A}{\partial x}\bigg|_{x=0} = -c_{A0}\mathscr{D}_{AB}\left(-\frac{\partial}{\partial x}\operatorname{erf}\frac{x}{\sqrt{4\mathscr{D}_{AB}z/v_{\max}}}\right)\bigg|_{x=0} \\ &= +c_{A0}\mathscr{D}_{AB}\left(\frac{2}{\sqrt{\pi}}e^{-x^2/(4\mathscr{D}_{AB}z/v_{\max})}\cdot\frac{1}{\sqrt{4\mathscr{D}_{AB}z/v_{\max}}}\right)\bigg|_{x=0} \\ &= c_{A0}\sqrt{\frac{\mathscr{D}_{AB}v_{\max}}{\pi z}} \end{aligned} \tag{18.8-17}$$

---

[2]R. L. Pigford, PhD Thesis, University of Illinois (1941). **Robert Lamar Pigford** (1917–1988) was a leader in the area of diffusion and mass transfer. He was for many years chairman of the Chemical Engineering Department at the University of Delaware. Together with W. R. Marshall, Jr., he wrote *Applications of Differential Equations to Chemical Engineering Problems.*

[3]The solution is worked out in detail by the method of combination of variables (or similarity method) in Example 3.8-1.

[4]M. Abramowitz and I. A. Stegun, *Handbook of Mathematical Functions*, Dover, New York, 9th printing (1972), pp. 310 et seq.

To get this result, Eq. C.6-3 has been used. Then the total molar flow rate of A across the surface at $x = 0$ (i.e., being absorbed by a liquid film of length L and width W) is

$$\begin{aligned} W_A &= \int_0^W \int_0^L N_{Ax}|_{x=0}\, dz\, dy \\ &= W c_{A0} \sqrt{\frac{\mathscr{D}_{AB} v_{\max}}{\pi}} \int_0^L \frac{1}{\sqrt{z}}\, dz \\ &= W L c_{A0} \sqrt{\frac{4\mathscr{D}_{AB} v_{\max}}{\pi L}} \end{aligned} \tag{18.8-18}$$

This result can also be obtained by integrating the product $v_{\max} c_A$ over the flow cross section at $z = L$ (see Problem 18C.2).

Equation 18.8-18 shows that the mass-transfer rate is directly proportional to the square root of the diffusivity and inversely proportional to the square root of the "exposure time," $t_{\exp} = L/v_{\max}$. This approach for studying gas absorption was apparently first proposed by Higbie.[5]

The problem discussed in this section illustrates the "penetration model" of mass transfer. This model is discussed further in Chapters 21 and 22.

EXAMPLE 18.8-1

Gas Absorption from Rising Bubbles

Estimate the rate at which gas bubbles of A are absorbed by liquid B as the gas bubbles rise at their terminal velocity v_t through a clean quiescent liquid.

SOLUTION

Gas bubbles of moderate size, rising in liquids free of surface-active agents, undergo a toroidal (Rybczynski-Hadamard) circulation as shown in Fig. 18.8-2. The liquid moves downward relative to each rising bubble, enriched in species A near the interface in the manner of the falling film in Fig. 18.8-1. The depth of penetration of the dissolved gas into the liquid is slight over the major part of the bubble, because of the motion of the liquid relative to the bubble and because of the smallness of the liquid-phase diffusivity $\mathscr{D}_{AB}$. Thus, as a rough approximation, we can use Eq. 18.8-18 to estimate the rate of gas absorption, replacing the exposure time $t_{\exp} = L/v_{\max}$ for the falling film by D/v_t for the bubble, where D is the instantaneous bubble

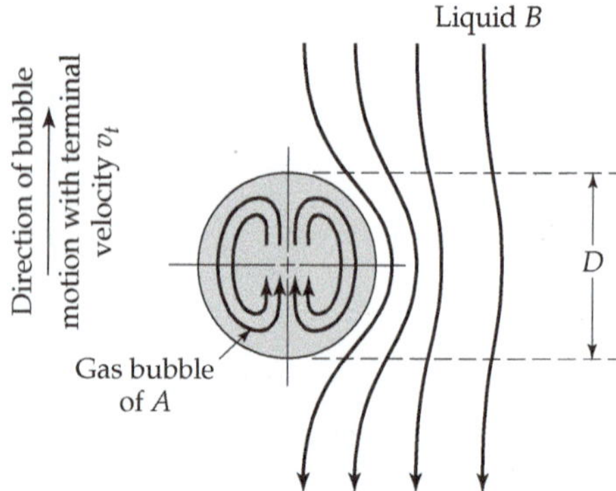

Fig. 18.8-2. Absorption of gas A into liquid B.

[5] R. Higbie, *Trans. AIChE*, **31**, 365–389 (1935). **Ralph Wilmarth Higbie** (1908–1941), a graduate of the University of Michigan, provided the basis for the "penetration model" of mass transfer; he worked at E. I. du Pont de Nemours & Co., Inc., and also at Eagle-Picher Lead Co.; then he taught at the University of Arkansas and the University of North Dakota.

diameter. This gives a rough estimate[5] of the molar absorption rate, averaged over the bubble surface, as

$$(N_{Ar})_{avg} = \sqrt{\frac{4\mathscr{D}_{AB}v_t}{\pi D}}\, c_{A0} \tag{18.8-19}$$

Here c_{A0} is the solubility of gas A in liquid B at the interfacial temperature and partial pressure of gas A. This equation has been approximately confirmed by Hammerton and Garner[6] for gas bubbles 0.3 cm to 0.5 cm in diameter rising through carefully purified water.

A more refined analysis can be obtained by using the creeping flow[7] solution for flow around a sphere, and the result is

$$(N_{Ar})_{avg} = \sqrt{\frac{4\mathscr{D}_{AB}v_t}{3\pi D}}\, c_{A0} \tag{18.8-20}$$

instead of Eq. 18.8-19.

Trace amounts of surface-active agents cause a marked decrease in absorption rates from small bubbles, by forming a "skin" around each bubble and thus effectively preventing internal circulation. The molar absorption rate in the small-diffusivity limit then becomes proportional to the $\frac{2}{3}$ power of the diffusivity, as for a solid sphere (see Eq. 22.2-10).

A similar approach has been used successfully for predicting mass-transfer rates during drop-formation at a capillary tip.[8]

§18.9 DIFFUSION INTO A FALLING LIQUID FILM (SOLID DISSOLUTION)[1]

We now turn to a problem very similar to that discussed in the previous section. Liquid B is flowing in laminar motion down a vertical wall as shown in Fig. 18.9-1. The film begins far enough up the wall so that v_z depends only on y for $z \geq 0$. For $0 < z < L$ the wall is coated with a species A that is slightly soluble in B.

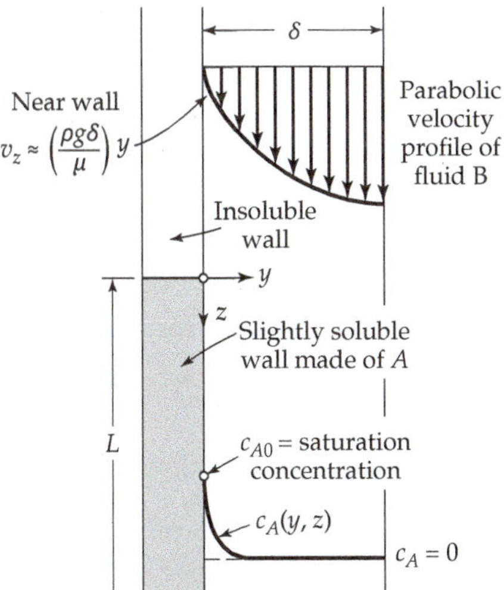

Fig. 18.9-1. Solid A dissolving into a falling film of liquid B, moving with a fully developed parabolic velocity profile.

[6]D. Hammerton and F. H. Garner, *Trans. Inst. Chem. Engrs. (London)*, **32**, S18–S24 (1954).

[7]V. G. Levich, *Physicochemical Hydrodynamics*, Prentice-Hall, Englewood Cliffs, NJ (1962), p. 408, Eq. 72.9. This reference gives many additional results, including liquid-liquid mass-transfer and surfactant effects.

[8]H. Groothuis and H. Kramers, *Chem. Eng. Sci.*, **4**, 17–25 (1955).

[1]H. Kramers and P. J. Kreyger, *Chem. Eng. Sci.*, **6**, 42–48 (1956); see also R. L. Pigford, *Chem. Eng. Prog. Symposium Series* No. 17, Vol. **51**, pp. 79–92 (1955) for the analogous heat-conduction problem.

For short distances downstream, species A will not diffuse very far into the falling film. That is, A will be present only in a very thin boundary layer near the solid surface. Therefore, the diffusing A molecules will experience a velocity distribution that is characteristic of the falling film right next to the wall, $y = 0$. The velocity distribution is given in Eq. 2.2-22. In the present situation $\cos\beta = 1$, and $x = \delta - y$, and

$$v_z(y) = \frac{\rho g \delta^2}{2\mu}\left[1 - \left(1 - \frac{y}{\delta}\right)^2\right] = \frac{\rho g \delta^2}{2\mu}\left[2\left(\frac{y}{\delta}\right) - \left(\frac{y}{\delta}\right)^2\right] \tag{18.9-1}$$

At and adjacent to the wall $(y/\delta)^2 \ll (y/\delta)$, so that for this problem the velocity is, to a very good approximation, a linear function of y: $v_z(y) = (\rho g \delta/\mu)y \equiv ay$. This means that Eq. 18.8-6, which is appropriate here, becomes for short distances downstream

$$ay\frac{\partial c_A}{\partial z} = \mathscr{D}_{AB}\frac{\partial^2 c_A}{\partial y^2} \tag{18.9-2}$$

where $a = \rho g \delta/\mu$. This equation is to be solved with the boundary conditions

B.C. 1: at $z = 0$, $c_A = 0$ (18.9-3)

B.C. 2: at $y = 0$, $c_A = c_{A0}$ (18.9-4)

B.C. 3: as $y \to \infty$, $c_A \to 0$ (18.9-5)

In the second boundary condition, c_{A0} is the solubility of A in B. The third boundary condition is used instead of the correct one ($\partial c_A/\partial y = 0$ at $y = \delta$), since for short contact times we feel intuitively that it will not make any difference. After all, since the molecules of A penetrate only slightly into the film, they cannot get far enough to "see" the outer edge of the boundary at $y = \delta$, and hence, they cannot distinguish between the true boundary condition and the approximate boundary condition in Eq. 18.9-5. The same kind of reasoning was encountered in Example 11.5-3.

The form of the boundary conditions in Eqs. 18.9-3 to 18.9-5—that is, $c_A \to 0$ both as $z \to 0$ and as $y \to \infty$—suggests the method of combination of variables, similar to that employed in Examples 3.8-1 and 11.5-3. Therefore, we try $c_A(y,z)/c_{A0} = f(\eta)$, where $\eta = y(a/9\mathscr{D}_{AB}z)^{1/3}$. This combination of the independent variables can be shown to be dimensionless, and the factor of "9" is included in order to make the solution look neater.

When this change of variables is made, the partial differential equation in Eq. 18.9-2 reduces to an ordinary differential equation

$$\frac{d^2f}{d\eta^2} + 3\eta^2\frac{df}{d\eta} = 0 \tag{18.9-6}$$

with the boundary conditions $f(0) = 1$ and $f(\infty) = 0$.

This second-order equation, which is of the form of Eq. C.1-9, has the solution

$$f = C_1\int_0^\eta e^{-\bar{\eta}^3}d\bar{\eta} + C_2 \tag{18.9-7}$$

The constants of integration can then be evaluated using the boundary conditions to give

$$C_2 = 1 \tag{18.9-8}$$

$$C_1 = -1\Big/\int_0^\infty e^{-\eta^3}d\eta = -1\Big/\tfrac{1}{3}\int_0^\infty e^{-x}x^{-2/3}dx = -1\Big/\tfrac{1}{3}\Gamma\left(\tfrac{1}{3}\right) = -1\Big/\Gamma\left(\tfrac{4}{3}\right) \tag{18.9-9}$$

by use of Eqs. C.4-1 and C.4-11. Therefore, we get

$$\frac{c_A(\eta)}{c_{A0}} = \frac{\int_\eta^\infty e^{-\bar{\eta}^3}d\bar{\eta}}{\int_0^\infty e^{-\bar{\eta}^3}d\bar{\eta}} = \frac{\int_\eta^\infty e^{-\bar{\eta}^3}d\bar{\eta}}{\Gamma\left(\frac{4}{3}\right)} \tag{18.9-10}$$

for the concentration profile, in which $\Gamma\left(\frac{4}{3}\right) = 0.8930\ldots$ is the gamma function of $\frac{4}{3}$ (see §C.4). Next the local molar flux at the wall can be obtained as follows

$$N_{Ay}|_{y=0} = -\mathscr{D}_{AB}\left.\frac{\partial c_A}{\partial y}\right|_{y=0} = -\mathscr{D}_{AB}c_{A0}\left.\left[\frac{d}{d\eta}\left(\frac{c_A}{c_{A0}}\right)\frac{\partial \eta}{\partial y}\right]\right|_{y=0}$$

$$= -\mathscr{D}_{AB}c_{A0}\left.\left[-\frac{\exp(-\eta^3)}{\Gamma\left(\frac{4}{3}\right)}\left(\frac{a}{9\mathscr{D}_{AB}z}\right)^{1/3}\right]\right|_{y=0} = +\frac{\mathscr{D}_{AB}c_{A0}}{\Gamma\left(\frac{4}{3}\right)}\left(\frac{a}{9\mathscr{D}_{AB}z}\right)^{1/3} \tag{18.9-11}$$

Then the molar flow of A across the entire mass-transfer surface at $y = 0$ is

$$W_A = \int_0^W \int_0^L N_{Ay}|_{y=0} dz\, dx = \frac{2\mathscr{D}_{AB}c_{A0}WL}{\Gamma\left(\frac{7}{3}\right)}\left(\frac{a}{9\mathscr{D}_{AB}L}\right)^{1/3} \tag{18.9-12}$$

where $\Gamma\left(\frac{7}{3}\right) = \frac{4}{3}\Gamma\left(\frac{4}{3}\right) = 1.1907\ldots$.

The problems discussed in this and the preceding section are examples of two types of asymptotic solutions that are encountered again in Chapter 22. It is therefore important that these two problems be thoroughly understood. Note that in §18.8, $W_A \propto (\mathscr{D}_{AB}L)^{1/2}$, whereas in this section $W_A \propto (\mathscr{D}_{AB}L)^{2/3}$. The differences in the exponents reflect the nature of the velocity gradient at the mass-transfer interface: in §18.8, the velocity gradient is zero, whereas in this section, the velocity gradient is nonzero.

§18.10 DIFFUSION AND CHEMICAL REACTION INSIDE A POROUS CATALYST

Up to this point we have discussed diffusion in gases and liquids in systems of simple geometry. We now wish to apply the shell mass balance method and Fick's first law to describe diffusion within a porous catalyst pellet. We make no attempt to describe the diffusion inside the tortuous void passages in the pellet. Instead, we describe the "averaged" diffusion of the reactant in terms of an "effective diffusivity."[1,2,3]

Specifically, we consider a spherical porous catalyst particle of radius R, as shown in Fig. 18.10-1. This particle is in a catalytic reactor, submerged in a fluid stream containing the reactant A and the product B. In the neighborhood of the surface of the particular catalyst particle under consideration, we presume that the concentration is c_{AR} moles of A per unit volume. Species A diffuses through the tortuous passages in the catalyst and

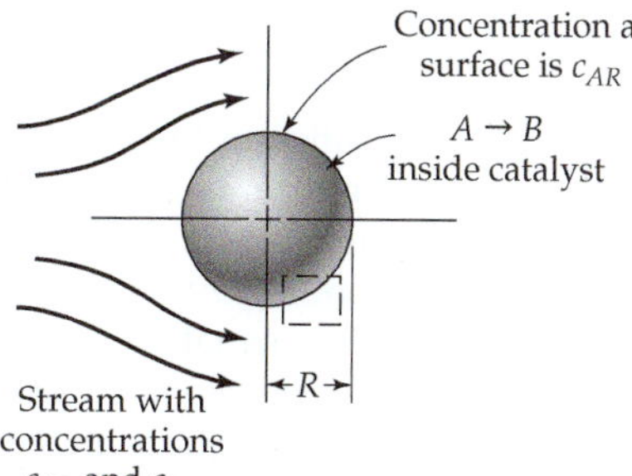

Fig. 18.10-1. A spherical catalyst that is porous. For a magnified version of the inset, see Fig. 18.10-2.

[1] E. W. Thiele, *Ind. Eng. Chem.*, **31**, 916–920 (1939).
[2] R. Aris, *Chem. Eng. Sci.*, **6**, 262–268 (1957).
[3] A. Wheeler, *Advances in Catalysis*, Academic Press, *New York* (1950), Vol. **3**, pp. 250–326.

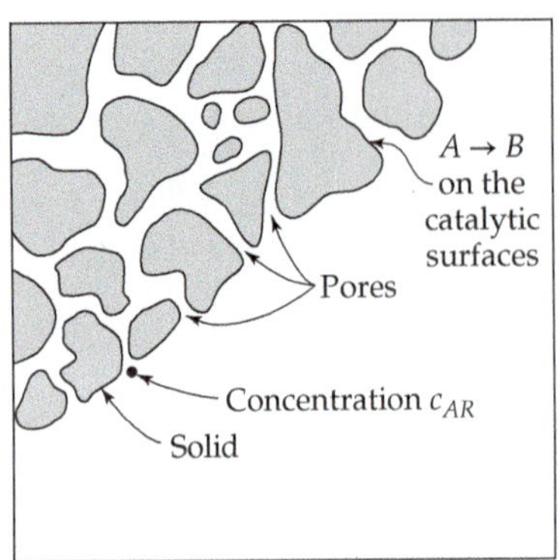

Fig. 18.10-2. Pores in the catalyst, in which diffusion and chemical reaction occur.

is converted to B on the catalytic surfaces, as sketched in Fig. 18.10-2. We assume that the concentration of species A, averaged over a volume large compared to the pore size but small compared to R, is a function of only r.

We start by making a molar shell balance for species A on a spherical shell of thickness Δr within a single catalyst particle

$$N_{Ar}|_r \cdot 4\pi r^2 - N_{Ar}|_{r+\Delta r} \cdot 4\pi(r+\Delta r)^2 + R_A \cdot 4\pi r^2 \Delta r = 0 \tag{18.10-1}$$

Here $N_{Ar}|_r$ is the number of moles per unit area of A passing in the r direction through an imaginary spherical surface at a distance r from the center of the sphere. The source term $R_A \cdot 4\pi r^2 \Delta r$ is the molar rate of production of A by chemical reaction in the shell of thickness Δr. Dividing by $4\pi\Delta r$ and letting $\Delta r \to 0$ gives

$$\lim_{\Delta r \to 0} \frac{(r^2 N_{Ar})|_{r+\Delta r} - (r^2 N_{Ar})|_r}{\Delta r} = r^2 R_A \tag{18.10-2}$$

or, using the definition of the first derivative,

$$\frac{d}{dr}(r^2 N_{Ar}) = r^2 R_A \tag{18.10-3}$$

This limiting process is clearly in conflict with the fact that the porous medium is granular rather than continuous. Consequently, in Eq. 18.10-3 the symbols N_{Ar} and R_A cannot be interpreted as quantities having a meaningful value at a point. Rather we have to regard them as quantities averaged over a small neighborhood of the point in question—a neighborhood small with respect to the dimension R, but large with respect to the dimensions of the passages within the porous particle.

We now *define* an "effective diffusivity" for species A in the porous medium by

$$N_{Ar} = -\mathscr{D}_{A,\text{eff}} \frac{dc_A}{dr} \tag{18.10-4}$$

in which c_A is the locally averaged concentration of the gas A contained within the pores. Note that we have neglected the convection term (see Eq. 18.0-1), because the bulk flow within porous catalyst particles is typically negligible. The effective diffusivity $\mathscr{D}_{A,\text{eff}}$ must be measured experimentally. It depends generally on pressure and temperature and also on the catalyst pore structure. The actual mechanism for diffusion in pores is complex, since the pore dimensions may be smaller than the mean-free path of the diffusing molecules. We do not belabor the question of mechanism here but simply *assume* that Eq. 18.10-4 can adequately represent the diffusion process.[4]

[4]See R. B. Bird, W. E. Stewart, and E. N. Lightfoot, *Transport Phenomena*, revised second edition, Wiley, New York (2007), Chapter 24.

When the preceding expression is inserted into Eq. 18.10-3, we get, for constant effective diffusivity,

$$\mathscr{D}_{A,\text{eff}}\frac{1}{r^2}\frac{d}{dr}\left(r^2\frac{dc_A}{dr}\right) = -R_A \tag{18.10-5}$$

We now consider the situation where species A disappears according to a first-order chemical reaction on the catalytic surfaces that form all or part of the "walls" of the winding passages. Let a be the available catalytic surface area per unit volume (of solids + voids). Then $R_A = -k_1''ac_A$, and Eq. 18.10-5 becomes

$$\mathscr{D}_{A,\text{eff}}\frac{1}{r^2}\frac{d}{dr}\left(r^2\frac{dc_A}{dr}\right) = k_1''ac_A \tag{18.10-6}$$

This equation is to be solved with the boundary conditions that $c_A = c_{AR}$ at $r = R$, and that c_A is finite at $r = 0$.

Equation 18.10-6 can be solved using Eqs. C.1-6a or C.1-6b. However, equations containing the operator $(1/r^2)(d/dr)[r^2(d/dr)]$ can frequently be solved by using a "standard trick"—namely, a change of variable $c_A(r)/c_{AR} = (1/r)f(r)$. The equation for $f(r)$ is then

$$\frac{d^2f}{dr^2} = \left(\frac{k_1''a}{\mathscr{D}_{A,\text{eff}}}\right)f \tag{18.10-7}$$

This is a standard second-order differential equation (see Eq. C.1-4a), which can be solved in terms of exponentials or hyperbolic functions. When it is solved and the result divided by r, we get the following solution of Eq. 18.10-6 in terms of hyperbolic functions (see §C.5):

$$\frac{c_A(r)}{c_{AR}} = \frac{C_1}{r}\cosh\sqrt{\frac{k_1''a}{\mathscr{D}_{A,\text{eff}}}}\,r + \frac{C_2}{r}\sinh\sqrt{\frac{k_1''a}{\mathscr{D}_{A,\text{eff}}}}\,r \tag{18.10-8}$$

Application of the boundary conditions gives $C_1 = 0$ (since otherwise c_A would be infinite at $r = 0$) and $C_2 = R/\sinh\sqrt{k_1''a/\mathscr{D}_{A,\text{eff}}}R$. Thus, we get finally

$$\boxed{\frac{c_A(r)}{c_{AR}} = \left(\frac{R}{r}\right)\frac{\sinh\sqrt{k_1''a/\mathscr{D}_{A,\text{eff}}}\,r}{\sinh\sqrt{k_1''a/\mathscr{D}_{A,\text{eff}}}\,R}} \tag{18.10-9}$$

for the concentration distribution inside the catalyst particle.

In studies on chemical kinetics and catalysis, one is frequently interested in the molar flux N_{AR} or the molar flow W_{AR} at the surface $r = R$:

$$W_{AR} = 4\pi R^2 N_{AR} = -4\pi R^2\mathscr{D}_{A,\text{eff}}\left.\frac{dc_A}{dr}\right|_{r=R} \tag{18.10-10}$$

When Eq. 18.10-9 is substituted into this expression, we get

$$W_{AR} = 4\pi R\mathscr{D}_{A,\text{eff}}c_{AR}\left(1 - \sqrt{\frac{k_1''a}{\mathscr{D}_{A,\text{eff}}}}R\coth\sqrt{\frac{k_1''a}{\mathscr{D}_{A,\text{eff}}}}R\right) \tag{18.10-11}$$

This result gives the rate of conversion (in moles/sec) of A to B in a single catalyst particle of radius R in terms of the parameters describing the diffusion and reaction processes.

If the catalytically active surface were all exposed to the stream of concentration c_{AR}, then the species A would not have to diffuse through the pores to a reaction site. The molar rate of conversion would then be given by the product of the available surface and the surface reaction rate:

$$W_{AR,0} = \left(\tfrac{4}{3}\pi R^3\right)(a)(-k_1''c_{AR}) \tag{18.10-12}$$

Taking the ratio of the last two equations, we get

$$\eta_A = \frac{W_{AR}}{W_{AR,0}} = \frac{3}{\phi^2}(\phi \coth \phi - 1) \tag{18.10-13}$$

in which $\phi = \sqrt{k_1''a/\mathscr{D}_{A,\text{eff}}}R$ is the *Thiele modulus*,[1] encountered in §18.4. The quantity η_A is called the *effectiveness factor*.[1–3,5] It is the quantity by which $W_{AR,0}$ has to be multiplied to account for the intraparticle diffusional resistance to the overall conversion process.

For nonspherical catalyst particles, the preceding results may be applied approximately by reinterpreting R. We note that, for a sphere of radius R, the ratio of volume to external surface is $R/3$. For nonspherical particles, we redefine R in Eq. 18.10-13 as

$$R_{\text{nonsph}} = 3\left(\frac{V_P}{S_P}\right) \tag{18.10-14}$$

where V_P and S_P are the volume and external surface area of a single catalyst particle. The absolute value of the conversion rate is then given approximately by

$$|W_{AR}| \approx V_P a k_1'' c_{AR} \eta_A \tag{18.10-15}$$

where

$$\eta_A = \frac{1}{3\Lambda^2}(3\Lambda \coth 3\Lambda - 1) \tag{18.10-16}$$

in which the quantity $\Lambda = \sqrt{k_1''a/\mathscr{D}_{A,\text{eff}}}(V_P/S_P)$ is a generalized modulus.[2,3]

The particular utility of the quantity Λ may be seen in Fig. 18.10-3. It is clear that when the exact theoretical expressions for η_A are plotted as a function of Λ, the curves have common asymptotes for large and small Λ and do not differ from one another very

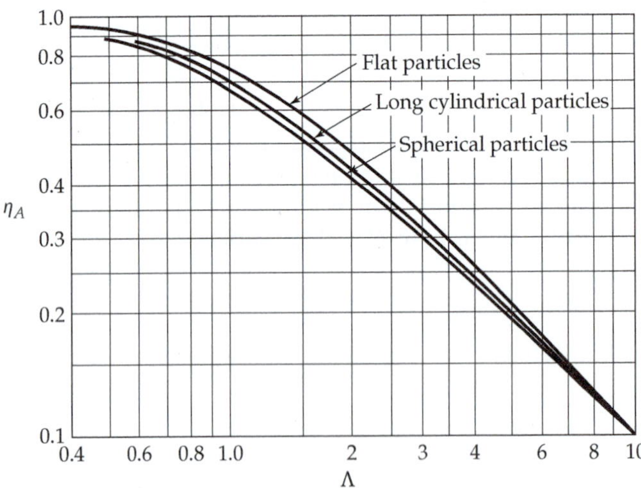

Fig. 18.10-3. Effectiveness factors for porous solid catalysts of various shapes. [R. Aris, *Chem. Eng. Sci.*, **6**, 262–268 (1957).]

[5]O. A. Hougen and K. M. Watson, *Chemical Process Principles*, Wiley, New York (1947), Part III, Chapter XIX. See also *CPP Charts*, by O. A. Hougen, K. M. Watson, and R. A. Ragatz, Wiley, New York (1960), Fig. E. See also C. G. Hill, *Introduction to Chemical Engineering Kinetics and Reactor Design*, Wiley, New York (1977); C. G. Hill and T. W. Root, *Introduction to Chemical Engineering Kinetics and Reactor Design*, 2nd edition, Wiley, New York (2013); J. B. Rawlings and J. G. Ekerdt, *Chemical Reactor Analysis and Design Fundamentals*, 2nd edition, Nob Hill Publishing, Madison, WI (2012).

18A.5 Determination of diffusivity for ether-air system. The following data on the evaporation of ethyl ether ($C_2H_5OC_2H_5$), with a liquid density of 0.712 g/cm^3, have been tabulated by Jost.[1] The data are for a tube of 6.16 mm diameter, a total pressure of 747 mm Hg, and a temperature of 22°C.

Decrease of the ether level (measured from the open end of the tube), in mm	Time, in seconds, required for the indicated decrease of level
from 9 to 11	590
from 14 to 16	895
from 19 to 21	1185
from 24 to 26	1480
from 34 to 36	2055
from 44 to 46	2655

The molecular weight of ethyl ether is 74.12, and its vapor pressure at 22°C is 480 mm Hg. It may be assumed that the ether concentration at the open end of the tube is zero. Jost has given a value of $\mathscr{D}_{AB}$ for the ether-air system of 0.0786 cm^2/s at 0°C and 760 mm Hg.

(a) Use the evaporation data to find $\mathscr{D}_{AB}$ at 747 mm Hg and 22°C, assuming that the arithmetic average gas-column lengths may be used for $z_2 - z_1$ in Fig. 18.6-1. Assume further that the ether-air mixture is ideal and that the diffusion can be regarded as binary.

(b) Convert the result to $\mathscr{D}_{AB}$ at 760 mm Hg and 0°C using Eq. 17.6-1.

18A.6 Mass flux from a circulating bubble.

(a) Use Eq. 18.8-20 to estimate the rate of absorption of CO_2 (component A) from a carbon dioxide bubble 0.5 cm in diameter rising through pure water (component B) at 18°C and at a pressure of 1 atm. The following data[2] may be used: $\mathscr{D}_{AB} = 1.46 \times 10^{-5}$ cm^2/s, $c_{A0} = 0.041$ g-mol/liter, $v_t = 22$ cm/s.

(b) Recalculate the rate of absorption, using the experimental results of Hammerton and Garner,[3] who obtained a surface-averaged k_c of 117 cm/hr (see Eq. 18.1-2).

Answers: **(a)** 1.17×10^{-6} g-mol/cm^{2}s; **(b)** 1.33×10^{-6} g-mol/cm^{2}s

18B.1 Diffusion through a stagnant film—alternate derivation. In §18.6 an expression for the evaporation rate was obtained in Eq. 18.6-16 by differentiating the concentration profile found a few lines before. Show that the same results may be derived without finding the concentration profile. Note that at steady state, N_{Az} is a constant according to Eq. 18.6-2. Then Eq. 18.6-3 can be integrated directly to get Eq. 18.6-16.

18B.2 Error in neglecting the convection term in evaporation.

(a) Rework the problem in the text in §18.6 by neglecting the convection term $x_A(N_{Az} + N_{Bz})$ in Eq. 18.0-1. Show that this leads to

$$N_{Az} = \frac{c\mathscr{D}_{AB}}{z_2 - z_1}(x_{A1} - x_{A2}) \tag{18B.2-1}$$

This is a useful approximation if A is present only in very low concentrations.

(b) Obtain the result in (a) from Eq. 18.6-16 by making the appropriate approximation.

(c) What error is made in the determination of $\mathscr{D}_{AB}$ in Example 18.6-2 if the result in (a) is used?

Answer: 2.2%

---

[1] W. Jost, *Diffusion*, Academic Press, New York (1952), pp. 411–413.

[2] G. Tammann and V. Jessen, *Z. anorg. allgem. Chem.*, **179**, 125–144 (1929); F. H. Garner and D. Hammerton, *Chem. Eng. Sci.*, **3**, 1–11 (1954).

[3] D. Hammerton and F. H. Garner, *Trans. Inst. Chem. Engrs.* (London), **32**, S18–S24 (1954).

18B.3 Effect of mass-transfer rate on the concentration profiles.

(a) Combine the result in Eq. 18.6-13 with that in Eq. 18.6-16 to get

$$\frac{1-x_A(z)}{1-x_{A1}} = \exp\left(\frac{N_{Az}(z-z_1)}{c\mathscr{D}_{AB}}\right) \tag{18B.3-1}$$

(b) Obtain the same result by integrating Eq. 18.6-3 directly, using the fact that N_{Az} is constant.

(c) Note what happens when the mass-transfer rate becomes small. Expand Eq. 18B.3-1 in a Taylor series and keep two terms only, as is appropriate for small N_{Az}. What happens to the slightly curved lines in Fig. 18.6-1 when N_{Az} is very small?

18B.4 Absorption with chemical reaction

(a) Rework the problem discussed in the text in §18.4, but take $z = 0$ to be the bottom of the beaker and $z = L$ at the gas–liquid interface.

(b) In solving Eq. 18.4-7, we took the solution to be of the sum of two hyperbolic functions. Try solving the problem by using the equally valid solution $\Gamma(\zeta) = C_1 \exp(\phi\zeta) + C_2 \exp(-\phi\zeta)$.

(c) In what way do the results in Eqs. 18.4-10 and 18.4-12 simplify for very large L? for very small L? Interpret the results physically.

18B.5 Absorption of chlorine by cyclohexene. Chlorine can be absorbed from Cl_2-air mixtures by olefins dissolved in CCl_4. It was found[4] that the reaction of Cl_2 with cyclohexene (C_6H_{10}) is second order with respect to Cl_2 and zero order with respect to C_6H_{10}. Hence, the rate of disappearance of Cl_2 per unit volume is $k_2'''c_A^2$ (where A designates Cl_2).

Rework the problem of §18.4 where B is a C_6H_{10} - CCl_4 mixture, assuming that the diffusion can be treated as pseudobinary. Assume that the air is essentially insoluble in the C_6H_{10} - CCl_4 mixture. Let the liquid phase be sufficiently deep that L can be taken to be infinite.

(a) Show that the concentration profile is given by

$$\frac{c_A(z)}{c_{A0}} = \left[1 + \sqrt{\frac{k_2'''c_{A0}}{6\mathscr{D}_{AB}}}\,z\right]^{-2} \tag{18B.5-1}$$

(b) Obtain an expression for the rate of absorption of Cl_2 by the liquid.

(c) Suppose that a substance A dissolves in and reacts with substance B so that the rate of disappearance of A per unit volume is some arbitrary function of the concentration, $f(c_A)$. Show that the rate of absorption of A is given by

$$N_{Az}|_{z=0} = \sqrt{2\mathscr{D}_{AB}\int_0^{c_{A0}} f(c_A)\,dc_A} \tag{18B.5-2}$$

Use this result to check the result of (b).

18B.6 Oxygen uptake by a bacterial aggregate. Under suitable circumstances the rate of oxygen metabolism by bacterial cells is very nearly zero order with respect to oxygen concentration. We examine such a case here and focus our attention on a spherical aggregate of cells, which has a radius R. We wish to determine the total rate of oxygen uptake by the aggregate as a function of aggregate size, oxygen mass concentration ρ_0 at the aggregate surface, the metabolic activity of the cells, and the diffusional behavior of the oxygen. For simplicity we consider the aggregate to be homogeneous. We then approximate the metabolic rate by an effective volumetric reaction rate $r_{O_2} = -k_0'''$ and the diffusional behavior by Fick's law, with an effective pseudobinary diffusivity $\mathscr{D}_{O_2m}$. Because the solubility of oxygen is very low in this system, both convective oxygen transport and transient effects may be neglected.[5]

---

[4] G. H. Roper, *Chem. Eng. Sci.*, **2**, 18–31, 247–253 (1953).

[5] J. A. Mueller, W. C. Boyle, and E. N. Lightfoot, *Biotechnol. and Bioengr.*, **10**, 331–358 (1968).

(a) Show by means of a shell mass balance that the quasi-steady-state oxygen concentration profile is described by the differential equation

$$\frac{1}{\xi^2}\frac{d}{d\xi}\left(\xi^2\frac{d\chi}{d\xi}\right) = N \tag{18B.6-1}$$

where $\chi = \rho_{O_2}/\rho_0$, $\xi = r/R$, and $N = k_0'''R^2/\rho_0\mathscr{D}_{O_2m}$.

(b) There may be an oxygen-free core in the aggregate, such that $\chi = 0$ for $\xi < \xi_0$, if N is sufficiently large. Write appropriate boundary conditions to integrate Eq. 18B.6-1 for this situation. To do this, it must be recognized that both χ and $d\chi/d\xi$ are zero at $\xi = \xi_0$. What is the physical significance of this last statement?

(c) Perform the integration of Eq. 18B.6-1 and show how ξ_0 may be determined.

(d) Sketch the total oxygen uptake rate and ξ_0 as functions of N, and discuss the possibility that no oxygen-free core exists.

Answer: **(c)** $\chi = 1 - \frac{N}{6}(1-\xi^2) + \frac{N}{3}\xi_0^3\left(\frac{1}{\xi} - 1\right)$ for $\xi \geq \xi_0 \geq 0$, where ξ_0 is determined as a function of N from $\xi_0^3 - \frac{3}{2}\xi_0^2 + \left(\frac{1}{2} - \frac{3}{N}\right) = 0$

18B.7 Diffusion from a suspended droplet. A droplet of liquid A, of radius r_1, is suspended in a stream of gas B as shown in Fig. 18.6-4. We postulate that there is a spherical stagnant gas film of radius r_2 surrounding the droplet. The concentration of A in the gas phase is x_{A1} at $r = r_1$ and x_{A2} at the outer edge of the film, $r = r_2$.

(a) By a shell balance, show that for steady-state diffusion r^2N_{Ar} is a constant within the gas film, and set the constant equal to $r_1^2N_{Ar1}$, the value at the droplet surface.

(b) Show that Eq. 18.0-1 and the result in (a) lead to the following equation for x_A:

$$r_1^2N_{Ar1} = -\frac{c\mathscr{D}_{AB}}{1-x_A}r^2\frac{dx_A}{dr} \tag{18B.7-1}$$

(c) Integrate this equation between the limits r_1 and r_2 to get

$$N_{Ar1} = \frac{c\mathscr{D}_{AB}}{r_2 - r_1}\left(\frac{r_2}{r_1}\right)\ln\frac{x_{B2}}{x_{B1}} \tag{18B.7-2}$$

What is the limit of this expression as $r_2 \to \infty$?

18B.8 Rate of leaching. In studying the rate of leaching of a substance A from solid particles by a solvent B, we may postulate that the rate-controlling step is the diffusion of A from the particle surface through a stagnant liquid film of thickness δ out into the main stream (illustrated in Fig. 18B.8). The molar solubility of A in B is c_{A0}, and the concentration in the main stream is $c_{A\delta}$.

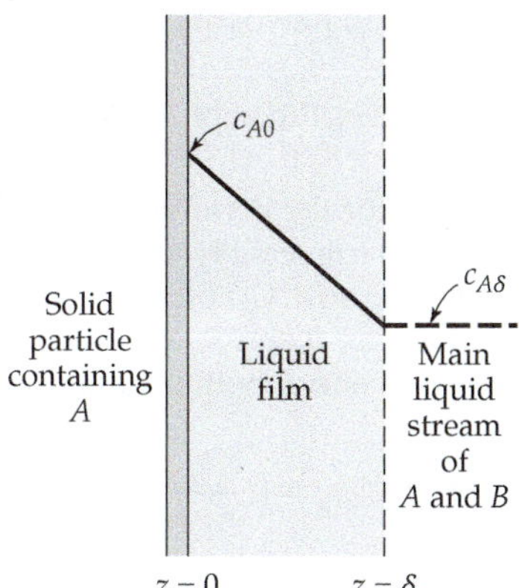

Fig. 18B.8 Leaching of A by diffusion into a stagnant liquid film of B.

(a) Obtain a differential equation for c_A as a function of z by making a mass balance on A over a thin slab of thickness Δz. Assume that $\mathscr{D}_{AB}$ is constant and that A is only slightly soluble in B. Neglect the curvature of the particle.

(b) Show that, in the absence of chemical reaction in the liquid phase, the concentration profile is linear.

(c) Show that the rate of leaching is given by

$$N_{Az} = \frac{\mathscr{D}_{AB}(c_{A0} - c_{A\delta})}{\delta} \tag{18B.8-1}$$

18B.9 Diffusion with homogeneous and heterogeneous reactions. Gas A diffuses through a stagnant film to a catalytic surface where A is instantaneously converted to B by a molecular rearrangement, as shown in Fig. 18B.9. When B leaves the surface, it begins to decompose very slowly by a first-order reaction, with rate constant k_1'''. Because the reaction is so slow, the mixture is composed almost exclusively of species A and B, and thus, we may treat this problem as one of equimolar counterdiffusion ($N_{Az} = -N_{Bz}$).

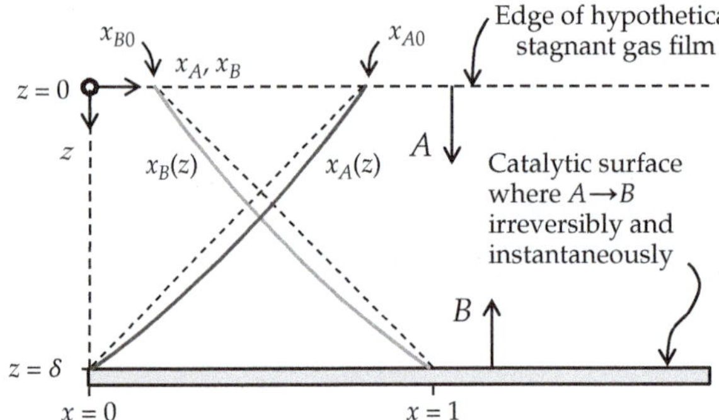

Fig. 18B.9 Schematic diagram of diffusion near a catalyst particle surface, where A is converted to B. Species B subsequently decomposes slowly via a homogeneous reaction. Solid curves for $x_A(z)$ and $x_B(z)$ are calculated for $x_{B0} = 0.2$ and $b = 1$; dashed lines are for $x_{B0} = 0.2$ and $b = 0$.

(a) Find the rate at which A is converted to B in this diffusion-plus-chemical-reaction process at steady state.

(b) What happens to the solution when $k_1''' \to 0$?

(c) Is the solution physically meaningful for all values of k_1'''?

Answer: **(a)** $N_{Az}|_{z=0} = \left(\frac{c\mathscr{D}_{AB}}{\delta}\right)\left(\frac{b}{\sinh b}\right)(1 - x_{B0}\cosh b)$ where $b = \sqrt{k_1'''\delta^2/\mathscr{D}_{AB}}$

18B.10 Diffusion with fast second-order reaction. In a steady-state, isothermal flow system, a solid A is dissolving in a flowing liquid stream S. Assume in accordance with the film model that the surface of A is covered with a stagnant liquid film of thickness δ and that the liquid outside the film is well mixed (see Fig. 18.6-2).

(a) Develop an expression for the rate of dissolution of A into the liquid if the concentration of A in the main liquid stream is negligible.

(b) Develop a corresponding expression for the dissolution rate if the liquid contains a substance B, which reacts instantaneously and irreversibly with A: $A + B \to P$. (An example of this system would be the dissolution of benzoic acid, C_6H_5COOH, in an aqueous NaOH solution.) The main liquid stream consists primarily of B and S, with B at a mole fraction of $x_{B\infty}$. (*Hint:* It is necessary to recognize that species A and B both diffuse toward a thin reaction zone as shown in Fig. 18B.10.)

Answers: **(a)** $N_{Az}|_{z=0} = \left(\frac{c\mathscr{D}_{AS}x_{A0}}{\delta}\right)$; **(b)** $N_{Az}|_{z=0} = \left(\frac{c\mathscr{D}_{AS}x_{A0}}{\delta}\right)\left(1 + \frac{x_{B\infty}\mathscr{D}_{BS}}{x_{A0}\mathscr{D}_{AS}}\right)$

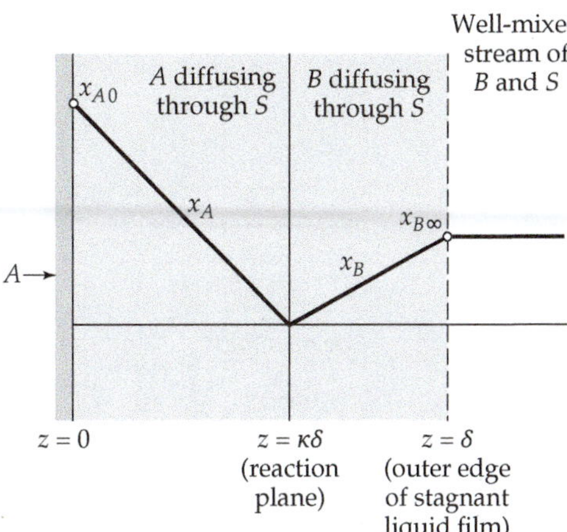

Fig. 18B.10 Concentration profiles for diffusion with rapid second-order reaction. The concentration of product P may be neglected.

18B.11 A sectioned-cell experiment[6] for measuring gas-phase diffusivity. Liquid A is allowed to evaporate through a stagnant gas B at 741 mm Hg total pressure and 25°C. At that temperature, the vapor pressure of A is known to be 600 mm Hg. After steady state has been attained, the cylindrical column of gas is divided into sections as shown in Fig. 18B.11. For a 4-section apparatus with a total height of 4.22 cm, the analysis of the gas samples thus obtained gives the following results:

Section	$(z - z_1)$ in cm: Bottom of section	$(z - z_1)$ in cm: Top of section	Mole fraction of A
I	0.10	1.10	0.757
II	1.10	2.10	0.641
III	2.10	3.10	0.469
IV	3.10	4.10	0.215

The measured evaporation rate of A at steady state is 0.0274 g-mol/hr. Ideal-gas behavior may be assumed.

(a) Verify the following expression for the concentration profile at steady state:

$$\ln \frac{x_{B2}}{x_B(z)} = \frac{N_{Az}(z_2 - z)}{c\mathscr{D}_{AB}} \tag{18B.11-1}$$

(b) Plot the natural log of the mole fraction x_B in each cell versus the value of z at the midplane of the cell. Is a straight line obtained? What are the intercepts at z_1 and z_2? Interpret these results.

(c) Use the concentration profile of Eq. 18B.11-1 to find analytical expressions for the average concentrations in each section of the tube.

(d) Find the best value of $\mathscr{D}_{AB}$ from this experiment.

Answer: **(d)** 0.154 cm^2/s

[6] E. J. Crosby, *Experiments in Transport Phenomena*, Wiley, New York (1961), Experiment 10.a.

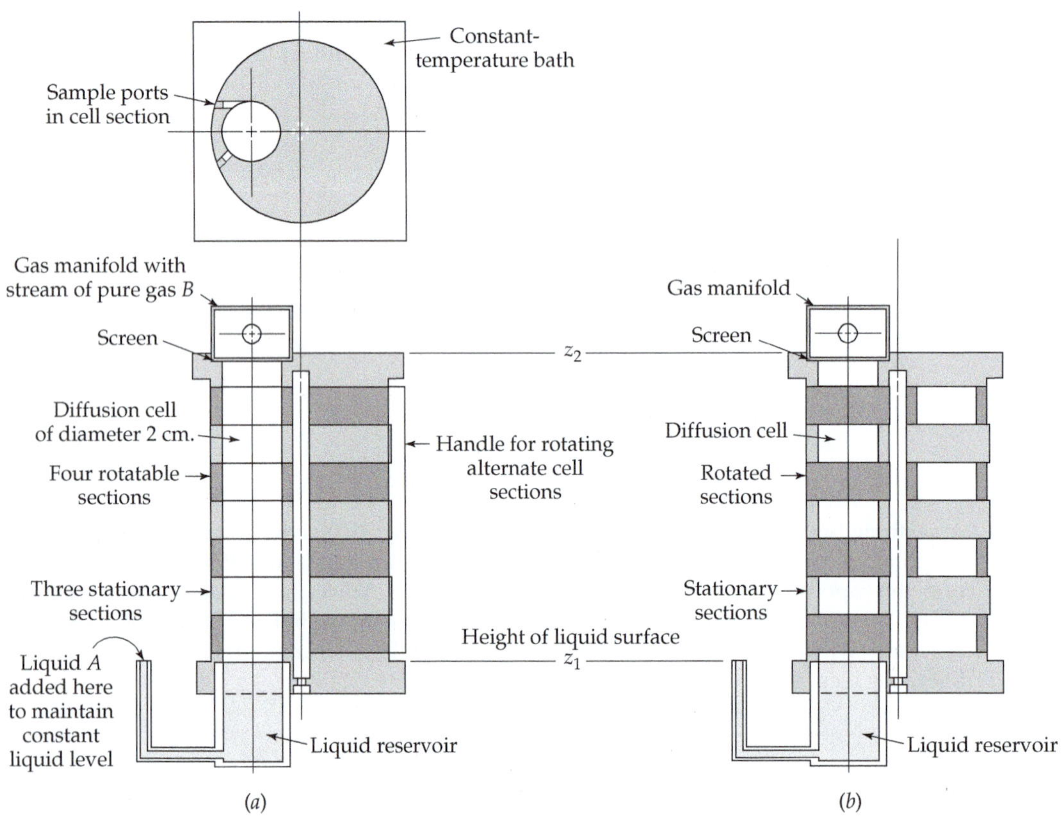

Fig. 18B.11 A sectioned-cell experiment for measuring gas diffusivities. (*a*) Cell configuration during the approach to steady state. (*b*) Cell configuration for gas sampling at the end of the experiment. [Adapted from E. J. Crosby, *Experiments in Transport Phenomena*, Wiley, New York (1961), Experiment 10.a.]

18B.12 Tarnishing of metal surfaces. In the oxidation of most metals (excluding the alkali and alkaline-earth metals) the volume of oxide produced is greater than that of the metal consumed. This oxide thus tends to form a compact film effectively insulating the oxygen and metal from each other. For the derivations that follow it may be assumed that

(a) For oxidation to proceed, oxygen must diffuse through the oxide film and that this diffusion follows Fick's law.

(b) The free surface of the oxide film is saturated with oxygen from the surrounding air.

(c) Once the film of oxide has become reasonably thick, the oxidation becomes diffusion controlled; that is, the dissolved oxygen concentration is essentially zero at the oxide-metal surface.

(d) The rate of change of dissolved oxygen content of the film is small compared to the rate of reaction. That is, quasi-steady-state conditions may be assumed.

(e) The reaction involved is $\frac{1}{2}xO_2 + M \rightarrow MO_x$.

We wish to develop an expression for rate of tarnishing in terms of oxygen diffusivity through the oxide film, the densities of the metal and its oxide, and the stoichiometry of the reaction. Let c_O be the solubility of O_2 in the film, c_f the molar density of the film, and z_f the thickness of the film. Show that the film thickness is

$$z_f = \sqrt{\frac{2\mathscr{D}_{O_2 - MO_x} t}{x} \frac{c_O}{c_f}} \tag{18B.12-1}$$

This result, the so-called "quadratic law," gives a satisfactory empirical correlation for a number of oxidation and other tarnishing reactions.[7] Most such reactions are, however, much more complex than the mechanism given above.[8]

18B.13 Effectiveness factors for thin disks. Consider porous catalyst particles in the shape of thin disks, such that the surface area of the edge of the disk is small in comparison with that of the two circular faces (see Fig. 18B.13). Apply the method of §18.10 to show that the steady-state concentration profile is

$$\frac{c_A(z)}{c_{As}} = \frac{\cosh\sqrt{k_1''a/\mathscr{D}_{A,\text{eff}}}\, z}{\cosh\sqrt{k_1''a/\mathscr{D}_{A,\text{eff}}}\, b} \tag{18B.13-1}$$

where c_{As} is the surface concentration at $z = \pm b$, and z and b are described in the figure. Show that the total mass-transfer rate at the surfaces $z = \pm b$ is

$$|W_A| = 2\pi R^2 c_{As} \mathscr{D}_{A,\text{eff}} \lambda \tanh \lambda b \tag{18B.13-2}$$

in which $\lambda = \sqrt{k_1''a/\mathscr{D}_{A,\text{eff}}}$. Show that, if the disk is sliced parallel to the xy plane into n slices, the total mass-transfer rate becomes

$$|W_A^{(n)}| = 2\pi R^2 c_{As} \mathscr{D}_{A,\text{eff}} \lambda n \tanh(\lambda b/n) \tag{18B.13-3}$$

Obtain the expression for the effectiveness factor by taking the limit

$$\eta_A = \lim_{n\to\infty} \frac{|W_A|}{|W_A^{(n)}|} = \frac{\tanh \lambda b}{\lambda b} \tag{18B.13-4}$$

Express this result in terms of the parameter Λ defined in §18.10.

Surface $z = +b$

Catalyst particle — $z = 0$ (midplane)

Surface $z = -b$

Fig. 18B.13 Side view of a disk-shaped catalyst particle.

18B.14 Diffusion and heterogeneous reaction in a slender cylindrical tube with a closed end. A slender cylindrical pore of length L, cross-sectional area S, and perimeter P, is in contact at its open end with a large body of well-mixed fluid, consisting of species A and B, as shown in Fig. 18B.14. Species A, a minor constituent of this fluid, disappears into the pore, diffuses in the z direction and reacts on its walls. The rate of this reaction may be expressed as $(\mathbf{n} \cdot \mathbf{n}_A)|_{\text{surface}} = f(\omega_{A0})$; that is, at the wall the mass flux normal to the surface is some

[7] G. Tammann, *Z. anorg. allgem. Chemie*, **124**, 25–35 (1922).

[8] W. Jost, *Diffusion*, Academic Press, New York (1952), Chapter IX. For a discussion of the oxidation of silicon, see R. Ghez, *A Primer of Diffusion Problems*, Wiley, New York (1988), §2.3.

function of the mass fraction, ω_{A0}, of A in the fluid adjacent to the solid surface. The mass fraction ω_{A0} depends on z, the distance from the inlet. Because A is present in low concentration, the fluid temperature and density may be considered constant, and the diffusion flux is adequately described by $\mathbf{j}_A = -\rho \mathscr{D}_{AB} \nabla \omega_A$, where the diffusivity may be regarded as a constant. Because the pore is long compared to his lateral dimension, concentration gradients in the lateral directions may be neglected. Note the similarity with the problem discussed in §10.5.

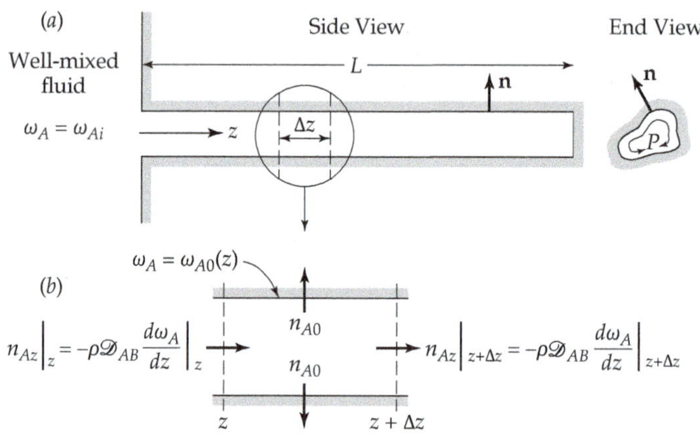

Fig. 18B.14 (*a*) Diffusion and heterogeneous reaction in a long, noncircular cylinder. (*b*) Region of thickness Δz over which the mass balance is made.

(a) Show by means of a shell balance that, at steady state,

$$-\frac{dn_{Az}}{dz} = \frac{P}{S} f(\omega_{A0}) \tag{18B.14-1}$$

(b) Show that the steady-state mass-average velocity v_z is approximately zero for this system.

(c) Substitute the appropriate form of Fick's law into Eq. 18B.14-1, and integrate the resulting differential equation for the special case that $f(\omega_{A0}) = k_1''\omega_{A0}$. To obtain a boundary condition at $z = L$, neglect the rate of reaction on the closed end of the cylinder; why is this a reasonable approximation?

(d) Develop an expression for the total rate w_A of disappearance of A in the cylinder.

(e) Compare the results of parts (c) and (d) with those of §10.5, both from the standpoint of the mathematical development and the nature of the assumptions made.

Answers: **(c)** $\dfrac{\omega_A(z)}{\omega_{Ai}} = \dfrac{\cosh N[1-(z/L)]}{\cosh N}$ where $N = \sqrt{\dfrac{PL^2k_1''}{S\rho\mathscr{D}_{AB}}}$; **(d)** $w_A = (S\rho\mathscr{D}_{AB}\omega_{Ai}/L)N \tanh N$

18B.15 Effect of temperature and pressure on evaporation rate.

(a) In §18.6, what is the effect of a change of temperature and pressure on the quantity x_{A1}?

(b) If the pressure is doubled, how is the evaporation rate in Eq. 18.6-16 affected?

(c) How does the evaporation rate change when the system temperature is raised from T to T'?

18B.16 Reaction rates in large and small particles.

(a) Obtain the following limits for Eq. 18.10-11:

$$R \to 0: \qquad W_{AR} = -\left(\tfrac{4}{3}\pi R^3\right)(k_1''a)c_{AR} \tag{18B.16-1}$$

$$R \to \infty: \qquad W_{AR} = -\left(4\pi R^2\right)\sqrt{k_1''a\mathscr{D}_{A,\text{eff}}}\,c_{AR} \tag{18B.16-2}$$

Interpret these results physically.

(b) Obtain the corresponding asymptotes for the system discussed in Problem 18B.13. Compare them with the results in (a).

18B.17 Evaporation rate for small mole fraction of the volatile liquid. In Eq. 18.6-17, expand

$$\frac{1}{(x_B)_{\ln}} = \left(\frac{1}{x_{A1} - x_{A2}}\right)\left(\ln\frac{1 - x_{A2}}{1 - x_{A1}}\right) \tag{18B.17-1}$$

in a Taylor series appropriate for small mole fractions of A. First rewrite the logarithm of the quotient as the difference of the logarithms. Then expand $\ln(1 - x_{A1})$ and $\ln(1 - x_{A2})$ in Taylor series about $x_{A1} = 0$ and $x_{A2} = 0$, respectively. Verify that Eq. 18.6-18 is correct.

18C.1 Diffusion and reaction in a partially impregnated catalyst. Consider a catalytic sphere like that in §18.10, except that the active ingredient of the catalyst is present only in the annular region between $r = \kappa R$ and $r = R$:

In Region I ($0 \leq r < \kappa R$), $\quad k_1''a = 0 \quad$ (18C.1-1)

In Region II ($\kappa R \leq r \leq R$), $\quad k_1''a = \text{constant} > 0 \quad$ (18C.1-2)

Such a situation may arise when the active ingredient is put on the particles after pelleting, as is done for many commercial catalysts.

(a) Integrate Eq. 18.10-6 separately for the active and inactive regions. Then apply the appropriate boundary conditions to evaluate the integration constants, and solve for the concentration profile in each region. Give qualitative sketches to illustrate the forms of the profiles.

(b) Evaluate W_{AR}, the total rate of reaction in a single particle.

18C.2 Absorption rate in a falling film. The result in Eq. 18.8-18 may be obtained by an alternative procedure.

(a) According to an overall mass balance on the film, the total moles of A transferred per unit time across the gas–liquid interface must be the same as the total molar rate of flow of A across the plane $z = L$. This is calculated as follows:

$$W_A = \lim_{\delta\to\infty}(W\delta v_{\max})\left(\frac{1}{\delta}\int_0^\delta c_A\big|_{z=L}dx\right) = Wv_{\max}\int_0^\infty c_A\big|_{z=L}dx \tag{18C.2-1}$$

Explain this procedure carefully.

(b) Insert the solution for c_A in Eq. 18.8-15 into the result of (a) to obtain:

$$\begin{aligned} W_A &= Wv_{\max}c_{A0}\frac{2}{\sqrt{\pi}}\int_0^\infty\left(\int_{x/\sqrt{4\mathscr{D}_{AB}L/v_{\max}}}^\infty \exp\left(-\xi^2\right)\,d\xi\right)dx \\ &= Wv_{\max}c_{A0}\frac{2}{\sqrt{\pi}}\sqrt{\frac{4\mathscr{D}_{AB}L}{v_{\max}}}\int_0^\infty\left(\int_u^\infty \exp\left(-\xi^2\right)\,d\xi\right)du \end{aligned} \tag{18C.2-2}$$

In the second line, the new variable $u = x/\sqrt{4\mathscr{D}_{AB}L/v_{\max}}$ has been introduced.

(c) Change the order of integration in the double integral, to get

$$W_A = WLc_{A0}\sqrt{\frac{4\mathscr{D}_{AB}v_{\max}}{\pi L}}\cdot 2\int_0^\infty \exp(-\xi^2)\left(\int_0^\xi du\right)\,d\xi \tag{18C.2-3}$$

Explain by means of a carefully drawn sketch how the limits are chosen for the integrals (see §C.7). The integrals may now be done analytically to get Eq. 18.8-18.

18C.3 Estimation of the required length of an isothermal reactor. Let a be the area of catalyst surface per unit volume of a packed-bed catalytic reactor (see Fig. 18.5-1), and S be the cross-sectional area of the reactor. Suppose that the rate of mass flow through the reactor is w (in lb_m/hr, for example).

(a) Show that a steady-state mass balance on substance A over a length dl of the reactor leads to

$$\frac{d\omega_{A0}}{dl} = -\frac{SaN_A M_A}{w} \tag{18C.3-1}$$

(b) Use the result of (a) and Eq. 18.5-17, with the assumptions of constant δ and $\mathscr{D}_{AB}$, to obtain an expression for the reactor length L needed to convert an inlet stream of composition $x_A(0)$ to an outlet stream of composition $x_A(L)$. (*Hint*: Equation (I) of Table 17.4-1 may be useful.)

Answer: **(b)** $L = \left(\dfrac{w\delta M_B}{2Sac\mathscr{D}_{AB}}\right)\displaystyle\int_{x_A(0)}^{x_A(L)} \frac{dx_{A0}}{[M_A x_{A0} + M_B(1 - x_{A0})]^2 \ln\left(1 - \frac{1}{2}x_{A0}\right)}$

Chapter 19

The Equations of Change for Binary Mixtures

In Chapter 18, problems in diffusion were formulated by making shell balances on one of the diffusing species. In this chapter, we make a mass balance for one of the chemical species in a binary mixture over an arbitrary, tiny element of volume fixed in space to establish the equations of continuity for each species. Insertion of the mass flux vector expressions then leads to the diffusion equations in a variety of forms. These diffusion equations can be used to set up any of the problems in Chapter 18 and more complicated ones as well.

Then we summarize all of the equations of change for binary mixtures: the equations of continuity, motion, and energy. These include all the equations of change that were given in Chapters 3 and 11. Next we summarize the expressions for the fluxes. All of these equations are given in general form, although for problem solving we generally use simplified versions of them.

The remainder of the chapter is devoted to solving problems involving diffusion in binary mixtures.

§19.1 THE EQUATIONS OF CONTINUITY FOR A BINARY MIXTURE

In this section we apply the law of conservation of mass to each chemical species in a mixture of A and B. The system we consider is a tiny volume element $\Delta x \Delta y \Delta z$ fixed in space through which the mixture is flowing (see Fig. 19.1-1). Within this mixture there may be a chemical reaction occurring, and we use the symbol r_A to indicate the rate at which species A is being produced, with dimensions of mass per unit volume per unit time.

The various contributions to the mass balance on A are

Rate of increase of mass of A in the volume element: $\qquad (\partial \rho_A/\partial t)\Delta x \Delta y \Delta z \qquad$ (19.1-1)

Rate of addition of mass of A across the face at x: $\qquad n_{Ax}|_x \Delta y \Delta z \qquad$ (19.1-2)

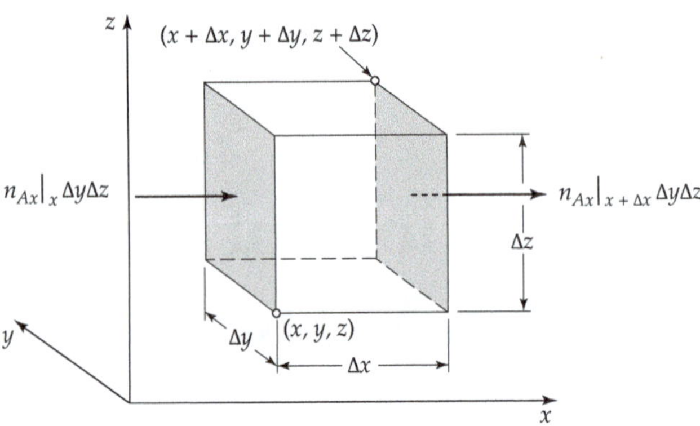

Fig. 19.1-1. Small fixed volume element through which species A is being transported by convection and/or diffusion. The arrows indicate the mass-transfer rates in and out of the volume element through the two shaded surfaces located at x and $x + \Delta x$.

Rate of removal of mass of A across the face at $x + \Delta x$: $\quad n_{Ax}|_{x+\Delta x}\Delta y\Delta z \qquad$ (19.1-3)

Rate of production of mass of A by chemical reaction: $\quad r_A \Delta x\Delta y\Delta z \qquad$ (19.1-4)

We also need to write down the addition and removal terms for the y and z directions, analogous to Eqs. 19.1-2 and 19.1-3. The components of the total mass flux vector, n_{Ax}, n_{Ay}, and n_{Az}, include both the convective flux and the diffusive flux contributions. When the entire mass balance is written down and divided by $\Delta x\Delta y\Delta z$, one obtains, after letting the size of the volume element decrease to zero,

$$\frac{\partial \rho_A}{\partial t} = -\left(\frac{\partial n_{Ax}}{\partial x} + \frac{\partial n_{Ay}}{\partial y} + \frac{\partial n_{Az}}{\partial z}\right) + r_A \tag{19.1-5}$$

This is the *equation of continuity for species A* in a binary mixture. By replacing A everywhere by B, we get the equation of continuity for species B. Equation 19.1-5 describes the change in mass concentration of species A with time at a fixed point in space by the convection and diffusion of A, as well as by chemical reactions that produce or consume species A.

Equation 19.1-5 can be written in vector notation as

$$\frac{\partial \rho_A}{\partial t} = -(\nabla \cdot \mathbf{n}_A) + r_A \tag{19.1-6}$$

Alternatively, we can use Eq. (B) of Table 17.4-1 to write

$$\frac{\partial \rho_A}{\partial t} = -(\nabla \cdot \rho_A \mathbf{v}) - (\nabla \cdot \mathbf{j}_A) + r_A \tag{19.1-7}$$

$\frac{\partial \rho_A}{\partial t}$ =	$-(\nabla \cdot \rho_A \mathbf{v})$	$-(\nabla \cdot \mathbf{j}_A)$	$+r_A$
rate of increase of mass of A per unit volume	net rate of addition of mass of A per unit volume by convection	net rate of addition of mass of A per unit volume by diffusion	rate of production of mass of A per unit volume by reaction

An analogous equation can be written for species B. When the equations for A and B are added, we get

$$\frac{\partial \rho}{\partial t} = -(\nabla \cdot \rho \mathbf{v}) \tag{19.1-8}$$

which is the *equation of continuity for the binary mixture*. This equation is identical to the equation of continuity for a pure fluid given in Eq. 3.1-4. In obtaining Eq. 19.1-8 we had to use Eq. (B) of Table 17.1-1 and also the fact that the law of conservation mass requires that $r_A + r_B = 0$. Finally we note that Eq. 19.1-8 becomes

$$(\nabla \cdot \mathbf{v}) = 0 \tag{19.1-9}$$

for a binary mixture of *constant mass density* ρ.

In the preceding discussion we used mass units, but a corresponding derivation is possible using molar units. The equation of continuity for species A in molar quantities, obtained by dividing Eq. 19.1-6 by the molecular weight of species A, is

$$\frac{\partial c_A}{\partial t} = -(\nabla \cdot \mathbf{N}_A) + R_A \tag{19.1-10}$$

where R_A is the molar rate of production of A per unit volume. By using Eq. (F) of Table 17.4-1, the above equation can be rewritten as

$$\frac{\partial c_A}{\partial t} = -\left(\nabla \cdot c_A\mathbf{v}^*\right) - \left(\nabla \cdot \mathbf{J}_A^*\right) + R_A \tag{19.1-11}$$

$\frac{\partial c_A}{\partial t}$	$-\left(\nabla \cdot c_A\mathbf{v}^*\right)$	$-\left(\nabla \cdot \mathbf{J}_A^*\right)$	$+R_A$
rate of increase of moles of A per unit volume	net rate of addition of moles of A per unit volume by convection	net rate of addition of moles of A per unit volume by diffusion	rate of production of moles of A per unit volume by reaction

with an analogous equation for species B. When the equations for the two species are added, we get

$$\frac{\partial c}{\partial t} = -(\nabla \cdot c\mathbf{v}^*) + (R_A + R_B) \tag{19.1-12}$$

where Eq. (H) of Table 17.3-1 has been used. The chemical reaction terms do not drop out, because the number of moles is not necessarily conserved in a chemical reaction. Finally we note that, in a *fluid mixture with constant molar density*, it is *not* true that $(\nabla \cdot \mathbf{v}^*) = 0$, but rather that

$$(\nabla \cdot \mathbf{v}^*) = \frac{1}{c}(R_A + R_B) \tag{19.1-13}$$

We have thus seen that the equation of continuity of species A in a binary mixture may be written in two forms, Eq. 19.1-7 and Eq. 19.1-11. When these equations are rewritten in terms of the mass fraction and mole fraction, respectively, we get

$$\frac{\partial}{\partial t}(\rho\omega_A) = -(\nabla \cdot \rho\omega_A\mathbf{v}) - (\nabla \cdot \mathbf{j}_A) + r_A \tag{19.1-14}$$

$$\frac{\partial}{\partial t}(cx_A) = -\left(\nabla \cdot cx_A\mathbf{v}^*\right) - \left(\nabla \cdot \mathbf{J}_A^*\right) + R_A \tag{19.1-15}$$

When the products are differentiated, these equations become

$$\rho\frac{\partial \omega_A}{\partial t} + \underline{\omega_A\frac{\partial \rho}{\partial t}} = -\rho(\mathbf{v} \cdot \nabla\omega_A) - \underline{\omega_A(\nabla \cdot \rho\mathbf{v})} - (\nabla \cdot \mathbf{j}_A) + r_A \tag{19.1-16}$$

$$c\frac{\partial x_A}{\partial t} + \underline{x_A\frac{\partial c}{\partial t}} = -c(\mathbf{v}^* \cdot \nabla x_A) - \underline{x_A(\nabla \cdot c\mathbf{v}^*)} - (\nabla \cdot \mathbf{J}_A^*) + R_A \tag{19.1-17}$$

Now it can be seen that when Eq. 19.1-8, multiplied by ω_A, is subtracted from Eq. 19.1-16, the dashed-underlined terms disappear and Eq. 19.1-18 below results. Similarly, when

Eq. 19.1-12, multiplied by x_A, is subtracted from Eq. 19.1-17, the dashed-underlined terms disappear, and a term $-x_A(R_A + R_B)$ appears on the right side, as in Eq. 19.1-19 below. Therefore, two additional equivalent forms of the species equation of continuity for A are

$$\boxed{\rho\left(\frac{\partial \omega_A}{\partial t} + (\mathbf{v}\cdot\nabla\omega_A)\right) = -(\nabla\cdot\mathbf{j}_A) + r_A} \tag{19.1-18}$$

$$\boxed{c\left(\frac{\partial x_A}{\partial t} + (\mathbf{v}^*\cdot\nabla x_A)\right) = -(\nabla\cdot\mathbf{J}_A^*) + R_A - x_A(R_A + R_B)} \tag{19.1-19}$$

These two equations contain the same physical content, but the first is written in mass quantities and the second in molar quantities. These equations are tabulated in Cartesian, cylindrical, and spherical coordinates in §B.10. To use these equations, we have to insert the appropriate expressions for the fluxes and the chemical reaction terms. We give three special simplifications of the equation of continuity below.

(i) Binary mixtures with constant $\rho\mathscr{D}_{AB}$

For this assumption, Eq. 19.1-18 becomes, after inserting Fick's law from Eq. (A) of Table 17.4-2,

$$\rho\left(\frac{\partial \omega_A}{\partial t} + (\mathbf{v}\cdot\nabla\omega_A)\right) = \rho\mathscr{D}_{AB}\nabla^2\omega_A + r_A \tag{19.1-20}$$

with a corresponding equation for species B. This equation is appropriate for describing the diffusion in *dilute liquid solutions* at constant temperature and pressure. The left side may also be written as $\rho D\omega_A/Dt$. Equation 19.1-20 without the reaction term r_A is of the same form as Eqs. 11.2-8 or 11.2-9. This similarity is quite important, because it is the basis for the analogies that are frequently used between heat and mass transport in flowing fluids with constant physical properties. Equation 19.1-20 is displayed in Cartesian, cylindrical, and spherical coordinates in §B.11.

(ii) Binary mixtures with constant $c\mathscr{D}_{AB}$

For this assumption, Eq. 19.19 becomes, after inserting Fick's law from Eq. (D) of Table 17.4-2,

$$c\left(\frac{\partial x_A}{\partial t} + (\mathbf{v}^*\cdot\nabla x_A)\right) = c\mathscr{D}_{AB}\nabla^2 x_A + (x_B R_A - x_A R_B) \tag{19.1-21}$$

with a corresponding equation for B. This equation is useful for *low-density gases* at constant temperature and pressure. The left side may *not* be written as cDx_A/Dt because of the appearance of $\mathbf{v}^*$ rather than $\mathbf{v}$. Equation 19.1-21 can be written in Cartesian, cylindrical, and spherical coordinates using §B.11.

(iii) Binary mixtures with zero velocity and no chemical reactions

In the absence of chemical reactions, with $\mathbf{v} = 0$ and $\rho =$ constant in Eq. 19.1-20 (or $\mathbf{v}^* = 0$ and $c =$ constant in Eq. 19.1-21), then we get

$$\frac{\partial c_A}{\partial t} = \mathscr{D}_{AB}\nabla^2 c_A \tag{19.1-22}$$

which is called *Fick's second law of diffusion*, or sometimes simply *the diffusion equation*. This equation is usually used for diffusion in *solids* or *stationary liquids* (that is, $\mathbf{v} = 0$ in Eq. 19.1-20) and for *equimolar counter-diffusion* in gases (that is, $\mathbf{v}^* = 0$ in Eq. 19.1-21). By equimolar counter-diffusion we mean that the net molar flux with respect to stationary coordinates is zero; in other words, that for every mole of A that moves, say, in the positive z direction, there is a mole of B that moves in the negative z direction. We had an example of this in the two-bulb experiment in §18.7. Equation 19.1-22 can be written in Cartesian, cylindrical, and spherical coordinates by simplifying the corresponding equations in §B.11.

Note that Eq. 19.1-22 has exactly the same form as the *heat-conduction equation* of Eq. 11.2-10. This similarity is the basis for analogies between many heat-conduction and diffusion problems in solids. Keep in mind that many hundreds of problems described by Eq. 11.2-10 or Eq. 19.1-22 have been tabulated in the monographs of Carslaw and Jaeger[1] and Crank.[2]

EXAMPLE 19.1-1

Convection, Diffusion, and Chemical Reaction[3]

In Fig. 19.1-2 we show a system in which a liquid, B, moves slowly upward through a slightly soluble porous plug of A. Some of the A dissolves in B and then slowly disappears by a first-order reaction. Assume that the velocity profile is flat across the tube. Assume further that c_{A0} is the solubility of unreacted A in B. Because A is only slightly soluble, we can ignore the presence of reaction products, and treat the system as a binary system (the *pseudo-binary* assumption). Neglect temperature effects associated with the heat of reaction.

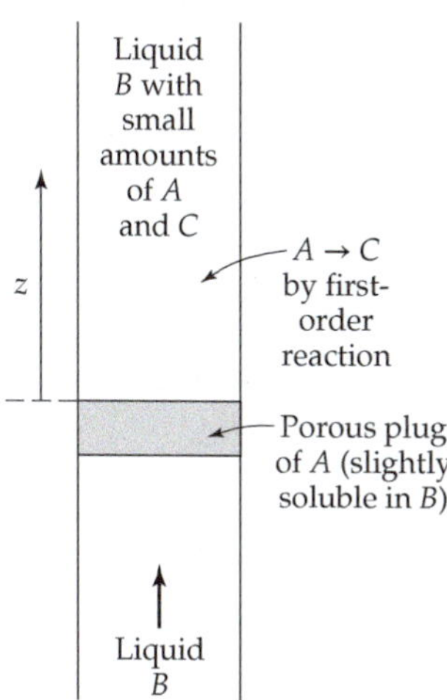

Fig. 19.1-2. Simultaneous diffusion, convection, and chemical reaction.

SOLUTION

Equation 19.1-20 is appropriate for dilute liquid solutions. Dividing this equation by the molecular weight M_A and specializing for the one-dimensional steady-state problem at hand, we get for constant ρ

$$v_0 \frac{dc_A}{dz} = \mathscr{D}_{AB} \frac{d^2 c_A}{dz^2} - k_1''' c_A \qquad (19.1\text{-}23)$$

which is to be solved with the boundary conditions that $c_A = c_{A0}$ at $z = 0$, and that $c_A \to 0$ as $z \to \infty$.

---

[1] H. S. Carslaw and J. C. Jaeger, *Conduction of Heat in Solids*, Oxford University Press, 2nd edition (1959).

[2] J. Crank, *The Mathematics of Diffusion*, Oxford University Press, 2nd edition (1975).

[3] W. Jost, *Diffusion*, Academic Press, New York (1952), pp. 58–59.

Equation 19.1-23 is expected to have a solution of the form of Eq. C.1-7a

$$c_A(z) = C_1 \exp(n_+ z) + C_2 \exp(n_- z) \tag{19.1-24}$$

where n_+ and n_- are the two solutions of the quadratic equation

$$\mathscr{D}_{AB} n^2 - v_0 n - k_1''' = 0 \tag{19.1-25}$$

which are

$$n_+ = \frac{v_0 + \sqrt{v_0^2 + 4\mathscr{D}_{AB} k_1'''}}{2\mathscr{D}_{AB}} > 0 \tag{19.1-26}$$

$$n_- = \frac{v_0 - \sqrt{v_0^2 + 4\mathscr{D}_{AB} k_1'''}}{2\mathscr{D}_{AB}} < 0 \tag{19.1-27}$$

The boundary condition $c_A \to 0$ as $z \to \infty$ requires that $C_1 = 0$. Applying the boundary condition that $c_A = c_{A0}$ at $z = 0$, we get $C_2 = c_{A0}$. Therefore, the solution to Eq. 19.1-23 is

$$\frac{c_A(z)}{c_{A0}} = \exp\left[-\left(\sqrt{1 + \frac{4\mathscr{D}_{AB} k'''}{v_0^2}} - 1\right)\left(\frac{v_0 z}{2\mathscr{D}_{AB}}\right)\right] \tag{19.1-28}$$

and the two boundary conditions are satisfied. This example illustrates the use of the equation of continuity of A for setting up a diffusion problem with convection and chemical reaction.

---

§19.2 SUMMARY OF THE BINARY MIXTURE CONSERVATION LAWS

In the three main parts of this book we have introduced the conservation laws. In Chapter 3, the conservation of mass and conservation of linear momentum for pure fluids were presented. In Chapter 11, we added the conservation of energy for pure fluids. In §19.1, we formulated the mass conservation equations for the chemical species in a binary mixture.

In Table 19.2-1, we summarize the equations of change for binary mixtures in terms of the total fluxes—the fluxes with respect to stationary axes. The equation numbers indicate where each equation first appeared. All these equations have the same form

$$\begin{Bmatrix} \text{rate of} \\ \text{increase} \\ \text{of } X \text{ per} \\ \text{unit} \\ \text{volume} \end{Bmatrix} = \begin{Bmatrix} \text{net rate} \\ \text{of addition} \\ \text{of } X \text{ per} \\ \text{unit} \\ \text{volume} \end{Bmatrix} + \begin{Bmatrix} \text{rate of} \\ \text{production} \\ \text{of } X \text{ per} \\ \text{unit} \\ \text{volume} \end{Bmatrix} \tag{19.2-1}$$

in which X refers to mass, momentum, and energy, respectively. In each equation, the net rate of addition of X per unit volume is the negative of a divergence term. The "rates of production" arise from a chemical reaction in the first equation and from the external force field $\mathbf{g}$ in the other two. Each equation is the statement of a *conservation law.*

The three *total fluxes* (the vector $\mathbf{n}_A$, the tensor $\boldsymbol{\phi}$, and the vector $\mathbf{e}$) appearing in Table 19.2-1 may be written as the sum of *convective fluxes* and *molecular fluxes*. These fluxes are displayed in Table 19.2-2, along with the equation numbers corresponding to their first appearance.

When the flux expressions of Table 19.2-2 are substituted into the conservation equations in Table 19.2-1 and then converted to the D/Dt form by means of Eqs. 3.5-6

Table 19.2-1. The Conservation Statements in Terms of the Total Fluxes

Mass of A:	$\dfrac{\partial}{\partial t}\rho\omega_A = -(\nabla \cdot \mathbf{n}_A) + r_A$	(A)[a] (Eq. 19.1-6)
Momentum:	$\dfrac{\partial}{\partial t}\rho\mathbf{v} = -[\nabla \cdot \boldsymbol{\phi}] + \rho\mathbf{g}$	(B)[b] (Eq. 3.2-8)
Energy:	$\dfrac{\partial}{\partial t}\rho\left(\dfrac{1}{2}v^2 + \hat{U}\right) = -(\nabla \cdot \mathbf{e}) + (\rho\mathbf{v} \cdot \mathbf{g})$	(C)[b] (Eq. 11.1-6)

[a]When the equations of continuity for species A and B are added, the equation of continuity for the binary mixture is obtained.

$$\frac{\partial}{\partial t}\rho = -(\nabla \cdot \rho\mathbf{v}) \qquad \text{(D) (Eq. 3.1-4)}$$

Here $\mathbf{v}$ is the mass-average velocity (see Eq. (D) of Table 17.4-1).

[b]If each species K is acted on by a different force per unit mass $\mathbf{g}_K$ (where K is either A or B), then in Eq. (B), replace $\rho\mathbf{g}$ by $(\rho_A\mathbf{g}_A + \rho_B\mathbf{g}_B)$, and in Eq. (C), replace $(\rho\mathbf{v} \cdot \mathbf{g})$ by $(\mathbf{n}_A \cdot \mathbf{g}_A + \mathbf{n}_B \cdot \mathbf{g}_B)$. These substitutions are required, for example, if the species are ions with different charges, and are acted upon by electric fields. Problems of this sort are considered in Chapter 24.

and 3.5-7, we get the equations of change obtained previously, which are summarized in Table 19.2-3.

In addition to the conservation equations, we need to have the expressions for the fluxes in terms of the gradients and the transport properties, the latter being functions of the temperature, density, and concentration. Finally, we need the *thermal equation of state* $p = p(\rho,T,x_A)$ and the *caloric equation of state* $\hat{U} = \hat{U}(\rho,T,x_A)$, and information about the rates of any homogeneous chemical reactions occurring.

The equations of motion and energy, given in Table 19.2-3, are not necessarily in their most convenient forms for solving some problems. Let us first consider the equation of motion. In §11.3, it was pointed out that the *equation of motion* as presented in Chapter 3

Table 19.2-2. The Total, Convective, and Molecular Fluxes for Binary Mixtures (All with the Same Sign Convention)

Entity	Total flux	=	Convective flux	+	Molecular flux	
Mass of A:	$\mathbf{n}_A$	=	$\rho\mathbf{v}\omega_A$	+	$\mathbf{j}_A$	(A)[a] (Eq. 17.4-1)
Momentum:	$\boldsymbol{\phi}$	=	$\rho\mathbf{v}\mathbf{v}$	+	$\boldsymbol{\pi}$	(B)[b] (Eq. 1.3-2)
Energy:	$\mathbf{e}$	=	$\rho\mathbf{v}\left(\frac{1}{2}v^2 + \hat{U}\right)$	+	$(\mathbf{q} + \mathbf{w})$	(C)[c] (Eq. 9.4-1)

[a]The velocity $\mathbf{v}$ is the mass-average velocity defined in Eq. (A) of Table 17.2-1.

[b]The molecular momentum-flux tensor is $\boldsymbol{\pi} = p\boldsymbol{\delta} + \boldsymbol{\tau}$ (see Eq. 1.2-15).

[c]The molecular energy-flux vector is $\mathbf{q} + \mathbf{w} = \mathbf{q} + [\boldsymbol{\pi} \cdot \mathbf{v}] = \mathbf{q} + p\mathbf{v} + [\boldsymbol{\tau} \cdot \mathbf{v}]$ (see Eq. 9.4-1), where, for mixtures, $\mathbf{q}$ is the conductive-plus-diffusive heat-flux vector and $\mathbf{w}$ is the work-flux vector. The conductive-plus-diffusive heat-flux vector contains both Fourier's law of heat conduction and terms for heat transport that accompanies diffusion, as described in §19.3 (see Eq. 19.3-3). The work-flux vector arises only in flow systems.

Table 19.2-3. Equations of Change, Based on Conservation Laws, for Binary Mixtures in Terms of the Molecular Fluxes

Total mass:	$\dfrac{D\rho}{Dt} = -\rho(\nabla \cdot \mathbf{v})$	(A) (Eq. (A) of Table 3.5-1)
Mass of A:	$\rho\dfrac{D\omega_A}{Dt} = -(\nabla \cdot \mathbf{j}_A) + r_A$	(B)[a] (Eq. 19.1-18)
Momentum:	$\rho\dfrac{D\mathbf{v}}{Dt} = -\nabla p - [\nabla \cdot \boldsymbol{\tau}] + \rho\mathbf{g}$	(C)[b] (Eq. (B.) of Table 3.5-1)
Energy:	$\rho\dfrac{D}{Dt}\left(\dfrac{1}{2}v^2 + \hat{U}\right) = -(\nabla \cdot \mathbf{q}) - (\nabla \cdot p\mathbf{v}) - (\nabla \cdot [\boldsymbol{\tau} \cdot \mathbf{v}]) + (\rho\mathbf{v} \cdot \mathbf{g})$	(D)[b] (Eq. (D) of Table 11.4-1)

[a] Analogous to the equation for the conservation of mass of A, there is an equation for the mass of B. When these two equations are added, we get the conservation of total mass (in Eq. A).

[b] See Note (b) of Table 19.2-1 on how the term $\rho\mathbf{g}$ in Eq. (C) and the term $(\rho\mathbf{v} \cdot \mathbf{g})$ in Eq. (D) need to be modified when each species is acted on by a different external force. As described in §19.3, for binary mixtures, the conductive-plus-diffusive heat-flux vector contains both Fourier's law of heat conduction and terms accounting for heat transport that accompanies diffusion (see Eq. 19.3-3).

[c] In Chapter 3, we also gave an equation of change for angular momentum (Eq. 3.4-1).

is suitable for setting up forced-convection problems, but that an alternative form (Eq. 11.3-2) is desirable for displaying explicitly the buoyant forces resulting from temperature inequalities in free-convection problems. In binary mixtures with concentration inequalities as well as temperature inequalities, we write the equation of motion as in Eq. (B) of Table 3.5-1 and use an approximate equation of state formed by making a double Taylor expansion of $\rho(T,\omega_A)$ about the values $\overline{T},\overline{\omega}_A$:

$$\rho(T,\omega_A) = \overline{\rho} + \left.\frac{\partial \rho}{\partial T}\right|_{\overline{T},\overline{\omega}_A}(T-\overline{T}) + \left.\frac{\partial \rho}{\partial \omega_A}\right|_{\overline{T},\overline{\omega}_A}(\omega_A - \overline{\omega}_A) + \cdots$$
$$\approx \overline{\rho} - \overline{\rho}\overline{\beta}(T-\overline{T}) - \overline{\rho}\overline{\zeta}(\omega_A - \overline{\omega}_A) \tag{19.2-2}$$

Here the coefficient $\overline{\zeta} = -(1/\overline{\rho})(\partial\rho/\partial\omega_A)_{T,p}$, evaluated at $\overline{T}$ and $\overline{\omega}_A$, relates the density to the composition. This coefficient is the mass-transfer analog of the coefficient $\overline{\beta} = -(1/\rho)(\partial\rho/\partial T)_{p,\omega_A}$ introduced in Eq. 11.3-1 in the discussion of free convection in pure fluids. When this approximate equation of state is substituted into the $\rho\mathbf{g}$ term of the equation of motion (but not into the $\rho D\mathbf{v}/Dt$ term, where we set $\rho = \overline{\rho}$), we get the *Boussinesq equation for a binary mixture*, with gravity as the only external force,

$$\rho\frac{D\mathbf{v}}{Dt} = (-\nabla p + \overline{\rho}\mathbf{g}) - [\nabla \cdot \boldsymbol{\tau}] - \overline{\rho}\mathbf{g}\overline{\beta}(T-\overline{T}) - \overline{\rho}\mathbf{g}\overline{\zeta}(\omega_A - \overline{\omega}_A) \tag{19.2-3}$$

The last two terms describe the buoyant force resulting from the temperature and composition variations within the fluid.

Next we turn to the *equation of energy*. Recall that in Table 11.4-1 the energy equation for pure fluids was given in a variety of forms. The same can be done for binary mixtures, and a few of the many possible forms for the energy equation are given in Table 19.2-4. Note that it is not necessary to add a term S_c (as we did in Chapter 10) to account for the thermal energy released by homogeneous chemical reactions. This information is implicitly included in the $\hat{U}$ and $\hat{H}$ functions, because the energies of formation and mixing of the various species must be taken into account. Furthermore, information about chemical reactions is explicitly included as $-(\overline{H}_A R_A + \overline{H}_B R_B)$ in Eq. (D) of Table 19.2-4.

Table 19.2-4. Various Forms of the Equation of Energy for a Binary Mixture, When Gravity Is the Only External Force [a,b]

$$\rho \frac{D}{Dt}\left(\frac{1}{2}v^2 + \hat{U} + \hat{\Phi}\right) = -(\nabla \cdot \mathbf{q}) - (\nabla \cdot p\mathbf{v}) - (\nabla \cdot [\boldsymbol{\tau} \cdot \mathbf{v}]) \quad \text{(A)}^c$$

$$\rho \frac{D}{Dt}\left(\frac{1}{2}v^2 + \hat{U}\right) = -(\nabla \cdot \mathbf{q}) - (\nabla \cdot p\mathbf{v}) - (\nabla \cdot [\boldsymbol{\tau} \cdot \mathbf{v}]) + (\rho\mathbf{v} \cdot \mathbf{g}) \quad \text{(B)}$$

$$\rho \frac{D\hat{U}}{Dt} = -(\nabla \cdot \mathbf{q}) - p(\nabla \cdot \mathbf{v}) - (\boldsymbol{\tau} : \nabla\mathbf{v}) \quad \text{(C)}$$

$$\rho \hat{C}_p \frac{DT}{Dt} = -(\nabla \cdot \mathbf{q}) - (\boldsymbol{\tau} : \nabla\mathbf{v}) + \left(\frac{\partial \ln \hat{V}}{\partial \ln T}\right)_{p,x_A} \frac{Dp}{Dt} + \overline{H}_A\left[\left(\nabla \cdot \frac{\mathbf{j}_A}{M_A}\right) - R_A\right] + \overline{H}_B\left[\left(\nabla \cdot \frac{\mathbf{j}_B}{M_B}\right) - R_B\right] \quad \text{(D)}^d$$

$$\frac{\partial}{\partial t}(c_A\overline{H}_A + c_B\overline{H}_B) + (\nabla \cdot (\mathbf{N}_A\overline{H}_A + \mathbf{N}_B\overline{H}_B)) = (\nabla \cdot k\nabla T) - (\boldsymbol{\tau} : \nabla\mathbf{v}) + \frac{Dp}{Dt} \quad \text{(E)}^e$$

[a] For binary mixtures, $\mathbf{q} = -k\nabla T + (\overline{H}_A/M_A)\mathbf{j}_A + (\overline{H}_B/M_B)\mathbf{j}_B + \mathbf{q}^{(x)}$, where $\mathbf{q}^{(x)}$ is a usually negligible term associated with the diffusion-thermo effect (see §19.3, and R. B. Bird, W. E. Stewart, and E. N. Lightfoot, *Transport Phenomena*, Wiley, New York, Revised 2nd Edition (2007), Eq. 24.2-6).

[b] The equations in this table are written for the situation that the same external force is acting on both species. If this is not the case, then $(\mathbf{j}_A \cdot \mathbf{g}_A) + (\mathbf{j}_B \cdot \mathbf{g}_B)$ must be added to the right sides of Eqs. (A), (C), (D), and (E); also the last term in Eq. (B) has to be replaced by $(\mathbf{n}_A \cdot \mathbf{g}_A) + (\mathbf{n}_B \cdot \mathbf{g}_B)$.

[c] Exact only if $\partial\hat{\Phi}/\partial t = 0$.

[d] Equation (D) was first published by L. B. Rothfeld, PhD Thesis, University of Wisconsin; see also Problem 19D.2 of R. B. Bird, W. E. Stewart, and E. N. Lightfoot, *op cit.*

[e] The contribution of $\mathbf{q}^{(x)}$ to the heat-flux vector has been omitted in this equation.

§19.3 SUMMARY OF THE BINARY MIXTURE MOLECULAR FLUXES

The equations of change have been given in terms of the fluxes of mass, momentum, and energy. To solve these equations, we have to write the fluxes in terms of the transport properties and the gradients of concentration, velocity, and temperature. Here we summarize the molecular flux expressions for binary mixtures:

Mass:
$$\mathbf{j}_A = -\rho\mathscr{D}_{AB}\nabla\omega_A \quad (19.3\text{-}1)$$

Momentum:
$$\boldsymbol{\tau} = -\mu[\nabla\mathbf{v} + (\nabla\mathbf{v})^\dagger] + \left(\frac{2}{3}\mu - \kappa\right)(\nabla \cdot \mathbf{v})\boldsymbol{\delta} \quad (19.3\text{-}2)$$

Energy:
$$\mathbf{q} = -k\nabla T + (\overline{H}_A/M_A)\mathbf{j}_A + (\overline{H}_B/M_B)\mathbf{j}_B \quad (19.3\text{-}3)$$

The limitations of these molecular flux expressions need to be summarized:

a. The *diffusive mass-flux vector* above is for binary systems only (*Fick's law*). The generalization to multicomponent systems has been made (the Maxwell-Stefan equations), but we do not discuss those equations here (see §24.6). Concentration gradients (or activity gradients) are not the only driving forces for diffusion. There are also "cross effects": temperature gradients can create a mass flux (*thermal diffusion* or the "Soret effect") and pressure gradients can create a mass flux (*pressure diffusion*). In addition, there is *forced diffusion*, caused by unequal forces acting on the constituent species. These other mechanisms are neglected in the remainder of this chapter, but they are discussed again briefly in Chapter 24.

b. The *molecular (viscous) momentum-flux tensor* for mixtures is the same as that for pure fluids (a generalization of *Newton's law of viscosity*). Once again we mention that the dilatational viscosity κ is seldom important. Of course, for polymers and other complex fluids, Eq. 19.3-2 has to be replaced by more complex equations (see Chapter 8).

c. The *conductive-plus-diffusive heat-flux vector* given in Eq. 19.3-3 contains a conduction term (*Fourier's law*) and a term that accounts for the heat transported by the interdiffusing species, which may be quite important. The heat transport by concentration gradients (the *diffusion thermo effect* or the "Dufour effect") $\mathbf{q}^{(x)}$ is often negligible. The thermal conductivity k for a mixture in Eq. 19.3-3 is defined as the proportionality constant between the heat flux and the temperature gradient in the absence of any mass fluxes.

We conclude this discussion with a few comments about the total energy-flux vector **e**. By substituting Eq. 19.3-3 into Eq. (C) of Table 19.2-2, we get after some rearranging

$$\begin{aligned}\mathbf{e} &= \rho\left(\frac{1}{2}v^2 + \hat{U}\right)\mathbf{v} + \mathbf{q} + p\mathbf{v} + [\boldsymbol{\tau}\cdot\mathbf{v}] \\ &= \rho\left(\frac{1}{2}v^2 + \hat{U}\right)\mathbf{v} - k\nabla T + \left(\frac{\overline{H}_A}{M_A}\mathbf{j}_A + \frac{\overline{H}_B}{M_B}\mathbf{j}_B\right) + p\mathbf{v} + [\boldsymbol{\tau}\cdot\mathbf{v}] \\ &= -k\nabla T + \left(\frac{\overline{H}_A}{M_A}\mathbf{j}_A + \frac{\overline{H}_B}{M_B}\mathbf{j}_B\right) + \rho(\hat{U} + p\hat{V})\mathbf{v} + \underline{\frac{1}{2}\rho v^2\mathbf{v} + [\boldsymbol{\tau}\cdot\mathbf{v}]} \end{aligned} \tag{19.3-4}$$

In some situations, notably in films and low-velocity boundary layers, the dashed underlined terms are negligible. When they may be discarded, we get

$$\begin{aligned}\mathbf{e} &= -k\nabla T + \left(\frac{\overline{H}_A}{M_A}\mathbf{j}_A + \frac{\overline{H}_B}{M_B}\mathbf{j}_B\right) + \rho\hat{H}\mathbf{v} \\ &= -k\nabla T + \left(\frac{\overline{H}_A}{M_A}\mathbf{j}_A + \frac{\overline{H}_B}{M_B}\mathbf{j}_B\right) + (c_A\overline{H}_A + c_B\overline{H}_B)\mathbf{v}\end{aligned} \tag{19.3-5}$$

or

$$\begin{aligned}\mathbf{e} &= -k\nabla T + \left(\frac{\overline{H}_A}{M_A}\mathbf{j}_A + \frac{\overline{H}_B}{M_B}\mathbf{j}_B\right) + \left(\rho_A\frac{\overline{H}_A}{M_A} + \rho_B\frac{\overline{H}_B}{M_B}\right)\mathbf{v} \\ &= -k\nabla T + \left(\frac{\overline{H}_A}{M_A}\mathbf{n}_A + \frac{\overline{H}_B}{M_B}\mathbf{n}_B\right) \\ &= -k\nabla T + (\overline{H}_A\mathbf{N}_A + \overline{H}_B\mathbf{N}_B)\end{aligned} \tag{19.3-6}$$

Finally, for ideal-gas mixtures, this expression can be further simplified by replacing the partial molar enthalpies (such as $\overline{H}_A$) by the molar enthalpies (such as $\tilde{H}_A$). Equation 19.3-6 is the standard starting point for solving one-dimensional problems in simultaneous heat and mass transfer.[1]

[1] T. K. Sherwood, R. L. Pigford, and C. R. Wilke, *Mass Transfer*, McGraw-Hill, New York (1975), Chapter 7. **Thomas Kilgore Sherwood** (1903–1976) was a professor at MIT for nearly 40 years, and then taught at the University of California in Berkeley. Because of his many contributions to the field of mass transfer, the Sherwood number (Sh) was named after him.

EXAMPLE 19.3-1

The Partial Molar Enthalpy

The partial molar enthalpy $\overline{H}_A$, which appears in Eqs. 19.3-3 and 19.3-6, is defined for a binary mixture as

$$\overline{H}_A = \left(\frac{\partial H}{\partial n_A}\right)_{T,p,n_B} \tag{19.3-7}$$

in which n_A is the number of moles of A in the mixture, and the subscript n_B indicates that the derivative is to be taken holding the number of moles of B constant. The enthalpy $H(n_A,n_B;T,p)$ is called an "extensive property," because, if the number of moles of each component is multiplied by a constant K, the enthalpy itself will be multiplied by K

$$H(Kn_A,Kn_B;T,p) = KH(n_A,n_B;T,p) \tag{19.3-8}$$

Mathematicians refer to this kind of function as being "homogeneous of degree 1." For such functions Euler's theorem[2] can be used to conclude that

$$H = n_A\overline{H}_A + n_B\overline{H}_B \tag{19.3-9}$$

(a) Prove that, for a binary mixture, the partial molar enthalpies at a given mole fraction can be determined by plotting the enthalpy per mole as a function of mole fraction, and then determining the intercepts of the tangent drawn at the mole fraction in question (see Fig. 19.3-1). This shows one way to get the partial molar enthalpy from data on the enthalpy of the mixture.

(b) How else could one get the partial molar enthalpy?

SOLUTION

(a) Throughout the remainder of this example, we omit the subscripts p and T, indicating that they are being held constant. First we write the expressions for the intercepts as follows:

$$\overline{H}_A = \tilde{H} - x_B\left(\frac{\partial \tilde{H}}{\partial x_B}\right)_n \quad \text{and} \quad \overline{H}_B = \tilde{H} + x_A\left(\frac{\partial \tilde{H}}{\partial x_B}\right)_n \tag{19.3-10,11}$$

in which $\tilde{H} = H/(n_A + n_B) = H/n$ is the enthalpy per mole.

To verify the correctness of Eq. 19.3-10, we rewrite it in terms of H

$$\overline{H}_A = \frac{H}{n} - \frac{x_B}{n}\left(\frac{\partial H}{\partial x_B}\right)_n \tag{19.3-12}$$

Now the expression $\overline{H}_A = (\partial H/\partial n_A)_{n_B}$ implies that H is a function of n_A and n_B, whereas $(\partial H/\partial x_A)_n$ implies that H is a function of x_A and n. The relation between these derivatives is given by the chain rule of partial differentiation. To apply this rule, we need the relations among the independent variables, which, in this example, are

$$n_A = (1 - x_B)n \quad \text{and} \quad n_B = x_B n \tag{19.3-13,14}$$

Therefore, we may write the chain rule as

$$\begin{aligned}\left(\frac{\partial H}{\partial x_B}\right)_n &= \left(\frac{\partial H}{\partial n_A}\right)_{n_B}\left(\frac{\partial n_A}{\partial x_B}\right)_n + \left(\frac{\partial H}{\partial n_B}\right)_{n_A}\left(\frac{\partial n_B}{\partial x_B}\right)_n \\ &= \overline{H}_A(-n) + \overline{H}_B(+n)\end{aligned} \tag{19.3-15}$$

---

[2]Euler's theorem for functions $f(x_1,x_2,\cdots x_n)$ that are homogeneous of degree k states that

$$\sum_{j=1}^{n} x_j \frac{\partial f}{\partial x_j} = kf \tag{19.3-8a}$$

See, for example, M. D. Greenberg, *Foundations of Applied Mathematics*, Prentice-Hall, Englewood Cliffs, NJ (1978), p. 128.

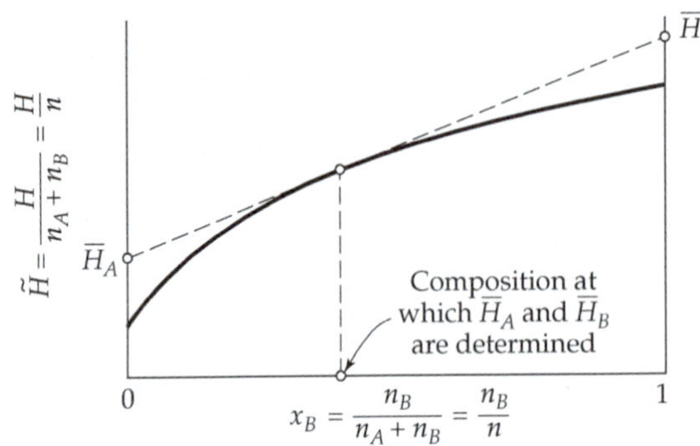

Fig. 19.3-1. The "method of intercepts" for determining partial molar quantities in a binary mixture.

Substitution of this into Eq. 19.3-12 and use of Euler's theorem ($H = n_A\bar{H}_A + n_B\bar{H}_B$) then shows that Eq. 19.3-12 is satisfied. This proves the validity of Eq. 19.3-10, and the correctness of Eq. 19.3-11 can be proved similarly.

(b) One can also get $\bar{H}_A$ by using the definition in Eq. 19.3-7 and measuring the slope of the curve of H versus n_A, holding n_B constant. One can also get $\bar{H}_A$ by measuring the enthalpy of mixing and using

$$H = n_A\bar{H}_A + n_B\bar{H}_B = n_A\tilde{H}_A + n_B\tilde{H}_B + \Delta H_{\text{mix}} \tag{19.3-16}$$

Often the enthalpy of mixing is neglected and the enthalpies of pure substances are given as $\tilde{H}_A \approx \tilde{C}_{pA}(T - T^0)$ and a similar expression for $\tilde{H}_B$. This is a standard assumption for gas mixtures at low to moderate pressures.

Other methods for evaluating partial molar quantities may be found in current textbooks on thermodynamics.

§19.4 THE EQUATIONS OF CHANGE AND SOLVING STEADY-STATE DIFFUSION PROBLEMS

The equations of change in §19.2 can be used to solve all the problems of Chapter 18 (see, for example, Problems 19B.1 through 19B.3) and more difficult ones as well. Unless the problems are idealized or simplified, transport phenomena problems for mixtures are quite complicated, and usually numerical techniques are required.[1] Here we solve a few introductory problems to illustrate the use of the equations of change.

EXAMPLE 19.4-1

Simultaneous Heat and Mass Transport[2]

A hot condensable vapor A is diffusing at steady state through a stagnant film of noncondensable gas B to a cold surface at $y = 0$, where A condenses. Develop expressions for the mole fraction profile $x_A(y)$ and the temperature profile $T(y)$ for the system pictured in Fig. 19.4-1, given the mole fractions and temperatures at both film boundaries ($y = 0$ and $y = \delta$). *Assume* ideal-gas behavior and uniform pressure. Furthermore *assume* the physical properties to be constant, evaluated at some mean temperature and composition. Neglect radiative heat transfer.

This example illustrates how energy and mass transport are combined in separation processes.

[1] W. E. Stewart and M. Caracotsios, *Computer-Aided Modeling of Reacting Systems*, Wiley-Interscience, New York (2008).

[2] A. P. Colburn and T. B. Drew, *Trans. Am. Inst. Chem. Engrs.*, **38**, 197–212 (1937).

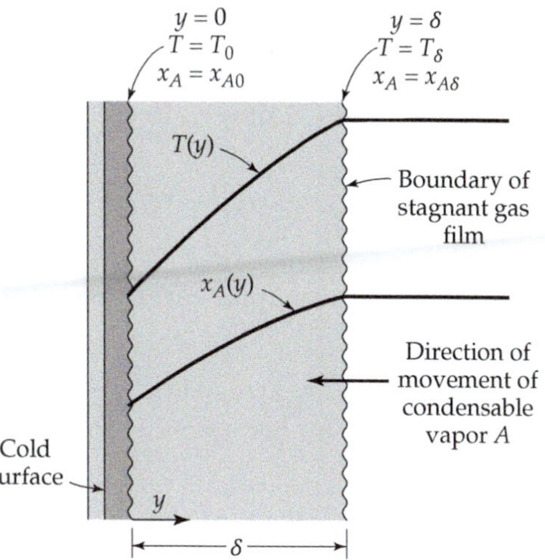

Fig. 19.4-1. Condensation of a hot vapor A on a cold surface in the presence of a noncondensable gas B.

SOLUTION

To determine the desired quantities, we have to solve the equations of continuity and energy for this system. Simplification of Eq. 19.1-10 and Eq. (C) of Table 19.2-1 for steady, one-dimensional transport, in the absence of chemical reaction and external forces gives, with help of Eq. A of Table A.7-1,

Continuity of A: $$\frac{dN_{Ay}}{dy} = 0 \tag{19.4-1}$$

Energy: $$\frac{de_y}{dy} = 0 \tag{19.4-2}$$

Therefore, both N_{Ay} and e_y are constant throughout the film.

Equation 19.3-6 suggests that, in order to get the energy flux, we need first to have information about the molar flux. Therefore, we begin by solving the equation of continuity to determine the *mole fraction profile*. The molar flux of A through stagnant B was given in Eq. 18.0-2 (or 18.6-3):

$$N_{Ay} = -\frac{c\mathscr{D}_{AB}}{1 - x_A}\frac{dx_A}{dy} \tag{19.4-3}$$

Substitution of Eq. 19.4-3 into Eq. 19.4-1 and integration of the resulting differential equation gives (see §18.6)

$$\frac{1 - x_A(y)}{1 - x_{A0}} = \left(\frac{1 - x_{A\delta}}{1 - x_{A0}}\right)^{y/\delta} \tag{19.4-4}$$

Here we have taken $c\mathscr{D}_{AB}$ to be constant at the value for the mean film temperature. We can then evaluate the constant flux N_{Ay} from the last two equations as

$$N_{Ay} = \frac{c\mathscr{D}_{AB}}{\delta}\ln\frac{1 - x_{A\delta}}{1 - x_{A0}} \tag{19.4-5}$$

Note that N_{Ay} is negative, because species A is condensing, after diffusing in the $-y$ direction. Equations 19.4-4 and 19.4-5 may be combined to put the concentration profiles into an alternative form

$$\frac{x_A(y) - x_{A0}}{x_{A\delta} - x_{A0}} = \frac{1 - \exp\left[\left(N_{Ay}/c\mathscr{D}_{AB}\right)y\right]}{1 - \exp\left[\left(N_{Ay}/c\mathscr{D}_{AB}\right)\delta\right]} \tag{19.4-6}$$

To get the *temperature profile*, we use the energy flux from Eq. 19.3-6 for an ideal gas along with Eq. 9.4-4, and the fact that $N_{By} = 0$,

$$e_y = -k\frac{dT}{dy} + \left(\tilde{H}_A N_{Ay} + \tilde{H}_B N_{By}\right)$$
$$= -k\frac{dT}{dy} + N_{Ay}\left(\tilde{H}_{A0} + \tilde{C}_{pA}(T - T_0)\right) \tag{19.4-7}$$

Here we have chosen the wall temperature T_0 as the reference temperature for the enthalpy. Insertion of this expression into Eq. 19.4-2 and integration between the limits $T = T_0$ at $y = 0$ and $T = T_\delta$ at $y = \delta$ gives

$$\frac{T(y) - T_0}{T_\delta - T_0} = \frac{1 - \exp\left[\left(N_{Ay}\tilde{C}_{pA}/k\right) y\right]}{1 - \exp\left[\left(N_{Ay}\tilde{C}_{pA}/k\right) \delta\right]} \tag{19.4-8}$$

It can be seen that the temperature profile is not linear for this system except in the limit as $N_{Ay}\tilde{C}_{pA}/k \to 0$. Note the similarity between Eq. 19.4-6 and Eq. 19.4-8.

The *conduction* energy flux at the wall, when accompanied by mass transport, is greater than that in the absence of mass transport. Thus, using a subscript "0" to indicate zero mass transport, we may write

$$\frac{-k(dT/dy)|_{y=0}}{-k(dT/dy)_0|_{y=0}} = \frac{-\left(N_{Ay}\tilde{C}_{pA}/k\right)\delta}{1 - \exp\left[\left(N_{Ay}\tilde{C}_{pA}/k\right)\delta\right]} \tag{19.4-9}$$

We see that the rate of heat transport is directly affected by the simultaneous mass transport, whereas the mass flux is not directly affected by the simultaneous heat transport. In applications at temperatures below the normal boiling point of species A, the quantity $N_{Ay}\tilde{C}_{pA}/k$ is small, and the right side of Eq. 19.4-9 is very nearly equal to unity (see Problem 19A.1).

EXAMPLE 19.4-2

Diffusion and Chemical Reaction in a Liquid

(a) A solid sphere (radius R) of substance A is suspended in a liquid B in which it is slightly soluble, and with which it undergoes a first-order chemical reaction with rate constant k_1'''. Find the quasi-steady-state concentration profile; that is, assume that R is constant, and determine $c_A(r)$. The molar solubility of A in B is c_{A0}.

(b) Next, relax the assumption of constant R, and perform a mass balance on A over the sphere volume to obtain an expression for the sphere radius as a function of time.

(c) How does the result in (a) simplify if there is no chemical reaction?

SOLUTION

(a) The differential equation for the steady-state diffusion from a sphere is obtained by simplifying Eq. B.11-3, using the postulate that $c_A = c_A(r)$. The result is

$$\mathscr{D}_{AB}\frac{1}{r^2}\frac{d}{dr}\left(r^2\frac{dc_A}{dr}\right) - k_1'''c_A = 0 \tag{19.4-10}$$

When we introduce the dimensionless quantities

$$\Gamma(\xi) = \frac{c_A(r)}{c_{A0}} \qquad \xi = \frac{r}{R} \qquad b = \sqrt{\frac{k_1'''R^2}{\mathscr{D}_{AB}}} \tag{19.4-11,12,13}$$

the diffusion equation becomes

$$\frac{1}{\xi^2}\frac{d}{d\xi}\left(\xi^2\frac{d\Gamma}{d\xi}\right) - b^2\Gamma = 0 \tag{19.4-14}$$

According to Eq. C.1-6b, this differential equation has the solution

$$\Gamma(\xi) = C_1 \frac{e^{-b\xi}}{\xi} + C_2 \frac{e^{+b\xi}}{\xi} \tag{19.4-15}$$

The boundary conditions that have to be applied are $\Gamma = 1$ at $\xi = 1$, and $\Gamma \to 0$ as $\xi \to \infty$. The constants of integration are then $C_1 = 1$ and $C_2 = 0$. Therefore, the solution to Eq. 19.4-14 is

$$\Gamma(\xi) = \frac{1}{\xi} \frac{e^{-b\xi}}{e^{-b}} \tag{19.4-16}$$

The molar flux of A at the sphere surface is then

$$N_{Ar}|_{r=R} = -\mathscr{D}_{AB} \left.\frac{dc_A}{dr}\right|_{r=R} = -\frac{c_{A0}\mathscr{D}_{AB}}{R} \left.\frac{d\Gamma}{d\xi}\right|_{\xi=1} = +\frac{c_{A0}\mathscr{D}_{AB}}{R}(1+b) \tag{19.4-17}$$

and the loss of A from the sphere in moles per unit time is

$$W_A = 4\pi R^2 \left(\frac{c_{A0}\mathscr{D}_{AB}}{R}\right)\left(1 + \sqrt{\frac{k_1''' R^2}{\mathscr{D}_{AB}}}\right) \tag{19.4-18}$$

Note that it has been assumed that, in the *quasi-steady-state* diffusion, the radius of the sphere is *not* changing significantly with time.

(b) We may now write an unsteady mass balance on the dissolving sphere. Since the volume of the sphere is changing very slowly, it is possible to use the *quasi-steady-state* result in Eq. 19.4-18 in setting up the equation for the mass balance. Thus, equating the negative of rate of change of mass of A within the sphere (i.e., the rate of mass *loss*) to the rate of transport of A away from the sphere surface, we obtain

$$-\frac{d}{dt}\left(\frac{4}{3}\pi R^3 \rho_{\text{sph}}\right) = 4\pi R^2 \left(\frac{c_{A0}\mathscr{D}_{AB}M_A}{R}\right)\left(1 + \sqrt{\frac{k_1''' R^2}{\mathscr{D}_{AB}}}\right) \tag{19.4-19}$$

First we divide through by $\frac{4}{3}\pi\rho_{\text{sph}}$ and then put all factors containing R on the left side of the equation to get

$$-\frac{R}{1 + \sqrt{k_1''' R^2/\mathscr{D}_{AB}}} \frac{dR}{dt} = \frac{c_{A0}\mathscr{D}_{AB}M_A}{\rho_{\text{sph}}} \tag{19.4-20}$$

Next we introduce the dimensionless variable $Y = \sqrt{k_1''' R^2/\mathscr{D}_{AB}}$ and then rewrite Eq. 19.4-20 as

$$-\frac{Y}{1+Y} \frac{dY}{dt} = \frac{k_1''' c_{A0} M_A}{\rho_{\text{sph}}} \tag{19.4-21}$$

Then we integrate from time t_0 to time t to get

$$-\int_{Y_0}^{Y} \frac{Y}{1+Y} dY = \frac{k_1''' c_{A0} M_A}{\rho_{\text{sph}}} \int_{t_0}^{t} dt \tag{19.4-22}$$

or

$$(Y(t) - Y_0) - \ln\left(\frac{1 + Y(t)}{1 + Y_0}\right) = -\frac{k_1''' c_{A0} M_A}{\rho_{\text{sph}}}(t - t_0) \tag{19.4-23}$$

Hence, the final expression for R as a function of t is

$$\sqrt{\frac{k_1'''}{\mathscr{D}_{AB}}}(R - R_0) - \ln\left(\frac{1 + \sqrt{k_1''' R^2/\mathscr{D}_{AB}}}{1 + \sqrt{k_1''' R_0^2/\mathscr{D}_{AB}}}\right) = -\frac{k_1''' c_{A0} M_A}{\rho_{\text{sph}}}(t - t_0) \tag{19.4-24}$$

This equation would have to be solved numerically to get the radius of the sphere as a function of time. For example, one can evaluate $t - t_0$ for various values of R, and then plot R vs. $t - t_0$.

(c) If there is no chemical reaction, $b = 0$, and the concentration profile and molar flux at the surface given by Eqs. 19.4-16 and 19.4-17 simplify to

$$\frac{c_A(r)}{c_{A0}} = \frac{R}{r} \qquad N_{Ar}|_{r=R} = \frac{c_{A0}\mathscr{D}_{AB}}{R} \tag{19.4-25,26}$$

if we assume that the concentration far from the sphere is still zero. Now if we define a mass-transfer coefficient by $N_{Ar}|_{r=R} = k_c(c_{A0} - c_{A\infty})$—analogously to the definition of the heat-transfer coefficient according the Newton's law of cooling (see Eq. 18.1-2)—we then get for a sphere of diameter D

$$k_c = \frac{\mathscr{D}_{AB}}{R} = \frac{2\mathscr{D}_{AB}}{D} \qquad \text{and} \qquad \text{Sh} = \frac{k_c D}{\mathscr{D}_{AB}} = 2 \tag{19.4-27,28}$$

Here Sh is the *Sherwood number* (analogous to the Nusselt number). The value Sh = 2 is the limiting value for this dimensionless group for the situation where the fluid surrounding the sphere has zero velocity. This limiting value will be encountered again in Chapter 22.

EXAMPLE 19.4-3

Concentration Profile in a Tubular Reactor

A catalytic tubular reactor is shown in Fig. 19.4-2. A dilute solution of solute A in a solvent S is in fully developed laminar flow in the region $z < 0$. When it encounters the catalytic wall in the region $0 \leq z \leq L$, solute A is instantaneously and irreversibly rearranged to an isomer B. Write the diffusion equation appropriate for this problem at steady state, and find the solution for short distances into the reactor. *Assume* that the flow is isothermal and neglect the presence of B.

SOLUTION

For the assumptions stated above, the flowing liquid will always be very nearly pure S. The quantities ρ and $\mathscr{D}_{AS}$ can be regarded as constant, and the diffusion of A in S can be described by the steady-state version of Eq. 19.1-20 (ignoring the presence of a small amount of the reaction product B—the *pseudobinary assumption*); equivalently, we can use Eq. B.11-2, which is in cylindrical coordinates, with constant ρ and $\mathscr{D}_{AS}$. Note that because the reaction occurs at the wall, $r_A = 0$ in the species balance. The relevant equations of change for the system are then (after dividing Eq. 19.1-20 by M_A)

Continuity of A:
$$v_z \frac{\partial c_A}{\partial z} = \mathscr{D}_{AS}\left[\frac{1}{r}\frac{\partial}{\partial r}\left(r\frac{\partial c_A}{\partial r}\right) + \underline{\underline{\frac{\partial^2 c_A}{\partial z^2}}}\right] \tag{19.4-29}$$

Motion:
$$0 = -\frac{d\mathscr{P}}{dz} + \mu\frac{1}{r}\frac{d}{dr}\left(r\frac{dv_z}{dr}\right) \tag{19.4-30}$$

We make the usual assumption that axial diffusion is negligible compared to axial convection, and therefore we omit the dashed-underlined term (compare with Eqs. 10.9-13 and 10.9-14). Equation 19.4-30 can be solved to give the parabolic velocity profile, $v_z(r) = v_{z,\max}[1 - (r/R)^2]$. When this result is substituted into Eq. 19.4-29, we get

$$v_{z,\max}\left(1 - \left(\frac{r}{R}\right)^2\right)\frac{\partial c_A}{\partial z} = \mathscr{D}_{AS}\frac{1}{r}\frac{\partial}{\partial r}\left(r\frac{\partial c_A}{\partial r}\right) \tag{19.4-31}$$

This is to be solved with the boundary conditions

B.C. 1: $\quad$ at $z = 0$, $\quad c_A = c_{A0}$ $\quad$ (19.4-32)

B.C. 2: $\quad$ at $r = R$, $\quad c_A = 0$ $\quad$ (19.4-33)

B.C. 3: $\quad$ at $r = 0$, $\quad c_A = \text{finite}$ $\quad$ (19.4-34)

For short distances z into the reactor, the concentration c_A differs from c_{A0} only near the wall, where the velocity is practically linear. Hence, we can introduce the variable $y = R - r$, neglect curvature terms, and replace B.C. 3 by a fictitious boundary condition at $y = \infty$ (see Example 11.5-3 for a detailed discussion of this method of treating the entrance region of the tube for the analogous heat-transfer problem).

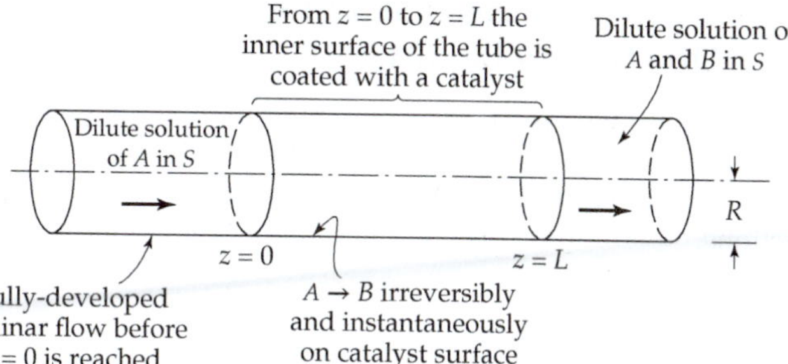

Fig. 19.4-2. Schematic diagram of a tubular reactor.

The reformulated problem statement is then

$$2v_{z,\max}\frac{y}{R}\frac{\partial c_A}{\partial z} = \mathscr{D}_{AS}\frac{\partial^2 c_A}{\partial y^2} \qquad (19.4\text{-}35)$$

with the boundary conditions

B.C. 1: at $z = 0$, $c_A = c_{A0}$ (19.4-36)

B.C. 2: at $y = 0$, $c_A = 0$ (19.4-37)

B.C. 3: as $y \rightarrow \infty$, $c_A \rightarrow c_{A0}$ (19.4-38)

As in Examples 3.8-1 and 11.5-3, and §18.9, this problem can be solved by the method of combination of independent variables by seeking a solution of the form $c_A(y,z)/c_{A0} = f(\eta)$, where $\eta = (y/R)(2v_{z,\max}R^2/9\mathscr{D}_{AS}z)^{1/3}$. One thus obtains an ordinary differential equation

$$f'' + 3\eta^2 f' = 0 \qquad (19.4\text{-}39)$$

with boundary conditions

B.C. 1: at $\eta = 0$, $f = 0$ (19.4-40)

B.C. 2: at $\eta = \infty$, $f = 1$ (19.4-41)

Solution of this problem gives (see Eq. C.1-9 and Eq. 18.9-10)

$$f(\eta) = \frac{c_A(y,z)}{c_{A0}} = \frac{\int_0^\eta \exp(-\bar{\eta}^3)d\bar{\eta}}{\int_0^\infty \exp(-\eta^3)d\eta} = \frac{\int_0^\eta \exp(-\bar{\eta}^3)d\bar{\eta}}{\Gamma\left(\frac{4}{3}\right)} \qquad (19.4\text{-}42)$$

This solution may be shown to satisfy Eqs. 19.4-39 to 19.4-41.

Experiments of the type described here have proven useful for obtaining mass-transfer data at high Schmidt numbers.[3] A particularly attractive reaction is the reduction of ferricyanide ions on metallic surfaces according to the reaction

$$\mathrm{Fe(CN)_6^{-3}} + \mathrm{e}^- \rightarrow \mathrm{Fe(CN)_6^{-4}} \qquad (19.4\text{-}43)$$

in which ferricyanide and ferrocyanide take the place of A and B, respectively, in the above development. The electrochemical reaction is quite rapid under properly chosen conditions. Furthermore, since it involves only electron transfer, the physical properties of the solution are almost entirely unaffected. The forced-diffusion effects neglected here (i.e., transport caused by body forces that differ among the molecules) may be suppressed by the addition of an indifferent electrolyte in excess.[4,5]

---

[3] D. W. Hubbard and E. N. Lightfoot, *Ind. Eng. Chem. Fundam.*, **5**, 370–379 (1966).

[4] J. S. Newman, *Electrochemical Systems*, Prentice-Hall, Englewood Cliffs, N.J., 2nd edition (1991), §1.10.

[5] J. R. Selman and C. W. Tobias, *Advances in Chemical Engineering*, **10**, Academic Press, New York, N.Y. (1978), pp. 212–318.

§19.5 THE EQUATIONS OF CHANGE AND SOLVING UNSTEADY-STATE DIFFUSION PROBLEMS

In this section we give two examples of time-dependent diffusion. The first deals with evaporation of a volatile liquid and illustrates deviations from Fick's second law that arise at high mass-transfer rates. The second deals with unsteady diffusion with chemical reactions.

EXAMPLE 19.5-1

Unsteady Evaporation of a Liquid (the "Arnold Problem")

It is desired to predict the rate at which a volatile liquid A evaporates into pure B in a tube of infinite length. The liquid level is maintained at $z = 0$ at all times. The temperature and pressure are assumed constant, and the vapors of A and B form an ideal-gas mixture. Hence, the molar density c is constant throughout the gas phase, and $\mathscr{D}_{AB}$ may be considered to be constant. It is further assumed that species B is insoluble in liquid A, and that the molar-average velocity in the gas phase does not depend on the radial coordinate.

The problem discussed in §18.6 differs from the one treated here in two respects: the former involves steady-state diffusion, and the latter is an unsteady-state problem; and the former considers a tube of finite length, whereas the latter has an infinitely long tube.

SOLUTION

For this system, the equation of continuity for the mixture, given in Eq. 19.1-13, becomes

$$\frac{\partial v_z^*}{\partial z} = 0 \tag{19.5-1}$$

in which v_z^* is the z component of the molar-average velocity. Integration with respect to z gives

$$v_z^* = v_{z0}^*(t) \tag{19.5-2}$$

which illustrates that v_z^* depends on time, but is independent of position. Here and elsewhere in this problem, the subscript "0" indicates a quantity evaluated at $z = 0$. According to Eq. (H) of Table 17.4-1, this velocity can be written in terms of the total molar fluxes of A and B as

$$v_z^*(t) = v_{z0}^*(t) = \frac{N_{Az0} + N_{Bz0}}{c} \tag{19.5-3}$$

However, N_{Bz0} is zero because of the insolubility of species B in liquid A. Then use of Eq. (E) of Table 17.4-2 gives finally

$$v_z^*(t) = -\frac{\mathscr{D}_{AB}}{1 - x_{A0}} \left.\frac{\partial x_A}{\partial z}\right|_{z=0} \tag{19.5-4}$$

in which x_{A0} is the interfacial gas-phase concentration, evaluated here on the assumption of interfacial equilibrium. For an ideal-gas mixture, this is just the vapor pressure of pure A divided by the total pressure.

The equation of continuity of Eq. 19.1-21 then becomes

$$\frac{\partial x_A}{\partial t} - \left(\frac{\mathscr{D}_{AB}}{1 - x_{A0}} \left.\frac{\partial x_A}{\partial z}\right|_{z=0} \right) \frac{\partial x_A}{\partial z} = \mathscr{D}_{AB} \frac{\partial^2 x_A}{\partial z^2} \tag{19.5-5}$$

This result can also be obtained using Eq. B.11-1. Equation 19.5-5 is to be solved with the initial and boundary conditions:

I.C.: at $t = 0$, $x_A = 0$ (19.5-6)

B.C. 1: at $z = 0$, $x_A = x_{A0}$ (19.5-7)

B.C. 2: as $z \to \infty$, $x_A \to 0$ (19.5-8)

We rewrite Eq. 19.5-5 in terms of $X(z,t) = x_A/x_{A0}$

$$\frac{\partial X}{\partial t} - \left(\frac{\mathscr{D}_{AB} x_{A0}}{1 - x_{A0}} \left.\frac{\partial X}{\partial z}\right|_{z=0} \right) \frac{\partial X}{\partial z} = \mathscr{D}_{AB} \frac{\partial^2 X}{\partial z^2} \tag{19.5-9}$$

Then based on our experience in Example 3.8-1 (and specifically Eq. 3.8-6), we now try a combination of variables of the form $X(z,t) = X(Z)$, where $Z = z/\sqrt{4\mathscr{D}_{AB}t}$. However, since Eqs. 19.5-5 and 19.5-9 contain the parameter x_{A0}, we can anticipate that X will depend not only on Z, but also parametrically on x_{A0}.

We first show how to transform each of the derivatives in Eq. 19.5-9 (compare with Eqs. 3.8-7 and 3.8-8):

$$\frac{\partial X}{\partial t} = \frac{dX}{dZ}\frac{\partial Z}{\partial t} = \frac{dX}{dZ}\frac{z}{\sqrt{4\mathscr{D}_{AB}}}\left(-\frac{1}{2}t^{3/2}\right) = \frac{dX}{dZ}\left(-\frac{Z}{2t}\right) \tag{19.5-10}$$

$$\frac{\partial X}{\partial z} = \frac{dX}{dZ}\frac{\partial Z}{\partial z} = \frac{dX}{dZ}\frac{1}{\sqrt{4\mathscr{D}_{AB}t}} \quad \text{and} \quad \frac{\partial^2 X}{\partial z^2} = \frac{d^2X}{dZ^2}\frac{1}{4\mathscr{D}_{AB}t} \tag{19.5-11,12}$$

$$\left.\frac{\partial X}{\partial z}\right|_{z=0} = \left.\frac{dX}{dZ}\right|_{Z=0}\frac{1}{\sqrt{4\mathscr{D}_{AB}t}} \tag{19.5-13}$$

Substituting Eqs. 19.5-10 to 19.5-13 into Eq. 19.5-9 and then multiplying the entire equation by $4t$ gives

$$\frac{d^2X}{dZ^2} + 2Z\frac{dX}{dZ} + \left[-2\cdot\left(-\frac{1}{2}\frac{x_{A0}}{1-x_{A0}}\left.\frac{dX}{dZ}\right|_{Z=0}\right)\right]\frac{dX}{dZ} = 0 \tag{19.5-14}$$

If, now, we define the constant quantity inside the parentheses as φ, we get finally

$$\frac{d^2X}{dZ^2} + 2(Z-\varphi)\frac{dX}{dZ} = 0 \quad \text{with} \quad \varphi(x_{A0}) = -\frac{1}{2}\frac{x_{A0}}{1-x_{A0}}\left.\frac{dX}{dZ}\right|_{Z=0} \tag{19.5-15,16}$$

We note that $\varphi(x_{A0})$ is a dimensionless molar-average velocity, $\varphi = v_z^*\sqrt{t/\mathscr{D}_{AB}}$, as can be seen by comparing Eqs. 19.5-16 and 19.5-4. Note that since φ depends only on x_{A0}, it follows that v_z^* is inversely proportional to $\sqrt{t}$. The initial and boundary conditions in Eqs. 19.5-6 to 19.5-8 now become

B.C. 1: at $Z = 0$ $\quad X = 1$ (19.5-17)

B.C. 2 and I.C.: as $Z \to \infty$, $\quad X \to 0$ (19.5-18)

Equation 19.5-15 can be solved by first letting $dX/dZ = Y$. This gives a first-order differential equation for Y that can be solved to obtain

$$Y(Z) = C_1 \exp\left[-(Z-\varphi)^2\right] \equiv \frac{dX}{dZ} \tag{19.5-19}$$

This is now a first-order differential equation for $X(Z)$, which gives on integration

$$X(Z) = C_1\int_0^Z \exp\left[-(\overline{Z}-\varphi)^2\right]d\overline{Z} + C_2 \tag{19.5-20}$$

The integration constants are determined by using Eqs. 19.5-17 and 19.5-18 to get

$$X(Z) = 1 - \frac{\int_0^Z \exp\left[-(\overline{Z}-\varphi)^2\right]d\overline{Z}}{\int_0^\infty \exp\left[-(\overline{Z}-\varphi)^2\right]d\overline{Z}} = 1 - \frac{\int_{-\varphi}^{Z-\varphi}\exp(-W^2)dW}{\int_{-\varphi}^{\infty}\exp(-W^2)dW} \tag{19.5-21}$$

where $W = \overline{Z} - \varphi$. Then we use the definition of the error function in §C.6 and some of the properties of this function, in particular, that $-\text{erf}(-\varphi) = \text{erf}\,\varphi$ and that $\text{erf}\,\infty = 1$. This leads to the final expression for the distribution of the mole fraction of A:[1]

$$X(Z) = 1 - \frac{\text{erf}\,(Z-\varphi) + \text{erf}\,\varphi}{\text{erf}\,\infty + \text{erf}\,\varphi} = \frac{1 - \text{erf}\,(Z-\varphi)}{1 + \text{erf}\,\varphi} \tag{19.5-22}$$

---

[1] J. H. Arnold, *Trans. AIChE*, **40**, 361–378 (1944). **Jerome Howard Arnold** (1907–1974) taught at MIT, the University of Minnesota, the University of North Dakota, and the University of Iowa; he worked for Standard Oil of California (1944–1948) and was the Director of the Contra Costa Transit District (1956–1960). See also V.-D. Dang and W. N. Gill, *AIChE Journal*, **16**, 793–802 (1970).

To get the function $\varphi(x_{A0})$, this mole-fraction distribution has to be substituted into Eq. 19.5-16. This gives

$$\varphi(x_{A0}) = \frac{1}{\sqrt{\pi}} \frac{x_{A0}}{1 - x_{A0}} \frac{\exp(-\varphi^2)}{1 + \text{erf}\,\varphi} \tag{19.5-23}$$

Rather than solving this by trial and error to get φ as a function of x_{A0}, it is easier to obtain x_{A0} as a function of φ:

$$x_{A0}(\varphi) = \frac{1}{1 + \left[\sqrt{\pi}\,(1 + \text{erf}\,\varphi)\,\varphi \exp \varphi^2\right]^{-1}} \tag{19.5-24}$$

A graph of the function $\varphi(x_{A0})$ is given in Fig. 19.5-1, and the concentration profiles are shown in Fig. 19.5-2.

Let V_A be the volume of A produced by evaporation up to time t. Then the rate of production of vapor volume from a surface of area S is

$$\frac{dV_A}{dt} = v_z^* S = \frac{N_{Az0} S}{c} = S\varphi \sqrt{\frac{\mathscr{D}_{AB}}{t}} \tag{19.5-25}$$

Here we have used Eq. 19.5-3. Integration with respect to t then gives

$$V_A(t) = S\varphi \sqrt{4\mathscr{D}_{AB} t} \tag{19.5-26}$$

This relation can be used for measuring the diffusivity from experimental data on the total rate of evaporation from a surface below an infinite stagnant film (see Problem 19A.2).

We can now assess the importance of including the convective transport of species A in the tube. If Fick's second law (Eq. 19.1-22) had been used to determine $X(Z)$ (i.e., neglecting convection), we would have obtained for the production of vapor volume

$$V_A^{\text{Fick}}(t) = S x_{A0} \sqrt{\frac{4\mathscr{D}_{AB} t}{\pi}} \tag{19.5-27}$$

Thus, we can rewrite Eq. 19.5-26 as

$$V_A(t) = S x_{A0} \sqrt{\frac{4\mathscr{D}_{AB} t}{\pi}} \cdot \psi \tag{19.5-28}$$

The factor $\psi = \varphi\sqrt{\pi}/x_{A0}$, given in Fig. 19.5-1, is a correction for the deviation from the Fick's-second-law results caused by the nonzero molar-average velocity. We see that the deviation becomes especially significant when x_{A0} is large, that is, for liquids with large volatility.

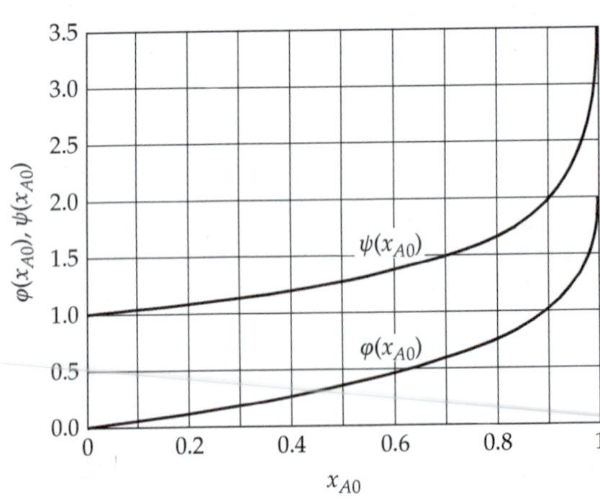

Fig. 19.5-1. Graph of $\varphi(x_{A0})$ and $\psi(x_{A0})$. Note that both φ and ψ approach ∞ as $x_{A0} \to 1$.

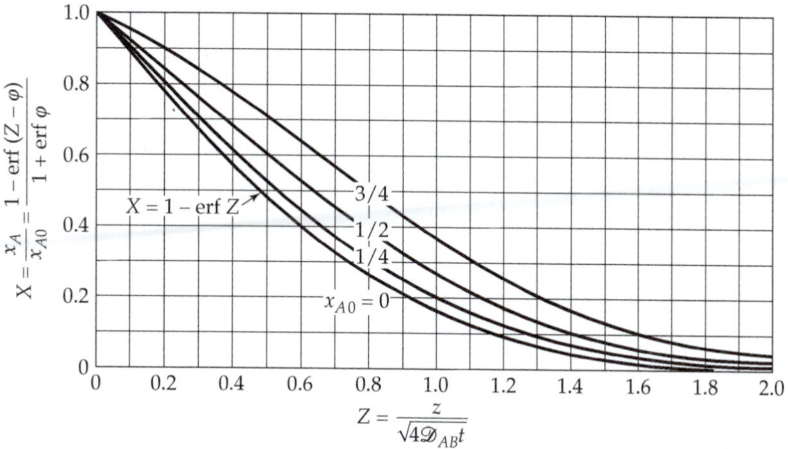

Fig. 19.5-2. Concentration profiles in time-dependent evaporation, showing that the deviation from Fick's law increases with the volatility of the evaporating liquid.

In the preceding analysis it is assumed that the system is isothermal. Actually, the interface will be cooled by the evaporation, particularly at large values of x_{A0}. This effect can be minimized by using a small-diameter tube made of a good thermal conductor. For application to other mass-transfer systems, however, the analysis given here needs to be extended by including the solution to the energy equation, so that the interfacial temperature and compositions can be calculated.

EXAMPLE 19.5-2

Gas Absorption with Rapid Reaction[2,3]

Gas A is absorbed by a stationary liquid solvent S, the latter containing solute B. Species A reacts with B in an instantaneous irreversible reaction according to the equation $aA + bB \rightarrow$ Products. It may be assumed that Fick's second law (Eq. 19.1-22) adequately describes the diffusion processes, since A, B, and the reaction products are present in S in low concentrations. Obtain expressions for the time-dependent concentration profiles.

SOLUTION

Because of the instantaneous reaction of A and B, there will be a plane parallel to the liquid-vapor interface at a distance z_R from it, that separates the region containing no A from that containing no B. The distance z_R is a function of t, since the boundary between A and B retreats as B is used up in the chemical reaction.

The differential equations for c_A and c_B are then

$$\frac{\partial c_A}{\partial t} = \mathscr{D}_{AS} \frac{\partial^2 c_A}{\partial z^2} \quad \text{for } 0 \leq z \leq z_R(t) \tag{19.5-29}$$

$$\frac{\partial c_B}{\partial t} = \mathscr{D}_{BS} \frac{\partial^2 c_B}{\partial z^2} \quad \text{for } z_R(t) \leq z < \infty \tag{19.5-30}$$

---

[2]T. K. Sherwood, R. L. Pigford, and C. R. Wilke, *Absorption and Extraction*, McGraw-Hill, New York, 3rd edition (1975), Chapter 8. See also G. Astarita, *Mass Transfer with Chemical Reaction*, Elsevier, Amsterdam (1967), Chapter 5.

[3]For related problems with moving boundaries associated with phase changes, see H. S. Carslaw and J. C. Jaeger, *Conduction of Heat in Solids*, Oxford University Press, 2nd edition (1959). See also S. G. Bankoff, *Advances in Chemical Engineering*, Academic Press, New York (1964), Vol. **5**, pp. 76–150; J. Crank, *Free and Moving Boundary Problems*, Oxford University Press (1984).

These are to be solved with the following initial and boundary conditions:

I.C.:	at $t = 0$,	$c_B = c_{B\infty}$ for $z > 0$	(19.5-31)
B.C. 1:	at $z = 0$,	$c_A = c_{A0}$	(19.5-32)
B.C. 2, 3:	at $z = z_R(t)$,	$c_A = 0$ and $c_B = 0$	(19.5-33)
B.C. 4:	at $z = z_R(t)$,	$-\frac{1}{a}\mathscr{D}_{AS}\frac{\partial c_A}{\partial z} = +\frac{1}{b}\mathscr{D}_{BS}\frac{\partial c_B}{\partial z}$	(19.5-34)
B.C. 5:	as $z \to \infty$,	$c_B \to c_{B\infty}$	(19.5-35)

Here c_{A0} is the interfacial concentration of A, and $c_{B\infty}$ is the original concentration of B. The fourth boundary condition is the stoichiometric requirement that a moles of A consume b moles of B (see Problem 19B.6).

The absence of a characteristic length in this problem, and the fact that $c_B = c_{B\infty}$ both at $t = 0$ and $z = \infty$ suggests trying a combination of variables. Experience has shown that in Example 3.8-1 and Example 19.5-1 (with $v_z^* = 0$), we could have assumed at the outset that the solution would be of the form $f = C_1 + C_2 \text{erf } Z$, where Z is some dimensionless combination of the independent variables; in Example 3.8-1, $Z = y/\sqrt{4\nu t}$, and in Example 19.5-1, $Z = z/\sqrt{4\mathscr{D}_{AB}t}$. Therefore, maybe it would not be altogether unreasonable to try the same idea here and propose a solution of the form

$$\frac{c_A}{c_{A0}} = C_1 + C_2 \text{erf}\frac{z}{\sqrt{4\mathscr{D}_{AS}t}} \quad \text{for } 0 \le z \le z_R(t) \tag{19.5-36}$$

$$\frac{c_B}{c_{B\infty}} = C_3 + C_4 \text{erf}\frac{z}{\sqrt{4\mathscr{D}_{BS}t}} \quad \text{for } z_R(t) \le z < \infty \tag{19.5-37}$$

These functions satisfy the differential equations, and *if* the constants of integration, C_1 to C_4, can be so chosen that the initial and boundary conditions are satisfied, we will have the complete solution to the problem.

Application of the initial condition and the first three boundary conditions permits the evaluation of the integration constants in terms of $z_R(t)$, thereby giving

From B.C. 1:
$$C_1 = 1 \tag{19.5-38}$$

From B.C. 2:
$$C_2 = -\frac{1}{\text{erf } z_R/\sqrt{4\mathscr{D}_{AS}t}} \tag{19.5-39}$$

From solving the I.C. and B.C. 3 simultaneously:

$$C_3 = -\frac{\text{erf } z_R/\sqrt{4\mathscr{D}_{BS}t}}{1 - \text{erf } z_R/\sqrt{4\mathscr{D}_{BS}t}} \tag{19.5-40}$$

$$C_4 = \frac{1}{1 - \text{erf } z_R/\sqrt{4\mathscr{D}_{BS}t}} \tag{19.5-41}$$

With these expressions for C_1, C_2, C_3, and C_4, we then get

$$\frac{c_A(z,t)}{c_{A0}} = 1 - \frac{\text{erf }(z/\sqrt{4\mathscr{D}_{AS}t})}{\text{erf }(z_R/\sqrt{4\mathscr{D}_{AS}t})} \quad \text{for } 0 \le z \le z_R(t) \tag{19.5-42}$$

$$\frac{c_B(z,t)}{c_{B\infty}} = 1 - \frac{1 - \text{erf }(z/\sqrt{4\mathscr{D}_{BS}t})}{1 - \text{erf }(z_R/\sqrt{4\mathscr{D}_{BS}t})} \quad \text{for } z_R(t) \le z < \infty \tag{19.5-43}$$

B.C. 5 is then automatically satisfied. Finally, insertion of these solutions into B.C. 4 gives the following implicit equation from which $z_R(t)$ can be obtained:

$$1 - \text{erf}\sqrt{\frac{\gamma}{\mathscr{D}_{BS}}} = \frac{ac_{B\infty}}{bc_{A0}}\sqrt{\frac{\mathscr{D}_{BS}}{\mathscr{D}_{AS}}}\text{erf}\sqrt{\frac{\gamma}{\mathscr{D}_{AS}}}\exp\left(\frac{\gamma}{\mathscr{D}_{AS}} - \frac{\gamma}{\mathscr{D}_{BS}}\right) \tag{19.5-44}$$

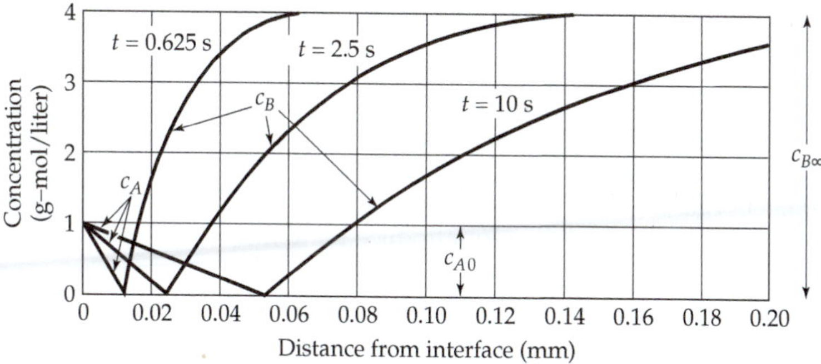

Fig. 19.5-3. Gas absorption with rapid chemical reaction, with concentration profiles given by Eqs. 19.5-42 to 19.5-44 (for $a = b$). This calculation was made for $\mathscr{D}_{AS} = 3.9 \times 10^{-5}$ ft^2/hr and $\mathscr{D}_{BS} = 1.95 \times 10^{-5}$ ft^2/hr. [T. K. Sherwood and R. L. Pigford, *Absorption and Extraction*, McGraw-Hill, New York (1952), p. 336.]

This equation can be solved for the constant γ, which is defined as $z_R^2/4t$. Thus $z_R(t)$, the location of the reaction plane, increases as $\sqrt{t}$.

To calculate the concentration profiles, one first solves Eq. 19.5-44 for $\sqrt{\gamma}$, and then inserts this value for $z_R/\sqrt{4t}$ in Eqs. 19.5-42 and 19.5-43. Some calculated concentration profiles are shown in Fig. 19.5-3, to illustrate the rate of movement of the reaction plane.

From the concentration profiles we can calculate the rate of mass transfer at the interface:

$$N_{Az0}(t) = -\mathscr{D}_{AS} \left.\frac{dc_A}{dz}\right|_{z=0} = \frac{c_{A0}}{\operatorname{erf}\sqrt{\gamma/\mathscr{D}_{AS}}}\sqrt{\frac{\mathscr{D}_{AS}}{\pi t}} \tag{19.5-45}$$

The average rate of absorption up to time t is then

$$N_{Az0,\text{avg}}(t) = \frac{1}{t}\int_0^t N_{Az0}(\bar{t})\,d\bar{t} = 2\frac{c_{A0}}{\operatorname{erf}\sqrt{\gamma/\mathscr{D}_{AS}}}\sqrt{\frac{\mathscr{D}_{AS}}{\pi t}} \tag{19.5-46}$$

Hence, the average rate up to time t is just twice the instantaneous rate.

§19.6 CONCLUDING COMMENTS

The main purpose of this chapter has been to summarize the principal equations of transport phenomena (see also the summary inside the front cover). The readers should by now be impressed with the unity of the subject, and the similarities—and differences— among the equations for mass, momentum, and energy. They should be able to point to each term in the equations of Tables 19.2-1, 19.2-2, and 19.2-3 and discuss its physical meaning.

The equations of conservation and the expressions for the fluxes have been given in great generality. In this textbook we have illustrated how the equations may be simplified in order to solve some straightforward problems for which idealized models can be used to approximate real systems. Of course, if one wants to use more realistic models, it will be necessary to keep more terms in the equations and then resort to numerical solutions.

The illustrative examples in this chapter have emphasized problems involving diffusion, with and without chemical reactions. The chemical kinetics involved have necessarily been trivially simple. If one wishes to treat more realistic chemical reactions, then it is necessary to be able to treat multicomponent systems. Although these have been extensively studied in the research literature, they are not appropriate for inclusion in an introductory textbook on transport phenomena. This subject is discussed very briefly in §24.6.

QUESTIONS FOR DISCUSSION

1. Under what conditions is it true that $(\nabla \cdot \mathbf{v}) = 0$? $(\nabla \cdot \mathbf{v}^*) = 0$?
2. Summarize the content of §19.1, using mass units. What are the main results?
3. Point to each of the terms in Table 19.2-1, and tell exactly what the meaning is. Do the same for Table 19.2-2.
4. In Eqs. 19.3-5 and 19.3-6, explain how each line follows from the preceding one.
5. How does one know to introduce the dimensionless variables in Eqs. 19.4-11, 19.4-12, and 19.4-13?
6. In Eq. 19.4-24, would it be easier to solve and get t as a function of R?
7. Explain how one goes from Eq. 19.4-31 and its boundary conditions to Eq. 19.4-35 and its boundary conditions.
8. In Example 19.5-1, does it make physical sense that v_z^* should be a function of t? What does this functionality turn out to be at the end of the example?
9. In Eqs. 19.5-10 to 19.5-13, what is being held constant in each of the partial derivatives?
10. Verify that Eq. 19.5-22 satisfies Eq. 19.5-5 and the boundary and initial conditions.
11. In Example 19.5-2, what do you think of the solution method, where one guesses the form of a solution and then sees if constants of integration can be found that satisfy the boundary and initial conditions?
12. What are some conditions for which the convection term in the continuity equation can be neglected (e.g., Eq. 19.1-22)? What are some consequences of neglecting the convection when it shouldn't be neglected?

PROBLEMS

19A.1 Dehumidification of air. For the system of Example 19.4-1, let the vapor be H_2O and the stagnant gas be air. Assume the following conditions (which are representative for air conditioning): (i) at $y = \delta$, $T = 80°F$ and $x_{H_2O} = 0.018$; (ii) at $y = 0$, $T = 50°F$ in Fig. 19.4-1.

(a) For $p = 1$ atm, calculate the right side of Eq. 19.4-9.

(b) Compare the conductive and diffusive heat flux at $y = 0$. What is the physical significance of your answer?

Answer: **(a)** 1.004

19A.2 Measurement of diffusivity by unsteady-state evaporation. Use the following data to determine the diffusivity of ethyl propionate (species A) into a mixture of 20 mole% air and 80 mole% hydrogen (this mixture being treated as a pure gas B).

Increase in vapor volume (cm^3)	$\sqrt{t}$ (s$^{1/2}$)
0.01	15.5
0.11	19.4
0.22	23.4
0.31	26.9
0.41	30.5
0.50	34.0
0.60	37.5
0.70	41.5

These data were obtained[1] by using a glass tube 200 cm long, with an inside diameter 1.043 cm; the temperature was 27.9°C and the pressure 761.2 mm Hg. The vapor pressure of ethyl propionate at this temperature is 41.5 mm Hg. Note that t is the actual time from the start of the evaporation, whereas the volume increase is measured from $t \approx 240$ s.

---

[1] D. F. Fairbanks and C. R. Wilke, *Ind. Eng. Chem.*, **42**, 471–475 (1950).

19A.3 Rate of evaporation of n-octane. At 20°C, how many grams of liquid n-octane will evaporate into N_2 in 24.5 hr in a system such as that studied in Example 19.5-1 at system pressures of **(a)** 1 atm, and **(b)** 2 atm? The area of the liquid surface is 1.29 cm^2, and the vapor pressure of n-octane at 20°C is 10.45 mm Hg.

Answer: **(a)** 6.8 mg

19A.4 Absorption with rapid second-order reaction. Make the following calculations for the reacting system depicted in Fig. 19.5-3:

(a) Verify the location of the reaction zone, using Eq. 19.5-44.

(b) Calculate N_{A0} at $t = 2.5$ s.

19A.5 Measurement of diffusivity by the point-source method. It is desired to design a flow system to utilize the results of Problem 19C.1 for the measurement of $\mathscr{D}_{AB}$ (see Fig. 19C.1).[2] The approaching stream of pure B will be directed vertically upward, and the gas composition will be measured at several points along the z axis.

(a) Calculate the gas-injection rate W_A, in g-mol/s required to produce a mole fraction $x_A \approx 0.01$ at a point 1 cm downstream of the source, in an ideal gaseous system at 1 atm and 800°C, if $v_0 = 50$ cm/s and $\mathscr{D}_{AB} \approx 5$ cm^2/s.

(b) What is the maximum permissible error in the radial position of the gas-sampling probe, if the measured composition x_A is to be within 1% of the centerline value?

19B.1 Steady evaporation. Rework the problem solved in §18.6 for the system shown in Fig. 18.6-1, dealing with the evaporation of liquid A into gas B, starting from Eq. 19.1-21.

(a) First obtain an expression for $\mathbf{v}^*$ using Eq. (H) of Table 17.4-1, as well as Fick's law in the form of Eq. (E) of Table 17.4-2.

(b) Show that Eq. 19.1-21 then becomes the following nonlinear second-order differential equation

$$\frac{d^2x_A}{dz^2} + \frac{1}{1-x_A}\left(\frac{dx_A}{dz}\right)^2 = 0 \tag{19B.1-1}$$

(c) Solve this equation to get the mole-fraction profile given in Eq. 18.6-13. (*Hint*: let $w = dx_A/dz$.)

19B.2 Gas absorption with homogeneous chemical reaction. Rework the problem solved in §18.4 (see Fig. 18.4-1), by starting with Eq. 19.1-20. What assumptions do you have to make to get Eq. 18.4-4.

19B.3 Diffusion in falling films. Using the equations of Chapter 19, show how to set up and solve the problems discussed in §18.8 and §18.9.

19B.4 Time-dependent evaporation. In Example 19.5-1, obtain the first and second derivatives of the solution in Eq. 19.5-22, dX/dZ and d^2X/dZ^2. Then use these to verify that the solution does indeed satisfy the differential equation in Eq. 19.5-15.

Also verify that substitution of the first derivative dX/dZ into Eq. 19.5-15 gives Eq. 19.5-23. Finally, show how Eq. 19.5-24 is obtained.

19B.5 Fick's second law solution for unsteady evaporation of a liquid. Rework Example 19.5-1 using Fick's second law (Eq. 19.1-22) to obtain Eq. 19.5-27.

---

[2]This is a precise method for measurements of diffusivity at high temperatures. For a detailed description of the method, see R. E. Walker and A. A. Westenberg, *J. Chem. Phys.*, **29**, 1139–1146, 1147–1153 (1958). For a summary of measured values and comparisons with the Chapman-Enskog theory, see R. M. Fristrom and A. A. Westenberg, *Flame Structure*, McGraw-Hill, New York (1965), Chapter XIII.

19B.6 Stoichiometric boundary condition for rapid irreversible reaction. The reactant fluxes in Example 19.5-2 must satisfy the stoichiometric relation

$$\text{at } z = z_R(t), \quad \frac{1}{a}c_A(v_{Az} - v_R) = -\frac{1}{b}c_B(v_{Bz} - v_R) \tag{19B.6-1}$$

in which $v_R = dz_R/dt$. Show that this relation leads to Eq. 19.5-34 when use is made of Fick's first law, with the assumptions of constant c and instantaneous irreversible reaction.

19B.7 Time for a droplet to evaporate. A droplet of pure A of initial radius R is suspended in a large body of motionless gas B. The concentration of A in the gas phase is x_{AR} at $r = R$ and zero at an infinite distance from the droplet.

(a) Assuming that R is constant, show that at steady state

$$R^2 N_{Ar}|_{r=R} = -\frac{c\mathscr{D}_{AB}}{1 - x_A} r^2 \frac{dx_A}{dr} \tag{19B.7-1}$$

where $N_{Ar}|_{r=R}$ is the molar flux in the r direction at the droplet surface, c is the total molar concentration in the gas phase, and $\mathscr{D}_{AB}$ is the diffusivity in the gas phase. Assume constant temperature and pressure throughout. Show that integration of Eq. 19B.7-1 from the droplet surface to infinity gives

$$RN_{Ar}|_{r=R} = -c\mathscr{D}_{AB} \ln(1 - x_{AR}) \tag{19B.7-2}$$

(b) We now let the droplet radius R be a function of time, and treat the problem as a quasi-steady one. Then the rate of decrease of moles of A within the drop can be equated to the instantaneous rate of loss of mass across the liquid-gas interface

$$-\frac{d}{dt}\left(\frac{4}{3}\pi R^3 c_A^{(L)}\right) = 4\pi R^2 N_{Ar}|_{r=R} = -4\pi R c\mathscr{D}_{AB} \ln(1 - x_{AR}) \tag{19B.7-3}$$

where $c_A^{(L)}$ is the molar density of pure liquid A. Show that when this equation is integrated from $t = 0$ to $t = t_0$ (the time for complete evaporation of the droplet), one gets

$$t_0 = \frac{c_A^{(L)} R^2}{2c\mathscr{D}_{AB} \ln[1/(1 - x_{AR})]} \tag{19B.7-4}$$

Does this result look physically reasonable?

19B.8 Concentration-dependent diffusivity. A stationary liquid layer of liquid B is bounded by planes $z = 0$ (a solid wall) and $z = b$ (a gas-liquid interface). At these planes, the concentration of A is c_{A0} and c_{Ab}, respectively. The diffusivity $\mathscr{D}_{AB}$ is a function of the concentration of A.

(a) Starting from Eq. 19.1-5 derive a differential equation for the steady-state concentration distribution.

(b) Show that the concentration distribution is given by

$$\frac{\displaystyle\int_{c_A(z)}^{c_{A0}} \mathscr{D}_{AB}(c_A')dc_A'}{\displaystyle\int_{c_{Ab}}^{c_{A0}} \mathscr{D}_{AB}(c_A)dc_A} = \frac{z}{b} \tag{19B.8-1}$$

(c) Verify that the molar flux at the solid-liquid surface is

$$N_{Az}|_{z=0} = \frac{1}{b}\int_{c_{Ab}}^{c_{A0}} \mathscr{D}_{AB}(c_A)dc_A \tag{19B.8-2}$$

(d) Now assume that the diffusivity can be expressed as a Taylor series in the concentration

$$\mathscr{D}_{AB}(c_A) = \overline{\mathscr{D}}_{AB}\left[1 + \beta_1(c_A - \bar{c}_A) + \beta_2(c_A - \bar{c}_A)^2 + \cdots\right] \tag{19B.8-3}$$

in which $\bar{c}_A = \frac{1}{2}(c_{A0} + c_{Ab})$ and $\overline{\mathscr{D}}_{AB} = \mathscr{D}_{AB}(\bar{c}_A)$. Then show that

$$N_{Az}|_{z=0} = \frac{\overline{\mathscr{D}}_{AB}}{b}(c_{A0} - c_{Ab})\left[1 + \frac{1}{12}\beta_2(c_{A0} - c_{Ab})^2 + \cdots\right] \tag{19B.8-4}$$

(e) How does Eq. 19B.8-4 simplify if the diffusivity is a linear function of the concentration?

19B.9 Diffusion of a finite sphere of material. At time $t = 0$, a dilute solute A in solvent B has a uniform concentration c_{A0} inside a spherical region of radius R, and zero concentration for $R \leq r < \infty$. For $t > 0$, diffusion immediately begins to take place. It has been shown[3] that the concentration as a function of position and time is

$$\frac{c_A(r,t)}{c_{A0}} = \frac{1}{2}\left(\text{erf}\frac{R+r}{\sqrt{4\mathscr{D}_{AB}t}} + \text{erf}\frac{R-r}{\sqrt{4\mathscr{D}_{AB}t}}\right) - \frac{1}{r}\sqrt{\frac{\mathscr{D}_{AB}t}{\pi}}\left(\exp\left(-(R-r)^2/4\mathscr{D}_{AB}t\right) - \exp\left(-(R+r)^2/4\mathscr{D}_{AB}t\right)\right) \quad (19\text{B}.9\text{-}1)$$

The quantity c_{A0} may also be written as $n_A/\left(\frac{4}{3}\pi R^3\right)$, where n_A is the number of moles of A that are in the spherical region of radius R.

(a) Show that at $t = 0$, the concentration given by Eq. 19B.9-1 is $c_A(r,t) = c_{A0}$ for $r \leq R$.

(b) Show that, if $R \to 0$ with n_A constant, Eq. 19B.9-1 is the same as Eq. 19B.10-1. Solving this requires a lot of tenacity and the application of L'Hôpital's rule three times (lengthy!). Alternatively, try expanding Eq. 19B.9-1 about $R = 0$.

(c) Does Eq. 19B.9-1 satisfy the diffusion equation?

19B.10 Diffusion from an instantaneous point source. At time $t = 0$, n_A moles of A are injected into a large body of fluid B. Take the point of injection to be the origin of coordinates. The material A diffuses radially in all directions. The solution of the problem may be found in Carslaw and Jaeger:[4]

$$c_A(r,t) = \frac{n_A}{(4\pi\mathscr{D}_{AB}t)^{3/2}}e^{-r^2/4\mathscr{D}_{AB}t} \quad (19\text{B}.10\text{-}1)$$

(a) Verify that Eq. 19B.10-1 satisfies Fick's second law.

(b) Verify that Eq. 19B.10-1 satisfies the boundary conditions at $r = \infty$.

(c) Show that Eq. 19B.10-1, when integrated over all space, gives n_A, as it should.

(d) What happens to Eq. 19B.10-1 when $t \to 0$?

19B.11 Oxidation of silicon.[5] A slab of silicon is exposed to gaseous oxygen (species A) at pressure p, producing a layer of silicon dioxide (species B) as shown in Fig. 19B.11. The layer extends from the surface $z = 0$, where the oxygen dissolves with concentration $c_{A0} = Kp$, to the surface at $z = \delta(t)$, where the oxygen and silicon undergo a first-order reaction with rate constant k_1''. The thickness $\delta(t)$ of the growing oxide layer is to be predicted. A quasi-steady-state approach is useful here, inasmuch as the advancement of the reaction front is very slow.

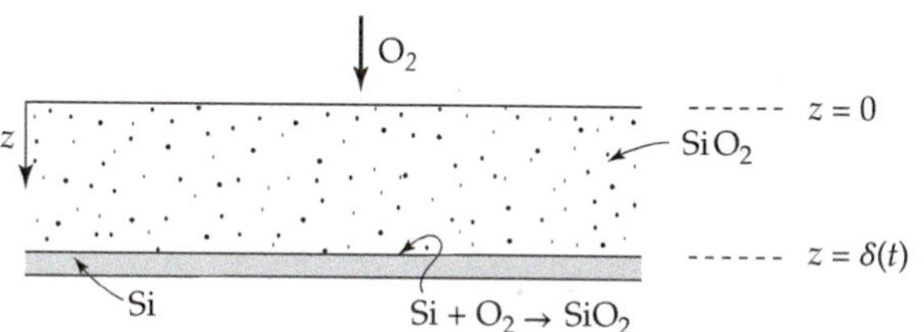

Fig. 19B.11 Oxidation of silicon.

[3] J. Crank, *The Mathematics of Diffusion*, Oxford University Press, 1st edition (1956), p. 27.

[4] H. S. Carslaw and J. C. Jaeger, *Conduction of Heat in Solids*, Oxford University Press, 2nd edition (1959), p. 257. See also, J. Crank, *The Mathematics of Diffusion*, Oxford University Press, 1st edition (1956), p. 27.

(a) First solve the diffusion equation in Eq. 19.1-22, with the term $\partial c_A/\partial t$ neglected, and apply the boundary conditions to obtain

$$c_A(z) = c_{A0} - (c_{A0} - c_{A\delta})\frac{z}{\delta} \tag{19B.11-1}$$

in which the concentration $c_{A\delta}$ at the reaction plane is as yet unknown.

(b) Next make a unsteady-state molar O_2 balance on the region $0 \leq z \leq \delta(t)$ to obtain, with the aid of the Leibniz formula of §C.3,

$$c_{A\delta}\frac{d\delta}{dt} = -\mathscr{D}_{AB}\frac{dc_A}{dz} - k_1''c_{A\delta} \tag{19B.11-2}$$

(c) Now write an unsteady-state molar balance on SiO_2 in the same region to obtain

$$+k_1''c_{A\delta} = \frac{1}{\tilde{V}_B}\frac{d\delta}{dt} \tag{19B.11-3}$$

(d) In Eq. 19B.11-2, evaluate $d\delta/dt$ by using Eq. 19B.11-3 and $\partial c_A/\partial z$ from Eq. 19B.11-1. This will yield an equation for $c_{A\delta}$

$$\frac{k_1''\delta\tilde{V}_B}{\mathscr{D}_{AB}}c_{A\delta}^2 + \left(1 + \frac{k_1''\delta}{\mathscr{D}_{AB}}\right)c_{A\delta} = c_{A0} \tag{19B.11-4}$$

Inserting numerical values into Eq. 19B.11-4 shows that the quadratic term can be safely neglected.[5]

(e) Combine Eqs. 19B.11-3 and 19B.11-4 (without the quadratic term) to get a differential equation for $\delta(t)$. Show that this leads to

$$\frac{\delta^2}{2\mathscr{D}_{AB}} + \frac{\delta}{k_1''} = \tilde{V}_B c_{A0} t \tag{19B.11-5}$$

which agrees with the experimental data.[5] Interpret the result.

19B.12 Manipulation of an equation of continuity of A. Show that for constant total density ρ, Eq. 19.1-20 can be rewritten in rectangular coordinates as

$$\frac{\partial c_A}{\partial t} = -\left(\frac{\partial}{\partial x}v_x c_A + \frac{\partial}{\partial y}v_y c_A + \frac{\partial}{\partial z}v_z c_A\right) + \mathscr{D}_{AB}\left(\frac{\partial^2 c_A}{\partial x^2} + \frac{\partial^2 c_A}{\partial y^2} + \frac{\partial^2 c_A}{\partial z^2}\right) + R_A \tag{19B.12-1}$$

This form will be used in Chapter 20.

19C.1 Diffusion from a point source in a moving stream. A stream of fluid B in laminar motion has a uniform velocity v_0 (as shown in Fig. 19C.1). At some point in the stream, taken to be the origin of coordinates, species A is injected at a small rate W_A g-mol/s. This rate is assumed to be sufficiently small that the mass-average velocity will not deviate appreciably from v_0. Species A is swept downstream (in the z direction), and at the same time it diffuses both axially and radially.

[5] R. Ghez, *A Primer of Diffusion Problems*, Wiley-Interscience, New York (1988), pp. 46–55; this book discusses a number of problems that arise in the microelectronics field.

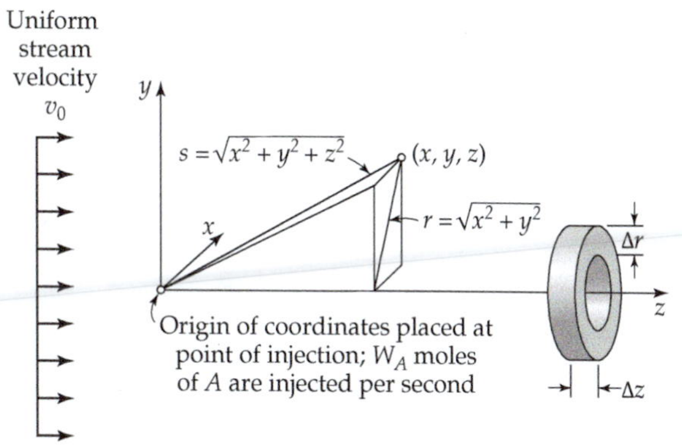

Uniform stream velocity v_0

$s = \sqrt{x^2 + y^2 + z^2}$ (x, y, z) $r = \sqrt{x^2 + y^2}$ Δr Δz

Origin of coordinates placed at point of injection; W_A moles of A are injected per second

Fig. 19C.1 Diffusion of A from a point source into a stream of B that moves with a uniform velocity.

(a) Show that Eq. 19.1-21 for c and $\mathscr{D}_{AB}$ assumed constant leads to the following partial differential equation:

$$v_0 \frac{\partial c_A}{\partial z} = \mathscr{D}_{AB} \left[\frac{1}{r} \frac{\partial}{\partial r} \left(r \frac{\partial c_A}{\partial r} \right) + \frac{\partial^2 c_A}{\partial z^2} \right] \tag{19C.1-1}$$

(b) Show that Eq. 19C.1-1 can also be written as

$$v_0 \left(\frac{z}{s} \frac{\partial c_A}{\partial s} + \frac{\partial c_A}{\partial z} \right) = \mathscr{D}_{AB} \left[\frac{1}{s^2} \frac{\partial}{\partial s} \left(s^2 \frac{\partial c_A}{\partial s} \right) + \frac{\partial^2 c_A}{\partial z^2} + 2 \frac{z}{s} \frac{\partial^2 c_A}{\partial s \partial z} \right] \tag{19C.1-2}$$

in which $s^2 = r^2 + z^2$.

(c) Verify (lengthy!) that the solution[6]

$$c_A = \frac{W_A}{4\pi \mathscr{D}_{AB} s} \exp \left[- (v_0 / 2\mathscr{D}_{AB}) (s - z) \right] \tag{19C.1-3}$$

satisfies the differential equation above.

(d) Show further that the following boundary conditions are also satisfied by Eq. 19C.1-3:

B.C. 1: $\quad$ as $s \to \infty, \quad c_A \to 0$ $\quad$ (19C.1-4)

B.C. 2: $\quad$ as $s \to 0, \quad -4\pi s^2 \mathscr{D}_{AB} \dfrac{\partial c_A}{\partial s} \to W_A$ $\quad$ (19C.1-5)

B.C. 3: $\quad$ at $r = 0, \quad \dfrac{\partial c_A}{\partial r} = 0$ $\quad$ (19C.1-6)

Explain the physical meaning of each of these boundary conditions.

(e) Show how data on $c_A(r,z)$ for given v_0 and $\mathscr{D}_{AB}$ may be plotted, when the preceding solution applies, to give a straight line with slope $v_0/2\mathscr{D}_{AB}$ and intercept $\ln \mathscr{D}_{AB}$.

(f) How is the solution to this problem related to that in Problem 19B.10?

19C.2 **Steady state diffusion from a rotating disk.**[7] A large disk (taken here to be infinite) is rotating with an angular velocity Ω in an infinite expanse of liquid B. Its surface is coated with a material A that is slightly soluble in B. The goal of this problem is to find the rate at which A dissolves in B. (Note: The solution to this problem can be applied to a disk of finite radius R with negligible error.) This system has been used for studying removal of behenic acid from stainless steel surfaces.[8]

---

[6] H. S. Carslaw and J. C. Jaeger, *Conduction of Heat in Solids*, Oxford University Press, 2nd edition (1959), pp. 266–267.

[7] V. G. Levich, *Physicochemical Hydrodynamics*, Prentice-Hall, Englewood Cliffs, New Jersey (1962), §11.

[8] C. S. Grant, A. T. Perka, W. D. Thomas, and R. Caton, *AIChE Journal*, **42**, 1465–1476 (1996).

The fluid dynamics of this problem was solved by von Kármán[9] and later corrected by Cochran.[10] It was found that the velocity components can be expressed, except near the disk edge, as

$$v_r(r,\zeta) = \Omega r F(\zeta); \quad v_\theta(r,\zeta) = \Omega r G(\zeta); \quad v_z(z) = \sqrt{\Omega \nu} H(\zeta) \tag{19C.2-1}$$

where $\zeta = z\sqrt{\Omega/\nu}$. The functions F, G, and H have the following expansions,[9] with $a = 0.510$ and $b = -0.616$:

$$F(\zeta) = a\zeta - \frac{1}{2}\zeta^2 - \frac{1}{3}b\zeta^3 - \frac{1}{12}b^2\zeta^4 - \cdots \tag{19C.2-2}$$

$$G(\zeta) = 1 + b\zeta + \frac{1}{2}a\zeta^3 + \frac{1}{12}(ab-1)\zeta^4 - \cdots \tag{19C.2-3}$$

$$H(\zeta) = -a\zeta^2 + \frac{1}{3}\zeta^3 + \frac{1}{6}\zeta^4 + \cdots \tag{19C.2-4}$$

Furthermore, in the limit as $\zeta \to \infty$, $H \to -0.886$, and F, G, F', and G' all approach zero. Also, it is known that the boundary-layer thickness is proportional to $\sqrt{\nu/\Omega}$, except near the edge of the disk.

For the diffusion problem, we have to solve Eq. 19.1-20 with $r_A = 0$ and the velocity field given by Eq. 19C.2-1. The boundary conditions are

B.C. 1	at $z = 0$, $\quad \rho_A = \rho_{A0}$	(19C.2-5)
B.C. 2	as $z \to \infty$, $\quad \rho_A \to 0$	(19C.2-6)
B.C. 3	at $r = 0$, $\quad \partial\rho_A/\partial r = 0$	(19C.2-7)
B.C. 4	as $r \to \infty$, $\quad \partial\rho_A/\partial r \to 0$	(19C.2-8)

Equation 19C.2-7 states that there is no diffusion mass flux at $r = 0$, and Eq. 19C.2-8 states that, at the outer edge of the disk (here taken to be infinite), there will be no mass flux by diffusion.

Since there can be but one solution to this linear problem, it may be seen that a solution of the form $\rho_A = \rho_A(z)$—i.e., the concentration of A is independent of r (!)—can be found that satisfies the differential equation and all the boundary conditions.

(a) Show that, for this problem, Eq. 19.1-20 becomes

$$v_z \frac{d\rho_A}{dz} = \mathscr{D}_{AB}\frac{d^2\rho_A}{dz^2} \quad \text{or} \quad H(\zeta)\frac{d\rho_A}{d\zeta} = \frac{1}{\text{Sc}}\frac{d^2\rho_A}{d\zeta^2} \tag{19C.2-9a,b}$$

This ordinary differential equation is solved by making the substitution $p = d\rho_A/d\zeta$, thereby getting a first-order separable equation $dp/d\zeta = \text{Sc}H(\zeta)p$.

(b) Show that the solution of Eq. 19C.2-9b is

$$\ln p = \text{Sc}\int_0^\zeta H(\bar{\zeta})\,d\bar{\zeta} + \ln C_1 \quad \text{or} \quad \frac{d\rho_A}{d\zeta} = C_1 \exp\left(\text{Sc}\int_0^\zeta H(\bar{\zeta})\,d\bar{\zeta}\right) \tag{19C.2-10a,b}$$

in which the overbars are used to indicate dummy variables of integration. Show that a further integration gives

$$\rho_A(\zeta) = C_1\int_0^\zeta \exp\left(\text{Sc}\int_0^{\bar{\zeta}} H(\bar{\bar{\zeta}})\,d\bar{\bar{\zeta}}\right)d\bar{\zeta} + C_2 \tag{19C.2-11}$$

Apply the boundary conditions that $\rho_A(0) = \rho_{A0}$ and $\rho_A(\infty) = 0$, and verify that Eq. 19C.2-11 then becomes

$$\frac{\rho_A(\zeta)}{\rho_{A0}} = 1 - \frac{\displaystyle\int_0^\zeta \exp\left(\text{Sc}\int_0^{\bar{\zeta}} H(\bar{\bar{\zeta}})\,d\bar{\bar{\zeta}}\right)d\bar{\zeta}}{\displaystyle\int_0^\infty \exp\left(\text{Sc}\int_0^{\bar{\zeta}} H(\bar{\bar{\zeta}})\,d\bar{\bar{\zeta}}\right)d\bar{\zeta}} \tag{19C.2-12}$$

---

[9]T. von Kármán, *Zeits. f. angew. Math. u. Mech.*, **1**, 244–247 (1921).
[10]W. G. Cochran, *Proc. Camb. Phil. Soc.*, **30**, 365–375 (1934).

(c) In the high Schmidt number limit, we need take only the first term in the expansion of $H(\zeta)$; show that this results in

$$\frac{\rho_A(\zeta)}{\rho_{A0}} = 1 - \frac{\int_0^{\zeta} \exp\left[\mathrm{Sc}\int_0^{\bar{\zeta}} \left(-a\bar{\bar{\zeta}}^2\right) d\bar{\bar{\zeta}}\right] d\bar{\zeta}}{\int_0^{\infty} \exp\left[\mathrm{Sc}\int_0^{\bar{\zeta}} \left(-a\bar{\bar{\zeta}}^2\right) d\bar{\bar{\zeta}}\right] d\bar{\zeta}} = 1 - \frac{\int_0^{\zeta} \exp\left(-\frac{1}{3}\mathrm{Sc}a\bar{\zeta}^3\right) d\bar{\zeta}}{\int_0^{\infty} \exp\left(-\frac{1}{3}\mathrm{Sc}a\bar{\zeta}^3\right) d\bar{\zeta}} \tag{19C.2-13}$$

(d) Show that the integral in the denominator can be evaluated analytically

$$\int_0^{\infty} \exp\left(-\frac{1}{3}\mathrm{Sc}a\bar{\zeta}^3\right) d\bar{\zeta} = \frac{\int_0^{\infty} \exp(-u^3)du}{\sqrt[3]{\frac{1}{3}\mathrm{Sc}a}} = \frac{\frac{1}{3}\Gamma\left(\frac{1}{3}\right)}{\sqrt[3]{\frac{1}{3}\mathrm{Sc}a}} = \frac{\Gamma\left(\frac{4}{3}\right)}{\sqrt[3]{\frac{1}{3}\mathrm{Sc}a}} \tag{19C.2-14}$$

and that the final expression for the concentration profile is

$$\frac{\rho_A(\zeta)}{\rho_{A0}} = 1 - \frac{\sqrt[3]{\frac{1}{3}\mathrm{Sc}a}}{\Gamma\left(\frac{4}{3}\right)} \int_0^{\infty} \exp\left(-\frac{1}{3}\mathrm{Sc}a\bar{\zeta}^3\right) d\bar{\zeta} \tag{19C.2-15}$$

(e) Then verify that the mass flux in the z direction is

$$j_{Az} = -\mathscr{D}_{AB}\frac{d\rho_A}{dz} = +\rho_{A0}\mathscr{D}_{AB}\frac{\sqrt[3]{\frac{1}{3}\mathrm{Sc}a}}{\Gamma\left(\frac{4}{3}\right)} \exp\left(-\frac{1}{3}\mathrm{Sc}a\zeta^3\right)\frac{d\zeta}{dz} \tag{19C.2-16}$$

in which $d\zeta/dz = \sqrt{\Omega/\nu}$.

(f) Finally, show that

$$j_{Az}|_{z=0} = \rho_{A0}\mathscr{D}_{AB}\frac{\sqrt[3]{\frac{1}{3}\mathrm{Sc}a}}{\Gamma\left(\frac{4}{3}\right)}\sqrt{\frac{\Omega}{\nu}} = \frac{\sqrt[3]{\frac{1}{3}(0.510)}}{\Gamma\left(\frac{4}{3}\right)}\rho_{A0}\mathscr{D}_{AB}\mathrm{Sc}^{1/3}\sqrt{\frac{\Omega}{\nu}}$$
$$= 0.620\rho_{A0}\mathscr{D}_{AB}\left(\frac{\nu}{\mathscr{D}_{AB}}\right)^{1/3}\sqrt{\frac{\Omega}{\nu}} \tag{19C.2-17}$$

is the mass flux on one side of the disk surface

Chapter 20

Concentration Distributions in Turbulent Flow

In the preceding chapters we have derived the equations for diffusion in a fluid or solid, and we have shown how one can obtain expressions for the concentration distribution, when no fluid turbulence is involved. In this chapter we turn our attention to mass transport in turbulent flow.

The discussion here is quite similar to that in Chapter 12, and much of that material can be taken over by analogy. Specifically, §12.4 and §12.5 can be taken over directly by replacing heat-transfer quantities by mass-transfer quantities. In fact, the problems discussed in those sections have been tested more meaningfully in mass transfer, since the range of experimentally accessible Schmidt numbers is considerably greater than that for Prandtl numbers.

We restrict ourselves here to isothermal binary systems, and make the assumption of constant mass density and diffusivity. Therefore, the partial differential equation describing diffusion in a flowing fluid (Eq. 19.1-20) is of the same form as that for heat conduction in a flowing fluid (Eq. 11.2-9), except for the inclusion of the chemical reaction term in the former.

The most apparent influence of turbulence on mass transport is the enhanced transport perpendicular to the main flow. Consider the flow of a fluid in catalytic tubular reactor, where species A must be transported to the wall in order for it to react (e.g., as in Example 19.4-3). If the flow is laminar with the only nonzero velocity component in the axial direction, then mass transport perpendicular to the tube wall can occur only by diffusion. This mode of mass transport can be very slow. On the other hand, if the flow is turbulent, there is a nonzero, fluctuating component of the velocity perpendicular to the tube walls, which results in the convective transport of mass from the bulk fluid to the tube wall. This mode of transport is typically much faster. This enhanced transport can also be interpreted as mixing—the turbulent motion mixes the fluid within the tube, creating a more uniform concentration profile than observed in laminar flow.

§20.1 CONCENTRATION FLUCTUATIONS AND THE TIME-SMOOTHED CONCENTRATION

The discussion in §12.1 about the temperature fluctuations and time-smoothing can be taken over by analogy for the molar concentration c_A. In a turbulent stream, c_A will be a rapidly oscillating function that can be written as the sum of a time-smoothed value $\bar{c}_A$ and a turbulent concentration fluctuation c'_A

$$c_A(t) = \bar{c}_A + c'_A(t) \tag{20.1-1}$$

which is analogous to Eq. 12.1-1 for the temperature. Of course, $\bar{c}_A$ will also depend on position, and may also vary slowly with time if the driving force for motion is not constant. By virtue of the definition of c'_A, we see that $\overline{c'_A} = 0$. However, quantities such as $\overline{v'_x c'_A}$, $\overline{v'_y c'_A}$, and $\overline{v'_z c'_A}$ are not zero, because the local fluctuations in concentration and velocity are not independent of one another.

The time-smoothed concentration profiles $\bar{c}_A(x,y,z,t)$ are those measured, for example, by the withdrawal of samples from the fluid stream at various points and various times. In tube flow with mass transfer at the wall, one expects that the time-smoothed concentration $\bar{c}_A$ will vary only slightly with position in the turbulent core, where the transport by turbulent eddies predominates. In the slowly moving region near the bounding surface, on the other hand, the concentration $\bar{c}_A$ will be expected to change within a small distance from its turbulent-core value to the wall value. The steep concentration gradient is then associated with the slow molecular diffusion process in the viscous sublayer, in contrast to the rapid eddy transport in the fully developed turbulent core.

§20.2 TIME-SMOOTHING OF THE EQUATION OF CONTINUITY OF SPECIES *A*

We begin with the equation of continuity for species A, which we presume is disappearing by an nth-order chemical reaction.[1] For the case of constant total density ρ, Equation 19.1-20 can be written in rectangular coordinates as (see Problem 19B.12)

$$\frac{\partial c_A}{\partial t} = -\left(\frac{\partial}{\partial x} v_x c_A + \frac{\partial}{\partial y} v_y c_A + \frac{\partial}{\partial z} v_z c_A\right) + \mathscr{D}_{AB}\left(\frac{\partial^2 c_A}{\partial x^2} + \frac{\partial^2 c_A}{\partial y^2} + \frac{\partial^2 c_A}{\partial z^2}\right) - k'''_n c_A^n \tag{20.2-1}$$

Here k'''_n is the reaction rate constant for the nth-order chemical reaction, and it is presumed to be independent of position. In subsequent equations we shall consider $n = 1$ and $n = 2$ in order to emphasize the difference between reactions of first and higher order.

When c_A is replaced by $\bar{c}_A + c'_A$, and v_i by $\bar{v}_i + v'_i$, we obtain after time-averaging

$$\frac{\partial \bar{c}_A}{\partial t} = -\left(\frac{\partial}{\partial x} \bar{v}_x \bar{c}_A + \frac{\partial}{\partial y} \bar{v}_y \bar{c}_A + \frac{\partial}{\partial z} \bar{v}_z \bar{c}_A\right) - \underline{\left(\frac{\partial}{\partial x} \overline{v'_x c'_A} + \frac{\partial}{\partial y} \overline{v'_y c'_A} + \frac{\partial}{\partial z} \overline{v'_z c'_A}\right)}$$
$$+ \mathscr{D}_{AB}\left(\frac{\partial^2 \bar{c}_A}{\partial x^2} + \frac{\partial^2 \bar{c}_A}{\partial y^2} + \frac{\partial^2 \bar{c}_A}{\partial z^2}\right) - \begin{cases} k'''_1 \bar{c}_A & \text{or} \\ k'''_2 \left(\bar{c}_A^2 + \underline{\overline{c'^2_A}}\right) \end{cases} \tag{20.2-2}$$

Comparison of this equation with Eq. 20.2-1 indicates that the time-smoothed equation differs in the appearance of some extra terms, marked here with dashed underlines. The terms containing $\overline{v'_i c'_A}$ describe the turbulent mass transport and we designate them

---

[1] S. Corrsin, *Physics of Fluids*, **1**, 42–47 (1958).

by $\bar{J}_{Ai}^{(t)}$, the ith component of the turbulent molar-flux vector.[2] We have now met the third of the turbulent fluxes, and we may now summarize their components thus:

Turbulent molar-flux vector components: $$\bar{J}_{Ai}^{(t)} = \overline{v_i' c_A'} \quad (20.2\text{-}3)$$

Turbulent momentum-flux vector components: $$\bar{\tau}_{ij}^{(t)} = \rho \overline{v_i' v_j'} \quad (20.2\text{-}4)$$

Turbulent energy-flux vector components: $$\bar{q}_i^{(t)} = \rho \hat{C}_p \overline{v_i' T'} \quad (20.2\text{-}5)$$

All of these are defined as fluxes with respect to the mass-average velocity **v**.

It is interesting to note that there is an essential difference between the behaviors of chemical reactions of different orders. The first-order reaction expression has the same form in the time-smoothed equation as in the original equation. The second-order reaction, on the other hand, contributes on time-smoothing an extra term $-k_2''' \overline{c_A'^2}$, this being the manifestation of the interaction between the chemical kinetics and the turbulent fluctuations.

We now summarize all three of the time-smoothed equations of change for turbulent flow of an isothermal, binary fluid mixture with constant ρ, $\mathscr{D}_{AB}$, and μ:

Continuity: $$(\nabla \cdot \bar{\mathbf{v}}) = 0 \quad (20.2\text{-}6)$$

Motion: $$\rho \frac{D\bar{\mathbf{v}}}{Dt} = -\nabla \bar{p} - [\nabla \cdot (\bar{\boldsymbol{\tau}}^{(v)} + \underline{\bar{\boldsymbol{\tau}}^{(t)}})] + \rho \mathbf{g} \quad (20.2\text{-}7)$$

Continuity of A: $$\frac{D\bar{c}_A}{Dt} = -\left(\nabla \cdot \left(\bar{\mathbf{J}}_A^{(l)} + \underline{\bar{\mathbf{J}}_A^{(t)}}\right)\right) - \begin{cases} k_1''' \bar{c}_A & \text{or} \\ k_2''' \left(\bar{c}_A^2 + \underline{\overline{c_A'^2}}\right) \end{cases} \quad (20.2\text{-}8)$$

Here $\bar{\mathbf{J}}_A^{(l)} = -\mathscr{D}_{AB} \nabla \bar{c}_A$, and it is understood that the D/Dt is to be written with the time-smoothed mass-average velocity $\bar{\mathbf{v}}$ in it.

§20.3 SEMIEMPIRICAL EXPRESSIONS FOR THE TURBULENT MASS FLUX

In the preceding section it was shown that the time-smoothing of the equation of continuity of A gives rise to a turbulent mass flux with components $\bar{J}_{Ai}^{(t)} = \overline{v_i' c_A'}$. To solve mass-transport problems in turbulent flow, it may be useful to postulate a relation between $\bar{J}_{Ai}^{(t)}$ and the time-smoothed concentration gradient. A number of empirical expressions can be found in the literature, but we present here only the two most popular ones.

a. Eddy diffusivity

By analogy with Fick's first law of diffusion, we may write

$$\bar{J}_{Ay}^{(t)} = -\mathscr{D}_{AB}^{(t)} \frac{d\bar{c}_A}{dy} \quad (20.3\text{-}1)$$

which is the defining equation for the *turbulent diffusivity* $\mathscr{D}_{AB}^{(t)}$, also called the *eddy diffusivity*. As is the case with the eddy viscosity and the eddy thermal conductivity, the

---

[2]The symbol $\mathbf{J}_A$ is used here to represent the *molar* flux of A with relative to the *mass*-average velocity **v**. This differs from the symbols $\mathbf{j}_A$ and $\mathbf{J}_A^*$ defined in Chapter 17 (these are the mass flux of A relative to the mass-average velocity, and the molar flux of A relative to the molar-average velocity, respectively). The term $\bar{\mathbf{J}}_A^{(t)}$ is then the time-averaged turbulent molar flux of A relative to the mass-average velocity. In general, the molar flux of A relative to the mass-average velocity can be written $\mathbf{J}_A = c_A(\mathbf{v}_A - \mathbf{v})$.

eddy diffusivity is not a physical property characteristic of the fluid, but depends on the position, direction, and the nature of the flow field.

The eddy diffusivity $\mathscr{D}_{AB}^{(t)}$ and the eddy kinematic viscosity $\nu^{(t)} = \mu^{(t)}/\rho$ have the same dimensions, namely length squared divided by time. Their ratio

$$\text{Sc}^{(t)} = \frac{\nu^{(t)}}{\mathscr{D}_{AB}^{(t)}} \tag{20.3-2}$$

is a dimensionless quantity, known as the *turbulent Schmidt number*. As is the case with the turbulent Prandtl number, the turbulent Schmidt number is of the order of unity (see the discussion in §12.3). Thus, the eddy diffusivity may be estimated by simply replacing it by the turbulent kinematic viscosity, about which a fair amount is known. This is done in §20.4, which follows.

b. The mixing-length expression of Prandtl and Taylor

According to the mixing-length theory of Prandtl, momentum, energy, and mass are all transported by the same mechanism. Hence, by analogy with Eqs. 4.4-4 and 12.3-3 we may write

$$\bar{j}_{Ay}^{(t)} = -l^2 \left|\frac{d\bar{v}_x}{dy}\right| \frac{d\bar{c}_A}{dy} \tag{20.3-3}$$

where l is the Prandtl mixing length introduced in Chapter 4. The quantity $l^2|d\bar{v}_x/dy|$ appearing here corresponds to $\mathscr{D}_{AB}^{(t)}$ as well as to $\nu^{(t)}$ and $\alpha^{(t)}$ in Eqs. 4.4-4 and 12.3-3. Thus, the mixing-length theory satisfies the *Reynolds analogy* $\nu^{(t)} = \alpha^{(t)} = \mathscr{D}_{AB}^{(t)}$, or $\text{Pr}^{(t)} = \text{Sc}^{(t)} = 1$.

§20.4 ENHANCEMENT OF MASS TRANSFER BY A FIRST-ORDER REACTION IN TURBULENT FLOW[1]

We now examine the effect of the chemical reaction term in the turbulent diffusion equation. Specifically we study the effect of the reaction on the rate of mass transfer at the wall for steadily driven turbulent flow in a tube, where the wall (of material A) is slightly soluble in the fluid (a liquid B) flowing through the tube. Material A dissolves in liquid B and then disappears by a first-order reaction. We shall be particularly interested in the behavior with high Schmidt numbers and large reaction rates.

The discussion is divided into two parts: (a) an analysis for infinitely fast reactions, and (b) an analysis for fast and slow reactions.

a. Infinitely fast reactions and $\bar{c}_A$ independent of z

For tube flow with axial symmetry and with $\bar{c}_A$ independent of the time, Eq. 20.2-8 becomes

$$\bar{v}_z \frac{\partial \bar{c}_A}{\partial z} = \frac{1}{r}\frac{\partial}{\partial r}\left(r\left(\mathscr{D}_{AB} + \mathscr{D}_{AB}^{(t)}\right)\frac{\partial \bar{c}_A}{\partial r}\right) - k_1''' \bar{c}_A \tag{20.4-1}$$

Here we have made the customary assumption that the axial transport by both molecular and turbulent diffusion can be neglected. We want to find the mass-transfer rate at the wall

$$+\mathscr{D}_{AB}\left.\frac{\partial \bar{c}_A}{\partial r}\right|_{r=R} = k_c(c_{A0} - \bar{c}_{A,\text{axis}}) \tag{20.4-2}$$

[1]O. T. Hanna, O. C. Sandall, and C. L. Wilson, *Ind. Eng. Chem. Research*, **26**, 2286–2290 (1987). An analogous problem dealing with falling films is given by O. C. Sandall, O. T. Hanna, and F. J. Valeri, *Chem. Eng. Communications*, **16**, 135–147 (1982).

where c_{A0} and $\bar{c}_{A,\text{axis}}$ are the concentrations of A at the wall and at the tube axis. As pointed out in the preceding section, the turbulent diffusivity is zero at the wall, and consequently does not appear in Eq. 20.4-2. The quantity k_c is a *mass-transfer coefficient*, based on a molar concentration driving force, which is the analog of the heat-transfer coefficient h. The coefficient h was discussed in Chapter 14, and mentioned in Chapters 9 and 10 in connection with "Newton's law of cooling." In the following development we take $\bar{c}_{A,\text{axis}}$ to be zero, assuming that the reaction rate is sufficiently rapid that the diffusing species never reaches the tube axis. Accordingly $\partial\bar{c}_A/\partial r$ must also be zero at the tube axis. After analyzing the system under this assumption, we will relax the assumption and give computations for a wider range of reaction rates.

We now define the dimensionless concentration $C = \bar{c}_A/c_{A0}$. Then we make the further assumption that, for large z, the concentration will be independent of z, Eq. 20.4-1 becomes

$$\frac{1}{r}\frac{\partial}{\partial r}\left(r\left(\mathscr{D}_{AB} + \mathscr{D}_{AB}^{(t)}\right)\frac{\partial C}{\partial r}\right) = k_1''' C \tag{20.4-3}$$

This equation may now be multiplied by r and integrated from an arbitrary position to the tube wall to give:

$$k_c R - r(\mathscr{D}_{AB} + \mathscr{D}_{AB}^{(t)})\frac{\partial C}{\partial r} = k_1''' \int_r^R \bar{r}C(\bar{r})d\bar{r} \tag{20.4-4}$$

Here the boundary condition at $r = R$ has been used, as well as the definition of the mass-transfer coefficient in Eq. 20.4-2. Then a second integration from $r = 0$ to $r = R$ gives

$$k_c R \int_0^R \frac{1}{r(\mathscr{D}_{AB} + \mathscr{D}_{AB}^{(t)})}dr - 1 = k_1''' \int_0^R \frac{1}{r(\mathscr{D}_{AB} + \mathscr{D}_{AB}^{(t)})}\left[\int_r^R \bar{r}C(\bar{r})\,d\bar{r}\right]dr \tag{20.4-5}$$

Here use has been made of the boundary conditions that $C = 0$ at $r = 0$, and that $C = 1$ at $r = R$.

Next we introduce the variable $y = R - r$, since the region of interest is right next to the wall. Then we get

$$k_c R \int_0^R \frac{1}{(R-y)(\mathscr{D}_{AB} + \mathscr{D}_{AB}^{(t)})}dy - 1 = k_1''' \int_0^R \frac{1}{(R-y)(\mathscr{D}_{AB} + \mathscr{D}_{AB}^{(t)})}\left[\int_0^y (R-\bar{y})\,C(\bar{y})d\bar{y}\right]dy \tag{20.4-6}$$

in which $C(\bar{y})$ is not the same function of $\bar{y}$ as $C(\bar{r})$ is of $\bar{r}$. For large Sc the integrands are important only in the region where $y \ll R$, so that $R - y$ may be safely approximated by R. Furthermore, we can use the fact that the turbulent diffusivity in the neighborhood of the wall is proportional to the third power of the distance from the wall (see Eq. 12.3-5). When the integrals are rewritten in terms of $\sigma = y/R$, we get the dimensionless equation

$$\frac{1}{2}\left(\frac{k_c D}{\mathscr{D}_{AB}}\right)\left(\frac{\mathscr{D}_{AB}}{\nu}\right)\int_0^1 \frac{1}{(\mathscr{D}_{AB}/\nu) + K\sigma^3}d\sigma - 1$$

$$= \left(\frac{k_1''' R^2}{\nu}\right)\int_0^1 \frac{1}{(\mathscr{D}_{AB}/\nu) + K\sigma^3}\left[\int_0^\sigma C(\bar{\sigma})\,d\bar{\sigma}\right]d\sigma \tag{20.4-7}$$

in which $K = \text{Sc}(Rv_*/14.5\nu)^3$, where $v_* = \sqrt{\tau_0/\rho}$ is the friction velocity introduced in §4.3. This equation contains several dimensionless groupings: the Schmidt number $\text{Sc} = \nu/\mathscr{D}_{AB}$, a dimensionless reaction-rate parameter $\text{Rx} = k_1''' R^2/\nu$, and a dimensionless mass-transfer coefficient $\text{Sh} = k_c D/\mathscr{D}_{AB}$, known as the Sherwood number ($D = 2R$ being the tube diameter).

In the limit that $\text{Rx} \to \infty$, the solution to Eq. 20.4-3 under the given boundary conditions is $C(\sigma) = \exp(-\text{Sh}\,\sigma/2)$. Substitution of this into Eq. 20.4-7 then gives after straightforward integration

$$\frac{1}{2}\frac{\text{Sh}}{\text{Sc}}I_0 - 1 = 2\frac{\text{Rx}}{\text{Sh}}I_0 - 2\frac{\text{Rx}}{\text{Sh}}I_1 \tag{20.4-8}$$

in which

$$I_0 = \int_0^1 \frac{1}{\text{Sc}^{-1} + K\sigma^3}d\sigma \tag{20.4-9}$$

$$I_1 = \int_0^1 \frac{\exp(-\text{Sh}\,\sigma/2)}{\text{Sc}^{-1} + K\sigma^3}d\sigma \tag{20.4-10}$$

Equation 20.4-8 can be solved[1] to give Sh as a function of Sc, Rx, and K, that is, the dimensionless mass-transfer rate at the wall in terms of the diffusivity and the reaction rate.

The foregoing solution of Eq. 20.4-3 is reasonable when Sc, Rx, and z are sufficiently large, and is an improvement over the result given by Vieth, Porter, and Sherwood.[2] However, in the absence of chemical reaction, Eq. 20.4-3 fails to describe the downstream increase of C caused by the transfer of species A into the fluid. Thus, the mass-transfer enhancement by the chemical reaction cannot be assessed realistically from the results of either Ref. 1 or Ref. 2.

b. Fast and slow reactions and $\bar{c}_A$ dependent on z

For a better analysis of the enhancement problem, we use Eq. 20.4-1 to get a more complete differential equation for C:

$$\bar{v}_z\frac{\partial C}{\partial z} = \frac{1}{r}\frac{\partial}{\partial r}\left(r\left(\mathscr{D}_{AB} + \mathscr{D}_{AB}^{(t)}\right)\frac{\partial C}{\partial r}\right) - k_1'''\,C \tag{20.4-11}$$

The assumption that $C = 0$ at $r = 0$ is then replaced by the zero-flux condition $\partial C/\partial r = 0$ there. We represent $\mathscr{D}_{AB}^{(t)}$ in this geometry as $l^2|d\bar{v}_z/dr|$ for fully developed flow, by use of a position-dependent mixing length as in Eq. 20.3-3. Introducing dimensionless notations $v^+ = \bar{v}_z/v_*$, $z^+ = zv_*/\nu$, $r^+ = rv_*/\nu$, and $l^+ = lv_*/\nu$, based on the friction velocity $v_* = \sqrt{\tau_0/\rho}$ of §4.3, we can then express Eq. 20.4-11 in the dimensionless form

$$\begin{aligned} v^+\frac{\partial C}{\partial z^+} &= \frac{1}{r^+}\frac{\partial}{\partial r^+}\left(r^+\left(\frac{\mathscr{D}_{AB} + \mathscr{D}_{AB}^{(t)}}{\nu}\right)\frac{\partial C}{\partial r^+}\right) - \left[\!\left[\frac{k_1'''\,\nu}{v_*^2}\right]\!\right]C \\ &= \frac{1}{r^+}\frac{\partial}{\partial r^+}\left(r^+\left(\frac{1}{\text{Sc}} + (l^+)^2\left|\frac{dv^+}{dr^+}\right|\right)\frac{\partial C}{\partial r^+}\right) - \text{Da}C \end{aligned} \tag{20.4-12}$$

in which a Damköhler number $\text{Da} = k_1'''\,\nu/v_*^2$ has been introduced.

We next have to insert into Eq. 20.4-12, an expression for the dimensionless velocity profile $v^+(r^+)$. To get this, we simplify Eq. 4.2-12 to obtain the equation of motion for steadily driven tube flow in the z direction

$$0 = \frac{\mathscr{P}_0 - \mathscr{P}_L}{L} - \frac{1}{r}\frac{d}{dr}\left(r\left(\bar{\tau}_{rz}^{(l)} + \bar{\tau}_{rz}^{(t)}\right)\right) \tag{20.4-13}$$

Next we let $\tau_0 = (\mathscr{P}_0 - \mathscr{P}_L)R/2L$ (the average wall shear stress), use Newton's law of viscosity for $\bar{\tau}_{rz}^{(l)}$ (Eq. B.1-13), and use the mixing-length expression for $\bar{\tau}_{rz}^{(t)}$ (Eq. 4.4-4) and write

$$-\rho l^2\left|\frac{d\bar{v}_z}{dr}\right|\frac{d\bar{v}_z}{dr} - \mu\frac{d\bar{v}_z}{dr} = \tau_0\frac{r}{R} \tag{20.4-14}$$

---

[2]W. R. Vieth, J. H. Porter, and T. K. Sherwood, *Ind. Eng. Chem. Fundam.*, **2**, 1–3 (1963).

When we introduce the distance from the wall, $y = R - r$, we can rewrite this equation as

$$\rho l^2 \left(\frac{d\bar{v}_z}{dy}\right)^2 + \mu \frac{d\bar{v}_z}{dy} = \tau_0 \left(1 - \frac{y}{R}\right) \tag{20.4-15}$$

We can now transform to dimensionless quantities, $v^+ = v_z/v_*$, $y^+ = yv_*/\nu$, $l^+ = lv_*/\nu$, and $R^+ = Rv_*/\nu$, where $v_* = \sqrt{\tau_0/\rho}$, and obtain

$$(l^+)^2 \left(\frac{dv^+}{dr^+}\right)^2 + \frac{dv^+}{dr^+} = 1 - \frac{y^+}{R^+} \qquad (\text{for } 0 \leq y^+ \leq R^+) \tag{20.4-16}$$

The quadratic formula may be applied to Eq. 20.4-16 to give

$$\frac{dv^+}{dy^+} = \begin{cases} \dfrac{-1 + \sqrt{1 + 4(l^+)^2[1 - (y^+/R^+)]}}{2(l^+)^2} & \text{if } y^+ > 0 \\ 1 & \text{if } y^+ = 0 \end{cases} \tag{20.4-17}$$

Finally we insert an expression for the mixing length as given by the Hanna-Sandall-Mazet modification[3] of the van Driest equation[4] (see Eq. 4.4-7), which in dimensionless form is

$$l^+ = \frac{lv_*}{\nu} = 0.4y^+ \frac{1 - \exp(-y^+/26)}{\sqrt{1 - \exp(-0.26y^+)}} \qquad \text{for } 0 \leq y^+ \leq R^+ \tag{20.4-18}$$

Then $v^+(y^+)$ is computable by numerically integrating Eq. 20.4-17. The resulting function $v^+(y^+)$ closely resembles the plotted curve in Fig. 4.5-3, with small changes near $y^+ = 30$ where the plotted curve has a slope discontinuity, and near the centerline where the calculated function $v^+(y^+)$ attains a maximum value dependent on the dimensionless wall radius R^+, whereas the curve in Fig. 4.5-3 incorrectly does not.

Equations 20.4-12, 20.4-17, and 20.4-18 were solved numerically[5] for fully developed flow of a fluid with kinematic viscosity $\nu = 0.6581 \text{cm}^2/\text{s}$ in a tube with 3 cm inner diameter, at Re = 10,000, Sc = 200, and various Damköhler numbers Da. These calculations were done with the software package Athena Visual Workbench.[6] The resulting Sherwood numbers, $\text{Sh} = k_c D/\mathscr{D}_{AB}$ based on k_c as defined in Eq. 20.4-2, are plotted in Fig. 20.4-1 as functions of z^+ for various values of the Damköhler number Da. These results lead to the following conclusions:

1. In the absence of reaction (that is, when Da = 0), the Sherwood number falls off rapidly with increasing distance into the mass-transfer region. This behavior is consistent with the results of Sleicher and Tribus[7] for a corresponding heat-transfer problem, and confirms that the convection term of Eq. 20.4-10 is essential for this system. This term was neglected in References 2 and 3 by regarding the concentration profiles as "fully developed."
2. In the presence of a pseudo-first-order homogeneous reaction of the solute (that is, when Da > 0) the Sherwood number falls off downstream less rapidly, and ultimately attains a constant asymptote that depends on the Damköhler number. Thus, the enhancement factor, defined as Sh (with reaction)/Sh (without reaction), can increase considerably with increasing distance into the mass-transfer region.

---

[3] O. T. Hanna, O. C. Sandall, and P. R. Mazet, *AIChE Journal*, **27**, 693–697 (1981).
[4] E. R. van Driest, *J. Aero. Sci.*, **23**, 1007–1011, 1036 (1956).
[5] M. Caracotsios, personal communication.
[6] Information on this package is available at www.athenavisual.com and from stewart_associates.msn.com.
[7] C. A. Sleicher and M. Tribus, *Trans. ASME*, **79**, 789–797 (1957).

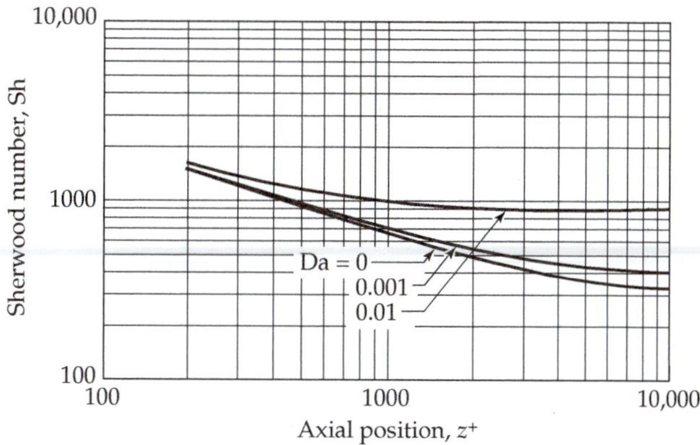

Fig. 20.4-1. Calculated Sherwood numbers, $\text{Sh} = k_c D/\mathscr{D}_{AB}$, for turbulent mass transfer from the wall of a tube, with and without homogeneous first-order chemical reaction. Results calculated at Re = 10,000 and Sc = 200, as functions axial position $z^+ = zv_*/\nu$ and Damköhler number $\text{Da} = k_1''' \nu/v_*^2$.

§20.5 CONCLUDING COMMENTS

In this, the third and last of the turbulence chapters, relatively little has been added that is really new in the subject. The notions of time-smoothed quantities, turbulent fluctuations, and correlations have been discussed in Chapters 4 and 12, and their reappearance here has not been surprising.

The only new item has been the occurrence of homogeneous chemical reactions in turbulent flows. We have given an example of the interaction of turbulence and a first-order homogeneous reaction in §20.4. It was seen that there can be a significant enhancement of the mass transfer resulting from the chemical reaction in the flow system. The interaction of turbulence and second-order homogeneous reactions has also been analyzed.[1]

The reader will have noted that a number of assumptions have gone into the above development. Most of these are probably defensible at high Schmidt numbers and very rapid reaction rates. Little can be said about their appropriateness otherwise.

QUESTIONS FOR DISCUSSION

1. Discuss the similarities and differences between turbulent heat and mass transport.
2. Discuss the behavior of first- and higher-order reactions in the time-smoothing of the equation of continuity for a given species. What are the consequences of this?
3. To what extent are the turbulent momentum flux, heat flux, and mass flux similar in form?
4. What empiricisms are available for describing the turbulent mass flux?
5. How can eddy diffusivities be measured, and on what do they depend?
6. In view of the assumptions made in §20.4, would you expect to get trustworthy results for mass transfer in turbulent tube flow without chemical reaction just by setting Rx = 0 in Eq. 20.4-8?

PROBLEMS

20A.1 Determination of eddy diffusivity. In Problem 19C.1 we gave the formula for the concentration profiles in diffusion from a point source in a moving stream (see Fig. 19C.1). In isotropic highly turbulent flow, Eq. 19C.1-3 may be modified by replacing $\mathscr{D}_{AB}$ by the eddy diffusivity $\mathscr{D}_{AB}^{(t)}$. This equation has been found to be useful for determining the eddy diffusivity. The molar flow rate of carbon dioxide is 1/1000 that of air.

(a) Show that if one plots $\ln sc_A$ versus $s - z$ the slope is $-v_0/2\mathscr{D}_{AB}^{(t)}$.

---

[1]R. B. Bird, W. E. Stewart, and E. N. Lightfoot, *Transport Phenomena*, Wiley, New York, Revised 2nd Edition (2007).

(b) Use the data on the diffusion of CO_2 from a point source in a turbulent air stream shown in Fig. 20A.1 to get $\mathscr{D}_{AB}^{(t)}$ for these conditions: pipe diameter, 15.24 cm; $v_0 = 1512$ cm/s.

(c) Compare the value of $\mathscr{D}_{AB}^{(t)}$ with the molecular diffusivity $\mathscr{D}_{AB}$ for the system CO_2-air.

(d) List all assumptions made in the calculations.

Answer: **(b)** $\mathscr{D}_{AB}^{(t)} = 20\ \text{cm}^2/\text{s}$

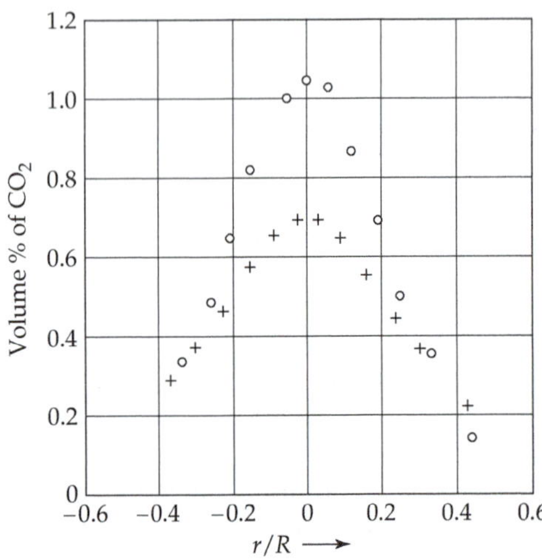

Fig. 20A.1 Concentration traverse data for CO_2 injected into a turbulent air stream with Re = 119,000 in a tube of diameter 15.24 cm. The circles are concentrations at a distance $z = 112.5$ cm downstream from the injection point, and the crosses are concentrations at $z = 152.7$ cm. [Experimental data are taken from W. L. Towle and T. K. Sherwood, *Ind. Eng. Chem.*, **31**, 457–462 (1939).]

20A.2 Heat- and mass-transfer analogy. Write the mass-transfer analog of Eq. 12.4-20. What are the limitations of the resulting equation?

20B.1 Wall mass flux for turbulent flow with no chemical reactions. Use the diffusional analog of Eq. 12.4-21 for turbulent flow in circular tubes, and the Blasius formula for the friction factor, to obtain the following expression for the Sherwood number,

$$\text{Sh} = 0.0160\text{Re}^{7/8}\text{Sc}^{1/3} \tag{20B.1-1}$$

valid for large Schmidt numbers.[1]

20B.2 Alternate expressions for the turbulent mass flux. Seek an asymptotic expression for the turbulent mass flux for long circular tubes with a boundary condition of constant wall mass flux. Assume that the net mass-transfer rate across the wall is small.

(a) Parallel the approach for laminar flow heat transfer in §10.9 to write

$$\Pi(\xi,\zeta) = \frac{\omega_A - \omega_{A1}}{j_{A0}D/\rho\mathscr{D}_{AB}} = C_1\zeta + \Pi_\infty(\xi) \tag{20B.2-1}$$

in which $\xi = r/D$, $\zeta = (z/D)/\text{ReSc}$, ω_{A1} is the inlet mass fraction of A, and j_{A0} is the interfacial mass flux of A into the fluid.

(b) Next use the equation of continuity for species A to obtain

$$4\frac{\bar{v}_z(\xi)}{\langle\bar{v}_z\rangle} = \frac{1}{\xi}\frac{d}{d\xi}\left[\left(1 + \frac{\text{Sc}}{\text{Sc}^{(t)}}\frac{\mu^{(t)}}{\mu}\right)\xi\frac{d\Pi_\infty}{d\xi}\right] \tag{20B.2-2}$$

[1] O. T. Hanna, O. C. Sandall, and C. R. Wilson, *Ind. Eng. Chem. Res.*, **28**, 2286–2290 (1987).

in which $\text{Sc}^{(t)} = \mu^{(t)}/\rho\mathscr{D}_{AB}^{(t)}$. This equation is to be integrated with the boundary conditions that Π_∞ is finite at $\xi = 0$ and $d\Pi_\infty/d\xi = -1$ at $\xi = \frac{1}{2}$.

(c) Integrate the above equation once with respect to ξ to obtain

$$\frac{d\Pi_\infty}{d\xi} = \frac{\frac{1}{2} - 4\int_\xi^{1/2} (\bar{v}_z(\bar{\xi})/\langle \bar{v}_z \rangle)\bar{\xi}d\bar{\xi}}{\xi[1 + (\text{Sc}/\text{Sc}^{(t)})(\mu^{(t)}/\mu)]} \tag{20B.2-3}$$

If one has information about the time-averaged velocity profile, Eq. 20B.2-3 can be integrated to obtain the concentration profile.

20B.3 An asymptotic expression for the turbulent mass flux.[2] Start with the final results of Problem 20B.2, and note that for sufficiently large Sc all curvature of the concentration profile will take place very near the wall where $\bar{v}_z(\xi)/\langle \bar{v}_z \rangle \approx 0$ and $\xi \approx \frac{1}{2}$. Assume that $\text{Sc}^{(t)} = 1$ and use Eq. 4.4-2 to obtain

$$\frac{d\Pi_\infty}{d\xi} = \frac{1}{[1 + \text{Sc}(\mu^{(t)}/\mu)]} = \frac{1}{[1 + \text{Sc}(yv^*/14.5\nu)^3]} \tag{20B.3-1}$$

Introduce the new coordinate $\eta = \text{Sc}^{1/3}(yv_*/14.5\nu)$ into Eq. 20B.3-1 to get an equation for $d\Pi_\infty/d\eta$ valid within the laminar sublayer. Then integrate from $\eta = 0$ (where $\bar{\omega}_A = \bar{\omega}_{A0}$) to $\eta = \infty$ (where $\bar{\omega}_A \approx \bar{\omega}_{Ab}$) to obtain an explicit relation for the wall mass flux j_{A0}. Compare with the analog of Eq. 12.4-21 obtained in Problem 20B.1.

20B.4 Deposition of silver from a turbulent stream. An approximately 0.1 *N* solution of KNO_3 containing 1.00×10^{-6} g-equiv. $AgNO_3$ per liter is flowing between parallel Ag plates, as shown in Fig. 20B.4(*a*). A small voltage is applied across the plates to produce a deposition of Ag on the cathode (lower plate) and to polarize the circuit completely (that is, to maintain the Ag^+ concentration at the cathode very nearly zero). Forced diffusion may be ignored, and the Ag^+ may be considered to be moving to the cathode by ordinary (that is, Fickian) diffusion and eddy diffusion only. Furthermore, this solution is sufficiently dilute that the effects of the other ionic species on the diffusion of Ag^+ are negligible.

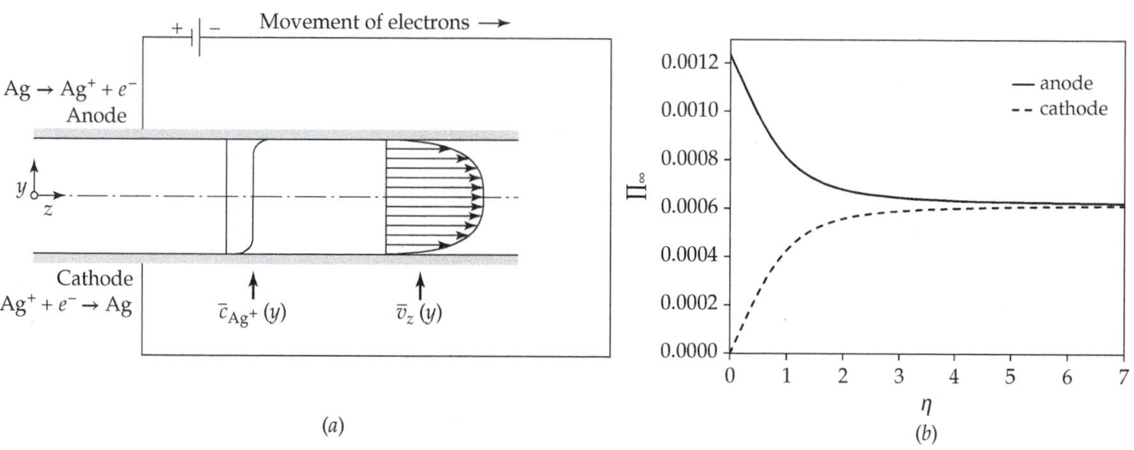

Fig. 20B.4 (*a*) Electrodeposition of Ag^+ from a turbulent stream flowing in the positive *z* direction between two parallel plates. (*b*) Concentration profiles in electrodeposition of Ag at an electrode (dimensionless quantities defined in Problem 20B.2 and 20B.3).

[2] C. S. Lin, R. W. Moulton, and G. L. Putnam, *Ind. Eng. Chem.*, **45**, 636 (1953).

(a) Calculate the Ag^+ concentration profile, assuming that (i) the effective binary diffusivity of Ag^+ through the water is 1.06×10^{-5} cm^2/s; (ii) the truncated Lin, Moulton, and Putnam expression of Eq. 4.4-2 for the turbulent velocity distribution in round tubes is valid for "slit flow" as well, if four times the hydraulic radius is substituted for the tube diameter; (iii) the plates are 1.27 cm apart, and $\sqrt{\tau_0/\rho}$ is 11.4 cm/s.

(b) Estimate the rate of deposition of Ag on the cathode, neglecting all other electrode reactions.

(c) Does the method of calculation in part (a) predict a discontinuous slope for the concentration profile at the center plane of the system? Explain.

Answers: **(a)** See Fig. 20B.4(*b*); **(b)** 6.76×10^{-12}equiv/$cm^2 \cdot s$

Chapter 21

Dimensional Analysis for Flowing Mixtures

This chapter on dimensional analysis is the mass-transfer analog of Chapter 5 (momentum transfer) and Chapter 13 (heat transfer). Here we dimensionally analyze the equations of change summarized in §19.2, using special cases of the flux expressions in §19.3. The aim is to identify the controlling dimensionless parameters of representative mass-transfer problems, and to provide an introduction to the mass-transfer correlations given in the next chapter. The focus here is on nonreacting systems, which we will exploit in the next chapter where we discuss correlations for mass-transfer coefficients.

§21.1 DIMENSIONAL ANALYSIS OF THE EQUATIONS OF CHANGE OF A BINARY MIXTURE

As in the previous chapters dealing with dimensional analysis, we restrict the discussion primarily to systems of constant physical properties. Here we are concerned with flowing, nonreacting binary mixtures, in which diffusion may be occurring. The equation of continuity for the mixture takes the familiar form

Continuity:
$$(\nabla \cdot \mathbf{v}) = 0 \tag{21.1-1}$$

The equation of motion may be written in the Boussinesq approximation (see §11.3) by putting Eq. 19.3-2 and Eq. 21.1-1 into Eq. 19.2-3, and replacing $-\nabla p + \bar{\rho}\mathbf{g}$ by $\nabla\mathscr{P}$. For a constant-viscosity Newtonian fluid this gives

Motion:
$$\bar{\rho}\frac{D\mathbf{v}}{Dt} = \mu\nabla^2\mathbf{v} - \nabla\mathscr{P} - \bar{\rho}\mathbf{g}\bar{\beta}(T-\bar{T}) - \bar{\rho}\mathbf{g}\bar{\zeta}(\omega_A - \bar{\omega}_A) \tag{21.1-2}$$

The energy equation—in the absence of chemical reactions, viscous dissipation, and external forces other than gravity—is obtained from Eq. (D) of Table 19.2-4, with Eq. 19.3-3. In using the latter we further neglect the diffusional transport of energy relative to the mass-average velocity. For constant thermal conductivity, k, this leads to

Energy:
$$\frac{DT}{Dt} = \alpha \nabla^2 T \tag{21.1-3}$$

in which $\alpha = k/\bar{\rho}\hat{C}_p$ is the thermal diffusivity. For nonreacting binary mixtures with constant ρ and $\mathscr{D}_{AB}$, Eq. 19.1-18 becomes

Continuity of A:
$$\frac{D\omega_A}{Dt} = \mathscr{D}_{AB} \nabla^2 \omega_A \tag{21.1-4}$$

For the assumptions that have been made, the analogy between Eqs. 21.1-3 and 21.1-4 is clear.

We now introduce the reference quantities l_0, v_0, and $\mathscr{P}_0$, used in §5.1 and §13.1, the reference temperatures T_0 and T_1, and the analogous reference mass fractions ω_{A0} and ω_{A1}. Then the dimensionless quantities that will arise are

$$\breve{x} = \frac{x}{l_0} \quad \breve{y} = \frac{y}{l_0} \quad \breve{z} = \frac{z}{l_0} \quad \breve{t} = \frac{v_0 t}{l_0} \tag{21.1-5}$$

$$\breve{\mathbf{v}} = \frac{\mathbf{v}}{v_0} \quad \breve{\nabla} = l_0 \nabla \quad \frac{D}{D\breve{t}} = \left(\frac{l_0}{v_0}\right)\frac{D}{Dt} \quad \breve{\mathscr{P}} = \frac{\mathscr{P} - \mathscr{P}_0}{\bar{\rho} v_0^2} \tag{21.1-6}$$

$$\breve{T} = \frac{T - T_0}{T_1 - T_0} \quad \breve{\omega}_A = \frac{\omega_A - \omega_{A0}}{\omega_{A1} - \omega_{A0}} \tag{21.1-7}$$

Here it is understood that $\mathbf{v}$ is the mass-average velocity of the mixture. It should be recognized that for some problems other choices of dimensionless variables may be preferable.

In terms of the dimensionless variables defined above, the equations of change may now be written as[1]

Continuity:
$$(\breve{\nabla} \cdot \breve{\mathbf{v}}) = 0 \tag{21.1-8}$$

Motion:
$$\frac{D\breve{\mathbf{v}}}{D\breve{t}} = \frac{1}{\mathrm{Re}} \breve{\nabla}^2 \breve{\mathbf{v}} - \breve{\nabla}\breve{\mathscr{P}} - \frac{\mathrm{Gr}}{\mathrm{Re}^2}\frac{\mathbf{g}}{g}(\breve{T} - \bar{\breve{T}}) - \frac{\mathrm{Gr}_\omega}{\mathrm{Re}^2}\frac{\mathbf{g}}{g}(\breve{\omega}_A - \bar{\breve{\omega}}_A) \tag{21.1-9}$$

Energy:
$$\frac{D\breve{T}}{D\breve{t}} = \frac{1}{\mathrm{RePr}} \breve{\nabla}^2 \breve{T} \tag{21.1-10}$$

Continuity of A:
$$\frac{D\breve{\omega}_A}{D\breve{t}} = \frac{1}{\mathrm{ReSc}} \breve{\nabla}^2 \breve{\omega}_A \tag{21.1-11}$$

The dimensionless Reynolds, Prandtl, and thermal Grashof numbers have been given in Table 13.1-1, but two additional groups have appeared here:

$$\mathrm{Sc} = \left[\!\left[\frac{\mu}{\rho \mathscr{D}_{AB}}\right]\!\right] = \left[\!\left[\frac{\nu}{\mathscr{D}_{AB}}\right]\!\right] = \text{Schmidt number} \tag{21.1-12}$$

$$\mathrm{Gr}_\omega = \left[\!\left[\frac{\bar{\rho}^2 g \bar{\zeta}(\omega_{A1} - \omega_{A0}) l_0^3}{\mu^2}\right]\!\right] = \text{diffusional Grashof number} \tag{21.1-13}$$

The Schmidt number is the ratio of momentum diffusivity (kinematic viscosity ν) to mass diffusivity and represents the relative ease of molecular momentum and mass transfer.

[1]The dimensionless equations of change written in terms of the mole fraction are equivalent to Eqs. 21.1-8 through 21.1-11, with $\breve{\omega}_A$ replaced by $\breve{x}_A$, and Gr_ω replaced by Gr_x. These two new dimensionless quantities are defined $\breve{x}_A = (x_A - x_{A0})/(x_{A1} - x_{A0})$, and $\mathrm{Gr}_x = [\![\bar{\rho}^2 g \bar{\xi}(x_{A1} - x_{A0}) l_0^3/\mu^2]\!]$, where $\bar{\xi} = -(1/\bar{\rho})(\partial\rho/\partial x_A)_{T,p}$.

It is analogous to the Prandtl number (Pr), which is the ratio of the momentum diffusivity (ν) to the thermal diffusivity (α). The diffusional Grashof number arises because of the buoyant force caused by the concentration inhomogeneities. The products RePr and ReSc in Eqs. 21.1-10 and 21.1-11 are known as Péclet numbers, Pé and Pé_{AB}, respectively.

The dimensional analysis of mass-transfer problems parallels that for heat-transfer problems. We illustrate the method by three examples: (i) The strong similarity between Eqs. 21.1-10 and 21.1-11 permits the solutions of many mass-transfer problems by analogy with previously solved heat-transfer problems; such an analogy is used in §21.2. (ii) Frequently the transfer of mass requires or releases energy, so that the heat and mass transfer must be considered simultaneously, as is illustrated in §21.3. (iii) Sometimes, as in many industrial mixing operations, diffusion plays a subordinate role in mass transfer and need not be given detailed consideration, as illustrated in §21.4. In §21.5, we illustrate the use of the Buckingham pi theorem in dimensional analysis of mass-transfer problems.

We shall see then that, just as for heat transfer, the use of dimensional analysis for the solution of practical mass-transfer problems is an art. This technique is normally most useful when at least some of the phenomena represented in the many dimensionless ratios can be neglected. Estimation of the relative importance of pertinent dimensionless groups normally requires considerable experience.

§21.2 CONCENTRATION DISTRIBUTION ABOUT A LONG CYLINDER

Here we consider the prediction of the concentration distribution near a very long, isothermal cylinder of volatile solid A, immersed in a gaseous stream of species B, which is insoluble in solid A. The system is similar to that pictured in Fig. 13.2-1, except that here we are considering the transfer of mass rather than heat. The vapor pressure of the solid is assumed to be small compared to the total pressure in the gas, so that the mass-transfer system is virtually isothermal. The question is: can the results of §13.2 be used to make the prediction?

We can answer the question in the affirmative, if it can be shown that suitably defined concentration profiles are identical to the dimensionless temperature profiles in the heat-transfer system:

$$\breve{\omega}_A(\breve{x},\breve{y},\breve{z}) = \breve{T}(\breve{x},\breve{y},\breve{z}) \tag{21.2-1}$$

This equality will be realized if the differential equations and boundary conditions for the two systems can be put into identical dimensionless form.

We therefore begin by choosing the same reference length, velocity, and pressure as in §13.2, and an analogous composition function $\breve{\omega}_A = (\omega_A - \omega_{A0})/(\omega_{A\infty} - \omega_{A0})$. Here ω_{A0} is the mass fraction of A in the gas adjacent to the interface, and $\omega_{A\infty}$ is the value far from the cylinder. We also specify that $\overline{\omega}_A = \omega_{A0}$, so that $\breve{\overline{\omega}}_A = 0$. The equations of change needed here are then Eqs. 21.1-8, 21.1-9, and 21.1-11. Thus, the differential equations here and in §13.2 are analogous, except for the appearance of the viscous heating term in Eq. 13.1-3.

As for the boundary conditions, we have here for the mass-transfer problem

B.C. 1: as $\breve{x}^2 + \breve{y}^2 \to \infty$, $\quad \breve{\mathbf{v}} \to \boldsymbol{\delta}_x \quad \breve{\omega}_A \to 1$ (21.2-2)

B.C. 2: at $\breve{y}^2 = \frac{1}{4}$, $\quad \breve{\mathbf{v}} = \dfrac{1}{\text{ReSc}}\dfrac{(\omega_{A0} - \omega_{A\infty})}{(1 - \omega_{A0})}\nabla\breve{\omega}_A \quad \breve{\omega}_A = 0$ (21.2-3)

B.C. 3: as $\breve{x}^2 + \breve{y}^2 \to \infty$ and $\breve{y} = 0$, $\quad \breve{\mathscr{P}} \to 0$ (21.2-4)

The second boundary condition on $\breve{\mathbf{v}}$, obtained with the help of Fick's first law, states that there is an interfacial radial velocity resulting from the sublimation of A.

If we compare the above description with that for heat transfer in §13.2, we see that there is no counterpart to the viscous dissipation term in the energy equation and no heat-transfer counterpart to the interfacial radial velocity component in the boundary condition of Eq. 21.2-3. The descriptions are otherwise analogous, however, with $\breve{\omega}_A$, Sc, and Gr_ω taking the places of $\breve{T}$, Pr, and Gr.

When the Brinkman number is sufficiently small, viscous dissipation will be unimportant, and that term in the energy equation can be neglected. Neglecting the Brinkman number term is appropriate, except for flows of very viscous fluids with large velocity gradients or in hypersonic boundary layers (§10.8). Similarly, in situations where $(1/\mathrm{ReSc})[(\omega_{A0} - \omega_{A\infty})/(1 - \omega_{A0})]$ is very small, it may be set equal to zero without introducing appreciable error. If these limiting conditions are met, analogous behavior will be obtained for heat and mass transfer. More precisely, the dimensionless concentration $\breve{\omega}_A$ will have the same dependence on $\breve{x}$, $\breve{y}$, $\breve{z}$, Re, Sc, and Gr_ω as the dimensionless temperature $\breve{T}$ will have on $\breve{x}$, $\breve{y}$, $\breve{z}$, Re, Pr, and Gr. The dimensionless concentration and temperature profiles will then be identical at a given Re, whenever Sc = Pr and $\mathrm{Gr}_\omega = \mathrm{Gr}$.

The thermal Grashof number can, at least in principle, be varied at will by changing $T_0 - T_\infty$. Hence, it is likely that the desired Grashof numbers can be obtained. However, it can be seen from Tables 9.5-1 and 17.5-1 that the Schmidt numbers for gases can vary over a considerably wider range than can the Prandtl numbers. Hence, it may be difficult to obtain a satisfactory thermal model of the mass-transfer process, except within a limited range of the Schmidt number.

Another possibly serious obstacle to achieving similar heat- and mass-transfer behavior is the possible nonuniformity of the surface temperature. The heat of sublimation must be obtained from the surrounding gas, and this in turn will cause the solid temperature to become lower than that of the gas. Hence, it is necessary to consider both heat and mass transfer simultaneously. A very simple analysis of simultaneous heat and mass transfer is discussed in the next section.

§21.3 FOG FORMATION DURING DEHUMIDIFICATION

A sketch showing the basic elements in a dehumidifier is shown in Fig. 21.3-1. Humid air enters a tube at temperature T_1 and humidity ω_{W1} (the mass fraction of water vapor). It leaves the tube at a lower temperature T_2 (below the dew point) and humidity ω_{W2}. The cooling is accomplished by an outer tube containing a liquid refrigerant that enters at temperature T_r and then boils off. The liquid refrigerant flow is countercurrent to the air flow. The cooling is sufficiently effective that the condensed water film can be assumed to be at T_r.

We wish to determine the range of refrigerant temperatures that may be used without the danger of fog formation. Fog is undesirable, because most of the tiny water droplets constituting the fog will pass through the tube along with the air, unless special collectors are provided. Fog can form if the wet air becomes supersaturated at any point in the system.

It is convenient to choose the dimensionless variables to be

$$\breve{T} = \frac{T - T_r}{T_1 - T_r} \qquad \breve{\omega}_W = \frac{\omega_W - \omega_{Wr}}{\omega_{W1} - \omega_{Wr}} \tag{21.3-1,2}$$

For the air (A)-water (W) system at moderate temperatures, the assumption of constant ρ and $\mathscr{D}_{AW}$ is reasonable, with air regarded as a single species. The heat capacities of water vapor and air are unequal, but the diffusional transport of energy is expected to be small. Hence, Eqs. 21.1-9 to 21.1-11 provide a reasonably reliable description of the dehumidification process. The boundary conditions needed to integrate the equations include $\breve{\omega}_W = \breve{T} = 1$ at the tube inlet, $\breve{\omega}_W = \breve{T} = 0$ at the gas–liquid boundary, and the no-slip boundary condition for the velocity $\breve{\mathbf{v}}$.

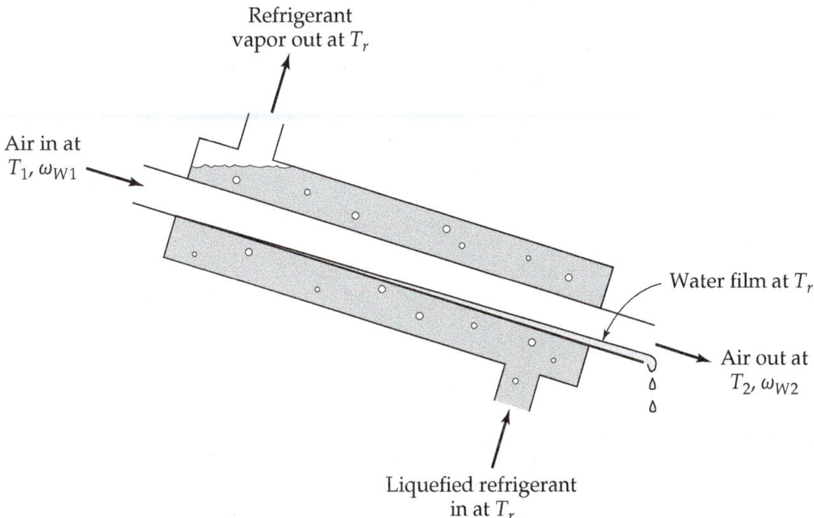

Fig. 21.3-1. Schematic representation of a dehumidifier. Air enters with temperature T_1 and humidity ω_{W1} (the mass fraction of water vapor). It leaves with outlet temperature T_2 and humidity ω_{W2}. Because the heat transfer to the refrigerant is very effective, the temperature of the air-condensate interface is close to the refrigerant temperature T_r.

We find then that the dimensionless profiles are related by

$$\check{\omega}_W(\check{x},\check{y},\check{z},\text{Re},\text{Gr}_\omega,\text{Gr},\text{Sc},\text{Pr}) = \check{T}(\check{x},\check{y},\check{z},\text{Re},\text{Gr},\text{Gr}_\omega,\text{Pr},\text{Sc}) \quad (21.3\text{-}3)$$

Thus, $\check{\omega}_W$ is the same function of its arguments as $\check{T}$ is of its arguments *in the exact order given*. Since in general Gr_ω is not equal to Gr and Sc is not equal to Pr, the two profiles are not similar. This general result is too complex to be of much value.

However, for the air-water system, at moderate temperatures and near-atmospheric pressure, Sc is about 0.6 and Pr is about 0.71.

If we assume for the moment that Sc and Pr are equal, the dimensional analysis becomes much simpler. For this special situation, the energy and species continuity equations are identical. Since the boundary conditions on $\check{\omega}_W$ and $\check{T}$ are also the same, the dimensionless concentration and temperature profiles are then identical. It should be noted that the equality of Gr_ω and Gr is not required. This is because the Grashof numbers affect the concentration and temperature profiles only by way of the velocity **v**, which appears in both the continuity equation and energy equation in the same way.

Therefore, with the assumption that Sc = Pr, we have

$$\check{\omega}_W = \check{T} \quad (21.3\text{-}4)$$

at each point in the system. This means, in turn, that *every* temperature-concentration pair in the tube lies on a straight line between (T_1, ω_{W1}) and (T_r, ω_{Wr}) on a psychrometric chart. This is shown graphically in Fig. 21.3-2 for a representative set of conditions. Note that (T_r, ω_{Wr}) must lie on the saturation curve, since equilibrium is very closely approximated.

It follows that there can be no fog formation if a straight line drawn between (T_1, ω_{W1}) and (T_r, ω_{Wr}) does not cross the saturation curve. Then the lowest refrigerant temperature that cannot produce fog is represented by the point of tangency of a straight line through (T_1, ω_{W1}) with the saturation curve.

It should be noted that *all* of the conditions along the line from the inlet (T_1, ω_{W1}) to (T_r, ω_{Wr}) will occur in the gas even though the bulk or cup-mixing conditions vary only

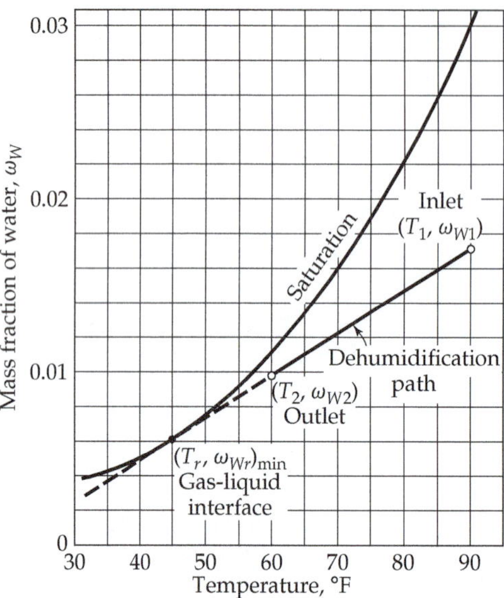

Fig. 21.3-2. A representative dehumidification path. The dehumidification path shown here corresponds to $T_{r,\min}$, the lowest refrigerant temperature ensuring the absence of fog. The dehumidification path for this situation is a tangent to the saturation curve through the point (T_1, ω_{W1}), representing the given inlet-air conditions. Calculated dehumidification paths for lower refrigerant temperatures would cross the saturation curve. Saturation water vapor concentrations would then be exceeded, making fog formation possible.

from (T_1, ω_{W1}) to (T_2, ω_{W2}). Thus, some fog can form even if saturation is not reached in the bulk of the flowing gas. For air entering at 90°F and 50% relative humidity, the minimum safe refrigerant temperature is about 45°F. It may also be seen from Fig. 21.3-2 that it is not necessary to bring all of the wet air to its dew point in order to dehumidify it. It is necessary only that the air be saturated at the cooling surface. The exit bulk conditions (T_2, ω_{W2}) can be anywhere along the dehumidification path between (T_1, ω_{W1}) and (T_r, ω_{Wr}), depending on the effectiveness of the apparatus used. Calculations based on the assumed equality of Sc and Pr have proven very useful for the air-water system.

In addition, it can be seen, by considering the physical significance of the Schmidt and Prandtl numbers, that the above-outlined calculation procedure is conservative. Since the Schmidt number is slightly smaller than the Prandtl number, dehumidification will proceed proportionally faster than cooling, and temperature-concentration pairs will be slightly below the dehumidification path drawn in Fig. 21.3-2. In condensing organic vapors from air, the reverse situation often occurs. Then the Schmidt numbers tend to be higher than the Prandtl numbers, and cooling proceeds faster than condensation. Conditions then lie above the straight line of Fig. 21.3-2, and the danger of fog formation is increased.

§21.4 BLENDING OF MISCIBLE FLUIDS

We want to develop by means of dimensional analysis the general form of a correlation for the time required to blend two miscible fluids in the agitated tank depicted in Fig. 21.4-1. Assume that the two fluids and their mixtures have essentially the same physical properties.

It will be assumed that the achievement of "equal degrees of blending" in any two mixing operations means obtaining the same dimensionless concentration profile in each. That is, the dimensionless solute concentration $\breve{\omega}_A$ is the same function of suitable dimensionless coordinates $(\breve{r},\theta,\breve{z})$ of the two systems when the degrees of blending are equal. These concentration profiles will depend on suitably defined dimensionless groups appearing in the pertinent conservation equations and their boundary conditions, and on a dimensionless time.

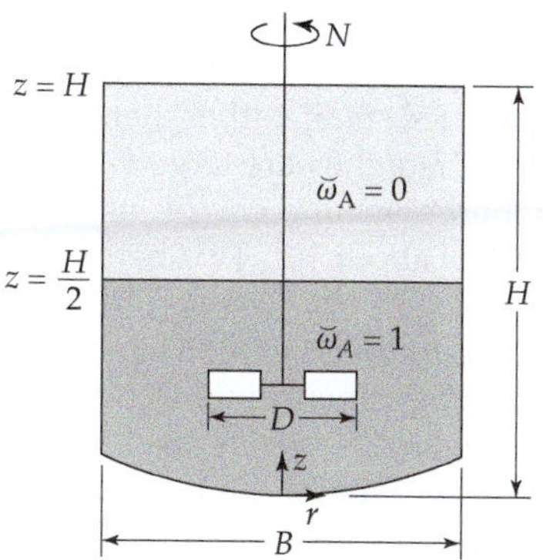

Fig. 21.4-1. Blending of miscible fluids. At zero time, the upper half of the tank is solute free, and the lower half contains a uniform distribution of solute at a dimensionless concentration of unity, and the fluid is motionless. The impeller is caused to turn at a constant rate of rotation N for all times greater than zero. Positions in the tank are given by the cylindrical coordinates r, θ, z, with r measured radially from the impeller axis, and z upward from the bottom of the tank.

In this problem we select the following definitions for the dimensionless variables:

$$\breve{r} = \frac{r}{D} \qquad \breve{z} = \frac{z}{D} \qquad \breve{\mathbf{v}} = \frac{\mathbf{v}}{ND} \qquad \breve{t} = Nt \qquad \breve{p} = \frac{p - p_0}{\rho N^2 D^2} \tag{21.4-1}$$

Here D is the impeller diameter, N is the rate of rotation of the impeller in revolutions per unit time, and p_0 is the prevailing atmospheric pressure. The dimensionless pressure $\breve{p}$ is used here rather than the quantity $\breve{\mathscr{P}}$ defined in §5.1; the formulation with $\breve{p}$ is simpler and gives equivalent results. Note that $\breve{t}$ is equal to the total number of turns of the impeller since the start of mixing.

The conservation equations describing the system are Eqs. 21.1-8, 21.1-9, and 21.1-11 with zero Grashof numbers. The dimensionless groups arising in these equations are Re, Fr, and Sc. The boundary conditions include the vanishing of $\mathbf{v}$ on the tank wall and of p on the free liquid surface. In addition we have to specify the initial conditions

$$\text{I.C. 1:} \qquad \text{at } \breve{t} \leq 0, \qquad \breve{\omega}_A = 0 \qquad \text{for } \frac{1}{2}\frac{H}{D} < \breve{z} \leq \frac{H}{D} \tag{21.4-2}$$

$$\text{I.C. 2:} \qquad \text{at } \breve{t} \leq 0, \qquad \breve{\omega}_A = 1 \qquad \text{for } 0 \leq \breve{z} < \frac{1}{2}\frac{H}{D} \tag{21.4-3}$$

$$\text{I.C. 3:} \qquad \text{at } \breve{t} \leq 0, \qquad \breve{\mathbf{v}} = \mathbf{0} \qquad \text{for } 0 \leq \breve{z} < \frac{H}{D} \text{ and } 0 \leq \breve{r} < \frac{1}{2}\frac{B}{D} \tag{21.4-4}$$

and the requirement of no slip on the impeller (see Eq. 5.3-4).

We find then that the dimensionless concentration profiles are functions of Re, Sc, Fr, the dimensionless time $\breve{t}$, the tank geometry ratios (via H/D and B/D), and the relative proportions of the two fluids. That is,

$$\breve{\omega}_A = f(\breve{r}, \breve{z}, \breve{t}, \text{Re, Fr, Sc, geometry, initial conditions}) \tag{21.4-5}$$

It is frequently possible to reduce the number of variables to be investigated.

It has been observed that, if the tank is properly baffled,[1] no vortices of importance occur; that is, the free liquid surface is effectively level. Under these circumstances, or in the absence of a free liquid surface, the Froude number does not appear in the system description, as we found in §5.3.

It is further found that in most operations on low-viscosity liquids, the rate-limiting step is the creation of a finely divided dispersion of one fluid in the other. In such a dispersion, the diffusional processes take place over very small distances. As a result, molecular diffusion is not rate limiting, and the Schmidt number Sc has little importance. It is

[1] A common and effective baffling arrangement for vertical cylindrical tanks with axially mounted impellers is a set of four evenly spaced strips along the tank wall, with their flat surfaces in planes through the tank axis, extending from the top to the bottom of the tank and at least two-tenths of the distance to the tank center.

further found that the effect of the Reynolds number (Re) is negligible under most commonly encountered conditions. This is because most of the mixing takes place in the interior of the tank where viscous effects are small, rather than in the boundary layers adjacent to the tank and impeller surfaces, where they are large.[2]

For most impeller-tank combinations in common use, the Reynolds number (Re) is unimportant when its value is above about 10^4. This behavior has been substantiated by a number of investigators.[3]

We thus arrive, *after extensive experimentation*, at a surprisingly simple result. When all of the assumptions above are valid, the dimensionless concentration profile depends only on $\breve{t}$. Hence, *the dimensionless time required to produce any desired degree of mixing is a constant* for a given system geometry. In other words, the total number of turns of the impeller during the mixing process determines the degree of blending, independently of Re, Fr, Sc, and tank size—provided, of course, that the tanks and impellers are geometrically similar.

For the same reasons, in a properly baffled tank, the dimensionless velocity distribution and the volumetric pumping efficiency of the impeller are nearly independent of the Froude number (Fr) and of the Reynolds number (Re), when $\text{Re}>10^4$.

§21.5 THE BUCKINGHAM PI THEOREM

It was pointed out in §5.5 and §13.5 that the Buckingham pi theorem can often be useful in suggesting relations among dimensionless groups of the quantities occurring in complicated problems. According to this theorem, if we are dealing with q quantities, $x_1, x_2, \ldots x_q$, involving d dimensions, then one can obtain a relation among $q - d$ different dimensionless groups.

Here we illustrate the use of this theorem for a transport problem involving a mixture.

EXAMPLE 21.5-1

Diffusion into a Falling Liquid Film

The problem of a gas absorbing into and diffusing within a falling liquid film was considered in §18.8. Gas A is absorbed into a laminar falling film of liquid B of constant thickness δ, width W and length L (see Fig. 18.8-1). The material A is only slightly soluble in B (with solubility c_{A0}) so that the viscosity of the liquid is unaffected and $\mathscr{D}_{AB}$ is constant. Since the flow is laminar, the velocity profile is $v_z(x) = v_{\max}[1 - (x/\delta)^2]$.

In §18.8, a solution to this problem was obtained in the limit of short contact times, or small penetration distances. The objective here is to use dimensional analysis and the Buckingham pi theorem to obtain relevant dimensionless groups that can be used to correlate data for the rate of absorption W_A, even when the assumption of small penetration distances is not valid.

SOLUTION

This problem contains seven quantities that contain three dimensions: δ [=] L, W [=] L, L [=] L, c_{A0} [=] moles/L^3, $\mathscr{D}_{AB}$ [=] L^2/t, $v_{\max}$ [=] L/t, and W_A [=] moles/t. We can therefore obtain four dimensionless groups, each of the form

$$\Pi = \delta^a W^b L^c c_{A0}^d \mathscr{D}_{AB}^e v_{\max}^f W_A^g \tag{21.5-1}$$

[2]The insensitivity of the required mixing time to the Reynolds number can be seen intuitively from the fact that the term $(1/\text{Re})\nabla^2\breve{\mathbf{v}}$ in Eq. 21.1-9 becomes small compared to the acceleration term $D\breve{\mathbf{v}}/D\breve{t}$ at large Re. Such intuitive arguments are dangerous, however, and the effect of Re is always important in the immediate neighborhood of solid surfaces. Here the amount of mixing taking place in the neighborhood of solid surfaces is small and can be neglected.

The insensitivity of the required mixing time to the Schmidt number can be seen from the time-averaged equation of continuity in Chapter 20. At large Re, the turbulent mass flux is much greater than that due to molecular diffusion, except in the immediate neighborhood of the solid surfaces.

[3]E. A. Fox and V. E. Gex, *AIChE Journal*, **2**, 539–544 (1956); H. Kramers, G. M. Baars, and W. H. Knoll, *Chem. Eng. Sci.*, **2**, 35–42 (1953); J. G. van de Vusse, *Chem. Eng. Sci.*, **4**, 178–200, 209–220 (1955).

By substituting the dimensions of the seven quantities into Eq. 21.5-1, we find that the dimensions of Π are

$$\Pi \ [=] \ L^{a+b+c-3d+2e+f} \text{moles}^{d+g} t^{-e-f-g} \tag{21.5-2}$$

The exponents are obtained by requiring that Π be dimensionless. This gives the system of equations

$$a + b + c - 3d + 2e + f = 0 \tag{21.5-3}$$

$$d + g = 0 \tag{21.5-4}$$

$$e + f + g = 0 \tag{21.5-5}$$

This is an underspecified system of three equations and seven unknowns, and thus we can select values for four of the unknowns a-g to obtain dimensionless groups (as long as the four selected values do not violate Eqs. 21.5-3 through 21.5-5). Because we seek a relationship between W_A and the other quantities, we will select $g = 1$ for one of the dimensionless groups and $g = 0$ for the other three dimensionless groups.

For the dimensionless group with $g = 1$, Eq. 21.5-4 gives $d = -1$. Equation 21.5-5 reduces to $e + f = -1$; we select $e = -1$, because we expect W_A to increase with $\mathscr{D}_{AB}$ from our knowledge of mass transfer. Equation 21.5-5 then gives $f = 0$. Substituting all of this information into Eq. 21.5-3 gives $a + b + c = -1$. We may select $a = b = 0$ to obtain $c = -1$. Thus, the first dimensionless group is

$$\Pi_1 = \frac{W_A}{c_{A0}\mathscr{D}_{AB}L} = \text{a dimensionless absorption rate} \tag{21.5-6}$$

For a second dimensionless group, we enforce $g = 0$, and thus Eq. 21.5-4 gives $d = 0$. We further let $f = 1$, and thus Eq. 21.5-5 gives $e = -1$. Therefore, Eq. 21.5-3 reduces to $a + b + c = 1$. We may select $a = 0$ and $b = 0$ to obtain $c = 1$. The second dimensionless group is thus

$$\Pi_2 = \frac{v_{\text{max}}L}{\mathscr{D}_{AB}} = \text{a dimensionless velocity} \tag{21.5-7}$$

This dimensionless group is related to other known dimensionless groups by $\Pi_2 = \frac{3}{8}\text{ReSc}(L/\delta)$, where $\text{Re} = 4\langle v_z\rangle\delta\rho/\mu = \frac{8}{3}v_{\text{max}}\delta\rho/\mu$, as defined in §2.2, and $\text{Sc} = \mu/\rho\mathscr{D}_{AB}$.

The remaining two dimensionless groups also use $g = 0$ and therefore $d = 0$. For both we also let $e = f = 0$, which reduces Eq. 21.5-3 to $a + b + c = 0$. Two independent solutions of this equation are $(a,b,c) = (1,0,-1)$ and $(0,1,-1)$, yielding the final two dimensionless groups

$$\Pi_3 = \frac{\delta}{L}, \qquad \text{and} \qquad \Pi_4 = \frac{W}{L} \tag{21.5-8,9}$$

These two groups could have been obtained by inspection since δ, W, and L have the same dimensions. We also note that some of the choices above are arbitrary; one could have obtained other dimensionless groups, such as $W_A/c_{A0}\mathscr{D}_{AB}\delta$, $v_{\text{max}}W/\mathscr{D}_{AB}$, L/δ, and W/δ.

In §18.8, the absorption rate was obtained assuming short penetration lengths. The result,

$$W_A = WLc_{A0}\sqrt{\frac{4\mathscr{D}_{AB}v_{\text{max}}}{\pi L}} \tag{21.5-10}$$

can be rewritten in dimensionless form as

$$\Pi_1 = \Pi_4\sqrt{\frac{4}{\pi}\Pi_2} \tag{21.5-11}$$

The dimensionless group Π_3 does not appear because the parameter δ does not enter into the analysis when the penetration length is small. When the penetration length is not small, Eq. 21.5-11 is no longer correct, but dimensional analysis still provides the relationship

$$\Pi_1 = f(\Pi_2, \Pi_3, \Pi_4) \tag{21.5-12}$$

The precise form of the function f is not obtained using dimensional analysis alone, but can be obtained empirically from experimental data.

§21.6 CONCLUDING COMMENTS

In this last chapter dealing with dimensional analysis, four illustrations have been given for systems with two chemical species. The reader should have been struck by several features of these examples.

a. The presence of more than one chemical species makes the dimensional analysis considerably more complex.

b. It was necessary to make more use of intuition than in the analyses done in Chapters 5 and 13 for pure fluids.

c. It was necessary to make more use of physical property data in some of the arguments regarding the relative magnitudes of Sc and Pr.

d. Experimental verification of some of the results had to be relied upon.

Nonetheless, the use of dimensional analysis can prove very useful for the description of the performance of some systems, and, at least in some cases, for obtaining useful, simple working equations.

We have also illustrated the use of the Buckingham pi theorem to obtain dimensionless groups that can be used to represent data. This approach requires even more insight to obtain convenient dimensionless groups, and is thus less desirable than dimensional analysis of the equations of change.

QUESTIONS FOR DISCUSSION

1. What are some of the advantages of writing the equations of change in dimensionless form?
2. Why is nondimensionalizing the equations of change preferred over the method illustrated in §21.5 for obtaining dimensionless groups? Under what conditions would the method in §21.5 be preferred?
3. Which dimensionless groups contain only physical properties? Which contain physical properties as well as information about the geometry, the strength of flow, or imposed temperatures?
4. How does one decide if a problem is one of forced convection, free convection, or of an intermediate case?
5. Can the Schmidt number be interpreted as a ratio of phenomena? Explain.
6. When listing the dimensions of the quantities that are important in a problem, what could be possible implications of a fundamental dimension appearing in only one of the quantities?
7. Given a set of q quantities with d dimensions, describe how the $q - d$ dimensionless group may be obtained from a system of d linear equations.

PROBLEMS

21A.1 Calculating Schmidt numbers. Calculate values for the Schmidt number, $\text{Sc} = \mu/\rho\mathscr{D}_{AB}$, for the following cases:

(a) $\mu = 10^{-3}$ Pa · s, $\rho = 10^3$ kg/m^3, $\mathscr{D}_{AB} = 10^{-9}$ m^2/s

(b) $\mu = 6.7 \times 10^{-4}$ lb$_m$/ft · s, $\rho = 62$ lb$_m$ /ft^3, $\mathscr{D}_{AB} = 1.08 \times 10^{-8}$ ft^2/s

(c) $\mu = 2.07 \times 10^{-5}$ Pa · s, $\rho = 0.0817$ lb$_m$/ft^3, $\mathscr{D}_{AB} = 0.181$ cm^2/s

21B.1 Verifying dimensionless groups. Verify that the dimensionless groups Π_1 and Π_2 in Example 21.5-1 are dimensionless.

21B.2 Alternate dimensional analysis. Rework the dimensional analysis of §21.1 using a dimensionless pressure defined $\breve{\mathscr{P}} = (\mathscr{P} - \mathscr{P}_0)/(\mu v_0/l_0)$ instead of $\breve{\mathscr{P}} = (\mathscr{P} - \mathscr{P}_0)/\overline{\rho} v_0^2$. Do new dimensionless groups appear?

21B.3 Dimensional analysis with a homogeneous reaction. Rework the dimensional analysis of §21.1 when a first order homogeneous reaction is also occurring. Specifically, nondimensionalize the equations of change when Eq. 21.1-4 is rewritten

$$\frac{D\omega_A}{Dt} = \mathscr{D}_{AB}\nabla^2\omega_A - k_1'''\omega_A \tag{21B.3-1}$$

where k_1''' is a constant. What new dimensionless groups appear?

21C.1 Dissolution into a falling film. Use the Buckingham pi theorem to obtain the dimensionless groups for the problem of a falling liquid film into which a solid is dissolving, as described in §18.9.

(a) Obtain a relationship between a dimensionless rate of dissolution and other relevant dimensionless groups when the penetration length is not necessarily small.

(b) Show that the result for W_A for short penetration lengths given in Eq. 18.9-12 can be rewritten in terms of the dimensionless groups you obtained.

Chapter 22

Interphase Transport in Nonisothermal Mixtures

Here we build on earlier discussions of binary diffusion to provide means for predicting the behavior of mass-transfer operations such as distillation, absorption, adsorption, extraction, drying, membrane filtrations, and heterogeneous chemical reactions. This chapter has many features in common with Chapters 6 and 14. It is particularly closely related to Chapter 14, because there are many situations where the analogies between heat and mass transfer can be regarded as exact.

There are, however, important differences between heat and mass transfer, and we will devote much of this chapter to exploring these differences. Since many mass-transfer operations involve fluid-fluid interfaces, we have to deal with distortions of the interfacial shape by viscous drag resulting from inhomogeneities in temperature and composition. In addition, there may be interactions between heat and mass transfer. Furthermore, at high mass-transfer rates, the temperature and concentration profiles may be distorted. These effects complicate and sometimes invalidate the neat analogy between heat and mass transfer that one might otherwise expect.

In Chapter 14 the interphase heat transfer involved the movement of heat to or from a solid surface, or the heat transfer between two fluids separated by a solid surface. Here we will encounter heat and mass transfer between two contiguous phases: fluid-fluid or fluid-solid. This raises the question as to how to account for the resistance to diffusion provided by the fluids on both sides of the interface.

We begin the chapter by defining, in §22.1, the mass- and heat-transfer coefficients for binary mixtures in one phase (liquid or gas). Then in §22.2 we show how analytical solutions to diffusion problems lead to explicit expressions for the mass-transfer coefficients. In that section we give some analytic expressions for mass-transfer coefficients at high Schmidt numbers for a number of relatively simple systems. We emphasize the different behavior of systems with fluid-fluid and solid-fluid interfaces.

In §22.3 we show how dimensional analysis leads to predictions involving the Sherwood number (Sh) and the Schmidt number (Sc), which are the analogs of the Nusselt number (Nu) and the Prandtl number (Pr) defined in Chapter 14. Here the emphasis is on the analogies between heat transfer in pure fluids and mass transfer in binary mixtures. Then in §22.4 we proceed to the definition of mass-transfer coefficients for systems with diffusion in two adjoining phases. We show there how to apply the information about mass transfer in single phases to the understanding of mass transfer between two phases.

In this chapter we have limited the discussion to a few key topics on mass transfer and transfer-coefficient correlations. Further information is available in specialized textbooks on these and related topics not covered here, such as mass transfer with chemical reactions, combined heat and mass transfer by free convection, effects of interfacial forces on heat and mass transfer, and transfer coefficients at high net mass-transfer rates.[1–5]

§22.1 DEFINITION OF MASS- AND HEAT-TRANSFER COEFFICIENTS IN ONE PHASE

In this chapter we relate the rates of mass transfer across phase boundaries to the relevant concentration differences. We consider binary systems only. These relations are analogous to the heat-transfer correlations of Chapter 14 and contain *mass-transfer coefficients* in place of the heat-transfer coefficients of that chapter. The system may have a true phase boundary, as in Fig. 22.1-1, Fig. 22.1-2, or Fig. 22.1-4, or an abrupt change in hydrodynamic properties, as in the system of Fig. 22.1-3, containing a porous solid. Figure 22.1-1 shows the

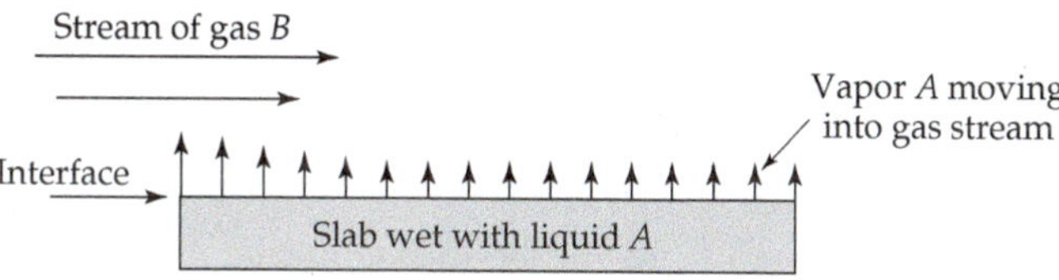

Fig. 22.1-1. Example of mass transfer across a plane boundary: drying of a saturated slab.

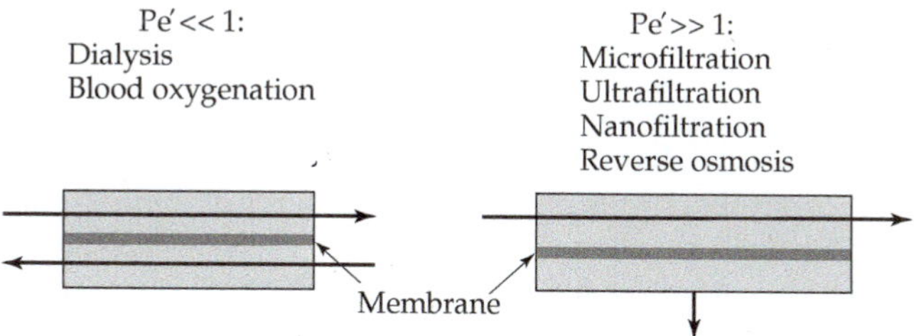

Fig. 22.1-2. Two rather typical kinds of membrane separators, classified here according to a Péclet number, $\text{Pé} = \delta v/\mathscr{D}_{\text{eff}}$, for the flow through the membrane. Here δ is the membrane thickness, v is the velocity at which solvent passes through the membrane, and $\mathscr{D}_{\text{eff}}$ is the effective solute diffusivity through the membrane. The heavy line represents the membrane, and the arrows represent the flow along or through the membrane.

---

[1] T. K. Sherwood, R. L. Pigford, and C. R. Wilke, *Mass Transfer*, McGraw-Hill, New York (1975).

[2] R. E. Treybal, *Mass Transfer Operations*, 3rd edition, McGraw-Hill, New York (1980).

[3] E. L. Cussler, *Diffusion: Mass Transfer in Fluid Systems*, 2nd edition, Cambridge University Press (1997).

[4] D. E. Rosner, *Transport Processes in Chemically Reacting Flow Systems* (Unabridged), Dover, New York (2000).

[5] R. B. Bird, W. E. Stewart, and E. N. Lightfoot, *Transport Phenomena*, Revised 2nd Edition, Wiley, New York (2007).

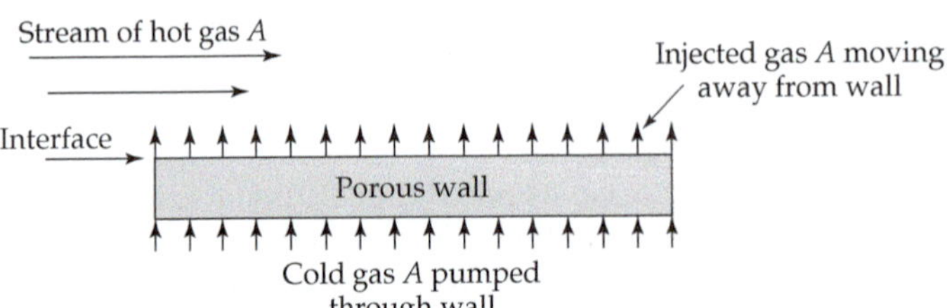

Fig. 22.1-3. Example of mass transfer through a porous wall: transpiration cooling.

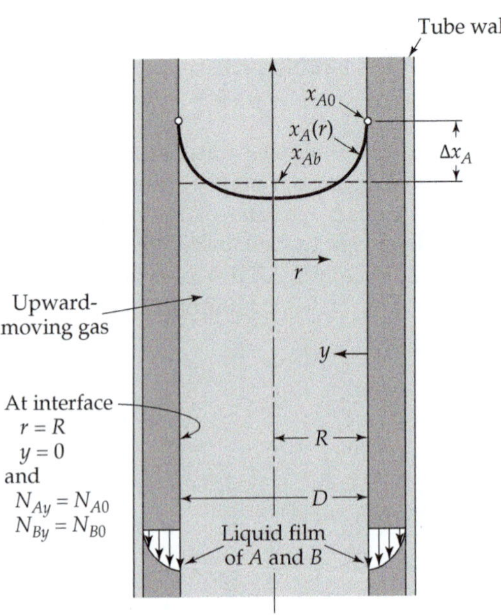

Fig. 22.1-4. Example of a gas-liquid contacting device: the wetted-wall column. Two chemical species *A* and *B* are moving from the downward-flowing liquid stream into the upward-flowing gas stream in a cylindrical tube.

evaporation of a volatile liquid, often used in experiments to develop mass-transfer correlations. Figure 22.1-2 shows a permselective membrane, in which a selectively permeable surface permits more effective transport of solvent than of a solute that is to be retained, as in ultrafiltration of protein solutions and the desalting of sea water. Figure 22.1-3 shows a macroscopically porous solid, which can serve as a mass-transfer surface or can provide sites for adsorption or reaction. Figure 22.1-4 shows an idealized liquid-vapor contactor where the mass-transfer interface may be distorted by viscous or surface-tension forces.

In each of these systems, there will be both heat and mass transfer at the interface, and each of these fluxes will have a molecular (diffusive) and a convective term (here we have moved the convective term to the left side of the equation):

$$N_{A0} - x_{A0}(N_{A0} + N_{B0}) = -\left(c\mathscr{D}_{AB}\frac{\partial x_A}{\partial y}\right)\bigg|_{y=0} \quad (22.1\text{-}1)$$

$$e_0 - (N_{A0}\overline{H}_{A0} + N_{B0}\overline{H}_{B0}) = -\left(k\frac{\partial T}{\partial y}\right)\bigg|_{y=0} \quad (22.1\text{-}2)$$

These equations are just Eq. 18.0-1 and Eq. 19.3-6 written at the mass-transfer interface ($y = 0$). They describe the interphase molar flux of species *A* and the interphase flux of energy (excluding the kinetic energy and the contribution from $[\boldsymbol{\tau} \cdot \mathbf{v}]$). Both N_{A0} and e_0 are defined as positive in a specific direction, as indicated in Figs. 22.1.1 through 22.1.4. Care must be taken to define the direction of positive flux when using interphase transport coefficients in problems.

In Chapter 14 we defined the heat-transfer coefficient in the absence of mass transfer by Eq. 14.1-1 ($Q = hA\Delta T$). For surfaces with mass and heat transfer, Eqs. 22.1-1 and 22.1-2 suggest that the following definitions are appropriate:

$$W_{A0} - x_{A0}(W_{A0} + W_{B0}) = k_{xA}A\Delta x_A \tag{22.1-3}$$

$$E_0 - (W_{A0}\overline{H}_{A0} + W_{B0}\overline{H}_{B0}) = hA\Delta T \tag{22.1-4}$$

Here W_{A0} is the number of moles of species A per unit time going through the transfer surface at $y = 0$, and E_0 is the total amount of energy going through the surface. The transfer coefficients k_{xA} and h are not defined until the area A and the driving forces Δx_A and ΔT have been specified. All the comments in Chapter 14 regarding these definitions may be taken over in this chapter, with the result that a subscript 1, ln, a, m, or loc can be added to make clear the type of driving force that is used. Throughout this chapter, however, we shall mainly use the local transfer coefficients, and occasionally the mean transfer coefficients. Also, in this chapter, molar fluxes of the species will be used, since in chemical engineering this is traditional. The relations between the mass-transfer expressions in molar and mass units are summarized in Table 22.2-1.

Local transfer coefficients are defined by writing Eqs. 22.1-3 and 22.1-4 for a differential area. Since $dW_{A0}/dA = N_{A0}$ and $dE_0/dA = e_0$, we get the definitions

$$N_{A0} - x_{A0}(N_{A0} + N_{B0}) = k_{xA,\text{loc}}\Delta x_A \tag{22.1-5}$$

$$e_0 - (N_{A0}\overline{H}_{A0} + N_{B0}\overline{H}_{B0}) = h_{\text{loc}}\Delta T \tag{22.1-6}$$

Next, we note that the left side of Eq. 22.1-5 is just J^*_{A0}, and that the left side of a similar equation written for species B is J^*_{B0}. But since $J^*_{A0} = -J^*_{B0}$ and $\Delta x_A = -\Delta x_B$, we find that $k_{xA,\text{loc}} = k_{xB,\text{loc}}$, and therefore, we can write both mass-transfer coefficients as $k_{x,\text{loc}}$, which has units of (moles)/(area)(time). Furthermore, if the heat of mixing is zero (as in ideal-gas mixtures), we can replace $\overline{H}_{A0}$ by $\tilde{C}_{pA,0}(T_0 - T^\circ)$, where T° is an arbitrarily chosen reference temperature, as explained in Example 19.3-1. A similar replacement may be made for $\overline{H}_{B0}$. With these changes we get

$$N_{A0} - x_{A0}(N_{A0} + N_{B0}) = k_{x,\text{loc}}\Delta x_A \tag{22.1-7}$$

$$e_0 - (N_{A0}\tilde{C}_{pA,0} + N_{B0}\tilde{C}_{pB,0})(T_0 - T^\circ) = h_{\text{loc}}\Delta T \tag{22.1-8}$$

We remind the reader that rapid mass transfer across phase boundaries can distort the velocity, temperature, and concentration profiles, as we have already seen in §18.6 and in Example 19.4-1. The correlations provided in §22.3, as well as their analogs in Chapters 6 and 14, are all for small mass-transfer rates, that is, for situations in which the convective terms in Eqs. 22.1-7 and 22.1-8 are negligible compared to the first term. Such situations are common, and most correlations in the literature suffer from the same limitation. Interphase transport with high mass-transfer rates is discussed extensively elsewhere.[1]

In much of the chemical engineering literature, the mass transfer coefficients are defined by

$$N_{A0} = k^0_{x,\text{loc}}\Delta x_A \tag{22.1-9}$$

The relation of this "apparent" mass-transfer coefficient to that defined by Eq. 22.1-7 is

$$k^0_{x,\text{loc}} = \frac{k_{x,\text{loc}}}{[1 - x_{A0}(1 + r)]} \tag{22.1-10}$$

in which $r = N_{B0}/N_{A0}$. Other widely used mass-transfer coefficients are defined by

$$N_{A0} = k^0_{c,\text{loc}}\Delta c_A \quad \text{and} \quad N_{A0} = k^0_{\rho,\text{loc}}\Delta \rho_A \tag{22.1-11}$$

---

[1]R. B. Bird, W. E. Stewart, and E. N. Lightfoot, *Transport Phenomena*, Revised 2nd Edition, Wiley, New York (2007), §22.8.

for liquids, and

$$N_{A0} = k^0_{p,\text{loc}} \Delta p_A \tag{22.1-12}$$

for gases. In the limit of low solute concentrations and low net mass-transfer rates, for which most correlations have been obtained

$$\lim_{x_{A0}(1+r)\to 0} \begin{Bmatrix} k^0_{x,\text{loc}} \\ ck^0_{c,\text{loc}} \\ pk^0_{p,\text{loc}} \\ \rho k^0_{\rho,\text{loc}} \end{Bmatrix} = k_{x,\text{loc}} \tag{22.1-13}$$

The superscript "0" indicates that these quantities are applicable only for *small mass-transfer rates* and *small mole fractions of species A*.

In many industrial contactors, the true interfacial area is not known. An example of such a system would be a column containing a random packing of irregular solid particles. In such a situation, one can define a volumetric mass-transfer coefficient, $k_x a$, incorporating the interfacial area for a differential region of the column. The rate at which moles of solute A are transferred to the interstitial fluid in a volume Sdz of the column is then given by

$$\begin{aligned} dW_{A0} &= (k_x a)(x_{A0} - x_{Ab})Sdz + x_{A0}(dW_{A0} + dW_{B0}) \\ &\approx (k^0_x a)(x_{A0} - x_{Ab})Sdz \end{aligned} \tag{22.1-14}$$

Here the interfacial area, a, per unit volume is combined with the mass-transfer coefficient, S is the total column cross section, and z is measured in the primary flow direction. Correlations for predicting values of these coefficients are available, but they should be used with caution. Rarely do they include all the important parameters, and as a result they cannot be safely extrapolated to new systems. Furthermore, although they are usually described as "local," they actually represent a poorly defined average over a wide range of interfacial conditions.[2–6]

We conclude this section by defining a dimensionless group widely used in the mass-transfer literature and in the remainder of this book:

$$\text{Sh} = \frac{k_x l_0}{c\mathscr{D}_{AB}} \tag{22.1-15}$$

which is called the Sherwood number, and which is based on the characteristic length l_0. This quantity can be "decorated" with subscripts 1, a, m, ln, and loc, in the same manner as k_x.

§22.2 ANALYTICAL EXPRESSIONS FOR MASS-TRANSFER COEFFICIENTS

The main value of mass-transfer coefficients is that they can be used to relate mass-transfer rates to concentration differences without requiring knowledge of the concentration profiles. Thus, the challenges of solving for concentration profiles can be avoided. However,

---

[2] J. Stichlmair and J. F. Fair, *Distillation Principles and Practice*, Wiley, New York (1998).

[3] H. Z. Kister, *Distillation Design*, McGraw-Hill, New York (1992).

[4] J. C. Godfrey and M. M. Slater, *Liquid-Liquid Extraction Equipment*, Wiley, New York (1994).

[5] R. H. Perry, D. W. Green, and J. O. Maloney, *Perry's Chemical Engineer's Handbook*, McGraw-Hill, New York, 7th edition (1997).

[6] J. E. Vivian and C. J. King, in *Modern Chemical Engineering* (A. Acrivos, ed.), Reinhold, New York (1963).

it is instructive to use concentration profiles obtained for idealized problems to predict mass-transfer coefficients, and to begin to understand the dependence of mass-transfer coefficients on other variables.

In the preceding chapters we have obtained a number of analytical solutions for concentration profiles and for the associated molar fluxes in idealized situations. From these solutions we can now derive the corresponding mass-transfer coefficients. These are usually presented in dimensionless form in terms of the Sherwood numbers. We summarize these analytical expressions here for use in later sections of this chapter. All of the results given in this section are for systems with a slightly soluble component A and small diffusivities $\mathscr{D}_{AB}$. It may be helpful to refer to Table 22.2-1 where the dimensionless groups for heat and mass transfer have been summarized.

a. Mass transfer in falling films on plane surfaces

The absorption of a slightly soluble gas A into a falling film of pure liquid B was analyzed in §18.8 in the limit of short contact times such that the dissolved species A does not "penetrate" very far into the film. The rate of absorption W_{A0} was determined in Eq.

Table 22.2-1. Analogies among Heat and Mass Transfer at Low Mass-Transfer Rates

	Heat-transfer quantities (pure fluids)	Binary mass-transfer quantities (isothermal fluids, molar units)	Binary mass-transfer quantities (isothermal fluids, mass units)
Profiles	T	x_A	ω_A
Diffusivity	$\alpha = k/\rho\hat{C}_p$	$\mathscr{D}_{AB}$	$\mathscr{D}_{AB}$
Effect of profiles on density	$\beta = -\frac{1}{\rho}\left(\frac{\partial\rho}{\partial T}\right)_p$	$\xi = -\frac{1}{\rho}\left(\frac{\partial\rho}{\partial x_A}\right)_{p,T}$	$\zeta = -\frac{1}{\rho}\left(\frac{\partial\rho}{\partial \omega_A}\right)_{p,T}$
Flux	$\mathbf{q}$	$\mathbf{J}_A^* = \mathbf{N}_A - x_A(\mathbf{N}_A + \mathbf{N}_B)$	$\mathbf{j}_A = \mathbf{n}_A - \omega_A(\mathbf{n}_A + \mathbf{n}_B)$
Transfer rate	Q	$W_{A0} - x_{A0}(W_{A0} + W_{B0})$	$w_{A0} - \omega_{A0}(w_{A0} + w_{B0})$
Transfer coefficient	$h = \dfrac{Q}{A\Delta T}$	$k_x = \dfrac{W_{A0} - x_{A0}(W_{A0} + W_{B0})}{A\,\Delta x_A}$	$k_\omega = \dfrac{w_{A0} - \omega_{A0}(w_{A0} + w_{B0})}{A\,\Delta\omega_A}$
Dimensionless groups common to all three correlations	$\text{Re} = l_0 v_0 \rho/\mu$ $\text{Fr} = v_0^2/g l_0$	$\text{Re} = l_0 v_0 \rho/\mu$ $\text{Fr} = v_0^2/g l_0$	$\text{Re} = l_0 v_0 \rho/\mu$ $\text{Fr} = v_0^2/g l_0$
Dimensionless groups that are different	$\text{Nu} = h l_0/k$ $\text{Pr} = \hat{C}_p\mu/k$ $\text{Gr} = l_0^3\bar{\rho}^2 g\beta\Delta T/\mu^2$ $\text{Pé} = \text{Re Pr} = \rho l_0 v_0 \hat{C}_p/k$	$\text{Sh} = k_x l_0/c\mathscr{D}_{AB}$ $\text{Sc} = \mu/\rho\mathscr{D}_{AB}$ $\text{Gr}_x = l_0^3\bar{\rho}^2 g\xi\Delta x_A/\mu^2$ $\text{Pé} = \text{ReSc} = l_0 v_0/\mathscr{D}_{AB}$	$\text{Sh} = k_\omega l_0/\rho\mathscr{D}_{AB}$ $\text{Sc} = \mu/\rho\mathscr{D}_{AB}$ $\text{Gr}_\omega = l_0^3\bar{\rho}^2 g\zeta\Delta\omega_A/\mu^2$ $\text{Pé} = \text{ReSc} = l_0 v_0/\mathscr{D}_{AB}$
Chilton-Colburn j-factors	$j_H = \text{NuRe}^{-1}\text{Pr}^{-1/3}$ $= \dfrac{h}{\rho\hat{C}_p v_0}\left(\dfrac{\hat{C}_p\mu}{k}\right)^{2/3}$	$j_D = \text{ShRe}^{-1}\text{Sc}^{-1/3}$ $= \dfrac{k_x}{c v_0}\left(\dfrac{\mu}{\rho\mathscr{D}_{AB}}\right)^{2/3}$	$j_D = \text{ShRe}^{-1}\text{Sc}^{-1/3}$ $= \dfrac{k_\omega}{\rho v_0}\left(\dfrac{\mu}{\rho\mathscr{D}_{AB}}\right)^{2/3}$

Note: (a) the subscript 0 on l_0 and v_0 indicates the characteristic length and velocity respectively, whereas the subscript 0 on the mole (or mass) fraction and molar (or mass) flux means "evaluated at the interface." (b) All three of these Grashof numbers can be written as $\text{Gr} = l_0^3\rho g\,\Delta\rho/\mu^2$, provided that the density change is caused only by a difference of temperature or composition.

18.8-18, which can be put into the form

$$W_{A0} = \left(\sqrt{\frac{4\mathscr{D}_{AB} v_{\max}}{\pi L}} \right) (WL)(c_{A0} - 0) \tag{22.2-1}$$

Equation 22.1-3 can be rewritten in terms of a mean mass-transfer coefficient $k^0_{c,m}$ (e.g., using the a mean mass-transfer coefficient version of Eq. 22.1-11) with molar concentration units and for small concentrations ($x_A \ll 1$) as

$$W_{A0} = k^0_{c,m} A \Delta c_A \tag{22.2-2}$$

where $A = WL$ is the contact area between the liquid and gas phases. Comparison of Eqs. 22.2-1 and 22.2-2 gives

$$k^0_{c,m} = \sqrt{\frac{4\mathscr{D}_{AB} v_{\max}}{\pi L}} \tag{22.2-3}$$

or

$$\text{Sh}_m = \frac{k^0_{c,m} L}{\mathscr{D}_{AB}} = \sqrt{\frac{4 L v_{\max}}{\pi \mathscr{D}_{AB}}} = \sqrt{\frac{4}{\pi} \left(\frac{L v_{\max} \rho}{\mu} \right) \left(\frac{\mu}{\rho \mathscr{D}_{AB}} \right)} = 1.128(\text{ReSc})^{1/2} \tag{22.2-4}$$

This shows that the Sherwood number (the dimensionless mass-transfer coefficient) depends on the Reynolds number and the Schmidt number, and that Re is defined in terms of the maximum velocity $v_{\max}$ in the film and the film length L. The Reynolds number could also be defined in terms of the average film velocity with a different numerical coefficient.

This "penetration model" for the mean Sherwood number can also be written

$$\text{Sh}_m = \sqrt{\frac{4L^2}{\pi \mathscr{D}_{AB} t_{\exp}}} \tag{22.2-5}$$

where $t_{\exp} = L/v_{\max}$ is the time of exposure of a fluid element on the surface of the moving liquid stream to the gas containing the absorbing species (see §18.8). It is possible to employ this model to estimate the mean Sherwood number in other geometries in which a liquid stream is exposed to a gas for a short period.

Similarly, for the dissolution of a slightly soluble material A from the wall into a falling liquid film of pure B, we can put Eq. 18.9-12 into the form of Eq. 22.2-2 as follows:

$$W_{A0} = \left(\frac{2\mathscr{D}_{AB}}{\Gamma\left(\frac{7}{3}\right)} \sqrt[3]{\frac{a}{9\mathscr{D}_{AB} L}} \right) (WL)(c_{A0} - 0) \equiv k^0_{c,m} A \Delta c_A \tag{22.2-6}$$

Then, using the definition of a given just after Eq. 18.9-1 and the expression for the maximum velocity in the film in Eq. 2.2-23, we find the Sherwood number as follows:

$$\begin{aligned} \text{Sh}_m = \frac{k^0_{c,m} L}{\mathscr{D}_{AB}} &= \frac{2}{\Gamma\left(\frac{7}{3}\right)} \sqrt[3]{\frac{(2v_{\max}/\delta)L^2}{9\mathscr{D}_{AB}}} \\ &= \frac{1}{\Gamma\left(\frac{7}{3}\right)} \sqrt[3]{\frac{16}{9} \left(\frac{L}{\delta} \right) \left(\frac{L v_{\max} \rho}{\mu} \right) \left(\frac{\mu}{\rho \mathbf{D}_{AB}} \right)} \\ &= 1.017 \sqrt[3]{\left(\frac{L}{\delta} \right)} (\text{ReSc})^{1/3} \end{aligned} \tag{22.2-7}$$

In this instance we have not only the Reynolds number and Schmidt number appearing, but also the ratio of the film length to the film thickness.

These two problems first introduced in Chapter 18—gas absorption by a falling film and the dissolution of a solid wall into a falling film—illustrate two important situations. In the first problem, there is no velocity gradient at the gas-liquid interface, and the quantity ReSc appears to the $\frac{1}{2}$-power in the expression for the Sherwood number. In the second problem, there is a velocity gradient at the solid-liquid interface, and the quantity ReSc appears to the $\frac{1}{3}$-power in the Sherwood number expression.

b. Mass transfer for flow around spheres

Next we consider the diffusion that occurs in the creeping flow around a spherical gas bubble and around a solid sphere. This pair of systems corresponds to the two systems discussed in the previous subsection.

For the gas absorption from a gas bubble surrounded by a liquid in creeping flow, we can put Eq. 18.8-20 in the form of Eq. 22.2-2 thus:

$$W_{A0,\text{avg}} = \sqrt{\frac{4}{3\pi}\frac{\mathscr{D}_{AB}v_\infty}{D}}A(c_{A0} - 0) \equiv k^0_{c,m}A\Delta c_A \tag{22.2-8}$$

where A is the surface area of the bubble. The Sherwood number is then

$$\begin{aligned}\text{Sh}_m &= \frac{k^0_{c,m}D}{\mathscr{D}_{AB}} = \sqrt{\frac{4}{3\pi}\frac{Dv_\infty}{\mathscr{D}_{AB}}} = \sqrt{\frac{4}{3\pi}\left(\frac{Dv_\infty\rho}{\mu}\right)\left(\frac{\mu}{\rho\mathscr{D}_{AB}}\right)}\\ &= 0.6515(\text{ReSc})^{1/2}\end{aligned} \tag{22.2-9}$$

Here the Reynolds number is defined using the approach velocity v_∞ of the fluid (or, alternatively, the terminal velocity of the rising bubble).

For the creeping flow around a solid sphere with a slightly soluble coating that dissolves into the approaching fluid, the analogous heat-transfer problem has been solved and the heat-transfer rate has been determined;[1] by analogy, the flux of dissolving material can be written

$$N_{A0,\text{avg}} = \frac{(3\pi)^{2/3}}{2^{7/3}\Gamma\left(\frac{4}{3}\right)}\sqrt[3]{\frac{\mathscr{D}^2_{AB}v_\infty}{D^2}}(c_{A0} - 0) \equiv k^0_{c,m}\Delta c_A \tag{22.2-10}$$

This result may be rewritten in terms of the Sherwood number as

$$\begin{aligned}\text{Sh}_m &= \frac{k^0_{c,m}D}{\mathscr{D}_{AB}} = \frac{(3\pi)^{2/3}}{2^{7/3}\Gamma\left(\frac{4}{3}\right)}\sqrt[3]{\frac{Dv_\infty}{\mathscr{D}_{AB}}} = \frac{(3\pi)^{2/3}}{2^{7/3}\Gamma\left(\frac{4}{3}\right)}\sqrt[3]{\left(\frac{Dv_\infty\rho}{\mu}\right)\left(\frac{\mu}{\rho\mathscr{D}_{AB}}\right)}\\ &= 0.991(\text{ReSc})^{1/3}\end{aligned} \tag{22.2-11}$$

As in subsection (*a*) we have ReSc to the $\frac{1}{2}$-power for the gas-liquid system and ReSc to the $\frac{1}{3}$-power for the liquid-solid system.

Both Eq. 22.2-9 and Eq. 22.2-11 are valid only for creeping flow. However, they are not valid in the limit that Re goes to zero. As we know from Problem 10B.1 and Eq. 14.4-5, if there is no flow past the solid sphere or the spherical bubble, $\text{Sh}_m = 2$. It has been found that a satisfactory description of the mass transfer all the way down to $\text{Re} = 0$ can be obtained by using the simple superpositions $\text{Sh}_m = 2 + 0.6515(\text{ReSc})^{1/2}$ for flow around gas bubbles and $\text{Sh}_m = 2 + 0.991(\text{ReSc})^{1/3}$ for flow around solid spheres, in lieu of Eqs. 22.2-9 and 22.2-11, respectively.

---

[1] R. B. Bird, W. E. Stewart, and E. N. Lightfoot, *Transport Phenomena*, Revised 2nd Edition, Wiley, New York (2007), Chapter 12, Eq. 12.4-34.

c. Mass transfer in the neighborhood of a rotating disk

For a disk of diameter D coated with a slightly soluble material A rotating with angular velocity Ω in a large region of liquid B, the mass flux at the surface of the disk is independent of position. According to Eq. 19C.2-17 we have

$$N_{A0} = 0.620 \left(\frac{\mathscr{D}_{AB}^{2/3}\Omega^{1/2}}{\nu^{1/6}} \right)(c_{A0} - 0) \equiv k_{c,m}^0 \Delta c_A \tag{22.2-12}$$

This may be expressed in terms of the Sherwood number as

$$\begin{aligned} \text{Sh}_m = \frac{k_{c,m}^0 D}{\mathscr{D}_{AB}} &= 0.620 \left(\frac{D\Omega^{1/2}\rho^{1/6}}{\mathscr{D}_{AB}^{1/3}\mu^{1/6}} \right) = 0.620\sqrt{\frac{D(D\Omega)\rho}{\mu}}\sqrt[3]{\frac{\mu}{\rho\mathscr{D}_{AB}}} \\ &= 0.620\text{Re}^{1/2}\text{Sc}^{1/3} \end{aligned} \tag{22.2-13}$$

Here the characteristic velocity in the Reynolds number is chosen to be $D\Omega$.

§22.3 EMPIRICAL CORRELATIONS FOR BINARY MASS- AND HEAT-TRANSFER COEFFICIENTS IN ONE PHASE

In this section we show that correlations for binary mass-transfer coefficients at low mass-transfer rates can be obtained directly from their heat-transfer analogs simply by a change of notation. These correspondences are quite useful, and many heat-transfer correlations have, in fact, been obtained from their mass-transfer analogs.

To illustrate the background of these useful analogies and the conditions under which they apply, we begin by presenting the diffusional analog of the dimensional analysis given in §14.3. Consider the steady, isothermal flow of a liquid solution of A in B, in the tube shown in Fig. 22.3-1. The fluid enters the tube at $z = 0$ with velocity uniform out to very near the wall and with a uniform inlet composition x_{A1}. From $z = 0$ to $z = L$, the tube wall is coated with a solid solution of A and B, which dissolves slowly and maintains the interfacial liquid composition constant at x_{A0}. For the moment we assume that the physical properties ρ, μ, c, and $\mathscr{D}_{AB}$ are constant.

The mass-transfer situation just described is mathematically analogous to the heat-transfer situation described at the beginning of §14.3. To emphasize the analogy, the equations for the two systems are presented together. Thus, the rate of heat addition by conduction between 1 and 2 in Fig. 14.3-1 and the molar rate of addition of species A by

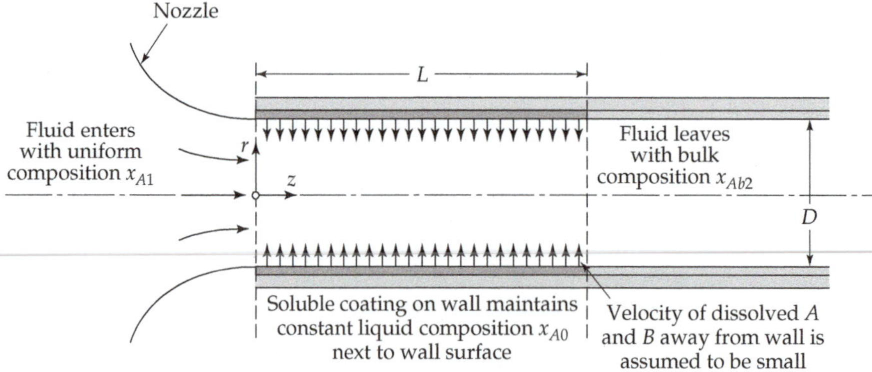

Fig. 22.3-1. Mass transfer in a pipe with a soluble wall.

diffusion between 1 and 2 in Fig. 22.3-1 are given by the following expressions, valid for either laminar or turbulent flow:

Heat transfer:
$$Q(t) = \int_0^L \int_0^{2\pi} \left(+k \left. \frac{\partial T}{\partial r} \right|_{r=R} \right) R \, d\theta \, dz \tag{22.3-1}$$

Mass transfer:
$$W_{A0}(t) - x_{A0}(W_{A0}(t) + W_{B0}(t)) = \int_0^L \int_0^{2\pi} \left(+c\mathscr{D}_{AB} \left. \frac{\partial x_A}{\partial r} \right|_{r=R} \right) R \, d\theta \, dz \tag{22.3-2}$$

Equating the left sides of these equations to $h_1(\pi DL)(T_0 - T_1)$ and $k_{x1}(\pi DL)(x_{A0} - x_{A1})$, respectively, we get for the transfer coefficients

Heat transfer:
$$h_1(t) = \frac{1}{\pi DL(T_0 - T_1)} \int_0^L \int_0^{2\pi} \left(+k \left. \frac{\partial T}{\partial r} \right|_{r=R} \right) R \, d\theta \, dz \tag{22.3-3}$$

Mass transfer:
$$k_{x1}(t) = \frac{1}{\pi DL(x_{A0} - x_{A1})} \int_0^L \int_0^{2\pi} \left(+c\mathscr{D}_{AB} \left. \frac{\partial x_A}{\partial r} \right|_{r=R} \right) R \, d\theta \, dz \tag{22.3-4}$$

We now introduce the dimensionless variables $\breve{r} = r/D$, $\breve{z} = z/D$, $\breve{T} = (T - T_0)/(T_1 - T_0)$, and $\breve{x}_A = (x_A - x_{A0})/(x_{A1} - x_{A0})$ and rearrange to obtain

Heat transfer:
$$\text{Nu}_1(t) = \frac{h_1 D}{k} = \frac{1}{2\pi L/D} \int_0^{L/D} \int_0^{2\pi} \left(- \left. \frac{\partial \breve{T}}{\partial \breve{r}} \right|_{\breve{r}=\frac{1}{2}} \right) d\theta \, d\breve{z} \tag{22.3-5}$$

Mass transfer:
$$\text{Sh}_1(t) = \frac{k_{x1} D}{c\mathscr{D}_{AB}} = \frac{1}{2\pi L/D} \int_0^{L/D} \int_0^{2\pi} \left(- \left. \frac{\partial \breve{x}_A}{\partial \breve{r}} \right|_{\breve{r}=\frac{1}{2}} \right) d\theta \, d\breve{z} \tag{22.3-6}$$

Here Nu is the Nusselt number for heat transfer without mass transfer, and Sh is the Sherwood number for isothermal mass transfer at small mass-transfer rates. The Nusselt number is a dimensionless temperature gradient integrated over the surface, and the Sherwood number is a dimensionless concentration gradient integrated over the surface.

These gradients can, in principle, be evaluated from Eqs. 13.1-7, 13.1-8, and 13.1-9 (for heat transfer) and Eqs. 21.1-8, 21.1-9, and 21.1-11 written in terms of mole fraction of species A (for mass transfer), under the following boundary conditions (with $\breve{\mathbf{v}}$ and $\breve{\mathscr{P}}$ defined as in §14.3 and with time averaging of the solutions if the flow field is turbulent):

Velocity and pressure:

$$\text{at } \breve{z} = 0, \qquad \breve{\mathbf{v}} = \boldsymbol{\delta}_z \qquad \text{for } 0 \leq \breve{r} < \frac{1}{2} \tag{22.3-7}$$
$$\text{at } \breve{r} = \frac{1}{2}, \qquad \breve{\mathbf{v}} = \mathbf{0} \qquad \text{for } \breve{z} > 0 \tag{22.3-8}$$
$$\text{at } \breve{r} = 0 \text{ and } \breve{z} = 0, \qquad \breve{\mathscr{P}} = 0 \tag{22.3-9}$$

Temperature:

$$\text{at } \breve{z} = 0, \qquad \breve{T} = 1 \qquad \text{for } 0 \leq \breve{r} < \frac{1}{2} \tag{22.3-10}$$
$$\text{at } \breve{r} = \frac{1}{2}, \qquad \breve{T} = 0 \qquad \text{for } 0 \leq \breve{z} \leq L/D \tag{22.3-11}$$

Concentration:

$$\text{at } \breve{z} = 0, \qquad \breve{x}_A = 1 \qquad \text{for } 0 \leq \breve{r} < \frac{1}{2} \tag{22.3-12}$$
$$\text{at } \breve{r} = \frac{1}{2}, \qquad \breve{x}_A = 0 \qquad \text{for } 0 \leq \breve{z} \leq L/D \tag{22.3-13}$$

The boundary condition in Eq. 22.3-8, on the velocity at the wall, is accurate for the heat-transfer system and also for the mass-transfer system, provided that $x_{A0}[W_{A0} + W_{B0}]$ is small; the latter criterion is discussed in §22.1. No boundary conditions are needed at the outlet plane, $\check{z} = L/D$, when we neglect the $\partial^2/\partial z^2$ terms of the conservation equations as discussed in §6.2 and §14.3.

If we can neglect the heat production by viscous dissipation in Eq. 13.1-9, and if there is no production of A by chemical reaction as in Eq. 21.1-11, then the differential equations for heat and mass transport are analogous, along with the boundary conditions. It follows then that the dimensionless profiles of temperature and concentration (time-smoothed, when necessary) are similar,

$$\check{T} = F(\check{r},\theta,\check{z};\text{Re},\text{Pr}); \qquad \check{x}_A = F(\check{r},\theta,\check{z};\text{Re},\text{Sc}) \qquad (22.3\text{-}14,15)$$

with the same form of F in both systems. Thus, to get the concentration profiles from the temperature profiles, one replaces $\check{T}$ by $\check{x}_A$ and Pr by Sc.

Finally, inserting the profiles into Eqs. 22.3-5 and 22.3-6 and performing the integrations and then time-averaging gives for *forced convection*

$$\text{Nu}_1 = G(\text{Re},\text{Pr},L/D); \qquad \text{Sh}_1 = G(\text{Re},\text{Sc},L/D) \qquad (22.3\text{-}16,17)$$

Here G is the same function in both equations. The same formal expression is obtained for Nu_a, Nu_{ln}, Nu_{loc} as well as for the corresponding Sherwood numbers. This important analogy permits one to write down a mass-transfer correlation from the corresponding heat-transfer correlation merely by replacing Nu by Sh, and Pr by Sc. The same can be done for any geometry, and for both laminar and turbulent flow. Note, however, that to get this analogy one has to assume (i) constant physical properties, (ii) small net mass-transfer rates, (iii) no chemical reactions, (iv) no viscous dissipation heating, (v) no absorption or emission of radiant energy, and (vi) no pressure diffusion, thermal diffusion, or forced diffusion. Some of these effects will be discussed in Chapter 24.

For *free convection* around objects of any given shape, a similar analysis shows that

$$\text{Nu}_m = H(\text{Gr},\text{Pr}); \qquad \text{Sh}_m = H(\text{Gr}_x,\text{Sc}) \qquad (22.3\text{-}18,19)$$

Here H is the same function in both cases, and the Grashof numbers are defined similarly for both processes (see Table 22.2-1 for a summary of the analogous quantities for heat and mass transfer).

To allow for the variation of physical properties in mass-transfer systems, we extend the procedures introduced in Chapter 14 for heat-transfer systems. That is, we generally evaluate the physical properties at some kind of mean film composition and temperature, except for the viscosity ratio μ_b/μ_0.

We now give three illustrations of how to "translate" from heat-transfer to mass-transfer correlations:

a. Forced convection around spheres

For forced convection around a solid sphere, Eq. 14.4-5 and its mass-transfer analog are:

$$\text{Nu}_m = 2 + 0.60\text{Re}^{1/2}\text{Pr}^{1/3}; \qquad \text{Sh}_m = 2 + 0.60\text{Re}^{1/2}\text{Sc}^{1/3} \qquad (22.3\text{-}20,21)$$

Equations 22.3-20 and 21 are valid for constant surface temperature and composition, respectively, and for small mass-transfer rates. They may be applied to simultaneous heat and mass transfer under restrictions (i)–(vi) given after Eq. 22.3-17.

b. Forced convection along a flat plate

As another illustration of the use of analogies, the Colburn analogy for momentum and energy transfer in the laminar boundary layer along a flat plate, can be extended to include mass transfer:

$$j_{H,\text{loc}} = j_{D,\text{loc}} = \frac{1}{2}f_{\text{loc}} = 0.332\text{Re}_x^{-1/2} \qquad (22.3\text{-}22)$$

The Chilton-Colburn "j-factors," one for heat transfer and one for diffusive mass transfer, are defined as[1]

$$j_{H,\text{loc}} = \frac{\text{Nu}_{\text{loc}}}{\text{RePr}^{1/3}} = \frac{h_{\text{loc}}}{\rho \hat{C}_p v_\infty}\left(\frac{\hat{C}_p \mu}{k}\right)^{2/3} \tag{22.3-23}$$

$$j_{D,\text{loc}} = \frac{\text{Sh}_{\text{loc}}}{\text{ReSc}^{1/3}} = \frac{k_{x,\text{loc}}}{c v_\infty}\left(\frac{\mu}{\rho \mathscr{D}_{AB}}\right)^{2/3} \tag{22.3-24}$$

This three-way analogy in Eq. 22.3-22 is accurate for Pr and Sc near unity within the limitations mentioned after Eq. 22.3-17. For flow around other objects, the friction factor part of the analogy is not valid because of the form drag, and even for flow in circular tubes the analogy with $\frac{1}{2}f_{\text{loc}}$ is only approximate (see §14.4).

c. *The Chilton-Colburn analogy*

The more widely applicable empirical analogy

$$j_H = j_D = \text{a function of Re, geometry, and boundary conditions} \tag{22.3-25}$$

has proven to be useful for transverse flow over around cylinders, flow through packed beds, and flow in tubes at high Reynolds numbers. For flow in ducts and packed beds, the "approach velocity" v_∞ has to be replaced by the average velocity or the superficial velocity. Equation 22.3-25 is the usual form of the *Chilton-Colburn analogy*. It is evident from Eqs. 22.3-20 and 22.3-21, however, that the analogy is valid for flow around spheres only when Nu and Sh are replaced by (Nu – 2) and (Sh – 2).

It would be very misleading to leave the impression that all mass-transfer coefficients can be obtained from the analogous heat-transfer coefficient correlations. For mass transfer we encounter a much wider variety of boundary conditions and other ranges of the relevant variables. The non-analogous behavior is addressed elsewhere.[2]

EXAMPLE 22.3-1

Evaporation from a Freely Falling Drop

A spherical drop of water, 0.05 cm in diameter, is falling at a velocity of 215 cm/s through dry, still air at 1 atm pressure with no internal circulation. Estimate the instantaneous rate of evaporation from the drop, when the drop surface is at $T_0 = 70°\text{F}$ and the air (far from the drop) is at $T_\infty = 140°\text{F}$. The vapor pressure of water at 70°F is 0.0247 atm. Assume quasi-steady-state conditions.

SOLUTION

Designate water as species A and air as species B. The solubility of air in water may be neglected, so that $W_{B0} = 0$. Then assuming that the evaporation rate is small, we may write Eq. 22.1-3 for the entire spherical surface as

$$W_{A0} = k_{xm}(\pi D^2)\frac{x_{A0} - x_{A\infty}}{1 - x_{A0}} \tag{22.3-26}$$

The mean mass-transfer coefficient, k_{xm}, may be predicted from Eq. 22.3-21.

The film conditions needed for estimating the physical properties are obtained as follows:

$$T_f = \frac{1}{2}(T_0 + T_\infty) = \frac{1}{2}(70°\text{F} + 140°\text{F}) = 105°\text{F} \tag{22.3-27}$$

$$x_{Af} = \frac{1}{2}(x_{A0} + x_{A\infty}) = \frac{1}{2}(0.0247 + 0) = 0.0124 \tag{22.3-28}$$

---

[1]T. H. Chilton and A. P. Colburn, *Ind. Eng. Chem.*, **26**, 1183–1187 (1934).

[2]R. B. Bird, W. E. Stewart, and E. N. Lightfoot, *Transport Phenomena*, Revised Second Edition, Wiley, New York (2007), §22.5-8.

In computing x_{Af}, we have assumed ideal-gas behavior, equilibrium at the interface, and complete insolubility of air in water. The mean mole fraction, x_{Af}, of the water vapor is sufficiently small that it can be safely neglected in evaluating the physical properties at the film conditions:

$$c = 3.88 \times 10^{-5} \text{ g-mol/cm}^3$$
$$\rho = 1.12 \times 10^{-3} \text{ g/cm}^3$$
$$\mu = 1.91 \times 10^{-4} \text{ g/cm} \cdot \text{s (from Table 1.4-1)}$$
$$\mathcal{D}_{AB} = 0.292 \text{ cm}^2/\text{s (from Eq. 17.6-1)}$$
$$\text{Sc} = \left(\frac{\mu}{\rho\mathcal{D}_{AB}}\right) = \frac{(1.91 \times 10^{-4} \text{ g/cm} \cdot \text{s})}{(1.12 \times 10^{-3} \text{ g/cm}^3)(0.292 \text{ cm}^2/\text{s})} = 0.58$$
$$\text{Re} = \left(\frac{Dv_\infty\rho}{\mu}\right) = \frac{(0.05 \text{ cm})(215 \text{ cm/s})(1.12 \times 10^{-3} \text{ g/cm}^3)}{(1.91 \times 10^{-4} \text{ g/cm} \cdot \text{s})} = 63$$

When these values are used in Eq. 22.3-21, we get

$$\text{Sh}_m = 2 + 0.60(63)^{1/2}(0.58)^{1/3} = 5.96 \tag{22.3-29}$$

and the mean mass-transfer coefficient is then

$$k_{xm} = \frac{c\mathcal{D}_{AB}}{D}\text{Sh}_m = \frac{(3.88 \times 10^{-5} \text{ g-mol/cm}^3)(0.292 \text{ cm}^2/\text{s})}{(0.05 \text{ cm})}(5.96)$$
$$= 1.35 \times 10^{-3} \text{ g-mol/s} \cdot \text{cm}^2 \tag{22.3-30}$$

Then from Eq. 22.3-26 the evaporation rate is found to be

$$W_{A0} = (1.35 \times 10^{-3} \text{ g-mol/s} \cdot \text{cm}^2)(\pi)(0.05 \text{ cm})^2 \frac{(0.0247 \text{ atm}/1 \text{ atm}) - 0}{1 - (0.0247 \text{ atm}/1 \text{ atm})}$$
$$= 2.70 \times 10^{-7} \text{ g-mol/s} \tag{22.3-31}$$

This result corresponds to a decrease of 1.23×10^{-3} cm/s in the drop diameter and indicates that a drop of this size will fall a considerable distance before it evaporates completely.

In this example, for simplicity, the velocity and surface temperature of the drop were given. In general, these conditions must be calculated from momentum and energy balances, as discussed in Problem 22B.1.

EXAMPLE 22.3-2

The Wet- and Dry-Bulb Psychrometer

We next turn to a problem where the analogy between heat and mass transfer leads to a surprisingly simple and useful, if approximate, result. The system, shown in Fig. 22.3-2, is a pair of thermometers, one of which is covered with a cylindrical wick kept saturated with water. The wick will cool by evaporation into the moving air stream, and for steady operation its temperature will approach an asymptotic value known as the *wet-bulb temperature*. The bare thermometer, on the other hand, will tend to approach the actual temperature of the approaching air, and this value is called the *dry-bulb temperature*. Develop an expression for determining the humidity of the air from the wet- and dry-bulb temperature readings neglecting radiation and assuming that the replacement of the evaporating water has no significant effect on the wet-bulb temperature measurement. In Problem 22B.2 it is shown how radiation can be taken into account.

SOLUTION

For simplicity, we assume that the fluid velocity is high enough that the thermometer readings are unaffected by radiation and by heat conduction along the thermometer stems, but not so high that viscous dissipation heating effects become significant. These assumptions are usually satisfactory for glass thermometers and for gas velocities of 30 to 100 ft/s. The dry-bulb temperature is then the same as the temperature T_∞ of the approaching gas, and the wet-bulb temperature is the same as the temperature T_0 of the outside of the wick.

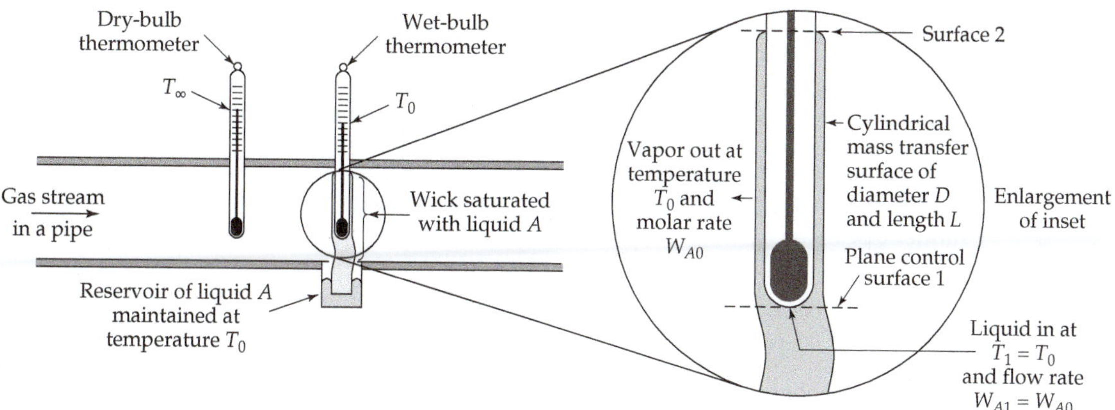

Fig. 22.3-2. Sketch of a wet-bulb and dry-bulb psychrometer installation. It is assumed that no heat or mass moves across plane 2.

Let species A be water and species B be air. An energy balance is made on a system that contains a length L of the wick (the distance between planes 1 and 2 in the figure). The rate of heat addition to the system by the gas stream is $h_m(\pi DL)(T_\infty - T_0)$. Enthalpy also enters via plane 1 at a rate $W_{A1}\overline{H}_{A1}$ in the liquid phase, and leaves at the mass-transfer surface at a rate $W_{A0}\overline{H}_{A0}$, both of these occurring at a temperature T_0. Hence, the energy balance gives

$$h_m(\pi DL)(T_\infty - T_0) = W_{A0}(\overline{H}_{A1} - \overline{H}_{A0}) \tag{22.3-32}$$

since the water enters the system at plane 1 at the same rate that it leaves as water vapor at the mass-transfer interface 0. To a very good approximation, $\overline{H}_{A1} - \overline{H}_{A0}$ may be replaced by $\Delta\tilde{H}_{\text{vap}}$, the molar heat of vaporization of water.

From the definition of the mass-transfer coefficient

$$W_{A0} - x_{A0}(W_{A0} + W_{B0}) = k_{xm}(\pi DL)(x_{A0} - x_{A\infty}) \tag{22.3-33}$$

in which $W_{B0} = 0$. Combination of Eqs. 22.2-32 and 22.2-33 gives then

$$\frac{(x_{A0} - x_{A\infty})}{(T_\infty - T_0)(1 - x_{A0})} = \frac{h_m}{k_{xm}\Delta\tilde{H}_{\text{vap}}} \tag{22.3-34}$$

Then using the definitions of Nu_m and Sh_m, and noting that $\rho\hat{C}_p = c\tilde{C}_p$, we may rewrite Eq. 22.3-34 as

$$\frac{(x_{A0} - x_{A\infty})}{(T_\infty - T_0)(1 - x_{A0})} = \frac{\text{Nu}_m}{\text{Sh}_m}\left(\frac{\text{Sc}}{\text{Pr}}\right)\frac{\tilde{C}_p}{\Delta\tilde{H}_{\text{vap}}} \tag{22.3-35}$$

Because of the analogy between heat and mass transfer, we can expect that the mean Nusselt and Sherwood numbers will be of the same form:

$$\text{Nu}_m = F(\text{Re})\text{Pr}^n; \qquad \text{Sh}_m = F(\text{Re})\text{Sc}^n \tag{22.3-36,37}$$

where F is the same function of Re in both expressions. Therefore, knowing the dry- and wet-bulb temperatures and the mole fraction of the water vapor adjacent to the wick (x_{A0}), we can calculate the concentration of the water vapor in the air stream from

$$\frac{(x_{A0} - x_{A\infty})}{(T_\infty - T_0)(1 - x_{A0})} = \left(\frac{\text{Sc}}{\text{Pr}}\right)^{1-n}\frac{\tilde{C}_p}{\Delta\tilde{H}_{\text{vap}}} \tag{22.3-38}$$

The exponent n depends to a slight extent on the geometry, but is not far from $\frac{1}{3}$, and the quantity $(\text{Sc}/\text{Pr})^{1-n}$ is not far from unity.[3] Furthermore, the wet-bulb temperature is seen to be independent of the Reynolds number as a result of the assumption introduced in Eqs. 22.3-36 and 22.3-37. This result would also have been obtained by using the Chilton-Colburn relations, which would give $n = \frac{1}{3}$ directly.

The interfacial gas composition x_{A0} can be accurately predicted, at low mass-transfer rates, by neglecting the heat- and mass-transfer resistance of the interface itself (see §22.4 for further discussion of this point). One can then represent x_{A0} by the vapor-liquid equilibrium relationship:

$$x_{A0} = x_{A0}(T_0, p) \tag{22.3-39}$$

A relation of this kind will hold for given species A and B if the liquid is pure A as assumed above. A commonly used approximation of this relationship is

$$x_{A0} = \frac{p_{A,\text{vap}}}{p} \tag{22.3-40}$$

in which $p_{A,\text{vap}}$ is the vapor pressure of pure A at temperature T_0. This relation assumes tacitly that the presence of B does not alter the partial pressure of A at the interface, and that A and B form an ideal-gas mixture.

If an air-water mixture at 1 atm pressure gives a wet-bulb temperature of 70°F and a dry-bulb temperature of 140°F, then

$$\begin{aligned}
p_{A,\text{vap}} &= 0.0247 \text{ atm (see Example 22.3-1)} \\
x_{A0} &= 0.0247, \text{ from Eq. 22.3-40} \\
\tilde{C}_p &= 6.98 \text{ Btu/lb-mol} \cdot {}^\circ\text{F at } 105^\circ\text{F, the film temperature}^4 \\
\Delta\tilde{H}_{\text{vap}} &= 18{,}900 \text{ Btu/lb-mol at } 70^\circ\text{F}^4 \\
\text{Sc} &= 0.58 \text{ (see Example 22.3-1)} \\
\text{Pr} &= 0.74, \text{ from Eq. 9.7-16}
\end{aligned}$$

Substitution into Eq. 22.3-38, with $n = \frac{1}{3}$, then gives

$$\frac{(0.0247 - x_{A\infty})}{(140^\circ\text{F} - 70^\circ\text{F})(1 - 0.0247)} = \left(\frac{0.58}{0.74}\right)^{2/3} \frac{(6.98 \text{ Btu/lb-mol} \cdot {}^\circ\text{F})}{(18{,}900 \text{ Btu/lb-mol})} \tag{22.3-41}$$

From this the mole fraction of water in the approaching air is

$$x_{A\infty} = 0.0033 \tag{22.3-42}$$

Since we assumed that the film concentration was $x_A = 0$ as a first approximation, we could go back and make a second approximation by using an average film concentration of $\frac{1}{2}(0.0247 + 0.0033) = 0.0140$ in the physical property calculations. The physical properties are not known accurately enough here to justify recalculation.

The calculated result in Eq. 22.3-42 is in only fair agreement with published humidity charts, because these are typically based on the adiabatic saturation temperature rather than the wet-bulb temperature.[4]

---

[3] A somewhat different equation, with $1 - n = 0.56$, was recommended for measurements in air by C. H. Bedingfield and T. B. Drew, *Ind. Eng. Chem.*, **42**, 1164–1173 (1950).

[4] O. A. Hougen, K. M. Watson, and R. A. Ragatz, *Chemical Process Principles*, 2nd edition, Wiley, New York, Part I (1954), p. 120.

EXAMPLE 22.3-3

Creeping-Flow Mass Transfer in Packed Beds

Many important adsorptive operations, from purification of proteins in modern biotechnology to the recovery of solvent vapor by dry-cleaning establishments, occur in dense particulate beds and are typically carried out in steady creeping flow, i.e., at $\text{Re} = D_p v_0 \rho/\mu < 20$. Here D_p is the effective particle diameter, and v_0 is the superficial velocity, defined as volumetric flow rate divided by the total cross section of the bed (see §6.4). It follows that the dimensionless velocity $\mathbf{v}/v_0$ will have a spatial distribution independent of the Reynolds number. Detailed information is available only for spherical packing particles.

Using the dimensional analysis discussion at the beginning of this section, predict the form of the steady-state mass-transfer coefficient correlation for creeping flow.

SOLUTION

The dimensional analysis procedure in §21.1 may be used, with D_p as the characteristic length and v_0 the characteristic velocity. Then, from Eq. 21.1-11, we see that the dimensionless concentration depends only on the product ReSc, in addition to the dimensionless position coordinates and the geometry of the bed.

The most extensive data are for creeping flow at large Péclet numbers. Experimental data on the dissolution of benzoic acid spheres in water[5] have yielded the result

$$\text{Sh}_m = \frac{1.09}{\varepsilon}(\text{ReSc})^{1/3} \qquad \text{ReSc} \gg 1 \tag{22.3-43}$$

where ε is the volume fraction of the bed occupied by the flowing fluid. Equation 22.3-43 is reasonably consistent with the relation

$$\text{Sh}_m = 2 + 0.991(\text{ReSc})^{1/3} \tag{22.3-44}$$

which incorporates the creeping-flow solution for flow around an isolated sphere[6] ($\varepsilon = 1$) (see §22.2(*b*)). This suggests that the flow pattern around an isolated sphere is not much different from that around a sphere surrounded by other spheres, particularly near the sphere surface where most of the mass transport takes place.

No reliable data are available for the limiting behavior at very low values of ReSc, but numerical calculations for a regular packing[7] predict that the Sherwood number asymptotically approaches a constant near 4.0 if based on a local difference between interfacial and bulk compositions.

Behavior within the solid phase is far more complex, and no simple approximation is wholly trustworthy. However, experiments[8] show that where intraparticle mass transport is described by Fick's second law, one can use the simpler approximation

$$\text{Sh}_m = \frac{k_{c,s} D_p}{\mathscr{D}_{AB}} \approx 10 \tag{22.3-45}$$

where $k_{c,s}$ is the effective mass-transfer coefficient within the solid phase, and $\mathscr{D}_{As}$ is the diffusivity of A in the solid phase. The equation is for "slow" changes in the solute concentration bathing the particle. This is an asymptotic solution for a linear change of surface concentration with time,[9] and has been justified[10] by calculations. For passage of a Gaussian (bell-shaped) concentration wave moving through the packed column, "slow" means that the passage time (temporal standard deviation) of the wave is long relative to the particle diffusional response

---

[5]E. J. Wilson and C. J. Geankopolis, *Ind. Eng. Chem. Fundamentals*, **5**, 9–14 (1966). See also J. R. Selman and C. W. Tobias, *Advances in Chemical Engineering*, **10**, 212–318 (1978), for an extensive summary of mass-transfer coefficient correlations obtained by electrochemical measurements.

[6]V. G. Levich, *Physicochemical Hydrodynamics*, Prentice-Hall, Englewood Cliffs, NJ (1962), §14.

[7]J. P. Sørensen and W. E. Stewart, *Chem. Eng. Sci.*, **29**, 811–837 (1974).

[8]A. M. Athalye, J. Gibbs, and E. N. Lightfoot, *J. Chromatography*, **589**, 71–85 (1992).

[9]H. S. Carslaw and J. C. Jaeger, *Conduction of Heat in Solids*, 2nd edition, Oxford University Press (1959), §9.3, Eqs. 10 and 11.

[10]J. F. Reis, E. N. Lightfoot, P. T. Noble, and A. S. Chiang, *Sep. Sci. Tech.*, **14**, 367–394 (1979).

time, which is of the order of $D_p^2/6\mathscr{D}_{As}$. Fick's second law must be solved with the detailed history of surface concentration when this inequality is not satisfied.

In packed beds, just as with tube flow, one must keep in mind the fact that there will be nonuniformities in the concentration as a function of the radial coordinate. This was discussed in §14.5.

EXAMPLE 22.3-4

Mass Transfer to Drops and Bubbles

In both gas-liquid[11] and liquid-liquid[12] contactors, sprays of liquid drops or clouds of bubbles are frequently encountered. Contrast their mass-transfer behavior with that of solid spheres.

SOLUTION

Many different types of behavior are encountered, and it is found that surface forces can play a very important role. These latter are discussed in some detail elswhere.[2] We consider here only some limiting cases and refer readers to the above-cited references.

Very small drops and bubbles behave like solid spheres and can be treated by the correlations in Example 22.3-3 and in Chapter 14. However, if both adjacent phases are free of surfactants and small particulate contaminants, the interior phase circulates and carries the adjacent regions of the exterior phase along. This stress-driven "Hadamard-Rybczinski circulation"[13] increases the mass-transfer rates markedly, often by an order of magnitude, and the rates can be estimated from the extensions[14–17] of the "penetration model" discussed in §18.8. Thus, for a spherical bubble of gas A with diameter D rising through a clean liquid B, the external Sherwood number (i.e., on the liquid side) lies in the range[17]

$$\sqrt{\frac{4}{3\pi}\frac{Dv_t}{\mathscr{D}_{AB}}} < \text{Sh}_m < \sqrt{\frac{4}{\pi}\frac{Dv_t}{\mathscr{D}_{AB}}} \tag{22.3-46}$$

where v_t is the terminal velocity (see Eqs. 18.8-19 and 18.8-20).

The size at which the transition from the solid-like behavior to circulation occurs depends on degree of surface contamination and is not easily predicted.

Very large drops or bubbles oscillate,[14] and both phases follow a modified penetration theory,

$$\text{Sh}_m \approx \sqrt{\frac{4.8D^2\omega}{\pi\mathscr{D}_{AB}}} \tag{22.3-47}$$

with angular frequency of oscillation[18]

$$\omega = \sqrt{\frac{192\sigma}{D^3(3\rho_D + 2\rho_C)}} \tag{22.3-48}$$

where σ is the interfacial tension, and ρ_D and ρ_C are the densities of the drops and the continuous medium.

The success of this model implies that the boundary layer is refreshed once every oscillation, but there is also a small effect of periodic stretching of the surface.

---

[11] J. Stichlmair and J. F. Fair, *Distillation Principles and Practice*, Wiley, New York (1998).

[12] J. C. Godfrey and M. M. Slater, *Liquid-Liquid Extraction Equipment*, Wiley, New York (1994).

[13] J. Happel and H. Brenner, *Low Reynolds Number Hydrodynamics*, Martinus Nijhoff, The Hague (1983).

[14] J. B. Angelo, E. N. Lightfoot, and D. W. Howard, *AIChE Journal*, **12**, 751–760 (1966).

[15] J. B. Angelo and E. N. Lightfoot, *AIChE Journal*, **14**, 531–540 (1968).

[16] W. E. Stewart, J. B. Angelo, and E. N. Lightfoot, *AIChE Journal*, **16**, 771–786 (1970).

[17] R. Higbie, *Trans. AIChE*, **31**, 365–389 (1935).

[18] R. R. Schroeder and R. C. Kintner, *AIChE Journal*, **11**, 5–8 (1965).

§22.4 DEFINITION OF MASS-TRANSFER COEFFICIENTS IN TWO PHASES

Recall that in §10.3 we introduced the concept of an overall heat-transfer coefficient, U, to describe the heat transfer between two streams separated from each other by a wall. This overall coefficient accounted for the thermal resistance of the wall itself, as well as the thermal resistance in the fluids on either side of the wall.

We now treat the analogous situation for mass transfer, except that here we are concerned with two fluids in intimate contact with one another, so that there is no wall resistance or interfacial resistance. This is the situation most commonly met in practice. Since the interface itself contains no significant mass, we may begin by assuming the continuity of the total mass flux at the interface for any species being transferred. Then for the system shown in Fig. 22.4-1 we write

$$N_{A0}|_{\text{gas}} = N_{A0}|_{\text{liquid}} = N_{A0} \tag{22.4-1}$$

for the interfacial fluxes toward the liquid phase. Then using the definition given in Eq. 22.1-9, we get

$$N_{A0} = k^0_{y,\text{loc}}(y_{Ab} - y_{A0}) = k^0_{x,\text{loc}}(x_{A0} - x_{Ab}) \tag{22.4-2}$$

in which we are now following the tradition of using x for mole fractions in the liquid phase and y for mole fractions in the gas phase. We now have to interrelate the interfacial compositions in the two phases.

In nearly all situations this can be done by assuming equilibrium across the interface, so that adjacent gas and liquid compositions lie on the equilibrium curve (see Fig. 22.4-2), which is regarded as known from solubility data:

$$y_{A0} = f(x_{A0}) \tag{22.4-3}$$

Exceptions to this are: (i) extremely high mass-transfer rates, observed for gas phases at high vacuum, where N_{A0} approaches $p_{A0}/\sqrt{2\pi M_A RT}$, the equilibrium rate at which gas molecules impinge on the interface; and (ii) interfaces contaminated with high concentrations of adsorbed particles or surfactant molecules. Situation (i) is quite rare, and situation (ii) normally acts indirectly by changing the flow behavior, rather than causing deviations from equilibrium. In extreme cases surface contamination can provide additional transport resistances.

To describe rates of interphase transport, one can either use Eqs. 22.4-2 and 22.4-3 to calculate interface concentrations and then proceed to use the single-phase coefficients, or

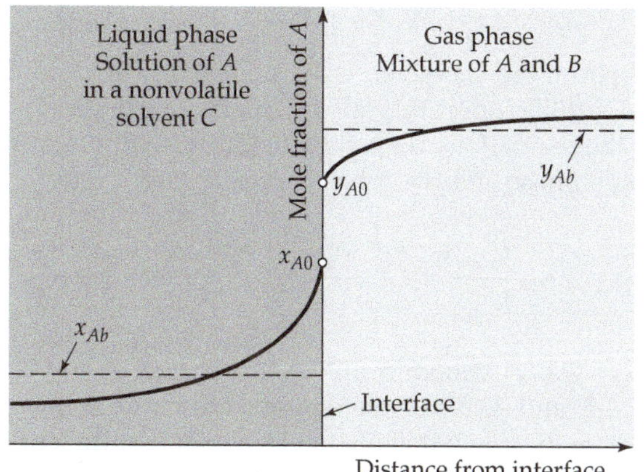

Fig. 22.4-1. Concentration profiles in the neighborhood of a gas-liquid interface.

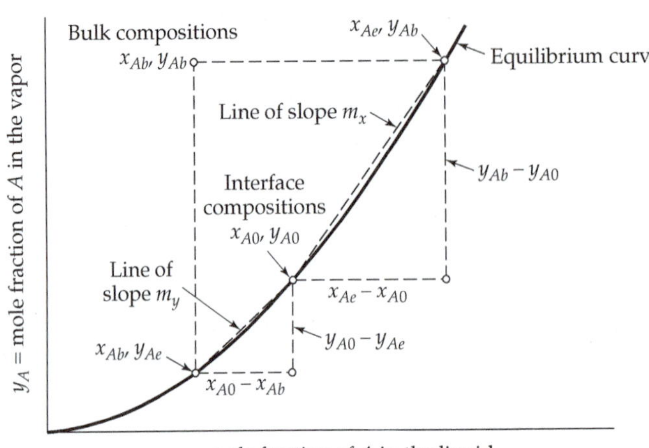

Fig. 22.4-2. Relations among gas- and liquid-phase compositions, and the graphical interpretations of m_x and m_y.

else work with overall mass-transfer coefficients

$$N_{A0} = K^0_{y,\text{loc}}(y_{Ab} - y_{Ae}) = K^0_{x,\text{loc}}(x_{Ae} - x_{Ab}) \tag{22.4-4}$$

Here y_{Ae} is the gas-phase composition in equilibrium with a liquid at composition x_{Ab}, and x_{Ae} is the liquid-phase composition in equilibrium with a gas at composition y_{Ab}. The quantity $K^0_{y,\text{loc}}$ is the overall mass-transfer coefficient "based on the gas phase," and $K^0_{x,\text{loc}}$ is the overall mass-transfer coefficient "based on the liquid phase." Here again molar flux N_{A0} is taken to be positive for transfer to the liquid phase.

Equating the quantities in Eqs. 22.4-2 and 22.4-4 gives two relations

$$K^0_{x,\text{loc}}(x_{Ae} - x_{Ab}) = k^0_{x,\text{loc}}(x_{A0} - x_{Ab}) \tag{22.4-5}$$

$$K^0_{y,\text{loc}}(y_{Ab} - y_{Ae}) = k^0_{y,\text{loc}}(y_{Ab} - y_{A0}) \tag{22.4-6}$$

connecting the two-phase coefficients with the single-phase coefficients.

The quantities x_{Ae} and y_{Ae} introduced in the above three relations may be used to define quantities m_x and m_y as follows:

$$m_x = \frac{y_{Ab} - y_{A0}}{x_{Ae} - x_{A0}}; \qquad m_y = \frac{y_{A0} - y_{Ae}}{x_{A0} - x_{Ab}} \tag{22.4-7,8}$$

As may be seen from Fig. 22.4-2, m_x is the slope of the line connecting points (x_{A0}, y_{A0}) and (x_{Ae}, y_{Ab}) on the equilibrium curve, and m_y is the slope of the line from (x_{Ab}, y_{Ae}) to (x_{A0}, y_{A0}).

From the above relations we can then eliminate the concentrations and get relations among the single-phase and two-phase mass-transfer coefficients:

$$\frac{k^0_{x,\text{loc}}}{K^0_{x,\text{loc}}} = 1 + \frac{k^0_{x,\text{loc}}}{m_x k^0_{y,\text{loc}}}; \qquad \frac{k^0_{y,\text{loc}}}{K^0_{y,\text{loc}}} = 1 + \frac{m_y k^0_{y,\text{loc}}}{k^0_{x,\text{loc}}} \tag{22.4-9,10}$$

The first of these was obtained from Eqs. 22.4-5, 22.4-2, and 22.4-7, and the second from Eqs. 22.4-6, 22.4-2, and 22.4-8. If the equilibrium curve is nearly linear over the range of interest, then $m_x = m_y = m$, which is the local slope of the curve at the interfacial conditions. We see, then, that the expressions in Eqs. 22.4-9 and 22.4-10 both contain a ratio of single-phase coefficients weighted with a quantity m. This quantity is of considerable importance:

(i) If $k^0_{x,\text{loc}}/mk^0_{y,\text{loc}} \ll 1$, the mass-transport resistance of the gas phase has little effect, and it is said that the mass transfer is *liquid-phase controlled*. In practice, this means that the system design should favor liquid-phase mass transfer.

(ii) If $k^0_{x,\text{loc}}/mk^0_{y,\text{loc}} \gg 1$, then the mass transfer is *gas-phase controlled*. In a practical situation, this means that the system design should favor gas-phase mass transfer.

(iii) If $0.1 < k^0_{x,\text{loc}}/mk^0_{y,\text{loc}} < 10$, roughly, one must be careful to consider the interactions of the two phases in calculating the two-phase transfer coefficients. Outside this range the interactions are usually unimportant. We return to this point in the example below.

The mean two-phase mass transfer coefficients must be defined carefully, and we consider here only the special case where bulk concentrations in the two adjacent phases do not change significantly over the total mass transfer surface S. We may then define K^0_{xm} by

$$(N_{A0})_m = \frac{1}{S}\int_S K^0_{x,\text{loc}}(x_{Ae} - x_{Ab})\, dS = K^0_{xm}(x_{Ae} - x_{Ab}) \tag{22.4-11}$$

so that, when Eq. 22.4-9 is used,

$$K^0_{xm} = \frac{1}{S}\int_S \frac{1}{(1/k^0_{x,\text{loc}}) + (1/m_x k^0_{y,\text{loc}})}\, dS \tag{22.4-12}$$

Frequently area mean overall mass-transfer coefficients are calculated from area mean coefficients for the two adjoining phases:

$$K^0_{x,\text{approx}} = \frac{1}{(1/k^0_{xm}) + (1/m_x k^0_{ym})} \tag{22.4-13}$$

The two mean values in Eqs. 22.4-12 and 22.4-13 can be significantly different.

EXAMPLE 22.4-1

Determination of the Controlling Resistance

Oxygen is to be removed from water using nitrogen gas at atmospheric pressure and 20°C in the form of bubbles exhibiting internal circulation, as shown in Fig. 22.4-3. Estimate the relative importance of the two mass-transfer coefficients $k^0_{x,\text{loc}}$ and $k^0_{y,\text{loc}}$. Let A stand for O_2, B for H_2O, and C for N_2.

SOLUTION

This can be done by assuming that the penetration model (see Eq. 18.8-19 or Eq. 22.2-5) holds in each phase, so that

$$k^0_{x,\text{loc}} \approx k_{x,\text{loc}} = c_l\sqrt{\frac{4\mathscr{D}_{AB}}{\pi t_{\text{exp}}}}; \qquad k^0_{y,\text{loc}} \approx k_{y,\text{loc}} = c_g\sqrt{\frac{4\mathscr{D}_{AC}}{\pi t_{\text{exp}}}} \tag{22.4-14}$$

where c_l and c_g are the total molar concentrations in the liquid and gas phases, respectively. The effective exposure time, t_{exp}, is the same for each of the phases.

The solubility of O_2 in water at 20°C is 1.38×10^{-3} g-mol/liter at an oxygen partial pressure of 760 mm Hg, the vapor pressure of water is 17.535 mm Hg, and the total pressure in the solubility measurements is 777.5 mm Hg. At 20°C, the diffusivity of O_2 in water is $\mathscr{D}_{AB} = 2.1 \times 10^{-5}\ \text{cm}^2/\text{s}$, and in the gas phase the diffusivity for $O_2 - N_2$ is $\mathscr{D}_{AC} = 0.2\ \text{cm}^2/\text{s}$. Assuming that $m_x = m_y = m$, we can then write (from Eq. 22.4-14)

$$\frac{k^0_{x,\text{loc}}}{mk^0_{y,\text{loc}}} = \frac{c_l}{c_g}\sqrt{\frac{\mathscr{D}_{AB}}{\mathscr{D}_{AC}}}\cdot\frac{1}{m} \tag{22.4-15}$$

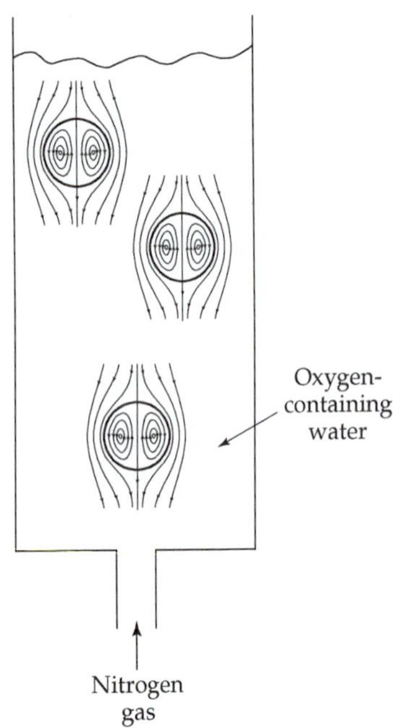

Fig. 22.4-3. Schematic diagram of an oxygen stripper, in which oxygen from the water diffuses into the nitrogen gas bubbles.

Into this we must substitute

$$\frac{c_l}{c_g} = \frac{c_l}{(p/RT)} = \frac{(55.56 \text{ g-mol/liter})}{(777.5 \text{ mm Hg}/760 \text{ mm Hg/atm})/(0.08206 \text{ liter} \cdot \text{atm/g-mol} \cdot \text{K})(293.15 \text{ K})} = 1306 \quad (22.4\text{-}16)$$

$$\sqrt{\frac{\mathscr{D}_{AB}}{\mathscr{D}_{AC}}} = \sqrt{\frac{2.1 \times 10^{-5} \text{ cm}^2/\text{s}}{0.2 \text{ cm}^2/\text{s}}} = 0.0102 \quad (22.4\text{-}17)$$

$$\frac{1}{m} = \frac{x_{Ae}}{y_{Ae}} = \frac{(1.38 \times 10^{-3} \text{ g-mol/liter})/(55.56 \text{ g-mol/liter})}{(760 \text{ mm Hg})/(777.5 \text{ mm Hg})} = 2.54 \times 10^{-5} \quad (22.4\text{-}18)$$

55.56 g-mol/liter being the molar concentration of pure water. Substitution of these numerical values into Eq. 22.4-15 then gives

$$\frac{k^0_{x,\text{loc}}}{mk^0_{y,\text{loc}}} = (1306)(0.0102)(2.54 \times 10^{-5}) = 3.38 \times 10^{-4} \quad (22.4\text{-}19)$$

This corresponds to case (i) below Eq. 22.4-10. Therefore, only the liquid-phase resistance is significant, and the assumption of penetration behavior in the gas phase is not critical to the determination of liquid-phase control. It may also be seen that the dominant factor is the low solubility of oxygen in water. One may generalize and state that absorption or desorption of sparingly soluble gases is almost always liquid-phase controlled. Correction of the gas-phase coefficient for net mass transfer is clearly not significant, and the correction for the liquid phase is negligible.

§22.5 CONCLUDING COMMENTS

We have seen in this chapter that mass-transfer coefficients can be calculated analytically in some special cases. However, values for a wider range of situations can be estimated using correlations. These correlations are identical to the analogous correlations for heat-transfer coefficients when written in dimensionless form. This very useful result arises because the dimensionless equations of change and the dimensionless interphase transfer coefficients are the same for energy and mass transport in a wide range of situations.

There are sometimes, of course, exceptions to the relatively simple results presented here. Interphase mass transfer can be complicated by a variety of phenomena, such as chemical reactions, combined heat and mass transfer by free convection, effects of interfacial forces, and high net mass-transfer rates. The interested reader is referred to references given in the introduction to learn more about these phenomena and their impact on mass–transfer coefficients.

QUESTIONS FOR DISCUSSION

1. Under what conditions can the analogies in §22.3 be applied? Why is the heat-transfer coefficient in Eq. 22.1-6 defined differently from that in Eq. 14.1-1—or is it?
2. Some of the mass-transfer coefficients in this chapter have a superscript "0." Explain carefully what this superscript denotes.
3. What conclusions can you draw from the analytical calculations of mass-transfer coefficients in §22.2?
4. What is the significance of the "2" in Eqs. 22.3-20 and 22.3-21?
5. What is the meaning of the subscripts "0," "*e*," and "*b*" in §22.4?

PROBLEMS

22A.1 Prediction of mass-transfer coefficients in closed channels. Estimate the gas-phase mass-transfer coefficients for water vapor evaporating into pure air at 2 atm and 25°C, and a mass flow rate of 1570 lb_m/hr, in the systems listed below. Take $\mathscr{D}_{AB} = 0.130\ cm^2/s$.

(a) A 6-in. i.d. vertical pipe with a falling film of water on the wall. Use the following correlation[1] for gases in a wetted-wall column:

$$\mathrm{Sh_{loc}} = 0.023\mathrm{Re}^{0.83}\mathrm{Sc}^{0.44} \quad (\mathrm{Re} > 2000) \tag{22A.1-1}$$

(b) A 6 in. diameter packed bed of water-saturated spheres, with $a = 100\ ft^{-1}$.

22A.2 Calculation of gas composition from psychrometric data. A stream of moist air has a wet-bulb temperature of 80°F and a dry-bulb temperature of 130°F, measured at 800 mm Hg total pressure and high air velocity. Compute the mole fraction of water vapor in the air stream. For simplicity, consider water as a trace component in estimating the film properties.

Answer: $x_{A\infty} = 0.0169$ (using $n = 0.44$ in Eq. 22.3-38)

22A.3 Calculating the inlet air temperature for drying in a fixed bed. A shallow bed of water-saturated granular solids is to be dried by blowing dry air through it at 1.1 atm pressure and a superficial velocity of 15 ft/s. What air temperature is required to keep the solids at a surface temperature of 60°F? Neglect radiation. See §14.5 for forced-convection heat-transfer coefficients in fixed beds.

22A.4 Rate of drying of granular solids in a fixed bed. Calculate the initial rate of water removal in the drying operation described in Problem 22A.3, if the solids are cylinders with $a = 180\ ft^{-1}$.

22B.1 Evaporation of a freely falling drop. A drop of water, 1.00 mm in diameter, is falling freely through dry, still air at pressure of 1 atm and a temperature of 100°F with no internal circulation. Assume quasi-steady-state behavior and a small mass-transfer rate to compute **(a)** the velocity of the falling drop, **(b)** the surface temperature of the drop, and **(c)** the rate of change of the drop diameter in cm/s. Assume that the film properties are those of dry air at 80°F.

Answers: **(a)** 395 cm/s; **(b)** 55.1°F; **(c)** -5.55×10^{-4} cm/s

---

[1] E. R. Gilliland and T. K. Sherwood, *Ind. Eng. Chem.* **26**, 516–523 (1934).

22B.2 Effect of radiation on psychrometric measurements. Suppose that a wet-bulb thermometer and a dry-bulb thermometer are installed in a long duct with constant inside surface temperature T_s and that the gas velocity is small. Then the dry-bulb temperature T_{db} and the wet-bulb temperature T_{wb} should be corrected for radiation effects. We assume, as in Example 22.3-2, that the thermometers are so installed that the heat conduction along the glass stems can be neglected.

(a) Make an energy balance on a unit area of the dry bulb to obtain an equation for the gas temperature T_∞ in terms of T_{db}, T_s, h_{db}, e_{db}, and a_{db} (these last two are the emissivity and absorptivity of the dry bulb).

(b) Make an energy balance on a unit area of the wet bulb and obtain an expression for the evaporation rate.

(c) Compute $x_{A\infty}$ for the pressure and thermometer readings of Example 22.3-2, with the additional information that $v_\infty = 15$ ft/s, $T_s = 130°\text{F}$, $e_{db} = a_{db} = e_{wb} = a_{wb} = 0.93$, dry-bulb diameter = 0.1 in., and wet-bulb diameter = 0.15 in. including the wick.

Answer: **(c)** $x_{A\infty} = 0.0023$

22B.3 Oxygen stripping. Calculate the rate at which oxygen transfers from quiescent oxygen-saturated water at 20°C to a bubble of pure nitrogen 1 mm in diameter, if the bubble acts as a rigid sphere. Note that it will first be necessary to determine the bubble velocity of rise through the water. Assume that the equilibrium concentration (solubility) of oxygen in water is given by[2]

$$c_{O_2} = (2.17 - 0.0507T + 5.604 \times 10^{-4}T^2)p \tag{22B.3-1}$$

where c_{O_2} [=] mmol/liter, T [=] °C, and p [=] atm. Also assume that the diffusivity of oxygen in water is given by[3]

$$\mathscr{D}_{O_2W} = (1.04 + 0.053T) \times 10^{-5} \tag{22B.3-2}$$

where $\mathscr{D}_{O_2W}$ [=] cm²/s and T [=] °C.

22B.4 Controlling diffusional resistance. Water drops 2 mm in diameter are being oxygenated by falling freely through pure oxygen at 20°C and a partial pressure of 1 atm. Do you need to know the gas-phase diffusivity to calculate the rate of oxygen transport? Why? The solubility of oxygen under these conditions is 1.39 mmol/liter, and the diffusivity is about 2.1×10^{-5} cm²/s.

---

[2]Obtained from regression of data in J. E. Bailey and D. F. Ollis, *Biochemical Engineering Fundamentals*, 2nd edition, McGraw-Hill, New York (1986); p. 463.

[3]Regression of data at 37°C by E. E. Spaeth and S. K. Friedlander, *Biophys. J.* (1967), p. 827, and data at 25°C reported in T. K. Sherwood, R. L. Pigford, and C. R. Wilke, *Mass Transfer*, McGraw-Hill, New York (1975).

Chapter 23

Macroscopic Balances for Multicomponent Systems

Applications of the laws of the conservation of mass, momentum, and energy to engineering flow systems have been discussed in Chapter 7 (isothermal systems) and Chapter 15 (nonisothermal systems). In this chapter we continue the discussion by introducing three additional factors not encountered in the earlier chapters: (a) the fluid in the system is composed of more than one chemical species; (b) chemical reactions may be occurring, along with changes of composition and the production or consumption of heat; and (c) mass may be entering the system through the bounding surfaces (that is, across surfaces other than planes 1 and 2). Various mechanisms by which mass may enter or leave through the bounding surfaces of the system are shown in Fig. 23.0-1.

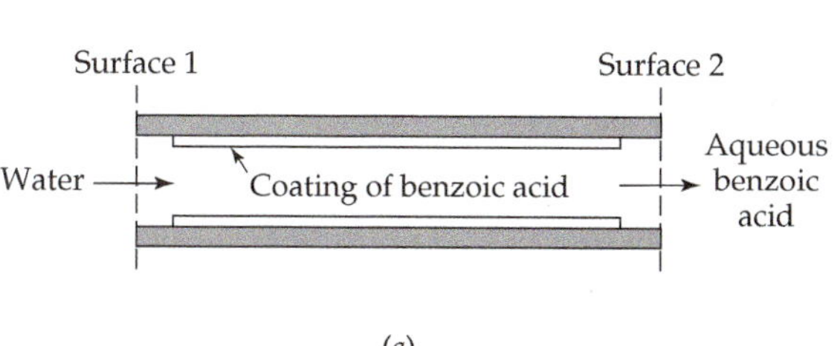

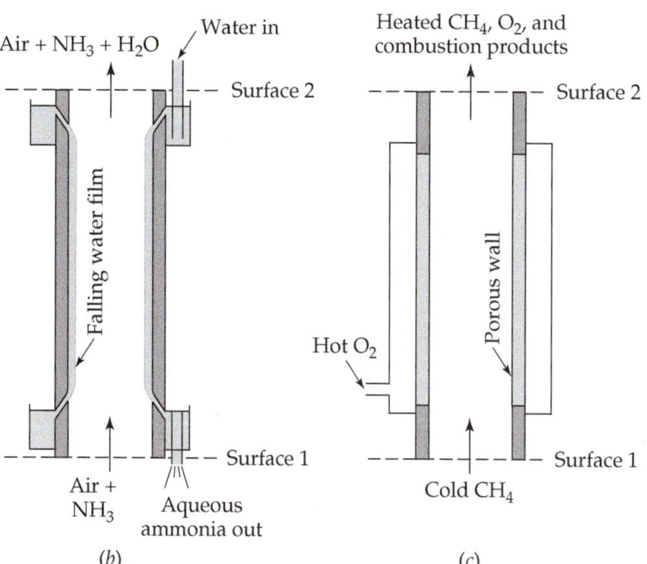

Fig. 23.0-1. Ways in which mass may enter or leave through boundary surfaces: (*a*) benzoic acid enters system by dissolution of the wall; (*b*) water vapor enters the system, defined as the gas phase, by evaporation, and ammonia vapor leaves by absorption; (*c*) oxygen enters the system by transpiration through a porous wall.

In this chapter we summarize the macroscopic balances for the more general situation described above. Each of these balances will now contain one extra term, to account for mass, momentum, or energy transport across the bounding surfaces. The balances thus obtained are capable of describing industrial mass-transfer processes, such as absorption, extraction, ion exchange, and selective adsorption. Inasmuch as entire treatises have been devoted to these topics, all we try to do here is to show how the material discussed in the preceding chapters paves the way for the study of the mass-transfer operations. The reader interested in pursuing the subject further should consult the available textbooks and treatises.[1–8]

The main emphasis in this chapter is on the mass balances for mixtures. For that reason, §23.1 is accompanied by three examples, which illustrate problems arising in environmental science, stagewise processes, and biomedical science. In §23.2 to §23.4 the other macroscopic balances are given. In Table 23.5-1 they are summarized for systems with multiple inlets and outlets. The last two sections of the chapter illustrate the applications of the macroscopic balances to more complex systems.

Although Part III of this book is concerned primarily with binary mixtures, the macroscopic balances can be stated for systems containing three or more chemical species with very little effort. Therefore, we go directly to multicomponent systems in this chapter, and treat binary systems as special cases.

§23.1 THE MACROSCOPIC MASS BALANCES

The statement of the law of conservation of mass of chemical species α in a macroscopic flow system containing N chemical species is

$$\boxed{\frac{dm_{\alpha,\text{tot}}}{dt} = -\Delta w_\alpha + w_{\alpha,0} + r_{\alpha,\text{tot}} \qquad \alpha = 1,2,3,\ldots, N} \tag{23.1-1}$$

This is a generalization of Eq. 7.1-3. Here $m_{\alpha,\text{tot}}$ is the instantaneous total mass of α in the system, and $-\Delta w_\alpha = w_{\alpha 1} - w_{\alpha 2} = \rho_{\alpha 1}\langle v_1\rangle S_1 - \rho_{\alpha 2}\langle v_2\rangle S_2$ is the difference between the mass rates of flow of species α across planes 1 and 2. The quantity $w_{\alpha,0}$ is the mass rate of addition of species α to the system by mass transfer across the bounding surface. Note that $w_{\alpha,0}$ is positive when mass is *added* to the system, just as Q and W_m are taken to be positive in the total energy balance when heat is added to the system and work is done on the system by moving parts. Finally, the symbol $r_{\alpha,\text{tot}}$ stands for the rate of production of species α by homogeneous and heterogeneous reactions within the system.[1]

---

[1] W. L. McCabe, J. C. Smith, and P. Harriot, *Unit Operations of Chemical Engineering*, McGraw-Hill, New York, 6th edition (2000).

[2] T. K. Sherwood, R. L. Pigford, and C. R. Wilke, *Mass Transfer*, McGraw-Hill, New York (1975).

[3] R. E. Treybal, *Mass Transfer Operations*, McGraw-Hill, New York, 3rd edition (1980).

[4] C. J. King, *Separation Processes*, McGraw-Hill, New York (1971).

[5] C. D. Holland, *Multicomponent Distillation*, McGraw-Hill, New York (1963).

[6] T. C. Lo, M. H. I. Baird, and C. Hanson, eds., *Handbook of Solvent Extraction*, Wiley-Interscience, New York (1983).

[7] R. T. Yang, *Gas Separations by Adsorption Processes*, Butterworth, Boston (1987).

[8] J. D. Seader and E. J. Henley, *Separation Process Principles*, Wiley, New York (1998).

[1] The quantities $m_{\alpha,\text{tot}}$, $w_{\alpha,0}$, and $r_{\alpha,\text{tot}}$ may be expressed as integrals:

$$m_{\alpha,\text{tot}} = \int_V \rho_\alpha\, dV; \quad w_{\alpha,0} = -\int_{S_0} (\mathbf{n}\cdot\rho_\alpha \mathbf{v}_\alpha)\, dS; \quad r_{\alpha,\text{tot}} = \int_V r_\alpha\, dV + \int_S r_\alpha^{(s)}\, dS \tag{23.1-1a,b,c}$$

in which $\mathbf{n}$ is the outwardly directed unit normal vector, and S_0 is that portion of the bounding surface over which mass transfer occurs, here assumed stationary. The integrands in $r_{\alpha,\text{tot}}$ are the homogeneous and heterogeneous reaction rates respectively.

Recall that, in Table 15.5-1, the molecular and eddy transport of momentum and energy across surfaces 1 and 2 in the direction of flow have been neglected with respect to the convective transport. The same is done everywhere in this chapter—in Eq. 23.1-1 and in the other macroscopic balances presented here.

If all N equations in Eq. 23.1-1 are summed, we get

$$\frac{dm_{\text{tot}}}{dt} = -\Delta w + w_0 \tag{23.1-2}$$

in which $w_0 = \sum_\alpha w_{\alpha,0}$, and use has been made of the law of conservation of mass in the form $\sum_\alpha r_{\alpha,\text{tot}} = 0$.

It is often convenient to write Eq. 23.1-1 in molar units:

$$\boxed{\frac{dM_{\alpha,\text{tot}}}{dt} = -\Delta W_\alpha + W_{\alpha,0} + R_{\alpha,\text{tot}} \qquad \alpha = 1,2,3,\ldots, N} \tag{23.1-3}$$

Here the capital letters represent the counterparts of the lowercase symbols in Eq. 23.1-1. When Eq. 23.1-3 is summed over all species, the result is

$$\frac{dM_{\text{tot}}}{dt} = -\Delta W + W_0 + \sum_{\alpha=1}^{N} R_{\alpha,\text{tot}} \tag{23.1-4}$$

Note that the last term is not in general zero, because the number of moles produced and the total number of moles consumed are not equal in many reaction systems.

In some applications, such as spatially continuous mass-transfer operations, it is customary to rewrite Eqs. 23.1-1 or 23.1-3 for a differential element of the system (that is, in the "d-form" discussed in §15.4). Then the differentials $dw_{\alpha,0}$ or $dW_{\alpha,0}$ can be expressed in terms of local mass-transfer coefficients.

EXAMPLE 23.1-1

Disposal of an Unstable Waste Product

A fluid stream emerges from a chemical plant with a constant mass flow rate $w = \rho Q$ and discharges into a river (Fig. 23.1-1(*a*)). It contains a waste material A at mass fraction ω_{A0}, which is unstable and decomposes at a rate proportional to its concentration according to the expression $r_A = -k_1''' \rho_A$—that is, by a first-order reaction.

To reduce pollution, it is decided to allow the effluent stream to pass through a holding tank of volume V, before discharging into the river (Fig. 23.1-1(*b*)). The tank is equipped with an efficient stirrer that keeps the fluid in the tank at very nearly uniform composition. At time $t = 0$ the fluid begins to flow into the empty tank. No liquid flows out until the tank has been filled up to the volume V.

Develop an expression for the concentration of the fluid in the tank as a function of time, both during the tank-filling process and after the tank has been completely filled.

SOLUTION

(a) We begin by considering the period during which the tank is being filled—that is, the period $t \leq \rho V/w = V/Q$, where ρ is the density and Q is the volume flow rate of the fluid mixture. We apply the macroscopic mass balance of Eq. 23.1-1 to the holding tank. The quantity $m_{A,\text{tot}}$ on the left side is $wt\omega_A$ at time t. The mass rate of flow entering the tank is $w\omega_{A0}$, and there is no outflow during the tank-filling stage. No A is entering or leaving through a mass-transfer interface. The rate of production of species is $r_{A,\text{tot}} = (wt/\rho)(-k_1''' \rho_A) = -k_1''' m_{A,\text{tot}}$. Therefore, the macroscopic mass balance for species A during the filling period is

$$\frac{d}{dt} m_{A,\text{tot}} = w\omega_{A0} - k_1''' m_{A,\text{tot}} \tag{23.1-5}$$

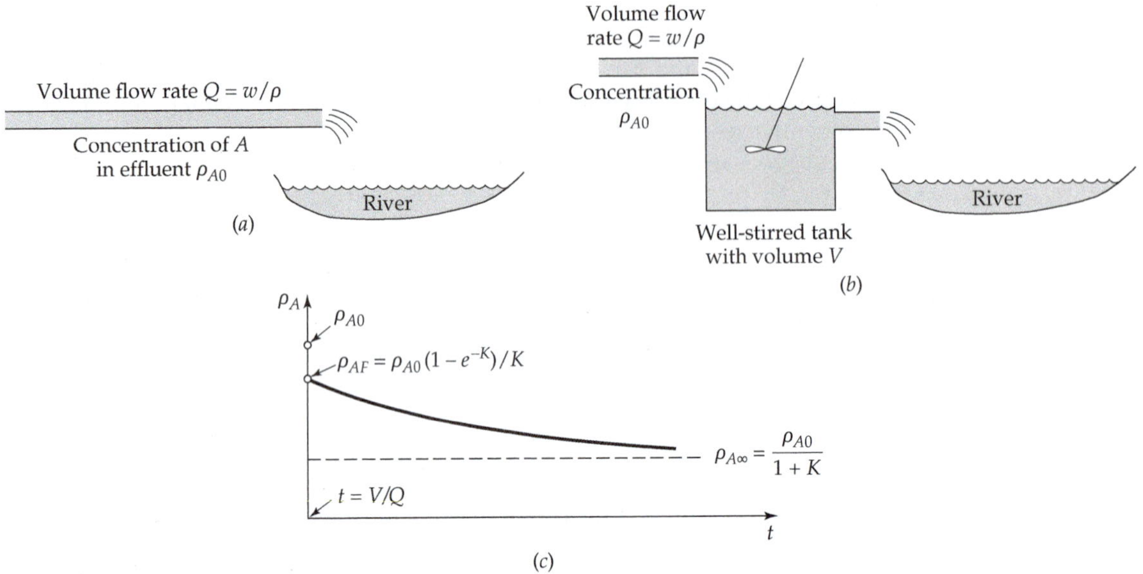

Fig. 23.1-1. (*a*) Waste stream with unstable pollutant emptying directly into a river. (*b*) Waste stream with a holding tank that allows the unstable pollutant to decay prior to going into the river. (*c*) Sketch showing the concentration of pollutant being discharged into the river after the holding tank has been filled (the dimensionless quantity K is $k_1''' V/Q$).

This first-order differential equation can be solved with the initial condition that $m_{A,\text{tot}} = 0$ at $t = 0$ to give

$$m_{A,\text{tot}} = \frac{w\omega_{A0}}{k'''}(1 - \exp(-k_1''' t)) \tag{23.1-6}$$

This may be written in terms of the instantaneous mass fraction of A in the tank by using the relation $m_{A,\text{tot}} = wt\omega_A$:

$$\frac{\omega_A}{\omega_{A0}} = \frac{1 - \exp(-k_1''' t)}{k_1''' t} \qquad \left(t \le \frac{\rho V}{w}\right) \tag{23.1-7}$$

The mass fraction of A at the instant when the tank is full, ω_{AF}, is then

$$\frac{\omega_{AF}}{\omega_{A0}} = \frac{1 - e^{-K}}{K} \tag{23.1-8}$$

in which $K = k_1'''\rho V/w = k_1''' V/Q$.

(b) The mass balance on the tank after it has been filled is

$$\frac{d}{dt}(\rho_A V) = w\omega_{A0} - w\omega_A - k_1'''\rho_A V \tag{23.1-9}$$

When we divide this equation by w and introduce the dimensionless time $\tau = (w/\rho V)t$, it becomes

$$\frac{d\omega_A}{d\tau} + (1 + K)\omega_A = \omega_{A0} \tag{23.1-10}$$

This first-order differential equation can be solved with the initial condition that $\omega_A = \omega_{AF}$ at $\tau = 1$ to give (see Eq. C.1-2)

$$\frac{\omega_A(\tau) - [\omega_{A0}/(1+K)]}{\omega_{AF} - [\omega_{A0}/(1+K)]} = e^{-(1+K)(\tau-1)} \qquad \left(t \ge \frac{\rho V}{w}\right) \tag{23.1-11}$$

This shows that as time progresses the mass fraction of the pollutant being discharged into the river decreases exponentially, with a limiting value of

$$\omega_{A\infty} = \frac{\omega_{A0}}{1+K} = \frac{\omega_{A0}}{1 + (k_1'''\rho V/w)} \tag{23.1-12}$$

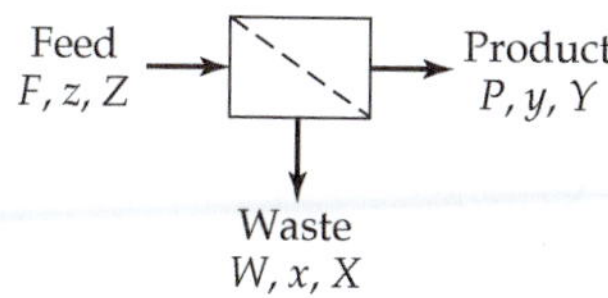

Fig. 23.1-2. Binary splitter, in which a feed stream F (a mixture of A and B) is split into a product stream P (with y being the mole fraction of A) and a waste stream W (with x being the mole fraction of A).

The curve for the mass concentration $\rho_A = \omega_A \rho$ as a function of time after the filling of the tank is shown in Fig. 23.1-1(*c*). This curve can be used to determine conditions such that the effluent concentration will be in the permitted range. Equation 23.1-12 can be used to decide on the volume V of the holding tank that is required.

EXAMPLE 23.1-2

Binary Splitters

Describe the operation of a binary splitter, one of the commonest and simplest separation devices (see Fig. 23.1-2). Here a binary mixture of A and B enters the apparatus in a feed stream at a molar rate F, and by some separation mechanism it is split into a product stream at a molar rate P and a waste stream with molar rate W. The mole fraction of A (the desired component) in the feed stream is z, and the mole fractions of A in the product and waste streams are y and x, respectively.

SOLUTION

We start by writing the steady-state macroscopic mass balances for component A and for the entire fluid as

$$zF = yP + xW \tag{23.1-13}$$

$$F = P + W \tag{23.1-14}$$

It is customary to define the ratio $\theta = P/F$ of the molar rates of the product and feed streams as the *cut*. Equation 23.1-13 then becomes, after eliminating W by use of Eq. 23.1-14,

$$z = \theta y + (1 - \theta)x \tag{23.1-15}$$

Normally the cut θ and the feed composition z are taken to be known.

We now need a relation between the feed and waste compositions, and it is conventional to write an equation relating the compositions of the two outgoing streams:

$$Y = \alpha X \tag{23.1-16}$$

Here α is known as the *separation factor*, also usually taken as known, and which characterizes the separation capability of the splitter. Here Y and X are the mole ratios defined by

$$Y = \frac{y}{1-y} \quad \text{and} \quad X = \frac{x}{1-x} \tag{23.1-17,18}$$

In terms of the mole fractions, Eq. 23.1-16 may be written as

$$y = \frac{\alpha x}{1 + (\alpha - 1)x} \quad \text{or} \quad x = \frac{y}{\alpha - (\alpha - 1)y} \tag{23.1-19,20}$$

Equations 23.1-15 and 23.1-19 (or 23.1-20) describe completely the splitter operation.

For vapor-liquid splitting—that is, equilibrium distillation—it is typical to define the *ideal* splitter in terms of an operation in which the product and waste streams are in equilibrium. For this situation, α is the *relative volatility*, and for thermodynamically ideal systems, it is just the ratio of the component vapor pressures. Even for nonideal systems, α changes relatively slowly with composition.

For *real* splitters one can then define α in terms of an empirical correction factor—for example, the *efficiency*—defined by

$$\alpha = E\alpha^* \tag{23.1-21}$$

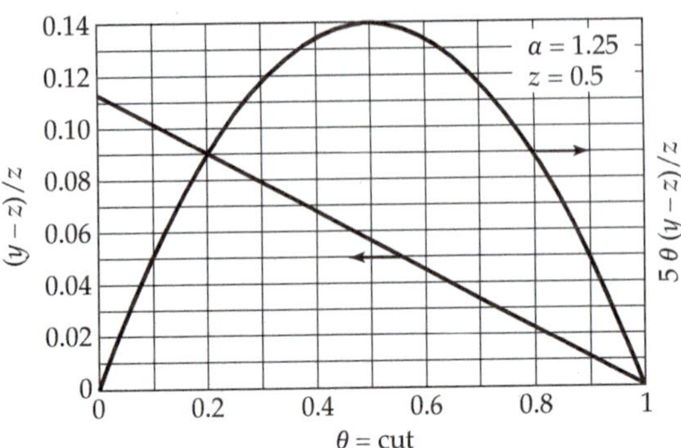

Fig. 23.1-3. Behavior of a binary splitter when the separation factor is $\alpha = 1.25$.

where α^* is the separation factor for the ideal model, and E is a correction factor that accounts for the failure of the actual system to meet the ideal behavior.

We thus find that, for a given feed composition, the *enrichment* $(y - z)/z$ produced by the splitter, is a function of the cut θ and the separation factor α. The enrichment can be calculated from the following equation, which is obtained by combining Eqs. 23.1-15 and 23.1-20:

$$z = \theta y + (1 - \theta)\frac{y}{\alpha - (\alpha - 1)y} \tag{23.1-22}$$

This is a quadratic equation for y, which can be solved when z and α are specified, and then the enrichment $(y - z)/z$ is obtained. An example is given in Fig. 23.1-3 where both $(y - z)/z$ and $5\theta(y - z)/z$ are plotted as functions of θ for $z = 0.5$ and $\alpha = 1.25$ (the latter being a reasonable value for many processes). It may be seen that, whereas the maximum enrichment $(y - z)/z$ is obtained for vanishingly small cuts, the product of enrichment and product rate peaks at an intermediate θ value. Finding an optimum θ value is a problem that must be addressed on economic grounds.

Simple splitters of the general type pictured in Fig. 23.1-2 are very widely used as building blocks in multistage separation processes. They include evaporators and crystallizers, which typically have a very high separation factor α per stage, and systems for distillation, gas absorption, and liquid extraction, where α can vary widely. All of these applications are well covered in standard texts on unit operations.

Membrane processes are rapidly increasing in importance, and many of the design principles were developed for the isotope fractionation industry.[2] Discussions of modern applications are also available.[3]

EXAMPLE 23.1-3

Compartmental Analysis

One of the simplest and most useful applications of the species macroscopic mass balance is *compartmental analysis*, in which a complex system is treated as a network of perfect mixers, each of constant volume, connected by ducts of negligible volume, with no dispersion occurring in the connecting ducts. Imagine mixing units, labeled $1, 2, 3, \ldots, n, \ldots, N$, containing various species (labeled with indices $\alpha, \beta, \gamma, \ldots$). Then the mass concentration $\rho_{\alpha n}$ of species α in unit n changes with time according to the equation

$$V_n \frac{d\rho_{\alpha n}}{dt} = \sum_{m=1}^{N} Q_{mn}(\rho_{\alpha m} - \rho_{\alpha n}) + V_n r_{\alpha n} \tag{23.1-23}$$

Here V_n is the volume of unit n, Q_{mn} is the volumetric flow rate of solvent flow from unit m to unit n, and $r_{\alpha n}$ is the rate of formation of species α per unit volume in unit n.

---

[2] E. Von Halle and J. Schacter, *Diffusion Separation Methods*, in Volume **8** of *Kirk-Othmer Encyclopedia of Chemical Technology* (M. Howe-Grant, ed.), 4th edition, Wiley, New York (1993), pp. 149–203.

[3] W. S. W. Ho and K. K. Sirkar, *Membrane Handbook*, Van Nostrand Reinhold, New York (1992), p. 954; R. D. Noble and S. A. Stern, *Membrane Separations Technology*, Elsevier, Amsterdam (1995), p. 718.

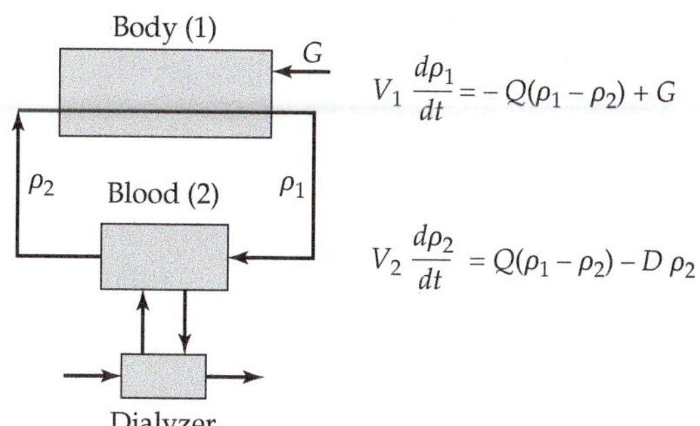

Fig. 23.1-4. Two-compartment model used to analyze the functioning of a dialyzer.

Show how such a model can be specialized to describe the removal of toxic metabolic products (that is, the toxic materials resulting from the human metabolism) from a patient by *hemodialysis*. Hemodialysis is the periodic removal of toxic metabolites achieved by contacting the blood and a dialysis fluid in countercurrent flow, separated by a cellophane membrane that is permeable to the metabolite.

SOLUTION

The simple two-compartment model of Fig. 23.1-4 has been found to be adequate for representing the hemodialysis system. Here the large block, or compartment 1 (labeled "body"), represents the combined body fluids, except for those in the blood, which are represented by compartment 2. The blood circulates via a branching system of vessels through compartment 1 at a volumetric rate Q, and in the process extracts solute across the vessel walls. This process is highly efficient, and a single solute is assumed to leave compartment 1 at concentration ρ_1, equal to the concentration throughout that compartment. At the same time, the solute is being formed within the body fluids at a constant rate G, and during dialysis it is being extracted from the blood by the dialyzer at a rate $D\rho_2$. The proportionality constant D is known as the "dialyzer clearance" and is fixed by the dialyzer design and operating conditions.

The very complex process actually taking place is modeled by the two equations

$$V_1 \frac{d\rho_1}{dt} = -Q(\rho_1 - \rho_2) + G \tag{23.1-24}$$

$$V_2 \frac{d\rho_2}{dt} = Q(\rho_1 - \rho_2) - D\rho_2 \tag{23.1-25}$$

with $D = 0$ between the dialysis periods. Because we are considering a single solute, the concentrations have only one subscript to indicate the compartment. We measure the time t from the start of a dialysis procedure, when the blood and body fluids are very nearly in equilibrium with each other, so that we may write the initial conditions as

I.C.: At $t = 0$, $$\rho_1 = \rho_2 = \rho_0 \tag{23.1-26}$$

where ρ_0 is a constant. We now want to get an explicit expression for the toxic metabolite concentration in the blood as a function of time.

We start by adding Eqs. 23.1-24 and 23.1-25 and solving for $d\rho_1/dt$. The latter is then substituted into the time derivative of Eq. 23.1-25 to obtain a differential equation for the metabolite concentration in the blood

$$\frac{d^2\rho_2}{dt^2} + \left(\frac{Q}{V_1} + \frac{Q}{V_2} + \frac{D}{V_2}\right)\frac{d\rho_2}{dt} + \frac{QD}{V_1V_2}\rho_2 = \frac{QG}{V_1V_2} \tag{23.1-27}$$

with

I.C.: At $t = 0$, $$\rho_2 = \rho_0 \quad \text{and} \quad \frac{d\rho_2}{dt} = -\frac{D\rho_0}{V_2} \tag{23.1-28}$$

The second initial condition is obtained from using Eqs. 23.1-25 and 23.1-26.

This equation is now to be solved for the following specific parameter values, which are typical for the removal of creatinine from a 70-kg adult human:

Quantity	V_1 (liters)	V_2 (liters)	Q (liters/min)	D (liters/min)	G (g/min)	ρ_0 (g/liter)
Magnitude	43	4.5	5.4	0.3	0.0024	0.140

The differential equation and initial conditions now take the form:

$$\frac{d^2\rho_2}{dt^2} + (1.3922\ \text{min}^{-1})\frac{d\rho_2}{dt} + (0.00837\ \text{min}^{-2})\rho_2 = 6.70 \times 10^{-5}\ \text{g liter}^{-1}\text{min}^{-2} \tag{23.1-29}$$

$$\text{I.C.:} \quad \text{At time } t = 0, \quad \rho_2 = 0.140\ \text{g liter}^{-1} \quad \text{and} \quad \frac{d\rho_2}{dt} = -0.00933\ \text{g liter}^{-1}\ \text{min}^{-1} \tag{23.1-30}$$

in which concentration is in grams per liter, and time is in minutes. The complementary function that satisfies the associated homogeneous equation is

$$\rho_{2,\text{cf}}(t) = C_1 \exp(-0.006043\ \text{min}^{-1})t + C_2 \exp(-1.386\ \text{min}^{-1})t \tag{23.1-31}$$

and the particular integral is

$$\rho_{2,\text{pi}} = 0.0080\ \text{g liter}^{-1} \tag{23.1-32}$$

The complete solution to the nonhomogeneous equation is given by the sum of the complementary function and the particular integral. When the constants of integration are determined from the initial conditions, we get

$$\begin{aligned}\rho_2(t) &= 0.1258\ \text{g liter}^{-1} \exp\left[(-0.006043\ \text{min}^{-1})t\right] \\ &\quad + 0.00621\ \text{g liter}^{-1} \exp\left[(-1.386\ \text{min}^{-1})t\right] + 0.0080\ \text{g liter}^{-1}\end{aligned} \tag{23.1-33}$$

$$\begin{aligned}\frac{d\rho_2}{dt} &= -(0.000760\ \text{g liter}^{-1}) \exp\left[(-0.006043\ \text{min}^{-1})t\right] \\ &\quad - (0.0086\ \text{g liter}^{-1}) \exp\left[(-1.386\ \text{min}^{-1})t\right]\end{aligned} \tag{23.1-34}$$

during the dialysis period.

For the recovery period following dialysis, we assume here that the patient has no kidney function, so that the dialyzer clearance D is zero. Equation 23.1-27 takes the simpler form

$$\frac{d^2\rho_2'}{dt'^2} + Q\left(\frac{V_1 + V_2}{V_1 V_2}\right)\frac{d\rho_2'}{dt'} = \frac{QG}{V_1 V_2} \tag{23.1-35}$$

where ρ' is the concentration during the recovery period. The complementary function and particular integral are

$$\rho_{2,\text{cf}}'(t') = C_3 \exp\left[-Q\left(\frac{V_1 + V_2}{V_1 V_2}\right)t'\right] + C_4 \tag{23.1-36}$$

$$\rho_{2,\text{pi}}'(t') = \frac{Gt'}{V_1 + V_2} \tag{23.1-37}$$

in which t' is the time measured from the start of the recovery period. Inserting the numerical values, we then get for the concentration during the recovery period and its time derivative

$$\rho_2'(t') = C_3 \exp\left[(-1.325\ \text{min}^{-1})t'\right] + (5.05 \times 10^{-5}\ \text{g liter}^{-1}\ \text{min}^{-1})t' + C_4 \tag{23.1-38}$$

$$\frac{d\rho_2'}{dt'} = -1.325 C_3 \exp\left[(-1.325\ \text{min}^{-1})t'\right] + (5.05 \times 10^{-5}\ \text{g liter}^{-1}\ \text{min}^{-1}) \tag{23.1-39}$$

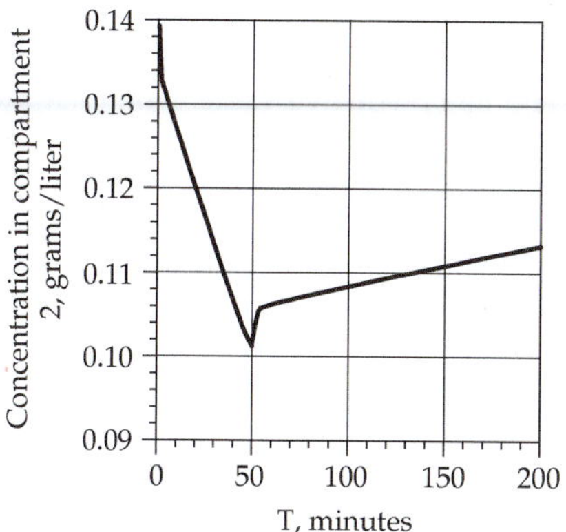

Fig. 23.1-5. Pharmacokinetics of dialysis: model prediction.

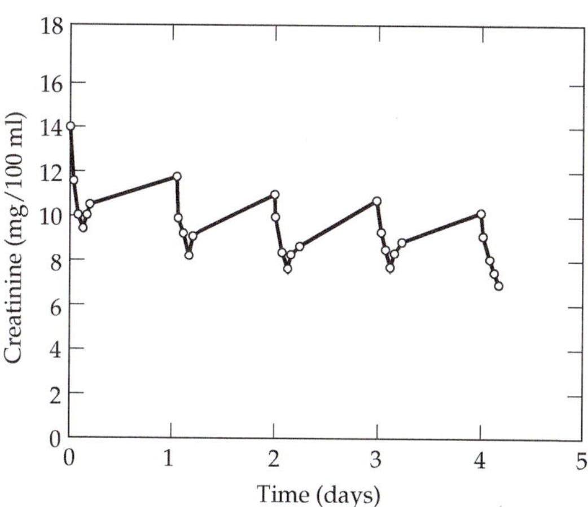

Fig. 23.1-6. Experimental (open circles) and simulated creatinine data (solid curve) for a dialysis patient. [R. L. Bell, K. Curtiss, and A. L. Babb, *Trans. Amer. Soc. Artificial Internal Organs*, **11**, 183 (1965).]

The integration constants are to be determined from the matching conditions at $t' = 0$,

At $t' = 0$, $$\rho_2' = \rho_2 \quad \text{and} \quad \rho_1' = \rho_1 \qquad (23.1\text{-}40,41)$$

We need a second initial condition for determining the integration constants in Eq. 23.1-38. This can be obtained from Eq. 23.1-25 and the corresponding equation for ρ_2' (i.e., with $D = 0$), combined with the two relations in Eqs. 23.1-40 and 23.1-41. This relation is

At $t' = 0$, $$\frac{d\rho_2'}{dt'} = \frac{d\rho_2}{dt} + \frac{D\rho_2}{V_2} \qquad (23.1\text{-}42)$$

For illustrative purposes, we shall end the dialysis at 50 minutes, for which

$$\rho_2(t = 50 \text{ min}) = 0.10099 \text{ g liter}^{-1} = \rho_2' \qquad (23.1\text{-}43)$$

$$\frac{d\rho_2}{dt}(t = 50 \text{ min}) = -0.005618 \text{ g liter}^{-1} \text{ min}^{-1} \qquad (23.1\text{-}44)$$

We now have enough information to determine the constants of integration, and therefore, we get for the concentration in the blood during the recovery period

$$\rho_2'(t') = 0.1053 \text{ g liter}^{-1} - (0.00424 \text{ g liter}^{-1}) \exp\left[(-1.325 \text{ min}^{-1})t'\right] + (5.05 \times 10^{-5} \text{ g liter}^{-1} \text{ min}^{-1})t' \qquad (23.1\text{-}45)$$

Equations 23.1-33 and 23.1-45 are plotted Fig. 23.1-5.

Of perhaps more interest is Fig. 23.1-6, which shows the application of Eqs. 23.1-24 and 23.1-25 to an actual patient. Here the points represent data, and the curves are the model predictions. Here only the dialyzer clearance and the creatinine concentrations are known, and the data of the first cycle are used to estimate the remaining parameters. The resulting model is then used to predict the next three cycles. It may be seen that this approach does an excellent job of correlating data and has predictive value. Note that the sudden rise in creatinine concentration at 50 min results from the fact that the dialyzer is no longer removing it from the blood. As a result, the disequilibrium between the blood and the rest of the body then becomes smaller.

Similar compartmental models have wide application in medicine, where they are referred to as *pharmacokinetic models*.[4] *A priori* pharmacokinetic modeling, where model parameters are determined independently of the process being modeled, was pioneered by Bischoff and Dedrick.[5]

§23.2 THE MACROSCOPIC MOMENTUM AND ANGULAR MOMENTUM BALANCES

The macroscopic statements of the laws of conservation of momentum and angular momentum for a fluid mixture, with gravity as the only external force, are

$$\frac{d\mathbf{P}_{\text{tot}}}{dt} = -\Delta\left(\frac{\langle v^2\rangle}{\langle v\rangle}w + pS\right)\mathbf{u} + \mathbf{F}_{s\to f} + \mathbf{F}_0 + m_{\text{tot}}\mathbf{g} \qquad (23.2\text{-}1)$$

$$\frac{d\mathbf{L}_{\text{tot}}}{dt} = -\Delta\left(\frac{\langle v^2\rangle}{\langle v\rangle}w + pS\right)[\mathbf{r}\times\mathbf{u}] + \mathbf{T}_{s\to f} + \mathbf{T}_0 + \mathbf{T}_{\text{ext}} \qquad (23.2\text{-}2)$$

These (seldom-used) equations are the same as Eqs. 7.2-2 and 7.3-2, except for the addition of the terms $\mathbf{F}_0$ and $\mathbf{T}_0$, which are the net influxes[1] of momentum and angular momentum into the system by mass transfer. For most mass-transfer processes these terms are so small that they can be safely neglected.

§23.3 THE MACROSCOPIC ENERGY BALANCE

For a fluid mixture, the statement of the law of conservation of energy is

$$\frac{d}{dt}(U_{\text{tot}} + K_{\text{tot}} + \Phi_{\text{tot}}) = -\Delta\left[\left(\hat{U} + p\hat{V} + \frac{1}{2}\frac{\langle v^3\rangle}{\langle v\rangle} + \hat{\Phi}\right)w\right] + Q_0 + Q + W_m \qquad (23.3\text{-}1)$$

[4] P. G. Welling, *Pharmacokinetics*, American Chemical Society (1997).

[5] K. B. Bischoff and R. L. Dedrick, *J. Pharm. Sci.*, **87**, 1347–1357 (1968); *AIChE Symposium Series*, **64**, 32–44 (1968).

[1] These terms may be written as integrals,

$$\mathbf{F}_0 = -\int_{S_0} [\mathbf{n}\cdot\rho\mathbf{v}\mathbf{v}]\,dS \qquad (23.2\text{-}1a)$$

$$\mathbf{T}_0 = -\int_{S_0} [\mathbf{n}\cdot\{\mathbf{r}\times\rho\mathbf{v}\mathbf{v}\}]\,dS \qquad (23.2\text{-}2a)$$

in which $\mathbf{n}$ is the outwardly directed unit normal vector.

This equation is the same as Eq. 15.1-2, except that an additional term[1] Q_0 has been added. This term accounts for the addition of energy to the system as a result of the mass transfer. It may be of considerable importance, particularly if material is entering through the bounding surface at a much higher or lower temperature than that of the fluid inside the flow system, or if it reacts chemically in the system.

When chemical reactions are occurring, considerable heat may be released or absorbed. This heat of reaction is automatically taken into account in the calculation of the enthalpies of the entering and leaving streams (see Example 23.5-1).

In some applications, in which the energy-transfer rates across the surface are functions of position, it is more convenient to rewrite Eq. 23.3-1 in the d-form, that is, over a differential portion of the flow system as described in §15.4. Then the increment of heat added, dQ, is expressible in terms of a local heat-transfer coefficient.

§23.4 THE MACROSCOPIC MECHANICAL ENERGY BALANCE

A careful examination of the derivation of the mechanical energy balance in §7.7 shows that the result obtained there applies to mixtures as well as to pure fluids. If we now include the mass-transfer boundary surface S_0, then we get

$$\frac{d}{dt}(K_{\text{tot}} + \Phi_{\text{tot}}) = -\Delta\left[\left(\frac{1}{2}\frac{\langle v^3\rangle}{\langle v\rangle} + \hat{\Phi} + \frac{p}{\rho}\right)w\right] + B_0 + W_m - E_c - E_v \tag{23.4-1}$$

This is the same as Eq. 7.4-2, except for the addition of the term B_0, which accounts for the mechanical energy transport across the mass-transfer boundary.[1] The use of this equation is illustrated in Example 23.5-4.

§23.5 USE OF THE MACROSCOPIC BALANCES TO SOLVE STEADY-STATE PROBLEMS

The macroscopic balances are summarized in Table 23.5-1 for systems with more than one entry and exit plane. The terms with subscript 0 describe the addition or removal of mass, momentum, angular momentum, energy, and mechanical energy at mass-transfer surfaces. Usually these balances are not used in their entirety for problem solving, but it is convenient to have a complete listing of them. For steady-state problems, the left sides of

---

[1]This term may be written as an integral:

$$Q_0 = -\int_{S_0}\left(\mathbf{n}\cdot\left\{\tfrac{1}{2}\rho v^2\mathbf{v} + \rho\hat{\Phi}\mathbf{v} + \sum_{\alpha=1}^{N} c_\alpha \overline{H}_\alpha \mathbf{v}_\alpha\right\}\right)dS \tag{23.3-1a}$$

in which $\mathbf{n}$ is the outwardly directed unit normal vector. The origin of this term may be seen by referring to Eq. 19.3-5.

[1]In terms of a surface integral, this term is given by

$$B_0 = -\int_{S_0}\left(\mathbf{n}\cdot\left[\tfrac{1}{2}\rho v^2 + \rho\hat{\Phi} + p\right]\mathbf{v}\right)dS \tag{23.4-1a}$$

Table 23.5-1. Unsteady-State Macroscopic Balances for Nonisothermal Multicomponent Systems ($\alpha = 1,2,3 \ldots N$)

Mass:

$$\frac{d}{dt}m_{\text{tot}} = \sum w_1 - \sum w_2 + w_0 = \sum \rho_1 \langle v_1 \rangle S_1 - \sum \rho_2 \langle v_2 \rangle S_2 + w_0 \quad \text{(A)}$$

Mass of species α:

$$\frac{d}{dt}m_{\alpha,\text{tot}} = \sum w_{\alpha 1} - \sum w_{\alpha 2} + w_{\alpha 0} + r_{\alpha,\text{tot}} \qquad \alpha = 1,2,3,\ldots N \quad \text{(B)}$$

Momentum:

$$\frac{d}{dt}\mathbf{P}_{\text{tot}} = \sum \left(\frac{\langle v_1^2 \rangle}{\langle v_1 \rangle} w_1 + p_1 S_1 \right) \mathbf{u}_1 - \sum \left(\frac{\langle v_2^2 \rangle}{\langle v_2 \rangle} w_2 + p_2 S_2 \right) \mathbf{u}_2 + m_{\text{tot}}\mathbf{g} + \mathbf{F}_0 - \mathbf{F}_{f \to s} \quad \text{(C)}$$

Angular momentum:

$$\frac{d}{dt}\mathbf{L}_{\text{tot}} = \sum \left(\frac{\langle v_1^2 \rangle}{\langle v_1 \rangle} w_1 + p_1 S_1 \right) [\mathbf{r}_1 \times \mathbf{u}_1] - \sum \left(\frac{\langle v_2^2 \rangle}{\langle v_2 \rangle} w_2 + p_2 S_2 \right) [\mathbf{r}_2 \times \mathbf{u}_2] + \mathbf{T}_{\text{ext}} + \mathbf{T}_0 - \mathbf{T}_{f \to s} \quad \text{(D)}$$

Mechanical energy:

$$\frac{d}{dt}(K_{\text{tot}} + \Phi_{\text{tot}}) = \sum \left(\frac{1}{2}\frac{\langle v_1^3 \rangle}{\langle v_1 \rangle} + gh_1 + \frac{p_1}{\rho_1} \right) w_1 - \sum \left(\frac{1}{2}\frac{\langle v_2^3 \rangle}{\langle v_2 \rangle} + gh_2 + \frac{p_2}{\rho_2} \right) w_2 + W_m + B_0 - E_c - E_v \quad \text{(E)}$$

(Total) energy:

$$\frac{d}{dt}(K_{\text{tot}} + \Phi_{\text{tot}} + U_{\text{tot}}) = \sum \left(\frac{1}{2}\frac{\langle v_1^3 \rangle}{\langle v_1 \rangle} + gh_1 + \hat{H}_1 \right) w_1 - \sum \left(\frac{1}{2}\frac{\langle v_2^3 \rangle}{\langle v_2 \rangle} + gh_2 + \hat{H}_2 \right) w_2 + W_m + Q_0 + Q \quad \text{(F)}$$

Notes:

[a] $\sum w_{\alpha 1} = w_{\alpha 1a} + w_{\alpha 1b} + w_{\alpha 1c} + \cdots$, where $w_{\alpha 1a} = \rho_{\alpha 1a} v_{1a} S_{1a}$, and so on; Equations (A) and (B) can be written in molar units by replacing the lowercase symbols by capital letters and adding to Eq. (A) the term $\Sigma_\alpha R_{\alpha,\text{tot}}$ to account for the fact that moles need not be conserved in a chemical reaction.

[b] h_1 and h_2 are elevations above an arbitrary datum plane.

[c] $\hat{H}_1$ and $\hat{H}_2$ are enthalpies per unit mass (for the mixture) relative to some arbitrarily chosen reference state; see Example 19.3-1.

[d] All equations are written for compressible flow; for incompressible flow, $E_c = 0$. The quantities E_c and E_v are defined in Eqs. 7.4-3 and 7.4-4.

[e] $\mathbf{u}_1$ and $\mathbf{u}_2$ are unit vectors in the direction of flow.

the equations may be omitted. As we saw in Chapters 7 and 15, considerable intuition is required in using the macroscopic balances, and sometimes it is necessary to supplement the equations with experimental observations.

EXAMPLE 23.5-1

Energy Balances for a Sulfur Dioxide Converter

Hot gases from a sulfur burner enter a converter, in which the sulfur dioxide present is to be oxidized catalytically to sulfur trioxide, according to the reaction $SO_2 + \frac{1}{2}O_2 \rightleftarrows SO_3$. How much heat must be removed from the converter per hour to permit a 95% conversion of the SO_2 for the conditions shown in Fig. 23.5-1? Assume that the converter is large enough for the components of the exit gas to be in thermodynamic equilibrium with one another. That is, the partial pressures of the exit gases are related by the equilibrium constraint

$$K_p = \frac{p_{SO_3}}{p_{SO_2} p_{O_2}^{1/2}} \quad (23.5\text{-}1)$$

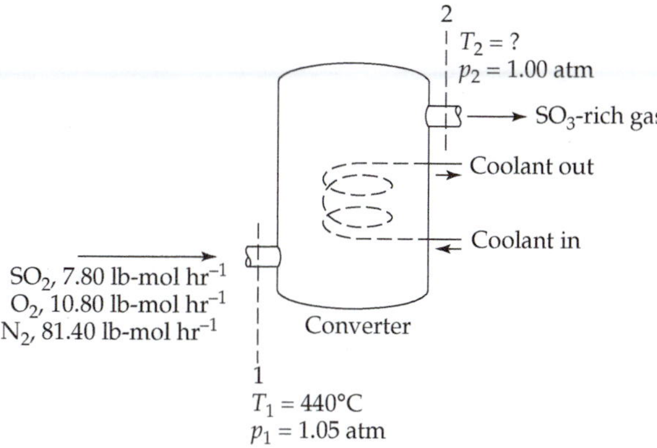

Fig. 23.5-1. Catalytic oxidation of sulfur dioxide.

Approximate values of K_p for this reaction are

T (K)	600	700	800	900
K_p (atm$^{-1/2}$)	9500	880	69.5	9.8

SOLUTION

It is convenient to divide this problem into two parts: **(a)** first we use the mass balance and equilibrium expression to find the desired exit temperature, and then **(b)** we use the energy balance to determine the required heat removal.

(a) ***Determination***[1] ***of T_2.*** We begin by writing the steady-state macroscopic mass balance, Eq. 23.1-3, for the various constituents in the two streams in the form:

$$W_{\alpha 2} = W_{\alpha 1} + R_{\alpha,\text{tot}} \tag{23.5-2}$$

In addition, we take advantage of the two stoichiometric relations

$$R_{SO_2,\text{tot}} = -R_{SO_3,\text{tot}} \tag{23.5-3}$$

$$R_{O_2,\text{tot}} = \tfrac{1}{2}R_{SO_2,\text{tot}} \tag{23.5-4}$$

We can now get the desired molar flow rates through surface 2:

$$W_{SO_2,2} = 7.80 \text{ lb-mol/hr} - (0.95)(7.80 \text{ lb-mol/hr}) = 0.38 \text{ lb-mol/hr} \tag{23.5-5}$$

$$W_{SO_3,2} = 0 + (0.95)(7.80 \text{ lb-mol/hr}) = 7.42 \text{ lb-mol/hr} \tag{23.5-6}$$

$$W_{O_2,2} = 10.80 \text{ lb-mol/hr} - \tfrac{1}{2}(0.95)(7.80 \text{ lb-mol/hr}) = 7.09 \text{ lb-mol/hr} \tag{23.5-7}$$

$$W_{N_2,2} = W_{N_2,1} = 81.40 \text{ lb-mol/hr} \tag{23.5-8}$$

$$W_2 = 0.38 + 7.42 + 7.09 + 81.40 = 96.29 \text{ lb-mol/hr} \tag{23.5-9}$$

Next, substituting numerical values into the equilibrium expression Eq. 23.5-1 gives

$$K_p = \frac{(7.42/96.29 \text{ atm})}{(0.38/96.29 \text{ atm})[(7.09/96.29 \text{ atm})]^{1/2}} = 72.0 \text{ atm}^{-1/2} \tag{23.5-10}$$

The ratio (7.42/96.29) is the mole fraction of SO_3 in the mixture at plane 2, where the total pressure is 1 atm; assuming an ideal-gas mixture, the partial pressure of SO_3 is then (7.42/96.29 atm). Similarly, the other ratios are the partial pressures of SO_2 and O_2. The value of K_p in Eq. 23.5-10

[1] See O. A. Hougen, K. M. Watson, and R. A. Ragatz, *Chemical Process Principles*, Part II, 2nd edition, Wiley, New York (1959), pp. 1017–1018.

corresponds to an exit temperature T_2 of about 510°C, according to the equilibrium data given above.

(b) ***Calculation of the required heat removal.*** As indicated by the results of Example 15.3-1, changes in kinetic and potential energy may be neglected here in comparison with changes in enthalpy. In addition, for the conditions of this example, we may assume ideal-gas behavior. Then, for each constituent, $\overline{H}_\alpha = \tilde{H}_\alpha(T)$. We may then write the macroscopic energy balance, Eq. 23.3-1, as

$$-Q = \sum_{\alpha=1}^{N} (W_\alpha \tilde{H}_\alpha)_1 - \sum_{\alpha=1}^{N} (W_\alpha \tilde{H}_\alpha)_2 \tag{23.5-11}$$

For each of the individual constituents we may write

$$\tilde{H}_\alpha = \tilde{H}_\alpha^\circ + (\tilde{C}_{p\alpha})_{\text{avg}}(T - T^\circ) \tag{23.5-12}$$

Here $\tilde{H}_\alpha^\circ$ is the standard enthalpy of formation[2] of species α from its constituent elements at the enthalpy reference temperature T°, and $(\tilde{C}_{p\alpha})_{\text{avg}}$ is the enthalpy-mean heat capacity[2] of the species between T and T°. For the conditions of this problem, we may use the following[2] numerical values for these physical properties (the last two columns are obtained from Eq. 23.5-12):

Species	$\tilde{H}_\alpha^\circ$ cal/g-mol at 25°C	$(\tilde{C}_{p\alpha})_{\text{avg}}$ [cal/g-mol · °C] from 25°C to 440°C	510°C	$(W_\alpha \tilde{H}_\alpha)_1$ Btu/hr	$(W_\alpha \tilde{H}_\alpha)_2$ Btu/hr
SO_2	−70,960	11.05	11.24	−931,900	−44,800
SO_3	−94,450	—	15.87	0	−1,158,700
O_2	0	7.45	7.53	60,100	46,600
N_2	0	7.12	7.17	433,000	509,500
			Totals	−438,800	−647,400

Substitution of the preceding values into Eq. 23.5-11 gives the required rate of heat removal:

$$-Q = (-438{,}800) - (-647{,}400) = 208{,}600 \text{ Btu/hr} \tag{23.5-13}$$

EXAMPLE 23.5-2

Height of a Packed-Tower Absorber[3]

It is desired to remove a soluble gas A from a mixture of A and an insoluble gas B by contacting the mixture with a nonvolatile liquid solvent L in the apparatus shown in Fig. 23.5-2. The apparatus consists essentially of a vertical pipe filled with a randomly arranged packing of small rings of a chemically inert material. The liquid L is sprayed evenly over the top of the packing and trickles over the surfaces of these small rings. In so doing, it is intimately contacted with the gas mixture that is passing up the tower. This direct contacting between the two streams permits the transfer of A from the gas to the liquid.

The gas and liquid streams enter the apparatus at molar rates of $-W_G$ and W_L, respectively, on an A-free basis. Note that the gas rate is negative, because the gas stream is flowing from plane 2 to plane 1 in this problem. The molar ratio of A to G in the entering gas stream is $Y_{A2} = y_{A2}/(1 - y_{A2})$, and the molar ratio of A to L in the entering liquid stream is $X_{A1} = x_{A1}/(1 - x_{A1})$. Develop an expression for the tower height Z required to reduce the molar ratio of A in the gas stream from Y_{A2} to Y_{A1}, in terms of the mass-transfer coefficients in the two streams and the stream rates and compositions.

[2]See, for example, O. A. Hougen, K. M. Watson, and R. A. Ragatz, *Chemical Process Principles*, Part I, 2nd edition, Wiley, New York (1959), pp. 257, 296.

[3]J. D. Seader and E. J. Henley, *Separation Process Principles*, Wiley, New York (1988).

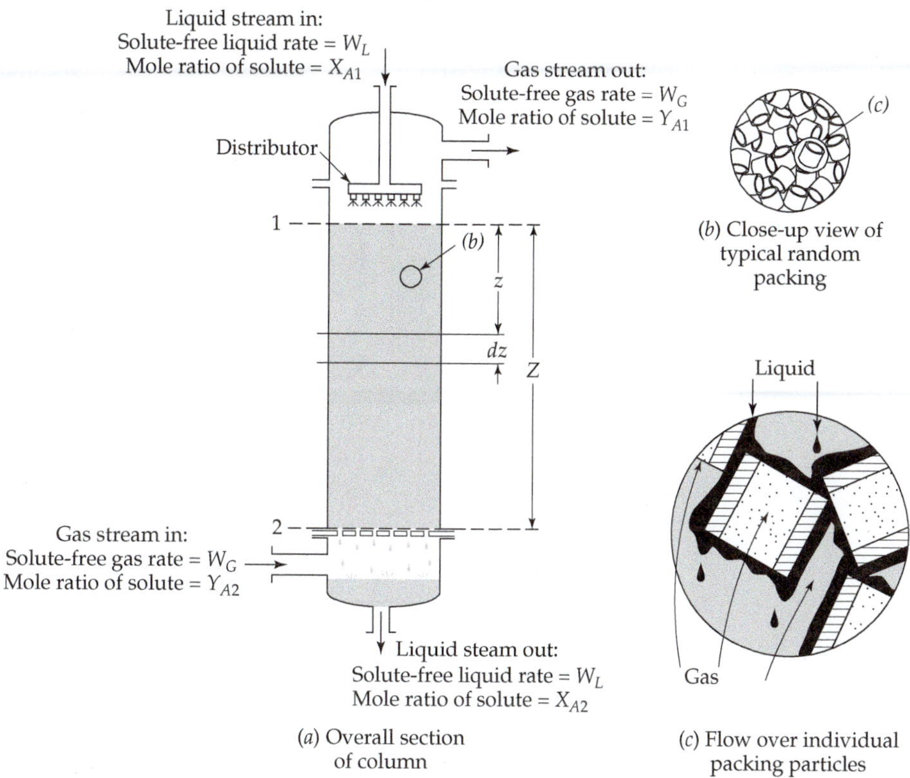

Fig. 23.5-2. A packed-column mass-transfer apparatus in which the descending phase is dispersed. Note that in this drawing W_G is negative; that is, the gas is flowing from 2 toward 1.

Assume that the concentration of A is always small in both streams, so that the operation may be considered isothermal and so that the high mass-transfer rate corrections to the mass-transfer coefficients are not needed. That is, the mass-transfer coefficients, k_x^0 and k_y^0, defined in the second line of Eq. 22.1-14, can be used.

SOLUTION

Since the behavior of a packed tower is quite complex, we replace the true system by a hypothetical model. We consider the system to be equivalent to two streams flowing side-by-side with no back-mixing, as shown in Fig. 23.5-3, and in contact with one another across an interfacial area a per unit volume of the packed column (see Eq. 22.1-14).

We further assume that the fluid velocity and composition of each stream are uniform over the tower cross section and neglect both eddy and molecular transport in the flow direction. We also consider the concentration profiles in the direction of flow to be continuous curves, not appreciably affected by the placement of the individual packing particles.

The model resulting from these simplifying assumptions is probably not a very satisfactory description of a packed tower. The neglect of back-mixing and fluid-velocity nonuniformity are probably particularly serious. However, the presently available correlations for mass-transfer coefficients have been calculated on the basis of this model, which should therefore be employed when these correlations are used.

We are now in a position to develop an expression for the column height, and we do this in two stages: **(a)** First we use the overall macroscopic mass balance to determine the exit liquid-phase composition and the relation between bulk compositions of the two phases at each

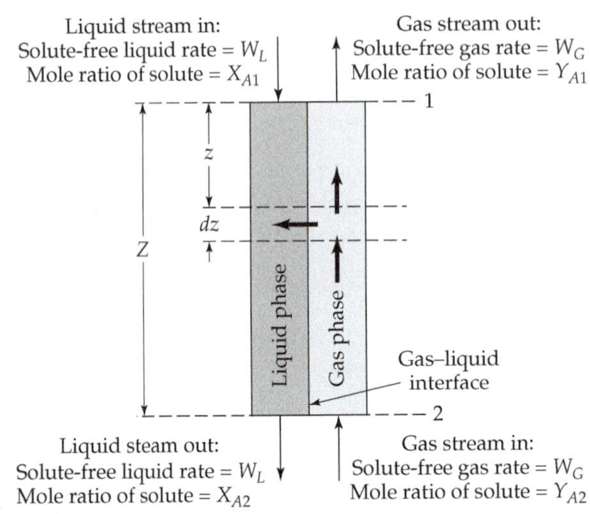

Fig. 23.5-3. Schematic representation of a packed-tower absorber, showing a differential element on which a mass balance is made.

point in the tower. **(b)** We then use these results along with the differential form of the macroscopic mass balance to determine the interfacial conditions and the required tower height.

(a) ***Overall macroscopic mass balances.*** For the solute A we write the macroscopic mass balance of Eq. 23.1-3 for each stream of the system between planes 1 and 2 as

Liquid stream
$$W_{Al2} - W_{Al1} = W_{Al,0} \tag{23.5-14}$$
Gas stream
$$W_{Ag2} - W_{Ag1} = W_{Ag,0} \tag{23.5-15}$$

Here the subscripts Al and Ag refer to the solute A in the liquid and gas streams, respectively. Since the number of moles leaving the liquid stream must enter the gas stream across the interface, $W_{Al,0} = -W_{Ag,0}$, and Eqs. 23.5-14 and 23.5-15 may be combined to give

$$W_{Al2} - W_{Al1} = -(W_{Ag2} - W_{Ag1}) \tag{23.5-16}$$

This can now be rewritten in terms of the compositions of the entering and leaving streams by setting $W_{Al2} = W_L X_{A2}$, and so on, and then rearrangement gives

$$X_{A2} = X_{A1} - \frac{W_G}{W_L}(Y_{A2} - Y_{A1}) \tag{23.5-17}$$

Thus, we have found the concentration of A in the exit liquid stream.

By replacing plane 2 by a plane at a distance z down the column, Eq. 23.5-17 may be used to obtain an expression relating bulk stream compositions at any point in the tower:

$$X_A = X_{A1} - \frac{W_G}{W_L}(Y_A - Y_{A1}) \tag{23.5-18}$$

Equation 23.5-18 (the "operating line") is shown in Fig. 23.5-4 along with the equilibrium distribution for the conditions of Problem 23A.2.

(b) ***Application of the macroscopic balances in the d-form.*** We now apply Eq. 23.1-3 to a small length dz of the tower, first to estimate the interfacial conditions and then determine the tower height required for a given separation.

(i) ***Determination of interfacial conditions.*** Because only A is transferred across the interface, we may write, according to the second line of Eq. 22.1-14 (which presumes low concentrations of A and small mass-transfer rates):

$$dW_{Al,0} = (k_x^0 a)(x_{A0} - x_A)S dz \tag{23.5-19}$$

$$dW_{Ag,0} = (k_y^0 a)(y_{A0} - y_A)S dz \tag{23.5-20}$$

Here a is the interfacial area per unit volume of the packed-bed tower, S is the cross-sectional area of the tower, x_{A0} and y_{A0} are the interfacial mole fractions of A in the liquid and gas phases, respectively, and x_A and y_A are the corresponding bulk concentrations (the index b is being omitted here, so that x_A, y_A, X_A, and Y_A are all bulk compositions).

Then, since $x_A = X_A/(X_A + 1) \approx X_A$ and $y_A = Y_A/(Y_A + 1) \approx Y_A$, Eqs. 23.5-19 and 23.5-20 may be combined to give

$$\frac{Y_A - Y_{A0}}{X_A - X_{A0}} = -\frac{(k_x^0 a)}{(k_y^0 a)} \tag{23.5-21}$$

This equation enables us to determine Y_{A0} as a function of Y_A. For any Y_A, one may locate X_A on the operating line (mass balance). One then draws a straight line of slope $-(k_x^0 a)/(k_y^0 a)$ through the point (Y_A, X_A), as shown in Fig. 23.5-4. The intersection of this line with the equilibrium curve then gives the local interfacial compositions (Y_{A0}, X_{A0}).

(ii) ***Determination of required column height.*** Application of Eq. 23.1-1 to the gas stream in a volume Sdz of the tower gives

$$W_G dY_A = dW_{Ag,0} \tag{23.5-22}$$

This expression may be combined with Eq. 23.5-20 for the dilute solutions being considered to obtain

$$-W_G dY_A = (k_y^0 a)(Y_A - Y_{A0})Sdz \tag{23.5-23}$$

This equation may now be rearranged and integrated from $z = 0$ to $z = Z$:

$$Z = -\frac{W_G}{S(k_y^0 a)} \int_{Y_{A1}}^{Y_{A2}} \frac{dY_A}{Y_A - Y_{A0}} \tag{23.5-24}$$

Equation 23.5-24 is the desired expression for the column height required to effect the specified separation. In writing Eq. 23.5-24 we have neglected the variation of

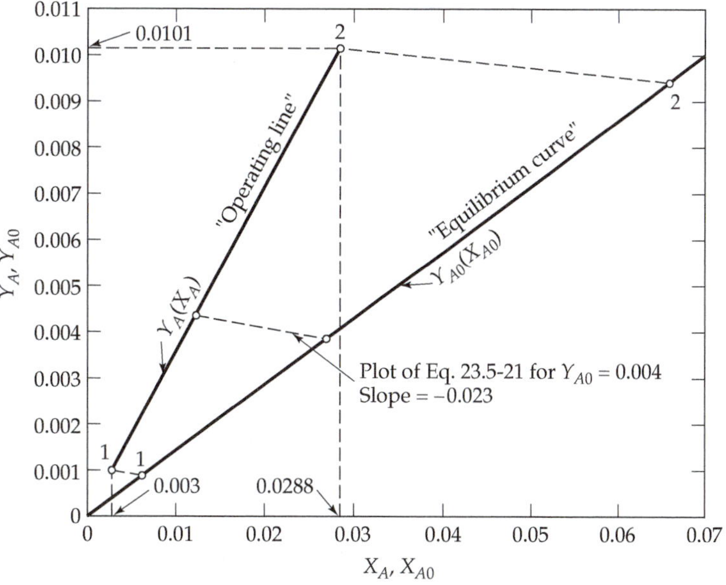

Fig. 23.5-4. Calculation of interfacial conditions in the absorption of cyclohexane from air in a packed column (see Problem 23A.2).

the mass-transfer coefficient k_y^0 with composition. This is usually permissible only for dilute solutions.

In general, Eq. 23.5-24 must be integrated by numerical or graphical procedures. However, for dilute solutions, it may frequently be assumed that the operating and equilibrium curves of Fig. 23.5-4 are straight lines. If, in addition, the ratio k_x^0/k_y^0 is constant, then $Y_A - Y_{A0}$ varies linearly with Y_A. We may then integrate Eq. 23.5-24 to obtain (see Problem 23B.1)

$$Z = \frac{W_G}{S(k_y^0 a)} \frac{Y_{A2} - Y_{A1}}{(Y_{A0} - Y_A)_{\ln}} \tag{23.5-25}$$

where

$$(Y_{A0} - Y_A)_{\ln} = \frac{(Y_{A0} - Y_A)_2 - (Y_{A0} - Y_A)_1}{\ln[(Y_{A0} - Y_A)_2/(Y_{A0} - Y_A)_1]} \tag{23.5-26}$$

Equation 23.5-25 can be rearranged to give

$$W_{Ag,0} = W_G(Y_{A2} - Y_{A1}) = (k_y^0 a)ZS(Y_{A0} - Y_A)_{\ln} \tag{23.5-27}$$

Comparison of Eq. 23.5-27 and Eq. 15.4-15 shows the close analogy between packed towers and simple heat exchangers. Expressions analogous to Eq. 23.5-24, but containing the overall mass-transfer coefficient K_y^0 may also be derived (see Problem 23B.1). Again, we may use the final results, Eqs. 23.5-25 or 23.5-27, for either cocurrent or countercurrent flow. Keep in mind, however, that the simplified model used to describe the packed tower is not so reliable as the corresponding one used for heat exchangers.

EXAMPLE 23.5-3

Linear Cascades

We saw in Example 23.1-2 that the degree of separation possible in a simple binary splitter can be quite limited, and it is therefore often desirable to combine individual splitters in a countercurrent *cascade* such as that shown in Fig. 23.5-5. Here the feed to any splitter stage is the sum of the waste stream from the splitter immediately above it and the product from the splitter immediately below.

Show how such an arrangement can increase the degree of separation relative to that obtained in a single splitter.

SOLUTION

For the system as a whole we can write a mass balance for the desired product and for the solution as a whole. That is, we treat the entire system as a splitter and write

$$z_F F = y_P P + x_W W; \qquad F = P + W \tag{23.5-28,29}$$

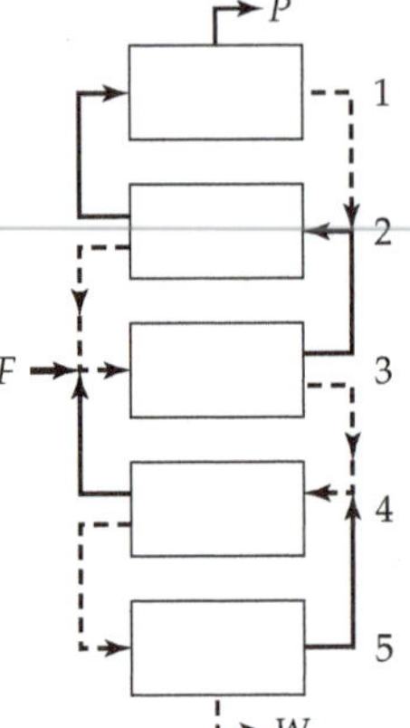

Fig. 23.5-5. A linear cascade. Upward flows are shown by solid lines, and downward flows by dashed lines.

It will be assumed here that all of quantities in these equations are given, so that the problem is specified insofar as the overall mass balances are concerned. It remains for us to determine the number of stages required to meet these conditions.

We begin by writing a set of mass balances over the top portion of the column, here the top two stages for illustrative purposes (see Fig. 23.5-5):

$$y_3 U_3 - x_2 D_2 = y_P P; \qquad U_3 - D_2 = P \tag{23.5-30,31}$$

Here U_n and D_n are the upflowing and downflowing streams from stage n, and the y_n and x_n are the corresponding mole fractions of the desired solute. When P is eliminated between Eqs. 23.5-30 and 23.5-31, we get

$$\frac{y_3 - y_P}{x_2 - y_P} = \frac{D_2}{U_3} \tag{23.5-32}$$

This equation gives the relation between the compositions of the downflowing and upflowing streams passing each other at any column cross section above the feed stage, in terms of the corresponding flow rates. This relation, when shown on an x-y plot (which is called a *McCabe-Thiele diagram*[3,4]) is known as the *operating line* for the system. We concentrate for the moment on compositions and return later to the problem of determining stream rate ratios.

The phase compositions in each stage are assumed to satisfy an equilibrium relation such as (see Eq. 23.1-19)

$$y_n = \frac{\alpha x_n}{1 + (\alpha - 1)x_n} \tag{23.5-33}$$

or, more generally, $y_n = f(x_n)$, where $f(x)$ is taken to be a known function.

Equations 23.5-32 and 23.5-33 (or its generalization) now permit determination of all compositions in the portion of the column above the feed point, usually known as the *rectifying section*, and similar calculations can be made for the *stripping section*, the portion below the feed point. We may then determine the number of stages required for the separation under consideration and the proper location of the feed stage.

First, however, we need to determine the stream rate ratios required in Eq. 23.5-32, and we consider three special cases here:

(a) *Total reflux.* This special mode of operation, in which P and W are zero, is important, as it provides the smallest possible number of stages that can yield the desired output concentrations. Here

$$U_n = D_{n-1} \tag{23.5-34}$$

for all n, and the operating line is given by

$$y_n = x_{n-1} \tag{23.5-35}$$

This simple relation holds for all cascades at total reflux. The stage compositions are plotted in Fig. 23.5-6 (for a product mole fraction of 0.9 and a waste mole fraction of 0.1), along with an equilibrium curve of the form of Eq. 23.5-33 with $\alpha = 2.5$.

The steplike lines between the equilibrium and operating lines in this figure suggest a graphical method of determining stage compositions: each "step" between the equilibrium and operating lines represents an increment one-component splitter or stage. The diagram thus suggests that five stages are required for this rather simple separation. However, for the situation of total reflux and constant relative volatility α, it is simplest to recognize that

$$Y_n = Y_{n-1}/\alpha \tag{23.5-36(a)}$$

so that

$$Y_N = Y_1/\alpha^{N-1} \tag{23.5-36(b)}$$

For the situation pictured in Fig. 23.5-6, we have then

$$\log\left(\frac{0.9/0.1}{0.1/0.9}\right) = (N-1)\log 2.5 \tag{23.5-37}$$

[4]W. L. McCabe and E. W. Thiele, *Ind. Eng. Chem.*, **17**, 605–611 (1925).

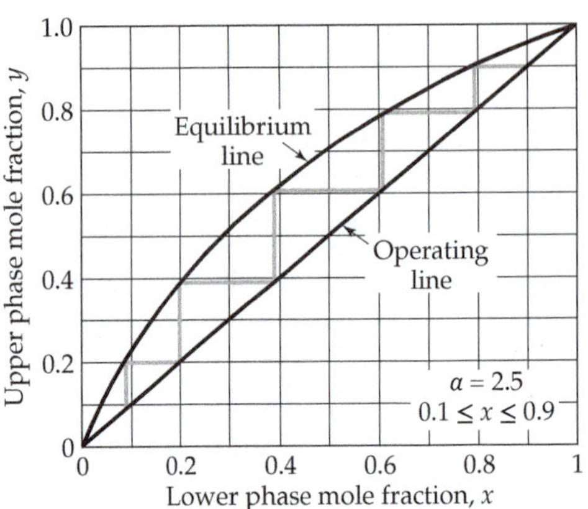

Fig. 23.5-6. McCabe-Thiele diagram for total reflux, with $\alpha = 2.5$ and $0.1 \leq x \leq 0.9$.

or

$$N = 1 + \frac{\log 81}{\log 2.5} = 5.796 \tag{23.5-38}$$

which is more accurate, but similar to the graphical estimate.

If products are to be withdrawn, it is necessary to calculate the stream-rate ratios, and the means for doing so vary with the specific operation under consideration.

(b) ***Thermodynamic constraints: adiabatic cascades and minimum reflux.*** For most of the common stagewise operations, stream ratios are determined by thermodynamic constraints, and these are thoroughly discussed in a wide variety of unit operations texts. We need not repeat this readily accessible information here, but we briefly consider distillation, the most widely used of all, by way of example. In principle, stream ratios in distillation are determined by assuming adiabatic columns and a set of energy balances.

However, it is very often permissible to assume equal molar heats of vaporization for the various species and to neglect "sensible heats" (i.e., the $\tilde{C}_p\Delta T$ contributions to $\Delta\tilde{H}$). With these simplifications the stream rates U_n and D_n are constants. We may then write for any position above the feed plate

$$U = D + P \quad \text{and} \quad y_{n+1}U = x_n D + y_P P \tag{23.5-39,40}$$

and below the feed plate

$$D = U + W \quad \text{and} \quad x_m D = y_{m+1}U + x_W W \tag{23.5-41,42}$$

Here the indices n and m refer, respectively, to the upper or rectifying section (above the feed point) and to the lower or stripping system of the column (below the feed point).

By way of example we consider the system in part (a) for a saturated liquid feed, equimolar in the two species involved, and operated at minimum reflux: the smallest amount of returning liquid from the top plate that can produce the desired separation. This situation will occur when the operating line touches the equilibrium curve, and in the system being considered, this "pinch" will occur first at the feed plate. The vapor composition on the feed plate is then given by

$$Y_F = 2.5X_F = 2.5 \quad \text{or} \quad y_F = \frac{Y_F}{1 + Y_F} = 0.7143 \tag{23.5-43,44}$$

The operating line then has two branches, one above and one below the feed plate, as shown in Fig. 23.5-7.

Any real column must operate between the limits of total and minimum reflux, but normal operation is just a few percent above the minimum. This is because the cost of individual plates tends to be much lower than the costs associated with increasing the reflux (the liquid returned to the column by condensation of vapor from the top plate): increasing the steam load required to return vapor from the liquid leaving the bottom plate, the condenser load to return

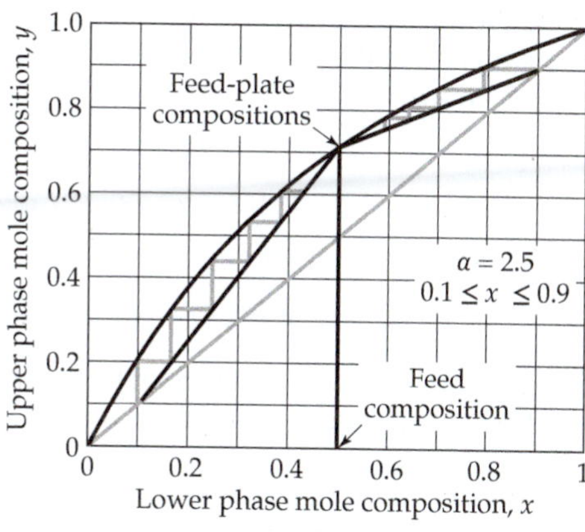

Fig. 23.5-7. McCabe-Thiele diagram for minimum reflux, with $\alpha = 2.5$ and $0.1 \leq x \leq 0.9$.

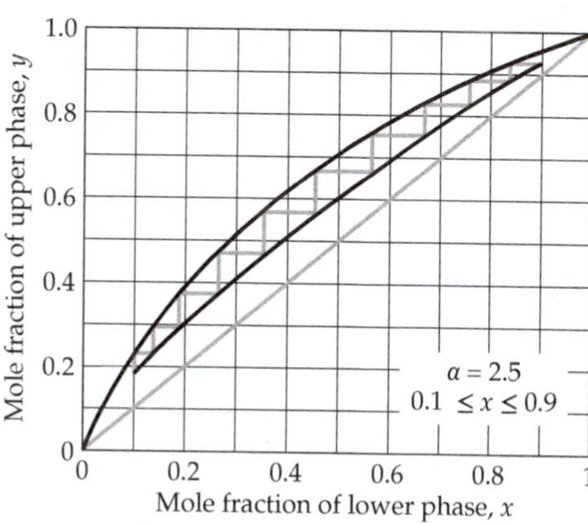

Fig. 23.5-8. McCabe-Thiele diagram for an ideal cascade, with $\alpha = 2.5$ and $0.1 \leq x \leq 0.9$.

the overhead vapor, and the capital costs of larger column diameter, larger reboiler, to return vapor at the bottom, and condenser, to return liquid at the top.

(c) ***Transport constraints and ideal cascades.*** For separation via selectively permeable membranes, the ratio of the product to waste streams is governed by the pressure exerted across the membrane, and the energy required to produce this pressure must be renewed for every stage of the cascade. This gives the designer an extra degree of freedom and has led to a wide variety of cascade configurations. First developed for isotopes,[5] membrane cascades have now been developed for industrial gas separations[6] and many other applications.

We consider here by example *ideal* cascades, which are those in which only streams of identical composition are mixed. In the terms of this example, that means

$$Y_{n+1} = X_{n-1} = \frac{Y_{n-1}}{\alpha} \tag{23.5-45}$$

and, by extension,

$$Y_{n+1} = \frac{Y_n}{\sqrt{\alpha}} \tag{23.5-46}$$

It follows that just twice as many stages are needed as at total reflux, and that the operating line lies half way between the "equilibrium" curve and the 45° line. As shown in Fig. 23.5-8, it has a continuous derivative across the feed stage.

---

[5]E. Von Halle and J. Schacter, *Diffusion Separation Methods*, in *Kirk-Othmer Encyclopedia of Chemical Technology*, Volume **8**, Wiley, New York (1993), pp. 149–203.

[6]R. Agrawal, *Ind. Eng. Chem. Research*, **35**, 3607–3617 (1996); R. Agrawal and J. Xu, *AIChE Journal*, **42**, 2141–2154 (1996).

Ideal cascades provide the smallest possible total stage stream flows, but the flows now vary with position: they are highest at the feed stage and decrease toward the ends of the cascade. For this reason they are known as *tapered cascades* (see Problem 23B.6).

EXAMPLE 23.5-4

Expansion of a Reactive Gas Mixture through a Frictionless Adiabatic Nozzle

An equimolar mixture of CO_2 and H_2 is confined at 1000K and 1.50 atm in the large insulated pressure tank shown in Fig. 15.5-6. Under these conditions the reaction

$$CO_2 + H_2 \Leftrightarrow CO + H_2O \tag{23.5-47}$$

may take place. After being stored in the tank long enough for the reaction to proceed to equilibrium, the gas is allowed to escape through the small converging nozzle shown to the ambient pressure of 1 atm.

Estimate the temperature and velocity of the escaping gas through the nozzle throat (**a**) assuming that no appreciable reaction takes place during passage of gas through the nozzle, and (**b**) assuming instant attainment of thermodynamic equilibrium at all points in the nozzle. In each case, assume that the expansion is adiabatic and frictionless.

SOLUTION

We begin by assuming quasi-steady-state operation, flat velocity profiles, and negligible changes in potential energy. We also assume constant heat capacities and ideal-gas behavior, and we neglect the diffusion in the direction of flow. We may then write the macroscopic energy balance, Eq. 23.3-1, in the form

$$\tfrac{1}{2}v_2^2 = \hat{H}_1 - \hat{H}_2 \tag{23.5-48}$$

Here the subscripts 1 and 2 refer to conditions in the tank and at the nozzle throat, respectively, and, as in Example 15.5-4, the fluid velocity in the tank is assumed to be zero.

To determine the enthalpy change, we equate $d(\frac{1}{2}v^2)$ from the d-form of the steady-state energy balance (Eq. 23.3-1) to $d(\frac{1}{2}v^2)$ of the d-form of the steady-state mechanical energy balance (Eq. 23.4-1) to get

$$\frac{1}{\rho}dp = d\hat{H} \tag{23.5-49}$$

In addition to Eq. 23.5-49, we use the ideal-gas law and an expression for $\hat{H}(T)$, obtained with the help of Eq. B of Table 17.1-2 generalized to multicomponent mixtures (i.e., $M = \sum_{\alpha=1}^{N} x_\alpha M_\alpha$), Eq. 19.3-16, and the relation $\rho\hat{H} = c\tilde{H}$, to get

$$p = cRT = \frac{\rho RT}{M} = \frac{\rho RT}{\sum_{\alpha=1}^{N} x_\alpha M_\alpha} \tag{23.5-50}$$

$$\hat{H} = \frac{\tilde{H}}{\rho/c} = \frac{\sum_{\alpha=1}^{N} x_\alpha \overline{H}_\alpha}{M} = \frac{\sum_{\alpha=1}^{N} x_\alpha[\tilde{H}_\alpha^{\circ} + \tilde{C}_{p\alpha}(T - T^{\circ})]}{\sum_{\alpha=1}^{N} x_\alpha M_\alpha} \tag{23.5-51}$$

Here x_α is the mole fraction of the species α at temperature T, and $\tilde{H}_\alpha^{\circ}$ is the molar enthalpy of species α at the reference temperature T°. The evaluation of $\hat{H}$ is discussed separately for the two approximations.

Approximation (a): Assumption of very slow chemical reaction. Here the x_α are constant at the equilibrium values for 1000K, and we may write Eq. 23.5-51 as

$$d\hat{H} = \left(\frac{\Sigma_\alpha x_\alpha \tilde{C}_{p\alpha}}{\Sigma_\alpha x_\alpha M_\alpha}\right) dT \tag{23.5-52}$$

Hence, we may write Eq. 23.5-49 in the form

$$d \ln p = \left(\frac{\Sigma_\alpha x_\alpha \tilde{C}_{p\alpha}}{R} \right) d \ln T \tag{23.5-53}$$

Since x_α and $\tilde{C}_{p\alpha}$ are assumed constant, this equation may be integrated from (p_1,T_1) to (p_2,T_2) to get

$$T_2 = T_1 \left(\frac{p_2}{p_1} \right)^{R/\Sigma_\alpha x_\alpha \tilde{C}_{p\alpha}} \tag{23.5-54}$$

We may now combine this expression with Eqs. 23.5-48 and 23.5-51 to obtained the desired expression for the gas velocity at plane 2:

$$v_2 = \left\{ 2T_1 \left[1 - \left(\frac{p_2}{p_1} \right)^{R/\Sigma_\alpha x_\alpha \tilde{C}_{p\alpha}} \right] \frac{\Sigma_\alpha x_\alpha \tilde{C}_{p\alpha}}{\Sigma_\alpha x_\alpha M_\alpha} \right\}^{1/2} \tag{23.5-55}$$

By substituting numerical values into Eqs. 23.5-54 and 23.5-55, we obtain (see Problem 23A.1) $T_2 = 919.1\text{K}$ and $v_2 = 1739$ ft/s. It may be seen that this treatment is very similar to that presented in Example 15.5-4. It is also subject to the restriction that the throat velocity must be subsonic; that is, the pressure in the nozzle throat cannot fall below that fraction of p_1 required to produce sonic velocity at the throat (see Eq. 15B.6-2). If the ambient pressure falls below this critical value of p_2, the throat pressure will remain at the critical value, and there will be a shock wave beyond the nozzle exit.

Approximation (b): Assumption of very rapid reaction. We may proceed here as in part (a), except that the mole fractions x_α must now be considered functions of the temperature defined by the equilibrium relation

$$\frac{(x_{H_2O})(x_{CO})}{(x_{H_2})(x_{CO_2})} = K_x(T) \tag{23.5-56}$$

and the stoichiometric relations

$$x_{H_2O} = x_{CO}; \quad x_{H_2} = x_{CO_2}; \quad \sum_{\alpha=1}^{4} x_\alpha = 1 \tag{23.5-57,58,59}$$

The quantity $K_x(T)$ in Eq. 23.5-56 is the known equilibrium constant for the reaction. It may be considered as a function only of temperature, because of the assumed ideal-gas behavior and because the number of moles present is not affected by the chemical reaction. Equations 23.5-57 and 23.5-58 follow from the stoichiometry of the reaction and the composition of the gas originally placed in the tank.

The expression for the final temperature is now considerably more complicated. For this reaction, where $\Sigma_\alpha x_\alpha M_\alpha$ is constant, Eqs. 23.5-49 and 23.5-50 may be combined to give

$$R \ln \frac{p_2}{p_1} = \int_{T_1}^{T_2} \left(\frac{d\tilde{H}}{dT} \right) d \ln T \tag{23.5-60}$$

where, with the heat capacities approximated as constants,

$$\frac{d\tilde{H}}{dT} = \sum_{\alpha=1}^{4} [\tilde{H}_\alpha^\circ + \tilde{C}_{p\alpha}(T - T^\circ)] \frac{dx_\alpha}{dT} + \sum_{\alpha=1}^{4} x_\alpha \tilde{C}_{p\alpha} \tag{23.5-61}$$

In general, the integral in Eq. 23.5-60 must be evaluated numerically, since the x_α and the dx_α/dT are all complicated functions of temperature governed by Eqs. 23.5-56 to 23.5-59. Once T_2 has been determined from Eq. 23.5-60, however, v_2 may be obtained by use of Eqs. 23.5-48 and 23.5-51. By substituting numerical values into these expressions, we obtain (see Problem 23B.2) $T_2 = 937$ K and $v_2 = 1748$ ft/s.

We find, then, that both the exit temperature and the velocity from the nozzle are greater when chemical equilibrium is maintained throughout the expansion. The reason for this is that the equilibrium shifts with decreasing temperature in such a way as to release heat of reaction to the system. Such a release of energy will occur with decreasing temperature in any system at chemical equilibrium, regardless of the reactions involved. This is one consequence of the famous rule of Le Châtelier. In this case, the reaction is endothermic as written and the equilibrium constant decreases with falling temperature. As a result, CO and H_2O are partially reconverted to H_2 and CO_2 on expansion, with a corresponding release of energy.

It is interesting that in rocket engines the exhaust velocity, hence the engine thrust, are also increased if rapid equilibration can be obtained, even though the combustion reactions are strongly exothermic. The reason for this is that the equilibrium constants for these reactions increase with falling temperature so that the heat of reaction is again released on expansion. This principle has been suggested as a method for improving the thrust of rocket engines. The increase in thrust potentially obtainable in this way is quite large.

This example has been chosen for its simplicity. Note in particular that if a change in the number of moles accompanies the chemical reaction, then the equilibrium constant, and hence the enthalpy, are functions of the pressure. In this case, which is quite common, the variables p and T implicit in Eq. 23.5-60 cannot be separated, and a numerical integration of this equation is required. Such integrations have been performed, for example, for the prediction of the behavior of supersonic wind tunnels and rocket engines, but the calculations involved are too lengthy for presentation here.

§23.6 USE OF THE MACROSCOPIC BALANCES TO SOLVE UNSTEADY-STATE PROBLEMS

In §23.5 the discussion was restricted to steady state. Here we move on to the transient behavior of multicomponent systems. Such behavior is important in a large number of practical operations, such as leaching and drying of solids, chromatographic separations, and chemical reactor operations. In many of these processes heats of reaction as well as mass transfer must be considered. A complete discussion of these topics is outside the scope of this text, and we restrict ourselves to two simple examples. More extensive discussions may be found elsewhere.[1]

EXAMPLE 23.6-1

Start-up of a Chemical Reactor

It is desired to produce a substance B from a raw material A in a chemical reactor of volume V equipped with a stirrer that is capable of keeping the entire contents of the reactor fairly homogeneous. The formation of B is reversible, and the forward and reverse reactions may be considered first order, with reaction-rate constants k_{1B}''' and k_{1A}''', respectively. In addition, B undergoes as irreversible first-order decomposition, with a reaction-rate constant k_{1C}''', to a third component C. The chemical reactions of interest may be represented as

$$A \rightleftarrows B \rightarrow C \tag{23.6-1}$$

At zero time, a solution of A at a concentration c_{A0} is introduced to the initially empty reactor at a constant mass flow rate w.

Develop an expression for the amount of B in the reactor, when it is just filled to its capacity V, assuming that there is no B in the feed solution and neglecting changes of fluid properties.

[1] W. R. Marshall, Jr., and R. L. Pigford, *The Application of Differential Equations to Chemical Engineering Problems*, University of Delaware Press, Newark, DE (1947); B. A. Ogunnaike and W. H. Ray, *Process Dynamics, Modeling, and Control*, Oxford University Press (1994).

SOLUTION

We begin by writing the unsteady-state macroscopic mass balances for species A and B. In molar units these may be expressed as:

$$\frac{dM_{A,\text{tot}}}{dt} = \frac{wc_{A0}}{\rho} - k'''_{1B}M_{A,\text{tot}} + k'''_{1A}M_{B,\text{tot}} \tag{23.6-2}$$

$$\frac{dM_{B,\text{tot}}}{dt} = -(k'''_{1A} + k'''_{1C})M_{B,\text{tot}} + k'''_{1B}M_{A,\text{tot}} \tag{23.6-3}$$

Next we eliminate $M_{A,\text{tot}}$ from Eq. 23.6-3. First we differentiate this equation with respect to t to get

$$\frac{d^2M_{B,\text{tot}}}{dt^2} = -(k'''_{1A} + k'''_{1C})\frac{dM_{B,\text{tot}}}{dt} + k'''_{1B}\frac{dM_{A,\text{tot}}}{dt} \tag{23.6-4}$$

In this equation, we replace $dM_{A,\text{tot}}/dt$ by the right side of Eq. 23.6-2, and then use Eq. 23.6-3 to eliminate $M_{A,\text{tot}}$. In this way we obtain a linear second-order differential equation for $M_{B,\text{tot}}$ as a function of time:

$$\frac{d^2M_{B,\text{tot}}}{dt^2} + (k'''_{1A} + k'''_{1B} + k'''_{1C})\frac{dM_{B,\text{tot}}}{dt} + k'''_{1B}k'''_{1C}M_{B,\text{tot}} = \frac{k'''_{1B}wc_{A0}}{\rho} \tag{23.6-5}$$

This equation is to be solved with the initial conditions

I.C. 1: at $t = 0$, $M_{B,\text{tot}} = 0$ (23.6-6)

I.C. 2: at $t = 0$, $\dfrac{dM_{B,\text{tot}}}{dt} = 0$ (23.6-7)

This equation can be integrated to give (see Eq. C.1-7a)

$$M_{B,\text{tot}}(t) = \frac{wc_{A0}}{\rho k'''_{1C}}\left(\frac{s_-}{s_+ - s_-}\exp(s_+t) - \frac{s_+}{s_+ - s_-}\exp(s_-t) + 1\right) \tag{23.6-8}$$

where

$$2s_\pm = -(k'''_{1A} + k'''_{1B} + k'''_{1C}) \pm \sqrt{(k'''_{1A} + k'''_{1B} + k'''_{1C})^2 - 4k'''_{1B}k'''_{1C}} \tag{23.6-9}$$

Equations 23.6-8 and 23.6-9 give the total mass of B in the reactor as a function of time, up to the time at which the reactor is completely filled. These expressions are very similar to the equations obtained for the temperature controller in Example 15.5-2. It can be shown, however, that s_+ and s_- are both real and negative, and therefore, $M_{B,\text{tot}}$ cannot oscillate (see Problem 23B.3).

EXAMPLE 23.6-2

Unsteady Operation of a Packed Column

There are many industrially important processes in which mass transfer takes place between a fluid and a granular porous solid: for example, recovery of organic vapors by adsorption on charcoal, extraction of caffeine from coffee beans, and separation of aromatic and aliphatic hydrocarbons by selective adsorption on silica gel. Ordinarily, the solid is held fixed, as indicated in Fig. 23.6-1, and the fluid is allowed to percolate through it. The operation is thus inherently unsteady, and the solid must be periodically replaced or "regenerated," that is, returned to its original condition by heating or other treatment. To illustrate the behavior of such fixed-bed mass-transfer operations, we consider as a physically simple case, the removal of a solute from a solution by passage through an adsorbent bed.

In this operation, a solution containing a single solute A at mole fraction x_{A1} in a solvent B is passed at a constant volumetric flow rate w/ρ through a packed tower. The tower packing consists of a granular solid capable of adsorbing A from the solution. At the start of the percolation, the interstices of the bed are filled with pure liquid B, and the solid is free of A. The percolating fluid displaces this solvent evenly so that the solution concentration of A is always uniform over any cross section. For simplicity, it is assumed that the equilibrium concentration of A adsorbed on the solid is proportional to the local concentration of A in the solution. It is

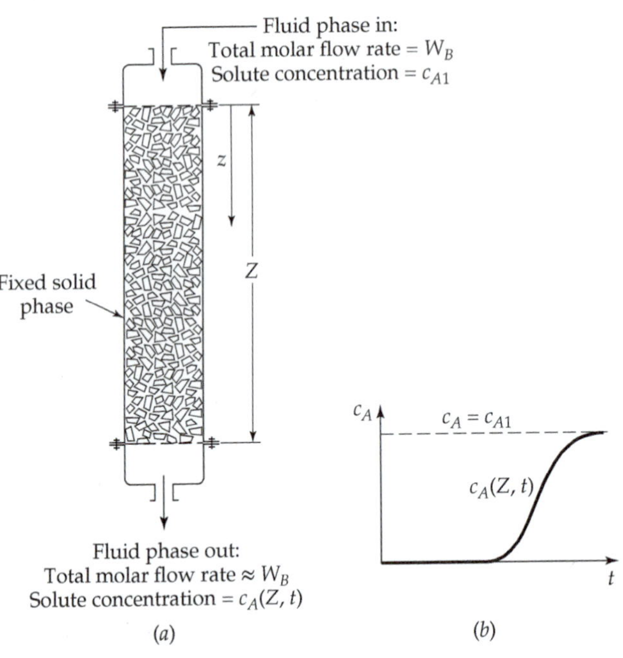

Fig. 23.6-1. A fixed-bed absorber: (*a*) pictorial representation of equipment; (*b*) a typical effluent curve.

also assumed that the concentration of A in the percolating solution is always small and that the resistance of the porous solid to intraparticle mass transfer is negligible.

Develop an expression for the concentration of A in the column as a function of time and distance down the column.

SOLUTION

Paralleling the treatment of the gas absorber in Example 23.5-2, we think of the two phases as being continuous and existing side by side as pictured in Fig. 23.6-2. We again define the contact area per unit packed volume of column as a. Now, however, one of the phases is stationary, and unsteady-state conditions prevail. Because of this locally unsteady behavior, the macroscopic mass balances are applied locally, over a small increment of the column of height Δz. We may

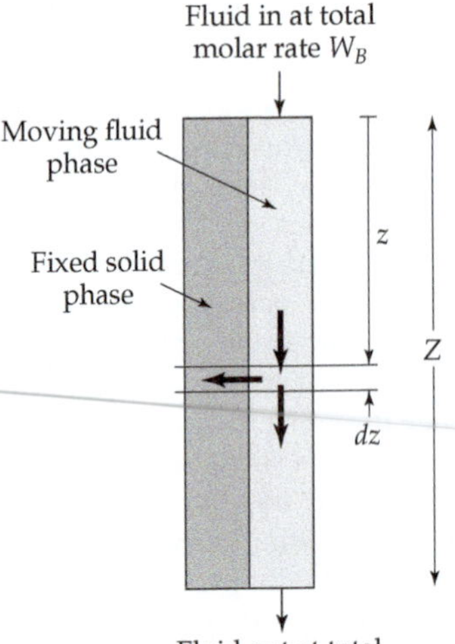

Fig. 23.6-2. Schematic model for a fixed-bed absorber, showing a differential element over which a mass balance is made.

use Eq. 23.1-3 and the assumption of dilute solutions to state that the molar rate of flow of solvent, W_B, is essentially constant over the length of the column and the time of operation. We now proceed to use Eq. 23.1-3 to write the conservation of mass relations for species A in each phase of a section the column of height Δz.

For the *solid phase* in this increment of column we may apply Eq. 23.1-3 locally, keeping in mind that now $M_{A,\text{tot}}$ depends on both z and t:

$$\frac{\partial M_{A,\text{tot}}}{\partial t} = W_{A0} \tag{23.6-10}$$

or

$$(1-\varepsilon)S\Delta z \frac{\partial c_{As}}{\partial t} = (k_x^0 a)(x_A - x_{A0})S\Delta z \tag{23.6-11}$$

where use has been made of Eq. 22.1-14, and the symbols have the following meanings:

ε = volume fraction of column occupied by the liquid
S = cross-sectional area of (empty) column
c_{As} = moles of adsorbed A per unit volume of the solid phase
x_A = bulk mole fraction of A in the fluid phase
x_{A0} = interfacial mole fraction of A in the fluid phase, assumed to be in equilibrium with c_{As}
k_x^0 = fluid-phase mass-transfer coefficient, defined in Eq. 22.1-14, for small mass-transfer rates

Note that, in writing Eq. 23.6-11, we have neglected convective mass transfer through the solid-fluid interface. This is reasonable if x_{A0} is much smaller than unity. We have also assumed that the particles are small enough so that the concentration of the solution surrounding any given particle is essentially constant over the particle surface.

For the *fluid phase*, in the column increment under consideration, Eq. 23.1-3 becomes

$$\frac{\partial M_{A,\text{tot}}}{\partial t} = -\Delta W_A + W_{A0} \tag{23.6-12}$$

or

$$\varepsilon c S\Delta z \frac{\partial x_A}{\partial t} = -W_B \Delta z \frac{\partial x_A}{\partial z} - (k_x^0 a)(x_A - x_{A0})S\Delta z \tag{23.6-13}$$

Here c is the total molar concentration of the liquid. Equation 23.6-13 may be rewritten by the introduction of a modified time variable, defined by

$$t' = t - \left(\frac{\varepsilon c S}{W_B}\right) z \tag{23.6-14}$$

It may be seen that, for any position in the column, t' is the time measured from the instant that the percolating solvent "front" has reached the position in question. By rewriting Eqs. 23.6-13 and 23.6-11 in terms of t', we get

$$\left(\frac{\partial x_A}{\partial z}\right)_{t'} = -\frac{(k_x^0 a)S}{W_B}(x_A - x_{A0}) \tag{23.6-15}$$

$$\left(\frac{\partial c_{As}}{\partial t'}\right)_z = \frac{(k_x^0 a)}{(1-\varepsilon)}(x_A - x_{A0}) \tag{23.6-16}$$

Equations 23.6-15 and 23.6-16 combine the equations of conservation of mass for each phase with the assumed mass-transfer rate expression. These two equations are to be solved simultaneously along with the interphase equilibrium distribution, $x_{A0} = mc_{As}$, in which m is a constant. The boundary conditions are

B.C. 1:	at $t' = 0$,	$c_{As} = 0$	for all $z > 0$	(23.6-17)
B.C. 2:	at $z = 0$,	$x_A = x_{A1}$	for all $t' > 0$	(23.6-18)

Before solving these equations, it is convenient to rewrite them in terms of the following dimensionless variables:

$$X(\zeta,\tau) = \frac{x_A}{x_{A1}} \qquad Y(\zeta,\tau) = \frac{mc_{As}}{x_{A1}} \qquad \zeta = \frac{(k_x^0 a)S}{W_B}z \qquad \tau = \frac{(k_x^0 a)m}{(1-\varepsilon)}t' \qquad (23.6\text{-}19,20,21,22)$$

In terms of these variables, the differential equations and boundary conditions take the form

$$\frac{\partial X}{\partial \zeta} = -(X - Y); \qquad \frac{\partial Y}{\partial \tau} = +(X - Y) \qquad (23.6\text{-}23,24)$$

with the boundary conditions $Y(\zeta,0) = 0$ and $X(0,\tau) = 1$.

The solution[2] to Eqs. 23.6-23 and 23.6-24 for these boundary conditions is

$$X(\zeta,\tau) = 1 - \int_0^{\zeta} e^{-(\tau+\bar{\zeta})} J_0(i\sqrt{4\tau\bar{\zeta}})d\bar{\zeta} \qquad (23.6\text{-}25)$$

Here $J_0(ix)$ is a zero-order Bessel function of the first kind. The Taylor series of $J_0(ix)$ is $1 + \frac{1}{4}x^2 + \frac{1}{64}x^4 + \cdots$, and hence, $J_0(ix)$ is a real quantity. This solution is presented graphically in several available references.[3]

§23.7 CONCLUDING COMMENTS

In this the third chapter on the macroscopic balances, there are really no new concepts introduced. The balances given in Chapter 15 are amended to take into account the presence of more than one chemical species. They are easily restricted to binary systems by letting the sums on α go from A to B.

The applications of these macroscopic balances lead right into topics treated in chemical engineering under such labels as "material and energy balances," "process control," and "unit operations."

QUESTIONS FOR DISCUSSION

1. How are the macroscopic balances for multicomponent mixtures derived? How are they related to the equations of change?
2. In Eq. 23.1-1, how are homogeneous and heterogeneous reactions accounted for? What is the physical meaning of $w_{\alpha,0}$?
3. Give a specific example of a system in which the last term in Eq. 23.1-4 is zero.
4. In using Table 23.5-1 one normally specifies the directions of the streams (that is, whether they are input or output streams). How could one proceed if the flow directions change with time?
5. Summarize the calculation procedures for the enthalpy per unit mass, $\hat{H} = \hat{U} + p\hat{V}$, in Eq. 23.3-1, and the partial molar enthalpy in Eq. 23.3-1a. What are these quantities for ideal-gas mixtures?
6. Review the derivation of the mechanical energy balance in §7.7. What would have to be changed in that derivation, if one wishes to apply it to a nonisothermal, reacting mixture in a flow system with no mass–transfer surfaces?
7. To what extent does this chapter provide the background for studying the unit operations, such as absorption, extraction, distillation, and crystallization?
8. What changes would have to be made in this chapter to describe processes in a space ship or on the surface of the moon?

---

[2]This result was first obtained by A. Anzelius, *Z. angew. Math. u. Mech.*, **6**, 291–294 (1926), for the analogous problem in heat transfer. See also H. Bateman, *Partial Differential Equations of Mathematical Physics*, Dover, New York (1944), pp. 123–125.

[3]See, for example, O. A. Hougen and K. M. Watson, *Chemical Process Principles*, Part III, Wiley, New York (1947), p. 1086. Their y/y_0, $b\tau$, and aZ correspond to our X, τ, and ζ.

PROBLEMS

23A.1 Expansion of a gas mixture: very slow reaction rate. Estimate the temperature and velocity of the water-gas mixture at the discharge end of the nozzle in Example 23.5-4 if the reaction rate is very slow. Use the following data: $\log_{10} K_x = -0.15$, $\tilde{C}_{p,H_2} = 7.217$, $\tilde{C}_{p,CO_2} = 12.995$, $\tilde{C}_{p,H_2O} = 9.861$, $\tilde{C}_{p,CO} = 7.932$ (all heat capacities are in Btu/lb-mol · °F). Is the nozzle exit pressure equal to the ambient pressure?

Answers: 919.1 K, 1739 ft/s; yes, the nozzle flow is subsonic

23A.2 Height of a packed-tower absorber. A packed tower of the type described in Example 23.5-2 is to be used for removing 90% of the cyclohexane from a cyclohexane-air mixture by absorption into a nonvolatile light oil. The gas stream enters the bottom of the tower at a volumetric rate of 363 ft^3/min, at 30°C, and at 1.05 atm pressure. It contains 1% cyclohexane by volume. The oil enters the top of the tower at a rate of 20 lb-mol/hr, also at 30°C, and it contains 0.3% cyclohexane on a molar basis. The vapor pressure of cyclohexane at 30°C is 121 mm Hg, and solutions of it in the oil may be considered to follow Raoult's law.

(a) Construct the operating line for the column.

(b) Construct an equilibrium curve for the range of operation encountered here. Assume the operation to be isothermal and isobaric.

(c) Determine the interfacial conditions at each end of the column.

(d) Determine the required tower height using Eq. 23.5-24 if $k_x^0 a = 0.32$ lb-mol/hr · ft^3, $k_y^0 a = 14.2$ lb-mol/hr · ft^3, and the tower cross section S is 2.00 ft^2.

(e) Repeat part (d), using Eq. 23.5-25.

Answer: **(d)** ca. 62 ft; **(e)** 60 ft

23B.1 Effective average driving forces in a gas absorber. Consider a packed-tower gas absorber of the type discussed in Example 23.5-2. Assume that the solute concentration is always low and that the equilibrium and operating lines are both very nearly straight. Under these conditions, both $k_y^0 a$ and $k_x^0 a$ may be considered constant over the mass-transfer surface.

(a) Show that $(Y_A - Y_{Ae})$ varies linearly with Y_A. Note that Y_A is the bulk mole ratio of A in the gas phase and Y_{Ae} is the equilibrium gas-phase mole ratio over a liquid of bulk composition X_A (see Fig. 22.4.2).

(b) Repeat part (a) for $(Y_A - Y_{A0})$.

(c) Use the results of parts (a) and (b) to show that

$$W_{Ag,0} = (k_y^0 a) ZS (Y_{A0} - Y_A)_{\ln} \tag{23B.1-1}$$

$$W_{Ag,0} = (K_y^0 a) ZS (Y_{Ae} - Y_A)_{\ln} \tag{23B.1-2}$$

The overall mass-transfer coefficient K_y^0 is defined by Eq. 22.4-4. Note that this part of the problem may be solved by analogy with the development in Example 15.4-1.

23B.2 Expansion of a gas mixture: very fast reaction rate. Estimate the temperature and velocity of the water-gas mixture at the discharge end of the nozzle in Example 23.5-4 if the reaction rate may be considered infinitely fast. Use the data supplied in Problem 23A.1 as well as the following: at 900K, $\log_{10} K_x = -0.34$; $\tilde{H}_{H_2} = +6340$, $\tilde{H}_{H_2O}(g) = -49{,}378$, $\tilde{H}_{CO} = -16{,}636$, $\tilde{H}_{CO_2} = -83{,}242$ (all enthalpies are given in cal/g-mol). For simplicity, neglect the effect of temperature on heat capacity, and assume that $\log_{10} K_x$ varies linearly with temperature between 900K and 1000K. The following simplified procedure is recommended:

(a) It may be seen in advance that T_2 will be higher than for slow reaction rates, and hence greater than 920K (see Problem 23A.1). Show that, over the temperature range to be encountered, $\tilde{H}$ varies very nearly linearly with the temperature according to the expression $(d\tilde{H}/dT)_{\text{avg}} \approx 12.40$ cal/g-mol · K.

(b) Substitute the result in (a) into Eq. 23.5-60 to show that $T_2 \approx 937$K.

(c) Calculate $\tilde{H}_1$ and $\tilde{H}_2$, and show by use of Eq. 23.5-48 that $v_2 = 1750$ ft/s.

23B.3 Start-up of a chemical reactor.

(a) Solve Eq. 23.6-5 along with the given initial conditions to show that Eq. 23.6-8 correctly describes $M_{B,\text{tot}}$ as a function of time.

(b) Show that s_+ and s_- in Eq. 23.6-9 are area real and negative. *Hint*: show that

$$(k'''_{1A} + k'''_{1B} + k'''_{1C})^2 - 4k'''_{1B}k'''_{1C} = (k'''_{1A} - k'''_{1B} + k'''_{1C})^2 + 4k'''_{1A}k'''_{1B} \tag{23B.3-1}$$

(c) Obtain expressions for $M_{A,\text{tot}}$ and $M_{C,\text{tot}}$ as functions of time.

23B.4 Irreversible first-order reaction in a continuous reactor. A well-stirred reactor of volume V is initially completely filled with a solution of solute A in a solvent S at concentration c_{A0}. At time $t = 0$, an identical solution of A in S is introduced at a constant mass flow rate w. A small constant stream of dissolved catalyst is introduced at the same time, causing A to disappear according to an irreversible first-order reaction with rate constant k'''_1 s^{-1}. The rate constant may be assumed independent of composition and time. Show that the concentration of A in the reactor (assumed isothermal) at any time is

$$\frac{c_A(t)}{c_{A0}} = \left(1 - \frac{wt_0}{\rho V}\right)e^{-t/t_0} + \frac{wt_0}{\rho V} \tag{23B.4-1}$$

in which $t_0 = [(w/\rho V) + k'''_1]^{-1}$.

23B.5 Mass and energy balances in an adiabatic splitter. One hundred pounds of 40% by mass of superheated aqueous ammonia with a specific enthalpy of 420 Btu/lb$_m$ is to be flashed adiabatically to a pressure of 10 atm. Calculate the compositions and masses of the liquid and vapor produced. For the purposes of this problem you may assume that at thermodynamic equilibrium

$$\log_{10} Y_{NH_3} = 1.4 + 1.53 \log_{10} X_{NH_3} \tag{23B.5-1}$$

where Y_{NH_3} and X_{NH_3} are the *mass* ratios of ammonia to water. The enthalpies of saturated liquid and vapor at 10 atm may be assumed to be

$$\hat{H} = 1210 - 465y_{NH_3} - 115y_{NH_3}^{12} \tag{23B.5-2}$$

Btu/lb$_m$ of saturated vapor, and

$$\hat{h} = 330 - 950x_{NH_3} + 740x_{NH_3}^2 \tag{23B.5-3}$$

Btu/lb$_m$ of saturated liquid. Here x_{NH_3} and y_{NH_3} are *mass* fractions of ammonia.

Answer: $P = 36.4$ lb$_m$, $y_P = 0.750$, $\hat{H}_P = 858$ Btu/lb$_m$, $W = 63.6$ lb$_m$, $x_W = 0.20$, $\hat{h}_W = 140$ Btu/lb$_m$

23B.6 Flow distribution in an ideal cascade. Determine the upflowing and downflowing stream flows of individual stages for the ideal cascade described in Example 23.5-3. Express your results as fractions of the feed rate, and start from the bottom of the cascade ($n = 0$). Use 9 stages as the closest integer providing the desired separation. Begin by calculating the upflowing and downflowing stream compositions and then use the mass balances

$$D_{n+1} = U_n + W; \quad x_{n+1}D_{n+1} = y_nU_n + x_WW \tag{23B.6-1}$$

below the feed plate, and the corresponding balances above it. The bottom's composition (W) corresponds to mole fractions of $x_0 = x_W = 0.1$.

23C.1 Irreversible second-order reaction in an agitated tank. Consider a system similar to that discussed in Problem 23B.4, except that the solute disappears according to a second-order reaction; that is, $R_{A,\text{tot}} = -k'''_2 Vc_A^2$. Develop an expression for c_A as a function of time by the following method:

(a) Use a macroscopic mass balance on solute A over the tank to obtain a differential equation describing the evolution of c_A with time.

(b) Rewrite the differential equation and the accompanying initial condition in terms of the variable

$$u(t) = c_A(t) + \frac{w}{2\rho V k_2'''}\left(1 + \sqrt{1 + \frac{4\rho V k_2''' c_{A0}}{w}}\right) \tag{23C.1-1}$$

The nonlinear differential equation obtained in this way is a *Bernoulli differential equation*.

(c) Now put $v = 1/u$ and perform the integration. Then rewrite the result in terms of the original variable c_A.

23C.2 Protein purification. It is desired to purify a binary protein mixture using an ideal cascade of individual ultrafiltration units of the type shown in Fig. 23C.2. The larger of the two membrane units is the source of separation and each protein flux across the membrane is expressed by

$$N_i = c_i v S_i \tag{23C.2-1}$$

where N_i is the transmembrane protein flux of species i, c_i is its concentration in the upstream solution (assumed to be well mixed), v is the transmembrane superficial velocity, and S_i is a protein-specific *sieving factor*. The smaller membrane unit is solely to maintain a solvent balance and can be ignored for the purposes of this problem.

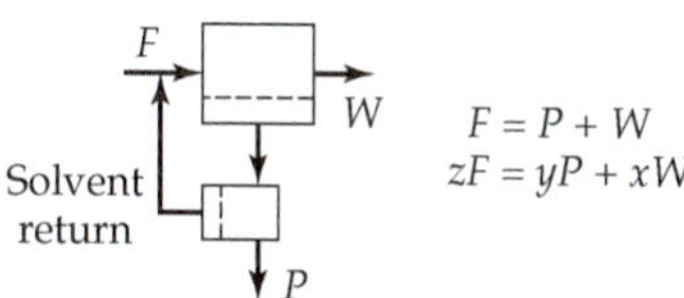

Fig. 23C.2 A membrane-based binary splitter.

(a) Show that the enrichment of protein 1 relative to protein 2 is given by

$$Y_1 = \alpha_{12} X_1 \tag{23C.2-2}$$

where Y_1 and X_1 are the mole ratios of protein 1 to protein 2 in the product and waste streams, respectively, and $\alpha_{12} = S_1/S_2$.

(b) Determine the number of stages required in an ideal cascade to produce 99% pure protein 1 from a 90% feed in 95% yield as a function of α_{12}. It is suggested that α_{12} be considered as having a range of 2 to 200.

(c) Calculate the output concentrations, yield, and stream flow rates for a three-state cascade, with $\alpha_{12} = 40$, and with a feed of 90% purity to the middle stage.

Chapter 24

Other Mechanisms for Mass Transport

In Part I, Chapters 1 to 7 were concerned with Newton's "law" of viscosity, but Chapter 8 dealt with "non-Newtonian fluids." In Part II, Chapters 9 to 15 described the consequences of Fourier's "law" of heat conduction, but Chapter 16 was about an entirely different mechanism, namely "radiation." In Part III, Chapters 17 to 23 concerned themselves with the movement of masses of chemical species according to Fick's "law" of diffusion. It was mentioned in Chapter 19 that mass could also be transported by a temperature gradient, by a pressure gradient, and by external forces acting on the various species. This chapter describes these mechanisms of mass transport. We conclude the chapter with a brief discussion of multicomponent systems.

§24.1 NONEQUILIBRIUM THERMODYNAMICS

The topics to be discussed here have been organized under the general heading of *nonequilibrium thermodynamics*. This subject is a generalization of classical thermodynamics, the latter being restricted to systems at equilibrium. It was Onsager[1] who first developed this subject, which has found many applications in physics, chemistry, and biology. Onsager's theory is, however, restricted to "linear laws," such as Newton's law of viscosity, Fourier's law of heat conduction, and Fick's (first) law of diffusion. If one wishes to consider nonlinear relationships (non-Newtonian fluids being a good example), then one has to use some kind of more general theory. At the moment, the general formalism of Öttinger is perhaps

[1]Nobel Laureate **Lars Onsager** (1903–1976) studied chemical engineering at the Technical University of Norway in Trondheim. After working with Peter Debye in Zürich for two years, he held teaching positions at several universities before moving on to Yale University. His famous work on the thermodynamics of irreversible processes, which led to the Nobel Prize, can be found in L. Onsager, *Phys. Rev.*, **37**, 405–426 (1931); **38**, 2265–2279 (1931).

the best approach.[2] In this chapter we consider only the Onsager theory, which has been discussed elsewhere for binary systems[3] and multicomponent systems.[4,5,6]

In the nonequilibrium thermodynamics approach, one starts by writing down the *equation of change for entropy*

$$\rho \frac{D\hat{S}}{Dt} = -(\nabla \cdot \mathbf{s}) + g_S \tag{24.1-1}$$

in which $\hat{S}$ is the entropy per unit mass of a multicomponent fluid, $\mathbf{s}$ is the entropy flux vector, and g_S is the rate of entropy production per unit volume. It can be shown that, for a binary system,[3]

$$\mathbf{s} = \frac{1}{T}\mathbf{q}^{(h)} + \left(\frac{\overline{S}_A}{M_A} - \frac{\overline{S}_B}{M_B}\right)\mathbf{j}_A \tag{24.1-2}$$

$$Tg_S = -(\mathbf{q}^{(h)} \cdot \nabla \ln T) - \frac{cRT}{\rho\omega_A\omega_B}(\mathbf{j}_A \cdot \mathbf{d}_A) - (\boldsymbol{\tau} : \nabla\mathbf{v}) - \left(\frac{\overline{G}_A}{M_A} - \frac{\overline{G}_B}{M_B}\right)r_A \tag{24.1-3}$$

In these equations $\mathbf{q}^{(h)}$ is the heat flux with the diffusional enthalpy flux (which involves no temperature gradient) subtracted off

$$\mathbf{q}^{(h)} = \mathbf{q} - \left(\frac{\overline{H}_A}{M_A} - \frac{\overline{H}_B}{M_B}\right)\mathbf{j}_A \tag{24.1-4}$$

The vector $\mathbf{d}_A$ is the generalized driving force for the diffusion of component A, which consists of an *activity gradient, a pressure gradient, and a difference between the external forces acting on the two species*

$$\mathbf{d}_A = x_A \nabla \ln a_A + \frac{1}{cRT}(\phi_A - \omega_A)\nabla p + \frac{\rho\omega_A\omega_B}{cRT}(\mathbf{g}_B - \mathbf{g}_A) \tag{24.1-5}$$

Here a_A is the activity of species A, $\phi_A = c_A\overline{V}_A$ is the volume fraction of species A, and $\mathbf{g}_A$ is the external force per unit mass acting on species A. In the operation $\nabla \ln a_A$, it is understood that the derivative is to be taken at constant T and p.

The entropy production term in Eq. 24.1-3 is made up of the products of fluxes and driving forces. The first two terms involve vector fluxes and vector forces. The following term involves a tensor flux and a tensor driving force, and the last term has a scalar driving force and a scalar flux. According to *Curie's law*, there will be coupling between the fluxes and forces of the same tensorial order, or between the fluxes and forces of orders that differ by two. That means that the first two fluxes will have the form

$$\mathbf{q}^{(h)} = -\alpha_{00}\nabla \ln T - \alpha_{0A}\mathbf{d}_A \tag{24.1-6}$$

$$\mathbf{j}_A = -\alpha_{A0}\nabla \ln T - \alpha_{AB}\mathbf{d}_A \tag{24.1-7}$$

According to *Onsager's reciprocal relations* the off-diagonal terms of the 2 x 2 α-matrix are related ($\alpha_{0A} = (cRT/\rho_A)\alpha_{A0}$).

---

[2]H. C. Öttinger, *Beyond Equilibrium Thermodynamics*, Wiley, New York (2005).

[3]L. Landau and E. M. Lifshitz, *Fluid Mechanics*, 2nd edition, Pergamon Press, Oxford (1987).

[4]J. O. Hirschfelder, C. F. Curtiss, and R. B. Bird, *Molecular Theory of Gases and Liquids*, second corrected printing, Wiley, New York (1964), Chapter 11; C. F. Curtiss and R. B. Bird, *Ind. Eng. Chem. Research*, **38**, 2515–2522 (1999), errata **40**, 1791 (2001); R. B. Bird, W. E. Stewart, and E. N. Lightfoot, *Transport Phenomena*, revised 2nd edition, Wiley, New York (2007), Chapter 24;

[5]S. Chapman and T. G. Cowling, *The Mathematical Theory of Non-Uniform Gases*, 3rd edition, Cambridge University Press (1970), p. 274.

[6]R. B. Bird and D. J. Klingenberg, *Advances in Water Resources*, **62**, 238–242 (2013); *errata*: in Eq. 3.15, $\omega_\alpha\omega_\beta$ on the right side should be in the denominator instead of the numerator; also, in the first sentence of Sec. 4, the reference should be to Eq. 3.15 rather than Eq. 3.13.

The expression for the mass flux becomes

$$\mathbf{j}_A = -D_A^T \nabla \ln T - \rho \frac{\omega_A \omega_B}{x_A x_B} Đ_{AB} \mathbf{d}_A \tag{24.1-8}$$

Here $Đ_{AB}$ is the *Maxwell-Stefan diffusivity* (different from the Fick diffusivity $\mathscr{D}_{AB}$) and D_A^T is the *thermal diffusion coefficient*. Then, after some manipulation, one can arrive at the following expressions for the heat and mass flux vectors for a binary system:

$$\mathbf{j}_A = -\rho \frac{\omega_A \omega_B}{x_A x_B} Đ_{AB} \begin{bmatrix} x_A (\nabla \ln a_A)_{p,T} + \dfrac{1}{cRT}(\phi_A - \omega_A)\nabla p \\ + \dfrac{\rho \omega_A \omega_B}{cRT}(\mathbf{g}_B - \mathbf{g}_A) + k_T \nabla \ln T \end{bmatrix} \tag{24.1-9}$$

$$\mathbf{q} = -k\nabla T + \left(\frac{\overline{H}_A}{M_A} - \frac{\overline{H}_B}{M_B} \right) \mathbf{j}_A + 2 \frac{cRT}{\rho} \frac{x_A x_B}{\omega_A \omega_B} \frac{D_A^T}{Đ_{AB}} \mathbf{j}_A \tag{24.1-10}$$

Note that $\mathbf{J}_A^*$ can also be obtained using $\mathbf{J}_A^*/cx_A x_B = \mathbf{j}_A/\rho\omega_A\omega_B$ (see Table 17.3-1). Equation 24.1-9 states that a flux of mass of species A can result from a gradient in the

Table 24.1-1. Experimental Thermal Diffusion Ratios for Liquids and Low-Density Gas Mixtures

Liquids:[a]

Components A–B	T(K)	x_A	k_T
$C_2H_2Cl_4$–n-C_6H_{14}	298	0.5	1.08
$C_2H_4Br_2$–$C_2H_4Cl_2$	298	0.5	0.225
$C_2H_2Cl_4$–CCl_4	298	0.5	0.060
CBr_4–CCl_4	298	0.09	0.129
CCl_4–CH_3OH	313	0.5	1.23
CH_3OH–H_2O	313	0.5	−0.137
cyclo-C_6H_{12}–C_6H_6	313	0.5	0.100

Gases:

Components A–B	T(K)	x_A	k_T
Ne–He[b]	330	0.80	0.0531
		0.40	0.1004
N_2–H_2[c]	264	0.706	0.0548
		0.225	0.0663
D_2–H_2[d]	327	0.90	0.0145
		0.50	0.0432
		0.10	0.0166

[a] R. L. Saxton, E. L. Dougherty, and H. G. Drickamer, *J. Chem. Phys.*, **22**, 1166–1168 (1954); R. L. Saxton and H. G. Drickamer, *J. Chem. Phys.*, **22**, 1287–1288 (1954); L. J. Tichacek, W. S. Kmak, and H. G. Drickamer, *J. Phys. Chem.*, **60**, 660–665 (1956).

[b] B. E. Atkins, R. E. Bastick, and T. L. Ibbs, *Proc. Roy. Soc. (London)*, **A172**, 142–158 (1939).

[c] T. L. Ibbs, K. E. Grew, and A. A. Hirst, *Proc. Roy. Soc. (London)*, **A173**, 543–554 (1939).

[d] H. R. Heath, T. L. Ibbs, and N. E. Wild, *Proc. Roy. Soc. (London)*, **A178**, 380–389 (1941).

activity of A (ordinary diffusion), from a pressure gradient (pressure diffusion), from external force inequalities (forced diffusion), and from a gradient in the temperature (the latter is the *thermal diffusion effect* or the *Soret effect*). Equation 24.1-10 states that a heat flux can result from a temperature gradient (thermal conduction), from heat transport associated with the mass transport, and finally a presumably small contribution due to the *diffusion thermo effect* or the *Dufour effect*. The thermal diffusion coefficient D_A^T is usually replaced by the *thermal diffusion ratio* k_T, defined by

$$k_T = \frac{D_A^T}{\rho Đ_{AB}} \frac{x_A x_B}{\omega_A \omega_B} \tag{24.1-11}$$

When k_T is positive, species A moves toward the colder region, and when it is negative, species A moves toward the warmer region. Some sample values of k_T for gases and liquids are given in Table 24.1-1. Finally, we note that Eq. 24.1-10 is the origin of Eq. 19.3-3.

For binary mixtures of dilute gases, it is found by experiment that the species with the larger molecular weight usually goes to the colder region. If the molecular weights are about equal, then usually the species with the largest diameter moves to the colder region. In some instances there is a change in the sign of the thermal diffusion ratio as the temperature is lowered.[5]

In the next few sections we give some examples of applications of Eq. 24.1-9.

§24.2 CONCENTRATION DIFFUSION AND DRIVING FORCES

In Chapter 17 we wrote Fick's first law by stating that the mass (or molar) flux is proportional to the gradient of the mass (or mole) fraction. This is summarized in Table 17.3-1.

On the other hand, in Eq. 24.1-9, it appears that the thermodynamics of irreversible processes dictates using the *activity gradient* as the driving force for concentration diffusion. In this section we show that either the activity gradient or the mass (or mole) fraction gradient driving force may be used, but that each choice requires a different diffusivity. These two diffusivities are related, and we illustrate this for a binary mixture.

When we drop the pressure-, thermal-, and forced-diffusion terms from Eq. 24.1-9, we get in molar units (using Eq. (I) of Table 17.3-1)

$$\mathbf{J}_A^* = -cĐ_{AB} x_A \nabla \ln a_A \tag{24.2-1}$$

This may be rewritten by making use of the fact that the activity is a function of the concentration to obtain

$$\mathbf{J}_A^* = -cĐ_{AB} x_A \left(\frac{\partial \ln a_A}{\partial \ln x_A}\right)_{T,p} \nabla \ln x_A = -cĐ_{AB} \left(\frac{\partial \ln a_A}{\partial \ln x_A}\right)_{T,p} \nabla x_A \tag{24.2-2}$$

The activity may be written as the product of the *activity coefficient* and the mole fraction, $a_A = \gamma_A x_A$, so that

$$\mathbf{J}_A^* = -cĐ_{AB} \left[1 + \left(\frac{\partial \ln \gamma_A}{\partial \ln x_A}\right)_{T,p}\right] \nabla x_A \tag{24.2-3}$$

If the mixture is "ideal," then the activity coefficient is equal to unity, and Eq. 24.2-1 becomes the same as Eq. (G) of Table 17.3-1, and $Đ_{AB} = \mathscr{D}_{AB}$.

If the mixture is "nonideal," one can define the binary diffusivity $\mathscr{D}_{AB}$ as

$$\mathscr{D}_{AB} = Đ_{AB} \left(\frac{\partial \ln a_A}{\partial \ln x_A}\right)_{T,p} = Đ_{AB} \left[1 + \left(\frac{\partial \ln \gamma_A}{\partial \ln x_A}\right)_{T,p}\right] \tag{24.2-4}$$

Then Eqs. 24.2-2 and 24.2-3 become

$$\mathbf{J}_A^* = -c\mathscr{D}_{AB} \nabla x_A \tag{24.2-5}$$

which is one of the forms of Fick's law (see Eq. (G) of Table 17.3-1). In order to measure $Ð_{AB}$, one has to have measurements of the activity as a function of concentration, and for this reason $Ð_{AB}$ has not been popular.

For ideal mixtures $\mathscr{D}_{AB}$ and $Ð_{AB}$ are identical, and they are linear functions of the mole fraction as shown in Fig. 24.2-1. For nonideal mixtures $\mathscr{D}_{AB}$ and $Ð_{AB}$ are different nonlinear functions of the mole fraction; an example is shown in Fig. 24.2-2. However, it has been observed that the product $\mu Ð_{AB}$ has been found for *some* nonideal mixtures

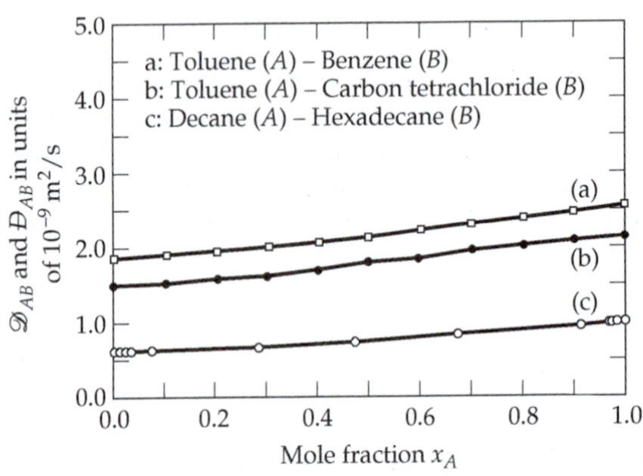

Fig. 24.2-1. Diffusivity in ideal liquid mixtures at 25°C. [P. W. M. Rutten, *Diffusion in Liquids*, Delft University Press (1992), p. 31.]

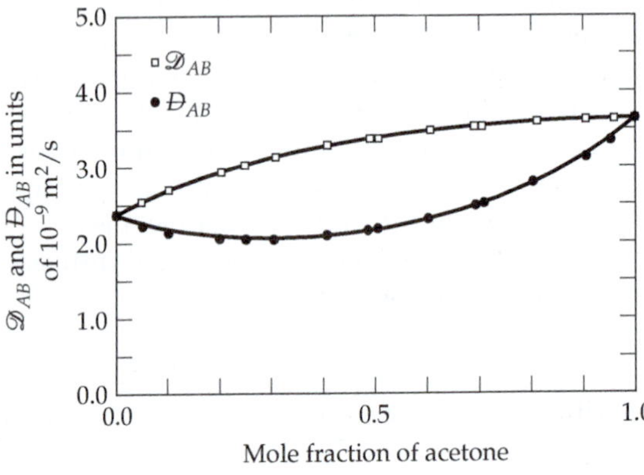

Fig. 24.2-2. Diffusivity in a nonideal liquid mixture (acetone–chloroform at 25°C). [P. W. M. Rutten, *Diffusion in Liquids*, Delft University Press (1992), p. 32.]

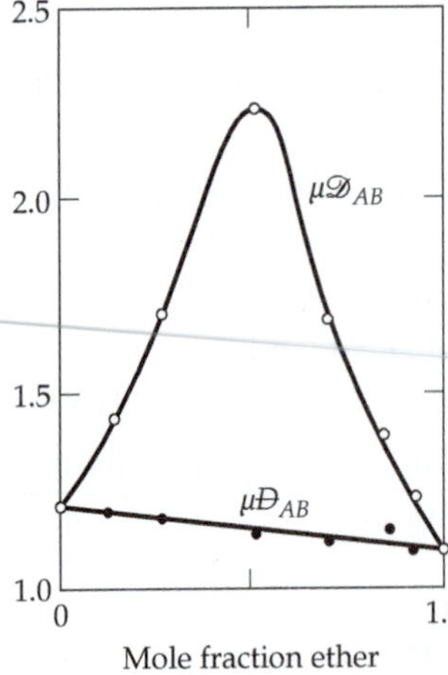

Fig. 24.2-3. Effect of activity on the product of viscosity and diffusivity for liquid mixtures of chloroform and diethyl ether. [R. E. Powell, W. E. Roseveare, and H. Eyring, *Ind. Eng. Chem.*, **33**, 430–435 (1941).]

to be very nearly linear in the mole fraction, whereas $\mu\mathscr{D}_{AB}$ is not (see Fig. 24.2-3). There is no compelling reason to prefer one diffusivity over the other. Most of the diffusivities reported in the literature are $\mathscr{D}_{AB}$, and not $Ð_{AB}$.

§24.3 THERMAL DIFFUSION AND THE CLUSIUS-DICKEL COLUMN

In this section we discuss the diffusion of species under the influence of a *temperature gradient*. To illustrate the phenomenon, we consider the system shown in Fig. 24.3-1, two bulbs joined together by an insulated tube of small diameter and filled with a mixture of ideal gases A and B. The bulbs are maintained at constant temperatures of T_1 and T_2, respectively, and the diameter of the insulated tube is small enough to eliminate convection currents substantially. Ultimately the system arrives at a steady state, with gas A enriched at one end of the tube and depleted at the other. We want to find an expression for $x_{A2} - x_{A1}$, the difference of the mole fractions at the two ends of the tube.

After steady state has been achieved, there is no net motion of either A or B, so that $\mathbf{J}_A^* = 0$. If we take the tube axis to be in the z direction, then from Eq. 24.1-9 we get

$$\frac{dx_A}{dz} + \frac{k_T}{T}\frac{dT}{dz} = 0 \tag{24.3-1}$$

Here the activity a_A has been replaced by the mole fraction x_A, as is appropriate for an ideal-gas mixture. Usually the degree of separation in an apparatus of the kind being considered here is small. We may therefore ignore the position dependence of k_T and integrate this equation to get

$$x_{A2} - x_{A1} = -\int_{T_1}^{T_2} \frac{k_T}{T}\, dT \tag{24.3-2}$$

Because the dependence of k_T on T is rather complicated, it is customary to assume k_T constant at the value for some mean temperature T_m. Performing the integral in Eq. 24.3-2 then gives (approximately)

$$x_{A2} - x_{A1} = -k_T(T_m)\ln\frac{T_2}{T_1} \tag{24.3-3}$$

The recommended[1] mean temperature is

$$T_m = \frac{T_1 T_2}{T_2 - T_1}\ln\frac{T_2}{T_1} \tag{24.3-4}$$

Equations 24.3-3 and 24.3-4 are useful for estimating the order of magnitude of thermal diffusion effects.

Unless the temperature gradient is very large, the separation will normally be quite small. Therefore, it has been advantageous to combine the thermal diffusion effect with free convection between two vertical walls, one heated and the other cooled. The heated stream then ascends, and the cooled one descends. The upward stream will be richer in one

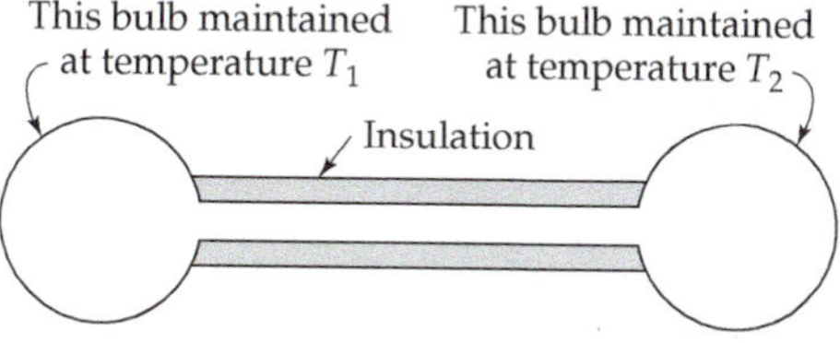

Fig. 24.3-1. Steady-state binary thermal diffusion in a two-bulb apparatus. The mixture of gases A and B tends to separate under the influence of the thermal gradient.

[1] H. Brown, *Phys. Rev.*, **58**, 661–662 (1940).

of the components, say A, and the downward stream will be richer in B. This is the principle of the operation of the *Clusius-Dickel column*.[2–4] By coupling many of these columns together in a "cascade," it is possible to perform a separation. During World War II this was one of the methods used for separating the uranium isotopes by using uranium hexafluoride gas. The method has also been used with some success in the separation of organic mixtures, where the components have very nearly the same boiling points, so that distillation is not an option.

The thermal diffusion ratio can also be obtained from the Dufour (diffusion-thermo) effect, but the analysis of the experiment is fraught with problems and experimental errors difficult to avoid.[5]

§24.4 PRESSURE DIFFUSION AND THE ULTRACENTRIFUGE

Next we examine the diffusion produced by a *pressure gradient*. If a sufficiently large pressure gradient can be established, then a measurable separation can be effected. One example of this is the ultracentrifuge, which has been used to separate enzymes and proteins. In Fig. 24.4-1 we show a small cylindrical cell in a very high-speed centrifuge. The length of the cell, L, is short with respect to the radius of rotation R_0, the solution density may be considered a function of composition only, and the temperature is constant. It is desired to determine the distribution of the two components at steady state in terms of their partial molar volumes, position in the cell, the pressure gradient, and the angular velocity of rotation of the rotating element, Ω. The pressure gradient is obtained from the equation of motion as

$$\frac{dp}{dz} = \rho g_\Omega \approx -\rho\Omega^2 R_0 \tag{24.4-1}$$

For simplicity, we assume that the partial molar volumes and the activity coefficients are constant over the range of compositions existing in the cell.

At steady state $\mathbf{J}_A^* = 0$, and hence the relevant terms in Eq. 24.1-9 give for species A

$$\frac{dx_A}{dz} + \frac{M_A x_A}{RT}\left(\frac{\overline{V}_A}{M_A} - \frac{1}{\rho}\right)\frac{dp}{dz} = 0 \tag{24.4-2}$$

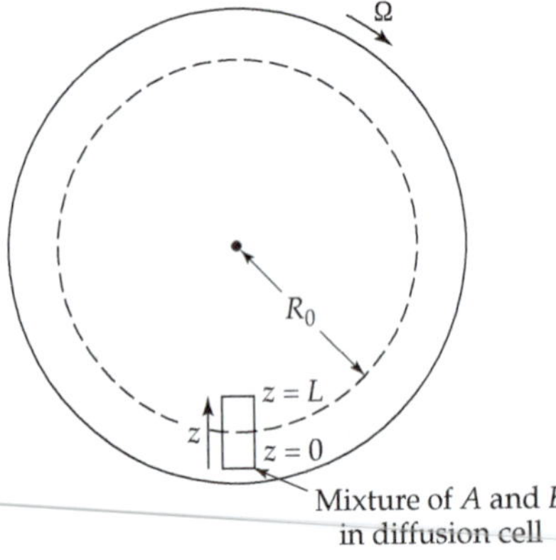

Fig. 24.4-1. Steady-state pressure diffusion in a centrifuge. The mixture in the diffusion cell tends to separate by virtue of the pressure gradient produced in the centrifuge.

[2] K. Clusius and G. Dickel, *Z. Phys. Chem.*, **B44**, 397–450, 451–473 (1939).

[3] K. E. Grew and T. L. Ibbs, *Thermal Diffusion in Gases*, Cambridge University Press (1952); K. E. Grew, in *Transport Phenomena in Fluids* (H. J. M. Hanley, ed.), Marcel Dekker, New York (1969), Chapter 10.

[4] R. B. Bird, *Advances in Chemical Engineering*, **1**, 155–239 (1956), §4.D.2; *errata*, 2, 325 (1958).

[5] S. Chapman and T. G. Cowling, *The Mathematical Theory of Nonuniform Gases*, 3rd edition, Cambridge University Press (1970), pp. 268–271.

Inserting the appropriate expression for the pressure gradient, and then multiplying by $(\overline{V}_B/x_A)\,dz$, we get for species A

$$\overline{V}_B \frac{dx_A}{x_A} = -\overline{V}_B \frac{g_\Omega}{RT}(\rho\overline{V}_A - M_A)dz \tag{24.4-3}$$

Then we write a similar equation for species B, which is

$$\overline{V}_A \frac{dx_B}{x_B} = -\overline{V}_A \frac{g_\Omega}{RT}(\rho\overline{V}_B - M_B)dz \tag{24.4-4}$$

Subtracting Eq. 24.4-4 from Eq. 24.4-3 we get

$$\overline{V}_B \frac{dx_A}{x_A} - \overline{V}_A \frac{dx_B}{x_B} = \frac{g_\Omega}{RT}(M_A\overline{V}_B - M_B\overline{V}_A)dz \tag{24.4-5}$$

We now integrate this equation from $z = 0$ to some arbitrary value of z, taking account of the fact that the mole fractions of A and B at $z = 0$ are x_{A0} and x_{B0}, respectively. This gives

$$\overline{V}_B \int_{x_{A0}}^{x_A} \frac{dx_A}{x_A} - \overline{V}_A \int_{x_{B0}}^{x_B} \frac{dx_B}{x_B} = \frac{M_A\overline{V}_B - M_B\overline{V}_A}{RT} \int_0^z g_\Omega\,dz \tag{24.4-6}$$

If g_Ω is constant over the range of integration (which it normally would be), then we get

$$\overline{V}_B \ln \frac{x_A}{x_{A0}} - \overline{V}_A \ln \frac{x_B}{x_{B0}} = \frac{M_A\overline{V}_B - M_B\overline{V}_A}{RT} g_\Omega z \tag{24.4-7}$$

Then we take the exponential of both sides to find

$$\left(\frac{x_A}{x_{A0}}\right)^{\overline{V}_B} \left(\frac{x_{B0}}{x_B}\right)^{\overline{V}_A} = \exp\left[\left(M_B\overline{V}_A - M_A\overline{V}_B\right)\left(\frac{R_0\Omega^2 z}{RT}\right)\right] \tag{24.4-8}$$

This describes the steady-state concentration distribution for a binary system in a constant centrifugal force field. Notice that, since this result contains no transport coefficients at all, the same result can be obtained by an equilibrium thermodynamics analysis.[1] However, if one wishes to analyze the time-dependent behavior of a centrifugation, then the diffusivity for the mixture A-B will appear in the result, and the problem cannot be solved by equilibrium thermodynamics.

§24.5 ION FLUXES AND THE NERNST-PLANCK EQUATION

We next consider the diffusion of electrolytes in a solution. This brings us immediately into a multicomponent problem. Here, however, we consider just a very dilute solution, in which one species (A) is an ion, and the only other species is the solvent (B). Then we can omit the term $\mathbf{g}_B$ and set ω_B equal to unity.

Then, in the absence of pressure diffusion and thermal diffusion, Eq. 24.1-9 becomes for ionic species A

$$\mathbf{j}_A = -\rho \frac{\omega_A \omega_B}{x_A x_B} Ð_{AB} \left[x_A(\nabla \ln a_A)_{p,T} - \frac{\rho\omega_A}{cRT}\mathbf{g}_A\right] \tag{24.5-1}$$

Since the ionic species A is quite dilute, we may make the assumption that a_A is not very different from x_A so that

$$\mathbf{J}_A^* = -c\mathcal{D}_{AB}\left[\nabla x_A - \frac{\rho_A}{cRT}\mathbf{g}_A\right] \tag{24.5-2}$$

---

[1]E. A. Guggenheim, *Thermodynamics*, North Holland, Amsterdam (1950), pp. 356–360.

The force on the ion may be given as $\mathbf{g}_A = -(z_A F/M_A)\nabla\phi$, where z_A is the valence of the ion (positive or negative), $F = 96845$ abs.-coulombs/g-equivalent is the *Faraday constant*, M_A is the molecular weight of the ion, and ϕ is the electric potential. When this is inserted into Eq. 24.5-2, we get

$$\begin{aligned}\mathbf{J}_A^* &= -c\mathscr{D}_{AB}\left[\nabla x_A + \frac{\rho_A}{cRT}\left(\frac{z_A F}{M_A}\right)\nabla\phi\right] \\ &= -c\mathscr{D}_{AB}\left[\nabla x_A + \frac{x_A}{RT} z_A F \nabla\phi\right]\end{aligned} \tag{24.5-3}$$

Finally, since c is very nearly a constant in this dilute solution, the expression for the molar flux can be rewritten as

$$\mathbf{J}_A^* = -\mathscr{D}_{AB}\left[\nabla c_A + c_A z_A\left(\frac{F}{RT}\right)\nabla\phi\right] \tag{24.5-4}$$

which is called the *Nernst-Planck* equation. This can be regarded as a modification of Fick's first law for an ionic species in a dilute solution of electrolyte.

§24.6 MULTICOMPONENT SYSTEMS; THE MAXWELL-STEFAN EQUATIONS

For systems with N chemical species, the story is very similar to that in §24.1, but considerably more complicated. Here we content ourselves with giving only the main results. Detailed treatments are given elsewhere.[1,2]

The expressions for the entropy flux $\mathbf{s}$ and entropy production rate per unit volume g_S are generalizations of Eqs. 24.1-2 and 24.1-3:

$$\mathbf{s} = \frac{1}{T}\left(\mathbf{q}^{(h)} - \sum_{\alpha=1}^{N}\frac{T\overline{S}_\alpha}{M_\alpha}\mathbf{j}_\alpha\right) \tag{24.6-1}$$

$$Tg_S = -(\mathbf{q}^{(h)}\cdot\nabla\ln T) - \sum_{\alpha=1}^{N}\left(\frac{\mathbf{j}_\alpha}{\rho_\alpha}\cdot cRT\mathbf{d}_\alpha\right) - (\boldsymbol{\tau}:\nabla\mathbf{v}) - \sum_{\alpha=1}^{N}\frac{\overline{G}_\alpha}{M_\alpha}r_\alpha \tag{24.6-2}$$

where $\mathbf{q}^{(h)} = \mathbf{q} - \sum_{\alpha=1}^{N}(\overline{H}_\alpha/M_\alpha)\mathbf{j}_\alpha$, and the generalized driving force is

$$cRT\mathbf{d}_\alpha = c_\alpha RT\sum_{\beta=1}^{N-1}\left(\frac{\partial\ln a_\beta}{\partial\omega_\beta}\right)_{T,p,\omega}\nabla\omega_\beta + (\phi_\alpha - \omega_\alpha)\nabla p - \rho_\alpha\mathbf{g}_\alpha + \omega_\alpha\sum_{\beta=1}^{N}\rho_\beta\mathbf{g}_\beta \tag{24.6-3}$$

We next write down the appropriate generalization of Eqs. 24.1-6 and 24.1-7 and apply the Onsager reciprocal relations. This gives the fluxes in terms of the driving forces

$$\mathbf{j}_\alpha = -D_\alpha^T\nabla\ln T + \rho_\alpha\sum_{\beta=1}^{N}\mathbb{D}_{\alpha\beta}\mathbf{d}_\beta \qquad \alpha = 1,2,3,\ldots,N \tag{24.6-4}$$

These are the *generalized Fick equations*, and the $\mathbb{D}_{\alpha\beta}$ are the *multicomponent Fick diffusivities* (which are not the same as the binary diffusivities). These equations can be inverted to

---

[1] J. O. Hirschfelder, C. F. Curtiss, and R. B. Bird, *Molecular Theory of Gases and Liquids*, Wiley, New York (1954), second corrected printing (1964), Chapter 11; S. R. de Groot and P. Mazur, *Non-Equilibrium Thermodynamics*, North Holland, Amsterdam (1962); R. B. Bird, W. E. Stewart, and E. N. Lightfoot, *Transport Phenomena*, revised 2nd edition, Wiley, New York (2007), Chapter 24.

[2] R. B. Bird and D. J. Klingenberg, *Advances in Water Resources*, **62**, 238–242 (2013); *errata*: in Eq. 3.15, $\omega_\alpha\omega_\beta$ on the right side should be in the denominator instead of the numerator; also, in the first sentence of Sec. 4, the reference should be to Eq. 3.15 rather than Eq. 3.13.

give the driving forces in terms of the fluxes[3]

$$\mathbf{d}_\alpha = -\sum_{\beta \neq \alpha}^{N} \frac{x_\alpha x_\beta}{\text{Đ}_{\alpha\beta}} \left(\frac{D_\alpha^T}{\rho_\alpha} - \frac{D_\beta^T}{\rho_\beta} \right) \nabla \ln T - \sum_{\beta \neq \alpha}^{N} \frac{x_\alpha x_\beta}{\text{Đ}_{\alpha\beta}} \left(\frac{\mathbf{j}_\alpha}{\rho_\alpha} - \frac{\mathbf{j}_\beta}{\rho_\beta} \right) \qquad \alpha = 1,2,3,\ldots, N \qquad (24.6\text{-}5)$$

These are the *generalized Maxwell-Stefan equations*, and the $\text{Đ}_{\alpha\beta}$ are the *multicomponent Maxwell-Stefan diffusivities*. Equations 24.6-4 and 24.6-5 are the generalizations of Eqs. (A) and (C) of Table 17.4-2. The $\mathbb{D}_{\alpha\beta}$ and $\text{Đ}_{\alpha\beta}$ are interrelated by the Curtiss-Bird equation[4]

$$\text{Đ}_{\alpha\beta} = \frac{x_\alpha x_\beta}{\omega_\alpha \omega_\beta} \frac{\sum_{\gamma \neq \alpha} \mathbb{D}_{\alpha\gamma} (\text{adj } B_\alpha)_{\gamma\beta}}{\sum_{\gamma \neq \alpha} (\text{adj } B_\alpha)_{\gamma\beta}} \qquad \alpha,\beta = 1,2,3,\ldots, N \qquad (24.6\text{-}6)$$

in which $(B_\alpha)_{\beta\gamma} = -\mathbb{D}_{\beta\gamma} + \mathbb{D}_{\alpha\gamma}$—that is, the $\beta\gamma$ component of a matrix called B_α, which is of order $(N-1) \times (N-1)$ with the $\beta = \alpha$ row and the $\gamma = \alpha$ column being excluded—and adj B_α is the matrix adjoint to B_α.

For an isothermal mixture of low-density gases, the $\text{Đ}_{\alpha\beta}$ are, to a very good approximation, the same as $\mathscr{D}_{\alpha\beta}$. If there is no thermal, pressure, or forced diffusion, then Eq. 24.6-5 reduces to

$$\nabla x_\alpha = -\sum_{\beta=1}^{N} \frac{x_\alpha x_\beta}{\mathscr{D}_{\alpha\beta}} (\mathbf{v}_\alpha - \mathbf{v}_\beta) = -\sum_{\beta=1}^{N} \frac{1}{c\mathscr{D}_{\alpha\beta}} (x_\beta \mathbf{N}_\alpha - x_\alpha \mathbf{N}_\beta) \qquad \alpha = 1,2,3,\ldots, N \qquad (24.6\text{-}7)$$

Here the $\mathscr{D}_{\alpha\beta}$ are the *binary* diffusivities given in Chapter 17. Equations 24.6-7 are referred to as the Maxwell-Stefan equations, since Maxwell[5] suggested them for binary mixtures, and Stefan[6] proposed extending the binary equation to multicomponent mixtures.

Because of the difficulty in getting analytical solutions to the Maxwell-Stefan equations, various approximate methods have been suggested. One such method, first suggested by Hougen and Watson,[7] is the following: for the diffusion of species i in the multicomponent mixture, we write

$$\mathbf{N}_i = -c\mathscr{D}_{im} \nabla x_i + x_i \sum_{\alpha=1}^{N} \mathbf{N}_\alpha \qquad (24.6\text{-}8)$$

by analogy with the binary expression in Eq. (E) of Table 17.4-2. Then we solve Eq. 24.6-8 for ∇x_i and equate the result to ∇x_i from Eq. 24.6-7 in the Maxwell-Stefan equations to get for *collinear molar fluxes* $\mathbf{N}_\alpha$,

$$\frac{1}{c\mathscr{D}_{im}} = \frac{\sum_{\alpha=1}^{N} (1/c\mathscr{D}_{i\alpha})(x_\alpha \mathbf{N}_i - x_i \mathbf{N}_\alpha)}{\mathbf{N}_i - x_i \sum_{\alpha=1}^{N} \mathbf{N}_\alpha} \qquad (24.6\text{-}9)$$

In general, the $\mathscr{D}_{im}$ are functions of position. For situations in which this dependence is slight, we may generalize the binary diffusion formulas and mass-transfer coefficient

---

[3]This kind of inversion was first performed by H. J. Merk, *Appl. Sci. Res.*, **A8**, 73–99 (1959), when he was a graduate student in engineering physics at the Technical University of Delft.

[4]C. F. Curtiss and R. B. Bird, *Ind. Eng. Chem. Research*, **38**, 2515–2522 (1999); errata **40**, 1791 (2001).

[5]J. C. Maxwell, *Phil. Mag.*, **XIX**, 19–32 (1860), **XX**, 21–32, 33–36 (1868).

[6]J. Stefan, *Sitzungsber. Kais. Akad. Wiss. Wien*, **LXIII**(2), 63–124 (1871), **LXV**(2), 323–363 (1872).

[7]O. A. Hougen and K. M. Watson, *Chemical Process Principles*, Vol. **III**, Wiley, New York (1947), pp. 977–979. Methods for evaluating $\mathscr{D}_{im}$ for special cases have been developed by C. R. Wilke, *Chem. Engr. Prog.*, **46**, 95–104 (1950) and W. E. Stewart, *NACA Tech. Note 3208* (1954).

correlations by simply replacing $\mathscr{D}_{AB}$ by $\mathscr{D}_{im}$. For some special kinds of diffusing systems, Eq. 24.6-9 becomes particularly simple:

a. For trace components 2,3 ..., N in nearly pure 1,

$$\mathscr{D}_{im} = \mathscr{D}_{i1} \tag{24.6-10}$$

b. For systems in which the $\mathscr{D}_{i\alpha}$ are all the same,

$$\mathscr{D}_{im} = \mathscr{D}_{i\alpha} \tag{24.6-11}$$

c. For systems in which 2,3,..., N move with the same velocity or are stationary,

$$\frac{1-x_1}{\mathscr{D}_{1m}} = \sum_{\alpha=2}^{N} \frac{x_\alpha}{\mathscr{D}_{1\alpha}} \tag{24.6-12}$$

For systems in which the variation of $\mathscr{D}_{im}$ is considerable, the assumption of linear variation with composition or distance has proven useful.[8] The $\mathscr{D}_{im}$ approach to solving multicomponent problems seems to give pretty good results for calculating mass-transfer rates, but a less satisfactory quantitative description of concentration profiles.

EXAMPLE 24.6-1

Diffusion in a Three-Component Gas System

To illustrate the setting up of multicomponent diffusion problems for gases, we rework the problem of §18.6 when liquid water (species 1) is evaporating into air, regarded as a binary mixture of nitrogen (2) and oxygen (3) at 1 atm and 352K. We take the air-water interface to be at $z = 0$ and the top end of the diffusion tube to be at $z = L$. We consider the vapor pressure of water to be known, so that x_1 is known at $z = 0$ (that is, $x_{10} = 341$ mmHg/760 mmHg $= 0.449$), and the mole fractions of all three gases are known at $z = L$: $x_{1L} = 0.10$, $x_{2L} = 0.75$, $x_{3L} = 0.15$. The diffusion tube has a length $L = 11.2$ cm.

The conservation of mass leads, as in §18.6, to the following expressions:

$$\frac{dN_{\alpha z}}{dz} = 0 \quad \alpha = 1,2,3 \tag{24.6-13}$$

From this it may be concluded that the molar fluxes of the three species are all constants at steady state. Since species 2 and 3 are not moving, we conclude that N_{2z} and N_{3z} are both zero.

Next we need the expressions for the molar fluxes from Eq. 24.6-7. Since $x_1 + x_2 + x_3 = 1$, we need only two of the three available equations, and we select the equations for species 2 and 3. Since $N_{2z} = 0$ and $N_{3z} = 0$, these equations simplify considerably:

$$\frac{dx_2}{dz} = \frac{N_{1z}}{c\mathscr{D}_{12}} x_2; \qquad \frac{dx_3}{dz} = \frac{N_{1z}}{c\mathscr{D}_{13}} x_3 \tag{24.6-14,15}$$

Note that the diffusivity $\mathscr{D}_{23}$ does not appear here, because there is no relative motion of 2 and 3. These equations can be integrated from an arbitrary height z to the top of the tube at L, to give for constant $c\mathscr{D}_{\alpha\beta}$

$$\int_{x_2}^{x_{2L}} \frac{d\bar{x}_2}{\bar{x}_2} = \frac{N_{1z}}{c\mathscr{D}_{12}} \int_z^L d\bar{z}; \qquad \int_{x_3}^{x_{3L}} \frac{d\bar{x}_3}{\bar{x}_3} = \frac{N_{1z}}{c\mathscr{D}_{13}} \int_z^L d\bar{z} \tag{24.6-16,17}$$

Integration then gives

$$\frac{x_2}{x_{2L}} = \exp\left(-\frac{N_{1z}(L-z)}{c\mathscr{D}_{12}}\right); \qquad \frac{x_3}{x_{3L}} = \exp\left(-\frac{N_{1z}(L-z)}{c\mathscr{D}_{13}}\right) \tag{24.6-18,19}$$

and the mole fraction profile of water vapor in the diffusion column will be

$$x_1 = 1 - x_{2L} \exp\left(-\frac{N_{1z}(L-z)}{c\mathscr{D}_{12}}\right) - x_{3L} \exp\left(-\frac{N_{1z}(L-z)}{c\mathscr{D}_{13}}\right) \tag{24.6-20}$$

[8] H. W. Hsu and R. B. Bird, *AIChE Journal*, **6**, 551–553 (1960).

When we apply the boundary condition at $z = 0$, we get

$$x_{10} = 1 - x_{2L} \exp\left(-\frac{N_{1z}L}{c\mathscr{D}_{12}}\right) - x_{3L} \exp\left(-\frac{N_{1z}L}{c\mathscr{D}_{13}}\right) \tag{24.6-21}$$

which is a transcendental equation for N_{1Z}.

According to Reid, Prausnitz, and Poling,[9] $\mathscr{D}_{12} = 0.364\ \text{cm}^2/\text{s}$ and $\mathscr{D}_{13} = 0.357\ \text{cm}^2/\text{s}$ at 352 K and 1 atm. At these conditions $c = 3.46 \times 10^{-5}$ g-mol/cm^3. To get a quick solution to Eq. 24.6-21, we take both diffusivities to be equal[10] to 0.36 cm^2/s. Then we get

$$0.449 = 1 - 0.90 \exp\left(-\frac{N_{1z}\,(11.2)}{(3.462 \times 10^{-5})(0.36)}\right) \tag{24.6-22}$$

from which it is found that $N_{1z} = 5.523 \times 10^{-7}$g-mol/cm$^2 \cdot$ s. This can be used as a first guess in solving Eq. 24.6-21 more exactly, if desired. Then the entire profiles can be calculated from Eqs. 24.6-18 to 24.6-20.

§24.7 CONCLUDING COMMENTS

This chapter has been a very brief, introductory discussion of the use of the thermodynamics of irreversible processes to describe the coupling of the heat and mass flux vectors. Most of our attention has been focused on the mass flux vector and the various types of driving forces that can produce a species mass flux. According to Curie's law, there is a coupling between the momentum-flux tensor and the chemical reaction rate, but no discussion of this has been given.

If one really wants to delve into the thermodynamic approach to transport phenomena, multicomponent systems have to be dealt with. This is particularly true for electrolyte and colloidal systems, where the systems are necessarily multicomponent in nature. However, that is not a subject for an introductory text such as this.

This chapter has concerned itself only with the basic ideas of the subject. The reader should now be aware that thermal, pressure, and forced diffusion are part of the subject of mass transport. For multicomponent systems, there is a set of equations—the *Maxwell-Stefan* equations—that describe the diffusion in systems of three and more chemical species. If there are n chemical species, there will be $\frac{1}{2}n(n-1)$ diffusivities $Ð_{\alpha\beta}$. These multicomponent equations admit analytical solutions only for a few simple problems.

QUESTIONS FOR DISCUSSION

1. Name the four kinds of mass transport that are discussed in this chapter.
2. Do any new transport properties have to be defined in the chapter?
3. Discuss briefly the operation of a Clusius-Dickel column.
4. Discuss the operation of an ultracentrifuge. What kinds of materials may be fractionated in this kind of instrument?
5. What is the difference between ordinary (equilibrium) thermodynamics and nonequilibrium thermodynamics?
6. Discuss the Maxwell-Stefan equations, their origin, their use, and the methods of solving them.

---

[9] R. C. Reid, J. M. Prausnitz, and B. E. Poling, *The Properties of Gases and Liquids*, 4th edition, McGraw-Hill, New York (1987), p. 591.

[10] The solution to ternary diffusion problems in which two of the binary diffusivities are equal was discussed by H. L. Toor, *AIChE Journal*, **3**, 198–207 (1957).

PROBLEMS

24A.1 Thermal diffusion.

(a) Estimate the steady-state separation of H_2 and D_2 occurring in the simple thermal diffusion apparatus shown in Fig. 24.3-1 under the following conditions: T_1 is 200 K, T_2 is 600 K, the mole fraction of D_2 is initially 0.10, and the effective average k_T is 0.0166.

(b) At what temperature should this average k_T have been evaluated?

Answers: **(a)** The mole fraction of H_2 is higher by 0.0183 in the hot bulb; **(b)** 330 K

24A.2 Ultracentrifugation of proteins. Estimate the steady-state concentration profile when a typical albumin solution is subjected to a centrifugal field 50,000 times the gravitational acceleration under the following conditions:

Cell length = 1.0 cm
Molecular weight of albumin = 45,000 g/g-mol
Apparent density of albumin in solution = $M_A/\overline{V}_A = 1.34$ g/cm^3
Mole fraction of albumin (at $z = 0$), $x_{A0} = 5 \times 10^{-6}$
Apparent density of water in the solution = 1.00 g/cm^3
Temperature = 75°F

Answer: $x_A = 5 \times 10^{-6} \exp(-22.7z)$, with z in cm

24B.1 Catalytic oxidation of carbon monoxide. Figure 24B.1 shows schematically how oxygen and carbon monoxide combine at a catalytic surface (palladium) to make carbon dioxide, according to the technologically important reaction[1]

$$O_2 + 2CO \rightarrow 2CO_2 \tag{24B.1-1}$$

Assume that the reaction occurs instantaneously and irreversibly at the catalytic surface. The gas composition at the outer edge of the film (at $z = 0$) is presumed known, and the catalyst surface is at $z = \delta$. The temperature and pressure are assumed to be independent of position throughout the film. Label the species as follows: $O_2 = 1$, $CO = 2$, and $CO_2 = 3$.

(a) Show that the molar fluxes are constant across the film, and that the stoichiometry dictates that $N_{1z} = \frac{1}{2}N_{2z} = -\frac{1}{2}N_{3z}$.

(b) Show that the Maxwell-Stefan equations give

$$\frac{dx_3}{dz} = -\frac{N_{3z}}{c\mathscr{D}_{13}}\left(1 + \tfrac{1}{2}x_3\right) \tag{24B.1-2}$$

$$\frac{dx_1}{dz} = \frac{N_{3z}}{2c\mathscr{D}_{12}}(1 - 3x_1 - x_3) + \frac{N_{3z}}{2c\mathscr{D}_{13}}(2x_1 + x_3) \tag{24B.1-3}$$

In Eq. 24B.1-2 we have made the approximation that $\mathscr{D}_{23} \approx \mathscr{D}_{13}$; justify this by getting σ_{13}, σ_{23}, ε_{13}, and ε_{23} from the table of Lennard-Jones potential parameters.

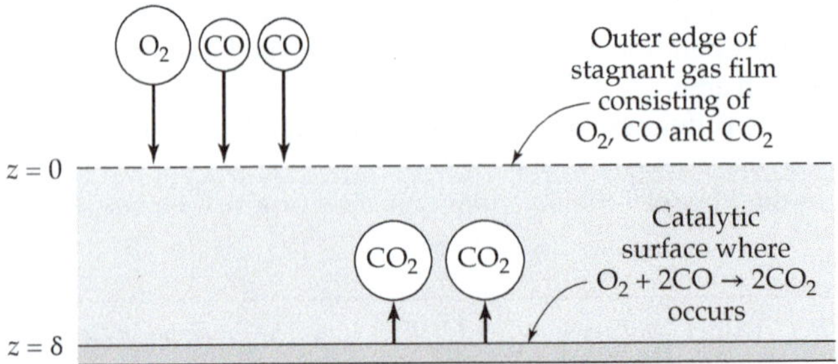

Fig. 24B.1 Three-component system with a catalytic chemical reaction.

[1] B. C. Gates, *Catalytic Chemistry*, Wiley, New York (1992), pp. 356–362; C. N. Satterfield, *Heterogeneous Catalysis in Industrial Practice*, 2nd edition, McGraw-Hill, New York (1991), Chapter 8.

(c) Integrate[2] Eq. 24B.1-2, and then combine the result with Eq. 24B.1-3. Integrate the resulting equation to get $x_1(z)$.

(d) From $x_1(z)$ and a similar result for $x_2(z)$ obtain x_3 at $z = \delta$.

(e) Finally, solve the equation for $x_3(\delta)$ to get $N_{3z}(\delta)$, which is the rate of production of carbon dioxide at the catalytic surface.

24C.1 The Lightfoot form for the Maxwell-Stefan equations. We want to develop an expression for the mass flux alternative to that given in Eq. 24.6-5, which may be rewritten as

$$\sum_{\substack{\beta=1\\ \beta\neq\alpha}}^{N} \frac{x_\alpha x_\beta}{Đ_{\alpha\beta}}(\mathbf{v}_\alpha - \mathbf{v}_\beta) = -x_\alpha(\nabla \ln x_\alpha)_{T,p} - \frac{1}{cRT}\left[(\phi_\alpha - \omega_\alpha)\nabla p - \rho_\alpha \mathbf{g}_\alpha + \omega_\alpha \sum_{\beta=1}^{N} \rho_\beta \mathbf{g}_\beta\right] \tag{24C.1-1}$$

First, verify that we may rewrite the left side of this equation as

$$\sum_{\substack{\beta=1\\ \beta\neq\alpha}}^{N} \frac{x_\alpha x_\beta}{Đ_{\alpha\beta}}(\mathbf{v}_\alpha - \mathbf{v}_\beta) = \sum_{\substack{\beta=1\\ \beta\neq\alpha}}^{N} \frac{x_\alpha x_\beta}{Đ_{\alpha\beta}}(\mathbf{v}_\gamma - \mathbf{v}_\beta) + x_\alpha(\mathbf{v}_\alpha - \mathbf{v}_\gamma)\sum_{\substack{\beta=1\\ \beta\neq\alpha}}^{N} \frac{x_\beta}{Đ_{\alpha\beta}} \tag{24C.1-2}$$

Next write the sum in the second term as a sum over all β, and then add a term to compensate for the error introduced to get

$$\sum_{\substack{\beta=1\\ \beta\neq\alpha}}^{N} \frac{x_\beta}{Đ_{\alpha\beta}} = \sum_{\substack{\beta=1\\ \text{all } \beta}}^{N} \frac{x_\beta}{Đ_{\alpha\beta}} - \frac{x_\alpha}{Đ_{\alpha\alpha}} \tag{24C.1-3}$$

Note that this change has introduced into the sum over all β a term containing $Đ_{\alpha\alpha}$, which has never been defined, because it was not needed. Therefore, we are free to define $Đ_{\alpha\alpha}$ in any way we choose. Let us choose the following definition:

$$\frac{x_\alpha}{Đ_{\alpha\alpha}} = -\sum_{\substack{\beta=1\\ \beta\neq\alpha}}^{N} \frac{x_\beta}{Đ_{\alpha\beta}} \qquad \text{or} \qquad \sum_{\substack{\beta=1\\ \text{all } \beta}}^{N} \frac{x_\beta}{Đ_{\alpha\beta}} = 0 \tag{24C.1-4a,4b}$$

Then verify that Eq. 24C.1-1 becomes

$$\sum_{\substack{\beta=1\\ \text{all } \beta}}^{N} \frac{x_\alpha x_\beta}{Đ_{\alpha\beta}}(\mathbf{v}_\gamma - \mathbf{v}_\beta) = -x_\alpha(\nabla \ln x_\alpha)_{T,p} - \frac{1}{cRT}\left[(\phi_\alpha - \omega_\alpha)\nabla p - \rho_\alpha \mathbf{g}_\alpha + \omega_\alpha \sum_{\beta=1}^{N} \rho_\beta \mathbf{g}_\beta\right] \tag{24C.1-5}$$

with the auxiliary relation Eq. 24C.1-4b. This expression was first proposed by Lightfoot.[3]

---

[2]Three-component problems with two diffusivities equal have been discussed by H. L. Toor, *AIChE Journal*, **3**, 198–207 (1957).

[3]E. N. Lightfoot, *Transport Phenomena in Living Systems*, Wiley, New York (1974), p. 164.

Postface

Of all the messages we have tried to convey throughout this book, the most important is the role of the *equations of change*. We have shown how these equations may be solved to obtain the velocity, temperature, and concentration profiles. We have also seen that they form the starting point for the study of the transport phenomena in turbulent systems. Also, when they are integrated over large systems, they give the *macroscopic balances*.

No introductory text can possibly meet the needs of every reader. Therefore, we have tried to present a *solid basis in the fundamentals* necessary for tackling problems presently unforeseen. We have also attempted to give literature citations to works where additional material may be found.

We have also provided examples to illustrate how to solve simple, straightforward problems. Problems at the ends of the chapters enable the readers to test their command of the material. No attempt has been made to discuss *numerical methods*, inasmuch as there are many books available for learning that material, as well as software for performing such calculations. The fundamentals provided herein are necessary for setting up such approaches, troubleshooting implementations, and interpreting and analyzing the results.

It should be recognized that great advances are being made in the molecular theory of transport phenomena. *Molecular simulation techniques* are proving to be very powerful techniques for the understanding of flow in complex systems, flow in porous materials, behavior of thin films, and the motions of polymeric materials in flow systems.

Simple models of *turbulent flow* have been discussed in this book, but these simplified treatments are but a feeble introduction to an enormous research area. Here, numerical methods have proven exceptionally useful. Applications include the forces on aircraft, internal combustion engines, mixer performance, and interpretation of meteorological observations.

Complex problems, for which analytical solutions are not possible, can also be approached by *dimensional analysis*. It is important to understand that the equations of change play a very important role in this kind of approach.

Of increasing importance is the solving of *non-Newtonian flow problems*. This includes polymers, suspensions, emulsions, and biological fluids. Such problems are now being studied by molecular theories, which require considerable background in nonequilibrium statistical mechanics.

No music or oral communication would be possible without *compressible flow*. This subject is needed for the study of shock waves, space vehicle reentry, weather phenomena, and the awesome destructive power of tornadoes.

Some problems involving *flow with chemical reactions* have been presented in this book, but we have given only the simplest possible reactions in our examples. For studying combustion, flame propagation, and explosion phenomena, more complex chemical reactions must be invoked. The same is true in biological problems and transport in the human body, where much has yet to be done.

In this book, we have said very little about *electrical and magnetic effects* in transport phenomena. The transport in ionic systems is quite important in biological systems. Magnetic effects appear in situations ranging from semi-active damping systems in automobiles to various phenomena observed in interstellar space.

No problem in engineering or biotechnology can be solved entirely by reliance on transport phenomena, thermodynamics, and chemical kinetics, although these topics are essential to the fundamental understanding of the problems. Those in the business of solving complex problems may have to resort to *heuristic approaches* to supplement reliance on basic theory. And, above all, the ultimate test of any approach will be in *experimentation*. As the Dutch Nobel Laureate, H. Kamerlingh Onnes, said: "meten is weten," which means "to measure is to know."

RBB
WES
ENL
DJK

Appendix A

Vector and Tensor Notation[1]

The physical quantities encountered in transport phenomena fall into three categories: *scalars*, such as temperature, pressure, volume, and time; *vectors*, such as velocity, momentum, and force; and (second-order) *tensors*, such as the stress, momentum flux, and velocity gradient tensors. We distinguish among these quantities thus:

s = scalar (lightface Italic)

$\mathbf{v}$ = vector (boldface Roman)

$\boldsymbol{\tau}$ = second-order tensor (boldface Greek)

In addition, boldface Greek letters with one directional subscript (such as $\boldsymbol{\delta}_i$) are vectors.

For vectors and tensors, several different kinds of multiplication are possible. Some of these require the use of special multiplication signs to be defined later: the single dot ($\cdot$), the double dot ($:$), and the cross ($\times$). We put these special multiplications, or sums thereof, in different kinds of enclosures to indicate the type of result produced:

parentheses ( ) = scalar

brackets [ ] = vector

braces { } = second-order tensor

No special significance is attached to the kind of parentheses if the only operations enclosed are addition and subtraction, or a multiplication in which $\cdot$, $:$, and $\times$ do not appear. Hence $(\mathbf{v} \cdot \mathbf{w})$ and $(\boldsymbol{\tau} : \nabla \mathbf{v})$ are scalars, $[\nabla \times \mathbf{v}]$ and $[\boldsymbol{\tau} \cdot \mathbf{v}]$ are vectors, and $\{\mathbf{v} \cdot \nabla \boldsymbol{\tau}\}$

[1] This appendix is very similar to Appendix A of R. B. Bird, R. C. Armstrong, and O. Hassager, *Dynamics of Polymeric Liquids, Vol. 1, Fluid Mechanics*, 2nd edition, Wiley-Interscience, New York (1987). There, in §8, a discussion of nonorthogonal coordinates is given. Also in Table A.7-4, there is a summary of the del operations for bipolar coordinates.

and $\{\boldsymbol{\sigma} \cdot \boldsymbol{\tau} + \boldsymbol{\tau} \cdot \boldsymbol{\sigma}\}$ are second-order tensors. On the other hand, $\mathbf{v} - \mathbf{w}$ may be written as $(\mathbf{v} - \mathbf{w})$, $[\mathbf{v} - \mathbf{w}]$, or $\{\mathbf{v} - \mathbf{w}\}$, since no dot or cross operations appear. Similarly $\mathbf{vw}$, $(\mathbf{vw})$, $[\mathbf{vw}]$, and $\{\mathbf{vw}\}$ are all equivalent.

Actually, scalars can be regarded as zero-order tensors and vectors as first-order tensors. The multiplication signs may be interpreted thus:

Multiplication Sign	Order of Result
None	Σ
$\times$	$\Sigma - 1$
$\cdot$	$\Sigma - 2$
$:$	$\Sigma - 4$

in which Σ represents the sum of the orders of the quantities being multiplied. For example, $s\boldsymbol{\tau}$ is of the order $0 + 2 = 2$, $\mathbf{vw}$ is of the order $1 + 1 = 2$, $\boldsymbol{\delta}_1\boldsymbol{\delta}_2$ is of the order $1 + 1 = 2$, $[\mathbf{v} \times \mathbf{w}]$ is of the order $1 + 1 - 1 = 1$, $(\boldsymbol{\sigma} : \boldsymbol{\tau})$ is of the order $2 + 2 - 4 = 0$, and $\{\boldsymbol{\sigma} \cdot \boldsymbol{\tau}\}$ is of the order $2 + 2 - 2 = 2$.

The basic operations that can be performed on scalar quantities need not be elaborated on here. However, the laws for the algebra of scalars may be used to illustrate three terms that arise in the subsequent discussion of vector operations:

a. For the multiplication of two scalars, r and s, the order of multiplication is immaterial so that the *commutative* law is valid: $rs = sr$.

b. For the successive multiplication of three scalars, q, r, and s, the order in which the multiplications are performed is immaterial, so that the *associative* law is valid: $(qr)s = q(rs)$.

c. For the multiplication of a scalar s by the sum of scalars p, q, and r, it is immaterial whether the addition or multiplication is performed first, so that the *distributive* law is valid: $s(p + q + r) = sp + sq + sr$.

These laws are *not* generally valid for the analogous vector and tensor operations described in the following paragraphs.

§A.1 VECTOR OPERATIONS FROM A GEOMETRICAL VIEWPOINT

In elementary physics courses, one is introduced to vectors from a geometrical standpoint. In this section we extend this approach to include the operations of vector multiplication. In §A.2 we give a parallel analytic treatment.

Definition of a Vector and Its Magnitude

A vector $\mathbf{v}$ is defined as a quantity of a given magnitude and direction. The magnitude of the vector is designated by $|\mathbf{v}|$ or simply by the corresponding lightface symbol v. Two vectors $\mathbf{v}$ and $\mathbf{w}$ are equal when their magnitudes are equal and when they point in the same direction; they do not have to be collinear or have the same point of origin. If $\mathbf{v}$ and $\mathbf{w}$ have the same magnitude but point in opposite directions, then $\mathbf{v} = -\mathbf{w}$.

Addition and Subtraction of Vectors

The addition of two vectors can be accomplished by the familiar parallelogram construction, as indicated by Fig. A.1-1a. Vector addition obeys the following laws:

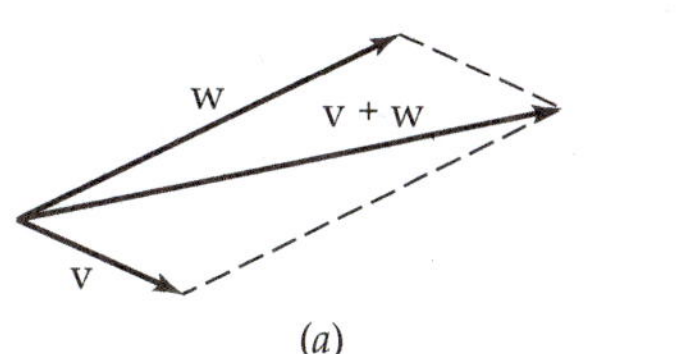

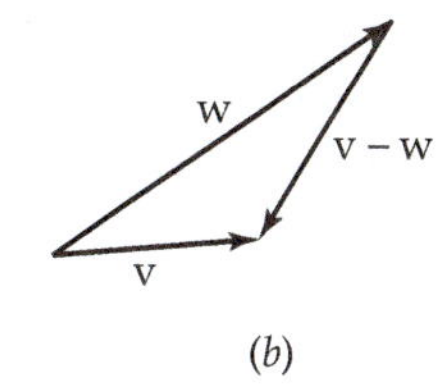

Fig. A.1-1. (*a*) Addition of vectors; (*b*) subtraction of vectors.

Commutative: $(\mathbf{v} + \mathbf{w}) = (\mathbf{w} + \mathbf{v})$ (A.1-1)

Associative: $(\mathbf{v} + \mathbf{w}) + \mathbf{u} = \mathbf{v} + (\mathbf{w} + \mathbf{u})$ (A.1-2)

Vector subtraction is performed by reversing the sign of one vector and adding; thus $\mathbf{v} - \mathbf{w} = \mathbf{v} + (-\mathbf{w})$. The geometrical construction for this is shown in Fig. A.1-1b.

Multiplication of a Vector by a Scalar

When a vector is multiplied by a scalar, the magnitude of the vector is altered but its direction is not. The following laws are applicable

Commutative: $s\mathbf{v} = \mathbf{v}s$ (A.1-3)

Associative: $r(s\mathbf{v}) = (rs)\mathbf{v}$ (A.1-4)

Distributive $(q + r + s)\mathbf{v} = q\mathbf{v} + r\mathbf{v} + s\mathbf{v}$ (A.1-5)

Scalar Product (or Dot Product) of Two Vectors

The scalar product of two vectors **v** and **w** is a scalar quantity defined by

$$(\mathbf{v} \cdot \mathbf{w}) = vw \cos \phi_{\mathbf{vw}} \quad (0 \leq \phi_{\mathbf{vw}} < \pi) \tag{A.1-6}$$

in which $\phi_{\mathbf{vw}}$ is the angle between the vectors **v** and **w**. The scalar product is then the magnitude of **w** multiplied by the projection of **v** on **w**, or vice versa (Fig. A.1-2a). Note that the scalar product of a vector with itself is just the square of the magnitude of the vector

$$(\mathbf{v} \cdot \mathbf{v}) = |\mathbf{v}|^2 = v^2 \tag{A.1-7}$$

The rules governing scalar products are as follows:

Commutative: $(\mathbf{u} \cdot \mathbf{v}) = (\mathbf{v} \cdot \mathbf{u})$ (A.1-8)

Not Associative: $(\mathbf{u} \cdot \mathbf{v})\mathbf{w} \neq \mathbf{u}(\mathbf{v} \cdot \mathbf{w})$ (A.1-9)

Distributive: $(\mathbf{u} \cdot \{\mathbf{v} + \mathbf{w}\}) = (\mathbf{u} \cdot \mathbf{v}) + (\mathbf{u} \cdot \mathbf{w})$ (A.1-10)

Vector Product (or Cross Product) of Two Vectors

The vector product of two vectors **v** and **w** is a vector defined by

$$[\mathbf{v} \times \mathbf{w}] = \{vw \sin \phi_{\mathbf{vw}}\}\mathbf{n}_{\mathbf{vw}} \tag{A.1-11}$$

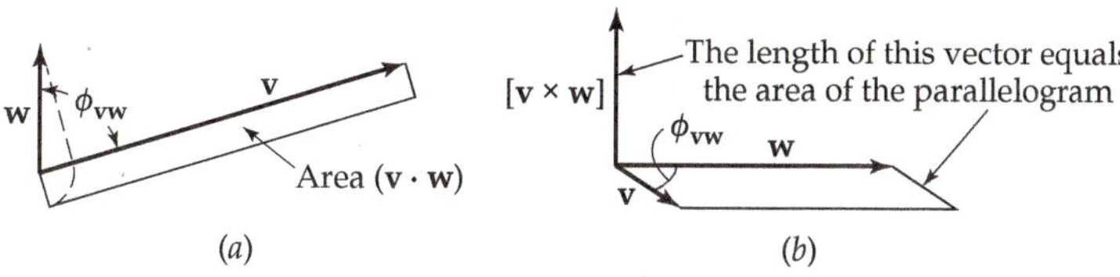

Fig. A.1-2. Products of two vectors: (*a*) the scalar product; (*b*) the vector product.

in which $\mathbf{n}_{vw}$ is a vector of unit length (a "unit vector") perpendicular to both $\mathbf{v}$ and $\mathbf{w}$ and pointing in the direction that a right-handed screw will move if turned from $\mathbf{v}$ toward $\mathbf{w}$ through the angle ϕ_{vw}. The vector product is illustrated in Fig. A.1-2b. The magnitude of the vector product is just the area of the parallelogram defined by the vectors $\mathbf{v}$ and $\mathbf{w}$. It follows from the definition of the vector product that

$$[\mathbf{v} \times \mathbf{v}] = \mathbf{0} \tag{A.1-12}$$

Note the following summary of laws governing the vector product operation:

Not Commutative: $$[\mathbf{v} \times \mathbf{w}] = -[\mathbf{w} \times \mathbf{v}] \tag{A.1-13}$$

Not Associative: $$[\mathbf{u} \times [\mathbf{v} \times \mathbf{w}]] \neq [[\mathbf{u} \times \mathbf{v}] \times \mathbf{w}] \tag{A.1-14}$$

Distributive: $$[\{\mathbf{u} + \mathbf{v}\} \times \mathbf{w}] = [\mathbf{u} \times \mathbf{w}] + [\mathbf{v} \times \mathbf{w}] \tag{A.1-15}$$

Multiple Products of Vectors

Somewhat more complicated are multiple products formed by combinations of the multiplication processes just described:

(a) $rs\mathbf{v}$ (b) $s(\mathbf{v} \cdot \mathbf{w})$ (c) $s[\mathbf{v} \times \mathbf{w}]$
(d) $(\mathbf{u} \cdot [\mathbf{v} \times \mathbf{w}])$ (e) $[\mathbf{u} \times [\mathbf{v} \times \mathbf{w}]]$ (f) $([\mathbf{u} \times \mathbf{v}] \cdot [\mathbf{w} \times \mathbf{z}])$
(g) $[[\mathbf{u} \times \mathbf{v}] \times [\mathbf{w} \times \mathbf{z}]]$

The geometrical interpretations of the first three of these are straightforward. The magnitude of $(\mathbf{u} \cdot [\mathbf{v} \times \mathbf{w}])$ can be shown to represent the volume of a parallelepiped with edges defined by the vectors $\mathbf{u}$, $\mathbf{v}$, and $\mathbf{w}$.

EXERCISES

1. What are the "orders" of the following quantities: $(\mathbf{v} \cdot \mathbf{w})$, $(\mathbf{v} - \mathbf{u})\mathbf{w}$, $(\mathbf{ab} : \mathbf{cd})$, $[\mathbf{v} \cdot \rho\mathbf{wu}]$, $[[\mathbf{a} \times \mathbf{f}] \times [\mathbf{b} \times \mathbf{g}]]$?

2. Draw a sketch to illustrate the inequality in Eq. A.1-9. Are there any special cases for which it becomes an equality?

3. A mathematical plane surface of area S has an orientation given by a unit normal vector $\mathbf{n}$, pointing downstream of the surface. A fluid of density ρ flows through this surface with a velocity $\mathbf{v}$. Show that the mass rate of flow through the surface is $w = \rho(\mathbf{n} \cdot \mathbf{v})S$.

4. A constant force $\mathbf{F}$ acts on a body moving with a velocity $\mathbf{v}$, which is not necessarily collinear with $\mathbf{F}$. Show that the rate at which $\mathbf{F}$ does work on the body is $W = (\mathbf{F} \cdot \mathbf{v})$.

5. The angular velocity $\mathbf{W}$ of a rotating solid body is a vector whose magnitude is the rate of angular displacement (radians per second) and whose direction is that in which a right-handed screw would advance if turned in the same direction. The position vector $\mathbf{r}$ of a point is the vector from the origin of coordinates to the point. Show that the velocity of any point in a rotating solid body is $\mathbf{v} = [\mathbf{W} \times \mathbf{r}]$, relative to an origin located on the axis of rotation.

§A.2 VECTOR OPERATIONS IN TERMS OF COMPONENTS

In this section a parallel analytical treatment is given to each of the topics presented geometrically in §A.1. In the discussion here we restrict ourselves to Cartesian coordinates and label the axes as 1, 2, 3 corresponding to the usual notation of x, y, z of Cartesian coordinates; only right-handed coordinates are used. Except when noted, the formulas given in this and the following section are also valid in cylindrical coordinates (where 1, 2, 3 correspond to r, θ, z) and to spherical coordinates (where 1, 2, 3 correspond to r, θ, ϕ). Cylindrical and spherical coordinate systems are illustrated in Fig. A.6-1.

Many formulas can be expressed compactly in terms of the *Kronecker delta* δ_{ij} and the *permutation symbol* ε_{ijk}. These quantities are defined thus:

$$\begin{cases} \delta_{ij} = +1, & \text{if } i = j \quad \text{(A.2-1)} \\ \delta_{ij} = 0, & \text{if } i \neq j \quad \text{(A.2-2)} \end{cases}$$

$$\begin{cases} \varepsilon_{ijk} = +1, & \text{if } ijk = 123, 231, \text{ or } 312 \quad \text{(A.2-3)} \\ \varepsilon_{ijk} = -1, & \text{if } ijk = 321, 132, \text{ or } 213 \quad \text{(A.2-4)} \\ \varepsilon_{ijk} = 0, & \text{if any two indices are alike} \quad \text{(A.2-5)} \end{cases}$$

Note also that $\varepsilon_{ijk} = (1/2)(i - j)(j - k)(k - i)$.

Several relations involving these quantities are useful in proving some vector and tensor identities

$$\sum_{j=1}^{3} \sum_{k=1}^{3} \varepsilon_{ijk}\varepsilon_{hjk} = 2\delta_{ih} \tag{A.2-6}$$

$$\sum_{k=1}^{3} \varepsilon_{ijk}\varepsilon_{mnk} = \delta_{im}\delta_{jn} - \delta_{in}\delta_{jm} \tag{A.2-7}$$

Note that a three-by-three determinant may be written in terms of the ε_{ijk}

$$\begin{vmatrix} a_{11} & a_{12} & a_{13} \\ a_{21} & a_{22} & a_{23} \\ a_{31} & a_{32} & a_{33} \end{vmatrix} = \sum_{i=1}^{3} \sum_{j=1}^{3} \sum_{k=1}^{3} \varepsilon_{ijk} a_{1i} a_{2j} a_{3k} \tag{A.2-8}$$

The quantity ε_{ijk} thus selects the necessary terms that appear in the determinant and affixes the proper sign to each term.

The Unit Vectors

Let $\boldsymbol{\delta}_1, \boldsymbol{\delta}_2, \boldsymbol{\delta}_3$ be the "unit vectors" (i.e., vectors of unit magnitude) in the direction of the 1, 2, 3 axes[1] (Fig. A.2-1). We can use the definitions of the scalar and vector products to tabulate all possible products of each type

$$\begin{cases} (\boldsymbol{\delta}_1 \cdot \boldsymbol{\delta}_1) = (\boldsymbol{\delta}_2 \cdot \boldsymbol{\delta}_2) = (\boldsymbol{\delta}_3 \cdot \boldsymbol{\delta}_3) = 1 \quad \text{(A.2-9)} \\ (\boldsymbol{\delta}_1 \cdot \boldsymbol{\delta}_2) = (\boldsymbol{\delta}_2 \cdot \boldsymbol{\delta}_3) = (\boldsymbol{\delta}_3 \cdot \boldsymbol{\delta}_1) = 0 \quad \text{(A.2-10)} \end{cases}$$

$$\begin{cases} [\boldsymbol{\delta}_1 \times \boldsymbol{\delta}_1] = [\boldsymbol{\delta}_2 \times \boldsymbol{\delta}_2] = [\boldsymbol{\delta}_3 \times \boldsymbol{\delta}_3] = \mathbf{0} & & \text{(A.2-11)} \\ [\boldsymbol{\delta}_1 \times \boldsymbol{\delta}_2] = \boldsymbol{\delta}_3; \quad [\boldsymbol{\delta}_2 \times \boldsymbol{\delta}_3] = \boldsymbol{\delta}_1; \quad [\boldsymbol{\delta}_3 \times \boldsymbol{\delta}_1] = \boldsymbol{\delta}_2 & & \text{(A.2-12)} \\ [\boldsymbol{\delta}_2 \times \boldsymbol{\delta}_1] = -\boldsymbol{\delta}_3; \quad [\boldsymbol{\delta}_3 \times \boldsymbol{\delta}_2] = -\boldsymbol{\delta}_1; \quad [\boldsymbol{\delta}_1 \times \boldsymbol{\delta}_3] = -\boldsymbol{\delta}_2 & & \text{(A.2-13)} \end{cases}$$

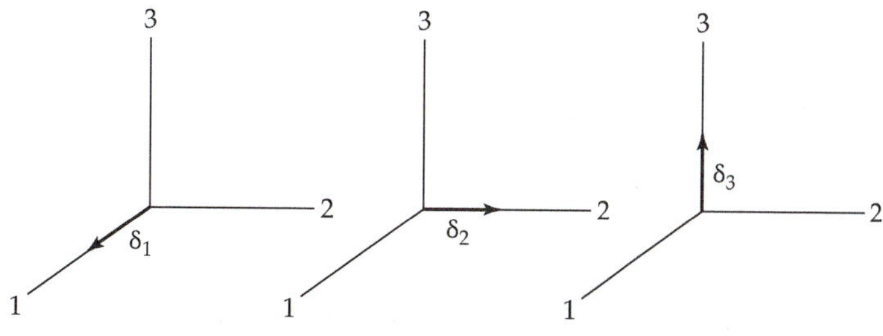

Fig. A.2-1. The unit vectors $\boldsymbol{\delta}_i$; each vector is of unit magnitude and points in the ith direction.

---

[1]In most elementary texts the unit vectors are called $\boldsymbol{i}, \boldsymbol{j}, \boldsymbol{k}$. We prefer to use $\boldsymbol{\delta}_1, \boldsymbol{\delta}_2, \boldsymbol{\delta}_3$ because the components of these vectors are given by the Kronecker delta. That is, the component of $\boldsymbol{\delta}_1$ in the 1-direction is δ_{11} or unity; the component of $\boldsymbol{\delta}_1$ in the 2-direction is δ_{12} or zero.

All of these relations may be summarized by the following two relations:

$$\boxed{(\boldsymbol{\delta}_i \cdot \boldsymbol{\delta}_j) = \delta_{ij}} \tag{A.2-14}$$

$$\boxed{[\boldsymbol{\delta}_i \times \boldsymbol{\delta}_j] = \sum_{k=1}^{3} \varepsilon_{ijk}\boldsymbol{\delta}_k} \tag{A.2-15}$$

in which δ_{ij} is the Kronecker delta, and ε_{ijk} is the permutation symbol defined in the introduction to this section. These two relations enable us to develop analytic expressions for all the common dot and cross operations. In the remainder of this section and in the next section, in developing expressions for vector and tensor operations all we do is to break all vectors up into components and then apply Eqs. A.2-14 and A.2-15.

Expansion of a Vector in Terms of its Components

Any vector **v** can be completely specified by giving the values of its projections v_1, v_2, v_3, on the coordinate axes 1, 2, 3 (Fig. A.2-2). The vector can be constructed by adding vectorially the components multiplied by their corresponding unit vectors:

$$\mathbf{v} = \boldsymbol{\delta}_1 v_1 + \boldsymbol{\delta}_2 v_2 + \boldsymbol{\delta}_3 v_3 = \sum_{i=1}^{3} \boldsymbol{\delta}_i v_i \tag{A.2-16}$$

Note that a *vector associates a scalar with each coordinate direction.*[2] The v_i are called the "components of the vector **v**" and they are scalars, whereas the $\boldsymbol{\delta}_i v_i$ are vectors, which when added together vectorially, give **v**.

The magnitude of a vector is given by

$$|\mathbf{v}| = v = \sqrt{v_1^2 + v_2^2 + v_3^2} = \sqrt{\sum_i v_i^2} \tag{A.2-17}$$

Two vectors **v** and **w** are equal if their components are equal: $v_1 = w_1$, $v_2 = w_2$, and $v_3 = w_3$. Also $\mathbf{v} = -\mathbf{w}$, if $v_1 = -w_1$, and so on.

Addition and Subtraction of Vectors

The sum or difference of vectors **v** and **w** may be written in terms of components as

$$\mathbf{v} \pm \mathbf{w} = \sum_i \boldsymbol{\delta}_i v_i \pm \sum_i \boldsymbol{\delta}_i w_i = \sum_i \boldsymbol{\delta}_i (v_i \pm w_i) \tag{A.2-18}$$

Geometrically, this corresponds to adding up the projections of **v** and **w** on each individual axis and then constructing a vector with these new components. Three or more vectors may be added in exactly the same fashion.

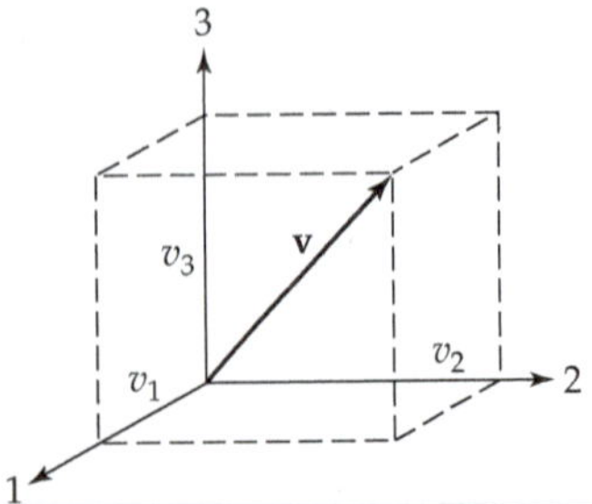

Fig. A.2-2. The components v_i of the vector **v** are the projections of the vector on the coordinate axes 1, 2, and 3.

[2]For a discussion of the relation of this definition of a vector to the definition in terms of the rules for transformation of coordinates, see W. Prager, *Mechanics of Continua*, Ginn, Boston (1961).

Multiplication of a Vector by a Scalar

Multiplication of a vector by a scalar corresponds to multiplying each component of the vector by the scalar:

$$s\mathbf{v} = s\left\{\sum_i \boldsymbol{\delta}_i v_i\right\} = \sum_i \boldsymbol{\delta}_i \{sv_i\} \qquad \text{(A.2-19)}$$

Scalar Product (or Dot Product) of Two Vectors

The scalar product of two vectors **v** and **w** is obtained by writing each vector in terms of components according to Eq. A.2-16 and then performing the scalar-product operations on the unit vectors, using Eq. A.2-14

$$\begin{aligned}(\mathbf{v}\cdot\mathbf{w}) &= \left(\left\{\sum_i \boldsymbol{\delta}_i v_i\right\}\cdot\left\{\sum_j \boldsymbol{\delta}_j w_j\right\}\right) = \sum_i\sum_j (\boldsymbol{\delta}_i\cdot\boldsymbol{\delta}_j) v_i w_j \\ &= \sum_i\sum_j \delta_{ij} v_i w_j = \sum_i v_i w_i \end{aligned} \qquad \text{(A.2-20)}$$

Hence the scalar product of two vectors is obtained by summing the products of the corresponding components of the two vectors. Note that $(\mathbf{v}\cdot\mathbf{v})$ (sometimes written as $\mathbf{v}^2$ or as v^2) is a scalar representing the square of the magnitude of **v**.

Vector Product (or Cross Product) of Two Vectors

The vector product of two vectors **v** and **w** may be worked out by using Eqs. A.2-16 and A.2-15:

$$\begin{aligned}[\mathbf{v}\times\mathbf{w}] &= \left[\left\{\sum_j \boldsymbol{\delta}_j v_j\right\}\times\left\{\sum_k \boldsymbol{\delta}_k w_k\right\}\right] \\ &= \sum_j\sum_k [\boldsymbol{\delta}_j\times\boldsymbol{\delta}_k] v_j w_k = \sum_i\sum_j\sum_k \varepsilon_{ijk}\boldsymbol{\delta}_i v_j w_k \\ &= \begin{vmatrix}\boldsymbol{\delta}_1 & \boldsymbol{\delta}_2 & \boldsymbol{\delta}_3 \\ v_1 & v_2 & v_3 \\ w_1 & w_2 & w_3\end{vmatrix}\end{aligned} \qquad \text{(A.2-21)}$$

Here we have made use of Eq. A.2-8. Note that the ith-component of $[\mathbf{v}\times\mathbf{w}]$ is given by $\sum_j\sum_k \varepsilon_{ijk} v_j w_k$; this result is often used in proving vector identities.

Multiple Vector Products

Expressions for the multiple products mentioned in §A.1 can be obtained by using the preceding analytical expressions for the scalar and vector products. For example, the product $(\mathbf{u}\cdot[\mathbf{v}\times\mathbf{w}])$ may be written

$$(\mathbf{u}\cdot[\mathbf{v}\times\mathbf{w}]) = \sum_i u_i[\mathbf{v}\times\mathbf{w}]_i = \sum_i\sum_j\sum_k \varepsilon_{ijk} u_i v_j w_k \qquad \text{(A.2-22)}$$

Then, from Eq. A.2-8, we obtain

$$(\mathbf{u}\cdot[\mathbf{v}\times\mathbf{w}]) = \begin{vmatrix} u_1 & u_2 & u_3 \\ v_1 & v_2 & v_3 \\ w_1 & w_2 & w_3\end{vmatrix} \qquad \text{(A.2-23)}$$

The magnitude of $(\mathbf{u}\cdot[\mathbf{v}\times\mathbf{w}])$ is the volume of a parellelepiped defined by the vectors **u**, **v**, **w** drawn from a common origin. Furthermore, the vanishing of the determinant is a necessary and sufficient condition that the vectors **u**, **v**, and **w** be coplanar.

The Position Vector

The usual symbol for the position vector—that is, the vector specifying the location of a point in space—is **r**. *In Cartesian coordinates*, the components of **r** are then x_1, x_2, and x_3, so that

$$\mathbf{r} = \sum_i \boldsymbol{\delta}_i x_i \tag{A.2-24}$$

This is an irregularity in the notation, since the components have a symbol different from that for the vector. The magnitude of **r** is usually called $r = \sqrt{x_1^2 + x_2^2 + x_3^2}$, and this r is the radial coordinate in spherical coordinates (see Fig. A.6-1 for how to write the position vector in cylindrical and spherical coordinates).

EXAMPLE A.2-1

Proof of a Vector Identity

The analytical expressions for dot and cross products may be used to prove vector identities; for example, verify the relation

$$[\mathbf{u} \times [\mathbf{v} \times \mathbf{w}]] = \mathbf{v}(\mathbf{u} \cdot \mathbf{w}) - \mathbf{w}(\mathbf{u} \cdot \mathbf{v}) \tag{A.2-25}$$

SOLUTION

The i-component of the expression on the left side can be expanded as

$$\begin{aligned}[\mathbf{u} \times [\mathbf{v} \times \mathbf{w}]]_i &= \sum_j \sum_k \varepsilon_{ijk} u_j [\mathbf{v} \times \mathbf{w}]_k = \sum_j \sum_k \varepsilon_{ijk} u_j \left\{ \sum_l \sum_m \varepsilon_{klm} v_l w_m \right\} \\ &= \sum_j \sum_k \sum_l \sum_m \varepsilon_{ijk} \varepsilon_{klm} u_j v_l w_m = \sum_j \sum_k \sum_l \sum_m \varepsilon_{ijk} \varepsilon_{lmk} u_j v_l w_m \end{aligned} \tag{A.2-26}$$

We may now use Eq. A.2-7 to complete the proof

$$\begin{aligned}[\mathbf{u} \times [\mathbf{v} \times \mathbf{w}]]_i &= \sum_j \sum_l \sum_m (\delta_{il}\delta_{jm} - \delta_{im}\delta_{jl}) u_j v_l w_m = v_i \sum_j \sum_m \delta_{jm} u_j w_m - w_i \sum_j \sum_l \delta_{jl} u_j v_l \\ &= v_i \sum_j u_j w_j - w_i \sum_j u_j v_j = v_i (\mathbf{u} \cdot \mathbf{w}) - w_i (\mathbf{u} \cdot \mathbf{v}) \end{aligned} \tag{A.2-27}$$

which is just the i-component of the right side of Eq. A.2-25. In a similar way one may verify such identities as

$$(\mathbf{u} \cdot [\mathbf{v} \times \mathbf{w}]) = (\mathbf{v} \cdot [\mathbf{w} \times \mathbf{u}]) \tag{A.2-28}$$

$$([\mathbf{u} \times \mathbf{v}] \cdot [\mathbf{w} \times \mathbf{z}]) = (\mathbf{u} \cdot \mathbf{w})(\mathbf{v} \cdot \mathbf{z}) - (\mathbf{u} \cdot \mathbf{z})(\mathbf{v} \cdot \mathbf{w}) \tag{A.2-29}$$

$$[[\mathbf{u} \times \mathbf{v}] \times [\mathbf{w} \times \mathbf{z}]] = ([\mathbf{u} \times \mathbf{v}] \cdot \mathbf{z})\mathbf{w} - ([\mathbf{u} \times \mathbf{v}] \cdot \mathbf{w})\mathbf{z} \tag{A.2-30}$$

EXERCISES

1. Write out the following summations:

   (a) $\sum_{k=1}^{3} k^2$ **(b)** $\sum_{k=1}^{3} a_k^2$ **(c)** $\sum_{j=1}^{3} \sum_{k=1}^{3} a_{jk} b_{kj}$ **(d)** $\left(\sum_{j=1}^{3} a_j \right)^2 = \sum_{j=1}^{3} \sum_{k=1}^{3} a_j a_k$

2. A vector **v** has components $v_x = 1, v_y = 2, v_z = -5$. A vector **w** has components $w_x = 3, w_y = -1, w_z = 1$. Evaluate:

   (a) $(\mathbf{v} \cdot \mathbf{w})$

   (b) $[\mathbf{v} \times \mathbf{w}]$

   (c) The length of **v**

   (d) $(\boldsymbol{\delta}_x \cdot \mathbf{v})$

   (e) $[\boldsymbol{\delta}_x \times \mathbf{w}]$

   (f) $\phi_{\mathbf{vw}}$

   (g) $[\mathbf{r} \times \mathbf{v}]$, where **r** is the position vector.

3. Evaluate:
   (a) $([\boldsymbol{\delta}_1 \times \boldsymbol{\delta}_2] \cdot \boldsymbol{\delta}_3)$ (b) $[[\boldsymbol{\delta}_2 \times \boldsymbol{\delta}_3] \times [\boldsymbol{\delta}_1 \times \boldsymbol{\delta}_3]]$.

4. Show that Eq. A.2-6 is valid for the particular case $i = 1, h = 2$.
   Show that Eq. A.2-7 is valid for the particular case $i = j = m = 1, n = 2$.

5. Verify that $\sum_{j=1}^{3} \sum_{k=1}^{3} \varepsilon_{ijk} a_{jk} = 0$ if $a_{jk} = a_{kj}$.

6. Explain carefully the statement after Eq. A.2-21 that the ith component of $[\mathbf{v} \times \mathbf{w}]$ is $\sum_j \sum_k \varepsilon_{ijk} v_j w_k$.

7. Verify that $([\mathbf{v} \times \mathbf{w}] \cdot [\mathbf{v} \times \mathbf{w}]) + (\mathbf{v} \cdot \mathbf{w})^2 = v^2 w^2$ (the "identity of Lagrange").

§A.3 TENSOR OPERATIONS IN TERMS OF COMPONENTS

In the last section we saw that expressions could be developed for all common dot and cross operations for vectors by knowing how to write a vector $\mathbf{v}$ as a sum $\sum_i \boldsymbol{\delta}_i v_i$, and by knowing how to manipulate the unit vectors $\boldsymbol{\delta}_i$. In this section we follow a parallel procedure. We write a tensor $\boldsymbol{\tau}$ as a sum $\sum_i \sum_j \boldsymbol{\delta}_i \boldsymbol{\delta}_j \tau_{ij}$, and give formulas for the manipulation of the unit dyads $\boldsymbol{\delta}_i \boldsymbol{\delta}_j$; in this way, expressions are developed for the commonly occurring dot and cross operations for tensors.

The Unit Dyads

The unit vectors $\boldsymbol{\delta}_i$ were defined in the preceding discussion and then the *scalar products* $(\boldsymbol{\delta}_i \cdot \boldsymbol{\delta}_j)$ and *vector products* $[\boldsymbol{\delta}_i \times \boldsymbol{\delta}_j]$ were given. There is a third kind of product that can be formed with the unit vectors—namely, the *dyadic products* $\boldsymbol{\delta}_i \boldsymbol{\delta}_j$ (written without any multiplication symbol). According to the rules of notation given in the introduction to Appendix A, the products $\boldsymbol{\delta}_i \boldsymbol{\delta}_j$ are tensors of the second order. Since $\boldsymbol{\delta}_i$ and $\boldsymbol{\delta}_j$ are of unit magnitude, we will refer to the products $\boldsymbol{\delta}_i \boldsymbol{\delta}_j$ as *unit dyads*. Whereas each unit vector in Fig. A.2-1 represents a single coordinate direction, the unit dyads in Fig. A.3-1 represent *ordered* pairs of coordinate directions.

(In physical problems we often work with quantities that require the simultaneous specification of two directions. For example, the flux of x-momentum across a unit area of surface perpendicular to the y direction is a quantity of this type. Since this quantity is sometimes not the same as the flux of y-momentum perpendicular to the x direction, it is evident that specifying the two directions is not sufficient; we must also agree on the order in which the directions are given.)

The dot and cross products of unit vectors were introduced by means of the geometrical definitions of these operations. The analogous operations for the unit dyads are introduced formally by relating them to the operations for unit vectors

$$(\boldsymbol{\delta}_i \boldsymbol{\delta}_j : \boldsymbol{\delta}_k \boldsymbol{\delta}_l) = (\boldsymbol{\delta}_j \cdot \boldsymbol{\delta}_k)(\boldsymbol{\delta}_i \cdot \boldsymbol{\delta}_l) = \delta_{jk} \delta_{il} \tag{A.3-1}$$

$$[\boldsymbol{\delta}_i \boldsymbol{\delta}_j \cdot \boldsymbol{\delta}_k] = \boldsymbol{\delta}_i (\boldsymbol{\delta}_j \cdot \boldsymbol{\delta}_k) = \boldsymbol{\delta}_i \delta_{jk} \tag{A.3-2}$$

$$[\boldsymbol{\delta}_i \cdot \boldsymbol{\delta}_j \boldsymbol{\delta}_k] = (\boldsymbol{\delta}_i \cdot \boldsymbol{\delta}_j) \boldsymbol{\delta}_k = \delta_{ij} \boldsymbol{\delta}_k \tag{A.3-3}$$

$$\{\boldsymbol{\delta}_i \boldsymbol{\delta}_j \cdot \boldsymbol{\delta}_k \boldsymbol{\delta}_l\} = \boldsymbol{\delta}_i (\boldsymbol{\delta}_j \cdot \boldsymbol{\delta}_k) \boldsymbol{\delta}_l = \delta_{jk} \boldsymbol{\delta}_i \boldsymbol{\delta}_l \tag{A.3-4}$$

$$\{\boldsymbol{\delta}_i \boldsymbol{\delta}_j \times \boldsymbol{\delta}_k\} = \boldsymbol{\delta}_i [\boldsymbol{\delta}_j \times \boldsymbol{\delta}_k] = \sum_{l=1}^{3} \varepsilon_{jkl} \boldsymbol{\delta}_i \boldsymbol{\delta}_l \tag{A.3-5}$$

$$\{\boldsymbol{\delta}_i \times \boldsymbol{\delta}_j \boldsymbol{\delta}_k\} = [\boldsymbol{\delta}_i \times \boldsymbol{\delta}_j] \boldsymbol{\delta}_k = \sum_{l=1}^{3} \varepsilon_{ijl} \boldsymbol{\delta}_l \boldsymbol{\delta}_k \tag{A.3-6}$$

These results are easy to remember: one simply takes the dot (or cross) product of the nearest unit vectors on either side of the dot (or cross); in Eq. A.3-1 two such operations are performed.

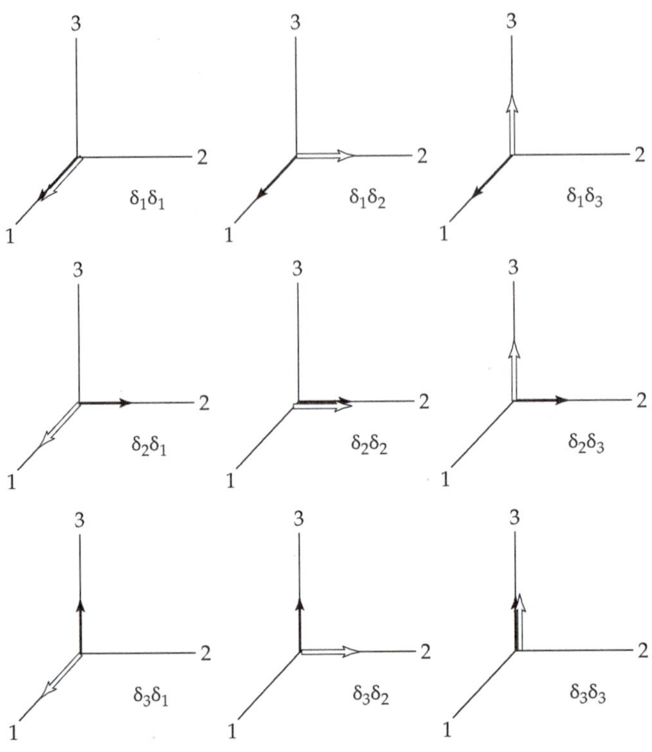

Fig. A.3-1. The unit dyads $\boldsymbol{\delta}_i\boldsymbol{\delta}_j$. The solid arrows represent the first unit vector in the dyadic product, and the hollow vectors the second. Note that $\boldsymbol{\delta}_1\boldsymbol{\delta}_2$ is not the same as $\boldsymbol{\delta}_2\boldsymbol{\delta}_1$.

Expansion of a Tensor in Terms of Its Components

In Eq. A.2-16 we expanded a vector in terms of its components, each component being multiplied by the appropriate unit vector. Here we extend this idea and define[1] a (second-order) tensor as *a quantity that associates a scalar with each ordered pair of coordinate directions* in the following sense:

$$\begin{aligned}\boldsymbol{\tau} &= \boldsymbol{\delta}_1\boldsymbol{\delta}_1\tau_{11} + \boldsymbol{\delta}_1\boldsymbol{\delta}_2\tau_{12} + \boldsymbol{\delta}_1\boldsymbol{\delta}_3\tau_{13} \\ &\quad + \boldsymbol{\delta}_2\boldsymbol{\delta}_1\tau_{21} + \boldsymbol{\delta}_2\boldsymbol{\delta}_2\tau_{22} + \boldsymbol{\delta}_2\boldsymbol{\delta}_3\tau_{23} \\ &\quad + \boldsymbol{\delta}_3\boldsymbol{\delta}_1\tau_{31} + \boldsymbol{\delta}_3\boldsymbol{\delta}_2\tau_{32} + \boldsymbol{\delta}_3\boldsymbol{\delta}_3\tau_{33} \\ &= \sum_{i=1}^{3}\sum_{j=1}^{3} \boldsymbol{\delta}_i\boldsymbol{\delta}_j\tau_{ij}\end{aligned} \tag{A.3-7}$$

The scalars τ_{ij} are referred to as the "components of the tensor $\boldsymbol{\tau}$."

There are several special kinds of second-order tensors worth noting:

1. If $\tau_{ij} = \tau_{ji}$, the tensor is said to be *symmetric*.
2. If $\tau_{ij} = -\tau_{ji}$, the tensor is said to be *antisymmetric*.
3. If the components of a tensor are taken to be the components of $\boldsymbol{\tau}$, but with the indices transposed, the resulting tensor is called the *transpose* of $\boldsymbol{\tau}$ and given the symbol $\boldsymbol{\tau}^{\dagger}$:

$$\boldsymbol{\tau}^{\dagger} = \sum_i\sum_j \boldsymbol{\delta}_i\boldsymbol{\delta}_j\tau_{ji} \tag{A.3-8}$$

[1]Tensors are often defined in terms of the transformation rules; the connections between such a definition and that given above is discussed by W. Prager, *Mechanics of Continua*, Ginn, Boston (1961).

4. If the components of the tensor are formed by ordered pairs of the components of two vectors **v** and **w**, the resulting tensor is called the *dyadic product of* **v** and **w** and given the symbol **vw**:

$$\mathbf{vw} = \sum_i \sum_j \boldsymbol{\delta}_i \boldsymbol{\delta}_j v_i w_j \tag{A.3-9}$$

   Note that $\mathbf{vw} \neq \mathbf{wv}$, but that $(\mathbf{vw})^\dagger = \mathbf{wv}$.

5. If the components of the tensor are given by the Kronecker delta δ_{ij}, the resulting tensor is called the *unit tensor* and given the symbol $\boldsymbol{\delta}$:

$$\boldsymbol{\delta} = \sum_i \sum_j \boldsymbol{\delta}_i \boldsymbol{\delta}_j \delta_{ij} \tag{A.3-10}$$

The magnitude of a tensor is defined by

$$\begin{aligned} |\boldsymbol{\tau}| = \tau &= \sqrt{\frac{1}{2}(\boldsymbol{\tau} : \boldsymbol{\tau}^\dagger)} \\ &= \sqrt{\frac{1}{2}\sum_i \sum_j \tau_{ij}^2} \end{aligned} \tag{A.3-11}$$

Addition of Tensors and Dyadic Products

Two tensors are added thus:

$$\boldsymbol{\sigma} + \boldsymbol{\tau} = \sum_i \sum_j \boldsymbol{\delta}_i \boldsymbol{\delta}_j \sigma_{ij} + \sum_i \sum_j \boldsymbol{\delta}_i \boldsymbol{\delta}_j \tau_{ij} = \sum_i \sum_j \boldsymbol{\delta}_i \boldsymbol{\delta}_j (\sigma_{ij} + \tau_{ij}) \tag{A.3-12}$$

That is, the sum of two tensors is that tensor whose components are the sums of the corresponding components of the two tensors. The same is true for dyadic products.

Multiplication of a Tensor by a Scalar

Multiplication of a tensor by a scalar corresponds to multiplying each component of the tensor by the scalar:

$$s\boldsymbol{\tau} = s\left\{\sum_i \sum_j \boldsymbol{\delta}_i \boldsymbol{\delta}_j \tau_{ij}\right\} = \sum_i \sum_j \boldsymbol{\delta}_i \boldsymbol{\delta}_j \{s\tau_{ij}\} \tag{A.3-13}$$

The same is true for dyadic products.

The Scalar Product (or Double Dot Product) of Two Tensors

Two tensors may be multiplied according to the double dot operation

$$\begin{aligned} (\boldsymbol{\sigma} : \boldsymbol{\tau}) &= \left(\left\{\sum_i \sum_j \boldsymbol{\delta}_i \boldsymbol{\delta}_j \sigma_{ij}\right\} : \left\{\sum_k \sum_l \boldsymbol{\delta}_k \boldsymbol{\delta}_l \tau_{kl}\right\}\right) = \sum_i \sum_j \sum_k \sum_l (\boldsymbol{\delta}_i \boldsymbol{\delta}_j : \boldsymbol{\delta}_k \boldsymbol{\delta}_l) \sigma_{ij} \tau_{kl} \\ &= \sum_i \sum_j \sum_k \sum_l \delta_{il} \delta_{jk} \sigma_{ij} \tau_{kl} = \sum_i \sum_j \sigma_{ij} \tau_{ji} \end{aligned} \tag{A.3-14}$$

in which Eq. A.3-1 has been used. Similarly, we may show that

$$(\boldsymbol{\tau} : \mathbf{vw}) = \sum_i \sum_j \tau_{ij} v_j w_i \tag{A.3-15}$$

$$(\mathbf{uv} : \mathbf{wz}) = \sum_i \sum_j u_i v_j w_j z_i \tag{A.3-16}$$

The Tensor Product (the Single Dot Product) of Two Tensors

Two tensors may also be multiplied according to the single dot operation

$$\{\boldsymbol{\sigma}\cdot\boldsymbol{\tau}\} = \left\{\left(\sum_i\sum_j \boldsymbol{\delta}_i\boldsymbol{\delta}_j\sigma_{ij}\right)\cdot\left(\sum_k\sum_l \boldsymbol{\delta}_k\boldsymbol{\delta}_l\tau_{kl}\right)\right\} = \sum_i\sum_j\sum_k\sum_l\{\boldsymbol{\delta}_i\boldsymbol{\delta}_j\cdot\boldsymbol{\delta}_k\boldsymbol{\delta}_l\}\sigma_{ij}\tau_{kl}$$

$$= \sum_i\sum_j\sum_k\sum_l \delta_{jk}\boldsymbol{\delta}_i\boldsymbol{\delta}_l\sigma_{ij}\tau_{kl} = \sum_i\sum_l \boldsymbol{\delta}_i\boldsymbol{\delta}_l\left(\sum_j \sigma_{ij}\tau_{jl}\right) \quad \text{(A.3-17)}$$

That is, the il-component of $\{\boldsymbol{\sigma}\cdot\boldsymbol{\tau}\}$ is $\sum_j\sigma_{ij}\tau_{jl}$. Similar operations may be performed with dyadic products. It is common practice to write $\{\boldsymbol{\sigma}\cdot\boldsymbol{\sigma}\}$ as $\boldsymbol{\sigma}^2$, $\{\boldsymbol{\sigma}\cdot\boldsymbol{\sigma}^2\}$ as $\boldsymbol{\sigma}^3$, and so on.

The Vector Product (or Dot Product) of a Tensor with a Vector

When a tensor is dotted into a vector, we get a vector

$$[\boldsymbol{\tau}\cdot\mathbf{v}] = \left[\left\{\sum_i\sum_j \boldsymbol{\delta}_i\boldsymbol{\delta}_j\tau_{ij}\right\}\cdot\left\{\sum_k \boldsymbol{\delta}_k v_k\right\}\right] = \sum_i\sum_j\sum_k[\boldsymbol{\delta}_i\boldsymbol{\delta}_j\cdot\boldsymbol{\delta}_k]\tau_{ij}v_k$$

$$= \sum_i\sum_j\sum_k \boldsymbol{\delta}_i\delta_{jk}\tau_{ij}v_k = \sum_i \boldsymbol{\delta}_i\left\{\sum_j \tau_{ij}v_j\right\} \quad \text{(A.3-18)}$$

That is, the ith component of $[\boldsymbol{\tau}\cdot\mathbf{v}]$ is $\sum_j\tau_{ij}v_j$. Similarly, the ith component of $[\mathbf{v}\cdot\boldsymbol{\tau}]$ is $\sum_j v_j\tau_{ji}$. Note that $[\boldsymbol{\tau}\cdot\mathbf{v}] \neq [\mathbf{v}\cdot\boldsymbol{\tau}]$ unless $\boldsymbol{\tau}$ is symmetric.

Recall that when a vector $\mathbf{v}$ is multiplied by a scalar s, the resultant vector $s\mathbf{v}$ points in the same direction as $\mathbf{v}$ but has a different length. However, when $\boldsymbol{\tau}$ is dotted into $\mathbf{v}$, the resultant vector $[\boldsymbol{\tau}\cdot\mathbf{v}]$ differs from $\mathbf{v}$ in *both* length and direction; that is, the tensor $\boldsymbol{\tau}$ "deflects" or "twists" the vector $\mathbf{v}$ to form a new vector pointing in a different direction.

The Tensor Product (or Cross Product) of a Tensor with a Vector

When a tensor is crossed with a vector, we get a tensor:

$$\{\boldsymbol{\tau}\times\mathbf{v}\} = \left\{\left(\sum_i\sum_j \boldsymbol{\delta}_i\boldsymbol{\delta}_j\tau_{ij}\right)\times\left(\sum_k \boldsymbol{\delta}_k v_k\right)\right\} = \sum_i\sum_j\sum_k[\boldsymbol{\delta}_i\boldsymbol{\delta}_j\times\boldsymbol{\delta}_k]\tau_{ij}v_k$$

$$= \sum_i\sum_j\sum_k\sum_l \varepsilon_{jkl}\boldsymbol{\delta}_i\boldsymbol{\delta}_l\tau_{ij}v_k = \sum_i\sum_l \boldsymbol{\delta}_i\boldsymbol{\delta}_l\left\{\sum_j\sum_k \varepsilon_{jkl}\tau_{ij}v_k\right\} \quad \text{(A.3-19)}$$

Hence, the il-component of $\{\boldsymbol{\tau}\times\mathbf{v}\}$ is $\sum_j\sum_k\varepsilon_{jkl}\tau_{ij}v_k$. Similarly the lk-component of $\{\mathbf{v}\times\boldsymbol{\tau}\}$ is $\sum_i\sum_j\varepsilon_{ijl}v_i\tau_{jk}$.

Other Operations

From the preceding results, it is not difficult to prove the following identities:

$$[\boldsymbol{\delta}\cdot\mathbf{v}] = [\mathbf{v}\cdot\boldsymbol{\delta}] = \mathbf{v} \quad \text{(A.3-20)}$$

$$[\mathbf{uv}\cdot\mathbf{w}] = \mathbf{u}(\mathbf{v}\cdot\mathbf{w}) \quad \text{(A.3-21)}$$

$$[\mathbf{w}\cdot\mathbf{uv}] = (\mathbf{w}\cdot\mathbf{u})\mathbf{v} \quad \text{(A.3-22)}$$

$$(\mathbf{uv}:\mathbf{wz}) = (\mathbf{uw}:\mathbf{vz}) = (\mathbf{u}\cdot\mathbf{z})(\mathbf{v}\cdot\mathbf{w}) \quad \text{(A.3-23)}$$

$$(\boldsymbol{\tau}:\mathbf{uv}) = ([\boldsymbol{\tau}\cdot\mathbf{u}]\cdot\mathbf{v}) \quad \text{(A.3-24)}$$

$$(\mathbf{uv}:\boldsymbol{\tau}) = (\mathbf{u}\cdot[\mathbf{v}\cdot\boldsymbol{\tau}]) \quad \text{(A.3-25)}$$

EXERCISES

1. The components of a symmetric tensor $\boldsymbol{\tau}$ are

$$\begin{array}{lll} \tau_{xx} = 3 & \tau_{xy} = 2 & \tau_{xz} = -1 \\ \tau_{yx} = 2 & \tau_{yy} = 2 & \tau_{yz} = 1 \\ \tau_{zx} = -1 & \tau_{zy} = 1 & \tau_{zz} = 4 \end{array}$$

The components of a vector **v** are

$$v_x = 5 \qquad v_y = 3 \qquad v_z = -2$$

Evaluate
(a) $[\boldsymbol{\tau} \cdot \mathbf{v}]$ **(c)** $(\boldsymbol{\tau} : \boldsymbol{\tau})$ **(e)** $\mathbf{vv}$
(b) $[\mathbf{v} \cdot \boldsymbol{\tau}]$ **(d)** $(\mathbf{v} \cdot [\boldsymbol{\tau} \cdot \mathbf{v}])$ **(f)** $[\boldsymbol{\tau} \cdot \boldsymbol{\delta}_x]$

2. Evaluate
(a) $[[\boldsymbol{\delta}_1\boldsymbol{\delta}_2 \cdot \boldsymbol{\delta}_2] \times \boldsymbol{\delta}_1]$ **(c)** $(\boldsymbol{\delta} : \boldsymbol{\delta})$
(b) $(\boldsymbol{\delta} : \boldsymbol{\delta}_1\boldsymbol{\delta}_2)$ **(d)** $\{\boldsymbol{\delta} \cdot \boldsymbol{\delta}\}$

3. If $\boldsymbol{\alpha}$ is symmetrical and $\boldsymbol{\beta}$ is antisymmetrical, show that $(\boldsymbol{\alpha} : \boldsymbol{\beta}) = 0$.

4. Explain carefully the statement after Eq. A.3-17 that the *il*-component of $\{\boldsymbol{\sigma} \cdot \boldsymbol{\tau}\}$ is $\sum_j \sigma_{ij}\tau_{jl}$.

5. Consider a rigid structure composed of point particles joined by massless rods. The particles are numbered 1,2,3,..., N, and the particle masses are m_ν ($\nu = 1,2,\ldots, N$). The locations of the particles with respect to the center of mass are $\mathbf{R}_\nu$. The entire structure rotates on an axis passing through the center of mass with an angular velocity **W**. Show that the angular momentum with respect to the center of mass is

$$\mathbf{L} = \sum_\nu m_\nu[\mathbf{R}_\nu \times [\mathbf{W} \times \mathbf{R}_\nu]] \tag{A.3-26}$$

Then show that the latter expression may be rewritten as

$$\mathbf{L} = [\boldsymbol{\Phi} \cdot \mathbf{W}] \tag{A.3-27}$$

where

$$\boldsymbol{\Phi} = \sum_\nu m_\nu\{(\mathbf{R}_\nu \cdot \mathbf{R}_\nu)\boldsymbol{\delta} - \mathbf{R}_\nu\mathbf{R}_\nu\} \tag{A.3-28}$$

is the *moment-of-inertia tensor*.

6. The kinetic energy of rotation of the rigid structure in Exercise 5 is

$$K = \sum_\nu \frac{1}{2}m_\nu(\dot{\mathbf{R}}_\nu \cdot \dot{\mathbf{R}}_\nu) \tag{A.3-29}$$

where $\dot{\mathbf{R}}_\nu = [\mathbf{W} \times \mathbf{R}_\nu]$ is the velocity of the νth particle. Show that

$$K = \frac{1}{2}(\boldsymbol{\Phi} : \mathbf{WW}) \tag{A.3-30}$$

§A.4 VECTOR AND TENSOR DIFFERENTIAL OPERATIONS

The vector differential operator ∇, known as "nabla" or "del," is defined in rectangular coordinates as

$$\nabla = \boldsymbol{\delta}_1\frac{\partial}{\partial x_1} + \boldsymbol{\delta}_2\frac{\partial}{\partial x_2} + \boldsymbol{\delta}_3\frac{\partial}{\partial x_3} = \sum_i \boldsymbol{\delta}_i\frac{\partial}{\partial x_i} \tag{A.4-1}$$

in which the $\boldsymbol{\delta}_i$ are the unit vectors and the x_i are the variables associated with the 1, 2, 3 axes (i.e., the x_1, x_2, x_3 are the Cartesian coordinates normally referred to as x, y, z). The symbol ∇ is a vector-operator—it has components like a vector but it cannot stand alone;

it must operate on a scalar, vector, or tensor function. In this section we summarize the various operations of ∇ on scalars, vectors, and tensors. As in §A.2 and §A.3, we decompose vectors and tensors into their components and then use Eqs. A.2-14 and A.2-15, and Eqs. A.3-1 to 6. Keep in mind that in this section equations written out in component form are valid only for Cartesian coordinates, for which the unit vectors $\boldsymbol{\delta}_i$ are constants; curvilinear coordinates are discussed in §A.6 and §A.7.

The Gradient of a Scalar Field

If s is a scalar function of the variables x_1, x_2, x_3, then the operation of ∇ on s is

$$\nabla s = \boldsymbol{\delta}_1 \frac{\partial s}{\partial x_1} + \boldsymbol{\delta}_2 \frac{\partial s}{\partial x_2} + \boldsymbol{\delta}_3 \frac{\partial s}{\partial x_3} = \sum_i \boldsymbol{\delta}_i \frac{\partial s}{\partial x_i} \tag{A.4-2}$$

The vector thus constructed from the derivatives of s is designated by ∇s (or grad s) and is called the *gradient* of the scalar field s. The following properties of the gradient operation should be noted.

Not Commutative: $$\nabla s \neq s\nabla \tag{A.4-3}$$

Not Associative: $$(\nabla r)s \neq \nabla(rs) \tag{A.4-4}$$

Distributive: $$\nabla(r+s) = \nabla r + \nabla s \tag{A.4-5}$$

The Divergence of a Vector Field

If the vector $\mathbf{v}$ is a function of the space variables x_1, x_2, x_3, then a scalar product may be formed with the operator ∇; in obtaining the final form, we use Eq. A.2-14:

$$\begin{aligned}(\nabla \cdot \mathbf{v}) &= \left(\left\{\sum_i \boldsymbol{\delta}_i \frac{\partial}{\partial x_i}\right\} \cdot \left\{\sum_j \boldsymbol{\delta}_j v_j\right\}\right) = \sum_i \sum_j (\boldsymbol{\delta}_i \cdot \boldsymbol{\delta}_j) \frac{\partial}{\partial x_i} v_j \\ &= \sum_i \sum_j \delta_{ij} \frac{\partial}{\partial x_i} v_j = \sum_i \frac{\partial v_i}{\partial x_i}\end{aligned} \tag{A.4-6}$$

This collection of derivatives of the components of the vector $\mathbf{v}$ is called the *divergence* of $\mathbf{v}$ (sometimes abbreviated div $\mathbf{v}$). Some properties of the divergence operation should be noted

Not Commutative: $$(\nabla \cdot \mathbf{v}) \neq (\mathbf{v} \cdot \nabla) \tag{A.4-7}$$

Not Associative: $$(\nabla \cdot s\mathbf{v}) \neq (\nabla s \cdot \mathbf{v}) \tag{A.4-8}$$

Distributive: $$(\nabla \cdot \{\mathbf{v} + \mathbf{w}\}) = (\nabla \cdot \mathbf{v}) + (\nabla \cdot \mathbf{w}) \tag{A.4-9}$$

The Curl of a Vector Field

A cross product may also be formed between the ∇ operator and the vector $\mathbf{v}$, which is a function of the three space variables. This cross product may be simplified by using Eq. A.2-15 and written in a variety of forms

$$\begin{aligned}[\nabla \times \mathbf{v}] &= \left[\left\{\sum_j \boldsymbol{\delta}_j \frac{\partial}{\partial x_j}\right\} \times \left\{\sum_k \boldsymbol{\delta}_k v_k\right\}\right] \\ &= \sum_j \sum_k [\boldsymbol{\delta}_j \times \boldsymbol{\delta}_k] \frac{\partial}{\partial x_j} v_k = \sum_i \sum_j \sum_k \varepsilon_{ijk} \boldsymbol{\delta}_i \frac{\partial}{\partial x_j} v_k \\ &= \begin{vmatrix} \boldsymbol{\delta}_1 & \boldsymbol{\delta}_2 & \boldsymbol{\delta}_3 \\ \dfrac{\partial}{\partial x_1} & \dfrac{\partial}{\partial x_2} & \dfrac{\partial}{\partial x_3} \\ v_1 & v_2 & v_3 \end{vmatrix} \\ &= \boldsymbol{\delta}_1 \left\{\frac{\partial v_3}{\partial x_2} - \frac{\partial v_2}{\partial x_3}\right\} + \boldsymbol{\delta}_2 \left\{\frac{\partial v_1}{\partial x_3} - \frac{\partial v_3}{\partial x_1}\right\} + \boldsymbol{\delta}_3 \left\{\frac{\partial v_2}{\partial x_1} - \frac{\partial v_1}{\partial x_2}\right\}\end{aligned} \tag{A.4-10}$$

The vector thus constructed is called the *curl* of **v**. Other notations for $[\nabla \times \mathbf{v}]$ are curl **v** and rot **v**, the latter being common in the German literature. The curl operation, like the divergence, is distributive but not commutative or associative. Note that the ith component of $[\nabla \times \mathbf{v}]$ is $\sum_j \sum_k \varepsilon_{ijk}(\partial/\partial x_j)v_k$.

The Gradient of a Vector Field

In addition to the scalar product $(\nabla \cdot \mathbf{v})$ and the vector product $[\nabla \times \mathbf{v}]$ one may also form the dyadic product $\nabla\mathbf{v}$:

$$\nabla\mathbf{v} = \left\{ \sum_i \boldsymbol{\delta}_i \frac{\partial}{\partial x_i} \right\} \left\{ \sum_j \boldsymbol{\delta}_j v_j \right\} = \sum_i \sum_j \boldsymbol{\delta}_i \boldsymbol{\delta}_j \frac{\partial}{\partial x_i} v_j \tag{A.4-11}$$

This is called the *gradient* of the vector **v** and is sometimes written grad **v**. It is a second-order tensor whose ij-component[1] is $(\partial/\partial x_i)v_j$. Its transpose is

$$(\nabla\mathbf{v})^\dagger = \sum_i \sum_j \boldsymbol{\delta}_i \boldsymbol{\delta}_j \frac{\partial}{\partial x_j} v_i \tag{A.4-12}$$

whose ij-component is $(\partial/\partial x_j)v_i$. Note that $\nabla\mathbf{v} \neq \mathbf{v}\nabla$ and $(\nabla\mathbf{v})^\dagger \neq \mathbf{v}\nabla$.

The Divergence of a Tensor Field

If the tensor $\boldsymbol{\tau}$ is a function of the space variables x_1, x_2, x_3, then a vector product may be formed with operator ∇; in obtaining the final form we use Eq. A.3-3:

$$\begin{aligned} [\nabla \cdot \boldsymbol{\tau}] &= \left[\left\{ \sum_i \boldsymbol{\delta}_i \frac{\partial}{\partial x_i} \right\} \cdot \left\{ \sum_j \sum_k \boldsymbol{\delta}_j \boldsymbol{\delta}_k \tau_{jk} \right\} \right] = \sum_i \sum_j \sum_k [\boldsymbol{\delta}_i \cdot \boldsymbol{\delta}_j \boldsymbol{\delta}_k] \frac{\partial}{\partial x_i} \tau_{jk} \\ &= \sum_i \sum_j \sum_k \delta_{ij} \boldsymbol{\delta}_k \frac{\partial}{\partial x_i} \tau_{jk} = \sum_k \boldsymbol{\delta}_k \left\{ \sum_i \frac{\partial}{\partial x_i} \tau_{ik} \right\} \end{aligned} \tag{A.4-13}$$

This is called the *divergence* of the tensor $\boldsymbol{\tau}$, and is sometimes written div $\boldsymbol{\tau}$. The kth component of $[\nabla \cdot \boldsymbol{\tau}]$ is $\sum_i (\partial/\partial x_i)\tau_{ik})$. If $\boldsymbol{\tau}$ is the product $s\mathbf{vw}$, then

$$[\nabla \cdot s\mathbf{vw}] = \sum_k \boldsymbol{\delta}_k \left\{ \sum_i \frac{\partial}{\partial x_i} (s v_i w_k) \right\} \tag{A.4-14}$$

The Laplacian of a Scalar Field

If we take the divergence of a gradient of the scalar function s, we obtain

$$\begin{aligned} (\nabla \cdot \nabla s) &= \left(\left\{ \sum_i \boldsymbol{\delta}_i \frac{\partial}{\partial x_i} \right\} \cdot \left\{ \sum_j \boldsymbol{\delta}_j \frac{\partial s}{\partial x_j} \right\} \right) \\ &= \sum_i \sum_j \delta_{ij} \frac{\partial}{\partial x_i} \frac{\partial s}{\partial x_j} = \left\{ \sum_i \frac{\partial^2}{\partial x_i^2} s \right\} \end{aligned} \tag{A.4-15}$$

The collection of differential operators operating on s in the last line is given the symbol ∇^2; hence in rectangular coordinates

$$(\nabla \cdot \nabla) = \nabla^2 = \frac{\partial^2}{\partial x_1^2} + \frac{\partial^2}{\partial x_2^2} + \frac{\partial^2}{\partial x_3^2} \tag{A.4-16}$$

This is called the *Laplacian* operator. (Some authors use the symbol Δ for the Laplacian operator, particularly in the older German literature; hence $(\nabla \cdot \nabla s)$, $(\nabla \cdot \nabla)s$, $\nabla^2 s$, and Δs

---

[1] *Caution:* Some authors define the ij-component of $\nabla\mathbf{v}$ to be $(\partial/\partial x_j)v_i$.

are all equivalent quantities.) The Laplacian operator has only the distributive property, as do the gradient, divergence, and curl.

The Laplacian of a Vector Field

If we take the divergence of the gradient of the vector function **v**, we obtain

$$\begin{aligned}[\nabla\cdot\nabla\mathbf{v}] &= \left[\left\{\sum_i \boldsymbol{\delta}_i\frac{\partial}{\partial x_i}\right\}\cdot\left\{\sum_j\sum_k \boldsymbol{\delta}_j\boldsymbol{\delta}_k\frac{\partial}{\partial x_j}v_k\right\}\right]\\ &= \sum_i\sum_j\sum_k[\boldsymbol{\delta}_i\cdot\boldsymbol{\delta}_j\boldsymbol{\delta}_k]\frac{\partial}{\partial x_i}\frac{\partial}{\partial x_j}v_k\\ &= \sum_i\sum_j\sum_k \delta_{ij}\boldsymbol{\delta}_k\frac{\partial}{\partial x_i}\frac{\partial}{\partial x_j}v_k = \sum_k\boldsymbol{\delta}_k\left(\sum_i\frac{\partial^2}{\partial x_i^2}v_k\right)\end{aligned} \tag{A.4-17}$$

That is, the kth component of $[\nabla\cdot\nabla\mathbf{v}]$ is, in Cartesian coordinates, just $\nabla^2 v_k$. Alternative notations for $[\nabla\cdot\nabla\mathbf{v}]$ are $(\nabla\cdot\nabla)\mathbf{v}$ and $\nabla^2\mathbf{v}$.

Other Differential Relations

Numerous identities can be proved using the definitions just given:

$$\nabla rs = r\nabla s + s\nabla r \tag{A.4-18}$$

$$(\nabla\cdot s\mathbf{v}) = (\nabla s\cdot\mathbf{v}) + s(\nabla\cdot\mathbf{v}) \tag{A.4-19}$$

$$(\nabla\cdot[\mathbf{v}\times\mathbf{w}]) = (\mathbf{w}\cdot[\nabla\times\mathbf{v}]) - (\mathbf{v}\cdot[\nabla\times\mathbf{w}]) \tag{A.4-20}$$

$$[\nabla\times s\mathbf{v}] = [\nabla s\times\mathbf{v}] + s[\nabla\times\mathbf{v}] \tag{A.4-21}$$

$$[\nabla\cdot\nabla\mathbf{v}] = \nabla(\nabla\cdot\mathbf{v}) - [\nabla\times[\nabla\times\mathbf{v}]] \tag{A.4-22}$$

$$[\mathbf{v}\cdot\nabla\mathbf{v}] = \frac{1}{2}\nabla(\mathbf{v}\cdot\mathbf{v}) - [\mathbf{v}\times[\nabla\times\mathbf{v}]] \tag{A.4-23}$$

$$[\nabla\cdot\mathbf{v}\mathbf{w}] = [\mathbf{v}\cdot\nabla\mathbf{w}] + \mathbf{w}(\nabla\cdot\mathbf{v}) \tag{A.4-24}$$

$$(s\boldsymbol{\delta}:\nabla\mathbf{v}) = s(\nabla\cdot\mathbf{v}) \tag{A.4-25}$$

$$[\nabla\cdot s\boldsymbol{\delta}] = \nabla s \tag{A.4-26}$$

$$[\nabla\cdot s\boldsymbol{\tau}] = [\nabla s\cdot\boldsymbol{\tau}] + s[\nabla\cdot\boldsymbol{\tau}] \tag{A.4-27}$$

$$\nabla(\mathbf{v}\cdot\mathbf{w}) = [(\nabla\mathbf{v})\cdot\mathbf{w}] + [(\nabla\mathbf{w})\cdot\mathbf{v}] \tag{A.4-28}$$

EXAMPLE A.4-1

Proof of a Tensor Identity

Prove that for *symmetric* $\boldsymbol{\tau}$:

$$(\boldsymbol{\tau}:\nabla\mathbf{v}) = (\nabla\cdot[\boldsymbol{\tau}\cdot\mathbf{v}]) - (\mathbf{v}\cdot[\nabla\cdot\boldsymbol{\tau}]) \tag{A.4-29}$$

SOLUTION

First we write out the right side in terms of components:

$$(\nabla\cdot[\boldsymbol{\tau}\cdot\mathbf{v}]) = \sum_i\frac{\partial}{\partial x_i}[\boldsymbol{\tau}\cdot\mathbf{v}]_i = \sum_i\sum_j\frac{\partial}{\partial x_i}\tau_{ij}v_j \tag{A.4-30}$$

$$(\mathbf{v}\cdot[\nabla\cdot\boldsymbol{\tau}]) = \sum_j v_j[\nabla\cdot\boldsymbol{\tau}]_j = \sum_j\sum_i v_j\frac{\partial}{\partial x_i}\tau_{ij} \tag{A.4-31}$$

The left side may be written as

$$(\boldsymbol{\tau}:\nabla\mathbf{v}) = \sum_i\sum_j\tau_{ji}\frac{\partial}{\partial x_i}v_j = \sum_i\sum_j\tau_{ij}\frac{\partial}{\partial x_i}v_j \tag{A.4-32}$$

the second form resulting from the symmetry of $\boldsymbol{\tau}$. Subtraction of Eq. A.4-31 from Eq. A.4-30 will give Eq. A.4-32.

Now that we have given all the vector and tensor operations, including the various ∇ operations, we want to point out that the dot and double dot operations can be written down at once by using the following simple rule: *a dot implies a summation on adjacent indices*. We illustrate the rule with several examples.

To interpret $(\mathbf{v} \cdot \mathbf{w})$, we note that $\mathbf{v}$ and $\mathbf{w}$ are vectors, whose components have one index. Since both symbols are adjacent to the dot, we make the indices for both of them the same and then sum on them: $(\mathbf{v} \cdot \mathbf{w}) = \Sigma_i v_i w_i$. For double dot operations such as $(\boldsymbol{\tau} : \nabla\mathbf{v})$, we proceed as follows. We note that $\boldsymbol{\tau}$, being a tensor, has two subscripts, whereas ∇ and $\mathbf{v}$ each have one. We therefore set the second subscript of $\boldsymbol{\tau}$ equal to the subscript on ∇ and sum; then we set the first subscript of $\boldsymbol{\tau}$ equal to the subscript on $\mathbf{v}$ and sum. Hence we get $(\boldsymbol{\tau} : \nabla\mathbf{v}) = \Sigma_i \Sigma_j \tau_{ji} (\partial/\partial x_i) v_j$. Similarly, $(\mathbf{v} \cdot [\nabla \cdot \boldsymbol{\tau}])$ can be written down at once as $\Sigma_i \Sigma_j v_j (\partial/\partial x_i) \tau_{ij}$ by performing the operation in the inner enclosure (the brackets) before the outer (the parentheses).

To get the ith component of a vector quantity, we proceed in exactly the same way. To evaluate $[\boldsymbol{\tau} \cdot \mathbf{v}]_i$ we set the second index of the tensor $\boldsymbol{\tau}$ equal to the index on $\mathbf{v}$ and sum to get $\Sigma_j \tau_{ij} v_j$. Similarly, the ith component of $[\nabla \cdot \rho\mathbf{v}\mathbf{v}]$ is obtained as $\Sigma_j (\partial/\partial x_j)(\rho v_j v_i)$. Becoming skilled with this method can save a great deal of time in interpreting the dot and double dot operations in Cartesian coordinates.

EXERCISES

1. Perform all the operations in Eq. A.4-6 by writing out all the summations instead of using the $\sum$ notation.

2. A field $\mathbf{v}(x,y,z)$ is said to be *irrotational* if $[\nabla \times \mathbf{v}] = \mathbf{0}$. Which of the following fields are irrotational?
   (a) $v_x = by \qquad v_y = 0 \qquad v_z = 0$
   (b) $v_x = bx \qquad v_y = 0 \qquad v_z = 0$
   (c) $v_x = by \qquad v_y = bx \qquad v_z = 0$
   (d) $v_x = -by \qquad v_y = bx \qquad v_z = 0$

3. Evaluate $(\nabla \cdot \mathbf{v})$, $\nabla\mathbf{v}$, and $[\nabla \cdot \mathbf{v}\mathbf{v}]$ for the four fields in Exercise 2.

4. A vector $\mathbf{v}$ has components
$$v_i = \sum_{j=1}^{3} a_{ij} x_j$$
with $a_{ij} = a_{ji}$ and $\sum_{i=1}^{3} a_{ii} = 0$; the a_{ij} are constants. Evaluate $(\nabla \cdot \mathbf{v})$, $[\nabla \times \mathbf{v}]$, $\nabla\mathbf{v}$, $(\nabla\mathbf{v})^{\dagger}$, and $[\nabla \cdot \mathbf{v}\mathbf{v}]$. (*Hint:* In connection with evaluating $[\nabla \times \mathbf{v}]$, see Exercise 5 in §A.2.)

5. Verify that $\nabla^2(\nabla \cdot \mathbf{v}) = (\nabla \cdot (\nabla^2\mathbf{v}))$, and that $[\nabla \cdot (\nabla\mathbf{v})^{\dagger}] = \nabla \cdot (\nabla \cdot \mathbf{v})$.

6. Verify that $(\nabla \cdot [\nabla \times \mathbf{v}]) = 0$ and $[\nabla \times \nabla s] = \mathbf{0}$.

7. If $\mathbf{r}$ is the position vector (with components x_1, x_2, x_3) and $\mathbf{v}$ is any vector, show that
   (a) $(\nabla \cdot \mathbf{r}) = 3$
   (b) $[\nabla \times \mathbf{r}] = \mathbf{0}$
   (c) $[\mathbf{r} \times [\nabla \cdot \mathbf{v}\mathbf{v}]] = [\nabla \cdot \mathbf{v}[\mathbf{r} \times \mathbf{v}]]$ (where $\mathbf{v}$ is a function of position)

8. Develop an alternative expression for $[\nabla \times [\nabla \cdot s\mathbf{v}\mathbf{v}]]$.

9. If $\mathbf{r}$ is the position vector and r is its magnitude, verify that
   (a) $\nabla \dfrac{1}{r} = -\dfrac{\mathbf{r}}{r^3}$
   (b) $\nabla f(r) = \dfrac{1}{r}\dfrac{df}{dr}\mathbf{r}$
   (c) $\nabla (\mathbf{a} \cdot \mathbf{r}) = \mathbf{a} \qquad$ if $\mathbf{a}$ is a constant vector

10. Write out in full in Cartesian coordinates

(a) $\frac{\partial}{\partial t}\rho\mathbf{v} = -[\nabla \cdot \rho\mathbf{v}\mathbf{v}] - \nabla p - [\nabla \cdot \boldsymbol{\tau}] + \rho\mathbf{g}$

(b) $\boldsymbol{\tau} = -\mu\{\nabla\mathbf{v} + (\nabla\mathbf{v})^{\dagger} - \frac{2}{3}(\nabla \cdot \mathbf{v})\boldsymbol{\delta}\}$

§A.5 VECTOR AND TENSOR INTEGRAL THEOREMS

For performing general proofs in continuum physics, several integral theorems are extremely useful.

The Gauss–Ostrogradskii Divergence Theorem

If V is a closed region in space enclosed by a surface S, then

$$\int_V (\nabla \cdot \mathbf{v})dV = \int_S (\mathbf{n} \cdot \mathbf{v})dS \tag{A.5-1}$$

in which $\boldsymbol{n}$ is the outwardly directed unit normal vector. This is known as the *divergence theorem* of Gauss and Ostrogradskii. Two closely allied theorems for scalars and tensors are

$$\int_V \nabla s \, dV = \int_S \mathbf{n}s \, dS \tag{A.5-2}$$

$$\int_V [\nabla \cdot \boldsymbol{\tau}] \, dV = \int_S [\mathbf{n} \cdot \boldsymbol{\tau}] \, dS \qquad \text{(A.5-3)}^1$$

The last relation is also valid for dyadic products **vw**. Note that, in all three equations, ∇ in the volume integral is just replaced by **n** in the surface integral.

The Leibniz Formula for Differentiating a Volume Integral[1]

Let V be a closed moving region in space enclosed by a surface S; let the velocity of any surface element be $\mathbf{v}_S$. Then, if $s(x,y,z,t)$ is a scalar function of position and time,

$$\frac{d}{dt}\int_V s dV = \int_V \frac{\partial s}{\partial t}dV + \int_S s(\mathbf{v}_S \cdot \mathbf{n})dS \tag{A.5-4}$$

This is an extension of the *Leibniz formula* for differentiating a single integral (see Eq. C.3-2); keep in mind that $V = V(t)$ and $S = S(t)$. Equation A.5-4 also applies to vectors and tensors.

If the integral is over a volume, the surface of which is moving with the local fluid velocity (so that $\mathbf{v}_S = \mathbf{v}$), then use of the equation of continuity leads to the additional useful result:

$$\frac{d}{dt}\int_V \rho s dV = \int_V \rho\frac{Ds}{Dt}dV \tag{A.5-5}$$

in which ρ is the fluid density. Equation A.5-5 is sometimes called the *Reynolds transport theorem*.

EXERCISES

1. Consider the vector field

$$\mathbf{v} = \boldsymbol{\delta}_1 x_1 + \boldsymbol{\delta}_2 x_3 + \boldsymbol{\delta}_3 x_2$$

Evaluate both sides of Eq. A.5-1 over the region bounded by the planes $x_1 = 0$, $x_1 = 1$; $x_2 = 0$, $x_2 = 2$; $x_3 = 0$, $x_3 = 4$.

[1] M. D. Greenberg, *Foundations of Applied Mathematics*, Prentice-Hall, Englewood Cliffs, N.J. (1978), pp. 163–164.

2. Consider the time-dependent scalar function:

$$s = x + y + zt$$

Evaluate both sides of Eq. A.5-4 over the volume bounded by the planes: $x = 0$, $x = t$; $y = 0$, $y = 2t$; $z = 0$, $z = 4t$. The quantities x, y, z, t are dimensionless.

3. Evaluate both sides of Eq. A.5-2 for the function $s(x, y, z) = x^2 + y^2 + z^2$. The volume V is the triangular prism lying between the two triangles whose vertices are (2, 0, 0), (2, 1, 0), (2, 0, 3), and (−2, 0, 0), (−2, 1, 0), (−2, 0, 3).

§A.6 VECTOR AND TENSOR ALGEBRA IN CURVILINEAR COORDINATES

Thus far we have considered only Cartesian coordinates x, y, and z. Although formal derivations are usually made in Cartesian coordinates, for working problems it is often more natural to use curvilinear coordinates. The two most common curvilinear coordinate systems are the *cylindrical* and the *spherical*. In the following we discuss only these two systems, but the method can also be applied to all *orthogonal* coordinate systems—that is, those in which the three families of coordinate surfaces are mutually perpendicular.

We are primarily interested in knowing how to write various differential operations, such as ∇s, $[\nabla \times \mathbf{v}]$, and $(\boldsymbol{\tau} : \nabla \mathbf{v})$ in curvilinear coordinates. It turns out that we can do this in a straightforward way if we know, for the coordinate system being used, two things: (a) the expression for ∇ in curvilinear coordinates; and (b) the spatial derivatives of the unit vectors in curvilinear coordinates. Hence, we want to focus our attention on these two points.

Cylindrical Coordinates

In cylindrical coordinates, instead of designating the coordinates of a point by x, y, z, we locate the point by giving the values of r, θ, z. These coordinates[1] are shown in Fig. A.6-1*a*. They are related to the Cartesian coordinates by

$$x = r\cos\theta \qquad \text{(A.6-1)} \qquad\qquad r = +\sqrt{x^2 + y^2} \qquad \text{(A.6-4)}$$

$$y = r\sin\theta \qquad \text{(A.6-2)} \qquad\qquad \theta = \arctan\left(y/x\right) \qquad \text{(A.6-5)}$$

$$z = z \qquad \text{(A.6-3)} \qquad\qquad z = z \qquad \text{(A.6-6)}$$

To convert derivatives of scalars with respect to x, y, z into derivatives with respect to r, θ, z, the "chain rule" of partial differentiation[2] is used. The derivative operators are readily found to be related thus:

---

[1]*Caution:* We have chosen to use the familiar r,θ,z-notation for cylindrical coordinates rather than to switch to some less familiar symbols, even though there are two situations in which confusion can arise: (a) occasionally one has to use cylindrical and spherical coordinates in the same problem, and the symbols r and θ have different meanings in the two systems; (b) occasionally one deals with the position vector $\mathbf{r}$ in problems involving cylindrical coordinates, but then the magnitude of $\mathbf{r}$ is not the same as the coordinate r, but rather $\sqrt{r^2 + z^2}$. In such situations, as in Fig. A.6-1, we can use overbars for the cylindrical coordinates and write $\bar{r}, \bar{\theta}, \bar{z}$. For most discussions bars will not be needed.

[2]For example, for a scalar function $\chi(x,y,z) = \psi(r,\theta,z)$:

$$\left(\frac{\partial \chi}{\partial x}\right)_{y,z} = \left(\frac{\partial r}{\partial x}\right)_{y,z}\left(\frac{\partial \psi}{\partial r}\right)_{\theta,z} + \left(\frac{\partial \theta}{\partial x}\right)_{y,z}\left(\frac{\partial \psi}{\partial \theta}\right)_{r,z} + \left(\frac{\partial z}{\partial x}\right)_{y,z}\left(\frac{\partial \psi}{\partial z}\right)_{r,\theta}$$

Note that we are careful to use different symbols χ and ψ, since χ is a different function of x, y, z than ψ is of r, θ, and z!

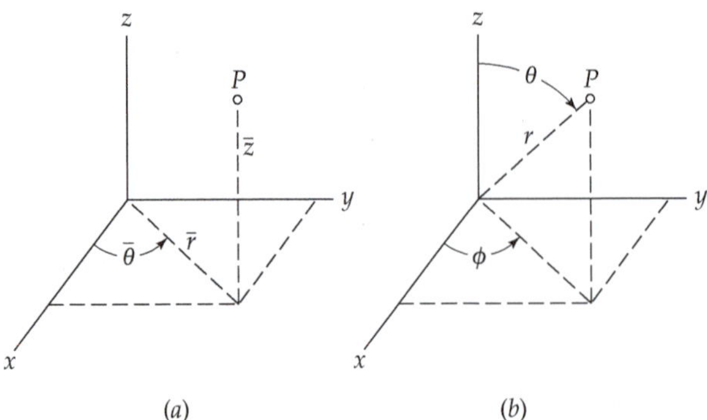

Fig. A.6-1. (*a*) Cylindrical coordinates[1] with $0 \leq \bar{r} < \infty, 0 \leq \bar{\theta} < 2\pi, -\infty < \bar{z} < \infty$. (*b*) Spherical coordinates with $0 \leq r < \infty, 0 \leq \theta \leq \pi, 0 \leq \phi < 2\pi$. Note that $\bar{r}$ and $\bar{\theta}$ in cylindrical coordinates are *not* the same as r and θ in spherical coordinates. Note carefully how the position vector **r** and its length r are written in the three coordinate systems:

Cartesian:	$\mathbf{r} = \boldsymbol{\delta}_x x + \boldsymbol{\delta}_y y + \boldsymbol{\delta}_z z;$	$r = \sqrt{x^2 + y^2 + z^2}$
Cylindrical:	$\mathbf{r} = \boldsymbol{\delta}_r \bar{r} + \boldsymbol{\delta}_z \bar{z};$	$r = \sqrt{\bar{r}^2 + \bar{z}^2}$
Spherical:	$\mathbf{r} = \boldsymbol{\delta}_r r;$	$r = r$

$$\begin{cases} \dfrac{\partial}{\partial x} = (\cos\,\theta)\dfrac{\partial}{\partial r} + \left(-\dfrac{\sin\,\theta}{r}\right)\dfrac{\partial}{\partial \theta} + (0)\dfrac{\partial}{\partial z} & \text{(A.6-7)} \\ \dfrac{\partial}{\partial y} = (\sin\,\theta)\dfrac{\partial}{\partial r} + \left(\dfrac{\cos\,\theta}{r}\right)\dfrac{\partial}{\partial \theta} + (0)\dfrac{\partial}{\partial z} & \text{(A.6-8)} \\ \dfrac{\partial}{\partial z} = (0)\dfrac{\partial}{\partial r} + (0)\dfrac{\partial}{\partial \theta} + (1)\dfrac{\partial}{\partial z} & \text{(A.6-9)} \end{cases}$$

With these relations, derivatives of any scalar functions (including, of course, components of vectors and tensors) with respect to x, y, and z can be expressed in terms of derivatives with respect to r, θ, and z.

Having discussed the interrelationship of the coordinates and derivatives in the two coordinate systems, we now turn to the relation between the unit vectors. We begin by noting that the unit vectors $\boldsymbol{\delta}_x$, $\boldsymbol{\delta}_y$, $\boldsymbol{\delta}_z$ (or $\boldsymbol{\delta}_1$, $\boldsymbol{\delta}_2$, $\boldsymbol{\delta}_3$ as we have been calling them) are independent of position—that is, independent of x, y, z. In cylindrical coordinates the unit vectors $\boldsymbol{\delta}_r$ and $\boldsymbol{\delta}_\theta$ will depend on position, as we can see in Fig. A.6-2. The unit vector $\boldsymbol{\delta}_r$ is a vector of unit length in the direction of increasing r; the unit vector $\boldsymbol{\delta}_\theta$ is a vector of unit length in the direction of increasing θ. Clearly as the point P is moved around on the xy-plane, the directions of $\boldsymbol{\delta}_r$ and $\boldsymbol{\delta}_\theta$ change. Trigonometrical arguments lead to the following relations:

$$\begin{cases} \boldsymbol{\delta}_r = (\cos\,\theta)\,\boldsymbol{\delta}_x + (\sin\,\theta)\boldsymbol{\delta}_y + (0)\boldsymbol{\delta}_z & \text{(A.6-10)} \\ \boldsymbol{\delta}_\theta = (-\sin\,\theta)\boldsymbol{\delta}_x + (\cos\,\theta)\boldsymbol{\delta}_y + (0)\boldsymbol{\delta}_z & \text{(A.6-11)} \\ \boldsymbol{\delta}_z = (0)\boldsymbol{\delta}_x + (0)\boldsymbol{\delta}_y + (1)\boldsymbol{\delta}_z & \text{(A.6-12)} \end{cases}$$

These may be solved for $\boldsymbol{\delta}_x$, $\boldsymbol{\delta}_y$, and $\boldsymbol{\delta}_z$ to give

$$\begin{cases} \boldsymbol{\delta}_x = (\cos\,\theta)\,\boldsymbol{\delta}_r + (-\sin\,\theta)\boldsymbol{\delta}_\theta + (0)\boldsymbol{\delta}_z & \text{(A.6-13)} \\ \boldsymbol{\delta}_y = (\sin\,\theta)\boldsymbol{\delta}_r + (\cos\,\theta)\boldsymbol{\delta}_\theta + (0)\boldsymbol{\delta}_z & \text{(A.6-14)} \\ \boldsymbol{\delta}_z = (0)\boldsymbol{\delta}_r + (0)\boldsymbol{\delta}_\theta + (1)\boldsymbol{\delta}_z & \text{(A.6-15)} \end{cases}$$

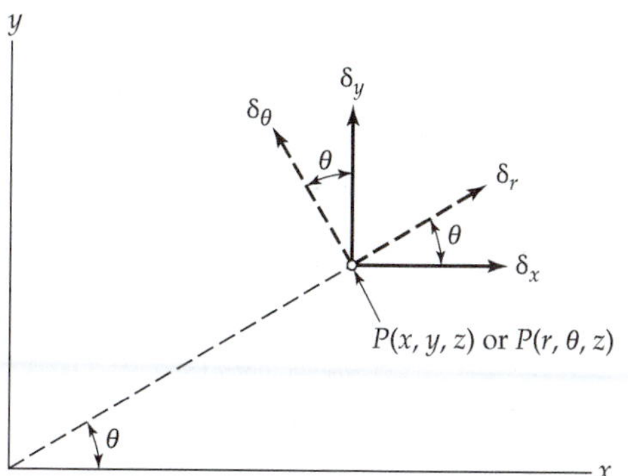

Fig. A.6-2. Unit vectors in Cartesian and cylindrical coordinates. The z-axis and the unit vector $\boldsymbol{\delta}_z$ have been omitted for simplicity.

The utility of these two sets of relations will be made clear in the next section.

Vectors and tensors can be decomposed into components with respect to cylindrical coordinates as was done for Cartesian coordinates in Eqs. A.2-16 and A.3-7 (i.e., $\mathbf{v} = \boldsymbol{\delta}_r v_r + \boldsymbol{\delta}_\theta v_\theta + \boldsymbol{\delta}_z v_z$). Also, the multiplication rules for the unit vectors and unit dyads are the same as in Eqs. A.2-14 and A.2-15 and A.3-1 to A.3-6. Consequently the various dot and cross product operations (but *not* the differential operations!) are performed as described in §A.2 and §A.3. For example,

$$(\mathbf{v}\cdot\mathbf{w}) = v_r w_r + v_\theta w_\theta + v_z w_z \tag{A.6-16}$$

$$[\mathbf{v}\times\mathbf{w}] = \boldsymbol{\delta}_r(v_\theta w_z - v_z w_\theta) + \boldsymbol{\delta}_\theta(v_z w_r - v_r w_z) + \boldsymbol{\delta}_z(v_r w_\theta - v_\theta w_r) \tag{A.6-17}$$

$$\begin{aligned}\{\boldsymbol{\sigma}\cdot\boldsymbol{\tau}\} &= \boldsymbol{\delta}_r\boldsymbol{\delta}_r(\sigma_{rr}\tau_{rr} + \sigma_{r\theta}\tau_{\theta r} + \sigma_{rz}\tau_{zr}) \\ &\quad + \boldsymbol{\delta}_r\boldsymbol{\delta}_\theta(\sigma_{rr}\tau_{r\theta} + \sigma_{r\theta}\tau_{\theta\theta} + \sigma_{rz}\tau_{z\theta}) \\ &\quad + \boldsymbol{\delta}_r\boldsymbol{\delta}_z(\sigma_{rr}\tau_{rz} + \sigma_{r\theta}\tau_{\theta z} + \sigma_{rz}\tau_{zz}) \\ &\quad + \text{etc.}\end{aligned} \tag{A.6-18}$$

Spherical Coordinates

We now tabulate for reference the same kind of information for spherical coordinates r, θ, ϕ. These coordinates are shown in Figure A.6-1*b*. They are related to the Cartesian coordinates by

$$x = r\sin\theta\cos\phi \tag{A.6-19}$$
$$y = r\sin\theta\sin\phi \tag{A.6-20}$$
$$z = r\cos\theta \tag{A.6-21}$$

$$r = +\sqrt{x^2+y^2+z^2} \tag{A.6-22}$$
$$\theta = \arctan\left(\sqrt{x^2+y^2}/z\right) \tag{A.6-23}$$
$$\phi = \arctan(y/x) \tag{A.6-24}$$

For the spherical coordinates we have the following relations for the derivative operators:

$$\frac{\partial}{\partial x} = (\sin\theta\cos\phi)\frac{\partial}{\partial r} + \left(\frac{\cos\theta\cos\phi}{r}\right)\frac{\partial}{\partial\theta} + \left(-\frac{\sin\phi}{r\sin\theta}\right)\frac{\partial}{\partial\phi} \tag{A.6-25}$$

$$\frac{\partial}{\partial y} = (\sin\theta\sin\phi)\frac{\partial}{\partial r} + \left(\frac{\cos\theta\sin\phi}{r}\right)\frac{\partial}{\partial\theta} + \left(\frac{\cos\phi}{r\sin\theta}\right)\frac{\partial}{\partial\phi} \tag{A.6-26}$$

$$\frac{\partial}{\partial z} = (\cos\theta)\frac{\partial}{\partial r} + \left(-\frac{\sin\theta}{r}\right)\frac{\partial}{\partial\theta} + (0)\frac{\partial}{\partial\phi} \tag{A.6-27}$$

The relations between the unit vectors are

$$\boldsymbol{\delta}_r = (\sin\theta\cos\phi)\boldsymbol{\delta}_x + (\sin\theta\sin\phi)\boldsymbol{\delta}_y + (\cos\theta)\boldsymbol{\delta}_z \tag{A.6-28}$$

$$\boldsymbol{\delta}_\theta = (\cos\theta\cos\phi)\boldsymbol{\delta}_x + (\cos\theta\sin\phi)\boldsymbol{\delta}_y + (-\sin\theta)\boldsymbol{\delta}_z \tag{A.6-29}$$

$$\boldsymbol{\delta}_\phi = (-\sin\phi)\boldsymbol{\delta}_x + (\cos\phi)\boldsymbol{\delta}_y + (0)\boldsymbol{\delta}_z \tag{A.6-30}$$

and

$$\begin{cases} \boldsymbol{\delta}_x = (\sin\theta\cos\phi)\,\boldsymbol{\delta}_r + (\cos\theta\cos\phi)\boldsymbol{\delta}_\theta + (-\sin\phi)\boldsymbol{\delta}_\phi & \text{(A.6-31)} \\ \boldsymbol{\delta}_y = (\sin\theta\sin\phi)\boldsymbol{\delta}_r + (\cos\theta\sin\phi)\boldsymbol{\delta}_\theta + (\cos\phi)\boldsymbol{\delta}_\phi & \text{(A.6-32)} \\ \boldsymbol{\delta}_z = (\cos\theta)\boldsymbol{\delta}_r + (-\sin\theta)\boldsymbol{\delta}_\theta + (0)\boldsymbol{\delta}_\phi & \text{(A.6-33)} \end{cases}$$

And, finally, some sample operations in spherical coordinates are

$$\begin{aligned}(\boldsymbol{\sigma}:\boldsymbol{\tau}) = {} & \sigma_{rr}\tau_{rr} + \sigma_{r\theta}\tau_{\theta r} + \sigma_{r\phi}\tau_{\phi r} \\ & + \sigma_{\theta r}\tau_{r\theta} + \sigma_{\theta\theta}\tau_{\theta\theta} + \sigma_{\theta\phi}\tau_{\phi\theta} \\ & + \sigma_{\phi r}\tau_{r\phi} + \sigma_{\phi\theta}\tau_{\theta\phi} + \sigma_{\phi\phi}\tau_{\phi\phi} \end{aligned} \tag{A.6-34}$$

$$(\mathbf{u}\cdot[\mathbf{v}\times\mathbf{w}]) = \begin{vmatrix} u_r & u_\theta & u_\phi \\ v_r & v_\theta & v_\phi \\ w_r & w_\theta & w_\phi \end{vmatrix} \tag{A.6-35}$$

That is, the relations (not involving ∇!) given in §A.2 and §A.3 can be written directly in terms of spherical components.

EXERCISES

1. Show that

$$\int_0^{2\pi}\int_0^{\pi} \boldsymbol{\delta}_r \sin\theta \, d\theta \, d\phi = \mathbf{0}$$

$$\int_0^{2\pi}\int_0^{\pi} \boldsymbol{\delta}_r\boldsymbol{\delta}_r \sin\theta \, d\theta \, d\phi = \frac{4}{3}\pi\boldsymbol{\delta}$$

where $\boldsymbol{\delta}_r$ is the unit vector in the r direction in spherical coordinates.

2. Verify that in spherical coordinates $\boldsymbol{\delta} = \boldsymbol{\delta}_r\boldsymbol{\delta}_r + \boldsymbol{\delta}_\theta\boldsymbol{\delta}_\theta + \boldsymbol{\delta}_\phi\boldsymbol{\delta}_\phi$.

§A.7 DIFFERENTIAL OPERATIONS IN CURVILINEAR COORDINATES

We now turn to the use of the ∇-operator in curvilinear coordinates. As in the previous section, we work out in detail the results for cylindrical and spherical coordinates.

Cylindrical Coordinates

From Eqs. A.6-10, A.6-11, and A.6-12 we can obtain expressions for the spatial derivatives of the unit vectors $\boldsymbol{\delta}_r$, $\boldsymbol{\delta}_\theta$, and $\boldsymbol{\delta}_z$:

$$\boxed{\frac{\partial}{\partial r}\boldsymbol{\delta}_r = \mathbf{0} \qquad \frac{\partial}{\partial r}\boldsymbol{\delta}_\theta = \mathbf{0} \qquad \frac{\partial}{\partial r}\boldsymbol{\delta}_z = \mathbf{0}} \tag{A.7-1}$$

$$\boxed{\frac{\partial}{\partial \theta}\boldsymbol{\delta}_r = \boldsymbol{\delta}_\theta \qquad \frac{\partial}{\partial \theta}\boldsymbol{\delta}_\theta = -\boldsymbol{\delta}_r \qquad \frac{\partial}{\partial \theta}\boldsymbol{\delta}_z = \mathbf{0}} \tag{A.7-2}$$

$$\boxed{\frac{\partial}{\partial z}\boldsymbol{\delta}_r = \mathbf{0} \qquad \frac{\partial}{\partial z}\boldsymbol{\delta}_\theta = \mathbf{0} \qquad \frac{\partial}{\partial z}\boldsymbol{\delta}_z = \mathbf{0}} \tag{A.7-3}$$

The reader would do well to interpret these derivatives geometrically by considering the way $\boldsymbol{\delta}_r$, $\boldsymbol{\delta}_\theta$, $\boldsymbol{\delta}_z$ change as the location of P is changed in Fig. A.6-2.

We now use the definition of the ∇-operator in Eq. A.4-1, the expressions in Eqs. A.6-13, A.6-14, and A.6-15, and the derivative operators in Eqs. A.6-7, A.6-8, and A.6-9 to obtain the formula for ∇ in cylindrical coordinates

$$\nabla = \boldsymbol{\delta}_x \frac{\partial}{\partial x} + \boldsymbol{\delta}_y \frac{\partial}{\partial y} + \boldsymbol{\delta}_z \frac{\partial}{\partial z}$$
$$= (\boldsymbol{\delta}_r \cos\theta - \boldsymbol{\delta}_\theta \sin\theta)\left(\cos\theta \frac{\partial}{\partial r} - \frac{\sin\theta}{r}\frac{\partial}{\partial \theta}\right)$$
$$+ (\boldsymbol{\delta}_r \sin\theta + \boldsymbol{\delta}_\theta \cos\theta)\left(\sin\theta \frac{\partial}{\partial r} + \frac{\cos\theta}{r}\frac{\partial}{\partial \theta}\right) + \boldsymbol{\delta}_z \frac{\partial}{\partial z} \quad \text{(A.7-4)}$$

When this is multiplied out, there is considerable simplification, and we get

$$\boxed{\nabla = \boldsymbol{\delta}_r \frac{\partial}{\partial r} + \boldsymbol{\delta}_\theta \frac{1}{r}\frac{\partial}{\partial \theta} + \boldsymbol{\delta}_z \frac{\partial}{\partial z}} \quad \text{(A.7-5)}$$

for *cylindrical* coordinates. This may be used for obtaining all differential operations in cylindrical coordinates, provided that Eqs. A.7-1, A.7-2, and A.7-3 are used to differentiate any unit vectors on which ∇ operates. This point will be made clear in the subsequent illustrative example.

Spherical Coordinates

The spatial derivatives of $\boldsymbol{\delta}_r$, $\boldsymbol{\delta}_\theta$, and $\boldsymbol{\delta}_\phi$ are obtained by differentiating Eqs. A.6-28, A.6-29, and A.6-30:

$$\boxed{\frac{\partial}{\partial r}\boldsymbol{\delta}_r = \mathbf{0} \qquad \frac{\partial}{\partial r}\boldsymbol{\delta}_\theta = \mathbf{0} \qquad \frac{\partial}{\partial r}\boldsymbol{\delta}_\phi = \mathbf{0}} \quad \text{(A.7-6)}$$

$$\boxed{\frac{\partial}{\partial \theta}\boldsymbol{\delta}_r = \boldsymbol{\delta}_\theta \qquad \frac{\partial}{\partial \theta}\boldsymbol{\delta}_\theta = -\boldsymbol{\delta}_r \qquad \frac{\partial}{\partial \theta}\boldsymbol{\delta}_\phi = \mathbf{0}} \quad \text{(A.7-7)}$$

$$\boxed{\frac{\partial}{\partial \phi}\boldsymbol{\delta}_r = \boldsymbol{\delta}_\phi \sin\theta \qquad \frac{\partial}{\partial \phi}\boldsymbol{\delta}_\theta = \boldsymbol{\delta}_\phi \cos\theta \qquad \frac{\partial}{\partial \phi}\boldsymbol{\delta}_\phi = -\boldsymbol{\delta}_r \sin\theta - \boldsymbol{\delta}_\theta \cos\theta} \quad \text{(A.7-8)}$$

Use of Eqs. A.6-31, A.6-32, and A.6-33 and Eqs. A.6-25, A.6-26, and A.6-27 in Eq. A.4-1 gives the following expression for the ∇-operator:

$$\boxed{\nabla = \boldsymbol{\delta}_r \frac{\partial}{\partial r} + \boldsymbol{\delta}_\theta \frac{1}{r}\frac{\partial}{\partial \theta} + \boldsymbol{\delta}_\phi \frac{1}{r\sin\theta}\frac{\partial}{\partial \phi}} \quad \text{(A.7-9)}$$

in *spherical* coordinates. This expression may be used for obtaining differential operations in spherical coordinates, provided that Eqs. A.7-6, 7, and 8 are used for differentiating the unit vectors.

In Table A.7-1, Table A.7-2, and Table A.7-3 we summarize the differential operations most commonly encountered in Cartesian, cylindrical, and spherical coordinates.[1] The curvilinear coordinate expressions given can be obtained by the method illustrated in the following two examples.

EXAMPLE A.7-1

Differential Operations in Cylindrical Coordinates

Derive expressions for $(\nabla \cdot \mathbf{v})$ and $\nabla\mathbf{v}$ in cylindrical coordinates.

SOLUTION

(a) We begin by writing ∇ in cylindrical coordinates and decomposing $\mathbf{v}$ into its components

$$(\nabla \cdot \mathbf{v}) = \left(\left\{\boldsymbol{\delta}_r \frac{\partial}{\partial r} + \boldsymbol{\delta}_\theta \frac{1}{r}\frac{\partial}{\partial \theta} + \boldsymbol{\delta}_z \frac{\partial}{\partial z}\right\} \cdot \{\boldsymbol{\delta}_r v_r + \boldsymbol{\delta}_\theta v_\theta + \boldsymbol{\delta}_z v_z\}\right) \quad \text{(A.7-10)}$$

---

[1]For other coordinate systems see the extensive compilation of P. Moon and D. E. Spencer, *Field Theory Handbook*, Springer, Berlin (1961). In addition, an orthogonal coordinate system is available in which one of the three sets of coordinate surfaces is made up of coaxial cones (but with noncoincident apexes); all of the $\boldsymbol{\nabla}$-operations have been tabulated by the originators of this coordinate system, J. F. Dijksman and E. P. W. Savenije, *Rheol. Acta*, **24**, 105–118 (1985).

Expanding, we get

$$(\nabla \cdot \mathbf{v}) = \left(\boldsymbol{\delta}_r \cdot \frac{\partial}{\partial r}\boldsymbol{\delta}_r v_r\right) + \left(\boldsymbol{\delta}_r \cdot \frac{\partial}{\partial r}\boldsymbol{\delta}_\theta v_\theta\right) + \left(\boldsymbol{\delta}_r \cdot \frac{\partial}{\partial r}\boldsymbol{\delta}_z v_z\right)$$
$$+ \left(\boldsymbol{\delta}_\theta \cdot \frac{1}{r}\frac{\partial}{\partial \theta}\boldsymbol{\delta}_r v_r\right) + \left(\boldsymbol{\delta}_\theta \cdot \frac{1}{r}\frac{\partial}{\partial \theta}\boldsymbol{\delta}_\theta v_\theta\right) + \left(\boldsymbol{\delta}_\theta \cdot \frac{1}{r}\frac{\partial}{\partial \theta}\boldsymbol{\delta}_z v_z\right)$$
$$+ \left(\boldsymbol{\delta}_z \cdot \frac{\partial}{\partial z}\boldsymbol{\delta}_r v_r\right) + \left(\boldsymbol{\delta}_z \cdot \frac{\partial}{\partial z}\boldsymbol{\delta}_\theta v_\theta\right) + \left(\boldsymbol{\delta}_z \cdot \frac{\partial}{\partial z}\boldsymbol{\delta}_z v_z\right) \quad \text{(A.7-11)}$$

We now use the relations given in Eqs. A.7-1, A.7-2, and A.7-3 to evaluate the derivatives of the unit vectors. This gives

$$(\nabla \cdot \mathbf{v}) = (\boldsymbol{\delta}_r \cdot \boldsymbol{\delta}_r)\frac{\partial v_r}{\partial r} + (\boldsymbol{\delta}_r \cdot \boldsymbol{\delta}_\theta)\frac{\partial v_\theta}{\partial r} + (\boldsymbol{\delta}_r \cdot \boldsymbol{\delta}_z)\frac{\partial v_z}{\partial r} + (\boldsymbol{\delta}_\theta \cdot \boldsymbol{\delta}_r)\frac{1}{r}\frac{\partial v_r}{\partial \theta}$$
$$+ (\boldsymbol{\delta}_\theta \cdot \boldsymbol{\delta}_\theta)\frac{1}{r}\frac{\partial v_\theta}{\partial \theta} + (\boldsymbol{\delta}_\theta \cdot \boldsymbol{\delta}_z)\frac{1}{r}\frac{\partial v_z}{\partial \theta} + \frac{v_r}{r}(\boldsymbol{\delta}_\theta \cdot \boldsymbol{\delta}_\theta) + \frac{v_\theta}{r}(\boldsymbol{\delta}_\theta \cdot \{-\boldsymbol{\delta}_r\})$$
$$+ (\boldsymbol{\delta}_z \cdot \boldsymbol{\delta}_r)\frac{\partial v_r}{\partial z} + (\boldsymbol{\delta}_z \cdot \boldsymbol{\delta}_\theta)\frac{\partial v_\theta}{\partial z} + (\boldsymbol{\delta}_z \cdot \boldsymbol{\delta}_z)\frac{\partial v_z}{\partial z} \quad \text{(A.7-12)}$$

Since $(\boldsymbol{\delta}_r \cdot \boldsymbol{\delta}_r) = 1$, $(\boldsymbol{\delta}_r \cdot \boldsymbol{\delta}_\theta) = 0$, and so on, the latter simplifies to

$$(\nabla \cdot \mathbf{v}) = \frac{\partial v_r}{\partial r} + \frac{1}{r}\frac{\partial v_\theta}{\partial \theta} + \frac{v_r}{r} + \frac{\partial v_z}{\partial z} \quad \text{(A.7-13)}$$

which is the same as Eq. A of Table A.7-2. The procedure is a bit tedious, but it *is* straightforward.

(b) Next we examine the dyadic product $\nabla\mathbf{v}$:

$$\nabla\mathbf{v} = \left\{\boldsymbol{\delta}_r\frac{\partial}{\partial r} + \boldsymbol{\delta}_\theta\frac{1}{r}\frac{\partial}{\partial \theta} + \boldsymbol{\delta}_z\frac{\partial}{\partial z}\right\}\{\boldsymbol{\delta}_r v_r + \boldsymbol{\delta}_\theta v_\theta + \boldsymbol{\delta}_z v_z\}$$
$$= \boldsymbol{\delta}_r\boldsymbol{\delta}_r\frac{\partial v_r}{\partial r} + \boldsymbol{\delta}_r\boldsymbol{\delta}_\theta\frac{\partial v_\theta}{\partial r} + \boldsymbol{\delta}_r\boldsymbol{\delta}_z\frac{\partial v_z}{\partial r} + \boldsymbol{\delta}_\theta\boldsymbol{\delta}_r\frac{1}{r}\frac{\partial v_r}{\partial \theta} + \boldsymbol{\delta}_\theta\boldsymbol{\delta}_\theta\frac{1}{r}\frac{\partial v_\theta}{\partial \theta} + \boldsymbol{\delta}_\theta\boldsymbol{\delta}_z\frac{1}{r}\frac{\partial v_z}{\partial \theta}$$
$$+ \boldsymbol{\delta}_\theta\boldsymbol{\delta}_\theta\frac{v_r}{r} - \boldsymbol{\delta}_\theta\boldsymbol{\delta}_r\frac{v_\theta}{r} + \boldsymbol{\delta}_z\boldsymbol{\delta}_r\frac{\partial v_r}{\partial z} + \boldsymbol{\delta}_z\boldsymbol{\delta}_\theta\frac{\partial v_\theta}{\partial z} + \boldsymbol{\delta}_z\boldsymbol{\delta}_z\frac{\partial v_z}{\partial z}$$
$$= \boldsymbol{\delta}_r\boldsymbol{\delta}_r\frac{\partial v_r}{\partial r} + \boldsymbol{\delta}_r\boldsymbol{\delta}_\theta\frac{\partial v_\theta}{\partial r} + \boldsymbol{\delta}_r\boldsymbol{\delta}_z\frac{\partial v_z}{\partial r} + \boldsymbol{\delta}_\theta\boldsymbol{\delta}_r\left(\frac{1}{r}\frac{\partial v_r}{\partial \theta} - \frac{v_\theta}{r}\right) + \boldsymbol{\delta}_\theta\boldsymbol{\delta}_\theta\left(\frac{1}{r}\frac{\partial v_\theta}{\partial \theta} + \frac{v_r}{r}\right)$$
$$+ \boldsymbol{\delta}_\theta\boldsymbol{\delta}_z\frac{1}{r}\frac{\partial v_z}{\partial \theta} + \boldsymbol{\delta}_z\boldsymbol{\delta}_r\frac{\partial v_r}{\partial z} + \boldsymbol{\delta}_z\boldsymbol{\delta}_\theta\frac{\partial v_\theta}{\partial z} + \boldsymbol{\delta}_z\boldsymbol{\delta}_z\frac{\partial v_z}{\partial z} \quad \text{(A.7-14)}$$

Hence, the rr-component is $\partial v_r/\partial r$, the $r\theta$-component is $\partial v_\theta/\partial r$, and so on, as given in Table A.7-2.

EXAMPLE A.7-2

Differential Operations in Spherical Coordinates

Find the r-component of $[\nabla \cdot \boldsymbol{\tau}]$ in spherical coordinates.

SOLUTION

Using Eq. A.7-9 we have

$$[\nabla \cdot \boldsymbol{\tau}]_r = \left[\left\{\boldsymbol{\delta}_r\frac{\partial}{\partial r} + \boldsymbol{\delta}_\theta\frac{1}{r}\frac{\partial}{\partial \theta} + \boldsymbol{\delta}_\phi\frac{1}{r\sin\theta}\frac{\partial}{\partial \phi}\right\} \cdot \{\boldsymbol{\delta}_r\boldsymbol{\delta}_r\tau_{rr} + \boldsymbol{\delta}_r\boldsymbol{\delta}_\theta\tau_{r\theta} + \boldsymbol{\delta}_r\boldsymbol{\delta}_\phi\tau_{r\phi}\right.$$
$$\left.+ \boldsymbol{\delta}_\theta\boldsymbol{\delta}_r\tau_{\theta r} + \boldsymbol{\delta}_\theta\boldsymbol{\delta}_\theta\tau_{\theta\theta} + \boldsymbol{\delta}_\theta\boldsymbol{\delta}_\phi\tau_{\theta\phi} + \boldsymbol{\delta}_\phi\boldsymbol{\delta}_r\tau_{\phi r} + \boldsymbol{\delta}_\phi\boldsymbol{\delta}_\theta\tau_{\phi\theta} + \boldsymbol{\delta}_\phi\boldsymbol{\delta}_\phi\tau_{\phi\phi}\}\right]_r \quad \text{(A.7-15)}$$

We now use Eqs. A.7-6, A.7-7, A.7-8 and Eq. A.3-3. Since we want only the r-component, we select only those terms that contribute to the coefficient of $\boldsymbol{\delta}_r$:

$$\left[\boldsymbol{\delta}_r\frac{\partial}{\partial r} \cdot \boldsymbol{\delta}_r\boldsymbol{\delta}_r\tau_{rr}\right] = [\boldsymbol{\delta}_r \cdot \boldsymbol{\delta}_r\boldsymbol{\delta}_r]\frac{\partial \tau_{rr}}{\partial r} = \boldsymbol{\delta}_r\frac{\partial \tau_{rr}}{\partial r} \quad \text{(A.7-16)}$$

$$\left[\boldsymbol{\delta}_\theta\frac{1}{r}\frac{\partial}{\partial \theta} \cdot \boldsymbol{\delta}_\theta\boldsymbol{\delta}_r\tau_{\theta r}\right] = [\boldsymbol{\delta}_\theta \cdot \boldsymbol{\delta}_\theta\boldsymbol{\delta}_r]\frac{1}{r}\frac{\partial}{\partial \theta}\tau_{\theta r} + \text{other term} \quad \text{(A.7-17)}$$

$$\left[\boldsymbol{\delta}_\phi\frac{1}{r\sin\theta}\frac{\partial}{\partial \phi} \cdot \boldsymbol{\delta}_\phi\boldsymbol{\delta}_r\tau_{\phi r}\right] = [\boldsymbol{\delta}_\phi \cdot \boldsymbol{\delta}_\phi\boldsymbol{\delta}_r]\frac{1}{r\sin\theta}\frac{\partial}{\partial \phi}\tau_{\phi r} + \text{other term} \quad \text{(A.7-18)}$$

$$\left[\boldsymbol{\delta}_\theta \frac{1}{r}\frac{\partial}{\partial \theta} \cdot \boldsymbol{\delta}_r\boldsymbol{\delta}_r\tau_{rr}\right] = \frac{\tau_{rr}}{r}\left[\boldsymbol{\delta}_\theta \cdot \left\{\frac{\partial}{\partial \theta}\boldsymbol{\delta}_r\right\}\boldsymbol{\delta}_r\right] + \frac{\tau_{rr}}{r}\left[\boldsymbol{\delta}_\theta \cdot \boldsymbol{\delta}_r\left\{\frac{\partial}{\partial \theta}\boldsymbol{\delta}_r\right\}\right]$$

$$= \frac{\tau_{rr}}{r}[\boldsymbol{\delta}_\theta \cdot \boldsymbol{\delta}_\theta\boldsymbol{\delta}_r] = \boldsymbol{\delta}_r\frac{\tau_{rr}}{r} \tag{A.7-19}$$

$$\left[\boldsymbol{\delta}_\phi \frac{1}{r\sin\theta}\frac{\partial}{\partial \phi} \cdot \boldsymbol{\delta}_r\boldsymbol{\delta}_r\tau_{rr}\right] = \frac{\tau_{rr}}{r\sin\theta}\left[\boldsymbol{\delta}_\phi \cdot \left\{\frac{\partial}{\partial \phi}\boldsymbol{\delta}_r\right\}\boldsymbol{\delta}_r\right]$$

$$= \frac{\tau_{rr}}{r\sin\theta}[\boldsymbol{\delta}_\phi \cdot \boldsymbol{\delta}_\phi \sin\theta\,\boldsymbol{\delta}_r] = \boldsymbol{\delta}_r\frac{\tau_{rr}}{r} \tag{A.7-20}$$

$$\left[\boldsymbol{\delta}_\theta \frac{1}{r}\frac{\partial}{\partial \theta} \cdot \boldsymbol{\delta}_\theta\boldsymbol{\delta}_\theta\tau_{\theta\theta}\right] = \boldsymbol{\delta}_r\left(-\frac{\tau_{\theta\theta}}{r}\right) + \text{other term} \tag{A.7-21}$$

$$\left[\boldsymbol{\delta}_\phi \frac{1}{r\sin\theta}\frac{\partial}{\partial \phi} \cdot \boldsymbol{\delta}_\theta\boldsymbol{\delta}_r\tau_{\theta r}\right] = \boldsymbol{\delta}_r\frac{\tau_{\theta r}\cos\theta}{r\sin\theta} \tag{A.7-22}$$

$$\left[\boldsymbol{\delta}_\phi \frac{1}{r\sin\theta}\frac{\partial}{\partial \phi} \cdot \boldsymbol{\delta}_\phi\boldsymbol{\delta}_\phi\tau_{\phi\phi}\right] = \boldsymbol{\delta}_r\left(\frac{-\tau_{\phi\phi}}{r}\right) + \text{other terms} \tag{A.7-23}$$

Combining the above results we get

$$[\nabla \cdot \boldsymbol{\tau}]_r = \frac{1}{r^2}\frac{\partial}{\partial r}(r^2\tau_{rr}) + \frac{\tau_{\theta r}}{r}\cot\theta + \frac{1}{r}\frac{\partial}{\partial \theta}\tau_{\theta r} + \frac{1}{r\sin\theta}\frac{\partial \tau_{\phi r}}{\partial \phi} - \frac{\tau_{\theta\theta} + \tau_{\phi\phi}}{r} \tag{A.7-24}$$

Note that this expression is correct whether or not $\boldsymbol{\tau}$ is symmetric.

EXERCISES

1. If $\mathbf{r}$ is the instantaneous position vector for a particle, show that the velocity and acceleration of the particle are given by (use Eq. A.7-2):

$$\mathbf{v} = \frac{d}{dt}\mathbf{r} = \boldsymbol{\delta}_r\dot{r} + \boldsymbol{\delta}_\theta r\dot{\theta} + \boldsymbol{\delta}_z\dot{z} \tag{A.7-25}$$

$$\mathbf{a} = \boldsymbol{\delta}_r(\ddot{r} - r\dot{\theta}^2) + \boldsymbol{\delta}_\theta(r\ddot{\theta} + 2\dot{r}\dot{\theta}) + \boldsymbol{\delta}_z\ddot{z} \tag{A.7-26}$$

in cylindrical coordinates. The dots indicate time derivatives of the coordinates.

2. Obtain $(\nabla \cdot \mathbf{v})$, $[\nabla \times \mathbf{v}]$, and $\nabla\mathbf{v}$ in spherical coordinates, and $[\nabla \cdot \boldsymbol{\tau}]$ in cylindrical coordinates.

3. Use Table A.7-2 to write down directly the following quantities in cylindrical coordinates:

(a) $(\nabla \cdot \rho\mathbf{v})$, where ρ is a scalar
(b) $[\nabla \cdot \rho\mathbf{v}\mathbf{v}]_r$, where ρ is a scalar
(c) $[\nabla \cdot p\boldsymbol{\delta}]_\theta$, where p is a scalar
(d) $(\nabla \cdot [\boldsymbol{\tau} \cdot \mathbf{v}])$
(e) $[\mathbf{v} \cdot \nabla\mathbf{v}]_\theta$
(f) $\nabla\mathbf{v} + (\nabla\mathbf{v})^\dagger$

4. Verify that the entries for $\nabla^2\mathbf{v}$ in Table A.7-2 can be obtained by any one of the following methods:

(a) First verify that, in cylindrical coordinates the operator $(\nabla \cdot \nabla)$ is

$$(\nabla \cdot \nabla) = \frac{\partial^2}{\partial r^2} + \frac{1}{r}\frac{\partial}{\partial r} + \frac{1}{r^2}\frac{\partial^2}{\partial \theta^2} + \frac{\partial^2}{\partial z^2} \tag{A.7-27}$$

and then apply the operator to $\mathbf{v}$.

(b) Use the expression for $[\nabla \cdot \boldsymbol{\tau}]$ in Table A.7-2, but substitute the components for $\nabla\mathbf{v}$ in place of the components of $\boldsymbol{\tau}$, so as to obtain $[\nabla \cdot \nabla\mathbf{v}]$.

(c) Use Eq. A.4-22:

$$\nabla^2\mathbf{v} \equiv [\nabla \cdot \nabla\mathbf{v}] = \nabla(\nabla \cdot \mathbf{v}) - [\nabla \times [\nabla \times \mathbf{v}]] \tag{A.7-28}$$

and use the gradient, divergence, and curl operations in Table A.7-2 to evaluate the operations on the right side.

Table A.7-1. Summary of Differential Operations Involving the ∇-Operator in Cartesian Coordinates (x,y,z)

$$(\nabla \cdot \mathbf{v}) = \frac{\partial v_x}{\partial x} + \frac{\partial v_y}{\partial y} + \frac{\partial v_z}{\partial z} \quad \text{(A)}$$

$$\nabla^2 s \equiv (\nabla \cdot \nabla s) = \frac{\partial^2 s}{\partial x^2} + \frac{\partial^2 s}{\partial y^2} + \frac{\partial^2 s}{\partial z^2} \quad \text{(B)}$$

$$\begin{aligned}(\boldsymbol{\tau} : \nabla \mathbf{v}) = {} & \tau_{xx}\left(\frac{\partial v_x}{\partial x}\right) + \tau_{xy}\left(\frac{\partial v_x}{\partial y}\right) + \tau_{xz}\left(\frac{\partial v_x}{\partial z}\right) \\ & + \tau_{yx}\left(\frac{\partial v_y}{\partial x}\right) + \tau_{yy}\left(\frac{\partial v_y}{\partial y}\right) + \tau_{yz}\left(\frac{\partial v_y}{\partial z}\right) \\ & + \tau_{zx}\left(\frac{\partial v_z}{\partial x}\right) + \tau_{zy}\left(\frac{\partial v_z}{\partial y}\right) + \tau_{zz}\left(\frac{\partial v_z}{\partial z}\right) \end{aligned} \quad \text{(C)}$$

$$[\nabla s]_x = \frac{\partial s}{\partial x} \quad \text{(D)}$$

$$[\nabla s]_y = \frac{\partial s}{\partial y} \quad \text{(E)}$$

$$[\nabla s]_z = \frac{\partial s}{\partial z} \quad \text{(F)}$$

$$[\nabla \times \mathbf{v}]_x = \frac{\partial v_z}{\partial y} - \frac{\partial v_y}{\partial z} \quad \text{(G)}$$

$$[\nabla \times \mathbf{v}]_y = \frac{\partial v_x}{\partial z} - \frac{\partial v_z}{\partial x} \quad \text{(H)}$$

$$[\nabla \times \mathbf{v}]_z = \frac{\partial v_y}{\partial x} - \frac{\partial v_x}{\partial y} \quad \text{(I)}$$

$$[\nabla \cdot \boldsymbol{\tau}]_x = \frac{\partial \tau_{xx}}{\partial x} + \frac{\partial \tau_{yx}}{\partial y} + \frac{\partial \tau_{zx}}{\partial z} \quad \text{(J)}$$

$$[\nabla \cdot \boldsymbol{\tau}]_y = \frac{\partial \tau_{xy}}{\partial x} + \frac{\partial \tau_{yy}}{\partial y} + \frac{\partial \tau_{zy}}{\partial z} \quad \text{(K)}$$

$$[\nabla \cdot \boldsymbol{\tau}]_z = \frac{\partial \tau_{xz}}{\partial x} + \frac{\partial \tau_{yz}}{\partial y} + \frac{\partial \tau_{zz}}{\partial z} \quad \text{(L)}$$

$$[\nabla^2 \mathbf{v}]_x = \frac{\partial^2 v_x}{\partial x^2} + \frac{\partial^2 v_x}{\partial y^2} + \frac{\partial^2 v_x}{\partial z^2} \quad \text{(M)}$$

$$[\nabla^2 \mathbf{v}]_y = \frac{\partial^2 v_y}{\partial x^2} + \frac{\partial^2 v_y}{\partial y^2} + \frac{\partial^2 v_y}{\partial z^2} \quad \text{(N)}$$

$$[\nabla^2 \mathbf{v}]_z = \frac{\partial^2 v_z}{\partial x^2} + \frac{\partial^2 v_z}{\partial y^2} + \frac{\partial^2 v_z}{\partial z^2} \quad \text{(O)}$$

$$[\mathbf{v} \cdot \nabla \mathbf{w}]_x = v_x\left(\frac{\partial w_x}{\partial x}\right) + v_y\left(\frac{\partial w_x}{\partial y}\right) + v_z\left(\frac{\partial w_x}{\partial z}\right) \quad \text{(P)}$$

$$[\mathbf{v} \cdot \nabla \mathbf{w}]_y = v_x\left(\frac{\partial w_y}{\partial x}\right) + v_y\left(\frac{\partial w_y}{\partial y}\right) + v_z\left(\frac{\partial w_y}{\partial z}\right) \quad \text{(Q)}$$

$$[\mathbf{v} \cdot \nabla \mathbf{w}]_z = v_x\left(\frac{\partial w_z}{\partial x}\right) + v_y\left(\frac{\partial w_z}{\partial y}\right) + v_z\left(\frac{\partial w_z}{\partial z}\right) \quad \text{(R)}$$

Table A.7-1. *(Continued)*

$$\{\nabla \mathbf{v}\}_{xx} = \frac{\partial v_x}{\partial x} \quad \text{(S)}$$

$$\{\nabla \mathbf{v}\}_{xy} = \frac{\partial v_y}{\partial x} \quad \text{(T)}$$

$$\{\nabla \mathbf{v}\}_{xz} = \frac{\partial v_z}{\partial x} \quad \text{(U)}$$

$$\{\nabla \mathbf{v}\}_{yx} = \frac{\partial v_x}{\partial y} \quad \text{(V)}$$

$$\{\nabla \mathbf{v}\}_{yy} = \frac{\partial v_y}{\partial y} \quad \text{(W)}$$

$$\{\nabla \mathbf{v}\}_{yz} = \frac{\partial v_z}{\partial y} \quad \text{(X)}$$

$$\{\nabla \mathbf{v}\}_{zx} = \frac{\partial v_x}{\partial z} \quad \text{(Y)}$$

$$\{\nabla \mathbf{v}\}_{zy} = \frac{\partial v_y}{\partial z} \quad \text{(Z)}$$

$$\{\nabla \mathbf{v}\}_{zz} = \frac{\partial v_z}{\partial z} \quad \text{(AA)}$$

$$\{\mathbf{v} \cdot \boldsymbol{\nabla} \boldsymbol{\tau}\}_{xx} = (\mathbf{v} \cdot \boldsymbol{\nabla})\tau_{xx} \quad \text{(BB)}$$

$$\{\mathbf{v} \cdot \boldsymbol{\nabla} \boldsymbol{\tau}\}_{xy} = (\mathbf{v} \cdot \boldsymbol{\nabla})\tau_{xy} \quad \text{(CC)}$$

$$\{\mathbf{v} \cdot \boldsymbol{\nabla} \boldsymbol{\tau}\}_{xz} = (\mathbf{v} \cdot \boldsymbol{\nabla})\tau_{xz} \quad \text{(DD)}$$

$$\{\mathbf{v} \cdot \boldsymbol{\nabla} \boldsymbol{\tau}\}_{yx} = (\mathbf{v} \cdot \boldsymbol{\nabla})\tau_{yx} \quad \text{(EE)}$$

$$\{\mathbf{v} \cdot \boldsymbol{\nabla} \boldsymbol{\tau}\}_{yy} = (\mathbf{v} \cdot \boldsymbol{\nabla})\tau_{yy} \quad \text{(FF)}$$

$$\{\mathbf{v} \cdot \boldsymbol{\nabla} \boldsymbol{\tau}\}_{yz} = (\mathbf{v} \cdot \boldsymbol{\nabla})\tau_{yz} \quad \text{(GG)}$$

$$\{\mathbf{v} \cdot \boldsymbol{\nabla} \boldsymbol{\tau}\}_{zx} = (\mathbf{v} \cdot \boldsymbol{\nabla})\tau_{zx} \quad \text{(HH)}$$

$$\{\mathbf{v} \cdot \boldsymbol{\nabla} \boldsymbol{\tau}\}_{zy} = (\mathbf{v} \cdot \boldsymbol{\nabla})\tau_{zy} \quad \text{(II)}$$

$$\{\mathbf{v} \cdot \boldsymbol{\nabla} \boldsymbol{\tau}\}_{zz} = (\mathbf{v} \cdot \boldsymbol{\nabla})\tau_{zz} \quad \text{(JJ)}$$

where the operator $(\mathbf{v} \cdot \nabla) = v_x \frac{\partial}{\partial x} + v_y \frac{\partial}{\partial y} + v_z \frac{\partial}{\partial z}$

Table A.7-2. Summary of Differential Operations Involving the ∇-Operator in Cylindrical Coordinates (r,θ,z)

$$(\nabla \cdot \mathbf{v}) = \frac{1}{r}\frac{\partial}{\partial r}(rv_r) + \frac{1}{r}\frac{\partial v_\theta}{\partial \theta} + \frac{\partial v_z}{\partial z} \qquad \text{(A)}$$

$$\nabla^2 s \equiv (\nabla \cdot \nabla s) = \frac{1}{r}\frac{\partial}{\partial r}\left(r\frac{\partial s}{\partial r}\right) + \frac{1}{r^2}\frac{\partial^2 s}{\partial \theta^2} + \frac{\partial^2 s}{\partial z^2} \qquad \text{(B)}$$

$$\begin{aligned}(\boldsymbol{\tau} : \nabla \mathbf{v}) = {} & \tau_{rr}\left(\frac{\partial v_r}{\partial r}\right) + \tau_{r\theta}\left(\frac{1}{r}\frac{\partial v_r}{\partial \theta} - \frac{v_\theta}{r}\right) + \tau_{rz}\left(\frac{\partial v_r}{\partial z}\right) \\ & + \tau_{\theta r}\left(\frac{\partial v_\theta}{\partial r}\right) + \tau_{\theta\theta}\left(\frac{1}{r}\frac{\partial v_\theta}{\partial \theta} + \frac{v_r}{r}\right) + \tau_{\theta z}\left(\frac{\partial v_\theta}{\partial z}\right) \\ & + \tau_{zr}\left(\frac{\partial v_z}{\partial r}\right) + \tau_{z\theta}\left(\frac{1}{r}\frac{\partial v_z}{\partial \theta}\right) + \tau_{zz}\left(\frac{\partial v_z}{\partial z}\right) \qquad \text{(C)}\end{aligned}$$

$$[\nabla s]_r = \frac{\partial s}{\partial r} \qquad \text{(D)}$$

$$[\nabla s]_\theta = \frac{1}{r}\frac{\partial s}{\partial \theta} \qquad \text{(E)}$$

$$[\nabla s]_z = \frac{\partial s}{\partial z} \qquad \text{(F)}$$

$$[\nabla \times \mathbf{v}]_r = \frac{1}{r}\frac{\partial v_z}{\partial \theta} - \frac{\partial v_\theta}{\partial z} \qquad \text{(G)}$$

$$[\nabla \times \mathbf{v}]_\theta = \frac{\partial v_r}{\partial z} - \frac{\partial v_z}{\partial r} \qquad \text{(H)}$$

$$[\nabla \times \mathbf{v}]_z = \frac{1}{r}\frac{\partial}{\partial r}(rv_\theta) - \frac{1}{r}\frac{\partial v_r}{\partial \theta} \qquad \text{(I)}$$

$$[\nabla \cdot \boldsymbol{\tau}]_r = \frac{1}{r}\frac{\partial}{\partial r}(r\tau_{rr}) + \frac{1}{r}\frac{\partial}{\partial \theta}\tau_{\theta r} + \frac{\partial}{\partial z}\tau_{zr} - \frac{\tau_{\theta\theta}}{r} \qquad \text{(J)}$$

$$[\nabla \cdot \boldsymbol{\tau}]_\theta = \frac{1}{r^2}\frac{\partial}{\partial r}(r^2\tau_{r\theta}) + \frac{1}{r}\frac{\partial}{\partial \theta}\tau_{\theta\theta} + \frac{\partial}{\partial z}\tau_{z\theta} + \frac{\tau_{\theta r} - \tau_{r\theta}}{r} \qquad \text{(K)}$$

$$[\nabla \cdot \boldsymbol{\tau}]_z = \frac{1}{r}\frac{\partial}{\partial r}(r\tau_{rz}) + \frac{1}{r}\frac{\partial}{\partial \theta}\tau_{\theta z} + \frac{\partial}{\partial z}\tau_{zz} \qquad \text{(L)}$$

$$[\nabla^2 \mathbf{v}]_r = \frac{\partial}{\partial r}\left(\frac{1}{r}\frac{\partial}{\partial r}(rv_r)\right) + \frac{1}{r^2}\frac{\partial^2 v_r}{\partial \theta^2} + \frac{\partial^2 v_r}{\partial z^2} - \frac{2}{r^2}\frac{\partial v_\theta}{\partial \theta} \qquad \text{(M)}$$

$$[\nabla^2 \mathbf{v}]_\theta = \frac{\partial}{\partial r}\left(\frac{1}{r}\frac{\partial}{\partial r}(rv_\theta)\right) + \frac{1}{r^2}\frac{\partial^2 v_\theta}{\partial \theta^2} + \frac{\partial^2 v_\theta}{\partial z^2} + \frac{2}{r^2}\frac{\partial v_r}{\partial \theta} \qquad \text{(N)}$$

$$[\nabla^2 \mathbf{v}]_z = \frac{1}{r}\frac{\partial}{\partial r}\left(r\frac{\partial v_z}{\partial r}\right) + \frac{1}{r^2}\frac{\partial^2 v_z}{\partial \theta^2} + \frac{\partial^2 v_z}{\partial z^2} \qquad \text{(O)}$$

$$[\mathbf{v} \cdot \nabla \mathbf{w}]_r = v_r\left(\frac{\partial w_r}{\partial r}\right) + v_\theta\left(\frac{1}{r}\frac{\partial w_r}{\partial \theta} - \frac{w_\theta}{r}\right) + v_z\left(\frac{\partial w_r}{\partial z}\right) \qquad \text{(P)}$$

$$[\mathbf{v} \cdot \nabla \mathbf{w}]_\theta = v_r\left(\frac{\partial w_\theta}{\partial r}\right) + v_\theta\left(\frac{1}{r}\frac{\partial w_\theta}{\partial \theta} + \frac{w_r}{r}\right) + v_z\left(\frac{\partial w_\theta}{\partial z}\right) \qquad \text{(Q)}$$

$$[\mathbf{v} \cdot \nabla \mathbf{w}]_z = v_r\left(\frac{\partial w_z}{\partial r}\right) + v_\theta\left(\frac{1}{r}\frac{\partial w_z}{\partial \theta}\right) + v_z\left(\frac{\partial w_z}{\partial z}\right) \qquad \text{(R)}$$

Table A.7-2. *(Continued)*

$$\{\nabla \mathbf{v}\}_{rr} = \frac{\partial v_r}{\partial r} \tag{S}$$

$$\{\nabla \mathbf{v}\}_{r\theta} = \frac{\partial v_\theta}{\partial r} \tag{T}$$

$$\{\nabla \mathbf{v}\}_{rz} = \frac{\partial v_z}{\partial r} \tag{U}$$

$$\{\nabla \mathbf{v}\}_{\theta r} = \frac{1}{r}\frac{\partial v_r}{\partial \theta} - \frac{v_\theta}{r} \tag{V}$$

$$\{\nabla \mathbf{v}\}_{\theta\theta} = \frac{1}{r}\frac{\partial v_\theta}{\partial \theta} + \frac{v_r}{r} \tag{W}$$

$$\{\nabla \mathbf{v}\}_{\theta z} = \frac{1}{r}\frac{\partial v_z}{\partial \theta} \tag{X}$$

$$\{\nabla \mathbf{v}\}_{zr} = \frac{\partial v_r}{\partial z} \tag{Y}$$

$$\{\nabla \mathbf{v}\}_{z\theta} = \frac{\partial v_\theta}{\partial z} \tag{Z}$$

$$\{\nabla \mathbf{v}\}_{zz} = \frac{\partial v_z}{\partial z} \tag{AA}$$

$$\{\mathbf{v}\cdot\nabla\boldsymbol{\tau}\}_{rr} = (\mathbf{v}\cdot\nabla)\tau_{rr} - \frac{v_\theta}{r}(\tau_{r\theta} + \tau_{\theta r}) \tag{BB}$$

$$\{\mathbf{v}\cdot\nabla\boldsymbol{\tau}\}_{r\theta} = (\mathbf{v}\cdot\nabla)\tau_{r\theta} + \frac{v_\theta}{r}(\tau_{rr} - \tau_{\theta\theta}) \tag{CC}$$

$$\{\mathbf{v}\cdot\nabla\boldsymbol{\tau}\}_{rz} = (\mathbf{v}\cdot\nabla)\tau_{rz} - \frac{v_\theta}{r}\tau_{\theta z} \tag{DD}$$

$$\{\mathbf{v}\cdot\nabla\boldsymbol{\tau}\}_{\theta r} = (\mathbf{v}\cdot\nabla)\tau_{\theta r} + \frac{v_\theta}{r}(\tau_{rr} - \tau_{\theta\theta}) \tag{EE}$$

$$\{\mathbf{v}\cdot\nabla\boldsymbol{\tau}\}_{\theta\theta} = (\mathbf{v}\cdot\nabla)\tau_{\theta\theta} + \frac{v_\theta}{r}(\tau_{r\theta} + \tau_{\theta r}) \tag{FF}$$

$$\{\mathbf{v}\cdot\nabla\boldsymbol{\tau}\}_{\theta z} = (\mathbf{v}\cdot\nabla)\tau_{\theta z} + \frac{v_\theta}{r}\tau_{rz} \tag{GG}$$

$$\{\mathbf{v}\cdot\nabla\boldsymbol{\tau}\}_{zr} = (\mathbf{v}\cdot\nabla)\tau_{zr} - \frac{v_\theta}{r}\tau_{z\theta} \tag{HH}$$

$$\{\mathbf{v}\cdot\nabla\boldsymbol{\tau}\}_{z\theta} = (\mathbf{v}\cdot\nabla)\tau_{z\theta} + \frac{v_\theta}{r}\tau_{zr} \tag{II}$$

$$\{\mathbf{v}\cdot\nabla\boldsymbol{\tau}\}_{zz} = (\mathbf{v}\cdot\nabla)\tau_{zz} \tag{JJ}$$

where the operator $(\mathbf{v}\cdot\nabla) = v_r\frac{\partial}{\partial r} + \frac{v_\theta}{r}\frac{\partial}{\partial \theta} + v_z\frac{\partial}{\partial z}$

Table A.7-3. Summary of Differential Operations Involving the ∇-Operator in Spherical Coordinates (r,θ,ϕ)

$$(\nabla\cdot\mathbf{v}) = \frac{1}{r^2}\frac{\partial}{\partial r}(r^2 v_r) + \frac{1}{r\sin\theta}\frac{\partial}{\partial\theta}(v_\theta\sin\theta) + \frac{1}{r\sin\theta}\frac{\partial v_\phi}{\partial\phi} \tag{A}$$

$$\nabla^2 s \equiv (\nabla\cdot\nabla s) = \frac{1}{r^2}\frac{\partial}{\partial r}\left(r^2\frac{\partial s}{\partial r}\right) + \frac{1}{r^2\sin\theta}\frac{\partial}{\partial\theta}\left(\sin\theta\frac{\partial s}{\partial\theta}\right) + \frac{1}{r^2\sin^2\theta}\frac{\partial^2 s}{\partial\phi^2} \tag{B}$$

$$\begin{aligned}(\boldsymbol{\tau}:\nabla\mathbf{v}) = {}& \tau_{rr}\left(\frac{\partial v_r}{\partial r}\right) + \tau_{r\theta}\left(\frac{1}{r}\frac{\partial v_r}{\partial\theta} - \frac{v_\theta}{r}\right) + \tau_{r\phi}\left(\frac{1}{r\sin\theta}\frac{\partial v_r}{\partial\phi} - \frac{v_\phi}{r}\right)\\ &+ \tau_{\theta r}\left(\frac{\partial v_\theta}{\partial r}\right) + \tau_{\theta\theta}\left(\frac{1}{r}\frac{\partial v_\theta}{\partial\theta} + \frac{v_r}{r}\right) + \tau_{\theta\phi}\left(\frac{1}{r\sin\theta}\frac{\partial v_\theta}{\partial\phi} - \frac{v_\phi}{r}\cot\theta\right)\\ &+ \tau_{\phi r}\left(\frac{\partial v_\phi}{\partial r}\right) + \tau_{\phi\theta}\left(\frac{1}{r}\frac{\partial v_\phi}{\partial\theta}\right) + \tau_{\phi\phi}\left(\frac{1}{r\sin\theta}\frac{\partial v_\phi}{\partial\phi} + \frac{v_r}{r} + \frac{v_\theta}{r}\cot\theta\right)\end{aligned} \tag{C}$$

$$[\nabla s]_r = \frac{\partial s}{\partial r} \tag{D}$$

$$[\nabla s]_\theta = \frac{1}{r}\frac{\partial s}{\partial\theta} \tag{E}$$

$$[\nabla s]_\phi = \frac{1}{r\sin\theta}\frac{\partial s}{\partial\phi} \tag{F}$$

$$[\nabla\times\mathbf{v}]_r = \frac{1}{r\sin\theta}\frac{\partial}{\partial\theta}(v_\phi\sin\theta) - \frac{1}{r\sin\theta}\frac{\partial v_\theta}{\partial\phi} \tag{G}$$

$$[\nabla\times\mathbf{v}]_\theta = \frac{1}{r\sin\theta}\frac{\partial v_r}{\partial\phi} - \frac{1}{r}\frac{\partial}{\partial r}(r v_\phi) \tag{H}$$

$$[\nabla\times\mathbf{v}]_\phi = \frac{1}{r}\frac{\partial}{\partial r}(r v_\theta) - \frac{1}{r}\frac{\partial v_r}{\partial\theta} \tag{I}$$

$$[\nabla\cdot\boldsymbol{\tau}]_r = \frac{1}{r^2}\frac{\partial}{\partial r}(r^2\tau_{rr}) + \frac{1}{r\sin\theta}\frac{\partial}{\partial\theta}(\tau_{\theta r}\sin\theta) + \frac{1}{r\sin\theta}\frac{\partial}{\partial\phi}\tau_{\phi r} - \frac{\tau_{\theta\theta}+\tau_{\phi\phi}}{r} \tag{J}$$

$$[\nabla\cdot\boldsymbol{\tau}]_\theta = \frac{1}{r^3}\frac{\partial}{\partial r}(r^3\tau_{r\theta}) + \frac{1}{r\sin\theta}\frac{\partial}{\partial\theta}(\tau_{\theta\theta}\sin\theta) + \frac{1}{r\sin\theta}\frac{\partial}{\partial\phi}\tau_{\phi\theta} + \frac{(\tau_{\theta r}-\tau_{r\theta}) - \tau_{\phi\phi}\cot\theta}{r} \tag{K}$$

$$[\nabla\cdot\boldsymbol{\tau}]_\phi = \frac{1}{r^3}\frac{\partial}{\partial r}(r^3\tau_{r\phi}) + \frac{1}{r\sin\theta}\frac{\partial}{\partial\theta}(\tau_{\theta\phi}\sin\theta) + \frac{1}{r\sin\theta}\frac{\partial}{\partial\phi}\tau_{\phi\phi} + \frac{(\tau_{\phi r}-\tau_{r\phi}) + \tau_{\phi\theta}\cot\theta}{r} \tag{L}$$

$$\begin{aligned}[\nabla^2\mathbf{v}]_r = {}& \frac{\partial}{\partial r}\left(\frac{1}{r^2}\frac{\partial}{\partial r}(r^2 v_r)\right) + \frac{1}{r^2\sin\theta}\frac{\partial}{\partial\theta}\left(\sin\theta\frac{\partial v_r}{\partial\theta}\right) + \frac{1}{r^2\sin^2\theta}\frac{\partial^2 v_r}{\partial\phi^2}\\ &- \frac{2}{r^2\sin\theta}\frac{\partial}{\partial\theta}(v_\theta\sin\theta) - \frac{2}{r^2\sin\theta}\frac{\partial v_\phi}{\partial\phi}\end{aligned} \tag{M}$$

$$\begin{aligned}[\nabla^2\mathbf{v}]_\theta = {}& \frac{1}{r^2}\frac{\partial}{\partial r}\left(r^2\frac{\partial v_\theta}{\partial r}\right) + \frac{1}{r^2}\frac{\partial}{\partial\theta}\left(\frac{1}{\sin\theta}\frac{\partial}{\partial\theta}(v_\theta\sin\theta)\right)\\ &+ \frac{1}{r^2\sin^2\theta}\frac{\partial^2 v_\theta}{\partial\phi^2} + \frac{2}{r^2}\frac{\partial v_r}{\partial\theta} - \frac{2\cot\theta}{r^2\sin\theta}\frac{\partial v_\phi}{\partial\phi}\end{aligned} \tag{N}$$

$$\begin{aligned}[\nabla^2\mathbf{v}]_\phi = {}& \frac{1}{r^2}\frac{\partial}{\partial r}\left(r^2\frac{\partial v_\phi}{\partial r}\right) + \frac{1}{r^2}\frac{\partial}{\partial\theta}\left(\frac{1}{\sin\theta}\frac{\partial}{\partial\theta}(v_\phi\sin\theta)\right) + \frac{1}{r^2\sin^2\theta}\frac{\partial^2 v_\phi}{\partial\phi^2}\\ &+ \frac{2}{r^2\sin\theta}\frac{\partial v_r}{\partial\phi} + \frac{2\cot\theta}{r^2\sin\theta}\frac{\partial v_\theta}{\partial\phi}\end{aligned} \tag{O}$$

$$[\mathbf{v}\cdot\nabla\mathbf{w}]_r = v_r\left(\frac{\partial w_r}{\partial r}\right) + v_\theta\left(\frac{1}{r}\frac{\partial w_r}{\partial\theta} - \frac{w_\theta}{r}\right) + v_\phi\left(\frac{1}{r\sin\theta}\frac{\partial w_r}{\partial\phi} - \frac{w_\phi}{r}\right) \tag{P}$$

Table A.7-3. *(Continued)*

$$[\mathbf{v}\cdot\nabla\mathbf{w}]_\theta = v_r\left(\frac{\partial w_\theta}{\partial r}\right) + v_\theta\left(\frac{1}{r}\frac{\partial w_\theta}{\partial \theta} + \frac{w_r}{r}\right) + v_\phi\left(\frac{1}{r\sin\theta}\frac{\partial w_\theta}{\partial\phi} - \frac{w_\phi}{r}\cot\theta\right) \qquad \text{(Q)}$$

$$[\mathbf{v}\cdot\nabla\mathbf{w}]_\phi = v_r\left(\frac{\partial w_\phi}{\partial r}\right) + v_\theta\left(\frac{1}{r}\frac{\partial w_\phi}{\partial \theta}\right) + v_\phi\left(\frac{1}{r\sin\theta}\frac{\partial w_\phi}{\partial\phi} + \frac{w_r}{r} + \frac{w_\theta}{r}\cot\theta\right) \qquad \text{(R)}$$

$$\{\nabla\mathbf{v}\}_{rr} = \frac{\partial v_r}{\partial r} \qquad \text{(S)}$$

$$\{\nabla\mathbf{v}\}_{r\theta} = \frac{\partial v_\theta}{\partial r} \qquad \text{(T)}$$

$$\{\nabla\mathbf{v}\}_{r\phi} = \frac{\partial v_\phi}{\partial r} \qquad \text{(U)}$$

$$\{\nabla\mathbf{v}\}_{\theta r} = \frac{1}{r}\frac{\partial v_r}{\partial\theta} - \frac{v_\theta}{r} \qquad \text{(V)}$$

$$\{\nabla\mathbf{v}\}_{\theta\theta} = \frac{1}{r}\frac{\partial v_\theta}{\partial\theta} + \frac{v_r}{r} \qquad \text{(W)}$$

$$\{\nabla\mathbf{v}\}_{\theta\phi} = \frac{1}{r}\frac{\partial v_\phi}{\partial\theta} \qquad \text{(X)}$$

$$\{\nabla\mathbf{v}\}_{\phi r} = \frac{1}{r\sin\theta}\frac{\partial v_r}{\partial\phi} - \frac{v_\phi}{r} \qquad \text{(Y)}$$

$$\{\nabla\mathbf{v}\}_{\phi\theta} = \frac{1}{r\sin\theta}\frac{\partial v_\theta}{\partial\phi} - \frac{v_\phi}{r}\cot\theta \qquad \text{(Z)}$$

$$\{\nabla\mathbf{v}\}_{\phi\phi} = \frac{1}{r\sin\theta}\frac{\partial v_\phi}{\partial\phi} + \frac{v_r}{r} + \frac{v_\theta}{r}\cot\theta \qquad \text{(AA)}$$

$$\{\mathbf{v}\cdot\nabla\boldsymbol{\tau}\}_{rr} = (\mathbf{v}\cdot\nabla)\tau_{rr} - \left(\frac{v_\theta}{r}\right)(\tau_{r\theta}+\tau_{\theta r}) - \left(\frac{v_\phi}{r}\right)(\tau_{r\phi}+\tau_{\phi r}) \qquad \text{(BB)}$$

$$\{\mathbf{v}\cdot\nabla\boldsymbol{\tau}\}_{r\theta} = (\mathbf{v}\cdot\nabla)\tau_{r\theta} + \left(\frac{v_\theta}{r}\right)(\tau_{rr}-\tau_{\theta\theta}) - \left(\frac{v_\phi}{r}\right)(\tau_{\phi\theta}+\tau_{r\phi}\cot\theta) \qquad \text{(CC)}$$

$$\{\mathbf{v}\cdot\nabla\boldsymbol{\tau}\}_{r\phi} = (\mathbf{v}\cdot\nabla)\tau_{r\phi} - \left(\frac{v_\theta}{r}\right)\tau_{\theta\phi} + \left(\frac{v_\phi}{r}\right)[(\tau_{rr}-\tau_{\phi\phi}) + \tau_{r\theta}\cot\theta] \qquad \text{(DD)}$$

$$\{\mathbf{v}\cdot\nabla\boldsymbol{\tau}\}_{\theta r} = (\mathbf{v}\cdot\nabla)\tau_{\theta r} + \left(\frac{v_\theta}{r}\right)(\tau_{rr}-\tau_{\theta\theta}) - \left(\frac{v_\phi}{r}\right)(\tau_{\theta\phi}+\tau_{\phi r}\cot\theta) \qquad \text{(EE)}$$

$$\{\mathbf{v}\cdot\nabla\boldsymbol{\tau}\}_{\theta\theta} = (\mathbf{v}\cdot\nabla)\tau_{\theta\theta} + \left(\frac{v_\theta}{r}\right)(\tau_{r\theta}+\tau_{\theta r}) - \left(\frac{v_\phi}{r}\right)(\tau_{\theta\phi}+\tau_{\phi\theta})\cot\theta \qquad \text{(FF)}$$

$$\{\mathbf{v}\cdot\nabla\boldsymbol{\tau}\}_{\theta\phi} = (\mathbf{v}\cdot\nabla)\tau_{\theta\phi} + \left(\frac{v_\theta}{r}\right)\tau_{r\phi} + \left(\frac{v_\phi}{r}\right)[\tau_{\theta r} + (\tau_{\theta\theta}-\tau_{\phi\phi})\cot\theta] \qquad \text{(GG)}$$

$$\{\mathbf{v}\cdot\nabla\boldsymbol{\tau}\}_{\phi r} = (\mathbf{v}\cdot\nabla)\tau_{\phi r} - \left(\frac{v_\theta}{r}\right)\tau_{\phi\theta} + \left(\frac{v_\phi}{r}\right)[(\tau_{rr}-\tau_{\phi\phi}) + \tau_{\theta r}\cot\theta] \qquad \text{(HH)}$$

$$\{\mathbf{v}\cdot\nabla\boldsymbol{\tau}\}_{\phi\theta} = (\mathbf{v}\cdot\nabla)\tau_{\phi\theta} + \left(\frac{v_\theta}{r}\right)\tau_{\phi r} + \left(\frac{v_\phi}{r}\right)[\tau_{r\theta} + (\tau_{\theta\theta}-\tau_{\phi\phi})\cot\theta] \qquad \text{(II)}$$

$$\{\mathbf{v}\cdot\nabla\boldsymbol{\tau}\}_{\phi\phi} = (\mathbf{v}\cdot\nabla)\tau_{\phi\phi} + \left(\frac{v_\phi}{r}\right)[(\tau_{r\phi}+\tau_{\phi r}) + (\tau_{\theta\phi}+\tau_{\phi\theta})\cot\theta] \qquad \text{(JJ)}$$

where the operator $(\mathbf{v}\cdot\nabla) = v_r\frac{\partial}{\partial r} + \frac{v_\theta}{r}\frac{\partial}{\partial\theta} + \frac{v_\phi}{r\sin\theta}\frac{\partial}{\partial\phi}$

§A.8 INTEGRAL OPERATIONS IN CURVILINEAR COORDINATES

In performing the integrations of §A.5 in curvilinear coordinates, it is important to understand the construction of the volume elements, as is shown for cylindrical coordinates in Fig. A.8-1 and for spherical coordinates in Fig. A.8-2.

In doing *volume integrals*, the simplest situations are those in which the bounding surfaces are surfaces of the coordinate system. For *cylindrical coordinates*, a typical volume integral of a function $f(r,\theta,z)$ would be of the form

$$\int_{z_1}^{z_2}\int_{\theta_1}^{\theta_2}\int_{r_1}^{r_2} f(r,\theta,z)r\,dr\,d\theta\,dz \tag{A.8-1}$$

and for *spherical coordinates* a typical volume integral of a function $g(r,\theta,\phi)$ would be

$$\int_{\phi_1}^{\phi_2}\int_{\theta_1}^{\theta_2}\int_{r_1}^{r_2} g(r,\theta,\phi)r^2dr\,\sin\,\theta\,d\theta\,d\phi \tag{A.8-2}$$

Since the limits in these integrals ($r_1,r_2,\theta_1,\theta_2$,etc.) are constants, the order of the integration is immaterial.

In doing *surface integrals*, the simplest situations are those in which the integration is performed on one of the surfaces of the coordinate system. For *cylindrical coordinates* there are three possibilities:

On the surface $r = r_0$:
$$\int_{z_1}^{z_2}\int_{\theta_1}^{\theta_2} f(r_0,\theta,z)r_0\,d\theta\,dz \tag{A.8-3}$$

On the surface $\theta = \theta_0$:
$$\int_{z_1}^{z_2}\int_{r_1}^{r_2} f(r,\theta_0,z)\,dr\,dz \tag{A.8-4}$$

On the surface $z = z_0$:
$$\int_{\theta_2}^{\theta_2}\int_{r_1}^{r_2} f(r,\theta,z_0)r\,dr\,d\theta \tag{A.8-5}$$

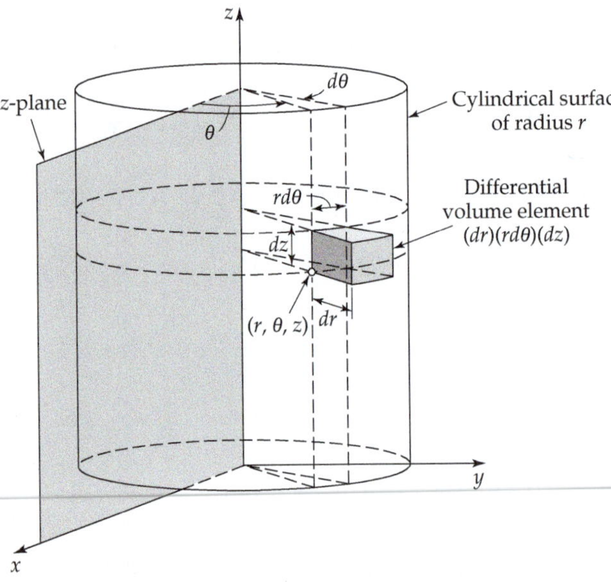

Fig. A.8-1. Differential volume element $r\,dr\,d\theta\,dz$ in cylindrical coordinates, and differential line elements dr, $r\,d\theta$, and dz. The differential surface elements are: $(r\,d\theta)(dz)$ perpendicular to the r direction (intermediate shading); $(dz)(dr)$ perpendicular to the θ direction (darkest shading); and $(dr)(r\,d\theta)$ perpendicular to the z direction (lightest shading).

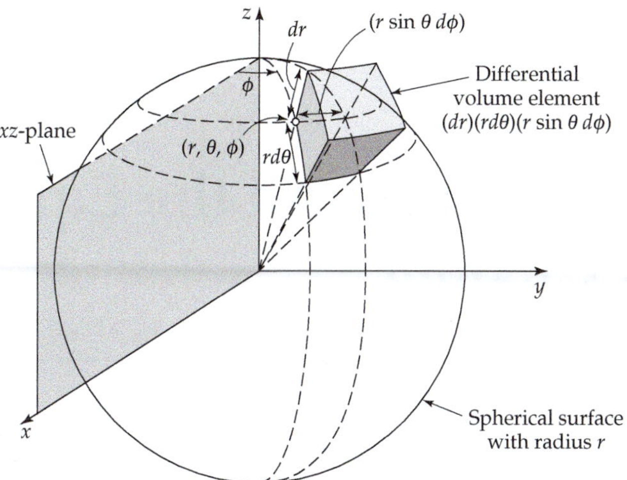

Fig. A.8-2. Differential volume element $r^2 \sin\theta\, dr\, d\theta\, d\phi$ in spherical coordinates, and the differential line elements dr, $r\,d\theta$, and $r\sin\theta\,d\phi$. The angle θ is measured from the z axis, that is, downward from the top of the sphere (in contrast to angles of latitude, which are measured upward from the equator). The differential surface elements are: $(r\,d\theta)(r\sin\theta\,d\phi)$ perpendicular to the r direction (lightest shading); $(r\sin\theta\,d\phi)(dr)$ perpendicular to the θ direction (darkest shading); and $(dr)(r\,d\theta)$ perpendicular to the ϕ direction (intermediate shading).

Similarly, for *spherical coordinates*:

On the surface $r = r_0$: $$\int_{\phi_1}^{\phi_2}\int_{\theta_1}^{\theta_2} g(r_0,\theta,\phi) r_0^2 \sin\theta\, d\theta\, d\phi \tag{A.8-6}$$

On the surface $\theta = \theta_0$: $$\int_{\phi_1}^{\phi_2}\int_{r_1}^{r_2} g(r,\theta_0,\phi) \sin\theta_0\, r\, dr\, d\phi \tag{A.8-7}$$

On the surface $\phi = \phi_0$: $$\int_{\theta_1}^{\theta_2}\int_{r_1}^{r_2} g(r,\theta,\phi_0) r\, dr\, d\theta \tag{A.8-8}$$

The reader should try making sketches to show exactly what areas are described by each of the above six surface integrals.

If the area of integration in a surface integral is not one of the surfaces of the coordinate system, then a book on differential and integral calculus should be consulted.

Appendix B

The Fluxes and the Equations of Change

§B.1 NEWTON'S LAW OF VISCOSITY

$$[\boldsymbol{\tau} = -\mu(\nabla\mathbf{v} + (\nabla\mathbf{v})^\dagger) + (\tfrac{2}{3}\mu - \kappa)(\nabla \cdot \mathbf{v})\boldsymbol{\delta}]$$

Cartesian coordinates (x,y,z):

$$\tau_{xx} = -\mu\left[2\frac{\partial v_x}{\partial x}\right] + (\tfrac{2}{3}\mu - \kappa)(\nabla \cdot \mathbf{v}) \quad \text{(B.1-1)}^a$$

$$\tau_{yy} = -\mu\left[2\frac{\partial v_y}{\partial y}\right] + (\tfrac{2}{3}\mu - \kappa)(\nabla \cdot \mathbf{v}) \quad \text{(B.1-2)}^a$$

$$\tau_{zz} = -\mu\left[2\frac{\partial v_z}{\partial z}\right] + (\tfrac{2}{3}\mu - \kappa)(\nabla \cdot \mathbf{v}) \quad \text{(B.1-3)}^a$$

$$\tau_{xy} = \tau_{yx} = -\mu\left[\frac{\partial v_y}{\partial x} + \frac{\partial v_x}{\partial y}\right] \quad \text{(B.1-4)}$$

$$\tau_{yz} = \tau_{zy} = -\mu\left[\frac{\partial v_z}{\partial y} + \frac{\partial v_y}{\partial z}\right] \quad \text{(B.1-5)}$$

$$\tau_{zx} = \tau_{xz} = -\mu\left[\frac{\partial v_x}{\partial z} + \frac{\partial v_z}{\partial x}\right] \quad \text{(B.1-6)}$$

in which

$$(\nabla \cdot \mathbf{v}) = \frac{\partial v_x}{\partial x} + \frac{\partial v_y}{\partial y} + \frac{\partial v_z}{\partial z} \quad \text{(B.1-7)}$$

[a] When the fluid is assumed to have constant density, the term containing $(\nabla \cdot \mathbf{v})$ may be omitted. For monatomic gases at low density, the dilatational viscosity κ is zero.

§B.1 NEWTON'S LAW OF VISCOSITY (continued)

Cylindrical coordinates (r,θ,z):

$$\tau_{rr} = -\mu\left[2\frac{\partial v_r}{\partial r}\right] + (\tfrac{2}{3}\mu - \kappa)(\nabla \cdot \mathbf{v}) \quad \text{(B.1-8)}^a$$

$$\tau_{\theta\theta} = -\mu\left[2\left(\frac{1}{r}\frac{\partial v_\theta}{\partial \theta} + \frac{v_r}{r}\right)\right] + (\tfrac{2}{3}\mu - \kappa)(\nabla \cdot \mathbf{v}) \quad \text{(B.1-9)}^a$$

$$\tau_{zz} = -\mu\left[2\frac{\partial v_z}{\partial z}\right] + (\tfrac{2}{3}\mu - \kappa)(\nabla \cdot \mathbf{v}) \quad \text{(B.1-10)}^a$$

$$\tau_{r\theta} = \tau_{\theta r} = -\mu\left[r\frac{\partial}{\partial r}\left(\frac{v_\theta}{r}\right) + \frac{1}{r}\frac{\partial v_r}{\partial \theta}\right] \quad \text{(B.1-11)}$$

$$\tau_{\theta z} = \tau_{z\theta} = -\mu\left[\frac{1}{r}\frac{\partial v_z}{\partial \theta} + \frac{\partial v_\theta}{\partial z}\right] \quad \text{(B.1-12)}$$

$$\tau_{zr} = \tau_{rz} = -\mu\left[\frac{\partial v_r}{\partial z} + \frac{\partial v_z}{\partial r}\right] \quad \text{(B.1-13)}$$

in which

$$(\nabla \cdot \mathbf{v}) = \frac{1}{r}\frac{\partial}{\partial r}(rv_r) + \frac{1}{r}\frac{\partial v_\theta}{\partial \theta} + \frac{\partial v_z}{\partial z} \quad \text{(B.1-14)}$$

---

[a] When the fluid is assumed to have constant density, the term containing $(\nabla \cdot \mathbf{v})$ may be omitted. For monatomic gases at low density, the dilatational viscosity κ is zero.

Spherical coordinates (r,θ,ϕ):

$$\tau_{rr} = -\mu\left[2\frac{\partial v_r}{\partial r}\right] + (\tfrac{2}{3}\mu - \kappa)(\nabla \cdot \mathbf{v}) \quad \text{(B.1-15)}^a$$

$$\tau_{\theta\theta} = -\mu\left[2\left(\frac{1}{r}\frac{\partial v_\theta}{\partial \theta} + \frac{v_r}{r}\right)\right] + (\tfrac{2}{3}\mu - \kappa)(\nabla \cdot \mathbf{v}) \quad \text{(B.1-16)}^a$$

$$\tau_{\phi\phi} = -\mu\left[2\left(\frac{1}{r\sin\theta}\frac{\partial v_\phi}{\partial \phi} + \frac{v_r + v_\theta\cot\theta}{r}\right)\right] + (\tfrac{2}{3}\mu - \kappa)(\nabla \cdot \mathbf{v}) \quad \text{(B.1-17)}^a$$

$$\tau_{r\theta} = \tau_{\theta r} = -\mu\left[r\frac{\partial}{\partial r}\left(\frac{v_\theta}{r}\right) + \frac{1}{r}\frac{\partial v_r}{\partial \theta}\right] \quad \text{(B.1-18)}$$

$$\tau_{\theta\phi} = \tau_{\phi\theta} = -\mu\left[\frac{\sin\theta}{r}\frac{\partial}{\partial \theta}\left(\frac{v_\phi}{\sin\theta}\right) + \frac{1}{r\sin\theta}\frac{\partial v_\theta}{\partial \phi}\right] \quad \text{(B.1-19)}$$

$$\tau_{\phi r} = \tau_{r\phi} = -\mu\left[\frac{1}{r\sin\theta}\frac{\partial v_r}{\partial \phi} + r\frac{\partial}{\partial r}\left(\frac{v_\phi}{r}\right)\right] \quad \text{(B.1-20)}$$

in which

$$(\nabla \cdot \mathbf{v}) = \frac{1}{r^2}\frac{\partial}{\partial r}(r^2 v_r) + \frac{1}{r\sin\theta}\frac{\partial}{\partial \theta}(v_\theta\sin\theta) + \frac{1}{r\sin\theta}\frac{\partial v_\phi}{\partial \phi} \quad \text{(B.1-21)}$$

---

[a] When the fluid is assumed to have constant density, the term containing $(\nabla \cdot \mathbf{v})$ may be omitted. For monatomic gases at low density, the dilatational viscosity κ is zero.

§B.2 FOURIER'S LAW OF HEAT CONDUCTION[a]

$$[\mathbf{q} = -k\nabla T]$$

Cartesian coordinates (x,y,z):

$$q_x = -k\frac{\partial T}{\partial x} \quad \text{(B.2-1)}$$

$$q_y = -k\frac{\partial T}{\partial y} \quad \text{(B.2-2)}$$

$$q_z = -k\frac{\partial T}{\partial z} \quad \text{(B.2-3)}$$

Cylindrical coordinates (r,θ,z):

$$q_r = -k\frac{\partial T}{\partial r} \quad \text{(B.2-4)}$$

$$q_\theta = -k\frac{1}{r}\frac{\partial T}{\partial \theta} \quad \text{(B.2-5)}$$

$$q_z = -k\frac{\partial T}{\partial z} \quad \text{(B.2-6)}$$

Spherical coordinates (r,θ,ϕ):

$$q_r = -k\frac{\partial T}{\partial r} \quad \text{(B.2-7)}$$

$$q_\theta = -k\frac{1}{r}\frac{\partial T}{\partial \theta} \quad \text{(B.2-8)}$$

$$q_\phi = -k\frac{1}{r\sin\theta}\frac{\partial T}{\partial \phi} \quad \text{(B.2-9)}$$

[a]For mixtures, the term $\Sigma_\alpha(\overline{H}_\alpha/M_\alpha)\mathbf{j}_\alpha$ must be added to $-k\nabla T$ (see Eq. 19.3-3).

§B.3 FICK'S (FIRST) LAW OF BINARY DIFFUSION[a]

$$[\mathbf{j}_A = -\rho \mathscr{D}_{AB} \nabla \omega_A]$$

Cartesian coordinates (x,y,z):

$$j_{Ax} = -\rho \mathscr{D}_{AB} \frac{\partial \omega_A}{\partial x} \quad \text{(B.3-1)}$$

$$j_{Ay} = -\rho \mathscr{D}_{AB} \frac{\partial \omega_A}{\partial y} \quad \text{(B.3-2)}$$

$$j_{Az} = -\rho \mathscr{D}_{AB} \frac{\partial \omega_A}{\partial z} \quad \text{(B.3-3)}$$

Cylindrical coordinates (r,θ,z):

$$j_{Ar} = -\rho \mathscr{D}_{AB} \frac{\partial \omega_A}{\partial r} \quad \text{(B.3-4)}$$

$$j_{A\theta} = -\rho \mathscr{D}_{AB} \frac{1}{r} \frac{\partial \omega_A}{\partial \theta} \quad \text{(B.3-5)}$$

$$j_{Az} = -\rho \mathscr{D}_{AB} \frac{\partial \omega_A}{\partial z} \quad \text{(B.3-6)}$$

Spherical coordinates (r,θ,φ):

$$j_{Ar} = -\rho \mathscr{D}_{AB} \frac{\partial \omega_A}{\partial r} \quad \text{(B.3-7)}$$

$$j_{A\theta} = -\rho \mathscr{D}_{AB} \frac{1}{r} \frac{\partial \omega_A}{\partial \theta} \quad \text{(B.3-8)}$$

$$j_{A\phi} = -\rho \mathscr{D}_{AB} \frac{1}{r \sin \theta} \frac{\partial \omega_A}{\partial \phi} \quad \text{(B.3-9)}$$

[a]To get the molar fluxes with respect to the molar average velocity, replace $\mathbf{j}_A$, ρ, and ω_A by $\mathbf{J}_A^*$, c, and x_A.

§B.4 THE EQUATION OF CONTINUITY[a]

$$[\partial \rho / \partial t + (\nabla \cdot \rho \mathbf{v}) = 0] \quad \text{(B.4-1)}$$

Cartesian coordinates (x,y,z):

$$\frac{\partial \rho}{\partial t} + \frac{\partial}{\partial x}(\rho v_x) + \frac{\partial}{\partial y}(\rho v_y) + \frac{\partial}{\partial z}(\rho v_z) = 0 \quad \text{(B.4-2)}$$

Cylindrical coordinates (r,θ,z):

$$\frac{\partial \rho}{\partial t} + \frac{1}{r}\frac{\partial}{\partial r}(\rho r v_r) + \frac{1}{r}\frac{\partial}{\partial \theta}(\rho v_\theta) + \frac{\partial}{\partial z}(\rho v_z) = 0 \quad \text{(B.4-3)}$$

Spherical coordinates (r,θ,φ):

$$\frac{\partial \rho}{\partial t} + \frac{1}{r^2}\frac{\partial}{\partial r}(\rho r^2 v_r) + \frac{1}{r \sin \theta}\frac{\partial}{\partial \theta}(\rho v_\theta \sin \theta) + \frac{1}{r \sin \theta}\frac{\partial}{\partial \phi}(\rho v_\phi) = 0 \quad \text{(B.4-4)}$$

[a]When the fluid is assumed to have constant mass density ρ, the equation simplifies to $(\nabla \cdot \mathbf{v}) = 0$.

§B.5 THE EQUATION OF MOTION IN TERMS OF τ

$$[\rho D\mathbf{v}/Dt = -\nabla p - [\nabla \cdot \boldsymbol{\tau}] + \rho \mathbf{g}]$$

Cartesian coordinates (x,y,z):[a]

$$\rho\left(\frac{\partial v_x}{\partial t} + v_x\frac{\partial v_x}{\partial x} + v_y\frac{\partial v_x}{\partial y} + v_z\frac{\partial v_x}{\partial z}\right) = -\frac{\partial p}{\partial x} - \left[\frac{\partial}{\partial x}\tau_{xx} + \frac{\partial}{\partial y}\tau_{yx} + \frac{\partial}{\partial z}\tau_{zx}\right] + \rho g_x \quad \text{(B.5-1)}$$

$$\rho\left(\frac{\partial v_y}{\partial t} + v_x\frac{\partial v_y}{\partial x} + v_y\frac{\partial v_y}{\partial y} + v_z\frac{\partial v_y}{\partial z}\right) = -\frac{\partial p}{\partial y} - \left[\frac{\partial}{\partial x}\tau_{xy} + \frac{\partial}{\partial y}\tau_{yy} + \frac{\partial}{\partial z}\tau_{zy}\right] + \rho g_y \quad \text{(B.5-2)}$$

$$\rho\left(\frac{\partial v_z}{\partial t} + v_x\frac{\partial v_z}{\partial x} + v_y\frac{\partial v_z}{\partial y} + v_z\frac{\partial v_z}{\partial z}\right) = -\frac{\partial p}{\partial z} - \left[\frac{\partial}{\partial x}\tau_{xz} + \frac{\partial}{\partial y}\tau_{yz} + \frac{\partial}{\partial z}\tau_{zz}\right] + \rho g_z \quad \text{(B.5-3)}$$

[a]These equations have been written without making the assumption that **τ** is symmetric. This means, for example, that when the usual assumption is made that the stress tensor is symmetric, τ_{xy} and τ_{yx} may be interchanged.

Cylindrical coordinates (r,θ,z):[b]

$$\rho\left(\frac{\partial v_r}{\partial t} + v_r\frac{\partial v_r}{\partial r} + \frac{v_\theta}{r}\frac{\partial v_r}{\partial \theta} + v_z\frac{\partial v_r}{\partial z} - \frac{v_\theta^2}{r}\right) = -\frac{\partial p}{\partial r} - \left[\frac{1}{r}\frac{\partial}{\partial r}(r\tau_{rr}) + \frac{1}{r}\frac{\partial}{\partial \theta}\tau_{\theta r} + \frac{\partial}{\partial z}\tau_{zr} - \frac{\tau_{\theta\theta}}{r}\right] + \rho g_r \quad \text{(B.5-4)}$$

$$\rho\left(\frac{\partial v_\theta}{\partial t} + v_r\frac{\partial v_\theta}{\partial r} + \frac{v_\theta}{r}\frac{\partial v_\theta}{\partial \theta} + v_z\frac{\partial v_\theta}{\partial z} + \frac{v_r v_\theta}{r}\right) = -\frac{1}{r}\frac{\partial p}{\partial \theta} - \left[\frac{1}{r^2}\frac{\partial}{\partial r}(r^2\tau_{r\theta}) + \frac{1}{r}\frac{\partial}{\partial \theta}\tau_{\theta\theta} + \frac{\partial}{\partial z}\tau_{z\theta} + \frac{\tau_{\theta r} - \tau_{r\theta}}{r}\right] + \rho g_\theta \quad \text{(B.5-5)}$$

$$\rho\left(\frac{\partial v_z}{\partial t} + v_r\frac{\partial v_z}{\partial r} + \frac{v_\theta}{r}\frac{\partial v_z}{\partial \theta} + v_z\frac{\partial v_z}{\partial z}\right) = -\frac{\partial p}{\partial z} - \left[\frac{1}{r}\frac{\partial}{\partial r}(r\tau_{rz}) + \frac{1}{r}\frac{\partial}{\partial \theta}\tau_{\theta z} + \frac{\partial}{\partial z}\tau_{zz}\right] + \rho g_z \quad \text{(B.5-6)}$$

[b]These equations have been written without making the assumption that **τ** is symmetric. This means, for example, that when the usual assumption is made that the stress tensor is symmetric, $\tau_{r\theta} - \tau_{\theta r} = 0$.

Spherical coordinates (r,θ,φ):[c]

$$\rho\left(\frac{\partial v_r}{\partial t} + v_r\frac{\partial v_r}{\partial r} + \frac{v_\theta}{r}\frac{\partial v_r}{\partial \theta} + \frac{v_\phi}{r\sin\theta}\frac{\partial v_r}{\partial \phi} - \frac{v_\theta^2 + v_\phi^2}{r}\right) = -\frac{\partial p}{\partial r}$$
$$- \left[\frac{1}{r^2}\frac{\partial}{\partial r}(r^2\tau_{rr}) + \frac{1}{r\sin\theta}\frac{\partial}{\partial \theta}(\tau_{\theta r}\sin\theta) + \frac{1}{r\sin\theta}\frac{\partial}{\partial \phi}\tau_{\phi r} - \frac{\tau_{\theta\theta} + \tau_{\phi\phi}}{r}\right] + \rho g_r \quad \text{(B.5-7)}$$

$$\rho\left(\frac{\partial v_\theta}{\partial t} + v_r\frac{\partial v_\theta}{\partial r} + \frac{v_\theta}{r}\frac{\partial v_\theta}{\partial \theta} + \frac{v_\phi}{r\sin\theta}\frac{\partial v_\theta}{\partial \phi} + \frac{v_r v_\theta - v_\phi^2\cot\theta}{r}\right) = -\frac{1}{r}\frac{\partial p}{\partial \theta}$$
$$- \left[\frac{1}{r^3}\frac{\partial}{\partial r}(r^3\tau_{r\theta}) + \frac{1}{r\sin\theta}\frac{\partial}{\partial \theta}(\tau_{\theta\theta}\sin\theta) + \frac{1}{r\sin\theta}\frac{\partial}{\partial \phi}\tau_{\phi\theta} + \frac{(\tau_{\theta r} - \tau_{r\theta}) - \tau_{\phi\phi}\cot\theta}{r}\right] + \rho g_\theta \quad \text{(B.5-8)}$$

$$\rho\left(\frac{\partial v_\phi}{\partial t} + v_r\frac{\partial v_\phi}{\partial r} + \frac{v_\theta}{r}\frac{\partial v_\phi}{\partial \theta} + \frac{v_\phi}{r\sin\theta}\frac{\partial v_\phi}{\partial \phi} + \frac{v_\phi v_r + v_\theta v_\phi\cot\theta}{r}\right) = -\frac{1}{r\sin\theta}\frac{\partial p}{\partial \phi}$$
$$- \left[\frac{1}{r^3}\frac{\partial}{\partial r}(r^3\tau_{r\phi}) + \frac{1}{r\sin\theta}\frac{\partial}{\partial \theta}(\tau_{\theta\phi}\sin\theta) + \frac{1}{r\sin\theta}\frac{\partial}{\partial \phi}\tau_{\phi\phi} + \frac{(\tau_{\phi r} - \tau_{r\phi}) + \tau_{\phi\theta}\cot\theta}{r}\right] + \rho g_\phi \quad \text{(B.5-9)}$$

[c]These equations have been written without making the assumption that **τ** is symmetric. This means, for example, that when the usual assumption is made that the stress tensor is symmetric, $\tau_{r\theta} - \tau_{\theta r} = 0$.

§B.6 THE EQUATION OF MOTION FOR A NEWTONIAN FLUID WITH CONSTANT ρ AND μ

$$[\rho D\mathbf{v}/Dt = -\nabla p + \mu\nabla^2\mathbf{v} + \rho\mathbf{g}]$$

Cartesian coordinates (x,y,z):

$$\rho\left(\frac{\partial v_x}{\partial t} + v_x\frac{\partial v_x}{\partial x} + v_y\frac{\partial v_x}{\partial y} + v_z\frac{\partial v_x}{\partial z}\right) = -\frac{\partial p}{\partial x} + \mu\left[\frac{\partial^2 v_x}{\partial x^2} + \frac{\partial^2 v_x}{\partial y^2} + \frac{\partial^2 v_x}{\partial z^2}\right] + \rho g_x \quad \text{(B.6-1)}$$

$$\rho\left(\frac{\partial v_y}{\partial t} + v_x\frac{\partial v_y}{\partial x} + v_y\frac{\partial v_y}{\partial y} + v_z\frac{\partial v_y}{\partial z}\right) = -\frac{\partial p}{\partial y} + \mu\left[\frac{\partial^2 v_y}{\partial x^2} + \frac{\partial^2 v_y}{\partial y^2} + \frac{\partial^2 v_y}{\partial z^2}\right] + \rho g_y \quad \text{(B.6-2)}$$

$$\rho\left(\frac{\partial v_z}{\partial t} + v_x\frac{\partial v_z}{\partial x} + v_y\frac{\partial v_z}{\partial y} + v_z\frac{\partial v_z}{\partial z}\right) = -\frac{\partial p}{\partial z} + \mu\left[\frac{\partial^2 v_z}{\partial x^2} + \frac{\partial^2 v_z}{\partial y^2} + \frac{\partial^2 v_z}{\partial z^2}\right] + \rho g_z \quad \text{(B.6-3)}$$

Cylindrical coordinates (r,θ,z):

$$\rho\left(\frac{\partial v_r}{\partial t} + v_r\frac{\partial v_r}{\partial r} + \frac{v_\theta}{r}\frac{\partial v_r}{\partial \theta} + v_z\frac{\partial v_r}{\partial z} - \frac{v_\theta^2}{r}\right) = -\frac{\partial p}{\partial r} + \mu\left[\frac{\partial}{\partial r}\left(\frac{1}{r}\frac{\partial}{\partial r}(rv_r)\right) + \frac{1}{r^2}\frac{\partial^2 v_r}{\partial \theta^2} + \frac{\partial^2 v_r}{\partial z^2} - \frac{2}{r^2}\frac{\partial v_\theta}{\partial \theta}\right] + \rho g_r \quad \text{(B.6-4)}$$

$$\rho\left(\frac{\partial v_\theta}{\partial t} + v_r\frac{\partial v_\theta}{\partial r} + \frac{v_\theta}{r}\frac{\partial v_\theta}{\partial \theta} + v_z\frac{\partial v_\theta}{\partial z} + \frac{v_r v_\theta}{r}\right) = -\frac{1}{r}\frac{\partial p}{\partial \theta} + \mu\left[\frac{\partial}{\partial r}\left(\frac{1}{r}\frac{\partial}{\partial r}(rv_\theta)\right) + \frac{1}{r^2}\frac{\partial^2 v_\theta}{\partial \theta^2} + \frac{\partial^2 v_\theta}{\partial z^2} + \frac{2}{r^2}\frac{\partial v_r}{\partial \theta}\right] + \rho g_\theta \quad \text{(B.6-5)}$$

$$\rho\left(\frac{\partial v_z}{\partial t} + v_r\frac{\partial v_z}{\partial r} + \frac{v_\theta}{r}\frac{\partial v_z}{\partial \theta} + v_z\frac{\partial v_z}{\partial z}\right) = -\frac{\partial p}{\partial z} + \mu\left[\frac{1}{r}\frac{\partial}{\partial r}\left(r\frac{\partial v_z}{\partial r}\right) + \frac{1}{r^2}\frac{\partial^2 v_z}{\partial \theta^2} + \frac{\partial^2 v_z}{\partial z^2}\right] + \rho g_z \quad \text{(B.6-6)}$$

Spherical coordinates (r,θ,ϕ):

$$\rho\left(\frac{\partial v_r}{\partial t} + v_r\frac{\partial v_r}{\partial r} + \frac{v_\theta}{r}\frac{\partial v_r}{\partial \theta} + \frac{v_\phi}{r\sin\theta}\frac{\partial v_r}{\partial \phi} - \frac{v_\theta^2 + v_\phi^2}{r}\right) = -\frac{\partial p}{\partial r}$$
$$+\mu\left[\frac{1}{r^2}\frac{\partial^2}{\partial r^2}(r^2 v_r) + \frac{1}{r^2\sin\theta}\frac{\partial}{\partial \theta}\left(\sin\theta\frac{\partial v_r}{\partial \theta}\right) + \frac{1}{r^2\sin^2\theta}\frac{\partial^2 v_r}{\partial \phi^2}\right] + \rho g_r \quad \text{(B.6-7)}^a$$

$$\rho\left(\frac{\partial v_\theta}{\partial t} + v_r\frac{\partial v_\theta}{\partial r} + \frac{v_\theta}{r}\frac{\partial v_\theta}{\partial \theta} + \frac{v_\phi}{r\sin\theta}\frac{\partial v_\theta}{\partial \phi} + \frac{v_r v_\theta - v_\phi^2\cot\theta}{r}\right) = -\frac{1}{r}\frac{\partial p}{\partial \theta}$$
$$+\mu\left[\frac{1}{r^2}\frac{\partial}{\partial r}\left(r^2\frac{\partial v_\theta}{\partial r}\right) + \frac{1}{r^2}\frac{\partial}{\partial \theta}\left(\frac{1}{\sin\theta}\frac{\partial}{\partial \theta}(v_\theta\sin\theta)\right) + \frac{1}{r^2\sin^2\theta}\frac{\partial^2 v_\theta}{\partial \phi^2} + \frac{2}{r^2}\frac{\partial v_r}{\partial \theta} - \frac{2\cot\theta}{r^2\sin\theta}\frac{\partial v_\phi}{\partial \phi}\right] + \rho g_\theta \quad \text{(B.6-8)}$$

$$\rho\left(\frac{\partial v_\phi}{\partial t} + v_r\frac{\partial v_\phi}{\partial r} + \frac{v_\theta}{r}\frac{\partial v_\phi}{\partial \theta} + \frac{v_\phi}{r\sin\theta}\frac{\partial v_\phi}{\partial \phi} + \frac{v_\phi v_r + v_\theta v_\phi\cot\theta}{r}\right) = -\frac{1}{r\sin\theta}\frac{\partial p}{\partial \phi}$$
$$+\mu\left[\frac{1}{r^2}\frac{\partial}{\partial r}\left(r^2\frac{\partial v_\phi}{\partial r}\right) + \frac{1}{r^2}\frac{\partial}{\partial \theta}\left(\frac{1}{\sin\theta}\frac{\partial}{\partial \theta}(v_\phi\sin\theta)\right) + \frac{1}{r^2\sin^2\theta}\frac{\partial^2 v_\phi}{\partial \phi^2} + \frac{2}{r^2\sin\theta}\frac{\partial v_r}{\partial \phi} + \frac{2\cot\theta}{r^2\sin\theta}\frac{\partial v_\theta}{\partial \phi}\right] + \rho g_\phi \quad \text{(B.6-9)}$$

[a] The quantity in the brackets in Eq. B.6-7 is *not* what one would expect from Eq. (M) for $[\nabla \cdot \nabla\mathbf{v}]$ in Table A.7-3, because we have added to Eq. (M) the expression for $(2/r)(\nabla \cdot \mathbf{v})$, which is zero for fluids with constant ρ. This gives a much simpler equation.

§B.7 THE DISSIPATION FUNCTION Φ_v FOR NEWTONIAN FLUIDS (SEE EQ. 3.3-3)

Cartesian coordinates (x,y,z):

$$\Phi_v = 2\left[\left(\frac{\partial v_x}{\partial x}\right)^2 + \left(\frac{\partial v_y}{\partial y}\right)^2 + \left(\frac{\partial v_z}{\partial z}\right)^2\right] + \left[\frac{\partial v_y}{\partial x} + \frac{\partial v_x}{\partial y}\right]^2 + \left[\frac{\partial v_z}{\partial y} + \frac{\partial v_y}{\partial z}\right]^2 + \left[\frac{\partial v_x}{\partial z} + \frac{\partial v_z}{\partial x}\right]^2 - \frac{2}{3}\left[\frac{\partial v_x}{\partial x} + \frac{\partial v_y}{\partial y} + \frac{\partial v_z}{\partial z}\right]^2 \tag{B.7-1}$$

Cylindrical coordinates (r,θ,z):

$$\Phi_v = 2\left[\left(\frac{\partial v_r}{\partial r}\right)^2 + \left(\frac{1}{r}\frac{\partial v_\theta}{\partial \theta} + \frac{v_r}{r}\right)^2 + \left(\frac{\partial v_z}{\partial z}\right)^2\right] + \left[r\frac{\partial}{\partial r}\left(\frac{v_\theta}{r}\right) + \frac{1}{r}\frac{\partial v_r}{\partial \theta}\right]^2 + \left[\frac{1}{r}\frac{\partial v_z}{\partial \theta} + \frac{\partial v_\theta}{\partial z}\right]^2 + \left[\frac{\partial v_r}{\partial z} + \frac{\partial v_z}{\partial r}\right]^2$$
$$- \frac{2}{3}\left[\frac{1}{r}\frac{\partial}{\partial r}(r v_r) + \frac{1}{r}\frac{\partial v_\theta}{\partial \theta} + \frac{\partial v_z}{\partial z}\right]^2 \tag{B.7-2}$$

Spherical coordinates (r,θ,ϕ):

$$\Phi_v = 2\left[\left(\frac{\partial v_r}{\partial r}\right)^2 + \left(\frac{1}{r}\frac{\partial v_\theta}{\partial \theta} + \frac{v_r}{r}\right)^2 + \left(\frac{1}{r \sin\theta}\frac{\partial v_\phi}{\partial \phi} + \frac{v_r + v_\theta \cot\theta}{r}\right)^2\right]$$
$$+ \left[r\frac{\partial}{\partial r}\left(\frac{v_\theta}{r}\right) + \frac{1}{r}\frac{\partial v_r}{\partial \theta}\right]^2 + \left[\frac{\sin\theta}{r}\frac{\partial}{\partial \theta}\left(\frac{v_\phi}{\sin\theta}\right) + \frac{1}{r\sin\theta}\frac{\partial v_\theta}{\partial \phi}\right]^2 + \left[\frac{1}{r\sin\theta}\frac{\partial v_r}{\partial \phi} + r\frac{\partial}{\partial r}\left(\frac{v_\phi}{r}\right)\right]^2$$
$$- \frac{2}{3}\left[\frac{1}{r^2}\frac{\partial}{\partial r}(r^2 v_r) + \frac{1}{r\sin\theta}\frac{\partial}{\partial \theta}(v_\theta \sin\theta) + \frac{1}{r\sin\theta}\frac{\partial v_\phi}{\partial \phi}\right]^2 \tag{B.7-3}$$

§B.8 THE EQUATION OF ENERGY IN TERMS OF q

$$[\rho \hat{C}_p DT/Dt = -(\nabla \cdot \mathbf{q}) - (\partial \ln \rho/\partial \ln T)_p Dp/Dt - (\boldsymbol{\tau} : \nabla \mathbf{v})]$$

Cartesian coordinates (x,y,z):

$$\rho \hat{C}_p\left(\frac{\partial T}{\partial t} + v_x\frac{\partial T}{\partial x} + v_y\frac{\partial T}{\partial y} + v_z\frac{\partial T}{\partial z}\right) = -\left[\frac{\partial q_x}{\partial x} + \frac{\partial q_y}{\partial y} + \frac{\partial q_z}{\partial z}\right] - \left(\frac{\partial \ln \rho}{\partial \ln T}\right)_p \frac{Dp}{Dt} - (\boldsymbol{\tau} : \nabla \mathbf{v}) \tag{B.8-1}$$ [a]

Cylindrical coordinates (r,θ,z):

$$\rho \hat{C}_p\left(\frac{\partial T}{\partial t} + v_r\frac{\partial T}{\partial r} + \frac{v_\theta}{r}\frac{\partial T}{\partial \theta} + v_z\frac{\partial T}{\partial z}\right) = -\left[\frac{1}{r}\frac{\partial}{\partial r}(r q_r) + \frac{1}{r}\frac{\partial q_\theta}{\partial \theta} + \frac{\partial q_z}{\partial z}\right] - \left(\frac{\partial \ln \rho}{\partial \ln T}\right)_p \frac{Dp}{Dt} - (\boldsymbol{\tau} : \nabla \mathbf{v}) \tag{B.8-2}$$ [a]

Spherical coordinates (r,θ,ϕ):

$$\rho \hat{C}_p\left(\frac{\partial T}{\partial t} + v_r\frac{\partial T}{\partial r} + \frac{v_\theta}{r}\frac{\partial T}{\partial \theta} + \frac{v_\phi}{r\sin\theta}\frac{\partial T}{\partial \phi}\right) = -\left[\frac{1}{r^2}\frac{\partial}{\partial r}(r^2 q_r) + \frac{1}{r\sin\theta}\frac{\partial}{\partial \theta}(q_\theta \sin\theta) + \frac{1}{r\sin\theta}\frac{\partial q_\phi}{\partial \phi}\right]$$
$$- \left(\frac{\partial \ln \rho}{\partial \ln T}\right)_p \frac{Dp}{Dt} - (\boldsymbol{\tau} : \nabla \mathbf{v}) \tag{B.8-3}$$ [a]

[a] The viscous dissipation term, $-(\boldsymbol{\tau} : \nabla \mathbf{v})$, is given in Appendix A, Tables A.7-1, A.7-2, and A.7-3. This term may usually be neglected, except for systems with very large velocity gradients. The term containing $(\partial \ln \rho/\partial \ln T)_p$ is zero for fluids with constant ρ.

§B.9 THE EQUATION OF ENERGY FOR PURE NEWTONIAN FLUIDS WITH CONSTANT[a] ρ AND k

$$[\rho \hat{C}_p DT/Dt = k\nabla^2 T + \mu\Phi_v]$$

Cartesian coordinates (x,y,z):

$$\rho\hat{C}_p\left(\frac{\partial T}{\partial t} + v_x\frac{\partial T}{\partial x} + v_y\frac{\partial T}{\partial y} + v_z\frac{\partial T}{\partial z}\right) = k\left[\frac{\partial^2 T}{\partial x^2} + \frac{\partial^2 T}{\partial y^2} + \frac{\partial^2 T}{\partial z^2}\right] + \mu\Phi_v \qquad \text{(B.9-1)}^b$$

Cylindrical coordinates (r,θ,z):

$$\rho\hat{C}_p\left(\frac{\partial T}{\partial t} + v_r\frac{\partial T}{\partial r} + \frac{v_\theta}{r}\frac{\partial T}{\partial \theta} + v_z\frac{\partial T}{\partial z}\right) = k\left[\frac{1}{r}\frac{\partial}{\partial r}\left(r\frac{\partial T}{\partial r}\right) + \frac{1}{r^2}\frac{\partial^2 T}{\partial \theta^2} + \frac{\partial^2 T}{\partial z^2}\right] + \mu\Phi_v \qquad \text{(B.9-2)}^b$$

Spherical coordinates (r,θ,ϕ):

$$\rho\hat{C}_p\left(\frac{\partial T}{\partial t} + v_r\frac{\partial T}{\partial r} + \frac{v_\theta}{r}\frac{\partial T}{\partial \theta} + \frac{v_\phi}{r\sin\theta}\frac{\partial T}{\partial \phi}\right) = k\left[\frac{1}{r^2}\frac{\partial}{\partial r}\left(r^2\frac{\partial T}{\partial r}\right) + \frac{1}{r^2\sin\theta}\frac{\partial}{\partial \theta}\left(\sin\theta\frac{\partial T}{\partial \theta}\right) + \frac{1}{r^2\sin^2\theta}\frac{\partial^2 T}{\partial \phi^2}\right] + \mu\Phi_v \qquad \text{(B.9-3)}^b$$

[a]This form of the energy equation is also valid under the less stringent assumptions k = constant and $(\partial \ln \rho/\partial \ln T)_p Dp/Dt = 0$. The assumption ρ = constant is given in the table heading because it is the assumption more often made.

[b]The function Φ_v is given in §B.7. The term $\mu\Phi_v$ is usually negligible, except in systems with large velocity gradients.

§B.10 THE EQUATION OF CONTINUITY FOR SPECIES α IN TERMS[a] OF $\mathbf{j}_\alpha$

$$[\rho D\omega_\alpha/Dt = -(\nabla\cdot\mathbf{j}_\alpha) + r_\alpha]$$

Cartesian coordinates (x,y,z):

$$\rho\left(\frac{\partial \omega_\alpha}{\partial t} + v_x\frac{\partial \omega_\alpha}{\partial x} + v_y\frac{\partial \omega_\alpha}{\partial y} + v_z\frac{\partial \omega_\alpha}{\partial z}\right) = -\left[\frac{\partial j_{\alpha x}}{\partial x} + \frac{\partial j_{\alpha y}}{\partial y} + \frac{\partial j_{\alpha z}}{\partial z}\right] + r_\alpha \qquad \text{(B.10-1)}$$

Cylindrical coordinates (r,θ,z):

$$\rho\left(\frac{\partial \omega_\alpha}{\partial t} + v_r\frac{\partial \omega_\alpha}{\partial r} + \frac{v_\theta}{r}\frac{\partial \omega_\alpha}{\partial \theta} + v_z\frac{\partial \omega_\alpha}{\partial z}\right) = -\left[\frac{1}{r}\frac{\partial}{\partial r}(r j_{\alpha r}) + \frac{1}{r}\frac{\partial j_{\alpha\theta}}{\partial \theta} + \frac{\partial j_{\alpha z}}{\partial z}\right] + r_\alpha \qquad \text{(B.10-2)}$$

Spherical coordinates (r,θ,ϕ):

$$\rho\left(\frac{\partial \omega_\alpha}{\partial t} + v_r\frac{\partial \omega_\alpha}{\partial r} + \frac{v_\theta}{r}\frac{\partial \omega_\alpha}{\partial \theta} + \frac{v_\phi}{r\sin\theta}\frac{\partial \omega_\alpha}{\partial \phi}\right) = -\left[\frac{1}{r^2}\frac{\partial}{\partial r}(r^2 j_{\alpha r}) + \frac{1}{r\sin\theta}\frac{\partial}{\partial \theta}(j_{\alpha\theta}\sin\theta) + \frac{1}{r\sin\theta}\frac{\partial j_{\alpha\phi}}{\partial \phi}\right] + r_\alpha \qquad \text{(B.10-3)}$$

[a]To obtain the corresponding equations in terms of $\mathbf{J}_\alpha^*$ make the following replacements:

Replace	ρ	ω_α	$\mathbf{j}_\alpha$	$\mathbf{v}$	r_α
by	c	x_α	$\mathbf{J}_\alpha^*$	$\mathbf{v}^*$	$R_\alpha - x_\alpha\sum_{\beta=1}^{N} R_\beta$

§B.11 THE EQUATION OF CONTINUITY FOR SPECIES A IN TERMS OF ω_A FOR CONSTANT[a] $\rho\mathscr{D}_{AB}$

$$[\rho D\omega_A/Dt = \rho\mathscr{D}_{AB}\nabla^2\omega_A + r_A]$$

Cartesian coordinates (x,y,z):

$$\rho\left(\frac{\partial\omega_A}{\partial t} + v_x\frac{\partial\omega_A}{\partial x} + v_y\frac{\partial\omega_A}{\partial y} + v_z\frac{\partial\omega_A}{\partial z}\right) = \rho\mathscr{D}_{AB}\left[\frac{\partial^2\omega_A}{\partial x^2} + \frac{\partial^2\omega_A}{\partial y^2} + \frac{\partial^2\omega_A}{\partial z^2}\right] + r_A \quad \text{(B.11-1)}$$

Cylindrical coordinates (r,θ,z):

$$\rho\left(\frac{\partial\omega_A}{\partial t} + v_r\frac{\partial\omega_A}{\partial r} + \frac{v_\theta}{r}\frac{\partial\omega_A}{\partial\theta} + v_z\frac{\partial\omega_A}{\partial z}\right) = \rho\mathscr{D}_{AB}\left[\frac{1}{r}\frac{\partial}{\partial r}\left(r\frac{\partial\omega_A}{\partial r}\right) + \frac{1}{r^2}\frac{\partial^2\omega_A}{\partial\theta^2} + \frac{\partial^2\omega_A}{\partial z^2}\right] + r_A \quad \text{(B.11-2)}$$

Spherical coordinates (r,θ,ϕ):

$$\rho\left(\frac{\partial\omega_A}{\partial t} + v_r\frac{\partial\omega_A}{\partial r} + \frac{v_\theta}{r}\frac{\partial\omega_A}{\partial\theta} + \frac{v_\phi}{r\sin\theta}\frac{\partial\omega_A}{\partial\phi}\right)$$
$$= \rho\mathscr{D}_{AB}\left[\frac{1}{r^2}\frac{\partial}{\partial r}\left(r^2\frac{\partial\omega_A}{\partial r}\right) + \frac{1}{r^2\sin\theta}\frac{\partial}{\partial\theta}\left(\sin\theta\frac{\partial\omega_A}{\partial\theta}\right) + \frac{1}{r^2\sin^2\theta}\frac{\partial^2\omega_A}{\partial\phi^2}\right] + r_A \quad \text{(B.11-3)}$$

[a] To obtain the corresponding equations in terms of x_A, make the following replacements:

Replace	ρ	ω_A	$\mathbf{v}$	r_A
by	c	x_A	$\mathbf{v}^*$	$x_B R_A - x_A R_B$

Appendix C

Mathematical Topics

In this appendix we summarize information on mathematical topics (other than vectors and tensors) that are useful in the study of transport phenomena.[1]

§C.1 SOME ORDINARY DIFFERENTIAL EQUATIONS AND THEIR SOLUTIONS

We assemble here a short list of differential equations that arise frequently in transport phenomena. The reader is assumed to be familiar with these equations and how to solve them. The quantities a, b, and c are real constants, and f, g, and h are functions of x or y. The C's are constants of integration.

Equation	Solution	
$\dfrac{dy}{dx} = \dfrac{f(x)}{g(y)}$	$\int g\,dy = \int f\,dx + C_1$	(C.1-1)
$\dfrac{dy}{dx} + f(x)y = g(x)$	$y = e^{-\int f\,dx}\left(\int e^{\int f\,dx} g\,dx + C_1\right)$	(C.1-2)
$\dfrac{d^2y}{dx^2} + a^2y = 0$	$y = C_1 \cos ax + C_2 \sin ax$	(C.1-3)
$\dfrac{d^2y}{dx^2} - a^2y = 0$	$y = C_1 \cosh ax + C_2 \sinh ax$ or	(C.1-4a)
	$y = C_3 e^{+ax} + C_4 e^{-ax}$	(C.1-4b)
$\dfrac{1}{x^2}\dfrac{d}{dx}\left(x^2\dfrac{dy}{dx}\right) + a^2y = 0$	$y = \dfrac{C_1}{x}\cos ax + \dfrac{C_2}{x}\sin ax$	(C.1-5)

[1]Some useful reference books on applied mathematics are: M. Abramowitz and I. A. Stegun, *Handbook of Mathematical Functions*, Dover, New York, 9th printing (1973); G. M. Murphy, *Ordinary Differential Equations and Their Solutions*, Van Nostrand, Princeton, NJ (1960); J. J. Tuma, *Engineering Mathematics Handbook*, 3rd edition, McGraw-Hill, New York (1987); M. D. Graham and J. B. Rawlings, *Modeling and Analysis Principles for Chemical and Biological Engineers*, Nob Hill Publishing, Madison, WI (2013).

$\frac{1}{x^2}\frac{d}{dx}\left(x^2\frac{dy}{dx}\right) - a^2y = 0$ $\quad y = \frac{C_1}{x}\cosh ax + \frac{C_2}{x}\sinh ax$ (C.1-6a)

$y = \frac{C_3}{x}e^{+ax} + \frac{C_4}{x}e^{-ax}$ (C.1-6b)

$\frac{d^2y}{dx^2} + a\frac{dy}{dx} + by = 0$ Solve the equation $n^2 + an + b = 0$, and get the roots $n = n_+$ and $n = n_-$. Then

(a) if n_+ and n_- are real and unequal, (C.1-7a)
$y = C_1e^{n_+x} + C_2e^{n_-x}$

(b) if n_+ and n_- are real and equal to n,
$y = e^{nx}(C_1x + C_2)$ (C.1-7b)

(c) if n_+ and n_- are complex: $n_\pm = p \pm iq$,
$y = e^{px}(C_1 \cos qx + C_2 \sin qx)$ (C.1-7c)

$\frac{d^2y}{dx^2} + 2x\frac{dy}{dx} = 0$ $\quad y = C_1\int_0^x e^{-\bar{x}^2}d\bar{x} + C_2$ (C.1-8)

$\frac{d^2y}{dx^2} + 3x^2\frac{dy}{dx} = 0$ $\quad y = C_1\int_0^x e^{-\bar{x}^3}d\bar{x} + C_2$ (C.1-9)

$\frac{d^2y}{dx^2} = f(x)$ $\quad y = \int_0^x\int_0^{\bar{x}} f(\bar{\bar{x}})d\bar{\bar{x}}\,d\bar{x} + C_1x + C_2$ (C.1-10)

$\frac{1}{x}\frac{d}{dx}\left(x\frac{dy}{dx}\right) = f(x)$ $\quad y = \int_0^x \frac{1}{\bar{x}}\int_0^{\bar{x}} \bar{\bar{x}}f(\bar{\bar{x}})\,d\bar{\bar{x}}\,d\bar{x} + C_1 \ln x + C_2$ (C.1-11)

$\frac{1}{x^2}\frac{d}{dx}\left(x^2\frac{dy}{dx}\right) = f(x)$ $\quad y = \int_0^x \frac{1}{\bar{x}^2}\int_0^{\bar{x}} \bar{\bar{x}}^2 f(\bar{\bar{x}})\,d\bar{\bar{x}}\,d\bar{x} - \frac{C_1}{x} + C_2$ (C.1-12)

$\frac{d^2y}{dx^2} = h(y)$ $\quad x = \int_0^y \frac{d\bar{y}}{\sqrt{2\int_0^{\bar{y}} h(\bar{\bar{y}})d\bar{\bar{y}} + C_1}} + C_2$ (C.1-13)

$x^3\frac{d^3y}{dx^3} + ax^2\frac{d^2y}{dx^2} + bx\frac{dy}{dx} + cy = 0$ (C.1-14)

$y = C_1x^{n_1} + C_2x^{n_2} + C_3x^{n_3}$, where the the n_k are the roots of the equation $n(n-1)(n-2) + an(n-1) + bn + c = 0$, provided that all roots are distinct.

---

[a]In Eqs. C.1-4 and C.1-6 the decisions as to whether to use the exponential forms or the trigonometric (or hyperbolic) functions are usually made on the basis of the boundary conditions on the problem or the symmetry properties of the solution.

[b]Equations C.1-5 and C.1-6 are solved by making the substitution $y(x) = u(x)/x$ and then solving the resulting equation for $u(x)$.

[c]In Eqs. C.1-8 to C.1-13, it may be convenient or necessary to change the lower limits of the integrals to some value other than zero.

§C.2 EXPANSIONS OF FUNCTIONS IN TAYLOR SERIES

In physical problems we often need to describe a function $y(x)$ in the neighborhood of some point $x = x_0$. Then we expand the function $y(x)$ in a "Taylor series about the point $x = x_0$":

$$y(x) = y|_{x=x_0} + \frac{1}{1!}\left(\left.\frac{dy}{dx}\right|_{x=x_0}\right)(x - x_0) + \frac{1}{2!}\left(\left.\frac{d^2y}{dx^2}\right|_{x=x_0}\right)(x - x_0)^2 + \frac{1}{3!}\left(\left.\frac{d^3y}{dx^3}\right|_{x=x_0}\right)(x - x_0)^3 + \cdots \tag{C.2-1}$$

The first term just gives the value of the function at $x = x_0$. The first two terms give a straight-line fit of the curve at $x = x_0$. The first three terms give a parabolic fit of the curve at $x = x_0$, and so on. The Taylor series is often used when only the first several terms are needed to describe the function adequately.

Here are a few Taylor series expansions of standard functions about $x = 0$:

$$e^{\pm x} = 1 \pm \frac{x}{1!} + \frac{x^2}{2!} \pm \frac{x^3}{3!} + \cdots \tag{C.2-2}$$

$$\ln(1 + x) = x - \frac{x^2}{2} + \frac{x^3}{3} - \frac{x^4}{4} + \cdots \tag{C.2-3}$$

$$\operatorname{erf} x = \frac{2x}{\sqrt{\pi}}\left(1 - \frac{x^2}{1!3} + \frac{x^4}{2!5} - \frac{x^6}{3!7} + \cdots\right) \tag{C.2-4}$$

$$\sqrt{1 \pm x} = 1 \pm \frac{1}{2}x - \frac{1 \cdot 1}{2 \cdot 4}x^2 \pm \frac{1 \cdot 1 \cdot 3}{2 \cdot 4 \cdot 6}x^3 - \cdots \tag{C.2-5}$$

$$\sin x = x - \frac{x^3}{3!} + \frac{x^5}{5!} - \frac{x^7}{7!} + \cdots \tag{C.2-6}$$

$$\cos x = 1 - \frac{x^2}{2!} + \frac{x^4}{4!} - \frac{x^6}{6!} + \cdots \tag{C.2-7}$$

$$\sinh x = x + \frac{x^3}{3!} + \frac{x^5}{5!} + \frac{x^7}{7!} + \cdots \tag{C.2-8}$$

$$\cosh x = 1 + \frac{x^2}{2!} + \frac{x^4}{4!} + \frac{x^6}{6!} + \cdots \tag{C.2-9}$$

Further examples may be found in calculus textbooks and handbooks. Taylor series can also be written for two and more independent variables.

§C.3 DIFFERENTIATION OF INTEGRALS (THE LEIBNIZ FORMULA)

Suppose we have a function $f(x,t)$ that depends on a space variable x and the time t. Then we can form the integral

$$I(t) = \int_{\alpha(t)}^{\beta(t)} f(x,t)\, dx \tag{C.3-1}$$

which is a function of t (see Fig. C.3-1(a)). If we want to differentiate this function with respect to t without evaluating the integral, we can use the Leibniz formula

$$\frac{d}{dt}\int_{\alpha(t)}^{\beta(t)} f(x,t)\, dx = \underset{(1)}{\int_{\alpha(t)}^{\beta(t)} \frac{\partial}{\partial t} f(x,t)\, dx} + \left(\underset{(2)}{f(\beta,t)\frac{d\beta}{dt}} \underset{(3)}{- f(\alpha,t)\frac{d\alpha}{dt}}\right) \tag{C.3-2}$$

Figure C.3-1(b) shows the meaning of the operations performed here: the first term on the right side gives the change in the integral because the function itself is changing with time; the second term accounts for the gain in area as the upper limit is moved to the right; and the third term shows the loss in area as the lower limit is moved to the right. This formula finds many uses in science and engineering.[1] The three-dimensional analog is given in Eq. A5.4.

---

[1] The Nobel laureate Richard P. Feynman, in his book *Surely You're Joking Mr. Feynman*, Bantam Books, New York (1985), p. 72 and p. 93, tells how he found the Leibniz formula useful on many occasions.

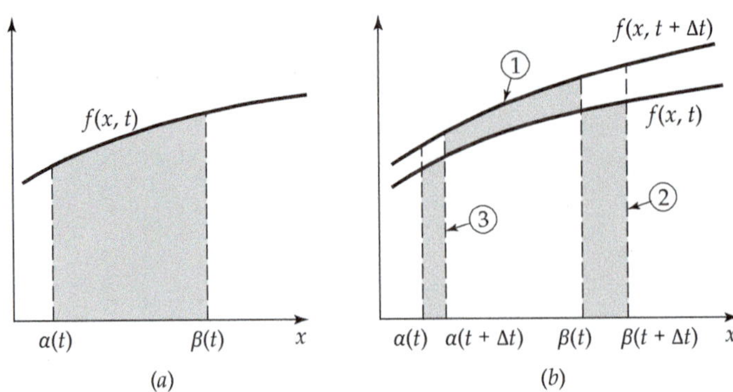

Fig. C.3-1. (*a*) The shaded area represents $I(t) = \int_{\alpha(t)}^{\beta(t)} f(x,t)\,dx$ at an instant t (Eq. C.3-1). (*b*) To get dI/dt, we form the difference $I(t+\Delta t) - I(t)$, divide by Δt, and then let $\Delta t \to 0$. The three numbered shaded areas correspond to the three numbered terms on the right side of Eq. C.3-2.

§C.4 THE GAMMA FUNCTION

The gamma function appears frequently as the result of integrations:

$$\Gamma(n) = \int_0^\infty x^{n-1} e^{-x} dx \tag{C.4-1}$$

$$\Gamma(n) = \int_0^1 \left(\ln \frac{1}{x} \right)^{n-1} dx \tag{C.4-2}$$

$$\Gamma(n+1) = \int_0^\infty \exp(-x^{1/n})\,dx \tag{C.4-3}$$

Several formulas for gamma functions are important:

$$\Gamma(n+1) = n\Gamma(n) \quad \text{(used to define } \Gamma(n) \text{ for negative } n\text{)} \tag{C.4-4}$$

$$\Gamma(n) = (n-1)! \quad \text{(when } n \text{ is an integer greater than 0)} \tag{C.4-5}$$

Some special values of the gamma function are:

$$\Gamma(1) = \Gamma(2) = 1 \tag{C.4-6}$$

$$\Gamma\left(\tfrac{1}{2}\right) = \sqrt{\pi} = 1.77245\ldots \tag{C.4-7}$$

$$\Gamma\left(\tfrac{3}{2}\right) = \tfrac{1}{2}\Gamma\left(\tfrac{1}{2}\right) = \tfrac{1}{2}\sqrt{\pi} = 0.88622\ldots \tag{C.4-8}$$

$$\Gamma\left(\tfrac{1}{3}\right) = 2.67893\ldots \tag{C.4-9}$$

$$\Gamma\left(\tfrac{2}{3}\right) = 1.35412\ldots \tag{C.4-10}$$

$$\Gamma\left(\tfrac{4}{3}\right) = \tfrac{1}{3}\Gamma\left(\tfrac{1}{3}\right) = 0.89297\ldots \tag{C.4-11}$$

$$\Gamma\left(\tfrac{7}{3}\right) = \tfrac{4}{3}\Gamma\left(\tfrac{4}{3}\right) = 1.19063\ldots \tag{C.4-12}$$

$$\Gamma(n+1) \approx n^n e^{-n} \sqrt{2\pi n} \left[1 + \frac{1}{12n} + \frac{1}{288n^2} - \frac{139}{51840n^3} - \cdots \right] \tag{C.4-13}$$

This last equation is *Stirling's formula* for large n. The gamma function is displayed in Fig. C.4-1.

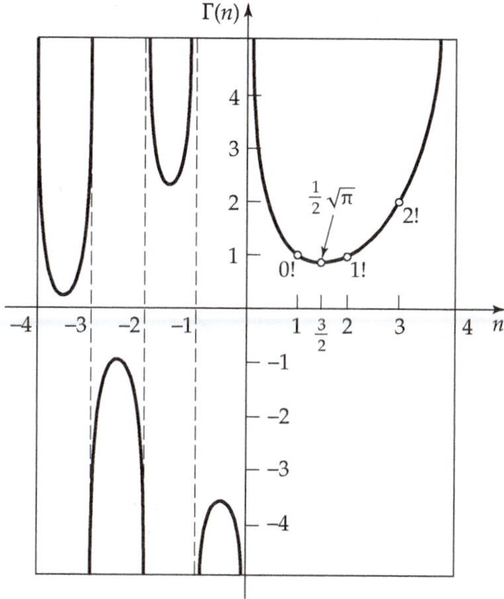

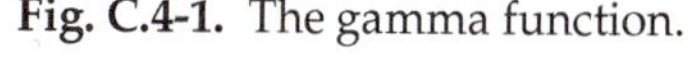
Fig. C.4-1. The gamma function.

Note that

$$\int_0^\infty e^{-x^3}dx = \frac{1}{3}\int_0^\infty e^{-y}y^{-2/3}dy = \frac{1}{3}\int_0^\infty e^{-y}y^{(1/3)-1}dy = \frac{1}{3}\Gamma\left(\frac{1}{3}\right) = \Gamma\left(\frac{4}{3}\right) \tag{C.4-14}$$

by performing the change of variables $x^3 = y$.

The function defined in Eqs. C.4-1, C.4-2, and C.4-3, above, is sometimes called the *complete gamma function* to distinguish it from the (upper) *incomplete gamma function* defined as follows:

$$\Gamma(n,z) = \int_z^\infty e^{-x}x^{n-1}dx \tag{C.4-15}$$

If the integral has limits 0 to z, it is called the *lower incomplete gamma function.*

§C.5 THE HYPERBOLIC FUNCTIONS

The hyperbolic sine (sinh x), the hyperbolic cosine (cosh x), and the hyperbolic tangent (tanh x) arise frequently in science and engineering problems. They are related to the hyperbola in very much the same way that the circular functions are related to the circle (see Fig. C.5-1). The circular functions (sin x and cos x) are periodic, oscillating functions, whereas their hyperbolic analogs are not (see Fig. C.5-2).

The hyperbolic functions are related to the exponential function as follows:

$$\cosh x = \frac{1}{2}(e^x + e^{-x}); \qquad \sinh x = \frac{1}{2}(e^x - e^{-x}) \tag{C.5-1,2}$$

The corresponding relations for the circular functions are:

$$\cos x = \frac{1}{2}(e^{ix} + e^{-ix}); \qquad \sin x = \frac{1}{2i}(e^{ix} - e^{-ix}) \tag{C.5-3,4}$$

Then one can derive a variety of standard relations for the hyperbolic functions, such as

$$\cosh^2 x - \sinh^2 x = 1 \tag{C.5-5}$$

$$\cosh(x \pm y) = \cosh x \cosh y \pm \sinh x \sinh y \tag{C.5-6}$$

$$\sinh(x \pm y) = \sinh x \cosh y \pm \cosh x \sinh y \tag{C.5-7}$$

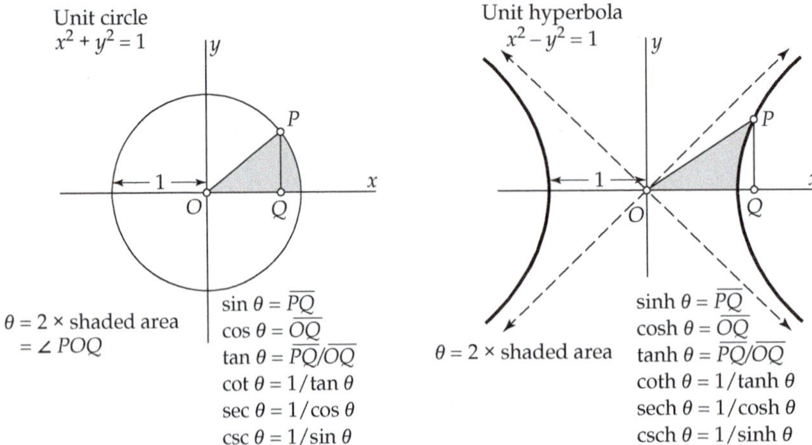

Fig. C.5-1. Comparison of circular and hyperbolic functions.

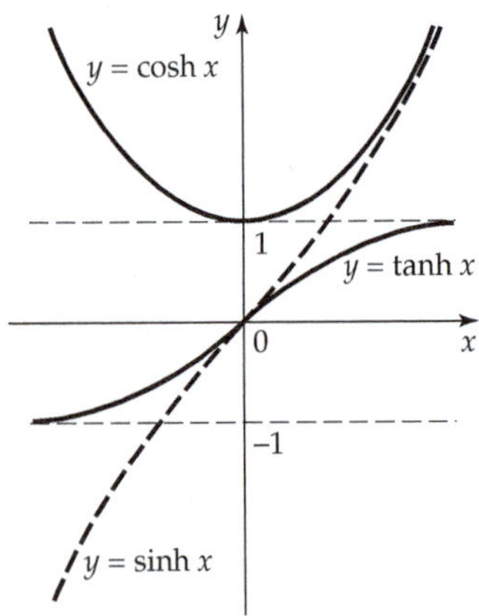

Fig. C.5-2. Comparison of the shapes of the hyperbolic functions.

$$\cosh ix = \cos x; \qquad \sinh ix = i \sin x \tag{C.5-8,9}$$

$$\frac{d \cosh x}{dx} = \sinh x; \qquad \frac{d \sinh x}{dx} = \cosh x \tag{C.5-10,11}$$

$$\int \cosh x \, dx = \sinh x; \qquad \int \sinh x \, dx = \cosh x \tag{C.5-12,13}$$

It should be kept in mind that $\cosh x$ and $\cos x$ are both even functions of x, whereas $\sinh x$ and $\sin x$ are odd functions of x. In Eqs. C.5-12,13 and C.5-12,13, an arbitrary constant of integration can be added to the right side.

§C.6 THE ERROR FUNCTION

The error function is defined as

$$\text{erf}\, x = \frac{\int_0^x \exp(-\bar{x}^2)d\bar{x}}{\int_0^\infty \exp(-\bar{x}^2)d\bar{x}} = \frac{2}{\sqrt{\pi}} \int_0^x \exp(-\bar{x}^2)d\bar{x} \tag{C.6-1}$$

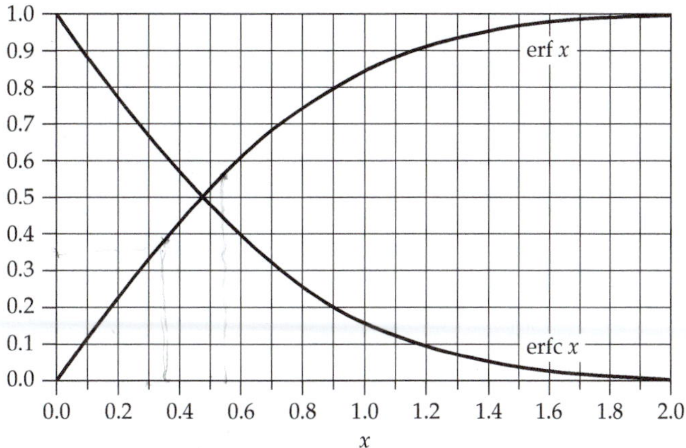

Fig. C.6-1. The error function.

This function, which arises naturally in numerous transport phenomena problems, is monotone increasing, going from erf $0 = 0$ to erf $\infty = 1$, and has the value of 0.99 at about $x = 2$. This function is plotted in Fig. C.6-1.

The Taylor series expression for the error function about $x = 0$ is given in Eq. C.2-4. The Taylor series about $1/x = 0$ is

$$\operatorname{erf} x = 1 - \frac{e^{-x^2}}{x\sqrt{\pi}}\left(1 - \frac{1}{2x^2} + \frac{1 \cdot 3}{2^2 x^4} - \frac{1 \cdot 3 \cdot 5}{2^3 x^6} + \cdots\right) \tag{C.6-2}$$

It is also worth noting that erf $(-x) = -$ erf x, and that

$$\frac{d}{dx}\operatorname{erf} u = \frac{2}{\sqrt{\pi}} e^{-u^2} \frac{du}{dx} \tag{C.6-3}$$

by applying the Leibniz formula to Eq. C.6-1.

The closely related function erfc $x = 1 -$ erf x is called the "complementary error function." The complementary error function is also plotted in Fig. C.6-1.

§C.7 CHANGING THE ORDER OF INTEGRATION

If we have a double integral of the form

$$\int_0^1 \int_0^x f(x,y)\, dy\, dx \tag{C.7-1}$$

we can picture the region of integration as in Fig. C.7-1(*a*). In the inner integral, we integrate $f(x,y)$ over the area of a thin vertical "strip" from the horizontal line $y = 0$ axis up to the diagonal line $y = x$. Then we add up the integrals over all of the "strips" of width dx from $x = 0$ to $x = 1$.

Alternatively, we may want to change the order of integration (Fig. C.7-1(*b*)) and write

$$\int_0^1 \int_y^1 f(x,y)\, dx\, dy \tag{C.7-2}$$

That is, we now integrate $f(x,y)$ over the area of a thin horizontal "strip" that goes from the diagonal line $x = y$ over to the vertical line $x = 1$. Then we add up the integrals over all of the "strips" of width dy going from $y = 0$ to $y = 1$.

That is, in both Eq. C.7-1(*a*) and Eq. C.7-1(*b*), we have to choose the limits of integration in such a way that we describe exactly the same (shaded) integration area. Since the area

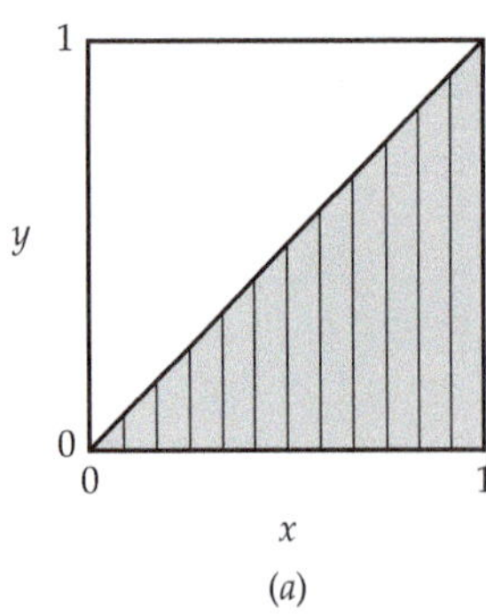

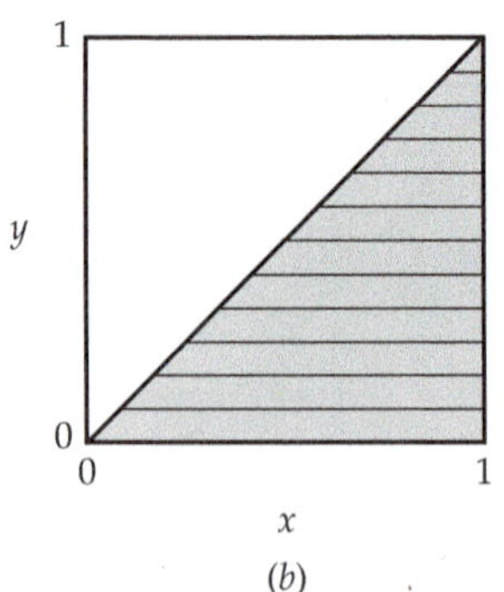

Fig. C.7-1. Interchanging order of integration.

of integration in this illustration is bounded by straight lines, it is relatively easy to select the limits for the integrals; if the area of integration is bounded by curves, selecting the limits for the integrals may be somewhat more complicated.

The principal reason for wanting to change the order of integration is that, whereas it may be impossible to evaluate the inner integral over y in Eq. C.7-1 analytically, it may be relatively easy to do the inner integral over x in Eq. C.7-2 analytically.

§C.8 L'HÔPITAL'S RULE

It often happens that it is desired to take the limit of the ratio of two functions $f(x)/g(x)$ as $x \to a$. If $f(a) = 0$ and $g(a) = 0$, then we obtain 0/0, which is indeterminate. However, sometimes the limit really does exist and it can be found by using L'Hôpital's rule, which states that

$$\lim_{x\to a}\frac{f(x)}{g(x)} = \lim_{x\to a}\frac{df/dx}{dg/dx} \quad \text{if the limit exists} \tag{C.8-1}$$

For example, although sin x and x both go to zero as x goes to zero,

$$\lim_{x\to 0}\frac{\sin x}{x} = \lim_{x\to 0}\frac{d(\sin x)/dx}{d(x)/x} = \lim_{x\to 0}\frac{\cos x}{1} = 1 \tag{C.8-2}$$

As a further illustration,

$$\lim_{x\to 1}\frac{2\ \ln x}{x-1} = \lim_{x\to 1}\frac{2d(\ln x)/dx}{d(x-1)/dx} = \lim_{x\to 1}\frac{2/x}{1} = 2 \tag{C.8-3}$$

The same procedure may be followed if $f(a) = g(a) = +\infty$ or $-\infty$.

$$\lim_{x\to \infty}\frac{3\ \ln x}{x-1} = \lim_{x\to \infty}\frac{3\,d(\ln x)/dx}{d(x-1)/dx} = \lim_{x\to \infty}\frac{3/x}{1} = 0 \tag{C.8-4}$$

Appendix D

Tables for Prediction of Transport Properties

Table D.1. Lennard-Jones (6–12) Potential Parameters and Critical Properties

		Lennard-Jones parameters			Critical properties[g, h]				
Substance	Molecular Weight M	σ (Å)	ε/κ (K)	Ref.	T_c (K)	p_c (atm)	$\tilde{V}_c$ (cm^3/g-mol)	$\mu_c \times 10^6$ (g/cm · s)	$k_c \times 10^6$ (cal/cm · s · K)
Light gases:									
H_2	2.016	2.915	38.0	*a*	33.3	12.80	65.0	34.7	—
He	4.003	2.576	10.2	*a*	5.26	2.26	57.8	25.4	—
Noble gases:									
Ne	20.180	2.789	35.7	*a*	44.5	26.9	41.7	156.	79.2
Ar	39.948	3.432	122.4	*b*	150.7	48.0	75.2	264.	71.0
Kr	83.80	3.675	170.0	*b*	209.4	54.3	92.2	396.	49.4
Xe	131.29	4.009	234.7	*b*	289.8	58.0	118.8	490.	40.2
Simple polyatomic gases:									
Air	28.964[i]	3.617	97.0	*a*	132.4[i]	37.0[i]	86.7[i]	193.	90.8
N_2	28.013	3.667	99.8	*b*	126.2	33.5	90.1	180.	86.8
O_2	31.999	3.433	113.	*a*	154.4	49.7	74.4	250.	105.3
CO	28.010	3.590	110.	*a*	132.9	34.5	93.1	190.	86.5
CO_2	44.010	3.996	190.	*a*	304.2	72.8	94.1	343.	122.
NO	30.006	3.470	119.	*a*	180.	64.	57.	258.	118.2
N_2O	44.012	3.879	220.	*a*	309.7	71.7	96.3	332.	131.
SO_2	64.065	4.026	363.	*c*	430.7	77.8	122.	411.	98.6
F_2	37.997	3.653	112.	*a*	—	—	—	—	—
Cl_2	70.905	4.115	357.	*a*	417.	76.1	124.	420.	97.0
Br_2	159.808	4.268	520.	*a*	584.	102.	144.	—	—
I_2	253.809	4.982	550.	*a*	800.	—	—	—	—
Hydrocarbons:									
CH_4	16.04	3.780	154.	*b*	190.7	45.8	98.7	159.	158.
$CH \equiv CH$	26.04	4.114	212.	*d*	308.7	61.6	112.9	237.	—
$CH_2 = CH_2$	28.05	4.228	216.	*b*	282.4	50.0	124.	215.	—
C_2H_6	30.07	4.388	232.	*b*	305.4	48.2	148.	210.	203.
$CH_3C \equiv CH$	40.06	4.742	261.	*d*	394.8	—	—	—	—
$CH_3CH = CH_2$	42.08	4.766	275.	*b*	365.0	45.5	181.	233.	—
C_3H_8	44.10	4.934	273.	*b*	369.8	41.9	200.	228.	—
n–C_4H_{10}	58.12	5.604	304.	*b*	425.2	37.5	255.	239.	—
i–C_4H_{10}	58.12	5.393	295.	*b*	408.1	36.0	263.	239.	—

n–C_5H_{12}	72.15	5.850	326.	*b*	469.5	33.2	311.	238.	—
i–C_5H_{12}	72.15	5.812	327.	*b*	460.4	33.7	306.	—	—
$C(CH_3)_4$	72.15	5.759	312.	*b*	433.8	31.6	303.	—	—
n–C_6H_{14}	86.18	6.264	342.	*b*	507.3	29.7	370.	248.	—
n–C_7H_{16}	100.20	6.663	352.	*b*	540.1	27.0	432.	254.	—
n–C_8H_{18}	114.23	7.035	361.	*b*	568.7	24.5	492	259.	—
n–C_9H_{20}	128.26	7.463	351.	*b*	594.6	22.6	548.	265.	—
Cyclohexane	84.16	6.143	313.	*d*	553.	40.0	308.	284.	—
Benzene	78.11	5.443	387.	*b*	562.6	48.6	260.	312.	—
Other organic compounds:									
CH_3Cl	50.49	4.151	355.	*c*	416.3	65.9	143.	338.	—
CH_2Cl_2	84.93	4.748	398.	*c*	510.	60.	—	—	—
$CHCl_3$	119.38	5.389	340.	*e*	536.6	54.	240.	410.	—
CCl_4	153.82	5.947	323.	*e*	556.4	45.0	276.	413.	—
C_2N_2	52.034	4.361	349.	*e*	400.	59.	—	—	—
COS	60.076	4.130	336.	*e*	378.	61.	—	—	—
CS_2	76.143	4.483	467.	*e*	552.	78.	170.	404.	—
CCl_2F_2	120.91	5.116	280.	*b*	384.7	39.6	218.	—	—

[a] J. O. Hirschfelder, C. F. Curtiss, and R. B. Bird, *Molecular Theory of Gases and Liquids*, corrected printing with notes added, Wiley, New York (1964).

[b] L. S. Tee, S. Gotoh, and W. E. Stewart, *Ind. Eng. Chem. Fundamentals*, **5**, 356–363 (1966). The values for benzene are from viscosity data on that substance. The values for other substances are computed from Correlation (iii) of the paper.

[c] L. Monchick and E. A. Mason, *J. Chem.Phys.*, **35**, 1676–1697 (1961); parameters obtained from viscosity.

[d] L. W. Flynn and G. Thodos, *AIChE Journal*, **8**, 362–365 (1962); parameters obtained from viscosity.

[e] R. A. Svehla, *NASA Tech. Report R-132* (1962); parameters obtained from viscosity. This report provides extensive tables of Lennard-Jones parameters, heat capacities, and calculated transport properties.

[f] Values of the critical constants for the pure substances are selected from K. A. Kobe and R. E. Lynn, Jr., *Chem. Rev.*, **52**, 117–236 (1962); *Amer. Petroleum Inst. Research Proj.* **44**, Thermodynamics Research Center, Texas A&M University, College Station, Texas (1966); and *Thermodynamic Functions of Gases*, F. Din (editor), Vols. **1–3**, Butterworths, London (1956, 1961, 1962).

[g] Values of the critical viscosity are from O. A. Hougen and K. M. Watson, *Chemical Process Principles*, Vol. **3**, Wiley, New York (1947), p. 873.

[h] Values of the critical thermal conductivity are from E. J. Owens and G. Thodos, *AIChE Journal*, **3**, 454–461 (1957).

[i] For air, the molecular weight M and the pseudocritical properties have been computed from the average composition of dry air as given in COESA, *U.S. Standard Atmosphere 1976*, U.S. Government Printing Office, Washington, D.C. (1976).

Table D.2. Collision Integrals for Use with the Lennard-Jones (6–12) Potential for the Prediction of Transport Properties of Gases at Low Densities[a,b,c]

$\kappa T/\varepsilon$ or $\kappa T/\varepsilon_{AB}$	$\Omega_\mu = \Omega_k$ (for viscosity and thermal conductivity)	$\Omega_{\mathscr{D},AB}$ (for diffusivity)	$\kappa T/\varepsilon$ or $\kappa T/\varepsilon_{AB}$	$\Omega_\mu = \Omega_k$ (for viscosity and thermal conductivity)	$\Omega_{\mathscr{D},AB}$ (for diffusivity)
0.30	2.840	2.649	2.7	1.0691	0.9782
0.35	2.676	2.468	2.8	1.0583	0.9682
0.40	2.531	2.314	2.9	1.0482	0.9588
0.45	2.401	2.182	3.0	1.0388	0.9500
0.50	2.284	2.066	3.1	1.0300	0.9418
0.55	2.178	1.965	3.2	1.0217	0.9340
0.60	2.084	1.877	3.3	1.0139	0.9267
0.65	1.999	1.799	3.4	1.0066	0.9197
0.70	1.922	1.729	3.5	0.9996	0.9131
0.75	1.853	1.667	3.6	0.9931	0.9068
0.80	1.790	1.612	3.7	0.9868	0.9008
0.85	1.734	1.562	3.8	0.9809	0.8952
0.90	1.682	1.517	3.9	0.9753	0.8897
0.95	1.636	1.477	4.0	0.9699	0.8845
1.00	1.593	1.440	4.1	0.9647	0.8796
1.05	1.554	1.406	4.2	0.9598	0.8748
1.10	1.518	1.375	4.3	0.9551	0.8703
1.15	1.485	1.347	4.4	0.9506	0.8659
1.20	1.455	1.320	4.5	0.9462	0.8617
1.25	1.427	1.296	4.6	0.9420	0.8576
1.30	1.401	1.274	4.7	0.9380	0.8537
1.35	1.377	1.253	4.8	0.9341	0.8499
1.40	1.355	1.234	4.9	0.9304	0.8463
1.45	1.334	1.216	5.0	0.9268	0.8428
1.50	1.315	1.199	6.0	0.8962	0.8129
1.55	1.297	1.183	7.0	0.8727	0.7898
1.60	1.280	1.168	8.0	0.8538	0.7711
1.65	1.264	1.154	9.0	0.8380	0.7555
1.70	1.249	1.141	10.0	0.8244	0.7422
1.75	1.235	1.128	12.0	0.8018	0.7202
1.80	1.222	1.117	14.0	0.7836	0.7025
1.85	1.209	1.105	16.0	0.7683	0.6878
1.90	1.198	1.095	18.0	0.7552	0.6751
1.95	1.186	1.085	20.0	0.7436	0.6640
2.00	1.176	1.075	25.0	0.7198	0.6414
2.10	1.156	1.058	30.0	0.7010	0.6235
2.20	1.138	1.042	35.0	0.6854	0.6088
2.30	1.122	1.027	40.0	0.6723	0.5964
2.40	1.107	1.013	50.0	0.6510	0.5763
2.50	1.0933	1.0006	75.0	0.6140	0.5415
2.60	1.0807	0.9890	100.0	0.5887	0.5180

[a]The values in this table, applicable for the Lennard-Jones (6–12) potential, are interpolated from the results of L. Monchick and E. A. Mason, *J. Chem. Phys.*, **35**, 1676–1697 (1961). The Monchick–Mason table is believed to be slightly better than the earlier table by J. O. Hirschfelder, R. B. Bird, and E. L. Spotz, *J. Chem. Phys.*, **16**, 968–981 (1948).

[b]This table has been extended to lower temperatures by C. F. Curtiss, *J. Chem. Phys.*, **97**, 7679–7686 (1992). Curtiss showed that at low temperatures, the Boltzmann equation needs to be modified to take into account "orbiting pairs" of molecules. Only by making this modification is it possible to get a smooth transition from quantum to classical behavior. The deviations are appreciable below dimensionless temperatures of 0.30.

[c]The collision integrals have been curve-fitted by P. D. Neufeld, A. R. Janzen, and R. A. Aziz, *J. Chem. Phys.*, **57**, 1100–1102 (1972), as follows:

$$\Omega_\mu = \Omega_k = \frac{1.16145}{T^{*0.14874}} + \frac{0.52487}{\exp(0.77320T^*)} + \frac{2.16178}{\exp(2.43787T^*)} \quad \text{(D.2-1)}$$

$$\Omega_{\mathscr{D},AB} = \frac{1.06036}{T^{*0.15610}} + \frac{0.19300}{\exp(0.47635T^*)} + \frac{1.03587}{\exp(1.52996T^*)} + \frac{1.76474}{\exp(3.89411T^*)} \quad \text{(D.2-2)}$$

where $T^* = \kappa T/\varepsilon$.

Appendix E

Constants and Conversion Factors

§E.1 MATHEMATICAL CONSTANTS

$$\pi = 3.14159\ldots$$
$$e = 2.71828\ldots$$
$$\ln\ 10 = 2.30259\ldots$$

§E.2 PHYSICAL CONSTANTS[1]

Gas law constant (R)	8.31451	J/g-mol · K
	8.31451×10^3	kg · m²/s² · kg-mol · K
	8.31451×10^7	g · cm²/s² · g-mol · K
	1.98721	cal/g-mol · K
	82.0578	cm³ atm/g-mol · K
	4.9686×10^4	lb_m ft²/s² · lb-mol · °R
	1.5443×10^3	ft · lb_f/lb-mol · °R
	10.731	ft³ · psia/lb-mol · °R
Standard acceleration of gravity (g_0)	9.80665	m/s²
	980.665	cm/s²
	32.1740	ft/s²
Joule's constant (J_c) (mechanical equivalent of heat)	4.1840	J/cal
	4.1840×10^7	erg/cal
	778.16	ft · lb_f/Btu
Avogadro's number ($\tilde{N}$)	6.02214×10^{23}	$(\text{g-mol})^{-1}$
Boltzmann's constant ($\kappa = R/\tilde{N}$)	1.38066×10^{-23}	J/K
	1.38066×10^{-16}	erg/K
Faraday's constant (F)	96485.3	abs. Coulomb/g-equivalent
Planck's constant (h)	6.62608×10^{-34}	J · s
	6.62608×10^{-27}	erg · s

[1] E. R. Cohen and B. N. Taylor, Physics Today (August 1996), pp. BG9–BG13; R. A. Nelson, Physics Today (August 1996), pp. BG15–BG16.

Stefan–Boltzmann constant (σ)	5.67051×10^{-8}	$W/m^2 \cdot K^4$
	1.3553×10^{-12}	$cal/s \cdot cm^2 \cdot K^4$
	1.7124×10^{-9}	$Btu/hr \cdot ft^2 \cdot R^4$
Electron charge (e)	1.60218×10^{-19}	Coulomb = $A \cdot s$
Speed of light in a vacuum (c)	2.99792×10^8	m/s

§E.3 CONVERSION FACTORS

In the tables that follow, to convert any physical quantity from one set of units to another, multiply it by the appropriate table entry. For example, suppose that p is given as 10 $lb_f/in.^2$ (psia), and we wish to have p in poundals/ft^2. From Table E.3-2 the result is

$$p = (10)(4.6330 \times 10^3) = 4.6330 \times 10^4 \text{ poundals/ft}^2$$

The entries in the shaded rows and columns are those that are needed for converting from and to SI units.

In addition to the tables, we give a few of the commonly used conversion factors here:

Given a quantity in these units:	Multiply by:	To get quantity in these units:
Pounds (mass)	453.59	Grams
Kilograms	2.2046	Pounds (mass)
Inches	2.5400	Centimeters
Meters	39.370	Inches
Gallons (U.S.)	3.7853	Liters
Gallons (U.S.)	231.00	Cubic inches
Gallons (U.S.)	0.13368	Cubic feet
Cubic feet	28.316	Liters
Kelvins	1.800000	Degrees Rankine (°R = °F + 459.67)
Degrees Rankine	0.555556	Kelvins (K = °C + 273.15)

Table E.3-1. Conversion Factors for Quantities Having Dimensions of F or ML/t^2

Given a quantity in these units ↓	Multiply by table value to convert to these units →	$N = kg \cdot m/s^2$ (Newtons)	$g \cdot cm/s^2$ (dynes)	$lb_m \cdot ft/s^2$ (poundals)	lb_f
$N = kg \cdot m/s^2$	(Newtons)	1	10^5	7.2330	2.24881×10^{-1}
$g \cdot cm/s^2$	(dynes)	10^{-5}	1	7.2330×10^{-5}	2.24881×10^{-6}
$lb_m \cdot ft/s^2$	(poundals)	1.3826×10^{-1}	1.3826×10^4	1	3.1081×10^{-2}
lb_f		4.4482	4.4482×10^5	32.1740	1

Table E.3-2. Conversion Factors for Quantities Having Dimensions of F/L^2 or M/Lt^2 (pressure, momentum flux)

Given a quantity in these units ↓ / Multiply by table value to convert to these units →	Pa (N/m^2) ($kg/m \cdot s^2$)	$dyne/cm^2$ ($g/cm \cdot s^2$)	$poundals/ft^2$ ($lb_m/ft \cdot s^2$)	lb_f/ft^2	$lb_f/in.^2$ (psia)[a]	atm	mm Hg	in. Hg
$Pa = N/m^2 = kg/m \cdot s^2$	1	10	6.7197×10^{-1}	2.0886×10^{-2}	1.4504×10^{-4}	9.8692×10^{-6}	7.5006×10^{-3}	2.9530×10^{-4}
$dyne/cm^2 = g/cm \cdot s^2$	10^{-1}	1	6.7197×10^{-2}	2.0886×10^{-3}	1.4504×10^{-5}	9.8692×10^{-7}	7.5006×10^{-4}	2.9530×10^{-5}
$poundals/ft^2 = lb_m/ft \cdot s^2$	1.4882	1.4882×10^{1}	1	3.1081×10^{-2}	2.1584×10^{-4}	1.4687×10^{-5}	1.1162×10^{-2}	4.3945×10^{-4}
lb_f/ft^2	4.7880×10^{1}	4.7880×10^{2}	32.1740	1	6.9444×10^{-3}	4.7254×10^{-4}	3.5913×10^{-1}	1.4139×10^{-2}
$lb_f/in.^2$ (psia)[a]	6.8947×10^{3}	6.8947×10^{4}	4.6330×10^{3}	144	1	6.8046×10^{-2}	5.1715×10^{1}	2.0360
atm	1.0133×10^{5}	1.0133×10^{6}	6.8087×10^{4}	2.1162×10^{3}	14.696	1	760	29.921
mm Hg	1.3332×10^{2}	1.3332×10^{3}	8.9588×10^{1}	2.7845	1.9337×10^{-2}	1.3158×10^{-3}	1	3.9370×10^{-2}
in. Hg	3.3864×10^{3}	3.3864×10^{4}	2.2756×10^{3}	7.0727×10^{1}	4.9116×10^{-1}	3.3421×10^{-2}	25.400	1

[a]This unit is preferably abbreviated "psia" (pounds per square inch absolute) or "psig" (pounds per square inch gage). Gage pressure is absolute pressure minus the prevailing barometric pressure. Sometimes the pressure is reported in "bars"; to convert from bars to pascals, multiply by 10^5, and to convert from bars to atmospheres, multiply by 0.98692.

Table E.3-3. Conversion Factors for Quantities Having Dimensions of FL or ML^2/t^2 (energy, work, torque)

Given a quantity in these units ↓ / Multiply by table value to convert to these units →	J ($kg \cdot m^2/s^2$)	ergs ($g \cdot cm^2/s^2$)	foot poundals $lb_m\ ft^2/s^2$	$ft \cdot lb_f$	cal[a]	Btu	hp-hr	kw-hr
$J = kg \cdot m^2/s^2$	1	10^{7}	2.3730×10^{1}	7.3756×10^{-1}	2.3901×10^{-1}	9.4783×10^{-4}	3.7251×10^{-7}	2.7778×10^{-7}
$ergs = g \cdot cm^2/s^2$	10^{-7}	1	2.3730×10^{-6}	7.3756×10^{-8}	2.3901×10^{-8}	9.4783×10^{-11}	3.7251×10^{-14}	2.7778×10^{-14}
foot poundals $= lb_m\ ft^2/s^2$	4.2140×10^{-2}	4.2140×10^{5}	1	3.1081×10^{-2}	1.0072×10^{-2}	3.9942×10^{-5}	1.5698×10^{-8}	1.1706×10^{-8}
$ft \cdot lb_f$	1.3558	1.3558×10^{7}	32.1740	1	3.2405×10^{-1}	1.2851×10^{-3}	5.0505×10^{-7}	3.7662×10^{-7}
thermochemical calories[a]	4.1840	4.1840×10^{7}	9.9287×10^{1}	3.0860	1	3.9657×10^{-3}	1.5586×10^{-6}	1.1622×10^{-6}
British thermal units	1.0550×10^{3}	1.0550×10^{10}	2.5036×10^{4}	778.16	2.5216×10^{2}	1	3.9301×10^{-4}	2.9307×10^{-4}
Horsepower hours	2.6845×10^{6}	2.6845×10^{13}	6.3705×10^{7}	1.9800×10^{6}	6.4162×10^{5}	2.5445×10^{3}	1	7.4570×10^{-1}
kilowatt hours	3.6000×10^{6}	3.6000×10^{13}	8.5429×10^{7}	2.6552×10^{6}	8.6042×10^{5}	3.4122×10^{3}	1.3410	1

[a]This unit, abbreviated "cal," is used in some chemical thermodynamic tables. To convert quantities expressed in International Steam Table calories (abbreviated "I. T. cal") to this unit, multiply by 1.000654.

Table E.3-4. Conversion Factors for Quantities Having dimensions[a] of M/Lt or Ft/L^2 (viscosity, density times diffusivity)

Given a quantity in these units ↓ / Multiply by table value to convert to these units →	Pa · s (kg/m · s)	g/cm · s (poises)	centipoises	lb_m/ft · s	lb_m/ft · hr	$lb_f \cdot s/ft^2$
Pa · s = kg/m · s	1	10	10^3	6.7197×10^{-1}	2.4191×10^3	2.0886×10^{-2}
g/cm · s = (poises)	10^{-1}	1	10^2	6.7197×10^{-2}	2.4191×10^2	2.0886×10^{-3}
centipoises	10^{-3}	10^{-2}	1	6.7197×10^{-4}	2.4191	2.0886×10^{-5}
lb_m/ft · s	1.4882	1.4882×10^1	1.4882×10^3	1	3600	3.1081×10^{-2}
lb_m/ft · hr	4.1338×10^{-4}	4.1338×10^{-3}	4.1338×10^{-1}	2.7778×10^{-4}	1	8.6336×10^{-6}
$lb_f \cdot s/ft^2$	4.7880×10^1	4.7880×10^2	4.7880×10^4	32.1740	1.1583×10^5	1

[a]When moles appear in the given and the desired units, the conversion factor is the same as for the corresponding mass units.

Table E.3-5. Conversion Factors for Quantities Having Dimensions of ML/t^3T or F/tT (thermal conductivity)

Given a quantity in these units ↓ / Multiply by table value to convert to these units →	W/m · K or kg · m/s³ · K	g · cm/s³ · K or erg/s · cm · K	lb_m ft/s³ °F	lb_f/s · °F	cal/s · cm · K	Btu/hr · ft · °F
W/m · K = kg · m/s³ · K	1	10^5	4.0183	1.2489×10^{-1}	2.3901×10^{-3}	5.7780×10^{-1}
g · cm/s³ · K	10^{-5}	1	4.0183×10^{-5}	1.2489×10^{-6}	2.3901×10^{-8}	5.7780×10^{-6}
lb_m ft/s³ °F	2.4886×10^{-1}	2.4886×10^4	1	3.1081×10^{-2}	5.9479×10^{-4}	1.4379×10^{-1}
lb_f/s · °F	8.0068	8.0068×10^5	3.2174×10^1	1	1.9137×10^{-2}	4.6263
cal/s · cm · K	4.1840×10^2	4.1840×10^7	1.6813×10^3	5.2256×10^1	1	2.4175×10^2
Btu/hr · ft · °F	1.7307	1.7307×10^5	6.9546	2.1616×10^{-1}	4.1365×10^{-3}	1

Table E.3-6. Conversion Factors for Quantities Having Dimensions of L^2/t (momentum diffusivity, thermal diffusivity, molecular diffusivity)

Given a quantity in these units ↓ / Multiply by table value to convert to these units →	m²/s	cm²/s	ft²/hr	centistokes
m²/s	1	10^4	3.8750×10^4	10^6
cm²/s	10^{-4}	1	3.8750	10^2
ft²/hr	2.5807×10^{-5}	2.5807×10^{-1}	1	2.5807×10^1
centistokes	10^{-6}	10^{-2}	3.8750×10^{-2}	1

Table E.3-7. Conversion Factors for Quantities Having Dimensions of M/t^3T or F/LtT (heat transfer coefficients)

Given a quantity in these units ↓	Multiply by table value to convert to these units →	W/m² K (J/m² s · K) kg/s³ K	W/cm² K	g/s³ K	lb_m/s^3 °F	lb_f/ft · s · °F	cal/cm² s · K	Btu/ ft² hr · °F
W/m² K =	kg/s³ K	1	10^{-4}	10^3	1.2248	3.8068×10^{-2}	2.3901×10^{-5}	1.7611×10^{-1}
W/cm² K		10^4	1	10^7	1.2248×10^4	3.8068×10^2	2.3901×10^{-1}	1.7611×10^3
g/s³ K		10^{-3}	10^{-7}	1	1.2248×10^{-3}	3.8068×10^{-5}	2.3901×10^{-8}	1.7611×10^{-4}
lb_m/s^3 °F		8.1647×10^{-1}	8.1647×10^{-5}	8.1647×10^2	1	3.1081×10^{-2}	1.9514×10^{-5}	1.4379×10^{-1}
lb_f/ft · s · °F		2.6269×10^1	2.6269×10^{-3}	2.6269×10^4	32.1740	1	6.2784×10^{-4}	4.6263
cal/cm² s · K		4.1840×10^4	4.1840	4.1840×10^7	5.1245×10^4	1.5928×10^3	1	7.3686×10^3
Btu/ft² hr · °F		5.6782	5.6782×10^{-4}	5.6782×10^3	6.9546	2.1616×10^{-1}	1.3571×10^{-4}	1

Table E.3-8. Conversion Factors for Quantities Having Dimensions[a] of M/L^2t or Ft/L^3 (mass transfer coefficients k_x or k_ω)

Given a quantity in these units ↓	Multiply by table value to convert to these units →	kg/m² s	g/cm² s	lb_m/ft² s	lb_m/ft² hr	lb_f s/ft³
kg/m² s		1	10^{-1}	2.0482×10^{-1}	7.3734×10^2	6.3659×10^{-3}
g/cm² s		10^1	1	2.0482	7.3734×10^3	6.3659×10^{-2}
lb_m/ft² s		4.8824	4.8824×10^{-1}	1	3600	3.1081×10^{-2}
lb_m/ft² hr		1.3562×10^{-3}	1.3562×10^{-4}	2.7778×10^{-4}	1	8.6336×10^{-6}
lb_f s/ft³		1.5709×10^2	1.5709×10^1	32.1740	1.1583×10^5	1

[a]When moles appear in the given and the desired units, the conversion factor is the same as for the corresponding mass units.

Notation

Numbers in parentheses refer to equations, sections, or tables in which the symbols are defined or first used. Dimensions are given in terms of mass (M), length (L), time (t), temperature (T), moles (mol), current (I), and dimensionless (---). Boldface symbols are vectors or tensors (see Appendix A). Symbols that appear infrequently are not listed.

A = area, L^2
a = absorptivity (16.2-1), ---
a = interfacial area per unit volume of packed bed (6.4-4), L^{-1}
a_A = activity of species A (24.1-5), ---
C_p = heat capacity at constant pressure (9.2-7), ML^2/t^2T
C_V = heat capacity at constant volume (9.7-6), ML^2/t^2T
c = speed of light (16.1-1), L/t
c = total molar concentration (Table 17.1-1), mol$/L^3$
c_A = molar concentration of species A (Table 17.1-1), mol$/L^3$
D = diameter of cylinder or sphere, L
D_p = particle diameter in packed bed (6.4-3), L
$\mathscr{D}_{AB}$ = binary diffusivity for system A-B (17.3-1), L^2/t
$\mathscr{D}_{\alpha\beta}$ = binary diffusivity for the pair α-β in a multicomponent system (24.6-7), L^2/t
$Ð_{\alpha\beta}$ = Maxwell-Stefan multicomponent diffusivity (24.6-5), L^2/t
$\mathbb{D}_{\alpha\beta}$ = Fick multicomponent diffusivity (24.6-4), L^2/t
D_A^T = thermal diffusion coefficient (24.1-8), M/Lt
d = molecular diameter (1.6-3), L
$\mathbf{d}_\alpha$ = diffusional driving force for species α (24.1-5), L^{-1}
E_{tot} = $U_{tot} + K_{tot} + \Phi_{tot}$ = total energy in a macroscopic system (15.1-2), ML^2/t^2
E_c = compression term in macroscopic mechanical energy balance (7.4-3), ML^2/t^3
E_v = viscous dissipation term in macroscopic mechanical energy balance (7.4-4), ML^2/t^3
e = 2.71828...
e = emissivity (16.2-3), ---
$\mathbf{e}$ = total energy flux vector (9.4-1, 19.3-4), M/t^3
$F_{12}, \overline{F}_{12}$ = direct, indirect view factor (16.4-9, 16.5-10), ---
$\mathbf{F}_{s\to f}$ = force exerted by the solid on the fluid (7.2-1) ML/t^2
f = friction factor (or drag coefficient) (6.1-1), ---
G = elastic modulus (8.4-2), M/Lt^2
G = $H - TS$ = Gibbs free energy (24.1-3), ML^2/t^2
G = $\langle\rho v\rangle$ = mass velocity (6.4-8), M/L^2t
$\mathbf{g}$ = gravitational acceleration (3.2-8), L/t^2
$\mathbf{g}_A$ = body force per unit mass acting on species A (Table 19.2-1), L/t^2
g_S = entropy production rate (24.1-1), M/Lt^3T

$H = U + pV$ = enthalpy (9.4-2), ML^2/t^2
H_A = enthalpy of species A (Table 19.2-4), ML^2/t^2
h = Planck's constant (16.1-2), ML^2/t
h = elevation (2.3-11), L
$h, h_1, h_{\text{ln}}, h_{\text{loc}}, h_a, h_m$ = heat-transfer coefficients (10.1-2, 14.1-1 to 6), M/t^3T
I = current density (10.6-1), I/L^2
$i = \sqrt{-1}$ (8.4-8), ---
$\mathbf{j}_A^{(c)}$ = convective mass flux vector of species A (17.2-5), M/L^2t
$\mathbf{j}_A$ = diffusive mass flux vector of species A (17.3-5), M/L^2t
$\mathbf{J}_A^{*(c)}$ = convective molar flux vector of species A (17.2-8), mol/L^2t
$\mathbf{J}_A^*$ = diffusive molar flux vector of species A (17.3-11), mol/L^2t
j_H, j_D = Chilton-Colburn j-factors (14.3-19, Table 22.2-1), ---
K_{tot} = total kinetic energy in a macroscopic system (7.4-1, 15.1-1), ML^2/t^2
K_x, K_y = two-phase mass-transfer coefficients (22.4-4), mol/L^2t
$\kappa = R/\tilde{N}$ = Boltzmann's constant (1.6-1), ML^2/t^2T
k = thermal conductivity (9.2-1, 9.2-6), ML/t^3T
k_x = single-phase mass-transfer coefficient (22.1-3), mol/L^2t; (for related quantities k_c, k_ρ, k_p, see 22.1-11, 22.1-12)
k_T = thermal diffusion ratio (24.1-11), ---
k_e = electrical conductivity (10.6-1), I^2t^3/ML^3
k_n'' = heterogeneous chemical reaction rate constant (18.0-6), $\text{mol}^{1-n}/L^{2-3n}t$
k_n''' = homogeneous chemical reaction rate constant (18.0-5), $\text{mol}^{1-n}/L^{3-3n}t$
L = length of film, tube, or slit (2.2-1, 2.3-3), L
$\mathbf{L}_{\text{tot}}$ = total angular momentum within a macroscopic system (7.3-1), ML^2/t
l = mixing length (in turbulence) (4.4-4), L
l_0 = characteristic length in dimensional analysis (5.1-3), L
M = molar mean molecular weight (Table 17.1-2), M/mol
M_A = molecular weight of species A (Table 17.1-2), M/mol
$M_{\alpha,\text{tot}}$ = total number of moles of species α in macroscopic system (23.1-3), mol
$m_{\alpha,\text{tot}}$ = total mass of species α in macroscopic system (23.1-1), M
m_{tot} = total mass in macroscopic system (7.1-1), M
m = mass of a molecule (1.6-1), M
m, n = parameters in power-law non-Newtonian viscosity model (8.3-3), M/Lt^{2-n}, ---
N = rate of shaft rotation (5.3-4), t^{-1}
$\tilde{N}$ = Avogadro's number, $(\text{g-mol})^{-1}$
$\mathbf{N}_A$ = total molar flux vector for species A (17.4-3), mol/L^2t
$\mathbf{n}_A$ = total mass flux vector for species A (17.4-1), M/L^2t
$\mathbf{n}$ = unit normal vector (7.7-2, A.5-1), ---
n = molecular concentration, or number density (1.6-2), L^{-3}
$\mathbf{P}_{\text{tot}}$ = total amount of momentum in a macroscopic flow system (7.2-1), ML/t

$\mathscr{P} = p + \rho g h$ = modified pressure (2.3-11), M/Lt^2
$\mathscr{P}_0$ = characteristic modified pressure used in dimensional analysis (5.1-4), M/Lt^2
p = fluid pressure, M/Lt^2
Q = volume rate of fluid flow across a surface (2.2-24), L^3/t
Q = rate of heat flow across a surface (9.2-1), ML^2/t^3
$Q_{\vec{12}}$ = radiant energy flow from surface 1 to surface 2 (16.4-5), ML^2/t^3
Q_{12} = net radiant energy interchange between surface 1 and surface 2 (16.4-8), ML^2/t^3
$\mathbf{q}$ = conductive heat-flux vector (9.2-2, 9.2-6), M/t^3
$\mathbf{q}^{(c)}$ = convective energy-flux vector (9.1-3), M/t^3
R = gas constant (in $p\tilde{V} = RT$) (9.7-14) $ML^2/t^2T\text{mol}$
R = radius of a cylinder or a sphere, L
R_h = mean hydraulic radius (6.2-15), L
R_A = molar rate of production of species A per unit volume by chemical reaction (18.0-5), mol/L^3t
r_A = mass rate of production of species A per unit volume by chemical reaction (19.1-4), M/L^3t
$\mathbf{r}$ = position vector (3.4-1), L
$r = \sqrt{x^2 + y^2}$ = radial coordinate in cylindrical coordinates (Fig. A.6-1), L
$r = \sqrt{x^2 + y^2 + z^2}$ = radial coordinate in spherical coordinates (Fig. A.6-1), L
S_1, S_2 = cross-sectional area at planes 1 and 2 (7.1-1), L^2
S = entropy (24.1-1), ML^2/t^2T
$\mathbf{s}$ = entropy-flux vector (24.1-1), M/t^3T
T = absolute temperature, T
$T_1 - T_0$ = characteristic temperature difference used in dimensional analysis (13.1-5), T
$\mathbf{T}_{s\to f}$ = torque exerted by a solid boundary on the fluid (7.3-1), ML^2/t^2
$\mathbf{T}_{\text{ext}}$ = external torque acting on system (7.3-1), ML^2/t^2
t = time, t
U = internal energy (9.1-1), ML^2/t^2
U = overall heat-transfer coefficient (10.3-15), M/t^3T
$\bar{u}$ = arithmetic mean molecular speed (1.6-1), L/t
$\mathbf{u}$ = unit vector in direction of flow (7.2-1), ---
$\mathbf{u}$ = displacement vector (8.4-2), ---
V = volume, L^3
$\mathbf{v}$ = mass-average velocity (17.2-1), L/t
$\mathbf{v}^\star$ = molar-average velocity (17.2-2), L/t
$\mathbf{v}_A$ = velocity of species A (17.2-1, Table 17.2-1), L/t
v_0 = characteristic velocity in dimensional analysis (5.1-4), L/t
v_s = speed of sound(9.8-2, 11C.1-4), L/t
$v_\star = \sqrt{\tau_0/\rho}$ friction velocity (4.3-6), L/t
W_m = rate of doing work on the system by the surroundings via moving parts (7.4-1), ML^2/t^3
W = width (of film, slit) (2.2-1), L

W = molar rate of flow across a surface (23.1-4), mol/t
W_α = molar rate of flow of species α across a surface (23.1-3), mol/t
w = mass rate of flow across a surface (2.2-26), M/t
w_α = mass flow rate of species α across a surface (23.1-1), M/t
$\mathbf{w}$ = work-flux vector (9.3-5), M/t^3
x_A = mole fraction of species A (1.5-2, Table 17.1-1), ---
y_A = mole fraction of species A (22.4-2), ---
x, y, z = Cartesian coordinates
y = distance from wall (in turbulence) (4.3-1), L
Z = wall collision frequency (1.6-2), $L^{-2}t^{-1}$

alpha
$\alpha = k/\rho\hat{C}_p$ = thermal diffusivity (9.2-7), L^2/t

beta
β = thermal coefficient of volume expansion (10.10-6), T^{-1}

gamma
$\gamma = C_p/C_V$ = heat capacity ratio (11.4-40), ---
$\dot{\boldsymbol{\gamma}} = \nabla\mathbf{v} + (\nabla\mathbf{v})^\dagger$ = rate-of-deformation tensor (8.3-1), t^{-1}

delta
$\Delta X = X_2 - X_1$ = difference between value of X at exit (plane 2) and value at entry (plane 1) (7.1-3)
δ = falling film thickness (2.2-21), boundary-layer thickness (after 3.8-16), L
$\boldsymbol{\delta}$ = unit tensor (1.2-13, A.3-10), ---
$\boldsymbol{\delta}_i$ = unit vector in the i-direction (A.2-9), ---
δ_{ij} = Kronecker delta (1.2-13, A.2-1), ---

epsilon
ε = fractional void space (6.4-3), ---
$\varepsilon, \varepsilon_{AB}$ = maximum attractive energy between two molecules (1.6-10, 17.7-13), ML^2/t^2
ε_{ijk} = permutation symbol (A.2-3), ---

zeta
ζ = concentration coefficient of volume expansion (19.2-2 and Table 22.2-1), ---

eta
η = non-Newtonian viscosity (8.2-1), M/Lt
η_0, η_∞ = zero-shear-rate viscosity, infinite-shear-rate viscosity (8.3-4), M/Lt
$\eta', -\eta''$ = components of the complex viscosity (8.2-4), M/Lt
$\bar{\eta}$ = elongational viscosity (8.2-5), M/Lt

theta
Θ = dimensionless temperature difference (10.4-1), ---
$\theta = \arctan(y/x)$ = angle in cylindrical coordinates (A.6-5), ---
$\theta = \arctan(\sqrt{x^2+y^2}/z)$ = angle in spherical coordinates (A.6-23), ---

kappa
κ = dilatational viscosity (1.2-7), M/Lt
κ = radius ratio for coaxial cylinders or concentric spheres (2.4-7, 11.4-30), ---

lambda
λ = wavelength of electromagnetic radiation (16.1-1), L
λ = mean-free path (1.6-3), L
λ = time constant in viscoelastic rheological models (8.4-3, 8.5-2), t

mu
μ = viscosity (1.2-1), M/Lt
$\mu^{(t)}$ = turbulent (eddy) viscosity (4.4-1), M/Lt

nu
$\nu = \mu/\rho$ = kinematic viscosity (1.2-3), L^2/t
ν = frequency of electromagnetic radiation (16.1-1), t^{-1}

xi — ξ = concentration coefficient of volume expansion (Table 22.2-1), ---

pi — π = 3.14159...

$\boldsymbol{\pi}^{(c)}$ = convective momentum-flux tensor (1.1-2, 1.1-6), M/Lt^2

$\boldsymbol{\pi} = \boldsymbol{\tau} + p\boldsymbol{\delta}$ = molecular momentum-flux tensor, molecular stress tensor (1.2-15), M/Lt^2

rho — ρ = density, M/L^3

ρ_A = mass of species A per unit volume of mixture (Table 17.1-1), M/L^3

$\rho^{(A)}$ = density of pure species A (Table 17.1-1), M/L^3

sigma — σ = Stefan-Boltzmann constant (16.2-10), M/t^3T^4

σ, σ_{AB} = collision diameter (1.6-10, 17.7-11), L

tau — $\boldsymbol{\tau}$ = (viscous) momentum-flux tensor, (viscous) stress tensor (1.2-2, 1.2-13), M/Lt^2

phi — Φ = potential energy (3.3-2, 7.4-1), ML^2/t^2

Φ_v = viscous dissipation function (3.3-3), t^{-2}

$\boldsymbol{\phi} = \rho\mathbf{vv} + \boldsymbol{\pi} = \rho\mathbf{vv} + p\boldsymbol{\delta} + \boldsymbol{\tau}$ = total momentum-flux tensor (1.3-1, 1.3-2), M/Lt^2

$\phi = \arctan y/x$ = angle in spherical coordinates (A.6-24), ---

φ = intermolecular potential energy (1.6-10), ML^2/t^2

psi — Ψ_1, Ψ_2 = first, second normal-stress coefficient (8.2-2 and 3), M/L

Ψ_v = viscous dissipation function (3.3-3), t^{-2}

omega — Ω = angular velocity (3.7-29), t^{-1}

$\Omega_\mu, \Omega_k, \Omega_{\mathscr{D}}$ = collision integrals (1.6-14, 9.7-13, 17.7-11), ---

ω_A = mass fraction of species A (Table 17.1-1), ---

$\omega_{A1} - \omega_{A0}$ = characteristic mass-fraction difference used in dimensional analysis (21.1-7), ---

Overlines

$\tilde{X}$ = per mole (9.7-6)

$\hat{X}$ = per unit mass (7.4-7, 9.1-1, 9.2-7)

$\overline{X}$ = partial molar (19.3-3, 19.3-7, 24.1-2)

$\overline{X}$ = time-smoothed (4.1-4, 4.2-1)

$\breve{X}$ = dimensionless (5.1-3)

Brackets

$\langle X \rangle$ = average value over a flow cross section

$(X), [X], \{X\}$ = used in vector-tensor operations when the brackets enclose dot or cross operations (Appendix A)

[=] = has the dimensions of, has units of

Superscripts

$X^\dagger$ = transpose of a tensor (1.2-13, A.3-8)

$X^{(c)}$ = convective flux (1.1-2, 1.1-6)

$X^{(t)}$ = turbulent (4.2-8)

$X^{(v)}$ = viscous (4.2-9)

X' = fluctuating quantity (4.2-1)

Subscripts

A, B = species A and B in binary systems

$\alpha, \beta, \ldots$ = species in multicomponent systems

a = arithmetic-mean driving force or associated transfer coefficient (14.1-3)
b = bulk or "cup-mixing" value for an enclosed stream (10.9-35, 14.1-2)
c = evaluated at the critical point (1.5-1)
ln = logarithmic-mean driving force or associated transfer coefficient (14.1-4)
loc = local driving force or associated transfer coefficient (14.1-5)
m = mean transfer coefficient for a submerged object (14.1-6)
r = reduced, relative to critical value (§1.5)
tot = total amount of quantity in a macroscopic system (7.1-1)
0 = quantity evaluated at a surface (23.1-1)
1, 2 = quantity evaluated at cross sections 1 and 2 (7.1-1)

Named dimensionless groups designated with two letters

Bi = Biot number (10.2-20)
Br = Brinkman number (10.8-9, Table 13.1-2)
Da = Damköhler number (18.5-24)
Ec = Eckert number (Table 13.1-2)
Fr = Froude number (5.1-10, Table 13.1-2)
Gr = Grashof number (10.10-17, Table 13.1-2)
Gr_ω = Diffusional Grashof number (21.1-13, Table 22.2-1)
Le = Lewis number (17.3-15)
Ma = Mach number (11.4-52, Table 13.1-2)
Nu = Nusselt number (10.9-37; 14.2-3 and 4)
Pé = Péclet number (9.2-8, Table 13.1-2)
Pr = Prandtl number (9.2-8, Table 13.1-2)
Ra = Rayleigh number (Table 13.1-2)
Re = Reynolds number (5.1-9, Table 13.1-2)
Sc = Schmidt number (17.3-14)
Sh = Sherwood number (22.2-4)
We = Weber number (5.1-11)

Mathematical operations

D/Dt = substantial derivative (3.5-4), t^{-1}
$\mathscr{D}/\mathscr{D}t$ = corotational derivative (8.5-1), t^{-1}
∇ = del operator (3.1-4, A.4-1), L^{-1}
$\ln x = 2.303 \log_{10} x$ = logarithm of x to the base e
$\log x$ = logarithm of x to the base 10
$\exp x = e^x$ = the exponential function of x
$\operatorname{erf} x = \dfrac{2}{\sqrt{\pi}}\displaystyle\int_0^x \exp(-t^2)\,dt$ = error function (3.8-15, §C.6)
$\operatorname{erfc} x = 1 - \operatorname{erf} x$ = complementary error function (3.8-15, §C.6)
$\Gamma(x) = \displaystyle\int_0^\infty t^{x-1}e^{-t}dt$ = the (complete) gamma function (11.5-58, §C.4)
$\Gamma(x,u) = \displaystyle\int_0^u t^{x-1}e^{-t}dt$ = the incomplete gamma function (11.5-62, §C.4)

Author Index

D

E

F

G

H

I

J

K

L

X

Y

Z

Subject Index

(*Bold entries indicate biographical information*)

C

G

H

N

S

T

U

V

W

Y

ALGEBRAIC OPERATIONS FOR VECTORS AND TENSORS IN CARTESIAN COORDINATES

(s is a scalar, $\mathbf{v}$ is a vector, and $\boldsymbol{\tau}$ is a tensor; dot or cross product operations enclosed within parentheses (...) are scalars, those enclosed within brackets [...] are vectors)

$$(\mathbf{v} \cdot \mathbf{w}) = v_x w_x + v_y w_y + v_z w_z = (\mathbf{w} \cdot \mathbf{v})$$

$$[\mathbf{v} \times \mathbf{w}]_x = v_y w_z - v_z w_y = -[\mathbf{w} \times \mathbf{v}]_x$$
$$[\mathbf{v} \times \mathbf{w}]_y = v_z w_x - v_x w_z = -[\mathbf{w} \times \mathbf{v}]_y$$
$$[\mathbf{v} \times \mathbf{w}]_z = v_x w_y - v_y w_x = -[\mathbf{w} \times \mathbf{v}]_z$$

$$[\boldsymbol{\tau} \cdot \mathbf{v}]_x = \tau_{xx} v_x + \tau_{xy} v_y + \tau_{xz} v_z \qquad [\mathbf{v} \cdot \boldsymbol{\tau}]_x = v_x \tau_{xx} + v_y \tau_{yx} + v_z \tau_{zx}$$
$$[\boldsymbol{\tau} \cdot \mathbf{v}]_y = \tau_{yx} v_x + \tau_{yy} v_y + \tau_{yz} v_z \qquad [\mathbf{v} \cdot \boldsymbol{\tau}]_y = v_x \tau_{xy} + v_y \tau_{yy} + v_z \tau_{zy}$$
$$[\boldsymbol{\tau} \cdot \mathbf{v}]_z = \tau_{zx} v_x + \tau_{zy} v_y + \tau_{zz} v_z \qquad [\mathbf{v} \cdot \boldsymbol{\tau}]_z = v_x \tau_{xz} + v_y \tau_{yz} + v_z \tau_{zz}$$

The above may be generalized to cylindrical coordinates by replacing x, y, z by r, θ, z, and to spherical coordinates by replacing x, y, z by r, θ, ϕ (see Figures 1.2-2, A.6-1, A.8-1, and A.8-2).

DIFFERENTIAL OPERATIONS FOR SCALARS, VECTORS, AND TENSORS IN CARTESIAN COORDINATES

$$(\nabla s)_x = \frac{\partial s}{\partial x} \qquad (\nabla s)_y = \frac{\partial s}{\partial y} \qquad (\nabla s)_z = \frac{\partial s}{\partial z}$$

$$[\nabla \times \mathbf{v}]_x = \frac{\partial v_z}{\partial y} - \frac{\partial v_y}{\partial z} \qquad [\nabla \times \mathbf{v}]_y = \frac{\partial v_x}{\partial z} - \frac{\partial v_z}{\partial x} \qquad [\nabla \times \mathbf{v}]_z = \frac{\partial v_y}{\partial x} - \frac{\partial v_x}{\partial y}$$

$$(\nabla \cdot \mathbf{v}) = \frac{\partial v_x}{\partial x} + \frac{\partial v_y}{\partial y} + \frac{\partial v_z}{\partial z}$$

$$(\mathbf{v} \cdot \nabla s) = v_x \frac{\partial s}{\partial x} + v_y \frac{\partial s}{\partial y} + v_z \frac{\partial s}{\partial z}$$

$$\nabla^2 s \equiv (\nabla \cdot \nabla s) = \frac{\partial^2 s}{\partial x^2} + \frac{\partial^2 s}{\partial y^2} + \frac{\partial^2 s}{\partial z^2}$$